GENERAL AND MOLECULAR PHARMACOLOGY

GENERAL AND MOLECULAR PHARMACOLOGY

Principles of Drug Action

Edited By

FRANCESCO CLEMENTI AND GUIDO FUMAGALLI

Co-editors

CHRISTIANO CHIAMULERA, EMILIO CLEMENTI,
RICCARDO FESCE, DIEGO FORNASARI, AND CECILIA GOTTI

WILEY

Translated and modified by Francesco Clementi and Guido Fumagalli. Originally published in Italian under the title "Farmacologia Generale e Molecolare. Il meccanismo d'azione dei farmaci", 4th edition, by Francesco Clementi and Guido Fumagalli, © UTET SpA – Unione Tipografico-Editrice Torinese, Torino, Italy (2012).

Copyright © 2015 by John Wiley & Sons, Inc. All rights reserved

Published by John Wiley & Sons, Inc., Hoboken, New Jersey
Published simultaneously in Canada

No part of this publication may be reproduced, stored in a retrieval system, or transmitted in any form or by any means, electronic, mechanical, photocopying, recording, scanning, or otherwise, except as permitted under Section 107 or 108 of the 1976 United States Copyright Act, without either the prior written permission of the Publisher, or authorization through payment of the appropriate per-copy fee to the Copyright Clearance Center, Inc., 222 Rosewood Drive, Danvers, MA 01923, (978) 750-8400, fax (978) 750-4470, or on the web at www.copyright.com. Requests to the Publisher for permission should be addressed to the Permissions Department, John Wiley & Sons, Inc., 111 River Street, Hoboken, NJ 07030, (201) 748-6011, fax (201) 748-6008, or online at http://www.wiley.com/go/permissions.

Limit of Liability/Disclaimer of Warranty: While the publisher and author have used their best efforts in preparing this book, they make no representations or warranties with respect to the accuracy or completeness of the contents of this book and specifically disclaim any implied warranties of merchantability or fitness for a particular purpose. No warranty may be created or extended by sales representatives or written sales materials. The advice and strategies contained herein may not be suitable for your situation. You should consult with a professional where appropriate. Neither the publisher nor author shall be liable for any loss of profit or any other commercial damages, including but not limited to special, incidental, consequential, or other damages.

For general information on our other products and services or for technical support, please contact our Customer Care Department within the United States at (800) 762-2974, outside the United States at (317) 572-3993 or fax (317) 572-4002.

Wiley also publishes its books in a variety of electronic formats and by print-on-demand. Not all content that is available in standard print versions of this book may appear or be packaged in all book formats. If you have purchased a version of this book that did not include media that is referenced by or accompanies a standard print version, you may request this media by visiting http://booksupport.wiley.com. For more information about Wiley products, visit us at www.wiley.com.

Library of Congress Cataloging-in-Publication Data

General and molecular pharmacology : principles of drug action / Francesco Clementi, Guido Fumagalli, editors ; Christiano Chiamulera, Emilio Clementi, Riccardo Fesce, Diego Fornasari, Cecilia Gotti, co-editors.
   1 online resource.
   Includes bibliographical references and index.
   Description based on print version record and CIP data provided by publisher; resource not viewed.
   ISBN 978-1-118-76859-4 (pdf) – ISBN 978-1-118-76868-6 (epub) – ISBN 978-1-118-76857-0 (cloth : alk. paper)
I. Clementi, Francesco, editor. II. Fumagalli, Guido, editor.
[DNLM: 1. Chemistry, Pharmaceutical. 2. Molecular Biology. 3. Pharmacological Phenomena. QV 744]
   RS403
   615.1′9–dc23

2015008591

CONTENTS

LIST OF CONTRIBUTORS	xlvi
PREFACE	xlix
SECTION 1 INTRODUCTION TO PHARMACOLOGY	1

1 Essential Lexicon of Pharmacology — 3
Francesco Clementi and Guido Fumagalli

The Social Impact of Pharmacology, 3
Essential Lexicon, 4
   Active Substances, 4
   Pharmacological Disciplines, 5
   Drug–Receptor Interactions, 6
   Measure of the Clinical Response, 7
Take-Home Message, 7

2 A Short History of Pharmacology — 8
Vittorio A. Sironi

Birth and Historical Developments of Pharmacology, 8
   From Magical and Natural Remedies of Ancient Medicine to Arabic Alchemy, 8
   From Monastic Medicine to Botanical Gardens, 9
   From Anatomical Renaissance to the "Experienz": Paracelsus' Spagyric, 10
   From Iatrochemistry to the Age of Enlightenment, 11
   From the Search of the Active Principle to the Discovery of Alkaloids
      and Glucosides, 12
   The Drug Synthesis Revolution: From Handmade to Industrial Production, 12
Modern Pharmacology, 13
   Ehrlich and Chemotherapy: The Concept of Receptor, 13
   The Birth of Modern Pharmacology, 14
The Biotechnology Era and the Pharmacology in the Third Millennium, 15
   The Impact of New Biotechnologies: Molecular Biology, Bioinformatics,
      and Combinatorial Chemistry, 15
   Biological Drugs and Pharmacology Perspectives, 16

Personalized Therapies and New Sceneries in Pharmaceutical Industry, 17
Take-Home Message, 18
Further Reading, 18

SECTION 2 GETTING THE DRUG TO ITS SITE OF ACTION   19

3 Cellular Basis of Pharmacokinetics   21
Riccardo Fesce and Guido Fumagalli

A Quick Journey with the Drug in the Body, 21
   Absorption, 21
   Distribution, 21
   Drug Elimination, 22
Crossing Cell Membranes, 23
   Passive Diffusion across Cell Membranes, 24
   Drug Transport across Cell Membranes, 25
   Endocytosis, 25
Drug Diffusion to Organs and Tissues, 27
   Properties of the Most Important Cell Barriers, 27
Take-Home Message, 30
Further Reading, 30

4 Drug Absorption and Administration Routes   31
Riccardo Fesce and Guido Fumagalli

General Rules About Drug Absorption Rate, 32
   Partition Coefficient, 32
   Drug Dispersion, 32
   Extension of the Absorbing Surface, 32
   Permeability of the Absorbing Surface, 32
   Vascularization, 33
Enteral Routes of Administration, 33
   Oral Route, 33
   Sublingual and Rectal Routes, 35
Systemic Parenteral Routes of Administration, 35
   The Intravascular Route, 35
   i.m. Injection, 36
   Subcutaneous and Intradermal Injections, 36
Other Routes of Drug Administration, 36
   Inhalation Route, 36
   Topical/Regional Routes, 37
   Intracavity Routes, 37
   Dermal or Transcutaneous Route, 37
   Mucosal Routes, 38
Absorption Kinetics, 38
   General Rules, 38
   Interrelation between Gene Therapy and Drug Delivery Techniques, 43
Take-Home Message, 44
Further Reading, 44

5 Drug Distribution and Elimination   45
Riccardo Fesce and Guido Fumagalli

Distribution, 46
   Tissues and Avidity for Drugs, 46
   The Apparent Distribution Volume, 48

Drug Binding to Plasma Proteins, 50
Factors That Determine the Distribution Rate of Drugs to the Various
  Compartments, 51
Elimination, 53
  The Concept of Half-Life, 53
  The Concept of Clearance, 54
Renal Excretion of Drugs, 55
  Glomerular Filtration, 56
  Tubular Functions and Pharmacokinetics, 56
  Active Transport of Organic Anions and Cations, 56
  Factors Determining Renal Clearance of Drugs, 57
Hepatic Excretion and Enterohepatic Cycle, 58
  Perfusion, Binding to Plasma Proteins, Enzymatic Activity,
    and Hepatic Clearance, 59
Take-Home Message, 59
Further Reading, 60

6 Drug Metabolism 61

Enzo Chiesara, Laura Marabini, and Sonia Radice

Metabolic Modification of Drug Activity, 61
Two Phases of Drug Metabolism, 62
  Phase I Reactions, 62
  Phase II Enzymatic Reactions, 66
Extrahepatic Biotransformations, 68
  Biotransformation by the Intestinal Flora, 69
Pharmacometabolic Induction and Inhibition, 69
  Induction of Drug Metabolism, 69
  Inhibition of Drug Metabolism, 71
Take-Home Message, 72
Further Reading, 72

7 Control of Drug Plasma Concentration 73

Riccardo Fesce and Guido Fumagalli

Time Course of Drug Plasma Concentration Following a Single
  Administration, 73
Drugs Distribute to Organs and Tissues and then are Eliminated, 74
  Description of Drug Plasma Concentration Time Course Following a Single
    Administration, 74
  Area under the Plasma Concentration Curve (AUC), 74
  The Plasma Concentration Peak, 75
Drug Plasma Concentration Time Course During Repetitive Administrations, 75
  During Repetitive Administrations, the Drug Plasma Concentration Time
    Course Is Given by the Sum of the Time Courses of the Single Doses, 75
  In a Chronic Therapy at Steady State, Each New Dose Replaces the Drug
    Amount that has been Eliminated Since the Last Administration, 77
  The Time to Reach the Steady State Depends on the Drug Half-Life, 77
  Plasma Concentration at Steady State, 78
  The Single Dose to Administer is Computed as a Function of the Interval
    between Successive Administrations, 78
  Fluctuations of Drug Plasma Concentration at Steady State, 79
  Absorption Kinetics Influence the Amplitude of Oscillations in Plasma
    Concentration at Steady State, 80
  Loading (Attack) Doses to Rapidly Attain Steady-State Concentration, 80
Multicompartmental Kinetics, 81
  Drug Binding to Plasma Proteins and Tissue Equilibration with Plasma, 81

The Particular Case of the Nephron, 82
Drugs Redistribution among Compartments, 83
Corrections of the Therapeutic Regimen, 83
Normally Available Pharmacokinetic Data Are Average Values, 84
Varying Dosage as a Function of Body Weight and Physical Constitution, 84
Varying Dosage as a Function of Age, 84
Dosage Correction in the Presence of Hepatic Pathologies, 86
Dosage Correction in the Presence of Renal Pathologies, 86
Take-Home Message, 87
Further Reading, 89

SECTION 3   RECEPTORS AND SIGNAL TRANSDUCTION   91

8 Drug–Receptor Interactions: Quantitative and Qualitative Aspects   93
Gian Enrico Rovati and Valérie Capra

General Properties of Drug Receptors, 93
Drug Receptors Are Molecules Relevant for Cellular Functions, 93
Not All Drugs Interact with a Receptor, 94
Drug Activity Follows to Drug–Receptor Complex Formation, 94
Drug–Receptor Interaction Is Mostly Mediated by Weak Chemical Bonds, 94
Reversible or Irreversible Drug–Receptor Interactions, 95
Characteristics of Drug–Receptor Interaction, 95
The Relationship between Drug Concentration and Drug–Receptor Complex Is Similar to the Michaelis–Menten Equation, 97
The Binding Isotherm and Its Linear Transformations Allow to Obtain the Parameters of the Drug–Receptor Interaction, 97
Receptors Can Be Heterogeneous, 99
Drug Competition for a Same Receptor Binding Site, 99
Quantitative Aspects of Drug Effects: Dose–Response Curves, 99
Potency and Efficacy, 100
From Drug–Receptor Interaction to Drug Response, 101
Occupancy Theory, 101
Modifications to the Occupancy Assumption, 102
Efficacy Theory, 104
Nonlinear Function between Receptor Occupancy and Tissue Response: EC_{50} Different from K_d, 106
Constitutively Active Receptors and Inverse Agonists, 107
Two-State Receptor Model and Beyond: Multiple Receptor States and "Biased" Signaling, 108
Take-Home Message, 108
Further Reading, 108

9 Receptors and Modulation of Their Response   109
Francesco Clementi and Guido Fumagalli

Classes of Receptors and Strategies of Signal Transduction, 109
Intracellular/Intranuclear Receptors, 110
Membrane Receptors, 110
Control of Receptor Localization in the Cell Membrane, 116
Intracellular Traffic of Cell Receptors, 117
How Receptors Reach the Cell Membrane and how Their Number is Regulated, 117
Modulation of Receptor Responses, 117
Receptor Modulation By Drugs, 118

Take-Home Message, 119
Further Reading, 119

10 Adaptation to Drug Response and Drug Dependence   **121**
Cristiano Chiamulera

Molecular, Cellular, and Systemic Adaptation, 122
   Cellular Adaptation, 122
   Effects of Repeated Drug Exposure, 122
Drug Addiction as a Paradigm of Allostatic Adaptation, 123
   Adaptation and Stages of Drug Addiction, 125
   Research on Drug Addiction, 127
Therapy for Drug Dependence, 128
Take-Home Message, 129
Further Reading, 129

11 Pharmacological Modulation of Posttranslational Modifications   **130**
Monica Di Luca, Flavia Valtorta, and Fabrizio Gardoni

Protein Phosphorylation, 130
   Protein Kinases, 131
   Protein Phosphatases, 132
SUMOylation, 133
Ubiquitination, 134
Glycosylation, 135
Acetylation, 135
Hydroxylation, 135
Carboxylation, 136
Methylation, 136
S-Nitrosylation, 136
Disulfide Bonds, 136
Lipid Modifications, 136
   Myristoylation, 136
   Prenylation, 136
   Palmitoylation, 137
   Attachment of Glycosylphosphatidylinositols, 137
Pharmacological Modulation of Post Traslational Modifications, 137
   Deacetylase Inhibitors, 137
   Glycosylation Inhibitors, 138
Take-Home Message, 138
Further Reading, 138

12 Calcium Homeostasis Within the Cells   **139**
Jacopo Meldolesi and Guido Fumagalli

The Cytosol: A Crossroad of Ca^{2+} Fluxes, 139
   Free Ca^{2+} in the Cytosol and Total Cell Calcium, 139
   The High-Affinity Buffering of Cytosolic Proteins, 140
The Plasma Membrane: Channels, Pumps, and Transporters, 140
   Surface Channels Permeable to Ca^{2+}, 140
   Surface Pumps and Transporters, 141
Ca^{2+} in Intracellular Organelles, 142
   The ER: A Rapidly Exchanging Ca^{2+} Pool, 142
   Mitochondria as Local Buffers of $[Ca^{2+}]_i$, 143
   ER and Mitochondria Allow Rapid Changes of $[Ca^{2+}]_i$ within Cells, 144
   $[Ca^{2+}]_i$ Control in Other Intracellular Structures, 145
   Local Relevance of Organelle Calcium Pools, 145

Ca^{2+} in Cell Pathology, 145
Take-Home Message, 146
Further Reading, 146

13 Pharmacology of Map Kinases — 147
Lucia Vicentini and Maria Grazia Cattaneo

The MAPK Family and the Activation Mechanism, 147
    The ERK Family, 147
    The JNK Family, 148
    The p38 Family, 150
    ERK5, 150
MAPK Specificity, 150
    Duration of Action, 150
    Multiple Enzyme Isoforms, 151
    Subcellular Localization and Interaction with Scaffold Proteins, 151
    miRNAs, 151
Pharmacological Inhibition of MAPK, 151
    Inhibition of the Ras/RAF/MEK/ERK Cascade, 151
    MEK, 152
    RAF, 152
    JNKs, 152
    p38, 153
Take-Home Message, 153
Further Reading, 153

14 Small G Proteins — 154
Erzsébet Ligeti and Thomas Wieland

Structure and Function of SMGs, 154
Physiological Roles of Components of Major SMG Families, 155
Posttranslational Modification and Subcellular Localization of SMGs, 157
Regulatory Proteins, 157
    GEFs, 158
    GAPs, 158
    GDIs, 159
Modulation of SMG Signaling By Bacterial Toxins and Drugs, 159
    Bacterial Toxins, 159
    Bacterial Toxins Modifying Host RhoGTPases, 160
    Bacterial Toxins Acting as GEFs or GAPs for Host GTPases, 162
    Modulation of the Prenylation Process, 163
    Modulation of GEFs, 163
    Modulation of GAPs, 164
Future Perspectives, 165
Take-Home Message, 165
Further Reading, 165

15 Integration in Intracellular Transduction of Receptor Signals — 166
Jacopo Meldolesi

Dualism of Receptors in the Nucleus and In the Cytoplasm, 166
Heterogeneity of Receptor Assembly, 167
Transduction Cascades Depend On Crosstalk and Complementarity Among Receptors, 168
    A Comprehensive Scenario of Signal Transduction: From GSKβ3 to AKT and mTOR, 169

Development of New Drugs and Therapies, 169
Take-Home Message, 170
Further Reading, 171

SECTION 4 RECEPTOR CLASSES   173

16 Ligand-Gated Ion Channels   175
Cecilia Gotti and Francesco Clementi

Tissue Distribution and Subcellular Localization of LGICs, 175
Molecular Organization of LGICs, 176
  Classification of LGICs, 176
Topology of LGICs, 179
  The Binding Sites for Endogenous Ligands, 179
  Structure and Localization of the Channel, 182
  Functions of the Cytoplasmic Domain, 183
Modulation of LGIC Activity, 184
  Desensitization, 185
  Subunit Composition and Biophysical and Pharmacological Properties
    of LGICs, 185
Crosstalk with Other Receptor Systems, 186
Mechanisms of Action of Drugs That Modulate LGICs, 187
Take-Home Message, 187
Further Reading, 188

17 G-Protein-Coupled Receptors   189
Lucia Vallar, Maria Pia Abbracchio, and Lucia Vicentini

Molecular Organization of GPCRs, 189
  The Ligand Binding Site, 191
  Dimerization, 192
  GPCR Mutations and Human Diseases, 192
Molecular Organization and Function of G Protein, 192
  Signaling by G Proteins, 193
  Structural and Functional Properties of α-Subunits, 194
  The βγ Complex, 194
  Regulation of Multiple Effectors Through G Proteins, 195
Effector Pathways of GPCRs, 195
  The Adenylyl Cyclase System, 195
  The Phospholipase C System, 196
Interaction of GPCRs with Other Proteins, 198
  GPCR-Interacting Proteins: Control of Receptor Function, Intracellular
    Trafficking, and Localization, 198
  G-Protein-Independent Signaling, 199
Perspectives, 200
Take-Home Message, 200
Further Reading, 200

18 Growth Factor Receptors   202
Silvia Giordano, Carla Boccaccio, and Paolo M. Comoglio

Molecular Structure of Growth Factor Receptors, 204
  Growth Factor Receptors are Tyrosine Kinases with Modular Structure, 204
  Modulatory Functions of the Transmembrane and Juxtamembrane Domains
    of Growth Factor Receptors, 204

Receptor Activation and Signal Transduction, 205
  Receptor Dimerization and Activation, 205
  Signal Transducers Binding to Phosphorylated Tyrosines, 207
  RAS-Dependent Transduction Pathway in Cell Proliferation and Neoplastic Transformation, 209
  Activation of Enzymes Generating Lipid Second Messengers, 210
  Cytoplasmic Tyrosine Kinases, 211
  Tyrosine Phosphatases Modulating Tyrosine Kinase Activity, 212
  STATs, 212
Pharmacological Approaches to the Control of Growth Factor Receptor Activity, 213
  mAbs and Small Kinase Inhibitors, 213
  Growth Factor Receptors as Targets for Anticancer Drugs, 215
Take-Home Message, 216
Further Reading, 216

19 Cytokine Receptors 217
Massimo Locati

Classification of Cytokines and Their Receptors, 217
  Hematopoietin Receptors, 217
  Tyrosine Kinase Receptors, 218
  The IL-1/IL-18 Receptor Family, 218
  The TNFR Family (Jelly Roll Motif Cytokines), 219
  Chemokine Receptors, 219
Cytokines in Their Biological Settings, 220
  Hematopoietic Cytokines, 220
  Cytokines in Innate Immunity, 220
  Cytokines of Adaptive Immunity, 222
  Anti-inflammatory Cytokines, 223
Pharmacology of Cytokines and Their Receptors, 223
Take-Home Message, 224
Further Reading, 224

20 Adhesion Molecule Receptors 225
Giorgio Berton and Carlo Laudanna

Adhesion Receptors, 225
  Classification, 225
  Functions, 226
Signal Transduction By Adhesion Receptors, 229
  FAK and Src-Family Kinases in Integrin Signal Transduction, 229
  Signal Transduction by Cadherins, 232
  Signal Transduction by CD44, 232
Adhesion Receptors as Drug Targets, 233
Take-Home Message, 234
Further Reading, 234

21 Soluble Cytokine Receptors and Monoclonal Antibodies in Pathophysiology and Therapy 235
Alberto Mantovani and Annunciata Vecchi

Soluble Receptors, 235
  The Paradigm of Viral Receptors for Cytokines, 236
  "Decoy" Receptors: Molecular Traps for Agonists, 236
  Soluble Receptors: Mechanisms of Generation and Action, 237
  Soluble Receptors: Antagonist Effects, 237

Soluble Receptors: Agonist Effects, 238
Soluble Receptors as Pharmacological Agents, 238
Monoclonal Antibodies, 239
Take-Home Message, 239
Further Reading, 239

SECTION 5 MODULATION OF GENE ESPRESSION 241

22 Pharmacology of Transcription 243
Roberta Benfante and Diego Fornasari

Introduction to the Mechanisms of Transcriptional Regulation, 243
  General Transcription Factors, 243
  The Promoter: A Multifunctional Region with Positive and Negative Regulatory Elements, 244
  Additional Factors Required to Stabilize PIC Structure, 246
  Mediator, 246
  Control of Specificity and Inducibility of Gene Expression, 246
  Coregulatory Complexes and Covalent Modification of Histones, 248
  Toward a Unified Theory of Transcriptional Regulation, 250
Transcriptional Regulation By Extracellular Stimuli, 251
  Three Classes of Inducible Transcription Factors, 251
From the Pharmacology of Transcription to Pharmacoepigenomics, 252
Take-Home Message, 255
Further Reading, 255

23 Pharmacogenetics and Personalized Therapy 256
Diego Fornasari

Variation of Drug Responses and the Definition of Pharmacogenetics, 256
Genetic Basis of Variability in Drug Response, 256
  Genes and Pharmacogenetics, 256
  Allelic Variants of Genes Involved in Drug Response, 257
  Copy Number Variation, 257
Genetic Polymorphisms and Drug Metabolism, 258
Genetic Polymorphisms in Genes Encoding Phase I Enzymes, 258
  Polymorphisms of *CYP2D6,* 258
  Allelic Variants in the CYP2C Subfamily, 260
  Role of Phase I Enzymes in Prodrug Activation, 260
  Thiopurine Methyltransferase, 262
  UGT1A1 and Irinotecan: A Case Already under Evaluation by the Regulatory Agencies, 263
Genetic Polymorphisms in Genes Encoding Transporters Involved in Drug Absorption, Distribution, and Excretion, 263
  ABCB1 and Multidrug Resistance, 263
  The OAT1B1 Transporter and Statin-Induced Myopathy, 264
Genetic Polymorphisms in Genes Encoding Molecular Targets of Drug Action, 264
  β1 Receptor Polymorphism and the Response to β-Blockers in Heart Failure, 264
  β2 Receptor Polymorphism and the Response to Antiasthma Therapy, 265
Methods of Pharmacogenetic Studies, 265
The Future of Pharmacogenetic, 266
Take-Home Message, 266
Further Reading, 267

24 Intracellular Receptors — 268
Adriana Maggi and Elisabetta Vegeto

Structural Features of Intracellular Receptors, 268
  Intracellular Receptor Classification, 268
  Molecular Organization and Functional Domains, 272
Intracellular Receptors as Ligand-Regulated Transcription Factors, 274
  Classification of Ligands, 274
  Ligand-Dependent Transcriptional Activation, 274
  Ligand-Dependent Transrepression, 274
  Transcriptional Activation of Intracellular Receptors in the Absence of Ligand, 275
  Temporal Oscillations of DNA Binding of Intracellular Receptors, 277
Physiological Activities and Pharmacological Control of Intracellular Receptors, 278
  Specificity of Action of Homologous Receptors, 278
  Tissue Specificity of Nuclear Receptor Activity, 278
  Receptor Agonists and Antagonists, 278
  CAR and SXR Receptors, 283
Take-Home Message, 283
Further Reading, 283

25 RNA Molecule as a Drug: From RNA Interference to Aptamers — 284
Valerio Fulci and Giuseppe Macino

Mechanisms of Action of RNA Drugs, 284
  RNA Interference, 285
  Inhibition of miRNA Biological Activity by Complementary Oligonucleotides, 286
  Antisense Oligonucleotide to Modify the Splicing Patterns, 286
  Aptamers, 286
Delivery, 287
General Chemical Structure, 288
  Phosphorothioates, 289
  Phosphoramidates, 289
  2'-O-Alkyl-Ribonucleotides, 289
  LNAs, 289
  PNA, 289
  Modifications of Purine and Pyrimidine Bases, 289
Pharmacotoxicology, 289
  Cellular Pharmacokinetics and Toxicology, 289
  Systemic Pharmacokinetics and Toxicology, 290
Present Use and Future Perspectives, 290
Take-Home Message, 292
Further Reading, 292

SECTION 6 REGENERATIVE MEDICINE — 293

26 Regenerative Medicine and Gene Therapy — 295
Luciano Conti and Elena Cattaneo

Principles of Regenerative Medicine, 296
Definition, Classification, and Features of Stem Cells, 296

Pluripotent Stem Cells, 298
Multipotent (or Adult) Stem Cells, 299
Stem Cell-Based Drugs, 299
Cell Therapy and Regenerative Medicine, 299
Regenerative Medicine Approaches to Epithelial Lesions, 300
Regenerative Medicine to Treat Cardiac Dysfunctions, 301
Stem Cell-Based Therapies for Skeletal Muscle Diseases, 301
Stem Cell-Based Therapies for Brain Diseases, 302
Gene Therapy, 303
Protocols for Gene Therapy, 303
Gene Therapy for Monogenic Inherited Diseases, 304
Future Perspectives, 306
Take-Home Message, 306
Further Reading, 306

SECTION 7   PHARMACOLOGICAL CONTROL OF MEMBRANE TRANSPORT   309

27 Ion Channels   311

Maurizio Taglialatela and Enzo Wanke

Ion Channels and Transporters, 311
Characterization and Function of Ion Channels, 312
Channel Classification According to Permeating Ions and Gating Mechanisms, 312
Permeation and Concentration Gradients, 312
Transmembrane Voltage Triggers Conformational Changes, 315
Current–Voltage Relationships and The Rectification Process, 315
Structural Organization of Ion Channels, 316
The Voltage Sensor of VGICs, 316
Inactivation, 319
Ion Selectivity, 319
Drugs and Ion Channels, 319
Drugs Interacting Directly with Ion Channels, 319
Modulation of Ion Channel *Activity* by Drugs Acting on Receptors Functionally Coupled to Ion Channels, 321
Sodium Channels, 322
Molecular Structure and Modulation, 322
Cellular Localization of VGSCs, 322
Pharmacology of VGSCs, 324
Calcium Channels, 326
Localization and Physiological Functions of VGCCs, 326
Structural Organization of VGCC, 328
VGCC Pharmacology, 328
Potassium Channels, 331
Structural Organization of Potassium Channels, 332
Nonselective Channels, Anionic Channel, and Others, 339
Cationic Channels Modulated by Cyclic Nucleotides, 339
Take-Home Message, 343
References, 343
Further Reading, 344

28 Membrane Transporters — 345
Lucio Annunziato, Giuseppe Pignataro, and Gianfranco Di Renzo

Transporter Classification, 345
  ATP-Dependent Active Transporters, 345
  ATP-Independent Transporters, 346
Na^+/K^+-ATPase, 346
  Structure, 348
  Biophysical Properties and Involvement in Intracellular Signaling, 348
  Pharmacology of Na^+/K^+-ATPase, 348
H^+/K^+-ATPase, 349
  Structure, Distribution, and Function, 350
  Regulation of the Gastric Proton Pump, 350
  Pharmacology of the Gastric Proton Pump, 351
Plasma Membrane Ca^{2+}-ATPase, 351
Sarcoplasmic/Endoplasmic Reticulum Ca^{2+}-ATPase, 352
  Structure, Distribution, and Regulatory Mechanisms of SERCA, 352
  Physiological Properties and Pharmacological Modulation, 353
Na^+/Ca^{2+} Exchanger, 353
  Structure and Distribution, 353
  Biophysical Properties and Physiological Role, 354
  NCX in Pathologies and Pharmacological Modulation, 354
Na^+/H^+ Exchanger, 355
  Structure, Distribution, and Functional Properties, 355
  Functional Significance and Pharmacological Modulation, 356
$Na^+/K^+/Cl^-$ Cotransporter, 356
  Structure and Distribution, 356
  Functional Properties, 357
  Role in Cell Physiology and Pathology, 357
  Pharmacology, 358
Take-Home Message, 358
Further Reading, 358

29 Neurotransmitter Transporters — 359
Gaetano Di Chiara

Neurotransmitter Transporters and Synaptic Function, 359
  Regulation of Transporter Activity and Traffic, 360
Neurotransmitter Transporter Families, 361
  Na^+/K^+-Dependent Transporters for Excitatory Amino Acids, 361
Na^+/Cl^--Dependent Plasma Membrane Transporters, 364
  Molecular Mechanism of Transport, 366
  GABA Transporters, 367
  The Serotonin Transporter, 369
  The Dopamine Transporter, 371
  The Noradrenaline and Adrenaline Transporters, 371
  H^+-Dependent Vesicular Transporters, 372
  The Vesicular Monoamine Transporters, 373
  The Vesicular Acetylcholine Transporter, 374
  The Vesicular Transporters for Excitatory Amino Acids, GABA, and Glycine, 374
Take-Home Message, 374
Further Reading, 374

SECTION 8   CONTROL OF PROTEOLYSIS   377

30 Intracellular Proteolysis   379
  Fabio Di Lisa and Edon Melloni

  General Characteristics of Proteases, 379
    Classification, 379
  Characteristics and Regulation of Intracellular Proteolysis, 381
    Function and Pharmacological Modulation of the Main Intracellular
      Proteolytic Systems, 382
  Lysosomal Proteases, 382
    Types and Families of Lysosomal Enzymes, 383
  Compartmentalized Proteases with Specific Functions, 385
    Signal Proteases, 385
    Maturation Proteases, 385
    Plasma Membrane Proteases, 386
  Cytoplasmic Proteases, 387
    The Proteasome, 387
  Proteasome Inhibitors as Anti-Inflammatory and
    Antitumoral Drugs, 388
    Other Serine Proteases for Protein Quality Control, 388
    Caspases: Initiators and Executors of Apoptosis, 389
    Calpain, 391
  Exogenous Intracellular Proteases, 392
    HIV and Inhibitors of Viral Proteases, 392
  Take-Home Message, 393
  Further Reading, 393

31 Extracellular Proteolysis   395
  Francesco Blasi

  The Extracellular Matrix Proteolytic Degradation Systems, 395
  Plasminogen System and its Activators, 396
    PAs as Fibrinolytic Agents, 396
    Localized Activation of uPA by Interaction with a Specific Receptor, 397
    Function and Pharmacological Modulation of the PA System, 398
    Pharmacological Inhibition of the PA System in Neoplastic Processes, 399
  Matrix Metalloproteases, 399
    Modular Structure of MMPs, 399
    MMP Activation at the Plasma Membrane, 399
    Function and Pharmacological Modulation of MMP, 400
  Take-Home Message, 400
  Further Reading, 401

**SECTION 9   CONTROL OF CELL CYCLE AND CELLULAR
PROLIFERATION**   403

32 Cell Cycle and Cell Death   405
  Marco Corazzari and Mauro Piacentini

  Cell Cycle, 405
    Timely Regulated Expression of Cyclins and Cell Cycle Progression, 405

Role of Retinoblastoma Binding Protein during G1–S Transition, 406
Cdk Inhibitors and Cell Cycle "Checkpoints", 407
Cell Death, 407
Programmed Cell Death Or Apoptosis, 407
    Apoptotic Bodies, 407
    Apoptotic Biochemical Pathways, 408
Necrosis, 412
    Morphological Features of Necrosis, 412
    Role of Mitochondria in the Necrotic Process, 413
Drugs and Apoptosis, 413
    Proapoptotic Drugs, 413
    Drugs That Inhibit Apoptosis, 413
Take-Home Message, 414
Further Reading, 414

33 Mechanisms of Action of Antitumor Drugs   415
Giovanni Luca Beretta, Laura Gatti, and Paola Perego

Conventional Antitumor Drugs, 417
    Alkylating Agents, 417
    Platinum Compounds, 418
    Antimetabolites, 418
    Drugs Acting on Microtubules, 420
    DNA Topoisomerase Inhibitors, 421
Target-Specific Antitumor Drugs, 423
    Inhibitors of Survival Factors, 423
    Proteasome Inhibitors, 425
    Epigenetic Factors as Drug Targets, 426
    Telomers and Telomerase, 426
Monoclonal Antibodies in Clinical Use, 427
    Mechanism of Action, 427
    Trastuzumab, 427
    Cetuximab, 428
    Bevacizumab, 429
Take-Home Message, 429
Further Reading, 429

SECTION 10   CONTROL OF CELLULAR METABOLISM   431

34 Mitochondria, Oxidative Stress, and Cell Damage: Pharmacological Perspectives   433
Clara De Palma, Orazio Cantoni, and Fabio Di Lisa

Reactive Oxygen Species (ROS), 433
    Mitochondrial Formation of ROS, 434
    Physiological Role of ROS, 434
    Pathophysiological Role of ROS in Mitochondrial Ca^{2+} Homeostasis, 435
The Mitochondrial PTP, 436
Drugs and Mitochondria, 437
    Antioxidants Directed to Mitochondria, 437
    Drugs Acting on Mitochondrial Channels to Prevent Mitochondrial Dysfunction, 437
    Drugs Promoting Mitochondrial Dysfunction for Possible Antineoplastic Treatments, 437

Drugs Acting on the Mitochondrial Metabolism, 437
Take-Home Message, 438
Further Reading, 438

35 Pharmacological Control of Lipid Synthesis 439
Lorenzo Arnaboldi, Alberto Corsini, and Nicola Ferri

Cholesterol Biosynthesis, 439
    Statins: Inhibitors of the HMG-CoA Reductase, 441
    Farnesyl Pyrophosphate Synthase Inhibitors (Nitrogen-Containing Bisphosphonates), 441
    Squalene Synthase Inhibitors, 442
Biosynthesis of Fatty Acids, 443
    Fatty Acid Synthase and Its Inhibitors, 443
    Fatty Acid Desaturases and Their Inhibitors, 444
Triglyceride Biosynthesis, 444
    Inhibitors of Diacylglycerol Acyltransferase, 445
Transcriptional Control of Genes Involved in Lipid Metabolism, 446
    Liver X Receptors and Liver X Receptor Synthetic Ligands, 446
    Peroxisome Proliferator-Activated Receptors and Their Pharmacological Modulation, 447
Lipid Transfer Proteins, 448
    ACAT and Its Inhibitors, 448
    CETP and Its Inhibitors, 449
    The MTP and Its Inhibitors, 450
Take-Home Message, 451
Further Reading, 451

36 Glucose Transport and Pharmacological Control of Glucose Metabolism 452
Paolo Moghetti and Giacomo Zoppini

Mechanisms of Glycemic Control, 452
    Modulation of GLUT4 Transporter Function by Insulin and Physical Activity, 456
    Insulin Receptors and Signal Transduction Pathway, 456
Pharmacology of Glycemic Control, 457
    Pharmacological Stimulation of β-Cell Activity, 458
    Modulation of Insulin Signaling, 459
    Modulation of Intestinal Glucose Absorption, 460
Insulin Resistance and New Therapeutic Perspectives, 461
Take-Home Message, 461
Further Reading, 461

SECTION 11   INTERCELLULAR COMMUNICATION 463

37 Pharmacological Regulation of Synaptic Function 465
Michela Matteoli, Elisabetta Menna, Costanza Capuano, and Claudia Verderio

The Synapse, 465
    Synaptic Organization Complexity and Synaptopathies, 466
The Presynaptic Compartment: Neurotransmitter Release, 468
    Chemical Messengers: Neurotransmitters and Neuropeptides, 468
    Secretory Granules and SVs, 469
    Neurotransmitter Fate, 471

xx   CONTENTS

The Postsynaptic Compartment: Signal Reception, 472
Synapse Formation, Maintenance, and Plasticity, 472
The Pharmacology of Neurosecretion, 474
    Drugs Interfering with Secretory Vesicle Transport, 474
    Drugs Interfering with Neurotransmitter Loading into Vesicles, 475
    Drugs and Toxins Interfering with Late Steps of Neuroexocytosis, 475
Take-Home Message, 476
Further Reading, 476

38 Catecholaminergic Transmission   477

Pier Franco Spano, Maurizio Memo, M. Cristina Missale, Marina Pizzi, and Sandra Sigala

The Catecholaminergic System in the Autonomic Nervous System, 477
    Anatomical Organization of the Sympathetic System, 478
    Cardiovascular Effects of the Sympathetic System, 479
    Other Noncardiovascular Effects of Catecholamines, 479
The Catecholaminergic Systems in the CNS, 480
    Distribution and Functions of the Adrenergic and Noradrenergic Systems, 480
    Distribution of the Dopaminergic System, 481
    Functions of the Dopaminergic Systems in the CNS, 482
Synthesis of Catecholamines, 483
Vesicular Storage and Release of Catecholamines, 484
Catabolism and Reuptake of Catecholamines, 485
    Catecholamine Reuptake, 486
Catecholamine Receptors, 487
    Adrenergic Receptors, 487
    DA Receptors, 489
Principles of Drug Action on Catecholaminergic Receptors, 490
    Drugs Acting on α-Adrenergic Receptors, 490
    Drugs Acting on β-Adrenergic Receptors, 491
    Drugs Acting on DA Receptors, 493
Take-Home Message, 495
Further Reading, 495

39 Cholinergic Transmission   496

Giancarlo Pepeu

Distribution and Function of the Cholinergic Systems, 496
    Cholinergic Transmission in the Peripheral Nervous System, 496
    The Cholinergic Transmission in the Central Nervous System, 497
    Role of Brain Cholinergic System in Learning, Memory, and Movement, 498
ACh Synthesis and Metabolism, 498
    ACh Precursors, 499
    ACh Hydrolysis by ChEs, 499
ACh Intracellular Storage and Release, 500
    ACh Release, 501
Cholinergic Receptors, 501
    Nicotinic Receptors, 501
    Muscarinic Receptors, 503
Drugs Acting on Cholinergic Receptors, 504
    Drugs Active on Nicotinic Receptors, 504

Drugs Active on Muscarinic Receptors, 504
Take-Home Message, 507
Further Reading, 508

40 The Serotonergic Transmission — 509
Maurizio Popoli, Laura Musazzi, and Giorgio Racagni

Functions and Distribution of the Serotonergic System in the Body, 509
   5-HT in the Nervous System, 509
   5-HT and the Cardiovascular System, 511
   5-HT in Gastrointestinal and Genitourinary Systems, 513
   5-HT in Metabolism and Endocrine System, 513
Synthesis and Metabolism of Serotonin, 513
Vesicular Storage, Release, and Extracellular Clearance of Serotonin, 514
   The 5-HT Reuptake System, 515
Classification of Serotonin Receptors, 515
Drugs Acting on Serotonin Receptors, 516
   Pharmacology of the $5-HT_1$ Receptors, 516
   Pharmacology of the $5-HT_2$ Receptors, 516
   Pharmacology of the $5-HT_3$ Receptor, 517
   Pharmacology of the $5-HT_4$ Receptors, 518
   Pharmacology of the $5-HT_6$, 518
   Pharmacology of the $5-HT_7$ Receptors, 518
Take-Home Message, 518
Further Reading, 518

41 Histaminergic Transmission — 520
Emanuela Masini and Laura Lucarini

Distribution and Function of the Histaminergic System, 521
   Histaminergic Neurons in the CNS, 521
   Functions of the Histaminergic System, 522
   Histamine in the CNS, 522
   Histamine in the Cardiovascular System, 523
   Histaminergic System in the Stomach, 523
   Histamine Effects on Smooth Muscles, 524
   Histamine and the Immune Response, 524
Synthesis and Metabolism of Histamine, 524
   Histamine Metabolism, 525
Storage and Release of Histamine, 525
   Pharmacological Modulation of Histamine Metabolism and Release, 526
Histamine Receptors and Their Pharmacological Modulation, 526
   Drugs Active on Histamine Receptors, 526
Take-Home Message, 528
Further Reading, 528

42 GABAergic Transmission — 529
Mariangela Serra, Enrico Sanna, and Giovanni Biggio

GABA Distribution, Synthesis, and Metabolism, 529
GABA Release and Reuptake, 530
GABA Receptor Classification, 531
$GABA_A$ Receptors, 531
   Drug Binding Sites on $GABA_A$ Receptors, 531
   Subunit Composition of Native Receptors, 533

Extrasynaptic $GABA_A$ Receptors: Phasic and Tonic
   Inhibition, 533
  Pharmacology of $GABA_A$ Receptors, 534
$GABA_B$ Receptors, 537
  Pharmacology of $GABA_B$ Receptors, 539
Schematic Overview of the Main Pharmacological Interventions on Gabaergic
   Synapses, 539
Take-Home Message, 540
Reference, 540
Further Reading, 540

43 Glutamate-Mediated Neurotransmission 541
Flavio Moroni

Glutamate Synthesis and Metabolism, 541
Glutamate Transporters, Vesicular Accumulation, and Signal
   Inactivation, 542
  Membrane Transporters, 542
  Vesicular Transporters, 544
  The Cystine–Glutamate Exchanger, 544
Glutamate Receptors, 544
  Glutamate Ionotropic Receptors, 544
  Metabotropic Receptors, 546
Glutamate Neurotransmission in Physiology and Pathology, 547
  Glutamate and Excitotoxicity, 547
  Glutamate and Depression, 549
Drugs and Excitatory Neurotransmission, 549
Glutamate-Mediated Neurotransmission in Brief, 551
Take-Home Message, 551
Further Reading, 551

44 Purinergic Transmission 553
Stefania Ceruti, Flaminio Cattabeni, and Maria Pia Abbracchio

Purines as Intercellular Transmitters, 554
  Source, Metabolism, and Release of Purines, 555
  Generation of Active Metabolites by ATP Hydrolysis, 555
Receptors for Purines, 557
  P1 Adenosine Receptors, 557
  P2 Receptors for ATP, 557
  P2X Ion Channel Receptors, 557
  P2Y GPCRs, 559
Biological Roles of Purines, 560
  Effects on the Cardiovascular System, 560
  Effects on the CNS and Peripheral Nervous System, 561
  Effects on the Respiratory System, 562
  Effects on Other Systems, 563
  Adenosine Effects Independent of Receptor Activation, 563
  Purines and Antitumor Therapy, 564
Take-Home Message, 564
Further Reading, 564

45 Neuropeptides 565
Lucia Negri and Roberta Lattanzi

Neuropeptide Synthesis, 566
  Propeptide Processing, 566

Storage and Secretion of Neuropeptides, 567
Peptidergic Transmission, 568
  Neuropeptide Receptors, 568
Neuropeptide Functions and Therapeutic Potential, 569
Take-Home Message, 569
Further Reading, 571

46 The Opioid System — 572
Patrizia Romualdi and Sanzio Candeletti

Endogenous Opioid Peptides, 572
  Opioid System Distribution, 574
Opioid Receptors, 575
  Signal Transduction, 575
Opioid Receptor Distribution and Effects, 576
  Modulation of Nociceptive Transmission, 576
  Respiratory Depression, 579
  Cardiovascular Effects, 579
  Effects on the GI Tract and Other Smooth Muscles, 579
  Effects on Food Intake and Body Temperature, 580
  Effects on the Immune System, 580
Tolerance and Physical Dependence to Opiates, 580
  Molecular Mechanisms of Cellular Adaptation to Chronic Exposure to Opiates, 580
  Molecular Mechanisms of Withdrawal, 582
  Regulation of Gene Transcription by Opioids, 582
Addiction to Opioids, 583
Take-Home Message, 584
Further Reading, 584

47 The Endocannabinoid System — 585
Daniela Parolaro and Tiziana Rubino

Cannabinoid Receptors, 585
  CB1R, 585
  CB2 Receptors, 586
  GPR55 Receptors, 586
  TRPV1 Receptors, 586
  PPARs, 586
Endocannabinoids, 586
  Anandamide, 586
  2-Arachidonoylglycerol, 587
  The Transporter, 588
Biological Functions of Endocannabinoid System, 588
  EC-Mediated Synaptic Plasticity, 589
  Other Biological Functions, 589
Drugs Affecting the ECS, 593
  Direct Agonists, 593
  Antagonists, 593
Take-Home Message, 595
Further Reading, 595

48 Pharmacology of Nitric Oxide — 596
Emilio Clementi

Chemistry and Biosynthesis of NO, 596
Biosynthesis of NO, 597
  NO Synthases, 597

Biochemistry of NO, 599
  NO and Activation of Guanylate Cyclase, 599
  NO and Inhibition of Cytochrome c Oxidase, 600
  Nitrosylation of Thiols, 601
  NO and MicroRNAs, 602
Systemic and Organ Effects of NO, 602
  Effects in the Cardiovascular System, 602
  NO and the Respiratory System, 603
  NO and Metabolic Diseases, 603
  NO in Central and Peripheral Nervous Systems, 603
  NO and Skeletal Muscle, 604
  NO and the Immune System, 605
Pharmacology of NO, 606
  Nitrate Vasodilators, 606
  Nitric Esters of Known Drugs, 606
  Stimulators of cGMP Action, 606
  NOS Inhibitors, 607
Take-Home Message, 607
Further Reading, 607

49 Arachidonic Acid Metabolism 608
Carlo Patrono and Paola Patrignani

Arachidonic Acid Release from Membrane Lipids, 609
Enzymatic Metabolism of Arachidonic Acid, 610
  PGH-Synthase Pathway, 610
  Cyclic Endoperoxide Metabolism, 613
  The Lipoxygenase Pathway, 615
  LTA_4 Metabolism, 616
  Transcellular Metabolism of PGH_2 and LTA_4, 617
Nonenzymatic Metabolism of Arachidonic Acid, 617
Eicosanoid Receptors, 618
  Prostanoid Receptors, 618
  Leukotriene Receptors, 620
Take-Home Message, 620
Further Reading, 621

SECTION 12 PHARMACOLOGY OF DEFENCE PROCESSES 623

50 Pharmacological Modulation of the Immune System 625
Carlo Riccardi and Graziella Migliorati

The Immune Response, 625
  The Concept of Autoimmunity, 626
Immunosuppressive Drugs, 626
  Anticancer Chemotherapeutic Agents, 626
  Immunosuppressive Drugs with Higher Specificity, 629
  Thalidomide, 630
  Calcineurin Inhibitors, 630
  Antibodies as Selective Immunosuppresants, 635
  Other Biological Drugs with Immunomodulatory
    Activity, 636
  Issues Associated with the Use of Antibodies as Pharmacological
    Agents, 637

Immunostimulant Drugs, 637
  Cytokines, 637
  Cytokines as Immunoadjuvants, 639
  New Molecular Targets: Phosphodiesterase 4 and p38 MAP Kinases Inhibitors, 639
Take-Home Message, 639
Further Reading, 639

51 Mechanism of Action of Anti-Infective Drugs 641
Francesco Scaglione

Antibacterial Drugs and Their Mechanisms of Action, 641
  Inhibitors of Peptidoglycan Synthesis, 643
  Inhibitors of Peptidoglycan Polymerization, 645
  Transcription Inhibitors, 646
  Translation Inhibitors, 646
  Inhibitors of DNA Synthesis and Replication, 649
  Inhibitors of Cytoplasmic Membrane Functions, 649
  Mechanisms of Resistance to Antibacterial Drugs, 650
Antifungal Drugs, 651
  Drugs Acting on the Cell Wall, 651
  Drugs Acting on the Cytoplasmic Membrane, 652
  Inhibitors of DNA and Protein Synthesis, 653
  Inhibitors of Enzymatic Metabolic Pathways, 653
  Mitotic Inhibitors, 653
  Mechanisms of Resistance to Antifungal Drugs, 653
Antiviral Drugs, 654
  Mechanisms of Action of Antiviral Drugs, 654
  Inhibitors of Viral DNA Replication, 654
  Drugs Active against Influenza Virus, 656
  Drugs Mainly Active on Hepatitis C Viruses (HCV), 657
  New HCV drugs, 658
  Other Drugs Active on HCV, 658
  Drugs Active against Hepatitis B virus (HBV), 658
  Drugs Active against Immunodeficiency virus (HIV), 659
  Reverse Transcriptase Inhibitors, 659
  HIV Protease Inhibitors, 660
  Inhibitors of Virus Entry into the Host Cells, 660
  Integrase Inhibitors, 660
Take-Home Message, 660
Further Reading, 661

SECTION 13 TOXICOLOGY AND DRUG INTERACTIONS 663

52 Introduction to Toxicology 665
Helmut Greim

Components of Risk Assessment, 666
  Hazard Identification, 666
  Dose–Response and Toxic Potency, 666
  Exposure Assessment, 667
  Risk Characterization, 668

Toxicological Evaluation of New and Existing Chemicals, 669
  General Requirements for Hazard Identification and Risk Assessment, 669
  Acute Toxicity, Subchronic Toxicity, and Chronic Toxicity, 669
  Irritation and Phototoxicity, 670
  Sensitization and Photosensitization, 670
  Genotoxicity, 670
  Carcinogenicity, 670
  Toxicity for Reproduction, 671
  Toxicokinetics, 671
  Mode and/or Mechanism of Action, 672
Test Guidelines, 672
  Alternatives to Animal Experiments, 672
General Approach for Hazard Identification and Risk Assessment, 673
  Evaluation of Mixtures, 674
  Evaluation of Uncertainties, 674
Toxicological Issues Related to Specific Chemical Classes, 674
Classification and Labelling (C&L) of Chemicals, 675
The Threshold of Toxicological Concern Concept, 675
The Precautionary Principle, 676
Conclusions, 676
Take-Home Message, 676
References, 676
Further Reading, 677

53 Drug Interactions 678
Achille P. Caputi, Giuseppina Fava, and Angela De Sarro

Pharmacokinetic Drug Interactions, 678
  Interactions during Drug Oral Absorption, 679
  Interactions during Drug Distribution, 681
  Interactions during Drug Biotransformation, 681
  Interaction during Drug Excretion, 683
Pharmacodynamic Drug Interactions, 684
Chemical Antagonism, 684
  Antidotes, 684
  Pharmaceutical Interactions and Incompatibility, 685
Interactions Between Herbal Remedies and Drugs, 685
Interactions Between Dietary Supplements and Drugs, 685
Take-Home Message, 686
Further Reading, 686

SECTION 14 DRUG DEVELOPMENT 687

54 Preclinical Research and Development of New Drugs 689
Ennio Ongini

Technological Innovation and Scientific Knowledge in Current Pharmaceutical
  Research, 689
The Research Strategies, 691
  Patent as a Driving Force for Innovation, 692
The Research Stages, 692
  Start-Up of a New Project and Identification of the Lead Compound, 692
  Pharmaceutical Chemistry: From the Mainstream Approach to the Molecular
    Modeling, 692

Drug Selection: *In Vitro* Assays and Experimental Models, 693
Biological Drugs, 694
Impact of Genomics Studies on Drug Research, 694
Pharmacokinetic Studies, 694
The Developmental Stages, 696
The Role of National and Supranational Legislation, 696
Scaling Up from Laboratory to Industrial Drug Preparation, 697
Choice of the Pharmaceutical Form, 697
Toxicology Studies, 697
The Drugs in 2020, 698
Take-Home Message, 698
Further Reading, 699
Web Sites, 699

55 Role of Drug Metabolism and Pharmacokinetics in Drug Development 700
Simone Braggio and Mario Pellegatti

DMPK in Drug Discovery, 700
Evolution of DMPK in Drug Discovery, 701
DMPK in the Regulatory Phase of Drug Development, 702
Safety of Metabolites, 703
PK in Clinical Trials, 704
Interindividual Variability, 705
DDI, 705
PK in Drug Formulation of Generic Equivalent Drugs, 706
New Perspectives, 706
Take-Home Message, 706
Further Reading, 707

56 Clinical Development of a New Drug and Methodology of Drug Trials 708
Carlo Patrono

Clinical Development of a New Drug, 708
Principles and Rules of Clinical Investigation, 708
The Path of Clinical Development of a New Drug, 709
Observational Studies and Randomized Clinical Trials, 710
The Key Role of the Primary Hypothesis, 711
The Choice of the Primary End Point, 711
Sample Size Calculation, 712
A Look At the Future, 712
Take-Home Message, 713
Further Reading, 713

INDEX 714

SUPPLEMENTARY CONTENTS (FOR ONLINE)

(The supplementary materials are correlated with the textbook and the relevance of each supplement to the content of the chapters is indicated by the symbol ☞ in the proper chapters)

E 1.1 Alternative or Nonconventional Therapies
Francesco Clementi

    Frequency of Use
    Phytotheraphy or Medicine that Makes Use of Plants
        Plant as Drugs
        Herboristic Preparations
        Aromatotherapy
        Toxicity of Natural Preparations
        Therapeutic Validity
    Homeopathy
        Principles
        Similia Similibus
        Doses
        Clinical Validity
        Is the Lack of Clear Data Related to a Lack of Research Funds?
    Possible Reasons for the Success of Nonconventional Medicine
        Ethical Problems Posed by Homeopathy
    References

E 1.2 The Placebo
Francesco Clementi

    Definition of Placebo
    Placebo Effect
    How to Study the Placebo Effect
    Biological Properties of Placebo and Nocebo
    Use of Placebo in Clinical Trials
    Ethical Problems Posed by the Use of Placebo
    The Placebo in Medical Practice
    References

E 2.1 The Drugs of the Ancients
Vittorio A. Sironi

Ancient Medicinal Preparations
The Treacle
Further Reading

E 2.2 Ehrlich's Chemotherapy: Receptor Theory and Clinical Experimentation
Vittorio A. Sironi

Methodology of Pharmacological Research and Beginning of Clinical
    Experimentation in Medicine
Birth and Evolution of the Receptor Theory
Further Reading

E 2.3 The Thalidomide Case
Vittorio A. Sironi

End of an Era and Birth of Pharmacovigilance
Molecules Never Die: The Return of the "Damned Drug" and Its New
    Therapeutic Options
Further Reading

E 3 Pharmacokinetics
Riccardo Fesce and Guido Fumagalli

A Rapid Journey with the Drug in the Organism
    The Drug Must Be Absorbed to Reach Its Target Organ
    The Drug Reaches Its Target Organ but It Is also Distributed to the
        Various Tissues in the Body
    The Drug Is Eliminated
Passage across Cellular Membranes
    Passive Diffusion across Cell Membranes
    Drug Transport across Cell Membranes
Diffusion of Drugs to Organs and Tissues
    Properties of the Most Important Cell Barriers
Suggested Reading
Drug Absorption and Administration Routes
    General Rules about Drug Absorption Speed
    Oral Route: Drug Absorption along the Gastrointestinal Tract
    Buccal (Sublingual) and Rectal Routes
    Systemic Parenteral Routes
    Further Routes
    Absorption Kinetics
    Drug Delivery: Formulations to Regulate the Timing and Site of Drug
        Release
Suggested Reading
Distribution
    Tissues Display Different Avidities for the Drug
    Apparent Distribution Volume
    Drug Binding to Plasma Proteins
    The Factors that Determine the Distribution Rate of Drugs to the Various
        Compartments
    Drug Elimination from the Organism

The Concept of Half-Life
The Concept of Clearance
Renal Excretion of Drugs
Hepatic Excretion and Enterohepatic Cycle
The Relation Between Metabolism and Excretion
Suggested Reading
Control of Drug Plasma Concentration
  The Time Course of Drug Plasma Concentration Following a Single Administration
  Time Course of Drug Plasma Concentration during Repetitive Administration
Multicompartmental Kinetics
  Equilibrium Concentrations Can Be Different Among Tissues
  The Rate of Equilibration of a Tissue with Plasma Depends on Local Plasma Flow and the Tissue Apparent Distribution Volume, V_d
  The Velocity of Equilibration of Tissues with Plasma Depends on Drug Binding to Plasma Proteins
  The Particular Case of the Nephron
  Drugs Can Redistribute among Compartments
  The Distribution of Drugs Administered by Inhalation
Corrections of the Therapeutic Regimen
  Normally Available Pharmacokinetic Data Are Average Values
  Varying Dosage as a Function of Body Weight and Physical Constitution
  Varying Dosage as a Function of Age
  Dosage Correction in the Presence of Hepatic Pathologies
  Dosage Correction in the Presence of Renal Pathologies
Suggested Reading

E 6.1 Factors that Modify Drug Metabolism
Enzo Chiesara, Laura Marabini, and Sonia Radice

Physiological Factors
  Differences Based on Species
  Differences Within Species
  Differences Related to Age or Gender
Pathological Factors
External Factors
  Diet
  Environment

E 6.2 Induction of Drug Metabolism
Enzo Chiesara, Laura Marabini, and Sonia Radice

Morphological and Biochemical Aspects of the Induction of Drug Metabolism
Molecular Mechanisms

E 6.3 Modulations of Efficacy Due to Interaction between Synthetic and Herbal Drugs
Laura Marabini, Sonia Radice, and Enzo Chiesara

Pharmacodynamic Interactions
Pharmacokinetic Interactions
Further Reading

E 7.1 Therapeutic Drug Monitoring
Dario Cattaneo

Optimization of Drug Therapy Guided by Pharmacokinetics
The Importance of AUC (Area Under the Plasma Concentration Curve)
Use of TDM through Single Sampling Strategies
Further Reading

E 7.2 The Distribution of Drugs Administered by Inhalation
Riccardo Fesce and Guido Fumagalli

E 8.1 Methods for Receptor Investigations
Gian Enrico Rovati and Valérie Capra

Concentration-Response (*In Vitro*) or Dose-Response (*In Vivo*) Curves
  Advantages
  Disadvantages
Ligand Binding
  Advantages
  Disadvantages
Molecular Biology
  Advantages
  Disadvantages
Label-Free Techniques
  Advantages
  Disadvantages
Further Reading

E 8.2 Receptor Binding Assays
Gian Enrico Rovati and Valérie Capra

E 9.1 Birth and Evolution of the Receptor Theory
Francesco Clementi, Guido Fumagalli, and Vittorio A. Sironi

References

E 9.2 Regulation of Receptor Response
Francesco Clementi and Guido Fumagalli

Desensitization
  Homologous and Heterologous Desensitization
  Desensitization of Ligand-Gated Ion Channels
  Desensitization of G Protein-Coupled Receptors
  Desensitization of Tyrosine Kinase Receptors
  Desensitization of Receptors with Intrinsic Guanylcyclase Activity
  Desensitization of Intracellular Receptors
Receptor Down-Regulation
  Early Downregulation
  Late Downregulation
  Receptor Endocytosis and Signal Transduction
  Desensitization Can Occur via Modulation of G Protein Activity
  Differences in Downregulation of Receptors Coupled to Phospholipase C Activation
  Downregulation of Tyrosine Kinase Receptors
Sensitization or Receptor Upregulation
  Drugs May Induce Expression of New Receptors
References

E 9.3 Intracellular Trafficking of Receptors
Francesco Clementi and Guido Fumagalli

Intracellular Traffic of Receptors
Receptor Localization to Specific Plasma Membrane Domains
Function of Lipid Rafts and Submembrane Matrix
Receptor Removal from the Membrane
    Endocytosis or Lateral Diffusion?
    An Example of Receptor Plasticity in Non-Nervous Tissues:
        β-Adrenergic Receptors in the Heart
    Pathologies Associated with Defects in Receptor Trafficking
References

E 10.1 Types of Memory
Cristiano Chiamulera

Learning
Priming
Conditioning
Operant Conditioning and Reinforcement

E 10.2 A Modern Definition of Memory
Cristiano Chiamulera

E 10.3 Conditioning as the Core of Addictive Behavior
Cristiano Chiamulera

E 10.4 Visualization Techniques
Cristiano Chiamulera

E 12.1 Cytosolic Proteins that Bind Ca^{2+} with High Affinity
Jacopo Meldolesi and Guido Fumagalli

References

E 12.2 Surface Ca^{2+} Channels: TRP, ORAI, SOCC Channels and their Functions
Jacopo Meldolesi and Guido Fumagalli

References

E 12.3 Ca^{2+} Pumps of Plasma Membrane and Intracellular Rapidly Exchanging Ca^{2+} Stores
Jacopo Meldolesi and Guido Fumagalli

References

E 12.4 Ca^{2+}-Binding Proteins of the ER Lumen
Jacopo Meldolesi and Guido Fumagalli

References

E 12.5 ER Channels: IP_3 and Ryanodine Receptors
Jacopo Meldolesi and Guido Fumagalli

References

E 12.6   **Local Ca^{2+} Spikes Can Evolve Into Oscillations And Waves**
Jacopo Meldolesi and Guido Fumagalli

   References

E 12.7   **Mitochondria: Semiautonomous Organelles that Need the ER to Operate**
Jacopo Meldolesi and Guido Fumagalli

   References

E 12.8   **The Nucleus is also Operative in Ca^{2+} Homeostasis**
Jacopo Meldolesi and Guido Fumagalli

   References

E 13.1   **Phosphatases with Dual Specificity as Regulators of MAPK Activity**
Lucia Vicentini and Maria Grazia Cattaneo

   Further Reading

E 13.2   **Role of MAPKs in Memory and Learning**
Lucia Vicentini and Maria Grazia Cattaneo

   Mutations in Genes Encoding ERK Cascade Components and Central Nervous System Diseases
   Further Reading

E 13.3   **ERK Activation by G Protein–Coupled Receptors**
Lucia Vicentini and Maria Grazia Cattaneo

E 15.1   **Neurotrophins**
Jacopo Meldolesi

   References

E 15.2   **Wnt and Hedgehog Signaling Pathways: Two Future Drug Targets**
Jacopo Meldolesi

   References

E 16.1   **Evolution of Pentameric Ligand–Gated Ion Channels**
Cecilia Gotti

   Models and Approaches Used to Study pLGIC Structure and Function
   Structure of Pentameric Ligand–Gated Ion Channels
   Agonist Binding and Channel Opening
      Role of Loop C in Channel Opening
      Channel Opening
   The Channel Gate
   pLGIC Mutations Causing Channelopathies
   References

E 16.2   **How to Identify the Amino Acids that Make Up the Inner Wall of Ligand-Gated Ion Channels**
Cecilia Gotti

   Reference

E 17.1 Orphan Receptors
Lucia Vallar, Maria Pia Abbracchio, and Lucia Vicentini

GPCR "Deorphanization," Implications for the Drug Discovery Process
Deorphanization of Receptors other than GPCRs: Example of Receptor Channels
Some Unsolved Issues and Future Directions
Further Reading

E 17.2 Structure and Conformation Modifications of GPCRs
Lucia Vallar, Maria Pia Abbracchio, and Lucia Vicentini

Further Reading

E 17.3 Drugs Active on Phosphodiesterases
Lucia Vallar, Maria Pia Abbracchio, and Lucia Vicentini

E 20.1 Adhesion and Platelet Activation
Giorgio Berton and Carlo Laudanna

Further Reading

E 20.2 Adhesion and Leukocyte Recruitment
Giorgio Berton and Carlo Laudanna

Further Reading

E 22.1 Pharmacoepigenomics
Diego Fornasari and Roberta Benfante

Epigenetic Modifications
    DNA Methylation
    Covalent Modifications of Histones and Chromatin Function
    Transcription Factors and the Recruitment of Histone Modifying Enzymes
Epigenetic Alterations and Human Pathologies
Epigenomic Drugs
    Inhibitors of DNA Methylation
    Inhibitors of DNA Demethylation
    Inhibitors of Histone Deacetylase
    Epigenomic Drugs and Brain
    Epigenomics and Adverse Drug Reactions
Future Perspectives
Further Reading

E 23.1 Genetic Polymorphisms
Diego Fornasari

Single Nucleotide Polymorphisms (SNPs)
    SNPs in Coding Regions
    Perigenic SNPs (pSNPs)
    Random SNPs
Copy Number Variations

E 24.1 Ligands of Intracellular Receptors
Adriana Maggi and Elisabetta Vegeto

Chemical Structure of Ligands and their Interactions with the Intracellular Receptors
Intracellular Receptor Genomic Interactions
Pharmacological Properties

E 26.1 History and Development of iPS Cell Research
Luciano Conti and Elena Cattaneo

Further Reading

E 26.2 Use of Blood Stem Cells in Hematology
Luciano Conti and Elena Cattaneo

Allogeneic Applications
Autologous Applications
Further Reading

E 26.3 Viral Vectors for Gene Therapy in Somatic Cells
Luciano Conti and Elena Cattaneo

E 26.4 Gene Therapy for Cystic Fibrosis and Tumors
Luciano Conti and Elena Cattaneo

Cystic Fibrosis
Tumors
Further Reading

E 27.1 How to Observe Ion Channel Currents in Real Time
Maurizio Taglialatela and Enzo Wanke

E 27.2 How to Study Interactions between Drugs and Ion Channels
Maurizio Taglialatela and Enzo Wanke

Heterologous Expression Systems
Functional Assays of Ion Channel Activity
   Electrophysiology-Based Assays
   Binding Studies
   Flux Assays
Optical Technologies
   Membrane Potential Indicators
   Ion-Selective Optical Probes
   Fluorescence Resonance Energy Transfer

E 27.3 Natural Peptide Toxins
Maurizio Taglialatela and Enzo Wanke

E 27.4 Physiopathology and Pharmacology of Muscular Contraction
Maurizio Taglialatela and Enzo Wanke

Striatal Muscle Contraction
Smooth Muscle Contraction
Role of Ryanodine Receptors in Muscular Contraction

E 27.5 Physiopathology of VGCCs: Genetic Studies in Animal Models And Humans
Maurizio Taglialatela and Enzo Wanke

    Knockout Mice Models of Ca_v1 Genes
    Knockout Mice Models of Ca_v2 Genes
    Knockout Mice Models of Ca_v3 Genes
    Perspectives

E 27.6 Drug-Induced Long QT Syndrome
Maurizio Taglialatela and Enzo Wanke

    Reference

E 27.7 TRP Channels
Maurizio Taglialatela and Enzo Wanke

    Further Reading

E 27.8 Nonvoltage-Dependent Na^+ Channels and their Roles as Mechanosensitive and Acid-Sensitive Transducers
Maurizio Taglialatela and Enzo Wanke

    ENaCs in Nephron
    ASICs Role in PNS: Pain and Sense of Touch
    ASICs at Synapses
    Further Reading

E 27.9 Anion Channels
Maurizio Taglialatela and Enzo Wanke

    Further Reading

E 27.10 Water Channels: Aquaporins
Maurizio Taglialatela and Enzo Wanke

E 27.11 Voltage-Independent Ca^{2+} Channels Activated by Store Depletion
Maurizio Taglialatela and Enzo Wanke

    References

E 27.12 Pharmacological Modulation of gap Junction Channels and Electrical Synapses
Maurizio Taglialatela, Enzo Wanke, and Francesco Clementi

    Structure and Function of Gap Junctions
    Physio-Pharmacological Control of Gap Junctions
    Electrical Synapses
    References

E 28.1 Systems for Drug Extrusion
Lucio Annunziato, Giuseppe Pignataro, and Gianfranco Di Renzo

E 28.2 Cardiac Glycosides
Lucio Annunziato, Giuseppe Pignataro, and Gianfranco Di Renzo

    Pharmacokinetics
    Mechanism of Action

Clinical Use
Toxicity
Endogenous Glycoside-Like Substances

E 28.3 Diuretic Drugs
Lucio Annunziato, Giuseppe Pignataro, and Gianfranco Di Renzo

Indirect Diuretics
Direct Diuretics
   Diuretics Acting on the Ascending Henle's Loop
   Diuretics of the Distal Convoluted Tubule
   Diuretics Acting on Collecting Ducts

E 28.4 The Na^+/Ca^{2+} Exchanger as a New Molecular Target for the Development of Drugs to Treat Cerebral Ischemia
Lucio Annunziato, Giuseppe Pignataro, and Gianfranco Di Renzo

E 29.1 Therapeutic Properties of GABA Transporter Inhibitors
Gaetano Di Chiara

E 29.2 Mechanism of Action of Amphetamine
Gaetano Di Chiara

E 29.3 Inhibitors of Amine Transporters and Antidepressant Drugs
Gaetano Di Chiara

Further Reading

E 30.1 Protease Classification and Nomenclature
Fabio Di Lisa and Edon Melloni

E 30.2 Autophagy
Fabio Di Lisa and Edon Melloni

Macroautophagy
Chaperone-Mediated Autophagy
Complementarity between Different Forms of Autophagy and Antagonism
   with Apoptosis
Pharmacological Modulation
Physiopathological and Therapeutic Relevance
   Oxidative Stress and Aging
   Neurodegeneration
   Cancer
Further Reading

E 30.3 Inhibitors of Angiotensin Converting Enzymes and their Action in Cardiovascular Pathologies
Francesco Clementi and Guido Fumagalli

Further Reading

E 30.4 Ubiquitin and Proteasome
Fabio Di Lisa and Edon Melloni

E 31.1 Plasminogen Activators and Cardiovascular Diseases
Francesco Blasi

Plasminogen Activators in Clinical Practice
Further Reading

E 31.2 Plasminogen Activators and Tumor Malignancy
Francesco Blasi

Genes Involved in the Metastatic Phenotype
Evidences of uPA Involvement in Tumor Invasiveness
uPA Activity and uPA/uPAR Interaction as Targets for Antimetastatic Therapy

E 31.3 uPAR, Cancer and Stem Cells
Francesco Blasi

References

E 32.1 p53: Regulator of Mitotic Cycle Progression and Apoptosis Inducer
Marco Corazzari and Mauro Piacentini

E 32.2 Mitotic Catastrophe
Marco Corazzari and Mauro Piacentini

E 32.3 Intrinsic Apoptotic Pathway Induced by Endoplasmic Reticulum
Marco Corazzari and Mauro Piacentini

Sensors of ER Stress
  PERK
  ATF6
  IRE1
ER Stress-Induced Cell Death
Further Reading

E 33.1 Mechanisms of Drug Resistance in Cancer Chemotherapy
Giovanni Luca Beretta, Laura Gatti, and Paola Perego

Mechanisms of Resistance to Conventional Antitumor Agents
  Alkylating Agents
  Platinum Compounds
  Antimetabolites
  Antimitotic Agents
  DNA Topoisomerase Inhibitors
Mechanisms of Resistance to Target-Specific Drugs
Additional Clinically Relevant Drugs
Further Reading

E 34.1 Structure, Organization, and Dynamics of Mitochondria
Clara De Palma

E 35.1 Pharmacology of the Mevalonate Pathway
Lorenzo Arnaboldi, Alberto Corsini, and Nicola Ferri

Mechanism of Action of Statins
  Pleiotropic Effects of Statins

Other Pharmacological Targets of the MVA Pathway
- Squalene Synthase
- Squalene Epoxidase
- Lanosterol Synthase
- Lanosterol 14-α-Demethylase
- Sterol-Δ-14 Reductase
- The Sterol 4,4-Demethylase Complex
- Sterol Δ-8-Δ-7 Isomerase
- Sterol-C5-Desaturase and Sterol-Δ-7-Desaturase
- Sterol Δ-24 Reductase

E 35.2 Pharmacology of the Biosynthesis of Fatty Acids
Lorenzo Arnaboldi, Alberto Corsini, and Nicola Ferri

Acetyl-CoA Carboxylase
Fatty Acid Synthase
Elongation and Desaturation of Fatty Acids
Fatty Acid Desaturases
FAS, SCD-1, and Tumors

E 35.3 Pharmacology of the Biosynthesis of Triglycerides
Lorenzo Arnaboldi, Alberto Corsini, and Nicola Ferri

Diacylglycerol Acyltransferase (DGAT)
Polyunsaturated ω-3 Fatty Acids
Niacin (Nicotinic Acid)

E 35.4 Role of LXR Receptors in Lipid Metabolism and Their Pharmacological Role
Lorenzo Arnaboldi, Alberto Corsini, and Nicola Ferri

Role of Peroxisome Proliferator-Activated Receptors (PPARs) in Lipid Metabolism

E 35.5 The Lipoproteins
Lorenzo Arnaboldi, Alberto Corsini, and Nicola Ferri

E 35.6 Lipid Transfer Proteins
Lorenzo Arnaboldi, Alberto Corsini, and Nicola Ferri

LCAT and Cholesteryl Esters
ACAT
ACAT and Alzheimer's Disease (AD)
Cholesteryl Ester Transfer Protein (CETP)
Inhibitors of Cholesteryl Ester Transfer Protein

E 36.1 The Controinsular Hormones
Paolo Moghetti and Giacomo Zoppini

E 37.1 Role of Astrocytes in Synaptic Transmission
Michela Matteoli, Elisabetta Menna, Costanza Capuano, and Claudia Verderio

Further Reading

E 37.2 Dynamic Organization of Synaptic Vesicle Pools
Michela Matteoli, Elisabetta Menna, Costanza Capuano, and Claudia Verderio

Pools of Synaptic Vesicles
The Biophysical Steps of the Release of the Neurotransmitter
   Vesicle Release from the Cytoskeleton
   Mobilization of Vesicles toward the Active Zone
   Vesicle Priming (Hemifusion)
   Fusion with the Presynaptic Membrane
   The Fusion Pore
Further Reading

E 37.3 Endocytotic Process of Synaptic Vesicles
Michela Matteoli, Elisabetta Menna, Costanza Capuano, and Claudia Verderio

Generalized Endocytosis
Clathrin-Mediated Endocytosis
Further Reading

E 37.4 Neuropeptide Secretion
Michela Matteoli, Elisabetta Menna, Costanza Capuano, and Claudia Verderio

Further Reading

E 37.5 Role of Lipids in the Exo-Endocytotic Cycle of Synaptic Vesicles
Michela Matteoli, Elisabetta Menna, Costanza Capuano, and Claudia Verderio

Further Reading

E 38.1 Dopamine and Parkinson's Disease
Pier Franco Spano, Maurizio Memo, M. Cristina Missale, Marina Pizzi, and Sandra Sigala

E 38.2 Brain Reward Circuits
Pier Franco Spano, Maurizio Memo, M. Cristina Missale, Marina Pizzi, and Sandra Sigala

E 38.3 Modulation of the Functional Properties of Dopamine Receptors by Interaction with Other Proteins
Pier Franco Spano, Maurizio Memo, M. Cristina Missale, Marina Pizzi, and Sandra Sigala

E 39.1 Methods for Investigating the Role of the Brain Cholinergic System in Learning and Memory
Giancarlo Pepeu

Approach 1: The Drugs
Approach 2: Knockout Mice and Lesions of the Cholinergic Neurons
The Methods
   Analysis of Cerebral Electrical Activity
   Cognitive Tests
   Measurement of ACh Release from Discrete Brain Areas during Behavioral Performances Using Microdialysis and Electrochemical Recording
References
Further Reading

E 39.2 Cholinesterases and Cholinesterase Inhibitors
Giancarlo Pepeu

Further Reading

E 39.3 Neuromuscular Blocking Agents
Giancarlo Pepeu

Further Reading

E 40.1 Drugs Acting on Serotonergic System
Maurizio Popoli, Laura Musazzi, and Giorgio Racagni

Antidepressant Drugs
Atypical Antipsychotics
Anxiolytic Drugs
Drugs for Migraine: Triptans
Antiemetic Drugs

E 40.2 Serotonergic System and Modulation of Pain Perception
Maurizio Popoli, Laura Musazzi, and Giorgio Racagni

Neurophysiology of Pain
Pain and Serotonin
Antidepressants for the Treatment of Chronic Neuropathic Pain
Triptans in Migraine Therapy

E 40.3 Pineal Gland, Melatonin, Serotonin, and Regulation of Circadian Rhythms
Maurizio Popoli, Laura Musazzi, and Giorgio Racagni

E 41.1 Histamine Receptors and Their Signal Transduction Pathways
Emanuela Masini and Laura Lucarini

H_1 Receptor
H_2 Receptor
H_3 Receptor
H_4 Receptor

E 42.1 Antiepileptic Drugs with a Gabaergic Mechanism of Action
Mariangela Serra, Enrico Sanna, and Giovanni Biggio

Benzodiazepines
Barbiturates and Deoxy-Barbiturates
Valproic Acid
Vigabatrin
Tiagabine

E 42.2 Physiological and Pharmacological Modulation of $GABA_A$ Receptor Gene Expression
Mariangela Serra, Enrico Sanna, and Giovanni Biggio

Further Reading

E 43.1 Long-Term Potentiation (LTP) and Long-Term Depression (LTD) of Excitatory Synaptic Transmission
Flavio Moroni

Possible Mechanisms Underlying LTP Induction and Maintenance

E 44.1 Purines and Ischemia
Stefania Ceruti, Flaminio Cattabeni, and Maria Pia Abbracchio

E 44.2 Drugs Acting on Purinergic System
Stefania Ceruti, Flaminio Cattabeni, and Maria Pia Abbracchio

$P2Y_2$ and A_3 Receptor Agonists to Treat Diseases with Altered Ion and Fluid Secretion
Adenosine A_{2A} Receptor Antagonists as Anti-Parkinson Agents
Adenosine A_3 Receptor Agonists as Anti-Inflammatory and Anticancer Agents
New Platelet $P2Y_{12}$ Receptor Antagonists as Antiplatelet and Antithrombotic Agents
Further Reading
  Sitology

E 45.1 Characteristics of Some Neuropeptides
Lucia Negri and Roberta Lattanzi

Melanocortins
Oxytocin and Vasopressin
Cholecystokinin
Somatostatin
NPY
Calcitonin Gene-Related Peptide
Bombesin
Vasoactive Intestinal Peptide
Corticotropin-Releasing Factor
Tachykinins
Neurotensin
Galanin
Melanin-Concentrating Hormone
Orexins
Cocaine and Amphetamine-Regulated Transcript
Agouti-Related Peptide
Prokineticins
Hypothalamic Factors That Control Hormonal Secretion
Neuropeptide W
Ghrelin
Leptin

E 45.2 Hypothalamus and Neuropeptides
Lucia Negri and Roberta Lattanzi

E 45.3 Neuropeptides and Nociception
Lucia Negri and Roberta Lattanzi

E 46.1 Opiate Drugs
Patrizia Romualdi and Sanzio Candeletti

Morphine
Codeine
Hydromorphone

Oxycodone
Heroin or Diacetylmorphine
Tramadol
Tapentadol
Levorphanol
Dextromethorphan
Pentazocine
Nalbuphine and Butorphanol
Meperidine
Diphenoxylate and Loperamide
Fentanyl
Sufentanil, Remifentanil, and Alfentanil
Methadone
Dextropropoxyphene
Buprenorphine
Naloxone and Naltrexone

E 46.2 Nociceptin
Patrizia Romualdi and Sanzio Candeletti

Motor Activity
Nociception
Addiction
Stress, Anxiety, and Depression
Feeding Behavior and Other Effects

E 46.3 Opioid Receptor Distribution
Patrizia Romualdi and Sanzio Candeletti

E 46.4 Neurobiological Basis of Acute and Chronic Pain
Patrizia Romualdi and Sanzio Candeletti

E 46.5 Opiate Addiction
Patrizia Romualdi and Sanzio Candeletti

The Addiction as a Disease: First Clinical Evidence
Natural History of Drug Addiction
Factors Involved in Addiction Development
Neurobiological Changes in Addiction
The Genetic of Addiction
Further Reading

E 47.1 Chemical Structure of Drugs Acting on the Endocannabinoid System
Daniela Parolaro and Tiziana Rubino

CB1/CB2 Agonists
CB1-Selective Agonists
CB2-Selective Agonists
CB1 Antagonists/Inverse Agonists
CB2 Antagonists/Partial Agonists
FAAH Inhibitors
MAGL Inhibitors
AEA Uptake Inhibitors
Endocannabinoid Biosynthesis Inhibitors

E 47.2 Cannabinoids and Drug Dependence
Daniela Parolaro and Tiziana Rubino

E 48.1 Chemistry of Nitric Oxide
Emilio Clementi

Interactions with Iron-Heme-Containing Proteins
Interactions with O_2
Interactions with Superoxide Anion (O_2^-) and Formation of Peroxynitrite
Interactions with Proteins Containing Nonheme Iron
Interactions with Ozone
S-Nitrosylation of Protein Cysteine Residues
Interactions with Amines and Tyrosines

E 48.2 Catalytic Activity, Molecular Features, and Regulation of NO Synthase
Emilio Clementi

Cofactors and Prosthetic Groups Responsible for NOS Catalytic Activity
Molecular Specificities and Regulation of NO Synthase Activity
    NOS I
    NOS II
    NOS III

E 48.3 Nitric Oxide and Control of Cell Death
Emilio Clementi

E 48.4 Role of NO in Inflammation and in Tumor Pathology
Emilio Clementi

NO, Chronic Inflammation and Autoimmune Diseases
NO, Leukocyte Adhesion and Chemotaxis
NO and Tumor Growth

E 49.1 Mechanism of Action of Glucocorticoids as Anti-Inflammatory Drugs
Carlo Patrono and Paola Patrignani

E 49.2 Mechanism of Action of Aspirin as Antithrombotic Drug
Carlo Patrono and Paola Patrignani

E 51.1 Mechanisms of Antibiotic Resistance in Bacteria
Francesco Scaglione

Multiple and Cross-Resistances
Mechanisms of Antibiotic Resistance
    Reduced Antibiotic Entry into Bacterial Cells
    Increased Antibiotic Extrusion Mediated by Efflux Pumps
    Antibiotic Inactivation
    Target Modifications
    Development of Alternative Metabolic Pathways

E 53.1 Interactions between Drugs and Grapefruit Juice
Achillle P. Caputi, Giuseppina Fava, and Angela De Sarro

Further Reading

E 54.1 **The Significant Contribution Given by Small Research Groups (Start-Ups and Spin-Offs)**
Ennio Ongini

    The Research Centers
    Industry–Academia Interplays
    Researchers Are a Key Factor for Success

E 54.2 **High-Throughput Screening**
Ennio Ongini

E 54.3 **The Birth of a Project on a New Drug**
Ennio Ongini

    The Medical Need
    Costs and Financial Issues
    Number of Patients and Orphan Drugs
    Scientific Knowledge and Technologies
    Competition
    Human Factor

E 54.4 **The Patent**
Ennio Ongini

    Product Patents
    Process Patents
    Use Patent
    Generics

E 54.5 **Biotech Drugs**
Ennio Ongini

E 54.6 **Toxicology Tests**
Ennio Ongini

    The Established Program
        Acute Toxicity
        Mutagenesis
        Toxicity Out of Repeated Administrations
        Reproductive Toxicity
        Cancerogenesis
        Special Toxicity
        Immunotoxicology

E 54.7 **Safety Pharmacology: How to Evaluate Whether a New Molecule Will Cause Side Effects on Vital Functions**
Ennio Ongini

LIST OF CONTRIBUTORS

Maria Pia Abbracchio, Professor of Pharmacology, University of Milano, Milan, Italy

Lucio Annunziato, Professor of Pharmacology, University "Federico II", Naples, Italy

Lorenzo Arnaboldi, Assistant Professor in Pharmacology, University of Milano, Milan, Italy

Roberta Benfante, Researcher, CNR Institute of Neuroscience, Milan, Italy

Giovanni Luca Beretta, Scientist in Molecular Pharmacology, Istituto Nazionale dei Tumori, Milan, Italy

Giorgio Berton, Professor of General Pathology, University of Verona, Verona, Italy

Giovanni Biggio, Professor di NeuropsicoPharmacology, University of Cagliari, Cagliari, Italy

Francesco Blasi, FIRC Institute of Molecular Oncology (IFOM), Milan, Italy

Carla Boccaccio, Professor of Histology, University of Torino, Torino, Italy

Simone Braggio, Researcher in Pharmacokinetics and Metabolism, Aptuit, Verona, Italy

Sanzio Candeletti, Assistant Professor of Pharmacology, University of Bologna, Bologna, Italy

Orazio Cantoni, Professor of Pharmacology, University "Carlo Bo", Urbino PU, Italy

Valérie Capra, Assistant Professor in Pharmacology, University of Milano, Milan, Italy

Costanza Capuano, Researcher in Neuroscience, Humanitas Research Center, Milan, Italy

Achille P. Caputi, Professor of Pharmacology, University of Messina, Messina, Italy

Flaminio Cattabeni, Professor of Pharmacology, University of Milano, Milan, Italy

Dario Cattaneo, Senior Scientist, Unit of Clinical Pharmacology, Luigi Sacco Hospital, Milan, Italy

Elena Cattaneo, Professor of Pharmacology, University of Milano, Milan, Italy

Maria Grazia Cattaneo, Assistant Professor of Pharmacology, University of Milano, Milan, Italy

Stefania Ceruti, Assistant Professor in Pharmacology, University of Milano, Milan, Italy

Cristiano Chiamulera, Professor of Pharmacology, University of Verona, Verona, Italy

Enzo Chiesara, Professor of Pharmacology, University of Milano and CNR Institute of Neuroscience, Milan, Italy

Emilio Clementi, Professor of Pharmacology, University of Milano, Milan, Italy

Francesco Clementi, Professor Emeritus of Pharmacology, University of Milano and CNR Institute of Neueroscince, Milan, Italy

Paolo M. Comoglio, Professor of Histology, University of Torino, Torino, Italy

Luciano Conti, Assistant Professor of Applied Biology, University of Trento, Trento, Italy

Marco Corazzari, Assistant Professor in Biology, University "Tor Vergata", Rome, Italy

LIST OF CONTRIBUTORS xlvii

Alberto Corsini, Professor of Pharmacology, University of Milano, Milan, Italy

Clara De Palma, Assistant Professor of Pharmacology, Luigi Sacco Hospital, Milan, Italy

Angela De Sarro, Professor of Pharmacology, University of Messina, Messina, Italy

Gaetano Di Chiara, Professor of Pharmacology, University of Cagliari, Cagliari, Italy

Fabio Di Lisa, Professor of Biochemistry, University of Padova, Padova, Italy

Monica Di Luca, Professor of Pharmacology, University of Milano, Milan, Italy

Gianfranco Di Renzo, Professor of Pharmacology, University "Federico II", Naples, Italy

Giuseppina Fava, Assistant Professor in Pharmacology, University of Messina, Messina, Italy

Nicola Ferri, Assistant Professor in Pharmacology, University of Milano, Milan, Italy

Riccardo Fesce, Professor of Physiology, University of Insubria, Busto Arsizio (Varese), Italy

Diego Fornasari, Professor of Pharmacology, University of Milano, Milan, Italy

Valerio Fulci, Assistant Professor in Biology, University "Sapienza", Rome, Italy

Guido Fumagalli, Professor of Pharmacology, University of Verona, Verona, Italy

Fabrizio Gardoni, Assistant Professor of Pharmacology, University of Milano, Milan, Italy

Laura Gatti, Researcher in Pharmacology, Istituto Nazionale dei Tumori, Milan, Italy

Silvia Giordano, Professor of Histology, University of Torino, Torino, Italy

Cecilia Gotti, Research Director, CNR Institute of Neuroscience, Milan, Italy

Helmut Greim, Professor of Toxicology, Technical University, Munich, Germany

Roberta Lattanzi, Assistant Professor of Pharmacology, University "Sapienza", Rome, Italy

Carlo Laudanna, Professor of General Pathology, University of Verona, Verona, Italy

Erzsebét Ligeti, Professor of Physiology, Semmelweiss University, Budapest, Hungary

Massimo Locati, Professor of Immunology, University of Milano, Milan, Italy

Laura Lucarini, Associate Professor in Pharmacology, University of Firenze, Florence, Italy

Giuseppe Macino, Professor di Biology, University "Sapienza", Rome, Italy

Adriana Maggi, Professor of Biotechnoloy, University of Milano, Milan, Italy

Alberto Mantovani, IRCCS Humanitas Clinical and Research Center, Humanitas University, Milan, Italy

Laura Marabini, Assistant Professor of Pharmacology, University of Milano, Milan, Italy

Emanuela Masini, Professor of Pharmacology, University of Firenze, Florence, Italy

Michela Matteoli, Professor of Pharmacology, Humanitas University and Director of CNR Institute of Neuroscience, Milan, Italy

Jacopo Meldolesi, Professor Emeritus of General Pharmacology, University Vita-Salute San Raffaele, Milan, Italy

Edon Melloni, Professor of Biochemistry, University of Genova, Genova, Italy

Maurizio Memo, Professor of Pharmacology, University of Brescia, Brescia, Italy

Elisabetta Menna, Researcher, CNR Institute of Neuroscience, Milan, Italy

Graziella Migliorati, Professor of Pharmacology, University of Perugia, Perugia, Italy

M. Cristina Missale, Professor of Pharmacology, University of Brescia, Brescia, Italy

Paolo Moghetti, Professor of Endocrinology, University of Verona, Verona, Italy

Flavio Moroni, Professor of Pharmacology, University of Firenze, Florence, Italy

Laura Musazzi, Assistant Professor in Pharmacology, University of Milano, Milan, Italy

Lucia Negri, Professor of Pharmacology, University "Sapienza", Rome, Italy

Ennio Ongini, Vice President, Nicox, Milan, Italy

Daniela Parolaro, Professor of Cellular and Molecular Pharmacology, University of Insubria, Varese, Italy

Paola Patrignani, Professor of Pharmacology, University "G. d'Annunzio", Chieti, Italy

Carlo Patrono, Professor of Pharmacology, Catholic University, Rome, Italy

Mario Pellegatti, Independent Pharmaceutical Scientist, Verona, Italy

Giancarlo Pepeu, Professor Emeritus of Pharmacology, University of Firenze, Florence, Italy

Paola Perego, Senior Scientist in Molecular Pharmacology, Istituto Nazionale dei Tumori, Milan, Italy

Mauro Piacentini, Professor of Cell Biology, University "Tor Vergata", Rome, Italy,

Giuseppe Pignataro, Professor of Pharmacology, University "Federico II", Naples, Italy

Marina Pizzi, Professor of Pharmacology, University of Brescia, Brescia, Italy,

Maurizio Popoli, Professor of Pharmacology, University of Milano, Milan, Italy

Giorgio Racagni, Professor of Pharmacology, University of Milano, Milan, Italy

Sonia Radice, Senior Scientist, Unit of Clinical Pharmacology, Luigi Sacco Hospital, Milan, Italy

Carlo Riccardi, Professor of Pharmacology, University of Perugia, Perugia, Italy

Patrizia Romualdi, Professor of Pharmacology, University of Bologna, Bologna, Italy

Gian Enrico Rovati, Professor of Pharmacology, University of Milano, Milan, Italy

Tiziana Rubino, Assistant Professor of Pharmacology, University of Insubria, Varese, Italy

Enrico Sanna, Professor of Pharmacology, University of Cagliari, Cagliari, Italy

Francesco Scaglione, Professor of Pharmacology, University of Milano, Milan, Italy

Mariangela Serra, Professor of Pharmacology, University of Cagliari, Cagliari, Italy

Sandra Sigala, Professor of Pharmacology, University of Brescia, Brescia, Italy

Vittorio A. Sironi, Professor of History of Medicine, University of Milano-Bicocca, Milan, Italy

Pier Franco Spano, Professor Emeritus of Pharmacology, University of Brescia, Brescia, Italy

Maurizio Taglialatela, Professor of Pharmacology, University of Molise, Campobasso, Italy

Lucia Vallar, Professor of Pharmacology, University of Milano, Milan, Italy

Flavia Valtorta, Professor of Pharmacology, University Vita-Salute San Raffaele, Milan, Italy

Annunciata Vecchi, Senior Scientist, Humanitas Clinical and Research Center, Milan, Italy

Elisabetta Vegeto, Assistant Professor of Pharmacology, University of Milano, Milan, Italy

Claudia Verderio, Senior Scientist, CNR Institute of Neuroscience, Milan, Italy

Lucia Vicentini, Professor of Pharmacology, University of Milano, Milan, Italy

Enzo Wanke, Professor Emeritus of Physiology, University of Milano-Bicocca, Milan, Italy

Thomas Wieland, Professor of Pharmacology, University of Heidelberg, Heidelberg, Germany

Giacomo Zoppini, Assistant Professor of Endocrinology, University of Verona, Verona, Italy

PREFACE

General and Molecular Pharmacology: Principles of Drug Actions introduces the reader to the basic and general pharmacology from a cellular and molecular perspective and provides a comprehensive description of the complexity of the drug actions in tissues, organs, and whole organism. In writing this book, we have followed a target-oriented rather than a drug-oriented approach. The aim is to point out how multidisciplinary investigations into drug targets allow researchers not only to develop novel therapeutic agents but also to uncover the physiological mechanisms underlying normal and abnormal functions of the human body.

This target-oriented approach to the study of pharmacology is intended to change the common perception of drugs as chemical entities merely producing effects on the organisms, and to highlight the interdisciplinary and dynamic nature of pharmacological research. With this approach, we hope to introduce young pharmacologists to a more integrate view of pharmacology and to foster a holistic approach to the analysis and treatments of diseases in which drugs are regarded as critical tools to discover new cell functions, mechanisms, and therapies.

To emphasize the critical role played by molecular and cellular studies in the development of new drugs and therapies, attention has been given to "nonclassical" drugs such as antibodies and decoy receptors, and to novel therapeutic tools and approaches such as nanoparticles, RNA pharmacology, stem cells, pharmacogenetics, and strategies of personalized therapy.

Emphasis has also been given to pharmacokinetics and to the processes of drug discovery and clinical experimentation and to the principles of toxicology. The medical and therapeutic effects of drugs have been considered on the basis of their mechanism of action.

The book consists of a printed and an electronic part[*]. The former includes the general and most codified aspects of pharmacology and is particularly suitable for students at their first encounter with the discipline; the latter consists of short essays addressing more specific subjects and methodologies and is intended for everyone having a good understanding of the basic principles and wishing to have more in-depth knowledge of specific topics (advanced students and teachers). The interrelations between printed and electronic parts of the book are clearly marked in the index and in the text.

Briefly, the book is organized as follows:

- Section 1 offers an introduction to the glossary and the history of pharmacology.
- Section 2 provides a detailed analysis of the pharmacokinetic principles that govern drug journey and fate in the human body; these represent essential concepts and tools to define patient-specific drug doses.
- Section 3 focuses on drug targets and introduces the reader to the different receptor classes and their signal transduction pathways, with an initial discussion of the mechanisms that modulate receptor adaptation and response.
- Section 4 presents a detailed discussion of individual classes of plasma membrane receptors and their transduction pathways.
- Sections 5 and 6 represent a novelty for a pharmacological textbook. In fact, Section 5 illustrates how gene expression mechanisms can serve as drug targets and how pharmacogenetic is leading to a personalized approach to therapy, while Section 6 illustrates the

---

[*]The electronic supplemental content to support use of this text is available online at http://booksupport.wiley.com.

huge therapeutic potential of stem cells (cells as drugs), discussing also current issues and future research directions in the field.
- Sections 7 through 10 offer a description of cellular processes and functions that play relevant roles in pathologies, and the mechanisms underlying the therapeutic and toxic effects of drugs acting on them. Also here, as throughout this book, functions, cell structures, and components are discussed as targets of already-established or newly discovered or not-yet identified drugs.
- Sections 11 and 12 cover intercellular communication, mainly at the level of nervous system, and defense mechanisms, respectively, describing the relevant targets for pharmacological intervention.
- Sections 13 and 14 illustrate the principles of modern toxicology and the strategies for drug development.

Based on our experience with the Italian editions of the book, we believe that *General and Molecular Pharmacology: Principles of Drug Actions* is suitable for graduate students in pharmacy, basic science, and medical and veterinary courses, for more advanced students in biology and biotechnology courses, and for PhD students in pharmacology, cell biology, molecular biology, and biochemistry. Moreover, it may represent a useful consultation book for post doctoral research scientists who seek a more in-depth understanding of the molecular interactions between drugs and their cellular and molecular targets, or for those who wish to understand strategies and issues in modern drug research and development. As a wise master builder, we laid a foundation, and another builds on it.

We are very grateful to all the authors who contributed to *General and Molecular Pharmacology: Principles of Drug Actions*. Being all active scientists, they have offered an updated and critical overview of their research fields bringing a flavor of the laboratory life to this book.

We are indebted to UTET, the publisher of the Italian version of this textbook, *Farmacologia generale e molecolare*, for helping and supporting us in the preparation of this English edition. We also wish to thank the Fondazione Emilio Trabucchi for supporting this work and Susanna Terzano for her serious and passionate editorial work on this English edition.

FRANCESCO CLEMENTI AND
GUIDO FUMAGALLI

SECTION 1

INTRODUCTION TO PHARMACOLOGY

1

ESSENTIAL LEXICON OF PHARMACOLOGY

FRANCESCO CLEMENTI AND GUIDO FUMAGALLI

By reading this chapter, you will:

- Become familiar with the pharmacological vocabulary

Pharmacology studies drugs and their interactions with living organisms. In a broad sense, a drug is any substance that induces functional changes in an organism through a chemical or physical action, regardless of whether the resulting effect is beneficial or detrimental to the health of the receiving organism. In a more strict medical sense, a drug is a substance used in the diagnosis, treatment, or prevention of diseases or as a component of a medication.

Thus, the fields of interest in pharmacology are multiple, including chemistry, molecular and cellular biology, pathology, clinic, and toxicology. Studying a drug may mean investigating the chemical properties underlying its biological activity, or the way to synthesize or extract it, or its effects on cells, tissues, and organs or on healthy and diseased human subjects. Pharmacology bridges basic and clinical research, as it provides tools for a rational therapy of known pathologies as well as empirical approaches for diagnosis and cure of obscure diseases (cure *ex adjuvantibus*).

As Leonardo da Vinci said, "The medical doctor will use drugs in an appropriate way when he knows what a human being is, what life and complexion are and what health is." In this sense, a pharmacologist is a physiologist, as she/he deals with substances and methods that modify physiological functions. However, she/he is also a pathologist capable of understanding chemical and molecular causes of pathologies, a biochemist and a molecular and cellular biologist using tools and methods of these disciplines to test drugs or vice versa employing drugs to investigate cellular and molecular processes, and a clinician using drugs to cure patients.

THE SOCIAL IMPACT OF PHARMACOLOGY

Using drugs to treat diseases and conditions has changed the life of humanity (see Chapter 2). Several factors have contributed to the remarkable improvements in population health that have occurred over the last century in developed countries: improved hygienic conditions, more widespread culture of prevention and health, richer and more balanced diet, and less physically wearing working conditions. However, such effects on life quality and duration would not have been so significant without the development of drugs and evidence-based approaches to human health care.

Drugs have contributed to the disappearance of many serious diseases. As an example, consider the number of daily deaths from bacterial pneumonia during winter months in general hospitals before World War II as compared to the almost total absence of this disease in current death records. Drugs have also reduced hospitalization, lowered incidence and gravity of several disease-induced permanent damages, and allowed to manage many serious and degenerative diseases. The most striking example of the role played by drugs in the control of serious diseases is the case of acquired immune deficiency syndrome (AIDS): in just a few years, specific antiviral drugs have turned the disease from "plague of the century" to a chronic illness compatible with an almost normal life. Pharmacology has also had an impact on surgery, as, for example, H2 histamine receptor antagonists and proton pump inhibitors have almost eliminated gastric and duodenal

---

General and Molecular Pharmacology: Principles of Drug Action, First Edition. Edited by Francesco Clementi and Guido Fumagalli.
© 2015 John Wiley & Sons, Inc. Published 2015 by John Wiley & Sons, Inc.

ulcers and the need for surgical gastrectomy and antibiotics have significantly reduced the number of appendectomies.

Drug success has been so impressive to induce part of the lay public to believe that there could be a "pill" for every disease or any uncomfortable symptom, causing an excessive "medicalization" of society and often the inability of people to accept that not all diseases are treatable and not all patients can be cured. Unfortunately, when such huge expectation on medicine collides with a failure, it may lead to a general distrust and irrational refusal of medical treatments pushing people toward the so-called alternative medicines. Based on the ethical principle that drug usage must be supported by scientific demonstrations of its effectiveness in patients, pharmacology has created strict rules that preclinical and clinical drug investigations (see Chapters 54, 55, and 56) have to comply with before drugs can be approved for clinical use. These rules are rigorously applied in drug development and use but are often ignored by producers and supporters of "alternative medicines." This lack of severe preclinical and clinical experimentations implies that "alternative" products and procedures often have not undergone a careful evaluation of benefits (if any) and risks. In the best cases, the evidence supporting these practices is anecdotal, reported by minor scientific journals often lacking serious review procedures (see Supplement E 1.1, "Alternative or Nonconventional Therapies").

Pharmacology has also a social impact. Consider, for example, drugs controlling fertility, emotions, and mental and physical performances or drugs affecting the individual's perception of the external world or the social plague of drug addiction.

Pharmacological treatment of diseases is a duty for social health systems; it must be guaranteed to the population but it also has costs. This has important implications, both positive and negative, for the relationships between drug companies and public health systems. All these scientific, economic, ethical aspects make pharmacology a discipline with very broad and loosen borders in which technical aspects are associated with equally important ethical issues.

ESSENTIAL LEXICON

For some of the readers, this book will be the first contact with pharmacology. For this reason, we include here a list of the terms that are typical of this discipline and frequently found in the text. For a more extended definition of each term, the reader should refer to individual chapters of the volume.

Active Substances

Drugs A drug is any substance that induces functional changes in an organism by chemical or physical action. In a biomedical context, a drug is a substance used in the diagnosis, treatment, or prevention of a disease or as a component of a medication. From a regulatory point of view, drug is any substance recognized in the Official National Pharmacopoeia or formulary. Nutrients or food constituents are not considered drugs.

From a historical point of view, the terms used to indicate drugs have undergone numerous conceptual and lexical changes. In the Middle Ages, the term *semplici* (simple, from the Latin *medicina simplex*) was used to indicate medicinal herbs (cultivated in *horto dei semplici*, the first one built in Salerno, Italy), minerals, and substances of animal origin with supposed therapeutic activity; the term "drug" was introduced in the eighteenth century to indicate a mixture of biologically active substances present in vegetal or animal organisms or parts of these organisms containing these substances (e.g., digitalis was a preparation made from foxglove *Digitalis lanata* leaves). The term derives probably from the Dutch word *droog* that indicated the barrels containing plants and dried spices imported from India, from which drugs were extracted.

Drug Name Drugs may be indicated by (i) the chemical name, which identifies the chemical and molecular composition; (ii) the generic or official name, meaning the name under which the drug is licensed by the manufacturer and internationally identified, with the initial in lower case; and (iii) the trade name or proprietary/brand name (patented) indicating the medicine containing the drug, with the initial in large upper case. Two examples:

1. Chemical name, 1-(1-methylethylamino)-3-(1-naphthyloxy) propan-2-ol HCl; generic name, propranolol; proprietary name, Bedranol, Inderal, InnoPran, Tesnol, Syprol, and so on
2. Chemical name, acetylsalicylic acid; generic name, aspirin; proprietary name, Aspirin, Ascriptin, Aspro, Disprin, and so on

Medicinal Preparation Specialty indicates the formulation labeled with the trade name under which an active ingredient is sold. Active ingredients can be prepared in different formulations: tablet, syrup, ointment, and so on. The active ingredient is generally associated with inert components called excipient. The excipient should not alter the pharmacological properties of the active ingredient, but may modify its pharmacokinetics. The preparation can be carried out on an industrial scale, according to the Official National Pharmacopoeias or in accordance with standards approved by the competent authorities. In this latter case, it is named "medicinal preparation" or "pharmaceutical specialty." When the patient needs an individual drug preparation, this can be prepared by the pharmacist, a compounding pharmacist in the United States.

Generic Drugs   It is a drug/substance whose patent is expired (usually 15–20 years after its approval). A generic drug is subjected to simplified authorization procedures by the competent authorities, which require the demonstration that it is pharmaceutically equivalent in terms of purity, stability, formulation and route(s) of administration, and its pharmacokinetic profile is comparable (±20%) to the product of reference. The price of generic drugs is lower, given the very limited cost of research and development. Currently, in developed countries, prescriptions for generic drugs are in the range of 50–75%.

Biological Drug and Biosimilar Medicine   It is a medicine containing macromolecules (proteins, glycoproteins, polysaccharides) made by or derived from living organisms. The term includes antibodies, interleukins, vaccines, and even cells. The active substance in biological drugs is usually more complex, larger, and usually more target specific than in classical drugs. Biological drugs are patented and since some patents have already expired, "biosimilar medicines" are now emerging. For biosimilars, comparison with the reference medicine is difficult because of the great variability and complexity of the preparation and purification procedures. Therefore, national legislations for the approval of biosimilars are more demanding than for chemical drugs (similarity does not equal identity) and involve stringent criteria for quality, safety, specificity, and effectiveness.

Medicinal Herbs or Herbal Remedies   These are plants or parts of them that have (or are deemed to have) healing and beneficial properties. Many active ingredients are contained in plants or animals or parts of them. Throughout the centuries, a series of *semplici* (simple) have been selected that—cultivated, processed, and stored in an appropriate manner—contain mixtures of active substances that are relatively reproducible in terms of quantity, composition, kinetics, bioavailability, and clinical effect. Production, preparation, and marketing of these preparations have been subjected to strict regulations in most countries (Chapter 2).

Many issues are associated with the use of medicinal herbs. Very often, their composition is not known and may vary depending on cultivar chosen and culture conditions. The stability of the active components during preparation and storage is generally not known either. In addition, contamination by bacteria, molds, or toxic compound may change their initial composition. The presence of heavy metals, in particular in some Indian medicine preparations, is responsible for a number of long-term intoxications. Frauds and adulterations are very common but difficult to uncover.

In recent years, complex sociocultural factors, rather than evidence-based criteria, have encouraged the spreading of the personalized use of "herbs" (see Supplement E 1.1).

Homeopathic Remedies   Homeopathy has been proposed by German physician C.F. Hahnemann in the early nineteenth century on the basis of a philosophical interpretation of nature and medicine that could be summarized with these words: "uses in sickness a medicine that is able to reproduce another artificial disease as close as possible to the previous one so that it will be healed: *similia similibus curantur*" and "homeopathic medicine should be healthier as the dose is reduced." In homeopathy, dose can be diluted up to 10^{-60} mol/L and above. These drug concentrations are not consistent with current scientific knowledge on the action mechanism of drugs and with scientific evidences of efficacy (see Supplement E 1.1).

Placebo   It is an inert substance devoid of significant biological effects on the organism given to "please" the patient. It is often used in controlled clinical trials to distinguish the drug effects from those due to the context in which the drug is administered. It can also have a negative effect, called nocebo (see Supplement E 1.2, "The Placebo").

Pharmacological Disciplines

Pharmacology is the branch of biomedical science that studies drugs and their interactions with living organisms. Due to the diversity of objectives, methodologies, and tools, pharmacology is divided into specialized branches.

General and Cellular and Molecular Pharmacology   It is the branch that analyzes the general mechanisms underlying drug action in the body and in particular at molecular and cellular level. As said by Benet in the introduction to the ninth edition of Goodman and Gilman, general pharmacology can be divided into pharmacodynamics, namely, what that the drug does to the body, and pharmacokinetics, namely, what the body does to the drug. This branch provides the knowledge on mechanisms of action of drugs and the rationale for their therapeutic use and individual adjustment of the dose.

Pharmacognosy   It is the study of drugs deriving from plant and animal sources.

Pharmacokinetics   It is the study of the events underlying drug absorption, distribution, metabolism, and elimination. Manipulation of these aspects allows to adjust drug dosage to individual subjects.

Clinical Pharmacology   It is the study of drug effectiveness and safety in humans. It provides established and validated scientific methods to design and evaluate clinical studies.

Pharmacogenetics   It investigates the genetic differences that can affect the individual or population responses to drugs. It is expected to provide important insight into the action mechanism of drugs and the etiology of diseases. A new branch is pharmacoepigenetics, which studies drugs capable of affecting chromatin and DNA modifications that regulate genomic functions involved in drug responses.

Pharmacogenomics It is the study of how genes affect the response to a drug. It combines pharmacology and genomics to establish drug therapies tailored on individual subjects and discovery of novel therapeutic targets. The terms "pharmacogenomics" and "pharmacogenetics" are often used interchangeably to describe studies investigating how genes affect individual responses to drugs.

Chemotherapy Literally, it is a therapy with a drug of chemical synthesis. Actually, it is a therapy (made of synthetic or extracted substances) aimed at destroying harmful living agents such as infectious agents or cancer cells.

Toxicology It is the study of the harmful effects of drugs or exogenous substances. Toxicological studies contribute to drug profiling.

Pharmacovigilance It is the discipline monitoring the occurrence of toxic effects after drug commercialization. It allows detection of rare toxic events and evaluates the appropriateness of prescriptions dispensed in clinical practice.

Pharmacoeconomics It is the study of drug cost/benefit from the economic point of view. This branch is becoming increasingly important given the difficulties encountered by countries to bear the costs of medical expenses. It also studies costs of production, fairness of sale price, and economic impact of drug use. It is required for proper management of health policies.

Drug–Receptor Interactions

Receptor It is the molecule whose function is modulated by interaction with a drug. The concept of receptor refers to the principle introduced by Paul Ehrlich: *corpora non agunt nisi fixata* (substances do not act if not bound). The substance that binds to its receptor in general is called "ligand." In biology, the term receptor also indicates a molecule used by cells to bind a diffusible substance released in the extracellular space by other cells for communication (i.e., first messengers, such as neurotransmitters, hormones, autacoids). The ligand can bind its receptor at orthosteric and/or allosteric sites. Occupation of orthosteric sites modifies the receptor function, usually activating it; ligand occupation of allosteric sites does not activate/inhibit receptor function, but modifies the receptor response to its physiological ligand.

Affinity It is the measure of drug ability to bind to its receptor. It is inversely correlated with the dissociation constant of the drug–receptor complex. Operatively, it is the negative logarithm (base 10) of the concentration required to bind 50% of the binding sites. The higher the affinity, the lower the drug concentration needed to bind the receptor to a given extent.

Competition It is the condition occurring when two drugs recognize the same binding site. Reciprocal hindrance between two drugs reduces their apparent affinity for the binding site. If one of two drugs binds irreversibly to the binding site, the interaction is called noncompetitive.

Agonist It is a drug that increases the probability of a biological response by activating or stabilizing the receptor active state.

Antagonist It is a drug that binds to a receptor and does not activate it but prevents the agonist from interacting with the binding site. The term should not to be confused with "functional antagonism," which specifically indicates cases in which two drugs act on different receptors mediating opposite responses (e.g., the effects of drugs active on the sympathetic system are often functionally antagonized by drugs active on the parasympathetic system).

Partial Agonist It is a drug that only partially activates a receptor.

Inverse Agonist It is a drug binding to a receptor that is active also in the absence of a ligand; an inverse agonist reduces the probability of the receptor to be in its active state and, in this case, induces a pharmacological response opposite to that of an agonist.

Allosteric Agonist or Antagonist It is a drug devoid of intrinsic activity (i.e., not able to activate or inhibit a receptor) that changes the receptor efficacy and/or potency upon agonist binding.

Effectiveness It is the magnitude of the effect generated. The term was coined by Stephenson to compare agonists that, upon binding to the same extent to a receptor population, produced effects of different entity. In evaluating the effectiveness of different drugs, the measure should relate to the same biological effect. Effectiveness should not be confused with "therapeutic utility."

Potency It is the dose (or concentration) required to produce an effect of a given entity (e.g., 50% of the maximal effect). It is often considered the least important property of a drug since low potency can be overcome by increasing the dose.

Selectivity At molecular level, it is the ability of a drug to interact with a limited number (preferably only one) of macromolecules. At the level of organism, it is the ability to generate responses attributable to the modulation of a single functional system. Selectivity may decrease by increasing drug concentration, since at high concentrations a drug may bind to low affinity targets.

Desensitization It is a reduction in the capacity of a receptor system to generate a response or to transduce a signal upon prolonged exposure to an agonist.

Downregulation It is a decrease in receptor number generally occurring upon prolonged activation by an agonist. It is different from desensitization that indicates a decreased capacity of signal transduction of the receptor molecule.

Tolerance It is the reduced ability of an organism to respond to repeated administrations of a drug. Tolerance can be consequence of receptor desensitization and/or downregulation. It should not be confused with "addiction." A tolerance that develops rapidly is called tachyphylaxis. A drug can induce tolerance and tachyphylaxis to another drug.

Sensibilization It is a gradual increase in response following repeated drug administrations.

Addiction It is the compulsive use of a psychoactive drug as a result of its repeated and continuous use. Abrupt discontinuation of the drug causes a withdrawal syndrome.

Side Effects These are effects of a drug different from those required for therapeutic purpose. Harmful side effects are called adverse drug reactions. They may result from the same mechanism of action producing the therapeutic effect (e.g., tachycardia for anticholinergic drugs used for intestinal spasms) or from a totally different mechanism. They can occur at therapeutic doses (e.g., orthostatic hypotension for organic nitrates used for cardiac angina), but they are more commonly observed at higher doses. For the most severe side effects, the term "toxic effects" or "adverse reactions" is more appropriate.

Measure of the Clinical Response

Risk/Benefit Ratio It is the ratio between patient's health risk (due to side or toxic effects) and beneficial effects produced by a drug. It is a relative evaluation, often subjective, that must take into account severity of the treated disease, benefits expected, and quality of life as perceived by the treated patient.

Therapeutic Index It is the ratio between the dose (or concentration) that can produce useful effects (ED: Effective Dose) and the dose (or concentration) responsible for side/toxic effects (e.g., toxic dose (TD) or lethal dose (LD)).

Meta-analysis It is a statistical method to compare and combine results of different studies and experiments with the aim of extrapolating statistically significant conclusions on a particular therapeutic effect of a drug treatment.

TAKE-HOME MESSAGE

- Pharmacology is a quantitative discipline with specific terminologies.
- Pharmacology has a number of interests ranging from molecular biology to economics.

2

A SHORT HISTORY OF PHARMACOLOGY

VITTORIO A. SIRONI

By reading this chapter, you will:

- Gain an overview of the history of mankind's search for remedies to treat diseases
- Learn how pharmacology began, how it has evolved over the centuries, and its current directions

BIRTH AND HISTORICAL DEVELOPMENTS OF PHARMACOLOGY

From Magical and Natural Remedies of Ancient Medicine to Arabic Alchemy

Humans have been searching for remedies to relieve pain and cure diseases since ancient times. The initial approach was empirical, simply based on the casual discovery of the beneficial properties of a procedure or an herb. Subsequently, it became speculative, following the belief that illnesses were caused by foreign entities—or evil spirits—taking possession of an individual's body and/or mind. For that reason, healing was supposed to require evil spirits to be driven away through rituals consisting of exorcisms and magical practices.

Egyptians knew plant and animal derivatives with therapeutic properties (castor oil, senna, pomegranate, tannin, opium, aloe, and mint are among the substances described in the *Papyrus Ebers*, the most extensive text of ancient medicine, dating from 1550 B.C.), but they used to accompany their preparation and administration with magic spells and exorcisms to chase away demons causing sickness. Similarly, Hebrews developed practices to prevent and cure diseases based on precise hygiene rules as well as on several therapeutically active plants, such as sycamore, hyssop, cumin, and pine tree, but at the same time ascribed to religious elements typical of their lifestyle a significant therapeutic role.

The humoral theory, developed by Hippocrates (460–337 A.C.) and further elaborated by Galen (129–199 D.C.), introduced the idea that diseases were due to an imbalance of four humors making up the body: blood, phlegm, yellow bile, and black bile. Each humor corresponded to one of the four primordial elements (earth, water, air, and fire), as well as to one of the fundamental qualities (cold, warm, moist, and dry), one of the seasons (spring, summer, autumn, and winter), and one of the life stages (childhood, youth, maturity, and old age). In ancient Greece, pharmacy (the Greek word for drug is *pharmakon*, which means both remedy and poison) was initially developed in the context of the dietetic practice as a collection of indications correlating properties of selected medicaments to the four temperaments (sanguine, phlegmatic, choleric, and melancholic) in ill organisms.

In the *Corpus Hippocraticum*, for the first time, we find a systematic description of the rules to collect "simple" plants (hellebore, opium, belladonna, veratrum, rue, and mint), the recipes to prepare medicaments (classified as laxatives, narcotics, diuretics, and emetics), and the indications for their use to treat different diseases.

Orderliness and naturalistic rationalism characterize also the "biology" of both Aristotle (384–282 A.C.) and his pupil Theophrastus (370–286 A.C.). Theophrastus wrote *Historia plantarum*, which is not only a neat and analytic attempt to taxonomically classify plants but also a real treatise on phytotherapy where he accurately describes the effects of poppy, hemlock, mandrake, and hellebore. Indeed, *Historia plantarum* can be regarded as a text anticipating the treatises on the pharmacology of medicinal plants that starting from the Middle Age will create the foundations of the apothecary profession.

General and Molecular Pharmacology: Principles of Drug Action, First Edition. Edited by Francesco Clementi and Guido Fumagalli.
© 2015 John Wiley & Sons, Inc. Published 2015 by John Wiley & Sons, Inc.

In ancient Rome, medicine and pharmacology proceeded along the path traced by Greeks. In their writings, Aulus Cornelius Celsus, Scribonius Largus, and Plinius the Elder summarized the knowledge gained over time. However, the one who revised and took it further, leaving a mark bound to last for centuries, was Dioscorides Pedanius, a Greek military surgeon and later a Roman citizen in the first century A.D. Dioscorides wrote a five-volume book, universally known as *De materia medica* (the title of the first Latin translation published in 1478 with annotations by Pietro Andrea Mattioli), that remained as a reference book in pharmacology until the eighteenth century. His opus, subdivided in 827 chapters, reports all known medicaments of vegetal (650), animal (85), and mineral (50) origin.

Similarly, Galen's *De simplicium medicamentis et facultatibus*, a book concerning therapy, had a universal and long-lasting influence. In his book, Galen listed 473 medicaments of plant origin that were going to represent the foundations of medicine throughout Europe for more than one and a half century. According to Galen and the humoral theory he embraced, curing a disease means administering treatments with actions and effects opposite to those of the disease, during all the different stages of the illness. Thus, he indicates "warming" agents to treat "cold" diseases or disease stages, and vice versa, "cooling" drugs to treat "hot" disease or stages. The maxim *Contraria contrariis curantur* (which means "cure the opposite with the opposite") represents the fundamental principle of Galen's pharmacotherapy, which also recommends the doctor to prepare the medicament (which is still called "galenical," after his name) himself and to administer it carefully and thriftily.

Among the multiple and various types of medicamentous preparations used at that time, we remember *catapozi* (simple pills); plasters; liniments; *trocisci* (powder poultice with vinegar and wine); *terra rara* ("rare clays," also called *terra sigillata* "sealed clays" because they were stamped with the seal indicating their origin), clays that were credited with special therapeutic properties; and the *triaca* or *teriaca*, a mixture of several components (mainly minced vipers) used to cure intoxications and poisonings. Its recipe was kept secret and its preparation was considered an initiation practice, revealed only to few authorized persons (see Supplement E 2.1, "The Drugs of the Ancients").

The fall of the Western Roman Empire, in 476 A.D., marks the beginning of the Middle Age, with the emerging Arab civilization gaining the supremacy in medicine and pharmacology knowledge.

Arabic medicine was based on the use of medicaments. As recorded by Albucasis in his writings, Avicenna's and Avenzoar's medicine avoided bloody practices typical of the "ferramentaria" (using iron and fire) surgery, in favor of procedures belonging to the "medicamentaria" (using medicaments) surgery (where surgery has the meaning of "handling").

Arab culture spread in the Western world "alchemy" (literally meaning "mixing"), an empirical practice that helped to extend the knowledge on some protochemical procedures (such as distillation, heating, and "bain-marie") suitable for obtaining compounds (also some new, like ammonia) and at the same time prompted the pursuit of utopistic goals, fruits of imagination, or dreams, such as seeking magic-natural answers to health problems (long-life elixir) and poverty (philosophical stone).

From Monastic Medicine to Botanical Gardens

During the Medieval barbarization, the conservative spirit of the monastic orders was pivotal to the preservation of traditions, transmission of knowledge, and development of new ideas. Monks preserved and copied ancient manuscripts, pondered their contents, and elaborated new models. Medical assistance provided in monasteries to *pauperes infirmi* represented a model for future hospitals, and monastic herboristery based on the cultivation of simple herbs further enriched pharmacotherapy.

Under the influence of monastic medicine, in the eleventh and twelfth centuries, the Schola Medica Salernitana (Salernitan Medical School) was founded, the first center of secular medicine, at the crossroads of Greco–Roman traditions and Arabic elaboration. Three books belonging to that school, the *Regimen Sanitatis Salernitanum* (a collection of poetry providing advice to preserve health and live long), the *Antidotarium* by Nicolò Salernitano, and the *Ars medendi* by Cofone, influenced European medicine for a long time. In particular, the *Antidotarium* listed methods to prepare medicaments and established the criteria for their proper, not aleatory, use, representing the starting point for all subsequent pharmacopoeias.

Shortly afterward, professional corporations began to appear, and medic and apothecary became distinct and juridically defined professions. The *Ordinationes*, dictated in 1240 by the Emperor Frederick II (1191–1250), King of Sicily, provided a novel institutional as well as scientific definition of medicine and pharmacy.

The birth of the universities, as places where established scientific knowledge was taught and learned, as well as the invention of the printing press (1450), highly accelerated the spread of new ideas and discoveries, speeding up the development of medicine and pharmacy.

Medical and pharmaceutical knowledge expanded rapidly and enormously. The fifteenth century saw a renewed interest in Plinius the Elder's *Naturalis historia* (printed in Venice in 1469) and, as mentioned before, the Latin translation of Dioscorides' writing (by Pietro Andrea Mattioli in 1478). Moreover, the discovery of the New World (1462) brought new plants of therapeutic interest (such as jailap, ipecacuanha, and guaiac) to the Old World. These new medicinal plants can be found depicted and described in several *Hortus sanitatis* (books on medicinal plants) printed during Renaissance alongside herbs already known in the Middle Age.

In 1498, in Florence, the first pharmacopoeia, entitled *Ricettario Fiorentino*, is printed in vernacular. This is a handbook for apothecaries describing all drugs and medicaments and their properties, as well as the procedure to prepare them. This renaissance of pharmacology culminated in 1533 with the foundation of the first botanical garden by Francesco Bonafede, in Padua.

From Anatomical Renaissance to the "Experienz": Paracelsus' Spagyric

In the history of science, 1543 is a revolutionary year: the seven books of the *De humani corporis fabrica* by Andreas Vesalius (1514–1564) were published in Basel, while in Norimberga, the six books of the *De revolutionibus orbium coelestium* by Nicolaus Copernicus (1473–1543) were printed. Vesalius' book started the "anatomic revolution," reinterpreting human microcosm, whereas Copernicus initiated the "astronomical revolution," reinterpreting universe macrocosm.

In the middle of the sixteenth century also, the science of drugs underwent a process of revision and innovation, with the beginning of new studies. The "pharmacological revolution" is strictly related to the new interpretation of medicine proposed by a reformist physician, Theophrastus Paracelsus (Fig. 2.1) (1493–1541), who culturally formed in the midst of the antiauthoritarian protest accompanying the Protestant Reformation.

Theophrastus Bombast von Hohenheim (the full name of Paracelsus), also called "the Luther of medicine," criticized and challenged the classical medicine tradition of Aristotle, Galen, and Avicenna, which he considered a source of bookish knowledge in contrast to the knowledge that could be gained by reading the "book of nature." According to Paracelsus, the real teaching is not found in textbooks but in the direct experience (*Experienz*) of diseases and their remedies, which spring from the bowels of "Mother Earth" in the form of thermal and mineral waters (*aqua vivimus*) and in the metals (*metalla*) they contain.

In the "health workshops," representing the first laboratories of the rising protochemistry, Paracelsus' alchimia or *chymia*—along with astronomy, philosophy, and virtue—is one of the pillars of real medicine. He called it "spagyric" (from the Greek words *spào*, to extract, and *agèiro*, to collect) to indicate the two main activities on which alchemy relies: analysis and synthesis. According to Paracelsus and his followers, true physicians have to focus on "extraction" just like miners and on "harvest" like farmers; medicine is a "farming of human beings" or a "metallurgical alchemy" producing effective drugs (Fig. 2.2).

The role of a pharmacist in Paracelsus' view is radically different from that of the traditional apothecary; he does not

FIGURE 2.1 Portrait of Philippus Aureolus Theophrastus Bombastus von Hohenheim, better known as Paracelsus (1493–1541). Unknown painter.

wear a long coat, as the latter does, but "a blacksmith leather apron," like a miner. In his laboratory, he pokes the fires under crucibles and alembics, melting, distilling, and sublimating metals, and from the "huge teaching deriving from metal transformation," he obtains new compounds, completely different from the traditional "simple" ones.

Paracelsus introduced new concepts. He was the first to use the word "alcohol" (from the Arabic word *al-kohl*, the matter of black coal) to indicated the "spirit" of the black wine. From the idea of *chaos*, he had the initial intuition that his follower Van Helmont further elaborated into the scientific concept of *gas*. From his personal experience with patients, whose suffering he soothed with *laudanum* (opium tincture), he derived the aphorism *similia similibus curantur* ("like cures like," reversing Galen's theory of opposites), which in Paracelsus' view had more an anthropological and curative meaning rather than a pharmacological and therapeutic one. For Paracelsus, "like cures like" means that infirms are cured by infirmary assistants, the poor by doctors of the poor, and farmers by country doctors. In this way, the concept "a good doctor is the first medicine" began to form, the first clue of the now known "placebo" effect.

FIGURE 2.2 The workshop of an alchemist. Painting of Giovanni Stradano, sixteenth century (Firenze, Palazzo Vecchio).

From Iatrochemistry to the Age of Enlightenment

The seventeenth was a century of transition and contradiction. In 1661, the publication of Boyle's *The Sceptical Chemist* marked the birth of chemistry as a science. However, those were also the times in which "plague spreaders"—people believed to deliberately create and spread "handcrafted" plague using chemical tools—were tried and executed.

Despite that, the way to the Age of Enlightenment, in which the supremacy of reason would be affirmed, was already paved. Following Galileo's doctrine, the iatromechanists, which were physicians and physicists at the same time, interpreted the organism as a machine. In 1628, William Harvey (1578–1657) described the mechanics of blood flow in his book *De motu cordis*. In 1680, Giovanni Alfonso Borelli (1608–1679) described muscular movement and locomotion in his *De motu animalium*. After Harvey's description of the macromachine pumping the blood, Marcello Malpighi (1682–1694), using the microscope, provided a description of two glomerular (kidney) and alveolar (lung) micromachines.

In parallel to this new mechanistic idea of organisms, a new discipline, called *iatrochemistry*, was developed, in particular in North Europe, as an extension of Paracelsus' thinking. Iatrochemists were at the same time physicians and chemists and considered the organism like an "alembic" in which chemical reactions occurred. In their view, the world, consisting of both living (organic) and nonliving (inorganic) matter, was made of tiny particles, like in the atomism of Democritus and Epicurus, rediscovered through the philosophical thought of Descartes and the quantitative method of Galileo Galilei.

However, these new theoretical perspectives did not bring any interesting practical novelty to pharmacy and pharmacology. Old theories were rediscovered and renewed: iatromechanics supported the solid parts of the organisms (solidism), whereas iatrochemistry supported the fluid parts (humoralism). Giovanni Battista Morgagni (1682–1771) developed a synthesis of new and old theories in his *De sedibus et causis morborum per anatomen indagatis* (published in Venice, 1761), a text of anatomic pathology and physiopathology revised in light of the new physical–chemical concepts.

At the time, medicaments adopted in medical practice included both new remedies, real or putative, such as the "triumphant" antimony, as well as old remedies, such as the "magical" *triaca*.

The eighteenth century sees the publication of the Linnaeus' monumental studies on the plant world. In his trilogy—*Systema naturae* (1735), *Philosophia botanica* (1750), and *Species plantarum* (1753)—Carl von Linné (1707–1778) lists, describes, and reclassifies all known plants. His most important contribution consists in the binomial nomenclature he adopted to univocally name and identify each plant species, which still represents the basis of modern classification (taxonomy) of living organisms. Moreover, he contributed to the relaunch of phytotherapy, which at the time was challenged and shadowed by the big development of chemical medicaments.

Around the same time, another Swedish, the German-born pharmacist Karl W. Scheele (1742–1786), founded the modern pharmaceutical chemistry. While working in the back of his pharmacy shop in Koping, near Stockholm, in 1774, he discovered oxygen (the air of fire). However, he was late in publishing his discovery and was preceded by the English chemist Joseph Priestley, who is now credited with it. In the following years, Scheele discovered many organic substances (such as formic, uric, lactic, citric, and malic acids and glycerine) subsequently employed in pharmacology during the nineteenth century, opening the way to studies in organic chemistry. Meanwhile, in the midst of the Age of the Enlightenment and the outbreak of the French revolution, Antoine-Laurent Lavoisier (1743–1794) refuted the old phlogiston theory and laid the foundations for the new quantitative chemistry.

From the Search of the Active Principle to the Discovery of Alkaloids and Glucosides

While the end of the eighteenth century saw all efforts directed toward the separation of fractions supposed to be therapeutically effective from medicaments, the beginning of the nineteenth century yielded the first successes of pharmaceutical chemistry in isolating active principles from medicinal plants (mainly cinchona tree, opium, and tobacco) with the aim of purifying and producing them in large amount.

Using increasingly refined techniques and supported by a gradually growing awareness, Louis-Charles Derosne (1803), Friedrich W. Serturner (1807), and Joseph Gay-Lussac (1817) succeeded in isolating morphine from opium. In 1809, Louis-Nicolas Vauquelin extracted nicotine from tobacco, and in 1820, Pierre-Joseph Pelletier and Joseph B. Caventou separated quinine from the cinchona bark, demonstrating that this new substance was responsible for the well-known antipyretic effect of the cinchona bark. In addition, they succeeded in extracting strychnine from *nux vomica* and caffeine from coffee beans.

These active substances, all containing nitrogen and behaving as bases (forming salts when combined with acids), were named alkaloids (1821, Wilhelm Meissner). Soon afterward, another class of active principles, the glucosides, was identified, which includes substances that do not behave as bases and upon hydrolysis release glucose as a secondary product. Two glucosides were soon to become crucial in pharmacology: digitalin, discovered in 1834 by Roger, and salicin, identified by Leroux in 1839.

The discovery of alkaloids and glucosides represents the continuation of the organic chemistry studies started, empirically, by Karl Scheele and taken forward in a more systematic way by Justus von Liebig (1803–1873), which culminated with the synthesis of urea carried out in 1831 by Friedrich Wöhler (1800–1882). It must be noted that the synthesis of urea, an organic compound, from inorganic substances (carbon and nitrogen), despite being now recognized as a revolutionary achievement, was initially neglected because urea was not seen as a product of vital processes but rather a waste product organisms dispose of through urine.

All these discoveries, together with the discovery of anesthetics, represent the major achievements of the new pharmacology of the nineteenth century. In pharmacology, the most significant figures of the time are physiologists/pharmacologists (such as François Magendie (1783–1855, College de France), Rudolf Buchheim (1820–1879, University of Giessen), Carl Binz (1932–1913, University of Bonn), Oswald Schmiedeberg (1838–1921, University of Strasbourg), Giovanni Semmola (1793–1865), and Arnaldo Cantani (1837–1893, University of Naples)) who studied the drug–organism interactions to evaluate both therapeutic and toxic effects of medicaments. Their work represents the application to pharmacology of the innovative ideas that in those years were stemming from medical and biological studies: the *médecine expérimentale* (experimental medicine) of Claude Bernard (1813–1878), the *cellular pathologie* (cellular pathology) of Rudolf Virchow (1821–1902), the germ theory of Louis Pasteur (1822–1895), and the microbial biology of Robert Koch (1843–1910).

The Drug Synthesis Revolution: From Handmade to Industrial Production

The need to obtain large amounts of therapeutically active plant extracts led the pharmacist Heinrich E. Merck (1794–1855) to establish in 1827 the first pharmaceutical factory in Germany. At that time, Germany was the European country, after the United Kingdom, in which the socioeconomical transformation triggered by the industrial revolution and the scientific culture was most advanced. Heinrich E. Merck transformed the apothecary shop owned by his family for several generations into a well-equipped artisan laboratory where substances such as cocaine and morphine were produced according to protoindustrial criteria.

During the nineteenth century, the real novelty in pharmacology consisted in the chemical synthesis of medicaments: medicaments extracted from plants or minerals began to be replaced by synthetic drugs, built in laboratory and capable of exerting selective therapeutic effects on patients. This represented a revolution in pharmacology that prompted the industrialization of the pharmaceutical production. Medicaments (*pharmaceutical specialties*) became not only innovative and easily accessible remedies with marked therapeutic properties but also products that could provide an economic return and therefore subject to the strict rules of the market.

The new procedures for drug production had to meet the new needs emerging from the rapid transformation of society. Industrialization and urbanization, with the consequent low-quality lifestyle, were resulting in increased spreading of infectious pathologies, thus increasing the demand of effective remedies to preserve the ability to work of town dwellers.

For several reasons (presence of a strong chemical industry, large funds, innovative approach to medical–biological research, and favorable entrepreneurial spirit), Germany and Switzerland are the countries in which the pharmaceutical industry developed at most, as a continuation or a derivation of the chemical dye industry. Bayer and Hoechst (1863), BASF (1865), and Schering (1871) in Germany and CIBA and Geigy (1884), Sandoz (1886), and Hoffmann-La Roche (1894) in Switzerland were the first and most important factories producing dyes and drugs in the second half of the nineteenth century.

In those years, pharmaceutical research focused mainly on looking for substances mimicking plant extracts in their ability of soothing pain or exerting antipyretic action.

Phenacetin, discovered in 1887 by Anton Kast and Karl Hinsberg, patented and launched on the market by Bayer in 1888, is the first real synthetic drug produced by a pharmaceutical industry and used in clinical practice. A decade later, in 1899, another compound representing a milestone in the history of pharmacology and pharmaceutical industry was released by Bayer: acetylsalicylic acid (best known as aspirin) discovered by Felix Hoffmann in 1898 (actually rediscovered, since the product had already been produced in 1853 by Charles F. Gerhardt) as a remedy for rheumatisms and pain is better tolerated than salicylic acid.

In Italy and France, pharmaceutical industry began with the transformation of several laboratories previously associated with apothecary stores. The most active—like those established by Giovanni Battista Schiapparelli (1795–1863) in Turin in 1824 and Carlo Erba (1811–1888) and Roberto Giorgio Lepetit (1842–1907) in Milan, respectively, in 1853 and 1868—turned into important pharmaceutical factories.

MODERN PHARMACOLOGY

Ehrlich and Chemotherapy: The Concept of Receptor

The new synthetic drugs produced by the emerging pharmaceutical industry were able to stop or reduce the most severe symptoms accompanying the diseases (such as fever or pain), but they were unable to reduce the incidence of infections, which represented the most common pathologies at the beginning of the twentieth century.

Sera and vaccines (in particular those against diphtheria and tetanus) and the increasing use of quinine (to treat malaria) significantly contributed to reduce the spread of some infections, but they were still not sufficient. To definitively eradicate infections (which scientists had already demonstrated to be due to pathogenic germs), new medicaments were required capable of selectively killing microbes without affecting the patient, thus achieving the *therapia sterilisans magna* (great sterilizing therapy), which was the ultimate, unreached goal of the medicine of the second half of the nineteenth century.

A new drug launched in 1910 by Hoechst to fight syphilis, arsphenamine, also known as Salvarsan or 606 (chemically known as dioxy-diamino-arsenobenzol, an arsenic compound), seemed to have the desired characteristics and to open a new era in the human struggle against infections. Salvarsan represents the final product of the studies carried out by a brilliant German scientist, Paul Ehrlich (1854–1915), who was a strenuous upholder of the importance of chemotherapy (therapy employing chemical substances) to eradicate infections. According to Ehrlich, effective therapies against infections require chemical compounds capable of recognizing pathogenic germs by chemical affinity, in order to destroy them without affecting the organism.

FIGURE 2.3 Daniel Bovet (1907–1992), the "father" of modern pharmacology.

The results obtained with arsphenamine in the treatment of syphilis were not as positive as expected. However, Ehrlich's ideas represented a turning point in the evolution of pharmacology, mainly for his clear intuition and definition of the concept of receptor (see Supplement E 2.2, "Ehrlich's Chemotherapy: Receptor Theory to Clinical Experimentation").

Ehrlich's chemotherapy represents also the first approach to the modern concept of experimental pharmacology, an essential premise to the therapeutic pharmacology developed a few decades later with the discovery of sulfonamides and antibiotics.

Despite the disappointing results obtained with arsphenamine, the concept of chemotherapy proposed by Ehrlich remained a fascinating working hypothesis in pharmacology. Following the direction indicated by Ehrlich, in 1932, Gerhard Domagk (1895–1964), a scientist working at Bayer, synthesized a dye, Prontosil, that soon proved to have a potent antibacterial activity. Oddly, the compound seemed to exert its action only *in vivo*, not *in vitro*. The reason for such unusual behavior was discovered in 1935 by a group of researchers headed by Daniel Bovet (1907–1992) (Fig. 2.3) at the Pasteur Institute of Paris. The antibacterial effect of

FIGURE 2.4 Alexander Fleming (1881–1955), who is credited with the discovery of penicillin.

Prontosil is associated with a portion of the compound that is released when the organism hydrolyzes the molecule. This discovery led to the generation of a new revolutionary class of drugs, the sulfonamides (or sulfa drugs), the first effective drugs to fight infections.

However, the real, definitive victory over infectious diseases occurred after 1942, during the Second World War, when the United States secretly began the industrial production and clinical application of penicillin, a substance with a powerful antibiotic activity derived from the *Penicillium* mold. Penicillin was discovered by accident in 1929 by Alexander Fleming (Fig. 2.4) (1881–1955) and stabilized by crystallization, in 1939, by Howard Florey (1898–1955) and Ernst Chain (1906–1979).

Penicillin and other antibiotics discovered during the following years (streptomycin in 1944, chloramphenicol and cephalosporin in 1947, neomycin in 1949, tetracyclines and erythromycin in 1952) had a huge impact on medicine and society, being able to significantly increase life expectancy and quality, and their discovery is regarded as one of the events that have changed the course of human civilization.

Sulfonamides and antibiotics represent the fulfillment of Ehrlich's idea of chemotherapy and the first drugs employed in experimental pharmacology to study organism–drug interactions. Moreover, they are the first products allowing modern pharmacology to develop, alongside the mainly "experimental" theoretical approach, a more applicative therapeutic one.

From the "medical matter" of the beginning of the nineteenth century, we have come to the pharmacotherapy of the second half of the twentieth century.

The Birth of Modern Pharmacology

Daniel Bovet (1907–1992), the 1957 Nobel Laureate in Medicine, is the scientist modern pharmacology owes the most. After discovering the antibacterial effects of sulfa drugs, in 1937, he began fundamental studies on histamine and its antagonists, the antihistamines. Subsequently, in 1947, he focused on gallamine opening the way to synthetic curare. In particular, his investigations on phenothiazines—substances with antihistamine activity that had been known for some time but never studied in depth—in 1952 led Jean Delay and Pierre Deniker to adopt chlorpromazine (Largactil) as a sedative in the clinical treatment of some psychosis, marking the official beginning of psychopharmacology.

The introduction of lithium salts into clinical practice, in 1949, and the discovery of antidepressant agents (anti-MAO and tricyclic antidepressants, TCAs, in 1952 and 1957, respectively) provided physicians with new tools to approach psychiatric disorders, offering hopes and freedom to thousands of individuals previously bound to be secluded, often for their entire life, in psychiatric hospital, the "insane jails."

In the 1960s, a new class of "minor sedatives," benzodiazepines (BZD), discovered by Leo Sternbach, proved very effective in controlling anxiety and reducing insomnia. Chlordiazepoxide (Librium, 1960) and diazepam (Valium, 1963) are the prototypes of a family of compounds that has deeply changed the lifestyle of millions of people around the world. Their introduction has significantly improved many people's life, but it has given rise to phenomena of drug abuse and severe and unjustified drug addiction.

A real pharmacotherapy explosion occurred in the 1950s, which radically changed medicine and pharmacology. After the beginning of psychopharmacology, new and more potent antibiotics were discovered and synthesized (rifampicin, 1957; ampicillin, 1961), some endowed with antimycotic activity (nystatin, 1950; griseofulvin, 1958), as well as substances active against tuberculosis (PAS, 1946; isoniazid, 1952), and more effective antispastic agents, analgesics, and nonsteroidal anti-inflammatory drugs (NSAIDs) offering a valid alternative to cortisone (introduced in clinical practice in 1949). New oral antidiabetics became available in 1955 for the treatment of diabetes mellitus, while the discovery and introduction of contraceptives, in 1956, marked the beginning of the era of birth control.

In the 1960s, the clinical management of devastating neurological conditions previously considered incurable, such as Parkinson's disease, was radically changed by the introduction of L-DOPA. Meanwhile, first diuretics and beta-blockers with hypotensive and cardioprotective action began to be used. The systematic adoption of anthelmintics, the discovery and prescription of new vitamins, the spreading of polio vaccines, and the application of chemotherapy to cancer treatment gradually led to significant improvements in the therapeutic approach to multiple pathological conditions.

This seemingly unstoppable progress of pharmacotherapy and the resulting optimism in the medical environment came to a sudden halt in the 1960s when the "thalidomide case" burst out. Thalidomide is a hypnotic drug that was marketed by a German company as absolutely safe but turned out to induce fetal malformations when administered to pregnant women, causing severe defects, mainly phocomelia, in newborns. Thalidomide was withdrawn from the market and the shock produced by its devastating effects prompted a critical and beneficial discussion among physicians and pharmacologists on use and safety of drugs. The thalidomide case was a clear indication that new legislations were needed to promote accurate pharmaceutical and clinical studies on drugs before marketing them, in order to guarantee the absence of severe adverse effects as well as the presence of a real therapeutic efficacy (see Chapter 52 and Supplement E 2.3, "The Thalidomide Case").

This fostered the rise of a new branch of pharmacology—alongside the traditional theoretical and experimental pharmacology—the clinical pharmacology, whose main task is to guarantee a rational use of drugs in clinical practice (see also Chapters 54, 55, and 56). Clinical pharmacology includes three main disciplines: pharmacovigilance, also known as drug safety (careful investigation of drug adverse effects), pharmacoepidemiology (quantitative evaluation of drug use and abuse), and pharmacoeconomics (economic evaluation of cost/benefit ratio of drug therapies and in particular their impact on public expenditure).

Several innovative drugs have been discovered and marketed in the last decades of the twentieth century: anti-H2 agents, such as cimetidine (1972), and proton pump inhibitors, such as omeprazole (1994) and its derivatives, for the treatment of peptic ulcers; immunosuppressants, such as cyclosporine, and antiviral drugs, such as acyclovir (1984); new calcium antagonists, angiotensin-converting enzyme (ACE) inhibitors, and angiotensin II receptor agonists for blood pressure control and cardioprotection; statins (1989) to treat familial hypercholesterolemia; agents for the treatment of prostatic hyperplasia, such as finasteride (1993); selective antimigraine drugs, such as triptans (1994); safer and more effective antidepressants, such as the selective serotonin reuptake inhibitors (SSRIs) (1992); and innovative antipsychotics, such as clozapine and clopenthixol (1996). A real success for the efficiency in therapeutic results, for individuals and society, took place in the field of chemotherapy against HIV and HCV infections, an example of the strong validity of close collaboration between industry and academic research.

These molecules, often prototypes of new classes of drugs, should (and we use the conditional mode) result from in-depth studies as well as from a careful assessment of risks and benefits.

Indeed, every drug displays not only beneficial but also adverse effects on the organism; therefore, its introduction in clinical practice has to be decided after a careful evaluation of the risk/benefit ratio, distinguishing pharmacological effects (drug–organism biochemical interactions, which not always are beneficial or curative) from therapeutic effects (the ability of drugs to act on the cause of the disease).

THE BIOTECHNOLOGY ERA AND THE PHARMACOLOGY IN THE THIRD MILLENNIUM

The Impact of New Biotechnologies: Molecular Biology, Bioinformatics, and Combinatorial Chemistry

In the last few decades, the use of bioinformatics and molecular biology techniques in medicine and pharmacology has offered new opportunities, both theoretical and practical, paving the way for new therapeutic approaches to diseases. The encounter between bioinformatics and biomedical technologies, on one side, and the traditional biological chemistry and pharmacology, on the other, have changed the way of physicians and pharmacologists think and work.

Until a few decades ago, the search for new drugs involved two main stages: (i) screening of large collections (libraries) of synthetic molecules, to test their affinity for specific biological targets, and (ii) evaluation of the therapeutic properties (at physiological and/or histological level) of the selected molecules in animal models affected by pathologies either spontaneous or induced by physical, chemical, or biological agents. The development of technologies to generate transgenic animals, carrying either a mutation in an endogenous gene or one or more heterologous genes, has provided new experimental models to investigate human pathologies (and to test the effectiveness of drugs designed to treat them) and also new biological sources of heterologous proteins of interest. From the 1970s, building on the rapidly increasing knowledge about enzymes and receptors provided by bioinformatics, traditional pharmacological approaches *in vivo* have been replaced by *in vitro* technologies, made even faster and more powerful by automated data processing systems.

Molecular biology has been the main trigger of a revolutionary change in medicine and pharmacology by offering tools to clone, study, modify, and recombine genes starting from their nucleotide sequences.

In 1973, recombinant DNA technologies opened up new therapeutic perspectives by allowing exploiting the biosynthetic machinery of living organisms, such as bacteria, yeast, and *in vitro* cultured cells, to obtain complex molecules like proteins (which in most cases would be impossible to synthesize using the traditional chemical procedures). Before the 1970s, all marketed drugs contained molecules either obtained by chemical synthesis (still representing most of the pharmacological compounds) or extracted from organs, tissues, and biological fluids of animals (such as insulin, extracted from pigs) or humans (such as the growth hormone, obtained from the hypophysis of dead individuals, or gonadotropins, purified from the urine of pregnant or postmenopausal women). These new "biological techniques" or biotechnologies have brought a revolution in the pharmacology field comparable to that triggered in the middle of the nineteenth century by the first synthetic drugs. The completion of the Human Genome Project in 2003, together with the development of new bioinformatics tools, has paved the way for new extraordinary pharmacological applications. Knowing the entire sequence of bases that make up the human genome means being potentially able to find the nucleotide sequence of all genes contained in the genome (but remember that the genome sequence in itself is not sufficient to identify genes; computer-assisted analysis may allow to predict which regions could correspond to putative protein-encoding genes, but a functional demonstration is required to confirm them as real genes) and to predict the amino acid sequence of the proteins they encode (most genes generate multiple transcripts as a result of alternative mRNA splicing and therefore multiple proteins). Furthermore, bioinformatics may help to predict the function of a protein starting from its amino acid sequence, by revealing sequence similarity or identity with other proteins already functionally characterized, and to identify the amino acid residues, which are critical for its activity. Bioinformatics tools are therefore invaluable for investigating the therapeutic potential of proteins and to devise the best pharmacological approach to target them.

In recent years, all steps in the drug development process—research, design, analysis, and production—have undergone drastic changes. In the past, the approach to drug discovery involved the analysis of natural products or the design and synthesis of molecules capable of binding to the biological target of interest. After an initial screening, the selected molecules were chemically modified to improve their affinity for the target and/or to enhance their therapeutic efficacy.

This traditional approach has been first assisted and then replaced by molecular modeling techniques. Using sophisticated softwares, today, it is possible to design and virtually analyze a drug (computer drug modeling or computer-aided drug design) before producing it. In this way, it is possible to optimize the chemical structure of compounds to improve their interaction with the specific target, simply starting from biological and chemical/physical information. Moreover, this approach allows to exclude candidate drugs (lead compounds) that with the traditional approach would probably be discarded at later stages because of their unsatisfying absorption, distribution, metabolism, excretion, or toxicity, thus allowing to save time and money (for more details, see Chapter 54). Algorithms have been developed to predict drug absorption, distribution, metabolism, excretion, and toxicity (ADMET bioactivity) that may help to identify problematic molecules before synthesizing them. Such virtual screenings are faster and cheaper than the traditional chemical procedures. Therefore, they are extremely interesting for the pharmaceutical industry, given the high costs of developing a therapeutic drug with the traditional approach.

Also, the advent of combinatorial chemistry has dramatically changed the way pharmacological research is carried out in both university and industry laboratories.

Combinatorial chemistry consists in a collection of chemical procedures to synthetize large amounts (libraries) of different variants of a starting chemical structure. The biological activities of the molecules obtained in this way are then systematically tested using rapid automated screening methods (high-throughput screening or HTPS) based on robots and data processing systems. The compounds with the desired biological activities are then further studied, whereas all the others are discarded. Such large-scale methods to produce and screen molecules highly increase the probability to identify products acting on specific targets.

This research approach has been exploited by a new kind of industries (such as Pharmacopeia, founded in 1993, and Affymax, founded in 1994) specialized in the production of large archives of molecules generated by combinatorial chemistry.

Drug virtual screening and rapid automated screening of compounds generated by combinatorial chemistry are the new frontiers in chemical bioinformatics, the discipline that, alongside genetic engineering, represent one of the most promising future directions of pharmacological research (see also Chapter 54).

Biological Drugs and Pharmacology Perspectives

The era of biological drugs officially started in 1982, when human insulin produced with DNA recombinant technologies by Genentech (the first biotech company founded, which soon became the main example of the new economic–industrial development in pharmaceutical production) was approved and registered in the United States as a drug.

Between the end of the 1980s and the beginning of the 1990s, efforts were mostly directed toward obtaining proteins for therapeutic purposes from genetically modified cells. Insulin, previously obtained from pig pancreas, was replaced by human insulin purified from recombinant bacteria carrying and expressing the insulin human gene. Gonadotropins, previously purified from the urine of pregnant or postmenopausal women, were replaced by recombinant proteins produced by mouse ovary cells carrying the human genes encoding the hormones. Recombinant interferon replaced the natural molecule previously extracted from fibroblasts.

Besides insulin, gonadotropins, and interferon, several other biological drugs have been produced using recombinant DNA technologies, such as growth hormone, erythropoietin, coagulation factor VIII, plasminogen inhibitors, interleukin-2, interferon-α and interferon-β, and antisense drugs.

From the 1990s onward, the increasing knowledge of the proteins encoded by the human genome has drawn the attention of researchers to secreted proteins (those released in the bloodstream in contrast to those building up the body) with the aim of identifying new therapeutically interesting molecules. This has determined a contrast between the universe of proteins that could be obtained with DNA technologies starting from the knowledge of gene functions and the universe of chemical structures that could be designed using the traditional chemical approach.

These new technological and scientific developments have opened the way to genomics and proteomics, two disciplines that have overturned the paradigm of pharmacology research. The traditional approach to drug discovery—from the pathology to the target, from the target to the screening, and from the screening to the drug—has been turned upside down: today, the discovery of a new drug starts with a gene and continues with the characterization of its protein functions to end up with the identification of a potential therapeutic application.

Future pharmacopoeia will include, alongside gene therapy, an increasing number of biological drugs such as monoclonal antibodies (produced using a technology devised in 1975 by Georges J. Kohler and César Milstein, which were awarded the Nobel Prize in 1984) to fight cancer and new vaccines.

Bioinformatics and genetics have provided tools also to investigate the clinical response to drugs.

Drug delivery techniques have undergone significant improvements too. Besides the traditional routes of drug administration (tablets, capsules, suppositories, intramuscular injections, and intravenous infusions), new methods for delivering medicines are now available: nasal sprays, transdermal patches, lyophilized oral compounds, delayed-release tablets (through osmotic pump systems), or controlled-release formulations (for instance, through the covalent attachment of PEG to the drug via a process called PEGylation or employing liposomes and nanoparticles as carriers). These methods allow a more precise and constant administration of the active principles, as confirmed by pharmacokinetics studies, for optimal therapeutic outcome.

Personalized Therapies and New Sceneries in Pharmaceutical Industry

Today, the ultimate ambitious goal of pharmacogenetics and pharmacogenomics is to predict the patient's response to a specific drug exploiting microarray-based technologies to evaluate possible resistance or adverse effects associated with his genotype and devise the best therapy for each subject (personalized therapy).

Men and women get ill in a different manner, as their different biology has a different impact on their health, the disease course, and their response to therapy. This observation has given rise to a new medical discipline, called gender-specific medicine that focuses on the effects of gender on diseases with the aim of finding more appropriate therapies specific for men or women. Consequently, also, a gender-specific pharmacology has been developed, to investigate the difference in the response of men and women to pharmacological treatments, with the aim of identifying gender-specific drugs to treat the same pathology, thus providing men and women with equally effective therapies.

This new pharmacology has changed and is still changing the rules of pharmaceutical industry and market. Over the last 20 years, the investments a pharmaceutical company has to make to develop a new drug have increased from 80 to 900 million dollars, whereas the time required for a new drug to reach the market has moved from 10 to 15 years. This is the main reason why many pharmaceutical companies have disappeared in the last few decades, because they are unable to keep up with progresses in the field or have merged, giving rise to a relatively small number of pharmaceutical giants (big pharma) that can control and influence pharmaceutical research and market worldwide. For instance, Novartis was established by the merger of Ciba-Geigy and Sandoz, Aventis by the merger of Hoechst Roussel and Rhone-Poulenc Rorer, and GSK by the merger of Glaxo Wellcome and SmithKline. However, alongside the big pharma, biotechnological progresses have prompted the creation and proliferation of small research laboratories or companies (biotech companies) focused on few but promising projects capable of attracting private investments. Initially, this phenomenon was restricted to the United States, but now, it has spread also in Europe. Some of them started as branches of bigger companies or research laboratories stemming from universities (spin-offs) or as small societies founded by groups of researchers and entrepreneurs to develop new projects (start-ups).

Hence, future drugs will be developed both in huge laboratories born from the merger of middle-big traditional pharmaceutical industries and in small dynamic companies that have benefited from the failures of many others, ditched by a market that does not allow mistakes.

Today, nanotechnologies and molecular drugs represent the ultimate frontiers in pharmacotherapy technology. Nanotechnologies concern the manipulation of matter at molecular and atomic level (in the scale from 1 to 100 nm) to obtain products in the macroscopic scale. In the nanotechnology world, no borders exist between chemistry, physics, engineering, mathematics, and biology; this means that it is a multidisciplinary research field. The properties of the matter in that scale make it very complex to build and test nanodevices. On the other side, nanotechnologies have a large number of potential applications in many different fields. Application of these technologies to medicine has given birth to nanomedicine, a new discipline that is expected to have a revolutionary impact on diagnostics and therapy. In particular, nanopharmacology is the research field focused on the therapeutic applications of nanotechnologies. It employs nanoparticles (vectors) to create nanodrugs with pharmacokinetic and therapeutic properties that would be impossible to achieve with drugs bigger than 1 μm.

Such nanodrugs act on targets in a specific and selective way (similarly to what occurs, for instance, in the pharmacology of RNA) and exert peculiar pharmacological effects. Nanoparticles allow the delivery of high amounts of drug specifically to the affected tissue or pathological cells (such as tumor cells). Moreover, given their solubility, they allow a prolonged exposure to the drug. They can release more drugs at the same time, with different and suitable pharmacokinetics. They allow drugs to cross the blood–brain barrier and cell membranes, to reach their correct therapeutic targets.

Different approaches (conventional or stabilized liposomes, PEGylated microspheres or nanoparticles, polymers or dendrimers, molecules with several "hooks" to mediate their anchorage to cells, microdevices containing microchips, MEM) are used to modify drug bioavailability in order to modulate the therapeutic response in accordance with the desired effects.

Another important feature of nanodrugs is that they become activated only in response to specific signals (such as a magnetic field). This property, called triggered response, allows nanoparticles to release the drug only when they reach the appropriate target.

Therefore, nanotechnologies offer the best tools to fulfill the personalized therapy, which is the ultimate goal of the medicine and pharmacology of the third millennium.

TAKE-HOME MESSAGE

- Started as a collection and description of natural remedies endowed with a magical aura, over the centuries, pharmacology has developed into a multidisciplinary approach to disease treatment aimed at the design of targeted drugs, based on "state-of-the-art" technologies and their industrial production.
- Over the last 100 years, scientific development and knowledge in pharmacology have been supported and promoted by technological advancements in biomedical sciences.
- Modern pharmacology relies on interaction between several independent yet correlated disciplines.

FURTHER READING

Bovet D., *Une chimie qui guérit. Histoire de la découverte des sulfamides*. Payot: Paris, 1988.

Burbaum J., Tobal G. M. (2002). Proteomics in drug discovery. *Current Opinion in Chemical Biology*, 6, 427–433.

Franconi F., Carru C., Malorni W., Vella S., Mercuro G. (2011). The effect of sex/gender on cardiovascular pharmacology. *Current Pharmaceutical Design*, 17, 1095–1107.

Freitas R.A. Jr. (2005a). What is nanomedicine? *Nanomedicine: Nanotechnology, Biology and Medicine*, 1, 2–9.

Freitas R.A. Jr. (2005b). Current status of nanomedicine and medical nanorobotics. *Journal of Computational and Theoretical Nanoscience*, 2, 1–25.

Gershell L. J., Atkins J. H. (2003). A brief history of novel drug discovery technologies. *Nature Reviews in Drug Discovery*, 2, 321–327.

Leake C. D., *An historical account of pharmacology to the 20th century*. Charles C. Thomas: Springfield, 1975.

Lindpaintner K. (2002). The impact of pharmacogenetics and pharmacogenomics of drug discovery. *Nature Reviews in Drug Discovery*, 1, 463–469.

Maehle A. H., Prull C. R., Halliwell R. F. (2002). The emergence of the drug receptor theory. *Nature Reviews in Drug Discovery*, 1, 637–641.

Nagle T., Berg C., Nassr R., Pang K. (2003). The further evolution of biotech. *Nature Reviews in Drug Discovery*, 2, 75–79.

Porter R., Telsh M., *Drugs and narcotics in history*. Cambridge University Press: Cambridge, 1929.

Sironi V. A., *Le officine della salute. Storia del farmaco e della sua industria in Italia*. Laterza: Roma-Bari, 1992.

Swinney D. C., Anthony J. (2011). How were new medicines discovered? *Nature Reviews. Drug Discovery*, 10, 507–519.

Watson G., *Theriac and mithridatum: a study in therapeutic*. Wellcome Historical Medical Library: London, 1966

SECTION 2

GETTING THE DRUG TO ITS SITE OF ACTION

3

CELLULAR BASIS OF PHARMACOKINETICS

RICCARDO FESCE AND GUIDO FUMAGALLI

> **By reading this chapter, you will:**
>
> - Learn the basic mechanisms of diffusion underlying drug absorption, distribution, metabolism, and excretion
> - Understand the implication of Fick's law on drug diffusion in biological compartments
> - Know the properties of cellular barriers and their effects on drug diffusion among body compartments

Therapeutic efficacy depends on a correct diagnosis and an appropriate choice of the drug. Both therapeutic and adverse responses to a drug are usually dose dependent. Once the appropriate drug has been selected, success of the therapy depends on the physician being able to induce appropriate concentrations in blood and tissues (including the target organ) by modulating drug dosage and administration protocol. This is a critical issue when the gap between effective and toxic levels (therapeutic window) of a drug is narrow. The critical questions a physician has to address are: How much drug? How frequently? By which route? To be able to answer these questions, it is essential to master the basic principles that govern drug absorption, distribution, and elimination. This field of pharmacology is named pharmacokinetics. A complete overview of pharmacokinetic principles and processes is offered at the website.

A QUICK JOURNEY WITH THE DRUG IN THE BODY

Absorption

A drug may occasionally be administered directly at the site of its action. This occurs, for example, with antiacid drugs, which are aimed at neutralizing the low pH in the gastric lumen without being absorbed, or with local anesthetics, which are directly administered in close proximity to the nerve trunks to be silenced.

However, in general, the target organ is distant from the site of drug administration. To cover this long distance, the drug must enter the systemic circulation. Thus, unless it is directly injected into blood vessels, the first step of the drug journey in the body is its absorption, meaning the series of processes that allow the drug to enter the bloodstream (absorption phase in Fig. 3.1). If the drug is injected in an interstitial space (i.e., intramuscular or subcutaneous administrations), the drug diffuses from the injection site and enters the capillary vessels, like many other endogenous solutes do. For any other administration route (e.g., oral, rectal, sublingual, and transcutaneous), the drug must also cross the epithelial barrier that separates the external world from the interstitial fluids and that may consist of single or multiple layers of cells.

Distribution

Once the drug has entered the organism and reached the bloodstream, it can diffuse to the whole organism (distribution phase in Fig. 3.1). Diffusion back to interstitial fluids takes place at capillaries; it is influenced by the capillary bed permeability, which may differ in the various districts (e.g., CNS and placenta) based on the anatomical and structural organization of the capillary network and the possible presence of pathological processes (e.g., inflammation). The basal lamina of the endothelium significantly hampers the diffusion of blood cells and substances with molecular weight similar to, or larger than, that of albumin (67 kDa). Thus, possible binding to plasma proteins or blood cells prevents drug diffusion to the tissues: only the unbound (free) fraction of the drug can reach tissue cells.

General and Molecular Pharmacology: Principles of Drug Action, First Edition. Edited by Francesco Clementi and Guido Fumagalli.
© 2015 John Wiley & Sons, Inc. Published 2015 by John Wiley & Sons, Inc.

22 CELLULAR BASIS OF PHARMACOKINETICS

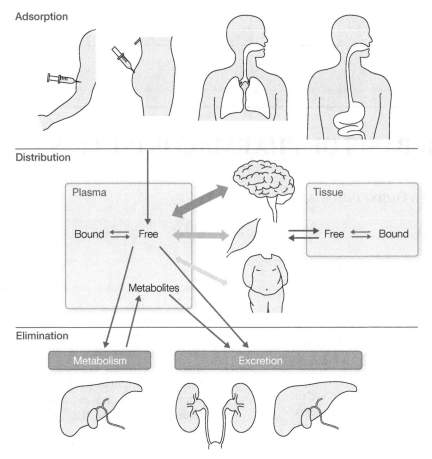

FIGURE 3.1 The journey of the drug in the body. There are three distinct but intermingled phases in pharmacokinetics: absorption, distribution, and elimination. The rate of absorption depends on the route of administration. Once the drug has been absorbed (i.e., has entered the bloodstream), it is distributed to the various organs and tissues by the cardiovascular system; the fraction that is not bound to large proteins or blood cells diffuses back and forth through the capillary wall to the interstitial spaces and reaches the cells. Elimination occurs by excretion and/or metabolism; excretion occurs mainly (but not exclusively) in the kidney, and metabolism mainly (but not exclusively) in the liver.

Through the systemic circulation, the drug is distributed to all organs of the body. However, this distribution process is not homogeneous. Indeed, the drug initially reaches the various body compartments with different velocities, mainly due to their different perfusion rates (arrows with different thickness in Fig. 3.1). Later on, its local concentration in each tissue may significantly depart from plasma concentration (drug "tropism" for specific organs or tissues). Organs that have large volume and/or capacity to bind drug, such as skin, fat, and bones, over time may accumulate substantial amounts of drug(s) even if they are poorly perfused.

Drug Elimination

About 4% of the whole plasma fluid is filtered every minute by renal glomeruli. The unbound drug contained in such volume is transferred to the preurine. This accounts for about 1/25 of the total free drug present in the blood every minute. All drug molecules that are not reabsorbed along the nephron will be eliminated (cleared) with the urine. The kidney is the most important (but not the only) organ for drug elimination from the body.

Some drugs are eliminated at the level of the pulmonary alveoli (i.e., anesthetic gases) or through other secretions such as sweat, milk, saliva, and bile. In this last case, a fraction of the drug can be reabsorbed in the intestine (enterohepatic recycling).

Metabolic enzymatic processing is another route of drug elimination. In principle, all cells and organs can metabolize drugs, but the liver is the most important organ for drug metabolism. Metabolic modifications change pharmacokinetic properties and/or efficacy of most drugs.

As the drug is eliminated, its plasma concentration decreases, and its effects may vanish. If the drug is repeatedly administered, each new dose adds to what remains of previous doses in patient's body. The kinetics of drug accumulation following repeated administrations will be discussed in detail in Chapter 7. Here, we would like to emphasize that pathological or functional alterations of the elimination processes (mainly in the kidney and liver) may modify both action duration and total quantity of the drug that accumulates in the body during a prolonged therapy, thus thwarting the therapy or determining toxic effects.

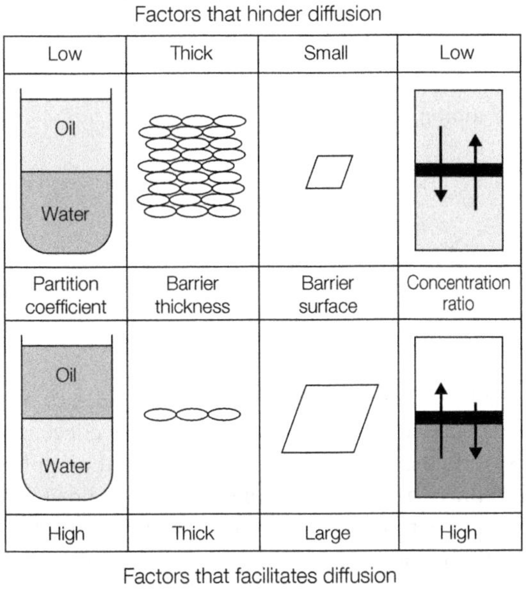

Therefore, pharmacokinetic aspects need to be considered in setting up the appropriate therapy.

CROSSING CELL MEMBRANES

In order to bind to its receptor and induce biological effects, a drug needs to diffuse from the site of administration, enter general circulation, exit at capillary level, and, in some cases, enter the cells. When administered by the oral route, the drug first has to cross the mucosal epithelium to reach the vascular bed; crossing of cellular barriers also occurs when other routes of administration are used. To act on its receptor, the drug then has to leave circulation and diffuse out of the capillaries into the interstitium and, in some cases, penetrate inside the cell. In the brain and placenta, a drug has to cross a cellular barrier to reach all cells. When transiting through the liver, it may enter into the hepatocytes to be metabolized by intracellular enzymes. When filtered at the renal glomerulus, the drug will be eliminated, totally or partially, depending on its capability of crossing the tubular epithelium to be reabsorbed. All these events indicate that absorption, distribution, and elimination depend on the capacity of a drug to cross cell membranes.

Three basic mechanisms sustain transcellular diffusion of a drug: passive diffusion, carrier-mediated transport, and endo-/exocytotic events (see Figs. 3.2 and 3.3). In most cases, passive diffusion governs the kinetics of drug transport across cell membranes.

FIGURE 3.2 Factors affecting passive diffusion across cell membranes. The upper part of the figure shows the conditions that slow down diffusion through cell membranes, and the lower, those that facilitate it.

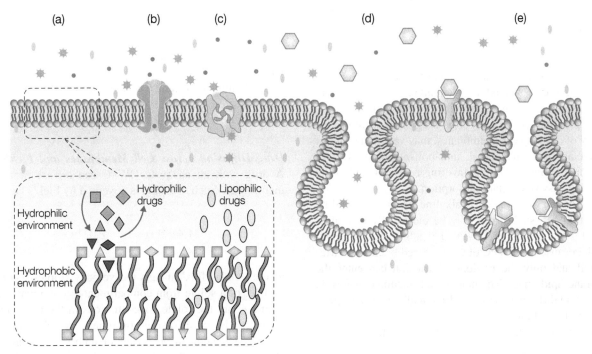

FIGURE 3.3 Cellular and molecular mechanisms supporting transmembrane diffusion of drugs. (a) Passive diffusion dependent on the drug oil–water partition coefficient (OWPC). (b) Transmembrane channel-mediated diffusion. (c) Carrier-mediated diffusion. (d) Fluid-phase endocytosis. (e) Receptor-mediated endocytosis.

Passive Diffusion across Cell Membranes

The capacity of a drug to cross cell membranes depends on its **partition coefficient**. The matrix of cell membranes consists of lipids (the lipid tails of phospholipids), whereas cytoplasm and extracellular spaces are aqueous solutions. In order to diffuse into a cell, a drug needs to be sufficiently water soluble to stay in solution in the extra- and intracellular aqueous solutions but also liposoluble enough to distribute in the lipidic environment of the membrane matrix. The degree of hydro-/lipophilicity of a compound can be determined by measuring its distribution in a volume containing water and oil: the ratio between its concentrations in the oily and aqueous phases is called oil–water partition coefficient (OWPC). When the OWPC is greater than 1, the compound is mostly lipophilic; when it approaches 0, the compound is highly hydrophilic. Since drugs must interact with both water and lipid environments, their OWPCs usually have intermediates values.

OWPC Properties The OWPC of a substance depends on its chemical and physical properties. Charged compounds are fully hydrophilic. The presence of groups capable of forming hydrogen bonds with water (carboxyl, alcohol, amine, aldehyde, and ketone moieties) also confers hydrophilicity. Molecules with a very low OWPC are excluded from the lipid phase and therefore have a negligible ability to penetrate and cross membranes (see Fig. 3.3a). A hydrophilic drug filtered in the preurine is poorly reabsorbed along the nephron and will be disposed of in the urine. By contrast, molecules with a high OWPC can freely cross cell barriers: they can be completely absorbed in the intestine, diffuse across barriers such as the blood–brain barrier (BBB), or even be absorbed through the skin (transcutaneous route). Finally, drugs with a particularly high OWPC do not easily diffuse across membranes because they tend to accumulate within the lipid matrix (as in the case of some antimycotic drugs).

The partition coefficient of a drug, and therefore its capacity of crossing cell membranes, may vary as a result of drug metabolism. In general, metabolism produces more hydrophilic compounds, thus favoring renal elimination by preventing passive tubular reabsorption.

Many drugs contain acidic or alkaline residues. Depending on solution pH, such drugs may be electrically neutral or charged. For these drugs, OWPC depends on the surrounding pH, because the OWPC of their ionized fraction is virtually null and only the nonionized fraction can enter the membrane lipid phase. The nonionized fraction depends on the absolute difference between drug acidity constant (pK_A) and pH of the solution.

The acidity constant of a molecule (K_A) is the ratio:

$$K_A = \frac{[\text{base}][H^+]}{[\text{acid}]}$$

where [acid] and [base] indicate the concentrations of the two species of the molecule. The negative decimal logarithm of K_A is indicated by pK_A; thus, we can also write

$$\frac{[\text{base}]}{[\text{acid}]} = \frac{K_A}{H^+} = 10^{(pH-pK_A)}$$

For a weak acid, the charged species is generally the base, whereas for a weak base, it is the acid. In such cases, the ionized/nonionized ratio is

$$\frac{[\text{base}]}{[\text{acid}]} = 10^{(pH-pK_A)} \text{ for a weak acid}$$

$$\frac{[\text{acid}]}{[\text{base}]} = 10^{(pK_A-pH)} \text{ for a weak base}$$

A marked pH difference between the two sides of a membrane may determine large differences in drug concentration, because only the nonionized species reaches the equilibrium between the two sides. Two examples are as follows: a weak acid such as aspirin ($pK_A = 3$) displays a 0.01 ratio between nonionized and ionized species in the gastric lumen (pH = 1; [base]/[acid] = $10^{(1-3)}$) and a 25,000 ratio in the plasma (pH = 7.4; [base]/[acid] = $10^{(7.4-3)}$). When the acid species (nonionized) reaches the equilibrium, the total concentrations will be more than 20,000-fold higher in the interstitial side than in the lumen. This may contribute to direct damage of the gastric mucosa.

With some drugs, the pH dependence of OWPC can be exploited for therapy. For example, in case of barbiturate intoxication (barbiturates are a class of weak acids endowed with hypnotic activity that are often used in suicide attempts), urine alkalinization can be used to increase the drug ionized quota in preurine and reduce its tubular reabsorption, thus accelerating its elimination (see Chapter 5).

Drug Diffusion across Cell Membranes and Fick's Law
A drug with an adequate OWPC can diffuse across cell membranes. Such diffusion is governed by Fick's law:

$$\text{Molar flux} = (c_1 - c_2) D \times \frac{A}{d}$$

where "molar flux" means the velocity (moles/second) of transfer of a solute from compartment 1 to compartment 2; c_1 and c_2 are the solute concentrations in the two compartments; D is the diffusion coefficient, which depends on the physicochemical properties of both solvent and solute (for cell membrane crossing, D mostly depends on the solute OWPC); A is the membrane surface area; and d is its thickness. For plasma membranes, d can be considered a constant;

for a tissue, d depends on the number of cellular layers that have to be crossed during the diffusion process (Fig. 3.2).

From this equation, we can conclude that:

1. The net molar flux of a drug across a biological membrane separating two compartments is proportional to the concentration difference between the two compartments; this implies that over time the flux decreases as the concentrations move toward equilibrium. As a consequence, the concentrations follow first-order kinetics (see Box 4.1).
2. Different drugs have distinct capacity of penetration, depending on their diffusion and partition coefficients.
3. The flux is directly proportional to the surface area of the barrier the drug has to cross; this is why drugs (and foods) taken by mouth are mostly absorbed in the small intestine, where the absorbing surface is huge ($>2000\,m^2$), as opposed to the gastric mucosa ($<0.5\,m^2$).
4. The thinner the barrier to cross, the more efficient the transfer: if its OWPC is adequate, a drug will be absorbed through the skin but at a lower rate than through a mucosa, due to the thickness of keratinized epithelium and derma. However, absorption may be extensive and fast if the skin is lesioned or de-epithelialized (reduced barrier thickness), and undesired toxic effects may appear.

Drug Transport across Cell Membranes

Drugs and substances, regardless of whether they can cross a membrane by passive diffusion, may be transported by specific mechanisms. These can be schematically grouped as follows: (i) transport by endo-, pino-, and transcytosis; (ii) active transport against concentration gradient (via transporter or exchanger pumps); (iii) carrier-mediated transport; and (iv) transport through pores of membrane channels (see Fig. 3.3)

Endocytosis

Hydrophilic drugs can cross cell membranes and barriers by endocytosis. During the process, the drug remains in aqueous solution. Vesicles can cross cells and release their content by exocytosis on the opposite side (transcytosis), a typical process occurring in capillary endothelium. In these cells, vesicles can fuse and form transcellular pores, allowing the diffusion of large molecules; this is especially relevant for anionic proteins.

Receptor-mediated endocytosis is a specialized form of endocytosis. Many receptor proteins tend to oligomerize when they bind their ligands (in some cases, also in absence of ligand) and to recruit specific proteins (adaptins) that guide the assembly of a protein coat on the cytoplasmic face of the plasma membrane. The main component of this coat is the protein clathrin, whose polymerization produces a local depression in the membrane (coated pit) that gradually turns into an endocytic coated vesicle. Afterward, coated vesicles shed their coat and fuse into large vacuolar structures (**endosomes**). Endosomes have an acidic lumen (pH 5), produced by a membrane proton pump (ATPase).

The internalized ligand and receptor can undergo different fates (Fig. 3.4):

1. In the acidic pH of the endosome, the internalized receptor–ligand complex dissociates; the receptor recycles to the cell surface, whereas the ligand is transported to the lysosome and degraded. A receptor may recycle several times before being degraded. This applies to LDL, insulin, LH, and other polypeptide hormones.

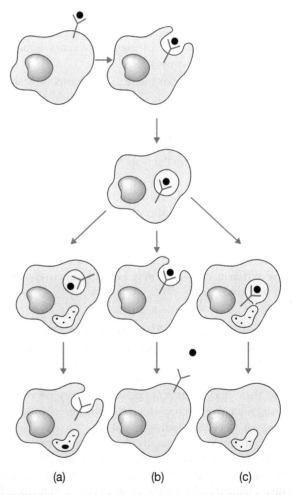

FIGURE 3.4 Possible fates of receptors and their ligands following receptor-mediated endocytosis. (a) The ligand remains in the endolysosomal compartment, while the receptor recycles back to the surface (e.g., LDL receptor). (b) Both receptor and ligand recycle back to the cell surface (e.g., transferrin system and intracellular uptake of iron). (c) Both receptor and ligand are degraded within the lysosome (e.g., EGF receptor).

2. Both receptor and ligand recycle to the plasma membrane. This is the case of transferrin and its receptor. When complexed with Fe^{3+} ions, transferrin binds to its receptor (expressed by all cells) with high affinity, and the complex is subsequently internalized. In the endosome, the low pH causes transferrin to release Fe^{3+} that can be used by the cell, whereas the transferrin–receptor complex is recycled to the cell membrane. Here, the complex dissociates because, in the absence of Fe^{3+}, transferrin has low affinity for its receptor in the neutral pH of the extracellular space. Transferrin is released in the circulation where it can bind other Fe^{3+} (mostly at enteric and hepatic levels), while the receptor is again available to bind iron-loaded transferrin molecules. This mechanism is necessary to deliver Fe^{3+} to cells, as free ions tend to generate insoluble ferric hydroxide at neutral pH.
3. Both receptor and ligand may be degraded, as in the case of the epidermal growth factor (EGF) and its receptor. Part of the EGF signaling occurs during the internalization pathway.
4. The ligand undergoes transcellular transport, whereas the receptor is recycled. This has been described in polarized (epithelial) cells and also for proteins such as IgA and IgM that are secreted in bile by the liver and in milk by breast cells and IgG that are transferred from the mother to the fetus in the placenta.

As described for other active transport mechanisms, receptor-mediated endocytosis is specific, energy dependent, and limited by the number of binding sites. On the other hand, it is more versatile and allows large molecules to enter (or cross) cells. In addition, the specific distribution of receptors among different cell types may help to selectively target drugs to specific cells.

Active Transport against Electrochemical Gradient The mechanisms of drug transport mediated by exchangers/cotransporters or membrane pumps are described in detail in Chapter 28. These mechanisms allow transport of solutes against their electrochemical gradients, as they are coupled to energy-supplying processes (ATP hydrolysis, flow of another ion species down its gradient).

Thus, transport of drugs by these mechanisms is only partially dependent on the drug concentration gradient, is saturable, and depends on the number of active transporter/pump molecules and their rate of exchange (cycles/sec). In the case of pumps, transport is generally unidirectional (because to revert the cycle the electrochemical gradient has to be greater than the energy released by ATP hydrolysis).

The mechanism of active transport brings substances into the cells, but it may also mediate extrusion, allowing cells to dispose of toxic compounds. Drugs that exert their action inside the target cells (e.g., some antineoplastic agents and antibiotics) may be actively extruded by cotransport systems and ATPase pumps (mostly belonging to the "ABC cassette" ATPase family). The complex system responsible for drug extrusion via active transport is usually defined multidrug resistance (MDR) because it underlies the property of tumor cells to be (or become) insensitive to specific therapeutic agents. Notably, certain tumor cell types overexpress detoxifying enzymes (as a result of the activation of xenobiotic-activated receptors, XARs) that can metabolize and inactivate antitumor drugs. In some cases, combined therapies including drugs capable of altering ion gradients or affecting the activity of extruding pumps or detoxicating enzymes have been used to increase the efficacy of chemotherapeutic agents.

Transporters and Facilitated Diffusion Drugs can cross membranes through transporters or carriers, which are membrane proteins that can alternately expose a substrate binding site on either of the two membrane sides. When ions and charged molecules bind membrane transporters, their charged groups are hidden in the complex, which therefore can cross the membrane and release the cargo on the other side. Some exogenous substances, such as the antibiotic valinomycin, can insert into biological membranes and act as carriers (selective for potassium ions, in the case of valinomycin). In general, carriers are quite selective, and many of them can transport classes of molecules, such as small organic anions or cations.

Similarly to passive diffusion, this type of transport (called **facilitated transport**) moves solutes down their concentration gradients. However, in contrast with passive diffusion, it is saturable and limited by the number of carriers and their turnover rate. This means that, above a critical drug concentration, carrier-mediated transport becomes constant (maximum transport), thereby showing mixed kinetics: zero order (independent of substrate concentration) at high substrate concentrations and first order (proportional to concentration) at lower concentrations.

Diffusion through Membrane Channels Hydrophilic or charged substances, such as ions and drugs, can cross the plasma membrane through specific channels. These are transmembrane proteins that can generate, within their own structure, a hydrophilic pore, or a cavity, lined by charged amino acid residues interconnecting the aqueous compartments on the two membrane sides. In most cases, these channels also contain residues that determine which charged species can move through them. In all cases, drug flow obeys the laws of diffusion: the solute flows across the membrane with a velocity proportional to its diffusion coefficient, to the membrane surface area, and to the density of pores; the net flux is proportional to the electrochemical gradient. Some drugs can insert into the lipid membrane and form ion-selective channels; this is the case of the antibiotic gramicidin that forms a cation-selective pore.

DRUG DIFFUSION TO ORGANS AND TISSUES

As discussed previously, a drug can move from a compartment to the next through passive diffusion or facilitated or active transport mechanisms. Depending on the mechanism, flow velocity may be determined by Fick's law (i.e., by solute concentration difference and barrier area/thickness) or, when a facilitated or an active transport is involved, by the density of transport molecules. In the last case, at high drug concentrations, the flow may be limited by saturation.

When we consider a drug entering a specific organ or tissue, also other factors come into play. To clarify, imagine a model of the cardiovascular system made of pipes and reservoirs in which water flows moved by a central pump. If we introduce some dye in proximity to the pump, most of it will move first to those reservoirs connected through the largest pipes (representing organs with the highest fluxes). Moreover, among the reservoirs with similar flux, the dye will reach the equilibrium first in those with the smallest volume. The critical parameter in this process is the flux-to-volume ratio, the so-called specific flux. For each reservoir, the time to equilibrium depends on its specific flux, and the time constant of the diffusion process is equal to the ratio volume/flux [ml/(ml/min)=min]. Similarly, each organ/tissue reaches the equilibrium with the plasma with a velocity proportional to its specific flux (milliliter of plasma per gram of tissue per minute). Note that, as discussed in Chapter 5, the volume to consider is not the real volume of the tissue/organ but rather its apparent distribution volume.

The considerations earlier may sound contradictory: on one hand, each organ/tissue should reach the equilibrium with the plasma based on the permeability of the cellular barriers that separate them; on the other hand, the determining factor seems to be the flux/volume ratio. Actually, here, a general rule applies: between two limiting factors, the most stringent prevails. The plasma reaching a tissue releases the drug at a rate that depends on the cell barrier permeability (and possibly on the transport system efficiency). However, the plasma reaching a tissue cannot leave more drug than that required to bring it to the equilibrium with the tissue. Thus, the equilibration rate cannot be higher than the specific flux, which therefore becomes the limiting factor. For example, a poorly lipophilic drug (e.g., β-blocker atenolol) hardly and slowly diffuses to the CNS, because of its slow diffusion through the BBB (see the following text), but it rapidly distributes to other tissues (e.g., skeletal and cardiac muscles). Vice versa, the diffusion of a highly lipophilic drug (e.g., the barbiturate Pentothal) is unaffected by the BBB but is limited by the specific perfusion. This means that the drug can reach high concentrations much more rapidly in the brain (highly perfused) than in the muscles or other tissues (see Chapter 5 for a more detailed discussion of these aspects).

The most relevant factors controlling drug (and solute in general) distribution to different body compartments are the following (see also Fig. 3.2):

1. The net drug flux across a barrier separating two compartments is proportional to the barrier surface area.
2. The flux is inversely proportional to the barrier thickness (e.g., several cell layers).
3. The flux is proportional to the density of pores or transporter molecules, or to the intensity of the endo-/exocytic activity, when these processes are involved.
4. The flux is proportional to the concentration difference across the membrane. This holds true when passive diffusion processes predominate or when carriers are involved but drug concentration is low. The flux becomes independent of drug concentration under conditions of saturation (of pumps, exchangers, endo-/exocytosis, etc.).
5. The equilibration rate between compartments is proportional to the ratio between the drug quantity that moves between compartments (every second) and the volume of the compartments.
6. The equilibration speed between a compartment and plasma, when not limited by diffusion/transport processes, is equal to the specific perfusion of the compartment (i.e., the ratio between hematic flux and compartment/tissue volume).

In the following pages, we will examine the structural and functional characteristics of the cell barriers that separate various compartments. Such characteristics determine not only the efficiency and velocity of drug penetration into various compartments but also the selectivity of the barrier, with reference to the physicochemical properties of different drugs and the possible presence of selective transport mechanisms.

Properties of the Most Important Cell Barriers

The Capillary Endothelium: A Very Weak Barrier for Low-Molecular-Weight Compounds The capillary endothelium represents a labile barrier for most drugs because of its very limited thickness, high endo-/exocytic and transcytotic activity, and the presence of fenestrae and membrane pores. The number of capillaries may vary from $50/mm^2$ in the skin to $2000/mm^2$ in the myocardial tissue; this offers a huge exchange surface area, allowing a very rapid diffusion of most drugs from the capillary bed to interstitial fluids.

Hydrophilic, polarized, and charged molecules, if not specifically transported, diffuse through pores and fenestrae, which have a molecular cutoff of approximately 60,000 Da (corresponding to albumin molecular weight). For lipophilic molecules capable of diffusing across cell membranes (including plasma-dissolved gases such as oxygen, nitrogen, carbon dioxide, and anesthetic gases), diffusion speed mostly depends on the partition coefficient (OWPC).

Therefore, (i) only the unbound (free in solution) drug molecules can cross the endothelial barrier: aggregation or binding to plasma proteins (see the following text) markedly decreases the speed at which drugs can leave the circulation; this phenomenon can be exploited to regulate drug availability through particular chemical formulations (see Table 4.10); (ii) changes in the morphofunctional organization of the capillary endothelium (e.g., inflammatory processes) may influence the velocity of drug diffusion from vascular bed to interstitial fluids at specific sites.

Different Permeabilities of Capillary Beds Capillary permeability varies among organs, because of differences in both endothelial cell thickness and pore number (Fig. 3.5). For example, the endothelium of renal glomeruli is highly permeable; in the liver, the sinusoid capillaries lack a basal lamina and the endothelium is discontinuous, so that plasma proteins can easily diffuse to hepatocytes.

Permeability properties of the capillary endothelium are modulated by specific physiopathological conditions and by drugs (e.g., histamine produces an increase in capillary permeability). In addition, perfusion of capillary beds may vary (due to arteriolar vasodilation or vasoconstriction), and this may influence the velocity of drug diffusion to certain tissue districts.

The Blood–Brain Barrier The endothelium of cerebral capillaries is particularly impermeable and sheathed by astrocytic projections to constitute the so-called BBB.

The BBB limits the penetration of hydrophilic drug into the CNS. Drugs can reach CNS cells through two paths: from the blood or from the cerebrospinal fluid (CSF). In both cases, the cell barriers they have to cross are noticeably more efficient and selective compared to normal capillary endothelia (Fig. 3.6).

Encephalic capillaries display very limited exo-/endocytic and transcytotic activity, lack the large pores that characterize the normal capillary endothelium, and are almost entirely sheathed by glial cells (see Fig. 3.5). Because of this particular morphological organization, in the CNS, the interstitial liquid composition is quite different from that of other compartments, being almost devoid of plasma proteins. In order to cross the BBB, a drug needs to be either actively transported or lipophilic.

Drugs with a high OWPC enter the CNS very rapidly, due to the high perfusion index of the encephalon (on average 0.5 ml/min/g of tissue, i.e., about 10-fold a resting skeletal muscle; up to 1.3 ml/min/g of cerebral gray matter). Moreover, given the very high quantity of lipids (abundance of cell membranes and myelin sheaths), lipophilic drugs tend to accumulate in the nervous tissue, especially in the white matter.

Hydrophilic drugs do not enter the CNS to a significant extent and may lack central effects unless they are directly administered in the CSF (an infrequent event) or the BBB is functionally impaired: this possibility must be considered in aged patients (an atherosclerotic insult may damage capillary walls) and in the presence of encephalic and meningeal inflammatory conditions or in cases of particularly elevated fever, especially in children.

In general, drug penetration in the CNS is strongly influenced by:

1. Binding to plasma proteins, as only the unbound fraction can diffuse across the BBB or in the liquor (CSF)
2. Ionization, as the ionized fraction is excluded from the CNS
3. Partition coefficient (OWPC) for the nonionized fraction

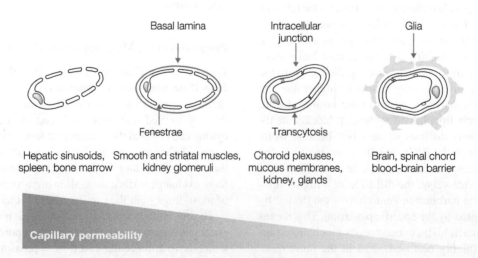

FIGURE 3.5 The morphological organization of capillary vessels varies in different districts resulting in local differences in permeability.

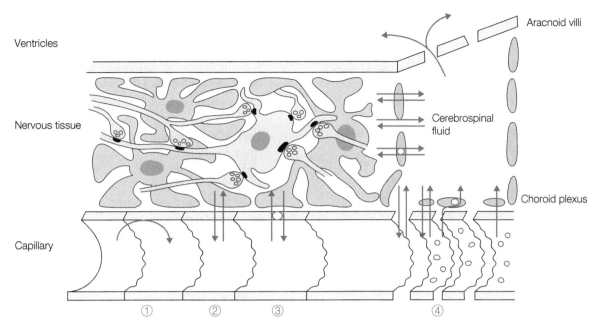

FIGURE 3.6 The blood–brain barrier (BBB). The interstitial fluid that surrounds nerve cells in the brain is in equilibrium with the cerebrospinal fluid (CSF), which is almost devoid of plasma proteins. The endothelium of cerebral vessels together with glial cell projections constitutes the BBB. This prevents the entry of any substance that cannot freely diffuse in lipid membranes into the cerebral interstitial fluid (1). Thus, only drugs with an adequate partition coefficient (2) or recognized by transport systems present in the BBB (3) can penetrate the CNS. The BBB is almost absent at the level of the choroid plexus (4) and other CNS regions (pituitary gland, median eminence, and area postrema on the floor of the fourth ventricle) where CSF is usually formed. At these sites, hydrophilic drugs can interact with the exposed brain tissue. Lack of BBB allows the chemoreceptor trigger zone, close to the area postrema, to sense toxic compounds or conditions in the blood and activate vomit.

The partition coefficient per se is not sufficient to predict the degree of penetration of a drug into the CNS; however, among drugs with similar chemical structure, a higher OWPC suggests a more efficient penetration.

Active transport systems are present in the BBB, which transfer amino acids, glucose, ions (e.g., Ca^{2+}, Mg^{2+}, other metals, and phosphates), small peptides with N-terminal tyrosine (e.g., enkephalins), and polypeptides (e.g., insulin, growth factors, and transferrin) into the CNS. Transport mechanisms can be exploited for therapeutic purposes, as in the case of L-DOPA therapy for Parkinson's disease; the compound, which is a precursor of the neurotransmitter dopamine, is hydrophilic, and its uptake in the brain is mediated by the amino acid transporter in the BBB.

In addition to capillaries, drugs can reach nerve cells in the brain through the CSF. This derives partly from the cerebral interstitial fluid, but mostly from an ultrafiltration process occurring at the choroid plexuses and similar circumventricular structures. At the plexuses, the endothelium has fenestrae and is more permeable than elsewhere in the CNS; however, its very limited extension (about 5000-fold less than the encephalic microcirculation) drastically limits the passage of hydrophilic drugs to the CNS.

The CNS areas close to choroid plexuses and some periventricular regions (pituitary gland, median eminence, and chemoreceptor trigger zone) are less "protected" by the BBB and therefore more sensitive to drug actions.

The Placental Barrier The placental barrier is less impermeable than the BBB. Placental villi are perfused by fetal capillary circulation and enclosed in the maternal hematic sinuses. Thus, the placental syncytium, interstitial space, and capillary endothelium of the villi separate fetal from maternal blood. The placental syncytium acts as a molecular filter in which all transport mechanisms discussed previously are particularly active: passive and facilitated (e.g., glucose) diffusion, active transport (for amino acids that are more concentrated in fetal than maternal blood and for calcium and phosphate), and receptor-mediated endocytosis (particularly efficient for transferrin, IgG, and other maternal peptides and proteins).

Passive diffusion from maternal to fetal blood and vice versa depends on the drug partition coefficient, the concentration difference, and binding to plasma proteins, as described for the BBB. Note that tissue layers separating fetal from maternal blood are about 25 μm thick in early pregnancy and only 2 μm at the end of pregnancy.

BBB and placental barrier differ in several respects. First, the placental barrier is less tight than BBB and hydrophilic drugs can cross it, being mainly limited by their molecular size. Second, maternal blood flows very slowly in the placenta, allowing drugs more time to move into fetal circulation. Thus, placental transfer is negligible for those drugs that are hydrophilic and large in size and are rapidly cleared from the blood of the mother.

Equilibrium kinetics are slow in the placenta also because of binding to fetal plasma proteins. In other compartments, interstitial fluids are essentially devoid of proteins capable of binding the drug; therefore, drug concentration rapidly equilibrates with the level of free drug in the plasma. In the placenta, fetal plasma proteins bind part of the drug, thus slowing down the rise of free drug level toward the equilibrium with the maternal blood. In other words, plasma proteins increase the apparent distribution volume, decrease the flux/volume ratio, and therefore slow down the kinetics (see also Box 7.1). Labor and delivery represent particular conditions, as the baby can suffer a marked acidosis. If the mother is administered acidic drugs (such as barbiturates) for anesthesia induction, these may significantly accumulate; given the reduced metabolic clearance of the newborn, this may produce persistent sedation for up to 12–24 h.

In general, it is appropriate to think that any drug given during pregnancy will reach the fetus as well, although slowly and possibly at lower concentrations. Available data suggest that equilibration between fetal and maternal blood takes at least 40 min in the most rapid cases; compounds that are large sized, hydrophilic, ionized, or highly bound to plasma proteins may take hours to equilibrate. Under these conditions, if the drug is rapidly eliminated by the mother, therapeutic doses can be administered with no significant effect on the fetus.

Similar considerations can be made about puerperium and breastfeeding, as virtually all compounds present in maternal blood at significant concentrations will also be found in milk.

TAKE-HOME MESSAGE

- Temporal changes in its local concentrations at different sites, including the therapeutic target, define the pharmacokinetic properties of a drug.
- The plasma concentration of a drug is the end result of its absorption, distribution, metabolism, and excretion.
- During its journey through the body, a drug crosses several cellular barriers, which may differ in permeability and in the variety of active transport mechanisms they are endowed with.
- After entering the bloodstream, drugs diffuse to all organs and tissues; along their journey within the body, they can be metabolized (mostly by the liver) or eliminated (mostly by the kidney).
- The capability of a drug to cross cellular barriers mostly depends on its physicochemical properties and in particular on its molecular size and oil–water partition coefficient (OWPC). Hydrophilic drugs mainly cross cellular barriers if transport mechanisms are present.
- Drug absorption, distribution, and elimination mostly depend on passive diffusion processes: in general, the net molar fluxes between compartments are proportional to concentration differences (first-order kinetics), and the rates of equilibration between plasma and tissues depend on the flux-to-volume ratio (perfusion index).

FURTHER READING

Birkett D.J., *Pharmacokinetic made easy*, McGraw-Hill Publishing Co: New York, 2003.

Clark R.D., Wolohan P.R. (2003). Molecular design and bioavailability. *Current Topics in Medicinal Chemistry*, 3, 1269–1288.

4

DRUG ABSORPTION AND ADMINISTRATION ROUTES

RICCARDO FESCE AND GUIDO FUMAGALLI

By reading this chapter, you will:

- Know the properties of the administration routes in terms of rate of absorption, drug transfer to the systemic circulation, and time course of plasma concentration
- Learn the advantages and limitations of different routes of administration and the concept of bioavailability
- Learn how to select the appropriate route of administration based on therapeutic needs and patient compliance
- Learn how some administration routes can be used to produce local and no systemic effects

---

TABLE 4.1 Drug administration routes

Enteral	Oral, sublingual, rectal	Per os
Parenteral systemic	Intravascular	Intravenous (i.v.; systemic), intracardiac, intra-arterial (regional)
	Intramuscular (i.m.)	i.m.
Others	Cutaneous	Subcutaneous, intradermal
	Organ	Intrathecal, intra-articular, inhalatory
	Intracavity	Intraperitoneal, intrapleural
	Transdermal	Both local and systemic
	Transmucosal	Ocular, vaginal, vesical

The term "systemic administration" indicates that the drug is administered in a way that allows it to enter the systemic circulation and diffuse to all organs and tissues including the targets. Routes of systemic administration are classified as enteral, if the site of absorption is in the digestive apparatus, and parenteral, if anywhere else. The term topic administration is used to indicate that the drug is applied directly where its effects are desired, whereas its absorption and diffusion to the rest of the body are usually (but not always) considered a side effect.

The different routes of drug administration are listed in Table 4.1. As shown in the table, the distinction between systemic and topic administration is not absolute, as a topic route can be used for controlled delivery of drugs in the systemic circulation and vice versa. For example, the subcutaneous route is generally used for systemic delivery of large molecules that are not adsorbed by oral ingestion (e.g., insulin) but is used by dentists for local application of anesthetics. Topic application on the skin is generally used for local skin treatments but is also used to obtain a slow and prolonged absorption of drugs that can cross cell barriers (e.g., transdermal patches of nitrates or antikinetotic drugs).

When choosing the route of administration, we need to consider the physicochemical properties of the drug and its therapeutic use. For example, highly hydrophilic drugs are usually poorly absorbed by the gastrointestinal apparatus and a parenteral injection is preferred; by contrast, lipophilic drugs are very well absorbed even when applied topically on the body surface (skin or mucosa).

The time course of drug plasma concentration varies depending on the route of administration. Rapid intravenous (i.v.) injection may induce high plasma concentration and toxic effects, whereas the same dose administered by intramuscular (i.m.) injection may result in lower peak values, no toxic effects, and longer action duration. For some drugs, slow- or multiphasic-release formulation is available. The choice of administration route may also be based on practical aspects and patient's compliance. For

---

General and Molecular Pharmacology: Principles of Drug Action, First Edition. Edited by Francesco Clementi and Guido Fumagalli.
© 2015 John Wiley & Sons, Inc. Published 2015 by John Wiley & Sons, Inc.

example, a chronic therapy should be based on the least traumatic procedure of administration (oral route), while with noncollaborating patients (comatose, newborn), the parenteral (endovenous) route is preferred. Psychological consideration should also be taken into account: in some cases, patients adhere more tightly to doctor's indications if the therapy is given by injection rather than pills.

GENERAL RULES ABOUT DRUG ABSORPTION RATE

In all routes of administration, except the intravascular injection, the drug has to cross several cellular barriers (or membranes) and diffuse in intercellular spaces to reach the bloodstream. Variables affecting the adsorption rates are listed in Table 4.2; some of them relate to properties of the substances, and others to those of the adsorbing tissue.

Partition Coefficient

The drug oil–water partition coefficient (OWPC; see Chapter 3) is relevant to absorption when the administration route involves crossing of cell barriers (parenteral, transdermal, inhalatory). The larger the OWPC and the smaller the drug molecular radius, the faster the crossing of membranes and cells. Since drugs have also to diffuse in aqueous interstitial fluids, they also need to have some degree of hydrophilicity. Drugs with low OWPC (hydrophilic), which hardly cross membranes, should be administered by parenteral injection.

Drug Dispersion

To be absorbed (i.e., to enter the bloodstream), a substance has to be fully dissolved. For drugs administered in solid formulation, the time required for the pharmaceutical preparation to be dissolved (e.g., in the gastrointestinal juices) may limit the absorption rate. To avoid delay caused by dispersion (solvation), some drugs are dissolved in water before administration, whereas formulations that dissolve slowly are used in slow-release (delayed) administration.

The dispersion rate of a drug formulation depends on its solubility, temperature, and in some cases solvent pH. A major role is played by the **excipient**, which is the pharmacologically inactive substance formulated with the active ingredient; the same active compound formulated with different excipient may show different absorption rates and/or bioavailability (Fig. 4.1). In the last few years, new technologies (including nanotechnology) have been used for excipient preparation to modulate drug absorption and delivery.

Extension of the Absorbing Surface

As discussed in Chapter 3, the larger the size of the absorbing area, the faster the absorption rate. Given its very large surface, the small intestine, rather than the gastric mucosa, is the main site of absorption of orally administered drugs. For the same reason, absorption of anesthetic gases occurs in lung alveoli.

Permeability of the Absorbing Surface

Permeability of epithelia is mostly dependent on their thickness and cornification (keratinization). Thus, drugs cross rapidly the monolayered epithelium of the small intestine, slowly the multilayered nonkeratinized pharyngeal epithelium, and very slowly the skin (multilayered keratinized epithelium). Monolayered epithelia may have different permeability depending on the number and function of their tight junctions. In addition, pathological conditions and loss of continuity (wounds) may cause rapid absorption of drugs.

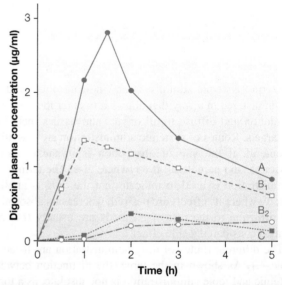

FIGURE 4.1 Pharmacokinetic variability. The graph shows the time course of digoxin plasma concentration following administration of four commercial formulations of the drug, produced by three distinct companies (A, B, and C), to the same subject. Trials were separated by 1-week intervals. In all cases, the subject received two digoxin pills (0.25 mg) with 100 ml water with empty stomach in the morning. Note the marked differences in bioavailability, also among preparations manufactured by the same company (B_1 and B_2). Different bioavailabilities lead to different plasma concentration peaks and time courses.

TABLE 4.2 Variables affecting rates of drug absorption

Variables dependent on the drug or its pharmaceutical composition	OWPC Dispersion/solvation
Variables dependent on the absorbing surface	Size Permeability

Drugs administered by injection into soft tissues rather than vascular structures are absorbed by diffusion from the injection site in the interstitial space to the closest capillaries. Large molecules can also enter the circulation by this route. The absorption rate depends on vascularization. Therefore, absorption is modulated by hydration and blood perfusion that may vary with age or in various pathophysiological conditions.

Vascularization

The absorption rate depends on vascularization and blood flow in the absorbing area. Absorption is faster following i.m. rather than subcutaneous injection because muscle tissue is more vascularized. Local inflammation or increased blood flux in response to physiological stimuli (exercise for muscle, heat exposure for skin) can modify the absorption rate of an area. For this reason, vasoconstrictors are often injected along with local anesthetics to slow down absorption and prolong their effect.

ENTERAL ROUTES OF ADMINISTRATION

Routes of enteral administration are oral, sublingual, and per rectum (or rectal). Administration by these routes is easy, is not painful, and does not require specialized personnel. Absorption occurs according to the pharmacokinetic principles described earlier and in the previous chapter. However, the properties of the gastrointestinal apparatus and its juices may change the chemical structure and activity of drugs. Therefore, not all drugs can be administered by this route. Drugs that are not absorbed in the gastrointestinal tract can be administered by the enteric route to exert local effects, as in the case of some antimicrobial agents.

The sublingual and the per rectum routes allow drugs to be delivered directly into the systemic circulation, thus avoiding the portal vein system and liver filtration. For these reasons, they are frequently used, instead of the oral route, for drugs that are extensively destroyed by the liver.

Oral Route

Oral Bioavailability With the oral route (per os), absorption occurs in the gastrointestinal tract, mostly by passive diffusion, and is therefore dependent on drug OWPC and possible level of ionization. Most drugs are significantly absorbed in the small intestine, which represents the portion of the gastrointestinal tract with the most extended surface and vascularization. To enter the systemic circulation, an orally administered drug first has to disperse and dissolve, then cross a cell barrier, diffuse to the vessels, and avoid liver filtration and metabolism. Each of these steps may reduce the amount of active drug that enters the systemic circulation and diffuses to the entire body. The ratio between the amount of drug reaching the systemic circulation and the amount administered is called oral bioavailability.

Oral bioavailability and rate of absorption may vary depending on the functional state of the digestive system, composition of its content, rate of transit in the stomach and small intestine, composition and activity of the intestinal flora, and drug sensitivity to the hepatic metabolic activity. When administering a drug by the oral route, all these variables have to be considered and taken into account.

Absorption along the Gastrointestinal Tract Absorption of drugs by oral route occurs mostly in the small intestine. The small intestine has a surface area of approximately $200 \, m^2$ and a blood flux of $1 \, l/min$; the corresponding values for the gastric mucosa are $1 \, m^2$ and $0.15 \, l/min$. In addition, the intestinal mucosa is permeable to small molecules even when their OWPC is relatively low. These features make the small intestine the major site of absorption of orally administered drugs; they also explain why absorption of a weak acid such as acetylsalicylic acid occurs in the intestine although the dissociated (charged) form prevails in the neutral pH of the intestine. In small bowel, the mucosa is permeable to many small-sized molecules, even though they may have a relatively low partition coefficient; thus, for many drugs, the absorption rate mostly depends on the intestinal blood flow.

For large or highly hydrophilic compounds, the absorption rate is slow and independent from the intestinal blood flow; for these substances, changes in functional integrity and/or motility of the intestinal mucosa may significant influence bioavailability. In this context, it is important to remind that transit time through the digestive system is usually 1–2 days, but through the small intestine is only 3 h, regardless of its content. Thus, changes in intestinal peristalsis (e.g., due to drugs) may have significant consequences on absorption time and significantly modify oral bioavailability of drugs that are slowly or incompletely absorbed. The extent of such changes is not always predictable; for this reason, it is common practice to avoid administration of drugs affecting intestinal motility when a long-term therapy is ongoing. A list of drugs that may interfere with oral absorption of other drugs is shown in Table 4.3.

Effects of Gastric Emptying on Oral Absorption of Drugs As stated earlier, absorption in the stomach is very limited; for this reason, the stomach should be considered as depot organ that releases drugs to be absorbed in the small intestine. Therefore, it is important to keep in mind factors and conditions that control gastric emptying as they determine when and how rapidly a drug reaches the intestine, thereby affecting beginning, intensity, and duration of drug effects.

TABLE 4.3 Modification of oral absorption by coadministration of other drugs

Drugs inducing the modification	Effect on absorption	Drugs whose absorption is modified
Anticholinergic drugs	Decreased	Paracetamol, tetracyclines, phenylbutazone, lithium (retard preparations)
	Increased	Digoxin
Opioids	Decreased	Alcohol, morphine
Tricyclic antidepressants	Decreased	Phenylbutazone
	Increased	4-Aminosalicylic acid
Metoclopramide	Decreased	Digoxin
	Increased	Aspirin, levodopa, paracetamol, lithium (retard preparations)

TABLE 4.4 Effects of food intake on absorption of some drugs administered per os

Increased absorption	Decreased absorption
Ampicillin, amoxicillin, rifampicin, aspirin, isoniazid, levodopa	Griseofulvin, carbamazepine, propranolol, metoprolol, spironolactone, hydralazine

TABLE 4.5 Chemical reactions occurring in the gastrointestinal tract that can modify drug oral bioavailability

Reaction	Drug	Effect
Complex formation	Tetracyclines	Insoluble complexes with polyvalent ions (Ca^{2+}, Al^{3+})
Conjugation		
Sulfation	Isoproterenol	Loss of pharmacological activity
Glucuronidation	Salicylamide	Loss of pharmacological activity
Decarboxylation	Levodopa	Loss of pharmacological activity
Hydrolysis		
Acidic	Penicillin G, erythromycin	Loss of pharmacological activity
Enzymatic	Aspirin	Formation of salicylic acid, active
Reduction	Sulfasalazine	Formation of 5-aminosalicylic acid, active

During fasting, the stomach cycles between contraction and relaxation phases of variable duration, which peak with an intense, ejective contraction about every 30 min. For drugs administered in liquid formulation, delivery to the small intestine depends on the occurrence of a propulsive contraction; this may cause some delay between administration and beginning of increase in blood concentration. For drugs administered in solid formulation, the time required for dissolution needs also to be considered since small solid debris may adhere to the gastric wall and be retained in the stomach.

The situation is more complex when the stomach is full. In this condition, the stomach shows continuous nonpropulsive contractions whose function is to mix food with gastric juices; only particles smaller than 2 mm can pass through the pylorus. Gastric emptying is usually slow; with a normal meal, it takes about 7 h, but it takes longer or less time with fat-rich foods and light meals, respectively. For drugs administered in solid form, a further delay may occur because of the time required for dissolution. Thus, for a fast absorption, a drug should be administered in liquid form with empty stomach; for a slow absorption, a solid formulation administered after a solid meal is preferred.

An increase in pH may accelerate gastric emptying in fasting condition; this probably explains why buffered and dissolved formulations of aspirin are absorbed so quickly. Changes in gastric pH can also be induced by treatments with antiacid drugs buffering or blocking H^+ secretion. Interestingly, composition of buffering antacids may influence gastric emptying: for example, Al^{3+}-containing salts delay whereas Mg^{2+}-containing salts accelerate gastric emptying.

Interference of Food with Drug Absorption Besides the reported effects of meals on gastric emptying and drug absorption, the presence of food in the intestine may modify both absorption rate and bioavailability of several drugs (Table 4.4). For most drugs (with the exception of large and highly hydrophilic compounds), the limiting factor for absorption is the intestinal blood flux; this may be significantly modified by the content of the upper digestive tract, with protein-rich meals causing an increase in blood flow and carbohydrate-rich meals producing the opposite effect. Pathological conditions may also modulate intestinal blood flow.

Changes in drug oral bioavailability may also occur as a result of meal-dependent modulation of the hepatic metabolic activity. This is the case of propranolol, a beta-adrenergic blocking agent that is usually extensively degraded by the liver. After a meal, the transporters mediating hepatic uptake of the drug are saturated by other substrates, and propranolol can escape degradation, thus resulting in higher plasma concentration when administered after or during meals.

Effects of Chemical Modifications of Drugs on Their Bioavailability Oral bioavailability of a drug can be modified by chemical reactions occurring in the gastrointestinal tract or in the liver. A partial list of the physicochemical events that may modify absorption and/or activity of some drugs is shown in Table 4.5. Aggregation and formation of insoluble aggregates with foods may occur, as in the case of tetracyclines, which interact with polyvalent cations, present in large amounts in dairy products or in some buffering antacids. Drug modifications may be due to chemicals, including gastric HCl, or to enzymes present in gastrointestinal fluids or intestinal flora. In most cases, drug biological activity is reduced.

Drugs absorbed in the gastrointestinal pathway may be cleared by the liver before entering the systemic circulation. Table 4.6 shows a list of drugs whose biotransformation during the first pass through the liver is so extensive to significantly reduce their bioavailable fraction. This phenomenon is called first-pass effect.

For some drugs, including the beta-adrenergic agonist isoproterenol and the antiangina trinitrine, the first-pass effect almost completely abolishes the therapeutic activity. In these circumstances, a change of route of administration (or drug) should be considered. For trinitrine, the sublingual route represents a useful alternative as it allows direct delivery to the systemic circulation via the superior vena cava, thus escaping liver capture and metabolism.

Sublingual and Rectal Routes

Both routes are used to deliver drugs that undergo extensive degradation in the digestive system or first-pass effect. By these routes, at least part of the drug escapes the liver first-pass effect. Indeed, blood from the lower and intermediate tracts of the rectum is drained by the corresponding hemorrhoidal plexuses and flows into the inferior vena cava (the blood drained from the upper third of the rectum enters the portal circulation), and the blood from the mouth is drained by the superior vena cava.

The sublingual route is the preferred for organic nitrate administration. These drugs are highly lipophilic and quickly cross the multilayered epithelium of the tongue; this muscular organ is highly vascularized, thus allowing fast absorption and rapid appearance of therapeutic effects.

TABLE 4.6 Drugs sensitive to first-pass effect

Amitriptyline	Tricyclic antidepressant
Desipramine	Tricyclic antidepressant
Diltiazem	Calcium channel blocker
5-Fluorouracil	Antineoplastic agent
Isoproterenol	Beta-adrenergic agonist
Labetalol	Beta-adrenergic antagonist
Lidocaine	Local anesthetic, antiarrhythmic agent
Mercaptopurine	Antineoplastic agent
Metoprolol	Beta-adrenergic antagonist
Morphine	Opioid analgesic
Neostigmine	Acetylcholinesterase inhibitor
Nifedipine	Calcium channel blocker
Nitroglycerin	Nitrovasodilator
Propranolol	Beta-adrenergic antagonist
Testosterone	Androgen, steroid hormone
Verapamil	Calcium channel blocker

The rectal route is indicated for patients with swallowing problems or when vomit is present. Absorption is usually slow, prolonged, and not always constant; for this reason, this route is used for maintaining active concentrations of certain NSAIDs and bronchodilators during the night hours.

SYSTEMIC PARENTERAL ROUTES OF ADMINISTRATION

The Intravascular Route

This is the route of choice when a rapid and/or controlled drug delivery is required since the intravascular route delivers the drug straight into the circulation. The injection is usually given into the veins. Arterial injections are limited to situations requiring drug delivery at high concentration in a specific vascular district, as in the case of some antineoplastic drugs or angiography. When using the intravascular route, the drug may be administered as a single-bolus injection or via continuous infusion (**drip injection**).

The i.v. route is also used to administer substances that are not adsorbed by other routes (e.g., because of their high molecular weight) and for irritating substances that cannot be administered intramuscularly or subcutaneously.

Risks Associated with i.v. Administration Most risks associated with i.v. injections can be limited simply applying good medical practice (see Table 4.7). The first risk is the inappropriate dosage; whereas in cases of excessive doses administered via the oral route vomit can be induced or adsorbing substances can be administered to wash out the drug excess from the gastric lumen, no treatment other than plasmapheresis is available for the i.v. route.

The second risk is associated with too rapid injections, for which the most sensitive organs are the highly perfused brain and heart. To appreciate the problem, consider the following example. For a good antiarrhythmic effect, a dose of 150 mg of lidocaine is injected in a 75 kg patient; since the drug distribution volume is 1 l/kg, after complete tissue distribution, the lidocaine plasma concentration will be 2 mg/l, which is the appropriate therapeutic concentration. A concentration as high as 10 mg/l may induce serious side

TABLE 4.7 Potential complication associated with i.v. administration of drug

Rapid toxic reactions	Due to fast injection
Infections	Nonsterile materials and/or compounds
Embolism	Injection of oily solution
	Injection of large volumes of gas or air
	Injection of solution of abnormal osmolarity
	Injection of solution containing precipitates

effects, especially in the CNS. If the dose is injected in 1 s, the entire dose will dilute in 0.1 l (the size of the cardiac output in 1 s), and lidocaine concentration will be, at least for a few seconds, higher than 1500 mg/l. Also, assuming that the injected dose will quickly dilute in the entire circulating blood (volume 5 l), the concentration will remain very high (150/5 = 30 mg/l), and serious toxic effects will affect the CNS. For this reason, it is good medical practice to give i.v. injections of drugs very slowly, using a time interval at least similar to the circulation time (2 min approximately). The slow procedure also allows detection of possible side effects before injecting the full dose.

Embolism and infections are other possible complications of i.v. injections. Embolism may be induced by (accidental) administration of oily suspensions intended for i.m. injection or by the presence of precipitates in the injected solution. Blood red cell aggregates may form following injection of hypertonic solutions, whereas rapid infusion of hypotonic solutions may produce hemolysis. Infections are now rare as disposable materials are commonly used. On the other hand, transmission of infection diseases via i.v. injections still represents a problem in some social communities, mostly of drug addicts.

i.m. Injection

The choice of the site for i.m. injection is based on the amount of solution to be infused (deltoid muscle for small volumes and gluteus or quadriceps for large). The deltoid muscle is typically used for vaccines or drugs that may cause hypersensitivity or adverse reactions since a tourniquet can be applied to reduce drug absorption in case of reaction. The most common site for i.m. injection is the gluteus muscle where volumes higher than 5 ml can be injected. Attention must be give to avoid injury of the sciatic nerve, especially in elderly people and in children where the muscle is smaller. Absorption of drugs dissolved in 5–10 ml of water solution is usually complete in 15–30 min. The absorption rate depends on the local circulation and is therefore affected by the level of activity of the muscle. OWPC, nature of the dissolving excipient, osmolarity, and volume of the injected solution also affect absorption time. Large molecules that do not enter the capillaries may reach the systemic circulation following lymphatic absorption from the muscle interstitial fluids.

Drugs that are insoluble at interstitial pH or suspended in oily solutions are absorbed significantly more slowly; this is exploited in "retard" formulations, which allow a delayed delivery of drugs. For example, penicillin G as sodium salt is absorbed in 2–3 h, whereas its procaine salt requires 24 h.

i.m. injection can be associated with pain (due to distension and inflammation of the tissue or to irritating substances), neural and vascular lesions (due to bad practice), and abscesses or necrosis (due to injection of necrotizing substances or to lack of sterility).

Subcutaneous and Intradermal Injections

An intradermal injection consists in the delivery of small volumes (0.1–0.2 ml) of substances to the dermal connective tissue. This route is generally used for testing allergens or sensitivity to drugs.

A subcutaneous injection is used for larger volumes (usually <2 ml), and the most common sites of administration are the ventral portion of the forearm and the lateral portions of the abdominal wall. Irritating substances should not be administered by this route.

The blood flow is usually smaller than in muscle and the absorption rate is accordingly slower. It can be further slowed down by the concomitant administration of vasoconstrictors, as in the case of local anesthesia. The subcutaneous tissue is ideal for implantation of devices or injection of "deposit" formulations for delayed and prolonged delivery of drugs to the systemic circulation (frequently used for steroidal hormone therapies). Interestingly, the absorption rate depends on the size and shape of the injected particles; for example, zinc–insulin preparations consist of homogeneous microcrystals of insulin that dissolve slowly in the subcutaneous tissue, thus allowing slow absorption and maintenance of appropriate plasma concentrations of the hormone for hours.

As in the case of i.m. injection, pain, abscess, and necrosis can be undesired effects of inappropriate subcutaneous injections.

OTHER ROUTES OF DRUG ADMINISTRATION

Inhalation Route

Pulmonary absorption of drugs occurs mostly at the alveolar level. Due to the large exchange area (about 70 m^2) and the thinness of the wall separating air from blood, absorption is very fast.

Drugs can be administered as gases or aerosols for topical (as in bronchial obstruction) or systemic (e.g., anesthetic gases) use. Anesthetic gases have high OWPC and rapidly diffuse through the cells of the alveolar epithelium and capillary endothelium. Onset and duration of the effects mostly depend on the blood/air partition coefficient; the pharmacokinetic of anesthetic gases is discussed in Chapter 7.

The inhalation route is adopted for topical use in the treatment of several bronchopulmonary diseases to deliver drugs such as bronchodilators, corticosteroids, antiallergics, mucolytics, and antibiotics. Although the inhalation route is used to obtain high local concentrations, a relevant proportion of the drug may still reach the circulation and induce systemic effects.

Deposition of Aerosol Particles along the Respiratory Tree Depends on Their Dimension Solid or liquid (nonvolatile) drugs can be dispersed in air or other gases and administered

TABLE 4.8 Percentages of retention of particles of different sizes along selective tracts of the airways

	Particle diameters							
	450 cm^3				1500 cm^3			
Inhalation volume	20	6	2	0.2	20	6	2	0.2
Upper airways	33	0	0	0	47	5	0	0
Large bronchi	54	22	5	1	51	47	11	1
Terminal bronchioles	6	30	11	7	1	12	5	4
Alveoli	0	30	25	8	0	16	28	12
Total retained	93	82	41	16	99	80	44	17

as aerosol, that is, in the form of small, slowly sedimenting particles that remain suspended in the air for a long time. Most aerosol particles adhere to the walls of the respiratory tract as a result of random impact; the fraction of "impacting" particles also depends on sedimentation and diffusion rates and on inertial precipitation (the term indicates the tendency of a particle to fly straight also when the flow changes direction, as in bronchial bifurcations).

Table 4.8 shows that the percentage of deposition in various portions of the respiratory tree mostly depends on particle diameter and inhaled volume. Particle sedimentation declines with their size; thus, large particles tend to be deposited along the upper airways and the first (high) portions of the tracheobronchial tree, whereas small particles reach the final ramifications (bronchioles and alveoli). Particles larger than 20 μm are useful to deliver drugs to the upper airways, and particles ranging from 1 to 10 μm are preferred to treat bronchiolar and alveolar pathologies. Particles <1 μm are mostly exhaled and very poorly retained. When the inhaled volume is small, the airflow is slower and deposition of intermediate particles occurs earlier and more extensively at bronchiolar and alveolar levels.

Deposition of aerosol particles is also influenced by airway clearance, and aerosol penetration in the respiratory tree is reduced in chronic obstructive pulmonary disease (COPD). Other critical factors are the phase of the respiratory cycle in which the aerosol is delivered, aerosol stability and superficial tension, and the technical features of the aerosol-generating device. For all these reasons, drug absorption by inhalation may be variable.

Drugs are often delivered by an aerosol spray that contains an inert propeller and dispenses a controlled quantity of drug (metered-dose inhaler). For patients unable to coordinate the release of the aerosol with inhalation, devices are available in which the inhaler is activated by the user's own breath-in.

Nebulizers are very common: a liquid drug preparation is nebulized by an air spurt, producing medium-/large-sized particles; these devices should only be used to treat pathologies of the upper airways and large bronchi.

Topical/Regional Routes

Intra-articular injections of anti-inflammatory drugs, local anesthetics, and antibiotics are used to treat disorders of the knee and other joints. Repeated and frequent applications may easily produce local injury.

Intrathecal injections allow to bypass the BBB and to obtain rapid effects at meningeal level and in spinal nerve roots. Drugs can be introduced in the spinal subarachnoid space (lumbar injection) or in the ventricular cavities. In general, prior the injection, an equal amount of cerebrospinal fluid (CSF) is withdrawn. This route can be used to treat cerebral/spinal acute infections or tumors and to inject contrast media for radiographic examinations. To obtain diffusion toward the most cephalic parts of the CNS, the drug can be dissolved in a solution less dense than the CSF and injected in the patient maintained in upright position. The intrathecal route can also be used to inject anesthetic drugs to obtain spinal anesthesia (injection in the subarachnoid space). More commonly, the anesthetic drug is injected in the epidural space. With epidural injection, spinal anesthesia develops more slowly (because the drug must diffuse across the dura mater as well). On the other hand, diffusion to other spinal levels is limited, side effects associated with orthosympathetic blockade (resulting in cardiocirculatory alterations) are less frequent, and dorsal roots are more efficiently blocked (at the intervertebral foramina), producing additional anesthetic effects.

Intracavity Routes

This term refers to drug injections directly into the peritoneal cavity (a systemic route) or into the pleural cavity (mostly to induce regional/local effects). Both cavities have a broad, highly vascularized surface area and rapidly absorb injected drugs. Care is required to avoid lesions to the internal organs.

Dermal or Transcutaneous Route

To obtain local effects, drugs can be applied directly on accessible sites such as the skin, eyes, orifices, and cavities of the respiratory, digestive, and urogenital apparatuses. Although not desired, systemic absorption and effects may occur; thus, knowing the factors that facilitate systemic absorption is important.

Preparations for topical cutaneous application are available in the form of oil, cream, unguent, paste, powder, lotion, spray, ointment, and liniment (assuasive).

Skin absorption may occur at three different sites: corneal layer, sweat glands (<0.1% of body surface), and hair follicles (~0.2% of body surface). Hair follicles represent the most rapid path, and for some drugs (e.g., steroid drugs) and toxic agents, they are the major absorption route. For some other drugs, the corneal layer may act as a depot organ releasing the drug slowly over a prolonged time period.

Factors Influencing Cutaneous Absorption Several factors influence skin absorption, the most important being skin integrity, thickness and hydration, and local blood flow. All these may vary among different regions of the same individual. The vehicle in which the drug is dissolved is also important.

Abrasions, ulcerative lesions, burnings, and other conditions may produce discontinuities in the corneal layer, thus interrupting the skin barrier and facilitating absorption by the systemic circulation. Penetration of hydrocortisone through intact skin has been demonstrated to remarkably increase (>50-fold) after stripping the skin with a dermal band-aid.

Inflammation, local or general elevation of body temperature, and association with irritating or rubefacient agents (generally used to treat pain or to increase perfusion of underlying tissues) can notably increase topical absorption of drugs. Cold, tourniquets, and simultaneous administration of vasoconstriction agents produce opposite effects: they can be used to reduce absorption of toxic substances.

Patient's age can influence skin absorption because infant skin is thinner and more hydrated than the adult. Moreover, since body volume in newborns and infants is small, skin absorption may results in high plasma concentrations.

Skin thickness and permeability vary in different body regions. Permeability is the lowest in the plantar region, is large in scrotum and posterior ear surface, and increases when we move from the forearm to the skull.

Changes in the hydration state of the corneal layer may significantly affect absorption. In occlusive bandages, corneal layer hydration may increase from the normal value, 10% up to 50%. Under these conditions, penetration of corticosteroids through the skin may increase up to 100-fold and produce significant systemic effects.

The vehicle used in the pharmaceutical preparation, as well as its evaporation or penetration, can significantly modify the ability of the active principle to penetrate the skin. Absorption obviously depends on the partition coefficient, but scarcely lipophilic drugs may penetrate much better when suspended in unguents or fatty creams. Propylene oxide, dimethyl sulfoxide, and similar hygroscopic solvents may modify absorption by altering the hydration state of the corneal layer.

Despite its low permeability, the cutaneous route can be used for systemic administration of some drugs, in particular those that would be extensively degraded if absorbed through the portal system: for example, ointments or transdermal patches of nitroglycerin are used for chronic therapy of angina pectoris. Dermal bands containing drugs for motion sickness are usually applied to the skin behind the ear to obtain reliable and steady absorption; indeed, the skin is particularly thin in that region and protected from external factors that may produce changes in local vascularization.

Mucosal Routes

Drugs may be administered topically on nasal, conjunctival, oropharyngeal, and vaginal mucosae. Due to the lack of corneal layer and the reduced epithelial stratification, absorption from these sites is generally extensive and may produce systemic effects.

For topically applied ophthalmic drugs, absorption occurs partly through the conjunctiva and partly through the nasal mucosa after drainage through the nasolacrimal ducts. Drugs may be delivered to the aqueous humor through the cornea and reach retinal vessels. Using this route, it is possible to visualize the retinal vascular network with fluorescent dyes. It is important to keep in mind that no more than 20 μl of liquid can be instilled into an eye and anything in excess is lost with tears or drained by the lacrimal duct.

The nasal route can be used for topical application of vasoconstrictors in the presence of nasal congestion. Prolonged use may lead to dysplasia of the mucosa and frank ulcerations, a common consequence of chronic cocaine abuse. The nasal route is used to administer peptides and proteic hormones (e.g., calcitonin) in spray form: here absorption is facilitated by the thinness of the nasal mucosa and the persistency of the applied product in the nasal cavity.

ABSORPTION KINETICS

General Rules

Intravascular injections aside, absorption requires diffusion of the drug from the administration compartment to the blood. The speed of this process depends on the variables listed in Table 4.2. In the following pages, we will examine how the administration route influences both absorption rate and amount of absorbed drug and as a result drug plasma levels.

Absorption of drugs through enteral or parenteral routes generally follows first-order kinetics (the flow is proportional to the concentration). This means that the quantity of drug absorbed per unit time is a constant percentage of the drug remaining to be absorbed.

First-order kinetics are described by a curve in a linear plot and by a straight line in a semilogarithmic plot (solid lines in Fig. 4.2). Kinetics are defined by an absorption rate constant, k_a (absorbed fraction per unit time), or by an absorption time constant $\tau = 1/k_a$. A useful parameter is the half-life ($t_{1/2}$), that is, the time needed to halve drug concentration in the absorbing compartment. The half-life is a fixed time, independent from the initial concentration (see Box 4.1). First-order kinetics generally indicate that transport occurs through diffusional mechanisms or unsaturated transport. In most cases, the absorption half-life ranges from 15 to 60 min.

For some drugs, absorption follows a zero-order kinetic, meaning that it proceeds at constant velocity (same amount absorbed per unit time). Zero-order kinetics are described by

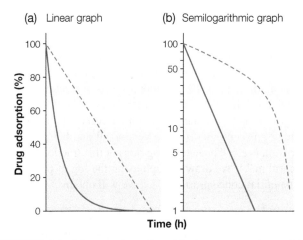

FIGURE 4.2 Linear and semilogarithmic representations of the time course of the drug quantity remaining to be absorbed in absorption processes characterized by first-order kinetics (solid lines) or zero-order kinetics (dashed lines).

a straight line in a linear plot and by a curve in a semilogarithmic plot (dashed lines in Fig. 4.2); they indicate that the drug is absorbed by saturated active transport mechanisms (transport velocity depends on number/activity of available transporters). In most cases, drug absorption follows first-order kinetics, which will be described in the following sections of this chapter.

Absorption Rate Determines Peak Plasma Concentration and Time to the Peak Drug plasma concentration is determined by the balance between absorption and elimination. Initially, absorption is maximal and elimination is null. As the drug is gradually absorbed, absorption slows down (drug concentration at the absorbing site decreases) and elimination increases (drug concentration at the clearance sites increases; for most drugs, elimination processes also follow first-order kinetics). At every moment, the variation of plasma concentration is the result of the difference between absorption and elimination fluxes. Thus, plasma concentration increases as long as more drug is absorbed than is eliminated; it stops rising (peak) when the two fluxes are equal and decreases when elimination processes prevail.

To appreciate how two different absorption rates can produce different concentration curves, consider the effects of an i.m. injection of the same dose in two muscles differently perfused (Fig. 4.3; in the a and b panels, dashed lines describe the absorption and solid lines the elimination fluxes). The peak of the plasma concentration curve corresponds to the time when the two fluxes—absorption and elimination—are equal (crossing of the solid and dashed lines in Fig. 4.3a and b). In the fast-absorption case (Fig. 4.3, panels a and c), a large fraction of the administered drug is absorbed before elimination becomes relevant, the peak is high, and plasma concentration begins to decrease early. In comparison, in the less perfused muscle (panels b and d in Fig. 4.3), the absorption rate is slower, the plasma concentration rises more slowly, and the concentration peak is lower and reached at a later time. Note also that the two elimination curves (solid line in a and b, Fig. 4.3) are different because of the different concentrations of drug in the blood. Given the participation of both absorption and elimination processes in determining plasma concentration time course, the time and height of the concentration peak will also be affected by any alteration in the elimination processes. This example suggests the following conclusions that hold true whenever the same drug dose is absorbed at different velocities: the slower the absorption, the lower and more delayed the peak in plasma concentration.

The Absorption Rate Varies Depending on the Administration Route Based on the previous statement, the peak concentration of a drug dose injected in the vein as a single bolus is higher than the peak concentration that can be achieved with any extravascular administration. As described in Table 4.2, the absorption rate depends on the properties of the drug and/or of the absorbing area. The absorption rate of a drug through any of the administration routes can only be determined experimentally; however, we can reasonably assume that, for drugs with intermediate partition coefficient, absorption rates will be increasingly slower for i.m., subcutaneous, and enteral routes. The differences are highlighted in Figure 4.4, where the same single dose is administered through different routes. In some cases, absorption may be delayed (e.g., using pills with an acid-resistant coat that dissolves only in the intestine) or slowed down (retard preparation).

A detailed analysis of the time course of plasma concentration will be undertaken in Chapter 7. Here, we can observe that with the extravascular administration, drug plasma concentration at peak level is the same achieved at that same moment when the same drug dose is injected directly in the bloodstream (see arrows in Fig. 4.4). However, in the case of extravascular administrations, part of the drug still remains to be absorbed, and thus, from the peak onward, plasma concentration is higher than that observed at comparable times after i.v. injection. Nevertheless, if with the extravascular route the bioavailability is low and only a small proportion of the administered drug is absorbed, the curve may remain constantly lower than the curve corresponding to the i.v. injection.

The events displayed in Figure 4.4 imply that using different administration routes may allow to avoid toxic plasma concentrations of the drug and to prolong therapeutic effects. In addition, it is important to note that, if the absorption rate is too slow, drug plasma concentration may not reach the minimum level within the therapeutic window.

If drug absorption rate is very slow (e.g., retard preparations), the drug may persist at effective levels for a much longer time period than that expected according to its normal elimination rate: in this case, pharmacological effect duration

BOX 4.1 FIRST-ORDER KINETICS

Imagine a pool containing 10 l of colored water that is continuously replaced by fresh water at a rate of 1 l/min. With the water, 1/10 of the color is removed; thus, supposing that the initial color concentration in the water is of 8.0 g/l, it becomes 7.2 g/l (8−0.8) after 1 min and 6.48 (7.2−0.72) after 2 min. To ease calculation, it is better to think in terms of fraction of the initial value (right axis of graph a); thus, starting with 1, after 1 min, what is left is 1−0.1=0.9, after 2 min 0.9−0.09=0.81, after 3 min 0.729, and after n min 0.9^n. In general, if a fraction k of the color is lost in the time unit (0.1 every minute in our example), what remains after t min is a fraction $(1-k)^t$ (a). The number e is such that if k is a very small number ($k \ll 1$), then $1-k \approx e^{-k}$; it follows that the fraction of the color remaining after t min is e^{-kt}. The natural log (log) of this fraction is $-kt$. For this reason, the number e is also named the "base to natural logarithm." If we draw a graph using the log of the color concentration (right axis of graph b) against the time, starting with the initial concentration of C_0, we will obtain

$$\log\left(C_0 \times e^{-kt}\right) = \log(C_0) - kt,$$

that is, a right line that starts from the log of the initial concentration of the color and decays with a slope k to the time (b); this allows to estimate k with a linear regression or as a ratio between the two sides (a/b) of the triangle in figure b. The parameter k is the fraction lost in the time unit and is named constant of elimination; its reverse, $\tau = 1/k$, is named elimination rate constant. This kinetic can be described by the equation

$$C = C_0 \times e^{-kt}$$

where C is the concentration at time t, C_0 is the initial concentration, k is the elimination rate constant, and e is the base to natural logarithm.

It is possible to demonstrate that, considering the time that each molecule remains in the pool and measuring the mean permanence in the pool or mean lifetime, this has the value of τ.

Another useful parameter to describe the temporal changes of the color concentration is the time required to halve the concentration; it can be appreciated from figures a and b, that the time required to change the concentration from 8 to 4, or from 4 to 2, or from 2 to 1 is always the same. This time (required to halve the color concentration) is named half-life and indicated by the symbol $t_{1/2}$. $t_{1/2}$ is the time when C is half of C_0, that is, $C/C_0 = 0.5$. Therefore,

$$e^{-kt} = \frac{1}{2} \rightarrow t_{1/2} = \frac{\log(2)}{k} = \tau \times 0.693$$

The half-life is the value that is sufficient to describe a first-order kinetic; in the example of the graph shown in the figure in the box, $t_{1/2}$ is 6.93 min. From $t_{1/2}$, we can estimate

$$\text{Mean lifetime}: \tau = 1.443 \times t_{1/2} \quad \text{elimination rate constant}: k = 0.693/t_{1/2}$$

Note that half-life and mean lifetime are proportional but not synonyms.

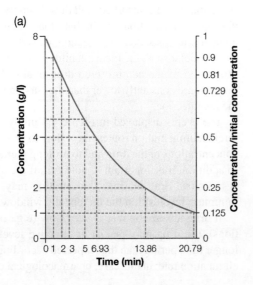
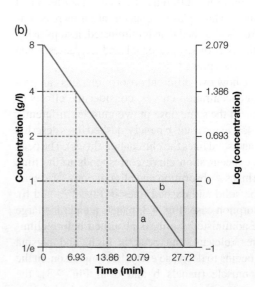

This type of kinetics is defined as first-order kinetic because the amount eliminated at each time is proportional to the concentration raised to the first power; if the amount eliminated is constant (proportional to C raised to the 0 power), as it occurs in a process mediated by a saturated transporter, the kinetic is of zero order.

First-order kinetics are more frequent in the events in which passive diffusion predominates. Note that a process mediated by a transporter follows a zero-order kinetic only when the transport is saturated. For cargo concentrations that are not saturating, the interaction with the transporters is governed by the same rules that describe ligand–receptor interaction, that is, the process is proportional to the ligand concentration and can be included into the more diffused category of natural events described by first-order kinetics.

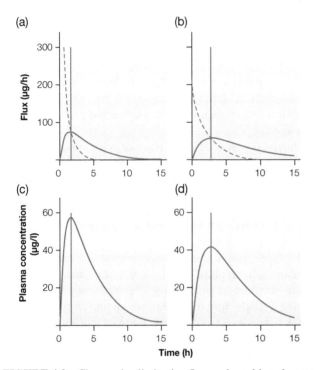

FIGURE 4.3 Changes in elimination flow and resulting changes in plasma concentration as consequences of absorption rate differences. A drug with 145 min half-life (elimination rate 25% per hour) is injected in two muscles differently perfused. The resulting absorption half-lives are 28 min (graphs a and c) or 84 min (graphs b and d), respectively. The different absorption rates (dashed lines in a and b) influence both the elimination flow (solid lines in a and b) and the plasma concentration (c and d). The peak in plasma concentration occurs when the absorption and elimination flows equal each other (vertical lines); it is delayed and has a lower value when absorption is slow. Following the peak, plasma concentration remains higher for a slower absorption rate (more drugs remain to be absorbed). (a and b) Absorption (dashed) and elimination flows (solid lines); c and d, plasma concentrations. Curves computed for $D = 350\,\mu g$, $V_d = 5\,l$, and clearance $= 1.25\,l/h$.

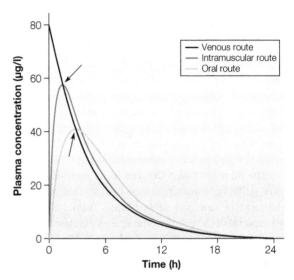

FIGURE 4.4 Time course of plasma concentrations of the same drug, after intravascular administration (maximum level at time $= 0$) or intramuscular or oral administration. Peaks induced by the extravascular administrations occur at later times and lower heights. They all cross the curve induced by the intravascular administration exactly at their peaking time, and from there onward, plasma concentrations induced by the extravascular administrations are higher than those obtained with the i.v. administration.

is determined by the absorption rate. For example, i.m. injection of 300,000 UI penicillin G produces a peak concentration of about 1 μg/ml by 1–3 h and a plasma concentration of about 0.1 μg/ml after 1 day and of 0.02 μg/ml after 2 days; by contrast, if 1,200,000 UI benzathine penicillin is administered, since the drug is very slowly absorbed, its plasma concentration is again 0.1 μg/ml after 1 day, but it reaches 0.02 μg/ml only after 14 days. When the absorption rate is the limiting factor, the drug quantity present in the patient's body (except at the administration site) is always a minor proportion of the administered dose.

Relationship between Absorbed Dose and Drug Plasma Concentrations At any moment, plasma drug concentration is determined by the difference between the amount of drug that has been absorbed and the amount that has been eliminated. Therefore, given the same administration route, plasma levels depend on the administered dose.

As previously stated, both absorption and elimination of most drugs follow first-order kinetics, meaning that the percentage of drugs absorbed and eliminated in a unit time (respectively, k_a and k_e) is constant and the quantity of drug

TABLE 4.9 Goals of drug delivery techniques

Stabilization and protection from degradation
Adaptation to oral ingestion or to inhalatory/transdermal administration of drugs usually administered by injection
Increase of half-life of drug in plasma
Targeting to selective cells/tissues/organ
Controlled release of drug by local factors or magnetic fields

---

being absorbed and eliminated is proportional to the quantity remaining to be absorbed or eliminated, respectively. Therefore, if we double the dose, also, the absorbed amount, eliminated amount, and plasma double.

Drug Delivery: Formulations to Control Absorption and Release Site Development of formulations to allow absorption of large (proteic) substances and/or release of drugs in the plasma with controlled kinetics has been a crucial aim of pharmaceutical research for many years.

From the initial idea of slowing down drug absorption by changing site and mode of administration as well as surface area of the administration devices (e.g., band-aids, subcutaneous pellets), research focus has shifted toward the development of new and sophisticated "vehicles" that may address several of the problems related to the classical aspects of drug pharmacokinetics (see Table 4.9). Development of novel technologies to control absorption characterizes a new field of pharmacology named "drug delivery."

Transdermal and Transmucosal Approaches Transdermal slow-release systems ("band-aids") have been used for several drugs including estrogens, nitroglycerin, and scopolamine. For the delivery of large molecules such as peptides or proteins, pellets or osmotic pumps implanted subcutaneously have been tested, with several limitations due to invasiveness and lack of molecular stability of the proteins. More recently, transdermal methods have been developed based on "biolistic" (microbullets) and ionophoretic approaches.

Two common approaches to deliver large (proteic) drugs are nasal application and pulmonary inhalation.

The nasal mucosa is highly permeable and richly vascularized. In addition, the nasal route allows drug to avoid the liver first-pass effect and is particularly suitable for autoadministration. The drug can be associated with bioadhesive polymers, which stick to the mucosa and release the drug for prolonged time and with protective substances that improve its absorption (biliary salts, surfactants, or cyclodextrins that facilitate transcellular passage).

Pulmonary inhalation also is very convenient; it avoids the use of needles and exploits a huge absorbing surface. This route is also used for systemic administration of peptide or protein drugs and hormones (narcotics, ACE inhibitors, insulin, calcitonin, heparin, antitrypsin, interferon, FSH, vaccines, and gene therapies). Compared to inhalation through traditional or dosed devices, inhalation of drug dry powder formulations (dry powder inhalation (DPI)) represents a promising innovation, as peptide and protein drugs are much more stable in dry formulations.

Polymers and Microparticles to Protect the Drug and Regulate Its Release In recent years, research in the field of polymers for drug delivery has been intense. Trapping the drug in a polymeric matrix (or capsule) protects it from degradation and allows its release at a controlled rate and, in some cases, its selective delivery to specific tissues or cell types.

After initial approaches with natural polymers, such as collagen or cellulose, polyanhydrides, polyesters, polyacrylates, and polyurethanes have been developed. With these compounds, the rate of matrix degradation (and drug release) can be determined based on the presence of heteroatoms (other than carbon) and on the polymerization conditions.

Polymers can be used to produce microspheres that act as small reservoirs hosting the drug or as matrices in which drug molecules are physically trapped in the polymer meshwork. An interesting development is the synthesis of biodegradable polymers that upon hydrolysis release the drug and turn into compounds (lactic and glycolic acids) that can be fully degraded to water and CO_2.

Heteropolymers also provide relevant advantages. By combining hydrophilic and lipophilic monomers, micellae can be obtained (as small as a few nanometers, possibly containing a single drug molecule), which protect the drug and can deliver it in a targeted way. For example, sugars bound to the micelle surface can target it to glycoreceptors present on cell membranes: this approach has been used with some success to specifically deliver the antitumor drug doxorubicin to solid tumors. Electroactive polymers have also been synthesized, in which content release is controlled by ion counterflow (dopants).

Even more versatile are hydrogels, hydrophilic polymers that resist to acidic environment. Indeed, they can release the trapped molecules upon swelling triggered by changes in pH, application of electric or magnetic fields, or ultrasounds, depending on their chemical composition. In this way, it is possible to actively control drug release and delivery at the desired site of action. Polymers with a polyacrylate skeleton are also superabsorbent and bioadhesive and can stick to mucous surfaces and slowly release their content. A brilliant example is the trapping of insulin in a hydrogel membrane to which the glucose oxidase enzyme has been bound: in the presence of high glucose concentrations, the enzyme produces gluconic acid that lowers hydrogel pH, thus causing insulin release.

Fullerene (a 60-carbon atom molecule organized as a sphere with hexagonal symmetry) represents a particular case, in that it has an internal cavity that can host metallic ions. Radioactive ions have been inserted for their selective delivery to tumor tissue, thereby reducing radiation exposure of other tissues.

Liposomes Liposomes are vesicles formed by a phospholipid double layer. In the last decades of the twentieth century, these microstructures were considered particularly promising as protective vehicles for drugs and as tools to selectively deliver them to specific tissues. Indeed, liposomes leave the bloodstream at actively proliferating vessels and at the leading edges of angiogenesis, which are abundant in neoplastic tissues. Some positive results were obtained with liposomes enclosing the antineoplastic drug doxorubicin. However, most attempts with first-generation liposomes failed because of their short life due to rapid binding by opsonins and degradation by mononuclear phagocytes of the reticuloendothelial system (RES). Practically, they proved useful only to deliver drugs to the RES itself.

Last-generation liposomes are sterically stabilized by a polyethylene glycol (PEG) coating that forms a hydrophilic halo and repels opsonins, thus eluding the RES. These liposomes have proved a good system for prolonged release and drug targeting. Delivering drugs to areas where neoangiogenesis occurs using liposomes has been used with promising results with stabilized liposomes. In anticancer therapy, such passive targeting allows a significant drug accumulation in solid tumors. For example, the antitumor drug doxorubicin encapsulated in liposomes can reach concentrations up to 10-fold higher in tumor infiltrate than in other sites. This approach is particularly effective in the therapy of Kaposi's sarcoma localized in the skin because liposome passive targeting tends to prefer cutaneous tissues.

Passive targeting properties of liposomes can be exploited also to treat infections; in this case, the increased permeability of the inflamed area leads to selective accumulation of liposomes and their pharmacological cargo.

However, the most ambitious goal of drug delivery approaches is the selective delivery to specific tissues or cell types. To this aim, research is focusing on the identification of specific and selective surface antigens on target cells that may interact with antibodies associated with liposomes or nanoparticles, thus allowing specific drug delivery. Although linking antibodies to stabilized liposomes is technically complex, promising experimental results have already been obtained. For example, liposomes carrying anti-CD34 antibodies can specifically deliver drugs to lymphoid cells expressing the CD34 surface antigen; similarly, liposomes carrying CC532 antibodies (CC532 is a rat colon carcinoma antigen) and loaded with the antitumor drug 5-fluorodeoxyuridine have provided encouraging therapeutic responses.

Liposomes and nanomedicine appear as the future methodological approaches for the treatment of tumor metastases, multiple tumors, leukemia, and myeloma forms for which effective therapies are not available today.

Interrelation between Gene Therapy and Drug Delivery Techniques

As a concluding remark, it is interesting to examine the twofold relation that links two important and novel fields of therapy: gene therapy and drug delivery (Table 4.10). On one hand, classical pharmacological knowledge can help to devise strategies to protect viruses and plasmids in the organism in order to deliver them to target cells, possibly in a selective way. To this aim, several of the techniques we have just described can be used for local or systemic delivery of gene sequences (regardless of the vector used): aerosol or dry powder inhalation; transdermal approaches, such as electroporation; oral administration targeting gastrointestinal mucosal cells; biodegradable depot formulations; and implantable pumps. On the other hand, gene therapy itself may become a form of "drug therapy" when the gene product (possibly transduced by the target tissue only) acts as therapeutic substance. Reproducibility, consistency, selectivity of expression, and number of cells

TABLE 4.10 Examples of therapeutic use of new "target delivery" techniques

Goal	Approach	Example
Targeting		
Organ	Bioadhesive polymers	Pathologies of the gastrointestinal mucosa
Tissue	Liposomes with cutaneous tropism	Tumors of the skin
Cells	Immunoliposomes against specific antigen	Treatment of solid or fluid tumors
Control of PK properties		
Protection from degradation	Nasal aerosol with adjuvants, dry powders for inhalation, polymers	Drugs sensitive to liver metabolism, proteins, peptides unstable when in solution
Slow/ delayed release	Systems for transdermal release (bands), polymers, microparticles, liposomes	Nitroglycerin, scopolamine, antineoplastic drugs
Controlled release	Implantable, systems pumps, polymers sensitive to pH or electric/magnetic fields and to counterions	Prolonged release of contraceptives, insulin-bound polymers that are degraded following enzymatic change in pH induced by increased concentration of blood glucose

expressing the therapeutic gene represent critical tasks to be addressed for efficient gene therapy.

TAKE-HOME MESSAGE

- Absorption is the transfer of the drug from the application/administration site to the bloodstream.
- The rate of absorption determines the speed at which drug plasma concentration (and the resulting effect) increases and the height of the plasmatic peak.
- Drugs can be administered directly in the bloodstream or through other routes. In the latter case, absorption rate may be governed by the presence of cellular barriers:
- The rate of absorption depends on the size and vascularization of the absorbing area. When cellular barriers are involved, thickness of the barriers and oil–water partition coefficient and molecular size of the drug also modulate its absorption rate.
- Differences in absorption rates can be exploited to calibrate velocity of onset and duration of the effects and to limit the peak in plasma concentration or even to produce only local effects (topical application); in this case, the possibility that systemic effects might also occur should always be considered.
- The use of enteral routes, which exploit the easiness of administration and the huge absorbing surface of the intestine, can be limited by drug metabolism in the liver (first-pass effect) and/or by metabolism in the intestinal flora before it can reach systemic circulation.
- Modern technologies of pharmaceutical formulation can be exploited to regulate the rate of drug absorption and/or to control its delivery to specific sites/cells.

FURTHER READING

Attar M., Lee V.H.H. (2003). Pharmacogenomic considerations in drug delivery. *Pharmacogenomics*, 4, 443–461.

Dressman J.B., Lennernas H., *Oral Drug Absorption: Protection and Assessment. Drugs and the Pharmaceutical Science*, Marcel Dekker: New York, 2000, vol. 106.

Harrington K.J., Syrigos K.N., Vile R.G. (2002). Liposomally targeted cytotoxic drugs for the treatment of cancer. *Journal of Pharmacy and Pharmacology*, 54, 11573–16600.

Mehvar R. (2003). Recent trends in the use of polysaccharides for improved delivery of therapeutic agents: pharmacokinetic and pharmacodynamic perspectives. *Current Pharmaceutical Biotechnology*, 4, 283–302.

Sloan K.B., Wasdo S. (2003). Designing for topical delivery: prodrugs can make the difference. *Medicinal Research Reviews*, 23, 763–793.

5

DRUG DISTRIBUTION AND ELIMINATION

RICCARDO FESCE AND GUIDO FUMAGALLI

By reading this chapter, you will:

- Learn how drugs move among plasma, organs, and tissues and understand the biological basis of the processes that mediate drug distribution and elimination
- Understand the concept of apparent distribution volume and the distribution kinetics of drugs in plasma and tissues
- Understand the principles of elimination kinetics, the meaning of rate and time constants, the importance of drug binding to plasma proteins, and the clinical application of operational concepts such as clearance and half-life
- Know the parameters and their use for quantitative description of pharmacokinetics of distribution and elimination
- Know and predict how physiological and pathological traits interfere with pharmacokinetics

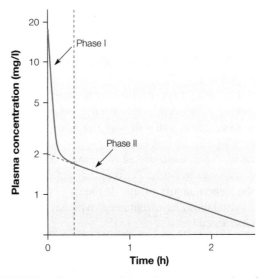

FIGURE 5.1 Time course of drug plasma concentration after i.v. administration. Semilogarithmic plot of the time course of lidocaine plasma concentration, following administration of a 150 mg bolus in vein. The initial decay phase is mostly due to drug distribution to tissues. The late phase reflects mainly the drug elimination process. Extrapolation to time 0 of the elimination phase yields an estimated (virtual) concentration at time 0 of 2 mg/l, corresponding to a $V_D = 150\,mg/2\,mg/l = 75\,l$.

The semilogarithmic plot in Figure 5.1 shows the time course of plasma concentration after intravenous (i.v.) injection of a 150 mg bolus of lidocaine, an antiarrhythmic drug. As you can see, two phases can be distinguished, characterized by a different decay rate: an initial rapid-decay phase (phase I, distribution phase) and a late slow-decay phase (phase II, elimination phase). During the distribution phase, the fall in plasma concentration is due to the drug leaving the vascular bed to distribute to the body tissues (including the heart, the therapeutic target), until blood and tissue concentrations reach the equilibrium. The slope in the distribution phase reflects the capacity of the drug to leave the blood vessels and the total volume in which it distributes, whereas in the elimination phase, it indicates the efficiency of the metabolic and excretion mechanisms responsible for its removal from the organism. The two processes actually superimpose in time and can only be distinguished if their rates are very different. However, it is useful to conceptually distinguish between the two phases, because they are governed by different phenomena.

General and Molecular Pharmacology: Principles of Drug Action, First Edition. Edited by Francesco Clementi and Guido Fumagalli.
© 2015 John Wiley & Sons, Inc. Published 2015 by John Wiley & Sons, Inc.

DISTRIBUTION

For a rapid grasping of some of the basic concepts describing this aspect of pharmacokinetics, see Box 5.1. For precisely predicting the drug action on a specific organ, one should be able to measure the drug concentration in that organ. This can be done in animal models but is almost impossible in humans where usually data can only be collected from blood and urine samples. However, blood concentration can allow estimating drug concentration in the compartment of interest when two parameters concerning the distribution processes are available: the equilibrium values (i.e., the concentrations attained in various compartments at the equilibrium) and rates (i.e., the velocities of drug distribution between blood and each tissue). Both these parameters vary depending on the drug and the compartment of interest.

In principle, for each tissue, the rate of equilibration with plasma is given by the ratio between hematic flux and tissue volume. However, a drug typically tends to reach different concentrations in different tissues and several factors may affect drug diffusion from capillaries to the interstice of specific tissues. In Box 5.2, drug distribution is illustrated using a hydraulic analogy.

Tissues and Avidity for Drugs

Drug concentration tends to be equal in the plasmatic and interstitial water; however, a fraction of the drug may be bound to proteins, lipids, macromolecules, and cell components and membranes, and this fraction may be quite variable among different tissues. Thus, the total drug concentrations in plasma, in tissue interstitial fluid, and in the tissue as a

---

BOX 5.1 EXAMPLES FOR GRASPING THE CONCEPT OF APPARENT VOLUME OF DISTRIBUTION

The concentration of a solute (C) in a reservoir is given by its amount (Q) divided by the volume of the reservoir (V), that is, $C = Q/V$. If such volume is not known, it can be calculated by adding a known amount of solute (Q) and measuring its concentration: $V = Q/C$.

Try this: take a 1 l bottle, fill it with water, and add 1 g of a dye. You will get a bottle of colored water, and you know that the resulting color corresponds to a dye concentration of 1 g/l. Now, take a flask of unknown volume. Add the dye until you get the same color: you will have the same concentration (C) having added a quantity (Q) of dye—the flask must have volume $V = Q/C$.

Second experiment: use an old, empty but uncleaned bottle (1 l) that contained olive oil; a thin layer of oil remains on the walls of the bottle. Perform the same measurement. If the dye you use is hydrophilic, you will get the same color with the same quantity of dye. If the dye is lipophilic, you will see a thin dark-colored film on the walls of the bottle, where oil residues have remained, whereas the water will be paler, as if the volume of the bottle were larger than 1 l. What happened?

Let us return to the reservoir of unknown volume. Suppose some oil residues are stuck to the reservoir walls: this will adsorb lipophilic substances. If the dye introduced into the reservoir has an intermediate partition coefficient, part of it will dissolve in the water and part in the oil residues. The resulting concentration in the reservoir water will depend on the oil–water partition coefficient (OWPC) of the solute. A highly hydrophilic solute will only dissolve in water, and its concentration will provide a reasonable estimation of the volume of the reservoir minus the volume occupied by the oil residues. An amphipathic solute (OWPC ≈ 1) will yield a more accurate estimate of the total volume. Conversely, a highly hydrophobic (lipophilic) solute will be largely captured by the oil residues, and only a minor fraction will dissolve in water, so that a lower concentration will be measured there. The resulting estimate of the reservoir volume (based on dye dissolved in water), $V = Q/C$, will be larger than the real volume (as C is smaller than expected). Such estimate is called apparent distribution volume and represents the volume of a "clean" reservoir (containing pure water only) equivalent to our "dirty" reservoir.

Such "apparent volume" does not need to have a physical correlate with specific compartments of the body. Still, it is an operational concept very useful to determine the amount of drug to be given to a patient in order to obtain a given plasma concentration. Body fluids (plasma, extracellular and intracellular fluids) are equivalent to the water in the reservoir of our example, while cell membranes, proteins, and any macromolecule or structure that may capture the drug are equivalent to the oil residues.

For a drug that cannot leave the circulation because of its large molecular size (e.g., dextrans), the distribution volume will be the volume of the plasma itself (~3 l). For a drug that diffuses to the extracellular spaces, the distribution volume will be that of the extracellular fluids (about 20% of body weight); for drugs that accumulate in specific compartments (lipophilic drugs in fat tissues and cell membranes, drugs sticking to the extracellular matrix components or actively captured by cells), the volume may be larger than the entire volume of the patient's body. If a drug is significantly bound to plasma proteins, its concentration in plasma is higher than in other water compartments of the body, and the resulting apparent distribution volume is smaller.

BOX 5.2 A HYDRAULIC MODEL OF PHARMACOKINETICS

The figure shows a schematic mechanical model of the cardiovascular system. The fluid is moved by a pump, flows in pipes of different calibers, and reaches reservoirs of different sizes.

A dye injected in the system will diffuse and eventually reach the same concentration everywhere. However, before reaching the equilibrium, the dye will have different concentration in the various reservoirs. Each reservoir exchanges an amount of water every minute (flux) that depends on the caliber of the connecting pipe, and it has to exchange all its water to equilibrate. Therefore, the time to equilibrium for each reservoir will depend on its flux to volume ratio (specific flux or specific perfusion). The larger the tube and the smaller the pool, the higher the specific flux and the shorter the time to equilibrate with the circulating fluid.

Drug distribution is governed by similar principles: each reservoir (also named compartment) represents a tissue or organ, and pipes represent vessels and the plasmatic compartment.

In highly perfused compartments, the time course of dye concentration may be biphasic (i.e., rise to high values and then decline). In fact, at early times (a in the figure), the dye has not significantly diffused to poorly perfused compartments. Therefore, the concentration in the circulating fluid and in well-perfused compartments is high and slowly declines as the dye distributes also to the large and poorly perfused compartments (b in the figure). Such redistribution phenomena constitute an important aspect in drug pharmacokinetics, especially for drugs acting on the CNS: a highly lipophilic drug may rapidly equilibrate with plasma in the brain (a highly perfused compartment), producing a major pharmacological effect. Such effect may vanish as a result of the redistribution phenomenon before the drug is significantly eliminated.

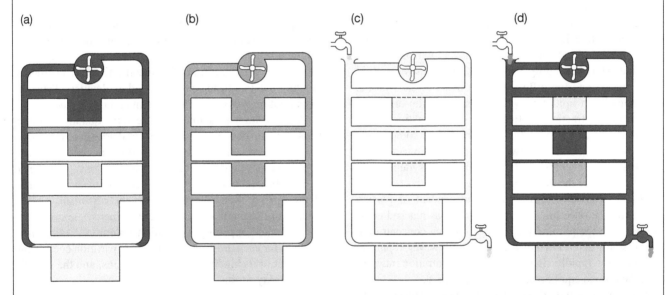

To make our hydraulic model more realistic, we introduce two faucets: one lets out a fixed fraction of the circulating fluid every minute, and the other replaces it with clean water. The two faucets simulate drug clearance: a constant fraction of plasma volume is "cleaned" every minute.

Dye concentration gradually declines, as it leaves the system through the outflow faucet. The rate of the elimination process depends again on the flux to volume ratio. Here, the flux equals the clearance (volume of colored fluid replaced by clean fluid every minute), while the volume is the total volume of the system. If the faucets are opened a bit more, this will increase the flux (clearance) and accelerate elimination.

Circulating blood actually is not in direct contact with interstitial fluids. The model should therefore include filters (depicted as dotted lines in the drawings) to represent diffusion barriers at the various capillary beds. A tight filter obviously hampers equilibration in highly perfused compartments, whereas it may be irrelevant in poorly perfused compartments. Consider the CNS: for a hydrophilic drug, the BBB behaves as a tight filter, and the CNS, though highly perfused, equilibrates quite slowly if at all; instead, for a lipophilic drug, the CNS behaves as a highly perfused, rapidly equilibrating tissue, and redistribution phenomena are observed.

It is important to recall here what we said about drug tropism and avidity of the various tissues for a drug. If a tissue tends to accumulate a drug to an equilibrium concentration twice the plasma concentration, at equilibrium it will contain a quantity of drug equal to that contained in a plasma volume equal to twice the tissue volume. Kinetically, the tissue behaves as if it had a volume (apparent distribution volume) twice its actual volume. Conversely, if a drug has tropism for plasma (it is captured by plasma proteins), the various compartments behave as if they had a volume smaller than their actual volume. Thus, in our hydraulic model, we should consider the volume of each reservoir not as the actual volume of the body compartment it represents, but as the apparent distribution volume of that compartment.

This model highlights the main principles to keep in mind to understand drug distribution in the organism: drug concentration in body compartments depends on (i) local plasmatic flux, (ii) physical–chemical properties of drug and tissue, (iii) permeability properties of capillary beds, and (iv) drug binding to plasma proteins. It also emphasizes that drug elimination rate is given by the clearance to apparent distribution volume ratio and explains how conditions that modify body composition and/or clearing organ function may significantly change drug pharmacokinetics and its concentration in plasma and tissues.

---

whole may be quite different. Furthermore, compartments with very slow equilibration rates (e.g., adipose tissue, very poorly perfused) may require a long time to equilibrate with plasma, and elimination may occur for a substantial amount of the dose before drug distribution equilibrium in the tissue is reached. For this reason, we will keep conceptually separated kinetic aspects from equilibrium values. We will consider the latter first.

The biochemical compositions of plasma, interstitial fluids, and intracellular fluids are quite different; some tissues are rich in lipids or macromolecules to which the drug can bind. Thus, each drug can partition in its own way between plasma and the different components of each tissue. For example, in lipid-enriched tissues, lipophilic drugs are present in two fractions: a free fraction in interstitial water and a larger fraction in membranes and lipidic components of the tissue. The ratio between total tissue concentration and plasma concentration (at equilibrium) is defined as tissue avidity for the drug (indicated as k_p).

Note that avidity is absolutely independent of plasma concentration; once the drug in the tissue has reached equilibrium with the plasma, the tissue to plasma concentration ratio is solely determined by the tissue k_p. However, if plasma concentration rapidly changes, tissue concentration may not have the time to equilibrate and will lag behind.

Even though free drug concentrations in plasma and tissue water have to be the same, the total drug concentration in the tissue at equilibrium may be quite different (avidity); tissues may accumulate drug as if they had a volume different from their actual one. Total drug concentration in plasma (plasma concentration) may also be quite different, because of drug binding to plasma proteins. These two concepts will be examined in detail.

The Apparent Distribution Volume

The organism can be considered as an ensemble of compartments with different functional properties. Among these, there are at least three main water compartments: plasma water, interstitial (extracellular) fluids, and intracellular fluids. In a 70 kg adult slim man, water accounts for about 60% of body weight (42 l): 4% of body weight (3 l) consists of plasma water, about 16% (11 l) of extracellular fluids, and 40% (28 l) of intracellular fluids. Actually, these volumes should be further subdivided, as distinct organs and tissues have different characteristics of perfusion, composition, and volume.

As seen earlier, tissue to plasma drug concentration ratio (at equilibrium) may be different from 1 and may vary among tissues. Indeed, in each tissue, the drug may be present in different amounts/concentrations due to its chemical–physical properties, affinity for specific tissue constituents, and binding to plasma proteins and molecules. Thus, its concentration will not be the same in all body tissues.

Concentration is defined by the ratio quantity/volume. Thus, the ratio between the administered drug dose (quantity) and the plasma concentration (quantity/liter) yields a volume of plasma in which the drug "appears" to have distributed at equilibrium. Obviously, this is not the actual volume of plasma—or water—contained in the organism, but the volume of plasma that is equivalent to the whole organism. This is called apparent distribution volume (V_D), and it is different for each drug, depending on its partition coefficient, the extent of its binding to plasma proteins, and the specific tissue avidity for it.

V_D Relates Plasma Concentration to the Total Drug Quantity in the Organism Knowing the apparent distribution volume for each drug is important to estimate its plasma concentration, once it has fully distributed to all tissues. The apparent distribution volume is given, by definition, by the following formula:

$$V_D = \frac{\text{amount of drug at distribution equilibrium}}{\text{drug plasma concentration}}$$

In general (but there are some important exceptions), the time needed to attain distribution equilibrium is much less than that needed to eliminate a substantial fraction of the absorbed drug. The V_D value can be estimated from a curve such as that in Figure 5.1: since during the elimination phase the drug is in equilibrium in all its apparent distribution volumes, linear extrapolation to time $t=0$ of the straight line representing the elimination phase in the semilogarithmic plot provides estimation of the drug concentration, $C_p(0)$, that would be obtained if the whole administered dose (D) instantaneously reached the distribution equilibrium. From $C_p(0)$, the apparent distribution volume is computed as follows:

$$V_D = \frac{D}{C_p(0)}$$

Each drug is characterized by its own V_D, determined by its physical–chemical properties, its partition coefficient, binding to plasma proteins, and the k_p (avidity) of each tissue for the drug.

Each Tissue and Organ Has Its Own V_d Each compartment in the organism gives its contribution to the total V_D of the organism. Given that k_p defines the ratio between tissue and plasma concentrations at equilibrium, each compartment has an apparent distribution volume equal to its actual volume multiplied by its avidity for the considered drug (k_p). Thus, the V_D of the whole organism is the sum of the plasma volume plus the V_d's of all compartments. As shown in the examples in Table 5.1, V_D varies enormously between drugs. If a drug V_D is very high (larger than the water volume in the body), it will accumulate in some organs ($k_p > 1$) that will behave as deposit organs. The reasons why an organ acts as a depot site are easy to understand in some cases, as in the case of adipose tissue and the brain for lipophilic drugs or for bones that tend to accumulate heavy metals in hydroxyapatite crystals. Other drugs may accumulate in specific organs: the antihelminthic and antiprotozoal agent quinacrine is 1000-fold more concentrated in the liver than plasma, and tetracyclines (antibiotics that can chelate calcium ions) accumulate in bone tissues and teeth.

The apparent distribution volume of drugs is conventionally expressed in units of volume per kilogram of body weight (Table 5.1). This reduces variability among people with different body weights. However, this conventional V_D does not take into account other important factors such as age, physical constitution (lean mass vs. fat mass), and some pathological conditions. In addition, some drugs may compete for binding sites at certain storage organs, resulting in reciprocal alteration of their V_D; this is an important factor to consider when performing multidrug treatments.

For each drug, the constant k_p and the apparent distribution volume of each compartment (=actual volume × k_p) determine the equilibrium concentration of the drug in the

TABLE 5.1 V_d **of various drugs**

Drug V_d	(l/kg)
Amiloride	17.0
Amiodarone	66.0
Amitriptyline	15.0
Bretylium	5.9
Caffeine	0.6
Carbamazepine	1.4
Chloramphenicol	0.9
Chloroquine	115.0
Chlorpromazine	21.0
Cimetidine	1.0
Clonidine	2.1
Codeine	2.6
Cyclosporine	1.3
Dexamethasone	0.8
Diazepam	1.1
Digitoxin	0.5
Diltiazem	3.1
Eryithromycin	0.7
Furosemide	0.1
Heparin	0.1
Imipramine	18.0
Kanamycin	0.3
Lidocaine	1.1
Methadone	3.8
Metoprolol	4.2
Minoxidil	2.7
Morphine	3.3
Naloxone	2.1
Neostigmine	0.7
Nicotine	2.6
Nifedipine	0.8
Nitroglycerin	3.3
Pentazocine	7.1
Phenobarbital	0.5
Prednisolone	1.5
Probenecid	0.2
Propranolol	4.3
Ranitidine	1.3
Thiopental	2.3
Tubocurarine	0.4
Valproic acid	0.2
Verapamil	5.0

tissue, but they also contribute, together with specific blood flux and capillary bed permeability, to determine the rate at which the equilibrium concentration is reached. In particular, the equilibrium concentration corresponds to k_p times the plasmatic concentration, and the equilibration rate corresponds to the blood flux over apparent volume ratio = flux/(volume × k_p) = (specific flux)/k_p, provided that the drug easily diffuses to the tissue. The rate is lower when diffusion is impaired because the drug does not easily cross cell membranes.

Drug Binding to Plasma Proteins

In blood, a drug can be free in the aqueous component or bound to plasma proteins or circulating cells. Thus, plasma can be considered as composed of two compartments: the protein phase, which can act as a circulating storage organ, and the aqueous phase, which equilibrates with all tissues and organs. Current analytical methods usually measure drug concentrations in plasma, as a whole, therefore water plus proteins. However, only the free fraction can actually leave the capillaries and equilibrate with tissues. Therefore, knowing drug plasma concentration may not be sufficient to accurately estimate its concentration in tissues. In particular, the free drug fraction in plasma may considerably vary among healthy individuals and in pathological conditions or during multidrug therapies.

Free and Bound Drug Concentration We need to know the free drug concentration in plasma not only to estimate the actual concentration at the target organ but also because the free (but not the bound) fraction is filtered at renal glomeruli and determines the drug elimination rate. For each drug, the degree of protein binding is generally expressed as the ratio between bound quantity to total quantity (or corresponding percentage). Theoretically, such ratio ranges from 0 to 1. A drug is said to be highly bound to plasma proteins when its ratio is above 0.9 (90%), whereas it is said not, or scarcely, bound when its ratio is below 0.2 (20%).

In terms of binding capacity for most endogenous or exogenous substances, albumin is the most important plasma protein. It represents about 50% of all plasma proteins and carries about a hundred charged groups, mostly negative at physiological pH. Lipophilic molecules and organic anions are the compounds most avidly bound by albumin: examples are free fatty acids, bilirubin, penicillin, sulfonamides, barbiturates, ascorbic acid (vitamin C), and also histamine, triiodothyronine, and thyroxine.

Another important carrier protein for small molecules is the acidic α_1-glycoprotein, an α_1-globulin with a high-affinity binding site for basic molecules. Steroids are bound by transcortin, an α_1-protein; α_2-, β_1-, and β_2-proteins transport lipids, cholesterol, liposoluble vitamins (A, K, D, E), hemoglobin (transported by a haptoglobin), zinc ions, copper ions (bound to ceruloplasmin), and iron ions (bound to transferrin).

Drug binding to plasma proteins follows the same principles that regulate drug–receptor interactions. As in that case, the bound fraction is determined by the ratio between binding and unbinding rates (i.e., the dissociation constant k that expresses the binding affinity). In a chemical reaction, equilibrium is attained when the reagents/products ratio equals the ratio between the velocities of the two reactions.

If $[B]$ is the bound drug concentration, $[F]$ the free drug concentration, n the number of binding sites on each protein molecule, and $[P]$ the protein concentration, the $n[P]-[B]$ difference corresponds to the concentration of free sites, and the equilibrium is attained for

$$\frac{(n[P]-[B])\times[F]}{[B]} = k \frac{[B]}{n\times[P]} = \frac{[F]}{k+[F]}$$

The amount of bound drug per mole of protein is

$$\frac{[B]}{[P]} = \frac{n\times[F]}{k+[F]}$$

This equation represents the typical drug–receptor binding curve, with a maximal asymptotic binding equal to n when the free concentration $[F]$ is much greater than k (see Chapter 8).

The fraction of drug bound to proteins is expressed by

$$\frac{[B]}{[B]+[F]} = \frac{n\times[P]}{k+n\times[P]+[F]}$$

This relation highlights several important facts:

1. The bound fraction (and vice versa the free fraction) depends on drug concentration, so that any statement such as "the drug is bound for more than 90% to plasma proteins" is incomplete, as this is true only for some values of drug concentration. When a statement like this is found, it usually implies "at the therapeutic concentrations usually employed": at higher concentrations, a minor fraction of the drug will be bound, as $[F]$ in the denominator increases, whereas at low concentrations, the bound fraction approaches the $n[P]/(k+n[P])$ ratio.

2. At very low drug concentrations (negligible $[F]$), the bound fraction approaches the limit $1/(1+k/(n[P]))$; in other words, it is determined by the ratio $k/(n[P])$, and therefore, if we consider a fixed plasma protein concentration $[P]$, it is determined by the dissociation constant, k, and by the number of binding sites on plasma proteins, n. The higher the affinity (low k), the higher the fraction bound at low concentrations: for very high affinities (very low k), the drug will be virtually totally bound when present at low concentrations.

3. A possible increase in binding sites ($n[P]$) leads to an increase in the drug bound fraction (the denominator decreases). By contrast, a decrease in binding sites (e.g., low plasma protein level or occupation of binding sites by other drugs) produces an increase in the free fraction.

4. When the drug is highly bound to plasma proteins (e.g., thyroid hormones), the total plasma concentration is higher than that in interstitial (and possibly intracellular) fluids. Thus, (i) plasma concentration is higher than that present at the receptor site—in this situation, the concepts of effective and toxic concentrations become ambiguous (concentration in plasma or at the receptor site?); (ii) the apparent distribution volume is much lower than the actual volume of all liquids reached by the drug; and (iii) when the free drug concentration reaches therapeutic values, the presence of other drugs or endogenous substances capable of displacing the bound fraction may cause free drug concentration to increase toward toxic levels.

The Relevant Parameter Is the Free Drug Concentration
The free drug concentration in the plasma aqueous phase is close to the concentration "seen" by its receptor. This should be kept in mind when the dosage has to be modified to maintain an effective plasma concentration in the presence of certain chronic pathologies (especially liver diseases). In these cases, it is important to consider the free drug concentration, because the total protein level or the serum protein profile may be altered or endogenous substances capable of interfering with drug binding to proteins may be present; under these conditions, the total plasma concentration may be strongly affected, and actually, it may not be necessary to change drug dosage.

For drugs binding plasma proteins with high affinity, the bound fraction may exceed 90%, and drug unbinding might be quite slow (as a general rule, high affinity, i.e., low k_d, implies slow dissociation rates). Plasma proteins can thus behave as large reservoirs for drug storage, determining two major effects:

1. Renal elimination may be scarce because only the free fraction is filtered by glomeruli.
2. Drug plasma concentration decays very slowly.

Both these effects imply that, after discontinuing administration, the drug may persist at effective levels in plasma for days or even weeks.

Despite these considerations, the use of total drug plasma concentration is frequent because it is generally much simpler to measure using standard laboratory techniques. Moreover, total plasma concentration takes into account the avidity of plasma with its protein content (plasma avidity, k_p, equals 1 by definition), so that the operative parameter V_D can be employed. This is very useful to define important parameters, including drug clearance. Moreover, it allows to estimate equilibration rates from the flux to volume ratio (plasma flux/apparent distribution volume of the tissue, V_d), to calculate the drug dose to be administered to obtain the desired plasma concentration, and to estimate the plasma concentration produced by a certain administration protocol (see following text).

Drugs Can Be Displaced from Plasma Proteins by Other Drugs or Endogenous Substances Drug binding to plasma proteins is not selective. Therefore, different molecules can compete for binding to the same site. Such competition may determine displacement and relevant alterations in drug free levels and effects. This is clinically relevant when the drug therapeutic index is narrow and the drug is largely bound to plasma proteins (>90%) and has a limited distribution volume. Under these conditions, transfer of the "freed" drug to tissues may determine a significant rise in drug concentration at its site of action and onset of toxic effects.

A typical example of dangerous consequences of drug displacement from plasma proteins is warfarin, an oral anticoagulant drug, often used in atrial fibrillation, postinfarction therapy, or after cardiac surgery. Patients treated with warfarin may suffer severe bleeding if concomitantly administered with drugs such as salicylates, phenylbutazone, or sulfonamides. Actually, the elevation in free drug concentration may be transient, as drug elimination mostly depends on free drug concentration (in general, as blood flows quite slowly in hepatic sinusoids and hepatocytes capture and metabolize the free drug, part of the bound drug is released and metabolized; however, since high binding to plasma proteins implies high affinity and slow unbinding, most bound drug may escape metabolism). The interaction of warfarin–phenylbutazone is even more complex, as the latter also inhibits hepatic metabolism of anticoagulant drug. Similar displacements occur for tolbutamide, an oral hypoglycemic agent, and phenytoin, an antiepileptic drug.

Displacement may affect endogenous products as well, as in the case of bilirubin and sulfonamides. In the past, this interaction used to be a cause of important iatrogenic complications in neonatal jaundice. Sulfonamides must not be given to pregnant women near term, because they cross the placenta and are secreted in milk. In the newborn, who has low capacity of conjugating bilirubin to glucuronide, plasma levels of unconjugated bilirubin (mostly bound to plasma proteins) rise during the first few days after birth, which diffuses to the skin and sclera causing neonatal jaundice; following displacement by sulfonamides, the increased free bilirubin concentration may lead to deposition at basal ganglia and subthalamic nuclei, producing the toxic encephalopathy referred to as kernicterus.

Factors That Determine the Distribution Rate of Drugs to the Various Compartments

The rate of drug distribution between circulating blood and a given tissue may be limited by perfusion or by local permeability. The factors influencing such permeability have been

TABLE 5.2 Hemodynamic parameters of some organs and tissues

Organs	Percent of body volume	Plasmatic flux (ml/min)	Percent of cardiac output	Perfusion rate (ml/min per ml of tissue)
Lung	0.7	2500	100	5
Kidney	0.4	650	22	2
Adrenal gland	0.03	12	0.2	0.6
Liver	2.3	650	27	0.4
Heart	0.5	100	4	0.3
Brain	2.0	350	14	0.25
Adipose tissue	10	100	4	0.01
Bones	16	125	5	0.01
Muscles (resting)	42	375	15	0.01

discussed in Chapter 3. If the drug is lipophilic and easily crosses membranes or if capillaries are highly permeable as in the muscles, skin, and kidney, the drug in the blood fully equilibrates with the tissue as it flows very slowly in the capillaries. In this case, hematic perfusion becomes the limiting factor to drug distribution between blood and tissue.

Hematic perfusion differs among various organs (Table 5.2). Blood perfusion values are usually expressed in ml/min per ml tissue (specific perfusion), that is, in min^{-1} (note that dimensionally this is a velocity, and it represents the rate constant of the blood–tissue exchange process). The value can vary under physiological or pathological conditions that modify total hematic flux (cardiac output) or produce a redistribution of flow (e.g., orthosympathetic activation, atheromas).

If the permeability is comparable, well-perfused tissues accumulate drugs more rapidly than less perfused ones. For example, upon i.v. administration, thiopental (a barbiturate used to induce anesthesia) accumulates very quickly in the highly perfused brain, where it produces its effect very rapidly. Then, as the drug distributes to other tissues having greater volumes but slower equilibration rates, its plasma level decreases (and the effect wanes) even though the drug has not yet been significantly metabolized or excreted (see Box 5.2).

Drug distribution to tissues generally follows first-order kinetics, that is, the concentration difference between plasma and tissue decreases by a constant fraction in a unit time. Such fraction is defined rate constant of the distribution process (k_T) and depends on hematic flux, tissue volume, and drug affinity for the tissue. In first-order kinetics, the time interval needed for the concentration difference to halve is fixed and is defined as half-life ($t_{1/2}$). Half-life can be computed as follows:

$$t_{1/2} = \frac{0.693}{k_T} = \frac{0.693 \times k_p \times V_T}{\Phi}$$

where k_p is the tissue to plasma drug concentration ratio at equilibrium (an index of the tissue avidity for the drug); V_T is the tissue volume, so that the product $k_p \times V_T$ is the apparent distribution volume of the tissue; Φ is the blood flux; and the number 0.693 is the natural logarithm of 2. The rate constant of drug distribution between blood and tissue is therefore defined by the ratio: (flux)/(apparent volume), that is, (specific flux)/(k_p).

If the arterial concentration of the drug is maintained constant through a continuous infusion (drip injection), tissue concentration will keep rising until equilibrium is reached (the same amount of drug enters and leaves the tissue). The time needed to reach equilibrium depends on the half-life of the phenomenon. Simple inspection of the above formulas indicates that the drug accumulation speed depends on hematic flux, and the higher the tissue volume and avidity for the drug, the longer the time needed to reach equilibrium. The consequences of this are illustrated in Figure 5.2a: for equal k_p, accumulation is faster in the kidney and brain (highly perfused) than in adipose tissue. However, in the same tissue, drugs with increasing k_p reach equilibrium at later times and at more elevated tissue concentrations (Fig. 5.2b).

All these considerations also hold true for the reverse process, that is, the drug release from tissue to blood. Thus, drug release is slower when the tissue is less perfused or has a greater avidity for the drug.

It is important to remind that the aforementioned considerations are valid when the drug can freely diffuse from blood to tissue and vice versa. Conditions may be different when diffusion is hampered by the histological organization of capillaries and surrounding tissue (the so-called barriers). In this case, the equilibration rates may be slower than predicted from the flux to volume ratio, being limited by the inability of venous blood to equilibrate with the tissue. Figure 5.3 shows how different drugs accumulate in the CNS with different rates. Here, the different kinetics are due to partition coefficient and ionization degree of the drugs. In general, tissue avidity for the drug also influences both magnitude and rate constant of tissue accumulation (and release). Note that such avidity does not depend exclusively on drug lipophilicity.

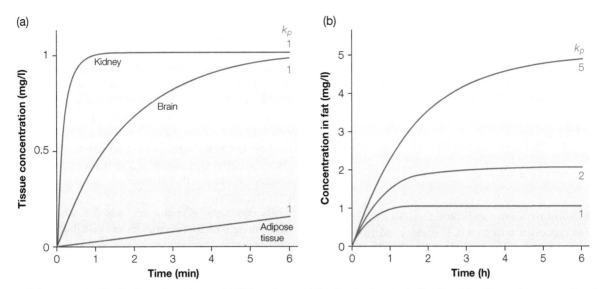

FIGURE 5.2 Drug distribution in various tissues. (a) When tissue avidity for the drug is similar ($k_p = 1$ in all cases), more perfused organs equilibrate faster with plasma. Over time, adipose tissue will also reach the equilibrium value (1 mg/ml) observed for the brain and kidney. (b) In the same tissue (here adipose tissue), different drugs may show different concentration time courses: if flux is constant, higher k_p values results in slower equilibration and higher final tissue concentrations.

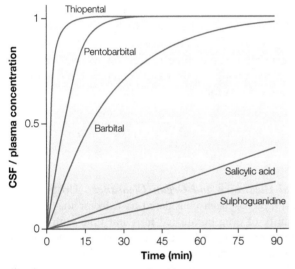

FIGURE 5.3 Differences among equilibration rates of drugs with cerebrospinal fluid (and CNS). Here, all drugs are assumed to have $k_p = 1$ and an equilibrium concentration equal to plasma concentration. In this case, the different rates are not determined by different V_D or flux/volume ratios, but by the different capacity of drugs to cross the BBB.

ELIMINATION

Drugs are eliminated by excretion and/or biotransformation. Biotransformation is described in detail in Chapter 6. The principal route of excretion is the kidney, but several drugs are excreted in significant amount with the bile and others by the lung (discussed in Chapter 7).

In general, most processes that lead to drug elimination follow first-order kinetics (see Box 4.1). However, it is important to remind that for some drugs the elimination kinetics may not be exclusively first order: for example, phenytoin and alcohol are mostly cleared by metabolic processes or active renal excretion, which are both saturable. In this case, elimination kinetics are zero order at high concentrations (a constant drug quantity is eliminated in the unit time) and only become first order at lower concentrations, when biotransformation or transport mechanisms are not saturated.

The Concept of Half-Life

First-order elimination kinetics are represented by a straight line in a semilogarithmic plot (see Fig. 4.2 and the elimination phase in Fig. 5.1). The graphs clearly show that the time interval that separates two points, in which one concentration is one-half of the other, is always the same and only depends on the slope of the line. The time interval needed for the concentration to halve its value is independent of the initial concentration and is said half-life or $t_{1/2}$. The equation that governs this kind of kinetics is

$$C(t) = C_0 \times e^{-kt} \quad \text{that is} \quad \frac{C(t)}{C_0} = e^{-kt} \quad (5.1)$$

where C_0 is the initial concentration, $C(t)$ is concentration at time t, and k is the rate constant of the process, that is, the fraction (%) by which drug concentration decays in the unit time. By definition, the half-life is the time needed for the concentration ratio $C(t)/C_0$ to fall to 0.5. Therefore, from Equation 5.1, we have

$$\exp(-kt_{1/2}) = 0.5; \quad k \times t_{1/2} = -\log(0.5) = \log(2) = 0.693 \tag{5.2}$$

Equation 5.2 leads to

$$t_{1/2} = \frac{0.693}{k} \text{ and } k = \frac{0.693}{t_{1/2}} \tag{5.3}$$

The half-life of a drug indicates the efficiency of the elimination processes acting on it. It is independent of drug concentration and solely dependent on the function of organs and systems eliminating the drug. Short half-life drugs (with high k) are quickly eliminated. Long half-life drugs persist for a long time in the organism. Each drug has its own half-life value that usually ranges from few minutes to weeks. It may vary among individuals, especially in the presence of pathologies that interfere with elimination processes. In general, drugs with a large distribution volume (one or more depot organs with large volume and high avidity) have long half-life, because as the drug is eliminated from plasma it is replaced by molecules accumulated in the storage sites.

Pathological alterations of the elimination organs generally produce an increase in drug half-life and prolong its effects (both beneficial and toxic). They also cause an increase of the peak value after administration (see Chapter 6). Changes in drug half-life require correction of the therapeutic regimen, especially for drugs with a narrow therapeutic window.

The Concept of Clearance

Once the drug has distributed to the tissues, the quantity present in the organism is given by the product $V_D \times C_p$ (apparent distribution volume times plasma concentration). Thus, the quantity eliminated in a unit time (Q_e) is given by the product $k_e \times V_D \times C_p$, which is the amount of drug contained in a volume of plasma equal to $k_e \times V_D$. In other words, this means that $k_e \times V_D$ liters of plasma are "cleared" per unit of time. Thus, the term clearance (Cl) indicates the volume of blood virtually cleared of the drug in a unit time by the elimination processes. Thus

$$Cl = k_e \times V_D \tag{5.4}$$

and the quantity eliminated in a unit time is equal to clearance times plasma concentration, that is,

$$Q_e = k_e \times V_D \times C_p = Cl \times C_p \tag{5.5}$$

Clearance is expressed in volume per unit time (typically, ml/min). In a given subject, k_e for a given drug is a constant, determined by the efficiency of the biological elimination processes; the apparent distribution volume is also a constant, dependent on the subject's physical constitution and the drug properties. Therefore, also, the clearance is a constant, independent of drug concentration (provided the elimination follows first-order kinetics; if zero-order kinetics occur, then the clearance varies with drug concentration, because a fixed quantity, eliminated in a unit time, corresponds to a volume of plasma that is inversely proportional to plasma concentration).

By combining Equations 5.3 and 5.4, we obtain the relation among half-life, clearance, and distribution volume:

$$t_{1/2} = \frac{0.693}{k_e} = \frac{0.693 \times V_D}{Cl} \tag{5.6}$$

This allows to correctly estimate the apparent distribution volume (V_D) by an alternative method that is more correct and precise than extrapolating to time zero the late decay phase of the plasma concentration curve. Plasma concentration and quantity of drug eliminated are measured over a given time period, t. Then the clearance is computed as the ratio (eliminated quantity/time)/(plasma concentration) (Eq. 5.5). Since $t_{1/2}$ can be easily derived from the plasma concentration curve (time needed for the concentration to halve its value), V_D is easily computed by rearranging Equation 5.6 as follows:

$$V_D = \frac{t_{1/2} \times Cl}{0.693} = 1.443 \times t_{1/2} \times Cl \tag{5.7}$$

Total Clearance and Organ Clearance Up to here, the clearance has been considered as a factor that relates the quantity of drug eliminated in a unit time with its plasma concentration ($Q_e = Cl \times C_p$). However, the clearance can also be examined in terms of drug elimination from a particular organ, which tells us about possible consequences of functional alterations of that organ, changes in binding to plasma proteins, or alterations in enzymatic or secretive activities. These considerations are useful to adapt the therapeutic regimen to specific physiopathological conditions of the patient (see Chapter 7).

The extraction rate of a drug passing through an organ is given by plasma flux (Φ) times the difference between arterial and venous blood concentrations ($C_A - C_V$); if such extraction rate is normalized to the rate of drug presentation ($\Phi \times C_A$), the extraction index (Eq. 5.8) is obtained:

$$\text{Extraction index (E.I.)} = \frac{C_A - C_V}{C_A} \tag{5.8}$$

If the extraction rate is instead examined in relation to arterial concentration, the value of organ clearance (Eq. 5.9) is obtained:

$$\text{Organ clearance} = \frac{\Phi \times (C_A - C_V)}{C_A} = \Phi \times \text{E.I.} \quad (5.9)$$

This is to say that the clearance of an organ is equal to the product of the plasma flux times the extraction index of the same organ.

The extraction index is obviously limited to a 0 to 1 range. Similarly, the clearance cannot exceed the plasma flux to the organ. The theoretical maximal clearance of an organ is its own plasma flux. For kidney and liver, plasma fluxes (and therefore maximal clearances) are 650 and 800 ml/min, respectively.

Total drug clearance can be seen as the sum of each organ clearance. In most cases, renal (Cl_R) and nonrenal clearances (Cl_{NR}) are considered. Most pharmacokinetic tables report total drug clearances, but they often indicate the percentage eliminated in the urines or by the liver, from which the values of renal (rCl) and nonrenal (nrCl) clearances can be estimated.

RENAL EXCRETION OF DRUGS

Kidneys function by filtering large volumes of plasma fluid in a poorly selective way, reabsorbing about 99% of the filtered water and part of the solutes in a highly selective way, and actively secreting organic acids and bases from plasma to tubular lumen.

The kidney is a highly perfused organ (≈ 1.3 l/min, i.e., about 25% of cardiac output). About 20% of the aqueous component (plasma flux, ≈ 650 ml/min) is filtered at the glomeruli, resulting in a glomerular filtration rate of about 130 ml/min (more than 150 l/day). The glomerulus is highly permeable to substances present in the plasma with molecular weight smaller than that of albumin, including drugs. It retains large plasma proteins and drugs bound to them.

Along the nephron, several mechanisms mediate reabsorption of 99% of the filtered water as well as ions and other substances. Figure 5.4 shows drug fates during their transit along the nephron. Drugs that can cross cell membranes are significantly reabsorbed in the tubules. Active processes may determine secretion or reabsorption of

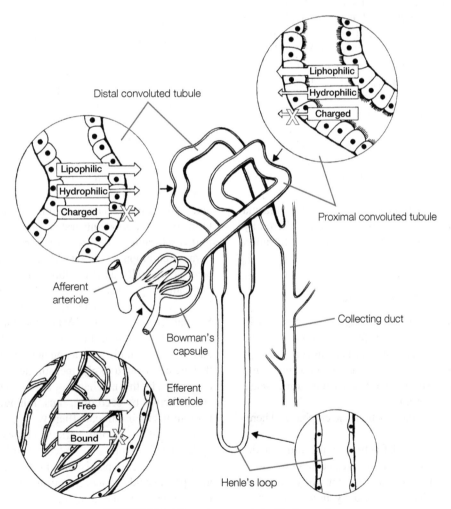

FIGURE 5.4 Drug movements inside the nephron.

compounds in the tubules, so that some compounds may be fully reabsorbed (e.g., glucose), while others are so efficiently secreted that virtually all plasma perfusing kidney parenchyma is deprived of them. In this case, the bound fraction of a drug may have the time to gradually equilibrate with the liquid phase during its flow in the capillaries, so that it is also significantly (sometimes completely) secreted and cleared.

Glomerular Filtration

Glomerular filtration rate is about 130 ml/min. A drug that is not reabsorbed or secreted in the tubules, and not bound to plasma proteins, has a renal clearance of 130 ml/min. If all the filtered drug is eliminated with the urine, a fraction $(0.130 \, l/V_D)$ of the free quota (q_f) contained in the organism is eliminated every minute, so that the renal elimination rate is $k_e = 0.13 \times q_f/V_D$.

Drugs not bound to plasma proteins $(q_f = 1)$ and not reabsorbed (e.g., very hydrosoluble and electrically charged drugs, such as some neuromuscular blocking agents that cannot cross membranes) distribute in plasma and interstitial spaces only (about 14 l in a 70 kg subject) and display a $k_e = 0.13/14 = 0.0093/min$, corresponding to an elimination time constant $\tau = 1/k_e = 108 \, min$ and a half-life of $0.693/k_e = 75 \, min$. This is a typical half-life value for very hydrophilic/charged drugs that are passively eliminated by the kidney (with no active tubular secretion).

The carbohydrate inulin, which is filtered at the glomeruli and is neither reabsorbed nor secreted in the tubule, is used to estimate the glomerular filtration rate. It has a half-life of 75 min with a V_D of about 14 l. From Equations 5.6 and 5.7, we know that $Cl = 0.693 \, V_D/t_{1/2}$; therefore, inulin clearance is $0.693 \times 14/75 = 129 \, ml/min$. Since all and only the inulin filtered at the glomeruli is eliminated, the quantity of plasma deprived of its inulin content equals the volume filtered by the glomeruli each minute, that is, the glomerular filtration rate (130 ml/min).

Tubular Functions and Pharmacokinetics

Figure 5.1 shows a schematic drawing of the nephron in which tubules are subdivided in different regions, each one characterized by distinct transport mechanisms for reabsorption and/or secretion of ion and larger compounds including drugs. For a detailed description of tubular functions, please refer to a physiology textbook.

Some aspects of the tubular functions are emphasized here:

1. Only the free, unbound fraction of the drug in the plasma is filtered by the glomerulus and can be found in the preurine. Drugs that are extensively bound to proteins are usually eliminated by nonrenal routes, mostly hepatic metabolism.

2. A drug that easily crosses membranes, in the preurine, is in equilibrium with the interstitial concentration; therefore, its concentration is comparable to the free drug concentration in plasma. Thus, for such a drug, renal clearance is close to: free fraction times urine volume (1–5% of filtrate volume). If the drug is highly bound to plasma proteins, its renal clearance is almost negligible.

3. Drugs with an intermediate degree of lipophilicity (partition coefficient) display a clearance ranging from negligible values (highly lipophilic drugs) to 130 ml/min (inulin clearance).

4. As mentioned earlier, pH influences the partition coefficient of weak bases and acids. Urinary pH is generally some units lower than plasma pH. Thus, compounds with $pK_A > 5$ are mostly reabsorbed if the acidic species is nonionized, whereas they are mostly disposed of if the basic species is nonionized. These aspects should be considered when acceleration of elimination is required, as in the case of intoxication by barbiturates (weak acids/hypnotics often used for suicide attempts). To alkalize the urine and increase barbiturate elimination, sodium bicarbonate and inhibitors of carbonic anhydrase are usually employed. Acidification can be used to accelerate amphetamine (weak base) elimination.

5. Two distinct transport mechanisms are responsible for active tubular secretion of endogenous compounds and drugs: they separately act on organic anions and cations and are both localized at the proximal tubule.

Active Transport of Organic Anions and Cations

This mechanism provides transcellular transport of carboxylated compounds and other organic anions, as well as compounds with partial electrical charges (e.g., sulfones and sulfonamides). The system is considered nonspecific in terms of transported substrates. In general, the anion transport is saturable, and different anions compete for transport. Para-aminohippuric acid (PAH), a derivative of hippuric acid, is generally used to study tubular anion transport. At not saturating concentration, PAH is fully eliminated and its clearance approximates renal plasma flow (650 ml/min). Several compounds (NSAIDs including salicylates and phenylbutazone; diuretics as furosemide, ethacrynic acid, and thiazides; antibiotics as penicillins and cephalosporins) that are conjugated to glycine, sulfate, or glucuronic acid are eliminated by this mechanism. Note that diuretics reach their site of action by secretion; thus, inhibition of such mechanism limits their clinical efficacy.

Competition among various anions for this transport system can be exploited to interfere with secretion of a drug, thus prolonging its persistence in the organism. The organic acid probenecid (an antigout drug) can be used to decrease penicillin secretion and maintain effective concentrations of the antibiotic for longer times.

The transport system for organic cations is saturable too. Endogenous compounds eliminated through this pathway include amine neurotransmitters (acetylcholine, dopamine, histamine, serotonin) and metabolites such as creatinine. Substrates for this transport system generally possess an amine group, positively charged at physiological pH. They include many natural alkaloids and their derivatives (e.g., the muscarinic blocker atropine and the anticholinesterase drug neostigmine), synthetic compounds (such as isoproterenol and hexamethonium), and opiate analgesics (such as morphine); interestingly, morphine can be conjugated to glucuronic acid in the liver and be thus excreted by the anion transport system.

An efficient transport mechanism is in charge of anion reabsorption; it is more selective than the tubular secretion system, so that most actively secreted anions are disposed of in the urine, but others are largely reabsorbed. This occurs, for example, to uric acid, actively secreted and reabsorbed in the tubule. Probenecid is another good substrate of the anion reabsorption system; for this reason, it can be employed in hyperuricemia and gout to compete with uric acid for tubular reabsorption, thereby increasing its elimination.

Factors Determining Renal Clearance of Drugs

Drug renal clearance is favored by absence of binding to plasma proteins, high hydrophilicity, ionization, and presence of active transport (secretion) systems in the tubule. As shown in Table 5.3, half-life values vary depending on distribution volume and renal clearance. A drug that is filtered, unable to cross membranes and not actively reabsorbed or secreted, will be found in the urine in the same amount (quantity, not concentration) present in the glomerular filtrate (filtration rate = 130 ml/min). For a drug that is reabsorbed, the clearance is lower, down to the limit of zero for total tubular reabsorption (e.g., glucose). Active secretion may instead produce an increase in drug clearance up to 650 ml/min (renal plasma flow) for complete active secretion (as observed with PAH).

Functional or pathologic modification of renal functions may significantly change drug pharmacokinetics. Changes are more pronounced for those drugs whose elimination occurs mostly through the kidney. Table 5.4 provides a list of drugs with different fractional renal elimination. Changes in renal function may induce significant changes in the half-life

TABLE 5.3 Half-life of drugs in relation to clearance and volume of distribution

	Drug distributed in		
Clearance (ml/min)	Plasma (3 l)	Extracellular fluids (41 l)	Total body water (42 l)
650 (active excretion)	$t_{1/2}=3$ min	$t_{1/2}=15$ min	$t_{1/2}=45$ min
130 (glomerular filtration)	$t_{1/2}=16$ min	$t_{1/2}=74$ min	$t_{1/2}=223$ min
30[a] (partial reabsorption)	$t_{1/2}=69$ min	$t_{1/2}=5$ h 22 min	$t_{1/2}=16$ h

[a] This is an example for reabsorbed drugs; clearance values range from 0 (complete reabsorption) to 130 ml/min (no reabsorption); higher values may be observed if active tubular secretion coexists.

TABLE 5.4 Renal and hepatic extraction indexes

	Low	Intermediate	High
	<0.3	0.3–0.7	>0.7
Hepatic extraction index[a]	Carbamazepine Diazepam Indomethacin Phenobarbital Theophylline Warfarin Valproic acid	Acetylsalicylic acid Quinidine Codeine Nortriptyline	Alprenolol Desimipramine Isoprenaline Lidocaine Morphine Nitroglycerin Pentazocine Propranolol
Renal extraction index[a]	Atenolol Digoxin Furosemide Gentamicin Lithium Tetracycline	Cimetidine Procainamide Cephalothin Penicillins (some)	Penicillins (some) Glucuronides (some)

[a] Drugs are considered for which the liver (or the kidney) accounts for >30% of total clearance.

values of those drugs that are substantially eliminated through the kidney. As discussed in detail in Chapter 7, this may require adjustments of the doses (or interval between doses) for a given drug. The correction is usually based on the change in creatine clearance, which represents a good comprehensive index of renal function and can be indirectly derived (using equations or tables) from creatinine plasma concentration. It should be considered that creatine clearance differs significantly in newborn and elderly as compared to normal adults (see Fig. 7.9). In addition, the fraction of water in the infant's body is larger, and the relationships between body weight and volume of distribution and between body weight and renal clearance are different compared to those of the adult. Thus, renal half-life values for these subjects may also be different.

HEPATIC EXCRETION AND ENTEROHEPATIC CYCLE

The liver performs two important functions on drugs present in the organism: metabolism and excretion. The two processes are strictly linked: phase II metabolism consists in the conjugation of lipophilic compounds and drugs with polar groups; this increases their molecular weight and polarity, making them more suitable for active biliary excretion.

The liver is especially equipped for processing both endogenous and xeno compounds since blood flows around the cells (Disse's space) without interference from the basal lamina and even compounds bound to plasma proteins can be exposed to the cell membrane of the hepatic cells. These also include phagocytes (Kupffer cells and mononuclear phagocytes of the reticuloendothelial systems) that can absorb and digest compounds arrived to the liver via the portal system and the hepatic artery. Drugs that do not cross cell membranes can enter hepatocytes via membrane systems capable of transporting large molecules such as bilirubin. Transporting proteins (ligandins and Z proteins) contribute to intracellular delivery of drugs to the microsomal system where their metabolic transformation occurs (see Chapter 6). Ligandin seems to possess also glutathione-S-transferase activity and to be able to conjugate activated substrates.

Drug excretion into the bile increases with their polarity and molecular weight. Ionization is augmented by addition of polarized anionic groups; the molecular weight is augmented by conjugation to glucuronic acid, glycine, glutathione, and sulfate groups. The molecular weight of the compound is relevant, as only molecules above 300–500 kDa are actively secreted in bile.

At least four distinct transport systems are involved in biliary secretion. They are specific for anions, bile acids, cations, and neutral substances, respectively. The transport is saturable and produces substrate concentration in the bile up to 50-folds higher than in the plasma. Competition for the same excretion transport system occurs between compounds of the same class but not between different classes. Among the organic acids excreted in bile, there are penicillin and the dye bromosulfophthalein, which is excreted as a glutathione conjugate; in the past, its rate of biliary excretion was used as an index of liver function. Although biliary acids are organic anions, they have a specific transport system. Cationic compounds are excreted through a third mechanism that requires the presence of a basic group (in general a charged amine) and nonpolar groups that make the compound amphipathic. The transport mechanism for neutral compounds is less well known: some compounds may be conjugated and transported by the anion transport system (these molecules are not excreted if the conjugation mechanisms are impaired or inhibited), whereas others are excreted even in the absence of any charged group. For some of these compounds (e.g., cardiac glycosides), excretion may be favored by the amphipathy and marked asymmetry of the molecule. Biliary excretion is the most important route for metal elimination, mostly through conjugation with reduced glutathione. Peptidic hormones may be excreted in bile via receptor-mediated transcytosis mechanisms.

Biliary excretion, as well as hepatocyte metabolic activity, is very important for orally administered drugs that reach the liver through the portal circulation, as they may undergo massive metabolism or excretion before entering the systemic circulation (presystemic elimination or first-pass effect). This may significantly reduce oral bioavailability of drugs. As a rule, presystemic elimination is saturable; drugs that undergo a marked first-pass effect, such as the β-blocker propranolol, may be administered by mouth if taken after a meal, when transport systems are saturated by food. However, drug bioavailability will be less predictable.

Once the drug has been secreted in bile, it is not necessarily eliminated from the organism: if its chemical–physical properties are compatible, it can be reabsorbed in the intestine. This mostly occurs with drugs that have been excreted after conjugation with glucuronic acid, as the conjugate can be easily hydrolyzed by the intestinal glucuronidase activity. In these cases, a continuous enterohepatic cycling takes place, which keeps the compound (endogenous or exogenous) in the organism until it is finally metabolized or excreted by the kidney. The enterohepatic cycle is essential to avoid continuous loss of endogenous substances such as biliary salts, vitamins D and B_{12}, folic acid, and estrogens.

Renal elimination of drugs is highly influenced by hydrophilicity and ionization; also in the liver, polar groups favor excretion. For most drugs, biotransformation mechanisms, especially in the liver, are aimed at increasing their hydrophilicity or polarization, thereby favoring their excretion.

In liver diseases, biliary excretion of drugs may be markedly affected; the situation is less clear than in the kidney, because functional impairment of hepatocytes, biliary tract, and vascularization may be quite variable and it is not easy to directly determine the levels of drug excretion in the bile.

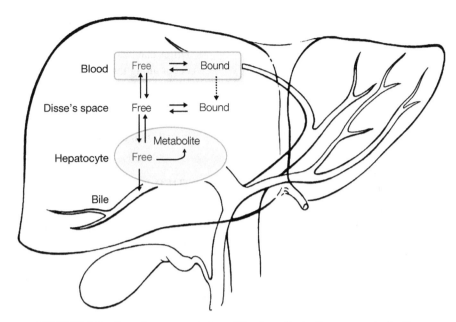

FIGURE 5.5 Stepwise representation of the drug elimination process in the liver.

Perfusion, Binding to Plasma Proteins, Enzymatic Activity, and Hepatic Clearance

The hepatic activities leading to drug clearance are schematically depicted in Figure 5.5. The drug enters the liver with the blood, partly free and partly bound to plasma proteins. The free fraction has the highest probability to enter the hepatocyte and be metabolized and possibly excreted with the bile. However, the anatomical organization of hepatic sinusoids allows also the absorption of plasma proteins and thus of the drugs bound to them. In addition, active uptake and metabolism of the drug by the hepatocytes produce a decrease in the drug free quota, so that part of the bound drug is released and then taken up by the hepatocytes, during the relatively long permanence of blood in the sinusoids.

Changes in liver blood flux may produce significant changes in drug hepatic clearance, particularly if the extraction index approaches 1. In this case, as the elimination processes involved are very efficient, drug clearance approaches the hepatic flux. Conversely, if the extraction index is low, drug venous concentration does not differ much from the arterial (by definition), and therefore, the effect of flux variation is minimal. Table 5.4 shows a list of drugs with their hepatic extraction index.

The effect of plasma protein binding is also related to the drug extraction index and is more marked when the index is low. In fact, in this case, the limiting factor for drug elimination is not the metabolism nor the secretion rate (nor the blood flux), but the drug ability to enter the hepatocyte by crossing the plasma membrane, a process that mostly involves the drug unbound fraction.

Pathological or iatrogenic changes in hepatic enzymatic activities (metabolic induction; see Chapter 6) and genetic alteration of hepatic enzymes are important when enzymatic activity is the limiting factor in drug elimination.

Since the liver is the most important organ in drug metabolism, in general a physician should be careful when administering drugs to patients affected by liver pathologies. However, available data suggest that a cautious attitude, although correct, is not always necessary. Variability exists since hepatic pathologies cannot be considered as a single entity for what concerns drug elimination. In terms of drug clearance, two main categories can be distinguished: chronic pathologies (cirrhosis in particular) and acute, reversible pathologies (e.g., viral hepatitis). In general, hepatic drug clearance is reduced in cirrhotic patients, whereas in acute pathologies and obstructive jaundice is reduced only for some drugs but not for all. Changes in hepatic clearance may also be due to individual factors, as described in detail in the next chapter. Finally, hepatic disorders may also induce substantial changes in plasma protein levels and profiles, thus changing the unbound fraction of drugs and their volumes of distribution and clearances. As predicted by Equation 5.6, this implies that hepatic disorders may modify drug half-lives.

TAKE-HOME MESSAGE

- Following absorption, drugs distribute to tissues via blood circulation. Drug concentrations in plasma and tissues may differ because of differential tissue avidity (ratio of drug concentration in tissue vs. plasma at equilibrium).

- Each tissue has an apparent distribution volume (actual volume × avidity) and thus contributes to the distribution volume of the organism, V_D.
- Drugs equilibrate between plasma and each tissue with a rate constant determined by tissue-specific perfusion and avidity; equilibration will be slower if diffusional barriers are present.
- Drugs can bind to plasma proteins; binding may influence V_D, elimination rate, and metabolism of a drug.
- For most drugs, elimination follows first-order kinetics.
- The elimination rate constant (fraction eliminated per unit time) equals the ratio between clearance (volume of plasma cleared in a time unit) and apparent distribution volume: $k_e = Cl/V_D$.
- Half-life is the time required for elimination of half of the drug present in the body; half-life is independent of the amount/concentration of drug and depends on the biological mechanisms of drug elimination.
- In the kidney, only the drug free quota is filtered at the glomerulus; for drugs that are neither reabsorbed nor excreted along the tubule, the clearance approximates the glomerular filtration rate (~130 ml/min). Higher clearance, up to total plasmatic renal flux (650 ml/min), is observed for drugs actively secreted along the tubule. Lipophilic drugs have low renal clearances because they are largely reabsorbed along the tubules.
- Drugs can be excreted in the bile and reabsorption may occur along the intestine (enterohepatic cycle).
- Metabolism usually makes drugs more hydrophilic, enhancing renal disposition; conjugation with hydrophilic moieties produces larger and amphipathic compounds that are more easily excreted in the bile.

FURTHER READING

Benet L.Z., Hoener B.A. (2002). Changes in plasma protein binding have little clinical relevance. *Clinical Pharmacology and Therapeutics*, 71, 115–121.

Port R.E., Wolf W. (2003). Non invasive methods to study drug distribution. *Investigational New Drugs*, 21, 157–168.

Roerts S.A. (2003). Drug metabolism and pharmacokinetics in drug discovery. *Current Opinion in Drug Discovery and Development*, 6, 66–80.

Smith Q.R. (2003). A review of blood-brain barrier transport techniques. *Methods in Molecular Medicine*, 89, 193–208.

Strolin Benedetti M., Baltes E.L. (2003). Drug metabolism and disposition in children. *Fundamental and Clinical Pharmacology*, 17, 281–299.

6

DRUG METABOLISM

ENZO CHIESARA, LAURA MARABINI, AND SONIA RADICE

> **By reading this chapter, you will:**
>
> - Understand the significance of metabolism in drug elimination and activity
> - Know the major biochemical reactions involved in drug biotransformation
> - Learn the genetic basis of drug biotransformation and the modulation by environmental factors

In order to be absorbed and distributed throughout all body tissues, drugs must have chemicophysical structures generally different from those facilitating their elimination. As shown in Chapter 5, lipophilicity and absence of electric charge facilitate cutaneous, pulmonary, and gastrointestinal absorption of drugs and allow them to cross cell membranes to reach their intracellular sites of action. In the absence of effective elimination systems, exposure to lipophilic drugs would result in their accumulation in the body.

However, there are elimination systems—urinary, biliary, and fecal excretion, pulmonary expiration, and transpiration—that are facilitated if substances acquire electric charges and are converted into hydrophilic chemical derivatives. Therefore, the organism needs to modify the structure of those substances. This biotransformation process is aimed at transforming foreign substances into more polar and water-soluble compounds inside the organism in order to increase their excretion and decrease their volume of distribution. The enzymes involved in biotransformation processes are called drug-metabolizing enzymes, although, besides drugs, they metabolize many other endogenous or exogenous compounds (xenobiotics) to which living beings are exposed. During evolution, differences and changes in diets have contributed to determine a marked variability and specific complexity of these enzymes among different species (see Supplement E 6.1, "Factors That Modify Drug Metabolism").

Nowadays, most living beings are exposed to xenobiotics from several sources, such as drugs, environmental pollutants, food (including additives and contaminants), cosmetics, and chemical products employed in agriculture.

METABOLIC MODIFICATION OF DRUG ACTIVITY

Metabolization, or more properly biotransformation, of a drug leads to the formation of more metabolites, different from each other, which either lack the pharmacological activity of the initial compound or are still pharmacologically active having a pharmacological spectrum at least partially similar to that of the original compound (e.g., heroin that is bioconverted into morphine). Other examples are the propranolol metabolites and some antidepressants that interfere with monoamine reuptake. Particularly important is the case of anxiolytic benzodiazepines, such as diazepam whose two main metabolites, nordiazepam and oxazepam, exhibit the same pharmacological characteristics of the original compound but have different half-lives (nordiazepam has two times the half-life of diazepam, whereas oxazepam has one-fourth of diazepam half-life). Thus, duration of diazepam action depends on which of the two compounds is generated by the enzymatic equipment of the subject.

Some drugs become active only after biotransformation. In this case, biotransformation gets the name of bioactivation, and the initial compound is called prodrug. Examples of prodrugs are some antidepressants, such as imipramine; some steroid hormones (prednisone and cortisone are activated to prednisolone and hydrocortisone, respectively); anticholinesterase parathion (used as a pesticide); and azathioprine, which is converted into mercaptopurine, a molecule capable of interfering with nucleic acid synthesis. Therefore, the pharmacological activity of prodrugs is dependent on the subject's metabolizing ability.

Biotransformation can also give rise to toxic metabolites, with no pharmacological activity; for example, the hepatotoxicity of nonsteroidal anti-inflammatory paracetamol is due to its hydroxylated metabolite reacting with other endocellular structures.

TWO PHASES OF DRUG METABOLISM

Drug biotransformations are enzymatic reactions and can be grouped into phase I (or functionalization) reactions and phase II (or conjugation) reactions. From a chemical perspective, oxidation, reduction, hydrolysis, hydration, dethioacetylation, and isomerization reactions belong to phase I. Phase I reactions are also called functionalization reactions because they modify substrates by introducing or exposing functional groups, such as –OH, –COOH, –SH, and –NH$_2$, suitable to phase II reactions, which use these functional groups as terminals to conjugate drugs with different molecules.

Many conjugation or synthetic reactions belong to phase II, the most important being glucuronidation, sulfation, methylation, acetylation, conjugation with amino acids, conjugation with glutathione, conjugation with fatty acids, and condensation.

Phase I reactions carry out the biological and chemical modification of xenobiotics, whereas phase II reactions increase their molecular weight and modify their liposolubility to convert them into water-soluble and highly ionized (at the physiological pH of the organism) compounds and facilitate their elimination.

Phase I Reactions

Phase I biotransformation reactions consist in a set of fundamental reactions: oxidation, reduction, hydrolysis, hydration, and less common mixed reactions.

Oxidations The oxidative enzyme system involved in drug metabolism is mainly found in the endoplasmic reticulum of the liver, kidney, lung, and intestine cells; it includes cytochrome P450 enzymes (CYPs), flavin-containing monooxygenases (FMOs), and epoxide hydrolases (EHs). The drug-metabolizing reactions carried out by this system are shown in Table 6.1.

Besides these microsomal reactions, there are other biotransformation reactions catalyzed by mixed-function oxidases mainly localized in mitochondria. These are highly substrate-selective enzymes that may be also involved in substrate synthesis, as in the case of endogenous steroids.

Cytochrome P450 The cytochrome P450 system is a superfamily of membrane-bound, heme-containing mixed-function oxygenases. Cytochrome P450 enzymes control different types of reactions, all based on the direct incorporation of a molecule of active oxygen into the substrate. Oxidative reactions involve large classes of endogenous compounds, such as cholesterol, as well as xenobiotics, like drugs (Table 6.1), carcinogens, pollutants, pesticides (Fig. 6.1), and herbal drugs.

Oxidative reactions require the presence of molecular oxygen, NADPH, and an oxidase system consisting of two coupled enzymes, NADPH–cytochrome P450 reductase and cytochrome P450. These enzymes are embedded in the phospholipid matrix of the endoplasmic reticulum (Fig. 6.2). Phospholipids are essential to allow reactions between the two enzymes and to allow the substrate to reach the active site.

Cytochrome P450 catalyzes a cyclic reaction that is based on the ability of its heminic iron to mediate cyclic oxidoreductase reactions on both the substrate and the molecular oxygen (Fig. 6.3). The substrate (RH) binds the oxidized form of cytochrome P450 (Fe^{3+}) to form an enzyme–substrate complex. Then cytochrome P450 is reduced (Fe^{2+}) through acquisition of one electron (e^-) supplied by NADPH–cytochrome P450 reductase. The reduced substrate–cytochrome complex incorporates a molecule of oxygen (O_2) and acquires another electron from NADPH

TABLE 6.1 Oxidative reactions

Reaction	Examples of substrate
Aromatic hydroxylation	Lignocaine
Aliphatic hydroxylation	Pentobarbital
Epoxidation	Benzo(a)pyrene
Oxidative deamination	Amphetamine
N-oxidation	2-acetylaminofluorene 3-methylpyridine
S-oxidation	Chlorpromazine
N-dealkylation	Diazepam
O-dealkylation	Codeine
S-dealkylation	6-methylthiopurine
Oxidative desulfurization	Parathion
Dehalogenation	Halothane
Alcohol oxidation	Ethanol

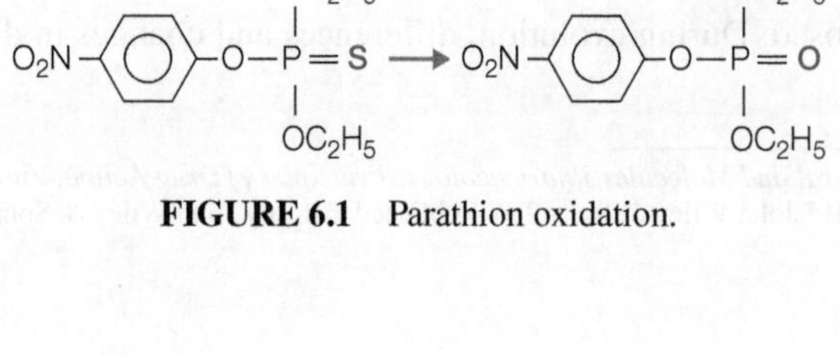

FIGURE 6.1 Parathion oxidation.

through NADPH–cytochrome P450 reductase. Metabolization of a substrate by cytochrome P450 requires one molecule of active oxygen to produce one molecule of oxidized substrate and water. In many cytochrome P450-dependent enzymatic reactions, depending on the substrate nature, an uncoupling of oxidative reactions occurs resulting in more O_2 being used than actually necessary to oxidize the substrate, with consequent formation of active oxygen (O_2^-) that is then transformed into water by a superoxide dismutase.

The different specificity, velocity, and affinity for various substrates have led first to hypothesize and then to demonstrate (by cloning P450 genes and identifying their allelic variants) that cytochrome P450 consists of many different families, subfamilies, and isoforms. These are characterized by an amino acid sequence identity ≥40% among family members and ≥59% among subfamilies members.

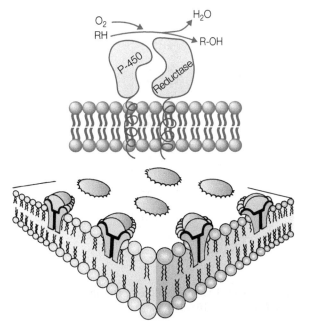

FIGURE 6.2 Localization of mixed-function oxidase system in the membrane of the smooth endoplasmic reticulum.

Current nomenclature of cytochrome P450 superfamily recommends the abbreviation CYP followed by an Arabic number (family), a capital letter (subfamily), and an Arabic number (isoform), as in CYP3A4.

By comparing the amino acid sequences of different members (and assuming a constant rate of evolutionary change), it has been possible to build a phylogenetic tree of cytochrome P450 evolution (see Fig. 6.4). Note the explosion of new genes in the last 400 million years, as well as the duplication of many genes within the CYP2 family. Such expansion of CYP genes can be explained by the so-called plant–animal warfare. The emergence of cytochrome P450 in plants, aimed at the synthesis of endogenous steroids and cholesterol in mitochondria (possibly as a defense against external adverse stimuli), was followed by the development of other biosynthetic systems producing additional defensive substances, such as phytoalexins. This was probably stimulated by the appearance of animal species, as the substances produced by those systems were able to make plants (the only source of food at that time) less desirable and digestible. As a "response," animals developed new P450 genes capable of demolishing that kind of substances. As animals spread throughout the world and met new plant forms for the first time, the "plant–animal warfare" became increasingly intense, favoring the explosion of new P450 genes. Since then, animal species have developed various forms of CYPs, mostly of CYP2 type (Table 6.2). This is obviously related to changes in eating habits and exposure (particularly humans) to a huge number of exogenous chemical substances (pollutants, drugs, and food). The large variety of CYP isoforms and the ability of each isoform to metabolize various substrates with different structures account for the drug interactions underlying the interindividual variability in the response to therapy and the adverse or even toxic effects of drugs.

Cytochrome P450 superfamily includes a group of "orphan" CYPs. This definition applies to a set of human CYP families, nearly one-fourth of all CYPs present in human genome, whose function, expression, and regulation are still unknown.

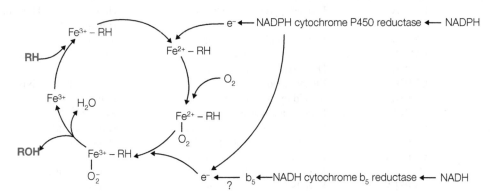

FIGURE 6.3 Diagram showing the catalytic cycle of cytochrome P450. RH, substrate; ROH, hydroxylated product.

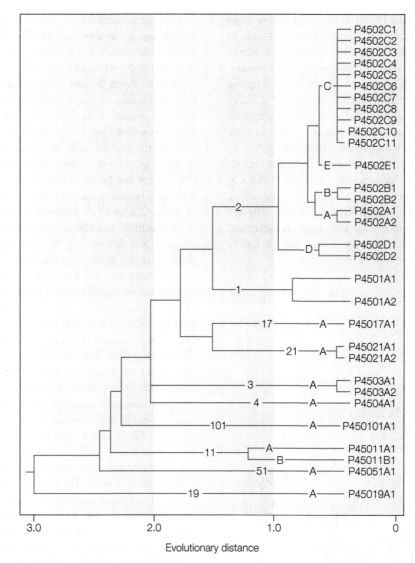

FIGURE 6.4 Evolution of the CYP superfamily. Adapted from Gonzales F.J., Gelboin H.V. (1994). Role of human cytochromes P450 in the metabolic activation of chemical carcinogens and toxins. *Drug Metabolism Reviews*, 26, 165–183.

TABLE 6.2 Some of principal cytochrome CYP families

Family	Subfamily	Reactions	Distribution
CYP1	CYP1A	Benzo(a)pyrene hydroxylation	Humans, rats, rabbits, mice
CYP2	CYP2A	Steroid hydroxylation	Humans, rats
	CYP2B	Benzphetamine demethylation	Rats, rabbits
	CYP2C	Steroid hydroxylation	Humans, rats, rabbits
	CYP2D	Debrisoquine hydroxylation	Humans, rats
	CYP2E	Ethanol oxidation	Humans, rats, rabbits
CYP3	CYP3A	Steroid hydroxylation	Rats
CYP4	CYP4A	Lauric acid hydroxylation	Rats
CYP11	CYP11A	Cholesterol side-chain cleavage	Humans, cattle
	CYP11B	Deoxycortisol 17β-hydroxylation	Humans, cattle
CYP17	CYP17A	Pregnenolone 17α-hydroxylation	Cattle
CYP21	CYP21A	Progesterone 21-hydroxylation	Humans, cattle, mice
CYP51	CYP51A	Phytohormone biosynthesis	Plants
CYP101	CYP101A	Camphor hydroxylation	Bacteria, yeasts

FMOs FMOs are phase I enzymes acting on hydrophobic substrates. They are mainly found in hepatocytes, localized in the endoplasmic reticulum membranes.

Six FMO families are known, with FMO 3 being the predominant in the liver. FMOs are only marginally involved in drug metabolism, and most times, they produce inert metabolites. Among the drugs metabolized by FMOs we remember are some H2 inhibitors (cimetidine and ranitidine), antipsychotics (clozapine), and antiemetics (itopride).

In contrast with CYPs, FMOs do not undergo induction and inhibition by other drugs used in the therapy (see Supplement E 6.2, "Induction of Drug Metabolism"); therefore, in the case of FMO-dependent biotransformation, the problem of interactions does not exist.

EHs EHs act on epoxides produced by cytochrome P450 enzymes and transform them in the corresponding dihydrodiols. EHs are classified as soluble EHs (sEHs) and membrane EHs (mEHs), in accordance with their localization in the soluble fraction or in the endoplasmic reticulum of hepatocytes. Epoxides are highly reactive electrophilic compounds that bind nucleophilic groups in proteins, DNA, and RNA and produce mutations and cell toxicity. EHs thus serve to inactivate potentially toxic metabolites generated by cytochrome P450, rather than drugs.

Cytochrome P450-Independent Mixed Oxidations and Mixed Oxidations Not Localized in the Endoplasmic Reticulum There are many other mixed-function oxidases, mainly localized in the soluble fraction of the liver, kidney, and lung cells. Although they are predominantly involved in the metabolism of endogenous compounds, they can oxidize some xenobiotics too. The most important of these oxidases are alcohol dehydrogenases and aldehyde dehydrogenases, xanthine oxidases, amino oxidases, aromatases, and alkyl hydrazine oxidases.

Alcohol dehydrogenases catalyze oxidation of alcohols into the corresponding aldehydes. Unlike the P450-dependent ethanol-oxidizing system typical of hepatic microsomes, these enzymes use NAD^+ as a cofactor and are dehydrogenases, which prevail in normal physiological situations. Cytochrome P450-mediated oxidation of ethyl alcohol is poorly relevant but plays a major role in the presence of induction even by ethanol itself (see Supplement E 6.2).

In this group of oxidation mechanisms, the enzymes called amino oxidases also play an important role. They are grouped into monoamine oxidases and diamine oxidases, which metabolize, respectively, catecholamines and endogenous diamines (such as histamine) into the corresponding aldehydes. Monoamine oxidases also metabolize exogenous amines from food (such as tyramine) and are mainly localized in nerve endings and in the liver.

Reductions In liver microsomes, also some reductive biotransformations take place. These reactions require NADPH, but, unlike those catalyzed by oxidases, are inhibited by oxygen. In this type of biotransformation, substrates to be metabolized can accept one or both electrons (e^-) coming from the P450 electron transfer chain, but not oxygen. Cytochrome P450 inhibitors also inhibit reductive reactions, either by competing with substrates for the binding sites or by forming complexes with heme iron and blocking electron transfer. The main substrates of these reactions are nitro and azo compounds (see two examples in Fig. 6.5). Reductive biotransformations involve both the flavoprotein NADPH–cytochrome P450 reductase and the terminal oxidase cytochrome P450.

These biotransformations play an important role in toxicology as they can produce toxic metabolites or reactive intermediates. For example, many nitro compounds are reduced to aminic derivatives that can be oxidized to N-hydroxylated metabolites with toxic characteristics. Carbon tetrachloride and halothane are two typical examples of reductive biotransformations resulting in the formation of toxic free radical.

Hydrolysis Esters, amides, hydrazides, and carbamates are hydrolyzed by different, specific or nonspecific, enzymes, mainly found in plasma as well as in endoplasmic reticulum and soluble fraction of hepatocytes.

The enzymatic activities controlling these reactions are esterases and amidases, which catalyze hydrolysis of ester- and amide-containing chemicals.

From a physiological perspective, the most important esterase is the acetylcholinesterase that hydrolyzes acetylcholine (see Chapter 39). Esterases may be nonspecific, such as plasma esterases reducing procaine into benzoic acid and diaminoethanol, or specific, such as those reducing meperidine in the liver.

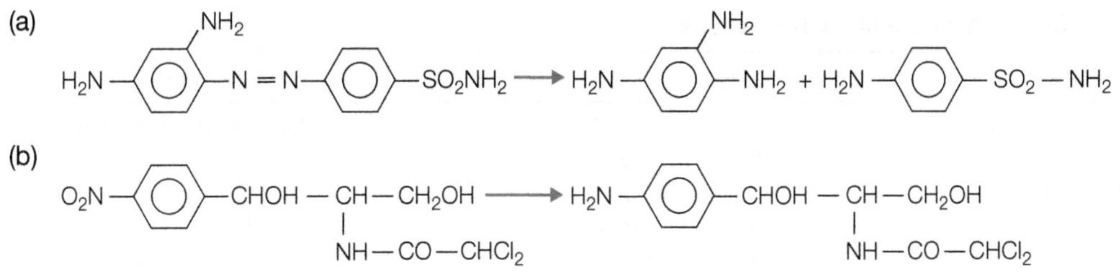

FIGURE 6.5 Reduction of (a) prontosil rubrum and (b) chloramphenicol.

The hydrolysis of amides is carried out by amidases, which are mainly found in plasma. These are characterized by a lower enzymatic rate compared to esterases and have a modest effect on drug metabolism. One of the most typical amidase-catalyzed reactions is the hydrolysis of the isoniazid hydrazine group.

Also, protein and peptide hydrolases exist, mainly in stomach secretions, but they are scarcely involved in drug metabolism.

Phase II Enzymatic Reactions

Phase II reactions are catalytic reactions that involve cofactors (such as uridine diphosphoglucuronic acid (UDP-GA)), transfer enzymes (such as UDP-glucuronosyltransferase(s) (UGT)), and substrate functional groups (in general, metabolites of phase I reactions) and produce conjugated products. These reactions require energy to activate cofactors (only seldom substrates) and convert them into high-energy intermediates. The activation is directly or indirectly mediated by ATP. The reactive functional groups –OH, –COOH, –NH, and –SH are mainly originated by phase I metabolism.

Phase II reactions are relatively numerous (Table 6.3) and involve different groups of enzymes. With few exceptions, these enzymes transform substrates into products that are more hydrophilic, more ionizable at physiological pH, and very often with a higher molecular weight. In other words, these reactions convert substrates into easily disposable and hardly resorbable products. All phase II reactions have a much higher reaction rate compared to phase I reactions and aim at blocking the biological activity of drugs.

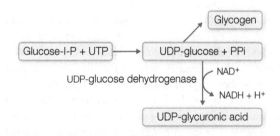

FIGURE 6.6 Synthesis of uridine diphosphoglucuronic acid.

Glucuronidations Because of the high number of different substrates and functional groups they involve, as well as the relatively abundant endogenous cofactor (the UDP-GA) they employ, glucuronidation reactions are the most important of the conjugation reactions.

Glucuronidation reactions consist in the transfer of glucuronic acid moieties from the high-energy cofactor, UDP-GA, to substrate functional groups. These reactions are catalyzed by UGTs. Unlike other phase II enzymes, which are mainly cytosolic, these enzymes are mostly localized in the endoplasmic reticulum of various tissues, such as the liver (the most enriched), kidney, intestine, skin, brain, and spleen. UDP-GA is one of the products of glycogen metabolism and is found in all body tissues; its synthesis, described in Figure 6.6, depends on cytosolic enzymes.

Glucuronides, the products of glucuronidation reactions, are generated from a large variety of substances, both exogenous and endogenous, including alcohols, phenols, hydroxylamines, carboxylic acids, amines, sulfonamides, and thiols. Because of their structure or the metabolic transformation they have undergone during phase I, these molecules expose –COOH, –OH, –NH$_2$, and –SH functional groups whose glucuronidation gives rise to oxygen-linked glucuronides (O-glucuronides), nitrogen-linked glucuronides (N-glucuronides), sulfur-linked glucuronides (S-glucuronides), and, although very rarely, carbon-linked glucuronides (C-glucuronides) in the presence of nucleophilic carbon atoms.

Glucuronide synthesis is mediated by different UGT isoforms. Nineteen UGT-encoding genes have been identified in the human genome; 9 are encoded by the UGT1 locus and 10 by the UGT2 family of genes. Four of them have been cloned, sequenced, and expressed in mammalian cell culture. Different isoforms respond to different inducers and show selectivity toward certain classes of substrates. The heterogeneity of these enzymes accounts not only for their different sensitivity to inducers and inhibitors but also for interspecies differences (see Supplement E 6.1), which are responsible for specific defect in glucuronidation of some classes of glucuronic acid acceptors.

All glucuronides can undergo subsequent cleavage by β-glucuronidases, enzymes that are mainly found in the intestinal microflora. These enzymes can hydrolyze *N*-, *O*-, and *S*-glucuronides with different efficiency, releasing an

TABLE 6.3 Principal phase II reactions

Reactions	Enzyme	Functional group
Glucuronidation	UDP-glucuronosyltransferase(s)	–OH, –COOH, –SH, –CH, –NH$_2$
Sulfation	Sulfotransferases	–OH, –NH$_2$, –SO$_2$NH$_2$
Methylation	Methyltransferases	–OH, –NH$_2$
N-acetylation	*N*-acetyltransferases	–OH, –NH$_2$, –SO$_2$NH$_2$
Glutathione conjugation	Glutathione-S-transferases	Epoxide, organic halides
Amino acid conjugation	*N*-acyltransferases	Aromatic –NH$_2$, –COOH

aglycone that can be reabsorbed and reenter enterohepatic circulation.

The compounds involved in this cycle tend to remain in the organism for long; they can undergo other biotransformation processes before being excreted and may therefore produce serious problems in therapies based on drugs with hepatotoxic activity (see following text; intestinal flora).

Sulfations Sulfations are another important type of conjugation reactions. These involve mainly phenols, but also alcohols, amines, and, to a lesser extent, thiols.

Sulfations are catalyzed by sulfotransferases (SULTs), soluble enzymes that are only found in the cytosol of liver, kidney, intestine, skin, and hormone-responsive cells.

Thirteen SULT human isoforms are known. These enzymes catalyze the transfer of inorganic sulfate groups to hydroxyl groups of phenols and aliphatic alcohols to produce esterified sulfates and ether sulfates. In these reactions, the cofactor is 3′-phosphoadenosine-5′-phosphosulfate (PAPS), synthesized in the cytosol from sulfate and ATP. Sulfate required for PAPS synthesis seems to be provided by cysteines through a relatively complex oxidation process.

Numerous SULTs with specific preferences for substrates have already been described. Phenol, alcohol, and arylamine SULTs are usually nonspecific and can metabolize a wide spectrum of drugs and xenobiotics. Conversely, steroid SULTs are specific, each one acting either on a single steroid or on selected groups of steroids.

Methylations Methylations are synthesis reactions mainly aimed at metabolizing endogenous compounds. However, some drugs can be methylated by unspecific methyltransferases found in the lungs or other tissues. Unlike other conjugation processes, methylations tend to disguise functional groups and to reduce substrate water solubility and ability to undergo further conjugation processes, thus not favoring their excretion.

These reactions require S-adenosylmethionine (SAM) as a cofactor. SAM is generated by a reaction between L-methionine and ATP catalyzed by L-methionine adenosyltransferase (Fig. 6.7). A methylation reaction consists in the transfer of a methyl group from SAM to the oxygen of alcohols, to the nitrogen of aminic groups, or to the thiol of sulfated substances, catalyzed by enzymes called methyltransferases. The products of these reactions are methylated substrates and S-adenosylhomocysteine.

Unlike other transfer enzymes, methyltransferases are relatively specific and are classified in accordance with substrate and corresponding methylated product. In humans, there are three kinds of methyltransferases: catechol-O-methyltransferase (COMT), phenol-O-methyltransferase (POMT), and thiopurine-S-methyltransferase (TSMT).

Acetylations Acetylations are among the most common biotransformations of drugs and environmental xenobiotics containing aromatic amines or hydrazine groups. The cofactor in these reactions is acetyl coenzyme A deriving from glycolysis or from direct interaction of acetate and coenzyme A mediated by CoA-S-acetyltransferase.

Transfer of acetyl groups from acetyl coenzyme A is mediated by *N*-acetyltransferases (NATs), cytosolic enzymes located in many tissues of several species. Primary aromatic amines, hydrazines, hydrazides, sulfonamides, and certain aliphatic primary amines are the most common substrates of NATs (Fig. 6.8). In humans, NATs are encoded by two genes, *NAT1* and *NAT2*.

A polymorphism in these enzymes affecting the acetylation of some substrates has been found in humans, mice, rabbits, and monkeys. This polymorphism has allowed classifying individuals into fast and slow acetylators on the basis of their rate of isoniazid acetylation. The polymorphism of NATs can explain difference in toxic effects produced by some substances in individuals of the same species; an example is the peripheral nervous damage induced by isoniazid in slow acetylators. Similarly, the different carcinogenic activity of aromatic amines may depend on the NAT polymorphism.

Acetylation reactions can sometimes disguise functional groups and lead to N-acetyl derivatives that are less water soluble than the original compounds, with possible toxicological consequences. Certain sulfamidic N-acetyl derivatives, less water soluble than the original drugs, can produce precipitates in renal tubules, thus causing renal damage.

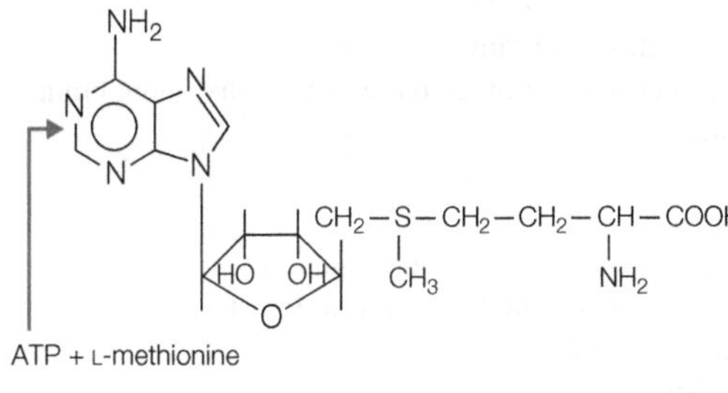

FIGURE 6.7 Synthesis of S-adenosylmethionine.

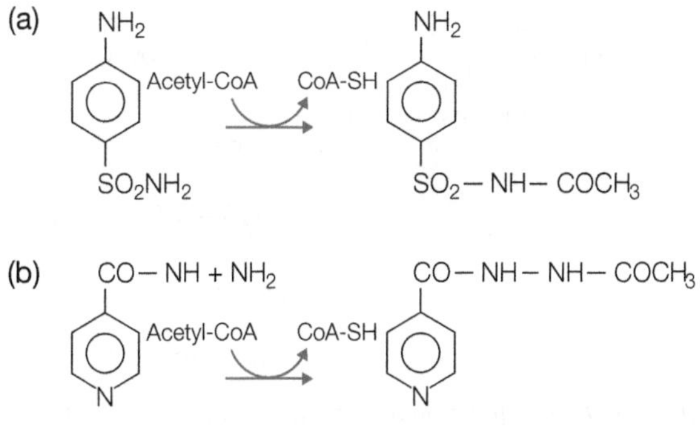

FIGURE 6.8 N-acetylation of (a) sulfonamide and (b) isoniazid.

Conjugations with Amino Acids Amino acids of exogenous origin can form derivatives with coenzyme A and react with endogenous amines to form conjugates; in this case, in contrast with other conjugations, it is the substrate to be activated, rather than the cofactor. The involved amino acids are normally glycine, glutamine, ornithine, arginine, and taurine.

Conjugations with glutamine prevail in humans, while conjugations with ornithine prevail in birds and reptiles. Taurine serves as an acyl acceptor in the conjugation of biliary acids. The reactions of peptide bond formation involve two sequential steps, catalyzed by different enzymes. The first step consists in the ATP-dependent activation of the acid into the thioester derivative of coenzyme A. In the second step, the coenzyme A thioester transfers its acyl group to the amine group of the receiving amino acid. These sequential modifications are catalyzed by ATP-dependent CoA ligase and N-acyltransferase, respectively. Both are soluble enzymes, even if similar activities are found in the matrix of hepatic and renal mitochondria. A representative example of this enzymatic mechanism is the conjugation of benzoic acid with glycine (Fig. 6.9).

Conjugations with Glutathione Glutathione is a tripeptide formed by glycine, γ-glutamic acid, and cysteine and is considered one of the most important compounds of the organism for its ability to remove potentially toxic electrophilic compounds. Given their high electrophilicity (also acquired through phase I reactions), various endogenous and exogenous substances, including drugs, can react with glutathione to form nontoxic conjugates. The substances that conjugate more easily with glutathione are epoxides, haloalkanes, nitroalkanes, alkenes, and azo- and nitro-aromatic compounds. Glutathione conjugates can either be excreted directly into urine and bile or undergo further metabolic processes.

Glutathione conjugations are mediated by glutathione-S-transferases (GSTs) that catalyze the transfer of glutathione to the electrophilic compound containing an electrophilic heteroatom (–O, –N, and –S).

More than 20 human GSTs have been identified up to now. These are grouped into two subfamilies, cytosolic and microsomal isoforms, characterized by different substrate selectivities and reaction rates (from 5 to 40 times higher in the cytosolic isoforms). Cytosolic GSTs catalyze the biotransformation of xenobiotics and drugs, while microsomal GSTs catalyze the biotransformation of endogenous compounds such as leukotrienes and prostaglandins. These enzymes are ubiquitous and mostly active in the blood, liver, intestine, kidney, and adrenal gland.

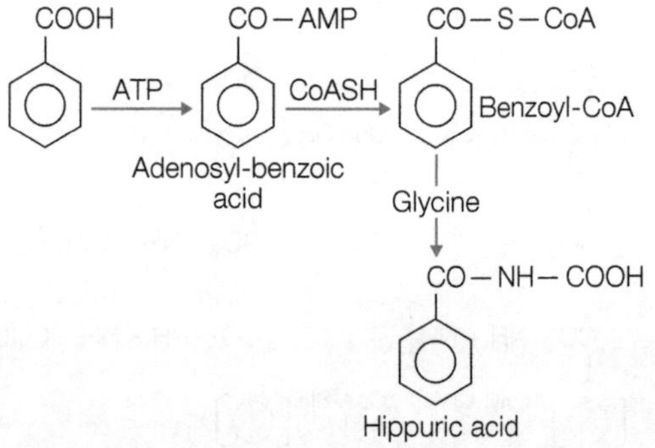

EXTRAHEPATIC BIOTRANSFORMATIONS

Although speed and capacity of extrahepatic biotransformations are not as high as those of the biotransformations occurring in the liver, they are important for both elimination and activation of many substances to which humans are exposed for a long time, even if at low levels. Extrahepatic tissues mostly containing biotransformation activities are those also involved in absorption or excretion of foreign substances, such as the lung, skin, gastrointestinal mucosa, and kidney (Table 6.4).

It is noticeable that, in contrast with liver where all parenchymal cells have biotransformation activity, in other tissues biotransformations are only found in certain cells, at least for what concerns phase I reactions. Enterocytes of small intestinal epithelium are important sites of phase I metabolism as they contain CYP3A. Therefore, after oral administration of drugs, various CYPs localized in the intestine and liver can sequentially regulate the amount of drug that reaches the bloodstream, thus modulating drug action. When this effect significantly affects the bioavailability of the drug, it is called "first-pass effect."

In general, subjects exhibit differences in intestinal and hepatic metabolism of both synthetic and natural drugs. These differences often contribute to the interindividual variability in the response to therapies and may cause adverse effects, therapeutic failures, and unexpected interactions

FIGURE 6.9 Conjugation of benzoic acid to glycine via benzoyl-coenzyme A.

TABLE 6.4 Distribution of most representative CYPs in extrahepatic tissues

CYP	Tissue
1A1	Lung, kidney, brain, GI tract, skin, placenta, lymphocytes
1B1	Lung, skin, kidney, testes, ovaries, fetal liver
2A6	Lung, nasal mucosa
2B6	GI tract, lung, brain
2C	GI tract (jejunal intestinal wall), larynx, lung, brain
2D6	Brain, GI tract
2E1	Lung, GI tract, placenta
2F1	Lung, placenta
2J2	Hearth, GI tract, nasal mucosa, lung
3A	GI tract, placenta, fetal liver, uterus, kidney, lung, brain
4B1	Lung, placenta
4A11	Kidney

leading to changes in therapeutic plans or emergence of toxic effects (see Chapter 53 and Supplement E 6.3, "Modulations of Efficacy due to the Interaction between Synthetic and Herbal Drugs").

Biotransformation by the Intestinal Flora

An important extrahepatic localization of biotransformation activities is represented by intestinal microorganisms. It has been estimated that intestinal microflora, or microbiota, has a biotransformation activity equivalent or higher than that of the liver, although very selective. Intestinal microflora has been estimated to encompass more than 400 bacterial species, which vary depending on the animal species, ages, diets, and pathological conditions. Because of the anaerobic condition of the intestinal tract, intestinal microbial biotransformations are mainly reductive reactions and produce more or less water-soluble metabolites. β-Glucuronidase and arylsulfatase reductase activities are also capable of regenerating drugs from their metabolites produced by hepatic enzymes and excreted in the gastrointestinal tract, allowing drugs to be absorbed again (this process is called enterohepatic circulation). This, in some cases, is responsible for drug-induced hepatopathies, such as those induced by sulfonamides or the "gray baby syndrome" caused by chloramphenicol.

PHARMACOMETABOLIC INDUCTION AND INHIBITION

Induction of Drug Metabolism

The activity of drug-metabolizing enzymes can be enhanced or inhibited by different agents and compounds through different mechanisms. This aspect of drug metabolism can have important therapeutic consequences. CYP superfamily is characterized by a basal expression that is relatively low in the absence of substrates and high in the presence of specific substrates or other compounds capable of inducing an increase in the enzymes themselves.

More than 50 years ago, it was observed for the first time that animals adapt to increased phenobarbital doses (tolerance) by increasing total concentration of CYPs, thus speeding up phenobarbital biotransformation.

Subsequently, an increased metabolic activity has been described for amino oxidases, and later the phenomenon has been recognized as a common feature of many CYP-dependent monooxygenases, including conjugative enzymes.

Furthermore, it has been subsequently shown that biotransformation enzymes, especially members of the CYP1A, CYP2B, CYP2C/H, CYP3A, and CYP4A gene subfamilies, display increased activity following administration of foreign substances, such as drugs, pesticides, industrial chemicals, and natural food products (such as ethanol). This increase in enzyme-metabolizing activity, called induction of drug metabolism, results from increased synthesis of biotransformation enzymes and in most case is tissue specific, dose dependent, and reversible upon removal of the inducer.

The increased metabolizing activity induced by a drug can accelerate the metabolism of the drug itself, as well as of other drugs nonfunctionally correlated. A good experimental example of this is the induction of the metabolism of strychnine and pentobarbital (a barbiturate) triggered by phenobarbital (another barbiturate) and other centrally active drugs. As shown in Tables 6.5 and 6.6, a 48 h pretreatment with different substances reduces the toxic action of both

TABLE 6.5 Effects of 48 h pretreatment with centrally acting drugs on strychnine toxicity

Pretreatment	No. of rats, total	No. of rats undergoing convulsion	No. of rats dead	Mortality (%)
Controls	45	36	34	76
Phenobarbital	24	1	0	0
Phenaglycodol	23	3	2	9
Glutethimide	24	4	2	8
Thiopental	16	5	3	18
Nikethamide	24	8	7	19
Primidone	16	6	4	25
Diphenylhydantoin	16	7	5	31
Urethane	16	9	7	44
Meprobamate	16	9	8	50
Carisoprodol	16	11	9	55
Pentobarbital	16	11	10	62

TABLE 6.6 Effects of 48 h pretreatment with centrally acting drugs on pentobarbital-induced sleep

Pretreatment	Sleep time (min) (means ± standard error)	Variation (%)
Controls	82 ± 2.8	100
Phenaglycodol	13 ± 3.1	−84
Glutethimide	26 ± 4.1	−68
Thiopental	22 ± 3.8	−73
Phenobarbital	28 ± 3.2	−66
Nikethamide	36 ± 3.4	−56
Meprobamate	46 ± 4.6	−44
Pentobarbital	47 ± 4.8	−43
Chlorbutol	51 ± 6.1	−37
Chlorpromazine	49 ± 4.5	−40
Fluopromazine	55 ± 4.6	−33
Urethane	52 ± 6.3	−37
Phenytoin	58 ± 5.4	−29
Hexobarbital	56 ± 5.8	−31
Primidone	62 ± 6.3	−26
Mephenesin dicarbamate	66 ± 5.3	−19
Carisoprodol	66 ± 4.5	−19

TABLE 6.7 Some inducers of hepatic biotransformations in animals

Classification	Example	Use/origin
Drugs	Phenobarbital	Antiseizure and sedative agent
	Rifampicin	Antibiotic
	Troleandomycin	Antibiotic
Insecticides	DDT, aldrin, lindane	Pesticides
Polycyclic aromatic	Benzo(a)pyrene	Environmental pollutants
Hydrocarbons	3-methylcholanthrene	Produced by industrial and domestic combustion
	Benz(a)anthracene	Petroleum processing
Alcohols	Ethanol	Beverages
Halogenated hydrocarbons	TCDD	Defoliant and herbicide contaminant
Food additives	Butylated hydroxyanisole	Food antioxidant and preservative

TABLE 6.8 Receptors, ligands, and enzymes involved in drug metabolism induction

Receptors	Ligands	Induced enzymes
Aryl hydrocarbon receptor (AhR)	Omeprazole	CYP1A1
Constitutive androstane receptor (CAR)	Phenobarbital	CYP1A2
Pregnane X receptor (PXR)	Rifampicin	CYP3A4
	Dexamethasone	CYP1B1
Farnesil X receptor (FXR)	Biliary acids	CYP2B6
Vitamin D receptor	Vitamin D	CYP2C9
Peroxisome proliferator-activated receptor (PPAR)	Fibrates	CYP4A
Retinoic acid receptor (RAR)	All-*trans* retinoic acid	Glutathione-S-transferase
		Glucuronyltransferase
		Sulfotransferase
Retinoid X receptors (RXR)	*cis*-Retinoic acid	Farnesyltransferase

strychnine and pentobarbital in rats. Besides phenobarbital and other barbiturate-like substances, polycyclic aromatic hydrocarbons (such as benzo(a)pyrene and 3-methylcholanthrene) represent another largely studied group of substances capable of metabolic induction.

Researches aimed at identifying possible inducers among drugs and chemical substances have shown that the majority of these compounds are lipophilic (Table 6.7). It is well known nowadays that foreign substances can affect enzyme content by activating the expression of genes encoding specific CYPs. This activation is mediated by a system of ligands and nuclear receptors that form complexes capable of activating transcription of specific target genes through RNA polymerase II (Table 6.8).

The discovery of high-affinity ligands, of inducible cell culture, and of aromatic hydrocarbon nuclear receptors (AhRs) with their binding mediators and nuclear translocators has allowed clarifying the molecular mechanism underlying CYP1A induction by aromatic polycyclic hydrocarbons (see Supplement E 6.2).

Subsequent identification of the peroxisome proliferator-activated receptor (PPAR) (see also Chapter 9) has revealed that this nuclear receptor plays a crucial role in the induction of CYP4As by proliferative substances such as fibrates (Table 6.8).

Moreover, studies into pregnane X receptor (PXR), which is activated by numerous drugs—including antibiotics (rifampicin and troleandomycin), Ca^{2+} antagonists (nifedipine), statins (mevastatin), oral antidiabetics, proteases inhibitors (ritonavir), and certain anticancer drugs (taxanes)—and constitutive androstane receptor (CAR), which is constitutively expressed, have confirmed the common mechanism of drug metabolism induction (Table 6.8).

Also, type 2 nuclear receptors are involved in drug metabolism and therapy. These belong to the steroid hormone receptor superfamily and are called orphan receptors because it is still unclear whether they recognize endogenous ligands or not.

Also in ligand–receptor relationship, there are species-specific differences similar to those observed in drug-metabolizing enzymes (see Supplement E 6.1). For example, rifampicin activates human PXR, but not mouse or rat PXR. This is another critical point that has to be taken into account when translating results from animal experimental models to humans.

On the contrary, some inducers (e.g., ethanol) that are active on CYP2E1 stabilize the enzymes at posttranscriptional level via a receptor-independent mechanism (see Supplement E 6.3, "Modulations of Efficacy Due to the Interaction between Synthetic and Herbal Drugs").

Inhibition of Drug Metabolism

Inhibition can be defined as a decrease in enzymatic activity due to mechanisms like direct inhibition of enzyme activity, destruction of enzyme, reduced supply of cofactors, and suppression of enzyme synthesis. In clinical practice, the first and second mechanisms are the most common. Competitive inhibition is due to the wide versatility of drug-metabolizing enzymes, whereby each CYP isoform can metabolize hundreds of different compounds, from drugs to pollutants, from endogenous compounds to nonnutritive plant components. This allows the elimination of a large number of compounds but also facilitates events of competitive inhibition whose extent depends on the relative affinity of xenobiotics for the enzyme binding sites. In humans, such phenomenon can occur during therapies based on concurrent administration of different drugs, because each CYP has only one allosteric binding site for specific substrates and therefore can bind a single molecule at a time (Tables 6.9 and 6.10). There are exceptions, such as CYP3A4, that can bind more than one molecule. If the enzyme inhibited is not the major responsible for the biotransformation of the inhibitor itself, competitive inhibition may not affect the inhibitory drug.

Inhibition can also result from competition for enzyme cofactors, blockade of transport components in multienzyme systems, or allosteric modifications of enzymatic structures. While *in vitro* inhibition is quite a simple and easy to observe phenomenon, it is worth noting that it is not as much *in vivo*. Many xenobiotics produce selective species-dependent inhibitory effects on certain CYP isoforms. The degree of inhibition varies over time, and some agents, such as SKF

TABLE 6.9 Substrates of the human CYP subfamilies mostly involved in drug metabolism

CYP1A1/2	CYP2C8/9	CYP2C19	CYP2D6	CYP3A4/5/7
Amitriptyline	Amitriptyline	Citalopram	Alprenolol	Alprazolam
Clozapine	Celecoxib	Clomipramine	Chlorpromazine	Amlodipine
Estradiol	Cerivastatin	Clopidogrel	Clomipramine	Atorvastatin
Fluvoxamine	Diclofenac	Imipramine	Desipramine	Cisapride
Haloperidol	Fluoxetine	Indomethacin	Fluoxetine	Clarithromycin
Naproxen	Glibenclamide	Lansoprazole	Haloperidol	Cyclosporine
Olanzapine	Glipizide	Nilutamide	Imipramine	Erythromycin
Ondansetron	Ibuprofen	Omeprazole	Paroxetine	Estradiol
Propranolol	Irbesartan	Pantoprazole	Propafenone	Indinavir
Theophylline	Losartan	Phenobarbitone	Timolol	Midazolam
Verapamil	Paclitaxel	Phenytoin	Amitriptyline	Nifedipine
Warfarin	Piroxicam	Proguanil	Risperidone	Quinidine
Zileuton	Repaglinide	Propranolol	Tamoxifen	Ritonavir
Zolmitriptan	Tolbutamide	Teniposide		Tacrolimus
	Torsemide	Ticlopidine		Triazolam

TABLE 6.10 Inhibitors of the human CYP subfamilies mostly involved in drug metabolism

CYP1A1/2	CYP2C8/9	CYP2C19	CYP2D6	CYP3A4/5/7
Amiodarone	Amiodarone	Chloramphenicol	Amiodarone	Amiodarone
Cimetidine	Fenofibrate	Cimetidine	Bupropion	Aprepitant
Ciprofloxacin	Fluconazole	Fluoxetine	Celecoxib	Cimetidine
Fluvoxamine	Fluvastatin	Ketoconazole	Cimetidine	Clarithromycin
Interferon	Gemfibrozil	Lansoprazole	Cinacalcet	Diltiazem
Quinolones	Glitazone	Modafinil	Citalopram	Erythromycin
	Isoniazid	Omeprazole	Clomipramine	Fluconazole
	Lovastatin	Pantoprazole	Flecainide	Indinavir
	Montelukast	Ticlopidine	Fluoxetine	Itraconazole
	Phenylbutazone	Topiramate	Methadone	Ketoconazole
	Sertraline		Ondansetron	Nefazodone
	Sulfamethoxazole		Paroxetine	Nelfinavir
	Tamoxifen		Quinidine	Progesterone
	Teniposide		Ranitidine	Ritonavir
	Trimethoprim		Sertraline	Saquinavir
	Zafirlukast		Ticlopidine	Verapamil

525A or piperonyl butoxide, initially inhibit CYPs and subsequently produce induction.

Inhibition due to enzyme inactivation occurs when a metabolite of a substrate binds tightly or irreversibly to an enzyme; in this case, reactivation can only occur by synthesis of new enzyme.

Many foreign and toxic substances and drugs can destroy liver CYPs through different mechanisms. One of them is known as suicide inhibition and is mediated by certain substances with porphyrogenic activity containing olefinic or acetylenic functions; these substances form a green pigment in the liver that has been recently identified as the alkylation product deriving from P450/substrate/heme interaction. Most of these substances are relatively inactive by themselves and require metabolic activation through CYPs; this mechanism explains why they are defined as "suicide" substrates for hemoprotein. It should be stressed that the toxic action of these metabolites seems to be extremely selective for CYPs. Among the substances that inhibit CYP-catalyzed reactions through suicide inactivations are certain halogenated alkanes (CCI4), alkenes (vinyl chloride, trichloroethylene), and therapeutic compounds containing allylic groups (e.g., secobarbital) or acetylenics (e.g., norethindrone acetate).

Conversely, other substances inhibit biotransformations by reducing the synthesis of either biotransformation enzymes or enzymes involved in cofactor production. Besides the most common inhibitors of protein synthesis, this group includes also some substances with a more selective mechanism of inhibition. For instance, 3-amino-1-2-3-triazole inhibits CYP synthesis by inhibiting porphyrin synthesis, while cobalt seems to reduce liver CYP synthesis by inhibiting heme synthesis.

Inhibition may also results from decreased availability of cofactors required for conjugation reactions. Examples of this mechanism of inhibition are L-methionine sulfoximine and buthionine sulfoximine that inhibit glutathione synthesis; dimethyl maleate, glycidol, and other similar substances that quickly reduce tissue glutathione stores by forming conjugates with it; galactosamine that inhibits UDP-GA synthesis by emptying the hepatic stores of uridine; and borneol and salicylamide that reduce UDP-GA availability by conjugating with it.

Several chemical substances can directly act on CYP oxidoreductive system. For instance, carbon monoxide and ethyl isocyanides behave as ligands for the reduced heme, thus competing with molecular oxygen. In this way, they become powerful inhibitors of oxidative reactions and, in the case of carbon monoxide, of reductive reactions as well. Finally, some classes of drugs or foreign substances can inhibit biotransformations by forming of inactive complexes with hemoprotein. These substances are CYP substrates and need to be biotransformed to perform their inhibitory effect as olefinic and acetylenic substances.

An interesting example of this type of inhibition and its clinical consequences concerns the treatment with triacetyloleandomycin, an antibiotic structurally similar to erythromycin that can produce liver cholestasis in the presence of oral contraceptives, ischemic episodes in the presence of ergotamine, and neurological impairment in the presence of carbamazepine and theophylline. These toxic interactions suggest that the antibiotic can reduce metabolism and therefore excretion of these drugs.

In conclusion, we report two examples that illustrate the relevance of metabolism induction and inhibition in therapy. (i) Consider the case of old patients who may be treated with up to 10 drugs at the same time. Each of them can interfere with the metabolism of the others, therefore modifying drug kinetics and properties, and this should be taken into account in designing the individual therapy. (ii) Drug metabolism inhibition can be purposely therapeutically exploited as in the case of antiviral therapy in HIV-infected patients, in which ritonavir, an inhibitor of P450 cytochromes, could be prescribed to boost the response to some antiviral agents.

TAKE-HOME MESSAGE

- Biotransformation modifies drug activity and pharmacokinetics.
- Phase I enzymes include various isoforms of cytochrome p450; the enzymes are identified by the acronym CYP followed by a number and a capital letter (family and subfamily identification) and by a number (specific gene/isoenzyme).
- Most drugs are metabolized by CYP3A3/4 and CYP2D6.
- Phase II reactions increase hydrophilicity and molecular weight of drugs, thus facilitating their elimination by the bile and/or kidney.
- Concomitant drugs, nutrients, chemicals, and diseases may enhance or reduce drug biotransformation by any CYP.

FURTHER READING

Daly A.K., Cholerton S., Gregory W., Idle J.R. (1993). Metabolic polymorphism. *Pharmacology & Therapeutics*, 57, 129–160.

Gibson G.G., Skett P. *Introduction to Drug Metabolism*. 2nd Ed. Blackie Academic and Professional: London, 1994.

Li J.H., Lu A.Y. (2001). Interindividual variability in inhibition and induction of cytochrome P450 enzymes. *Annual Review of Pharmacology and Toxicology*, 41, 535–567.

Stark K., Guengerich F.P. (2007). Characterization of orphan human cytochromes P450. *Drug Metabolism Reviews*, 39, 627–637.

Whitlock J.P. (1999). Induction of cytochrome P4501A1. *Annual Review of Pharmacology and Toxicology*, 39, 103–125.

Wilkinson G.R. (2005). Drug metabolism and variability among patients in drug response, *New England Journal of Medicine*, 352, 2211–2221.

7

CONTROL OF DRUG PLASMA CONCENTRATION

RICCARDO FESCE AND GUIDO FUMAGALLI

By reading this chapter, you will:

- Learn to predict drug plasma concentrations based on its pharmacokinetic properties
- Know differences in plasma concentration depending on dose and time interval between administrations
- Learn to use pharmacokinetic data to define the best therapeutic regimen
- Understand how to modify a drug therapy according to patient age and/or disease

After administration, drug plasma levels and their variations over time depend on the properties of absorption, distribution, and elimination processes and may be quite different, whether the drug is administered once or repetitively. Thus, to correctly design the therapeutic regimen for each patient is fundamental to know the general rules governing these processes.

TIME COURSE OF DRUG PLASMA CONCENTRATION FOLLOWING A SINGLE ADMINISTRATION

Drug absorption rate depends on the administration route, the functional properties of the absorption site, and the physical–chemical properties of the drug (in some cases, of excipients as well). Following administration of a single dose, drug plasma concentration can be predicted to rise with a velocity that depends on the number and structure of the cell barriers the drug has to cross and on the presence of transport mechanisms. This means that the same drug dose can produce different maximum plasma levels depending on the administration route (Fig. 7.1). The highest peak is generally attained upon intravenous (i.v.) administration and the lowest upon enteral administration; most of the parenteral routes generally yield intermediate values.

In general, drug entry in the bloodstream from an extravascular administration site is well described by first-order kinetics: a quantity of drug proportional to the concentration difference between absorbing compartment (e.g., intestinal

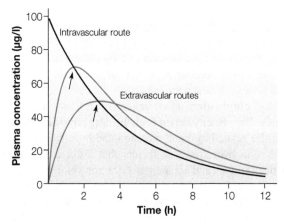

FIGURE 7.1 Single dose administration by different routes. Plasma concentration time courses following administration of the same drug by intravascular injection (maximum value at time=0) or via various extravascular routes (assuming complete bioavailability). All extravascular routes yield a delayed and lower peak. At the peak, the extravascular curves intersect the intravascular curve (arrows). From the peak time onward, the plasma concentrations produced by the extravascular routes are higher than those produced by the intravascular. The areas under all curves are equal.

General and Molecular Pharmacology: Principles of Drug Action, First Edition. Edited by Francesco Clementi and Guido Fumagalli.
© 2015 John Wiley & Sons, Inc. Published 2015 by John Wiley & Sons, Inc.

lumen) and plasma is absorbed in a unit time. The total absorbed quantity rises more rapidly at early times and gradually approaches the total value $D_0 \times F$, where D_0 is the administered dose and F is the absorbed fraction (bioavailability). The quantity remaining to be absorbed decreases exponentially, so that the absorbed quantity increases according to the function $Q_A = D \times (1 - \exp(-t/\tau_a))$; the absorption time constant, τ_a, ranges from a theoretical value of zero (i.v. injection) to some tens of minutes or more for oral administration. Time constants in the order of hours, days, or even weeks can be achieved with depot or slow-release formulations. This formula is a rough approximation, as it does not consider the interference exerted by distribution and elimination processes on drug plasma concentration time course.

DRUGS DISTRIBUTE TO ORGANS AND TISSUES AND THEN ARE ELIMINATED

Once the drug has entered the bloodstream, it distributes to the various areas of the organism and reaches equilibrium at concentration values that depend on tissue avidity for the drug. The velocity of the process depends on local drug flux (which may be limited either by local blood flux or by the permeability of cellular barriers) and the apparent distribution volume of the compartment. The speed at which the drug leaves the vascular bed to distribute to tissues also influences absorption (driven by the concentration difference between absorption site and plasma). Thus, the apparent distribution volume (V_D) of a drug may influence its absorption rate. In general, drugs with high V_D attain lower plasma concentrations (plasma concentration is inversely proportional to V_D) with longer absorption time constants (drug plasma concentration increases slowly because, as it gradually enters the circulation, it also leaves it to reach the tissues).

Drug elimination also generally follows first-order kinetics. The velocity depends on the drug physical–chemical properties related to renal excretion and hepatic metabolism but also on hydrodynamic factors that influence renal and hepatic clearance and apparent distribution volume. Thus, the time constant of drug elimination is given by the ratio V_D/Cl (Cl stands for clearance). The higher the V_D and the lower the Cl, the slower is the drug elimination process and the longer the drug action.

Description of Drug Plasma Concentration Time Course Following a Single Administration

After a single administration, drug concentration will obviously rise to a maximum and then decay as the elimination process proceeds. In general, the rising phase displays a time constant equal to the smallest one between absorption and elimination time constants, whereas the decay follows the other time constant. This implies that the more different the two time constants, the higher the peak value. Moreover, given the same absorption rate, the slower the absorption and the elimination, the longer is the time to the peak.

If we ignore the distribution kinetics, to a first approximation, after a single drug administration, plasma concentration will follow a two-exponential time course, well described by the following equation:

$$C_P(t) = \frac{D}{V_D} \times \frac{\tau_e}{\tau_e - \tau_a}\left(\exp\left(-\frac{t}{\tau_e}\right)\right) - \left(\exp\left(-\frac{t}{\tau_a}\right)\right)$$
$$= \frac{D}{\mathrm{Cl}} \times \frac{\exp(-t/\tau_e) - \exp(-t/\tau_a)}{\tau_e - \tau_a} \quad (7.1)$$

Here, $C_p(t)$ is the plasma concentration at time t after administration, D is the absorbed dose, V_D is the apparent distribution volume of the drug, τ_e and τ_a are the time constants of elimination and absorption, respectively, and Cl is the drug clearance, computed from the ratio V_D/τ_e. For simplicity, from here on, we assume that drugs are completely absorbed (100% bioavailability). For incomplete absorption (bioavailability < 100%), D must be corrected for drug bioavailability. Both bioavailability and τ_a depend, for each drug, on the administration route.

It is important to note that, in Equation 7.1, V_D is considered as a constant: thus, the equation is accurate only after the drug has distributed among body compartments (or if distribution is fast). The relations among administration route, absorption, and elimination rates, plasma concentration, and time to the peak are complex; to clarify this, we will introduce a particularly useful pharmacokinetic parameter, the area under the plasma concentration curve (AUC).

Area under the Plasma Concentration Curve (AUC)

In first-order elimination kinetics, a constant fraction of the drug quantity present in the organism is eliminated in a unit time. Such fraction is indicated as $k_e = 1/\tau_e$.

At any time (t), the quantity of drug in the organism equals plasma concentration times the apparent distribution volume, $Q(t) = C_p(t) \times V_D$. Correspondingly, a quantity $Q(t) \times k_e = C_p(t) \times V_D \times k_e$ is eliminated in a unit time. From the definition of clearance ($\mathrm{Cl} = V_D \times k_e$), we can derive that the drug quantity eliminated in a unit time equals $C_p(t) \times \mathrm{Cl}$.

Thus, each y-axis value of the curve of the plasma concentration time course, multiplied by Cl, yields the elimination rate at exactly that time (the corresponding x-axis value). Thus, if we multiply all values of the curve by Cl, the curve becomes the drug elimination curve. This means that the area under such curve (the sum of the quantities eliminated at each time) is equal to the total quantity of eliminated drug, which in turn is equal to the total quantity of absorbed drug (and administered, if bioavailability is complete). In other words, if a quantity of drug $Q = D \times F_A$ has been absorbed

(F_A = absorbed fraction), the AUC will be equal to $D \times F_A/Cl$, regardless of the administration route.

The AUC is useful to estimate the fraction absorbed through a given administration route: this can be obtained as the ratio between the AUC obtained using a specific route and that obtained by i.v. injection. Moreover, it can be employed to estimate drug clearance: this can be calculated as absorbed dose over AUC ratio. Once the clearance has been computed this way, we can derive drug half-life from the late phase of the plasma concentration curve and quite accurately calculate V_D as $V_D = 1.443 \times t_{1/2} \times Cl$.

An important consequence of the relation between absorbed quantity and AUC is that provided the bioavailability does not change, the area under the plasma concentration curve for the same drug and the same patient (same clearance) does not change when we change the administration route. This is clearly illustrated in Figure 7.1 showing the plasma concentration time courses obtained by administering the same dose of the same drug through different routes (with different absorption rates but with full bioavailability in all cases). Initially, the slower administration routes produce lower drug plasma concentrations, but these will necessarily increase at late times, so that the area under the curves is the same in all conditions.

The Plasma Concentration Peak

If a drug is administered through an intravascular bolus (absorption time constant $\tau_a \to 0$), Equation 7.1 becomes

$$C_P(t) = \frac{D}{V_D} \times \exp\left(\frac{-t}{\tau_e}\right) \qquad (7.2)$$

This equation assumes that the drug instantaneously distribute to all its V_D. Therefore, it does not accurately describe the early phases of the plasma concentration time course, but it provides a good description of the later phases. The integral of Equation 7.2 for t moving from 0 to ∞ has value AUC = $D \times \tau_e/V_D = D/Cl$, as expected. Equation 7.1 can be shown to yield exactly the same integral (AUC), confirming that whatever the value of τ_a (and the administration route), the AUC remains the same.

The peak in plasma concentration occurs after a time T_P that depends on both absorption and elimination rates, and in particular, the slower is the absorption (and the slower the elimination), the later it occurs. Analytically, it can be shown that $T_P = \tau_a \times \log(\tau_e/\tau_a)/(1 - \tau_a/\tau_e)$. At this particular time point, plasma concentration is given, according to Equation 7.1 (extravascular administration), by:

$$C_P(T_P) = \frac{D \times F_A}{V_d} \times \exp\left(\frac{-T_P}{\tau_e}\right), \qquad (7.3)$$

which equals the value of Equation 7.2 (i.v. administration) for $t = T_P$, for full bioavailability ($F_A = 1$). Thus, plasma concentration curves obtained after extravascular administration cross the curve obtained with bolus i.v. administration at the time of their peak. Regardless of the administration route, plasma concentration at peak time is the same that would be obtained via an i.v. bolus (Fig. 7.1); but for extravascular administrations, some drug still has to be absorbed at peak time. Therefore, from this time onward, plasma concentrations will remain higher than those produced by i.v. injection. A particular aspect of drug kinetics and distribution concerns drugs administered by inhalation that are discussed in Supplement E 7.1, "Therapeutic Drug Monitoring."

DRUG PLASMA CONCENTRATION TIME COURSE DURING REPETITIVE ADMINISTRATIONS

In clinical practice, it is often necessary to attain and maintain (possibly for several years) a given value of drug plasma concentration. In this case, we need to set up a chronic therapy and predict the effect of each administration on the drug plasma concentration produced by previous doses.

We consider the simple case of i.v. administration of a drug that rapidly equilibrates in its distribution volume (Fig. 7.2). Plasma concentration has an initial peak value given by the ratio D/V_D and then decays exponentially according to Equation 7.4:

$$C_t = C_0 \times \exp(-kt) \qquad (7.4)$$

Recalling the definition of half-life ($t_{1/2} = \ln(2)/k$), we can rewrite

$$C_t = C_0 \times \exp(-kt) = C_0 \times \exp\left(-\ln(2) \times \frac{t}{t_{1/2}}\right) = C_0 \times \left(\frac{1}{2}\right)^{t/t_{1/2}}$$

This means that at time $t = t_{1/2}$, C_t is one-half of C_0, or, in other words, after one half-life, 50% of the administered dose will still be present in the patient's body, after two half-lives $(1/2)^2 = 25\%$, and so on. After five half-lives, only about $(1/2)^5$ will be left in the body, which is a negligible fraction of the administered dose ($\approx 3\%$). Thus, if a second dose is given after five half-lives, the new dose simply replaces the previous one (Fig. 7.2a). However, if doses are administered at shorter intervals, each of them will add to what remains of the previous doses (Fig. 7.2b).

During Repetitive Administrations, the Drug Plasma Concentration Time Course Is Given by the Sum of the Time Courses of the Single Doses

In first-order kinetics, concentrations tend to change by a constant fraction (k) in a unit time. Suppose a patient is given 100 mg of a drug while 20 mg of the preceding dose are still present. His body will contain 120 mg of the drug and $k \times 120$ mg will be eliminated in a unit time. This is exactly

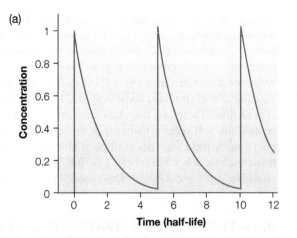
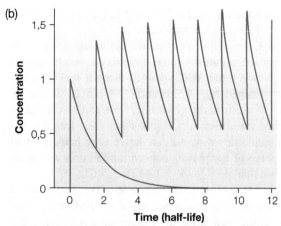

FIGURE 7.2 Repetitive drug administration. The *x*-axis shows time in "half-life" units. (a) When doses are administered at sufficiently long time intervals (five half-lives), no significant summation occurs between successive doses. (b) Administrations at shorter intervals produce drug accumulation; plasma concentration fluctuates around higher mean levels than those obtained with a single dose.

as if $k \times 20$ mg of the previous dose and $k \times 100$ mg of the new one were separately eliminated. After a half-life, 60 mg will be present, after two half-lives 30 mg, and so on, exactly as if 10 mg of the previous dose and 50 of the new one remained after one half-life, 5 mg of the old dose and 25 of the new one remained after two half-lives, and so on. In other words, it makes no difference whether the declines (or time courses) of the single doses are separately examined or the total quantity present in the organism is considered. As a consequence, a plasma concentration curve can be drawn for each administered dose, and the plasma concentration time course during repetitive administration will be deduced from the sum of the time courses of the single doses, as illustrated in Figure 7.3.

When a drug is repetitively administered at constant doses and at constant time intervals, drug plasma concentration rises as illustrated in Figure 7.3. As you can see, the superposition of the time courses produced by single doses (upper panel) gradually stabilizes on a constant pattern (after about five half-lives) and the total plasma concentration (lower panel) similarly stabilizes on a periodic, steady fluctuation (steady state). Before steady state is reached, the "tails" of previous administrations are lacking in the upper panel, whereas the concentration gradually increases toward equilibrium in the lower panel.

The dashed line in Figure 7.3b, which represents the smoothed time course of plasma concentration, highlights that the rising phase toward equilibrium displays the same shape as the plasma concentration decay after each dose: the ascent toward equilibrium has an exponential time course, with the same time constant as the decay due to elimination (τ_e). Though this may appear odd, it may be seen as a consequence of the fact that the "tails" of previous administrations (which have an exponential time course governed by τ_e) are missing from the total plasma concentration curve in the initial phases.

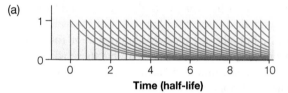
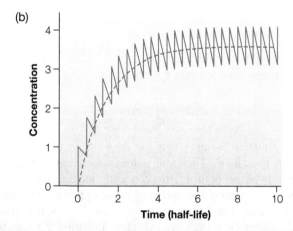

FIGURE 7.3 Drug accumulation and approach to equilibrium during repetitive drug administration. In both plots, the time unit is the drug half-life. (a) The plasma concentrations corresponding to each drug dose, separately considered and superimposed. Each dose produces a rapid rise and an exponential decay, with time constant determined by drug elimination processes. Note that after five half-lives, the plot becomes repetitive, periodic, and invariant. (b) The actual plasma concentration curve produced by this administration protocol. The curve is the sum of all curves in (a). The initial phase is characterized by a continuous increase in average plasma concentration (dashed line) toward equilibrium, whereas the late phase ("plateau," about five half-lives after the first dose) displays a stable average value around which plasma concentration fluctuates. In this phase, called administration equilibrium, each new dose replaces the quantity that has been eliminated in the interval between two successive administrations. The rising phase of average plasma concentration draws a monoexponential curve governed by the time constant of the elimination process.

Drug concentration does not keep rising indefinitely. After five half-lives, a negligible fraction of a given dose remains; a dose given five half-lives after the first one simply replaces it; the next dose will replace the second, and so on. Thus, plasma concentration will keep fluctuating around a value that depends on the quantity of drug that has been administered during the last five half-lives.

In a Chronic Therapy at Steady State, Each New Dose Replaces the Drug Amount that has been Eliminated Since the Last Administration

In the example shown in Figure 7.4, 200 mg of a drug that produces a plasma concentration of 10 mg/l ($V_D = 20 l$) is administered i.v. at intervals equal to one half-life. At the second administration, drug plasma concentration has decreased to 5 mg/ml and the new dose brings it to 15 mg/ml. At the third administration, drug plasma concentration is 7.5 mg/ml and the new dose brings it to 17.5 mg/ml. As shown in the figure (and as previously said), from the fifth dose on, plasma concentration steadily fluctuates between 10 and 20 mg/ml, as each dose simply replaces the drug amount eliminated in the interval between administrations. Immediately after each administration, the concentration is 20 mg/ml (the total quantity in the patient's body is 400 mg, i.e., two doses), and it declines to 10 mg/ml after one half-life (it halves, as does the total quantity in the body that declines to 200 mg), that is, when a new dose will be given. Such new dose will bring back the total quantity in the body to 400 mg and the concentration to 20 mg/ml.

The Time to Reach the Steady State Depends on the Drug Half-Life

The time needed to reach equilibrium does not change if we change the administration rate (provided doses and intervals remain the same), as it only depends on the time needed for the first dose to be completely eliminated and therefore on the drug $t_{1/2}$. This should be kept in mind when treating a patient. For example, if the drug used is digitoxin (a cardiac glycoside), whose half-life is 7 days, more than 1 month of treatment is needed for plasma concentration to reach the steady state; alternatively, the physician can use digoxin, which has a shorter half-life (about 36 h). Specific protocols can be devised in this kind of situations, using one or more loading doses (see the following).

Drug half-life determines both the rate of plasma concentration rise toward equilibrium and the rate of decay once the therapy is interrupted (Fig. 7.5). This aspect should be considered when administering drugs with long half-life and high toxicity or low therapeutic index. Since drug half-life depends on the elimination mechanisms and their efficiency, the physician should consider the possibility that individual factors or pathological conditions (often distinct

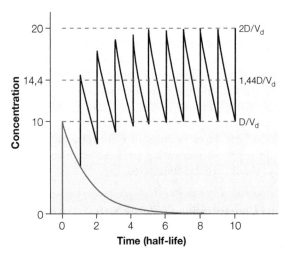

FIGURE 7.4 Quantitative aspects of repetitive drug administration. The drug is administered through intravascular route at regular intervals of one half-life; each dose is 200 mg, and the drug has a distribution volume of 20 l. Drug distribution is assumed to be much faster than elimination. Note that (i) each administration produces an increase in plasma concentration equal to the dose over distribution volume ratio, $D/V_D = 200\,mg/20\,l = 10\,mg/l$; (ii) the concentration at the time of each administration is one-half the maximum concentration produced by the previous administration; and (iii) the administration equilibrium is achieved with concentrations that fluctuate between a maximum of twice and a minimum of once the dose over distribution volume ratio. The average concentration at steady state C_{ss} is 1.44 times the D/V_D ratio.

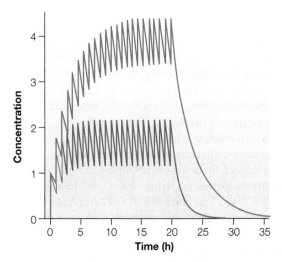

FIGURE 7.5 Effects of different elimination rates on plasma concentrations obtained with drug repetitive administration (same doses, same intervals). Comparison of the curves indicates that the same therapeutic regimen produces (i) a faster rise to steady state, (ii) a lower average steady-state concentration, and (iii) a faster decline of plasma concentration upon therapy interruption if the elimination is faster (lower curve). The magnitude of concentration fluctuations at steady state does not change under the two conditions, since it only depends on the dose over distribution volume ratio. However, the fractional fluctuation is much more marked in the situation in which the elimination is faster, and the average concentration is lower.

from the pathology targeted by the drug) may alter the efficiency of the drug elimination mechanisms. The average drug concentration at steady state depends on the total quantity of drug administered during a half-life and therefore on the half-life itself. As a consequence, functional or pathological alterations of drug $t_{1/2}$ may produce changes in steady-state drug plasma concentration, loss of efficacy, or toxic effects. Possible changes in drug clearance over time (e.g., in a patient recovering from an acute renal insufficiency) should also be considered.

Plasma Concentration at Steady State

The concept of equilibrium implies that, at steady state, the patient eliminates the drug at exactly the same rate at which the drug is administered. If we succeed in obtaining an effective plasma concentration at steady state, $(C_P)_{SS}$, the patient will eliminate an average drug amount equal to $(C_P)_{SS} \times Cl$ in a unit time. This is also the quantity of drug that has to be administered to maintain the steady state. If the clearance is expressed in ml/min and the effective plasma concentration is expressed in µg/ml, the resulting dose will be expressed in µg/min; the equivalent value in mg/h, or in another measuring unit, can be easily computed. For a drug with clearance Cl, administered at the dose D every time interval T (clearance, concentration, and time are expressed in compatible units, e.g., l/h, mg/l, and h, respectively), the reference equation is

$$D = (C_P)_{SS} \times Cl \times T \quad \text{and conversely,} \quad (C_P)_{SS} = \frac{D}{Cl \times T} \quad (7.5)$$

that predicts the average plasma concentration obtained at steady state by administering doses D at time intervals T. Equation 7.5 can be used to calculate the drug dose and the time interval between successive administrations needed to obtain the desired average C_P. For example, the α_2-adrenergic agonist clonidine, used to treat hypertension, has Cl = 3 ml/min/kg and an effective plasma concentration of 1 ng/ml. In a 70 kg subject, the dosage to obtain the desired concentration after five half-lives is 210 ng/min: $C_{eff} \times Cl = 1\,ng/ml \times 3\,ml/min/kg \times 70\,kg$. This corresponds to 12.6 µg/h, or 0.3 mg/die. If drug bioavailability is incomplete, the computed value needs to be modified accordingly.

Equation 7.5 can be rewritten, using the definition of clearance, $Cl = V_D \times k_e$, and the relation $k_e = 0.693/t_{1/2}$, as follows:

$$(C_P)_{SS} = 1.443 \times \frac{D}{V_D} \times \frac{t_{1/2}}{T} \quad (7.6)$$

Note that the ratio $t_{1/2}/T$ represents the number of administrations during a half-life. If $T = t_{1/2}$, we have one administration per each half-life. In simple words, the average plasma concentration at steady state, during repetitive administration of a drug at fixed doses and time intervals, equals 1.443 times the total dose administered during a half-life, divided by the apparent distribution volume.

In the case of clonidine, V_D is ≈ 3 l/kg and half-life is ≈ 12 h; using Equation 7.6, it is possible to calculate clonidine C_P at steady state during chronic administration of 0.5 mg/die:

$$(C_P)_{SS} = 1.443 \times \frac{500\,\mu g}{3 \times 70\,l} \times \frac{12\,h}{24\,h} = 1.7\,\mu g/l$$

Quantity of drug present in the body and drug plasma concentration are two correlated values, provided that drug distribution is fast. However, this does not apply when drug concentrations in various compartments are not at equilibrium. The interference of distribution with plasma concentration time course will be discussed in a later section.

Provided that doses (D) are constant, the quantity administered per unit time is inversely proportional to the interval between administrations. Figure 7.6 illustrates the result of administering the same drug dose with three different time intervals between administrations. As the administration frequency increases, the average steady-state concentration increases, but the time needed to attain the steady state does not change.

Equation 7.6 is particularly important because it defines that the average plasma concentration (and thus the average drug quantity in the organism) at steady state is not a function of the dose alone or the time interval between administrations alone, but is a function of both, that is, function of the drug quantity administered per unit time.

The Single Dose to Administer is Computed as a Function of the Interval between Successive Administrations

Equation 7.6 can be rewritten to calculate the dose D to be administered to obtain the desired average concentration C at steady state, as a function of the chosen administration interval:

$$D = (C_P)_{SS} \times 0.693 \times V_D \times \frac{T}{t_{1/2}} \quad (7.7)$$

The concept of bioavailability should always be kept in mind: the quantity of drug that effectively reaches the bloodstream may be only a fraction of the administered dose, D. In Equations 7.6 and 7.7, D refers to the drug quantity that effectively enters the body and is distributed to organs and tissues (absorbed dose).

Although oversimplified, Equation 7.7 can be used to design a therapeutic protocol on a rational basis. Consider the following example: we need to administer i.v. the antiarrhythmic drug lidocaine to a 70 kg patient to reach the effective plasma concentration of 4 mg/l. Lidocaine distribution volume V_D is 1.1 l/kg and its half-life is 1.8 h. When the drug is administered once every hour, the single dose D_h computed from Equation 7.7 is

$$D_h = 0.693 \times C_P \times V_D \times \frac{T}{t_{1/2}} = 0.693(4\,mg/l)(77l)\left(\frac{1}{1.8}\right)$$
$$= 119\,mg$$

that means about 2 mg/min (119/60), if we intend to set up a continuous infusion by drip injection. Note that this is the dose needed to have the desired drug plasma concentration once the steady state has been reached, that is, after about 9 h (5 × 1.8 h). If we need to obtain the desired plasma concentration more rapidly, then we have to change the treatment (see the following).

Equation 7.6 can also be used to roughly estimate the plasma concentration that a given protocol would produce in a given patient. For example, to estimate the plasma concentration of the cardiac glycoside digoxin in a patient treated for more than a month with 2 tablets/day per os, containing 0.2 mg of active principle, pharmacokinetic data should be collected from the manufacturer or a pharmacology textbook. For this drug, $V_D = 7\,l/kg$, the half-life is 1.7 days and the oral bioavailability is 0.75 (absorbed fraction). The distribution volume in a patient of 70 kg is 490 l (70 kg × 7 l/kg), and the absorbed daily dose is 0.3 mg (2 × 0.2 mg × 0.75). Since the drug quantity in the patient's body at steady state is 1.443 times the total dose administered over a time interval equal to one drug half-life (Eq. 7.6), such value is 0.736 mg (1.443 × 0.3 mg × 1.7), that is, 736 μg. Plasma concentration is obtained by dividing this value by the distribution volume: $C = 736\,\mu g/490\,l = 1.5\,\mu g/l$.

Fluctuations of Drug Plasma Concentration at Steady State

The ratio between maximum and minimum drug concentrations during oscillations at steady state strictly depends on the administration interval. This can be appreciated in the curves in Figure 7.6 showing that closer administrations induce larger $(C_P)_{SS}$, but the difference between consecutive administrations does not change (because the amount administered is the same). The concentration can be kept constant (and fluctuation reduced) at steady state by administering the drug continuously (e.g., via drip infusion). If administrations need to be separated in time for practical reasons, the daily dose computed according to Equation 7.7 will have to be subdivided into single doses maintaining constant ratio between doses and interval between doses. The shorter the administration interval, the smaller the single dose and, as consequence, the smaller is the resulting oscillations in plasma concentration at steady state. This is particularly important for drugs with a low therapeutic index. In such situations, the most relevant parameter is the tolerable percentage

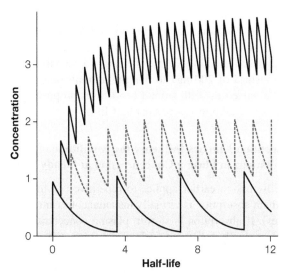

FIGURE 7.6 Effect of the administration interval on the plasma concentration time course. Inspection of the curves indicates that reducing the administration interval (for the same drug at the same dose): (i) higher average concentrations are attained at steady state; (ii) the time to reach steady state does not change; (iii) the absolute magnitude of the fluctuations in plasma concentration at steady state (maximum−minimum) is the same as it simply corresponds to a single dose divided by the distribution volume; and (iv) the fractional fluctuations (range/average concentration) become smaller.

fluctuation in plasma concentration at steady state. Once the steady state has been reached, each dose moves the plasma concentration from a minimum to a maximum of the fluctuation range. The amplitude of fluctuations at steady state is given by the ratio D/V_D. If this value is compared to the average steady-state concentration (from Eq. 7.6), the percentage fluctuation amounts to $(D/V_D)/C_{SS} = T/(1.443 \times t_{1/2})$. If we decide that the acceptable fluctuation must not be larger than a certain percentage P of the plasma concentration, then $T/(1443 \times t_{1/2}) < P$, and $T < 1.443 \times P \times t_{1/2}$. Thus, to limit fluctuations within 50% (about ±25%), the administration interval must be shorter than $(1.443 \times 0.5 = 0.72) \times t_{1/2}$. About three administrations per half-life will be needed to limit fluctuations to 20% (about ±10%): $1.443 \times 0.2 = 0.29 \times t_{1/2}$; seven administrations per half-life ($T = 0.14 \times t_{1/2}$) to limit fluctuations to 10% (±5%).

Given these considerations, it is possible to evaluate the fluctuation size in digoxin plasma levels in the case reported earlier (200 μg every 12 h, $V_D = 490\,l$, half-life = 41 h, absorbed fraction = 75%). The average plasma concentration at steady state is 1.5 μg/l, as seen earlier. It will fluctuate by a quantity equal to (absorbed fraction) × $D/V_D = 0.75 \times 200/490 = 0.3\,\mu g/l$, that is, between 1.65 and 1.35 μg/l. This corresponds to a fluctuation of about $100 \times 0.3/1.5 = 20\%$. If we wish to reduce the fluctuation to 10%, we have to change the single dose and the administration interval. Each dose needs to move plasma concentration from 0.95- to 1.05-fold the average plasma concentration, which corresponds to an absorbed dose equal to $(C_P)_{SS} \times V_D/10 = 1.5\,\mu g/l \times 490\,l/10 = 73.5\,\mu g$ and to the administration of single doses of about $100\,\mu g$ (bioavailability 0.75). The single dose will be halved (100 instead of 200 mg) so that the administration interval can be halved as well, from 12 to 6 h (as proposed earlier, about seven administrations per half-life).

Absorption Kinetics Influence the Amplitude of Oscillations in Plasma Concentration at Steady State

The discussion earlier applies to repetitive administrations in which absorption is virtually instantaneous (intravascular routes). If absorption is slower, plasma concentration rises more gradually following each administration, and concentration peaks are less prominent (Fig. 7.1; compare also with Fig. 7.7). The average plasma concentrations at steady state are the same for the three conditions illustrated in Figure 7.7, but maximum and minimum fluctuation levels differ.

Such a difference can be exploited when the drug has a narrow therapeutic window; oral administration smoothes fluctuations and may prevent plasma concentration from attaining levels associated with toxic effects.

When the absorption rate is low (e.g., retard preparations; see Chapter 4), and in particular when the absorption rate becomes slower than the elimination rate, since

$$\tau_a > \tau_e, \exp\left(\frac{-t}{\tau_a}\right) > \exp\left(\frac{-t}{\tau_e}\right),$$

we observe that both the numerator and the denominator in Equation 7.1 change their signs. The resulting equation, for plasma concentration, is

$$C_P(t) = \frac{D}{Cl} \times \frac{\exp(-t/\tau_a) - \exp(-t/\tau_e)}{\tau_a - \tau_e},$$

which is identical to Equation 7.1, except that τ_a and τ_e have been switched. This means that, if absorption is slower than elimination, the rising phase of plasma concentration, following each administration, will be governed by the elimination time constant, whereas the decay will be governed by the absorption time constant. Actually, this is not so surprising, if we consider a very slow absorption as something similar to a repetitive administration. In fact, we know that in that case the plasma level rises with a kinetic that depends solely on the elimination rate (and reaches steady state after five elimination half-lives). Since in case of slow absorption the drug quantity remaining to be absorbed, and consequently the drug quantity absorbed per unit time, slowly decreases (with a rate coincident with the absorption rate), the late decay is governed by the absorption rate constant.

Since for very slow absorption rates, absorption and elimination rates interchange their roles in determining the plasma concentration curve, repetitive administrations will

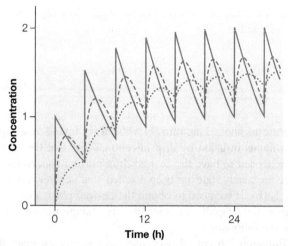

FIGURE 7.7 Effects of different administration routes on plasma concentration time course during repetitive administration. Full bioavailability and identical elimination rate are assumed for all conditions. Slower absorption rates (dashed curves) produce smoothed and narrower fluctuations at steady state. However, the time needed to reach steady state (which only depends on the elimination rate, that is, on drug half-life) and the average plasma concentration at steady state do not change. Average plasma concentration is given by the quantity of drug absorbed during a half-life, independently of absorption rate, multiplied by $1.443/V_D$.

produce a rise to plateau governed by the absorption rate constant, and when the treatment is interrupted, drug levels will decay with this same kinetic (recall that normally both the rise to plateau and the decay upon therapy interruption are governed by the elimination rate). In this case, the elimination rate only influences the shape of the oscillations, which will be less marked when the elimination rate is slower.

Loading (Attack) Doses to Rapidly Attain Steady-State Concentration

As we said, five half-lives are needed to reach steady state in a repetitive administration protocol. For most drugs, half-life has a value of a few hours, but some have half-life of several days. If it is important to rapidly attain the effective concentration (C_{eff}) with a drug having a particularly long half-life, the physician essentially has two possibilities: either using an equivalent drug with more favorable kinetic properties or administering one or more loading doses in a short time, in order to rapidly bring drug plasma concentration to the C_{eff} value.

The dose needed to rapidly attain C_{eff} is said to be the loading dose (D_L) and is computed as follows:

$$D_L = C_{eff} \times V_D \quad (7.8)$$

This dose, possibly corrected for incomplete bioavailability, can be given in a single administration or in fractioned doses administered at relatively short intervals. Protocols to administer loading doses need to take into account magnitude and danger of possible toxic phenomena that may be produced by the high plasma concentration determined by the loading dose. In fact, the actual volume in which the drug initially distributes may be quite smaller than the V_D attained at steady state and the resulting plasma concentration may persist at toxic levels for several hours.

The loading dose can be computed as a function of the chosen oral maintenance dose. The latter is $D = 0.693 \times V_D \times C_{eff} \times T/t_{1/2}$ (Eq. 7.7, where T is the administration interval); using the aforementioned definition $D_L = V_D \times C_{eff}$ (Eq. 7.8), we obtain:

$$D = 0.693 \times D_L \times \frac{T}{t_{1/2}} \quad \text{and} \quad D_L = 1.443 \times D \times \frac{t_{1/2}}{T} \quad (7.9)$$

Thus, the loading dose equals 1.443 times the dose administered at steady state during a drug half-life. For example, for digitoxin (7-day half-life) at 0.1 mg die, the loading dose would be $1.443 \times 0.1\,\text{mg} \times \times 7 \approx 1\,\text{mg}$.

MULTICOMPARTMENTAL KINETICS

Under many circumstances, it is important to estimate drug concentration in a particular organ or tissue. When more than one compartment is considered, two aspects need to be clearly identified: (i) the drug concentration ratio between the two compartments at equilibrium and (ii) the rate at which the two compartments proceed toward equilibrium.

We have seen in Chapter 5 that each organ (compartment) may display its own specific avidity (affinity) for the drug; thus, the drug quantity present in a given compartment at distribution equilibrium depends on the avidity (kp) and the volume of the compartment (V). The product $V \times kp$ is equivalent to a certain volume of plasma and is called apparent distribution volume of the compartment (V_d). The total V_D of the subject is given by the sum of the V_d of all compartments. Since the drug tends to bind to or concentrate in certain structures (plasma proteins, membranes, aqueous or lipidic phase of the tissue), concentrations at equilibrium may be different in any two compartments. In particular, the concentration ratio at equilibrium, between any two compartments "1" and "2," will be $C_1/C_2 = k_{p(1)}/k_{p(2)}$.

The rate of equilibration between two tissues is given by the ratio between the flux of drug (drug quantity that moves from one compartment to the other in a unit time) and the total drug quantity that has to move from one compartment to the other. This defines a rate constant, k, as the fractional decrease (of the distance from equilibrium) per unit time. One of the two compartments generally is plasma, as in general all tissues have to equilibrate with plasma.

In order to evaluate the velocity of the equilibration between a tissue and plasma, we need first to consider the venous blood that leaves the tissue. Unless the drug transfer to the interstice is impaired (binding to plasma proteins, cellular barriers, and BBB), the plasma equilibrates with the tissue during the slow passage through the capillary bed, so that the free drug concentration in venous plasma will be the same as in the tissue aqueous phase (C_T). In arterial plasma, the free drug concentration is $C_A \times F$, where C_A is the drug plasma concentration and F is the free drug fraction (not bound to plasma protein). Along the capillary bed in the tissue, free drug plasma concentration equilibrates with the tissue, moving from $C_A \times F$ to C_T; accordingly, the total drug plasma concentration (free + bound) moves from C_A to C_T/F. The quantity of drug released by plasma to the tissue in a unit time (drug flux) is $(C_A - C_T/F) \times \Phi$, where Φ stands for plasma flux (ml/min) to the tissue.

The total quantity of drug that is transferred from blood to tissue is given (to a first approximation) by the product of the apparent distribution volume of the tissue, V_T, times the difference in concentration between arterial blood and blood that is in equilibrium with the tissue, that is, $V_T \times (C_A - C_T/F)$.

The velocity of the equilibration process (ratio between quantity transferred in a unit time and total quantity to be transferred) is given by the ratio Φ/V_T. Remember that barriers that impair drug diffusion to the tissue slow down equilibration and the rate constant is limited by barriers rather than by the flux to volume ratio. For further details on multicompartmental kinetics, see also Box 7.1.

Drug Binding to Plasma Proteins and Tissue Equilibration with Plasma

Since high binding to plasma proteins determines a decrease in the apparent distribution volume of tissues, it will also accelerate their equilibration with blood. The kinetics may be limited by the presence of barriers: in such situation, plasma is unable to equilibrate with the tissue during its passage in the capillaries, and therefore, the drug equilibration rate is governed by the diffusion across the barrier.

Notwithstanding the approximations implied in these derivations, all this illustrates the operative handiness of the concept of apparent distribution volume; being defined as the ratio between total drug quantity in a compartment (or the whole body) and plasma concentration (total, not only free drug), the relation between apparent distribution volumes and distribution rate constants is simple and direct.

However, binding to plasma proteins introduces further complications: if drug affinity for plasma proteins is very high, dissociation may be very slow. In this case, only the free fraction of the drug has the time to equilibrate during the passage through the capillaries, and drug flux only is $(C_A \times F - C_T) \times \Phi$, which is a fraction F of what was computed earlier. In case of 99% binding, $F = 0.01$, drug flux to the tissue may be reduced by 100-folds, with marked slowing of the kinetics.

BOX 7.1 MULTICOMPARTMENTAL KINETICS

Drug redistribution phenomena clearly show that the simple difference between two exponentials governed by absorption and elimination rate constants (see Eq. 7.2 in text) is inadequate to properly describe drug plasma concentration time course. An accurate description of time courses in a multicompartmental system such as the human organism can only be obtained by simultaneously considering drug fluxes between plasma and each of the compartments (organs and tissues). Schematically, each of these fluxes arises from a compartment reaching equilibrium with plasma, with first-order kinetics and a time constant determined by the apparent volume-to-plasma flux ratio (except a possibly relevant slowdown produced by diffusion barriers). Thus, the overall time course of plasma concentration will be described by the sum of many exponentials, each with its own size coefficient and time constant. For each compartment, the time constant will be the volume/perfusion ratio, while the size coefficient will reflect the contribution of the compartment to overall kinetics and will essentially depend on the distribution volume of the compartment.

All compartments with a similar specific perfusion (flux/volume ratio) can be grouped, because they contribute to the same exponential component (same time constant), which will have a size coefficient reflecting the sum of the distribution volumes of the compartments belonging to the same group. By this procedure, we can distinguish, besides the hematic, the absorption, and the elimination compartments:

- The cerebral (highly perfused) compartment
- The muscular (high volume, ~50% body weight, low–moderate perfusion, good avidity for lipophilic drugs) compartment
- The visceral (low volume, high perfusion) compartment
- The highly perfused compartment (heart, kidney, adrenal, quite low volume but very high perfusion)
- The adipose compartment (variable volume, very low perfusion, high avidity for lipidic drugs)
- The poorly perfused tissues (such as the epidermis, several liters of volume, scarce perfusion)

Overall kinetics are dominated by the hematic and muscular/visceral compartments, and often, the approximation of a single compartment with volume corresponding to all body liquids accessible to the drug is acceptable.

The multicompartmental approach allows us to predict the drug concentration time course in the specific compartment where the drug is supposed to act (e.g., the CNS) and to account for sequestration, storage, and redistribution phenomena. Under specific physiological and pathological conditions, perfusion of some compartments may undergo remarkable changes, for example, in the "fight or flight" response, muscular perfusion is increased, while cutaneous and visceral perfusion may be markedly decreased: in the state of shock, visceral, cutaneous, and renal fluxes are profoundly reduced. Such changes may produce relevant alterations in drug distribution kinetics.

The Particular Case of the Nephron

The concept of equilibration between compartments can be applied to the nephron as well. Here, the situation is different, as the principal mechanism of drug transfer from plasma compartment to nephron consists in its passage together with plasma water during glomerular filtration. Thus, the quantity of drug that leaves the plasma during filtration is not proportional to the free and not the total plasma concentration. In contrast with what occurs during the diffusion from capillaries to tissues, in the nephron, drug diffusion does not change the free drug concentration, because it diffuses together with the solvent. Therefore, drug diffusion is not accompanied by its dissociation from plasma proteins to reequilibrate bound and free fractions. Apart from active secretion, in all cases (regardless of whether the drug is eliminated with urines or partly (or fully), actively or passively, reabsorbed), the excreted quantity is proportional to the free drug concentration in plasma. Therefore, renal clearance (the plasma volume cleared of drug every minute) is inversely proportional to drug binding to plasma proteins.

It is important to remind here that drug elimination rate is given by the clearance over V_D ratio. As we have just seen, renal clearance is proportional to the free drug fraction (free/total). A slightly different consideration applies to V_D drug concentration in tissues, C_T, is in equilibrium with the **free** drug plasma concentration. Therefore, the concentration ratio between tissues and plasma (and consequently the V_D) will change if free and bound fractions change. However, plasma necessarily contributes a fixed distribution volume of 3 l. Thus, part of the V_D (3 l) is insensitive to drug binding to plasma proteins; the V_D will change less than the clearance, especially when it has very low values (compared to the 3 l of plasma volume), that is, for very high binding to plasma proteins. This suggest that the clearance over V_D ratio, and therefore the drug elimination rate, may significantly decrease for drugs showing very high plasma protein binding.

If the drug is actively secreted in the tubules, the free drug concentration in the interstice decreases while it is excreted, and the drug bound to plasma proteins dissociates to maintain the equilibrium between bound and free fractions.

If dissociation from plasma proteins is rapid, drug flux (elimination rate) is proportional to total plasma concentration, the clearance is insensitive to protein binding and the velocity of the process (renal excretion in this case) is again inversely proportional to the distribution volume.

The placenta is another compartment where drug binding to plasma proteins is an important factor in slowing down distribution processes: here, equilibration kinetics are slow because the maternal flow is low compared to the large apparent distribution volume of the fetus. Fetal plasma proteins act as a reservoir for drugs highly bound to plasma proteins, thereby increasing the apparent distribution volume and slowing down transplacental equilibration kinetics.

Drugs Redistribution among Compartments

Highly perfused organs (with high hematic flux per unit volume) tend to equilibrate very rapidly with plasma. Low perfusion organs, in particular those having large distribution volumes, equilibrate very slowly. Consider a very lipophilic drug, such as the anxiolytic benzodiazepine diazepam or the hypnotic barbiturate thiopental. Following an i.v. administration, both drugs easily cross the blood–brain (BBB) and rapidly distribute to the CNS. As the hematic flow is much higher in the CNS than in several other organs and tissues, the CNS equilibrates with plasma before significant drug diffusion and equilibration has occurred at most of the other sites. Initially, the apparent distribution volume for the drugs is essentially constituted by the plasma, brain, liver, and a few more organs with small V_D, and effective concentrations of diazepam or thiopental are rapidly attained at these organs. Afterward, the drugs gradually equilibrate in others tissues and the distribution volumes increases; plasma concentration accordingly decreases, and the concentration in the brain—in rapid equilibrium with plasma—also decreases and may fall below the effective concentration levels. Drug effects vanish, so that their action duration seems to be very short. However, such short action duration is not related to drug metabolism or excretion, but to drug redistribution. If drug concentration in the CNS has not fall below the effective levels by the end of the redistribution phase, then its action will only terminate when its concentration further decline due to metabolism or excretion, and its action duration will be thus determined by drug half-life, rather than by its redistribution rate.

Figure 7.8 shows an example of computer-simulated time courses of brain, plasma, and muscle concentrations of 100 mg dose thiopental. Note that the redistribution rate (initial variations, highlighted in the panel a) is much faster than the elimination rate (late decline, better seen in the panel b). This is particularly important for highly lipophilic drugs because they usually have a final distribution volume much higher than the initial one, as the drug tends to widely distribute to muscles and to accumulate in adipose tissue. As

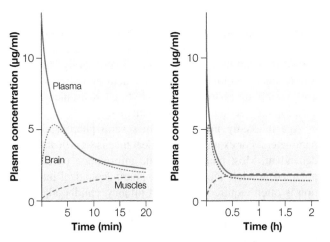

FIGURE 7.8 Computer-simulated time courses of plasma and tissue (brain and muscle) concentrations after intravenous administration of 100 mg thiopental to a 70 kg subject. Brain concentration rapidly equilibrates with plasma (left panel, time scale in minutes) and both rapidly decline during the first 10 min (redistribution phase); both plasma and brain concentration curves slowly stabilize while muscle concentration keeps rising; later on, all three concentrations move in parallel and very slowly decay (right panel, time scale in hours), as predicted by the quite low value of the drug clearance. The effect on the brain ensues rapidly and is short lived; however, its decline is not due to drug elimination, but to its redistribution to other tissues.

we saw, the elimination rate is determined by the clearance over V_D ratio. Thus, considering that the kidney is unable to excrete them (they are largely reabsorbed by passive diffusion in the tubule), and that metabolic clearance of lipophilic drugs is generally small with respect to the V_D, we can predict that their elimination rates are low and their half-lives are long. The paradox, which is clinically particularly important, is that drugs such as diazepam and thiopental, which appear to have very short action duration upon single administration, may become quite long-acting drugs if high doses are employed or if repetitively administered, so as to saturate the tissues.

CORRECTIONS OF THE THERAPEUTIC REGIMEN

Several factors, including genetic differences, diseases, age, body weight, concurrent use of other drugs, and environmental factors, make the population of patients affected by a given disease heterogeneous in terms of sensitivity to a given therapeutic regimen. This means that a regimen that in some patients is appropriate and effective to cure the disease may be toxic or insufficient in others. In most cases, pharmacokinetic parameters need to be reexamined and appropriately adjusted for each subject, before using them to calculate dosages with the general formulas.

The importance of genetic variability in determining wide interindividual differences in metabolic disposal of drugs or in receptor response or transduction mechanisms is discussed in detail in Chapters 6 and 23. Variability is mainly due to pharmacokinetic (e.g., isoniazid and slow vs. fast acetylators) or pharmacodynamic (e.g., resistance to the anticoagulant action of warfarin) factors.

Age markedly interferes with several pharmacokinetic parameters. For example, renal clearance is significantly age dependent. Most pharmacokinetic information on drugs is obtained in adult subjects, but a pharmacological intervention is often required in the first infancy, childhood, or old age. Some preliminary ideas about the criteria to be kept in mind when adapting dosages to single subjects are offered in the following pages.

Normally Available Pharmacokinetic Data Are Average Values

In order to correctly design a therapeutic protocol, it is necessary to know drug bioavailability (absorbed fraction F) and clearance (Cl), together with the apparent distribution volume (V_D) or drug half-life, $t_{1/2}$. Most pharmacology textbooks report these values in pharmacokinetic tables including the most commonly employed drugs; these can also be found on dedicated web sites. However, those provided are average values that are subjected to interindividual variability (standard deviation in the order of 20–50%). Though this may appear sufficient to adequately design a therapeutic protocol, one should consider that using the values reported in the tables, in 95% of cases, the achieved plasma concentration will be comprised between 35 and 270% of the desired concentration. For many drugs, such approximation may not be a problem, but it is not acceptable for drugs with a low therapeutic index; in these cases, the dosage needs to be defined based on individual estimates of Cl and V_D or by monitoring drug concentration in plasma. Even for drugs with a good therapeutic index, marked alterations in these parameters may require adjustments of the therapeutic regimen.

Varying Dosage as a Function of Body Weight and Physical Constitution

Body liquid volumes, muscular mass, hematic flux, and functionality of each organ have a strict correlation with body weight. This implies that distribution volume and clearance vary in parallel (note that a parallel change in V_D and Cl implies no change in elimination rate constant and half-life). These parameters are generally reported in units (l and ml/min, respectively) per kg body. Usually, correction to the standard dosage (generally referred to an average, male, adult, 70 kg subject) is applied when patients differ more than 30% from such average and especially in children.

Variations in body composition may be more relevant. For lipophilic drugs, the distribution volume may markedly change in obese people—more than estimated by simple correction for body weight—whereas, for polar drugs, patient's tallness and muscular structure are more relevant.

Varying Dosage as a Function of Age

Renal function varies during the course of life. For example, Figure 7.9 displays how creatinine clearance changes with age. Creatinine is distributed to all aqueous volumes in the body and is almost entirely present in free form both in blood and tissues. It is cleared by the kidney and has a clearance equal to the glomerular filtration rate. Creatinine clearance (per kg body weight) is low at birth, increases during the first months of life, is quite high in childhood, and gradually decreases with age. The latter aspect can be estimated using the following formula that relates creatinine clearance (Cl_{cr}) in a healthy individual (>20 years of age) with that of the "model" subject (70 kg, 55 years old):

$$Cl_{cr}(ml/min) = \frac{(140-age) \times weight(kg)}{70} \quad (7.10)$$

Several other factors, relevant to drug pharmacokinetic, remarkably decrease with age, such as total renal plasma flux, cardiac index, and maximum respiratory capacity. In general, the newborn and the elderly must be considered as patients with a reduced clearance and therefore subjected to possible drug accumulation. However, this is not the case for children after the 6 months as shown in Figure 7.9.

The most important aspects of this "pharmacokinetic immaturity" of the newborn are the reduced metabolic and excretion capacities and the different body composition (fluid compartment volumes). Some indicative differences

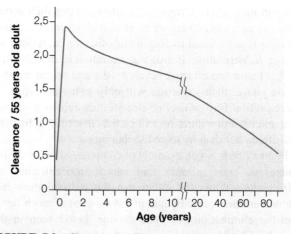

FIGURE 7.9 Changes in clearance per kg body weight as a function of age. Values are reported after normalization to the average clearance in a healthy, male, 70 kg body weight, 55-year-old subject.

TABLE 7.1 Comparison between pharmacokinetic parameters in children and adults

		Average newborn	Average adult
Body weight	(kg)	3.5	70
Body water	(%)	77	60
	(l)	2.7	42
Inulin clearance	(ml/min)	3	130
(glomerular filtration)	$t_{1/2}$ (min)	630	224
PAH clearance (tubular	(ml/min)	12	650
secretion)	$t_{1/2}$ (min)	160	45

in pharmacokinetic parameters between newborn and average adult are listed in Table 7.1. In newborn, renal clearance is much lower and drug half-life may be increased up to 3–4-folds, also due to the relatively larger water volume per kg body weight (the half-life is proportional to the ratio V_D over Cl, which is respectively larger and lower in the newborn, after normalization for body weight). In addition to a lower filtration rate, due to a lower renal flux to water volume ratio in the organism, active tubular secretion capacity is also lower, as evidenced by the comparison of inulin (filtered) with PAH (filtered and actively secreted) clearance.

Tubular secretion capacity rapidly increases in the first weeks after birth; maturation appears to be accelerated (especially for organic cation transport) by the presence of substrates. Indeed, in experimental animals, pups born of mothers treated with penicillin in late pregnancy display an almost complete development of tubular organic acid secretion systems at birth. Hepatic metabolism and excretion of drugs are also reduced in newborn.

Reduced clearance may represent a serious problem in case of chronic administration of drugs (especially lipophilic drugs) during puerperium and breastfeeding. During intrauterine life, the fetus can be considered as a further organ, in equilibrium with maternal blood, and therefore with a clearance capacity comparable to the maternal one. After birth, on one side, the infant can still be exposed to new drug doses through the milk, and on the other side, his/her excretion and metabolic capacities are reduced; thus, the drug may accumulate, especially if lipophilic, and give rise to toxic effects.

In children, merely reducing dosage as a function of body weight is generally inaccurate, because cardiac output, renal and hepatic blood flux, and glomerular filtration rate are all better correlated to body surface area than to body weight. Therefore, during a chronic therapy, it is advisable to correct drug dosage using the following equation:

$$\text{Dosage in children} = \left(\frac{\text{body weight}(m^2)}{1.8\,m^2}\right) \times \text{dosage in adults} \quad (7.11)$$

where $1.8\,m^2$ is the body surface area of a normal 70 kg adult subject. Body surface area can be estimated from body weight based on the observation that the former is roughly proportional to the latter elevated to 0.7. Thus, Equation 7.11 can be rewritten as

$$\text{Dosage in children} = \left(\frac{\text{body weight}(kg)}{70\,kg}\right)^{0.7} \times \text{dosage in adults} \quad (7.12)$$

To illustrate how Equation 7.12 can be used, we will consider the example of phenobarbital. The normal dosage for anticonvulsant therapy in the average adult is 100 mg die. In a 15 kg child, the correct dosage is $(15/70)^{0.7} \times 100$ mg die = 37 mg die. Note that the dosage normalized to body weight is 2.3 mg/kg in the child versus 1.43 mg/kg in the adult: the reason of such difference is that clearance is proportional to body surface area, so that clearance/kg increases as body weight decreases (the exponent is < 1).

Drug dosage correction as a function of body surface area (or weight) must not be performed in an automatic way because of two main reasons: (i) renal clearance widely varies during life and especially at its extremes, as clearly illustrated in Figure 7.9, and (ii) the distribution volume, V_D/kg, may vary quite remarkably due to the different body composition: for example, the percentage of water in body volume changes from 77% in the newborn to 60% in the adult (Table 7.1).

Consideration should also be given to the therapeutic regimen that we plan to use. If a single administration is considered, the appropriate dose will be mostly determined by the volume in which the drug will distribute: $D \approx C_{eff} \times V_D$. Thus, the dose per kg body weight may be larger than in the adult. Conversely, if repetitive administration is considered, drug half-life (and the ratio between V_D and clearance) becomes relevant: the maintenance dose should be $D \approx C_{eff} \times Cl$, and dosage (per kg) should be lower in newborns and larger in children than in adults. The reduced clearance of newborns is of little concern in acute drug administration, because it only extends drug persistence in the organism and the duration of its effects (which might even be advantageous). However, with repetitive administration, steady-state plasma concentration of the drug significantly increases if the drug half-life is noticeably increased. It should also be considered that hepatic metabolism and excretion are also remarkably reduced in infancy, introducing a further element of complexity in the determination of the optimal dosage for chronic administration.

In elderly patients, renal and hepatic metabolism and excretion are generally reduced, and during a chronic therapy, there is a risk of drug accumulation. Indeed, in elderly patients chronically treated with intermediate half-life drugs, accumulation phenomena often occur giving rise to toxic effects or prolongation of drug effects. For example, chronic treatment with benzodiazepines, used as sleep inducers, may result in sedation during the day, often associated with low reactivity or signs of cognitive impairment that may even be misdiagnosed as Alzheimer's disease or atherosclerotic dementia.

The following equation, which combines 7.11 and 7.12, can be used to determine and personalize dosage for drugs mostly cleared by the kidney with kinetics similar to creatinine:

$$\text{Dose} = \frac{(140 - \text{age})(\text{weight}(\text{kg}))^{0.7}}{1660} \times \text{standard dose} \quad (7.13)$$

This Equation 7.13 applies to any subject at least 20 years old, regardless of the body weight: the denominator is the product of renal clearance per kg in the "model" subject (55 years, 70 kg) times (70 kg) elevated to 0.7.

With respect to the elderly, a final important consideration concerns the possible impairment of the BBB, due to atherosclerotic insult. Hydrosoluble drugs that are safely administered to the average adult patient as they penetrate the CNS to a negligible extent should be administered with caution in these cases because they may produce unpredictable central effects.

Dosage Correction in the Presence of Hepatic Pathologies

In some hepatic pathologies, portal-systemic shunt is present. This may cause significant changes in bioavailability of orally administered drug normally undergoing massive first-pass effect. In most cases, the major problem for the physician is to evaluate and quantify the change in drug hepatic clearance (see Chapter 5, section "Hepatic Excretion and Enterohepatic Cycle"). In addition, changes in pharmacokinetic properties may widely vary among different drugs, and different liver pathologies may differently affect the same drug. Given such variability, it is not possible to establish general rules to rationally modify therapeutic regimens. The physician should in general refer to the available empirical data in the literature and carefully monitor the effects of the therapy on the individual patient.

A common consequence of severe liver pathologies is a decreased protein synthesis resulting in hypoproteinemia.

In contrast with drug displacement from plasma protein binding, which may determine an increase (even quite significant) in the free drug quantity in the organism and possible associated toxic effects, hypoproteinemia generally ensues very slowly and gradually, the storage compartment constituted by plasma proteins simply disappears, whereas all other compartments maintain their normal kinetic properties. As a result, free drug concentrations in plasma and tissues remain unchanged. Under steady-state conditions, the appropriate dosage produces the correct (free drug) plasma concentration: though the reservoir of plasma proteins is missing, free drug concentration is unchanged and the same elimination rate matches the same free drug concentration—thus, no change needs be introduced in therapeutic regimen.

As an example, consider a patient treated with digitoxin, with a dosage regimen of 100 μg die, which produces a total plasma concentration of 20 ng/ml. As the drug is 97% bound to plasma proteins, the free drug concentration in plasma (and in the aqueous component of all tissues) is 0.6 ng/ml (Table 7.2). If hypoproteinemia ensues and plasma protein binding falls to 90%, the same dose will produce the same free plasma concentration (if the rate of administration is unchanged, and therefore also the rate of elimination, which is proportional to the free drug concentration). The clearance will parallel the free/total drug ratio and the V_D will change roughly to the same extent (apart from the 3 l of plasma that is unaffected). The resulting slight change in the V_D/Cl ratio will be reflected in a slight decrease in drug half-life, which is however irrelevant for therapy, as the free drug concentration (and the concentration at the site of action) is totally unaffected.

It should be noted that total plasma concentration is markedly reduced by a decrease in plasma protein binding; therefore, in such cases, it is recommended to obtain information from the chemical laboratory on the free, rather than total, drug plasma concentration.

Dosage Correction in the Presence of Renal Pathologies

Drug renal clearance may display wide variability in normal population (standard deviation up to 50%). In pharmacokinetic tables, for drugs with particularly variable renal clearance, a relation is usually reported that allows to approximately estimating drug clearance based on creatinine clearance.

TABLE 7.2 Example of changes in digoxin pharmacokinetic parameters in the presence of hypoproteinemia

Dose (mg/day)	$t_{1/2}$ (day)	Dose (mg/day)	Percentage binding	V_D (l)	Cl (l/h)	C_p (ng/ml)	C_{free} (ng/ml)
0.1	7	0.7	97	50	0.21	20	0.6
0.1	6.7	0.67	90	160	0.7	6	0.6

As V_D and Cl represent plasma liters and liters/hour, respectively, they inversely vary with plasma concentration, for equal free drug concentration and equal tissue concentration.

Creatinine clearance is estimated from plasma creatinine concentration using the relation

$$Cl_{cr} = \frac{(140-\text{age}) \times (\text{weight (kg)})^{0.7}}{72 \times \text{serum creatinine (mg/100 ml)}} \text{ ml/min} \quad (7.14)$$

For women, the denominator value 72 is replaced with 85.

Evaluating renal drug clearance is obviously very important for drugs mostly excreted by this route, in subjects affected by renal pathologies and impairment of renal function in general. Correcting the dosage is necessary especially for drugs with a narrow therapeutic window.

Since a reduction in drug elimination implies a prolonged drug half-life, the results are

a. an increase in the time needed to reach steady state during repetitive administration, together with
b. a decrease in the relative magnitude of plasma concentration fluctuations between successive administrations.
c. an increase in average plasma concentration at steady state if the dosage is not modified.

The last inconvenience can be overcome by modifying the administered dose per unit time by a fraction equal to the estimated reduction in creatinine clearance if the drug is entirely eliminated by the kidney. For drugs that are eliminated also by other routes, the dosage modification should take into account the ratio between renal and nonrenal clearances. For example, a drug with renal clearance $Cl_R = 100$ ml/min and nonrenal clearance $Cl_{NR} = 80$ ml/min has total clearance $100 + 80 = 180$ ml/min. If in a patient creatinine clearance is reduced by 50%, Cl_R drops to 50 ml/min, while Cl_{NR} remains unchanged, and total clearance is $Cl_T = 50 + 80 = 130$ ml/min. The variation in total Cl, calculated as $Cl_T(\text{pt})/Cl_T(\text{ctl})$, in this example is $130/180 = 0.72$.

The dose needed to maintain a given plasma concentration is directly proportional to Cl_T and will have to be modified to the same extent. The following equations may prove useful (ctl and pt stand for healthy control and patient, respectively):

$$\text{Dose}_{pt} = \text{standard dose} \times \frac{Cl_T(\text{pt})}{Cl_T(\text{ctl})} \quad (7.15)$$

where the ratio between the total Cl of the patient and that of the control is derived from the creatinine clearance according to the following relation:

$$\frac{Cl_T(\text{pt})}{Cl_T(\text{ctl})} = \frac{(Cl_{cr}(\text{pt})/Cl_{cr}(\text{ctl})) \times Cl_R + Cl_{NR}}{Cl_T} \quad (7.16)$$

Consider the following example of administration correction because of reduced renal clearance. For the antiarrhythmic drug diisopyramide, pharmacokinetic data are $t_{1/2} = 6$ h; $V_D = 0.6$ l/kg. To attain the concentration of 3 mg/l in a 70 kg 40-year-old subject with normal serum creatinine level (1 mg/100 ml), a correct dosage of 0.5 g die can be computed using Equation 7.7. Fifty five percent of the drug is eliminated by the kidney. When renal insufficiency ensues in our patient, serum creatinine level rises to 2 mg/100 ml. From Equation 7.14, we can estimate that kidney function is reduced by about 50%. The dosage modification required to maintain diisopyramide plasma concentration unaltered can be calculated with Equations 7.15 and 7.16. The correct dosage is $(0.5 \text{ g die}) \times (55/2 + 45)/100 = 0.36$ g die.

For an overview of the relevant aspects that should be considered for defining the therapeutic protocols, see Box 7.2.

TAKE-HOME MESSAGE

- The dosage of a single administration is determined by the product of the effective concentration (C_{eff}) by the apparent distribution volume of the drug (V_D). V_D is generally proportional to body weight; it is larger in children (higher body water fraction) and may vary depending on the lean/fat mass ratio for lipophilic drugs.
- When using extravascular administration routes, plasma concentration rises with the absorption time constant and decays with the elimination time constant (vice versa for absorption slower than elimination). Absorption constant varies with the route of administration; the slower the absorption rate, the lower the peak value of plasma concentration.
- The area under the plasma concentration curve (AUC) times drug clearance equals the total quantity of drug eliminated (dose × absorption fraction). It is independent of the administration route (for equal bioavailability). Drug clearance can be computed from the ratio dose/AUC.
- The drug distributes to tissues with speed proportional to the specific plasmatic flux (flux/distribution volume), unless diffusion is hampered (e.g., BBB); drug action on highly perfused tissues may be terminated by redistribution before drug elimination.
- Stable drug plasma concentrations can be achieved by repetitive administrations. Four to five half-lives are needed to reach steady state.
- At steady state, the amount eliminated between two repeated doses equals the amount administered with a single dose. Thus, the appropriate dosage to reach desired C_{eff} is given by $C_{eff} \times$ clearance = eliminated quantity per unit time. Plasma concentration oscillates over a range = (single dose)/V_D.
- For drugs eliminated with first-order kinetics, the dose (D) administered at time interval (T) to reach C_{eff} at steady state can also be calculated as a function of V_D and half-life: $D/T = 0.693 \times C_{eff} \times V_D/t_{1/2}$.

BOX 7.2   DEFINING THE THERAPEUTIC PROTOCOL

Here, we recapitulate the relevant relations useful to define a therapeutic protocol. A brief guide to the optimization of drug therapy through pharmacokinetics can be found in Supplement E 7.2, "The Distribution of Drugs Administered by Inhalation."

Suppose we wish to maintain an effective concentration of a given drug in our patient.

First of all, we need to look for the following parameters in the pharmacokinetic tables:

- Bioavailability, B, relevant for oral administration
- Effective concentration, C_{eff}, (µg/ml or mg/l)
- Clearance per kg, Cl_k (ml/min/kg)
- Apparent distribution volume per kg, V_k (l/kg)

Then, we need to determine the patient's V_D (liters) = $V_k \times$ [body weight] and the clearance (l/h), $Cl = Cl_k \times 60 \times$ [body weight]/1000.

To define the daily dose, D_{die}, we use Equation 7.5 for $T = 24$ (h): Daily dose (mg) = $D_{die} = 24 \times C_{eff} \times Cl$

Once we have defined the administration interval (T, hours), we can calculate the single dose as

$$\text{Single dose (mg)} = D = \frac{D_{die} \times T}{24}$$

If drug half-life ($t_{1/2}$) is available, we can use Equation 7.6 to estimate average plasma concentration at steady state (mg/l)

$$C_{eq} = \frac{1.44 \times D \times B \times t_{1/2}}{T \times V_D}$$

for D in mg and using the same time unit to express T and $t_{1/2}$.

If the half-life is not available, we can calculate it reminding that $t_{1/2} = 0.693 \times V_D/Cl$. This parameter is useful to define the most appropriate administration interval. If we accept ±50% fluctuations in drug plasma concentration at steady state, then the administration interval can be $T = t_{1/2}$. If we wish fluctuation not to exceed ±20%, T should be in the order of $t_{1/2}/3$. A value of T about $t_{1/2}/7$ yields ±10% fluctuations. Note that slow absorption will reduce fluctuations with respect to these estimates.

The time to steady state (loading time) is ~$3 \times V_D/Cl \sim 5 t_{1/2}$.

To accelerate the loading phase, a loading dose (single or fractionated) can be employed, based on the effective concentration and the distribution volume:

$$\text{Loading dose } D_L = C_{eff} \times V_D$$

or, as a function of the maintenance single dose (D),

$$\text{Loading dose} (D_L) = \frac{D \times V_D}{Cl \times T} = 1.44 \frac{D \times t_{1/2}}{T}$$

It is important to keep in mind that following administration, the drug may take a relevant time to distribute to all its V_D, so that loading doses may produce plasma concentrations much higher than predicted from the previous relations and may determine toxic effects.

In case of decreased renal function, dosages of drugs cleared by the kidney and with low therapeutic index have to be modified. Patient's creatinine clearance (Cl_{cr}) can be used as an indicator. For drugs cleared by both renal (Cl_R) and extrarenal (Cl_{NR}) routes, the latter probably are not modified. The decrease in total clearance (Cl_T) can be considered and the dosage adjusted as follows:

$$\text{Dose adjustment factor} = \frac{(Cl_{cr}(pt)/Cl_{cr}(ctl)) \times Cl_R + Cl_{NR}}{Cl_T} \quad (\text{pt} = \text{patient}, \text{ctl} = \text{control})$$

In children, the dosage should be reduced considering body surface area, using

$$\text{Dosage in children} = \left(\frac{\text{body weight (kg)}}{70 \text{ kg}}\right)^{0.7} \times \text{dosage in adults}$$

Let's see how these equations can be applied to design a therapeutic protocol. Consider to use atenolol in a 75 kg patient.

B 56±30%
C_{eff} 0.5–1.0 mg/l
Cl_k 2.4±0.2 ml/min/kg
V_k 0.95±0.15 l/kg
$t_{1/2}$ 6.1±2.0 ore

In our patient, $V_D = 0.95 \times 75 = 71.25$ l. To obtain a $C_{eq} = 0.75$ mg/l, by administering single doses at 6 h intervals, the single dose required is

$$D = \frac{C_{eq} \times T \times V_D \times 0.693}{B \times t_{1/2}} = \frac{0.75\,\text{mg/l} \times 6(\text{h}) \times 71.25(\text{l}) \times 0.693}{0.56 \times 6.1(\text{h})} = 65\,\text{mg}$$

The difference between minimum and maximum concentrations at steady state can be estimated from the ratio $D/V_D = 65\,\text{mg}/71.25\,\text{l} = 0.91\,\text{mg/l}$.

The drug is 95% eliminated by renal excretion. In case renal insufficiency ensues with a halved Cl_{cr}, the same dose (65 mg) will have to be administered at 12 h intervals, or the single dose will have to be halved (this will also reduce plasma concentration fluctuations).

- Drug clearance is generally proportional to body surface (~weight to 0.7 power); it is lower in infants, higher in children (per kg) compared to adults, and reduced in the elderly and possibly when kidney and/or liver diseases are present. Altered clearances require dosage adjustment.

FURTHER READING

Alcorn J., Mcnamara P.J.J. (2003). Pharmacokinetics in the newborn. *Advanced Drug Delivery Reviews*, 55, 667–686.

Gandhi M., Aweeka F., Greenblatt R.M., Blaschke T.F.F. (2004). Sex differences in pharmacokinetics and pharmacodynamics. *Annual Review of Pharmacology Toxicology*, 44, 499–523.

Leahy D.E. (2003). Progress in simulation modelling for pharmacokinetics. *Current Topics in Medicinal Chemistry*, 3, 1257–1268.

Mangoni A.A., Jackson S.M. (2004). Age-related changes in pharmacokinetics and pharmacodynamics basic principles and practical applications. *British Journal of Clinical Pharmacology*, 57, 6–14.

Pichette V., Leblond F.A.A. (2003). Drug metabolism in chronic renal failure. *Current Drug Metabolism*, 4, 91–103.

9

RECEPTORS AND MODULATION OF THEIR RESPONSE

FRANCESCO CLEMENTI AND GUIDO FUMAGALLI

By reading this chapter, you will:

- Learn the classification of receptors for endogenous ligands based on molecular structure and the relationship between molecular structure and signal transduction mechanisms
- Learn the molecular and cellular strategies that control receptor signaling and the consequences of modulation of receptor responses on therapies

---

Organization and functioning of multicellular organisms is based on intercellular communication mediated by production of messages on one side and presence of receptive molecules on the others. Messages are usually in the form of diffusible molecules (mediators or transmitters or hormones) released in the extracellular space; from the site of release, they reach, bind, and activate receptor molecules present on the surface of cells or inside them. Upon ligand binding, the receptor undergoes a conformational modification that triggers a cascade of events leading to the cellular response (signal transduction).

In pharmacology, the term "receptor" identifies also molecules that bind drugs (ligands) even if they are not involved in cell-to-cell communication. Thus, an enzyme, a pump, or a DNA can serve as receptors for drugs. In this and most of the following chapters, the term "receptor" will be used to indicate cell molecules that bind a mediator and generate a response signal.

Several of the currently available drugs act by interacting with receptors for endogenous mediators; their activity is dependent on number, biological properties, functional state, tissue distribution of the receptor they act on, and changes induced by pathological events. As described in Chapter 8, drugs may act by stabilizing a receptor in its active state (agonists) or in its resting state (antagonists, drugs that prevent activation of the receptor by the physiological ligand). Some receptors display high activity even when at rest; for these receptors, drugs that decrease the basal activity are named inverse agonists.

Knowledge of the molecular and cellular mechanisms underlying receptor response is crucial for understanding modern pharmacology. Molecular biology applied to pharmacology has allowed the identification of a number of receptor subtypes for the same mediator previously undetected by classical pharmacological approaches. The abundance of receptor subtypes for the same mediator emphasizes the plasticity of cellular communication systems, as cell responses generated by a same mediator may vary depending on the receptors expressed in the cell type involved. The abundance and tissue specificity of receptor subtypes open new and interesting perspectives for a highly selective pharmacology. Finally, the recent demonstration that members of the same receptor family may differ for small amino acid sequences (even a single amino acid) and that these minimal differences result in unique structural and functional properties may help explain interindividual differences in drug response observed in the population (see Chapter 23).

CLASSES OF RECEPTORS AND STRATEGIES OF SIGNAL TRANSDUCTION

In this chapter, we consider receptors (i.e., molecules or supramolecular complexes) that upon binding of one or more endogenous mediators undergo conformational changes that give rise to a biological effect. Proteins that bind endogenous ligands without directly generating biological signals are not considered "receptors." Examples of nonreceptor binding proteins are plasma steroid-binding proteins to which steroid hormones bind when present in the plasma (see Supplement E 9.1, "Birth and Evolution of the Receptor Theory").

Two major classes of receptors are recognized based on their subcellular localization: membrane receptors and intracellular/nuclear receptors. Membrane receptors are found on the cell surface, and their ligands are hydrophilic molecules that hardly cross the plasma membrane (e.g., classical and peptidergic neurotransmitters, growth factors, cytokines). Intracellular/nuclear receptors are localized in cytoplasm and/or nucleus, and their mediators are lipophilic substances (e.g., steroid and thyroid hormones, retinoic acid, vitamin D) that easily cross the cell membrane and reach their receptors.

The two classes of receptors also differ in the mechanism of signal transduction. Membrane receptors transduce the signal by modifying intracellular ion concentrations or generating second messengers or stimulating formation of biologically active macromolecular complexes; intracellular receptors are transcription factors that interact with specific DNA sequences and change gene expression and therefore cell protein composition.

In some cases, the same mediator can activate more than one class of receptors. For example, estrogen can activate a slow cellular response by binding to intracellular receptors (ERα and ERβ), and a fast response by binding to membrane receptors (GPER1/GPR30). Often, activation of membrane receptors, such as receptors for cytokines, tyrosine kinase receptors (TKRs), and G-protein-coupled receptors (GPCRs), leads to a long-term response through activation of transcription factors and thus modification of gene transcription.

Intracellular/Intranuclear Receptors

Intracellular receptors bind DNA and modulate gene transcription. So far, 48 potential receptors for lipophilic ligands have been identified and grouped in six families. About half of these potential receptors are actually "orphan receptors" (i.e., their ligand and/or function are unknown; their inclusion in this category of receptors is based on sequence homology). The remaining half includes receptors for sex hormones, glucocorticoids, mineralocorticoids, thyroid hormones, vitamins A and D, peroxisome proliferators (PPAR), retinoic acids, and other lipids. They are ligand-activated transcription factors that bind to specific DNA consensus sequences in the promoters of specific genes, regulating their transcription.

Members of this family are monomeric proteins in which three regions can be recognized: a carboxy-terminal domain, where the specific binding site for the lipophilic ligand is localized; a central domain, containing the sequence-specific DNA recognition site; and an amino-terminal domain, required for gene transactivation. In the absence of ligand, the receptor is present in cytoplasm or nucleus in an inactive

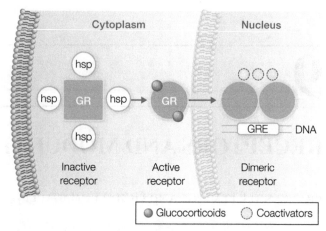

FIGURE 9.1 Mechanism of signal transduction of steroid hormone receptors. Hormones and lipophilic molecules can cross the cell membrane and activate intracellular receptors. The monomeric and inactive glucocorticoid receptor (GR) is found in the cytoplasm associated with heat shock proteins (hsp). Binding of glucocorticoids causes release of hsp, GR dimerization, and transfer into the nucleus. GR dimers interact with specific glucocorticoid-responsive elements (GREs; specific DNA sequences) present in the promoter of genes whose expression is modulated by glucocorticoids. GR recruits coactivator or corepressor proteins, and the resulting receptor complex, respectively, activates or inhibits gene transcription. Some intracellular receptors are resident in the nucleus.

form; the quiescent state is maintained by interaction with specific inhibitory proteins, members of the heat shock protein (hsp) family. Mediator (or agonist) binding induces a conformational change in the receptor that results in its dissociation from hsps and dimerization. If the receptor is in the cytoplasm, dimerization allows its transfer into the nucleus. In the nucleus, activated receptor complexes associate to specific DNA sequences (responsive elements, RE) in the promoters of genes whose transcription will be thus modulated (Fig. 9.1). The activity of these receptors is also modulated by binding of coactivators or corepressors. Agonists and antagonists act by interfering with activation or inactivation of these receptors (for a more detailed description, see Chapters 22 and 24).

Membrane Receptors

Membrane receptors mediate information transfer across the plasma membrane. Sequencing of the human genome has led to the identification of more than one thousand different receptors expressed on the plasma membrane of human cells. On the basis of their function and structure, membrane receptors can be grouped into 20 families. Only the nine most relevant families will be analyzed (Fig. 9.2), which are (i) ligand-gated ion channels (LGICs), (ii) GPCRs, (iii) receptors with intrinsic tyrosine kinase activity (TKRs), (iv) receptors with intrinsic guanylate cyclase

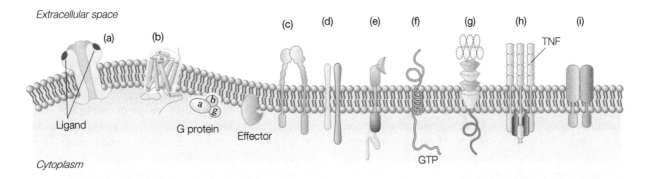

FIGURE 9.2 Families of membrane receptors. (a) Ligand-gated ion channels. (b) G-protein-coupled receptors with their seven transmembrane regions, associated trimeric G protein, and effector. (c) Receptors for extracellular matrix (integrins). (d) Cytokine receptors. (e) Receptors with intrinsic kinase activity (black cylinder). (f) Receptors with intrinsic guanylate cyclase activity. (g) Receptors for lipoproteins. (h) Receptors for TNF or death receptors. (i) "Toll-like" receptors.

activity, (v) receptors for cell adhesion and movement, (vi) cytokine receptors, (vii) receptors for tumor necrosis factor (TNFR), (viii) "Toll-like" receptors, and (ix) lipoproteins receptors. For many of the remaining families, endogenous ligands or active drugs are not known and thereby are classified as orphan receptors.

In general, LGIC activation produces transmembrane ion fluxes, thus generating rapid and short-lasting responses; by contrast, activation of a GPCR or any of the other receptor types generates responses that are usually slow and long lasting. Therefore, knowing the receptor types present in a cell is useful to predict the type of response to a given drug. On the other hand, a same mediator may activate different receptor types (see Table 9.1); for example, the neurotransmitter glutamate may produce responses that differ in terms of quality (type of response) and kinetics (speed of onset and duration). Furthermore, a single cell is often targeted by different stimuli and/or transmitters, and its final response will depend on the type and number of receptors present in the membrane, their activity, interactions with other receptors (crosstalk), and signal transduction systems activated (see Chapter 15). An additional level of complexity has emerged in recent years, since numerous data indicate that often receptor proteins do not act as "single" but as "complex," being organized in homodimers or homopolymers (a fact that may have important effects on parameters such as affinity for ligands and efficiency of signal transduction) or even in heterodimers/heteropolymers (i.e., constituted by different receptor proteins that may have different ligand specificity and signal transduction mechanisms) with functional and pharmacological properties very different from the monomeric parent receptors. This strategy is widely used in nature, for example, by chemokine receptors, to broaden anti-inflammatory response and adapt it to different needs.

In this chapter, we will summarize the most relevant properties of the nine major receptor families; the details are discussed in the following chapters.

TABLE 9.1 Neurotransmitters and their receptors with fast and slow signal transduction

Neurotransmitter	Fast response	Slow response
Acetylcholine	Nicotinic	Muscarinic
GABA	$GABA_A$	$GABA_B$
Glutamate	Ionotropic (AMPA, kainic, NMDA)	Metabotropic $(mGluR_1–mGluR_8)$
Serotonin	$5-HT_3$	$5-HT_1, 5-HT_2, 5-HT_4, 5-HT_5$
ATP	P2X	P2Y

LGIC Ligand-gated channels allow selective flow of ions according to their electrochemical gradient. LGICs are polymeric structures with three to five subunits arranged to form a transmembrane ion channel whose opening is regulated by binding of a neurotransmitter or an agonist. Opening of the receptor channel allows selected ions to flow in/out the cell along their electrochemical gradient, thus altering transmembrane electric potential. To this family belong nicotinic cholinergic receptors, $GABA_A$ and $GABA_C$ receptors, glycine receptors, ionotropic receptors for glutamate, $5-HT_3$ serotonin receptors, P2X purine receptors, and proton-activated channels. Most of these receptors (but not all) are concentrated at the interneuronal synapses (i.e., the nicotinic cholinergic receptor are in synapses at the neuromuscular junction, but extrasynaptic in epithelial cells).

Three morphofunctional regions can be recognized in the supramolecular receptor complex:

i. An extended extracellular portion where one or more neurotransmitter/ligand-binding sites are localized and that delimits a large funnel where ions are concentrated. Usually, binding sites involve two adjacent subunits, but their properties are often defined by other subunits as well. Allosteric binding sites are often present in the receptor extracellular portion.

ii. An intramembrane portion situated at the end of the funnel that forms the transmembrane ionic pore. The pore selects ions based on their size and charge; charged amino acids arranged in several rings along the inner surface of the channel determine ion charge selectivity. Agonist-induced opening of the channel is due to conformational modification of one or more subunits changing channel caliber and allowing ion flow. Drugs that bind this portion of the receptor interfere, by hindrance or by charge, with the kinetics of ion channel opening and closing.

iii. A cytoplasmic portion containing phosphorylation sites involved in the regulation of the kinetics of ion channel opening and closing and binding sites for adaptor proteins that link the receptor to the submembrane cytoskeleton and ensure its stability and correct localization in the cell membrane.

Drugs active on this class of receptors may target the orthosteric site (the binding site for the natural agonist, e.g., acetylcholine for the nicotinic receptor) or allosteric sites (e.g., benzodiazepines at $GABA_A$ receptors). Finally, some drugs may interfere with the biophysical and functional properties of the channel by binding to sites located in its lumen (e.g., hexamethonium for ganglionic nicotinic receptors or barbiturates for $GABA_A$ receptors).

This class of receptor also includes the transient receptor potential (TRP) channels. These are ligand-activated ion channels that respond to second messengers such as Ca^{2+}, IP_3, and anandamide (endogenous cannabinoid). Receptors for vanilloids and capsaicin (the irritant ingredient of hot peppers) are also LGICs; they are expressed mainly by sensory neurons of the trigeminal and dorsal root ganglia of the spinal cord where they act as identifiers of thermal and chemical noxious stimulus. Their molecular structure and (modest) voltage dependence make these receptors more similar to voltage-gated ion channels; for this reason, they will be described in Chapter 27.

GPCRs A common feature of this class of receptors is the generation of second messengers. Upon ligand binding, GPCRs transduce their signal by activating guanine nucleotide-binding proteins (G proteins), a family of heterotrimeric proteins that bind GTP and are endowed of intrinsic GTPase activity. This receptor family is the most abundant (about 800 members have been identified and their genes constitute approximately 2% of the genome) and is the target of several drugs used for therapy. Extracellular signals activating GPCRs include photons, neurotransmitters, odorants, nucleotides, amines, lipids, steroids, amino acids, peptides, glycoproteins, hormones, chemokines, and virus.

GPCRs are grouped in different families on the basis of molecular structure, ligand specificity, and G proteins they associate with. They are formed by a single polypeptide chain that crosses the plasma membrane seven times in correspondence of hydrophobic domains; for this reason, they are often called seven-transmembrane receptors (7TMRs) (Fig. 9.2). The receptor has the shape of a transmembrane globular particle with the ligand-binding site located in a hydrophilic pocket present in the extracellular or in the transmembrane portions of the molecule. The cytoplasmatic loop located between the transmembrane domains 5 and 6 is important for recognition and binding of specific G proteins; therefore, this site is responsible for the specificity of the effect triggered by receptor activation.

When an agonist binds a GPCR, the receptor undergoes a conformational change that increases its affinity for a specific G protein and determines its activation. The G protein is a trimer composed of α-, β-, and γ-subunits; activation involves binding of a GTP molecule to the α-subunit and dissociation of the trimeric complex. α-Subunit and β/γ complex modulate the activity of effectors that can be enzymes (e.g., adenylate cyclase and phospholipases) or ion channels. The active state persists until the α-subunit, which is endowed with GTPase activity, dephosphorylates GTP to GDP; then the trimeric inactive complex is reassembled. Effector activation by the G protein results in a transient increase in the intracellular concentration of the second messenger (Fig. 9.3).

The term second messenger identifies substances or ions (e.g., cAMP, cGMP, IP_3, and Ca^{2+}) whose concentration transiently varies in response to receptor activation. Second messengers mediate the conversion of signals, carried by extracellular first messengers and captured by receptors, into cellular response. Along the transfer, the message is amplified: in fact, a single G protein may activate multiple effector molecules, which in turn produce several molecules of second messengers, which in turn activate effectors (in most cases kinases) that modify the function of numerous substrates (Fig. 9.3). The duration of the effects induced by GPCR activation is independent of the duration of receptor–ligand interaction: indeed, it depends on the efficiency of the GTPase activity that stops G-protein activation, on the efficiency of the mechanisms that reestablish second messenger resting concentrations, and on the activity of the mechanisms that abolish posttranslational modifications of final effectors. The signal generated by brief stimulation of a GPCR may last up to several minutes.

GPCRs can form homodimers (in some cases homo-oligomers) or heterodimers (see Chapter 17). Increasing evidence indicates that the dimeric form is predominant; dimerization appears to be important during receptor synthesis, intracellular trafficking, and transfer and localization at the cell membrane. Dimers, in particular heterodimers, have pharmacological properties and signal transduction mechanisms different from those displayed by the monomeric form of the same receptor. They can also be tissue and disease specific; their presence explains some of the drug effects that are otherwise difficult to understand.

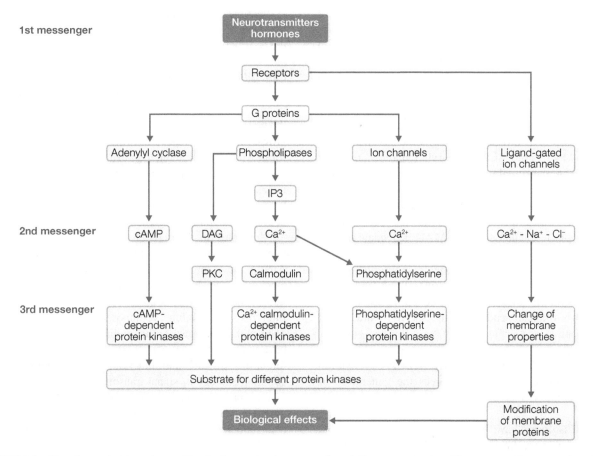

FIGURE 9.3 Signal transduction triggered by the interaction between a ligand (first messenger) and its receptor. Two examples of signal transduction pathways are shown: from GPCR (center) and from LGIC (right side). In the case of ligand–GPCR interaction, a G protein is activated (for simplicity, only one G protein is shown) that, in turn, can activate effectors (e.g., enzymes or ion channels). Effectors induce intracellular increase of a second messenger (e.g., cAMP, IP_3, and Ca^{2+}), which in turn can activate protein kinases leading to cellular events (e.g., contraction, secretion). Drugs are available for positive or negative modulation of every step of this cascade of signal transduction. In the case of LGIC, ligand binding results in opening of the receptor channel, ion flux (in or out), and rapid changes in intracellular ion concentrations. Depending on the ion involved, depolarization (cations) or hyperpolarization (Cl⁻) may ensue and trigger cellular responses (e.g., contraction, secretion, enzyme, and transporter activities). Ca^{2+} influx through LGIC may have little consequences on transmembrane potential but may activate important cellular consequences as Ca^{2+} is a second messenger.

Receptors with Intrinsic Kinase Activity Transmembrane receptor kinases play important roles in the control of survival, differentiation, and proliferation in diverse organs and cell types.

These receptors consist of a single polypeptide chain with an extracellular domain, containing the ligand-binding site, a single transmembrane region, and an intracellular domain with intrinsic kinase activity. Upon ligand binding, these receptors dimerize and this results in activation of their intrinsic kinase moiety that phosphorylates either tyrosine or serine/threonine residues. The best studied are the TKRs. Agonists of this large receptor superfamily are mostly growth factors (e.g., insulin, IGF-1, NGF, EGF, and FGF) and cytokines. The kinase, activated by receptor dimerization, selectively phosphorylates tyrosine residues present in the same receptor molecule (autophosphorylation) and in a number of other substrates (Fig. 9.2). Autophosphorylation leads to association with specific cytoplasmic proteins (adaptors) that, in most cases, contain Src homology 2 (SH2) domains. The signaling pathway downstream of the receptor depends on the type of adaptor proteins recruited and their particular location in the cell. Adaptors are also important in regulating TKR trafficking in and out of the membrane and their turnover. Many oncogenes are TKR-encoding genes carrying mutations that abolish the physiological regulation of receptor activity, resulting in uncontrolled cell proliferation. Tyrosine kinase inhibitors (small molecules or monoclonal antibodies) specific for different TKRs are innovative and often very effective therapeutic tools for the treatment of some cancers and their metastasis and for the regulation of immune and inflammatory responses.

An emerging group of receptors with intrinsic kinase activity phosphorylates serine/threonine residues. Agonists for these receptors are growth factors belonging to the TGF-β superfamily, including activin, bone morphogenic protein (BMP), and TGF-β. These proteins signal by contacting two distantly related transmembrane serine/threonine kinases called receptors I and II. The agonist binds directly to receptor II, which is a constitutively active kinase, and recruits receptor I into the complex, which is then phosphorylated by receptor II. Phosphorylation allows receptor I to propagate the signal to downstream signal transduction pathway controlled by small mother against decapentaplegic (SMAD) (from *Drosophila* SMAD) proteins. SMADs function as transcriptional comodulators; both SMAD translocations to the nucleus and SMAD-dependent regulation of target genes often require input from other signaling pathways. Thus, the transcriptional response to TGF-β family ligands is modulated by crosstalking with other signals received by the cell; indeed, the biological response to TGF-β is often dependent on the specific extracellular environment.

Receptors with Guanylate Cyclase Activity Activation of these membrane receptors induces the generation of the second messenger cyclic guanosine monophosphate (cGMP).

Several isoforms have been identified in mammals, but specific ligands are known for only few receptors. These include atrial natriuretic peptide (ANP), brain natriuretic peptide (BNP), intestinal natriuretic peptide (CNP), and an *E. coli* heat-stable toxin whose natural ligand is guanylin/uroguanylin.

The receptors are made of a single polypeptide chain with an extracellular domain, containing the hormone binding site, a hydrophobic transmembrane region, and an intracellular domain with intrinsic guanylate cyclase activity. The cytoplasmic domain also contains a protein kinase-like domain that controls receptor aggregation after activation, receptor efficiency, and ligand recognition. Upon receptor activation, cGMP intracellular concentration increases; the second messenger cGMP regulates the function of many end effectors (mainly kinases). Peptides and their receptors are poorly expressed but appear to play important roles in the cardiovascular and intestinal systems. Indeed, ANP gene knockout mice show hypertension resistant to sodium depletion, cardiac hypertrophy, changes in the extracellular matrix, and cardiac fibrosis. High levels of serum BNP are associated with increased risk of heart disease; CNP receptor is involved in secretion of intestinal fluids and in communication between the intestine and kidney in the control of sodium balance; CNP receptor gene knockout protects animals from diarrhea caused by *E. coli* enterotoxin.

Pharmacological modulation of these receptors is expected to have an impact in the regulation of blood pressure and in the control of diarrhea caused by bacterial toxins.

Cell Adhesion Receptors These are receptors mediating cell-to-cell and cell-to-microenvironment interactions. They transduce signals originating from adjacent cells and from the matrix in a series of intracellular events that modulate cell growth, motility, shape, and differentiation: they are crucial for instructing cells about their exact position in the organism both in physiological and pathological conditions (e.g., metastasis, cardiovascular diseases, tissue regeneration) and help delimiting specific membrane domains as in the case of neuronal and immunological synapses.

Molecules mediating cell adhesion are grouped in a number of families with different structures and ligands. Among these, integrins, cadherins, and selectins act both as mechanical anchorages between cells and extracellular matrix and as signal transducers after ligand interaction. Integrins are heterodimers composed of various combinations of α- and β-subunits; the heterodimers are cell specific and have individual selectivity/affinity for ligands. Although different in structure and function, these receptors share a common strategy of signal transduction based on formation of multiprotein complexes (in which cytoskeletal elements are often involved) that recruit transducers used also by other cell signaling pathways. For example, both integrins and CAMs activate tyrosine kinases (as the focal adhesion-associated kinase FAK); CAM signaling may involve G-protein-associated pathways, and integrins may activate MAP kinase cascade. In addition, these receptors often cooperate with other receptors and ion channels in a complex and integrated processing of extracellular signals.

Many drugs, now in clinical trial, target different integrins; examples are monoclonal antibodies, peptides, and small molecules tested as anticancer agents or immunomodulators.

Cytokine Receptors This is a large family of receptors controlling function of immune and hematopoietic cells. Cytokines are pleiotropic regulatory factors that control different functions, primarily in the immune and hematopoietic systems. Their receptors belong to families different for structure and signal transduction pathways that can be schematically classified in type 1 and type 2. Type 1 receptors bind many hematopoietic growth factors, GH, prolactin, and a number of interleukins; type 2 receptors bind interferons, interleukin-10 and interleukin-22, and blood coagulation factor VII.

Type 1 receptors are heterogeneous in terms of molecular structure; they are composed of two or more subunits and are divided into subgroups according to the presence of a common subunit responsible for signal transduction. For example, receptors for IL-3, IL-5, and CM-CSF, which belong to the interleukin-3 receptor family, each have a subunit capable of recognizing its own ligand, associated with a common subunit γc; members of the interleukin-6 receptor family have GP130 as the common subunit; members of the interleukin-2 family have IL-2Rα as common subunit. Receptor activation, initiated by binding of a cytokine, induces homologous or heterologous receptor oligomerization, followed by activation of JAK that in turn activate the transcription factor signal transducer and activator of transcription (STAT). At the same time, adaptors are recruited and can activate the MAP kinase and PI3 kinase pathways. The final effect of a given cytokine depends on types of receptor recruited and signaling pathways activated. Therefore, the response can vary depending on cell type and functional status: for example, the same interleukin may be involved in the generation of either immunological response or tolerance.

Type 2 receptors are formed by two similar but distinct transmembrane proteins containing, in their extracellular region, three fibronectin-like domains and two immunoglobulin-like domains that form the ligand recognition site. Receptor dimerization occurs upon ligand binding and activates the JAK pathway. A peculiarity of this receptor family is the production (by alternative splicing or proteolysis) of truncated receptors that are either secreted in the extracellular spaces or membrane bound. These truncated receptors bind the ligand but do not transduce the signal and therefore compete with signal-competent receptors for ligand binding, leading to fine-tuning of the cytokine effect.

The pharmacology of the cytokine system is very recent and has revolutionized the treatment of many autoimmune and inflammatory diseases. It is mostly based on specific monoclonal antibodies against different cytokines or their receptors, on soluble receptors chelating cytokines, and on small molecules directed against components of the signal transduction pathways (proteins and peptides have been identified that selectively inhibit JAK).

Receptors for Tumor Necrosis Factor or Death Receptors To this family belong the receptor for tumor necrosis factor alpha (TNFα) and the p75 neurotrophin receptor. TNFα is the main mediator of apoptosis and is implicated in the control of inflammation and immunity and in the pathogenesis of several chronic degenerative diseases. The p75 receptor for neurotrophins has a main role in inducing apoptosis.

These receptors are monomeric, with an extracellular region of about 40 amino acids containing one or two cysteine-rich domains, a transmembrane region, and a cytoplasmic domain called death domain (DD). Ligand binding induces receptor trimerization, followed by recruitment of specific adaptor proteins and assembly of a multiprotein signaling complex. Depending on the adaptor recruited, these receptors induce different effects. Two receptor types are known: TNFR1 recruits TNFR-associated DD (TRADD) that eventually activates NF-κB and mediates proinflammatory and immunostimulant activities. TNFR2 recruits Fas-associated DD (FADD) leading to activation of caspase-8 and caspase-10, thereby triggering the apoptotic program. Mutual interactions occur between the signaling cascades activated by the two receptors and the final effect results from the balance between the two pathways. Truncated forms of these receptors, secreted or membrane bound, do not activate signal transduction pathways but compete for ligand binding, thus finely modulating TNF effect (see Section "Decoy Receptors" and Chapters 19 and 21).

Receptor antagonists (mostly monoclonal antibodies) act as decoy receptors or as antagonists of membrane receptors; they have contributed to substantially modify the therapeutic outcome of several severe inflammatory diseases.

Toll-Like Receptors These receptors are involved in microorganism recognition and regulation of the inflammatory response. These receptors are a large family of membrane receptors present in all species, from insects to humans. Initially identified as regulators of embryonic development in *Drosophila*, they were later found in most species. Their name derives from the German "toll" (great!, important!), an exclamation pronounced by scientists the first time they saw a *Drosophila* mutant in the *Toll* gene. These receptors recognize molecular complexes and much conserved motifs present in prokaryotic and eukaryotic cells, and their activation leads to production of cytokines and chemokines. They are expressed by cells involved in the immune response to microorganisms, including macrophages, antigen-presenting cells, neutrophils, endothelial cells, cells of the digestive and pulmonary systems, and B and T lymphocytes.

Toll-like receptors consist of a peptide chain with a single transmembrane region, an extracellular region containing several leucine-rich motives and a cytoplasmic region homologous to the corresponding region of the IL-1 receptor. Receptor activation leads to NF-κB activation mediated by an adaptor protein, MyoD88 (the same used by the IL-1 receptor), IRAQ kinase, and TRAF6. They can also activate AP1 through a MAP kinase cascade or induce apoptosis by recruiting FADD and caspase-8. The signaling pathway depends on the type of receptor present on the cell surface.

At least 11 members of this receptor family have been identified in man, each specifically binding different ligands (e.g., TRL1 and TLR2 bind bacterial lipoproteins peptidoglycans and fibronectin; TLR4 binds LPS and gram-negative bacteria, the fusion protein of respiratory syncytial virus, and Taxol; TLR5 binds bacterial flagellin; TRL7 and TRL8 bind single-stranded RNAs; TLR9 binds unmethylated bacterial DNA). Type 3, 7, 8, and 9 receptors are located in endosomes, and the others on the plasma membrane. The receptors can form heterodimers with different affinity and selectivity toward ligands; in addition, their expression and their efficiency and specificity of ligand recognition are modulated by other cellular membrane proteins. All these properties contribute to the great plasticity of the TLR response.

A number of receptor agonists are already in clinical trials as adjuvants in vaccines, in viral and bacterial infections, and in cancer therapy. The major concern is that overstimulation of the immune system may lead to development of autoimmune diseases. Antagonists for receptors 4 and 7–9 are in clinical development for the therapy of autoimmune diseases with preliminary positive results.

Lipoproteins Receptors These receptors recognize plasma lipoproteins and many other ligands. They form a small family of 14 members (LDRR, VLDLR, and LRP 1–12) divided into four groups with slightly different structures. Their functions have not been fully elucidated, but the most important is related to regulation of endocytosis. The classical ligands are plasma lipoproteins and other 30 ligands with very different activities (e.g., TPA; blood coagulation factors 7, 8, and 9; complement; thrombospondin; antitrypsin; lactoferrin; and rhinoviruses). LRP 5 and LRP 6 are also coreceptors for Wnt ligands that stimulate signal transduction via Wnt-β-catenin. These receptors have a very large extracellular region containing LDL ligand-binding repeats, EGF domains, and other not well-characterized domains. The cytoplasmic portion contains motives that regulate internalization and other domains that interact with several adaptor proteins.

The most investigated function of these receptors is linked to lipid metabolism: they are involved in lipoprotein internalization, thus contributing to their removal from plasma and their transport into cells (see Chapter 35). In addition, they also have a classical receptor function activated by ligand binding. For example, binding of plasminogen activator and other serine proteases and metalloproteases induces a series of responses such as increase in intracellular Ca^{+2}, IP_3, and activation of ERK. Some members of this family of receptors have a role in modulation of synaptic transmission and development of the nervous system, probably mediated by the ability to modulate the NMDA glutamate receptor via binding to the PSD95 protein.

Decoy Receptors Some membrane receptors, typically those that bind cytokines and growth factors, can also exist in a truncated form lacking the intracellular domain; thus, they are able to bind ligands but unable to transduce an extracellular signal. They may be localized in the cell membrane and/or be secreted in the extracellular space and compete with their respective integral receptors for binding the same ligands.

Soluble receptors, or molecules that mimic their function, have been exploited as drugs. They have been studied in details in humans for the regulation of the immune and inflammatory systems, but other functions may be modulated by decoy receptors. For example, one of the oldest form of decoy receptor is the acetylcholine-binding protein (ABP) that is secreted in the cholinergic synapses of some snails and helps to modulate cholinergic synapses (see Chapters 16 and 39).

CONTROL OF RECEPTOR LOCALIZATION IN THE CELL MEMBRANE

The control of receptor number and localization in the cell membrane is essential for the correct and cell-specific response to a ligand. Examples of the relevance of this control are in the function of the neuronal and the immunological synapses. The "quality control" of cellular responses is assured by the presence of mechanisms that control localization and subcellular distribution of receptors and other molecules important for signal transduction. These mechanisms are necessary to ensure in every physiological condition the most appropriate receptor number and their correct localization especially in complex structures; consider, for example, the relevance of these controls in neuronal and immunological synapses (see Chapters 20 and 37).

In the neuromuscular junction, the nicotinic cholinergic receptor is localized at high density (10,000 molecules/$micron^2$) at the top of the postsynaptic membrane folds and at low density in the remaining adjacent segments. This particular location is strategic, as it places the receptor very close to the sites where presynaptic release of the neurotransmitter acetylcholine occurs. The high receptor density on the postsynaptic membrane (associated with the presence of similar mechanisms controlling the localization of voltage-dependent channels) also ensures that the ionic charges flowing through the activated receptors can reach high concentrations and generate a postsynaptic potential that propagate along the muscle plasma membrane. A similar control of receptor position also exists at CNS synapses. In other cells and for other neurotransmission systems, receptors are diffused and placed along dendrites and cell body. In such position, they can be activated by neurotransmitters or neuropeptides that, after release by nerve endings, diffuse in the intracellular space even for long distances (volume transmission).

Similarly, receptors activated by antigens are restricted to the immunological synapse, whereas receptors activated by growth factors or cytokines are more widely diffused throughout the cell membrane.

The control of receptor localization is also important for its interaction with other proteins essential for signal transduction. In many cases, the flow of information generated by receptor activation requires the assembly of a supramolecular complex containing receptor, adaptor proteins, signal transducers, and effector proteins. As an example of this supramolecular complex, consider the glutamatergic synapse where two types of LGICs are present: the AMPA and the NMDA receptors (see Chapter 43). Activation of AMPA receptor is needed for NMDA receptor activation. It has been observed that both receptors interact with proteins present in the postsynaptic matrix (PSD95–SAP90, Shank and Homer) and enriched in PDZ domains (domains of about 90 amino acids mediating protein–protein interactions). These proteins form a multiprotein network that links together transmembrane receptors, adaptor, and effector proteins for signal transduction and cytoskeletal elements. A similar situation occurs in GABAergic synapses, where the receptors for GABA and glycine are tied together by a hexagonal mesh made of gephyrin, a protein that mediates the link between receptors and cytoskeletal proteins. The supramolecular complexes and cytoskeleton represents the biochemical basis of the peculiar electron-dense aspect of postsynaptic membranes of CNS synapses, the so-called postsynaptic density.

To guarantee efficiency of the assembly mechanism, the different proteins of the receptor complex must be very close to each other, embedded in a favorable microenvironment that prevents their dispersion by diffusion. Lipid rafts represent a special environment where the assembly occurs; these are cell membrane areas enriched in cholesterol and selected proteins that can form, move, and disperse under the control of dynamic links with the cytoskeleton.

The link between receptors and cytoskeletal matrix is dynamic and can vary depending on cell function, especially along nerve cell plasma membrane and in pre- or postsynaptic sites. This has been shown by experimentally dissociating these macromolecular complexes via (i) intracellular injection of soluble synthetic peptides containing PDZ domains that displace receptors and signaling proteins from the postsynaptic matrix, (ii) antisense oligonucleotides that inhibit the synthesis of proteins containing PDZ domains, and (iii) compounds, such as the methyl-cyclodextrin, which destroys lipid rafts depleting them of cholesterol (see Chapter 9 and Supplements E 9.2, "Regulation of Receptor Response," and E 9.3, "Intracellular Trafficking of Receptors," for detailed discussion).

INTRACELLULAR TRAFFIC OF CELL RECEPTORS

How Receptors Reach the Cell Membrane and how Their Number is Regulated

Regardless of the nature of their ligand (e.g., neurotransmitter, cytokine, hormone) and their functional significance, membrane receptors are proteins synthesized in the endoplasmic reticulum (ER) in the cell body (or in the subsynaptic areas in neurons). Receptors made of subunits are assembled in the ER and then transported to the Golgi apparatus where they undergo posttranslational modifications. From there, mature receptor molecules are delivered to the plasma membrane and inserted into proper sites (synthesis of intracellular receptors is controlled by different pathways). The quality control of the process occurs before cell membrane insertion, so that only properly assembled receptors are transported to the cell surface. During synthesis, assembly, and transport, the receptor is normally protected from degradation by interaction with one or more chaperones (usually members of the hsp family). Receptors are synthesized in excess and those supernumerary (and not correctly assembled) are degraded by lysosomes or by the ubiquitin–proteasome system. This level of control is very complex and regulated by a number of biochemical reactions that can be enhanced or inhibited by physiological or pathological events or by drugs. For example, nicotine increases the number of neuronal nicotinic receptors present on the cell surface by facilitating the process of receptor assembly and maturation.

MODULATION OF RECEPTOR RESPONSES

The ability of cells to respond to external stimuli is modulated by intensity and duration of the stimulus. The strategies implemented to exert this control are variable and include mechanisms of receptor synthesis, transport and removal from the plasma membrane (control of receptor number), and modulation of the capacity of receptors and effectors to respond to the mediator (control of efficacy). Alterations in these controls underlie many pathological conditions. Examples are (i) myasthenia gravis (reduction of the number of acetylcholine receptor molecules in postsynaptic membrane of neuromuscular junction), (ii) testicular feminization syndrome (lack of androgen receptors due to genetic defects), and (iii) some forms of insulin-resistant diabetes (reduction in number of insulin receptors). In other cases, the alteration of receptors number is secondary to alterations of the mechanisms regulating their expression and is responsible, at least in part, of the disease symptoms (e.g., some of the cardiovascular symptoms of hyperthyroidism are due to increased expression of β-adrenergic receptors whose gene expression is positively controlled by thyroid hormones). Pathologies causing prolonged stimulation of a receptor signaling cascade can lead to changes in receptor number and/or in their ability to transduce the signal. A classical example is the adrenergic system in cardiac pathologies where a prolonged activation of the sympathetic system, such as in heart failure, induces at the same time an increase in number of β-adrenergic receptors in myocardial cells and a reduction of their ability to transduce the signal. Finally, a well-known phenomenon is the denervation-induced hypersensitivity to the neurotransmitter occurring at the neuromuscular junction. The hypersensitivity is due to increased expression of acetylcholine receptors as consequence of lack of receptor stimulation; in fact, it can be induced by chronic inhibition of acetylcholine release (e.g., poisoning by botulinum toxin) or chronic receptor blockade (with neuromuscular blocking agents or snake toxins). Denervation hypersensitivity is typically observed in cells forming synaptic contacts and, to a lesser extent, in other excitable tissues, such as smooth muscle tissues and glands. The functional significance of these adaptation phenomena is clear: in the absence of a signal, the number of receptors is increased to amplify cell sensitivity, whereas overstimulation is counteracted through a reduction of cell capacity to respond to the signal.

RECEPTOR MODULATION BY DRUGS

In analogy with the endogenous examples shown earlier, receptor response can be modulated by duration and intensity of exposure to agonists or antagonists. Receptor adaptation to pharmacological stimulation is bidirectional; chronic treatment with agonists generally leads to reduction of responses, whereas chronic treatment with receptor antagonists can induce an increased response.

Desensitization and Downregulation Overstimulation by an agonist can lead to desensitization and/or downregulation. Desensitization is the reduction in cell response to an agonist due to decreased ability of its receptor to transduce the signal. Downregulation is the reduction in cell response due to a decrease in number of functional receptors. These phenomena underlie drug tolerance, best described for psychoactive drugs (see Chapter 10).

Loss of the ability to respond to an agonist can be restricted to the specific receptor type it binds (homologous desensitization). However, prolonged activation of a receptor system may also induce cross-desensitization of other receptors sharing the same signal transduction pathway or the same effectors (heterologous desensitization). Heterologous desensitization is relatively common between GPCRs; it can be due to modifications of the receptor molecule (e.g., phosphorylation) or to modulation of G proteins and effector systems shared by different receptors (e.g., cyclases, phospholipases, ion channels). In theory, desensitization can occur at any level (e.g., reduced affinity for the ligand and inefficiency in transducing the signal). The biochemical basis and the extent of the desensitization can be unique for each receptor even inside the same cell; in addition, the desensitization affecting a given receptor may vary between different tissues. Despite this, it is possible to identify common mechanisms and strategies of desensitization within some of the largest receptor superfamilies.

Desensitization of LGICs primarily occurs by loss of ability to regulate the kinetics of ion channel opening. For these receptors, as well as for other types of ion channel, desensitization is an intrinsic functional property connected to specific portions of the amino acid sequence. The rate of desensitization can be modulated by phosphorylation of the receptor. In some cases, the phosphorylating kinase is activated by stimulation of other receptors; for example, the rate of desensitization of the muscle nicotinic acetylcholine receptor is increased by PKA-dependent phosphorylation, and PKA can be activated by the receptor for the neuropeptide CGRP coreleased by the same acetylcholine-secreting nerve terminal.

GPCRs are modulated by modifications affecting both affinity (reduced ability to bind the agonist) and efficiency in signal transduction (reduced ability to couple with G protein). These two events (which usually start rather quickly) are often followed by an increase in the rate of receptor removal from the plasma membrane (internalization), which leads to a significant reduction in the number of receptors expressed on the cell surface (downregulation). Thus, for most GPCRs, desensitization is associated and followed by internalization and downregulation. Internalized receptors can either remain inside the cell (as reservoir) and recycle to the cell surface, or be degraded. These processes are controlled via receptor phosphorylation by specific kinases (the β-adrenergic receptor kinase, BARK, is the first and best studied example) and by assembly of specific macromolecular complexes (involving arrestins) that mediate both receptor internalization and G-protein uncoupling.

In growth factor receptors (TKRs), internalization and receptor downregulation are part of the intrinsic mechanism of receptor activation. (See Supplement E 9.2 for a detailed description of the receptor desensitization, internalization, and recycling processes.)

Receptor Upregulation Chronic blockade of a receptor usually increases cell sensitivity to an agonist via mechanisms of receptor upregulation. In most cases, upregulation is associated with an increase in the number of receptors present on the cell surface mostly as a result of an increased expression of the receptor gene. These adaptation phenomena have clinical significance: the increased receptor number induced by chronic treatment with an antagonist results in tissue hypersensitivity when the treatment is discontinued, and in this condition, the natural agonist may induce extreme responses. Since the recovery to the original state (readaptation) takes time, a gradual discontinuation is usually recommended at the end of a chronic treatment with an antagonist. For analogy, a readaptation period is often required at the end of a chronic treatment with an agonist (see Supplement E 9.2 for more details).

TAKE-HOME MESSAGE

- Communications between cells involve ligand–receptor interaction. In most but not all cases, the ligand is a diffusible compound and activation of the receptor generates a cellular response.
- According to the subcellular localization, receptors can be classified as intracellular or plasma membrane.
- Receptors for endogenous messengers can be grouped according to their molecular structure; members of a same receptor family share common strategies of signal transduction.
- Receptor responses are modulated in space and time.
- Most drugs act by binding to orthosteric sites of one or more receptors stabilizing them in active, inactive, or desensitized states or to allosteric sites modifying the receptor response to ligands.
- Drugs may modulate receptor responses by interfering with receptor activation and intracellular signal transduction pathways or by modifying traffic, localization, and number of receptors at the cell membrane.

FURTHER READING

General

Alexander S. P. et al. (2013). The Concise Guide to Pharmacology 2013/14: Overview. *British Journal of Pharmacology*, 170, 1449–1458.

Groves J.T., Kuriyan J. (2010). Molecular mechanisms in signal transduction at the membrane. *Nature Structural and Molecular Biology*, 17, 659–665.

Haugh J.M. (2002). Localisation of receptor-mediated signal transduction pathways: the inside story. *Molecular Interventions*, 2, 293–307.

Rosenfelt R., Devi L.A. (2010). Receptor heteromerization and drug discovery. *Trends in Pharmacological Sciences*, 31, 124–130.

Intracellular Receptors

Gemain P., Staels B., Dacquet C., Spedding M., Laudet V. (2006). Overview of nomenclature of nuclear receptors. *Pharmacological Reviews*, 58, 685–704.

Mor A., White M.A., Fontoura B.M. (2014). Nuclear trafficking in health and disease. *Current Opinion in Cell Biology*, 28C, 28–35. doi: 10.1016/j.ceb.2014.01.007.

The British Pharmacological Society (2009). Nuclear receptors. *British Journal of Pharmacology*, 158, S157–S167.

Ligand-gated Ion Channels

Calimet N., Simoes M., Changeux J.P., Karplus M., Taly A., Cecchini M. (2013). A gating mechanism of pentameric ligand-gated ion channels. *Proceeding of the National Academy of Sciences USA*, 110, 3987–3996.

Changeaux J.P., Edelstein S.J. (1998). Allosteric receptors after 30 years. *Neuron*, 27, 959–980.

Lodge D. (2009). The history of the pharmacology and cloning of ionotropic glutamate receptors and the development of idiosyncratic nomenclature. *Neuropharmacology*, 56, 6–21.

Millar N.S., Gotti C. (2009). Diversity of vertebrate nicotinic acetylcholine receptors. *Neuropharmacology*, 56, 237–246.

The British Pharmacological Society (2009). LGIC. *British Journal of Pharmacology*, 158, S103–S121. doi: 10.1111/j.1476-5381.2009.00502.

G Protein-Coupled Receptors

Lagerstrom M.C., Schioth H.B. (2008). Structural diversity of G-protein-coupled receptors and significance for drug discovery. *Nature Reviews in Drug Discovery*, 7, 339–357.

Pierce K.L., Premont R.T., Lefkowitz R.J. (2002). Seven-transmembrane receptors. *Nature Reviews Molecular Cellular Biology*, 3, 639–649.

The British Pharmacological Society (2009). 7TM receptors. *British Journal of Pharmacology*, 158, S5–S101.

Catalytic Receptors

Garbers D.L., Chrisman T.D., Wiegn P., Katafuchi T., et al. (2006). Membrane guanylyl cyclase receptors: an update. *Trends in Endocrinology and Metabolism*, 17, 251–258.

Lemmon M.A., Schlessinger J. (2010). Cell signaling by receptor tyrosine kinases. *Cell*, 141, 1118–1134.

Massacesi C., Di Tomaso E., Fretault N., Hirawat S. (2013). Challenges in the clinical development of PI3K inhibitors. *Annals of the New York Academy of Sciences*, 1280, 19–23.

The British Pharmacological Society (2009). Catalytic receptors. *British Journal of Pharmacology*, 158, S169–S181.

Chemokine Receptors

Anders H.J., Romagnani P., Mantovani A. (2014). Pathomechanisms: homeostatic chemokines in health, tissue regeneration, and progressive diseases. *Trends in Molecular Medicine*, 20, 154–165.

Lazzennec G., Richmond A. (2010). Chemokines and chemokine receptors: new insights into cancer-related inflammation. *Trends in Molecular Medicine*, 16, 133–143

Schwarz M., Wells T. (2002). New therapeutics that modulate chemokine networks. *Nature Reviews Drug Discovery*, 1, 347–358.

Zweemer A.J., Toraskar J., Heitman L.H., Ijzerman A.P. (2014). Bias in chemokine receptor signalling. *Trends in Immunology*, 35, 243–252. 2014 S1471-4906(14)00029-5. doi: 10.1016/j.it.2014.02.004.

Toll-Like Receptors

Lester S.N., Li K. (2014). Toll-like receptors in antiviral innate immunity. *Journal of Molecular Biology*, 426, 1246–1264.

Parkinson T. (2008). The future of toll-like receptors therapeutics. *Current Opinion in Molecular Therapeutics*, 10, 21–31.

Zuany-Amorim C., Hastewell J., Walker C. (2002). Toll-like receptors as potential targets for multiple diseases. *Nature Reviews in Drug Discovery*, 1, 797–807.

Death Receptors

Dekkers M.P., Nikoletopoulou V., Barde Y.A. (2013). Cell biology in neuroscience. Death of developing neurons: new insights and implications for connectivity, *Journal of Cell Biology*, 203, 385–393.

Wilson N.S., Dixit V., Ashkenazi A. (2009). Death receptor signal transducers: nodes of coordination in immune signaling networks. *Nature Immunology*, 10, 348–355.

10

ADAPTATION TO DRUG RESPONSE AND DRUG DEPENDENCE

Cristiano Chiamulera

> **By reading this chapter, you will:**
>
> - Understand the definition of adaptation to drug effect, in particular for those known to induce drug dependence
> - Know the molecular, cellular, and systemic mechanisms underlying drug dependence phenomena (e.g., tolerance) and symptoms (e.g., compulsive relapse)
> - Learn recent research findings that may lead to clinical advancements of dependence treatment
> - Know addiction-related psychological and clinical concepts in the context of major drugs of abuse and their mechanisms of action and pharmacotherapy

Adaptation is an essential feature of the biological world. It is the property that allows biological systems to change in response to a changing environment. Therefore, adaptation identifies the set of events that throughout life, including the prenatal period, allows body development and maintenance in the presence of changing conditions. This can occur thanks to molecular mechanisms capable of implementing compensatory responses at various levels. Adaptation has particular interest when results from exposure to abnormal or pathological conditions (such as chronic stress, infections, substances, toxins, or food). In such situations, a new dynamic equilibrium state, named allostasis (allos = different), is established. Allostasis can be defined as the process that allows the body to preserve its physiological stability by changing its internal parameters in accordance with the environmental requirements (in contrast with homeostasis that is the state in which all physiological parameters are kept within normal values).

Among all systems, the central nervous system (CNS) is the one showing the highest degree of adaptation. Neuroplasticity characterizes CNS development and phenomena such as learning, memory, resilience, responses to drugs, and substances. Neuroscience research has shown that most individual's experiences are based on neuroplasticity: learning how to comb hair in an automatic way, acquiring a new language, recognizing facial expressions, etc. The discovery of common mechanisms underlying neuroplasticity has thus expanded the concept of "memory" that is no longer restricted to the action of recalling experiences, names, and concepts but includes also learning information concerning the body (see Supplement E 10.1, "Types of Memory").

The repeated pharmacological stimulation of a receptor, and the consequent responses, can act as inducers of adaptation. A paradigmatic phenomenon of drug-induced adaptation is drug dependence. Drug dependence is the loss of control over the use of a substance, resulting in compulsive drug seeking and drug taking, in spite of the adverse long-term consequences, including high relapse risk. Therefore, to understand and treat drug dependence is important to know the molecular and cellular effects of drugs of abuse and how these effects induce persistent changes in the body.

This chapter addresses the phenomenon of adaptation to substances and in particular to drugs of abuse. We will describe the common molecular and cellular events underlying adaptation that have provided the basis for understanding drug dependence phenomena (e.g., tolerance) and symptoms (e.g., compulsive relapse). Recent advances in research will be outlined in relation to clinical treatment of drug dependence. The online supplements provide details on the fundamental concepts of psychological and clinical relevance and on the main drugs of abuse, their mechanisms of action, and drugs used in cessation therapy.

MOLECULAR, CELLULAR, AND SYSTEMIC ADAPTATION

In the CNS, adaptation is established and maintained at molecular and cellular level along with modifications of neural circuits and brain processes. All cells have structures apt to receive information from other cells and the surrounding extracellular space. A neuron receives stimuli from other neurons via neurochemical transmission (endogenous neurotransmitters) through synaptic connections or from exogenous substances coming from the bloodstream (such as drugs). Synaptic plasticity can occur in the presence of an anomalous stimulus such as (i) increased or decreased synaptic concentration of the ligand, (ii) increased or decreased permanence of the ligand in the synapsis, and (iii) increased frequency of stimulation. These anomalies cause chronic adaptive responses. Neuroadaptation is a result of an integration of different synaptic plasticities and can be described in two different forms: one, defined as transient neuroplasticity, implies neuronal and synaptic changes lasting from hours to weeks; the other, named stable neuroplasticity, may persist for weeks and even result in permanent changes. An analogy exists between neuroadaptation and memory. According to the modern definition, the memory we retain of a word heard several times reflects adaptive changes in specific neural circuits, which have been repeatedly exposed to the same stimulus (the word). Thus, the activity of remembering is underlain by a persistent modification of synaptic and neural functions. Similar structural/functional changes occur following repeated exposure to a medicine or a drug of abuse. Transient neuroplasticity corresponds to the series of dynamic processes accompanying the transition from the original state to a new state associated with the abnormal stimulus (in analogy with the learning process). Stable neuroplasticity corresponds to the situation in which a new allostatic state has been finally established (in analogy with the storage of consolidated memories). This modern definition of cellular memory, extended also to pharmacological stimulation, has allowed the identification of common molecular mechanisms underlying the various forms of neuroplasticity (see Supplement E 10.2, "A Modern Definition of Memory").

Cellular Adaptation

Neuroadaptive responses upon chronic exposure to drugs of abuse may involve changes in (i) receptor conformation and allosteric state, (ii) cascade of second messengers and intracellular calcium levels, (iii) protein kinases, and (iv) transcription factors and regulation of gene expression. Changes in nuclear functions and alterations of transcription factors controlling the expression of specific genes can perturb intracellular transduction pathways, leading to functional changes in neurons and circuits involved in some psychological and behavioral processes. A limited number of transcription factors appear to play a key role in mediating adaptation to drugs. In addition, changes can occur also at posttranslational level; indeed, drugs can induce alterations in mRNA translation, protein degradation, and receptor targeting within the cell.

The new allostatic status persists as long as the exposure to the drug continues and even afterward. In case of abrupt drug cessation and abstinence, when the stimulus that has caused the adaptive response suddenly disappears, the allostatic state may manifest as an imbalance and thus as a potential disorder.

Effects of Repeated Drug Exposure

Many therapeutic treatments require more than a single acute administration of drug. The reasons may be different and related to the drug pharmacokinetic and pharmacodynamic profile or to the time course of development of the pathology. From the pharmacokinetic point of view, a repeated (several administrations of a single dosage, e.g., more pills per day) or constant (protracted administration, e.g., slow-release formulations) treatment is needed to reach appropriate drug plasma levels and maintain them within the therapeutic range. The dosing rationale may also be based on the pharmacodynamic properties of the drug and their modulation over time.

It is important to define the concept of chronic treatment. We define "chronic" a drug administration that is protracted in time and does not require drug plasma levels to remain constant. Chronic effects of a therapeutic drug or a drug of abuse can also be observed upon repeated administrations occurring only once a day or even once a week. This is particularly relevant in drug addiction: for example, ecstasy or alcohol addiction can ensue even upon limited administration occurring only during weekend. While chronic exposure to drugs induce the adaptive modifications necessary for the onset of clinical efficacy, it can also induce a series of side or toxic effects, such as reduced or absent clinical response and occurrence of adverse events and toxicity.

Two phenomena are caused by chronic exposure to drugs: tolerance and sensitization (or reverse tolerance). Conversely, the phenomena associated with the cessation of chronic exposure are physical and psychological withdrawal syndromes and abrupt cessation.

Tolerance Tolerance is the reduction or loss of the biological response induced by the drug occurring after repeated administrations (except reductions due to exacerbation of the disease under treatment). This gradual loss of therapeutic effect could be restored by increasing the dosage.

However, this is not easily practicable as pharmacological tolerance does not necessarily involve all the effects induced by the drug. Thus, increasing the dose to counteract the tolerance may lead to drug plasma levels that increase other effects not subjected to tolerance, which may be unwanted, toxic, or even lethal.

Tolerance is often described as an equal and opposite adaptive response to drug effect. The first level at which tolerance manifests involves receptors and transduction signals, with implications at cellular and systemic level. For example, one of the biological mechanisms that underlie tolerance to opiates is the downregulation of opioid receptors. Other mechanisms include (i) reduced synaptic activity, (ii) reduced number of synaptic buttons, (iii) reduced neurotransmitter synthesis, and (iv) modified transport and localization of receptors.

Sensitization Sensitization is an increased biological response to a chronic drug administration. The conceptualization of this phenomenon is not intuitive, as one may expect the body to develop adaptive responses opposite to the drug action, rather than additive or synergistic responses. In fact, whereas tolerance is a "defensive" adaptive mechanism, sensitization enhances the biological response pushing it toward a form of hypersensitivity. A sensitized response occurs when a pharmacological response is attained with doses lower than those normally administered, thus doses that were ineffective before the repeated exposure to the drug. In recent years, sensitization has been proposed as one of the main features of the psychological addiction to drugs of abuse. Like tolerance, sensitization may involve changes at various levels, from receptors to gene expression, from morphological changes to various psychobehavioral responses.

Chronic treatment with amphetamine, cocaine, nicotine, and other substances of abuse induces an increase in neuronal processes (dendrites) and number of synapses in the brain of laboratory animals. It is important to point out that sensitization is a form of allostatic adaptation that can be "recalled" even after a long time. In fact, psychopharmacological studies have revealed that the neurochemical changes underlying sensitization can persist for a long time even after complete cessation of addiction. In fact, it is well known that ex-alcoholics should refrain from taking substances containing even small quantities of alcohol, for example, liqueur chocolates, in order to avoid euphoric effects that can lead to relapse. Studies have shown that this occurs because chronic exposure to ethanol increases the number of NMDA glutamate receptors and such upregulation mediates the ability of low concentrations of ethanol (such as those in the alcoholic chocolates) to stimulate alcohol seeking and taking behaviors.

Withdrawal For many classes of drugs, withdrawal syndromes due to interruption of the treatment have been described. These syndromes vary from class to class and are observed when a chronic treatment is abruptly discontinued. The molecular mechanisms underlying withdrawal symptoms are all attributable to the unmasking of an adaptation that is not manifest until the drug is present in the body. Symptoms can be both physical and psychological, with differences from drug to drug. In general, these symptoms appear as forms of adaptation that are evident as effects opposite to the effects induced by the treatment (e.g., excitation after prolonged sedative treatment) or identical to the symptoms under treatment (e.g., insomnia after prolonged treatment with benzodiazepine sleeping pills).

For example, abstinence from the hypnotic–sedative opioid heroin induces physical withdrawal symptoms characterized by generalized excitation. In contrast, cessation of psychostimulant cocaine induces a syndrome characterized by mood and affectivity disorders. Nicotine withdrawal induces both physical and psychological symptoms, with the main physical symptom being weight gain. Despite the difference in clinical severity between heroin and nicotine withdrawal symptoms (e.g., cramps and pain in the former but not in the latter), they can both lead to relapse in drug addiction (e.g., to reduce pain in heroin addicts and for weight control in smokers).

DRUG ADDICTION AS A PARADIGM OF ALLOSTATIC ADAPTATION

The different phenomena of adaptation of the organism to drug-induced biological response are the current paradigm of drug addiction. Drug addiction is a behavioral disorder due to chronic intake of drugs of abuse. The term of drugs of abuse applies to all substances (e.g., alcoholic beverages, opium) that contain one or more molecules able to induce dependence (respectively ethanol and morphine) (Table 10.1 and Box 10.1). A drug of abuse induces dependence if it owns reinforcing properties, meaning that it induces pleasurable effects that can increase the probability that its assumption is repeated. When a substance acts as a reinforcer, it means that it induces and maintains an operant conditioning. Operant conditioning is the learning process that ensures repetition of those behaviors that lead to attainment of natural pleasures, such as courtship, approach, search for food, and drinking. Drugs of abuse that act as reinforcers use the same physiological mechanisms underlying motivated behaviors. The difference lies in the fact that whereas natural reinforcers act on brain mechanisms in a physiological manner, drugs of abuse induce so powerful effects as to induce neuroadaptive changes (see Supplement E 10.3, "Conditioning as the Core of Addictive Behavior").

TABLE 10.1 Substance of abuse and drug classes

Drug class	Drugs and substances of abuse	Clinical effects
CNS depressants	Alcoholic beverages (ethanol)	Euphoric effects at low doses
		Motor incoordination, perception, and cognitive disturbances at higher doses.
		Hypnotic effects, respiratory depression, and coma at intoxicating doses
	Opiates, morphine, heroin, methadone	Sedation and hypnosis at therapeutic doses
		Respiratory depression and coma at intoxicating doses
Psychostimulants	Coca leaves (cocaine), marijuana (cannabinoids), tobacco (nicotine), amphetamine, methamphetamine	Psychomotostimulant effects
		Increase performance and euphoria
		Craving and loss of control for drug intake
Hallucinogens	Lysergic acid (LSD), mescaline, psilocybin, cactus, magic mushrooms	Altered perception and thoughts
		Risk of psychotic effects
	Phencyclidine, ketamine	Altered perception and behavior.
		"Out of the body" experience
		Risk of perception
Empathogens	Ecstasy (methylendioxy-amphetamine; MDMA) and derivatives	Facilitation of affective states and emotional responses
		Effects often followed by dysphoria and depressive state

Drugs are substances from the most varied origins. They can be extracts and/or parts of plants, fungi, beverages, and synthetic compounds. All drugs contain chemical molecules with a psychoactive action, that is, able to modify psychological and behavioral processes. The main classes of drugs of abuse are classified according to the effect at the psycho-behavioral level.

BOX 10.1 CLASSES OF DRUGS OF ABUSE

Drugs of abuse are substances of very different origin: they can be extracts and/or parts of plants, fungi, beverages, and synthetic compounds. All drugs contain chemical molecules with a psychoactive action, that is, able to modify psychological and behavioral processes.

Drugs of abuse are classified according to their effect:

Sedative Hypnotics

- Opiates (morphine and derivatives, heroin, methadone) exert sedative effects that become hypnotics at higher doses. At even higher doses, they induce coma and death by inhibition of the respiratory center, the brain area responsible for breathing control.
- Alcoholic beverages (ethanol) at low doses induce euphoric effects, while at high doses they cause disturbance of motor coordination, perception and cognitive function, sedation, and hypnotic actions. Intoxication can lead to coma and death.

Psychostimulants

- Cocaine, amphetamines, methamphetamine, nicotine, cannabinoids induce a range of effects that lead to psychological dependence: reduced appetite, motor stimulation with reduced fatigue, mental excitement with euphoria, strong desire and craving for the substance, general state of activation and increased attention to stimuli associated with the drug, inability to control behavior, obsessive thoughts, and difficulty in maintaining abstinence.

Hallucinogens

- Lysergic acid (LSD), mescaline, psilocybin, cacti, and hallucinogenic mushrooms. These drugs are characterized by actions that lead to enhanced or altered perceptions, sometimes psychotic symptoms. Cannabinoids can be considered hallucinogens too.
- Phencyclidine (known as "angel dust") and ketamine are dissociative anesthetic that induces hallucinations, altered perception, and psychosis.

Entactogens

- Ecstasy (methylenedioxymethamphetamine (MDMA)) and derivatives are defined empathetic because they facilitate empathy and emotional openness at interpersonal level. Even if the use is not daily but generally limited to the weekend, the acute action is followed by a latent phase of depressive nature, which, depending on the subject, can last for days and is often the reason that leads to continued use of ecstasy.

Diagnostic criteria for drug dependence are (i) presence of a substance with reinforcing properties in the drug, (ii) establishment of drug-taking behavior, and (iii) occurrence of withdrawal symptoms. These main criteria define the general framework of psychological dependence on a drug of abuse. The addict is psychologically dependent on the substance as shown by compulsive drug-taking behavior. The addict may also need to take the drug to counteract the physical adverse effects due to abstinence; this is only observed in some forms of addiction, such as those to opiates and alcohol, and therefore is not a sufficient criterion. Another criterion used in the past to define drug dependence was the phenomenon of tolerance. However, not all addictive substances induce tolerance. Since the end of the 1980s, the definition of addiction has been widening, to include phenomena and symptoms such as craving and cue reactivity (i.e., hypersensitivity to environmental stimuli previously associated with the drug). The inclusion of these phenomena has expanded the definition of drug addiction to psychological dimensions such as cognition and affective processes. Drug addiction is not only defined as the cycle of "repeated administration of the substance to experience pleasurable effects" but involves also loss of conscious control and memory of these effects, even after long-term cessation, with a high risk of relapse. Thus, relapse prevention is the main therapeutic target of cessation. It is therefore important to understand mechanisms, processes, and factors that can lead to relapse even when substance assumption has been interrupted for a long time (Table 10.2).

Adaptation and Stages of Drug Addiction

All addictive drugs induce psychological dependence and contain molecules that have reinforcing properties. Despite these molecules own chemical diversity and act at different receptor targets, a common mechanism of action underlies their ability to act as reinforcers. A neuroanatomical pathway has been identified where all reinforcing drugs act by increasing the release of the neurotransmitter dopamine. Dopaminergic neurons in the area of the midbrain called the ventral tegmental area (VTA) project their axons to the ventral striatum in an area called the nucleus accumbens (NAc) and to the prefrontal cortex (Fig. 10.1). An increase in the frequency of action potentials in the VTA neurons leads to dopamine release in these target brain areas. This neuroanatomical pathway is the mesocorticolimbic dopaminergic pathway (Box 10.2).

The mesocorticolimbic dopaminergic pathway has a well-defined evolutionary role. It regulates behavioral responses to natural reinforcers (food, water, sex) in response to stimuli that directly act on the body as the primary reinforcers or indirectly as the myriad of stimuli associated with primary reinforcers that signal their availability (secondary reinforcers). Dopamine release does not mediate physical pleasure, but signals the pleasurable value of a given stimulus thus facilitating the reinforcement learning and regulating the motivated behavior. In this way, dopamine couples motivational relevance ("salience") to an enjoyable event, allowing learning to recognize an important stimulus for survival behavior (eating, drinking, mating, etc.).

TABLE 10.2 Definition of clinical phenomena and symptoms in drug addiction

Clinical phenomena and symptoms	Definition
Abuse	The recurrent use of a substance, for non-medicinal purposes, with resulting clinical impairment or distress
Compulsive relapse	The uncontrolled occurrence of substance-taking behavior
Craving	A subjective state of desire for substance effects
Cue reactivity	The pattern of physiological, psychological (e.g., the craving), and behavioral (e.g., relapse) responses that the former addict may present when exposed to stimuli and situations previously associated with substance use
Drug addiction	The complex pattern of phenomena and symptoms that are associated with the abuse of a substance
Drug of abuse	A substance with psychoactive properties with abuse liability
Physical dependence	The physical adaptation to substance use/abuse manifested by somatic withdrawal symptoms
Psychological dependence	The psycho-behavioral adaptation to substance use/abuse manifested primarily by compulsive behavior, difficulty in quitting, and difficulty to remain abstinent
Recreational or social use	A controlled pattern of substance use under certain situations
Reinforcing properties	The ability to induce a pleasant pharmacological effect, and therefore to increase the probability to repeat the use of the substance itself
Withdrawal syndrome	The pattern of signs and symptoms that occur at the abrupt cessation of substance use

For more detailed definitions, please refer to the revised chapter of "Substance-Related and Addictive Disorders" in the *Diagnostic and Statistical Manual of Mental Disorders* (DSM-5), American Association of Psychiatry, 2013.

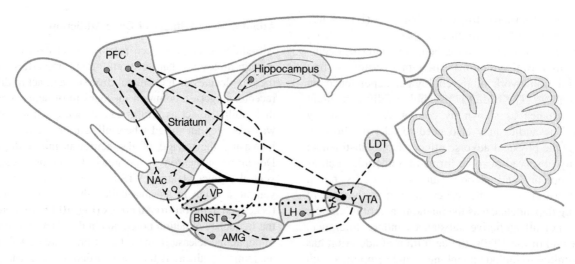

FIGURE 10.1 Mesocorticolimbic dopaminergic pathways. Schematic diagram of a sagittal section of rat brain with (solid lines) axonal projections from dopaminergic neuronal bodies in the ventral tegmental area (VTA) to terminals in the nucleus accumbens (NAc) and in the prefrontal cortex (PFC). Other brain areas involved are the amygdala (AMG), hippocampus, ventral pallidum (VP), bed nucleus of the stria terminalis (BNST), and lateral hypothalamus (LH). Excitatory glutamatergic (dashed lines) and GABAergic (dotted lines) pathways are also shown. From Kauer J.A., Malenka R.C. (2007). Synaptic plasticity and addiction. *Nature Reviews Neuroscience*, 8, 844–858.

BOX 10.2   MECHANISMS OF ACTION OF DRUGS WITH REINFORCING PROPERTIES

This box summarizes the neurochemical effects induced by the most common drugs of abuse. All these substances have in common the ability to increase dopamine levels in the mesocorticolimbic pathway, specifically in the nucleus accumbens and prefrontal cortex (see text). However, different drugs of abuse exert this stimulatory effect on the dopaminergic system via different mechanisms:

- Opiates: agonists at endogenous opioid receptors associated with G proteins, respectively referred to as μ, δ, and κ, induce inhibition of GABA interneurons in the VTA, thus disinhibiting VTA dopaminergic neurons. They also directly stimulate μ receptors on neurons of the nucleus accumbens stimulated by dopamine.
- Cocaine: inhibits dopamine reuptake by acting as a ligand on presynaptic dopamine transporter and increases synaptic levels of dopamine in the nucleus accumbens and prefrontal cortex.
- Amphetamines: stimulate dopamine release and increase synaptic levels of dopamine in the nucleus accumbens and prefrontal cortex.
- Ethanol: positive modulator of GABA receptor, induces inhibition of GABA interneurons in VTA and disinhibits dopaminergic neurons in VTA; exerts inhibitory modulation of glutamate NMDA receptor thereby reducing glutamatergic inhibition in the nucleus accumbens.
- Nicotine: agonist at nicotinic acetylcholine receptors expressed in the mesocorticolimbic system and in other brain areas; stimulation of nicotinic receptors on dopaminergic neurons in VTA enhances dopamine release; nicotinic receptors present on glutamatergic terminals favor glutamatergic activation in VTA.
- Cannabinoids: agonists at CB1 and CB2 cannabinoid receptors; activate CB1 receptors on neurons of the nucleus accumbens stimulated by dopamine.
- Phencyclidine and ketamine: antagonists at glutamate NMDA receptors thereby reducing glutamatergic inhibition in the nucleus accumbens.

Drugs of abuse induce effects similar to natural rewards. The major abnormal effect of reinforcing drugs is the way they stimulate dopamine release. Natural stimuli with motivational significance cause dopamine release by activating the mesocorticolimbic dopaminergic pathway in a controlled and homeostatic manner. Drugs act in a more intense and protracted fashion, by changing the homeostasis of the dopaminergic response and inducing allostatic adaptation. From the quantitative point of view, synaptic levels of dopamine can be higher (or persist for longer) than those sufficient to

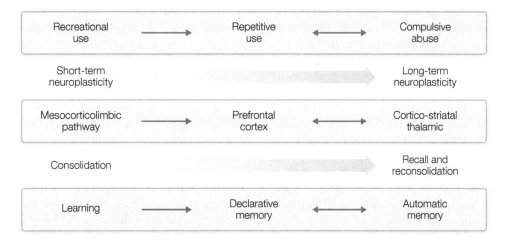

FIGURE 10.2 Schematic diagram showing the analogy between neuroplasticity induced by chronic drugs of abuse (different behavioral stages and major brain areas) and learning and memory processes. Modified by Kalivas P.W., O'Brien C. (2008). Drug addiction as a pathology of staged neuroplasticity. *Neuropsychopharmacology*, 33, 166–180.

stimulate dopamine receptors in the NAc and prefrontal cortex. This abnormal neurochemical response explains the strong reinforcing effect, and after repeated treatments, the consequent adaptation.

The adaptation induces modification of molecular and cellular processes of reinforcement. The observed phenomena vary from drug to drug and may consists in tolerance (gradual increase of effective dosage), psychological dependence (alterations of emotions and mood), and sensitization (increased sensitivity to the stimulus and high risk of relapse). These phenomena consequently are also mediated by other neuroanatomical structures involved in regulation of the mesocorticolimbic dopaminergic pathway, such as the amygdala, hippocampus, and cerebral cortex. Thus, drug addiction involves changes not only in motivational mechanisms but also in affective and cognitive processes. Neurochemical mechanisms modified by drugs of abuse suggest the formation of a kind of memory of the pleasurable effects of the drug (to be sought) and the unpleasant effects of abstinence (to be avoided).

In analogy with memory processes, first experiences of drug effect become learned behaviors. This learning phase is controllable, voluntary, meaning that search and intake of the substance can be adjusted and controlled. However, repeated exposure to drugs of abuse gradually (and increasingly) involves neural mechanisms controlling memory stabilization. An increasing involvement of glutamatergic transmission from the prefrontal cortex to the striatum and thalamus (the so-called corticostriatal–thalamic loop) gradually leads to transition from an executive, conscious process to an unconscious, compulsive, and automatic one. We can compare this process of adaptation to learning to do a task: initially, we operate in a conscious way, thinking about it, and then we do it "without thinking." From the neuroanatomical standpoint, this form of memory is stabilized in the corticostriatal–thalamic loop (Fig. 10.2).

Once dopamine has efficiently signaled to learn a motivated behavior and this behavior has been stabilized in the memory, dopamine release is no longer necessary to reinforce the drug intake behavior, but instead continues to indicate the presence of stimuli associated with it. This means that the dopaminergic mechanisms can still be activated even in the absence of the drug, and such stimuli may trigger relapse to drug-seeking behavior.

In summary, similarly to other types of memory, neuronal adaptation to chronic exposure to drugs of abuse initially occurs in brain areas involved in motivated behaviors but gradually involves other areas controlling affective and cognitive processes. In drug addiction, we can distinguish a "testing" phase, characterized by a controlled use of the drug that is compatible with other social activities and work. In this learning phase, the behavior of drug use is reversible. However, repeated use and abuse results in a stabilized form of behavior with loss of control and difficulty to change. Cessation attempts are often unsuccessful because of the difficulty to overcome physical and psychological symptoms of withdrawal, and the risk of relapse. This is due not only to the conditions of the addicted subject but also to a strong memory of past drug intake.

Research on Drug Addiction

In recent years, research has confirmed the adaptive nature of drug addiction at different levels. The complexity of the disorder, the diversity from substance to substance, and the strong role played by the individual's characteristics (genetic factors associated with drug addiction have been discovered) have led to the development of several lines of research based on multidisciplinary approaches. The use of biotechnology in addiction research has allowed to thoroughly investigating the role of changes in gene expression in the mechanism underlying neuroadaptation. A current focus of research in this field is the role of epigenetic processes in chronic drug effects. The development of animal models with high predictive validity is continuously contributing to translate results of basic research into clinical approaches. The development of noninvasive brain imaging techniques has revolutionized the research on brain processes of drug addiction. Indeed, it has allowed demonstrating the multidimensional complexity and temporal progression of brain neuroadaptation at various stages of addiction. The possibility to visualize brain activity has allowed studying many cerebral functions, psychological and behavioral responses, neurological damage, and functional abnormalities underlying drug addiction. Combinations of different assessment technologies are currently being developed to obtain data with high spatial resolution (to visualize the smallest area) and high temporal resolution (to see the fastest events) (see Supplement E 10.4, "Visualization Techniques"). Changes in cerebral blood flow after inhalation of nicotine, as measured by PET, have revealed that the drug induces changes similar to those induced by other drugs of abuse. Similar studies have shown that nicotine exerts significant effects on brain areas responsible for cognitive and affective changes in the smoker.

THERAPY FOR DRUG DEPENDENCE

Drug addiction is a chronic and relapsing behavioral disorder. Investigations into the neurobiological mechanisms underlying the different phenomena and stages of drug addiction have revealed its (mal)adaptive or allostatic nature. Being a complex and multifactorial disorder, it requires a therapeutic strategy integrating pharmacological, psychological, and social interventions. To date, no single-drug therapy is available to treat drug addiction: the pharmacological efficacy of the different treatments available is not statistically significant compared to that obtained with placebo. The effectiveness of pharmacological treatments increases when associated with psychological therapies, such as motivational, cognitive–behavioral, and group-based therapies. The main problem is the prevention of relapse, which can occur even after long-term cessation, usually due to major factors such as stress, drug reexposure, and occasional presence of conditioned stimuli.

Drug therapies currently approved for drug dependence are (i) replacement, (ii) aversive, and (iii) anticraving therapies. The aim of a replacement therapy is to limit abused drug intake rather than to cure dependence. It consists in the administration of a substance that fully or partially mimics the abused drug. Typically, these medicaments are partial agonists of the receptor to which the abused drug binds. Sometimes, it consists of the same addictive substance, for example, nicotine gums for smoking cessation. Replacement therapy has proven effective mainly at reducing the severity of abstinence symptoms. However, after the initial, critical days of abstinence, it is essential to associate the replacement therapy with a psychological intervention. The aversive therapy consists in the administration of a substance that induces unpleasant effects in case of relapse in drug taking. It is defined aversive because such unpleasant effects reduce the motivation to take the drug again. An example is disulfiram that, by inhibiting ethanol degradation, causes accumulation of acetaldehyde in the body, resulting in a series of unpleasant effects. The anticraving therapy consists in the use of drugs specifically developed to prevent relapse that act directly on the neurochemical mechanisms of dependence. These drugs have different mechanisms of action and their efficacy profile acts on different symptoms (see Table 10.3 and Box 10.3).

The pharmacological treatment of drug addiction has many limitations. The effectiveness of the pharmacological treatment has been widely demonstrated; however, it is effective in preventing relapse only in some patients. No medication is a "magic pill" by itself, and integration with social and psychological therapy is essential. Numerous clinical trials have shown how integrated treatments can increase effectiveness of pharmacotherapy.

The biology of drug addiction is not yet fully exploited in a therapeutically effective way. For example, we still do not know drugs capable of reversing the stable neuroplasticity induced by chronic exposure to drugs of abuse. Similarly, we do not know yet whether psychosocial treatments are effective in inducing changes in the brain. Finally, situational conditions, such as stimuli and environments with a risk of relapse to which the ex-addict is exposed after leaving the cessation unit, always represent unpredictable factors.

TABLE 10.3 Pharmacotherapies for treatment and detoxification of drug addiction

Pharmacotherapy	Clinical indication
Acamprosate	Relapse prevention in ex-alcoholics
Buprenorphine	Opioid detoxification therapy
Bupropion	Anticraving drug. Relapse prevention in ex-smokers
Disulfiram	Prevention of relapse to alcohol use
Methadone	Opioid substitution therapy
Naltrexone	Drinking reduction and prevention of relapse
Nicotine replacement therapies	Tobacco smoking substitution therapy
Varenicline	Relapse prevention in ex-smokers

U.S. Food and Drug Administration (FDA)-approved drugs (chronological order of approval).

BOX 10.3 DRUGS APPROVED FOR THE TREATMENT OF DRUG ADDICTION

Drugs approved by the US Food and Drug Administration for the treatment of drug addiction are listed below (in chronological order of approval). For simplicity, drug formulations and dosages are not shown. However, it is important to emphasize that the choice of the appropriate formulation is important for the individualization of treatment based on clinical responses, tolerance, and acceptability. Constant monitoring of outcomes can improve the overall clinical efficacy and patient's adherence to the pharmacological treatment within an integrated therapeutic intervention including individual and group psychotherapy and social support:

- Disulfiram. Inhibitor of acetaldehyde dehydrogenase. It induces aversive effects if the subject drinks alcohol because it increases acetaldehyde plasma levels. Prevent relapse in alcoholics.
- Methadone. It acts as a partial agonist of endogenous opioid μ receptors. Employed in opioid substitution therapy. Its long duration of action after oral administration facilitates the detoxification treatment of opiate addict.
- Nicotine replacement. Present in different formulations (e.g., gums, patches, tablets) that provide variables of plasma levels of nicotine depending on the type of smoker. Reduce withdrawal symptoms.
- Naltrexone. It possesses properties of endogenous opioid receptor antagonist, with greater affinity for the type μ. Reduces alcohol consumption and prevents relapse in alcoholics. Its effectiveness increases when associated with the integrated therapy.
- Bupropion. Increases synaptic levels of noradrenaline and dopamine by blocking their synaptic reuptake pumps. It possesses anticraving properties and prevents relapse in smokers. Its effectiveness increases when associated with the integrated therapy.
- Buprenorphine. It acts as a partial agonist at μ and κ opioid receptors and endogenous antagonist of type δ. Opioid substitution therapy. Effective in maintenance treatment of opiate addiction. It can be associated with naloxone, the endogenous opioid receptor antagonist, to reduce risk of abuse.
- Acamprosate. It acts as a partial agonist of NMDA receptors and indirectly on metabotropic glutamate receptors. It prevents relapse in ex-alcoholics.
- Varenicline. It acts as a partial agonist of the nicotinic receptor subtypes alpha4–beta2, known to mediate nicotine reinforcing effects. It prevents relapse in smokers.

In conclusion, drug addiction is a disorder that reflects an adaptation to drugs of abuse, and a joint effort of research and therapy is required to develop approaches for an effective and safe clinical intervention.

TAKE-HOME MESSAGE

- Adaptation is the property of biological systems that allows the achievement of a dynamic state of equilibrium, named allostasis, in response to changing environmental conditions. Allostasis is also the adaptive response to drugs and substances of abuse.
- Chronic exposure to drug of abuse can cause persistent changes in receptors, signal transduction mechanisms, regulation of gene expression, and posttranslational modifications.
- Pharmacological treatment of addiction is limited to a few drugs that help overcome some symptoms and phenomena of addiction but are not high efficient in preventing relapse.
- Best outcomes in relapse prevention are obtained by combining drugs with psychological and social therapies.

FURTHER READING

Di Chiara G., Bassareo V. (2007). Reward system and addiction: what dopamine does and doesn't do. *Current Opinions in Pharmacology*, 7, 69–76.

Everitt B.J., Robbins T.W. (2013). From the ventral to the dorsal striatum: devolving views of their roles in drug addiction. *Neuroscience Biobehavioral Reviews*, 37, 1946–1954.

Kalivas P.W., O'Brien C. (2008). Drug addiction as a pathology of staged neuroplasticity. *Neuropsychopharmacology*, 33, 166–180.

Koob G.F., Lloyd G.K., Mason B.J. (2009). Development of pharmacotherapies for drug addiction: a Rosetta Stone approach. *Nature Reviews in Drug Discovery*, 8, 500–514.

Robinson T.E., Berridge K.C. (2000). The psychology and neurobiology of addiction: an incentive-sensitization view. *Addiction*, 95(Suppl 2), S91–S117.

Robison A.J., Nestler E.J. (2011). Transcriptional and epigenetic mechanisms of addiction. *Nature Reviews Neuroscience*, 12, 623–637.

11

PHARMACOLOGICAL MODULATION OF POSTTRANSLATIONAL MODIFICATIONS

MONICA DI LUCA, FLAVIA VALTORTA, AND FABRIZIO GARDONI

By reading this chapter, you will:

- Gain a basic knowledge about the fine modulation of protein biological properties by posttranslational modifications (PTMs)
- Know different types of PTMs and their main function and properties
- Know how PTMs can be modulated by drugs

Cells undergo continuous modifications to respond to environmental stimuli. In eukaryotic cells, responses to different stimuli are complex and dynamic and are guaranteed by several diversified, often interconnected, mechanisms. A major mechanism is based on the so-called posttranslational modifications (PTMs) of proteins. PTMs are highly dynamic processes and consist in the addition of functional entities to one or more amino acid residues of a protein. By doing so, PTMs modify the biological properties of a protein through covalent modifications, which occur after mRNA translation, often at very last stages of their ribosomal biosynthetic pathway.

Nowadays, more than 200 PTMs are known. On the one hand, some PTMs as glycosylation, lipid addition, or disulfide bond formation are stable and essential for maturation and folding of newly synthesized proteins as they contribute to stabilization of protein secondary and tertiary structure. On the other hand, PTMs such as phosphorylation and SUMOylation are reversible and transient and play an important role in dynamic responses to stimuli.

One or more PTMs can take place with a well-defined temporal pattern on the same or different amino acid residues of the same protein. In cells, the interaction between proteins with specific PTM and other cellular components through their recognition domains induces a highly dynamic association/dissociation of macromolecular complexes. This process is often responsible for decoding cellular responses to different stimuli.

In eukaryotic cells, the great complexity and dynamism of the processes related to PTMs offer a great possibility of pharmacological targeting.

PROTEIN PHOSPHORYLATION

Protein phosphorylation represents the most common PTM, and it is known to modulate the biological function of a vast array of cellular proteins. The addition of a phosphate (PO_4^{3-}) group to a protein induces a conformational change in its structure that leads to a modification of its activation state (e.g., in enzymes or channels) or of its ability to interact with other cellular proteins or molecules. Notably, protein phosphorylation is a rapidly reversible process and therefore one of the most dynamic and versatile PTMs.

Protein phosphorylation is a widespread and common regulatory mechanism that takes place in both prokaryotic and eukaryotic organisms. Kinases and phosphatases are the two types of enzyme responsible for phosphorylation and dephosphorylation, respectively. In most cases, each kinase or phosphatase controls the phosphorylation of different substrate proteins; in this way, modulation of an individual kinase or phosphatase activity may result in activation/inhibition of several molecular pathways. On the other hand, many enzymes can phosphorylate the same substrate, thus

---

General and Molecular Pharmacology: Principles of Drug Action, First Edition. Edited by Francesco Clementi and Guido Fumagalli.
© 2015 John Wiley & Sons, Inc. Published 2015 by John Wiley & Sons, Inc.

leading to the convergence of various cellular stimuli on the same target molecule or molecular pathway.

Because the function of protein kinases and phosphatases has profound effects on a cell, their activity is highly regulated. Different types of input are able to modify the activity of kinases and phosphatases, thus inducing a modification of cellular phosphoproteome. These events can be mediated by direct mechanisms, for example, receptors bearing an enzymatic kinase activity, or indirect ones, for example, through activation of intracellular signaling pathways. Activation of G-protein-coupled receptors results in modulation of G-protein activity and consequent activation/inactivation of associated enzymes. These events ultimately lead to production of intracellular second messengers that are responsible for protein kinase activation. In some cases, the phosphorylation process is mediated by a kinase cascade, in which several protein kinases phosphorylate and activate each other, thus allowing amplification, convergence, or integration of different intracellular signaling pathways.

Protein Kinases

Protein kinases are important regulators of several cell functions and constitute one of the largest and most diverse gene families. By adding phosphate groups to substrate proteins, they modulate the overall function of many cellular proteins, thus orchestrating the activation of almost all cellular processes. Eukaryotic protein kinases comprise one of the largest superfamilies of homologous proteins and genes; of the 518 human protein kinases, 478 belong to a single superfamily whose catalytic domains are related in sequence. These can be clustered into groups, families, and subfamilies increasingly similar in sequence and biochemical function.

Protein kinases use the γ-phosphate of ATP (or GTP) to generate phosphate monoesters using protein alcohol groups (on Ser and Thr) and/or protein phenolic groups (on Tyr) as phosphate acceptors. Most kinases act either on both serine and threonine (serine/threonine kinases) or on tyrosine (tyrosine kinases). The amount of phosphate groups incorporated into tyrosine residues is far less than phosphate groups incorporated into serine/threonine residues of substrate proteins, with a ratio of about 1:100–1000. However, tyrosine phosphorylation plays a fundamental role in several cellular processes such as cell growth and differentiation. In addition, protein kinase families differ for their structure, activation mechanisms, and subcellular distribution. Most of these enzymes are soluble and mainly localized in the cell cytosolic compartment, whereas few of them are integral membrane components or are found attached to cellular membranes through lipid modifications. In few cases, kinase activation can also induce variation of its subcellular localization. For instance, protein kinase C is distributed in cell cytosol at resting state, and following calcium-dependent activation, it becomes transiently associated with the plasma membrane and other intracellular membranes.

Several cellular processes finely modulate the activity of different protein kinases depending on cell needs. Loss of these control mechanisms, due to mutations within specific domains, confers these proteins malignant transformation ability (see Chapters 9 and 18). In physiological conditions, the basal catalytic activity of protein kinases is limited by autoinhibitory mechanisms; usually, a specific protein domain interacts with the catalytic domain preventing its interaction with substrate. Protein kinase activation can be regulated by direct interaction with second messengers; calcium-dependent protein kinases require also the presence of a cofactor such as calmodulin (this protein binds calcium and then interacts with the protein kinase named calcium/calmodulin dependent; see Fig. 11.1b) or lipid molecules such as diacylglycerol. Another important second messenger is cAMP that controls PKA activation. In resting conditions, the catalytic subunit of PKA is complexed with the regulatory subunit, and as a result, the enzyme is inactive. When cAMP concentration rises, cAMP binds the regulatory subunit, and the catalytic subunit is released and can phosphorylate its substrates (see Fig. 11.2).

Activation of protein kinases is triggered or accompanied by their own phosphorylation or dephosphorylation. Kinase phosphorylation often occurs by autophosphorylation (the kinase phosphorylates itself), but it may also be catalyzed by different kinases. For example, calcium/calmodulin-dependent protein kinase II (CaMKII) phosphorylates itself at a specific threonine residue (Thr^{286}); this determines a conformational change of the enzyme itself that unmasks the catalytic site. This modification confers the enzyme the property to be active also in the absence of calcium ions and calmodulin (see Fig. 11.1b), thus prolonging the temporal effects of intracellular Ca^{2+} signaling even when Ca^{2+} concentration returns to basal levels.

Deactivation of protein kinases can occur in different ways: it can be induced either by a decrease of the cytoplasmic concentration of the second messenger responsible for kinase activation or by phosphorylation/dephosphorylation of the enzyme itself.

One of the most important features of protein kinases is their capability to phosphorylate only a limited number of substrates. In this way, upon activation of a transduction pathway, not all cellular proteins become phosphorylated, but only those involved in the transduction of that specific signal. The substrates of one specific kinase may differ in their amino acid sequences, but they all share a specific sequence surrounding the phosphorylation site that is called phosphorylation consensus domain.

As described earlier, one substrate can also be phosphorylated by different protein kinases activated by diverse intracellular signals, with either cooperative or antagonist effects.

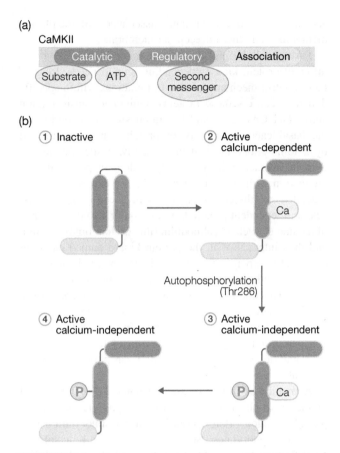

FIGURE 11.1 Structure of protein kinases and mechanism of activation of CaMKII. (a) The catalytic domain contains the ATP binding site and the substrate binding site. In addition, there is a regulatory domain containing an inhibitory sequence that blocks enzyme activity by interacting with the substrate binding site and thereby preventing kinase–substrate interaction. The regulatory domain contains binding sites for second messengers. (b) The inactive kinase (1) is depicted, with the catalytic region in its inactive status because of the regulatory region acting as pseudosubstrate. When the intracellular Ca^{2+} concentration increases, Ca^{2+} ions form a complex with calmodulin, and the complex binds to the regulatory region (2). In the presence of ATP, kinase molecules bound to the Ca^{2+}/calmodulin complex are able to phosphorylate molecules close to the same kinase (intermolecular autophosphorylation) (3), making the active kinase calcium independent. In this phase, detachment of the Ca^{2+} ion does not change the activation state of the kinase (4).

Protein Phosphatases

In every moment of a cell life, the level of phosphate incorporated into a specific protein results from the balance between the activity of protein kinases catalyzing the addition of phosphate groups and protein phosphatase catalyzing their removal.

Protein phosphatases are enzymes that catalyze hydrolysis of phosphoester bond at phosphoserine, phosphothreonine, and phosphotyrosine residues, leading to removal of phosphate groups. Most phosphatases consist of multiple domains containing one or more catalytic regions and other

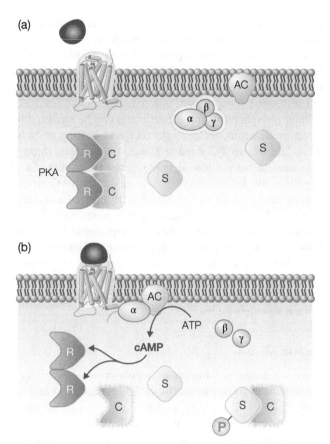

FIGURE 11.2 Activation of cAMP-dependent protein kinase (PKA). PKA consists of two subunits formed by separate polypeptides. In resting conditions (a), the catalytic subunit C forms a complex with the regulatory subunit R that keeps the enzyme in an inactive state. When cAMP concentration increases in the cell (b), cAMP binds the regulatory subunit, determining its detachment from the catalytic subunit. Thus, the inhibition is removed and the catalytic subunit can phosphorylate the substrates, S.

regions of variable size. Like protein kinases, also protein phosphatases can be divided into serine/threonine protein phosphatases and tyrosine protein phosphatases, depending on the amino acid residue they recognize in their substrates.

Tyrosine phosphatases bind only the phosphorylated form of their substrates: the recognized sequence includes the phosphorylated tyrosine as well as some adjacent residues. Therefore, the *in vivo* selectivity of this class of phosphatases primarily depends on the catalytic domain.

Serine/threonine phosphatases have an extremely different structure compared to tyrosine phosphatase; they are holoenzymes, consisting of a catalytic subunit associated with one or more additional subunits. These can have a regulatory function or can target the protein toward specific subcellular structures, such as glycogen or myofibrils. The catalytic domains of serine/threonine phosphatases are highly conserved, with a similar three-dimensional structure, and use the same mechanism of action.

Moreover, there are mixed-function phosphatases that are able to dephosphorylate phosphotyrosine, phosphoserine, and phosphothreonine residues, even if each enzyme has a preferential phosphatase activity. These phosphatases share conserved catalytic domains as well, but they interact with substrates through a site not containing the phosphorylated residue. Therefore, the specificity of this class of phosphatases is controlled by protein–protein interactions not involving the catalytic domain.

Protein phosphatases seem to have lower specificity compared to kinases, as they are generally able to dephosphorylate a large number of substrates. Furthermore, there seems to be no dephosphorylation consensus sequence for phosphatases; therefore, it is difficult to predict a priori which phosphatase dephosphorylates a specific substrate. Because of this lack of correspondence between groups of substrates phosphorylated by a specific kinase and groups of substrates dephosphorylated by a specific phosphatase, the protein kinase–phosphatase substrate networks are considerably complex to unravel.

Phosphatase activity can have multiple consequences. Dephosphorylation of proteins may affect their tertiary structure as well as their interactions with other cellular components. In particular, dephosphorylation of proteins involved in an intracellular signaling pathway can switch on/off that pathway. Most kinases are activated through phosphorylation of residues within their activation domain. Therefore, dephosphorylation of these residues by phosphatases inhibits their catalytic activity. For example, MAP kinases (MAPKs) are activated by MAP kinase kinase (MAPKK) through phosphorylation of a tyrosine and a threonine in their activation domains. Dephosphorylation of the phosphothreonine residue by phosphatase 2A (PP2A) or of both residues by mixed MKPs reduces MAPK activity (see also Chapter 13).

SUMOylation

SUMOylation is a PTM and a major regulator of the function of several proteins. Small ubiquitin-related modifier (SUMO) is a 97-amino-acid protein that can be covalently attached to specific lysine residues on target proteins. In mammalian cells, there are four SUMO paralogues, designated SUMO-1, SUMO-2, SUMO-3, and SUMO-4. All SUMO proteins are synthesized as inactive precursors that first undergo a C-terminal cleavage mediated by the family of sentrin/SUMO-specific protease (SENP) enzymes. This cleavage exposes a diglycine motif that allows SUMO to be conjugated to lysine residues in target proteins.

During each conjugation cycle (see Fig. 11.3), mature SUMO proteins are first activated in an ATP-dependent manner by the E1-activating enzyme; the activation requires the formation of heterodimers of SUMO-activating enzyme E1 (SAE1) and SAE2. This process leads to the formation of

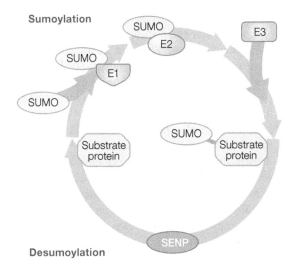

FIGURE 11.3 SUMOylation. SUMO proteins are synthesized in an immature form, cut at the C-terminus by the SENP enzymes, activated by conjugation to the active site of the E1 enzyme complex, and finally transferred to the active site of the E2 complex (Ubc9). The next step is SUMO conjugation, mediated by E2 in cooperation with E3, to the lysine residue of the protein substrate. Deconjugation of SUMOylated proteins is mediated by the enzyme SENP.

a thioester bond between a cysteine residue in the active site of E1 and the diglycine motif of SUMO. Next, activated SUMO proteins are translocated to the active site of the E2 SUMO-specific conjugating enzyme, named ubiquitin-conjugating 9 (Ubc9). Ubc9, either alone or in combination with a SUMO E3 enzyme, catalyzes the conjugation of SUMO to a specific lysine residue in the target protein. Very often (75% of cases), SUMO conjugation occurs at the ψKxD/E consensus sequence in target proteins, where ψ represents a large hydrophobic residue, K is the target lysine, and D/E are acidic residues. However, SUMOylation can also occur at lysine residues outside this motif and not all ψKxD/E motifs are SUMOylated.

Despite being a covalent protein modification, SUMOylation is a dynamic and reversible process that is regulated by deconjugation pathways involving the protease activity of deSUMOylation-specific enzymes, SENPs (the same SENP enzymes that are required for the maturation of SUMO precursors). Accordingly, the equilibrium between SUMO conjugation and SENP-mediated deconjugation regulates the SUMOylation degree of a specific protein. The identification of the molecular mechanisms involved in the regulation of these events is a very active area of current research.

Characterization of Ubc9 knockout mice has recently confirmed the critical importance of SUMOylation in eukaryotic cells. Homozygous Ubc9 knockout mice are not viable and die during embryogenesis because of developmental defects. Similarly, removal of Ubc9 gene in cell lines results in abnormalities ultimately leading to cell death mainly by apoptosis. SUMOylation is a major regulator of protein function in a wide range of cellular processes, such as genomic integrity, cell cycle, subcellular and nuclear localization of target proteins, modulation of protein–protein interaction, repression of transcription, and regulation of proteasomal degradation through a complex interaction with the ubiquitin system.

The majority of SUMO target proteins identified so far are either localized in the nucleus or involved in nuclear trafficking of cytosolic proteins. The functional consequences of SUMOylation are strictly correlated to the physiological function of target proteins and cell type. However, in many cases, the cellular consequences of protein SUMOylation remain poorly understood. Notably, recent studies indicate a direct role for protein SUMOylation in the pathogenesis of several human disorders such as cardiovascular and neuronal diseases.

UBIQUITINATION

In all organisms, proteins are continuously synthesized and degraded. Cellular protein content depends on the balance between protein synthesis and degradation. Studies on protein turnover rates in eukaryotic cells have shown that there are short-lived proteins (minutes) and long-lived ones (days). Abnormal or misfolded proteins are those most rapidly degraded. In eukaryotic cells, protein degradation is very often mediated by a specific PTM known as ubiquitination. This process consists in the addition of a ubiquitous 76-amino-acid regulatory protein named ubiquitin to a lysine residue of the target protein through an isopeptide bond.

The ubiquitin chain transfer onto target proteins is mediated by a cascade of enzymatic reactions involving specific enzymes, such as the ubiquitin-activating enzymes (E1), the ubiquitin-conjugating enzymes (E2), and the ubiquitin-protein ligases (E3) (Fig. 11.4). Specifically, ubiquitin is activated through the creation of a high-energy bond between the carboxyl group of its terminal glycine and a cysteine residue of the E1 enzyme. This bond is created using the energy provided by one ATP molecule. Afterward, ubiquitin is transferred onto another cysteine of the active site of an E2 enzyme (of which there are different types) by a trans(thio)esterification reaction. The last step requires a ubiquitin-protein ligase enzyme (or E3) that is able to interact with E2 and specifically with the substrate to be degraded. Hundreds of E3 enzymes exist; their variability and coordinated action with E2 proteins guarantee the high specificity to the entire degradation process.

Ubiquitination can be reversed by a specific family of proteases called deubiquitinases.

The main function of protein ubiquitination is the degradation of damaged, malfunctioning, or misfolded proteins in order to maintain cellular homeostasis. Ubiquitinated proteins are preferentially degraded by the proteasome system (for further information and details, refer to Chapter 30). However, ubiquitination can also be involved in regulation of many signal transduction pathways.

Malfunction of the ubiquitin system and consequent accumulation of ubiquitinated products in the cells can be related to the pathogenesis of some diseases such as Parkinson's disease.

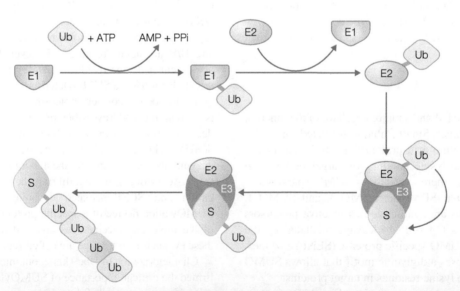

FIGURE 11.4 Ubiquitination. The chain transfer of ubiquitin (Ub) to a protein substrate occurs through a cascade of enzymatic reactions involving specific enzymes, such as Ub-activating enzymes (E1), enzymes capable of combining Ub (E2), and specific ligase (E3). In the first phase of the process, Ub is activated, with consumption of one molecule of ATP, through the creation of a high-energy bond with the enzyme E1. The Ub is then transferred to the active site of an enzyme E2 with a reaction of trans(thio)esterification, and finally, through the intervention of an enzyme E3 able to interact with the substrate to be degraded and E2, Ub is transferred to the substrate (S).

GLYCOSYLATION

Glycosylation is a process that consists in the addition of carbohydrates to the polypeptide chain of a protein. It is one of the most common PTMs and it has strong functional implications since it is also involved in subcellular localization of membrane proteins. Moreover, glycosylation allows proteins to reach their proper folding and therefore to exert their functions. Glycosylation protects proteins from protease attack; therefore, it influences the speed of protein degradation and its turnover. Finally, glycosylation is involved in protein "quality control" or, in other words, in cellular processes that aim at discarding any misfolded protein.

The main carbohydrates present in glycoproteins are glucose, galactose, mannose, fucose, and N-acetylglucosamine. These can be covalently bound to the asparagine (N-glycans) or to the serine/threonine (O-glycans) residues of a protein. The structural complexity of carbohydrates attached to glycoproteins requires an elaborate glycosylation process, which occurs through a precise sequence of reactions with an elevated energy cost.

N-Glycosylation begins in the rough endoplasmic reticulum (ER), while the protein is still being translated. The first step in the N-glycosylation process consists in the transfer of 14 carbohydrates (2 molecules of N-acetylglucosamine, 3 of glucose, and 9 of mannose) to a lateral asparagine residue of the peptidic chain. This oligosaccharide is assembled in the ER from single carbohydrates and is transferred to the protein by the enzyme glycosyltransferase. The glucidic chain can be transferred to any asparagine present in the protein primary structure provided that this amino acid is included in the consensus sequence Asn-X-Ser or Asn-X-Thr, where X can be any amino acid but a proline. The second step takes place also in the rough ER and is represented by the removal of three residues of glucose and one of mannose from the newly transferred glucidic chain. After that, proteins are delivered through vesicles to the Golgi apparatus. Here, each protein undergoes a precise and specific sequence of modifications depending to its future specific function. Removal or addition of single carbohydrates or of longer glucidic chains from a protein can occur at this stage. The specificity of the glucidic chain attached to proteins represents the mechanism through which cells are able to deliver them to their final subcellular compartment.

O-Glycosylation is a highly specific process, which does not involve sequential addition of carbohydrates to proteins. O-Glycosylation takes place entirely in the Golgi apparatus where carbohydrates are bound to the oxygen of serine or threonine residues in the polypeptide chain. Carbohydrates are added one at a time. The final number of carbohydrates bound to the protein by this process is usually limited to a few glucidic residues.

Carbohydrates often can modulate protein function. This is the case of the polysialic acid bound to the neural adhesion molecule N-CAM. Glycoproteins present on cellular surface play a crucial role in cell communication, maintenance of cell structure, and immune system recognition. Therefore, alterations of these glycoproteins can result in profound effects. For example, formation of erythrocyte aggregates, such as those observed in dyserythropoietic anemia type II, results from cleavage of glycoproteins on the surface of these blood cells. Furthermore, several viruses, bacteria, and parasites exploit the presence of carbohydrates attached to proteins on the cell surface to enter target cells. Vice versa, carbohydrate residues of serum glycoproteins and pituitary glycoprotein hormones are involved in the clearance of molecules from blood occurring in kidneys, as well as in targeting hormones to their target organs.

Moreover, defective degradation of glycoproteins by specific lysosomal hydrolases (glycosidases) is involved in some genetic diseases (e.g., mucolipidosis), which are caused by abnormal accumulation of partially degraded products. Instead, severe disorders, like mucolipidosis II and III, are related to defective glycoprotein transport to lysosomes.

ACETYLATION

Acetylation is a PTM that consists in the transfer of an acetyl group from acetyl-CoA to the ε-amino group of a lysine residue located in the substrate. This event neutralizes lysine positive charge and therefore alters the biological features of substrate. Histones are the first identified substrates of the enzymes catalyzing the addition or removal of acetyl groups that for this reason have been named, respectively, histone acetyltransferase (HAT) and histone deacetylase (HDAC). But histones are not the only substrates of these enzymes. The acetylation/deacetylation reactions involve different substrates such as transcription factors, cytoskeletal proteins, and proteins involved in cellular metabolism. So far, more than hundred proteins that are subject to acetylation have been identified; many of these are not nuclear proteins and are often localized at mitochondrial level. As for transcription factors, such as p53 and E2F1, acetylation increases their capability to bind target DNA sequences; several transcription factors contain a 60-amino-acid domain called bromodomain that binds acetylated lysine residues.

HYDROXYLATION

Hydroxylation (β-hydroxylation) consists in the addition a hydroxyl group to a proline residue and is mainly observed in proteins involved in oxidative metabolism. Hydroxylation of proline residues is catalyzed by monooxygenases, known as hydroxylases. These enzymes require NADPH and oxygen; one atom of O_2 is incorporated in the substrate and the other is reduced to water. The enzymatic complex includes also cytochrome P450 that modulates a respiratory chain in which electrons can flow from NADPH to oxygen. Examples of hydroxylation can be found in glucocorticoid synthesis and in the detoxification reactions of certain drugs. Neither enzymatic reactions responsible for removal of hydroxyl groups nor enzymes capable of mediating this type of modification have been described yet.

CARBOXYLATION

Carboxylation converts a glutamate residue within a target protein into γ-carboxyglutamate. This modification is found, for instance, in several clotting factors and other proteins of the coagulation cascade, where it modulates affinity of substrate proteins for calcium ions, a feature that is essential for correct functioning of clotting factors VII, IX, and X.

METHYLATION

Methylation of histone lysine or arginine residues has emerged as a critical event in regulation of epigenetic mechanisms associated with chromatin structural modification and in modulation of chromatin binding of nonhistonic proteins. As previously described for other PTMs, two different types of enzymes, methyltransferases and demethylases, are responsible for the dynamic regulation of the methylation status of substrate proteins. Given the increasing number of studies demonstrating a direct role for protein methylation in a variety of physiological processes as well as in cancer onset and progression, methyltransferases and demethylases have been suggested as possible novel targets for pharmacological intervention. However, this research area is still in the early phases of development, and so far, only few putative pharmacological agents have been identified and tested.

S-NITROSYLATION

S-Nitrosylation, the covalent attachment of a nitrogen monoxide group (NO) to the thiol side chain of a cysteine, represents an important mechanism for posttranslational regulation of several types of proteins. S-Nitrosylation conveys a large part of NO action on signal transduction and provides a key mechanism for redox-based physiological regulation. On the other hand, dysregulation of protein S-nitrosylation is associated with onset and symptoms of a growing list of pathological conditions, including cardiovascular diseases, neurodegenerative disorders of the CNS, and cystic fibrosis.

DISULFIDE BONDS

Disulfide bonds in proteins are formed between the thiol groups of cysteine residues. Disulfide bonds typically play a key role in folding and stability of proteins located in the extracellular medium. Interestingly, several proteins largely used in pharmacology, such as interferon, insulin, and antibodies, contain disulfide bonds. Considering that formation of a disulfide bond involves an oxidation reaction, disulfide bonds are usually unstable in cellular compartment, such as cytosol, characterized by a reducing environment. For this reason, in eukaryotic cells, disulfide bonds are formed in the lumen of the rough ER but not in the cytosol.

LIPID MODIFICATIONS

Lipid modifications are widespread and functionally relevant reactions in eukaryotic cells. In most cases, the lipid moiety is crucial to protein function as it allows an otherwise water-soluble protein to interact strongly with cell membranes.

Myristoylation

It is an irreversible modification catalyzed by the enzyme N-myristoyltransferase (NMT) in which a myristoyl group is covalently attached to an N-terminal amino acid of a nascent protein. Myristoylation most commonly occurs on glycine residues exposed during cotranslational removal of the N-terminal methionine. Myristoylation may also occur posttranslationally, for example, when previously internal glycine residues become exposed after caspase cleavage. Myristoylation is very common in proteins bound to the cytosolic side of the plasma membrane.

Prenylation

Prenylated proteins represent approximately 0.5–2% of all proteins. Prenylation involves covalent addition of a farnesyl (C15) or geranylgeranyl (C20) isoprenoid via a thioether linkage to conserved cysteine residues at or near protein C-terminus. Three known enzymes catalyze isoprenoid addition to proteins: protein farnesyltransferase (FTase), protein geranylgeranyltransferase type I (GGTase-I), and protein geranylgeranyltransferase type II (GGTase-II). Following prenylation, the last three amino acid residues are cleaved by an endoprotease and the carboxyl group of the modified cysteine is methylated by a specific methyl transferase. Known prenylated proteins include several membrane-bound proteins as well as proteins and enzymes involved in cell growth and differentiation. Among these, it is worth mentioning Ras and Ras-related GTP binding proteins (G proteins) and the gamma-subunit of trimeric G proteins. Prenylation mediates membrane attachment of most of these proteins.

Both isoprenoid chains, geranylgeranyl pyrophosphate and farnesyl pyrophosphate, are products of the hydroxymethylglutaryl-CoA (HMG-CoA) reductase pathway. This means that statins, which inhibit HMG-CoA reductase, prevent production of isoprenoids and consequently protein prenylation. Farnesyltransferase and geranylgeranyltransferase recognize the CaaX box at the C-terminus of the target protein, where C is the cysteine that is prenylated, a is an aliphatic amino acid, and X is any amino acid. Compounds mimicking the CaaX domain have been developed and tested as potent and selective inhibitors of prenylation enzymes.

Oncogenic mutations that permanently activate Ras are found in about 25–30% of all human tumors. Accordingly, identification of the molecular mechanisms whereby farnesylation regulates Ras function has led to the hypothesis that inhibitors of farnesyltransferase activity could represent a putative relevant target for anticancer therapy.

Palmitoylation

Palmitoylation represents one of the most frequent and versatile PTMs. It consists in addition of palmitate (C16:0) to cysteine residues through thioester linkage. The reversibility of palmitoylation makes it an attractive mechanism for regulating protein activity, and this feature has prompted intensive investigation on this modification. In addition, the family of proteins modified through thioester-linked palmitate is large and diverse. It includes transmembrane proteins and proteins that are synthesized on soluble ribosomes. The diverse contexts and functions of palmitoylation demonstrate the versatility of this modification. In most cases, palmitoylation increases protein affinity for membranes and thereby affects protein localization and function.

Attachment of Glycosylphosphatidylinositols

The oligosaccharide is attached to the carboxy-terminus of the polypeptide chain as a bridge between the protein and the phospholipid; for this reason, it is called glycosylphosphatidylinositol (GPI) anchor.

Newly synthesized proteins are attached to GPIs in the luminal leaflet of the ER by a GPI–protein transamidase complex. Protein substrates for GPI anchoring contain an N-terminal signal sequence for targeting to the ER and a C-terminal signal sequence responsible for GPI anchoring. The N-terminal signal sequence is cleaved by signal peptidase during or after translocation of the nascent polypeptide into the ER lumen.

Among the proteins to which GPI is added, there are acetylcholinesterase, alkaline phosphatase, the antigen lymphocyte Thy-1, and decay-accelerating factor (DAF), a factor that prevents complement-mediated lysis of erythrocytes. Lack of GPI addition prevents the superficial localization of DAF causing a pathological hemolysis, as observed in paroxysmal nocturnal hemoglobinuria.

In *Trypanosoma brucei* and mammalian cells, production of GPI anchor is inhibited by mannosamine, a derivative of mannose that is not recognized as a substrate by α2-mannosyltransferase, an enzyme involved in the assembly of the glucidic branches of GPI anchors.

A second class of inhibitors is specific for *Trypanosoma* that unlike mammalian cells incorporates myristic acid in GPI anchor. Homologs of myristic acid have been identified as potential drugs for the treatment of infections by this protozoan, as they are toxic for *Trypanosoma* but not for mammalian cells.

PHARMACOLOGICAL MODULATION OF POST TRANSLATIONAL MODIFICATIONS

PTMs in eukaryotic cells are particularly interesting from a pharmacological point of view because of their great molecular heterogeneity and dynamics. In particular, investigations into PTMs carried out in the last 10–15 years have led to a deeper understanding of their functions but also, and more importantly, have allowed the identification of some PTMs as new possible drug target. New classes of PTM inhibitors, some currently undergoing phase I or phase II clinical trials, allow highly selective inhibition of specific enzymes involved in modulation of most important PTMs.

Deacetylase Inhibitors

The biological role of protein acetylation both in physiological and pathological conditions is still not entirely clear, and it represents one of the focuses of current research. Recent studies have led to the identification and characterization of a large number of enzymes involved in deacetylation reactions, whereas much less is known about acetylases.

The high number of substrates of acetylation/deacetylation reactions suggests that these processes may have a key role in cell survival similarly to the phosphorylation/dephosphorylation processes. In addition, in the last few years, several clinical trials have been started to test drugs affecting the regulation of acetylation/deacetylation to be used as a novel therapeutic approach to human disorders such as neurodegenerative diseases and cancer. However, the mechanisms of action of most deacetylase inhibitors are still not completely understood and represent an active subject of current pharmacology research. In this context, it has been demonstrated that deacetylase inhibitors affect cell growth of several types of cancer cells. These molecules have been shown to be able to inhibit cell proliferation to induce differentiation and apoptosis of tumor cells with a low toxicity profile. Although most studies concerning the mechanism of action of deacetylase inhibitors have been confined to cell cultures and animal models, promising results have already been obtained with deacetylase inhibitors in ongoing clinical trials for the treatment of different types of tumors, such as neuroblastoma, melanoma, leukemia, as well as breast, prostatic, lung, ovarian, and colon cancers. Seven different classes of deacetylase inhibitors have been classified according to their structure: four of these classes of inhibitors are actually under clinical evaluation. Among them, vorinostat has been the first inhibitor approved by the FDA in 2006 for the treatment of cutaneous T cell lymphoma. In addition, romidepsin has been approved by the FDA with the same therapeutic indication.

Glycosylation Inhibitors

Metabolic inhibitors interfere with the metabolism of common precursors of the glycosylation processes. For example, 6-diazo-5-oxo-1-norleucine, a glutamine analog, inhibits glucosamine production by antagonizing glutamine.

Tunicamycin is a nucleoside antibiotic that inhibits N-glycosylation in eukaryotic cells by competitively blocking N-acetylglucosamine phosphotransferase, the enzyme responsible for transferring N-acetylglucosamine residues to dolichol phosphate.

Plant alkaloids block N-glycosylation by inhibiting glycosidases involved in the first steps of oligosaccharide processing. These agents do not interfere with glycoprotein folding, but rather, they alter the chemical properties of the saccharidic moiety. Inhibitors belonging to this class share a polyhydroxylated ring structure mimicking the orientation of hydroxyl residues in natural substrates.

Castanospermine and australine inhibit alpha-glucosidases involved in the initial processing of N-glycans.

Swainsonine inhibits alpha-mannosidase II leading to accumulation of mannose-containing sugars. In fact, consumption of *Swainsona canescens* and *Rhizoctonia leguminicola*, both highly enriched in swainsonine, causes locoism, a disease characterized by glycoprotein accumulation in lymph nodes.

TAKE-HOME MESSAGE

- Investigations into PTMs have led to a deeper understanding of their functions and roles in physiology and pathology and to the identification of drugs acting on/interfering with PTMs.
- In most cases, PTMs are reversible and both their generation and removal are finely modulated.
- PTMs may modulate protein maturation, function, and/or interaction.
- PTMs catalyzed by kinase and phosphates play major roles in cell signaling and receptor transduction and may be modulated by second messenger.

FURTHER READING

Alonzi D.S., Butter T.D. (2011). Therapeutic targets for inhibitors of glycosylation. *Chimia*, 65, 35–39.

Babst M. (2014).Quality control: quality control at the plasma membrane: one mechanism does not fit all. *Journal of Cell Biology*, 205, 11–20.

Beltrao P., Bork P., Krogan N.J., van Noort V. (2013). Evolution and functional cross-talk of protein post-translational modifications. *Molecular Systems Biology*, 9, 714.

Berndsen C.E., Wolberger C. (2014). New insights into ubiquitin E3 ligase mechanism. *Nature Structural & Molecular Biology*, 21, 301–307.

Prince H.M., Bishton M.J., Harrison S.J. (2009). Clinical studies of histone deacetylase inhibitors. *Clinical Cancer Research*, 15, 3958–3969.

Richon V.M., Garcia-Vargas J., Hardwick J.S. (2009). Development of vorinostat: current applications and future perspectives for cancer therapy. *Cancer Letters*, 280, 201–210.

Salaun C., Greaves J., Chamberlain L.H. (2010). The intracellular dynamic of protein palmitoylation. *The Journal of Cell Biology*, 191, 1229–1238.

Schedin-Weiss S., Winblad B., Tjernberg L.O. (2014). The role of protein glycosylation in Alzheimer disease. *The FEBS Journal*, 281, 46–62.

Tan J., Cang S., Ma Y., Petrillo R.L., Liu D. (2010). Novel histone deacetylase inhibitors in clinical trials as anti-cancer agents. *Journal of Hematology and Oncology*, 3, 5.

Wilkinson K.A., Henley J.M. (2010). Mechanisms, regulation and consequences of protein SUMOylation. *Biochemical Journal*, 428, 133–145.

12

CALCIUM HOMEOSTASIS WITHIN THE CELLS

JACOPO MELDOLESI AND GUIDO FUMAGALLI

> **By reading this chapter, you will:**
>
> - Know the mechanisms controlling the intracellular concentration of calcium ion, $[Ca^{2+}]_i$, at rest and upon stimulation
> - Know the structure/function of the key proteins controlling $[Ca^{2+}]_i$ and their possible pharmacological modulation
> - Understand the role of intracellular organelles in $[Ca^{2+}]_i$ control and the mechanisms regulating Ca^{2+} concentration within organelles
> - Know the consequences of the altered $[Ca^{2+}]_i$ control occurring in pathological conditions

In living cells, free calcium ions play a crucial role in signaling and regulation processes. Indeed, they contribute to membrane potential and to the control of a myriad of functions both in the cytosol and in the lumen of most intracellular organelles. In this chapter, we will deal with the regulation of calcium homeostasis in the cytosol and various cellular compartments. The indication Ca^{2+} will be used for free Ca^{2+}, regardless of its localization. The total and bound element will be indicated as calcium. In cell compartments, the free/bound equilibrium depends on several factors: Ca^{2+} concentration in the cytosol ($[Ca^{2+}]_i$), in the extracellular space $[Ca^{2+}]_o$, in the mitochondrial matrix $[Ca^{2+}]mit$, or in a compartment of interest and concentration of Ca^{2+}-binding sites and their affinity. The higher the number and the affinity of Ca^{2+}-binding sites, the lower the fraction of calcium that remains free, and vice versa.

THE CYTOSOL: A CROSSROAD OF Ca^{2+} FLUXES

Free Ca^{2+} in the Cytosol and Total Cell Calcium

With the exception of erythrocytes, total calcium in resting cells varies in the range of approximately 0.5–5 mmol/l, similar to the total calcium in the extracellular space (~2 mmol/l). However, the situation is very different when we consider the free Ca^{2+}: in the extracellular space, in fact, the concentration of free Ca^{2+}, $[Ca^{2+}]_o$, is approximately 1 mM, corresponding to approximately 50% of total calcium; in the cytosol, the concentration "in resting condition" varies from 50 to 100 nM, that is, approximately 10,000-fold smaller than the total cell content. Thus, most intracellular calcium is bound to molecules in both cytosol and intracellular structures. Given its high buffering capacity and the equilibrium with adjacent compartments (extracellular space, endoplasmic reticulum (ER), and mitochondria), cytosol is the site of numerous functions regulated by Ca^{2+}.

Increase in $[Ca^{2+}]_i$ may be due to Ca^{2+} influx through numerous ion channels distributed in the plasma membrane and activated by various mechanisms and/or to Ca^{2+} release from intracellular stores or other mechanisms. $[Ca^{2+}]_i$ increase is counteracted by enhanced Ca^{2+} extrusion at the plasma membrane through pumps and exchangers and by sequestration within intracellular organelles, due to pumps and exchangers molecularly and functionally distinct from those in the plasma membrane. Therefore, in general, $[Ca^{2+}]_i$

General and Molecular Pharmacology: Principles of Drug Action, First Edition. Edited by Francesco Clementi and Guido Fumagalli.
© 2015 John Wiley & Sons, Inc. Published 2015 by John Wiley & Sons, Inc.

increases are transient, because of the variable equilibrium among the processes of influx and release on one side and extrusion and segregation on the other side.

The High-Affinity Buffering of Cytosolic Proteins

In addition to the dynamic processes discussed so far, $[Ca^{2+}]_i$ is known to depend on a large number of cytosolic molecules, mostly proteins, that bind Ca^{2+} with high affinity (see Supplement E 12.1, "Cytosolic Proteins That Bind Ca^{2+} with High Affinity"). At rest, the buffering capacity of the cytosol is far from saturation. As a consequence, when Ca^{2+} ions move to the cytosol, they equilibrate rapidly with the binding molecules. Therefore, any increase in $[Ca^{2+}]_i$, for example, from 100 to 500 nM, is actually sustained by the transfer to the cytosol of much larger (up to 100-fold) amounts of the cation.

Buffering plays a major role in cytosol Ca^{2+} homeostasis. On one hand, it attenuates $[Ca^{2+}]_i$ increases and delays $[Ca^{2+}]_i$ decreases; on the other, it affects Ca^{2+} distribution and thus the distribution of processes that depend on $[Ca^{2+}]_i$. In water, diffusion of Ca^{2+} is fast, around $350\,\mu m^2/s$. In the cytosol, however, Ca^{2+} diffusion is much slower (15–50 $\mu m^2/s$). For this reason, Ca^{2+} ions released in the cytosol by channels of the cell surface and intracellular organelles rapidly saturate the buffering capacity of the area surrounding the site of release. In that area, therefore, $[Ca^{2+}]$ can reach for a few seconds values as high as 10 and possibly 100 μM, whereas the average $[Ca^{2+}]_i$ in the cell rises only moderately (e.g., to 1 μM). Even in resting cells, due to spontaneous activation of Ca^{2+} channels, the layer of cytosol adjacent to the plasma membrane exhibits a $[Ca^{2+}]_i$ approximately 10-fold higher than the rest of the cytoplasm. As a consequence, some processes that cannot occur in the middle of the cytoplasm can take place near the cell surface. Moreover, such dynamic cytosolic microdomains are not distributed at random but at peculiar sites often coinciding with sites of great importance for cells. For example, exocytosis of synaptic vesicles is a rapid (<100 μs), low-affinity, Ca^{2+}-dependent process that takes place at one of such cytosolic microdomains, the so-called active zone, that is enriched in voltage-gated Ca^{2+} channels. By contrast, exocytosis of granules of nonneural cells, a much slower process, takes place at $[Ca^{2+}]_i$ in the 1–10 μM range, sustained by release of Ca^{2+} from intracellular stores. Microdomains with high $[Ca^{2+}]_i$ do not occur only near the cell surface but also in the depth of the cytoplasm, for example, in the space between the ER and mitochondria (see also Supplements E 12.5–E 12.7).

To reconstruct the processes that govern $[Ca^{2+}]_i$, both at rest and upon stimulation, we will analyze separately Ca^{2+} movements across various cellular membranes. These processes are briefly presented in the following sections and discussed in more detail in Supplements E 12.2–E 12.8.

THE PLASMA MEMBRANE: CHANNELS, PUMPS, AND TRANSPORTERS

Surface Channels Permeable to Ca^{2+}

Because of the huge difference in Ca^{2+} concentration between extracellular space and cytosol, and the negative charge of the plasma membrane cytosolic face, Ca^{2+} electrochemical potential is higher than that of other ions. Therefore, passive Ca^{2+} flow across the plasma membrane is always an influx. A variety of channels characterized by distinct properties and regulated by various mechanisms support Ca^{2+} influx in all cells. These channels are presented here according to their different mechanisms of activation.

The voltage-activated or voltage-operated Ca^{2+} channels (VOCCs) (Fig. 12.1) are discussed in detail in Chapter 27. VOCCs are targets of numerous drugs and toxins. Some of these drugs, such as Ca^{2+} channel blockers, have important clinical uses. VOCCs are multifarious and are also expressed by nonexcitable cells, although at lower level.

The receptor channels (RCCs) are described in detail in Chapter 16. In these channels, opening is induced by the agonist binding (Fig. 12.1b, c). Ca^{2+} permeability is quite variable among different RCCs. For example, the NMDA glutamate receptors are highly permeable to Ca^{2+}, whereas muscle-type nicotinic cholinergic receptors are much more permeable to Na^+. However, other types of nicotinic receptors are highly permeable to Ca^{2+}, and their activation increases $[Ca^{2+}]_i$.

The third group of Ca^{2+} channels is highly complex and heterogeneous. From a molecular point of view, these channels belong to six families collectively indicated with the acronym TRP (transient receptor potential). However, only a subset of the 30 channels belonging to these families participates in regulation of Ca^{2+} homeostasis. Receptor-operated Ca^{2+} channels (ROCCs) are a subgroup of TRP channels that is most commonly present in neurons and excitable cells and is directly associated with G-protein-coupled receptors (by G_q and also G_i and G_o) and tyrosine kinase receptors. In cells expressing these channels, activation of specific receptors also induces ROCC opening and thereby Ca^{2+} influx (Fig. 12.1d). Therefore, ROCCs participate in receptor signaling and are often localized at critical sites of the plasma membrane. Another subgroup of the TRP family includes the store-operated channels (SOCCs), which are particularly important in nonexcitable cells and are activated following depletion of intracellular ER stores (Fig. 12.1e and Supplement E 12.2, "Surface Ca^{2+} Channels: TRP, ORAI, SOCC Channels and Their Functions"). The logic underlying the activity of these channels is clear. When Ca^{2+} stores are discharged, cells tend to lose Ca^{2+} through surface pumps and transporters. To counterbalance this loss and preserve cell homeostasis, ORAI1 channel (a member of the SOCC subgroup) is activated in the plasma membrane. ORAI1 activation is

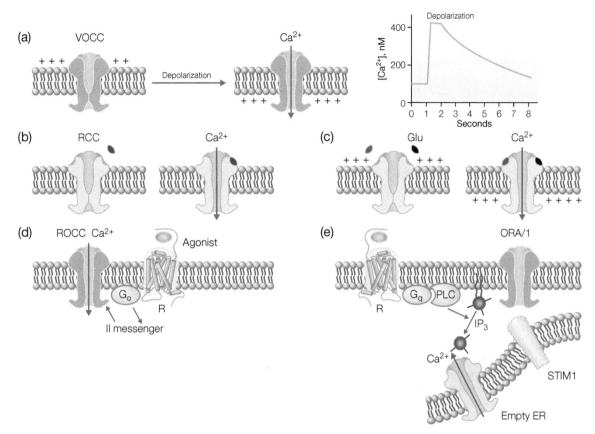

FIGURE 12.1 Ca^{2+} channels. (a) Increases in $[Ca^{2+}]_i$ induced by Ca^{2+} ion influxes through VOCCs. Length of $[Ca^{2+}]_i$ transients in the cytosol depend on both the kinetics of VOCC activation/inactivation (see Chapter 27) and the activity of plasma membrane pumps and transporters (see Chapter 28). (b) Ca^{2+} influx through RCCs depends on activation of these channels upon agonist binding. (c) At NMDA receptors, in addition to activation by glutamate, an important role is played also by depolarization and binding of allosteric regulators, such as D-serine (see Chapter 43). (d) In many cells, especially neurons and other neural cells, activation of some GPCRs coupled to IP_3 production results in parallel activation of ROCC channels mediated by specific G proteins indicated here as G_o. (e) GPCRs, acting via G_q and PLCβ activation, induce IP_3-mediated emptying of intracellular ER stores. As a consequence, the ER membrane protein STIM1 dimerizes and redistributes to ER areas close to the plasma membrane. There, STIM1 binds SOCCs of the Orai1 type (ORAI1), giving rise to activation of Ca^{2+} influx across the plasma membrane.

induced by its direct binding to STIM1, an ER protein that accumulates in ER regions closely associated with the plasma membrane when ER luminal $[Ca^{2+}]$ decreases.

Surface Pumps and Transporters

Surface pumps and transporters are integral membrane proteins that extrude Ca^{2+} through the plasma membrane, thus maintaining (or recovering after activation) the low $[Ca^{2+}]_i$ typical of resting cells.

The plasma membrane Ca^{2+} ATPase (PMCAs) bind cytosolic Ca^{2+} ions with high affinity (~1 μM) and transfer them to the extracellular space (Fig. 12.2a, Chapter 28, and Supplement E 12.3, "Ca^{2+} Pumps of Plasma Membrane and Intracellular Rapidly Exchanging Ca^{2+} Stores"). When $[Ca^{2+}]$ in the cytosolic layer proximal to the plasma membrane increases, Ca^{2+} associates with the high-affinity cytosolic protein calmodulin, and Ca^{2+}/calmodulin complex binds the regulatory site of PMCA increasing its V_{max}. PMCA continuously monitors the submembrane $[Ca^{2+}]_i$ and adjusts its own activity accordingly.

In neurons and cardiomyocytes, when $[Ca^{2+}]_i$ increases to a great extent, PMCA transport is coupled to activation of the Ca^{2+}/Na^+ antiporter (see Fig. 12.2b and Chapter 28). This antiporter is characterized by lower Ca^{2+} affinity (>5 μM) and higher capacity compared to PMCAs. For this reason, the antiporter activity is relevant for lowering $[Ca^{2+}]_i$ in cells exhibiting large and frequent elevations of $[Ca^{2+}]_i$ following stimulation. The antiporter is an indirect target of the cardioactive glycosides. Indeed, by inhibiting the Na^+/K^+ pump, these drugs cause a progressive Na^+ accumulation that in turn reduces Ca^{2+}/Na^+ antiporter activity, resulting in higher accumulation of Ca^{2+} in the sarcoplasmic reticulum. Release of this extra Ca^{2+} into the cytosol upon depolarization increases contraction strength (positive inotropic effect of cardiac glycosides).

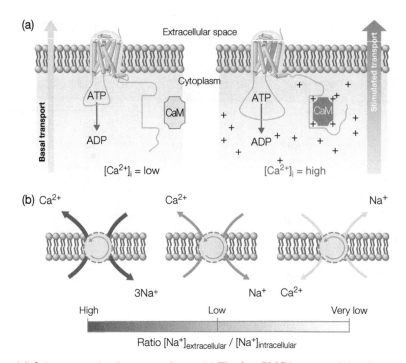

FIGURE 12.2 Extrusion of Ca^{2+} ion across the plasma membrane. (a) The four PMCA pumps of the plasma membrane are Ca^{2+} ATPases that use 1 mol of ATP to transport 1 mol of Ca^{2+} ions. Their activity is stimulated (thicker arrow) by the Ca^{2+}/calmodulin complex, which assembles when $[Ca^{2+}]_i$ increases. When ATP decreases, as in ischemia, PMCA activity is affected. (b) Ca^{2+}/Na^+ transporter is characterized by large capacitance and low affinity for Ca^{2+}. This transporter, therefore, is very active only when $[Ca^{2+}]_i$ is high. The activity of the transporter, indicated by the intensity of the arrow staining, depends on Na^+ gradient. When extracellular Na^+ is very low, the direction of transport can reverse. A moderate reduction of the gradient explains the positive inotropic effect of cardiac glycosides.

Ca^{2+} IN INTRACELLULAR ORGANELLES

Like in the cytoplasm, Ca^{2+} segregated within organelle lumen is partly bound and partly free, and the two pools are in equilibrium. However, in organelles, luminal proteins bind Ca^{2+} with a much lower affinity compared to most cytosolic proteins. As a consequence, Ca^{2+}-free pools within the organelles are proportionally much larger than the cytosolic pool, and Ca^{2+} flux from organelles to cytosol can be fast.

The ER: A Rapidly Exchanging Ca^{2+} Pool

The ER is a continuous and ubiquitous membrane system. Its surface varies from approximately $10 m^2/g$ in pancreatic acinar cells to a few cm^2/g in lymphocytes and many other cell types. Traditionally, the ER is divided in two sections, smooth and rough ER, both involved in Ca^{2+} homeostasis. Their distribution is nonrandom, being more concentrated in specialized areas of the cytoplasm where dynamics of $[Ca^2]_i$ are particularly important. The best known of these areas is the cytosol adjacent to the sarcoplasmic reticulum of striated (skeletal and heart) muscle. In this area, a specialized component of the ER is found, composed of two portions continuous with each other: the terminal and the longitudinal cisternae. Terminal cisternae are characterized by high concentrations of membrane channels and luminal low-affinity Ca^{2+}-binding proteins. Therefore, they rapidly release luminal Ca^{2+} accumulated by pumps in longitudinal cisternae. ER heterogeneity is a property not only of muscle but of all cells; indeed, pumps, luminal Ca^{2+}-binding proteins, and channels tend to concentrate in distinct portions of the endomembrane system (Fig. 12.3).

ER pumps and channels differ, both molecularly and functionally, from the corresponding plasma membrane proteins. When ER channels open and discharge Ca^{2+} segregated within its lumen, ER pumps are activated by the increased $[Ca^{2+}]_i$, resulting in rapid refilling of cisternae, which can therefore undergo a new discharge cycle. These processes also take place in *cis*-Golgi. Pumps in ER/*cis*-Golgi, referred to with the acronym SERCA (sarcoplasmic–endoplasmic reticulum Ca^{2+} ATPase), are quite evenly distributed in their membranes, and Ca^{2+} uptake occurs over a large part of their surface (for details, see Supplement E 12.3).

Luminal Ca^{2+}-binding proteins belong to different families, the best known being calsequestrins expressed by muscle fibers. Several of these low-affinity proteins bind Ca^{2+} with high capacity (up to 50 ions per molecule), and altogether, they bind considerable amounts of calcium segregated in ER lumen. This precludes precipitation of insoluble calcium salts, keeping calcium available for rapid release

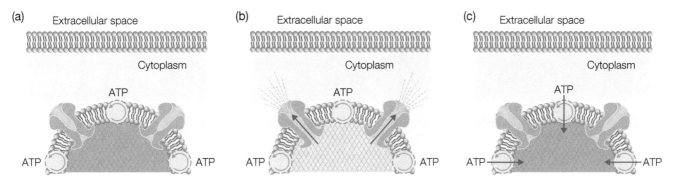

FIGURE 12.3 Intracellular rapidly exchanging Ca^{2+} stores. (a) The rapidly exchanging Ca^{2+} stores are organelles (ER, *cis*-Golgi) expressing Ca^{2+} pumps of the SERCA family (luminal proteins that bind Ca^{2+} with low affinity and high capacity) and channels (IP_3 and RY receptors) characterized by regulated activation. (b) Channels opening induces rapid discharge of Ca^{2+} from the organelle lumen into the cytosol, resulting in increased $[Ca^{2+}]_i$. (c) Subsequently, SERCAs reaccumulate calcium in the organelles, and $[Ca^{2+}]_i$ returns to resting level.

when channels are activated (Supplement E 12.4, "Ca^{2+}-Binding Proteins of the ER Lumen").

There are two classes of channels that directly connect ER lumen with the surrounding cytosol. The first takes its name from a drug, ryanodine (RY), a plant alkaloid not employed in therapy, and the second from its second messenger activator, the inositol 1,4,5-trisphosphate (IP_3).

Three RY channels subtypes have been identified. RYR1 channels are specific of muscle sarcoplasmic reticulum. They are coupled to VOCCs in the T tubules of the plasma membrane (Fig. 12.4a) and distributed transversally at fixed distance along the fibers. Fiber depolarization leads eventually to activation of these VOCCs. The ensuing activation of RY receptors (RYRs) releases Ca^{2+} from the reticulum into the cytosol, triggering coordinated contraction of the fiber. RYR2 is abundant in the heart, whereas RYR3 is more widely expressed but mostly in neurons. RYR2 and RYR3 activation depends largely on increases of local Ca^{2+} concentration. The ensuing process, named Ca^{2+}-induced Ca^{2+} release (CICR), determines a further increase in $[Ca^{2+}]_i$. Caffeine at very high mM concentration favors RYR channel activation, whereas RY prevents their inactivation.

Three forms of IP_3 receptors (IP_3R) are known, with different kinetic properties and differentially expressed in tissues. Their activation is triggered by IP_3 generated by receptors coupled to the G_o protein or endowed with tyrosine kinase activity. IP_3Rs are modulated by many ligands including Ca^{2+} itself and by interaction with proteins, both in the cytosol and in ER lumen. Further details about RY and IP_3 receptors can be found in Supplement E 12.5, "ER Channels: IP_3 and RY Receptors."

RYRs and IP_3Rs can coexist in the same ER membrane or be concentrated in specialized areas. The increased $[Ca^{2+}]_i$ induced by IP_3 stimulates CICR of RYR resulting in potentiation of the Ca^{2+} release response. Together with pumps, mitochondrial uptake and release processes, and cytosolic high-affinity binding proteins, IP_3Rs and RYRs sustain coordinate processes occurring in the cytoplasm of many cell types, oscillations (Fig. 12.4c), and waves of $[Ca^{2+}]_i$. Each oscillation/wave originates from a "pacemaker," a site from which Ca^{2+} release occurs preferentially, in general because of its high concentration of intracellular channels. Diffusion of the increased $[Ca^{2+}]_i$ to adjacent areas of the cytosol facilitates activation of RYR and of IP_3R, making the release process autoregenerative. When restricted to the "pacemaker" area, the process develops into periodic oscillations; when it moves because of activation of adjacent channels, it develops into waves along specific pathways. For example, in neuronal dendrites, waves have been described running both centrifugally and centripetally (see Supplement E 12.6, "Local Ca^{2+} Spikes Can Evolve into Oscillations and Waves").

Mitochondria as Local Buffers of $[Ca^{2+}]_i$

In the past, mitochondria were believed to be the most important intracellular calcium stores. But in the 1970s, it was demonstrated that the affinity of the major mitochondrial Ca^{2+} uptake system, the Ca^{2+} uniporter MCU, is relatively low ($K_d > 5\,\mu M$), suggesting that the physiological role of mitochondria in Ca^{2+} homeostasis was marginal. In the 1990s, however, it became clear that, in terms of Ca^{2+}, mitochondria operate coordinately with the ER: the transient "clouds" of $[Ca^{2+}]_i$ generated locally in the cytosol by activation of IP_3R and RYRs were shown in fact to be high enough to activate MCU in adjacent mitochondria. Recently, it has become clear that areas of the mitochondrial outer membrane are directly bound to and maintained at small distance from the ER surface. These conditions explain the observation that at rest mitochondrial calcium content is low, and when Ca^{2+} release from the ER is activated, mitochondria are rapidly but transiently loaded. Ca^{2+} accumulated, in fact, is released by antiporters that exchange Ca^{2+} with Na^+ or H^+ (Fig. 12.4d). Mitochondria, therefore, prevent cytosol to reach very high $[Ca^{2+}]_i$ levels in proximity of the ER. They work as local buffers allowing cytosolic signals to be protracted. In some cases, mitochondria participate in oscillations and

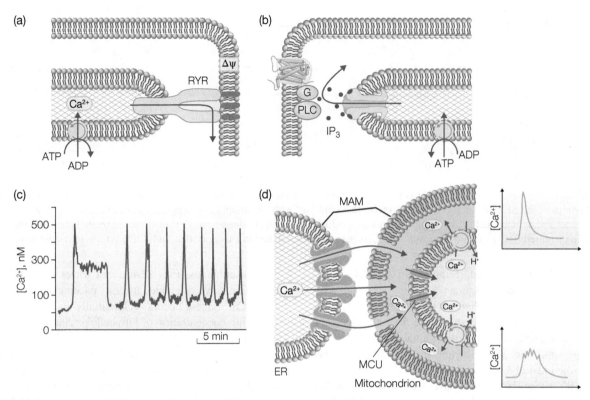

FIGURE 12.4 Processes of Ca^{2+} release from intracellular stores and their coordination. (a) In striated skeletal muscle fibers, RY receptors (RYRs) are concentrated in the external face of the sarcoplasmic reticulum terminal cisternae, directly attached to L-type VOCCs concentrated in the plasma membrane of the T tubules in the triads. Activation of RYRs occurs via direct coupling with VOCCs activated by drop in membrane potential ($\Delta\psi$). VOCCs, therefore, serve as voltage sensors triggering $[Ca^{2+}]_i$ increases. (b) Activation of IP_3 receptors occurs by binding of cytosolic IP_3 produced upon activation of PLCβ by activated GPCRs in the plasma membrane. IP_3 receptor sensitivity to IP_3 is modulated by Ca^{2+} itself, by ATP and by luminal and cytosolic proteins. (c) $[Ca^{2+}]_i$ oscillations generated in cells upon moderate stimulation, analogous to those taking place in physiological conditions. (d) Coordination of Ca^{2+} fluxes in the ER and mitochondria. Activation of IP_3 receptors in specialized ER areas called mitochondria-associated membranes (MAMs), close to the mitochondrial outer membrane, leads to rapid saturation of buffering proteins in the space between the ER and mitochondria, resulting in local $[Ca^{2+}]_i$ increase. The high levels reached by $[Ca^{2+}]_i$ are sufficient to activate MCU transporter in the inner mitochondrial membrane. This induces an increase in $[Ca^{2+}]$mit, which is then dissipated by Ca^{2+}/Na^+ and Ca^{2+}/H^+ transporters. The two subpanels on the right illustrate differences in $[Ca^{2+}]_i$ traces induced by (top) IP_3R activation and (bottom) by ER–mitochondria interaction. Details can be found in Supplements E 12.6 and E 12.7.

waves; in others, they block them, by affecting the spatiotemporal coordination of the ER and cytosol (see also Supplement E 12.7, "Mitochondria: Semiautonomous Organelles That Need the ER to Operate").

ER and Mitochondria Allow Rapid Changes of $[Ca^{2+}]_i$ within Cells

As already discussed in a previous section, Ca^{2+} diffusion in a highly buffered milieu such as the cytosol is slow. Activation of plasma membrane channels, in fact, induces rapid changes in $[Ca^{2+}]_i$ restricted to the superficial layer of the cytosol. The deep regions would be affected only after a delay (of seconds!), and thus, cellular activities would not take place coordinately. In the cytoplasm, however, diffusion can be fast, provided that the ER and mitochondria are involved. In fact, diffusion of IP_3 is approximately $250\,\mu m^2/s$. IP_3 generated at the plasma membrane can therefore assure the quasisimultaneous activation of all ER areas, independently of their localization within the cell, and shortly thereafter also of the associated mitochondria. This coordination is facilitated by the fact that Ca^{2+} release through IP_3Rs and RYRs is modulated positively by appropriate increases in $[Ca^{2+}]_i$. The ER areas bearing IP_3R and RYR and the adjacent mitochondria do therefore operate coordinately.

In striated muscle, such coordination is particularly sophisticated because RYRs are coupled directly to VOCCs in the plasma membrane. In this case, therefore, the main process activated by surface channels increase $[Ca^{2+}]_i$ rapidly in whole cytoplasm not by influx but by release from the sarcoplasmic reticulum. Also fascinating is the situation in neurons. Cooperation between surface channels and stores occurs also in these cells, but via different mechanisms. For example, [Ca^{2+}]$_i$ oscillations triggered by activation of VOCCs and receptors can remain restricted to postsynaptic spines, with no propagation to the dendritic shaft, contributing significantly to the complexity of signaling in neural networks.

[Ca^{2+}]$_i$ Control in Other Intracellular Structures

Other cellular structures appear to contribute to Ca^{2+} homeostasis but differently from the ER and mitochondria in terms of kinetics and function. Both at rest and upon stimulation, nuclear [Ca^{2+}] appears to change in relation to cytoplasmic [Ca^{2+}]$_i$. This may be explained assuming that also the nuclear envelope, which is continuous with the ER, functions as a rapidly exchanging Ca^{2+} store. Moreover, the envelope pores are fully permeable to IP$_3$ and Ca^{2+}. Finally, the buffering capacity of the nuclear matrix appears to be much weaker than that of the cytosol. In the nucleus, the role of Ca^{2+} is considerable. Several transcription factors, such as CREB/CBP, depend in fact on calmodulin for their function. Other factors are activated in the cytosol by Ca^{2+}-dependent processes such as phosphorylation and migrate to the nucleus upon activation. Details about Ca^{2+} in the nucleus are given in Supplement E 12.8, "The Nucleus Is also Operative in Ca^{2+} Homeostasis."

A structure that depends on Ca^{2+} for its functioning is the *trans*-Golgi network (TGN). High levels of the cation, accumulated by secretory pathway Ca^{2+} ATPase 1 (SPCA1), a pump different from SERCAs, are present in TGN cisternae and are then segregated within secretory granules. The quantitative data available are quite impressive. In chromaffin granules, which contain Ca^{2+}-binding proteins such as chromogranins and secretogranins, calcium is contained at approximately 60 mmol/l; in zymogen granules of pancreatic and salivary acinar cells, the content is approximately 30 mmol/l. Also, synaptic vesicles contain calcium in considerable amount. The function of these pools is still debated. A fraction of the stored calcium might be released in stimulated cells. It is clear, however, that the majority of this calcium is simply discharged by exocytosis together with the rest of the granule content.

Other intracellular organelles that contain calcium are endosomes, which trap quanta of extracellular medium in their lumen, and lysosomes, which might receive it from TGN just like secretory granules. Also in these cases, the functional significance of these segregated Ca^{2+} pools is obscure.

Local Relevance of Organelle Calcium Pools

Intracellular organelles use their internal pool not only to contribute to cytoplasmic calcium homeostasis but also for their own function. For example, in the ER, Ca^{2+} ions participate in folding and assembly of newly synthesized proteins and contribute to their localization within the cell. In the TGN, they contribute to granule matrix assembly, which, at least in some cells, has a viscoelastic consistency; in the nucleus, they have a role in transcription of many genes and, indirectly, also in other processes, such as DNA duplication. In mitochondria, Ca^{2+} has a role in energy metabolism: three dehydrogenases of the Krebs cycle are in fact Ca^{2+} dependent and their activity increases with [Ca^{2+}] mit. This is an additional reason why close proximity between ER and mitochondria is so important. In this regard, it should be noted that the low affinity of the mitochondrial transporter allows its activation only when cell stimulation is considerable. Accumulation of Ca^{2+} in mitochondria thus occurs only in pathologic conditions and leads to alteration of mitochondrial membrane potential, with ensuing defects of oxidative phosphorylation and stress, as it occurs in some pathologic conditions.

Ca^{2+} IN CELL PATHOLOGY

The spectrum of Ca^{2+} biological functions is much larger than that of other ions. Indeed, Ca^{2+} oscillations take place in eggs just after fertilization, and uncontrolled Ca^{2+} increases occur in cells undergoing degradation and death. Therefore, Ca^{2+} operates in cells from A to Z.

Changes in Ca^{2+} homeostasis take place in a variety of cell pathologies, especially in the nervous system and heart. Prolonged, large increases affect especially the mitochondria (depolarization, precipitation of calcium phosphate), cytoskeleton (depolymerization, disorganization), and membranes. Part of these effects is triggered by activation of varieties of Ca^{2+}-dependent enzymes such as proteases, phospholipases, and endonucleases.

Cytosolic Ca^{2+}-dependent proteases have been extensively investigated. The family of calpains is ubiquitous in cells where the enzymes operate in equilibrium with their inhibitor, calpastatin. Large and prolonged increases of [Ca^{2+}]$_i$ induce dissociation of calpains from calpastatin, with ensuing irreversible activation of the enzymes due to autodegradation of the sites mediating their binding to the inhibitor (see Chapter 30).

Ca^{2+}-dependent phospholipases are abundant in cell membranes. Among phospholipases A2, some are calmodulin dependent. Their physiological role is multifarious and includes the elimination of numerous toxic substances. Excessive activation, however, can lead to generation of metabolites with cytotoxic activity and ensuing membrane lysis.

In brain undergoing stroke, lesions are initially limited to the area not oxygenated in which neurons undergo rapid necrosis. In the subsequent hours, lesions expand in the so-called penumbra where neuronal death occurs by apoptosis.

The role of Ca^{2+} increases in these processes, and in particular, its participation in the slow neuronal death triggered by overactivation of NMDA glutamate receptors (a process often referred to as excitotoxicity) has been extensively discussed. At the moment, it is clear that $[Ca^{2+}]_i$ increases comparable to those induced by NMDA receptor activation have much lower toxic effects when induced by different mechanisms, such as prolonged cell depolarization caused by K^+ channel blockers like 4-aminopyridine. Moreover, the excitotoxic effect of glutamate has been shown to require, together with NMDA receptor activation, also activation of G_o-protein-coupled glutamatergic receptors. Finally, it should be emphasized that, in neurons, prolonged high $[Ca^{2+}]_i$ increases are accompanied by many additional effects, ranging from considerable changes in gene transcription to generation of toxic substances and metabolites. The complexity of the pathology occurring in those conditions is confirmed by pharmacological results obtained using drugs known to affect Ca^{2+} homeostasis such as VOCC blockers (also known as Ca^{2+} antagonists) and antagonists of glutamatergic RCCs. Drugs of both these families have been shown to exert some protective effect on neurons, for example, in case of stroke. The main problem with these drugs, however, is that they need to be administered concomitantly, or at least shortly after the pathology onset. Once the $[Ca^{2+}]_i$ increase has occurred, drugs become ineffective. We can conclude, therefore, that $[Ca^{2+}]_i$ increase appears as an important but often not sufficient condition to trigger slow excitotoxic neuronal death.

TAKE-HOME MESSAGE

- In resting cells, cytosolic concentration of free calcium ions is up to 10,000-fold lower than in the extracellular medium.
- This gradient is sustained by multiple mechanisms: cytosolic high-affinity/capacity buffering proteins, specific plasma membrane pumps and transporters, and calcium-sequestering organelles.
- Transient increases in cytosolic free Ca^{2+} concentration can be due to ion influx from extracellular medium and/or to Ca^{2+} release from intracellular storage organelles. In the former situation, Ca^{2+} increase is localized in the subplasma membrane region, whereas in the latter, it occurs at discrete intracellular sites.
- Ca^{2+} concentration is finely tuned also in the nucleus and organelles (ER, mitochondria) to meet precise functional needs of the cell.
- In a variety of pathological conditions, Ca^{2+} homeostasis is altered and significantly contributes to cell alterations, necrosis, and apoptosis.

FURTHER READING

Specific aspects of Ca^{2+} homeostasis and related problems are discussed in detail in supplementary materials E 12.1–E 12.8, where general references are also reported.

13

PHARMACOLOGY OF MAP KINASES

Lucia Vicentini and Maria Grazia Cattaneo

> **By reading this chapter, you will:**
>
> - Know the structure and function of different mitogen-activated protein kinase (MAPK) family members and their involvement in human diseases
> - Understand the functional significance of MAPK systems as components of signal transduction pathways
> - Know about novel pharmacological agents targeting MAPKs and/or their regulatory proteins

Mitogen-activated protein kinases (MAPKs) are a family of protein kinases that play an essential role in signal transduction and intracellular communication. Their expression and function have been highly conserved through evolution, from plants, fungi, and nematodes to mammals. MAPK activation in response to changes in the cellular environment leads to phosphorylation of serine and threonine residues on target proteins that can be either stimulated or inhibited as a consequence of this modification. The huge variety of these target proteins (including other protein kinases, transcription factors, enzymes, and cytoskeletal proteins) explains how MAPK pathways potentially regulate every aspect of cell biology. The pleiotropy of MAPK biological effects makes these proteins valuable targets for the development of innovative pharmacological agents for the treatment of human pathologies.

THE MAPK FAMILY AND THE ACTIVATION MECHANISM

In humans, four MAPK subfamilies have been identified: extracellular signal-regulated kinases (ERK1 and ERK2), c-Jun NH_2-terminal kinases (JNK1, JNK2, and JNK3), p38s (p38α, p38β, p38γ, and p38δ), and ERK5. Each group of MAPKs is activated by selective extracellular stimuli and controls specific cellular functions through phosphorylation of distinct sets of substrates.

Irrespective of their differences, MAPKs follow a common scheme of activation consisting of a cascade of phosphorylations catalyzed by sequentially activated kinases (Fig. 13.1). An extracellular stimulus activates a MAPK kinase kinase (MAPKKK) that phosphorylates specific serine and/or threonine residues of a MAPK kinase (MAPKK) that in turn activates MAPK itself via both tyrosine and threonine phosphorylation. Finally, activated MAPKs phosphorylate their nuclear, cytosolic, and/or membrane-associated target proteins, leading to an appropriate cellular response to the extracellular stimulus. MAPK activity is terminated through removal of phosphate groups by specific phosphatases that crucially control intensity and duration of cell responses. Phosphatases involved in regulation of MAPK activity are discussed in Supplement E 13.1, "Phosphatases with Dual Specificity as Regulators of MAPK Activity."

The ERK Family

The ERK family includes ERK1 and ERK2 and is mainly involved in the control of cell proliferation and differentiation. Human ERK1 and ERK2 are ubiquitously expressed and are encoded by two distinct genes. They show 84% identity in their amino acid sequences, similar substrate specificity and functions, and their parallel activation is required for cell proliferation.

The most typical ERK activators are growth factor tyrosine kinase receptors. Upon ligand binding, growth factor receptors undergo activation by autophosphorylation of selective tyrosine residues, which become recruitment sites

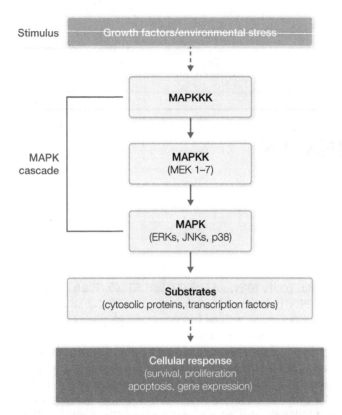

FIGURE 13.1 MAPK activation cascade. The MAPK cascade consists in the sequential activation of three protein kinases: MAPKK kinase (MAPKKK), MAPK kinase (MAPKK or MEK), and MAPK (ERKs, JNKs, p38). Activated MAPKs phosphorylate substrates such as cytosolic proteins and transcription factors. These effector molecules are essential for MAPK-mediated modulation of cellular responses.

for Grb2 and Shc adaptor proteins (see Chapter 18). Grb2 is bound to the guanine nucleotide exchange factor son of sevenless (SOS), and the Grb2–SOS complex moves to the plasma membrane, recruited to the activated receptor. Once at the membrane, Grb2–SOS exchange activity leads to activation of monomeric G protein Ras (see Chapter 14). RasGTP in turn activates RAF serine/threonine kinase (a MAPKKK), which phosphorylates MEK (a MAPKK) that in turn activates ERK via phosphorylation of specific tyrosine and threonine residues (Fig. 13.2). Activation of this canonical pathway allows a significant amplification of the signal; indeed, phosphorylation of 5% of MEKs results in activation of about 60% of endogenous ERKs.

ERK proteins can also be activated in response to G-protein-coupled receptor stimulation. The mechanisms involved, quite heterogeneous and still under study, are described in Supplement E 13.3, "ERK Activation by G-Protein-Coupled Receptors."

The biological effects mediated by ERK activation are typically long-term effects and depend on nuclear translocation of the activated enzyme. Indeed, in basal conditions, ERK is found in the cytoplasm associated to MEK or to scaffold or regulatory proteins and translocates to the nucleus only after growth factor stimulation. ERK nuclear translocation occurs upon autophosphorylation of a specific sequence allowing its interaction with importin-7, a protein that regulates ERK transfer through the nuclear pores. In the nucleus, ERK regulates gene expression via both direct and indirect mechanisms involving phosphorylation of transcription factors, coactivators, and chromatin-associated proteins, such as histones. Some genes, called immediate early genes (e.g., c-*FOS*), are typically activated within a few minutes after growth factor stimulation. These genes regulate transcription of late genes that modify cell genetic program.

ERK activity is also required for cell cycle progression through G1 phase (see Chapter 32). Its activation by mitogens stimulates cyclin D1 synthesis and association with cyclin-dependent kinase CDK4. Cyclin D1/CDK4 complex phosphorylates and inactivates the retinoblastoma (Rb) protein, thereby promoting activation of transcription factor E2F1, which controls expression of proteins required for DNA and chromosomal replication. Additionally, cyclin D1 promotes degradation of cell cycle inhibitor proteins such as p27.

Other mechanisms participate in ERK-mediated control of cell growth outside the nucleus. ERK is responsible for the activating phosphorylation of carbamoyl phosphate synthetase 2, a rate-limiting enzyme in pyrimidine synthesis. Another cytosolic substrate of ERK is the mitogen-activated protein kinase-interacting kinase (MNK) that phosphorylates eIF4E, a translation initiating factor, thus favoring ribosome recruitment to mRNAs and translation apparatus assembly and thereby protein synthesis.

Cell apoptosis is another crucial biological process modulated by ERKs through multiple mechanisms. ERK activation inhibits expression of proapoptotic proteins belonging to the Bcl-2 family. In addition, ERKs directly induce phosphorylation of FOXO3a, a transcription factor responsible for proapoptotic gene expression, and BIM, a proapoptotic Bcl-2 family member. ERK-mediated phosphorylation accelerates FOXO3a and BIM degradation, thus increasing cell resistance to apoptosis. Therefore, mutations in the Ras/Raf/MEK/ERK cascade resulting in upregulation of this pathway, which have been found in several human tumors (Fig. 13.3), may contribute to tumor development and chemotherapy resistance not only by promoting uncontrolled cell growth but also by increasing resistance to cell death.

Finally, ERKs also play an important role in the central nervous system (CNS), especially in neuronal differentiation and plasticity processes, as discussed in Supplement E 13.2, "Role of MAPKs in Memory and Learning."

The JNK Family

JNK proteins are classified as protein kinases activated by cellular stress (e.g., osmotic shock and UV exposure).

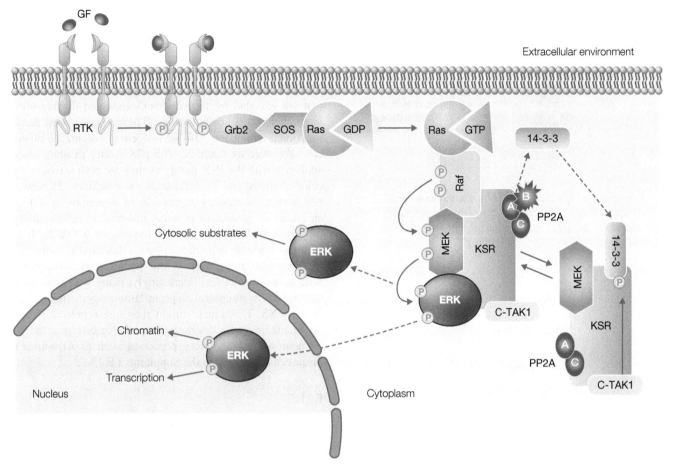

FIGURE 13.2 ERK activation through growth factor receptors. Growth factor (GF) binding induces dimerization and cross-phosphorylation of its receptor tyrosine kinase (RTK). Activated RTKs recruit Ras guanine nucleotide exchange factors, such as son of sevenless (SOS), through the growth factor receptor-bound-2 (Grb2) adaptor protein. Ras guanine nucleotide exchange factors generate RasGTP that activates its downstream kinases, Raf and MEK, leading to ERK phosphorylation and activation. Once activated, ERK phosphorylates cytosolic and nuclear targets. Ras activation is accompanied by membrane translocation of the scaffold protein kinase suppressor of Ras (KSR). In resting cells, inactive KSR is sequestered in the cytosol by binding to 14-3-3 protein that interacts with KSR-phosphorylated serine 392 (S392). RasGTP determines KSR S392 dephosphorylation by causing association of phosphatase-2A (PP2A) B subunit to PP2A A and C subunits, which are constitutively associated with KSR. This results in KSR release and translocation to the plasma membrane. KSR inactivation involves rephosphorylation of S392 by Cdc25-associated kinase-1 (C-TAK1).

However, they can be also stimulated by chemokines and growth factors and are involved in control of apoptosis. JNK proteins are encoded by three different genes that give rise to at least 10 different isoforms by alternative splicing. Among them, the best studied are JNK1, JNK2, and JNK3. JNK1 and JNK2 are ubiquitously expressed, while JNK3 is selectively present in the CNS and heart. Importantly, studies in knockout animals have shown that each isoform plays specific roles and its activity cannot be taken over by the other isoforms.

JNK stimulation induces activation of a variety of transcription factors, the main one being c-Jun, a component of the AP-1 transcription complex. AP-1 regulates the expression of a wide number of genes (see Chapter 22) and in particular stimulates transcription of genes encoding inflammatory cytokines. The expression of these genes typically increases following environmental stress and radiation, which are the conditions in which JNK activation occurs. In addition, activated JNKs can induce the expression of enzymes, such as collagenases and metalloproteases, which drive degradation of cell matrix proteins and contribute to joint and bone destruction occurring in rheumatoid arthritis and more in general to tissue damage occurring in autoimmune diseases. These observations suggest that JNKs may play a critical role in the pathogenesis of chronic inflammatory and autoimmune disorders.

Among JNK targets, there are also proteins like Bcl-2 and p53 that are master regulators of apoptosis (see Chapter 32). JNK activation commonly promotes apoptosis by triggering release of proapoptotic molecules from mitochondria and

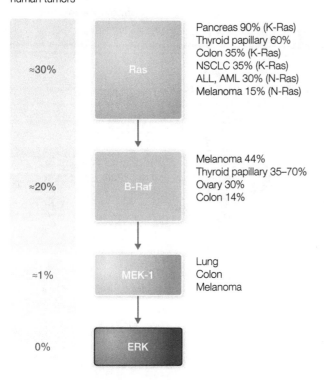

FIGURE 13.3 Mutations in Ras/Raf/MEK/ERK cascade identified in human tumors. Numbers in the left column indicate the frequency of Ras, B-Raf, MEK-1, and ERK mutations in all human tumors. In the right column are indicated the main tumors carrying these mutations and their frequency. ALL, acute lymphoblastic leukemia; AML, acute myeloid leukemia; NSCLC, non-small cell lung cancer.

expression of proapoptotic Bcl-2 family members required for execution of the apoptotic program. Interestingly, JNK activation has been observed in neurons during ischemic events and has been associated with apoptotic processes causing neuronal death. Knockout mice lacking the neurospecific JNK3 isoform show increased resistance to neuronal damage induced by excitatory agents or ischemia. These findings indicate JNK3 as a potential target for pharmacological neuroprotection in stroke.

JNK1 has been suggested to be involved in the pathogenesis of obesity and insulin-resistant diabetes. In particular, genetically obese mice, the so-called *ob/ob* mice, exhibit high JNK1 activity in the liver, skeletal muscles, and adipose tissue. Knocking down JNK1 in these animals results in reduced tendency to obesity and insulin resistance as well as in more efficient signal transduction through insulin receptors. On the contrary, JNK2 seems to be involved in the pathogenesis of insulin-dependent diabetes. Indeed, knocking out JNK2 in mice affected by type 1 diabetes results in reduced beta-pancreatic cell autoimmune destruction and slowing of disease progression. Therefore, JNK2 may control the autoimmune response typical of this type of diabetes.

The p38 Family

The p38 family consists of four members (α, β, γ, and δ) that are encoded by four distinct genes and display 60% amino acid sequence homology. These kinases show some functional redundancy; if one isoform is absent, the others can take over its function. The p38 family exhibits some analogies with the JNK family as they are both activated by cellular stress and proinflammatory cytokines. However, p38 activation induces expression of numerous cytokines and more in general of proteins involved in inflammatory processes such as type 2 cyclooxygenase (COX-2). It is therefore possible to predict a role for this MAPK subfamily (in particular for p38α) in the pathogenesis of chronic inflammatory diseases. Interestingly, many studies suggest that p38 exerts its control on proinflammatory cytokines also in the CNS. This kinase might therefore represent a new druggable target for degenerative pathologies characterized by neuronal inflammatory processes such as Alzheimer's disease, as discussed in the Supplement E 13.2.

ERK5

ERK5, also called big MAP kinase 1(BMK1), is larger than any other known MAPKs (≈80 kDa) and is distantly related to ERK1/2 as it contains a similar Thr–Glu–Tyr motif in its activation loop. ERK5 physiological function is still unclear, but cannot be taken over by ERK1/2. ERK5 knockout mice are embryonic lethal mainly because of defects in cardiovascular system development. Consistently, it has been shown that ERK5 is necessary for the activity of vascular endothelial growth factor (VEGF), the main growth factor regulating blood vessel formation.

MAPK SPECIFICITY

As described earlier, each MAPK is involved in the control of different and sometimes opposing cellular events. Considering that phosphorylation sites and substrate-interacting domains are very similar among different MAPKs, a crucial issue is the identification of the mechanisms conferring signal specificity. Factors controlling signal specificity include (i) duration of action, (ii) multiple enzyme isoforms, (iii) subcellular localization and interaction with scaffold proteins, and (iv) microRNAs (miRNAs).

Duration of Action

The first mechanism proposed to explain the specificity of MAPK action concerns signal duration and intensity. This hypothesis has been proposed based on results showing that in PC12 rat pheochromocytoma cells, both epidermal growth factor (EGF) and nerve growth factor (NGF) activate ERK via their specific tyrosine kinase receptors, but they induce opposite cell responses. Indeed, EGF acts as a growth inducer and NGF as a differentiation agent (see also Chapter 15). The ability to trigger cell proliferation or differentiation seems to depend on duration of ERK activity. Both EGF and NGF induce transient ERK activation via the canonical Ras/Raf/MEK/ERK pathway. However, NGF can also induce prolonged ERK activation through a mechanism involving stimulation of monomeric G protein Rap1 (which in PC12 cells is stimulated by NGF but not by EGF). It is important to note that this does not represent a general case. In other cell types, such as fibroblasts, transient ERK activation is also induced by nonproliferative stimuli, whereas mitogens trigger a long-lasting ERK stimulation (up to 6 h). Therefore, cellular background is a key factor in determining biological responses.

Multiple Enzyme Isoforms

Multiple isoforms of all MAPK cascade components have been described. For example, MAPKKKs include A-Raf, B-Raf, Raf-1, c-Mos, TPL2, and MEKK1. Each isoform can be activated by different extracellular stimuli and in turn stimulates different enzymes, thus increasing multiplicity and specificity of action of MAPKs in various physiological settings.

Subcellular Localization and Interaction with Scaffold Proteins

Primary signals activating MAPKs are generated at the plasma membrane where specific receptors sensing environmental stimuli are localized. In resting conditions, MAPK cascade components are located in the cytoplasm, but upon receptor stimulation, they change their cellular localization moving to other compartments where they interact with specific substrates activating specific cellular responses. For example, in resting conditions, ERKs are found in the cytosol bound to anchor proteins. Such binding is reversible, and after growth factor stimulation, about 60–70% of ERK molecules translocate to the nucleus where they modulate gene transcription. Therefore, regulation of subcellular localization of MAPK cascade components is an important cell strategy to selectively regulate biological responses following MAPK activation.

Scaffold proteins play a crucial role in the control of subcellular localization as they allow the formation of multimolecular complexes essential for proper spatial and temporal regulation of MAPK activity. Scaffold proteins bind various components of the MAPK cascade keeping them in close proximity, thus favoring efficient signal propagation. Moreover, they induce the recruitment of other target or regulatory proteins and contribute to signal stabilization by protecting MAPK from phosphatases. Among the scaffold proteins involved in MAPK regulation, kinase suppressor of Ras (KSR) is one of the best characterized. MEK constitutively binds KSR, whereas ERK binds it only after activation (Fig. 13.2). Importantly, it has been observed that expression of a Ras mutant form in KSR knockout mice does not induce skin tumors, as expected in wild-type animals, suggesting that scaffold proteins could represent an innovative and attractive pharmacological target for cancer therapy.

miRNAs

Recent studies have revealed that miRNAs are key modulators of signal propagation. miRNAs are noncoding RNAs that posttranscriptionally regulate mRNA stability, thus acting as genetic switches finely tuning cellular response to extracellular signals. Thus, mRNAs are tightly controlled by synthesis or turnover of specific regulatory miRNAs. For example, Let-7 miRNAs functionally inhibit Ras mRNA, and their expression is reduced in a subset of non-small cell lung cancer (NSCLC) patients characterized by poor prognosis. Similarly, miRNA-143/miRNA-145 is frequently lost in K-Ras mutant pancreatic cancers, and restoration of their expression has been shown to abrogate tumorigenesis.

PHARMACOLOGICAL INHIBITION OF MAPK

The central role played by MAPKs in the regulation of multiple cellular responses makes these enzymes a promising pharmacological target for human pathologies, in particular tumors and chronic inflammatory disorders. In the last few years, a number of selective inhibitors of various MAPK family members have been developed. However, up to now, the use of these inhibitors is mainly restricted to research purposes, and only few of them have been approved for clinical use.

Inhibition of the Ras/RAF/MEK/ERK Cascade

The canonical Ras/RAF/MEK/ERK cascade is the signal transduction pathway most frequently altered in human cancers. Mutational activation of Ras is found in almost 30% of all human cancers. More recently, B-RAF mutations have been described in about 20% of all tumors, and the percentage rises to 40% in melanoma (Fig. 13.3). Ras and RAF mutations induce persistent activation of ERK pathway, which results in deregulated cell growth and survival. ERK pathway upregulation keeps tumor cells in a constantly active state, thus driving cancer development and progression. Therefore, Ras and RAF inhibitors can be proposed as antitumoral agents.

MEK

Despite intensive effort, to date, inhibitors of mutated Ras have not shown any significant effect in clinical trials. For this reason, researchers have tried to block Ras indirectly, interfering with other proteins of the Ras signaling pathways. One such protein is MEK, which acts downstream Ras in MAPK pathway. Recent findings suggest that blocking MEK might be an effective way to attack tumors with mutated Ras/RAF proteins. The MEK inhibitor trametinib was the first drug of this class to be approved by the Food and Drug Administration (FDA) on May 2013 for the treatment of unresectable or metastatic melanoma positive for the B-RAF^{V600E} mutation, the most common mutated form of B-RAF. Some other MEK inhibitors are currently under clinical trials, alone or in combination with other drugs (Table 13.1). A crucial issue possibly explaining the modest clinical activity of ERK cascade inhibitors is the simultaneous dysregulation of multiple signaling pathways occurring in cancer cells. This means that inhibition of a single pathway may not be sufficient to promote growth arrest and apoptosis. Combination of ERK cascade inhibitors with agents targeting other signaling pathways (e.g., AKT and growth factor receptor inhibitors) may prove more effective compared to treatment with single pharmacological agents. Combinations with classical cytotoxic agents might also be efficacious. For example, clinical trials are currently being carried out to evaluate the MEK inhibitor selumetinib alone or in combination with various targeted and conventional chemotherapy agents (Table 13.1). It must be noted that MEK inhibitors show some severe side effects, including serious skin rashes, diarrhea, and dangerous drops in white blood cell counts. Also, a number of combination therapies have shown severe toxicity. Therefore, future efforts will be focused on finding the way to gain the most benefit from these drugs while minimizing their undesirable effects.

RAF

Promising results have also been obtained with RAF inhibitors. However, the beneficial effects of sorafenib, approved by the FDA for the treatment of kidney cancer and unresectable hepatocellular carcinoma, may not be due to its anti-RAF activity but rather to its inhibitory action on tumoral angiogenesis. Specific RAF inhibitors have given promising results in the treatment of melanoma. Vemurafenib and dabrafenib are the first RAF inhibitors to reach the market. Indeed, they have been approved (in August 2011 and May 2013, respectively) for the therapy of unresectable or metastatic melanoma positive for the B-RAF^{V600E} mutation. Interestingly, these drugs are effective only in tumors carrying this mutation, suggesting that a more in-depth knowledge of gene alterations involved in tumor initiation and progression may help to predict clinical outcome and plan targeted therapies.

In the future, new antitumoral agents capable of interfering with ERK cascade might originate from studies on the so-called lethal interactions. The aim of these studies is the identification of targets that cause cell death when inhibited in the presence of Ras mutations. An example to clarify this innovative approach is the lethal relationship existing between mutated K-Ras and the cell cycle protein CDK4; it has been shown that in a lung cancer model bearing mutated K-Ras, concurrent pharmacological or genetic blockade of CDK4 can arrest tumor progression.

JNKs

Studies in knockout animals have shown that each enzyme belonging to the JNK family exerts specific and unique functions. Therefore, for pharmacological purposes, it will be necessary to synthesize selective inhibitors. For example, a JNK3-specific inhibitor could be useful in the treatment of cerebral ischemia, whereas a JNK1 inhibitor could be considered for prevention of type 2 diabetes. However, despite the promising results obtained in preclinical models

TABLE 13.1 Ongoing clinical trials evaluating the mek inhibitor selumetinib[a]

Drugs	Disease	Phase of the study
Selumetinib	Large B-cell lymphoma	Phase II
Selumetinib	Melanoma	Phase II
Selumetinib + MK-2206	B-RAF^{V600} melanoma unresponsive to vemurafenib or dabrafenib	Phase II
Selumetinib + MK-2206	Metastaic pancreatic cancer	Randomized phase II
Selumetinib + MK-2206	Colorectal carcinoma	Pilot study
Selumetinib + erlotinib	NSCLC	Randomized phase II
Selumetinib + erlotinib	Pancreatic cancer	Phase II
Selumetinib + docetaxel	Melanoma with wild-type B-RAF	Randomized phase II
Selumetinib + docetaxel	K-Ras mutated NSCLC	Randomized phase II
Selumetinib + temozolomide	Metastatic uveal melanoma	Randomized phase II

Source: clinicaltrials.gov—U.S. National Institutes of Health.
NSCLC, non-small cell lung cancer.

[a] Selumetinib (AZD6244) is currently under study as single agent for relapsed or refractory diffuse large B-cell lymphoma and for stage III/IV melanoma harboring B-RAF or N-Ras mutations. In some other trials, it has been combined with MK-2206 (an AKT inhibitor), with erlotinib (an EGF receptor blocker) or with cytotoxic drugs such as docetaxel or temozolomide.

of cerebral ischemia and diabetes, the therapeutic properties of JNK inhibitors have not yet been confirmed in clinical trials.

Traditional JNK inhibitors block kinase activity by binding the protein ATP binding site. New inhibitors have been now designed from the minimal inhibitory sequence of JIP1, a scaffold protein that inhibits JNK activity. The first member of this drug class, the peptide D-Junk-1, strongly interferes with JNK activation in pathological but not in physiological conditions and has shown interesting therapeutic properties both in neuronal and nonneuronal tissues. So far, its use has been restricted to *in vitro* studies.

p38

p38 involvement in inflammatory cytokine production and pathogenesis of chronic inflammatory diseases has stirred a great interest in the development of drugs capable of blocking this enzyme. However, most attempts to generate clinically useful p38 inhibitors have failed because of toxicity or inadequate efficacy. As a result, companies have discontinued clinical trials with p38 inhibitors in rheumatoid arthritis and correlated disorders. Nonetheless, one of the most recent and less toxic agents, ARRY-797, has shown unexpected efficacy in the treatment of acute and chronic pain, opening up new perspectives for p38 inhibitors as analgesic drugs.

TAKE-HOME MESSAGE

- Four subfamilies of MAPKs have been so far identified: (i) ERKs, involved in proliferation and differentiation; (ii) JNKs, activated by cellular stress, growth factors, and cytokines and also involved in apoptosis and inflammation; (iii) p38, similar to JNKs for function and mechanisms of stimulation; and (iv) ERK5, involved in development of the cardiovascular system.
- Specificity of MAPK effects in different contexts is determined by duration of action, presence of multiple isoforms, interaction with specific scaffold proteins, and subcellular localization.
- The therapeutic value of ERK inhibitors for the treatment of human chronic inflammatory diseases is under investigation.

FURTHER READING

Avraham R., Yarden Y. (2012). Regulation of signalling by microRNAs. *Biochemical Society Transactions*, 40, 26–30.

Brown M.D., Sacks D.B. (2009). Protein scaffolds in MAP kinase signalling. *Cell Signalling*, 21, 462–469.

Jang S., Atkins M.B. (2013). Which drug, and when, for patients with BRAF-mutant melanoma? *Lancet Oncology*, 14, e60–e69.

Keshet Y., Seger R. (2010). The MAP kinase signaling cascades: a system of hundreds of components regulates a diverse array of physiological functions. *Methods in Molecular Biology*, 661, 3–38.

Kiel C., Serrano L. (2012). Challenges ahead in signal transduction: MAPK as an example. *Current Opinion in Biotechnology*, 23, 305–314.

Kyriakis J.M., Avruch J. (2012). Mammalian MAPK signal transduction pathways activated by stress and inflammation: a 10-year update. *Physiological Reviews*, 92, 689–737.

Santarpia L., Lippman S.M., El-Naggar A.K. (2012). Targeting the MAPK-RAS-RAF signaling pathway in cancer therapy. *Expert Opinion on Therapeutic Targets*, 16, 103–119.

Yang S.H., Sharrocks A.D., Whitmarsh A.J. (2013). MAP kinase signalling cascades and transcriptional regulation. *Gene*, 513, 1–13.

14

SMALL G PROTEINS

ERZSÉBET LIGETI AND THOMAS WIELAND

By reading this chapter, you will:

- Acquire a knowledge on the structure of small GTPases and understand the mechanism of GTP hydrolysis
- Know the physiological function of different small GTPases
- Understand the physiological mechanisms affecting the activation state of small GTPases
- Understand the mechanism of action of bacterial products and drugs that influence small GTPases
- Have a view of possible future drug targets related to small GTPases

Small G proteins (SMGs) control virtually all aspects of cell life, such as proliferation, survival, apoptosis, cell movement and contractility, shape changes, polarity, secretion, superoxide production, contacts to neighboring cells and extracellular matrix (ECM), intracellular vesicular traffic, and transport via plasma and nuclear membrane. Cell-to-cell communication mediated by soluble (hormones, paracrine agents, or neurotransmitters) or cell surface molecules involves binding and activation of plasma membrane receptors. SMGs are key elements in the downstream signaling of most plasma membrane receptors, being mainly responsible for correct timing of events. Their role in signal transduction of receptor tyrosine kinases (RTKs), such as epidermal growth factor (EGF) or platelet-derived growth factor (PDGF) receptors, was the first to be discovered. However, later, they were found to be involved also in signal transmission from G-protein-coupled receptors (GPCRs), integrins, and even ionotropic receptors.

STRUCTURE AND FUNCTION OF SMGs

SMGs are composed of one single polypeptide chain of 20–25 kDa. The G domain responsible for guanine nucleotide binding is conserved in the vast majority of guanine nucleotide-binding proteins. The H-Ras G domain, a 166-amino-acid fragment, was the first to be described, and similarities or differences in other SMGs were later reported. Its structure includes six antiparallel β-sheets surrounded by five α-helices. The nucleotide-binding pocket is lined by four to five highly conserved amino acid stretches: the N/TKXD and SAK motives responsible for the specific interaction with the nucleotide base; the P-loop around the β,γ-phosphate bond; and the regions called "switch I" and "switch II," which form hydrogen bonds with the oxygen of the nucleotide terminal phosphate group (Fig. 14.1). These hydrogen bonds (involving Ras Thr35 of switch I and Gly60 of switch II) stabilize the molecule and allow interactions with specific effector (target) proteins. Hence, the GTP-bound state is regarded as active. Members of different subfamilies of the SMG superfamily contain additional domains at the N- or C-terminal ends (as in Arf and Ran, respectively) or a helical insertion (as in Rho).

SMGs do not merely bind but also hydrolyze GTP, although their rate is one or two orders of magnitude slower than that of heterotrimeric G-protein α-subunits. GTP hydrolysis depends on a few conserved amino acids in critical position. Mutation of these residues (typically Gly12 and Gln61 in Ras) results in drastic decrease in the rate of GTP hydrolysis and insensitivity to regulatory proteins (see later), thus prolonging SMG active state.

Since in the GDP-bound state the mobility of the two switch regions is increased, a significant conformational

---

General and Molecular Pharmacology: Principles of Drug Action, First Edition. Edited by Francesco Clementi and Guido Fumagalli.
© 2015 John Wiley & Sons, Inc. Published 2015 by John Wiley & Sons, Inc.

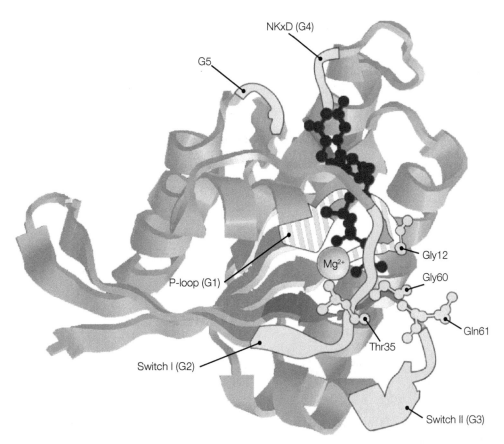

FIGURE 14.1 Structure of the highly conserved G domain. The guanine nucleotide-binding site in the G domain is formed by the flexible, catalytically important switch I/II regions, the phosphate-binding P-loop, and the NKxD motif, which controls nucleotide specificity. Mg^{2+} interacts with the nucleotide phosphate groups, and its position is determined by residues in switch I/II, P-loop, and DxxG motif (black). Binding of the guanine ring is further enhanced by direct and indirect interactions with the SAK (G5) region. Critical sequences (G1–G5) and amino acids are marked (Figure prepared by Dr. R. Csépányi-Kömi).

change accompanies GTP hydrolysis, thus terminating the interaction with the target protein. Because of their GTPase activity, G proteins are regarded as biological time switches playing an important role in controlling timing and temporal aspects of cellular processes.

PHYSIOLOGICAL ROLES OF COMPONENTS OF MAJOR SMG FAMILIES

The human genome contains approximately 150 SMGs (Ras superfamily) that are traditionally classified in five families, although some proteins cannot be assigned to any of these families. Four of the SMG families consist of many proteins with varying degree of homology, whereas the fifth family includes only one member (Table 14.1). However, the boundaries are not strict, and current research is revealing more and more interactions between members of different families (e.g., Ras and Rho or Rho and Rab). The major cellular functions in which classical members of the individual families are involved are indicated in Table 14.1. The molecular mechanism of action of SMGs involves either direct regulation of protein or lipid kinases or initiation of large molecular complex assembly; occasionally, both mechanisms are involved in the process.

Ras proteins, the first members of the entire superfamily, were discovered based on their homology to rat sarcoma virus genes. The three human Ras proteins (H-Ras, K-Ras, N-Ras) are key elements in the regulation of cell proliferation and survival. They are directly activated by different tyrosine kinase receptors (e.g., EGF, PDGF) but also by GPCRs via transactivation or second messengers. The first and best characterized direct effector of active Ras proteins is protein kinase Raf that initiates the activation of MAP kinase cascade leading to alteration of protein expression and eventually to enhanced proliferation. Another direct target is phosphatidylinositol-3 kinase (PI3K) that contributes to stimulation of protein kinase Akt. Akt itself controls protein synthesis and inhibits apoptosis, ultimately promoting cell survival. Active, GTP-bound Ras also directly interacts with PLCε and enhances generation of IP_3 and Ca^{2+} signals.

TABLE 14.1 Families of small GTPase Ras superfamily

Family	Ras	Rac/Rho	Rab	Arf	Ran
No. of SMGs	33	22	60	27	1
Representative member	H-Ras, K-Ras, N-Ras Rap1/2 Rheb	RhoA, RhoB, RhoC Rac1/2 Cdc42	Rab1–Rab60	Arf1 Arf6	Ran
Posttranslational modifications	Farnesylation (FTase) Palmitoylation	Geranylgeranylation (GGTase I)	Geranylgeranylation (GGTase II)	Myristoylation	
Localization	Plasma membrane (Golgi, ER)	Plasma membrane Cytosol	Endo- and exocytotic vesicles	Golgi ER (plasma membrane)	RanGTP: nucleus RanGDP: cytosol
No. of GEFs	29	~80	53	18	1
No. of GAPs	15	~70	38	31	1
No. of GDIs	0 (PDEδ)	3	2	0	0
Best-known direct targets	Raf kinase PI3 kinase mTOR	ROCK mDia WAVE P67phox WASP IRS53	Rabaptin Rabenosyn Early endosome antigen 1 (EEA1) PI3K	Coatomer	Importin Exportin
Main functions	Proliferation Death/survival Adhesion Protein synthesis	Cell shape Migration Contraction	Intracellular vesicular transport	Vesicle budding	Nuclear membrane transport

Active Ras proteins directly or indirectly control the function of other SMGs. Via direct interaction with relevant guanine nucleotide exchange factors (GEFs), they participate in Rac and Ral activation and indirectly control Cdc42 and Rheb activation state. Rheb is indirectly regulated by Akt, and it directly interacts and activates mTOR, the key regulator of protein synthesis.

Rap1 and Rap2 are also traditionally listed in the Ras family. These proteins have major roles in regulation of cell adhesion to ECM or to neighboring cells at adherent junctions. In cell–ECM communication, Rap1 has been shown to play a key role in "inside–out" signaling of integrins, by promoting assembly of the talin-based molecular complex that connects integrins to the actin cytoskeleton. Similarly, formation of molecular complexes that stabilize VE-cadherin and connect it to the cytoskeleton seems to be the main function of Rap1 in endothelial cells.

Members of the Ras homologous (Rho) family are primarily involved in organization of cell polarity, shape, and motility, mainly via regulation of the actin cytoskeleton. The functions of the three major members, Cdc42, Rac, and Rho, are interconnected, but Cdc42 seems to be responsible for determination of cell polarity, Rac for formation of branching actin chains of lamellipodia, and Rho for actin polymerization in form of stress fibers. Activation of Rho-family SMGs is initiated at the plasma membrane: the signaling pathway from both RTKs and GPCRs leads to several proteins regulating SMGs, activating GEFs, and/or inhibiting GAPs. The best-known direct target of Rho is Rho-kinase (ROCK) that phosphorylates and inhibits myosin light chain phosphatase, thereby promoting contraction of smooth muscle cells, but also directed movement of many other cell types allowing tail retraction. Another well-characterized direct target of Rho is mDia that is required for actin polymerization. The best-known direct targets of Rac include p21-activated kinase (PAK), involved in actin depolymerization; WAVE and IRS53, participating in the assembly of the molecular complex at the branching points of actin filaments; and p67phox, the central subunit of the superoxide-producing NADPH oxidase in phagocytes. Cdc42 directly interacts with WASP and IRS53, two proteins required for formation and bundling of actin filaments. Finally, via several intermediary interactions, both Rho and Rac have been shown to activate the Jun-kinase pathway that may contribute to metastasis formation.

Rab- and Arf-family proteins direct intracellular vesicular traffic. Rab proteins are present in the membrane of all endo- or exocytotic vesicles, and they mediate specific recognition of different vesicle populations. In their active form, they initiate the formation of large protein complexes that allow tethering of vesicles before docking and fusion. Rab proteins are also involved in coupling intracellular vesicles to motor proteins and in regulating vesicle motility. Cellular distribution of Arf proteins is more restricted: Arf1 is localized to the Golgi, whereas Arf6 is found mainly in the plasma membrane. At both sites, Arf proteins organize the formation of specific coats around budding vesicles. Besides promoting assembly of molecular complexes, both Rab and Arf SMGs modify the activity of several enzymes of the phospholipid metabolism, thereby creating favorable binding sites for members or regulators of coat assemblies.

Ras-like nuclear protein (Ran) is a key regulator of transport via the nuclear membrane in resting cells, whereas during cell division, it organizes the assembly of the mitotic spindle and reconstruction of the nuclear envelope.

POSTTRANSLATIONAL MODIFICATION AND SUBCELLULAR LOCALIZATION OF SMGs

SMGs are synthesized by cytosolic free ribosomes as hydrophilic proteins. Several steps of posttranslational modifications render them hydrophobic and allow their localization to membrane surfaces. The first step consists in the attachment of a 15-carbon (farnesyl) or 20-carbon (geranylgeranyl) prenyl group via thioester bond to a cysteine at the SMG C-terminal end.

Prenyltransferases recognize terminal CAAX or CXC motifs, where C stands for cysteine, A stands for any aliphatic amino acid, and X determines the specificity of the modifying enzyme. Farnesyltransferase (FTase) typically recognizes terminal serines, methionines, or glutamines and uses farnesylpyrophosphate (FPP) as carbon donor, whereas geranylgeranyltransferase I (GGTase I) recognizes preferably terminal leucines or phenylalanines and uses geranylgeranylpyrophosphate (GGPP) as carbon donor. These two enzymes share the same α-subunit and also recognize other amino acid sequences with low affinity. Geranylgeranyltransferase II (GGTase II) recognizes the terminal CC or CXC sequences of Rab proteins and adds two GG chains to SMGs.

Although prenyltransferases are soluble enzymes, they do not seem to react with soluble SMGs. Rather, nascent SMGs are captured by chaperone-like (scaffold, escort) proteins that regulate their interaction with prenyltransferase. In the case of SMGs containing a polybasic region (PBR) upstream of the CAAX box, this function is fulfilled by SmgGDS (GDP dissociation factor), whereas in the case of Rab proteins by Rab escort protein (Rep).

Prenylated SMGs become attached to the endoplasmic reticulum (ER), where the terminal amino acids are hydrolyzed by the endoprotease Rce1 and carboxymethylated by the isoprenylcysteine methyltransferase Icmt. Members of the Ras and Rho subfamilies without a PBR but containing another cysteine in their C-terminal region can also be palmitoylated by a Golgi-localized acyltransferase. Transfer to the plasma membrane occurs upon interaction with further regulatory proteins (see Section "GDIs").

Some of the last steps are reversible as a protein acylthioesterase can split the palmitoyl group and carboxylesterase 1 can remove the terminal methyl group. Both modifications weaken SMG membrane association.

Besides being modified at their C-terminal end, Arf proteins are typically myristoylated at their N-terminus. Prenylation, acylation, and methylation increase protein hydrophobicity. However, membrane association is also directed by electrostatic interactions between polybasic amino acid stretches, like in Ras or Rac proteins, and negatively charged phospholipids. Furthermore, also galectin-1, a protein with a hydrophobic pocket, has been shown to participate in the localization of oncogenic H-Ras to the plasma membrane inner leaflet. On the other hand, prenylation of SMGs not only influences membrane localization but also determines interaction with regulatory proteins, mainly guanine nucleotide dissociation inhibitors (GDIs, and functionally equivalent proteins), but also with selected GEFs and GTPase-activating proteins (GAPs) (see later).

Posttranslational modifications thus determine the characteristic localization of the different SMGs and are critical for their physiological function. Several mechanisms have been described in which physiological signals influence the activity of the modifying enzymes. In pancreatic β-cells, glucose has been shown to stimulate prenylation and carboxymethylation of several Rho-family GTPases by increasing FTase and GGTase I expression. These modifications accompany glucose-stimulated insulin secretion. On the other hand, insulin-activated FTase and GGTase I in several cell types significantly enhance the proportion of prenylated Ras and Rho proteins. Probably, phosphorylation of the common α-chain of the two enzymes via Shc and MAP kinases underlies this effect, which also increases tissue sensitivity to various growth factors. Elevated Ras farnesylation has been reported in hyperinsulinemic obese patients as well as in insulin receptor-expressing mammary tumors. More recently, activation of adenosine 2B receptors has been shown to induce phosphorylation of Rap1B, thereby decreasing the interaction of the SMG with SmgGDS and interfering with the prenylation step. In fact, in such conditions, a lower amount of Rap1B was found in the plasma membrane, resulting in a weakened function of adherent junctions and cell scattering. In neurons, both depolarization and brain-derived neurotrophic factor (BDNF) induce dendritic arborization via GGTase I activation and increased prenylation of Rho-family SMGs.

Pharmacological interference with prenylation, for example, with statins, has been shown to interrupt physiological regulation of insulin secretion or osteoclast activation.

REGULATORY PROTEINS

All five classical families of SMGs are controlled by two types of regulatory proteins: GEFs and GAPs. Members of the Rac/Rho and Rab families are also influenced by GDIs, whereas PDEδ plays an equivalent role in the case of Ras (Table 14.1).

GEFs

GDP dissociates very slowly from inactive monomeric GTPases. GEF binding to their GTPase substrates destabilizes GDP binding, thus promoting its dissociation. As cytosolic GTP concentration is usually about 10-fold higher than that of GDP, GTP binds to the open binding pocket. The role of a GEF is analogous to that exerted by agonist-bound GPCRs in the activation cycle of heterotrimeric G proteins; both GEFs and activated GPCRs destabilize GDP–GTPase interaction and stabilize G-protein nucleotide-free form. Next, GTP binding displaces the GEF, converting the GTPase in its active conformation.

GEFs are generally family specific and possess a characteristic domain of 20–30 kDa that is responsible for their activity. Several distinct GEF domains have been identified and structurally characterized. CDC25 and Sec7-like domains are found in GEFs interacting with the Ras and Arf families, respectively. For the Rho family, two structurally unrelated GEF domains exist in the animal kingdom. One is the DH domain, which is found in the Dbl-homology family of RhoGEFs always along with an adjacent pleckstrin homology (PH) domain, making up the DH–PH tandem motif typical of that GEF family. The other is the DHR2 catalytic domain, which is found in RhoGEFs of the DOCK family along with an upstream DHR1 domain required for membrane attachment. In addition to the critical GEF domain, GEF proteins contain a large variety of other domains, typically involved in protein–protein interactions or binding of phospholipids that they require for proper localization and regulation. In a few cases, the same protein is endowed with GEF activities for two different families, like RasGRF1 that has both Ras- and RacGEF activity and Trio that contains an N-terminal Rac-specific and a C-terminal Rho-specific DH–PH motif. The catalytic activity in many GEFs is under an autoinhibitory constraint due to intramolecular interactions. Processing of incoming signals and selective activation of proper downstream pathways occur via a common mechanism relieving such autoinhibition. Therefore, GEFs typically exert their actions in large molecular complex coupling, for example, plasma membrane receptors to activation of monomeric GTPases. In such complexes, relief of GEF autoinhibition can occur directly by protein–protein interaction or indirectly via production of second messengers (e.g., cAMP) or phosphorylation.

Here, below, we provide some examples to elucidate this principle:

1. After ligand binding, EGF receptor gets autophosphorylated on tyrosine residues. Via the SH2-domain-containing adapter protein Grb2, the RasGEF Sos translocates to the receptor complex and is activated by it, thus initiating the Ras-dependent Raf/Erk signaling cascade.
2. The NR2B subunit of NMDA glutamate receptors interacts directly with RasGRF1, coupling NMDA receptor stimulation to the Ras/Raf/Erk pathway.
3. Vasoconstricting mediators like endothelin-1, vasopressin, or thromboxane A_2 act on GPCRs coupled both to G_q/G_{11} and G_{12}/G_{13} heterotrimeric G proteins. The $G\alpha_{12}$- and $G\alpha_{13}$-subunits are able to bind members of the p115RhoGEF family (p115RhoGEF, PDZ-RhoGEF, and LARG), induce their translocation to the plasma membrane, and relieve their autoinhibition. $G\alpha_q$ forms a complex with and activates the membrane-attached p63RhoGEF. Thus, in vascular smooth muscle cells, more than one pathway links GPCR signal to Rho activation resulting in myosin phosphorylation and contraction.
4. GPCRs activating G_s-dependent adenylyl cyclases (e.g., β-adrenoceptors) stimulate cAMP synthesis. Besides activating its canonical effector protein kinase A, cAMP binds RapGEF EPAC and relieves autoinhibition by inducing a conformational change that allows free access of Rap to the Cdc25 domain. Thereby, activated Rap links G_s signaling to phospholipase Cε and protein kinase Cε activation.
5. The βγ-subunits of heterotrimeric G proteins activated by GPCRs (e.g., chemotactic receptors in neutrophil granulocytes) activate RacGEF P-Rex in cooperation with PI3K that produces phosphatidylinositol-3,4,5-phosphate, the suitable docking site for P-Rex. Activated Rac is a central regulator of both migration and superoxide production.
6. Activated SMGs, specifically RasGTP, are also able to bind GEFs as target proteins and induce activation of other small GTPases. For example, Tiam1 is a GEF with a Ras-binding domain. RasGTP stimulates Tiam1 RacGEF activity, resulting in Ras-induced activation of Rac and Rac-dependent cellular processes.

GAPs

These proteins enhance the intrinsic GTP hydrolytic activity of SMGs, thereby promoting their transition to the GDP-bound inactive conformation. In this way, they favor downregulation of SMG-mediated signaling, similarly to what regulator of G-protein signaling (RGS) proteins do on heterotrimeric G proteins. Whether GAPs inhibit or terminate the regulated intracellular process depends on the rate of its activation.

Similarly to GEFs, GAPs exert their action specifically on certain families of SMGs, but they may react with several members within the same family. The GAP activity is confined to a domain of 20–50 kDa, generally referred to as GAP domain. They possess a large variety of other domains allowing interactions with diverse proteins or lipids or regulatory mechanisms. The catalytic mechanism of the family-specific GAP domains shows some variations, but in most cases, it relies on a critical residue that in RasGAP and RhoGAPs is an arginine.

There are a few examples of dual-specificity GAPs. Such dual specificity may be due to two separate GAP domains, like in the case of ARAP1, which contains both an ArfGAP and a RhoGAP domain whose activities are independent. Alternatively, as in the case of SynGAP, the protein contains one canonical RasGAP domain and regions outside the GAP domain that contribute to RapGTP hydrolysis.

Substrate specificity of individual GAP molecules can be determined *in vitro*, or in cellular models. However, the results of the two approaches are not always consistent, as cellular regulatory processes may alter GAP substrate specificity. For example, association of p190RhoGAP to acidic phospholipids reversibly switches its substrate preference from Rho to Rac. Conversely, PKC-catalyzed phosphorylation of residues within the lipid-binding polybasic domain induces p190RhoGAP dissociation from the lipid bilayer and switches its substrate preference in the opposite direction, from Rac to Rho. Phosphorylation of MgcRacGAP by aurora B kinase extends its substrate specificity from Rac also to Rho.

In their native state, several GAPs are in a folded, autoinhibited conformation that limits accessibility of their catalytic site. Some GAPs also distinguish the small GTPase-prenylated form from the nonprenylated, restricting the reaction with the nonprenylated. Multidomain GAPs are subject to a wide variety of further regulatory mechanisms including phosphorylation, lipid binding, protein interactions, and degradation, which may affect SMG amount, activity, or localization.

Based on the GAP domain consensus sequence, about 150 potential GAPs have been identified in the human genome, although not all of them have been analyzed so far and some of those tested have shown no GAP activity. The number of potential GAPs specific for the Rho family largely exceeds the number of proteins belonging to this SMG family (Table 14.1).

Some experimental data suggest the involvement of specific GAPs in well-defined molecular complexes and processes, such as α2-chimaerin involvement in guidance receptor EphA4 signaling or oligophrenin-1 association to distinct glutamate receptor subunits. On the other hand, specific phenotypes resulting from mutation or lack of certain GAP molecules both in genetically modified animals and in human subjects suggest that replacement of one GAP by another of similar substrate specificity may be rather restricted.

Inhibition or deletion of certain GAPs has been shown to result in enhancement of specific functions (e.g., superoxide production and phagocytosis in neutrophil granulocytes, cell proliferation and tumor development), suggesting that constitutive GAP activity controls vital processes by downregulating SMG activities. However, there are examples showing that transient inhibition of GAP activity, either alone or along with GEF activity, is required for stimulated functions to appear.

Thus, GAPs are involved in cell activation, in spite of the fact that their main activity consists in downregulating SMG activity.

GDIs

GDIs are regulatory proteins acting only on members of the Rho and Rab families (Table 14.1). They preferentially interact with the GDP-bound form of the GTPase and prevent nucleotide exchange (their activity is similar to that exerted by the βγ dimers on the heterotrimeric G-protein activation cycle). Both RhoGDI and RabGDI bind SMGs in their prenylated form; consistently, the GDI crystal structure reveals a long lipid-binding pocket. RhoGDIs interact with several Rho-family members and are able to extract them from the membrane and sequester them in the cytosol.

Their main function seems to be the transfer of newly synthetized, posttranslationally modified prenylated SMGs from the ER/Golgi to the plasma membrane. However, the mechanism whereby SMGs dissociate from GDIs is not fully understood. Earlier studies indicated a role for phospholipids and phosphorylation, as well as for specialized GDI displacement factors. Recent data indicate that PDEδ, another protein with a deep lipid-binding pocket, plays probably a similar role in Ras protein shuttling to the plasma membrane.

MODULATION OF SMG SIGNALING BY BACTERIAL TOXINS AND DRUGS

Bacterial Toxins

As outlined before, the activity of monomeric GTPases is tightly controlled. This is also the case for the about 20 members of the Rho family. The most studied members of this family, RhoA, Cdc42, and Rac, are well known for their essential role in cytoskeleton regulation, which is very important for host–bacterial pathogen interactions. The Cytoskeleton is essential for epithelial and endothelial barrier functions, which prevent and limit pathogen invasion and dissemination within tissues. In addition, the immune cell cytoskeleton regulates migration, phagocytosis, and immune synapse formation. RhoGTPases are also involved in many of the signaling pathways required for effective immune responses. Indeed, they participate in development, activation, and function of B and T cells; leukocyte migration; phagocytosis; chemotactic gradient sensing; and regulation of NADPH oxidase activity in phagocytes.

Therefore, many bacterial pathogens try to overcome host defenses by interfering with host RhoGTPase activity. Many immune functions are blocked and some physiological processes of host cells are hijacked by the action of bacterial toxins. Many of them are exotoxins secreted by bacteria and taken up by host cells through their autonomous membrane translocation systems. Others are directly injected into host cells by cell-bound bacteria using type III or type IV secretion systems. Once in the eukaryotic cell, they cause modifications that affect RhoGTPase regulatory functions, thus interfering with the corresponding signaling pathways. By using such toxins, pathogens establish a "niche" for proliferation in a principally hostile environment.

Bacterial Toxins Modifying Host RhoGTPases

A growing number of bacterial protein toxins affect host RhoGTPase activity by catalyzing their covalent modifications (Table 14.2). The most common mechanisms underlying bacterial toxin entry in host cells are shown in Figure 14.2. The enzymatic activities of these toxins are diverse, and the modifications they catalyze include ADP-ribosylation, glycosylation, adenylylation, proteolysis, deamidation, and transglutamination. Some modifications (deamidation, transglutamination) of two important residues, Gln61 (Rac, Cdc42) or Gln63 (RhoA), inhibit GTPase activity resulting in RhoGTPase constitutive activity (Table 14.2, Fig. 14.3). RhoGTPase activation by Cnf1

TABLE 14.2 Bacterial toxins and effectors that modify signaling by monomeric GTPases

Toxins/effectors	Source organisms	Mode of action	Target GTPases
Activators			
Cnf1, Cnf2, Cnf3	Escherichia coli	A/B toxin, deamidation	Rho proteins, Rac proteins, Cdc42
Cnf	Yersinia pseudotuberculosis	A/B toxin, deamidation	RhoA
Dnt	Bordetella pertussis	A/B toxin, deamidation and transglutamination	Rho-family proteins
SopE and SopE2	Salmonella enterica subsp. enterica serovar Typhimurium	T3S, GEF	Rac proteins, Cdc42
BopE	Burkholderia pseudomallei	T3S, GEF	Rac proteins, Cdc42
IpgB2	Shigella flexneri	T3S, GEF	RhoA, Rac proteins, Cdc42
Map	Escherichia coli	T3S, GEF	Cdc42
EspM1, EspM2, EspM3	Escherichia coli Citrobacter rodentium	T3S, GEF	RhoA
EspT	Escherichia coli Citrobacter rodentium	T3S, GEF	Rac proteins, Cdc42
SifA and SifB	Salmonella typhimurium	T3S, GEF	Rho proteins?
DrrA	Legionella pneumophila	T4S, GEF, adenylylation	Rab1
EspG	Escherichia coli	Blocks ARFGAP binding	Arf1
Inhibitors			
SptP	Salmonella typhimurium	T3S, GAP	Rac proteins, Cdc42
YopE	Yersinia spp.	T3S, GAP	Rho proteins, Rac proteins, Cdc42
ExoS and ExoT	Pseudomonas aeruginosa	T3S, GAP	Rho proteins, Rac proteins, Cdc42
LepB	Legionella pneumophila	T4S, GAP	Rab1
C3	Clostridium botulinum	Exoprotein, ADP-ribosylation	RhoA, RhoB, RhoC
C3	Clostridium limosum	Exoprotein, ADP-ribosylation	RhoA, RhoB, RhoC
C3cer	Bacillus cereus	Exoprotein, ADP-ribosylation	RhoA, RhoB, RhoC
C3stau	Staphylococcus aureus	Exoprotein, ADP-ribosylation	RhoA, RhoB, RhoC
Toxins A and B	Clostridium difficile	A/B toxin, glucosylation	Rho-family proteins
Lethal toxin	Clostridium sordellii	A/B toxin, glucosylation	Ras-family proteins, Rac proteins
Hemorrhagic toxin	Clostridium sordellii	A/B toxin, glucosylation	Rho-family proteins
α-Toxin	Clostridium novyi	A/B toxin, N-acetylglucosamination	Rho-family proteins
TpeL	Clostridium perfringens	A/B toxin, N-acetylglucosamination	Ras, Ral, Rac, and Rap proteins
VopS	Vibrio parahaemolyticus	T3SS, adenylylation	Rho-family proteins
IbpA	Histophilus somni	T3SS, adenylylation	Rho-family proteins
YopT	Yersinia spp.	T3SS, proteolytic cleavage	Rho-family proteins

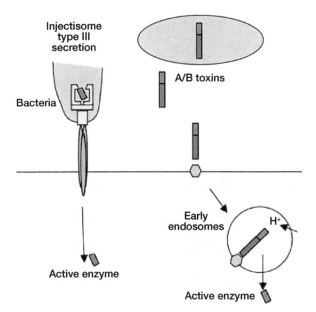

FIGURE 14.2 Modes of entry of bacterial toxins or effectors into host cell cytosol. Some bacteria (see examples in Table 14.2) have developed complex injection machineries (type III or type IV injectisomes) to directly insert virulence factors into host cell cytosol (see left portion of the figure). A/B toxins (see examples in Table 14.2) are secreted from bacteria and can cross cell membranes by receptor-mediated endocytosis. One component binds the surface receptor and stimulates vesicular endocytosis. Endosomes acidification triggers proteolytic processing and/or conformational change of toxins and allows the delivery of their catalytic domain to the cytosol (right portion of the figure).

deamidase causes formation of stress fibers, lamellipodia, and filopodia due to RhoA, Rac, and Cdc42 activation. In addition, Cnf1 induces multinucleate cell formation and uncoupling of S phase from mitosis. Although all Cnfs can deamidate a variety of RhoGTPases *in vitro*, they show a higher degree of specificity in living cells. Indeed, Cnf1 and Cnf2 from *Escherichia coli* activate several RhoGTPases. In contrast, *E. coli* Cnf3 and *Yersinia pseudotuberculosis* Cnf act specifically on RhoA. In addition, other modifications, like ADP-ribosylation of RhoA Asn41 residue, glycosylation, *N*-acetylglucosamination, or adenylylation of Thr35 (Rac, Cdc42) or Thr37 (Rho), lead to RhoGTPase inactivation. ADP-ribosylation of RhoA, RhoB, and RhoC Asn41 residue is catalyzed by C3-like inhibitory toxins of which at least seven are known sharing approximately 35–90% sequence identity (Table 14.1). All these toxins are small (~25 kDa) proteins and consist of only one ADP-ribosyltransferase domain. Although their cell accessibility is generally low, monocytes and macrophages take up efficiently the C3 toxin of *Clostridium botulinum*, suggesting that these cells are preferred targets for the toxin. The C3-like toxin of *Staphylococcus aureus* (C3stau) does not need to be taken up. The bacterium invades the host cell and secretes the toxin directly into the cytoplasm. ADP-ribosylated Rho proteins exhibit tight, high-affinity binding to GDI. This prevents RhoA translocation from the cytosol to the plasma membrane and thereby activation by GEFs. For example, loss in RhoA activity results in disassembly of stress fibers and redistribution of the actin cytoskeleton. Mono-O-glucosylation or N-acetylglucosamination of RhoA, RhoB, and RhoC Thr37 residue is another covalent modification carried out by bacterial toxins to inactivate these GTPases. This amino acid is involved in Mg^{2+} and GTP binding, and the modified proteins can no longer interact with GEFs and GAPs. Moreover, such covalent modification hinders the transition to the active conformation, even after GTP binding, and therefore prevents interaction with effectors. Thus, glucosylated or N-acetylglucosaminylated monomeric GTPases are uncoupled from cycling. They are bound to membranes and are not released into the cytosol. The prototypes of this toxin family are *Clostridium difficile* toxins A and B, which belong to the so-called A/B toxins. Such toxins bind with their C-terminus to their cell surface receptor, resulting in endocytosis of the toxin–receptor complex. The low pH of endosomes determines a structural change in these toxins, allowing their membrane insertion. After translocation of the toxin glucosyltransferase and protease domains into the cytosol, the cysteine protease domain autocatalytically cleaves the toxin and releases the glucosyltransferase domain into the cytosol. Toxin A and toxin B are the major virulence factors used by *C. difficile* to cause diarrhea and pseudomembranous colitis.

Clostridial glucosylating toxins display different preferences toward GTPases. *C. difficile* toxins glucosylate many members of the Rho subfamily (see Table 14.1). *Clostridium sordellii* lethal toxin specifically modifies Rac1 and Rac2 proteins. However, it additionally modifies small GTPases of the Ras family such as Ral, Rap, and Ras proteins. Ras-family proteins are also the main target of the *Clostridium perfringens* toxin TpeL. The highly conserved Thr37 in RhoA (Thr35 in Rac1 and Cdc42) can additionally be modified by toxin-catalyzed adenylylation, which consists in the addition of an AMP moiety derived from ATP. *Vibrio parahaemolyticus*, a bacterium causing food-borne gastroenteritis, secretes toxin VopS via a type III secretion system (T3SS). Once in the eukaryotic cell, VopS catalyzes RhoGTPase adenylylation, thereby inhibiting downstream signaling. Also, subspecies of *Yersinia*, for example, human pathogenic *Yersinia enterocolitica*, employ T3SS for injecting up to six effector proteins (*Yersinia* outer proteins (Yops)) into eukaryotic target cells. The outer protein YopT is a cysteine protease, which cleaves Rho-family proteins directly upstream the C-terminal cysteine residue carrying the isoprenyl moiety. Thus, GTPase membrane attachment is prevented and the protein is inactivated. Although RhoA is apparently the preferred substrate, YopT can also cleave RhoG and Rac proteins.

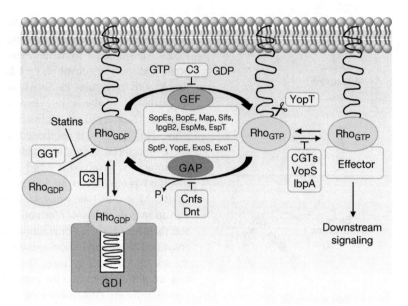

FIGURE 14.3 Bacterial toxins and effectors that modify the GTPase cycle of monomeric G proteins exemplified by Rho-family GTPases. Monomeric G proteins can adopt an active GTP-bound state and an inactive GDP-bound state. The choice between these two states is controlled by the relative activity of two proteins: guanine nucleotide exchange factors (GEFs) and GTPase-activating proteins (GAPs). While GEFs promote GDP release and thereby binding of intracellularly abundant GTP, GAPs increase the intrinsically slow GTPase activity of SMGs leading to rapid GTP hydrolysis and phosphate release. The bacterial effectors SopE, SopE2, BopE, Map, IpgB2, EspM1–EspM3, EspT, and SifA and SifB act as GEFs for eukaryotic RhoGTPases (see Table 14.2), whereas SptP, YopE, ExoS, and ExoT act as GAPs. Prenyltransferases, such as geranylgeranyltransferase (GGT), add lipophilic isoprenyl or fatty acid moieties to C-terminal cysteine residues. These lipophilic moieties serve as membrane anchors. Guanine nucleotide dissociation inhibitors (GDIs) of Rho and Rab GTPase families bind to the GDP-bound, lipid-modified form of the GTPase preventing nucleotide exchange. By hiding the GTPase lipophilic moiety in a long binding pocket, they are able to extract GTPases from the membrane and sequester them in the cytosol. ADP-ribosylation of Rho-family proteins by C3-like toxins (C3) causes GDIs to tightly bind GTPases and inhibits activation by GEFs. The toxin YopT is a protease that cleaves the C-terminal isoprenylated cysteine of Rho-family GTPases, leading to their release from the membrane and thereby inactivation. Clostridial glucosylating toxins (CGTs) attach glucose or N-acetylglucosamine to Rho-family proteins hindering their transition to the active conformation. The toxins VopS and IbpA catalyze adenylylation of Rho-family proteins. Adenylylated GTPases cannot interact with their cellular effectors. The cytotoxic necrotizing factors (Cnfs) and dermonecrotizing toxin (Dnt) cause deamidation and transglutamination of RhoGTPases, blocking GTP hydrolysis and hence making GTPases constitutively active.

Bacterial Toxins Acting as GEFs or GAPs for Host GTPases

Salmonella enterica subsp. *enterica* serovar Typhimurium injects two effectors, SopE and SopE2 (Table 14.2), into eukaryotic cells via a T3SS. Both effectors activate Rac proteins and Cdc42 in a GEF-like manner. The effector BopE from *Burkholderia pseudomallei* is a bacterial GEF very similar to SopE and SopE2 family. A family of so-called WXXXE effectors are produced by *Shigella flexneri* (IpgB1 andIpgB2), *E. coli* (Map and EspM2), *Citrobacter rodentium* (EspT), and *Salmonella typhimurium* (SifA and SifB). They share a common V-shaped three-dimensional (3D) structure with the SopE effectors and act as GEFs, too. The WXXXE motif is not directly involved in the catalysis of nucleotide exchange.

Remarkably, the structural changes induced at RhoGTPases by interaction with bacterial toxins of the WXXXE and SopE families resemble those that are induced by host cell GEFs belonging to the Dbl family, indicating a common molecular mechanism to catalyze nucleotide exchange. Also, monomeric GTPase Rab1A is targeted by a bacterial GEF. The type IV secretion system effector DrrA from *Legionella pneumophila* increases nucleotide exchange at Rab1 very efficiently. In addition, DrrA catalyzes Rab1 adenylylation and prevents its inactivation by GAPs (see following text).

EspG, a type III secretion effector from enterohemorrhagic *E. coli* (EHEC), also keeps ARF-family GTPases in the active state but via a different mechanism. It binds in a chaperone-like manner to ARFGTP, thereby inhibiting its interaction with host's ARFGAPs. ARFGTP hydrolysis and nucleotide exchange activities are necessary for vesicle transport in the Golgi network, which is therefore suppressed by EspG.

As outlined before, eukaryotic GAPs contain an arginine residue (often called arginine finger) that, together with a highly conserved glutamine residue in the SMG GTPase catalytic domain, is required for efficient GTP hydrolysis. The *S. typhimurium* 3TSS effector SptP also contains an arginine finger for rapid GTP hydrolysis by Rho-family proteins and thus acts as a RhoGAP. The exoenzymes ExoS and ExoT from *Pseudomonas aeruginosa* and YopE from *Y. pseudotuberculosis* facilitate GTP hydrolysis of Rho-family proteins by the same mechanism. A second type IV secretion system effector from *L. pneumophila*, LepB, has RabGAP activity, thus providing another way to disturb vesicle traffic in host cells.

Modulation of the Prenylation Process

Small-molecule FTase inhibitors have been developed, and several of them (such as tipifarnib, lonafarnib, BMS-214662, and L-778123) have been tested in many clinical studies targeting mainly oncogenic Ras proteins. However, no significant success has been reached up to now. The major reason is probably that several SMGs (most importantly K-Ras and N-Ras) escape inhibition via compensatory activity of GGTase I. Experiments on gene-deficient animals suggest that inhibition of both FTase and GGTase I could be more efficient.

An alternative approach to enzyme inhibition consists in the interference with prenyl substrate synthesis. Statins inhibit HMG-CoA reductase with consequent decrease of production of mevalonate, a common precursor in the synthesis of both cholesterol and FPP and GGPP. However, statins disrupt protein prenylation only at very high concentrations. Nitrogenous bisphosphonates, used in the treatment of osteoporosis, inhibit FPP synthase and deplete cells of both FPP and GGPP. Their high bone affinity provides a rationale for their use in bone tumors. Inhibitors of GGPP synthase are in development phase.

Inhibition of prenylation may have unexpected outcome: in one study, the applied statin inhibited prenylation of all major Rho-family members, but the amount of activated Rho, Rac, and Cdc42 and their biological effects were increased, due to impaired interaction with RhoGDI.

The most recent approaches include small molecules that interfere with K-Ras binding to PDEδ and attenuate shuttling of prenylated Ras to the plasma membrane and farnesyl-mimicking molecules that interfere with docking of prenylated Ras on galectin-1 in the plasma membrane.

Modulation of GEFs

Small-molecule chemicals, as a major structural class of drugs, are broadly pursued in targeting signaling pathways, and many of the drugs used today actually target GPCRs as surface proteins. However, heterotrimeric G proteins and monomeric GTPases are considered difficult to target as they are involved in too many parallel pathways. In addition, they exhibit globular structures with limited druggable surface areas. Therefore, as GEFs are pivotal regulators of monomeric GTPase activity and exhibit specific functions in distinct signaling pathways, they have been considered better potential drug targets for a control of the intracellular signal pathways. Still, GEFs are intracellular proteins and also require suitable hydrophobic pockets to be considered druggable. This significantly limits the scope of drug discovery efforts. Therefore, 3D structures of GEF–monomeric GTPase complexes have been used for structural mapping studies and virtual screening approaches to search for specific small-molecule inhibitors of monomeric GTPase–GEF interactions. The first substance identified using such approach is NSC23766, which blocks the interaction between Rac proteins and Tiam1 and Trio but not Vav GEFs (Table 14.2). Whether other Rac-specific GEFs are also inhibited has not been determined. NSC23766 is able to reach therapeutic intracellular concentrations both in cell culture and *in vivo*. Indeed, it has been studied with positive outcome in animal models for a variety of disorders, such as cardiac remodeling, cancer, and neuropathic pain, which are thought to benefit from inhibition of Rac activity. In addition, a more potent derivate of NSC23766, EHop-016, has been developed. Despite the beneficial profile reported by previous studies on NSC23766, a recent publication indicates a high potential for adverse effects. In fact, NSC23766 is not only an inhibitor of Rac1 GTPase activation by Tiam and Trio GEFs but also a nonselective, competitive antagonist at muscarinic acetylcholine receptors, within the same concentration range used for inhibiting Rac1 activation. In line with functional data, NSC23766 docking to the orthosteric binding pocket of muscarinic acetylcholine receptors has revealed that NSC23766 can be easily accommodated into it and can establish interactions with the receptor proteins. A variety of drugs currently used in clinical practice (e.g., oral antipsychotics) are also antagonists at muscarinic acetylcholine receptors and cause parasympatholytic side effects, such as blurred vision, dry mouth, and urinary retention, which often cause discontinuation of the treatment. Therefore, the unexpected antagonistic effect of NSC23766 at muscarinic acetylcholine receptors will clearly limit its potential as novel therapeutic agent. This highlights the need for proper pharmacological testing of such inhibitors before pursuing them further during the drug development process.

Nevertheless, virtual and high-throughput screenings for compounds capable of interfering with GEF–monomeric GTPase interaction have been successful for a variety of GEFs and GTPases (see Table 14.3), although the relevance of these novel compounds for future therapy still remains to be determined.

TABLE 14.3 Small molecules inhibiting activation of monomeric GTpases by guanine nucleotide exchange factors

Compound	Targeted GEFs	Targeted GTPase	Additional information
NSC23766 (N6-(2-((4-(diethylamino)-1-methylbutyl) amino)-6-methyl-4-pyrimidinyl)-2-methyl-4, 6-quinolinediamine)	Tiam1, Trio (N-terminal DH)	Rac	Also antagonist at muscarinic acetylcholine receptors
EHop-016 ((2-chloro-pyrimidin-4-yl)-(9-ethyl-9H-carbazol-3-yl)-amine)	Tiam1, Trio (N-terminal DH), Vav	Rac (Cdc42)	Derivative of NSC23766 with higher potency, inhibits also Cdc42 activation at higher concentrations
AZA1 (N*2*,N*4*-bis-(2-methyl-1H-indol-5-yl)-pyrimidine-2, 4-diamine)	Not determined	Rac, Cdc42	Derivative of NSC23766, only limited information available so far
ITX3 (2-((2,5-dimethyl-1-phenyl-1H-pyrrol-3-yl) methylene)-thiazolo(3,2-a) benzimidazol-3(2H)-one)	Trio (N-terminal DH)	Rac, RhoG	Seems to target a different interface compared to NSC23766
CPYPP (4-(3′-(2″-chlorophenyl)-2′-propen-1′-ylidene)-1-phenyl-3,5-pyrazolidinedione)	DOCK2, DOCK5, DOCK180	Rac	Only limited information available so far
ZCL278 (4-(3-(2-(4-bromo-2-chloro-phenoxy)-acetyl)-thioureido)-N-(4,6-dimethyl-pyrimidin-2-yl)-benzenesulfonamide)	Intersectin	Cdc42	Only limited information available so far
Y16 (4-(3-((3-methylbenzyl) oxy) benzylidene)-1-phenyl-3,5-pyrazolidinedione)	LARG, p115RhoGEF, PDZ-RhoGEF	RhoA (likely RhoB and RhoC)	Only limited information available so far
Brefeldin A	Sec7 domain GEFs	Arf	Allosteric inhibitor, highly cytotoxic possibly due to unrelated effects
Tyrphostin (AG1478)	Sec7 domain GEFs	Arf	Primarily described as an inhibitor of growth hormone receptor tyrosine kinase activity, apparently less cytotoxic
DCAI (4,6-dichloro-2-methyl-3-aminoethyl-indole)	Sos	Ras	Only limited information available so far
CE3F4 (6-fluoro-5,7-dibromo-2-methyl-1-formyl-1,2,3, 4-tetrahydroquinoline)	Epac	Rap	R-enantiomer is the most potent antagonist of cAMP binding, preferentially inhibits Epac1
ESI-09 (3-(5-tert-butyl-isoxazol-3-yl)-2-[(3-chlorophenyl)-hydrazono]-3-oxo-propionitrile)	Epac	Rap	Antagonist of cAMP binding

Modulation of GAPs

Small molecules capable of modifying the function of one or few specific GAPs are not known at present. A better characterization of the molecular complexes in which individual GAPs are involved and the effect of phosphorylation by definite kinases on their function or localization will offer new targets to modulate GAP activity in selected function.

In experimental settings, inhibition of GAP activity can be achieved by NaF. The fluoride complexes of aluminum and magnesium have been shown to mimic the GTP-bound state of the heterotrimeric G-protein α-subunit. In the case of SMGs, fluoride compounds allow formation of a stable complex between SMG and interacting GAP. Thus, addition of NaF (enough amounts of aluminum are present in all solutions and glassware) "titrates" out GAPs and diminishes their effect.

FUTURE PERSPECTIVES

Upregulation of SMG activity is most frequently the consequence of activating mutations, like the various Ras mutations found in many tumors. However, overexpression of GEFs (mainly RhoGEFs) or lack of GAPs (e.g., NF) has also been documented in the genesis of tumors. In these cases, existing and perspective therapy aims at inhibiting members of the enzyme cascade (e.g., Raf, PI3K, Akt) initiated by the overactivated SMG. Alternatively, downregulation of constitutively activated SMGs using siRNAs has also been attempted (see also Chapter 25).

TAKE-HOME MESSAGE

- Small GTPases regulate all aspects of cell life.
- Prevalence and duration of their GTP-bound active state are controlled by the activity of regulatory proteins (GEFs, GAPs, and GDIs) and factors controlling SMG prenylation state.
- Bacterial pathogens affect SMG activity by producing toxins directly inhibiting them or factors interfering with SMG regulatory proteins.
- Current pharmacological approaches target SMG prenylation and SMG–GEF complex formation.

FURTHER READING

Aktories K. (2011). Bacterial protein toxins that modify host regulatory GTPases. *Nature Reviews. Microbiology*, 16, 487–498.

Cherfils J., Zeghouf M. (2013). Regulation of small GTPases by GEFs, GAPs, and GDIs. *Physiological Reviews*, 93, 269–309.

Csépányi-Kömi R., Sáfár D., Grósz V., Tarján Z.L., Ligeti E. (2013). In silico tissue-distribution of human Rho family GTPase activating proteins. *Small GTPases*, 4, 90–101.

Gao Y., Dickerson J.B., Guo F., Zheng J., Zheng Y. (2004). Rational design and characterization of a Rac GTPase-specific small molecule inhibitor. *Proceedings of the National Academy of Sciences of the United States of America*, 101, 7618–7623.

Hong L., Kenney S.R., Phillips G.K., Simpson D., Schroeder C.E., Nöth J., Romero E., Swanson S., Waller A., Strouse J.J., Carter M., Chigaev A., Ursu O., Oprea T., Hjelle B., Golden J.E., Aubé J., Hudson L.G., Buranda T., Sklar L.A., Wandinger-Ness A. (2013). Characterization of a Cdc42 protein inhibitor and its use as a molecular probe. *Journal of Biological Chemistry*, 288, 8531–8543.

Levay M., Krobert K.A., Wittig K., Voigt N., Bermudez M., Wolber G., Dobrev D., Levy F.O., Wieland T. (2013). NSC23766, a widely used inhibitor of Rac1 activation, additionally acts as a competitive antagonist at muscarinic acetylcholine receptors. *Journal of Pharmacology and Experimental Therapeutics*, 347, 69–79.

Ligeti E., Welti S., Scheffzek K. (2012). Inhibition and termination of physiological responses by GTPase activating proteins. *Physiological Reviews*, 92, 237–272.

Shang X., Marchioni F., Evelyn C.R., Sipes N., Zhou X., Seibel W., Wortman M., Zheng Y. (2013). Small-molecule inhibitors targeting G-protein-coupled Rho guanine nucleotide exchange factors. *Proceedings of the National Academy of Sciences of the United States of America*, 110, 3155–3160.

Vetter I.R., Wittinghofer A. (2001). The guanine nucleotide-binding switch in three dimensions. *Science*, 294, 1299–1304.

Zverina E.A., Lamphear C.L., Wright E.N., Fierke C.A. (2012). Recent advances in protein prenyltransferases: substrate identification, regulation, and disease interventions. *Current Opinion in Chemical Biology*, 16, 544–552.

15

INTEGRATION IN INTRACELLULAR TRANSDUCTION OF RECEPTOR SIGNALS

JACOPO MELDOLESI

By reading this chapter, you will:

- Understand how *in vivo* cell homeostasis is governed by the simultaneous activity of multiple receptors often operating coordinately (crosstalk)
- Learn that any pharmacological treatment targeting receptors or intracellular signaling cascades can affect cell homeostasis not only at specific sites but also as a whole
- Understand that the effects of treatments specifically targeting receptors depend on transduction mechanisms that may vary from cell to cell

---

In pharmacology textbooks, intracellular signaling triggered by receptor activation is most often discussed in several separate chapters, each one focused on the specificity and peculiarities of the specific transduction pathway, such as rapidity for receptor channels, signal multiplicity for the G-protein-coupled receptors (GPCRs), or slowness for receptors inducing cell proliferation and/or gene transcription. When receptors are considered separately, these properties are correct. However, in living cells, signals do not occur separately. Rather, they are part of a complex scenario whose characteristics depend on a large variety of factors, such as multiplicity of receptors and intracellular signaling cascades, complexity of molecular architectures, and differential expression of targets, all working in an integrated fashion with also many other factors. Ideally, at least some of these properties should be taken into consideration when describing receptors and their signaling mechanisms. Practically, this is impossible. This chapter represents an attempt to drive readers' attention to these issues by presenting a series of specific examples in which signal integration has been found to be both biologically and therapeutically relevant.

DUALISM OF RECEPTORS IN THE NUCLEUS AND IN THE CYTOPLASM

The heterogeneity and complexity of the responses induced by individual agonists acting on different receptors on the same cell type (e.g., α- and β-adrenergic receptors in smooth muscles, channels and GPCRs activated by glutamate and GABA in neurons) have been known for many decades. In these cases, the receptors involved are all expressed at the cell surface. However, the possibility of a different type of heterogeneity, dependent on the coactivation of both surface and nuclear receptors, the latter working as transcription factors, started to be seriously investigated only about 10 years ago. The most convincing case of nuclear/plasma membrane coactivation is that of steroid receptors. Because of their high hydrophobicity, steroid hormones can easily cross the plasma membrane to reach their intracellular receptor. In the nucleus, active steroid receptors work as transcription factors (see Chapter 24), an activity that requires hours to become fully appreciable in terms of synthesis and accumulation of encoded proteins. Yet, already 30–40 years ago, many results obtained both in experimental endocrinology and in clinical practice were suggesting that in addition to these transcriptional effects and their delayed consequences, steroid hormones could also elicit rapid (1 or a few min) responses, including changes in cytosolic Ca^{2+} level, activation of tyrosine and serine/threonine protein kinases (e.g., Srk and MAP kinases (MAPKs)), and physiological processes such as glucose uptake and blood pressure changes.

The existence of nongenomic steroid receptors in the plasma membrane has been discussed for quite some time, but now is no longer questioned. These receptors appear to be of two types. The first type includes estrogen receptor 1 (ESR1), progesterone receptor membrane complex 1 (PGRMC1), membrane glucocorticoid receptor (mGR), and many others, all resembling their corresponding nuclear receptors and possibly encoded by the same genes. Localization of these receptors to the plasma membrane might result from mRNA alternative splicing or other post-transcriptional changes. The link between their activation and the aforementioned intracellular signals is still debated. The interest of these receptors is not limited to the rapid kinetics of their signaling, but extends also to other properties such as the specificity of their binding, which often differs in some respects from that of the corresponding nuclear receptors.

The second type is GPCRs, thereby profoundly different from the corresponding nuclear receptors also in terms of structure. The best known is G-protein estradiol receptor 1 (GPER1, formerly called GPR30), which has been shown to participate in key processes such as synaptic transmission in the hippocampus, heart contraction, and tumor cell proliferation. Interestingly, the specificity of this receptor appears not limited to estradiol. GPER1, in fact, is bound and activated also by aldosterone and possibly by other steroids.

The work in this field is very active, with hundreds of papers published each year. Although many new aspects still need to be clarified, what appears irreversibly acquired is the duality of the responses, nongenomic and genomic, supported by numerous experimental and clinical observations. This dualism appears not limited to steroid receptors. There are also other examples of hydrophobic agonists that appear as the opposite of steroids. This is the case of prostaglandins (see Chapter 49), derivatives of fatty acids known to activate GPCRs at the cell surface. In this case, there is evidence suggesting the existence of nuclear genomic receptors, which however have not been characterized in detail yet.

As a whole, the surface/nuclear dualism of steroid receptors may be considered similar to the surface receptor signaling that trigger changes in transcription through second messengers. Good examples are the tyrosine kinase (Trk) growth factor receptors (see Chapter 18). In addition to various cytoplasmic signals (activation of MAPK, Rho, and PI3 kinase (PI3K) cascades), these receptors phosphorylate STAT factors, inducing their nuclear translocation and their participation in transcription. Conversely, TGF-β receptor signaling was believed to consist only in serine/threonine phosphorylation of SMADs, another group of transcription factor, but in the last few years, it has been shown to be accompanied by activation of the cytoplasmic cascades typical of growth factor receptors. An effect on transcription occurs also following activation of hormone and transmitter receptors working via second messengers such as cAMP (Chapter 13) or Ca^{2+} (Chapter 12).

Whereas similarities exist between various types of receptors and steroid receptors in terms of duality of effects, they are profoundly different in terms of mechanisms. Specifically, in the case of steroid and possibly also prostaglandin receptors, activation of nuclear and surface receptors occurs simultaneously without the involvement of second messengers and thus with properties very different, both conceptually and operationally, from those of the other receptor types.

HETEROGENEITY OF RECEPTOR ASSEMBLY

The heterogeneity of responses can be due also to differences in receptor assembly. An example is $p75^{NTR}$, the first identified receptor of neurotrophins, expressed not only by neural but also by other cells (see Supplement E 15.1, "Neurotrophins"). In contrast with typical growth factor receptors, which are Trks (Chapter 14), $p75^{NTR}$ belongs to the TNF receptor family. In agreement with its structure, upon interaction with the sortilin coreceptor, $p75^{NTR}$ works as a death receptor. However, in cells expressing also one of the high-affinity neurotrophin receptors (the Trk receptors), $p75^{NTR}$ assembles preferentially with the latter and increases their neurotrophic effect. In other words, depending on whether it associates with sortilin or a Trk receptor, $p75^{NTR}$ contributes to either neuronal apoptosis or differentiation, respectively.

From a physiological point of view, both death and survival are essential for nervous system health, although at different times of the organism's life. However, the role played by $p75^{NTR}$ can be critical in pathological conditions. Recently, $p75^{NTR}$/sortilin complex has been found to be preferentially activated by neurotrophin precursors, whereas $p75^{NTR}$/Trk by mature neurotrophins. Therefore, cells releasing the receptor agonist take part in determining target cell fate. The proNGF/$p75^{NTR}$/sortilin complex has been shown to accumulate in neurons in the course of degenerative processes. For this reason, this complex is currently being investigated as a target of an innovative therapy for Parkinson's and Alzheimer's diseases.

Heterogeneity of assembly/response is not a bizarre property of $p75^{NTR}$ and neural cells. Similar results have been obtained in nonneural cells with classical cytokine receptors (see Chapter 19): for example, TNFR1 can induce either life or death depending on its transduction protein, TRAF2 or FADD. A special case is that of Wnt, a glycoprotein factor that acts though numerous GPCRs named Frizzleds. Wnt is of great importance during development, but it also plays essential roles in the physiology and pathology of various organs, especially the brain and heart (see Supplement E 15.2,

"Wnt and Hedgehog Signaling Pathways: Two Future Drug Targets?"). Wnt signaling is extremely complex as it involves the activation of at least three pathways: a canonical, β-catenin-dependent pathway mediated by specific complexes associated to Frizzled that undergo endocytosis and two noncanonical pathways mediated by either G proteins or small GTPases. Initially, such complexity was attributed to the heterogeneity of the ligands, which are 19, and the multiplicity of the Frizzled receptors, which are 10. Now, it is clear that canonical and noncanonical effects of Wnt can be induced by the activation of single receptors working via different transductional complexes (see Supplement E 15.2).

TRANSDUCTION CASCADES DEPEND ON CROSSTALK AND COMPLEMENTARITY AMONG RECEPTORS

In vivo, cells are continuously exposed to a large variety of signals of different intensity, from subliminal and weak to strong. Based on these signals, cells elaborate their responses and adjust the number and distribution of their receptors. Under physiological conditions, most responses do not result from the activation of a single receptor, but from the interaction among multiple receptors concomitantly activated and their ensuing signaling cascades. Interactions, or "crosstalk," among receptors significantly contribute to the overall signaling and play an important role in gene expression.

To exemplify how such crosstalk operates, we can consider GPCR activation a complex process involving multiple steps such as receptor homologous/heterologous dimerization, phosphorylation by GPCR kinases (GRKs), binding of arrestins, and others. Being common to all GPCRs, all these events are possible targets of their crosstalks. Other important examples of crosstalk are receptor phosphorylation by kinases activated by other receptor signaling (such as PKA, PKC, and CAMK II), collaboration between β- and α1-adrenergic receptors occurring in the heart through potentiation of voltage-operated Ca^{2+} channel (VOCC) type L and Ca^{2+} release from the endoplasmic reticulum, and collaboration between plasma membrane steroid receptors and growth factor receptors, taking place via activation of Srk. A final example of crosstalk involves two types of glutamatergic receptors, GPCRs operating via PIP2 hydrolysis, and channels, in particular the NMDA receptor that induces a large and prolonged Ca^{2+} influx upon activation. The cooperation between the two types of glutamatergic receptors is particularly important in postsynaptic spines where neurogranin, a protein that binds and keeps inactive calmodulin, is concentrated.

The PKC activation that follows glutamatergic GPCR activation results in neurogranin phosphorylation, which releases calmodulin. In the presence of the high $[Ca^{2+}]_i$ induced by NMDA activation, calmodulin induces activation of CAMK II, playing a key role in the process of long-term potentiation. Receptor crosstalk participates also in the response to drugs targeting single receptors. In fact, pharmacological blockade or stimulation of a single receptor can affect signaling cascades activated by other receptors collaborating with it. In the case of long-term responses dependent on crosstalk between multiple signaling cascades, interactions can be very complex. As an example, we will consider the process of neural cell differentiation, which has been extensively investigated in PC12 pheochromocytoma cell line that in many respects resembles adrenal medulla chromaffin cells. The neurotrophin NGF (see Supplement E 15.1) is used, and has been used for more than 40 years, to induce neuronal-like differentiation of PC12. This process includes outgrowth of numerous neurites that do not differentiate into axons and dendrites, block of proliferation, increase in cell volume, and increase in VOCCs, receptors, and other neurospecific proteins as well as in neuroendocrine dense-core vesicles and synaptic-like microvesicles. These phenotypic changes require time to occur: they become first appreciable after 12–24h and are complete 2–3 days later. By contrast, other growth factors (such as EGF) induce opposite effects, as they do not block but induce proliferation (see also Chapter 13). Surprisingly, activation of kinases (e.g., Srk) and GTPases (e.g., Ras) enhances both differentiation and proliferation. In addition, differentiation is not induced only by NGF but also by adhesion proteins (e.g., L1CAM) and peptides of the secretin family (e.g., PACAP pituitary peptide) whose receptors are coupled to adenylate cyclase. Agonists of other receptors with similar coupling or mere activation of PKA do not induce PC12 cell differentiation.

In order to explain these apparently contradictory results, we need to consider the process of PC12 differentiation from a more comprehensive viewpoint. Differentiation depends on integration of a variety of processes. Neurite outgrowth in PC12 (as well as axon outgrowth in neurons) requires enlargement of the cell surface sustained by exocytosis of various nonsecretory vesicles, associated with cytoskeleton reorganization and increase in number of channels, receptors, and many other proteins. These effects are the consequence of a change in gene expression, in particular in the transcription of many neuron-specific genes. To coordinate this type of responses, cells need to have specific properties (e.g., presence of clear and dense vesicles, cytoskeleton plasticity, specific mechanisms of gene expression control) and a battery of receptors capable of inducing responses of appropriate intensity and duration. Phosphorylation of the NGF tyrosine kinase receptor TrkA lasts much longer than EGF receptor phosphorylation. This is necessary for the activation of Rac1 and cdc42, the small GTPases that regulate exocytosis of specific vesicles and cytoskeleton reorganization. PACAP needs to activate (in a PKA-independent way) Rit, a small GTPase that allows several MAPKs to act in the nucleus. The differentiation processes induced by

NGF and PACAP are not identical. PC12 treated with NGF exhibit a larger number of neurites, whereas those treated with PACAP show a larger number of dense-core vesicles. Therefore, NGF and PACAP appear to operate via different signaling cascades. Their effects on PC12 differentiation are not identical but exhibit at least a partial degree of complementarity.

A Comprehensive Scenario of Signal Transduction: From GSKβ3 to AKT and mTOR

Now, we intend to reconstruct a comprehensive scenario including multiple signaling cascades. In fact, signaling cascades do not work separately in cells but are strictly connected to each other. Moreover, integration governs a number of key cell functions, including cell cycle, proliferation, differentiation, apoptosis, cell survival, metabolism, and migration. Therefore, a comprehensive approach to cell signaling has important implications in physiology, pathology, and pharmacology. From the pathology point of view, this approach is particularly important for cancer, inflammation, and neurodegeneration. We are going to focus on three enzymes, GSK3β, Akt, and mTOR, which are all serine/threonine kinases but with different properties. GSK3β, in contrast with most other kinases, is constitutively active, and it is inhibited, rather than activated, by various cell activation pathways. Therefore, phosphorylation by GSK3β most often results in inactivation of its targets. Akt (also called PKB) represents a critical step in the growth factor-activated PI3K cascade. In addition, Akt participates in the cascades triggered by adhesion proteins, cytokine receptors, and various GPCRs. mTOR participates in two distinct complexes, mTOR1 and mTOR2, which induce different and possibly opposite effects including cell proliferation and differentiation (Fig. 15.1).

Akt is activated by phosphoinositide-dependent kinase 1 (PDK1), which in turn becomes activated upon PIP3 production by PI3K. Activated Akt phosphorylates many important targets such as Bad and caspase-9 (involved in apoptosis), glucose transporters, transcription factors (e.g., CREB), and metabolic enzymes. Here, we will focus on two key targets, GSK3β and TSC2, which are inhibited by Akt phosphorylation. GSK3β is inhibited by at least three different pathways via Akt, MAPK, and Dishevelled, a messenger in Wnt pathways (see Supplement E 15.2). GSK3β inhibition attenuates the turnover of β-catenin, an important multifunctional protein. As a result, β-catenin can accumulate not only at cell junctions but also in the nucleus where it induces transcription of numerous genes including oncogenes such as c-Myc and cyclin D (see Supplement E 15.2). TSC2 (associated with TSC1) is a GAP protein that, among its functions, inhibits the Rheb small GTPase, the physiological activator of mTORC1. Thus, Akt activation induces mTORC1 activation, thereby resulting also in stimulation of protein synthesis. This, along with oncogene expression induced by β-catenin, stimulates cell growth.

On the contrary, TSC2 activation induces mTORC1 inhibition and GSK3β activation, resulting in decreased β-catenin accumulation in the nucleus and increased activation of the second target of rapamycin (TOR) complex, mTORC2. mTORC2 triggers cell differentiation via both cytoplasmic processes (mediated by small GTPases such as Rac1) and nuclear gene transcription (mediated by factors such as the forkhead box O, FOXO) (Fig. 15.1).

DEVELOPMENT OF NEW DRUGS AND THERAPIES

The examples illustrated highlight how receptor pharmacology is more complex than it usually appears when presented at the didactic level. These examples, and many others available, open several interesting perspectives in particular for the development of new drugs and the identification of drug associations that may be therapeutically useful. Research in this field is currently very active, and large investments have already been made or are expected in the near future. As always, this type of perspectives requires in-depth elaborations, both conceptual and operative, of new data coming from basic science. Here, we will consider a few developments expected in the next few years.

As we said previously, the dual activity of p75^{NTR}, the neurotrophin receptor that can induce either apoptosis or survival in neural cells, was first explained by demonstrating its ability to associate with either sortilin or Trks and then showing its different response to neurotrophin precursors (more active) and mature neurotrophins. Shortly thereafter, it was demonstrated that in neurons of Alzheimer's disease patients, p75^{NTR} is bound primarily to sortilin and contributes significantly to neurodegeneration, at least in some brain areas. These observations have opened up new perspectives. Whereas downregulation of sortilin, an abundant protein distributed in various subcellular compartments and involved in various functions, does not appear feasible, the development of antiprecursor drugs is currently being evaluated. If it is too early to make predictions, it is however clear that the pharmacology of neurodegenerative diseases in general, and Alzheimer's disease in particular, needs new directions and that resulting from p75^{NTR} studies could be one.

New directions for drug development have come also from studies on the signaling network. The reinterpretation of growth factor signaling cascade, recognizing the key role of Akt and mTOR, has attracted great interest. TOR is the acronym of "target of rapamycin," demonstrating the early involvement of pharmacology in these studies. For quite some time, rapamycin, as well as its analogs, and drugs with the same action were considered as antitumoral drugs. However, there is now ample evidence that these types of drugs are not

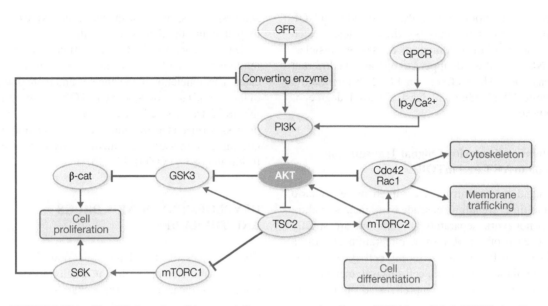

FIGURE 15.1 Simplified version of the growth factor receptor signaling and PI3 kinase/Akt/mTORC signaling.

very effective in most tumors. Today, the main focus of research in this field is the identification of specific associations having synergic effect, such as the association between rapamycin drugs and PI3K blockers.

GSK3β is also an important target of investigation. Given its involvement in numerous critical pathologies, the identification of drugs capable of modulating GSK3β could be of great interest, in particular for tumors and type 2 diabetes. So far, it has not been possible to dissociate its effects on insulin secretion from those on cell proliferation. Several drugs acting on GSK3β are currently being investigated.

A great interest exists also for Wnt and Hedgehog, another signaling pathway that shares some aspects with Wnt (see Supplement E 15.2). Initially, the two systems were investigated by molecular embryologists (Hedgehog in *Drosophila*); today, they are known to control oncogene expression and to contribute to the development of some important tumors. The work in this field aims at the development of new drugs and at the identification of effective associations and appears very promising.

Many cell signaling pathways operate under the control of feedback loops (see Fig. 15.1). Indeed, activation of many growth factor receptors induces activation of "converters" (e.g., IRS for the insulin receptor, GRAB2 and Shick for EGF and other receptors) that activate PI3K (also activated by GPCRs via IP3/Ca^{2+} pathway), resulting in PI3 generation and thereby Akt activation. Akt works as an inhibitory hub acting on a variety of targets and in particular TSC2, Rac1, and GSK3β (a Wnt target), which are involved in critical pathways. TSC2 inhibits mTORC1 and thus S6 kinase, which regulates protein synthesis and cell proliferation. TSC2 also activates mTORC2, which, besides stimulating cell differentiation, activates Akt and Rac1 and cdc42, two small GTPases

that control cell phenotype working on both cytoskeleton and membrane traffic to the plasma membrane. Another target of TSC2 is GSK3β, which increases turnover of β-catenin, a protein that stimulates proliferation by inducing oncogene transcription. To summarize, feedback inhibitions occur by at least three mechanisms: via Akt/TSC2/mTORC2/Rac1, via Akt/TSC2/GSK3β, and via S6 kinase, a component of the mTOR complex that operates on receptor converters. As a consequence, signaling is carefully controlled. For example, Akt activation by growth factors induces inhibition of TSC2, with ensuing mTORC1 activation and mTORC2 and GSK3β inhibition. The outcome of this sequence of events is the stimulation of proliferation and inhibition of differentiation. However, small declines in growth factor signaling tone, determined, for instance, by S6-induced feedback inhibition of receptor converters, can change the pathway and reverse the final outcome.

TAKE-HOME MESSAGE

- Heterogeneity of receptor assembly and crosstalks between signal transduction pathways may specify responses to a drug in different tissues.
- Hydrophobic agonists can interact with both intracellular and surface receptors, the former operating via transcription of specific genes and the latter via generation of second messengers. These two classes of receptors account, respectively, for the delayed and sustained effects and for the rapid effects.
- In the chemotherapy of cell proliferation, treatments targeting multiple strategic sites of important signaling cascades can induce synergic effects of increasing therapeutic interest.

FURTHER READING

Archbold H.C., Yang Y.X., Chen L., Cadigan K.M. (2012). How do they do Wnt they do?: regulation of transcription by the Wnt/β-catenin pathway. *Acta Physiologica*, 204, 74–109.

Bertrand F.E., Angus C.W., Partis W.J., Sigounas G. (2012). Developmental pathways in colon cancer: crosstalk between WNT, BMP, Hedgehog and Notch. *Cell Cycle*, 11, 4344–4351.

Chen L.W., Yung K.K., Chan Y.S., Shum D.K., Bolam J.P. (2008). The proNGF-p75NTR-sortilin signalling complex as new target for the therapeutic treatment of Parkinson's disease. *CNS and Neurological Disorders Drug Targets*, 7, 512–523.

Chen Y., Zeng J., Cen L., Chen Y., Wang X., Yao G., Wang W., Qui W., King K. (2009). Multiple roles of the p75 neurotrophin receptor in the nervous system. *Journal of International Medical Research*, 37, 281–288.

Dodge M.E., Lum L. (2011). Drugging the cancer stem cell compartment: lessons learned from the Hedgehog and Wnt signal transduction pathways. *Annual Review of Pharmacology and Toxicology*, 51, 289–310.

Ha H., Han D., Choi Y. (2009). TRAF-mediated TNFR-family signaling. *Current Protocols in Immunology*, 11, Unit 11.9D.

Hammes S.R., Levin E.R. (2012). Recent advances in extranuclear steroid receptor actions. *Endocrinology*, 152, 4489–4495.

Huang J., Manning B.D. (2009). A complex interplay between Akt, TSC2 and the two mTOR complexes. *Biochemical Society Transactions*, 37, 217–222.

Huang E.J., Reichardt L.F. (2003). Trk receptors: roles in neuronal signal transduction. *Annual Review of Biochemistry*, 72, 609–642.

Jin T., Fantus G.I., Sun J. (2008). Wnt and beyond Wnt: multiple mechanisms control the transcriptional property of beta-catenin. *Cellular Signalling*, 20, 1697–1704.

Kim W., Kim M., Jho E.H. (2013). Wnt/β-catenin signalling: from plasma membrane to nucleus. *Biochemistry Journal*, 450, 9–21.

Luine V.N., Frankfurt M. (2012). Estrogens facilitate memory processing through membrane mediated mechanisms and alterations in spine density. *Frontiers in Neuroendocrinology*, 33, 388–402.

Niehrs C. (2012). The complex world of WNT receptor signalling. *Nature Reviews. Molecular Cell Biology*, 13, 767–779.

O'Shea J., Holland S.M., Staud L.M. (2013). Jak and STATs in immunity, immunodeficiency, and cancer. *New England Journal of Medicine*, 368, 161–170.

Phukan S., Babu V.S., Kannoji A., Hariharan R., Balaji V.N. (2010). GSK3β: role in therapeutic landscape and development of modulators. *British Journal of Pharmacology*, 160, 1–19.

Rao T.P., Kuhl M. (2010). An updated overview on Wnt signalling pathways. *Circulation Research*, 106, 1798–1806.

Ravni A., Vaudry D., Gardin M.J., Eiden M.V., Falleuel-Morel A., Gonzalez B.J., Vaufry H., Eiden L.E. (2008). cAMP-dependent, protein kinase A-independent signaling pathway mediating neuritogenesis through Egr1 in PC12 cells. *Molecular Pharmacology*, 73, 1688–1708.

Schecterson L.C., Bothwell M. (2010). Neurotrophin receptors: old friends with new partners. *Developmental Neurobiology*, 70, 332–338.

Zhang Y.E. (2010). Non-Smad pathways of TGF-β signalling. *Cell Research*, 19, 128–139.

Zhong L., Gerges N.Z. (2010). Neurogranin and synaptic plasticity balance. *Communicative and Integrative Biology*, 3, 340–342.

SECTION 4

RECEPTOR CLASSES

16

LIGAND-GATED ION CHANNELS

CECILIA GOTTI AND FRANCESCO CLEMENTI

> **By reading this chapter, you will:**
>
> - Learn classifications and structures of ligand-gated ion channels (LGICs)
> - Understand structure–function relationship of LGIC family members
> - Learn the mechanisms of allosteric and orthosteric activation of LGICs and the molecular and cellular mechanisms whereby ligands activate LGICs and trigger intracellular signaling pathways
> - Learn the background for understanding the mechanism of action of drugs acting on LGICs and their potential significance as therapeutic agents

Ligand-gated ion channels (LGICs) are a superfamily of receptors mediating fast synaptic transmission, whose opening is modulated by specific endogenous neurotransmitters.

Neurotransmitter binding induces rapid conformational changes of these receptors resulting in the opening of an integral ion channel that allows specific ions to flow down their electrochemical gradients. This propagates neurotransmission and/or activates intracellular signaling pathways, thus beginning cellular responses to receptor activation.

LGICs are some of the oldest receptors in evolutionary terms. They are present in bacteria and plants and have undergone relatively minor structural and functional changes over time (see Supplement E 16.1, "Evolution of Pentameric Ligand-Gated Ion Channels").

LGICs are widely present in the central and peripheral nervous systems of animals and plants and are grouped into families on the basis of the pharmacology of their neurotransmitters. They include acetylcholine (ACh), GABA, glycine, glutamate and serotonin receptors, cGMP and cAMP cyclic nucleotide-activated receptors, and ATP receptors.

All these receptors are complexes of three to five subunits, either identical or different. On the basis of their subunit topology, LGICs can be divided into four classes:

1. The Cys-loop receptor superfamily including muscle and neuronal nicotinic receptors (AChR, nicotinic acetylcholine receptor (nAChR)), GABA receptors ($GABA_A R$), glycine receptors (Gly-R), and a subclass of serotonin receptors (5-HT_3)
2. Ionotropic glutamate receptors (NMDAR, AMPAR, and KAR)
3. Cyclic nucleotide-gated receptors (CNG-R)
4. Ionotropic ATP receptors (P2X-R)

Moreover, depending on their selective permeability, LGICs can be classified as cation selective or anion selective. Cation-selective LGICs depolarize (excite) neurons, whereas anion-selective LGICs, in most cases, hyperpolarize (inhibit) neurons (see Table 16.1)

TISSUE DISTRIBUTION AND SUBCELLULAR LOCALIZATION OF LGICs

LGICs are expressed in almost all cell types, where they are critical mediator of cell–cell communication. They are particularly important in communication between neurons and between neurons and nonneuronal cells. Communication between nerve cells mainly occurs at chemical synapses, specialized sites of cell–cell contact where electrical signals

---

General and Molecular Pharmacology: Principles of Drug Action, First Edition. Edited by Francesco Clementi and Guido Fumagalli.
© 2015 John Wiley & Sons, Inc. Published 2015 by John Wiley & Sons, Inc.

TABLE 16.1 List of ligand-gated ion channels and of their cloned subunits

Receptors	Abbreviations	Transmitters	Cloned subunits
Cation permeable receptors			
Muscle nicotinic receptor	AChR	Acetylcholine	$\alpha_1, \beta_1, \gamma, \delta, \epsilon$
Neuronal nicotinic receptor	nAChR	Acetylcholine	$\alpha_{2-10}, \beta_{2-4}$
Ionotropic glutamate receptors	AMPA-R, KA-R	Excitatory	GluA1–4, GluK1–5
	NMDA-R	Amino acids	GluN1, GluN2A–D, GluN3A–B
Serotonergic receptors	5-HT$_3$R	Serotonin	5-HT$_{3A-C}$
Purinergic receptors:	P2X-R	ATP	P2X1–7
Cyclic nucleotides-gated receptors	CNG-R	cAMP & cGMP	CNGA$_{1-4}$
			CNGB$_{1,3}$
Anion permeable receptors			
GABAergic receptor	GABA$_A$R	GABA	$\alpha_{1-6}, \beta_{1-3}, \gamma_{1-3}$
			$\delta, \epsilon, \pi, \theta$
			ρ_{1-3}
Glycinergic receptor	Gly-R	Glycine	α_{1-3}, β

trigger exocytotic release of neurotransmitters, which in turn activate postsynaptic receptors (see Chapter 37). The efficiency of transmission of such chemical signals is crucial for nervous system activity. LGICs may be located at pre- or postsynaptic sites on the cell body or dendritic spines and at the neuromuscular junction in the apical plasma membrane of the synaptic contact. Synaptic transmission through LGICs therefore relies on an intricate network of synaptic proteins involved in neurotransmitter release, channel and receptor activation, and modulation of signal transduction cascades. Recruitment and maintenance of LGICs at specific cell sites are essential for proper synaptic function and are mediated by intracellular scaffold proteins involved in receptor anchorage and intracellular signaling.

Rapsyn and gephyrin are two examples of scaffold proteins required for proper LGIC localization.

Rapsyn is essential for correct aggregation and localization of muscle nicotinic receptors (AChRs) at neuromuscular cholinergic synapses. It allows high-density AChR localization (10,000 molecules/μm^2) at the apex of the postsynaptic folds. Such strategic location, opposite the sites of neurotransmitter release, is critical for efficient synaptic transmission. Alterations of nAChR position contribute to the synaptic transmission failure that is typical of some types of myasthenia gravis.

Gephyrin is essential for the aggregation, localization, and function of Gly-R and some subtypes of GABA$_A$R in the brain and spinal cord inhibitory synapses.

Postsynaptic densities in the glutamatergic synapses have a very complex organization, and different glutamate receptor subtypes may be directly or indirectly associated with proteins involved in receptor anchoring, localization, signaling, and binding to the cytoskeleton (see Chapters 9 and 43). Moreover, translocation of different receptor subtypes at the postsynaptic membrane is regulated by neuronal electrical activity and/or activation of specific glutamate receptor subtypes, thus modulating the function of glutamatergic synapses.

Therefore, in muscle and CNS synapses, receptor localization and density in the postsynaptic membrane are controlled by specific receptor-associated proteins interacting with the cytoskeleton (Fig. 16.1).

However, it is important to note that LGICs may be present at extrasynaptic locations in the plasma membranes of skeletal muscle fibers and neurons, where they play different roles. Furthermore, in neurons, many LGICs are also expressed in the presynaptic membrane, where they modulate neurotransmitter release and regulate synaptic activity (Fig. 16.1 and Chapter 37). The mechanisms that maintain the LGIC localization in these membrane domains are still not completely clear.

MOLECULAR ORGANIZATION OF LGICs

Classification of LGICs

LGICs are oligomeric transmembrane proteins consisting of three to five either identical or different subunits, whose single rotational plane of symmetry is perpendicular to the plane of the membrane in which they are inserted. On the basis of the putative or known topology of their constituent subunits, they have been classified into four classes.

Class 1: Cys-Loop Receptors Cys-loop receptors take their name from a 13-amino-acid loop within the extracellular domain that is delimited by a pair of disulfide-bonded cysteine (Cys) residues (corresponding to Cys 128 and 142 of the α_1-subunit of the muscle nicotinic receptor) (Fig. 16.2a).

This class includes muscle nicotinic receptors (AChR), neuronal nicotinic receptors (nAChR), and GABA (GABA$_A$R), glycine (Gly-R), and serotonin receptors (5-HT$_3$R).

Cys-loop receptors consist of a ring of five subunits, all having a structure characterized by a large (~200-amino-acid-long) N-terminal extracellular domain, followed by four

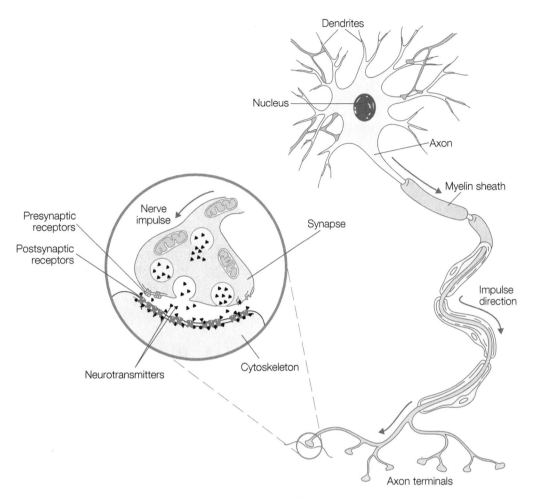

FIGURE 16.1 Information exchange at synapses. The inset shows a magnified synaptic contact. In the presynaptic terminal, vesicles containing neurotransmitters release their contents in the synaptic cleft. Neurotransmitter secretion is triggered by nerve impulses and modulated by presynaptic receptors. In a cholinergic synapse, for instance, presynaptic nAChRs bind the released ACh and modulate its secretion (autologous modulation by autoreceptors). In addition, presynaptic receptors for other neurotransmitters (e.g., α-adrenergic receptors activated by catecholamines secreted by neighboring terminals) may control synaptic function (heterologous modulation by heteroreceptors). Receptors on the postsynaptic membrane regulate cell responses to nervous impulses; their strategic position is controlled by postsynaptic proteins, involved in anchoring, scaffolding, and signaling.

transmembrane α-helical segments (TM1, TM2, TM3, TM4) interconnected by short intra- or extracellular hydrophilic loops. A long intracellular loop is present between TM3 and TM4, and a relatively short and variable extracellular COOH-terminal sequence is found after TM4 (Fig. 16.2a).

The subunits of this class of receptors show a high degree of amino acid sequence homology (between 20 and 80%) and a very similar distribution of hydrophobic transmembrane segments. Common to all subunits of this large family of LGICs is the presence in the first extracellular domain of a Cys-loop defined by two Cys that in the mammalian subunits are separated by 13 amino acids. Subunits are further classified into α- and non-α-subunits based on the presence of a Cys–Cys pair (residues 191–192 in *Torpedo* α1) near the beginning of TM1. A large number of α- and non-α-subunits have been cloned for all members of the superfamily.

Table 16.1 shows the list of subunits cloned for each of the receptors: AChR (α_1, β_1, γ, δ, ε), nAChR (α_{2-10}, β_{2-4}), GABA$_A$R (α_{1-6}, β_{1-3}, γ_{1-3}, δ, ε, π, θ, ρ_{1-3}), Gly-R (α_{1-3}, β), and 5-HT$_3$ (5-HT$_{3A-C}$).

Although some of Cys-loop LGICs can be formed by five identical subunits (homomeric receptors) (e.g., α_7 nicotinic neuronal receptor), the vast majority are heteropentamers consisting of two or more different subunits (heteromeric receptors) (e.g., $\alpha_4\beta_2$, $\alpha_6\alpha_4\beta_2$, $\alpha_3\beta_4$) (Fig. 16.2b, c).

In conclusion, class 1 receptors have a high degree of structural and functional homology, a common topology of the subunits that defines the agonist binding site, and the structure of the ion channel.

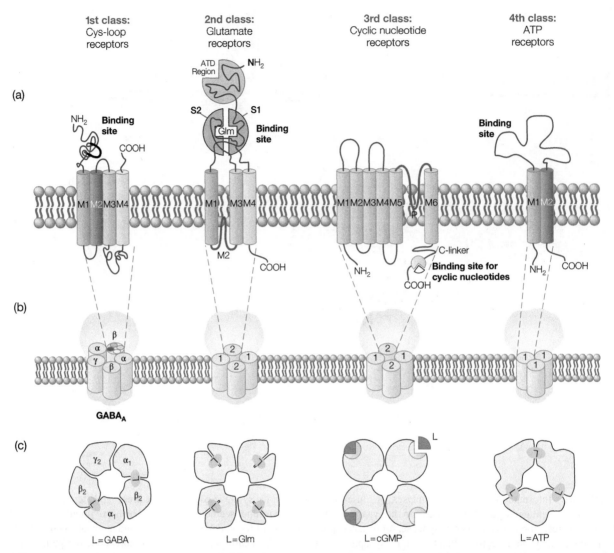

FIGURE 16.2 Structural comparison of LGICs. LGICs can be divided into four classes according to their amino acid sequence homology, subunit composition and stoichiometry, and quaternary structure as defined by X-ray crystallography. Class 1 includes the Cys-loop receptors for ACh, GABA, glycine, and serotonin; class 2 the ionotropic glutamate receptors; class 3 the cyclic nucleotide receptors; and class 4 the ATP receptors. (a) Proposed secondary structure of receptor subunits. See text for details. (b) Schematic subunit arrangement. (c) Cross sections of receptors showing subunit stoichiometry and ligand-binding sites. COOH, C-terminal region; Glm, glutamate; L, physiological ligand; NH_2, N-terminal region; TM1–TM6, transmembrane segments.

Class 2: Glutamate Receptors This class consists of ionotropic glutamate receptors, including α-amino-3-hydroxy-5-methyl-4-isoxazolepropionic acid receptor (AMPAR), kainate receptor (KAR), and .-methyl-D-aspartate receptor (NMDAR) (from the name of the agonists AMPA, kainate, and NMDA, which selectively and pharmacologically distinguish the three receptor subclasses) (see Chapter 43).

Glutamate receptors assemble as tetrameric complexes (dimers of dimers), formed by either identical (homomeric) or different (heteromeric) subunits of the same subclass (or AMPAR or KAR or NMDAR) (Fig. 16.2b).

Glutamate receptor subunits consist of a single polypeptide with a large N-terminal extracellular region, four hydrophobic transmembrane segments (TM1–TM4), and an intracellular C-terminal region (Fig. 16.2a). Only the TM1, TM3, and TM4 segments completely cross the lipid bilayer; segment TM2 forms a kink within the membrane and reenters the cytoplasm to form a P region similar to that found in the pore-forming domain of voltage-activated K^+ channels (Fig. 16.2a) (see Chapter 27).

Table 16.1 shows the list of subunits cloned for AMPAR (GluA1–4), KAR (GluK1–5), and NMDAR (GluN1, GluN2A–D, GluN3A–B).

Class 3: Receptors for cGMP and cAMP Cyclic Nucleotides This class includes channels gated by cGMP that are permeable to cations (Na$^+$, K$^+$, Ca^{2+}) and found in the cones of the retina, channels gated by cAMP and found in the olfactory epithelium, and channels gated by cyclic nucleotides and found in nonsensory tissues (hippocampus, heart, kidney, testis, intestine). Unlike the other LGICs, in these receptors, the ligand-binding site is intracellular. Structurally, they are homologous to the voltage-operated K$^+$ channel and consist of four subunits with intracellular N- and C-terminal regions, six transmembrane segments (TM1–TM6), and a segment (between TM5 and TM6) that forms a P region similar to the one found in voltage-dependent ion channels and glutamate receptors. Such P region and the TM6 transmembrane segment delimit the pore wall (Fig. 16.2). The subunits cloned for this class of receptors are called $CNGA_{1-4}$ and $CNGB_{1-3}$. *In vivo*, these receptors are hetero-oligomeric tetramers of CNGA and CNGB subunits, with different subunit stoichiometries depending on their localization (Fig. 16.2b, c).

Class 4: ATP P2X Ionotropic Receptors This class includes the ATP ionotropic receptors, which are oligomers consisting of three subunits having different topologies compared to other LGIC receptor subunits. Each subunit has N- and C-terminal intracellular regions, two transmembrane segments (TM1 and TM2), and a large extracellular region, about 280 amino acids long (Fig. 16.2a, b and Chapter 44). So far, seven subunits (P2X1–7) have been cloned for this class of receptors.

TOPOLOGY OF LGICs

nAChRs are considered the LGIC prototype as they were the first to be topographically reconstructed to show details of their molecular organization.

Figure 16.3 shows the overall structure of the muscle nicotinic receptor and its subunits. The five constituent subunits are symmetrically arranged in a precise order (α_1–γ–α_1–δ–β_1) to form a complex of about 300 kDa. The high-resolution electron microscopy data obtained by Unwin (2005) show that the nicotinic receptor has the shape of a 16 nm long cylinder with a diameter of 8 nm and consists of three parts: one extracellular part, one transmembrane pore, and a large exit region on the intracellular membrane surface. The extracellular part is funnel shaped, has an internal diameter of 2 nm, and protrudes about 6.5 nm into the extracellular space (Figs. 16.3a, c and 16.4a). The funnel seems to have the function of concentrating the ions flowing through the channel. The extracellular portion also contains binding sites for neurotransmitter and many drugs active on these receptors.

When it crosses the plasma membrane, the receptor narrows and forms the region that is critical for its permeability and selectivity. This region contains negatively charged amino acids, contributed by each of the five TM2 transmembrane segments (one for each subunit), which are arranged in three rings facing the inside of the channel and delimiting the cation-selective pore (Fig. 16.4). These rings select and concentrate the ions, ensuring that only the proper ions can pass through the pore. The negatively charged amino acids are found in cation-selective channels, such as the nicotinic receptors (Fig. 16.4b), but anion-selective channels, such as the Gly-R and GABA$_A$R, contain positive or neutral amino acids instead. This region also contains the ion "gate," which is normally closed but opens upon ACh binding.

In the cytoplasmic area, immediately after emerging from the membrane, the receptors widen and the ions entering or leaving the cytoplasm pass through narrow side openings (Fig. 16.4a). The cytoplasmic region also contains phosphorylation sites involved in the regulation of channel activity and binding sites for cytoskeletal proteins required for subcellular receptor localization.

The Binding Sites for Endogenous Ligands

The first step toward LGIC activation is the binding of the endogenous ligands. This causes the binding site to adopt a conformation(s) that stabilizes the channel in the open state. The binding sites for endogenous ligands are localized in the extracellular portion of the Cys-loop, ATP, and glutamate receptors, whereas they are found in the intracellular portion of the receptors activated by cyclic nucleotides (Fig. 16.2a).

Cys-Loop Receptors The most precise data on the Cys-loop receptor superfamily concern the muscle nicotinic receptor and have been obtained employing cross-linking compounds, specific drugs, and toxins. Relevant details have also come from the recent determination of the structure of the acetylcholine-binding protein (AChBP) purified from the marine snail *Lymnaea stagnalis*, at a resolution of 2.7 Å. AChBPs are a family of secreted soluble proteins consisting of five identical subunits that resemble the extracellular ligand-binding domain of AChRs (Fig. 16.5a) but lack the domains necessary to form a transmembrane ion channel. Alignment of the amino acid sequences of the ACh binding sites of the AChBPs and the neurotransmitter binding sites of the Cys-loop receptors showed a 15–20% sequence homology with the GABA$_A$R, Gly-R, and 5-HT3 and 25% homology with the nAChR. Almost all residues that are conserved in the Cys-loop receptor family members are present also in the AChBPs, including those required for ligand binding. This led to the hypothesis, and subsequent demonstration, that ligand-binding sites have a general common structure. The binding site for neurotransmitters (ACh, GABA, glycine, and serotonin) in the Cys-loop receptors lies at the interface between two adjacent subunits, one providing the main binding component (short noncontiguous sequences forming loops A, B, and C) and the other

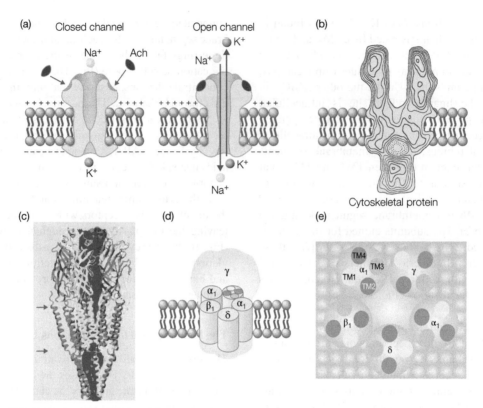

FIGURE 16.3 Muscle nicotinic cholinergic receptor channel and its molecular structure. (a) Upon binding of two ACh molecules to the two α_1/γ- and α_1/δ-subunit interfaces, a conformational change occurs leading to the opening of the intrinsic transmembrane pore of the receptor (see also Fig. 16.4c, d). (b) Reconstructed electron microscopic image of the nicotinic receptor from *Torpedo* electric organ membranes. The image was obtained by computer processing of the negatively stained receptors at a 4Å resolution. Note the long extracellular cavity forming the funnel in which ions to be translocated are collected, the narrowing of the lumen at the plasma membrane level (where the selective pore is found), and the cytoskeletal proteins controlling receptor localization in the plasma membrane. (c) Structural model of *Torpedo* nicotinic receptor obtained from X-ray crystallographic analysis. α_1-subunits are shown in white; non-α-subunits are shown in gray; and the channel is depicted dark. Arrows indicate the limits of the plasma membrane. Interestingly, nicotinic receptor reconstructions obtained by electron microscopy (b) and X-ray diffraction (c), two very different techniques, are extremely similar in shape and size. (d) Pentameric structure of the muscle nicotinic receptor. The five subunits are arranged in a circle in a precise order (α_1–γ–α_1–δ–β_1) to form the ion channel. Each subunit consists of a long extracellular N-terminal portion (*gray silhouette*) and four hydrophobic segments (TM1, TM2, TM3, TM4). (e) Cross-sectional view of the five subunits forming the pore. The relative positions of the transmembrane regions (TM1, TM3, TM4) and the TM2 region delimiting the channel walls are shown for each subunit.

contributing the complementary component formed by loops D, E, and F (see Fig. 16.5b and Supplement E 16.1).

In the muscle nicotinic receptors, ACh binds to the interfaces α_1/γ and α_1/δ, with the α-subunit contributing the major component and the γ- or δ-subunits contributing the complementary component. In the heteromeric Cys-loop receptors, not all the receptor subunits are involved in the formation of the binding sites; binding sites may be nonequivalent, and the number of binding sites may vary. For example, in the muscle nicotinic receptors, which have two nonequivalent binding sites at the α_1/γ and α_1/δ interfaces, receptor activation requires ACh binding to both sites, whereas in homomeric receptors, such as 5-HT$_{3A}$ or the α_7 nicotinic receptors, which have five identical binding sites, optimal activation requires binding to three binding sites only.

In muscle AChR, the conformational changes taking place after ACh binding have not been fully defined yet, but it is worth remembering that the binding sites located in the extracellular portion of the α-subunits are closely connected to the adjacent subunits. It is believed that binding of the cationic head of ACh to a subsite present in the α-subunit (loops A, B, C) and the simultaneous attraction of negatively charged amino acids of the loops E and F of δ- and γ-subunits close the two sites and create a slip movement between the subunits that leads to the opening of the channel (Fig. 16.4c, d) (for more details, see also Supplement E 16.1).

In frogs, the interaction of ACh with one molecule of AChR at the neuromuscular junction produces an influx of positive charges (mainly Na$^+$) at a speed of 10,000 ions/ms. The channel opens after ACh binding and remains in the open state for 1–4 ms, with a conductance of 50 picosiemens (pS).

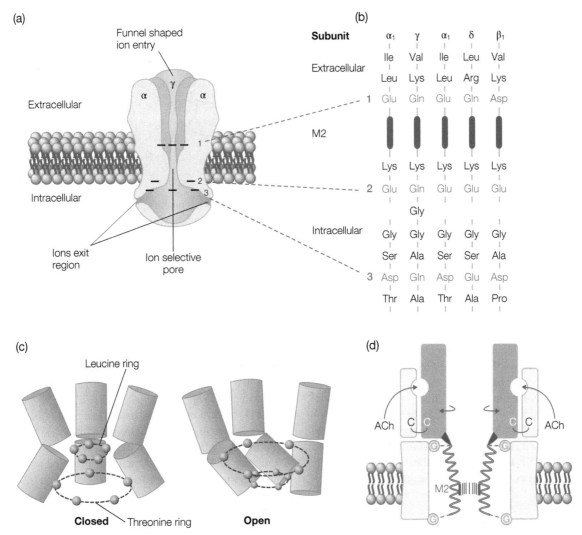

FIGURE 16.4 Structural and functional models of muscle nicotinic receptor. The reconstruction is based on Numa's and Unwin's findings. (a) Vertical section of a muscle nicotinic receptor showing the rings of negative charges that determine cation selectivity, the ion pore, and the ion entry and exit regions on the cytoplasmic side of the receptor. According to this model, three rings of negative charges around the pore are formed by negatively charged amino acids on the edges of the TM2 segments of each subunit. (b) Amino acid sequences of the TM2 flanking regions of each of the five subunits. The amino acids forming the rings shown in a (1, 2, and 3) are also indicated. (c) Arrangement of three TM2 segments of a muscle nicotinic receptor subunit in the closed (*left*) and open (*right*) state (for clarity, the fourth and fifth regions are not shown). When the channel is closed, the TM2 hemicylinders of the five subunits project toward the central axis of the channel, and five hydrophobic amino acids (leucines, one for each subunit) form a ring occluding the channel pore. When ACh binds to the receptor, the TM2 hemicylinders rotate inside the channel so that the leucine ring enlarges allowing ion flux. At the same time, the threonine ring shrinks and creates a selective ion filter. (d) Possible molecular interactions among nicotinic receptor domains. When ACh binds to the receptor, it induces a rotation in the α-subunits that is transmitted through the TM2 segments to the channel selective pore. Such rotation destabilizes the selective filter and causes the TM2 α-helices to adopt a new conformation, ion permeable.

The opening of a single channel depolarizes the postsynaptic membrane by 0.5 μV. The duration of channel opening depends on the agonist: carbachol opens the channel up to about one-third of the time observed with ACh, whereas succinylcholine causes the channel to remain open much longer.

The high degree of homology between both neurotransmitter binding sites and channel structures within the Cys-loop superfamily suggests that their coupling is similar in all Cys-loop receptors. This has been elegantly demonstrated using a chimeric receptor consisting of the extracellular N-terminal region of the $α_7$-subunit of the neuronal nicotinic receptor and the transmembrane region and C-terminus of the subunit of the 5-HT$_3$ receptor. This chimeric receptor is activated by ACh and nicotinic ligands but has the biophysical characteristics of the channel typical of the 5-HT$_3$ receptor.

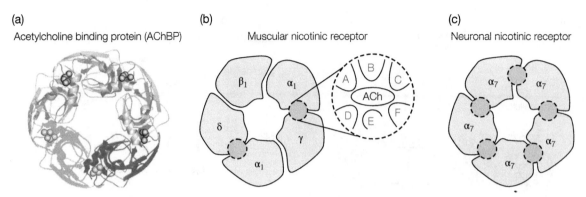

FIGURE 16.5 Subunit organization at the level of the ACh binding site in the acetylcholine-binding protein (AChBP) and two nAChR subtypes. (a) AChBP three-dimensional model obtained from crystallographic studies, showing α-helices and β sheets. Five identical subunits associate to form a structure with five ACh binding sites (*dots*). Subunits are depicted in order to facilitate their identification. (b) Subunit arrangement at ACh binding sites and (*enlarged*) domains involved in binding site formation in muscle nicotinic receptors. These receptors have two ACh binding sites at the interfaces between the α-subunit (main site, loops A, B, C) and the γ- or δ-subunit (complementary site, loops D, E, F). (c) Subunit arrangement at the five ACh binding sites in the homomeric α_7 nicotinic receptor. Binding site consists of the A, B, and C loops of an α_7-subunit (the primary binding site) and the D, E, and F loops of an adjacent α_7-subunit (complementary binding site).

Glutamate Receptors The binding site for the excitatory amino acids in the glutamatergic receptors is localized in the extracellular portion of the receptor. Crystallographic studies of the $GluA_2$ receptor have shown that each subunit is characterized by an N-terminal extracellular region (ATD region) of about 400 amino acids (Fig. 16.2a) and by the glutamate binding domain. This is highly conserved among different classes of glutamate receptors and is formed by two noncontiguous regions: S1, located between the ATD region and the beginning of the first transmembrane segment TM1, and S2, located between the TM3 and TM4 transmembrane segments (see Chapter 43). The ATD region and the glutamate binding site are organized as dimers of dimers in the four-subunit receptor molecule.

Crystallography, binding, and oocyte expression studies of chimeric receptors obtained by swapping the S1 and S2 regions of AMPAR and KAR and characterization of fusion proteins containing only S1 and S2 have shown that these regions are essential for glutamate binding and pharmacological specificity of the different receptor subtypes. In non-NMDA glutamate receptors, channel opening requires the binding of four glutamate molecules (one for each receptor subunit) (Fig. 16.2c), whereas NMDAR activation requires the binding of two distinct ligands: glutamate (the endogenous agonist) and glycine (the coagonist).

Cyclic Nucleotide and ATP Receptors In cyclic nucleotide-activated receptors, ligand-binding sites are localized in the C-terminal intracellular region of the subunits, within a sequence that is homologous to the cyclic nucleotide-binding regions of other proteins, such as cAMP- and cGMP-dependent protein kinases and transcription factors (Fig. 16.2a). The binding site is linked to the TM6 transmembrane region by an 80-amino-acid-long segment (C-linker) that triggers channel opening upon ligand binding. This linker region also contributes to the formation of contacts between channel subunits and favors channel tetramerization. These receptors are activated by the binding of two or four molecules of ligand per receptor (Fig. 16.2c).

ATP receptors consist of three subunits, and their activation requires the binding of three ATP molecules. Studies designed to identify the binding site have shown that amino acids located at different points in the large extracellular region, and preserved among the different P2X receptor subtypes, form a pocket between two adjacent subunits that consists of eight amino acids (four for each subunit) and binds ATP (Fig. 16.2). Through mutagenesis and heterologous expression studies, specific amino acids have been identified in the extracellular region that determine sensitivity to antagonist in different receptor subtypes. Knowledge of the molecular structure and conformation of neurotransmitter binding sites will advance the development of new drugs acting at this level.

Structure and Localization of the Channel

Cys-Loop Receptors

Most of the relevant data concerning the structure and localization of the ion channel in these receptors have come from structural and electrophysiological studies of the muscle nicotinic receptor employing the substituted-cysteine accessibility method (SCAM) and from the recent model of *Torpedo* AChR at 4Å resolution (see Supplement E 16.2, "How to Identify the Amino Acids That Make Up the Inner Wall of Ligand-Gated Ion Channels"). These studies indicate that the ion channel is a nearly symmetrical cylinder, with an inner ring made of five α-helices (TM2 segments). An outer ring of 15 α-helices (TM1, TM3, and TM4 segments) shields the inner ring from the membrane phospholipids (Fig. 16.2b). The TM2 segments are amphipathic α-helices containing a stretch of hydrophobic and uncharged amino acids that pair vertically relative to the plane of the membrane. Structural and electrophysiological studies have shown that the ion channel is mainly lined by the M2 domains of the five subunits (Fig. 16.3d, e).

As described previously, in the TM2 region, charged amino acids are arranged in such a way that when the five subunits assemble to form the pore, they create three rings of charges at different levels along the pore (Fig. 16.4a, b).

The first and third rings are those mainly responsible of the cation conductance, which can also be influenced by other amino acids in the TM2 segment. In the AChRs and 5-HT$_3$Rs (which are permeable to cations), the second ring consists of negatively charged amino acids, whereas in the GABA$_A$Rs and Gly-Rs (which are permeable to anions), it consists of uncharged amino acids. In the AChRs, it has been observed that amino acid changes in this ring severely alter receptor conductance, thus indicating that the ring is critical in determining ionic selectivity.

Electron microscopy studies have shown that the TM2 segment is not straight but it is made of two hemicylinders that may be arranged at different angles. When the channel is closed, the hemicylinders project toward the central axis of the channel, and five hydrophobic amino acids (one leucine for each subunit) form a ring that occludes the channel pore. A ring of hydrophilic amino acids (threonines) located below the leucine ring controls ion selectivity (Fig. 16.4c).

When ACh binds the receptor, it induces a rotation in the α-subunit that allows the hemicylinders of the M2 region to turn inside the channel, and the leucine ring widens. This destabilizes the selective filter and causes the α-helices of TM2 to adopt a new conformation that is permeable to ions (Fig. 16.4c, d).

These receptors do not easily discriminate different ionic species of the same charge: depending on the electrochemical gradient, both monovalent (Na$^+$ and K$^+$) and divalent ions (such as Ca^{2+}) can flow through the nAChR channel. However, subtypes of the same receptor may show partial selectivity: for example, Ca^{2+} permeability is low in muscle AChRs but high in some neuronal subtypes (particularly those containing the α$_7$-subunit). The amino acid composition of the M2 region forming the pore complex plays an important role in defining its ion selectivity. This is well exemplified by experiments involving the neuronal α$_7$ nicotinic receptor, in which substitution of a glutamic acid (charged) in the intermediate ring with an alanine (uncharged) leads to complete loss of Ca^{2+} permeability without affecting Na$^+$ and K$^+$ permeability.

Glutamate, Cyclic Nucleotides, and ATP Receptors The three different types of glutamate receptors have a common structure with the channel having an extracellular funnel-shaped vestibule that narrows down to the region of the selectivity pore. The pore walls are delimited by the P region, which functionally corresponds to the TM2 region. In some of the non-NMDAR subunits (see Chapter 43), the P regions contain, at a precise location, either a glutamine (Q) or an arginine (R). Receptor subtypes with Q-containing subunits are highly permeable to Na$^+$ and Ca^{2+}, whereas those with R-containing subunits are poorly permeable to Ca^{2+} ions. On the other hand, NMDARs contain an asparagine residue (N) in the same position that makes them highly permeable to Ca^{2+} and sensitive to magnesium blockade, two of the functional characteristics of this receptor (for details, see Chapter 43).

The channel pore region of the cyclic nucleotide receptors has been identified using the SCAM (see Supplement E 16.2). This region, called P region, is found between the TM5 and TM6 transmembrane segments and forms a loop with an α-helix structure that projects from the outside toward the central axis of the channel and then folds into an unwound structure toward the outside. The P region is highly homologous to the same region in voltage-gated K$^+$ channels, and besides forming the cation-selective filter, it also functions as a "gate" whose structural changes control the opening of the channel (Fig. 16.2a).

SCAM has also been used to identify the amino acid sequence of the pore in ATP receptors, which has three TM2 segments (one for each subunit) surrounded by three TM1 segments (Fig. 16.2a). X-ray studies of the crystallized P2X4 receptor have shown that the three subunits adopt a dolphin shape (Fig. 16.6a) and wrap themselves around each other to form the trimeric receptor (Fig. 16.6b). Moreover, these studies have shown that, in the closed state, the TM2 segments are steeply angled relative to the membrane plane, giving the pore an hourglass appearance with three different vestibules (top, middle, and extracellular) in the extracellular part and one in the intracellular part (Fig. 16.6c). The extracellular and central vestibules contain high levels of charged amino acids that have the function of concentrating the cations near the pore. The pore region is localized between the extracellular and the intracellular vestibules and, in the closed state, is inaccessible to ions coming from the outside; however, upon ATP binding, the three α-helices of the TM2 segments are rearranged to allow the opening of the pore.

Mutagenesis studies of the P2X2 receptors have shown that the same amino acid residues that occlude the pore in the closed state contribute to the formation of the cation-selective filter in the open state.

Functions of the Cytoplasmic Domain

The cytoplasmic area of the muscle AChRs mainly consist of long loops linking the TM3 and TM4 segments of the subunits and containing phosphorylation sites controlling the desensitization and the dynamics of receptor aggregation in

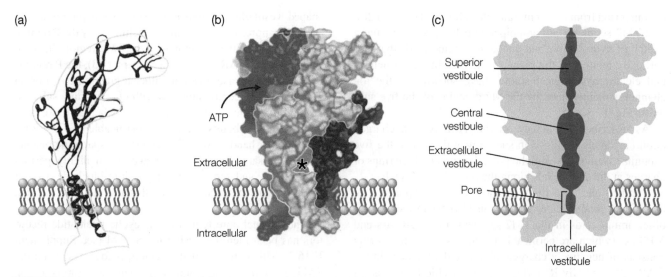

FIGURE 16.6 X-ray crystallography defined structure of the trimeric P2X receptor as follows: (a) "dolphin-shaped" silhouette of a single receptor subunit within a P2X receptor (*outline*). (b) Full surface of a P2X4 receptor with three identical subunits drawn in grays of different intensity; the ATP binding site is found at the interface between two adjacent subunits. The asterisk indicates one of the three fenestrations between subunits from which ions flow in/out the extracellular vestibule of the channel. (c) A schematic view P2X receptor ion channel showing the four vestibules and the selective ion pore. Comparison of this drawing with those in Figure 16.3b, c shows how a similar three-dimensional architecture can be built starting from very different molecular structures.

the postsynaptic membrane. These loops contain amino acid sequences that are targets of various kinases: for example, in the nicotinic receptor, protein kinase A (PKA) phosphorylates subunits δ and ε, protein kinase C (PKC) phosphorylates subunits α and δ, and a tyrosine kinase phosphorylates subunits β and δ. Phosphorylation can significantly modify receptor properties: PKA phosphorylation increases the number of spontaneous receptor openings, PKC phosphorylation increases the rate of receptor desensitization (see following text), and tyrosine kinase phosphorylation is important for ensuring high concentrations of AChRs at the postsynaptic membrane. In the case of $GABA_A R$, subunit phosphorylation has different effects on receptor function depending on the subtype: PKA negatively modulates the activity of the $β_1$-subunit-containing receptors but positively modulates the activity of those containing the $β_3$-subunit.

Kinases phosphorylating these receptors are activated by second messengers synthesized after stimulation of other receptors (see Chapters 9 and 13); this is an example of how different receptors with different signal transduction systems may influence each other.

The intracellular region of the subunits is also important for the localization of the different receptor subtypes. Chicken ciliary ganglion nicotinic receptors have two subtypes: the subtype containing the $α_3$-subunit has a synaptic localization, whereas the subtype containing the $α_7$-subunit has perisynaptic localization and is excluded from the synapse. When only the cytoplasmic loop between the TM3 and TM4 segments of the $α_7$-subunit is replaced with the homologous region of the $α_3$-subunit, a chimeric receptor is obtained displaying the same location as the $α_3$-containing receptor, indicating that the cytoplasmic loop of $α_3$ is sufficient to determine the synaptic localization of the receptor.

In $GABA_A Rs$, the intracellular region of the γ-subunit selectively binds a 17 kDa protein (GABARAP) that is involved in the $GABA_A$ receptor intracellular transport, localization, and accumulation at the level of the CNS inhibitory synapses (see Chapter 9).

MODULATION OF LGIC ACTIVITY

The extent of the cellular responses induced by LGIC activation can be modulated by at least two variables: the transmembrane potential and the duration of receptor stimulation.

In general, ion flow through the receptor channel varies in proportion to the membrane potential: that is, the current–voltage relation is linear. Only in excitatory amino acid receptors of NMDAR type the opening of the channel requires not only the occupation of the neurotransmitter binding sites but also a partial depolarization of the cell membrane (see Chapter 43).

It is important to remember that receptor channels are poorly selective and, consequently, Na^+ and Ca^{2+} ions can flow into the cell and K^+ ions can flow out through the nicotinic receptor channel. The Na^+ influx prevails at membrane potentials lower than the AChR reversal potential (~0 mV), whereas the K^+ outflow prevails at more positive

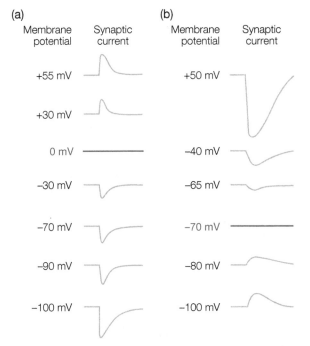

FIGURE 16.7 Direction and magnitude of currents flowing through LGICs. (a) Synaptic currents after nicotinic receptor activation by ACh (lines represent the sum of current variations at all activated channels). The AChR channel is equally permeable to Na^+ and K^+, which move following their electrochemical gradient. At the resting potential of $-90\,mV$, ACh activation of nicotinic receptors gives rise to a Na^+ depolarizing current. At more positive membrane potentials (-70, $-30\,mV$), the inward Na^+ current progressively decreases and the outward K^+ current increases. This leads to a decrease in the total synaptic current. At a membrane potential of $0\,mV$, the inward Na^+ current at each channel is balanced by an equal outward K^+ current, and the resulting synaptic current is equal to 0. At even more positive potentials ($+30$, $+50\,mV$), the outward K^+ current prevails and the direction of the synaptic current is reversed. (b) Variations in Cl^- currents at different membrane potentials. Cl^- currents vary following changes in membrane potential. At a membrane potential of $-65\,mV$, activation of LGICs permeable to Cl^- ($GABA_ARs$ and Gly-Rs) induces an inward hyperpolarizing current that increases in amplitude if the cell is gradually depolarized (-40, $+50\,mV$). At a membrane potential of $-70\,mV$ (equal to the Cl^- equilibrium potential), no current flows through the channel. At a lower potential (between -80 and $-100\,mV$), an outward depolarizing current returns cell potential to the Cl^- equilibrium potential. Currents may be different in cells having Cl^- equilibrium potential different from $-70\,mV$.

potential. Therefore, stimulation of a cation-permeable receptor has virtually no effect if the cell is already depolarized at a membrane potential equal to its reversal potential (see Fig. 16.7). For anion-permeable receptors, the influx of Cl^- ions (the only negative permeant ions) occurs until the membrane potential is higher than the Cl^- equilibrium potential (about $-70\,mV$); at lower potentials (between -80 and $-100\,mV$), the Cl^- ions flow out of the cell and restore membrane potential to the Cl^- equilibrium potential (Fig. 16.7). In any case, activation of these receptors inhibits synaptic activity because the membrane potential is maintained far from the threshold potential ($-55\,mV$).

Desensitization

The extent of the receptor response also depends on the duration of stimulation. In the case of the AChRs, treatment with nonhydrolyzable agonists (e.g., succinylcholine) or drugs that block ACh degradation (anticholinesterase agents) leads to continuous receptor stimulation. Under these conditions, each AChR fluctuates rapidly between a conductive and nonconductive state for several hundred milliseconds until it is stabilized in a nonconductive state not responding to the agonists. This is called desensitization. In the desensitized state, the receptor binds ACh with approximately 1000 times higher affinity, but it remains in a nonconductive state (see Chapter 9).

Desensitization is an intrinsic property of the receptor molecule, but as we saw previously, the desensitization rate and the number of receptor molecules involved are finely regulated by specific phosphorylation events induced by second messengers.

Subunit Composition and Biophysical and Pharmacological Properties of LGICs

We have previously seen that LGICs ion selectivity is determined by specific amino acid sequences in the TM2 region of their constituent subunits. Depending on their subunit content, functional and pharmacological properties of receptors such as conductance, channel opening time, sensitivity to desensitization, and ability to bind agonists and antagonists may vary. A typical example of the importance of subunit composition in defining LGIC properties is the nicotinic receptor. In embryonic muscle, this receptor consists of the α_1-, β_1-, γ-, and δ-subunits (with stoichiometry 2:1:1:1) and has a relatively long average opening time (4–11 ms) and a low level of conductance; in adult muscle receptors, the γ-subunit is replaced by an ϵ-subunit and the opening time is shorter (1–5 ms) and the conductance is high (see Fig. 16.8).

In the CNS, various nAChR isoforms are present that differ in terms of ion selectivity and sensitivity to agonists and antagonists. The nAChR is a heteropentamer consisting of two or three α-subunits and three or two β-subunits. Nine α-subunits (α_{2-10}) and three β-subunits (β_{2-4}) have so far been identified that are specific for nerve cells. The most common isoform in the CNS is a receptor containing the α_4- and β_2-subunits, which is sensitive to ACh and nicotine and is not blocked by the antagonist α-bungarotoxin. It has recently been shown that the same receptor can have other subunits (α_2, α_5, α_6, β_3) along with the α_4- and β_2-subunits, and these

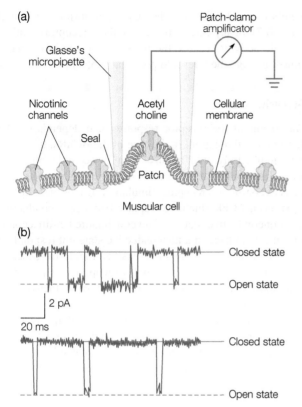

FIGURE 16.8 Changes in muscle AChR functional properties during development. (a) Schematic representation of a patch clamp unit used to determine the biophysical properties of single AChR channel. A glass micropipette with a 1 μm diameter is pressed against a skeletal muscle membrane after removal of the connective tissue. The micropipette is filled with saline solution and contains an electrode connected to an electrical circuit measuring currents flowing through the membrane at the tip of the pipette. (b) Single-channel conductance recorded in ACh-activated channels in fetal and adult muscle. In fetal muscles (*upper trace*), which express AChRs containing the α_1-, β_1-, γ-, and δ-subunits, receptor channels have a low conductance (40 pS) and a long open time (11 ms). In adult muscles (*lower trace*), which express AChRs containing the α_1-, β_1-, ε-, and δ-subunits, receptor channels have a higher conductance (59 pS) but a shorter open time (<6.6 ms).

modify its biophysical and pharmacological characteristics. Several CNS areas also contain α_7 homomeric receptors that, in contrast with other isoforms, are sensitive to α-bungarotoxin and display high Ca^{2+} permeability (see Chapter 39). Another example is the Gly-Rs: in the spinal cord areas, which are highly sensitive to strychnine, there is a prevalence of α_1-containing receptors, whereas in the more rostral areas, which are less sensitive to strychnine, α_2- and α_3-containing receptors are expressed.

In vitro, α- and β-subunits are sufficient to form a pentameric $GABA_AR$ r that opens upon GABA binding to the β-subunit; but sensitivity to benzodiazepines (whose binding site is on the α-subunit) is observed only in the presence of the γ-subunit, which is essential for the allosteric modulation of the receptor (see Chapter 42).

The cellular and molecular mechanisms controlling subunit heterogeneity of receptors are still unclear. The large number of receptors resulting from the combination of different subunits enables cells to vary parameters, such as subcellular localization (presynaptic or postsynaptic, diffuse or highly packed), rate of desensitization, and ion selectivity, which play a key role in the formation and maintenance of synaptic contacts and neuronal circuits. Furthermore, the heterogeneity among receptor subtypes may allow neuronal cells to modulate their own excitability and intensity of responses to environmental stimuli, which are important in learning and memory processes (see also Supplements E 9.2 and E 9.3).

CROSSTALK WITH OTHER RECEPTOR SYSTEMS

Each cell employs its own set of LGICs to modulate ion fluxes, integrate input signals, and (in nerve cells) propagate action potentials and release neurotransmitters. These receptors coexist with many other receptors and ion channels, and their activity can be modulated by second messengers, interaction with the βγ-subunits of G proteins, or posttransductional modifications; in addition, direct physical interactions with other receptors may modulate LGIC activity.

Indeed, it has recently been demonstrated that many LGICs have physical and/or functional interactions. For example, the P2X and $\alpha_3\beta_4$ nicotinic receptors coexpressed in enteric neurons control gastrointestinal functions, and their simultaneous activation produces an inward current that is less than the sum of the currents obtained when only one type of receptor is stimulated, thus indicating that the $\alpha_3\beta_4$ neuronal nicotinic receptor subtype inhibits the P2X receptor. Experimental evidence suggests that this might result from the physical interaction between the two receptors.

Direct interaction has also been demonstrated with GPCRs. In striatal neurons, the dopamine D_1 receptor interacts directly and selectively with the NR1 subunit of the NMDAR, and this modifies the intracellular trafficking of the D_1 receptor. It has also been shown that the D_2 receptor located in presynaptic terminals of mesencephalic dopaminergic neurons may physically interact with heteromeric nAChRs and this interaction modulates the release of dopamine in the striatum.

These studies clearly indicate that mechanisms by which neurotransmitters produce their effects in target cells are very complex, are poorly understood, and vary from cell to cell. The response of individual cells to a neurotransmitter depends on the levels of integration and coordination of different intracellular signaling pathways (see Chapter 15).

MECHANISMS OF ACTION OF DRUGS THAT MODULATE LGICs

Drugs competing with neurotransmitters for binding to the same orthosteric site on LGIC can act as agonists or antagonists, depending on their ability to stabilize the conformational changes required for opening or closing the intrinsic ion channel of the receptor. It is thus possible to use drugs to modulate the intensity and characteristics of the receptor response and the subsequent biological effects.

Receptor activity can also be modulated by drugs that bind to allosteric sites (Fig. 16.9). Occupation of these sites may have effects not only on the neurotransmitter binding kinetics but also on ion channel opening and closing and on the rate of transition to the desensitized state. In $GABA_A$, drugs such as benzodiazepines that bind to the α/γ-subunit interface modulate the binding of GABA to the orthosteric site at the α/β-subunit. Other drugs such as steroids, barbiturates, and picrotoxin control channel gating, whereas penicillin can occlude the open channel. Finally, benzodiazepines can reduce the rate of transition to the desensitized state.

Drugs that bind to allosteric sites and modulate nicotinic receptors are shown in Figure 16.9. The physiological significance of the allosteric sites in AChR is unknown, but it is likely that at least some are binding sites for as yet unknown endogenous substances. This has recently been demonstrated for a family of proteins secreted by neuronal cells (lynx), which share marked structural similarity with elapid snake venom proteins such as α-bungarotoxin (the first nicotinic receptor ligand to be discovered). In particular, when allosterically bound to the $α_4β_2$ nicotinic receptor, protein lynx1 inhibits receptor function by decreasing its affinity for ACh, accelerates receptor desensitization, and slows recovery from the desensitized state.

Drugs active at allosteric sites may have therapeutic value as a result of this mechanism of action (e.g., benzodiazepines); in the case of others, this mechanism can be considered secondary but still worth of attention. For example, many steroid hormones (dexamethasone, prednisolone, hydrocortisone, and progesterone) significantly inhibit muscle nicotinic receptor, and this may be relevant for therapeutic purposes. Some patients with myasthenia gravis (an autoimmune disease directed against AChRs and characterized by a decrease in the number of postsynaptic muscle AChRs) can receive corticosteroid immunosuppressive therapy, but especially at the beginning, they need to be closely monitored as the treatment may induce respiratory muscle weakness as a consequence of steroid blockade of the remaining nicotinic receptors. Physostigmine (Fig. 39.3), a potent cholinesterase inhibitor, sometimes used to treat myasthenia gravis, can enhance the activity of AChRs by binding to an allosteric site. The effect of this drug may be due to the fact that it synergistically slows neurotransmitter metabolism and directly activates the receptor.

In general, drugs that activate or inhibit LGICs through allosteric mechanisms can be pharmacologically very useful as they only modify receptors that are physiologically activated by neurotransmitters.

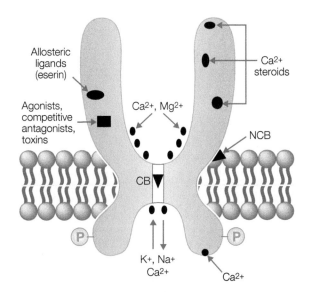

FIGURE 16.9 Orthosteric and allosteric agonist and antagonist binding sites modulating muscle AChR function. Orthosteric sites for agonists, competitive antagonists, and toxins are localized in the receptor extracellular domain. Allosteric sites that bind Ca^{2+}, Mg^{2+}, steroids, physostigmine, and galantamine, as well as competitive and noncompetitive channel blockers, are distributed on the outer part of the channel, in the channel itself, at the interface between the receptor and the membrane, and in the cytoplasmic part of the receptor. CB, channel blockers; NCB, noncompetitive blockers; P, phosphorylation site.

TAKE-HOME MESSAGE

- LGICs mediate rapid cell-to-cell communication and regulate intracellular ion concentrations and transmembrane potential.
- LGICs share similar structure and three-dimensional organization; they are multimeric proteins whose biophysical and pharmacological properties depend on their subunit composition.
- Four classes of LGICs have been recognized differing for number of subunits and modes of activation.
- Most LGICs are selective for ion charge and not for ion species.
- Activation/desensitization and channel properties can be modulated by posttranslational modification, allosteric modulators, and drugs binding to different sites of the receptor molecule.

FURTHER READING

Browne, L. E., Jiang, L. H., and North, R. A. (2010) New structure enlivens interest in P2X receptors. *Trends in Pharmacological Sciences*, 31, 229–237.

Cukkemane, A., Seifert, R., and Kaupp, U. B. (2011) Cooperative and uncooperative cyclic-nucleotide-gated ion channels. *Trends in Biochemical Sciences*, 36, 55–64.

Gotti, C., Zoli, M., and Clementi, F. (2006) Brain nicotinic acetylcholine receptors: native subtypes and their relevance. *Trends in Pharmacological Sciences*, 27, 482–491.

Hurst, R., Rollema, H., and Bertrand, D. (2013) Nicotinic acetylcholine receptors: from basic science to therapeutics. *Pharmacology & Therapeutics*, 137, 22–54.

Olsen, R. W., and Sieghart, W. (2008) International Union of Pharmacology. LXX. Subtypes of gamma-aminobutyric acid(A) receptors: classification on the basis of subunit composition, pharmacology, and function. Update. *Pharmacological Reviews*, 60, 243–260.

Sine, S. M., and Engel, A. G. (2006) Recent advances in Cys-loop receptor structure and function. *Nature*, 440, 448–455.

Smart, T. G., and Paoletti, P. (2012) Synaptic neurotransmitter-gated receptors. *Cold Spring Harbor Perspectives in Biology*, 4, pii: a009662. doi: 10.1101/cshperspect.a009662.

Stephen P. H. Alexander, Helen E. Benson, Elena Faccenda, Adam J. Pawson, Joanna L. Sharman, John C. McGrath, William A. Catterall, Michael Spedding, John A. Peters, Anthony J. Harmar and CGTP Collaborators (2013). The Concise Guide to Pharmacology: *Ligand-Gated Ion Channels*, 1582–1606.

Taly, A., Corringer, P. J., Guedin, D., Lestage, P., and Changeux, J. P. (2009) Nicotinic receptors: allosteric transitions and therapeutic targets in the nervous system. *Nature Reviews Drug Discovery*, 8, 733–750.

Traynelis, S. F., Wollmuth, L. P., McBain, C. J., Menniti, F. S., Vance, K. M., Ogden, K. K., Hansen, K. B., Yuan, H., Myers, S. J., and Dingledine, R. (2010) Glutamate receptor ion channels: structure, regulation, and function. *Pharmacological Reviews*, 62, 405–496.

Unwin, N. (2005) Refined structure of the nicotinic acetylcholine receptor at 4A resolution. *Journal of Molecular Biology*, 346, 967–989.

17

G-PROTEIN-COUPLED RECEPTORS

Lucia Vallar, Maria Pia Abbracchio, and Lucia Vicentini

By reading this chapter you will:

- Learn classifications and structures of G-protein-coupled receptors (GPCRs)
- Understand structure–function relationship of GPCR family members
- Learn the mechanisms of allosteric and orthosteric activation of GPCRs and the molecular and cellular mechanisms whereby ligands activate GPCRs and trigger intracellular signaling pathways
- Learn the background for understanding the mechanism of action of drugs acting on GPCRs and their potential significance as therapeutic agents.

A variety of hormones, neurotransmitters, and other chemical mediators induce their responses by coupling to G proteins, which in turn regulate specific effectors. G-protein-coupled receptors (GPCRs) represent the largest family of membrane receptors. In the human genome, about 800 genes encode for these receptors. GPCRs recognize ligands as different as catecholamines, serotonin, acetylcholine, GABA, glutamate, glycoprotein hormones, peptides, and molecules involved in cell adhesion. In addition, they are activated by light photons, odorants, and taste substances (Table 17.1). In response to these extracellular stimuli, GPCRs activate a great variety of intracellular signals, which control multiple functions. This family of receptors has therefore a great pharmacological impact. Indeed, about 30% of the currently used drugs target a GPCR.

This chapter will describe the structural and functional features of GPCRs, G proteins, and some of their effectors. The ability of GPCRs to interact with other intracellular proteins involved in receptor regulation or signal transduction will also be illustrated.

MOLECULAR ORGANIZATION OF GPCRs

GPCRs are characterized by significant differences in their amino acid sequence but share a common structural organization. Each receptor is formed by a single polypeptide chain that spans the plasma membrane seven times with extracellular N-terminus and intracellular C-terminus (Fig. 17.1). For this reason, GPCRs are also known as seven-transmembrane receptors (7TMRs). The highest number of conserved amino acid residues is found in the transmembrane segments, mostly in the regions closer to the cytoplasm, whereas the extracellular and cytoplasmic regions of the molecule may be greatly different in different receptors. Based on phylogenetic analysis, five main classes of GPCRs (classes A, B, C, D, and E) have been identified in humans (see Supplement E 17.2, "Structure and Conformation Modifications of GPCRs"). The rhodopsin or A family is by far the best characterized. In addition to rhodopsin, which is the photon-activated receptor expressed in the retina, this family includes receptors for catecholamines and serotonin, acetylcholine (muscarinic receptors), histamine, glycoprotein hormones, various peptide receptors, and many other ligands. All rhodopsin family members share common structural features.

TABLE 17.1 Examples of GPCRs

Receptor	G protein	Effector
Neurotransmitters		
Catecholamines		
Adrenergic β_1, β_2, β_3	G_s	↑ Adenylyl cyclase
Adrenergic α_1	G_q	↑ Phospholipase C
Adrenergic α_2	G_i/G_o	↓ Adenylyl cyclase
		↑ K⁺ channels
		↓ Ca²⁺ channels
Dopamine		
D_1	G_s	↑ Adenylyl cyclase
D_2	G_i/G_o	↓ Adenylyl cyclase
		↑ K⁺ channels
		↓ Ca²⁺ channels
Serotonin		
5-HT$_1$, 5-HT$_5$	G_i/G_o	↓ Adenylyl cyclase
		↑ K⁺ channels
		↓ Ca²⁺ channels
5-HT$_2$	G_q	↑ Phospholipase C
5-HT$_4$, 5-HT$_6$, 5-HT$_7$	G_s	↑ Adenylyl cyclase
Acetylcholine		
Muscarinic M$_1$, M$_3$, M$_5$	G_q	↑ Phospholipase C
Muscarinic M$_2$, M$_4$	G_i/G_o	↓ Adenylyl cyclase
		↑ K⁺ channels
		↓ Ca²⁺ channels
GABA		
GABA$_B$	G_i/G_o	↓ Adenylyl cyclase
		↑ K⁺ channels
		↓ Ca²⁺ channels
Glutamate		
mGlu$_1$, mGlu$_5$	G_q	↑ Phospholipase C
mGlu$_{2-4}$, mGlu$_{6-8}$	G_i/G_o	↓ Adenylyl cyclase
		↑ K⁺ channels
		↓ Ca²⁺ channels
Opioids	G_i/G_o	↓ Adenylyl cyclase
		↑ K⁺ channels
		↓ Ca²⁺ channels
Cannabinoids	G_i/G_o	↓ Adenylyl cyclase
		↑ K⁺ channels
		↓ Ca²⁺ channels
Hormones		
ACTH	G_s	↑ Adenylyl cyclase
LH	G_s	↑ Adenylyl cyclase
	G_q	↑ Phospholipase C
FSH	G_s	↑ Adenylyl cyclase
TSH	G_s	↑ Adenylyl cyclase
	G_q	↑ Phospholipase C
GHRH	G_s	↑ Adenylyl cyclase
CRH	G_s	↑ Adenylyl cyclase
GnRH	G_q	↑ Phospholipase C
TRH	G_q	↑ Phospholipase C
Somatostatin	G_i/G_o	↓ Adenylyl cyclase
		↑ K⁺ channels
		↓ Ca²⁺ channels
Vasopressin		
V_1	G_q	↑ Phospholipase C
V_2	G_s	↑ Adenylyl cyclase
Glucagon	G_s	↑ Adenylyl cyclase
Secretin	G_s	↑ Adenylyl cyclase
VIP	G_s	↑ Adenylyl cyclase
PTH	G_s	↑ Adenylyl cyclase
	G_q	↑ Phospholipase C
Calcitonin	G_s	↑ Adenylyl cyclase
	G_q	↑ Phospholipase C
Other mediators		
Histamine		
H_1	G_q	↑ Phospholipase C
H_2	G_s	↑ Adenylyl cyclase
H_3, H_4	G_i/G_o	↓ Adenylyl cyclase
		↓ Ca²⁺ channels
Adenosine		
A_1, A_3	G_i/G_o	↓ Adenylyl cyclase
		↑ K⁺ channels
		↓ Ca²⁺ channels
A_2	G_s	↑ Adenylyl cyclase
ATP		
P2Y$_{1-2}$, P2Y$_4$, P2Y$_6$, P2Y$_{11}$	G_q	↑ Phospholipase C
P2Y$_{12}$, P2Y$_{13}$, P2Y$_{14}$	G_i	↓ Adenylyl cyclase
Angiotensin II		
AT$_1$	G_q	↑ Phospholipase C
Bradykinin	G_q	↑ Phospholipase C
Thrombin		
PAR1	G_i	↓ Adenylyl cyclase
	G_q	↑ Phospholipase C
	G_{12}	↑ RhoGEF
PAR4	G_q	↑ Phospholipase C
	G_{12}	↑ RhoGEF
Lysophosphatidic acid		
LPA$_1$, LPA$_2$, LPA$_4$	G_i	↓ Adenylyl cyclase
	G_q	↑ Phospholipase C
	G_{12}	↑ RhoGEF
LPA$_3$	G_i	↓ Adenylyl cyclase
	G_q	↑ Phospholipase C
LPA$_5$	G_q	↑ Phospholipase C
	G_{12}	↑ RhoGEF
Eicosanoids		
Prostaglandin PGE$_2$		
EP$_1$	G_q	↑ Phospholipase C
EP$_2$, EP$_4$	G_s	↑ Adenylyl cyclase
EP$_3$	G_i	↓ Adenylyl cyclase
Prostaglandin PGF$_2\alpha$	G_q	↑ Phospholipase C
Prostacyclin	G_s	↑ Adenylyl cyclase
Thromboxane A$_2$	G_q	↑ Phospholipase C
	G_{12}	↑ RhoGEF
Photons		
Rhodopsin	G_t	↑ cGMP phosphodiesterase

A large variety of receptors transduce extracellular signals by coupling to G proteins. The table shows the main G proteins and the main effectors regulated by various receptors or receptor subtypes. A single receptor can activate more than one type of G protein and a single G protein can regulate multiple effectors.

ACTH, adrenocorticotropic hormone; CRH, corticotrophin-releasing hormone; FSH, follicle-stimulating hormone; GHRH, growth hormone-releasing hormone; GnRH, gonadotropin-releasing hormone; LH, luteinizing hormone; PTH, parathyroid hormone; TRH, thyrotropin-releasing hormone; TSH, thyroid-stimulating hormone.

↑, stimulation; ↓, inhibition.

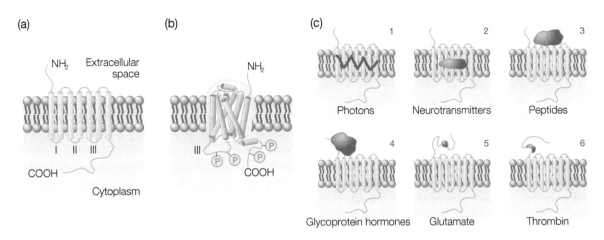

FIGURE 17.1 Molecular organization of GPCRs. GPCRs consist of a single polypeptide chain that spans the plasma membrane seven times. (a) The N-terminus of the receptor is extracellular and the C-terminus is intracellular. The seven transmembrane domains are connected by three extracellular regions and three cytoplasmic loops (I, II, and III). Loops II and III play an important role in interaction with G proteins. A schematic model of domain organization, derived from the crystal structures of some members of the rhodopsin family, is shown in (b). Phosphorylation sites present in the receptor intracellular loops and C-terminal tail are also indicated. (c) GPCRs use different strategies to interact with their ligands. In the case of rhodopsin (1), the inactive receptor is bound to *cis*-retinal through its transmembrane regions. Photons induce isomerization of *cis*-retinal to *trans*-retinal resulting in a conformational change that activates the receptor. (2) Many neurotransmitters and low-molecular-weight chemical mediators also bind a receptor "pocket" formed by the transmembrane domains. (3) Various peptides recognize both transmembrane and extracellular regions, whereas (4) glycoprotein hormones mainly interact with the receptor N-terminal portion. In the case of glutamate (5) and thrombin (6), activation is determined by the interaction between the N-terminus and specific extracellular regions of the receptor itself. Glutamate binding "forces" the N-terminal region to interact with receptor domains involved in activation. Thrombin cleaves the receptor, thus generating a new N-terminus capable of activating the receptor itself.

The Ligand Binding Site

To design molecules able to regulate receptor function, it is necessary to identify the receptor regions involved in ligand binding and those mediating the transition to an "active" conformation and interaction with G proteins. A variety of studies have shown that different types of endogenous ligands use different strategies to bind their specific receptors (Fig. 17.1). In the case of catecholamines, serotonin, histamine, acetylcholine, and other low-molecular-weight molecules, some or all the seven hydrophobic transmembrane regions contribute to the formation of a "pocket" within the plasma membrane that binds the ligand. In particular, studies on α- and β-adrenergic receptors have identified some amino acid residues crucial for catecholamine binding. In these receptors, the carboxyl group of an aspartic acid residue in the third transmembrane domain interacts with the aminic group of catecholamines, while two serine residues in the fifth transmembrane domain are responsible for binding the catechol moiety. On the contrary, receptor interaction with small peptides involves both transmembrane segments and some extracellular regions. The extracellular N-terminal region of the receptor represents the main binding site for large peptide hormones and for glycoprotein hormones, such as LH, FSH, and TSH. In some cases, it has been proposed that binding to this region "guides" the hormone toward a subsequent interaction with transmembrane segments and specific extracellular domains. Another interesting type of interaction is employed by glutamate and by thrombin and other proteases. In these cases, activation is due to interaction between the N-terminus and some extracellular regions of the receptor itself. Glutamate binding to the N-terminal end "forces" this part of the receptor to interact with extracellular domains. Thrombin and other proteases cleave the receptor N-terminus, thus generating a new N-terminal end that acts as a tethered ligand to induce receptor activation. In many cases, the physiological ligands for GPCRs are not known; these receptors are called orphan receptors (see Supplement E 17.1, "Orphan Receptors").

A great number of drugs act as either GPCR agonists or antagonists by interacting directly with the endogenous agonist binding site of the receptor ("orthosteric ligands"). However, more recent studies have identified numerous other agents that, by binding to a receptor site different from that used by the endogenous chemical mediator, can negatively or positively influence its action ("allosteric ligands"). These agents are believed to provide a more finely tuned regulation of receptor activity. Various allosteric modulators of different types of receptor are currently under investigation.

Irrespective of their binding mechanism, chemical mediators induce a conformational change in the receptor intracellular regions. At least in some receptors, the ligand causes a spatial reorganization of the transmembrane segments, which in turn modifies the receptor cytoplasmic surface, thus allowing G-protein binding (see Supplement E 17.2). The second and third cytoplasmic loops are particularly important for interaction with G proteins, which is crucial for further signal transmission.

Finally, various functionally important phosphorylation sites are present in the receptor cytoplasmic region, mainly in its C-terminal tail. These sites are involved in receptor phosphorylation by G-protein-coupled receptor kinases (GRKs) and in the consequent binding to β-arrestin (see Chapter 9). Formation of the receptor–β-arrestin complex is crucial not only for receptor desensitization but also for some signal transduction mechanisms, as illustrated later in this chapter.

Dimerization

Dimerization of receptor molecules is a crucial event for activation of several intracellular and membrane receptors (see Chapters 9, 18, and 24). Although it has long been thought that GPCRs only exist as monomers, it is now evident that also this class of receptors can form dimers or oligomers. Further complexity is added by the fact that dimers may be composed of identical receptors (homodimers) but also of different receptor subtypes or even receptors for different ligands (heterodimers). Examples of receptors that form dimers and/or heterodimers include $GABA_B$, glutamate (metabotropic), β-adrenergic, dopaminergic, opioid, adenosine, somatostatin, and vasopressin receptors.

Various receptors form dimers already in the endoplasmic reticulum. In some of these cases, it has been shown that dimerization is required for receptor transport to the plasma membrane. On the contrary, other receptors are present in the plasma membrane as monomers, and their dimerization appears to be a dynamic process often regulated by ligand interaction. Many receptors can bind the ligand and activate G proteins in their monomeric form. However, at least in some cases, dimerization is necessary for both ligand interaction and G-protein activation. A very well-characterized example of the importance of dimerization in GPCR trafficking and function comes from studies on the $GABA_B$ receptor. In this case, two different subtypes, $GABA_{B1}$ and $GABA_{B2}$, form a heterodimer. It has been shown that $GABA_{B1}$ receptor can reach the membrane only if associated with $GABA_{B2}$ receptor. In addition, the heterodimer represents the only functional form of the receptor. Indeed, $GABA_{B1}$ monomer is able to bind GABA but not to interact with G proteins, whereas $GABA_{B2}$ monomer does not recognize GABA but efficiently activate G proteins.

Although not necessarily required for receptor function, formation of homodimers and/or heterodimers can influence in multiple ways ligand binding, signaling through G proteins, and other receptor properties. In some dimers, activation of one receptor by its chemical mediator positively or negatively modifies, through an allosteric mechanism, the ability of the other receptor to interact with its ligand. For example, in the adenosine A_{2A}–dopamine D_2 receptor heterodimers found in the CNS, A_{2A} activation inhibits dopamine binding and consequent D_2 activation.

Further studies are certainly necessary to understand the physiological importance of GPCR homo- and heterodimerization. However, it is already possible to predict that the pharmacological impact will be great, since all the commercially available drugs interfering with GPCRs have been designed assuming the existence of monomeric receptors. It has been even hypothesized that some orphan GPCRs (see Supplement E 17.1) may work as such and may serve to regulate the function of other GPCRs by binding them and promoting formation of heterodimers with specific pharmacological properties.

GPCR Mutations and Human Diseases

A number of human diseases are associated with mutations of GPCRs. Mutations mainly affecting amino acidic residues in transmembrane segments or intracellular domains can cause constitutive activation of the receptor. These mutations have been identified in the TSH receptor in a high percentage of thyroid adenomas and in the LH receptor in familiar precocious puberty. Both TSH and LH receptors activate G_s and adenylyl cyclase with consequent increase in intracellular cAMP levels (see later in this chapter). cAMP initiates a cascade of intracellular events, which results in proliferation and differentiation of thyroid cells and testosterone production by Leydig cells. Receptor mutations cause G_s activation even in the absence of hormone and a persistent cAMP increase responsible for the development of the disease. Mutations inactivating the receptor have been identified in other diseases. For example, mutations in the vasopressin receptor have been found in patients affected by congenital diabetes insipidus. In most cases, these mutations cause receptor retention in the endoplasmic reticulum and loss of receptor function, thus preventing normal kidney response to vasopressin. Mutations that interfere with the function of rhodopsin, the photon-activated receptor present in retinal rods, are the most common cause of pigmentous retinitis.

In addition to the typical gene mutations described earlier, a variety of polymorphisms have been identified in all GPCR regions. These polymorphisms can influence receptor function and its response to drugs (see Chapter 23).

MOLECULAR ORGANIZATION AND FUNCTION OF G PROTEIN

G proteins are heterotrimers that bind and hydrolyze GTP. G proteins are associated with the cytoplasmic surface of the membrane and consist of three subunits designated,

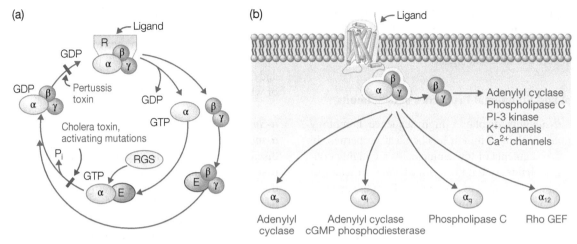

FIGURE 17.2 Signal transduction through G proteins. (a) G-protein cycle. G proteins are composed of α-, β-, and γ-subunits. When the receptor (R) is not activated by its ligand, the guanine nucleotide binding site of the α-subunit is occupied by GDP and α is associated with the βγ complex. Upon activation, the receptor interacts with the G protein and induces GDP–GTP exchange. As a result, the G protein dissociates from the receptor and α dissociates from βγ. Both α-GTP and βγ complex can bind and regulate specific effectors (E). GTP is then hydrolyzed by the α-subunit GTPase activity and α-GDP reassociates with βγ. RGS proteins greatly accelerate the rate of α-subunit GTP hydrolysis. Cholera toxin and pertussis toxin ADP-ribosylate some α-subunits altering G-protein cycle. Mutations identified in human tumors activate α-subunits inhibiting their GTPase activity. (b) Main G-protein effectors. Based on structural and functional properties, α chains have been divided in four classes: α_s, α_i, α_q, and α_{12}. α_s and α_i regulate adenylyl cyclase. The α_i group includes α_{i1}, α_{i2}, α_{i3}, α_o, and transducin, or α_t, which activates a cGMP phosphodiesterase in retinal cells. The α_q-subunits (α_q, α_{11}, α_{14}, and α_{16}) stimulate phospholipase C, whereas the α_{12} family, which includes α_{12} and α_{13}, regulates guanine nucleotide exchange factors for RhoGTPase (RhoGEF). For clarity, in the picture, G proteins, which physiologically are anchored to the plasma membrane, are represented in the cytoplasm.

in decreasing order of molecular weight, α, β, and γ. β- and γ-subunits are tightly bound together; therefore, from a functional point of view, a G protein can be considered a dimer formed by an α-subunit and a βγ complex. In humans, about 20 different α-subunits, 6 β-subunits, and 12 γ-subunits have been identified. G proteins are divided into different classes based on the identity of their α-subunits. The β- and γ-subunits can combine in multiple ways to produce a variety of βγ complexes. Whether and to what extent each βγ complex is selectively associated with defined α-subunits is not clear yet.

The G-protein α-subunits are able to bind the guanine nucleotide GTP. The name "G proteins," which is a short form for "GTP-binding proteins," refers to this essential feature of α chains. These subunits are also GTPases and hydrolyze the bound GTP to GDP. G proteins share their ability to bind and hydrolyze GTP with other families of proteins, often characterized by a monomeric structure (see Chapter 14).

Signaling by G Proteins

G proteins transfer signals from receptors to effectors through a GTP-regulated cycle. G proteins couple receptors to specific effectors, which can be enzymes, ion channels, or other molecules, via a cycle of activation/inactivation regulated by GTP binding and hydrolysis. This cycle, shown in Figure 17.2, can be summarized as follows: (i) when the hormone or neurotransmitter receptor is not activated by its specific agonist, the guanine nucleotide binding site in the α-subunit is occupied by GDP and α remains tightly associated with the βγ complex as an inactive heterodimer. (ii) Upon ligand binding, the activated receptor interacts with the G protein and induces GDP dissociation and subsequent GTP binding to the α-subunit. This event causes a conformational change in the α-subunit resulting in its dissociation from the βγ complex. Then, both α-GTP and free βγ interact with downstream effectors regulating their activity. (iii) The signal is terminated when the GTPase activity of the α-subunit hydrolyzes GTP to GDP and α-GDP associates with the βγ complex, forming again an inactive heterotrimer. Thus, G proteins transmit signals by continuously switching between an inactive GDP-bound conformation and an active GTP-bound conformation.

To ensure a rapid inactivation of G-protein signals, GTP hydrolysis is regulated by proteins capable of interacting with the α-subunits. In some systems, such as phospholipase C stimulation by G_q, the α-subunit effector itself accelerates GTP hydrolysis, thus acting like a GTPase-activating protein (GAP). In addition, a large family of proteins, named regulators of G-protein signaling or RGSs, interact with GTP-bound α-subunits and greatly accelerate (even more than 2000 times) the rate of GTP hydrolysis (Fig. 17.2). More than 30 RGS proteins have been identified. At least in some cases, specific RGS proteins bind to specific α-subunits. It is worth mentioning that RGS proteins interact also with many other intracellular molecules and appear to play multiple roles in cell function, including regulation of signal generation and integration between different signals.

Structural and Functional Properties of α-Subunits

G-protein α-subunits display a high sequence homology that accounts for their common biochemical properties. In particular, five segments of few amino acids are highly conserved in all α chains. In the three-dimensional structure of the molecule, these sequences associate to form the guanine nucleotide binding site. Other regions greatly vary in different α chains and confer specificity to the interaction with receptor and effector proteins. The C-terminal end is one of the most important regions mediating the communication between α chains and the intracellular portion of the receptor. Various regions within the α-subunit are involved in the interaction with effectors.

On the basis of sequence homology and common functional properties, G-protein α-subunits have been divided in four classes (Fig. 17.2). The first class is represented by α_s, which couples many receptors (e.g., β-adrenergic and TSH receptors) to adenylyl cyclase activation and thereby to cAMP production. The second class includes α_{i1}-, α_{i2}-, α_{i3}-, and α_o-subunits, which interact with many receptors for neurotransmitters and peptides (e.g., α_2-adrenergic and somatostatin receptors) and inhibit adenylyl cyclase. This class also includes transducin or α_t, which is expressed in retinal cells and couples rhodopsin to phosphodiesterase stimulation and cGMP hydrolysis. The third class includes α_q and other subunits with similar function that couple receptors (e.g., muscarinic M_1 and TRH receptor) to phospholipase C stimulation and therefore to generation of two second messengers, inositol 1,4,5-trisphosphate ($InsP_3$ or IP_3) and diacylglycerol (DAG). The fourth class is constituted by the α_{12}- and α_{13}-subunits, which interact with specific receptors (e.g., thrombin and thromboxane A_2 receptors) and stimulate guanine nucleotide exchange factors that activate the monomeric GTPase Rho, thereby regulating multiple cell functions.

Modification by Bacterial Toxins Some α-subunits are targets for bacterial toxins. Cholera toxin and pertussis toxin act by interfering with the function of some G-protein α-subunits. Both toxins are ADP-ribosyltransferases. Cholera toxin ADP-ribosylates an arginine in one of the α_s regions involved in GTP hydrolysis. Such modification results in inhibition of the GTPase activity of α_s that therefore remains blocked in its active form (see Fig. 17.2), inducing persistent adenylyl cyclase stimulation and continuous cAMP production and causing the massive loss of salts and water from the intestinal cells characteristic of cholera. Substrates of pertussis toxin are the α_i and α_o chains characterized by the presence of a cysteine in their C-terminus involved in their interaction with the receptor. ADP-ribosylation of this residue prevents G-protein activation by receptors (see Fig. 17.2), causing interruption of the signal originated by G_i and G_o. This is believed to contribute to the clinical symptoms of whooping cough.

α-Subunit Mutations and Human Diseases Mutations in α-subunit-encoding genes are present in some genetic diseases and endocrine tumors. Mutations in G-protein α-subunits are at the basis of a variety of pathological conditions. Type Ia pseudohypoparathyroidism (or Albright's hereditary osteodystrophy) is a dominant autosomal disease characterized by resistance to PTH and other hormones acting via receptors coupled to G_s. Most α_s mutations detected in this disease lead to reduced expression of the protein, thus preventing normal cell response. Activating mutations in the α chains are involved in the development of certain human tumors. Mutations causing constitutive activation of the α_s protein by inhibiting its GTPase activity have been identified in a high percentage of GH-secreting adenomas, in some thyroid tumors, and in the McCune–Albright syndrome, a genetic disease characterized by multiple endocrinopathies. These mutations in the α_s-encoding gene, which has been named *gsp* oncogene, induced growth and increased functional activity of certain cell types via persistent cAMP production, as previously said for mutations activating the TSH receptor. Analogous mutations have been detected in the α_{i2} gene in some ovarian and adrenal tumors and in the α_q gene in melanomas.

The βγ Complex

The βγ complex plays multiple roles in signal transmission via G proteins. βγ-Subunits are necessary for the interaction between G proteins and receptors; in fact, receptors bind with high-affinity αβγ-GDP but not α-GDP. Following receptor activation, the βγ complex acts as a signaling molecule that directly interacts with specific effectors, such as various enzymes and ion channels (Fig. 17.2). βγ-Subunits stimulate certain phospholipases C, stimulate or inhibit specific forms of adenylyl cyclase, and activate phosphatidylinositol 3-kinase (PI3K), which generates phosphatidylinositol-3,4,5-trisphosphates (PIP_3). In addition, the βγ complex inhibits Ca^{2+} channels and activates specific K^+ channels, including that expressed in cardiac pacemaker cells, which opens upon activation of the G_i-coupled acetylcholine M_2 muscarinic receptor. Opening of this channel causes K^+ efflux that counteracts the depolarizing effect of the *If* current characteristic of pacemaker cells, thereby inducing bradycardia. Finally, βγ-subunits recruit to the plasma membrane the GRKs that phosphorylate the receptor cytoplasmic regions (see preceding text and Chapter 9). Whether various combinations of β- and γ-subunits selectively interact with different receptors, effectors, or other molecules is not completely clear yet.

Regulation of Multiple Effectors Through G Proteins

By coupling to G proteins, a single receptor can regulate more than one effector, thus initiating a complex network of coordinated signaling events. Upon receptor activation, both the α-GTP and the βγ-subunits released from the G-protein heterotrimer are able to interact with effector proteins. In addition, both some α chains and the βγ complex regulate various effectors. Thus, by interacting with a single G-protein type, a receptor can activate multiple effector pathways. For example, G_i-coupled receptors such as dopamine D_2 and somatostatin receptors inhibit adenylyl cyclase through α chains and activate K^+ channels and inhibit Ca^{2+} channels through the βγ complex (Table 17.1). In many cases, a given receptor can also interact with more than one type of G protein. For example, TSH, LH, PTH, and calcitonin receptors are coupled to G_s and adenylyl cyclase stimulation, but they also induce phosphoinositide hydrolysis by interacting with G_q proteins (Table 17.1). PAR1 receptor activation by thrombin causes simultaneous activation of G_i, G_q, and G_{12} (Table 17.1).

It is believed that different active conformations of receptors mediate binding to different G proteins. Certain agonists cause a receptor to preferentially interact with a specific G protein. For example, some peptides acting on PAR1 receptor greatly favor activation of G_q over G_{12}. These interesting observations open the possibility to develop drugs selective or biased for a specific receptor signaling pathway (see Supplement E 17.2).

EFFECTOR PATHWAYS OF GPCRs

Activation of specific effectors translates the signal generated by receptor–G-protein interaction in a cellular message. In addition to the effectors described previously in this chapter, molecules controlled by G proteins include: phospholipase D, which leads to phosphatidic acid and DAG formation; phospholipase A_2, releasing arachidonic acid from membrane phospholipids (see Chapter 49); specific tyrosine kinases; and GAPs for the monomeric GTPase Rap (see Chapter 14). By controlling such a variety of effectors, G proteins can regulate multiple intracellular signaling pathways. In particular, it has been shown that both all α-subunits and the βγ complex can stimulate, by various mechanisms, the mitogen-activated protein kinase (MAPK) cascade, which is activated by growth factor receptors and other receptors and plays a crucial role in regulation of cell growth and differentiation (see Chapters 13 and 18). Here, we will describe the properties of the two best characterized enzymatic effectors of G proteins: adenylyl cyclase and phospholipase C. Ion channel regulation by G proteins is illustrated in Chapter 27.

The Adenylyl Cyclase System

Adenylyl cyclase is regulated by different G-protein subunits. Adenylyl cyclase is a ubiquitous enzyme that transforms ATP into cAMP. Adenylyl cyclase activity is regulated by a wide number of hormones, neurotransmitters, and other chemical mediators that interact with GPCRs. Some receptors stimulate adenylyl cyclase activity, whereas others inhibit it (Table 17.1). Nine main forms of adenylyl cyclase (named type 1–9) differentially expressed in different cell types have been identified. All these enzymes share a similar structure consisting of a short N-terminal segment and two large cytoplasmic domains (named C1 and C2) separated by two highly hydrophobic regions (M1 and M2) that span the plasma membrane six times (Fig. 17.3). All types of adenylyl cyclase display a high sequence homology in the two cytoplasmic regions, which are essential for catalytic

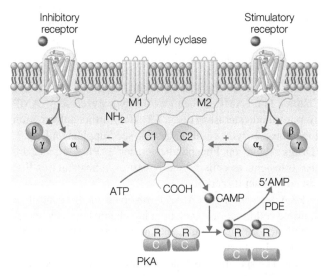

FIGURE 17.3 The adenylyl cyclase system. Adenylyl cyclase is characterized by two transmembrane regions (M1 and M2) and two large cytoplasmic regions (C1 and C2) and is regulated by stimulatory and inhibitory receptors. Agonists or antagonists of stimulatory receptors (e.g., $β_1$-adrenergic receptors) and inhibitory receptors (e.g., $α_2$-adrenergic and D_2 dopaminergic receptors) are largely used as drugs. Stimulation (+) and inhibition (−) of adenylyl cyclase are mediated by $α_s$-GTP and $α_i$-GTP, respectively. For clarity, in the picture, G-protein subunits are represented in the cytoplasm and their position with respect to adenylyl cyclase is arbitrary. Adenylyl cyclase stimulation leads to intracellular cAMP accumulation and consequent protein kinase A (PKA) activation. In its inactive form, PKA is composed of two regulatory subunits (R) and two catalytic subunits (C). cAMP binding causes release of the catalytic subunits, which phosphorylate a variety of cellular proteins. cAMP is hydrolyzed by phosphodiesterases (PDE).

activity, but greatly differ in other parts of the molecule. This heterogeneity reflects the different mechanisms of regulation of the enzymes.

Adenylyl cyclase stimulation occurs mainly via G_s activation by a variety of receptors. α_s-Subunits directly interact with C2 and the N-terminal portion of C1 in all types of adenylyl cyclases, thus stimulating enzyme activity (Fig. 17.3). On the contrary, G_i-coupled receptors inhibit adenylyl cyclase activity via the α_i-subunit (Fig. 17.3). Activated α_i chains bind C1 in certain types of enzymes and interfere with the conformational modification induced by α_s. Specific types of adenylyl cyclase can also be stimulated or inhibited by the G-protein βγ complex (Fig. 17.3).

The cAMP-Activated Protein Kinase A Adenylyl cyclase activity is dynamically controlled by a great variety of chemical mediators and is the main determinant of variations in intracellular cAMP levels. cAMP regulates many cellular proteins, including a specific protein kinase named protein kinase A (PKA). In the absence of cAMP, PKA is composed of two regulatory subunits and two catalytic subunits associated in an inactive tetrameric complex. When intracellular cAMP concentrations increase, the cyclic nucleotide binds to the regulatory subunits of the complex, causing dissociation of the catalytic subunits, which phosphorylate serine and threonine residues in specific proteins, thus modulating their activities (see Chapter 11). cAMP is then rapidly hydrolyzed to 5′-AMP by phosphodiesterases (Fig. 17.3) (for further information about these enzymes as drug targets, see Supplement E 17.3, "Drugs Active on Phosphodiesterases"). PKA effects are also transient, as cells normally express phosphatases that can counteract its activity.

PKA activation by cAMP is involved in the control of multiple cell functions, including ion channel activity, metabolic reactions, gene transcription, cardiac activity, smooth muscle cell relaxation, secretion, differentiation, and proliferation. The specific substrates phosphorylated by PKA and the consequent cell response are determined by the cellular localization of the enzyme. PKA localization is controlled by various members of the A-kinase-anchoring protein (AKAP) family, which bind the regulatory subunits of the inactive tetramer. There are more than 50 different AKAPs that localize PKA to specific regions in various cell types. AKAPs possess multiple domains involved in protein–protein interaction and favor the assembly of multimolecular complexes. These complexes can contain various components of the cAMP–PKA signaling pathway, like receptors, adenylyl cyclases, PKA-specific substrates, phosphodiesterases, and phosphatases. By tethering all these molecules, AKAPs control not only the intracellular localization but also the size and timing of PKA-mediated signals.

Finally, it is important to note that cAMP can also regulate cellular proteins independently of PKA. In some systems,

cAMP binds directly to membrane cation channels, thus triggering Na^+, K^+, and Ca^{2+} influx into the cell. The sinoatrial node. If current is one of the most important examples of this direct regulation. Noradrenaline-stimulated β-adrenergic receptors induce opening of the "f" channel by cAMP and thereby activation of the If depolarizing current, which has a major role in heart pacemaking and rate modulation. The drug ivabradine can selectively inhibit the If current and represents the first specific heart rate-lowering agent that has completed clinical development for stable angina pectoris and cardiac failure.

The Phospholipase C System

Phospholipase C hydrolyzes phosphatidylinositol 4,5-bisphosphate (PIP_2) and produces two second messengers. Phospholipase C is a phosphodiesterase activated by neurotransmitters, hormones, growth factors, and other chemical mediators, which catalyzes the hydrolysis of a specific membrane phospholipid, PIP_2. PIP_2 is found in the plasma membrane internal leaflet and derives from phosphorylation of phosphatidylinositol at positions 4 and 5 of the ring catalyzed by specific kinases (Fig. 17.4). Stearic acid and arachidonic acid occupy positions 1 and 2, respectively, of the phospholipid glycerol moiety. PIP_2 hydrolysis by phospholipase C gives rise to two second messengers, IP_3 and DAG, which regulate many cell functions such as metabolism, secretion, muscle cell contraction, neuronal activity, migration, and proliferation.

At least 13 phospholipase C isoforms have been identified. Based on their structural properties, these enzymes have been divided into six classes: β, γ, δ, ε, ζ, and η. Within each class, the different isoforms are indicated by numbers (e.g., β1 or γ2). Phospholipase C isoforms differ in both molecular weight and amino acid sequence. However, they all possess two regions, called X and Y, which share a certain degree of sequence homology and bear the catalytic activity. In other regions, the different types of enzymes contain distinct regulatory sequences. As it will be illustrated later in this Chapter, phospholipase C isoforms greatly differ in their mechanisms of regulation and are differentially expressed in various cell types.

In all isoforms, the catalytic activity measured *in vitro* is Ca^{2+} dependent. The hydrolysis is specific for inositol-containing phospholipids, with PIP_2 being the preferred substrate, followed by phosphatidylinositol 4-phosphate and phosphatidylinositol. In general, phospholipase C is located in the cytosol and translocates to the plasma membrane upon activation.

Mechanisms of Activation of Phospholipases C Phospholipases C can be activated by multiple mechanisms. The best characterized are those involved in phospholipase Cβ and Cγ activation. The four phospholipase Cβ isoforms are effectors of various G-protein subunits (Fig. 17.4). All

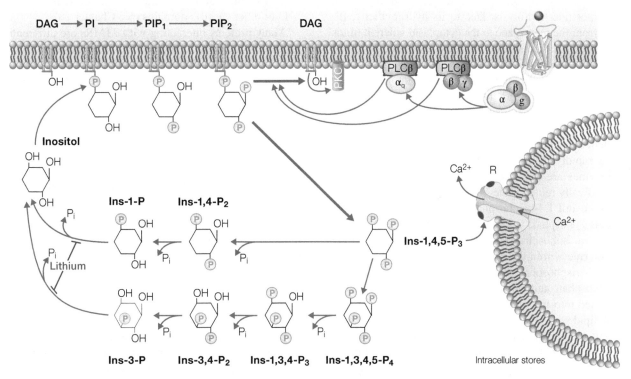

FIGURE 17.4 Phosphoinositide hydrolysis. Phosphatidylinositol (PI) is synthesized from inositol and diacylglycerol (DAG) and sequentially phosphorylated to phosphatidylinositol 4-phosphate (PIP_1) and phosphatidylinositol 4,5-bisphosphate (PIP_2). Phospholipase C (PLC) hydrolyzes PIP_2, thus generating diacylglycerol, which activates protein kinase C (PKC), and inositol 1,4,5-trisphosphate (Ins-1,4,5-P_3), which interacts with its specific receptor (R) to release Ca^{2+} from intracellular stores. By coupling to G proteins, 7TMRs activate type β phospholipases C. All these enzymes are effectors of the $α_q$-subunits. Some forms of phospholipase Cβ, such as phospholipase $Cβ_2$, are also stimulated by the βγ complex. The figure also illustrates the main metabolic pathways leading to the formation of inositol from Ins-1,4,5-P_3 and the site of action of lithium.

phospholipases Cβ are activated by G_q protein α-subunits. Some enzymes, such as phospholipase Cβ2, are also activated by the G-protein βγ complex. The Cβ isoforms interact with $α_q$ through their C-terminal regions, whereas phospholipase $Cβ_2$ interacts with the βγ complex via a sequence in its N-terminal domain. The interaction with $α_q$-subunit or βγ complex, which are both associated with the plasma membrane, involves phospholipase C translocation from the cytoplasm. The two phospholipase Cγ isoforms are activated by growth factor receptors (see Chapter 18). These receptors possess an intrinsic tyrosine kinase activity and, upon stimulation, undergo autophosphorylation. The receptor sequences containing the phosphorylated residues can then interact with one of the phospholipase Cγ SH2 domains located between the X and Y regions. As a consequence, the receptor tyrosine kinase phosphorylates phospholipase Cγ, stimulating its activity. Other types of receptors (e.g., antigen and immunoglobulin receptors as well as GPCRs) can activate phospholipase Cγ indirectly, by activating cytoplasmic tyrosine kinases that in turn phosphorylate the enzyme. Multiple and not completely understood mechanisms are involved in the activation of the other phospholipase C classes. These enzymes are often regulated by intracellular signaling molecules, such as Ca^{2+} ions or monomeric GTPases.

DAG and IP_3. DAG generated from phosphoinositide hydrolysis remains in the plasma membrane where it activates protein kinases C (PKCs) that phosphorylate serine and threonine residues in specific substrates (see Fig. 17.4 and Chapter 11). PKCs are a large family of enzymes, regulated by a variety of mechanisms and involved in many important cell functions. Seven PKC forms are activated by DAG. DAG binding causes PKC translocation from the cytosol to the plasma membrane, where the enzyme undergoes a conformational modification that activates the catalytic region. PKC activation requires the simultaneous interaction with DAG and phospholipids, and at least in some cases, it is Ca^{2+} dependent. DAG accumulation caused by phosphoinositide hydrolysis is transient. DAG is rapidly metabolized by a specific lipase or by a kinase that converts it into phosphatidic acid.

IP$_3$ is the other second messenger resulting from phosphoinositide hydrolysis. Due to its hydrophilicity, IP$_3$ diffuses from the membrane to the cytoplasm where it binds specific receptors located on vesicles associated with the endoplasmic reticulum that act as calcium stores (see Chapter 12). The IP$_3$ receptor is a channel that upon activation mediates calcium exit from intracellular stores, thus leading to transient increase in cytosolic calcium concentration (Fig. 17.4), which in turn activates multiple calcium-dependent enzymes and events.

IP$_3$ is rapidly metabolized in the cytoplasm (Fig. 17.4). Two enzymes are involved in this process: a phosphatase that specifically removes the 5-phosphate converting (1,4,5) IP$_3$ into inositol 1,4-bisphosphate and a kinase that specifically adds a phosphate to the 3 position, giving rise to inositol 1,3,4,5-tetrakisphosphate (IP$_4$). IP$_4$, which is believed to play a specific role within the cell, is then transformed into inositol 1,3,4-trisphosphate by IP$_3$/IP$_4$-5-phosphatase. Inositol 1,4-bisphosphate and inositol 1,3,4-trisphosphate are finally transformed into free inositol, which can be used again for phospholipid synthesis. Lithium, a drug used for the treatment of bipolar disorder, inhibits inositol monophosphatase, one of the phosphatases involved in inositol phosphate metabolism, thus causing inositol 1-phosphate accumulation and interruption of phospholipid resynthesis cycle (Fig. 17.4). Some studies have shown that this inhibition impairs phospholipase C-dependent signaling and modifies neuronal activity. It has been proposed that inhibition of inositol monophosphatase by lithium contributes, at least in part, to the therapeutic effects of the drug. This hypothesis is supported by the demonstration that inositol signaling disruptions are linked to memory impairment and depression. However, more recent studies have suggested that other mechanisms, like deactivation of GSK3β, an enzyme involved in regulation of various neuronal functions, may as well contribute to lithium effectiveness.

INTERACTION OF GPCRs WITH OTHER PROTEINS

A great variety of proteins interact with GPCRs. It is well known that 7TMRs interact via their cytoplasmic regions not only with G proteins but also with proteins involved in desensitization. A crucial role in this process is played by β-arrestin binding to activated receptors phosphorylated by GRKs (see Chapter 9). An increasing number of experimental data indicate that 7TMRs can also associate with many other proteins. In general, these proteins bind to sequences involved in protein–protein interaction located in the third cytoplasmic loop and in the C-terminus of the receptor. Many GPCRs contain in their C-terminal end a short sequence that recognizes PDZ domains in various cellular proteins. In several types of receptors, short proline-rich sequences mediate the interaction with proteins containing SH3 or other specific domains (see Chapter 18).

Many proteins interacting with 7TMRs are differentially expressed in distinct cell types. Some of these are constitutively associated with receptors, whereas others bind only activated receptors. In many cases, GPCR-interacting proteins can also bind other proteins, thus acting as scaffolds contributing to the assembly of multimolecular complexes associated with the receptor.

GPCR-Interacting Proteins: Control of Receptor Function, Intracellular Trafficking, and Localization

GPCR-associated proteins play a variety of important functions. In several cases, these proteins regulate receptor intracellular trafficking. Some proteins act as chaperones guiding newly synthesized receptors to the plasma membrane. Other control the destiny of activated receptors internalized by endocytosis, which can be either recycled back to the plasma membrane or targeted to lysosomes for degradation (see Chapter 9). For example, GASP proteins, which are associated with δ opioid, D$_2$ dopamine, and cannabinoid receptors, promote lysosomal degradation of receptors. In other cases, proteins associated with the receptor favor its recycling to the plasma membrane.

Another important function of GPCR-interacting proteins is the control of receptor localization in discrete regions of the plasma membrane. An example is Homer proteins, which bind a proline-rich sequence in the C-terminal end of mGluR1 and mGluR5 metabotropic glutamate receptors. Homer proteins interact with several other proteins, which in some cases are in turn associated with the cytoskeleton, thus promoting mGluR1 and mGluR5 anchoring to specific postsynaptic membrane areas of dendritic spines (see Chapter 43).

Proteins interacting with 7TMRs can also regulate signal transduction via G proteins. In some cases, they increase efficiency of G-protein-mediated responses by acting as scaffolds that bind various signaling molecules and tether them in close proximity to the receptor. mGluR1 and mGluR5 receptors associated with Homer proteins induce G$_q$ activation and consequent phospholipase C stimulation. Homer proteins can bind to a number of molecules involved in this signaling pathway, including IP$_3$ and ryanodine receptors, allowing efficient calcium release from intracellular stores upon receptor activation (see Chapter 12). Similarly, interaction between the PKA-binding proteins AKAPs and G$_s$-coupled β$_2$-adrenergic receptor enhances kinase activation by the receptor. In other cases, GPCR-associated proteins can reduce the efficiency of activation of certain signaling pathways via inhibition of receptor–G-protein interaction or recruitment of molecules that negatively regulate signal transduction. For example, the cytoplasmic regions of various receptors can bind directly or indirectly RGSs, proteins that accelerate GTPase activity of activated α-subunits.

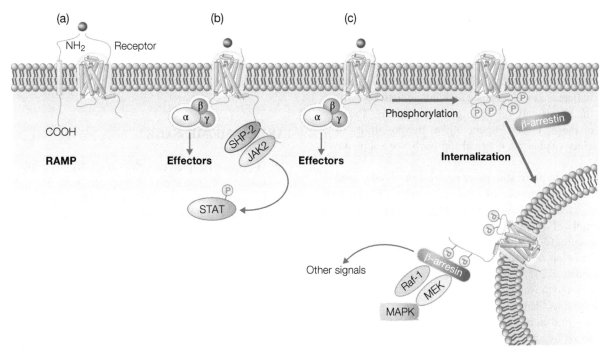

FIGURE 17.5 Signal transduction of GPCRs. 7TMRs interact with a variety of proteins and activate G-protein-independent signals. (a) Calcitonin receptor and calcitonin receptor-like receptor (CRLR) interact with RAMPs, which define the specificity of ligand–receptor interaction. (b) Many proteins interacting with the receptor intracellular regions are involved in signal transduction. Angiotensin II type 1 receptor interacts through its C-terminal tail with a complex containing SHP2 phosphatase and JAK2, resulting in JAK2 activation and STAT transcription factor phosphorylation. (c) Many receptors activate mitogen-activated protein kinases (MAPK) and other signals through β-arrestins. β-Arrestins bind activated receptors that have been phosphorylated by GRKs. β-Arrestin-dependent MAPK activation occurs in endosomes and involves assembly of a complex containing various signaling molecules, including Raf-1 and MEK kinases.

Therefore, using different mechanisms, GPCR-interacting proteins play an important role in regulating timing and intensity of G-protein signaling in each specific cell type.

In general, GPCR-associated proteins interact with intracellular regions of receptors and do not modify their pharmacological properties. An exception is represented by the transmembrane proteins called receptor activity-modifying protein 1 (RAMP1), RAMP2, and RAMP3 (Fig. 17.5). All RAMPs can interact with calcitonin receptor and calcitonin receptor-like receptor (CRLR) and define the specificity of receptor–ligand interaction. RAMP1-bound CRLR recognizes a peptide similar to calcitonin named calcitonin gene-related peptide (CGRP). CRLR association with RAMP2 or RAMP3 forms two different receptor subtypes that bind another peptide called adrenomedullin.

G-Protein-Independent Signaling

Over the past several years, it has been clarified that 7TMRs transduce signals not only via G proteins but also via other proteins interacting with the cytoplasmic regions of the receptor (Fig. 17.5). The best characterized G-protein-independent signals are those mediated by β-arrestins. Indeed, β-arrestins are regarded now as multifunctional adaptor molecules involved in receptor desensitization and trafficking as well as in signal transduction. It has been shown that receptor-bound β-arrestins can stimulate MAPK (see Chapter 13). In the case of β_2-adrenergic receptors, β-arrestins bind and activate Src kinase, causing the assembly of a multimolecular complex containing receptor, β-arrestin, Src, and various MAPK cascade components. With other receptors, such as angiotensin II type 1A receptor, β-arrestins stimulate MAPK by favoring formation of a complex containing receptor, β-arrestin, Raf, Mek kinases, and MAPK itself.

In several cases, β-arrestin-mediated MAPK activation originates from receptors internalized into endosomes. When compared to G-protein-mediated MAPK activation at the plasma membrane, the β-arrestin-dependent endosomal signal displays distinct spatiotemporal features and induces different cell responses. Indeed, the G-protein effect is rapid and transient and causes MAPK nuclear translocation and regulation of gene transcription, whereas β-arrestins induce a delayed and prolonged wave of MAPK stimulation, restricted to the cytoplasm. By using both endosomal and plasma membrane signals, 7TMRs can therefore mediate a complex regulation of the MAPK system. In addition to Src and components of the MAPK cascade, β-arrestins appear able to regulate various other signaling molecules, including PI3K and AKT, which are involved in regulation of cell survival.

Interestingly, upon receptor binding, chemical mediators activate both G-protein-dependent and β-arrestin-dependent signals, whereas some agonists and antagonists act preferentially or exclusively on one signaling pathway. The discovery of these "biased" agents opens the possibility of a highly selective pharmacological approach (see Supplement E 17.2).

In addition to β-arrestin, other proteins involved in signal transduction are capable of interacting with 7TMRs. For example, activation of angiotensin II type 1 receptors causes the assembly of a JAK2 tyrosine kinase-containing complex on the receptor C-terminal tail (Fig. 17.5), resulting in JAK2 phosphorylation and activation. Activated JAK2 can then phosphorylate STAT transcription factors, determining their translocation to the nucleus where they regulate gene expression. Several activated receptors, including $β_2$-adrenergic and parathyroid hormone receptors, interact with a protein called Na^+/H^+ exchanger regulatory factor 1 (NHERF1). NHERF1 binds the receptor C-terminal tail via a PDZ domain and is involved in various signaling pathways. For example, NHERF1–$β_2$-adrenergic receptor interaction blocks NHERF1 ability to inhibit Na^+/H^+ exchanger activity. Therefore, NHERF1 binding represents a G-protein-independent pathway used by $β_2$-adrenergic receptors to activate Na^+/H^+ exchanger in kidney cells.

PERSPECTIVES

During the last few years, our knowledge of 7TMR and G-protein structure has exponentially increased mostly through crystallography and biochemical and biophysical techniques such as nuclear magnetic resonance and mass spectrometry. Furthermore, the historical view that 7TMRs transduce their signal via a single specific G protein has been replaced by a more complex scenario. Now, we know that receptors can associate to form homodimers, heterodimers, or even more complex structures and that a receptor molecule can activate more than one G protein. We also know that, besides G proteins, 7TMRs bind also a variety of other proteins that often serve as scaffolds to assemble a receptor-associated multimolecular complex. These complexes can regulate receptor function and mediate G-protein-independent signals. All this knowledge can favor the development of new pharmacological approaches, including allosteric modulators of receptor function, agents that target receptor dimers, or agonists selective for specific G-protein-dependent or G-protein-independent signaling pathways. Future studies will certainly allow a more comprehensive view of how signal transduction by the various GPCRs is structurally and functionally organized in each specific cell type and will provide further basis for the development of new drugs controlling the function of this important class of receptors.

TAKE-HOME MESSAGE

- G-protein-coupled receptors (GPCRs) form the largest family of membrane receptors and mediate cell responses to chemical mediators, light photons, odorants, and taste molecules.
- All GPCRs consist of a single polypeptide chain that spans the plasma membrane seven times. Individual GPCR can form hetero-/homodimers sensitive to one or more ligands.
- G proteins are αβγ heterotrimers that bind and hydrolyze GTP. Upon activation by receptors, G proteins bind GTP and dissociate; α- and βγ-subunits bind to and regulate the activity of enzymes, ion channels, and other intracellular effectors. Activation is shut off by GTP hydrolysis by GTPase activity intrinsic to the α-subunit.
- Adenylyl cyclase catalyzes the formation of cAMP, which activates protein kinase A.
- Phosphoinositide hydrolysis by phospholipase C leads to the formation of two second messengers: inositol 1,4,5-trisphosphate (IP_3) and diacylglycerol.
- GPCRs interact with a variety of cellular proteins and can activate G-protein-independent signals.

FURTHER READING

Bradley S.J., Riaz S.A., Tobin A.B. (2014). Employing novel animal models in the design of clinically efficacious GPCR ligands. *Current Opinion in Cell Biology*, 27, 117–125.

Hanoune J., Defer N. (2001). Regulation and role of adenylyl cyclase isoforms. *Annual Review of Pharmacology and Toxicology*, 41, 145–174.

Lohse M.J. (2010). Dimerization in GPCR mobility and signaling. *Current Opinion in Pharmacology*, 10, 53–58.

Mujić-Delić A., de Wit R.H., Verkaar F., Smit M.J. (2014). GPCR-targeting nanobodies: attractive research tools, diagnostics, and therapeutics. *Trends in Pharmacological Sciences*, 35, 247–255.

Neves S.R., Ram P.T., Iyengar R. (2002). G protein pathways. *Science*, 296, 1636–1639.

Oldham W.M., Hamm H.E. (2008). Heterotrimeric G protein activation by G protein-coupled receptors. *Nature Reviews. Molecular Cell Biology*, 9, 60–71.

Ritter S.L., Hall R.A. (2009). Fine-tuning of GPCR activity by receptor-interacting proteins. *Nature Reviews. Molecular Cell Biology*, 10, 819–830.

Rosenbaum D.M., Rasmussen S.G.F., Kobilka B.K. (2009). The structure and function of G protein-coupled receptors. *Nature*, 459, 356–363.

Shukla A.K., Xiao K., Lefkowitz R.J. (2011). Emerging paradigms of β-arrestin-dependent seven transmembrane receptor signaling. *Trends in Biochemical Sciences*, 36, 457–469.

Suh P.G., Park J.I., Manzoli L., et al. (2008). Multiple roles of phosphoinositide-specific phospholipase C isozymes. *BMB Reports*, 41, 415–434.

Venkatakrishnan A.J., Depui X., Lebon G., et al. (2013). Molecular signatures of G-protein-coupled receptors. *Nature*, 494, 185–194.

Violin J.D., Crombie A.L., Soergel D.G., Lark M.W. (2014). Biased ligands at G-protein-coupled receptors: promise and progress. *Trends in Pharmacological Sciences*, 35, 308–316.

18

GROWTH FACTOR RECEPTORS

SILVIA GIORDANO, CARLA BOCCACCIO, AND PAOLO M. COMOGLIO

By reading this chapter, you will:

- Know the molecular organization and classification of growth factor receptors and their signal transduction
- Understand how these receptors are involved in cell proliferation
- Learn the strategies of pharmacological modulation of these receptors for therapeutic control of cancer

---

Communication among cells is crucial for development and survival of multicellular organisms. Over the last 50 years, several molecules, generally called growth factors, have been discovered, which are secreted in the extracellular environment and are able to stimulate proliferation. Growth factors control cell entry and progression through the cell cycle. They can be classified as competence and progression factors. The former recruit quiescent cells in the cell cycle; the latter are necessary for the transition from the G_1 to the S phase (see also Chapter 32). Due to historical reasons, growth factor nomenclature often suggests specificities that are not real. As a matter of fact, epidermal growth factor (EGF) also acts on fibroblasts and glial cells. Hepatocyte growth factor (HGF) acts as a mitogen not only for epithelial cells but also for endothelia and myoblasts. Nerve growth factor (NGF) acts not only on sympathetic ganglia cells but also on keratinocytes and lymphocytes. Furthermore, growth factors are not only involved in cell cycle control but also in control of differentiation, survival, motility, and some metabolic activities.

The biological activity of growth factors is mediated by high-affinity receptors located at the plasma membrane. Most of these receptors are tyrosine kinases, catalyzing the transfer of phosphate groups from ATP to tyrosine residues of protein substrates. Their enzymatic activity is tightly regulated: in the absence of the ligand, it is inhibited, and ligand binding to the extracellular part of the receptor removes the inhibition. In turn, kinase activity starts an intracellular cascade of biochemical reactions leading to functional responses either immediate and transitory or delayed and durable.

The physiology and pathology of growth factor receptors and other molecules involved in signal transduction are of great pharmacological interest because of their involvement in the pathogenesis of neoplasia and some metabolic defects. Growth factors, their receptors, and signal transducers are encoded by the so-called proto-oncogenes. In physiological conditions, many of these genes (at least a hundred in the human genome) control cell proliferation. Their altered counterparts, mutated or aberrantly expressed (oncogenes), can induce neoplastic transformation in experimental systems and are associated with the onset of human tumors. Mutations causing an increase in receptor expression or tyrosine kinase activity result in deregulated proliferative signaling (constitutive activation). On the contrary, a reduced activity of growth factor receptor may be responsible for morphogenetic defects or metabolic disorders: for example, alterations of the insulin receptor are involved in the pathogenesis of insulin-resistant diabetes.

General and Molecular Pharmacology: Principles of Drug Action, First Edition. Edited by Francesco Clementi and Guido Fumagalli.
© 2015 John Wiley & Sons, Inc. Published 2015 by John Wiley & Sons, Inc.

TABLE 18.1 Characteristics and biological activities of growth factors binding tyrosine kinase receptors

Growth factor	Structure	Activity	Receptor(s)
Epidermal growth factor (EGF)	Soluble 6 kDa protein deriving from proteolytic cleavage of a transmembrane precursor	Mitogenic for epithelial cells	EGFR (ErbB-1)
Transforming growth factor-α (TGF-α)	Soluble 6 kDa protein deriving from proteolytic cleavage of a 25 kDa precursor	Mitogenic for epithelial cells	EGFR (ErbB-1)
Amphiregulin	14–22 kDa glycoprotein	Mitogenic	EGFR (ErbB-1)
Heparin-binding epidermal growth factor (HB-EGF)	Soluble 14–20 kDa protein derived from a proteolytic cleavage of a transmembrane precursor	Mitogenic, chemotactic	EGFR/ErbB-4
Betacellulin	Soluble 32 kDa glycoprotein deriving from proteolytic cleavage of a membrane-bound precursor	Mitogenic for pancreatic and mammary cells	EGFR/ErbB-4
Neuregulin (Nrg) 1 and 2	NRG1 gene encodes at least 12 isoforms, deriving from alternative splicing; mature factors are 42–59 kDa glycoproteins deriving from proteolytic cleavage of a membrane-bound precursor	Morphogenesis of the mammary gland, development of the heart, glia, and nervous system	ErbB-3/ErbB-4
Insulin-like growth factor-1 (IGF-1), insulin	7.5 kDa peptides secreted as a single-chain form	Proliferation, differentiation, and cell survival	InsR/IGF-1R
Platelet-derived growth factor (PDGF)-AA, PDGF-AB, PDGF-BB	30 kDa homo-/heterodimers, bound by disulfide bonds formed by A (16–18 kDa) and/or B (12 kDa) chains	Chemotaxis, proliferation	PDGFRα/PDGRFβ
Stem cell growth factor (SCF)	31–36 kDa glycoprotein; transmembrane precursor cleaved in soluble forms	Proliferation, differentiation, migration, and survival of hematopoietic cells, melanocytes, and gametes	Kit
Colony-stimulating factor 1 (CSF-1)	70–90 kDa homodimeric glycoprotein; transmembrane precursor cleaved in soluble forms	Proliferation of monocyte–macrophage lineage; embryo implantation	Fms
Flt-3 L	Disulfide-linked 30 kDa heterodimer	Proliferation, survival of progenitor cells in the hematopoietic system	Flt-3
Fibroblast growth factor (FGF) 1–9	17–30 kDa single-chain glycoproteins; soluble, associated with proteoglycans	Proliferation and morphogenesis of endothelial cells	FGFR1–4
Vascular endothelial growth factor (VEGF) A–C	35–45 kDa homo-/heterodimeric glycoproteins	Mitogenic and morphogenic for endothelial cells	VEGFR1–3
Angiopoietin	70 kDa glycoprotein	Assembly of nonendothelial components of the vessel wall	Tie-2
Hepatocyte growth factor (HGF)	Secreted as a single-chain inactive precursor, transformed by proteolytic cleavage to disulfide-linked heterodimers (30 + 60 kDa)	Mitogenic, motogenic, and morphogenic for epithelial, endothelial, and mesenchymal cells	Met
Macrophage-stimulating protein (MSP)	Secreted as a single-chain inactive precursor, transformed by proteolytic cleavage to disulfide-linked heterodimers (25 + 45 kDa)	Mitogenic, motogenic, and morphogenic for epithelial, endothelial, and mesenchymal cells	Ron
Glial-derived neurotrophic factor (GDNF)	20 kDa homodimer	Survival factor for epithelial and neural cells	GDNFR/Ret complex
Nerve growth factor (NGF), brain-derived neurotrophic factor (BDNF), neurotrophin 3, 4, 5 (NT3, NT4, NT5)	15–25 kDa homodimers derived from inactive precursors	Mitogenic, differentiation, and survival factors	TrkA–C
Ephrin-A1–5	21–28 kDa proteins, soluble or GPI linked	Differentiation of the nervous system	Eph A1–8
Ephrin-B1–3	34 kDa transmembrane proteins	Differentiation of the nervous system	Eph B1–6
Gas6	75 kDa secreted protein	Mitogenic and survival factor for hematopoietic and nervous cells	Axl/Sky

MOLECULAR STRUCTURE OF GROWTH FACTOR RECEPTORS

Growth Factor Receptors are Tyrosine Kinases with Modular Structure

Growth factor receptors (Table 18.1) constitute a family of proteins sharing a common structure: they show an extracellular N-terminal portion, a single hydrophobic region crossing the plasma membrane, and a cytoplasmic domain containing the catalytic activity. Most of the known receptors consist of a single polypeptide chain. Exceptions are the insulin receptor (tetrameric), the HGF receptor (dimeric), and their homologs. Schematically, receptors are structured in five functional domains:

1. The extracellular domain includes the high-affinity binding site for the growth factor. It is formed by hundreds of amino acids and contains several glycosylation sites, cysteine residues (responsible for protein folding), and peculiar structural motives unique for each receptor.
2. The transmembrane domain crosses the lipid bilayer and is constituted by approximately 25 hydrophobic residues, generally followed, on the intracellular side, by several basic residues acting as a signal for membrane anchoring.
3. The intracellular juxtamembrane domain consists of about 50 amino acids and includes many regulatory residues.
4. The tyrosine kinase catalytic domain is formed by approximately 250 amino acids; in some cases, it is interrupted by a regulatory region.
5. The C-terminal tail contains on average 250 amino acids, but its length and functions are diverse in different receptors: it is the segment that mediates binding of signal transducer molecules to the activated receptor.

Tyrosine kinase receptors share a considerable homology in the transmembrane and catalytic domains. By contrast, extracellular domains are very divergent and exhibit different combinations of structural motives. Based on these differences, receptors can be subdivided in at least 15 subfamilies, each showing similar combinations of extracellular structural motives (Fig. 18.1). These motives may have two functions: some participate in the formation of complex tridimensional configurations that confer stability and specificity to the growth factor binding site, and others mediate receptor dimerization, a critical event for signal transduction. Usually, for each receptor family, there is a corresponding growth factor family. As a rule, every receptor binds a single growth factor with high affinity, but there are many exceptions: a receptor may bind several growth factors similar in structure, and a growth factor may bind to different members of the same receptor family (Table 18.1).

Modulatory Functions of the Transmembrane and Juxtamembrane Domains of Growth Factor Receptors

The receptor transmembrane region serves as a link between the extra- and intracellular domains. Signal reception and intracellular transduction depend on the nature of these two domains, while the transmembrane domain seems to have only a structural function. However, it has been demonstrated that a single point mutation in the hydrophobic transmembrane segment is sufficient to dramatically alter the receptor tyrosine kinase activity. For example, the activated form of the receptor encoded by the ERBB-2/neu oncogene, which has transforming activity, differs from the normal counterpart only for a glutamate residue replacing a valine in the transmembrane domain.

The juxtamembrane region (i.e., the amino acid sequence separating the transmembrane from the catalytic domain) is very well preserved among receptors of the same subfamily and is involved in the modulation of signal transduction by extrinsic stimuli. This process is called transmodulation. It has been demonstrated that in EGF and HGF receptors, this region contains threonine and serine residues, whose phosphorylation turns off the catalytic activity of the adjacent kinase domain. Phosphorylation of the juxtamembrane domain is mainly catalyzed by protein kinase C (PKC), which is activated by diacylglycerol (DAG) and Ca^{2+} ions. The concentration of these molecules is often increased upon stimulation of most growth factor receptors. In this way, receptor activation starts mechanisms of negative feedback regulation.

Moreover, in many receptors, the juxtamembrane domain contains tyrosine residues critical for ubiquitin-mediated receptor downregulation.

The Catalytic Domain Is Responsible for Signal Transduction
Unlike the extracellular domain, the intracellular domain is highly homologous among all growth factor receptors, reaching 90% among members of the same subfamily. The homology is concentrated in the tyrosine kinase catalytic domain, which is preceded, followed, and sometimes interrupted by regulatory sequences.

The resolution of the crystalline structure of fragments deriving from insulin and FGF receptors has allowed generating models of the general structure of catalytic domains. Figure 18.2 shows two lobes delimiting a recess, where the reaction that transfers phosphate groups from the ATP–Mg^{2+} complex to protein substrates takes place.

Comparative analysis of tyrosine kinase sequences has revealed 13-amino-acid residues that are conserved in corresponding positions in every receptor. These residues are essential for receptor catalytic function and interactions with substrates. Their mutation can affect either positively

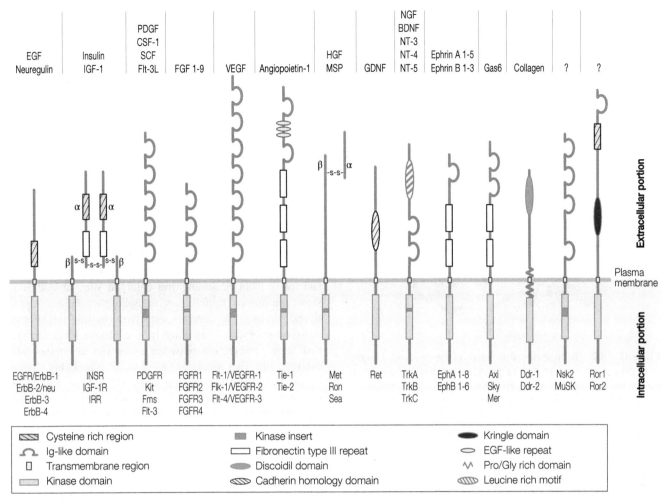

FIGURE 18.1 Subfamilies of tyrosine kinase receptors. Tyrosine kinase receptors are classified in subfamilies based on the structural characteristics of their extracellular domain. Representative members of each family are indicated, together with their corresponding ligands. The acronyms in block letters are fully described in Table 18.1.

or negatively the kinase function. The N-terminal lobe contains a glycine-rich motif (Gly-X-Gly-X-X-Gly) and a lysine residue that are critical for ATP binding. It has been demonstrated that mutation of this lysine abolishes kinase function. The C-terminal lobe contains the catalytic loop, characterized by the sequence His-Arg-Asp-Leu-Ala-Ala-Arg-Asn in all receptors. The Asp represents the catalytic residue, whereas the following amino acids allow access and identification of the substrate. The region between aspartate 1222 and glutamate 1253 in the C-terminal lobe contains the kinase activation loop. In most receptors, this loop includes one to three conserved tyrosines with regulatory function. Point mutations found in the receptors encoded by MET, RET, and KIT oncogenes and responsible for hereditary and nonhereditary tumors affect mainly the amino acids in the activation loop. These mutations increase kinase activity also in the absence of the receptor ligand.

RECEPTOR ACTIVATION AND SIGNAL TRANSDUCTION

Receptor Dimerization and Activation

Growth factor binding to the receptor induces activation of the cytoplasmic tyrosine kinase through a dimerization-based mechanism (Fig. 18.3). Receptors are free to move in the plasma membrane lipid bilayer and can form reversible dimers and oligomers. This process happens spontaneously, even in the absence of growth factors, as a result of casual multiple interactions among adjacent receptors. Usually, growth factors are single chained and bind a single receptor molecule. However, their binding to the receptor induces a conformational change in the extracellular domain that stabilizes receptor–receptor interactions and shifts the balance between monomeric and oligomeric structures in favor of the latter. Concerning the insulin receptor subfamily,

206 GROWTH FACTOR RECEPTORS

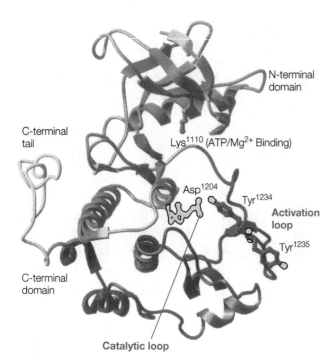

FIGURE 18.2 Tridimensional structure of the catalytic domain of tyrosine kinase receptors. The program "Modeler™" has been used to build a hypothetical model based on the crystal structure of the insulin receptor and the comparative sequence analysis of known receptors (by courtesy of Dr. L. Pugliese). Conserved amino acids, critical for enzymatic activity, are indicated. Their numbers refer to the sequence of the human HGF receptor. The N-terminal and C-terminal lobes of the catalytic domain delimit a pouch in which the two substrates—ATP (bound to the Lys^{1110}) and the tyrosine residue—interact. Phosphorylation of two tyrosines in the activation loop (Tyr^{1234} and Tyr^{1235}) increases the kinase V_{max} for the substrate.

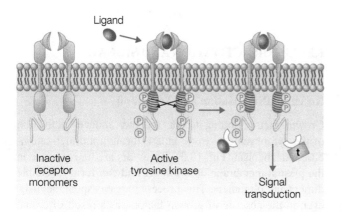

FIGURE 18.3 Model of tyrosine kinase receptor activation. Ligand binding to the extracellular domain induces receptor dimerization and allosteric modifications responsible for enzymatic activation. Reciprocal tyrosine phosphorylation (transphosphorylation) occurs between monomers of the same dimeric receptor. Upon SH2-mediated binding to phosphotyrosines (P in circles), signal transducers are activated and transduce the signal inside the cell.

the receptor unit exists as a covalent dimer even in the absence of the ligand. In the case of the PDGF receptor, the growth factor consists of two chains and stabilizes receptor dimers bridging two adjacent molecules. In some cases, receptor dimer formation requires complex interactions with other molecules connected to the membrane. For example, FGF-1 and FGF-2 induce dimerization of the receptor only in the presence of heparan sulfate proteoglycans.

Dimerization has two effects: it stabilizes the active status of the kinase and brings the intracellular domain of another receptor close to the catalytic domain, thus allowing receptor transphosphorylation at tyrosine residues located both within and outside the catalytic domain.

X-ray crystallography has demonstrated that, in insulin receptor, one of the conserved tyrosine residues in the kinase activation loop exerts a regulatory function. In the absence of the ligand, this tyrosine prevents exogenous protein substrates from accessing the catalytic site. In addition, steric hindrance inhibits its autophosphorylation. Upon ligand binding and receptor dimerization, this tyrosine is transphosphorylated resulting in a conformational change that allows interaction between exogenous substrates and catalytic site. This regulatory mechanism is probably shared by other receptor families, like the HGF, FGF, and NGF receptor families. In some receptor families, named **autocatalytic** (e.g., insulin and HGF receptor families), phosphorylation of tyrosine residues in the kinase domain also leads to an increase in the enzyme V_{max}.

Receptor trans-/autophosphorylation can also occur outside the catalytic domain, in the kinase insert (if present), in the juxtamembrane domain, or in the C-terminal end, creating high-affinity binding sites for cytoplasmic signal transducers containing specific domains (see later). Formation of these binding sites via trans-/autophosphorylation is a prerequisite for signal transduction. In the absence of trans-/autophosphorylation, tyrosine kinase activation is not sufficient to induce a biological effect. This has been demonstrated for the HGF receptor (encoded by the MET oncogene), which has a single multifunctional binding site for transducers, located in the C-terminal end, formed by a short amino acid sequence containing two tyrosines. Mutations turning these tyrosines into phenylalanines generate a receptor with perfectly preserved kinase activity and impaired biological activity as a consequence of its inability to bind and activate intracellular transducers. Transgenic mice bearing such mutations (MET knock-in mice) have the same phenotype of mice lacking the receptor-encoding gene (MET knockout mice).

The mechanism of receptor dimerization and activation can be altered in many pathological situations (Table 18.2). The presence of an excessive number of receptors, due to gene amplification and/or increased protein expression, augments the probability of casual interactions and dimer formation even in the absence of the ligand. Constitutive

TABLE 18.2 Tyrosine kinase receptors whose oncogenic activation is associated with onset of neoplasia or tumor angiogenesis

Receptor	Oncogenic activation	Neoplasia/associated neoplastic process
EGFR (ErbB-1)*	Overexpression (gene amplification); deletion of the extracellular domain	Breast, ovarian, lung, and other epithelial cancers; CNS tumors
ErbB-2 (HER2, Neu)	Overexpression (amplification), mutation	Breast, ovarian, gastric, lung, and intestinal cancers
ErbB-3 (HER3)	Overexpression; constitutive activation by heterodimerization with ErbB-2	Breast cancer
ErbB4 (HER4)	Overexpression	Breast and ovarian cancer
PDGFRα	Overexpression (amplification)	CNS tumors; ovarian cancer
PDGFRβ	Translocation (TEL–PDGFR); overexpression	TEL–PDGFR: chronic leukemia
CSF-1R (c-fms)	Point mutations; overexpression	Leukemia, CNS tumors
Kit (SCFR)	Germinal and somatic point mutations; small deletions; overexpression	Gastrointestinal stromal tumor, leukemia, and bone marrow preneoplastic alterations; testicular cancer
VEGFR1 (Flt1), VEGFR2 (Flk1), VEGFR3 (Flt4)	Expression; overexpression	Tumor angiogenesis; VEGFR3: also vascular cancer
FGFR1	Translocation (ZNF198–FGFR1); overexpression; point mutations	ZNF198–FGFR1: leukemia and lymphoma; deformities and skeletal dysplasia
FGFR2	Overexpression (amplification); truncation	Gastric, breast, prostate cancer
FGFR3	Translocation causing overexpression in plasma cells; germline point mutations	Translocation, multiple myeloma; germline mutations, achondroplasia
FGFR4	Overexpression (amplification)	Breast and ovarian cancer
TrkA	Translocations (Tpm–TrkA, Tpr–TrkA, Tfg–TrkA)	Thyroid cancer, pediatric CNS tumors
TrkC	Translocations (Tel–TrkC)	Pediatric sarcoma, leukemia
HGFR (Met)	Germinal and somatic point mutations; overexpression (amplification)	Mutations, kidney, gastric, and liver cancer; overexpression, intestinal, thyroid, liver, muscle tumors
RON	Overexpression	Liver and intestinal cancer
Eph A2, B2, B4	Overexpression	A2: cutaneous tumors (metastatic melanoma) B2, gastrointestinal carcinoma; B4, breast cancer
TIE1	Overexpression	Vascular and gastric tumors
Tl E2	Expression	Tumor angiogenesis
Ret	Translocations and inversions (PTC1, 2, etc.); germinal and somatic point mutations	PTC, thyroid cancer, especially associated with radiations; germinal point mutations, 2A and 2B multiple endocrine neoplasia (MEN); familial and sporadic thyroid cancer

activation of receptors belonging to the EGF and HGF subfamilies is found in many cases of carcinomas and in less frequent neoplastic pathologies. Another alteration responsible for constitutive activation of receptor kinases is the partial or total deletion of the extracellular domain, which removes the domain that usually inhibits receptor activation. Also, in this case, receptors are deregulated and exert transforming activity. Some examples are the truncated EGF receptor encoded by the viral oncogene v-erb-b1 (carried by the avian erythroblastosis virus) or those present in human glioblastoma multiforme.

Signal Transducers Binding to Phosphorylated Tyrosines

Intracytoplasmic signal transducers bind to phosphorylated receptors with high affinity through specific modules contained in their sequence. The most common is the so-called Src homology region 2 (SH2) domain, which recognizes and binds short amino acid sequences containing a phosphorylated tyrosine. The binding specificity is provided by the 3–5-amino-acid sequence downstream of the phosphorylated tyrosine. The specific association of SH2-containing transducers determines how signal transduction is activated.

The tridimensional structure of the SH2 domains and the modality of phosphotyrosine recruitment have been clarified through X-ray crystallography and nuclear magnetic resonance studies. SH2 domains are compact structures, formed by a hundred amino acids. Structural studies and analysis of phosphopeptide libraries have allowed to establish that the C-terminal portion of the SH2 domain contains three amino acids, immediately downstream the phosphotyrosine, that mediate high-affinity association between proteins containing different SH2 domains and specific phosphotyrosines in the receptor sequence. As a general rule, tyrosine kinase receptors contain many tyrosines that can be phosphorylated following receptor

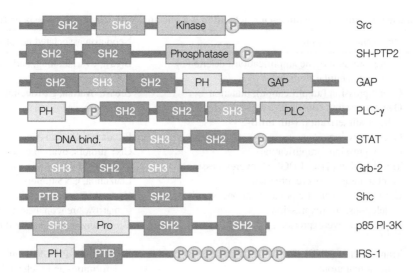

FIGURE 18.4 Structure of signal transducers that associate with active tyrosine kinase receptors through phosphotyrosine binding domains (SH2 or PTB). Some are endowed with enzymatic activity (in gray): Src (tyrosine kinase), SH-PTP2 (phosphatase), GAP (GTPase activator), and PLCγ (phospholipase). Some are transcription factors that bind DNA (STAT). Others are adaptors recruiting other signaling proteins to receptor and plasma membrane. Tyrosine phosphorylation sites (P) and conserved domain (SH3; Pro = proline-rich sequences; PH) critical for intermolecular interactions and signal transduction are indicated.

activation; they are located in the C-terminal end, in the juxtamembrane, and in the catalytic domain. Each phosphotyrosine is followed by a different consensus sequence and therefore preferentially binds one or few SH2-containing transducers. For example, in the PDGF receptor, the sequence pTyr-Val-Pro-Met specifically binds the SH2 domain of p85, the regulatory subunit of phosphatidylinositol 3-kinase (PI3K), whereas the sequence pTyr-Ile-Ile-Pro binds the SH2 domain of phospholipase Cγ (PLCγ).

Another peptide module involved in the recruitment of phosphotyrosine-containing sequences is the phosphotyrosine binding (PTB) domain. Less common than SH2, the PTB domain interacts with high affinity with phosphopeptides in which pTyr is immediately preceded by the Asn-Pro-X consensus sequence. Binding specificity is provided by a 5–8-hydrophobic-amino-acid sequence preceding the phosphorylated tyrosine. However, in some proteins, the PTB domain can also identify peptides in which the tyrosine is not phosphorylated. For this reason, the PTB domain is considered a motif identifying specific amino acid sequences, with different functions compared to the SH2. In fact, it can be found in adaptor proteins, like Shc and IRS1, which recruit transducers near tyrosine kinase receptors before their activation.

Localization of transducers at specific regions of the plasma membrane internal side is crucial for their association to the activated receptors. In some cases, localization is mediated by small lipid molecules bound to their N- or C-terminal domains; oncogenes *SRC* or *RAS* are an example. It has been established that pleckstrin homology (PH) domains can directly bind charged phosphoinositol heads in the plasma membrane, thus coupling the activity of inositol kinases and phosphatases with the regulation of intracellular signaling. Molecules of the GTPase family (dynamin), RAS regulators (SOS and GTPase-activating protein (GAP)), and cytoplasmic kinases (βARK) have PH domains.

Two classes of transducers contain SH2 domains (Fig. 18.4). One class includes proteins having their own enzymatic activity such as PLCγ; pp60src cytoplasmic tyrosine kinase, encoded by the proto-oncogene *c-SRC*; phosphotyrosine phosphatases SH-PTP1 and SH-PTP2; GTPase (GAP) activator; and ubiquitin ligase Cbl. This class includes also the signal transducer and activator of transcription (STAT) proteins, a more recently identified group of signal transducers that also act as gene transcription activators. The second class includes SH2-containing transducers devoid of intrinsic catalytic activity, which act either as signaling amplifier or as regulatory subunits of cytoplasmic enzyme molecules. Some examples are Grb2, Shc, and p85 (the PI3K noncatalytic subunit).

Binding of SH2-containing proteins to tyrosine kinase receptors may have many consequences such as:

1. It favors the interaction of the transducer with the receptor catalytic domain.
2. It induces conformational modifications in the transducer, independent from phosphorylation, modulating its activity, like in the case of PI3K.
3. It recruits the transducer to the plasma membrane, allowing its interaction with possible substrates associated with the plasma membrane, like lipid kinases, hydrolases, or the GTP/GDP exchange factor SoS.

Signal transduction from receptor-associated molecules to final effectors, either cytoplasmic proteins or factors controlling gene expression, involves enzyme cascades and production of second messengers. Some conserved domains allow transducers to assemble in supramolecular complexes: the most common one is the *SRC* homology region 3 (SH3) domain, which binds proline-rich sequences. The SH3 domain is formed by approximately 60 amino acids, organized in two main β sheets perpendicular to each other containing a group of conserved aromatic residues. Each SH3 domain binds polyproline sequences forming type II left-handed helices. These sequences are characterized by the repetition of PXXP motifs, in which the two prolines are critical for the high-affinity binding. The sequences are pseudosymmetric and bind the SH3 domain in both orientations. This extends the number of potential interactors and allows many spatial dispositions of the interacting proteins. Further binding specificity is provided by interactions between nonproline residues in the ligand and two variable domains flanking the SH3 hydrophobic β sheets. SH3 domains can be found in many SH2-containing signal transducers, as well as in many cytoskeletal proteins. While SH2-mediated interactions are regulated by tyrosine phosphorylation, which is a fast inducible and reversible process, SH3 binding to polyproline helices is constitutive and it does not seem to be affected by posttranslational modifications.

Many attempts have been made to relate a single receptor, or at least a specific signal transduction pathway, to a single and well-defined biological effect. However, experimental results indicate the opposite. Different receptors and transducers communicate and mutually affect each others, and each biological response is not the consequence of a single specific signal, but the result of more events able, altogether, to cross a threshold level.

RAS-Dependent Transduction Pathway in Cell Proliferation and Neoplastic Transformation

RAS GTPase, initially identified as the product of oncogenes carried by the Harvey and Kirsten murine sarcoma viruses, is mutated and constitutively activated in many human tumors (see also Chapter 14). RAS is a 21 kDa protein physiologically expressed in all the cells of the organism and anchored to the plasma membrane through a farnesyl group added to its C-terminal end. RAS works as a molecular switch, cycling between a GTP-bound active state and an inactive GDP-bound state. Since it has a much higher affinity for GDP than for GTP, the inactive Ras–GDP complex accumulates in the cell. Following activation of most tyrosine kinase receptors, RAS–GTP level increases rapidly and significantly. Transition from the GDP-bound to the GTP-bound form is stimulated by cytoplasmic proteins acting as guanine nucleotide-releasing factors (GNRF) or GDP/GTP exchange factors. The best characterized GNRF, named SoS, contains proline-rich sequences that bind with high affinity two SH3 domains in the Grb2 protein. Following receptor tyrosine phosphorylation, the Grb2–SoS complex, which is believed to be already formed in the cytoplasm, is recruited to the receptor, where SoS can interact with RAS. After binding GTP, RAS hydrolyzes it relatively slowly, returning to the inactive RAS–GDP state. RAS enzymatic activity is increased by GAPs that associate with phosphorylated receptors through SH2 domains. GAPs negatively modulate RAS pathway, but at the same time they behave as RAS effectors, binding and activating many intracellular proteins, some of which regulate cytoskeletal functions. For example, the loss of the NF1 gene, encoding a GAP called neurofibromin, is responsible for a familiar tumoral syndrome known as Recklinghausen neurofibromatosis.

Specific point mutations in the *RAS* gene are frequently found in many types of tumors. These are activating mutations that either reduce RAS GTPase activity or facilitate GDP/GTP exchange. In both cases, RAS stimulatory activity on proliferation becomes constitutive.

Signal transduction pathways downstream of RAS are mediated by proteins interacting with two short N-terminal domains of RAS (known as effector domain and switch region II). The most studied pathway is the one mediated by Raf, which consists in a cascade of serine/threonine kinase activations. Raf contains a N-terminal binding site (CR1), mediating its association with activated RAS and a serine/threonine kinase C-terminal domain. The mechanism of Raf activation is still not completely understood, but it seems to involve both its recruitment at the plasma membrane and serine/threonine phosphorylations. Activated Raf phosphorylates and activates MEK, a dual-specificity kinase that phosphorylates both tyrosine and threonine/serine residues. The most important MEK substrate is the serine/threonine kinase ERK/MAPK that phosphorylates many substrates including transcriptional factors of the TCF family. These factors induce the expression of the c-FOS proto-oncogene, whose product participates in the formation of AP1 transcriptional complex. The final result is the stimulation of the expression of genes that enable cells to overcome the G1–S restriction point of the cell cycle, leading to DNA replication.

Another kinase cascade parallel to that the Raf/MEK/ERK pathway is started by MEK-1 kinase. This kinase *in vivo* phosphorylates and activates a MEK-homologous kinase named SEK, which in turn phosphorylates and stimulates the SAPK/JNK (or Jun kinase), which activates the transcriptional factor encoded by the Jun proto-oncogene. Some evidence suggests that this signaling cascade may be activated by RAS via the G protein RAC, which seems necessary for Ras-induced transformation. However, it is not clear if the SAPK pathway is necessary to stimulate cell proliferation.

GROWTH FACTOR RECEPTORS

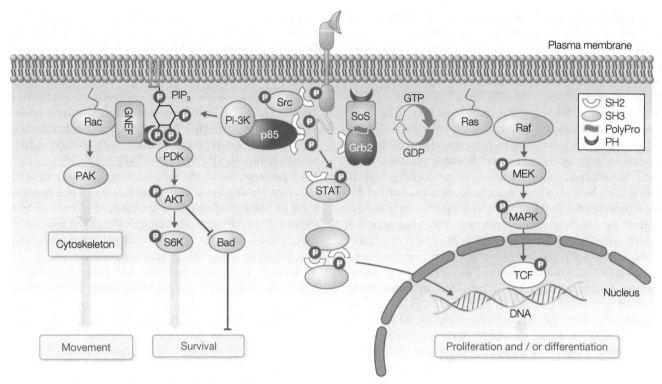

FIGURE 18.5 Main transduction pathways activated by tyrosine kinase receptors. Grb-2 binds phosphorylated tyrosines through its SH2 domain and SoS through its two SH3 domains. SoS activates the RAS-Raf-MEK-MAP kinase cascade that ends up with the activation of transcription factors regulating gene expression. STAT, that is tyrosine phosphorylated by the receptor kinase, forms dimers that translocate to the nucleus where they act as transcription factors. Both RAS and STAT pathways are involved in the control of cell proliferation and differentiation. P85, containing two SH2 domains, recruits PI3 kinase that phosphorylates phosphoinositides in D3. These bind PH-containing proteins like the PDK kinase, which activates AKT and other kinases. AKT activates S6K (ribosomal S6 kinase) and inactivates Bcl-2-associated death promoter (Bad, a pro-apoptotic protein that promotes apoptosis when non-phosphorylated). This pathway is involved in protection from apoptosis and cell survival. Another PH-containing molecule is the GTP/GDP exchanger (GNEF) that activates RAC. This acts on cytoskeletal proteins and regulates cell movement through kinases like PAK. The cytoplasmic tyrosine kinase Src is associated with the receptors that directly activate it. Src is involved in the control of cell movement, protection from apoptosis, and proliferation. The black arrows indicate that the transducer performs an enzymatic activity on its target, which leads to phosphorylation of amino acids—tyrosine, serine, or threonine—or inositols (circles with P).

Another RAS effector is PI3K (see later) whose catalytic subunit associates with RAS in a GTP-dependent manner (see Chapter 13).

Activation of Enzymes Generating Lipid Second Messengers

The most important among these enzymes are PLCγ and PI3K. PLCγ hydrolyzes phosphatidylinositol 4,5-bisphosphate (PIP_2) generating inositol 1,4,5-trisphosphate (IP_3) and DAG. PLCγ contains two SH2 groups and one SH3 group that allow its association with cytoskeletal proteins. After binding to the receptor, PLCγ undergoes tyrosine phosphorylation that increases its enzymatic activity. DAG is a strong activator of PKC, a serine/threonine kinase involved in the control of cell proliferation and transformation. PKC is the target of tumorigenic phorbol esters like TPA. IP_3 regulates calcium release from the rapid exchange intracellular deposits. The involvement of PLCγ in the signaling that leads to DNA replication and cell division is still a debated topic.

PI3K is a key regulator of mitosis and other activities correlated to neoplastic transformation and extracellular matrix invasion during tumoral progression. Besides, this enzyme plays a central role in the protection from apoptosis (Fig. 18.5). It has been found that a PI3K of retroviral origin causes avian sarcomas and transforms cultured cells. PI3K is the prototype of a superfamily of enzymes that phosphorylate phosphatidylinositols at D3 position. Three classes of PI3K have been identified based on their substrate specificity: class I preferably phosphorylates PIP_2, class II phosphorylates phosphatidylinositol 4-phosphate, and class III phosphorylates phosphatidylinositol. The activity of class I PI3Ks is mainly associated with receptor tyrosine kinases, the complex formed by the middle T-antigen (the principal transforming protein of polyomavirus), and pp60[Src]. Among the class I members, the best characterized is the heterodimer formed by the regulatory subunit p85 and the catalytic subunit p110. Following p85 association with the phosphorylated receptor, p110 translocates at the plasma membrane, where it phosphorylates phosphoinositides at D3 position. Tyrosine kinase receptors can activate PI3K both directly and indirectly. Recently, many targets of phosphatidylinositol-3-phosphate have been discovered; they are mostly involved in the control of cell survival, proliferation, and movement, and contain PH domains that bind phosphatidylinositols. The serine/threonine kinase Akt is directly activated by phosphatidylinositol 3,4-bisphosphate and via phosphorylation by PDK1, a kinase activated by phosphatidylinositol 3,4,5-trisphosphate. Akt, the cellular counterpart of the v-*akt* retroviral oncogene product, is responsible for the PI3K-mediated protection from apoptosis. Signal transduction downstream of Akt, although not entirely clear, involves phosphorylation of S6 kinase that plays an important role in G1–S transition of the cell cycle. Other PI3K targets are some members of the serine/threonine kinase C (PKC) family and in particular ε and λ. $PI(3,4,5)P_3$ can directly activate PKC or recruit its substrates to the plasma membrane, thus facilitating PKC-mediated phosphorylation. Finally, PI3K is involved in the control of cytoskeleton reorganization and extracellular matrix adhesion, which are both required for cell migration.

Cytoplasmic Tyrosine Kinases

Cytoplasmic tyrosine kinases are enzymes anchored to the plasma membrane through lipid groups. They are usually associated and activated by many transmembrane receptors, including most tyrosine kinase receptors, and regulate various cell events, such as proliferation, protection from apoptosis, and invasiveness, via mostly unknown mechanisms. They belong to a family that includes at least ten highly homologous members. Based on their expression pattern, they are divided into three groups: Src, Fyn, and Yes, expressed in almost all cell types; Blk, Fgr, Hck, Lck, and Lyn, mainly found in hematopoietic cells; and Frk group, expressed in epithelia. The product of the *SRC* proto-oncogene, pp60^{SRC}, is the prototype of this family.

Figure 18.6a shows the structure of these kinases taking Src as a model. Six different functional domains can be identified (starting from the N-terminus): a conserved domain (SH4), a "unique region," a SH3 domain, a SH2 domain, a catalytic domain, and a short C-terminal tail containing a tyrosine residue. The SH4 domain is a short sequence containing a myristoylation signal, allowing anchorage to the cell membrane. The "unique region" is the most variable among the Src family members. The SH3 and SH2 domains and the C-terminal end participate in a complex intramolecular regulation of Src catalytic activity. In resting state, the C-terminal tyrosine, which is phosphorylated by the Csk

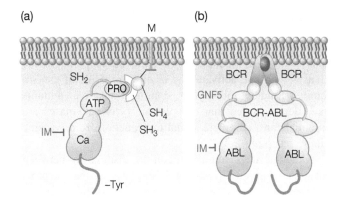

FIGURE 18.6 Molecular organization of non-receptor tyrosine kinases. (a) Non-receptor tyrosine kinases are formed by six functional domains. Starting from the N-terminus, there are a conserved domain (SH4), the "unique region," a SH3 domain, a SH2 domain, a catalytic domain, and a short C-terminal tail containing a tyrosine residue. The SH4 domain is a short amino acid sequence containing a myristoylation signal, which allows anchorage to the plasma membrane. SH3 and SH2 domains and the C-terminal end participate in a complex intramolecular regulation of Src catalytic activity. In resting state, the C-terminal tyrosine (Tyr) binds with high affinity the SH2 domain of the same molecule, favoring a closed and inactive conformation. Such conformation is further stabilized by interactions between SH3 domain, catalytic domain, and an adjacent region containing a proline-rich sequence (Pro). When the C-terminal tyrosine is dephosphorylated, the SH2 domain is free and the kinase turns into the active form. (b) In chronic myelogenous leukemia, the fusion between chromosomes 9 and 22 leads to the formation of a protein composed of part of BCR and part of ABL that spontaneously dimerizes and is constitutively active.

cytoplasmic kinase, establishes a high-affinity intramolecular interaction with the SH2 domain, favoring a closed and inactive conformation. Such conformation is further stabilized by interactions between SH3 domain, catalytic domain (Ca), and an adjacent region containing a proline-rich sequence (Pro). Binding to membrane PIP_2 plays an important role in this mechanism of negative regulation.

When the C-terminal tyrosine is dephosphorylated, through a still unclear mechanism, the SH2 domain is released and the kinase switches to the active form. Full enzymatic activity requires phosphorylation of a conserved tyrosine contained in the catalytic domain (Fig. 18.6a). Phosphotyrosines of activated tyrosine kinase receptors compete with the C-terminal end for binding to the Src SH2 domain, releasing the inhibitory intramolecular interaction. Association between Src and receptor tyrosine kinases results in phosphorylation of the tyrosine in Src catalytic domain, which has positive regulatory functions. The transforming oncogene v-src, isolated from the Rous sarcoma virus, lacks the C-terminal inhibitory region and therefore has a constitutive enzymatic activity.

Redundancy between members of the Src family present in the same cell and lack of selective inhibitors make difficult to define the specific biochemical and biological activity of these kinases. However, most receptor tyrosine kinases bind and activate one or more proteins of the Src family. Studies employing dominant-negative mutants have demonstrated that cytoplasmic tyrosine kinases are essential for a complete signal transduction from tyrosine kinase receptors.

Although it is often difficult to distinguish between substrates that are directly phosphorylated by the receptors and those phosphorylated by Src, many intracellular proteins are potential Src targets. Among them, focal adhesion kinase (FAK) is part of a complex transduction system associated with integrins (see Chapter 20), which are receptors controlling cell–extracellular matrix interactions. Activated FAK recruits Grb2 and stimulates the Ras/MAPK pathway. During cell movements, Src stimulates formation of lamellipodia and membrane ruffles. This effect seems to be mediated by PI3K and RAC activity, both elevated in cells transformed by Src. Activation of RAS/MAPK and PI3K is essential for Src-mediated cell movement, proliferation, and protection from apoptosis (see Chapters 13 and 20).

Chronic myelogenous leukemia, a massive clonal expansion of myelogenous cells, is caused by a chromosomal aberration called Philadelphia chromosome. This consists in the reciprocal translocation between the long arms of chromosomes 9 and 22 and causes the formation of a new gene, composed of part of the *BCR* gene and part of the *ABL* gene. The resulting chimeric protein BCR–ABL has a constitutive tyrosine kinase activity producing an intense and persistent intracellular signal that induces cell proliferation and resistance to apoptosis (Fig. 18.6b). Competitive inhibitors of this tyrosine kinase that bind the ATP site (Figs. 18.6 and 18.7) have been developed and are very effective in this pathology. Unfortunately, leukemic cells often become resistant to these drugs as they develop mutations affecting the site bound by the drugs in the ABL kinase. Recently, it has been observed that GNF-5, an allosteric inhibitor of the BCR–ABL kinase (Fig. 18.7), cooperates with classical competitive kinase inhibitors, blocking both the wild-type and the mutated sites, giving further therapeutic options.

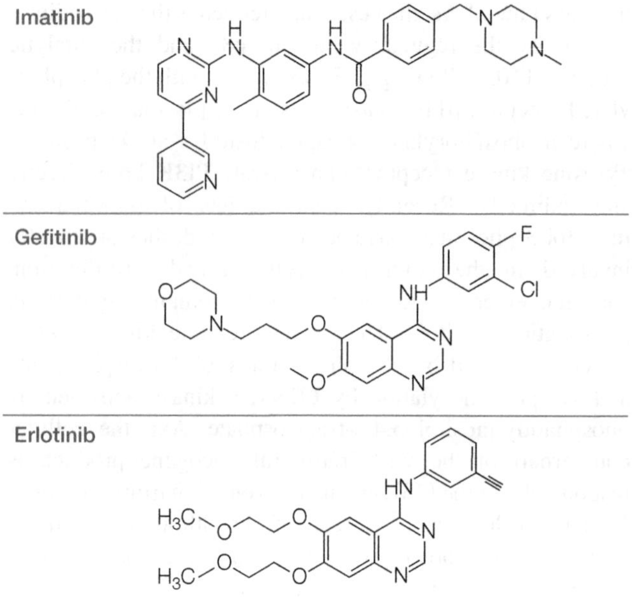

Tyrosine Phosphatases Modulating Tyrosine Kinase Activity

Activated receptor tyrosine kinases associate also SH2-containing enzymes endowed with phosphoprotein phosphatase activity. The two best characterized tyrosine phosphatases are SH-PTP1 and SH-PTP2. The first is also known as PTP-1C or SHP or HCP and is expressed almost exclusively in hematopoietic cells and some epithelial tissues. Conversely, SH-PTP2 (also known as Syp or

FIGURE 18.7 Structure of some tyrosine kinase competitive inhibitors.

PTP-2C) is present in most cell types. Both are associated with phosphorylated receptors and are activated by these receptors via phosphorylation. However, they seem to play opposite roles in signal transduction: SH-PTP1 dephosphorylates receptor tyrosine kinases and seems to be involved in negative regulation of signaling. On the contrary, SH-PTP2 seems to activate a still unknown positive signal transduction pathway that probably involves dephosphorylation of the phosphotyrosine in Src C-terminal end that exerts negative regulatory activity.

PTEN is a recently discovered member of the phosphatase family, encoded by a tumor suppressor gene, whose mutations are responsible for hereditary (Cowden syndrome) and sporadic tumors. This enzyme seems to play a very important role in the negative regulation of signal transduction: it can dephosphorylate phosphoserines, phosphotyrosines, and inositides phosphorylated in D3 by PI3K. It seems that, when mutated, PTEN is able to hydrolyze phosphatidylinositol trisphosphate, leading to hyperactivation of protein kinases and abnormal protection from apoptosis.

STATs

Receptor tyrosine kinases activate a signal transduction pathway that directly stimulates gene transcription. This pathway is mediated by molecules called STATs, a family of cytoplasmic proteins including at least six known isoforms. Initially identified as signal transducers of cytokine receptors in the hematopoietic system, isoforms 1, 3, and 5 are ubiquitously expressed and are involved in signal transduction of tyrosine kinase receptors. STAT proteins contain a DNA binding region at the N-terminus, followed by a SH3 domain, a SH2 domain, and a C-terminal sequence containing a tyrosine residue. Following tyrosine phosphorylation, the C-terminal sequence binds with high affinity the SH2 domain of another STAT molecule, forming homo- or heterodimers. These dimers move to the nucleus and stimulate transcription of genes involved in the control of cell cycle and differentiation. Among them, there are FOS, encoding a transcriptional factor, and WAF-1, whose product negatively regulates cell cycle progression.

Following cytokine receptor activation, STAT phosphorylation is mediated by cytoplasmic kinases of the JAK family. On the contrary, receptor tyrosine kinases can phosphorylate and directly activate STATs. The activity of these molecules is also modulated by serine phosphorylation by MAPK. In epithelial cells, STATs are critical transcriptional factors for terminal differentiation.

PHARMACOLOGICAL APPROACHES TO THE CONTROL OF GROWTH FACTOR RECEPTOR ACTIVITY

Over the past decades, advances in our knowledge of the mechanisms involved in cancer development have led to the introduction of new promising therapeutic strategies (called "molecular targeting") based on drugs designed to act on specific molecules critical for the maintenance of the neoplastic state. The development of inhibitors, especially monoclonal antibodies (mAbs) and small molecules, directed against activated oncogenes has been the most used approach. Among the oncogenes involved in human tumors, tyrosine kinases play a critical role (Table 18.2). This observation, together with the finding that cancer cell survival may depend on continuous expression of activated oncogenes (a concept defined as "oncogene addiction"), has made tyrosine kinases perfect targets for cancer targeted therapy.

Targeted therapies are based on the following principles: (i) the molecular target must be present and altered (qualitatively or quantitatively) in cancer cells; (ii) the target inhibition must be able to cause cancer cell death or at least to block their further growth; and (iii) the molecular drug must be poorly effective on normal cells and, hence, not very toxic for the patient. Therefore, patient's response to a targeted therapy depends on target activation state and role in development and tumor progression. In general, the most effective results can be observed when the altered molecules have a causal role in the maintenance of the transformed state. Conversely, minor effects are observed when the target is part of an important but not indispensable transduction pathway. It is important to highlight that aberrant activation of the target molecule—generally due to mutations, amplification, and/or hyperexpression—is often a predictor of response to the inhibitors. These observations should be taken into consideration not only during the selection of the target molecules in a specific type of tumor but also when deciding the treatment for a specific cancer patient. In other words, molecular therapies imply a personalization of the treatment, having as a prerequisite the demonstration that the altered molecule targeted by the drug is present in the tumor to be treated.

mAbs and Small Kinase Inhibitors

mAbs mAbs have been widely used in the clinic with promising results (Box 18.1). The main advantage of these molecules is their high specificity.

In general, mAbs have been used:

- For cancer immunotherapy. mAbs activate the immune system against cancer cells, thanks to their ability to promote complement-dependent cytotoxicity (CDC), antibody-dependent cell-mediated cytotoxicity (ADCC), and direct induction of apoptosis.
- To prevent interaction between receptors and their specific growth factors. In this context, mAbs can exert their therapeutic effects via different mechanisms: (i) binding to the ligand or the receptor, preventing their interaction, (ii) binding to the receptor and induction of receptor internalization, (iii) proteolytic cleavage and release of the receptor extracellular portion, and (iv) inhibition of receptor dimerization. Each mAb is believed to act through more than one mechanism.
- To deliver radioactive substances or drugs (conjugated to the antibody) specifically to the tumor cells.

In spite of the great results obtained in clinical practice, mAbs also present some disadvantages. For example, they are not always able to exert their effect on all cells within the tumor. This may be due to different factors: the increased interstitial pressure within the tumor prevents mAb through distribution; not all cancer cells express the antigen at the same level (heterogeneous antigen distribution); and cancer cells may undergo changes in the type of receptors they express. Another problem is the appearance of toxicity, often due to inhibition of the molecular target, essential for proper tissue functioning, in normal cells. An important example is the cardiotoxicity observed after the treatment of mammary carcinomas with trastuzumab.

Tyrosine Kinase Inhibitors Most small tyrosine kinase inhibitors developed so far are ATP competitors that interact with the ATP binding site "mimicking" the hydrogen bonds normally formed by ATP. Type 1 inhibitors bind to the enzyme active conformation, whereas type 2 inhibitors interact with the enzyme inactive form. Covalent inhibitors are considered stronger than competitive inhibitors.

BOX 18.1 FDA-APPROVED ANTINEOPLASTIC AGENTS

Drug	Target	Approved for the treatment of
Monoclonal antibodies		
Cetuximab (Erbitux®)	EGFR	Colon carcinoma, head and neck carcinoma
Panitumumab (Vectibix®)	EGFR	Colon carcinoma, head and neck carcinoma
Trastuzumab (Herceptin®)	HER2/Neu	Breast cancer
Bevacizumab	VEGF	Breast cancer, colon carcinoma, non-small cell lung cancer, glioblastoma multiforme
Tyrosine kinase (TK) receptor inhibitors		
Gefitinib (Iressa®)	EGFR	Non-small cell lung cancer
Erlotinib (Tarceva®)	EGFR	Non-small cell lung cancer, pancreatic carcinoma
Lapatinib (Tykerb)	EGFR, HER2/Neu	Breast cancer
Sunitinib (Sutent®)	PDGFR, VEGFR1–3; Kit; FLT3; CSF-1R; RET	Gastrointestinal cancer, kidney carcinoma
Pazopanib	VEGFR1–3	Kidney carcinoma
Agents directed against TKs		
Imatinib mesylate (Gleevec®)	Several TKs	Gastrointestinal stromal tumor, leukemia, dermatofibrosarcoma protuberans, myelodysplastic/myeloproliferative disorders, systemic mastocytosis
Lapatinib (Tykerb®)	HER2/EGFR	Advanced or metastatic breast cancer, gastric or gastroesophageal junction adenocarcinoma
Gefitinib (Iressa)	EGFR	Advanced non-small cell lung cancer
Erlotinib (Tarceva)	EGFR	Metastatic non-small cell lung cancer and pancreatic cancer
Vandetanib (Caprelsa®)	Several TKs	Metastatic medullary thyroid cancer
Crizotinib (Xalkori®)	ALK	ALK+ locally advanced or metastatic non-small cell lung cancer
Sorafenib (Nexavar®)	Several kinases	Renal cell carcinoma, hepatocellular carcinoma
Sunitinib (Sutent)	Several kinases	Metastatic renal cell carcinoma, gastrointestinal stromal tumors not responding to imatinib, or pancreatic neuroendocrine tumors
Pazopanib (Votrient®)	Several TKs	Advanced renal cell carcinoma and advanced soft tissue sarcoma
Regorafenib (Stivarga®)	Several TKs	Metastatic colorectal cancer
Cabozantinib (Cometriq™)	Several TKs	Metastatic medullary thyroid cancer
Trastuzumab (Herceptin)	HER2	Breast cancer
Pertuzumab (Perjeta™)	HER2	Used in combination with trastuzumab and docetaxel to treat metastatic breast cancer
Cetuximab (Erbitux)	EGFR	Squamous cell carcinoma of the head and neck or colorectal cancer
Panitumumab (Vectibix)	EGFR	Metastatic colon cancer
Bevacizumab (Avastin®)	VEGF	Glioblastoma, non-small cell lung cancer, metastatic colorectal cancer, and metastatic kidney cancer
Ziv-aflibercept (Zaltrap®)	Ziv-aflibercept consists of portions of two different VEGF receptors fused to a portion of an immune protein	Metastatic colorectal cancer

Despite being also ATP mimetic, they form an irreversible covalent bond with the kinase active site. Since in protein kinases ATP-binding site is highly homologous, both reversible and nonreversible inhibitors are often not specific for a single kinase and cross-react with other enzymes. Other inhibitors have their binding site outside the ATP pocket and can block the enzyme in its inactive conformation. These allosteric inhibitors show the highest selectivity, since their binding region has low sequence homology among protein kinases.

At the beginning of the "target therapy" era, it was believed that it would have been hard to identify inhibitors selective for a single molecular target and capable of providing a sufficient therapeutic index. Today, the design of kinase inhibitors can take advantage of sophisticated crystallographic and proteomic approaches and new screening tests. At the same time, many efforts have been made to identify and validate more appropriate molecular targets. To exemplify, in the following sections, we will describe the therapeutic strategies adopted to interfere with the activation of two paradigmatic classes of receptors: the epidermal growth factor receptor (EGFR) family, to directly inhibit cancer cell proliferation, and vascular endothelial growth factor receptor (VEGFR), to inhibit tumor neovascularization (see also Box 18.1).

Growth Factor Receptors as Targets for Anticancer Drugs

The EGFR Family The EGFR family is very important in the tumorigenic process. The EGFR family members are the first tyrosine kinase receptors for which a role in tumorigenesis has been demonstrated and they have been widely used as targets for antineoplastic therapies.

EGFR is activated by mutations or overexpression in lung, breast, stomach, head and neck, and colon carcinomas and in glioblastoma multiforme. Many strategies aimed at inhibiting EGFR are currently used or under study. These employ mAbs, tyrosine kinase inhibitors, ligand-conjugated toxins, and antisense molecules.

Cetuximab is a mAb directed against an EGFR extracellular epitope, which competitively inhibits binding of EGF and TGF-α ligands to the receptor. It has been approved as monotherapy or in association with radiotherapy or other conventional therapies for the treatment of colorectal metastatic carcinoma and head and neck squamous cell carcinoma.

Panitumumab was the first humanized mAb to be approved for patient treatment in association with chemotherapeutics. This antibody is interesting since, compared to cetuximab, it has a higher binding affinity, is a more powerful EGFR inhibitor, has a longer biological half-life, and is less immunogenic.

Clinical studies using small molecules that inhibit EGFR kinase activity in advanced lung carcinoma patients demonstrated that approximately 80% of patients with EGFR-activating mutations show rapid and enduring clinical responses, total or partial. It has been demonstrated that EGFR mutations identified in patients increase the sensitivity to EGFR inhibitors. Another lesson learned from clinical studies on EGFR inhibitors concerns the importance of patient selection. Indeed, almost only patients with tumors carrying EGFR mutations show positive responses to EGFR inhibitors in terms of global survival and disease-free survival. Based on many studies on different tumors, three EGFR inhibitors are now approved for antineoplastic therapy: gefitinib, erlotinib, and lapatinib.

HER2, an EGFR family receptor without a ligand, becomes active upon heterodimerization with another family member. HER2 gene amplification determines protein overexpression in approximately 25% of mammary carcinoma and correlates with an aggressive phenotype and a negative prognosis. Molecular therapies approved in mammary metastatic tumors with HER2 amplification are the mAb trastuzumab (directed against HER2 extracellular domain) and lapatinib, which is an inhibitor of EGFR and HER2 enzymatic activity. Recently, trastuzumab has been approved, combined to chemotherapy, in adjuvant therapy as well. New anti-HER2 mAbs (pertuzumab and ertumaxomab) are currently evaluated in clinical studies.

VEGFR VEGFR controls vessel formation and is essential for tumor development. The vascular system is formed through two processes: vasculogenesis and angiogenesis.

Vasculogenesis is the formation of the primary vascular system in the embryo, starting from progenitor cells, called angioblasts that differentiate around the third week after conception. Vascular endothelial growth factor (VEGF), via VEGFR2, regulates angioblast proliferation; via VEGFR1, it induces morphogenesis of primary capillaries. Ephrin receptors regulate differentiation in the arterial vascular areas (specifically expressing ephrin-B2 and its receptor) and in venous vascular areas (expressing ephrin-B4 and its receptor).

Angiogenesis is the development of new vessels starting from preexisting vessels and occurs both in the embryo and in postnatal life in physiological conditions (e.g., menstrual cycle, growth of the placenta), in defense and regeneration conditions (e.g., inflammation, healing), and in pathological conditions (e.g., alterations of retinal vascularization in diabetics, neoplasia, etc.). Angiogenesis requires degradation of the basal lamina in the parental vessel, followed by endothelial cell migration and proliferation, their organization in capillaries, and recruitment of periendothelial components of the vascular wall (smooth muscle cells). VEGF, through the receptors 1 and 2 present on endothelial cell surface, induces their proliferation and organization to form tubular structures; angiopoietin 1, through the Tie2 receptor, stimulates the endothelium to recruit smooth muscle cells. Angiopoietin 2, binding Tie2, causes destabilization of the vascular wall, followed by vessel regression and gemmation of a new vessel. By producing angiogenic growth factors, cancer cells are able to induce the formation of vessels necessary for oxygen and metabolite supply (a tumor cannot survive without vascularization when its diameter exceeds a few millimeters). Solid tumor vascularization or angiogenic switch is considered a critical event in tumor progression and involves some critical molecules, such as VEGF and its receptors (VEGFR and NP-1). Since VEGF is a key regulator in angiogenesis, many antiangiogenic therapies in use are directed against this factor or its receptor. The first antiangiogenic agent successfully used was bevacizumab, a monoclonal anti-VEGF antibody, which prevents VEGF interaction with its receptors. Bevacizumab is currently approved for patients with metastatic colorectal carcinoma (CRC) combined with chemotherapy, mammary carcinoma, lung cancer, and glioblastoma. Other anti-VEGF strategies currently underway involve soluble forms of VEGFR and VEGF antisense oligonucleotides. Among the approaches aimed at inhibiting VEGF receptors, the most developed is the use of anti-VEGFR tyrosine kinase inhibitors: at present, sunitinib and sorafenib are approved for clinical use.

TAKE-HOME MESSAGE

- Growth factor (GF) receptors are tyrosine kinases.
- Activation of GF receptors involves dimerization and assembly of cytoplasmic proteins endowed of specific phosphotyrosine binding domains, including SRC homology 2 (SH2) domains.
- The macromolecular complexes formed by receptor activation activate specific transducers, including Ras and PI3K, involved in the generation of signals of cell proliferation and neoplastic transformation.
- Cytoplasmic tyrosine kinases and tyrosine phosphatases modulate signal transduction by GFs.
- In cancer cells, deregulation of these receptors may result in the constitutive activation of the downstream signaling pathways, thus promoting uncontrolled proliferation.
- GF receptors can be inhibited by tyrosine kinase inhibitors that block their enzymatic activity or by monoclonal antibodies that prevent their interaction with the ligand or induce their degradation.

FURTHER READING

Arteaga C.L. and Engelman J.A. (2014). ERBB receptors: from oncogene discovery to basic science to mechanism-based cancer therapeutics. *Cancer Cell*, 17, 25, 282–303.

Barf T. and Kaptein A. (2012). Irreversible protein kinase inhibitors: balancing the benefits and risks. *Journal of Medicinal Chemistry*, 55, 6243–6262.

Blume-Jensen P. and Hunter, T. (2001). Oncogenic kinase signalling. *Nature*, 411, 355–365.

Chen T.W. and Bedard P.L. (2013). Personalized medicine for metastatic breast cancer. *Current Opinion in Oncology*, 25, 615–624.

Demoulin J.B. and Essaghir A. (2014). PDGF receptor signaling networks in normal and cancer cells. *Cytokine and Growth Factor Reviews*, 25, 273–283.

Downward J. (2003). Targeting RAS signalling pathways in cancer therapy. *Nature Reviews Cancer*, 3, 11–22.

Eigentler T.K., Meier F., Garbe C. (2013). Protein kinase inhibitors in melanoma. *Expert Opinion on Pharmacotherapy*, 14, 2195–2201.

Fabbro D., Parkinson D., Matter A. (2002). Protein tyrosine kinase inhibitors: new treatment modalities? *Current Opinion in Pharmacology*, 2, 374–381.

Goldman J.M. (2012). Ponatinib for chronic myeloid leukemia. *New England Journal of Medicine*, 367, 2148–2149.

Hanahan D. and Weinberg R.A. (2000). The hallmarks of cancer. *Cell*, 100, 57–70.

Hubbard S.R. and Till, J.H. (2000). Protein tyrosine kinase structure and function. *Annual Review of Biochemistry*, 69, 373–398.

Hunter T. (2000). Signaling – 2000 and beyond. *Cell*, 100, 113–127.

Kerbel R. and Folkman J. (2002). Clinical translation of angiogenesis inhibitors. *Nature Reviews Cancer*, 2, 727–739.

Lemmon M.A. and Schlessinger J. (2010). Cell Signaling by receptor tyrosine kinases. *Cell*, 141, 1117–1134.

Lieu C.H., Tan A.C., Leong S., Diamond J.R., Eckhardt S.G. (2013). From bench to bedside: lessons learned in translating preclinical studies in cancer drug development. *Journal of the National Cancer Institute*, 105, 1441–1456.

Matter A. (2001). Tumor angiogenesis as a therapeutic target. *Drug Discovery Today*, 6, 1005–1023.

Rozenfeld R. and Devi L.A. (2009). Receptor heteromerization and drug discovery. *Trends in Pharmacological Sciences*, 31, 124–130.

19

CYTOKINE RECEPTORS

MASSIMO LOCATI

> **By reading this chapter, you will:**
> - Know the main cytokine families and their receptors
> - Understand the role of the main cytokine families in the regulation of the immune response
> - Understand the possible strategies for the pharmacological control of cytokine networks

Cytokines are small glycoproteins with a broad structural and functional heterogeneity acting as soluble intercellular mediators. More strictly, the term cytokine is used to indicate molecules mainly acting on immune cells to regulate immune responses, thus excluding, for example, growth factors. In most cases, these mediators are produced by several different cell types and act in autocrine–paracrine fashion on a wide variety of cells and tissues. These properties distinguish cytokines from peptide hormones, which are usually produced by a single cell type, and act on a restricted number of different target cells in the body.

CLASSIFICATION OF CYTOKINES AND THEIR RECEPTORS

Considering its heterogeneity and complexity, an exhaustive and adequate classification of the cytokine system is hard to make. Here, we will adopt a hybrid classification based on the structural and functional properties of both cytokines and their receptors (Table 19.1).

Hematopoietin Receptors

Studies into the tridimensional structure of cytokines have allowed to identify a large family of molecules called hemopoietins, all containing one or more of four α-helical bundle motifs. A first large group of cytokines belonging to this family interacts with heterodimeric or more rarely heterotrimeric receptors composed of single membrane-spanning subunits. Each subunit typically contains one or more fibronectin-like type III domains in its extracellular portion and a highly conserved WSXWS sequence in its juxtamembrane region, required for receptor activation. Most of these receptors, called class I hemopoietin receptors, consist of two subunits: an α chain, which can bind selectively but with a low-affinity specific cytokine, and a β chain, required to create a high-affinity signaling receptor complex. As the β chain is shared by several cytokine receptors, we can identify a number of cytokine receptor subgroups. A first subgroup of class I hemopoietin receptors has CD131 as a common β chain and includes receptors for the interleukins IL-3 and IL-5 and granulocyte–macrophage colony-stimulating factor (GM-CSF); this subgroup stimulates different hematopoietic colonies in the bone marrow. A second subgroup shares gp130 as common β chain and includes receptors for IL-6, IL-11, ciliary neurotrophic factor (CNTF), oncostatin M (OSM), and leukemia inhibitory factor (LIF); these factors act synergistically with colony-stimulating factors (CSFs) and exert stimulatory effect on growth and differentiation of hematopoietic precursors and neurotrophic effects. A third subgroup of class I hemopoietin receptors shares the γ_c CD132 molecule as common β chain and includes receptors for IL-2, IL-4, IL-7, IL-9, and IL-15;

Table 19.1 Cytokine receptors

Type	Structural and signaling properties	Main ligands
Class I hemopoietin receptors	WSXWS sequence, no intrinsic kinase activity, signal transduction: JAK/STAT	IL-2, IL-3, IL-4, IL-5, IL-6, IL-7, IL-9, IL-11, IL-12, IL-13, IL-15, G-CSF, GM-CSF, CNTF, OSM, EPO, TPO, LIF
Class II hemopoietin receptors	No WSXWS sequence, no intrinsic kinase activity, signal transduction: JAK/STAT	IFNα, IFNβ, IFNγ, IL-10
Tyrosine kinase receptors	Intrinsic kinase activity, signal transduction: intracellular adaptors (SH2 domains)	EGF, PDGF, M-CSF, SCF, IGF-1, FGF
IL-1 receptor family	Intracellular TIR domains, signal transduction: intracellular adaptors (DD)	IL-1, IL-18
TNF receptor family	Intracellular DD, signal transduction: intracellular adaptors (DD)	TNFα, LTα, LTβ, FasL, TRAIL, RANKL
Chemokine receptors	Seven transmembrane domains, signal transduction: heterotrimeric G proteins and β-arrestin receptors	16 CXC chemokines, 27 CC chemokines, 1 CX3C chemokines, 1 C chemokine

Abbreviations are reported in the text.

this group has a prominent role in the regulation of lymphocyte differentiation and function.

Some class I hemopoietin receptors do not fit in this classification. These include IL-12 receptor, in which only the IL-12Rβ2 chain belong to this receptor class, and IL-13 receptor that consists of a selective α chain (IL-13Rα1 or IL-13Rα2) and the α chain (IL-4Rα chain) of the IL-4 receptor. It is worth noting that IL-4 and IL-13 cytokines share several biological properties including the ability to polarize the immune response.

Class I hemopoietin receptors include also the receptors for erythropoietin (EPO), thrombopoietin (TPO), and granulocyte colony-stimulating factor (G-CSF), which differ from those previously described because they are monomeric; for these receptors, activation of the signaling cascade seems to depend on ligand-dependent homodimerization.

Other groups of hemopoietins interact with heterotrimeric receptors, called class II hemopoietin receptors, which share with class I receptors the presence of fibronectin-like domains but lack the WSXWS sequence. These group of hemopoietins includes α and β (type I) interferons (IFN), playing a key role in the control of viral infections; IFNγ (type II), exerting a central role in the polarization of the immune response; and the anti-inflammatory cytokine IL-10.

Class I and II hematopoietin receptors lack enzymatic activity in their intracellular domain, and their signal transduction activity depends upon recruitment of Janus family tyrosine kinases (JAK1, JAK2, JAK3, and Tyk2). The receptor intracellular C-terminal tail constitutively interacts with the NH_2-terminal domain of the kinase. Upon ligand-dependent dimerization, transphosphorylation occurs between the JAK molecules associated with the two receptor chains, resulting in full activation of their enzymatic activity. Activated JAK phosphorylates itself and some tyrosine residues in the receptor, allowing its interaction with the src homology 2 (SH2) domains of signal transducer and activator of transcription (STAT) factors, which are thereby recruited to the signaling complex.

Tyrosine Kinase Receptors

A number of cytokines interact with single-transmembrane domain receptors endowed with tyrosine kinase activity in their intracellular domain. In most cases, these receptors act as monomers and have different types of domains in their extracellular portion. Several subgroups have been identified. Subgroup I includes epidermal growth factor (EGF) receptor and shows variable number of extracellular cysteine-rich domains. Subgroup II includes the insulin-like growth factor 1 (IGF-1) receptor and operates as an α2β2 heterodimer; similarly to subgroup I, these receptors have a variable number of cysteine-rich domains in their extracellular region. Subgroup III includes platelet-derived growth factor (PDGF) receptor, macrophage colony-stimulating factor (M-CSF) receptor, and stem cell factor (SCF) receptor. Subgroup IV includes fibroblast growth factor (FGF) receptor. Subgroup V includes the high-affinity nerve growth factor (NGF) receptor. Receptors belonging to the III, IV, and V subgroups contain a variable number of immunoglobulin-like domains in their extracellular region.

The signal transduction activity of these receptors requires ligand-dependent receptor dimerization, followed by autophosphorylation of tyrosine residues located in the intracellular domain, which are then recognized by SH2-containing adaptor proteins, including Grb–SOS (see also Chapter 18).

The IL-1/IL-18 Receptor Family

Cytokines belonging to the IL-1/IL-18 family are central to the development of inflammatory reactions and share a β trefoil structure and the ability to interact with a distinct group of receptors related to Toll-like receptors (TLRs), which are essential for pathogen recognition. These receptors are single-transmembrane domain molecules with three immunoglobulin-like domains in their extracellular region and one Toll-like/IL-1 receptors (TIR) domain in their intracellular region. IL-1RI and IL-18R can engage the ligands, but receptor signaling activity requires an accessory chain (AcP). AcP has no direct role in ligand recognition but recruits the signal transducers. The IL-1/IL-18 system also offers several examples of negative regulatory networks operating at receptor level. In the case of IL-1, besides the signaling-competent IL-1RI, there is a second IL-1 receptor, called IL-1RII and encoded by a different gene. IL-1RII is highly similar to IL-1RI in the extracellular region, but has no TIR domain in its intracellular region. Both receptors recruit AcP in a ligand-dependent fashion but with opposite functional outcomes. Furthermore, besides the agonistic cytokines IL-1α and IL-1β, a structurally related molecule exists, the IL-1 receptor antagonist (IL-1ra), that can engage IL-1RI and induce its blockade.

In the case of IL-18, the biological activity of the cytokine is controlled by a soluble binding molecule, the IL-18 binding protein (IL-18BP), that binds IL-18, preventing its interaction with IL-18R.

The signal transduction activity of these receptors is based on the protein–protein interaction TIR domain, which recruits the adaptor molecule MyD88. MyD88 contains two protein–protein interaction domains: a TIR domain that engages the corresponding TIR domain in the receptor tail and the death domain (DD, originally identified in the TNFα receptor) that allows MyD88 to recruit a class of kinases called IL-1 receptor-activated kinases (IRAKs). IRAKs phosphorylate the TNF receptor-associated factor 6 (TRAF6) adaptor, which in turn activates the serine/threonine kinase TAK1, leading to NF-κB activation, and an independent signaling pathway, resulting in activation of the stress kinases JNK and p38. These pathways activate the transcription of genes encoding inflammatory cytokines (IL-1, IL-6), chemokines (CXCL8, CCL2), endothelial adhesion molecules, enzymes that catalyze the production of effector molecules (inducible NO synthase, type 2 cyclooxygenase), and costimulatory molecules for the induction of adaptive immune response (CD80, CD86).

The TNFR Family (Jelly Roll Motif Cytokines)

Cytokines belonging to this family share a significant degree of structural homology (25–30%) and are expressed as homotrimeric type II transmembrane molecules released as soluble proteins after cleavage by membrane metalloproteases (see also Chapters 9 and 31). The founder molecule of this family, TNFα, owes its name to the necrotizing effect it exerts in some experimental tumors. However, the antitumoral activity is not its most relevant biological function. TNFα interacts with two single-transmembrane domain receptors, TNFR1 (or p55) and TNFR2 (or p75). The extracellular region of these receptors exhibits a conserved structural organization consisting of four cysteine-rich domains, whereas the intracellular regions mediating different biological properties are significantly different. TNFR1 is essentially a cell death receptor that activates an apoptotic program centered on the activation of the caspase cascade (see Chapter 32); TNFR2 is mainly involved in the inflammatory response and partially inhibits the cell death program induced by TNFR1. Upon interaction with the trimeric ligand, TNFR2 recruits the adaptor proteins TRAF1 and TRAF2, which activate JNK and p38. TRAF2 also recruits the receptor-interacting protein 1 (RIP1), which activates IKKβ, leading to inactivation of IκB and thereby induction of a proinflammatory NF-κB-dependent genetic program similar to the one described for the IL-1/IL-18R system. TNFR1 has no significant affinity for TRAF proteins and recruits the intracellular adaptor TNFR1-associated death domain (TRADD), which is conversely unable to bind TNFR2. The DD of TRADD interacts with the corresponding DD of a second adaptor protein, Fas-associated death domain (FADD), which recruits and activates caspase-8 initiating the cell death program (see Chapter 32). Besides TNFα, this family includes at least 10 other cytokines, including lymphotoxin α (LTα) and β (LTβ), and a set of ligands for receptors highly related to TNFR1 such as Fas (FasL), CD27 (CD27L), CD30 (CD30L), CD40 (CD40L), 4-1BB (4-1BBL), OX-40 (OX-40L), TNF-related apoptosis-inducing ligand (TRAIL), and receptor activator of NF-κB ligand (RANKL). Differently from TNFα, most of these molecules have no prominent role in the inflammatory response but are involved in the control of cell survival in different biological settings.

Chemokine Receptors

Chemokines represent the largest cytokine family, including 45 different members in humans. Their main function is the induction of directional cell migration along a ligand gradient, a process known as chemotaxis. Several chemokines are only produced upon induction of an immune/inflammatory reaction, while others are constitutively released and control lymphocyte differentiation and trafficking. Chemokine receptors belong to the large 7-transmembrane domain receptor superfamily and signal through heterotrimeric G proteins and arrestins.

Chemokines are usually positively charged and their interaction with the receptor involves two steps: the first step (docking) consists in the interaction with the negatively charged NH_2-terminal domain of a cognate receptor. This interaction does not activate the receptor, but it is essential for the second step, in which the chemokine induces structural changes in the transmembrane region of the receptor that are stabilized through interactions with the chemokine NH_2-terminal domain. In its active configuration, the receptor activates the α-subunit of heterotrimeric G proteins acting as a GDP–GTP exchange factor. The α- and β/γ-subunits then dissociate from the receptor and activate a panel of signal transducers (phospholipase $Cβ_2$, phospholipase A_2, phospholipase D, PI3 kinase, protein C kinase) inducing the cytoskeletal rearrangements required to sustain cell movement along the ligand gradient.

These receptors also induce signals, still not well characterized, that act on the intracytoplasmic portions of integrins, increasing their affinity for endothelial adhesion molecules (see Chapter 20). This allows the tight adhesion of leukocytes to the vascular endothelium and is essential for leukocyte extravasation.

Chemokine receptors also activate JNK, ERK1/ERK2, and p38 pathways, ultimately leading to activation of NF-κB and AP-1 transcription factors. Activated receptors are also recognized by members of a specific family of kinases called GRKs, which phosphorylate serine and threonine residues in the receptor COOH-terminal domain. These phosphorylated residues allow the recruitment of β-arrestins, which switch off the signaling cascade and control the receptor internalization via clathrin-coated pits (see also Chapters 9 and 17). Once the ligand–receptor complex is internalized in acidic compartments, the ligand is released and targeted for degradation, whereas the receptor is dephosphorylated and recycled to the plasma membrane.

CYTOKINES IN THEIR BIOLOGICAL SETTINGS

The structural classification followed in this chapter does not provide functional information on different cytokines. Although a detailed functional description of individual cytokines is beyond the scope of this text, in the next sections, we will provide a general description of groups of cytokines based on the main biological processes in which they are involved.

Hematopoietic Cytokines

The hematopoietic system produces eight cell lineages, consisting of terminally differentiated cells deriving from a pool of common stem cells via a number of hematopoietic precursors at different stages of differentiation. This process not only guarantees the production of many million cells per day under homeostatic conditions but allows to respond to "emergency situations" occurring in pathologic conditions by triggering the fast and coordinated expansion of selected lineages. The control of this process involves signals deriving from intercellular contacts, in particular between bone marrow stromal cells and hematopoietic precursors, and soluble signals mainly produced in the bone marrow microenvironment and acting in a paracrine fashion. These soluble mediators include about 20 cytokines involved in the control of different lineages, often acting in a synergistic way. Indeed, most of these mediators act preferentially on a restricted number of lineages but can also act synergistically with other mediators on other lineages (Fig. 19.1) and in several cases also on mature leukocytes. Hematopoietic cytokines include CSFs, which are specific for various lineages; IL-3, which acts as a multi-CSF; IL-5, a CSF of the eosinophil lineage; IL-7, a CSF specific for lymphoid cells; SCF; EPO, which is selectively active on the erythroid lineage; and TPO, the main regulator of thrombopoiesis. Some aspects of the hematopoietic activity are also controlled by cytokines mainly involved in other biological processes, including the proinflammatory cytokines IL-1 and IL-6, which stimulate progenitor cells, and the chemokines CCL3 and CXCL8, which, conversely, inhibit them.

Given their effects on bone marrow progenitors, CSFs are currently used in clinical practice. In particular, GM-CSF and G-CSF are used to reduce neutrophil nadir caused by pathological conditions or chemotherapeutic agents, thus limiting incidence and severity of incurring infections.

Cytokines in Innate Immunity

The inflammatory response is induced and controlled by a complex network of mediators including cytokines. These can be classified as primary inflammatory cytokines, produced and released locally at the beginning of the response (IL-1, TNFα, IL-6), and secondary inflammatory cytokines, induced by primary cytokines and mainly involved in leukocyte recruitment. Primary inflammatory cytokines have largely overlapping biological activities and act on a wide panel of target cells and tissues.

The IL-1 system includes two agonists, IL-1α and IL-1β, that are produced as procytokines and require processing by an enzyme, called caspase-1 or IL-1-converting enzyme (ICE), to exert their overlapping biological functions by interacting with the signaling receptor IL-1RI. The system includes also one antagonist, IL-1ra. This is selectively induced by anti-inflammatory cytokines in the same cells that produce IL-1α and IL-1β and compete with them for binding to IL-1RI.

TNFα is a another key mediator of inflammatory reactions. Similarly to IL-1, TNFα is synthesized as a membrane-bound precursor that is released upon processing by a metalloprotease called TNFα-converting enzyme (TACE). Both the membrane-bound and the soluble version of the cytokine interact with the receptors and are biologically active. TNFα and IL-1 are produced by a variety of cell types in response to different stimuli and set up a feed-forward loop by inducing each other. From a functional point of view, IL-1 and TNFα share most of their biological properties. At the site of inflammation, some of their major targets are

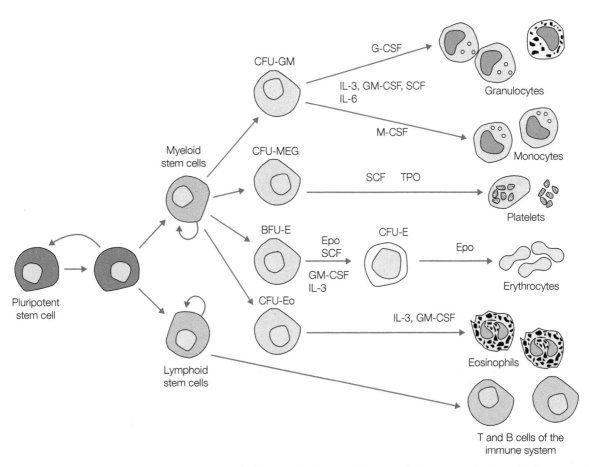

FIGURE 19.1 The role of cytokines in hematopoiesis. The hematopoietic process is responsible for the production of different cell lineages from pluripotent stem cell, which should on the one hand maintain a process of self-renewal and on the other allow progressive differentiation to the different cell types. This process is tightly controlled by different colony-stimulating factors, which support the differentiation of precursors to different cell lineage. The figure shows the most important of them in relation to cell differentiation.

endothelial cells, which respond to stimulation by activating a genetic proinflammatory and prothrombotic program that ends up with the induction of adhesion molecules that recruit inflammatory leukocytes, and enzymes that control the production of eicosanoids and NO with vasodilating activity. Moreover, IL-1 and TNFα induce the production of procoagulant tissue factor (TF) and inhibit the anticoagulant pathway supported by C protein and thrombomodulin. Another target of both IL-1 and TNFα is the connective tissue, including bone and cartilage cells and synoviocytes. Here, IL-1 and TNFα induce production of proteases and prostaglandins causing tissue damage and bone resorption. Finally, IL-1 and TNFα are major activators of infiltrating phagocytes. IL-1 and TNFα also have long-range effects, which are different for the two cytokines. IL-1 stimulates proliferation and differentiation of hematopoietic precursors, mostly through the induction of CSFs (GM-CSF in particular) and IL-6, whereas TNF inhibits hemopoietic precursors, an effect responsible for chronic disease anemia.

In the central nervous system, IL-1 (but not TNFα) acts as an endogenous pyrogen through the induction of IL-6 and prostaglandins, whereas both IL-1 and TNFα induce anorexia and asthenia and production of ACTH. In turn, ACTH induces a negative feedback based on the production of glucocorticoids, which are major anti-inflammatory mediators (see also Chapter 50). In the liver, IL-1, via IL-6, induces the production of acute-phase proteins, which amplify the innate immune response and support tissue remodeling. Besides its role in acute-phase response, IL-6 also acts as final effector of IL-1 in the bone marrow. As mentioned earlier, IL-6 interacts directly with a receptor complex constituted by a common chain, gp130, and a specific chain, IL-6Rα. However, IL-6 can activate cells lacking the membrane IL-6Rα chain through a soluble variant of IL-6Rα. This variant interacts with IL-6 in solution, and the complex, present in biological fluids, can recruit gp130 in a trans-signaling complex. This occurs in endothelial cells where IL-6 amplifies the expression of adhesion molecules and favors the production of chemokines acting on monocytes but not on neutrophils, thus contributing to the switch from acute to chronic inflammation.

In conclusion, at systemic level, the primary inflammatory cytokines IL-1 and TNFα act on the hematopoietic

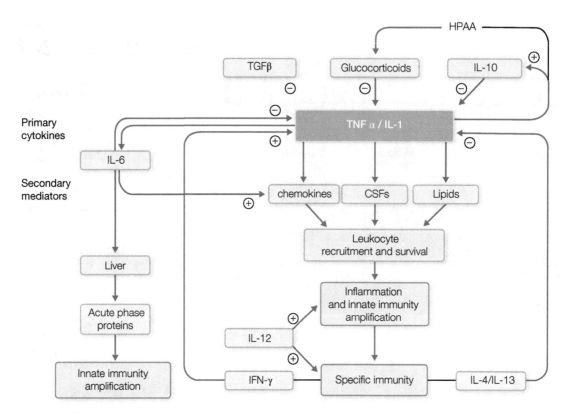

FIGURE 19.2 Cascade of inflammatory cytokines. Inflammatory cytokines activate and sustain the inflammatory response and the innate immune response. Inflammatory cytokines are classically divided into *primary*, prototypes of which are IL-1, TNF, and IL-6, acting on the liver that trigger the acute-phase response, and *secondary* chemokines, which amplify the recruitment of white blood cells and their survival in the tissues and prepare the scenario for the activation of specific immunity. The cascade of inflammatory cytokines is regulated by negative signals, represented in part by glucocorticoid hormones, derived from the activation of the hypothalamic–pituitary–adrenal axis (HPAA), and in part by anti-inflammatory cytokines such as IL-10 and TGF-β.

compartment, central nervous system, and liver and in several cases through IL-6 and overall coordinate the systemic inflammatory response. At the same time, at local level, these cytokines induce the production of adhesion molecules and chemokines, which amplify recruitment and survival of inflammatory leukocytes (Fig. 19.2) and set the stage for the activation of adaptive immunity.

Cytokines of Adaptive Immunity

Proliferation and differentiation of T and B lymphocytes are key steps in adaptive immunity and are both tightly controlled by a large panel of cytokines, including IL-12, IFNγ, IL-15, IL-4, and IL-2. In T lymphocytes, interaction of T cell receptor with its cognate antigen triggers a signaling cascade that activates the nuclear factor of activated T cells (NFAT), a transcription factor that induces the expression of the genes encoding IL-2 and IL-2R α chain. In the absence of further signals provided by costimulatory molecules, the IL-2 transcript is rapidly degraded and the lymphocyte anergized. On the contrary, in the presence of additional signals from costimulatory molecules, IL-2 production is significantly increased and the lymphocyte undergoes clonal expansion. The relevance of IL-2 in the induction of adaptive immunity is confirmed by the observation that the immunosuppressive drugs cyclosporin A and FK-506 act by suppressing IL-2 transcription.

Differentiation of $CD4^+$ T helper lymphocytes after antigen exposure is a second key step in adaptive immunity, also tightly controlled by complex cytokine-mediated networks (Fig. 19.3). In particular, IL-12, which is selectively produced by activated macrophages and dendritic cells, and IFNγ, a product of Th1 and NK cells, support Th1 and suppress Th2 differentiation, whereas IL-4, IL-5, and IL-13, which are typical Th2 products, perform the opposite function. Therefore, cytokines are at the same time key mediators and landmarks of polarized lymphocytes that activate different effector cells to coordinate the most appropriate immune response in different pathological settings. In particular, Th1-dependent production of IFNγ supports classical activation of macrophages, with increased cytotoxic activity against intracellular pathogens. Conversely, Th2-dependent production of IL-4 and IL-5 sustains alternative activation of macrophages and eosinophil

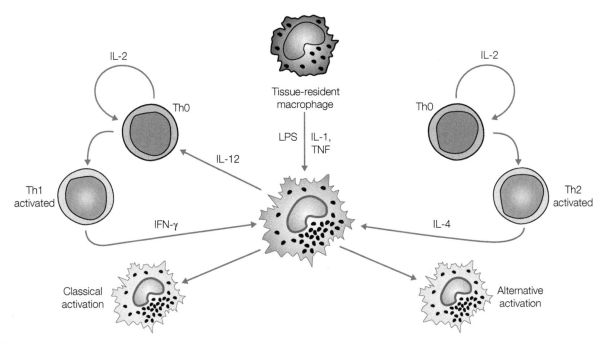

FIGURE 19.3 Role of cytokines in acquired immunity in the polarization of the immune response. The interaction of macrophages and phagocytic cells with pathogens (center of the figure) determines the types of cytokine produced and released and the consequent polarization of the specific immune response, that is, IFNγ for the polarization of type I and IL-12/IL-4 for the polarization of type II. The Th lymphocytes then further polarize the inflammatory response through the production of different specific cytokines.

recruitment that coordinate the most adequate immune response to extracellular pathogens and protozoa. Predominant cytokine profiles characterize noninfectious pathologic states: for example, allergic reactions are dominated by Th2 cytokines, whereas most autoimmune diseases are dominated by Th1 cytokines. This general observation correlates with the specific use of cytokines in therapeutic settings, such as the use of IFNγ in the Th2-dominated Omenn syndrome, an autosomal recessive immunodeficiency associated with mutations in genes affecting circulating levels of B and T cells.

Anti-inflammatory Cytokines

Cytokine classically operates in complex networks where different mediators reciprocally potentiate and inhibit each other, and in several cases, the same mediator exerts distinct effects on different aspects of the immune response. As an example, IL-6 has a primary role in acute-phase responses and in B cell activation but at the same time has some inhibitory effects on phagocytes. Similarly, IL-4 and IL-13 inhibit IFNγ-dependent classical activation of macrophages but coordinate type 2 immune responses and sustain alternative macrophage activation. Conversely, some cytokines mediate pure negative feedbacks. An example is IL-10, the best characterized anti-inflammatory cytokine. IL-10 is produced at late time points of inflammation by the same cells that produce proinflammatory cytokines (Fig. 19.4). It switches off the inflammatory response both by reducing the production of proinflammatory cytokines via interference with their NF-κB and STAT5-dependent transcription and destabilization of their coding transcripts and by inducing other anti-inflammatory mediators, such as IL-1ra. It is worth noting that some anti-inflammatory drugs, including steroids, operate at least in part by inducing IL-10.

PHARMACOLOGY OF CYTOKINES AND THEIR RECEPTORS

In some cases, cytokines themselves have been developed as pharmacological tools, such as for CSF or IL-1ra. However, most efforts have been focused on the development of strategies to inhibit their effects, using receptor inhibitors or blocking antibodies or by interfering with their receptor signaling activity.

The impressive efficacy of anti-TNFα antibodies in immunomediated diseases has raised a great interest in the identification of antibodies or soluble receptors able to bind and neutralize mediators known to be involved in the pathogenesis of a large number of human diseases, such as TNFα, IL-6, and IL-12/IL-23 (see also Chapter 21). Though cytokine receptors have been identified and structurally characterized in great detail, no specific small-molecule inhibitors have been developed so far. A noteworthy exception is represented by chemokine receptor inhibitors,

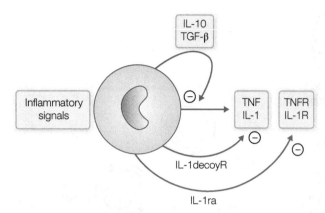

FIGURE 19.4 Circuits of negative regulation by primary inflammation cytokines. The inflammatory response, coordinated by the primary proinflammatory cytokines IL-1, TNF, and IL-6, is tightly controlled by anti-inflammatory cytokines. For example, IL-10 and TGF-β, produced at late time points of inflammation by the same cells that produce proinflammatory cytokines, switch off the inflammatory response by reducing the production of proinflammatory cytokines and inducing inhibitory molecules, including the receptor antagonist and/or the decoy IL-1 receptors.

which have been developed as a result of the great interest in these receptors as pharmacological targets because of their role as HIV-1 coreceptors. At present, inhibitors for CCR5 and CXCR4 have reached the market as important agents in the pharmacotherapy of HIV-1 infection, and other small compounds targeting other chemokine receptors are at advanced stages of development. As mentioned earlier, in several cases, different cytokines exert redundant biological effects because they trigger common signaling pathways controlling a restricted number of transcription factors. This consideration is at the basis of the strategy to target common signaling transducers used by different cytokines involved in a given biological process. A good example of this approach is represented by immunosuppressive drugs targeting the IL-2 system. Inhibitors of key signal transducers, such as the stress kinases JNK and ERK1/ERK2, as well as transcription factors, such as NF-κB, are in advanced development. Finally, an alternative approach is based on drugs capable of preventing cytokine production. The best example is represented by steroid drugs mimicking the anti-inflammatory effect of glucocorticoids in the inhibition of proinflammatory cytokines (IL-1, TNFα, IL-6, inflammatory chemokines) and adaptive immunity mediators (IFNγ, IL-4). Other examples are vitamin A, which inhibits IL-6 production, and vitamin D, which inhibits IL-1 and IL-12 production in macrophages.

TAKE-HOME MESSAGE

- Cytokines are a complex system of soluble mediators controlled by molecular networks.
- Cytokine receptors differ in molecular structures and mechanisms of signal transduction.
- The main steps of the innate and adaptive immune responses are controlled by different classes of cytokines that are specialized in different biological processes.
- Pharmacological approaches for controlling cytokine networks are based on production or removal of ligands or activation and inhibition of receptors and their signal transduction cascade.

FURTHER READING

Arisi G.M. (2014). Nervous and immune systems signals and connections: cytokines in hippocampus physiology and pathology. *Epilepsy & Behavior*, 38, 43–47.

Croft M. (2014). The TNF family in T cell differentiation and function – unanswered questions and future directions. *Seminars in Immunology*, 30, 205–218.

Gerard G., Rollins B.J. (2001). Chemokines and disease. *Nature Immunology*, 2, 108–115.

Gordon S. (2003). Alternative activation of macrophages. *Nature Reviews Immunology*, 3, 23–35.

Kane A., Deenick E.K., Ma C.S., Cook M.C., Uzel G., Tangye S.G. (2014). STAT3 is a central regulator of lymphocyte differentiation and function. *Current Opinion in Immunology*, 28C, 49–57.

Krumm B., Xiang Y., Deng J. (2014). Structural biology of the IL-1 superfamily: key cytokines in the regulation of immune and inflammatory responses. *Protein Science*, 23, 526–538.

Mantovani A., Dinarello C.A, Ghezzi P. *Pharmacology of cytokines*, Oxford University Press: London, 2000.

Mantovani A., Locati M., Vecchi A., Sozzani S., Allavena P. (2001). Decoy receptors as a strategy to regulate inflammatory cytokines and chemokines. *Trends in Immunology*, 22, 328–836.

Mantovani A., Sozzani S., Locati M., Allavena P., Sica A. (2002). Macrophage polarization: tumor-associated macrophages as a paradigm for polarized M2 mononuclear phagocytes. *Trends in Immunology*, 23, 549–555.

Sabat R., Ouyang W., Wolk K. (2014). Therapeutic opportunities of the IL-22-IL-22R1 system. *Nature Reviews Drug Discovery*, 13, 21–38.

Thomson A. *The cytokine handbook*, IV edition, Academic Press: San Diego, 2001.

Wolmald S., Hilton D.J. (2004). Inhibitors of cytokine signal transduction. *Journal of Biological Chemistry*, 279, 821–824.

20

ADHESION MOLECULE RECEPTORS

GIORGIO BERTON AND CARLO LAUDANNA

By reading this chapter, you will:

- Understand the basic aspects of adhesive interactions and regulation of adhesion receptor function
- Learn the mechanisms of signal transduction by adhesion receptors
- Know how several major cell responses are regulated by adhesion receptors in physiology and pathology
- Understand the role of adhesion receptors as drug targets

All multicellular organisms are characterized by the presence of mechanisms regulating cell–cell and cell–extracellular matrix interactions. In organisms endowed with a circulatory system, such adhesive interactions are also implicated in the recruitment of circulating cells into tissues and the repair of vessel wall damages. A great deal has been learned on molecules implicated in adhesive interactions, and recent studies have revealed that adhesive receptors can also transduce intracellular signals regulating cell migration, proliferation, and survival.

ADHESION RECEPTORS

Classification

Adhesion receptors may be broadly classified according to their structure in five distinct families (Fig. 20.1). Among these, some are implicated in adhesive interactions that can be defined as "stable," such as those involved in the organization of parenchymatous (e.g., liver, pancreas, lung) or lining (e.g., mucosal surface, endothelial layers) tissues. Cadherins are the main adhesive receptors mediating stable cell–cell interactions and integrins those implicated in stable cell–extracellular matrix interactions.

Cells migrating in the interstitium in both normal (i.e., during development or wound repair) and pathological (i.e., following malignant transformation) conditions are characterized by "dynamic" interactions wherein adhesive interactions required to anchor cells to the substratum are followed by deadhesive events allowing cells to crawl. In both hematopoietic and nonhematopoietic cells, the most important form of dynamic interactions is represented by integrin-mediated binding to extracellular matrix proteins.

A specialized and highly complex form of dynamic interaction is that occurring between leukocytes and vascular endothelium, which involves adhesion, crawling, and transmigration events and governs leukocyte recruitment into tissues and lymphocyte recirculation from the blood to the lymphatic circulation (see following text and Fig. 20.2). Leukocyte–endothelium interactions are mediated by different adhesion receptor families: selectins, integrins, mucins, and members of the immunoglobulin superfamily.

Another peculiar type of adhesive interaction is involved in the platelet-dependent phase of the hemostatic process. Platelets are anucleate cells circulating in the blood that, upon vessel wall damage, establish strong adhesive interactions with extracellular matrix proteins and form aggregates, forming a plug (platelet plug) that blocks blood leakage from the injured vessel (for more details on this process, see Supplement E 20.1, "Adhesion and Platelet Activation").

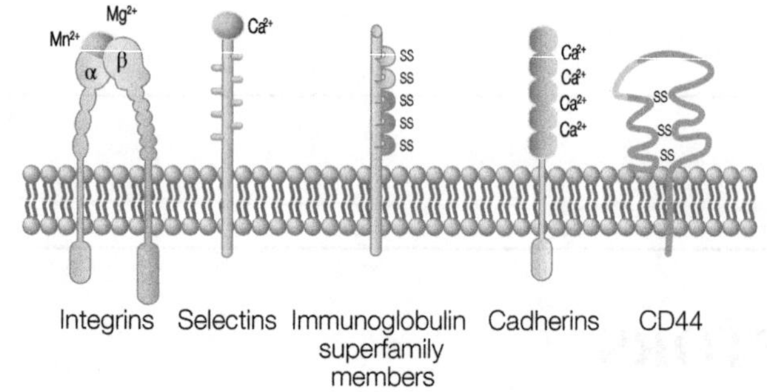

FIGURE 20.1 Simplified structure of integrins, selectins, immunoglobulin superfamily members, cadherins, and CD44. Protein domains involved in ligand binding are painted in gray. Integrin binding to their ligands depends on the divalent cations Mn^{2+} and Mg^{2+}. A list of main integrin ligands is reported in Table 20.1. Selectin binding to their ligand is Ca^{2+} dependent. Selectins recognize carbohydrate residues present in heavily glycosylated proteins (mucins). Polysaccharide residues recognized by selectins include sialyl-Lewis[x], sialyl-Lewis[A], and other sialylated and/or fucosylated oligosaccharides. Selectins are classified as L- (for leukocyte), P- (for platelet), and E- (for endothelial) selectins. Immunoglobulin superfamily members implicated in adhesion events include ICAM-1, ICAM-2, ICAM-3, VCAM-1, and Mad-CAM, which are mainly expressed by the vascular endothelium. Their expression is regulated via transcriptional mechanisms and is induced by proinflammatory cytokines. Cadherins are involved in homotypic adhesions through cadherin–cadherin Ca^{2+}-dependent interactions. Several alternative splicing isoforms of CD44 exist, resulting from the different combinations of 10 exons in the central part of the primary CD44 encoding mRNA. Such isoforms differ in their extracellular domains proximal to the plasma membrane; the figure shows the shortest isoform (called "s" for standard) that is ubiquitously expressed. The elective CD44 ligand is hyaluronic acid, a component of the extracellular matrix, but it can also bind other glycosaminoglycans and extracellular matrix proteins.

Functions

Cadherins in Nonhematopoietic Cell–Cell Interactions

Cadherins take part in the formation of the adherens junctions, which are sites of interactions between cells in solid organs and mucosal surfaces. This family of adhesion proteins (Fig. 20.1) is composed of different members named according to their tissue-specific expression: E-cadherin (epithelial), M-cadherin (muscular), P-cadherin (placental), VE-cadherin (endothelial), and N-cadherin (neuronal). They associate to form dimers that interact, in a calcium-sensitive manner, with cadherins of the same type expressed on adjacent cells. The cadherin intracellular cytoplasmic tail binds other proteins (catenins) and cytoskeletal components (vinculin, filamentous actin). Different isoforms of catenins (α, β, and γ or plakoglobin) may interact with each other or with cytoskeletal proteins. The molecular complex constituted by cadherins, catenins, and filamentous actin confers stiffness and solidity to adherens junctions. Additionally, it serves as a site of aggregation for other cytoplasmic proteins, including molecules involved in signal transduction (see following text and Fig. 20.6).

Integrin-Mediated Interactions between Cells and Extracellular Matrix Proteins and Cell Counterreceptors

Integrins are heterodimers resulting from the noncovalent association between an α chain and a β chain and are classified on the basis of the β chain structure (Fig. 20.1 and Table 20.1). Integrins are expressed in a wide range of cell types, can recognize a large number of ligands, and are implicated in a variety of biological functions. In mesenchymal and epithelial cells, integrin-dependent adhesion regulates differentiation, proliferation, survival, and migration. In addition, as mentioned earlier, integrins govern the process of leukocyte recruitment into tissues. Lastly, αIIbβ3, an integrin selectively expressed by platelets (Table 20.1), is essential for the formation of the platelet plug (see Supplement E 20.1). Among the diverse integrin functions, here, we will discuss those better characterized.

Two features are common to several integrins: on the one hand, their adhesive capacity is regulated, and on the other, they regulate cellular events by transducing intracellular signals upon binding appropriate ligands. Regulation of integrin adhesive capacity is referred to as inside-out signaling. This definition highlights the complex series of events leading to the generation of an intracellular signal (inside) by different surface receptors that eventually results in a conformational modification of integrin extracellular domains (out). This modification is essential to increase the integrin affinity for an appropriate ligand. Conversely, the signal transduction ability of integrins is referred to as outside-in signaling: when the appropriate ligand binds its extracellular domain (outside), the integrin behaves as a receptor transducing signals inside (in) the cell.

In the following section, we will discuss the molecular mechanisms underlying the inside-out and outside-in signaling events.

Integrins clustered in discrete domains of the plasma membrane and bound to their appropriate ligands serve as sites of assembly of protein complexes containing several cytoskeletal proteins, which interact with each other forming a sort of tridimensional spider web. Based on the complexity of their organization and the extent of their turnover, these structures are referred to as focal contacts or focal adhesions and contribute to the tightness of cell adhesion to a substratum. Different cytoskeletal proteins have been reported to bind to the β chain of integrins, including talin, kindlin, paxillin, α-actinin, and filamin. Filamin and α-actinin directly interact with filamentous actin, whereas talin and paxillin interact with vinculin, which indirectly bind filamentous actin by interacting with α-actinin. Notably, binding of talin and kindlin to the β chain cytoplasmic tail of β2 and β3 integrins has been recognized to be a central event in the

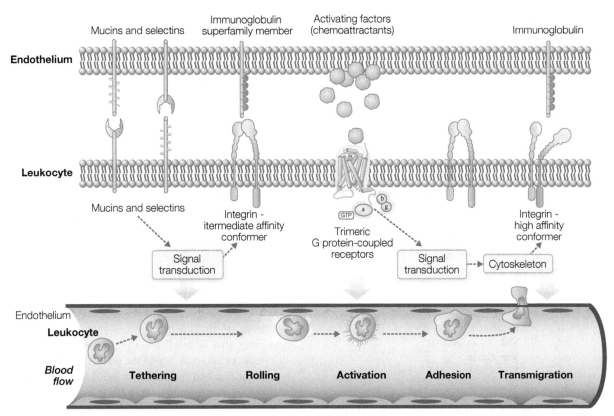

FIGURE 20.2 The widely accepted model describing the accumulation of various leukocyte types in different organ districts, under both physiological and pathological conditions, involves a sequence of stereotyped molecular and cellular events. During leukocyte recruitment, the coordinated action of tethering and rolling receptors (e.g., as mucins and selectins) and activating factors (e.g., chemotactic factors), along with arrest molecules (e.g., integrins and immunoglobulin superfamily members), generates a process finely regulated in time and space, which moves leukocytes outside the blood vessel. Complex intracellular signal transduction cascades control different aspects of the process. For example, the mucin PSGL-1 transduces intracellular signals that change integrin affinity from low to intermediate, thus favoring a "slow" rolling. By contrast, chemotactic factors, such as chemokines, transduce rapid intracellular signals that change integrin affinity from low to high, an event that critically contributes to the definitive arrest of leukocyte on the endothelium. This latter mechanism relies upon complex signaling cascades involving cytoskeletal proteins, such as talin-1 and FERMT3 (kindlin-3), which directly bind the integrin β chain cytoplasmic portion and determine the definitive integrin conformation transition to a high-affinity state. The model predicts that differences in leukocyte migration result from the combined action of adhesion molecules and activating factors differentially expressed on leukocyte surface and vascular endothelium. Indeed, diversity in leukocyte recruitment derives from the ability of specific combinations of adhesive molecules and activating factors to trigger the complete sequence of cellular events leading to extravasation of specific cells. For example, neutrophils express L-selectin (rolling receptor) and LFA-1 (arrest receptor) but are unable to migrate into the lymph nodes even if the latter express PNAD and ICAM-1 (endothelial ligands for L-selectin and αLβ2, respectively) on high endothelial venules (HEV). Indeed, neutrophils rolls on HEV but do not arrest nor migrate into the lymph node. This suggests that in normal lymph nodes, the factor required for activation of the integrin-dependent arrest of neutrophils is not present, and therefore, neutrophils move back to the bloodstream. Other examples are chemokines CCL21–CCL24–CCL26 (eotaxins), which are activating factor specific for eosinophils, controlling the final step of diapedesis. In this case, the specificity does not derive from the expression of adhesive molecules, since eosinophils and monocytes express the same selectins and integrins. However, tissues expressing high levels of eotaxin(s) direct the selective migration only of those cells (i.e., eosinophils) expressing the receptor specific for that chemokine. Similarly to the adhesion molecules, even activating factors show a consistent level of redundancy and promiscuity. For example, IL-8 is active on neutrophils, eosinophils, monocytes, and some lymphocyte subsets. Consequently, selective migration depends on the combined action of all the elements controlling the various steps in the process. So it is possible to theorize the existence of a "tissue area code" for every leukocyte type (similar to the ZIP code) resulting from proper combination of the right types of selectin + integrin + integrin-activating factor + endothelial ligand. Thus, in order to migrate in a certain tissue or organ, leukocytes need to recognize the complete code on the vascular district of the tissue. If this does not occur, leukocytes remain in the bloodstream.

TABLE 20.1 Integrins and their principal ligands[a]

β chain	α chain	Ligand
β1	α1	Collagen, laminin
	α2	Collagen, fibronectin, laminin, echovirus
	α3	Collagen, fibronectin, laminin
	α4	Fibronectin, VCAM
	α5	Fibronectin
	α6	Laminin
	α7	Laminin
	α8	QBRICK, nephronectin
	αv	
β2	αL	ICAM-1, ICAM-2, ICAM-3
	αM	ICAM-1, fibrinogen, factor X, C3bi, heparin and heparan sulfate, β-glucan, lipopolysaccharide (LPS)
	αX	Fibrinogen, C3bi, LPS
	αD	ICAM-3
β3	αIIb	Fibrinogen, fibronectin, von Willebrand factor
	αv	Vitronectin
β4	α6	Laminin
β5	αv	Vitronectin, HIV tat
β6	αv	Vitronectin
β7	α4	Mad-CAM, VCAM-1, fibronectin
β8	αv	Vitronectin

[a] Integrins are classified on the basis of their β chain sequence. The table reports the principal members of each subfamily. A few α chains (e.g., αv) can associate with different β chains. Some integrins (e.g., members of the β1 subfamily and αvβ3) are ubiquitously expressed; others are expressed only in a limited number of cell types. β2 integrins are also referred to as the leukocyte integrin subfamily because they are selectively expressed by leukocytes. The αIIbβ3 integrin is selectively expressed by platelets.

inside-out signaling mechanisms regulating integrin affinity for their ligand (see Supplement E 20.2, "Adhesion and Leukocyte Recruitment").

Integrins and Selectins in Adhesion and Recruitment of Leukocytes to the Vascular Endothelium To exert their role in biological defenses, white blood cells have the ability to migrate from the blood into the subendothelial interstitium every time infection or tissue damage occurs. Moreover, lymphocytes can migrate from the venules that pass through the lymph nodes in the lymph nodes themselves (a phenomenon called "lymphocyte recirculation") (Fig. 20.2).

Leukocyte migration occurs according to the following steps:

1. In the first phase, leukocytes interact with the endothelium (tethering) and roll on it; this event is mediated by the interaction between selectins and mucins.
2. During rolling, leukocytes come in contact with chemotactic factors, adsorbed on the endothelium surface, through specific receptors belonging to the large family of G-protein-coupled receptors (see Chapter 17). Interaction between these receptors and their appropriate ligands triggers the inside-out intracellular signals, leading to the third step responsible for leukocyte migration.
3. The third stage consists in a type of adhesion defined as "firm" that is mediated by the interaction between integrins of the β1 or β2 subfamilies and their ligands expressed on the endothelial cell surface, such as ICAM-1/ICAM-2 and VCAM-1 (see Fig. 20.1 and Table 20.1). Firm adhesion depends on a strengthening (activation) of the adhesive capacity of integrins induced in step 2. Integrin activation includes two distinct modalities: (i) conformational modification of the heterodimer, resulting in increased affinity (see preceding text and Figs. 20.2 and 20.3), and (ii) induction of lateral mobility of the heterodimer, resulting in cluster formation and valency increase; valency corresponds to the density of integrin heterodimers per area of plasma membrane involved in cell adhesion (Fig. 20.2).
4. Only after firm adhesion to the vascular endothelium, leukocytes can pass through the interendothelial junctions and migrate into the interstitium (for more details, see Supplement E 20.2).

Integrins and Other Receptors in Platelet Adhesion and Aggregation Vascular lesions are repaired through the coordinated action of platelets and a group of soluble blood proteins that constitute the coagulation system. The reparative process (hemostasis) ends up with the formation of a platelet plug stabilized by a mesh of insoluble, polymerized fibrin deriving from the proteolytic digestion of fibrinogen catalyzed by thrombin. Hemostasis is essential to avoid hemorrhages, but it is also responsible for the formation of thrombi, a common and often severe cause of blood vessel obstruction.

The role played by platelets in hemostasis depends on adhesive interactions between adhesive receptors expressed by platelets and (i) extracellular matrix proteins present beneath the endothelial surface and (ii) proteins present in the plasma or released following fusion of platelet α granules with the plasma membrane (details on the adhesion receptors implicated in the interaction between platelets and the subendothelial surface can be found in Supplement E 20.1).

A central feature of platelet aggregation (i.e., formation of a plug of platelets interacting with each other) is the activation of the integrin αIIbβ3 affinity for the plasma protein fibrinogen via inside-out signaling. Platelet adhesion to the subendothelium or binding of agonists (thromboxane A2, ADP, thrombin) to specific receptors trigger intracellular signals that lead to binding of talin and kindlin-3 to the αIIbβ3 integrin β chain. This enhances the affinity of αIIbβ3

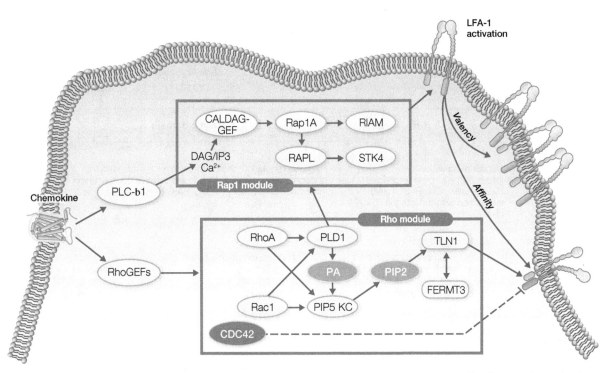

FIGURE 20.3 The intracellular signal transduction pathways regulating integrin activation are extraordinarily complex and only partially understood. To date, at least 68 molecules of the signal transduction machinery are known to modulate various aspects of integrin-dependent adhesiveness. However, the existing differences among cellular contexts, kinetics of activation, and types of proadhesive agonists suggest that there is probably a significant diversity and specificity in the signal transduction mechanisms regulating integrin activation. Although it is unlikely that a universal mechanism of integrin activation exists, there are two main signaling mechanisms known and highly studied: the signaling modules of Rho and Rap small GTPases. The figure shows the components of two signaling modules in the context of integrin activation by chemokines. In the Rho module, Rho GDP/GTP exchange factors activate RhoA and Rac20, which in turn activate phospholipase D1 (PLD1), which then activates phosphatidylinositol-4-phosphate 5-kinase (PIP5KC) through the synthesis of phosphatidic acid (PA). Furthermore, PIP5KC can be activated directly by RhoA and Rac20. PIP5KC generates phosphatidylinositol 4,5-bisphosphate (PIP2), which in turn activates talin-1 (TLN1) and, probably, FERMT3 (kindlin-3). These last two cytoskeletal proteins are crucial to induce integrin high-affinity state, thus mediating arrest of flowing leukocytes. In the Rho module, CDC42, although strongly homologous to RhoA and Rac20, serves as negative regulator. In the Rap module, accumulation of calcium and DAG, generated by phospholipase C beta 1 (PCL-β1), activates the GDP/GTP exchange factor CalDAG-GEF, which in turn activates Rap1A. Rap1A then activates RAPL, RIAM, and STK4. Although apparently not involved in the direct regulation of integrin affinity, the Rap module seems having a central role in the regulation of adhesion by adjusting the stabilization of integrin–ligand interaction. In the figure, all arrows indicate activation. Furthermore, the possibility that the Rap module could be regulated by the Rho module via PLD1 activation by RhoA is also indicated.

for fibrinogen that bridges platelets forming a relatively loose aggregate. The contemporary release of platelet α granule proteins (fibronectin, fibrinogen, von Willebrand factor, thrombospondin) allows multiple interactive events between these proteins and their specific receptors, thus enhancing the stability of the platelet aggregate (see Supplement E 20.1 for further details).

SIGNAL TRANSDUCTION BY ADHESION RECEPTORS

Adhesion receptors regulate several cell functions including migration, gene transcription, proliferation, and survival. Although different adhesion molecules (e.g., cadherins, selectins, CD44, and the mucin P-selectin glycoprotein ligand-1 (PSGL-1)) have been recognized as able to transduce intracellular signals, our discussion will focus on integrins, which are those that have been investigated in more detail.

FAK and Src-Family Kinases in Integrin Signal Transduction

Early studies on integrin signal transduction showed that compounds capable of inhibiting tyrosine kinase activities reduce integrin-dependent cell responses. In contrast with growth factor (GF) receptors (see Chapter 18), the cytoplasmic tails of integrin α and β chains lack any intrinsic tyrosine kinase activity. However, cytoplasmic tyrosine

ADHESION MOLECULE RECEPTORS

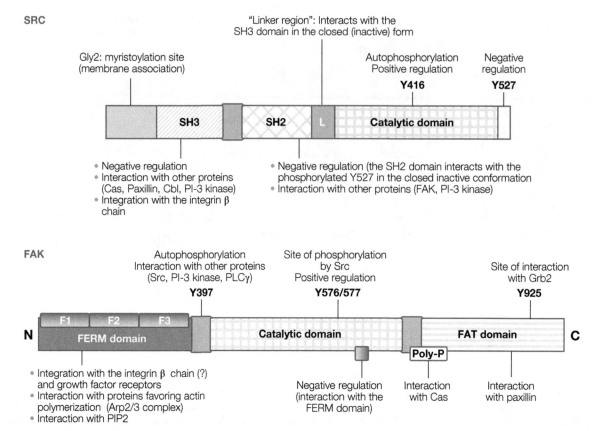

FIGURE 20.4 The figure summarizes Src and FAK domain organization. The Src family of cytoplasmic tyrosine kinases includes eight distinct members with differential patterns of expression: Src, Blk, Fgr, Fyn, Hck, Lck, Lyn, and Yes. Whereas Src, Fyn, and Yes are ubiquitously expressed, Blk, Fgr, Hck, and Lyn are mainly expressed in hematopoietic cells. The main Src domains are present also in other proteins. The SH2 (for "Src homology") domain mediates the interaction with short amino acid sequences containing a phosphorylated tyrosine; the SH3 domain binds proline-rich sequences. Src-family members can interact with several proteins, some of which indicated in the figure. Src-family members contain a glycine in position 2 of their sequence that dictates their posttranslational myristoylation (i.e., the covalent attachment site for the fatty acid myristic acid). With the exception of Src and Blk, the other family members have a cysteine in position 3 that dictates their palmitoylation. Fatty acylation of Src-family members favors their targeting to the phospholipid bilayer of cell membranes. In its phosphorylated state, residue 527 of Src (and analogous residues in the other members) binds to Src SH2 domain and favors another intramolecular interaction between the SH3 domain and a proline-rich sequence present in the linker (L) region. Src activation may results from dephosphorylation of phospho-Y527 or displacement of the intramolecular interactions involving the SH2 and SH3 domains by proteins containing phosphorylated tyrosines or proline-rich regions, respectively. Phosphorylation of the Y416 residue within the kinase domain results in Src full activation. Expression of FAK and its homolog Pyk2 is ubiquitous. The N- and C-terminals of FAK regulate its binding to several proteins. Mechanisms of FAK activation are similar to those of Src. An intramolecular interaction between the F2 subdomain of the four-point-one, ezrin, radixin, and moesin (FERM) domain maintains FAK in a closed, inactive conformation. Upon binding to talin bound to the integrin β chain, the lipid kinase phosphatidylinositol-4-phosphate 5-kinase type Iγ is activated and forms phosphatidylinositol 4,5-bisphosphate (PIP2). PIP2 binds to the FAK FERM domain disrupting its interaction with the kinase domain; this results in increased FAK autophosphorylation at the Y397 residue. Upon binding to phospho-Y397 through its SH2 domain, Src phosphorylates FAK at Y576/577, fully activating FAK kinase activity. Phosphorylation of different FAK tyrosine residues mediates its interaction with SH2-containing proteins. Additionally, a FAK proline-rich region mediates interaction with Cas.

kinases (i.e., focal adhesion kinase (FAK) and members of the Src family) are present in the multimolecular complexes organized by integrins (see preceding text and Figs. 20.4 and 20.5).

Both Src and FAK can associate with the integrin cytoplasmic tail and crosstalk to deliver downstream signals. FAK autophosphorylation at tyrosine residue 397 provides a binding site for Src SH2 domain. On the other hand, Src phosphorylates FAK at several tyrosine residues (Y576/577, Y861, Y925). Src binding to FAK Y397 residue enhances Src kinase activity, and Src phosphorylation of FAK Y576/577 residues increases FAK activity.

Several SH2 domain-containing adaptor or enzymatic proteins bind FAK and Src phosphorylated tyrosine residues

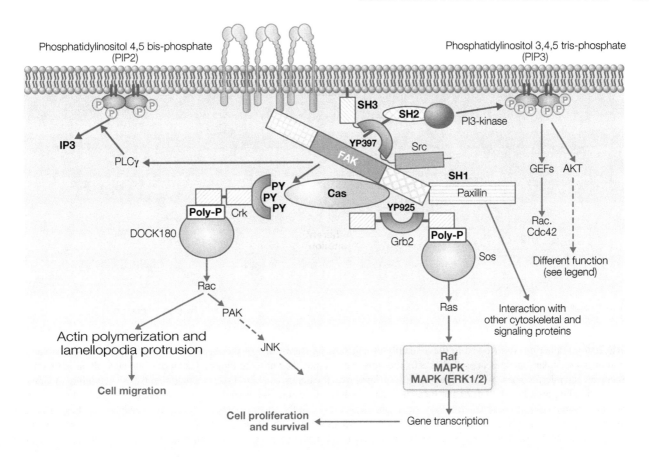

FIGURE 20.5 Model summarizing some of the mechanisms involved in signal transduction by integrins. Clustering of integrins bound to appropriate ligands results in the formation of multimolecular complexes containing cytoskeletal proteins (see text), FAK and Src. Phosphorylation of FAK Y397 allows binding of Src (see Fig. 20.4), phospholipase Cγ (PLCγ), and phosphatidylinositol 3-kinase (PI3 kinase). PLCγ triggers phosphoinositide turnover and generation of diacylglycerol and inositol 3-phosphate. PI3 kinase phosphorylates PIP2 at the 3 position of the inositol ring forming PIP3 that serves as interaction site with pleckstrin homology (PH) domain-containing proteins. Several guanine nucleotide exchange factors (GEFs), for example, Vav, have a PH domain. Another PH domain-containing protein implicated in the regulation of several cell responses is the serine/threonine (Ser/Thr) kinase Akt. PIP3-bound Akt becomes active upon phosphorylation by other kinases at specific Ser and Thr residues. Akt signal transduction regulates cell metabolism (glucose transport and metabolism, protein synthesis), apoptosis, and cell cycle. The protein Cas binds to FAK proline-rich sequences (see Fig. 20.4) and is phosphorylated by FAK and Src at several tyrosine residues. The SH2 domain-containing protein Crk binds to phosphorylated Cas and recruits the GEF for Rac DOCK180 via a SH3 domain–proline-rich region interaction. FAK phospho-Y925 binds the adaptor protein Grb2. Recruitment of Sos-1 (a RasGEF) by Grb2 results in the activation of one of the MAP kinase cascades; other MAP kinases may be activated via Rac (see Chapter 10). As a whole, FAK and Src signal transduction regulates cell migration, proliferation, survival, and gene transcription. Blue arrows indicate protein phosphorylated by FAK (or Src) and enzyme substrates; black arrows indicate activation of GTPases and other pathways.

(Fig. 20.4). Additionally, proline-rich sequences present in both FAK and Src can bind SH3 domain-containing proteins (Fig. 20.4) (for more details on FAK and Src signal transduction, see Supplement E 20.3, "Signal Transduction by FAK and Src").

Signal Transduction by Integrins and PSGL-1 in Leukocytes In leukocytes, adhesion receptors transduce intracellular signals via a specific module involving Src-family kinase members, adaptors containing an immunoreceptor tyrosine-based activation motif (ITAM), and the cytoplasmic tyrosine kinase Syk. This module was initially characterized in innate and adaptive immune cells in the context of the so-called immune receptors.

Antigen receptors in T and B lymphocytes (TCR and BCR, respectively) and receptors for immunoglobulin G or E (Fcγ or Fcε receptors, respectively), expressed in phagocytic cells or mast cells, activate members of the Src family of tyrosine kinases (see Fig. 20.4). These in turn phosphorylate two tyrosine residues in the ITAM present in the cytoplasmic tail of adaptor molecules (the heterodimers γ/ε or δ/ε or the homodimers ζ/ζ in T lymphocytes, the

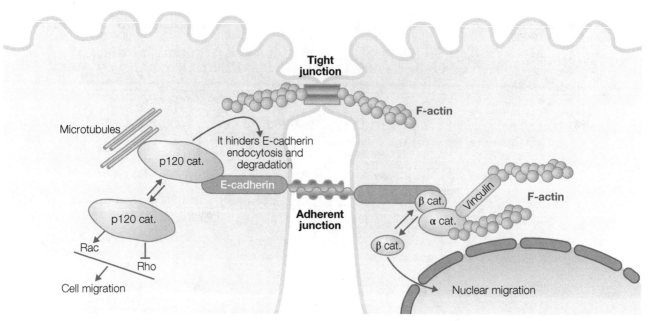

FIGURE 20.6 Cadherins are involved in the assembly of adherens junctions, one of the three molecular complexes responsible for the formation of intercellular junctions in epithelial cells (desmosomes are not shown in the figure). Cadherins interacts, albeit indirectly, with actin filaments and microtubules. α-Catenin, which is associated with β-catenin, can bind F-actin either directly or via binding to vinculin or other F-actin-binding proteins. Both p120 and β-catenin can switch from a cadherin-bound to a free cytosolic form. p120 catenin binding to cadherin cytoplasmic tail prevents cadherin endocytosis and degradation. In the free cytoplasmic form, p120 activates GTPases Cdc42 and Rac (which promote actin polymerization and lamellipodium formation) and inhibits Rho (which regulates stress fiber formation and cell arrest). Thus, as a result of Cdc42/Rac activation and Rho inhibition, p120 catenin promotes cell migration. β-Catenin detached from the cadherin cytoplasmic tail can migrate into the nucleus and act as a transcription factor regulating the expression of genes implicated in cell cycle regulation. A typical aspect of malignant cells is the so-called epithelial–mesenchymal transition. This phenomenon is thought to depend on a vicious cycle in which reduced expression of cadherins leads to an increase in the free cytoplasmic forms of p120 and β-catenin and consequent promotion of cell migration and cell proliferation.

heterodimers Igα/Igβ in B lymphocytes, or the homodimers γ/γ in innate immunity cells). In turn, the tyrosine-phosphorylated ITAMs interact with cytoplasmic tyrosine kinase ZAP-70 in T lymphocytes and Syk in B lymphocytes and innate immunity cells. ZAP-70/Syk binding to the phosphorylated ITAM tyrosine residues, and their tyrosine phosphorylation by a Src-family kinase member or another cytoplasmic tyrosine kinase (i.e., Abl), results in their full activation and phosphorylation of their downstream targets.

Recent studies have shown that the same module we have just described is also used by adhesion receptors regulating myeloid cell and platelet responses. For example, the mucin PSGL-1 (a P- and E-selectin ligand that regulates the rolling step in leukocyte recruitment; see Fig. 20.2) can activate a Src-family kinase, thus triggering phosphorylation of the ITAM present in the adaptor DAP12 or the Fcγ chain (see preceding text). This results in activation of the tyrosine kinase Syk. Also, signal transduction by β2 and β3 integrins in myeloid cells or platelets has been reported to involve a Src/ITAM-containing adaptors/Syk module.

Signal Transduction by Cadherins

Figure 20.6 summarizes some of the mechanisms mediating cadherin signal transduction. It is important to note that intracellular signals delivered by cadherins have been implicated in the stabilization of adherens junctions and other forms of cell–cell interactions (tight junctions and desmosomes). Hence, the concept has emerged that alterations of cadherin expression and/or signaling capacity are a distinctive feature of invasive, neoplastic cells.

Signal Transduction by CD44

CD44 (see Fig. 20.1) transduces intracellular signals via different direct or indirect mechanisms. For example, it can serve as a platform for the binding and proteolytic activation of GFs. Additionally, some CD44 splicing variants (see Fig. 20.1) may act as a GF coreceptors and thus favor GF receptor signaling, leading to increase proliferation and invasiveness. In analogy with integrins and cadherins, CD44 can organize multimolecular complexes by binding cytoskeletal proteins of ezrin, radixin, and moesin (ERM) family, filamentous actin, and other proteins.

ADHESION RECEPTORS AS DRUG TARGETS

Therapies have been developed to inhibit adhesive interactions that employ monoclonal antibodies or peptides capable of interacting with the integrin binding sites, thereby competing with the natural ligands. These therapies are used to inhibit leukocyte recruitment and platelet aggregation (see Table 20.1). In addition, drugs targeting components of the signal transduction pathways triggered by adhesion molecules have also been discovered. Among these, those targeting cytoplasmic tyrosine kinases have attracted a particular interest. Following the success of imatinib mesylate (Gleevec), an inhibitor of the cytoplasmic tyrosine kinase Abl used in the treatment of chronic myeloid leukemia (CML), second-generation inhibitors also targeting Src have been discovered (dasatinib, bosutinib). These drugs have started to be used in the therapy of human malignancies. Furthermore, FAK inhibitors are currently under study as potential antineoplastic drugs. Due to the role of Src-family tyrosine kinases Syk and Abl in the regulation of leukocyte

TABLE 20.2 Therapeutic targets and drug choice criteria

Main therapeutic target	Cellular effects	Clinical effects	Indications	Notes and choice criteria
α4β1 integrin[a]	Inhibition of α4β1 integrin–VCAM interaction	Reduced mononuclear leukocyte recruitment in autoimmune and inflammatory diseases	Multiple sclerosis	Natalizumab (Tysabri, Antegren): anti-α4 chain antibody
αLβ2	Inhibition of LFA-1 (CD11a/CD18)–ICAM interaction	Reduced leukocyte recruitment into the skin	Psoriasis	Efalizumab (Raptiva): anti-αL antibody
αIIbβ3 integrin[b]	Inhibition of αIIbβ3 integrin interaction with fibrinogen and other adhesion molecules	Reduced platelet adhesion and aggregation	Interventive coronary disease and acute coronary syndromes	Abciximab (ReoPro): anti-αIIbβ3 integrin antibody Tirofiban (Aggrastat): a nonpeptide molecule, chemically described as N-(butylsulfonyl)-O-[4-(4-piperidinyl)butyl]-L-tyrosine monohydrochloride monohydrate; antagonist of αIIbβ3 integrin Intrifiban (Integrilin, eptifibatide): cyclic heptapeptide containing 6 amino acids and 1-mercaptopropionyl (des-amino cysteinyl) residue; antagonist of αIIbβ3 integrin
α5β1[c]	Inhibition of α5β1–fibronectin interaction	Inhibition of neoplastic cell angiogenic activity and survival	Cancer: renal cell carcinoma, melanoma, lung and pancreatic cancers	Volociximab (M 200): anti-α5β1 antibody
αv and αvβ3[c]	Inhibition of αv-dependent adhesion	Inhibition of neoplastic cell angiogenic activity and survival	Cancer: melanoma, ovarian and prostatic cancers	Intetumumab (CNTO-95): anti-αv antibody Etaracizumab (Abegrin): anti- αvβ3 antibody
Selectins (L-, P-, and E-selectins)[d]	Inhibition of selectin-mediated interactions	Inhibition of leukocyte recruitment into inflammatory sites, inhibition of metastatic cell homing in different tissues	Autoimmune and inflammation-based pathologies; cancer	Cylexin (CY-1503), Bimosiamose (TBC-1269): glycomimetic antagonists

[a] Other inhibitory Ab or small molecules affecting α4β1/VCAM interaction are currently under study for the treatment of asthma, arthritis, and Crohn's disease.
[b] Other small inhibitory molecules affecting αIIbβ3 integrin function are under trial or have been approved for the treatment of thrombotic pathologies.
[c] Under clinical trial.
[d] Glycomimetic antagonists are under clinical trial.

adhesion, recruitment, and activation, it is predictable that tyrosine kinase inhibitors may find a role also in the treatment of autoimmune and inflammatory diseases (Table 20.2).

TAKE-HOME MESSAGE

- The adhesion between cells and between cells and extracellular matrix regulates important cellular functions such as building of tissues, recruitment and diapedesis of cells from vessels, and repair of vascular damage.
- Cell adhesion is mediated by several receptors differing for molecular structure and signal transduction.
- Adhesion mediated by integrins is modulated by "inside-out signaling" mechanisms that involves the generation of intracellular signals. Agonists induce a conformational change of the integrin that increases its affinity for the extracellular ligands.
- After interaction with the ligand, adhesion molecules, such as integrins, mucins, and cadherins, induce formation of supramolecular complexes with proteins involved in signal transduction and interaction with the cytoskeleton.
- Cellular signals generated by adhesion molecules may regulate proliferation, survival, and movement. Drugs interfering with integrin-dependent adhesion inhibit leukocyte recruitment in autoimmune diseases and platelet aggregation in thrombotic diseases.

FURTHER READING

Brieher W.M., Yap A.S. (2013). Cadherin junctions and their cytoskeleton(s). *Current Opinion in Cell Biology*, 25, 39–46.

Frame M.C., Patel H., Serrels, B., Lietha D., Eck, M.J. (2010). The FERM domain: organizing the structure and function of FAK. *Nature Reviews in Molecular Cell Biology*, 11, 802–814.

Legate K.R., Fässler, R. (2009). Mechanisms that regulate adaptor binding to beta-integrin cytoplasmic tails. *Journal of Cell Science*, 122, 187–198.

Ley K., Laudanna C., Cybulsky M.I., Nourshargh S. (2007). Getting to the site of inflammation: the leukocyte adhesion cascade updated. *Nature Reviews in Immunology*, 7, 678–689.

Moser M., Legate K.R., Fassler R. (2009). The tail of integrins, talin and kindlins. *Science*, 324, 895–899.

Perez-Moreno M., Jamora C., Fuchs E. (2003). Sticky business: orchestrating cellular signals at adherens junctions. *Cell*, 112, 535–548.

Provenzano P.P., Keely, P.J. (2009). The role of focal adhesion kinase in tumor initiation and progression. *Cell Adhesion and Migration*, 3, 347–350.

Weisberg E., Manley P.W., Cowan-Jacob S., Hochhaus A., Griffin, J.D. (2007). Second generation inhibitors of BCR-ABL for the treatment of imatinib-resistant chronic myeloid leukaemia. *Nature Reviews Cancer*, 7, 345–356.

Wolfenson H., Lavelin I., Geiger B. (2013). Dynamic regulation of the structure and functions of integrin adhesions. *Developmental Cell*, 24, 447–458.

21

SOLUBLE CYTOKINE RECEPTORS AND MONOCLONAL ANTIBODIES IN PATHOPHYSIOLOGY AND THERAPY

ALBERTO MANTOVANI AND ANNUNCIATA VECCHI

> **By reading this chapter, you will:**
>
> - Understand the concept of decoy receptor
> - Know how soluble receptors are generated
> - Understand the potential of monoclonal antibodies as therapeutic tools

Antibodies and soluble receptors constitute a strategy to capture and block the action of mediators, in particular polypeptide mediators such as cytokines and growth factors, before they interact with their endogenous receptors. This strategy is conceptually simple. However, with the exception of serum therapy against toxins and poisons, which pertain to a different sector from the one we are dealing with, blocking agonists have not received much attention in the classical pharmacological approaches. With the advent of the cytokine era, this picture has significantly changed. Blocking antibodies are the anticytokine strategy that has been most extensively used in human diseases, with important results, thus constituting a paradigm on which therapeutic strategies have been later compared and tested. Indeed, monoclonal antibodies directed against tumor necrosis factor (TNF) have formed a paradigm, demonstrating the therapeutic potential of these new pharmacological tools. Although the mechanism of action of blocking antibodies and soluble receptors is intuitively simple, reality may be more complex. In fact, antibodies and soluble receptors may have complex interactions with their agonist, acting as a delivery system or prolonging agonist half-life. The possibility to use these molecules as tools for pharmacological intervention has contributed to the development of better therapeutic agents and has allowed revealing unsuspected complexity in their interaction with specific agonists.

SOLUBLE RECEPTORS

G-protein-coupled receptors, being characterized by a structure with seven transmembrane domains, cannot be produced nor engineered as soluble molecules with good affinity for their agonists. Therefore, our discussion on soluble receptors is restricted to other receptor classes. In theory, soluble receptors for polypeptide factors such as cytokines should retain the ligand specificity and affinity typical of the plasma membrane-associated receptors. Moreover, they should not be recognized as foreign by the immune system and should not raise blocking antibodies. At the same time, always in theory, a major disadvantage of soluble receptors is their rapid clearance and their short half-life in circulating blood. To overcome this limitation, soluble receptors have been coupled with the Fc portion of IgG class (often IgG1) immunoglobulins that confer a longer half-life, in general of the order of many days. Moreover, fusion proteins between soluble receptors and immunoglobulins are divalent, thus having a greater avidity for the agonists even when expressed in transmembrane form (an example is TNF that before being secreted is expressed as a type 2 transmembrane protein).

Generation of soluble receptors is part of the homeostatic mechanisms of action of growth factors and cytokines. Therefore, understanding the pathophysiology of soluble receptors is essential to learn how they can be pharmacologically exploited.

The Paradigm of Viral Receptors for Cytokines

In their evolutionary relationship with organisms, viruses have developed strategies to regulate and control cell communication systems and in particular inflammatory cytokines. Indeed, viruses behave as "molecular pirates" that capture, adapt, and exploit cellular receptors to subvert cell communication and defense systems.

Viruses represent a paradigm of the *in vivo* ability of soluble cytokine receptors to modulate cytokine function. In the course of their evolution, many viruses, in particular those belonging to the poxvirus family, have captured cytokine receptor-encoding genes and have used them to subvert host inflammatory and immunological responses. Viruses have developed several strategies to gain control of cytokines and their receptors: for example, viruses may have in their genome genes encoding for immunosuppressive cytokines (e.g., IL-10 in Epstein–Barr virus) or soluble receptors for cytokines and cytokine binding molecules, as well as transmembrane cytokine receptors (Table 21.1). Thus, although experimental studies on soluble cytokine receptors suggest that these molecules may have a complex role that includes agonist presentation to signaling molecules that modulate agonist half-life and its protection, analysis of naturally occurring responses to viruses suggest that soluble cytokine receptors may constitute a mechanism to negatively regulate cytokine action, at least for cytokines such as interleukin-1 (IL-1), TNF, and interferons.

"Decoy" Receptors: Molecular Traps for Agonists

Decoy receptors constitute a general strategy to modulate cytokine and growth factor activity (Fig. 21.1). Ligand recognition, usually with high specificity and affinity, and response (via signal transduction) are two essential characteristics of receptors, according to the classical definition given by Langley in the late nineteenth century. Decoy receptors represent a deviation from this as they trap the ligand and/or essential components of the signal transduction cascade activated by the receptor, thereby preventing normal response. In other words, they are "baits and switches" (decoys) that serve as molecular traps for agonists and fundamental signal transduction components. Recent results suggest that these molecules perform both these functions. Decoy receptors differ from soluble receptors in that they do not mediate signaling after ligand binding and they are encoded by specific genes whose sole function is to constitute a negative control loop for cytokines and growth factors.

The paradigm of decoy receptors was originally formulated for the type II IL-1 receptor (Fig. 21.1). This receptor can be found either in a membrane-associated form or in a

TABLE 21.1 Cytokine receptors of viral origin

Protein	Virus	Specificity
B15R	Vaccinia	IL-1β, not IL-1α, IL-1ra
T2	Shape papilloma	TNFα and β
TNFR (rabbit)	Myxoma	TNFα and β
B28R	Vaccinia	TNFα and β
crmB	Cowpox	TNFα and β
G2R/G4R	Variola	TNFα and β
M-T7	Myxoma	IFNγ, chemokines
B8R	Vaccinia	IFNγ
B18R[a]	Vaccinia	IFNα/β
US28	CMV	CC chemokines[b]
ECRF3	Herpesvirus saimiri	CC chemokines
KSV-GPCR	KSV/HHV8	CC chemokines[c]

[a] Released and membrane.
[b] Membrane, transduces signal.
[c] Constitutively active, induces proliferation.

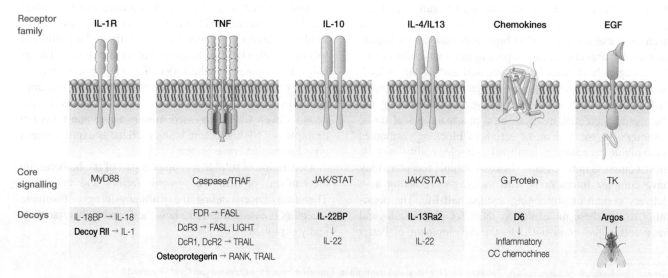

FIGURE 21.1 Decoy receptors belonging to different families of cytokines and growth factors (From Mantovani et al. (2001)).

free form released in biological fluids, either alone or complexed with an accessory protein (AcP). It binds the agonists IL-1β and IL-1α and prevents them from recognizing the signaling receptor formed by IL-1RI and AcP chains. Interestingly, this receptor (called IL-1 decoy) recognizes the two agonists but does not recognize IL-1 receptor antagonist (IL-1ra), the molecule with IL-1-antagonistic activity. It is worth mentioning that IL-1ra is now a drug approved for clinical use for the treatment of rheumatoid arthritis, an autoimmune disease in which IL-1 plays a key role.

In this way, the two systems negatively regulating IL-1 do not interfere with each other: one blocks the agonist molecules (the decoy receptor), and the other blocks the receptor (IL-1ra). Moreover, the IL-1 decoy receptor acts also as a dominant negative, as it forms a molecular complex with AcP and agonist, thereby subtracting them to the signaling receptors.

The paradigm of decoy receptors has been subsequently extended to other members of the IL-1 family and other cytokines and growth factors; indeed, in the IL-1 family, there is a decoy receptor for interleukin-18 (IL-18) (IL-18 binding protein). Furthermore, decoy receptors have been found in the TNF (e.g., osteoprotegerin), IL-10, IL-4, and IL-13 (IL-13Rα2) receptor families as well as in the TGFβ family. To summarize, decoy receptors differ from soluble receptors as they are genetically designed to serve solely as molecular traps and represent a general strategy for regulating inflammatory cytokine and growth factor functions.

More recently, it has been discovered that atypical seven-transmembrane domain receptors exist that do not mediate conventional activation signals. In particular, D6 receptor is a decoy and scavenger for inflammatory chemokines (Fig. 21.1). In addition, decoy receptors have been discovered in *Drosophila melanogaster*, indicating that decoy receptors constitute an evolutionary conserved strategy to modulate cytokine and growth factor activity. Therefore, the pharmacological use of soluble receptors is based on a pathophysiological control mechanism of cytokine and growth factor receptors.

Soluble Receptors: Mechanisms of Generation and Action

Soluble forms of signal-transducing receptors are produced through different mechanisms, including production of alternatively spliced mRNA transcripts and proteolytic cleavage or release of membrane receptors by phospholipase C. Table 21.2 summarizes the various mechanisms and molecules involved.

Alternative splicing gives rise to soluble receptor forms through different mechanisms. The simplest mechanism consists in the exclusion of exon encoding the receptor transmembrane portion; for example, this is the case of the GM-CSF receptor α chain. Another mechanism involves

TABLE 21.2 Mechanisms generating soluble cytokine receptors

Proteolytic cleavage	TNF, RI and RII
	IL-1 RII
	IL-2Rα
	IL-6Rα
	M-CSFR (fms)
	c-kit
Phospholipase	CNTFR
Alternative splicing	IL-4Rα
	IL-5Rα
	GM-CSFRα
	IL-7R
	IL-9
	LIF-R
	IFNα/βR
	c-kit

inclusion in the mature transcript of a "soluble exon" that causes arrest of polypeptide chain synthesis before the transmembrane region-encoding exon. This type of mechanism generates soluble receptors for IL-4, IL-5, and leukemia inhibitory factor (LIF).

Proteolytic enzymes are involved in the release of the soluble forms of TNF type 1 (or p55) and type 2 (or p75) receptors, IL-1 decoy receptor, and IL-2 receptor α chain. By contrast, the IL-6 receptor α chain can be released via both proteolytic cleavage and alternative splicing. Enzymes involved in shedding or proteolytic cleavage and release of membrane receptors have been only partially characterized. It is worth mentioning that TNF-converting enzyme (TACE), belonging to the ADAM family (ADAM17), causes membrane release of both TNF and receptor. In general, the enzymes belonging to the matrix metalloproteinase (MMP) family play a fundamental role in release of membrane receptors.

Soluble Receptors: Antagonist Effects

Virtually, all soluble receptors for cytokines and growth factors, including those having a relatively low affinity for ligands compared to membrane receptors, can inhibit *in vitro* binding and biological activity of cytokines and growth factors. In general, agonist binding to a soluble receptor prevents its binding to a functional membrane receptor, thereby blocking signaling and subsequent responses. From this point of view, soluble receptors block cytokine and growth factor action in a competitive manner, and their inhibitory effect depends on their relative concentrations and affinities. However, as discussed previously, at least some decoy receptors (IL-1, IL-18) behave as dominant negative, subtracting essential components to the signaling receptor complex (AcPs in these cases). Therefore, the effect of these molecules is higher than expected from simple competitors.

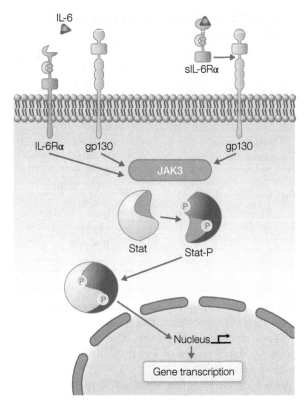

FIGURE 21.2 IL-6 and its receptor. IL-6 interacts with a membrane receptor consisting of two subunits, gp130 and IL-6Rα. Each subunit contains different molecular domains. gp130 is common to different receptors recognized by cytokines similar to IL-6. gp130 can also cooperate with the soluble form of IL-6Rα (sIL-6Rα) present in plasma and transduce signal in a process called "transsignaling." In this process, sIL-6Rα presents the agonist to the gp130 signaling chain. This is important in the activation of the vascular endothelium and in the progression from acute to chronic inflammation.

Soluble Receptors: Agonist Effects

Although soluble receptors generally behave as inhibitors, some important exceptions exist. The α chains of at least some IL-6 family members, even in soluble form, bind the agonist and present it to the gp130 chain that activates signal transduction (Fig. 21.2). This phenomenon is called transsignaling. Furthermore, at least in some experimental conditions, paradoxical phenomena of cytokine activity potentiation by their soluble receptors have been observed. This may result from different mechanisms, as prolongation of agonist half-life (e.g., TNF) or protection from proteolytic degradation. In conclusion, soluble receptors for cytokines and growth factors may have divergent effects on the biological activity of their ligands: on one hand, these receptors can prevent ligand interaction with signal-transducing receptors and behave as a dominant negative; on the other, in certain conditions, they can increase agonist stability and half-life. In general, the predominant effect is inhibitory.

However, the general concept that soluble receptors for cytokines or growth factors may have opposite effects on their ligands has implications for understanding their in vivo function and pharmacology.

Soluble Receptors as Pharmacological Agents

From a pharmacological point of view, soluble receptors can be used as such or as biotechnologically modified products. In addition, they can also represent a target for agents that increase their production. The IL-1 decoy receptor is an important example of this second possibility. IL-1 expression is strongly enhanced by signals such as inflammatory cytokines and in particular glucocorticoid hormones that have been shown, both in vitro and in vivo, to increase its production and release in animals and humans. It is likely that the increased production of IL-1 decoy receptor represents one of the mechanisms underlying glucocorticoid anti-inflammatory activity.

Soluble receptors for cytokines and growth factors are potential therapeutic agents to inhibit the activity of these endogenous mediators. When compared with blocking monoclonal antibodies, soluble receptors for cytokines have a similar ligand affinity but have the theoretical advantage of not being recognized as nonself by the immune system.

The main issue associated with the use of soluble receptors as drugs is their relatively short in vivo half-life, which usually does not exceed a few hours. By contrast, monoclonal antibodies have a half-life that is in the order of several days. Generation of fusion proteins between soluble receptors and the immunoglobulin Fc portion has circumvented this intrinsic limit of soluble receptors. The advantage of these hybrid molecules is constituted by an increased half-life, similar to that of antibodies. Furthermore, in principle, the dimeric nature of hybrid receptor constructs confers a greater avidity for agonists linked to the membrane, such as the 26kD transmembrane form of TNF, which is biologically active, or, more in general, for soluble multimeric agonists.

Soluble receptors for TNF have been the first molecules of this type to be translated into drugs for a human pathology. At the end of the 1980s, different groups discovered that human urine contains TNF inhibitory proteins. Using several hundred liters of human urine, the inhibitor molecule was purified and sequenced in Israel and shown to be a protein capable of binding TNF. This protein, called TNF-binding protein (TBP), was found to be the soluble form of type 1 and type 2 TNF membrane receptors (p55 and p75). Fusion proteins between soluble TNF receptors and the constant part of the immunoglobulin heavy chain were later generated, with the main purpose of extending the protein half-life in vivo. These fusion proteins exhibited anti-inflammatory and immunosuppressive activities in a variety of preclinical models of systemic and local inflammation. Today, Fc type

2–TNF receptor fusion proteins are approved for clinical use and, in particular, have proved effective in the treatment of rheumatoid arthritis, juvenile rheumatoid arthritis, and other immunoinflammatory diseases.

MONOCLONAL ANTIBODIES

At the end of the nineteenth century, Kitasato and von Behring conducted a pioneering work on the use of antisera against diphtheria toxin, thus establishing the principle that antibodies can be used as inhibitors of endogenous or exogenous toxic substances. Antibodies represent a classic and obvious strategy to inhibit toxic agents both exogenous and endogenous. The discovery of monoclonal antibody technology and more generally the advent of biotechnological techniques to engineer antibody molecules have opened up new avenues for the use of antibodies as blocking agents.

Monoclonal antibodies were discovered in the second half of the 1970s, but their translation into pharmacologically effective agents in humans occurred only in the 1990s. The major theoretical advantage of antibodies over, for example, natural soluble receptors is their long half-life in biological fluids. Moreover, high-affinity monoclonal antibodies can be obtained easily. The disadvantage of antibodies as blocking agents is that they can be antigenic and induce production of anti-antibodies inhibiting their activity. This limit is only partly overcome by humanization technologies, as production of antibodies against variable parts of antibodies inoculated in large quantities as drugs can always occur.

Different strategies of genetic engineering have been used to minimize antibody antigenicity. These include the construction of chimeric molecules consisting of the human immunoglobulin constant region fused to the Fv fragment of the mouse monoclonal antibody variable region. Furthermore, other techniques are available to humanize monoclonal antibodies, such as the transfer of complementarity-determining regions (CDRs) from mouse immunoglobulin genes into human immunoglobulin genes. The antibodies obtained in this way are 95% human as they contain only a small portion of murine sequences. In addition, it is possible to obtain fully human monoclonal antibodies using either classic cell technologies or molecular biology techniques to produce recombinant antibodies in bacteria. These are single-chain antibodies (single-chain variable fragment (scFv)). Finally, humanized monoclonal antibodies can also be produced in immunized, genetically modified mice.

At the time of writing, murine antibodies, hybrid antibodies, or humanized antibodies predominate in the field of therapeutic monoclonal antibodies.

A list of some monoclonal antibodies currently approved for pharmaceutical use in humans is provided in Table 21.3. As you can see, monoclonal antibodies have been applied as therapeutic agents in different areas such as cancer therapy, in particular lymphomas (e.g., anti-CD20 antibodies) and immunosuppression (anti-CD3 and anti-CD20 antibodies) via inflammatory cytokine blockade (in particular TNF, IL-6, and IL-1). The demonstration that anti-TNF antibodies can have a therapeutic effect in patients with rheumatoid arthritis and inflammatory bowel disease, such as Crohn's disease, has been crucial in promoting the development of anticytokine strategies based on soluble receptors or monoclonal antibodies. Therefore, therapeutic antibodies, initially developed to block exogenous toxins, are now undergoing a renaissance, in particular in those fields where they are employed to inhibit mediators (cytokines or growth factors) that in pathological conditions behave as endogenous "toxins."

TABLE 21.3 Selected Monoclonal Antibodies Currently in Clinical Use

Antibody	Target	Pathology
Alemtuzumab	CD52	B-CLL (2nd line)
Gemtuzumab	CD52	Acute myeloid leukemia
Rituximab	CD20	Non-Hodgkin lymphoma
Basiliximab	IL-2R	Graft rejection
Palivizumab	RSV	RS virus in children
Daclizumab	IL-2R	Graft rejection
Infliximab, adalimumab	TNF-α	Crohn's disease, rheumatoid arthritis
Nimotuzumab	EGFR	Squamous cell carcinoma, glioma
Bevacizumab	VEGF	Tumors and eye pathologies
Ipilimumab	CTLA-4	Metastatic melanoma

TAKE-HOME MESSAGE

- Decoy receptors uncouple the two basic properties of receptors: ligand recognition and signal transduction.
- Decoy receptors constitute a general strategy to modulate cytokine and growth factor activity.
- Decoy receptors trap ligands interfering with their binding to receptors.
- Decoy receptors can act as dominant negative subtracting accessory protein and agonists to the signaling receptors.
- Monoclonal antibodies are target specific and are used in the therapy of several autoimmune and tumor pathologies.

FURTHER READING

Hansel T.T., Kropshofer H., Singer T., Mitchell J.A., George A.J.T. (2010). The safety and side effects of monoclonal antibodies. *Nature Reviews Drug Discovery*, 9, 325–338.

Mantovani A., Bonecchi R., Locati M. (2006). Tuning inflammation and immunity by chemokine sequestration: decoys and more. *Nature Reviews in Immunology*, 6, 907–918.

Mantovani A., Dinarello C.A., Ghezzi P. *Pharmacology of cytokines*. Oxford University Press, Oxford, 2000.

Mantovani A., Locati M., Vecchi A., Sozzani S., Allavena P. (2001). Decoy receptors: a strategy to regulate inflammatory cytokines and chemokines. *Trends in Immunology*, 22, 328–336.

Mantovani A., Savino B., Locati M., Zammataro L., Allavena P., Bonecchi R. (2010). The chemokine system in cancer biology and therapy. *Cytokine and Growth Factor Reviews*, 21, 27–39.

Taylor P.C. (2010). Pharmacology of TNF blockade in rheumatoid arthritis and other chronic inflammatory diseases. *Current Opinion in Pharmacology*, 10, 308–315.

SECTION 5

MODULATION OF GENE ESPRESSION

22

PHARMACOLOGY OF TRANSCRIPTION

Roberta Benfante and Diego Fornasari

By reading this chapter, you will:

- Learn the molecular mechanisms of transcriptional regulation of gene expression
- Understand how gene expression can be modulated by drugs
- Learn how pharmacological modulation of transcription factors corrects altered gene expression in diseases

---

INTRODUCTION TO THE MECHANISMS OF TRANSCRIPTIONAL REGULATION

Transcription and translation are the biochemical events through which information encoded in genomic DNA sequence is converted into proteins. In multicellular organisms, all somatic cells contain the same genetic information but express only part of it. Indeed, some genes (such as those encoding enzymes of the main metabolic pathways) are expressed ubiquitously, whereas others are expressed only in specific cell types and tissues, such as globin in erytroid cells or albumin in the liver. In addition, some genes require appropriate extracellular stimuli to be expressed, such as interleukin-2 in T lymphocytes, or are expressed at different levels depending on cellular needs.

The large majority of molecular mechanisms that govern the level of protein expression operate by regulating gene transcription by RNA polymerase II (RNAP II), the enzyme that catalyzes DNA-dependent synthesis of messenger RNAs (mRNAs).

General Transcription Factors

General transcription factors (GTFs) mediate promoter recognition and proper positioning of RNAP II on gene regulatory regions. RNAP II is the multiprotein enzymatic complex that catalyzes polymerization of ribonucleotides using a single strand of DNA as a template (Fig. 22.1). Transcription initiation requires recruitment and correct positioning of RNAP II on specific sequences contained in the regulatory region (promoter) of the gene to be transcribed. RNAP II recruitment and positioning are assisted by protein factors, defined polymerase II GTFs, and identified by letters (e.g., transcription factor IID or TFIID). More precisely, the preinitiation complex (PIC) is composed of the DNA sequence to be transcribed, RNAP II, and five GTFs (TFIID, TFIIB, TFIIF, TFIIE, TFIIH), each with distinct functional properties and roles. TFIID plays a fundamental role since it starts the orderly sequence of events that culminate in PIC formation (see the following). TFIID is a multiprotein complex composed of the TATA-box binding protein (TBP) and other factors called TBP-associated factors (TAFs). As suggested by its name, TBP recognizes a specific DNA sequence (TATA box) rich in adenine and thymine localized approximately 30 nucleotides upstream the transcription start site of most eukaryotic genes. TAFs constitute a homogeneous family of proteins significantly similar both in sequence and three-dimensional structure to histones. TBP–TATA-box interaction is the first event in gene transcription. Subsequently, TFIIB establishes direct interactions with both TBP and DNA sequences immediately upstream and downstream the TATA box and recruits RNAP II to the promoter. Such recruitment occurs

PHARMACOLOGY OF TRANSCRIPTION

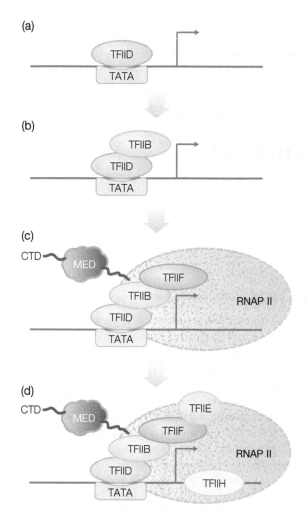

FIGURE 22.1 Preinitiation complex (PIC) formation. (a) The first event in PIC formation is the recognition of the TATA box by the general transcription factor TFIID, which consists of the TATA-box binding protein (TBP) and 12 TAFs (TBP-associated factors). (b) Interaction between TFIID and TFIIB determines recruitment and placement of RNAP II on the promoter. (c) The carboxy-terminal tail (CTD) of RNAP II binds Mediator (MED). (d) Entry of TFIIE and TFIIH completes PIC formation. CTD, carboxy-terminal domain; RNAP II, RNA polymerase II; TFII, transcription factor II.

primarily through interaction with TFIIF, which is associated with RNAP II. The final stage in PIC formation involves the recruitment of TFIIE and TFIIH. TFIIH is a nine-subunit complex and is the only GTF with defined enzymatic activities: (i) two helicase activities of opposite polarity, essential for unwinding and local denaturation of DNA, and (ii) a cyclin-dependent kinase (CDK) activity, responsible for the phosphorylation of serine residues in the carboxy-terminal portion of one of the RNAP II subunits, required for transition from transcription initiation to transcription elongation (Fig. 22.1).

The Promoter: A Multifunctional Region with Positive and Negative Regulatory Elements

Eukaryotic gene promoters contain specific DNA sequences (defined *cis*-acting element), usually, but not necessarily, located upstream the TATA box, which contribute to transcriptional regulation by binding transcription factors (defined *trans*-acting factor), other than GTFs, capable of either stimulating (activators) or inhibiting (repressors) transcription. Activators and repressors are able to directly bind DNA in a sequence-specific manner and to interact with PIC through general cofactors (TAFs and Mediator) and specific cofactors (see "Additional factors required to stabilize PIC structure"), promoting or inhibiting PIC formation and RNAP II activity.

The *cis*-acting elements present in its regulatory region determine how a given gene is regulated. Given the high number of activators and repressors existing, many different combinations of *cis*-acting elements are found within gene regulatory regions, each one determining the specific expression profile of the corresponding gene.

However, some common features can be found in the spatial and functional organization of promoters.

The Minimal Promoter The minimal (or core) promoter directs correct positioning of the PIC. It consists of sequences responsible for correct positioning of RNAP II and specific sites from which transcription starts (Fig. 22.2). Although the TATA box is important for TFIID binding, its presence has been observed only in 10–15% of tissue-specific gene promoters. Some genes use, as an alternative or in addition to the TATA box, different sequences known as initiator (Inr) and downstream promoter element (DPE). Promoters containing both these elements regulate transcription of highly expressed genes, whereas promoters lacking the TATA box but containing an Inr are associated with ubiquitously expressed genes, oncogenes, and transcription factor-encoding genes. DPE is typically found in promoters lacking TATA box (TATA-less) and requires for its function the presence of an Inr at a precise distance, critical for transcriptional activity. TBP is believed to play a fundamental role in PIC formation also in TATA-less promoter, although independently form direct DNA binding. Other elements are found upstream and downstream the TATA box, such as BRE^u and BRE^d, which are TFIIB binding sites, and downstream core element (DCE) (Fig. 22.2).

Other sequences contribute to the regulation of gene expression; these include both proximal and distal elements.

Proximal Promoter Elements Immediately upstream the core promoter (−40/−250 with respect to the transcription start site), usually, there are sequences binding ubiquitous activators, such as the GC box, which binds transcription

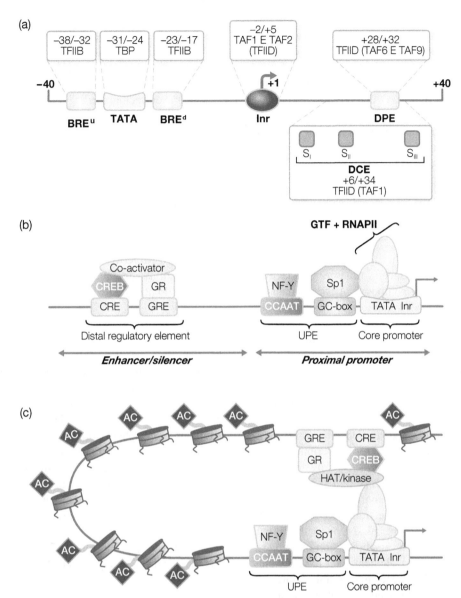

FIGURE 22.2 Structural organization of a gene regulatory sequence. (a) The core (minimal) promoter comprises DNA sequences responsible for correct positioning of RNA polymerase II and proper transcription initiation (+1, transcription start site). In many genes, it contains, besides the TATA box, other DNA elements capable of interacting with GTFs and TAFs. These elements can be found in some, but not all promoters, in different combinations. The properties of each minimal promoter are dictated by the presence/absence of these elements. TBP plays a fundamental role in PIC formation also in core promoter devoid of TATA box (TATA-less) by indirectly contacting DNA via TAFs. BRE^u, TFIIB recognition element upstream; BRE^d, TFIIB recognition element downstream; Inr, initiator; DCE, downstream core element; DPE, downstream promoter element; TAF, TBP-associated factors; TFIID, TFIIB, transcription factor II. (b and c) Promoter activity can be increased by ubiquitous transcription factors interacting with *cis*-acting DNA elements in the proximal regulatory region. These factors stabilize PIC by interacting with general cofactors (GTFs). Specificity and inducibility of gene expression is mediated by specific transcription factors binding to more distal region of the promoter (enhancer or silencer). These factors can recruit co-activators or co-repressor endowed with chromatin modifier activity that creates conditions suitable or unfavorable to transcription. For example, some co-activators with kinase activity stimulate transcription elongation by phosphorylating the C-terminal domain (CTD) of RNAP II. A widely accepted model to explain how interaction between factors bound to distal regions and RNAP II can occur suggests that the intervening sequences form a loop. CCAAT, CAAT box; CREB, cyclic AMP-responsive element-binding protein; GR, glucocorticoid receptor; GRE, glucocorticoid response element(s); HAT, histone acetyltransferase; UPE, upstream promoter elements.

factor Sp1, and the CAAT box, which can bind different transcription factors, including NF-Y and NF-1 (Fig. 22.2). All these sequences, often collectively called upstream promoter elements (UPE), along with the core promoter constitute the promoter itself. The GC box often represents the only element present in promoters of ubiquitously expressed and scarcely regulated genes; in these cases, more than one GC box is present allowing the binding of multiple Sp1 molecules, which synergistically stabilize PIC, thus ensuring a sufficient level of transcription. GC boxes can be found within GC-rich DNA regions called CpG islands, whose methylation status affects the expression level of the associated genes. Hypomethylation of CpG islands is associated with gene transcription, hypermethylation to gene silencing.

Distal Promoter Elements The human genome encodes at least 1700–1900 sequence-specific transcription factors. These proteins are generally composed of two domains: a DNA binding domain (DBD) that recognizes and binds specific DNA sequences and a regulatory domain whose primary function is to recruit co-factors, either having chromatin modification and remodeling activities or directly interacting with RNAP II.

Fundamental condition for transcription to take place is DNA accessibility to PIC-forming factors. Chromatin condensation prevents interaction between DNA and transcription factors; therefore, a decondensation process needs to occur for DNA to become available for transcription factor and RNAP II binding (see "Control of specificity and inducibility of gene expression"). Enhancers are DNA sequences that recruit transcription factors promoting chromatin decondensation. These factors allow the transition from a transcriptionally repressed state to a permissive one and facilitate assembly of the transcription machinery on promoters. Conversely, silencers are DNA elements that bind proteins and/or factors that inhibit gene expression acting on chromatin structure. Both enhancers and silencers can be found up to 100 kb away from the minimal promoter (Fig. 22.2). A widely accepted model to explain how such distant elements can interact with RNAP II suggests that DNA sequences interposed between them and promoters can form a loop (Fig. 22.2).

Additional Factors Required to Stabilize PIC Structure

Although RNAP II and GTFs are sufficient to promote basal transcription in *in vitro* reconstituted systems, PIC is a labile structure, which is believed to be only partially or rarely formed *in vivo* or with a very short half-life.

Indeed, one of the most consistent hypotheses about the molecular activity of activators suggests they stimulate gene transcription by stabilizing PIC. However, using *in vitro* reconstituted systems, it has been shown that activators are unable to stimulate basal transcription carried out by PIC, indicating that other factors must be involved. This has led to the discovery of the global or general co-factors.

Transcriptional co-factors are proteins that do not bind directly DNA, but form a bridge between activator(s) and PIC, often establishing contacts with TBP.

General co-factors can be grouped into two major classes: TAFs and proteins forming the Mediator complex.

At least 12 different TAF isoforms have been identified in yeast, *Drosophila*, and humans. Each of them can interact with TBP on one side and with a partially specific group of activators on the other (Fig. 22.3). Initially, TAFs were considered merely co-factors absolutely required for mediating the effects of activators. Subsequently, it was shown that TAFs can directly recognize DNA elements such as Inr or DPE, playing a key role in all TATA-less promoters (Fig. 22.2). Moreover, it is now believed that the role of TAFs is not as general as initially thought, since promoters exist that do not require them to respond to activators.

Mediator

Mediator is a multimodular complex resulting from the assembly of different proteins (in humans over 25, with a total mass of 1 MDa) in different combinations and stoichiometries. Mediator does not directly contact DNA, as expected for a cofactor. It physically interacts with RNAP II, stimulating phosphorylation of its carboxy-terminal domain (CTD), an event considered essential for transcription initiation and progression. Depending on its specific protein composition, Mediator may interact with both activators and repressors, mediating their effects on PIC and RNAP II (Fig. 22.3). However, Mediator can stimulate basal transcription also in the absence of activators, and for this reason, it is regarded as a constituent element of PIC, along with RNAP II and GTFs (Fig. 22.1).

Control of Specificity and Inducibility of Gene Expression

Transcription is a finely controlled process in terms of tissue specificity and/or ability to respond to extracellular stimuli. This is due to regulatory elements, such as enhancer and silencer, generally located upstream the promoter, which bind tissue-specific and/or inducible activators or repressors (Figs. 22.2 and 22.3). Often, the activity of these transcription factors is mediated by specific co-factors, either co-activators or co-repressors.

It is worth remembering that nuclear DNA is not "naked" and therefore freely accessible to general or promoter-specific transcription factors, but it is organized in a nucleoprotein structure known as chromatin. The basic unit of chromatin is the nucleosome that consists of 146 base pairs of DNA wrapped two times around a protein core, containing two copies of each of the four histone proteins: H2A, H2B,

INTRODUCTION TO THE MECHANISMS OF TRANSCRIPTIONAL REGULATION   247

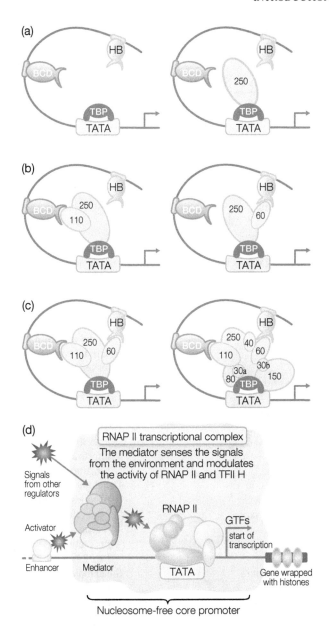

FIGURE 22.3 Functional role of Mediator and TAFs. The activators BCD and HB are unable to stimulate transcription of the target gene in the absence of specific TAFs. (a) TAFs, identified by numbers corresponding to their molecular weights, connect activators with TBP, allowing PIC stabilization. BCD and HB can stimulate transcription (b) separately or (c) together, in the latter case producing a synergistic effect. (d) Mediator is a multiprotein complex able to integrate different signals directed to RNAP II by physically interacting with it and by modulating the phosphorylation state of its CTD. Mediator also interacts with other PIC components.

H3, and H4. Transcription requires DNA to be accessible to GTFs, RNAP II, and sequence-specific activators but also to gradually denature, and subsequently renature, for the entire length of the region to be transcribed. The compact structure of chromatin prevents these processes, making transcription impossible. Therefore, a prerequisite for gene transcription to occur is chromatin reorganization, in particular at the level of its basic unit, the nucleosome. Co-factors can modify chromatin structure, thereby facilitating or inhibiting access of different transcriptional proteins to DNA.

Many of these specific co-factors have been identified while studying intracellular receptors that modulate gene expression in response to extracellular stimuli, often of hormonal nature (see Chapter 24 and the following).

To date, more than 300 specific co-factors (also called co-regulators) have been identified. They are involved in regulation of many physiological functions, and alteration of their activity has been associated with several pathological conditions, making them potential targets for the development of new therapies, in particular against tumors (Table 22.1; see Supplement E 22.1, "Pharmacoepigenomics"). Although initially co-factors seemed to lack common structural determinants, the analysis of their sequence has allowed identifying a large variety of enzymatic activities associated with these proteins, indicating that, besides bridging activators and repressors to PIC, they play an active role in the control of gene transcription. In addition, co-factors carry out their activities as part of large multiprotein complexes, in which more than one enzymatic activity may be present (Table 22.2). The ability to assemble multiprotein complexes makes these factors capable of transducing external signals in a precise and controlled manner, thereby increasing the complexity of the response through the temporal and tissue-specific regulation of specific genes. Co-factors can be classified into two groups, according to their activity: co-factors that covalently modify histones (acetylation/deacetylation, methylation/demethylation, phosphorylation/dephosphorylation, ubiquitination/deubiquitination, SUMOylation, and ADP-ribosylation) and co-factors that remodel chromatin in an ATP-dependent manner and control access of transcription factors and basal transcription machinery to promoters (see the following). Other factors play a fundamental role in assembly and recruitment or release of co-regulator complexes.

Among the first cloned co-factors, we remember steroid receptor co-activator (SRC)-1, nuclear receptor co-repressor (NCoR), and silencing mediator of retinoid and thyroid receptor (SMRT). Often, co-factors do not show specificity for a particular transcription factor, but have a pleiotropic influence on the transcription of a large number of genes. Their specificity of action is controlled by post-translational modifications, such as phosphorylation, methylation, and acetylation, taking place in response to different extracellular stimuli. Such modifications serve as a molecular code to control co-factor activity and their interaction with different transcription factors. Therefore, co-factors work as coordination centers of large transcriptional programs involved in growth, differentiation, and metabolic functions (e.g., PGC-1α in the metabolism of carbohydrates and lipids) by integrating signals from multiple transduction pathways.

TABLE 22.1 Co-factors are key regulatory elements of gene expression and modification of their activity and/or expression level leads to disease

Co-activator	Activity	Physiological function	Pathological defect	Syndrome	Therapy
CBP/p300	Histone acetyltransferase		Defects in DNA acetylation	Rubinstein–Taybi syndrome, mental retardation, growth defects, skeletal malformations, defects in long-term memory	HDAC inhibitors to counteract lack of histone acetyltransferase activity
HDAC5	Histone deacetylase		Role in depression		HDAC inhibitors
MECP2	Recruits HDACs			Rett syndrome	n.d.
SRC-3	Complex containing kinase, ubiquitin ligase, ATPase, methyltransferase, acetylase activities	Role in estrogen receptor controlled gene expression	Amplified in steroid-dependent tumors (breast cancer). It confers resistance to tamoxifen		n.d.
PGC-1 α	n.d. Acts as a platform to recruit co-factors such as HDACs of Sirtuin family	Role in energetic metabolism, mitochondria biogenesis, fatty acid oxidation, and stimulation of gluconeogenesis	Type 2 diabetes, polycystic ovary syndrome		Metabolic disorders are treated modulating Sirtuin activity with resveratrol
E6-AP co-activator			Angelman syndrome		
Sirtuin	HDAC		Metabolic syndrome		Resveratrol, a SIRT1 activator, is used to improve metabolic functions
p160	HAT	Development of reproductive organs, metabolism, growth	Aberrant expression in tumors (breast, prostate, and ovary cancer)		Development of peptides capable of interfering with receptor/co-factor interaction (see Box 22.1)
LSD1	Demethylase		Altered expression in tumors (e.g., prostate cancer)		Small molecules identified homolog to LSD1 and inhibiting monoamine and polyamine oxidases

CBP/p300: cAMP enhancer binding protein/p300; MECP2, methyl-CpG-binding protein.

For example, phosphorylation of NCoR and SMRT co-factors controls their intracellular localization, inducing their translocation from the nucleus to the cytoplasm, thereby preventing their repressive activity on transcription. By contrast, phosphorylation of SRC-3 (by kinases like Iκκα and/or CDK2) determines its association with other specific members of the complex (such as histone acetyltransferase (HAT) p300 and CBP or the methyltransferase CARM1) and affects its affinity for transcription factors, thus influencing expression of its target genes. Post-translational modifications of co-factors can also affect their stability. As a consequence, co-factor activity can be regulated also by proteasome degradation, via ubiquitin-dependent or ubiquitin-independent mechanisms. Many co-activators have an intrinsic E2 and E3 ubiquitin ligase activity (see Chapter 30).

Coregulatory Complexes and Covalent Modification of Histones

This class of complexes is composed of factors capable of covalently modifying the N-terminal portion of histones by acetylation, phosphorylation, and methylation. Acetylation is certainly the best characterized of these mechanisms; 40 years have passed since the discovery that hyperacetylation of lysine residues in histone N-terminal tails is involved in transcription activation.

TABLE 22.2 Coregulators complexes

Cofactor	Complex	Enzymatic activity/function
BRG1	SWI/SNF	ATPase
CoREST	CoREST	n.d.
CTBP1, CTBP2	CoREST	NAD-dependent dehydrogenase
HDAC1, HDAC2	SIN3, NuRD, CoREST	Deacetylation
HDAC3	NCoR, SMRT	Deacetylation
HDAC4–HDAC11	§	Deacetylation
LSD1	CoREST, NuRD	H3K4 demethylation
NCoR, SMRT	NCoR, SMRT	n.d.
SIN3	Sin3, CoREST	n.d.
SIRT1–SIRT5	§	NAD-dependent deacetylation

The complexes of the co-factors possess multiple enzymatic activities, allowing coordinated epigenetic modifications. NCoR and SMRT have high sequence homology at the protein level and their complexes possess similar activities and are always stably associated with HDAC3, responsible for deacetylase activity associated with events of repression of these complexes. n.d.: no specific enzymatic activity described; §: it has not been described association with any of the major co-repressor complexes. BRG1, Brahma-related gene-1; CoREST, RE1 silencing transcription factor/neural-restrictive silencing factor corepressor; CTBP1/2, C-terminal-binding protein 1/2; HDAC, histone deacetylase; LSD1, lysine specific demethylase 1; N-CoR, nuclear receptor corepressor; SIRT1-5, sirtuin 1-5; SMRT, silencing mediator of retinoid and thyroid receptor.

Acetylation has been suggested to promote transcription through multiple mechanisms. First of all, by neutralizing lysine charge, it probably destabilizes histone–DNA interaction, leading to exposure of nucleosomal sequences. Since nucleosomes interact with each other through histone tails, acetylation could help, or even anticipate, chromatin decondensation by contributing to the destruction of such superstructure. The acetylation status of a gene, and in particular of its promoter, is the result of the opposite activities of two complexes: HATs and histone deacetylases (HDACs). Nuclear HATs include three protein families: GNAT, MYST, and P300; GNAT and MYST families include more than one member. All nuclear HATs possess a highly conserved region for acetyl-CoA binding and are individually active on histones *in vitro*. However, *in vivo*, they operate as part of multiprotein complexes, cooperating with many accessory proteins. Based on their composition, distinct HAT complexes have been identified, each one showing different specificities for histones and different biological functions. In vertebrates, the HDAC family is composed of 11 members (HDAC1–HDAC11), subdivided into three subclasses: class I (HADC1, HDAC3, and HDAC8), class II (HDAC4, HDAC7, HDAC9, and HDAC10), and class IV (HDAC11). Class III consists of NAD-dependent deacetylases belonging to the Sirtuin family. HDAC1 and HDAC2 have been found in two multiprotein complexes, called Sin3 and NuRD, along with other proteins. Sin3 and NuRD have been

(a) Activator binding site

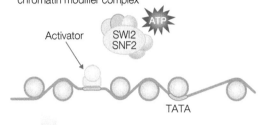

(b) The activator recruits the ATP-dependent chromatin modifier complex

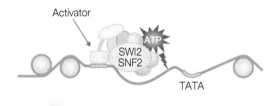

(c) The activator disassembles nucleosome structure by recruiting the ATP-dependent chromatin modifier complex

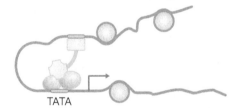

(d) The activator facilitates the assembly of general TFs

FIGURE 22.4 The ATP-dependent remodeling complexes regulate gene expression. (a) In basal conditions, the gene is not transcribed because its core promoter, and in particular its TATA box, is masked by nucleosomes. The binding site for the activator is however accessible. (b and c) Activator binding leads to recruitment of the ATP-dependent SWI2/SNF2 complex that causes nucleosome slippage, thereby exposing the TATA box and allowing PIC formation. (d) In this example, after binding, the activator makes contact with PIC. In other cases, the activator, through ATP-dependent remodeling complexes, mediates only chromatin remodeling.

frequently reported to be associated with transcriptional repressors, confirming that one of the mechanisms to silence or inhibit transcription involves stabilization of histone–DNA interaction via histone deacetylation.

Histone modifications represent a code (histone code) capable of directing specific and distinct transcriptional programs.

The ATP-Dependent Co-regulatory Complexes ATP-dependent complexes use energy derived from ATP hydrolysis to promote nucleosome "slippage" to expose or mask

PHARMACOLOGY OF TRANSCRIPTION

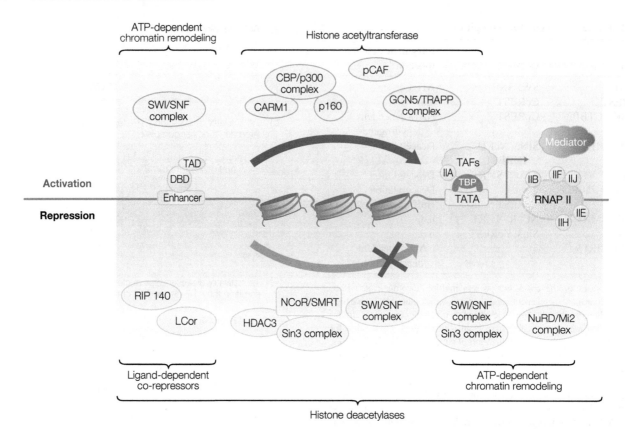

FIGURE 22.5 General model of gene expression regulation. Gene regulation results from integration of the activities of different factors. The recognition of specific sequences in the regulatory region of a gene by general and/or tissue-specific transcription factors is followed by recruitment of a large number of co-regulator complexes. Co-activator complexes include factors that contain chromatin remodeling activities, histone arginine methyltransferase, histone acetyltransferase, and components of the mediator complex mediating interaction with PIC. On the other hand, co-repressors include complexes with chromatin remodeling activity, basal co-repressors, such as NCoR and SMRT, which act as a platform for the recruitment of other complexes with histone deacetylase activities. Specific co-repressors, such as LCor (ligand-dependent nuclear receptor co-repressor) and RIP140 (receptor-interacting protein 140) for nuclear receptors, recruit other general co-repressors. Therefore, thorough and efficient regulation of gene expression in eukaryotes results from the action of two opposing molecular forces that determine activation or repression of specific set of genes depending on the extracellular signals. GCN5/TRAPP, general control 5 of amino-acid synthesis, yeast, homolog/transformation-transactivation domain-associated protein; RIP 140, receptor-interacting protein 140; NuRD/Mi2, nucleosome remodeling deacetylase complex; IIA, TFIIA; IIJ, TFIIJ.

specific DNA sequences. These complexes are multiprotein ATP-dependent structures, containing a central subunit endowed with ATPase activity and several accessory proteins. Three distinct families of ATP-dependent remodeling factors have been identified in humans and in other species (SWI2/SNF2, ISWI, and Mi-2) based on the structure of their central subunit, and several lines of evidence suggest that different complexes use distinct mechanisms to promote chromatin remodeling. Accessory proteins have been hypothesized to promote the localization of complexes at specific promoters, through interaction with specific activators. According to the current view, the mechanism would be as follows: a tissue-specific activator binds its DNA site localized in the region between two nucleosomes and is therefore accessible and recruits an ATP-dependent complex, which promotes PIC formation by unmasking nucleosomal sequences required for binding of TBP or ubiquitous activators (such as Sp1), leading to transcription of the gene (Fig. 22.4). This cascade of events has been demonstrated for MyoD, a transcriptional activator essential for skeletal muscle-specific gene expression, and for the entire differentiation process of skeletal muscle cells. Recruitment of the SWI/SNF complex has also been shown in the case of glucocorticoid and estrogen receptors, indicating that ATP-dependent chromatin remodeling is a control mechanism used also in transcriptional regulation by extracellular stimuli.

Toward a Unified Theory of Transcriptional Regulation

All regulatory processes described so far do not occur independently and separately from each other; a high degree of integration between them exists. For example, the two types of co-regulatory complexes frequently work together, acting in a different order depending on the gene considered. Peculiar is the case of the NuRD modifier complex. Indeed, NuRD contains HDAC1, HDAC2, but also Mi-2, which is the central subunit of an ATP-dependent complex. This means that NuRD can perform both activities, histone modification and chromatin remodeling, and belongs simultaneously to both types of complexes.

On the other hand, activators (and repressors) may regulate transcription simply by recruiting chromatin modifiers, thus exposing (or masking) DNA sequences for ubiquitous factors. Otherwise, they may contact PIC via either global co-factors or specific co-activators. Some coactivators are chromatin modifiers themselves: this is the case of CBP/p300, originally described as a co-activator and only subsequently characterized for its acetyltransferase activity (Fig. 22.5).

Therefore, there are many strategies of transcriptional regulation, all leading to the same final event: formation of a stable, properly positioned PIC enabling RNAP II to start mRNA synthesis.

TRANSCRIPTIONAL REGULATION BY EXTRACELLULAR STIMULI

Despite being an extremely complex process, transcription offers many opportunities for pharmacological intervention aiming at increasing or decreasing expression of specific genes involved in disease development. Although some cascades of transcriptional events seem to be poorly influenced by extracellular stimuli (this is the case of certain ontogenetic processes that seem to consist in the mere execution of an intrinsic genetic program), one of the main properties of the genome is its ability to react to information and requests coming from the extracellular environment by modifying transcription of specific genes. This can happen because the expression and/or function of many activators (and repressors) and co-factors is regulated by extracellular stimuli, making transcription of their target genes "sensitive" to extracellular cues and adjustable to cellular functional requirements.

Three Classes of Inducible Transcription Factors

Based on the molecular mechanisms responsible for their activation, we can distinguish at least three distinct classes of inducible transcription factors: (i) factors activated by ligand (i.e., receptors for steroid hormones; for an in-depth discussion of this class, see Chapter 24); (ii) factors activated by post-translational modifications, often consisting in phosphorylation of specific serine and threonine residues; and (iii) factors whose expression, and therefore activity, is regulated at transcriptional level (Fig. 22.6).

Serine and threonine residues are substrates of protein kinases activated by second messengers, such as PKA, PKC, calcium/calmodulin-dependent kinases (CaMK), MAP kinases, as well as phosphatases such as calcineurin (such as NF-AT). To the class of factors activated by post-translational modifications belongs cyclic AMP-responsive element-binding protein (CREB), whose activation makes transcription of some genes sensitive to changes in cAMP intracellular concentration (Fig. 22.7).

Given the high number of G-protein-coupled receptors that can activate adenylate cyclase, it is easy to understand how drugs blocking their activity ultimately interfere with genome activity by inhibiting CREB-dependent gene expression. This is exemplified by beta-blockers, drugs used in the treatment of heart failure that interfere with cardiac remodeling process by blocking CREB activity.

A particular subclass of transcriptional regulators activated by post-translational modifications is that formed by signal transducer and activator of transcription (STAT) proteins. These factors are found in latent form in the cytoplasm; when phosphorylated at specific tyrosine residues, they dimerize, via reciprocal recognition of phosphotyrosine residues through intrinsic SH2 domains, and migrate to the nucleus. Tyrosine residues can be directly phosphorylated by membrane receptors with intrinsic tyrosine kinase activity (e.g., EGFR) or by JAK, whose activation depends on stimulation of cytokine membrane receptors.

A particular case of transcription factor regulated by posttranslational modification is NF-κB. In unstimulated cells, NF-κB is found in the cytoplasm associated with IκB, an inhibitory protein masking NF-κB nuclear localization signal. NF-κB becomes activated when IκB is post-translationally modified by phosphorylation, resulting in NF-κB release and nuclear translocation.

As we said, the third class of inducible transcription factors includes proteins whose activity is regulated at transcriptional level. In contrast with transcription factors belonging to the first two classes that are always present in the cell (ready to be activated), factors belonging to the third class are not expressed in basal conditions, and their activation in response to appropriate stimuli occurs through induction of their gene transcription. Newly synthesized mRNAs migrate to the cytoplasm, where they are translated into proteins that are then transferred to the nucleus to regulate transcription of specific target genes. Transcriptional induction of transcription factors belonging to the third class is almost always mediated by transcription factors belonging to the second class.

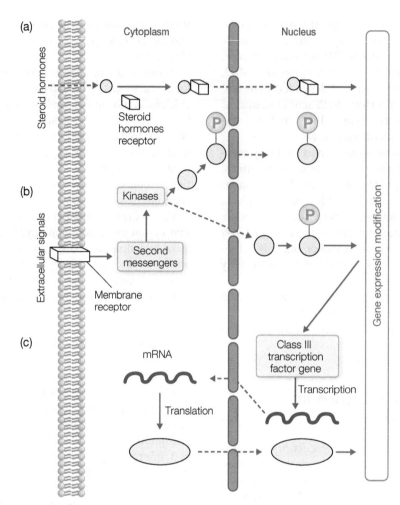

FIGURE 22.6 Mechanisms of transcriptional regulation by extracellular stimuli. Solid arrows indicate cause–effect relationships; dashed arrows indicate actual movements of molecules between different compartments. The three classes of transcription factors that respond to extracellular stimuli are represented. (a) Class I: factors whose activation depends on interaction with a ligand. This class includes steroid hormone receptors. The ligand–receptor interaction can take place in the cytoplasm, as shown, or directly in the nucleus where some receptors of this superfamily normally reside (e.g., thyroid hormone receptor). (b) Class II: transcription factors whose activity is controlled by phosphorylation of specific serine, threonine or tyrosine residues. In the figure, the role of phosphatases is not indicated, but also dephosphorylation can act as an activation signal. Some transcription factors are phosphorylated in the cytoplasm, and this allows them to migrate into the nucleus (such as STAT proteins). Others are phosphorylated in the nucleus, after migration of the specific kinase (e.g. CREB). Class II factors, besides regulating expression of numerous genes with the most diverse functions, also regulate expression of class III transcription factors (c).

Factors belonging to the third class are c-fos, fra-1, fra-2 (fos-related antigen (fra)), fos-B, c-Jun, Jun-B, Jun-D, and krox-20 and krox-24. Some of these, such as c-fos, are proto-oncogenes, proteins that when mutated or overexpressed participate in the processes of neoplastic cell transformation and immortalization. In physiological conditions, many of the extracellular stimuli determining expression of these transcription factors are growth factors. However, expression of these transcriptional proteins can be induced also by other stimuli (e.g., neurotransmitters) as shown in a variety of physiological and pathological situations, such as in cellular processes involved in learning and memory or in cerebral ischemic insults and seizures.

FROM THE PHARMACOLOGY OF TRANSCRIPTION TO PHARMACOEPIGENOMICS

From the first cell division, multicellular organisms need to regulate gene expression in a precise manner to ensure proper execution of the genetic programs underlying tissue development and differentiation, as well as maintenance of

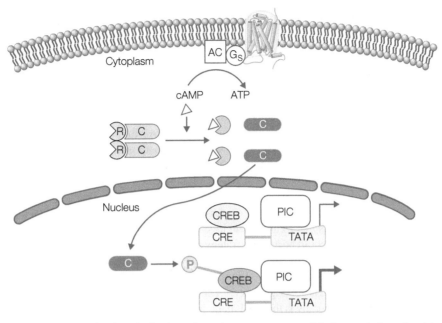

FIGURE 22.7 Mechanism of CREB activation. Activation of membrane receptors positively coupled to adenylate cyclase through stimulatory G proteins determines increase in cAMP intracellular concentrations. cAMP binds the regulatory subunits of protein kinase A, promoting their dissociation from the catalytic subunits, which then migrate into the nucleus and phosphorylate transcription factor cyclic AMP-responsive element-binding protein (CREB). CREB is bound to its specific DNA sequences even in basal conditions, but is not able to stimulate transcription. Its phosphorylation results in a conformational change that makes CREB competent to interact with the initiation complex and to stimulate transcription of target genes. AC, adenylate cyclase; C, catalytic subunit of protein kinase A; Gs, G-stimulatory protein; PIC, preinitiation complex; R, regulatory subunit of protein kinase A.

cellular homeostasis in response to a variety of extracellular stimuli. The precise regulation of gene transcription depends on promoter structure, cell cycle, and inductive signals. Each event that triggers transcriptional activation and/or repression results from the combined activity of many cis-regulatory elements as well as multiple protein factors, including RNAP II and GTFs required for PIC formation, specific transcriptional activators and repressors interacting with enhancers and silencers, and co-factors for remodeling and/or modifying chromatin. No one of these factors alone would be sufficient for correct and complete activation of the transcriptional program. As shown by studies investigating the transcriptome using RNA microarrays, the etiology of many diseases is associated with altered expression of multiple genes. This has given new impulse to the development of therapeutic strategies directly interfering with factors of the transcriptional machinery, with the aim of restoring a correct pattern of gene expression. In theory, this interference can be achieved at the level of transcription factors. Indeed, some drugs already used in clinical practice exert their action directly on transcription factors (e.g., drugs targeting intracellular receptors, cyclosporine and tacrolimus), or on co-factors involved in histone covalent modifications or in DNA methylation, that greatly influence expression of specific genes. Chromatin modifications and DNA methylation are at the basis of epigenetic regulation and are discussed in detail in Supplement E 22.1, "Pharmacoepigenomics." Although we have decided to discuss separately the two pharmacological approaches—the one targeting the epigenetic modifications and the other directly interfering with transcription factor activation—it is important to stress that their final goal is the same: restoring correct expression of genes involved in a given disease. Therefore, pharmacology of transcription and pharmacoepigenomics represent complementary aspects of the same problem and are becoming increasingly overlapping.

At present, the pharmacoepigenomics approach is still poorly specific, and it is based on agents that affect the general mechanisms controlling gene expression (acetylation, methylation, etc.). By contrast, the pharmacology of transcription, which largely coincides with the pharmacology of intracellular receptors, employs compounds with much higher specificity. An interesting intermediate situation is shown in Box 22.1, describing a class of synthetic modulators that are capable of interfering with a specific coactivator endowed with acetyltransferase activity. This example highlights how a better understanding of the mechanisms of gene regulation can provide cues for development of new drugs.

BOX 22.1 SYNTHETIC MODULATORS OF TRANSCRIPTION ACTIVATION: THE CASE OF CBP/p300

The activity of transcription factors can be modulated either by preventing their binding to specific DNA consensus sequence (via their DBD) or by blocking interaction between their transactivation domain (TAD) and other proteins, such as coactivators, using molecules structurally similar to their TAD. In the first case, many non-natural compounds capable of interfering with DBDs in *in vitro* systems and cell cultures have already been identified (Table 22.3). More difficult has been the discovery of molecules capable of inhibiting interactions with TADs. In cells, TAD activity is usually regulated by proteins that mask TAD site of interaction with co-activators and are removed/inactivated by inductive stimuli. Among the inhibitors identified so far, there are small molecules or peptides mimicking specific TADs thereby competing with them for coactivator binding. The most successful example concerns inhibitors of TADs that interact with coactivator CBP/p300, a HAT. CBP, and its homolog p300, interacts with more than 100 transcription factors, many of which are implicated in cancer. Modulation of CBP/p300 activity has assumed a role of important therapeutic interest. For example, a CBP/p300 interacting factor is HIF-1α, an intracellular receptor that regulates the response to hypoxia by inducing expression of genes such as VEGF and metalloproteinase and contributes to cancer progression and metastasis by promoting angiogenesis. HIF-1α activity depends on its interaction with CBP/p300; therefore, preventing this interaction could be a good strategy to block the process of metastasis and cancer development. A high-throughput screening has led to the identification of a natural product, chetomine, capable of inhibiting such interaction (Fig. Box 22.1, left). Although this compound blocks HIF-1α transcriptional activity in cellular models, it turned out to be toxic in animal models. A synthetic derivative, ETP-3 (Fig. Box 22.1, right), displays a reduced toxicity. To date, the main issue preventing the use of these compounds in therapy is that CBP/p300 is a general coactivator controlling a large number of genes; this means that blocking its action would affect the expression of 403 genes, 113 of which directly controlled by HIF-1α. Despite the pleiotropic activity of co-factors on gene expression regulation is a limitation at present, future investigation into the mechanisms of interaction between transcription factors and co-factors, as well as structural studies, will help to devise new therapeutic strategies to treat diseases whose etiogenesis is closely associated with altered regulation of transcription.

Chetomine   Dimeric ETP-3

FIGURE BOX 22.1

TABLE 22.3 Drugs that interfere with transcription factors binding to DNA

Gene	Molecule	Activity	IC$_{50}$	Biological effect
c-Myc	Mimetic peptide IIA6B17	Inhibitor of c-Myc/Max dimerization	28 μM	Inhibition of reporter gene transcription
c-Myc	28RH-NCN-1 e #764	Derivative of thiazolidinone 10058-F4 DNA binding inhibitor	29 and 4.6 μM	Inhibition of proliferation of HL60 leukemia cell line
HOXA13	Lactam carboxamide	DNA binding inhibitor	6.5 μM	Repression of transcription in *in vitro* reporter systems
E2F4	HLM006474	DNA binding inhibitor	29.8 μM	Stimulation of apoptosis; inhibition of proliferation and invasion of A375 melanocytes in dermic layer
STAT3	Galiellalactone	DNA binding inhibitor (natural molecule)	n.d	
STAT3	STA-21	DNA binding inhibitor	n.d.	
EWS–FLI1	NSC635473 and its derivative YK-4-279	Reduce binding to RNA helicase A (RHA) *in vitro*	10 μM	Inhibition of reporter gene transcription and antiproliferative activity on Ewing's sarcoma family tumor (ESFT) cells

EWS-FLI1, fusion protein involved in Ewing's sarcoma development; n.d., not determined; STAT, signal transducer and activator of transcription.

TAKE-HOME MESSAGE

- Gene transcription can be constitutive or inducible and is a finely regulated mechanism in terms of tissue specificity and time.
- Formation of a stable and properly positioned preinitiation complex (PIC) is essential for initiation of mRNA synthesis by RNAP II. PIC formation is mediated by transcription factors and general cofactors interacting with specific DNA sequences located in proximal or distal regulatory elements (enhancers/silencers).
- Activators and/or repressors regulate transcription by recruiting co-activators or co-repressors with chromatin modifier activity. Post-translational modification of histone proteins generates a "histone code" creating an environment favorable or unfavorable to transcription.
- Post-translational modifications of transcription factors, such as phosphorylation, dephosphorylation, and ubiquitination, allow cells to change their transducing machineries and transcriptomes to adapt to extracellular stimuli and/or preserve cellular homeostasis.
- Modification of chromatin and/or DNA mediates the epigenetic control of gene expression.
- Several drugs modulate transcription acting at different level.

FURTHER READING

Battle S., Maguire O., Campbell M.J. (2010). Transcription factor corepressors in cancer biology: roles and targeting. *International Journal of Cancer*, 126, 2511–2519.

Boland M.J., Nazor K.L., Loring J.F. (2014) Epigenetic regulation of pluripotency and differentiation. *Circulation Research*, 115:311–324.

Carrera I., Treisman J. (2008). Message in a nucleus: signaling to the transcriptional machinery. *Current Opinion in Genetics and Development*, 18, 397–403.

Helin K., Dhanak D. (2013) Chromatin proteins and modifications as drug targets. *Nature*, 502, 480–488.

Hsia E.Y., Goodson M.L., Zou J.X., Privalsky M.L., Chen H.W. (2010). Nuclear receptor coregulators as a new paradigm for therapeutic targeting. *Advance Drug Delivery Reviews*, 62, 1227–1237.

Juven-Gershon T., Kadonaga J.T. (2010). Regulation of gene expression via the core promoter and the basal transcriptional machinery. *Developmental Biology*, 339, 225–229.

Lee L.W., Mapp A.K. (2010). Transcriptional switches: chemical approaches to gene regulation. *The Journal of Biological Chemistry*, 285, 11033–11038.

Perissi V., Jepsen K., Glass C., Rosenfeld M.G. (2010). Deconstructing repression: evolving models of co-repressor action. *Nature Reviews Genetics*, 11, 109–123.

Tsukiyama T. (2002). The in vivo functions of ATP-dependent chromatin remodeling factors. *Nature Reviews Molecular Cell Biology*, 3, 422–429.

Weake V.M., Workman J.L. (2010). Gene Inducible expression: different regulatory mechanisms. *Nature Reviews Genetics*, 11, 426–437.

23

PHARMACOGENETICS AND PERSONALIZED THERAPY

DIEGO FORNASARI

By reading this chapter, you will:

- Understand the molecular consequences of polymorphisms in genes encoding proteins involved in drug response
- Learn the pharmacological and clinical consequences of polymorphisms and the role of pharmacogenetics in personalized medicine

---

VARIATION OF DRUG RESPONSES AND THE DEFINITION OF PHARMACOGENETICS

It is well known that the response to a given drug is not univocal but a wide spectrum of clinical responses exists as patients may show different combinations of therapeutic and adverse effects in response to the same treatment. In the past, such interpersonal variability was ascribed to nongenetic factors, such as age, kidney, hepatic health, and lifestyle (in particular diet, alcohol consumption, smoking, and simultaneous use of other drugs). Although all these factors may have a role and in some circumstances may entirely account for the differences in the way patients respond to the same drug, today, it is widely accepted that the individual's genetic background is always critical in determining the response to therapy and in some cases may be responsible for up to 95% of the variability in the clinical response to a given drug.

Variation in drug response represents one of the main problems in clinical practice, as it is responsible for therapeutic failures and adverse effects that may seriously harm patients, prolonging hospital stays and increasing sanitary costs. In the United States, it has been estimated that adverse effects occur in 6.7% of hospitalized patients, being fatal in 0.32% of cases. This means that drug adverse effects represent the fourth leading cause of death, corresponding to a social cost of almost 100 billion dollars a year.

According to the FDA, between 1998 and 2005, there has been a 2.6-fold increase in adverse drug reactions (ADR) reported, meaning four times higher than the increase in the total number of prescriptions over the same time interval.

Pharmacogenetics, a term coined by Vogel in 1959, is the discipline that studies how individual's genetic background affects drug activity, with the aim not only of predicting and preventing adverse reactions and therapeutic failures but also of allowing the identification of the most suitable drug and dose for each patient. The terms "pharmacogenetics" and "pharmacogenomics" are often used as synonyms, assuming pharmacogenomics as a genome-scale pharmacogenetics. In this chapter, we will use the term "pharmacogenetics" to indicate the study of interpersonal differences in DNA sequence affecting drug response.

GENETIC BASIS OF VARIABILITY IN DRUG RESPONSE

Genes and Pharmacogenetics

Genes involved in drug response can be classified in two groups: genes encoding primary therapeutic targets, such as receptors, ion channels, or enzymes, and genes encoding proteins involved in drug absorption, distribution, metabolism,

---

General and Molecular Pharmacology: Principles of Drug Action, First Edition. Edited by Francesco Clementi and Guido Fumagalli.
© 2015 John Wiley & Sons, Inc. Published 2015 by John Wiley & Sons, Inc.

and excretion. Polymorphisms in genes belonging to these two groups can modify both the pharmacokinetic and the pharmacodynamic of a certain drug causing adverse reactions or abolishing the expected clinical responses.

The existence of many variations in the genomic DNA of different individuals was already known when a paper accompanying the publication of the entire human genome sequence confirmed the presence of at least 1.42 million single nucleotide polymorphisms (SNPs). Since then, several projects have been launched to better estimate such variations (the HapMap and the ongoing 1000 Genomes projects). These have increased the provisional estimation of human genomic variations to several million polymorphisms, including all polymorphisms with a frequency greater than 5%. In general, it is believed that most human genes present variations among individuals, and several allelic variants of each gene exist in human populations. This means that even the same individual may carry two different allelic variants of a given gene (heterozygote). Several types of polymorphisms exist, but the most abundant are the SNPs. An SNP occurs when a specific position in the genomic DNA sequence may be occupied by two different nucleotides in the genome of two individuals (Fig. 23.1).

Allelic Variants of Genes Involved in Drug Response

SNPs are classified based on their position in the gene context: (i) SNPs falling within coding regions are called coding SNPs or cSNP; they are found in exons and can affect that amino acid sequence of the encoded proteins (see Supplement E 23.1, "Genetic Polymorphisms"). (ii) SNPs falling outside the coding regions are called perigenic SNPs or pSNPs and can be found within regulatory regions (promoters, enhancers), DNA region specifying the 5′ or 3′ untranslated regions (5′ or 3′ UTRs) of mRNAs, and introns and in particular in splicing junctions. pSNPs may affect not only the expression level of a protein (when they interfere with mRNA transcription, stability, and translatability) but also its amino acid sequence by interfering with its mRNA splicing (see Supplement E 23.1). pSNPs located in DNA regions specifying mRNA 3′ UTRs have recently attracted much interest since these mRNA regions are known to be targets of miRNAs (see Chapter 25) that regulate mRNA stability and/or translatability. Several examples of pSNPs affecting mRNA 3′ UTRs have already been shown to interfere with miRNA–mRNA pairing, causing altered expression of specific genes. For example, a pSNP in the 3′ UTR-specifying region of the dihydrofolate reductase (DHFR) gene has been demonstrated to prevent the binding of its mRNA to a specific miRNA, resulting in higher DHFR protein expression and resistance to methotrexate (an irreversible DHFR inhibitor used in cancer treatment and as immunosuppressant). (iii) Random SNPs (rSNPs) are SNPs located in intergenic regions (regions where no genes are found), which represent 98% of the human genome. rSNPs have no direct effect on gene expression, but they are very useful as diagnostic markers, as they may be linked (linkage disequilibrium) to genes involved in pathologies or in phenotypes relevant to pharmacogenetics and may contribute to their identification.

FIGURE 23.1 Genetic variation in the populations. In two individuals, the same genomic position can be occupied by two different nucleotides.

Copy Number Variation

Copy number variation (CNV) is another type of polymorphism of great importance in pharmacogenetics. A CNV is a structural variation in DNA characterized by the presence, on a given chromosome, of either no copy or more than one copy of a DNA segment having a length ranging from 1 kb to several megabases. This implies that if a CNV contains a gene, the subject carrying that CNV may have 0 or one or more than two copies of that gene, instead of the expected two copies (see Supplement E 23.1). Moreover, if the protein encoded by that gene is involved in the response to a given drug, the presence or absence of that CNV in the patient's genome will greatly affect his reaction to that drug, thus contributing to the variability in the drug clinical effects. As we will discuss in more detail later, an example of this is the *CYP2D6* gene of which up to 13 copies, all perfectly functioning, can be found in the genome.

GENETIC POLYMORPHISMS AND DRUG METABOLISM

Drug metabolism and, more generally, xenobiotic metabolism is a process occurring mainly (but not only) in the liver whose aim is to make more soluble the substances we introduce in the body, thus favoring their disposal through bile and urine. Drug metabolism (see Chapter 6 for details) is a multiphase process: phase I reactions are mostly functionalization reactions making substrates more reactive in order to favor the conjugation reactions occurring in phase II, which turn substrates into more hydrosoluble forms. Despite these chemical modifications, often, the disposal of xenobiotics is not a passive process, but additional mechanisms are required, which are collectively indicated as phase III.

SNPs have been found in several genes encoding enzymes involved in each step of drug metabolism, with variable effects on the activity of the gene product. For instance, completely inactive allelic variants of members of the cytochrome P450 superfamily, such as *CYP2D6*, *CYP2C19*, and *CYP2A6*, have been found. Some of the SNPs identified in these genes introduce stop codons, resulting in the production of truncated proteins, or cause amino acid substitutions inactivating the encoded proteins. In other cases, such as for phase II enzymes *N*-acetyltransferase 1 and 2 (NAT1 and NAT2), allelic variants have been identified reducing activity or half-life of the enzymes, in both cases resulting in a defective drug metabolism. It is worth remembering that each individual, being diploid, carries two copies (two alleles) of each autosomal locus, one inherited from the father and the other from the mother. If the two alleles are identical, we say that the individual is homozygous at that locus; if they are different, the individual is heterozygous. In general, as well as in the case of enzymes controlling xenobiotic metabolism, both alleles contribute to the production of the corresponding protein (but remember that there are exceptions to this rule, such as in the case of imprinted genes or in the allelic exclusion occurring in B lymphocytes). Therefore, a heterozygous subject carrying one active allele and one inactive at a given locus produces 50% of the amount of the encoded protein produced by a homozygous subject carrying two active alleles at the same locus. In contrast, a homozygous subject carrying two inactive alleles do not produce the corresponding protein at all. This implies that administration of a given drug may give rise to different clinical outcomes depending on the patient's genotype and in particular on what allelic forms of the enzymes involved in drug response he carries. It is important to point out that some drugs can be metabolized by more than one phase I or II enzyme, often with different effectiveness. Therefore, the individual's genotype relevant to pharmacogenetics includes all alleles at all loci involved in the response to a specific drug. This is a concept that we will discuss in more detail later: pharmacogenetic profiles often involve the concomitant actions of several genes.

GENETIC POLYMORPHISMS IN GENES ENCODING PHASE I ENZYMES

Polymorphisms of *CYP2D6*

The members of the cytochrome P450 superfamily are the most important enzymes in phase I reactions. To date, more than 60 different CYP450 isoforms have been identified, classified in at least 17 families and relative subfamilies. However, 72% of liver cytochromes consist of about 10 members belonging to seven different subfamilies (see Table 23.1). For each of these cytochromes, several allelic variants have been identified, and for many of them, precise associations with individual drug responses have been demonstrated. Although CYP2D6 represents only 1.5% of the total hepatic content of cytochromes, the relationships between its polymorphisms and the response to specific drugs have been extensively studied, providing one of the best examples of the huge clinical impact of pharmacogenetics. Indeed, the analysis of the pharmacological significance of CYP2D6 polymorphisms has provided information, methodological approaches, and novel concepts paving the way to subsequent studies on other members of the cytochrome P450 superfamily.

The first polymorphism in *CYP2D6* was discovered at the end of the 1970s when researchers were investigating the cause of the abnormal blood pressure response observed in some patients treated with debrisoquine, an antihypertensive drug. Since then, many other *CYP2D6* variants have been found (more than 50 in total), with SNPs that completely prevent the synthesis of the enzyme and others that give rise to a completely or partially inactive enzyme or to an enzyme with a different substrate specificity or even with an increased catalytic activity

TABLE 23.1 Percentage of liver cytochrome content

CYP isoform	%[a]
CYP1A2	12.7
CYP2A6	4.0
CYP2B6	0.2
CYP2C	18.2
CYP2D6	1.5
CYP2E1	6.6
CYP3A	28.8

[a]Average percentage of cytochrome content measured in liver microsome samples from 60 subjects (30 Caucasians, 30 Japanese).

TABLE 23.2 Molecular and clinical effects of polymorphisms in some genes encoding cytochrome P450 superfamily members

Enzyme	Main allelic variants	SNP molecular effect	SNP functional effect	Main drugs metabolized	Clinical outcome
CYP2C9	CYP2C9*2	Arg144Cys	Reduced activity	Warfarin Phenytoin	Bleeding
	CYP2C9*3	Ile359Leu	Altered substrate specificity	Tolbutamide NSAIDs	Hypoglycemia
CYP2C19	CYP2C19*2	Altered splicing	Inactive enzyme	Omeprazole	Improved clinical response and more effective eradication of *H. pylori*
	CYP2C19*3	Insertion of a stop codon	Inactive enzyme		
CYP2D6	CYP2D6*4	Altered splicing	Inactive enzyme	Antidepressants Antipsychotics β-blockers	Tardive dyskinesia with antipsychotics, cardiotoxicity with tricyclic antidepressants
	CYP2D6*10	Pro34Ser, Ser486Thr	Unstable enzyme		
	CYP2D6*17	Thr107Ile, Arg2965Cys, Ser486Thr	Reduced affinity for substrates		
	CYP2D6*2xN	Copy number variations	Increased enzymatic activity		Reduced pharmacological response to all drugs

TABLE 23.3 Frequency of some allelic variants of cytochrome P450 superfamily members in different human populations (%)

Allelic variants	Caucasians	Asians	Black Africans	Ethiopians and Saudi Arabians
CYP2C9*2	8–13	0	ND	ND
CYP2C9*3	6–9	2–3	ND	ND
CYP2C19*2	13	23–32	13	14–15
CYP2C19*3	0	6–10	ND	0–2
CYP2D6*4	12–21	1	2	1–4
CYP2D6*10	1–2	51	6	3–9
CYP2D6*17	0	ND	34	3–9
CYP2D6*2xN	1–5	0–2	2	10–16

ND - Not Determined.

(see Table 23.2). In addition, as mentioned earlier, *CYP2D6* is subject to CNV, meaning that individuals may carry from 0 to 13 functional copies of the gene. CNV of *CYP2D6* are particularly abundant in Ethiopia and in Saudi Arabia (with a prevalence up to 29% in some regions; see Table 23.3), but they are also present in Mediterranean countries including Italy, with a frequency of 5–10%. Such a huge number of allelic variants are reflected in the enormous variability in CYP2D6 activity observed *in vivo* in the populations, with individuals differing up to 1000 times in their ability to metabolize CYP2D6-specific substrates.

This concept is exemplified in Figure 23.2a, showing the results of a pharmacokinetic study on healthy voluntaries, subdivided into groups based on the number of functional copies of *CYP2D6* they carry. All subjects were administered a single and identical oral dose of nortriptyline and plasma concentration of the drug was measured over the next 72h. Nortriptyline is a tricyclic antidepressant having cardiotoxicity as a potential adverse effect, because its mechanism of action involves inhibition of catecholamine reuptake and antagonism of muscarinic and α1-adrenergic receptors. Such adverse effect mainly depends on the plasma concentration of the drug, which varies depending on the individual's *CYP2D6* genotype. As shown in the figure, subjects having no functional copies of *CYP2D6* displayed a higher peak of nortriptyline concentration in the blood and maintained a high plasma concentration of the drug for longer compared to the others (Fig. 23.2a). This effect is the consequence of reduced metabolism and disposal of nortriptyline prolonging its half-life. Looking at the graph, subjects unable to metabolize (nonmetabolizers) the drug show a larger area under the curve (AUC) compared to subjects carrying two functional copies of *CYP2D6*. In general, nonmetabolizers tend to develop adverse reactions to all drugs inactivated by CYP2D6. On the contrary, subjects carrying 13 functional copies of *CYP2D6* tend to be "resistant" to the same drugs, because they metabolize them too quickly hence preventing them from reaching their therapeutically effective blood concentrations.

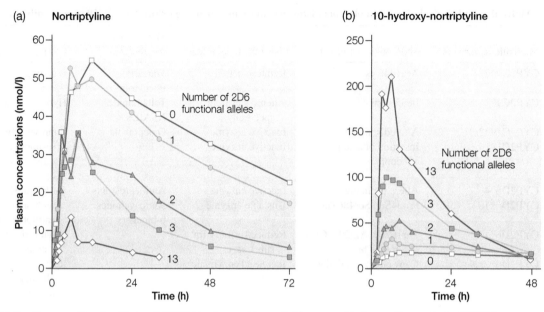

FIGURE 23.2 Pharmacokinetic effects of *CYP2D6* polymorphisms. Subjects stratified according to their *CYP2D6* genotype were treated with a single oral dose of nortriptyline and plasma concentrations of (a) nortriptyline and (b) the metabolite of nortriptyline were measured at different times after administration.

Allelic Variants in the CYP2C Subfamily

As mentioned before, the study of CYP2D6 has been rapidly followed by similar ones aimed at identifying allelic variants of other members of the cytochrome P450 superfamily. Particular attention has been devoted to the CYP2C subfamily, which is the second most expressed in the liver, after the CYP3A (Table 23.1).

As summarized in Table 23.2, two allelic variants of *CYP2C9* (*CYP2C9*2* and *CYP2C9*3*) have been indentified, encoding enzymes with reduced activity. Administration of standard doses of warfarin to an individual carrying these variants, especially if homozygous, is often followed by hemorrhage. Since the frequency of the *CYP2C9*2* and *CYP2C9*3* alleles among Caucasians is 8–13% and 6–9%, respectively (Table 23.3), this pharmacogenetic profile should be considered relatively common in Italy.

However, it must be noted that polymorphisms in genes encoding phase I enzymes not always cause adverse effects. On the contrary, some of them result in a better response to pharmacological treatments. An example of this is provided by some polymorphisms in the *CYP2C19* gene. As indicated in Table 23.2, two allelic variants of this cytochrome exist, named *CYP2C19*2* and *CYP2C19*3*, whose prevalence is highly variable among the Caucasian populations analyzed, whereas it is much higher in Asians, reaching a peak of 70% in one of the Melanesian islands (Table 23.3). Regardless of their frequency and the different genetic background in which are found, the clinical effects of these allelic variants are the same in Asians and Caucasians, as recently shown by an Italian study. When treated with gastric pump inhibitors, individuals homozygous for the normal allele, *CYP2C19*1*, recover less frequently from *H. pylori* infection compared to heterozygous subjects carrying one inactive allele or homozygous subjects with both allele inactive. Such difference in the clinical response depends on the pharmacokinetic properties of the drug: subjects carrying *CYP2C19*2* and *CYP2C19*3* alleles display a larger AUC compared to subjects homozygous for the normal allele, *CYP2C19*1*, meaning that in these individuals the drugs have an enhanced inhibitory effect on gastric secretion.

Role of Phase I Enzymes in Prodrug Activation

Drugs are not always administered in their active forms, that is, the forms capable of interacting with the molecular targets. Some drugs are introduced in the organism as prodrugs, inactive molecules that need to be metabolized by phase I enzymes to become active. With prodrugs, the therapeutic consequences of polymorphisms are the opposite of those observed with drugs.

Polymorphisms in CYP2D6 Gene and the Therapeutic Response to Some Oral Opioids

Some oral opioids used in pain therapy are administered as prodrugs and need to be demethylated at position 3 by CYP2D6 in order to become active and be able to bind the mu opioid receptors. Significant examples are tramadol and codeine. Figure 23.3 shows the pharmacokinetic effect of codeine on subjects carrying either two functional alleles of *CYP2D6* or no functional allele, as a result of a CNV. In this study, codeine was

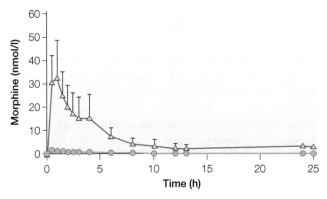

FIGURE 23.3 Effect of *CYP2D6* polymorphisms on the conversion of codeine into morphine. The graph shows the plasma concentration of morphine measured after administration of the same codeine dose in two subjects carrying either two functional copies (triangle) or no functional copy (circle) of *CYP2D6*. *x*-axis, time (h); *y*-axis, morphine plasma concentration.

administered as a single oral dose and blood concentration of morphine (the product resulting from codeine demethylation by *CYP2D6*) was then followed over time. As you can see, subjects carrying two functional copies of *CYP2D6* can activate the prodrug, releasing morphine in the blood, whereas those carrying the CNV do not show morphine in the blood and therefore do not respond to the therapy. What would you expect to occur in subjects carrying from 3 to 13 copies of *CYP2D6* under the same conditions? In those subjects, the conversion of codeine into morphine would be excessive and would trigger adverse effects ranging from nausea and vomiting, in subjects with only few copies of *CYP2D6*, to hallucinations, respiratory depression, and coma in subjects with high *CYP2D6* copy number. There is a large body of evidence indicating that these kind of adverse reactions to codeine and tramadol are much more frequent in subjects carrying the CNV than in subjects with two functional copies of *CYP2D6*.

Polymorphisms in CYP2D6 Gene and the Response to Tamoxifen Tamoxifen is a selective estrogen receptor modulator (SERM) used in the pharmacological treatment of breast cancer. It is a prodrug that needs to be processed by CYP2D6 to be converted in two active metabolites, 4-hydroxytamoxifen and endoxifen. It has been reported that women carrying either 0 or 1 functional copy of *CYP2D6* show a reduced disease-free survival compared to women with two or more *CYP2D6* copies (Fig. 23.4).

CYP2C19 and Antiplatelet Drugs: Risk of Fatal Therapeutic Failures Another important example of therapeutic failures due to lack of prodrug activation concerns clopidogrel, an antiplatelet drug often used in combination with low-dose aspirin, for instance, in subjects undergoing percutaneous coronary angioplasty to prevent stent occlusion.

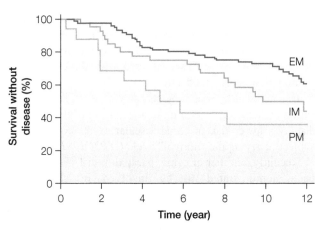

FIGURE 23.4 Kaplan–Meyer estimates of relapse-free survival based on metabolizer status. *x*-axis, time (years); *y*-axis, percentage of relapse-free survival patients. EM, extensive metabolizer (3 or more functional copies of *CYP2D6*); IM, intermediate metabolizer (2 functional copies of *CYP2D6*); PM, poor metabolizer (0–1 functional copy of *CYP2D6*).

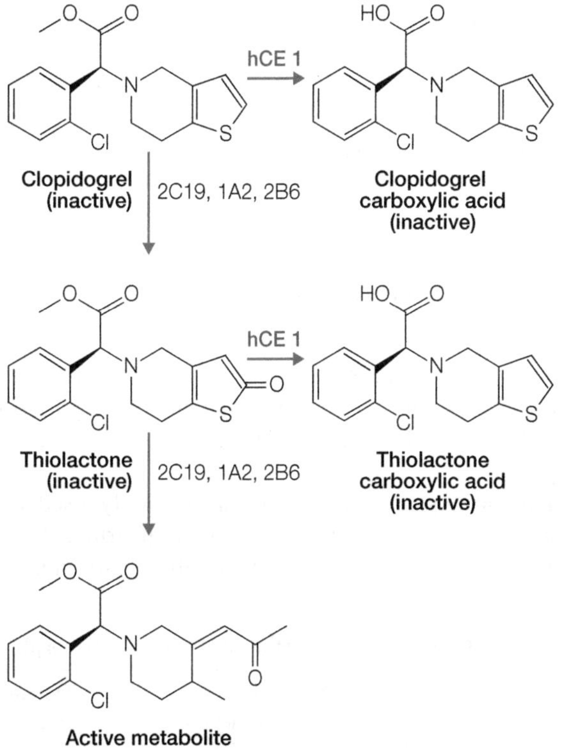

FIGURE 23.5 Metabolism of clopidogrel. hCE 1, human carboxylesterase 1.

Clopidogrel is an orally administered prodrug that is rapidly and largely (85%) inactivated by liver esterases. The remaining 15% needs to undergo two activation reactions involving CYPs, and in particular CYP2C19, as key players (Fig. 23.5). Several clinical studies have shown that subjects carrying the *CYP2C19*2* and *CYP2C19*3* allelic variants exhibit a reduced plasma concentration of the active metabolite and an increased platelet reactivity and are significantly more prone to serious cardiovascular events such as infarction, ictus, and death.

It is noteworthy that subjects carrying the *CYP2C19*17* variant, which codes for a more active form of the enzyme, display a lower risk of cardiovascular events compared to subjects carrying the "normal" allele *CYP2C19*1*.

Recently, another enzyme, paraoxonase 1 (PON1), has been demonstrated to be involved in clopidogrel activation; also for PON1, several inactive variants have been identified, associated with a reduced response to the drug. The exact contribution of CYP2C19 and POX1 to clopidogrel activation remains to be clarified. However, the genetic inability to respond to clopidogrel may soon be overcome by a new drug, prasugrel, another P2Y12 receptor antagonist but not requiring activation by CYP2C19.

Polymorphisms in Genes Encoding Phase II Enzymes The demonstration of the association existing between adverse reactions to isoniazid and inability to acetylate (and inactivate) the drug represents a milestone in the history of pharmacogenetics. Besides NAT2, whose genetic polymorphisms are responsible for interindividual differences in the ability to acetylate isoniazid, many phase II enzymes have multiple allelic variants present in the populations (Table 23.4).

Thiopurine Methyltransferase

Another enzyme that, similarly to CYP2D6, represents a paradigm in pharmacogenetics despite being involved in the metabolism of only a few drugs is thiopurine methyltransferase (TPMT), whose endogenous substrates are mostly unknown.

Thiopurines are among the most widely used antineoplastic drugs. They exert their pharmacological effects acting as purine antimetabolites. Administered as prodrugs, they are enzymatically converted by the enzyme hypoxanthine–guanine phosphoribosyltransferase (HPRT) in thioguanine nucleotides. When incorporated in the DNA, thioguanine nucleotides interfere with cell proliferation in several ways. However, the prodrug is not entirely converted in pharmacologically active molecules, because HPRT is in competition with TPMT, which in contrast transforms the prodrug in an inactive metabolite by *S*-methylation.

Thiopurine drugs include mercaptopurine, a first-line drug in the treatment of acute lymphoblastic leukemia; thioguanine, employed in myeloblastic leukemias; and azathioprine, an immunosuppressant commonly used in the therapy of autoimmune diseases but also in the prophylaxis of transplant rejection. The most severe adverse effect of thiopurines is strictly linked to their mechanism of action and consists in severe suppression of hematopoietic bone marrow activity. However, this event does not occur in all patients receiving the drugs, and its severity varies between subjects. Such interindividual variability is due to the existence of inactive allelic variants of the *TPMT* gene. Many SNPs have been identified in *TPMT*, but only three of them account for more than 90% of the clinically relevant allelic variants. These are cSNPs that introduce amino acid substitutions making the enzyme highly unstable and prone to rapid degradation by the proteasome, thereby resulting in lack of TPMT enzymatic activity. It has been estimated that about 10% of Caucasians carries one of the inactive TPMT alleles, and 1 every 300 is homozygous for them. Homozygous subjects show no TPMT activity in red cells, where the enzyme is mostly expressed (beside the liver), and when treated with thiopurines produce high amounts of thioguanine nucleotides, as HPRT activity is not counterbalanced by TPMT. These patients, often children in the case of acute lymphoblastic leukemia, almost inevitably undergo serious and potentially fatal myelosuppression. This often can be avoided simply finding the right dose of drug that can be reduce down to one-tenth of the normal dosage, in order to maintain its therapeutic efficacy in the absence of adverse effects.

TABLE 23.4 Molecular and clinical effects of polymorphisms in some genes encoding phase II enzymes

Enzyme	Molecular effects of most frequent SNPs	Functional effects of most frequent SNPs	Main drug metabolized	Clinical outcome
Dihydropyrimidine dehydrogenase	Altered splicing	Inactive enzyme	5-fluorouracile	Neurotoxicity, myelosuppression
NAT2		Reduced enzymatic activity	Isoniazid Hydralazine Procainamide	Neurotoxicity
Glucuronosyltransferase (UGT1A1)	Promoter alterations	Reduced expression	Irinotecan	Diarrhea, myelosuppression
TPMT	Amino acid substitutions	Unstable enzyme	Mercaptopurine Azathioprine	Myelosuppression

UGT1A1 and Irinotecan: A Case Already under Evaluation by the Regulatory Agencies

Another polymorphism that has already found application in clinical practice concerns the phase II enzyme UGT1A1. One of the substrate of this enzyme is the antitumoral drug irinotecan, employed in the treatment of several cancers such as colorectal cancer.

Irinotecan has a complex metabolism including activation and inactivation steps; moreover, its absorption and distribution involve several transporters. One crucial step, responsible for much of irinotecan toxicity, is the glucuronidation of its active metabolite SN-38 by the UGT1A1 hepatic enzyme, which favors its disposal in the bile. Polymorphisms in the promoter of the *UGT1A1* gene strongly reduce its transcription and expression, thus exposing the carriers to risk of adverse effects such as leukopenia and severe diarrhea.

Both FDA and EMA recommend to test patients' *UGT1A1* genotype prior to administration of irinotecan; homozygous subjects for variants reducing UGT1A1 expression should be treated with low initial doses of the drug or not treated with irinotecan at all.

GENETIC POLYMORPHISMS IN GENES ENCODING TRANSPORTERS INVOLVED IN DRUG ABSORPTION, DISTRIBUTION, AND EXCRETION

Drugs cross biological membranes according to precise physicochemical rules, but their permeation can be highly enhanced or inhibited by specific membrane transporters belonging to two superfamilies: solute carriers (SLCs) and ATP-binding cassette transporters (see also Chapter 28).

SLCs have been classified in 78 families and in humans are encoded by about 315 genes. They are molecules capable of moving endogenous compounds, ionic or nonionic, as well as drugs and xenobiotics, from one side of the membrane to the other. From the energetic point of view, these transporters mediate either facilitated or secondary active transports.

ABC transporters have been classified in 7 families including 49 members in total. Their substrates share a planar hydrophobic structure, containing some positively charged or neutral residues. This means that they can transport many different molecules, structurally and pharmacologically not correlated. ABC transporters mediate ATP-dependent primary active transports.

ABCB1 and Multidrug Resistance

ABCB1 (also called MDR1 or P-glycoprotein) is one of the members of the ATP transporter superfamily most extensively studied from both a pharmacological and pharmacogenetic point of view. ABCB1, also known as the molecule

TABLE 23.5 Drugs that are substrates of ABCB1 transporter

Drug class	Transported drugs
Antitumorals	Doxorubicin, imatinib, paclitaxel, vinblastine, vincristine
Calcium channel blockers	Diltiazem, verapamil
Beta-blockers	Bunitrolol, carvedilol
Inhibitors of HIV proteases	Indinavir, ritonavir
Antibiotics and antifungals	Erythromycin, levofloxacin, ketoconazole
Hormones	Testosterone, progesterone
Immunosuppressants	Cyclosporine, tacrolimus, sirolimus
Antiemetics	Ondansetron
Antihistamines	Terfenadine, fexofenadine
Miscellanea	Clopidogrel, digoxin, loperamide

responsible for multidrug resistance (MDR), is expressed on the apical membranes of enterocytes, brain capillary endothelial cells, and kidney tubules, on the biliary face of hepatocytes, and in the placenta. It mediates unidirectional transport, from the cytoplasm to the extracellular medium. Therefore, it acts as a barrier and a defense molecule that prevents xenobiotics from entering the body. In the intestine, it is expressed from the duodenum to the rectum along with members of the CYP3A family, which contribute to reducing the bioavailability of many drugs, such as those reported in Table 23.5.

More than 100 cSNPs have been identified in the *ABCB1* gene and their potential impact on drug response has been investigated. Their frequencies are largely unknown, except for a few of them. For example, the T variant of the SNP 3435C>3435T is present in about 52–57% of Caucasians and seems to determine a reduced expression of the transporter. The TT genotype has been shown to be associated with an increased plasma concentration of digoxin resulting in a higher probability to develop toxic effects. However, it is still unclear how this polymorphism affects the activity of the transporter. Indeed, 3435C>3435T is a silent polymorphism, meaning that it does not change the primary structure of the ABCB1 protein. It has been suggested that it might decrease mRNA stability, but so far, there is very little evidence supporting this idea. Interestingly, it has been pointed out that the new codon introduced by the SNP, despite specifying for the same amino acid, requires a very rare tRNA; this could slow down translation, allowing the nascent polypeptide to assume a different conformation possibly increasing the activity of the final protein. Regardless of the molecular mechanism, this polymorphism seems to affect the response to many drugs. For example, at the level of the blood–brain barrier, ABCB1 limits the uptake of some antidepressants, such as citalopram. Subjects carrying even only one copy of the T variant have a higher probability of remission compared to homozygous for the C variant, because higher amounts of the drug can enter the central nervous system.

The OAT1B1 Transporter and Statin-Induced Myopathy

OAT1B1 belongs to the SLC family and is involved in the bidirectional transport of many drugs, in particular many statins. A genome-wide study designed to identify genes whose allelic variants are associated with development of myopathy and rhabdomyolysis in subjects treated with 80 mg simvastatin revealed a cSNP in *OATB1B* tightly linked to the onset of the adverse effects. The transporter is localized in the hepatocyte external surface facing the bloodstream and pumps the drug coming from the intestine through the portal vein into the hepatocytes. The identified polymorphism seems to drastically reduce OATB1B activity, thereby allowing the drug to bypass hepatic filtration and to accumulate in skeletal muscles giving rise to the adverse effects.

GENETIC POLYMORPHISMS IN GENES ENCODING MOLECULAR TARGETS OF DRUG ACTION

The presence of polymorphisms in genes encoding molecular targets of drug action, such as receptors, ion channels, and enzymes, has a huge impact on patients' response to pharmacological treatments. Whereas polymorphisms in genes encoding metabolic enzymes seem to be responsible mainly for adverse effects (but not always; remember the cases of prodrugs and omeprazole), those in genes encoding therapeutic targets seem to affect mainly, but not only, the efficacy of the treatment. For example, depending on their localization, nonsilent cSNPs in genes encoding seven-membrane-spanning receptors can affect their affinity for agonists and antagonists, their interaction with G proteins thus changing their ability to transduce the signal, or their rate of agonist-induced internalization thereby changing their cell surface expression. To date, several examples of polymorphisms in drug targets have been described, some of which are indicated in Table 23.6.

One of the most interesting example concerns β-adrenergic receptors, which have a huge impact on the response to drugs used to treat cardiovascular and lung conditions.

β1 Receptor Polymorphism and the Response to β-Blockers in Heart Failure

Beta-blockers are first-line drugs in the therapy of chronic heart failure, but unfortunately, there is a significant interindividual variability in the outcome of the treatment. Some allelic variants of β1 receptors exist, two of which are due to a cSNP at position 389 where either a glycine (Gly) or an arginine (Arg) can be found. This polymorphism affects a region of the receptor that is essential for its interaction with the G_s protein. Indeed, *in vitro* functional studies have shown

TABLE 23.6 Molecular and clinical effects of polymorphisms in some genes encoding phase II enzymes

Clinical response caused by SNP	Primary targets	Drug
5-Lipoxygenase gene promoter	Zileuton	No drug response in asthma
Tardive dyskinesia	D2 and D3 dopaminergic receptors	Antipsychotics
ACE inhibitors	ACE	Evolution of diabetic nephropathy, left ventricular hypertrophy, endothelial dysfunction
Drug response in schizophrenia	5-HT2C receptor	Clozapine
Estrogen receptor alpha	Estrogens, SERM	Bone mineralization in postmenopausal women
β1- and β2-adrenergic receptors	β-blockers, bronchodilators	Reduced response, adverse effects

that the Arg389 variant is much more efficient at triggering cAMP production compared to the Gly389 variant.

In transgenic experimental models, the Arg389 variant determines myocardial hypercontractility, which rapidly evolves toward heart failure. Several studies demonstrate that heart failure patients with the Arg389Arg genotype (i.e., homozygous for the Arg389 allele) have a lower survival rate compared to Gly389Gly patients. In one of the first studies attempting to correlate the therapeutic response to β-blockers with the genotype, it was observed that out of a total of 224 heart failure patients treated with the β-blocker carvedilol, those carrying the Arg389Arg genotype responded better compared to Gly389Gly patients, as indicated by measurements of their left ventricular ejection fraction. However, it must be noted that this result has not been confirmed in all studies in which the two genotypes have been compared using as a criterion the left ventricular ejection fraction (an intermediate clinical outcome), possibly because of the different β-blockers employed.

On the contrary, most significant and consistent results have come from clinical studies in which patients with the same genotype (e.g., Arg389Arg) were either treated or untreated (placebo) with a given β-blocker and clinical outcomes such as death, hospitalization for heart failure, and need for heart transplant were considered. Such studies have clearly shown that β-blockers are very effective and beneficial in Arg389Arg patients but only poorly in Gly389Gly patients, suggesting that in Gly389Gly patients it may be preferable to use different drugs rather than β-blockers, given their poor effects and severe adverse effects.

Another fascinating hypothesis that could explain the better clinical response of Arg389Arg patients to carvedilol has come from studies employing a technique called fluorescent resonance energy transfer (FRET) to investigate the dynamics of activation and inactivation of the β1 receptor. These studies have shown that carvedilol acts as an inverse agonist on Arg389 β1 receptors and as an antagonist on Gly389 β1 receptors. In a clinical perspective, this means that in Arg389Arg subject's carvedilol is expected to block the basal activity of Arg389 β1 receptors, which in normal conditions is quite high, resulting in a better control of cardiac frequency and contractility.

β2 Receptor Polymorphism and the Response to Antiasthma Therapy

β_2-adrenergic receptors are mainly expressed in smooth muscles of blood vessels and bronchial airways, where in physiological conditions they determine vasodilation and bronchodilation in response to circulating catecholamines. This has been exploited in particular in the treatment of asthma of different etiology where β_2-adrenergic receptor is the main target of a large number of agonist drugs.

Three cSNPs have been identified in the β_2-adrenergic receptor gene that have been associated with the receptor altered expression, downregulation, or reduced coupling to G_s protein. In particular, two cSNPs have been studied: one cause the substitution of glycine for arginine at position 16, the other causes the substitution of glutamine for glutamate at position 27. A recent study reported that Arg16 homozygous individuals develop a complete desensitization of β_2 receptors, evaluated as progressive disappearance of vasodilation after continuous isoproterenol infusion. On the contrary, Gly16 homozygous individuals do not exhibit any change in vasodilation.

Also, polymorphisms affecting codon 27 are involved in the response to agonists. Glu27 homozygous subjects display a more pronounced vasodilation induced by isoproterenol compared to Gln27 homozygous individuals, regardless of codon 16 genotype. These results are consistent with others demonstrating that the therapeutic response in asthmatic patients undergoing a programmed and regular treatment with salbutamol or salmeterol was gradually reduced in Arg16 homozygous compared to Gly16 homozygous because of frequent exacerbations. Moreover, therapy interruption resulted in serious worsening in Arg16 homozygous but not in Gly16 homozygous. These observations seem to suggest that Arg16Arg genotype may be a useful marker to identify patients that should not receive β_2-adrenergic agonists on a regular basis. However, a more recent prospective study reports no variations in the response to salmeterol (either in combination with steroids or not) that could be explained by the genotype. Therefore, it still remains to be understood how pharmacogenetics contributes to the clinical response to bronchodilators and whether it could help to identify patients at risk of developing potentially fatal exacerbations when exposed to these drugs.

METHODS OF PHARMACOGENETIC STUDIES

In the last 15 years, a huge number of pharmacogenetics studies have been published, sometimes reporting conflicting results. In particular, there have been a number of studies suggesting opposite conclusions regarding the contribution of the same allelic variant to a specific drug response. In other cases, polymorphisms previously shown *in vitro* to affect the function or expression of a protein known to be involved in a specific drug response proved completely ineffective in subsequent clinical studies.

As expected, these discrepancies have raised concerns and suspicions on the real ability of pharmacogenetics to predict drug response from the individual genotype. Such inconsistencies are probably due both to the complexity of the genetic basis of many phenotypes analyzed and to the inadequacy of the methodological approaches available.

To understand whether a given SNP (allelic variant) or a given haplotype (a group of SNPs) is associated with a specific phenotype concerning a specific drug response, both direct and indirect studies can be carried out. Direct pharmacogenetics studies analyze polymorphisms in selected candidate genes (encoding, e.g., receptors or metabolic enzymes), which may be expected to be associated with drug response, based on pharmacological, pathological, and physiopathological considerations. These studies measure the frequency of specific allelic variants of the candidate genes in subjects exhibiting the same response to a given drug compared to a control group. For instance, this type of studies have led to the identification of the *CYP2D6* allelic variants responsible for adverse events or the *TPMT* variants associated with the toxic effects of purinic antimetabolites described earlier. The advantage of these studies is that the genes analyzed are already known; therefore, if a clear association is identified, the information can be easily applied for diagnostic purposes. However, these studies provide univocal results only if the pharmacogenetic trait is controlled only by the allelic variants of a single gene. Unfortunately, most pharmacogenetic phenotypes are determined by more than one gene and their inheritance often does not follow the Mendelian laws. The complexity of most pharmacogenetic traits can be explained using two different models (not mutually exclusive): (i) the polygenic model, whereby the pharmacogenetic trait depends on the additive effects of a large number of genes (and their relative allelic variants), each one having modest effects, and (ii) the epistatic model, whereby all the relevant allelic variants of the genes involved in the drug response have to be present for the pharmacogenetic phenotype to be expressed. Both models imply that the effect of each single gene on the pharmacogenetic trait is modest and the genotypic risk associated with each one is relatively low and difficult to assess.

The choice of a wrong genetic model to interpret a pharmacogenetic trait—a single-gene model instead of a polygenic model—is the reason why some pharmacogenetic studies end up with inconsistent results.

A more complex direct approach consists in the simultaneous analysis of the polymorphisms of a large number of genes already known to be involved in complex pharmacological pathways. For example, to explain the lack of therapeutic response to bronchodilators, we could analyze, at the same time, the allelic variants of the genes encoding β_2-adrenergic receptor, G_s protein, adenylate cyclase, arginase, NOS, and so on, in the attempt to associate a specific combination of allelic variants of all these genes with the pharmacogenetic trait. Some of these studies have already been carried out in different therapeutic areas, in most cases with significant, although often not definitive, results.

This approach is certainly more potent than a single candidate gene approach but has a clear limitation in that it is based on hypotheses supported by the information available, thus excluding all unknown or known genes that have not been associated with a given drug response yet.

Indirect studies consist in genome-wide analyses and do not need any previous knowledge on genes or allelic variants associated with a given phenotype. Some of these studies employ specific polymorphic markers, often functionally neutral because it is localized in intergenic regions, whose only property is to be very close to allelic variants involved in a given pharmacogenetic profile. When two loci—in this case a polymorphic site and an allele predisposing to a given drug response (or to a disease or any other phenotype)—are in close proximity, the probability that a recombination event would separate them during the meiotic process is extremely low. Therefore, they tend to be inherited together and to be found associated in the population. The phenomenon whereby two loci very rarely recombine is known as linkage disequilibrium and indicates that the two loci are extremely close to each other, on the same chromosome.

For the indirect approach to be informative, a large number of polymorphic sites are needed, evenly spaced throughout the human genome. According to an initial estimation, 500,000 polymorphic sites are required to explore the entire human genome exploiting linkage disequilibrium. The main limitations of these studies are their cost and the complexity of the data analysis.

A new technological revolution that promises to boost indirect studies in pharmacogenetics is the next generation sequencing. This consists in a collection of new methods for DNA sequencing devised to obtain the sequence of millions of bases in a short time and at relatively low cost. It has been predicted that in the next few years, these methods will allow to sequence the entire human genome in a few days at the cost of only 1000 dollars. These systems are already being used in the context of the "1000 Genomes" project, with impressive results. Using the new sequencing technology, in the future, it will be possible to identify all DNA polymorphisms present in the human genome and to carry out indirect pharmacogenetics studies in which all genes of several individuals would be analyzed simultaneously to establish their contribution to a given phenotype, either a disease or a specific drug response.

THE FUTURE OF PHARMACOGENETIC

To date, most pharmacogenetic data have come from direct studies, and in some cases, the results have been translated in therapeutic protocols (such as in the case of warfarin and irinotecan therapies). The next generation sequencing technology promises to give a boost to indirect studies, thereby offering the possibility to explain complex pharmacogenetic profiles through the identification of all genes, even the unsuspected ones, contributing to a given phenotype. This will allow establishing a clear association between complex genotypes and specific drug responses, thus providing the necessary information to design diagnostic tools to identify, in advance, patients that may not respond or may develop adverse effects in response to specific drugs, thereby reducing risks and costs resulting from non-effective therapies.

The possibility to prescribe drugs based on patient's genotype will have several benefits: a safer and more effective therapeutic intervention, a more rational selection of the dosage, a more focused and motivated search for alternative therapies to cure those patients that cannot receive a specific active principle, as well as a significant reduction of healthcare costs by eliminating ineffective treatments and hospitalization caused by drug adverse effects.

TAKE-HOME MESSAGE

- The enzymes involved in phase I and II of drug metabolism are key variables in drug response.
- Polymorphisms are present in genes coding for transporters involved in drug absorption, distribution, and elimination as well as in other drug targets exploited in therapy.
- Definition of the patient's genotype in relation to specific pharmacological treatments may allow reduction of adverse effects and therapeutic failures, with possible decrease of health costs and improvement of strategies for new drug development.

FURTHER READING

Crettol S., Petrovic N., Murray M. (2010). Pharmacogenetics of phase I and phase II drug metabolism. *Current Pharmaceutical Design*, 16, 204–219.

Crews K.R., Gaedigk A., Dunnenberger H.M., Leeder J.S., Klein T.E., Caudle K.E., Haidar C.E., Shen D.D., Callaghan J.T., Sadhasivam S., Prows C.A., Kharasch E.D., Skaar T.C.; Clinical Pharmacogenetics Implementation Consortium. (2014). Clinical pharmacogenetics implementation consortium guidelines for cytochrome P450 2D6 genotype and codeine therapy: 2014 update. *Clinical Pharmacology and Therapeutics*, 95, 376–82.

He Y., Hoskins J.M., McLeod H.L. (2011). Copy number variants in pharmacogenetic genes. *Trends in Molecular Medicine*, 17, 244–251.

Ivanov M., Barragan I., Ingelman-Sundberg M. (2014). Epigenetic mechanisms of importance for drug treatment. *Trends in Pharmacological Sciences*, 35(8), 384–396.

Johnson J.A., Liggett S.B. (2011). Cardiovascular pharmacogenomics of adrenergic receptor signaling: clinical implications and future directions. *Clinical Pharmacology and Therapeutics*, 89, 366–378.

Roses A.D., Saunders A.M., Lutz M.W., Zhang N., Hariri A.R., Asin K.E., Crenshaw D.G., Budur K., Burns D.K., Brannan S.K. (2014). New applications of disease genetics and pharmacogenetics to drug development. *Current Opinion in Pharmacology*, 14, 81–89.

Wang L., McLeod H.L., Weinshilboum R.M. (2011). Genomics and drug response. *New England Journal of Medicine*, 364, 1144–1153.

Zhang W., Dolan M.E. (2010). Impact of the 1000 genomes project on the next wave of pharmacogenomic discovery. *Pharmacogenomics*, 11, 249–256.

24

INTRACELLULAR RECEPTORS

ADRIANA MAGGI AND ELISABETTA VEGETO

> **By reading this chapter, you will:**
> - Know the general features of intracellular receptor structure and their classification
> - Understand the mechanisms underlying receptor specificity
> - Know the pharmacological profiles of the major classes of drugs acting on intracellular receptors

Intracellular receptors are one of the most important classes of transcription factors in the animal kingdom. In mammals, this large family includes proteins involved in embryo growth and development, as well as reproduction or metabolism in adulthood. Activation of these proteins occurs upon binding to lipophilic ligands that can readily diffuse through the cellular membrane.

Elwood V. Jensen was the first to propose the existence of intracellular receptors, when he observed that a radiolabeled steroid hormone administered *in vivo* rapidly concentrates in cell nuclei of target organs. Bert W. O'Malley and his collaborators elucidated the mechanism of action of steroid hormones using the chicken oviduct model system. They showed that hormone nuclear internalization occurring upon (^{3}H)-progesterone administration is followed by an increase in specific mRNAs and proteins (e.g., avidin). Moreover, O'Malley and coworkers demonstrated that progesterone could interact with an intracellular protein capable of binding DNA, leading to the hypothesis that such progesterone receptor protein could act as hormone-activated transcription factors. In the 1980s, glucocorticoid receptor was the first intracellular receptor to be cloned, shortly followed by estrogen and thyroid hormone receptors. All these proteins revealed a high structural homology suggesting their belonging to a family of closely related proteins also including receptors with unknown physiologic ligands. More recently, two groups headed by B.W. O'Malley and Ronald M. Evans, respectively, demonstrated the existence of coregulators and their essential role in controlling and enabling the transcriptional activity of intracellular receptors.

Today, we know that intracellular receptors control integrated, tightly interconnected transcriptional programs regulating key physiological functions. Thereby, a large number of drugs targeting these receptors are already in clinical use. Moreover, selected members of this family are currently under study as potential targets of innovative antitumoral therapies. A list of clinical conditions that can be modulated by drugs acting on intracellular receptors is provided in Box 24.1.

STRUCTURAL FEATURES OF INTRACELLULAR RECEPTORS

In humans, the intracellular receptor gene family includes 49 members, of which 24 are activated by direct binding to known endogenous ligands (hormones or metabolic products), whereas the remaining 25 are referred to as "orphans" as their activation mechanisms and ligands are currently unknown.

Intracellular Receptor Classification

Comparative analysis of intracellular receptor gene structures within and between species has identified six evolutionary conserved subfamilies with one or more receptor subtypes;

BOX 24.1 THERAPEUTIC TARGETS AND DRUG CHOICE CRITERIA

Main therapeutic target	Cellular effects	Clinical effects	Indications	Notes and choice criteria
Glucocorticoid receptor	Inhibition of gene expression and synthesis of inflammatory mediators	Immunosuppression	Allergy, asthma, acute inflammation, autoimmune and rheumatic diseases, prevention of cell and organ transplant rejection and graft-versus-host disease, COPD, inflammatory bowel disease[a]; collagen diseases	Side effect: increased risk of infection[b]
	Antiproliferative activity in eosinophils and lymphocytes	Lymphoid tumor regression; reduced pain and swelling associated with cancer	Acute and chronic lymphocytic leukemias, Hodgkins' and non-Hodgkin's lymphomas, multiple myeloma, and breast cancer	Administered in primary combination chemotherapy
	Stimulation of carbohydrate storage and protein catabolism; anti-insulin effects, thus raising blood glucose; aldosterone-like activity, increasing salt and water retention; enhanced gastric acid secretion; increased muscle and bone catabolism; effects on the nervous system			Several serious or mild side effects in chronic regimens: diabetes, osteoporosis, increase in blood glucose levels,[c] suppression of calcium absorption through various metabolic effects; nausea; abdominal or muscle pain; swelling of hands and feet; headache; arrhythmia; stunting of growth (in children); difficulty in sleeping; changes in behavior; unusual tiredness or weakness; rapid weight gain (due to increased appetite); wounds that will not heal; Cushing's syndrome[d]

Glucocorticoids are also called corticoids or steroids; besides the oral and intravenous routes of administration, inhaled, ocular, rectal, or topical pharmaceutical formulations of corticosteroids are available to treat diseases of the upper and lower airways, eye, intestine, and skin, respectively. One needs to gradually scale down the dosage to avoid serious side effects.

Main therapeutic target	Cellular effects	Clinical effects	Indications	Notes and choice criteria
Mineralocorticoid receptor	Sodium and water readsorption by kidney due to the increased transcription of the epithelial sodium channel (ENAC) gene in distal convoluted epithelial cells	Agonists: hormone replacement	Chronic adrenal insufficiency; hypoaldosteronism	Fludrocortisone
		Antagonists: diuretics, resulting in reduced risk of cardiovascular complications (arrhythmia, cerebrovascular events, ischemic heart disease) and reduced left ventricular mass. No potassium loss ("K-sparing diuretics")	Hypertension; primary aldosteronism[e]; bilateral adrenal hyperplasia; aldosterone-producing adenomas	Spironolactone side effects: gynecomastia, sexual dysfunction in man, and mastodynia and menstrual irregularities in women.[f] Eplerenone shows less side effects, yet spironolactone is more effective than eplerenone and the selective ENAC blocker, amiloride.

Aldosterone and cortisol have similar affinity for the MR; however, glucocorticoids circulate at roughly 100 times the level of mineralocorticoids. An enzyme exists in mineralocorticoid target tissues to prevent overstimulation by glucocorticoids. This enzyme, 11-beta hydroxysteroid dehydrogenase type II, catalyzes the deactivation of glucocorticoids to 11-dehydro metabolites; licorice is known to be an inhibitor of this enzyme; and chronic consumption can result in a condition known as pseudohyperaldosteronism.

BOX 24.1 (*Continued*)

Main therapeutic target	Cellular effects	Clinical effects	Indications	Notes and choice criteria
Estrogen receptor (ER)	Cell proliferation (cell cycle progression and cell survival)	Antagonists: inhibition of cancer progression	ER-positive breast cancer; endometrial cancer	Agonists are prescribed mainly for replacement therapies[g]
	Cell metabolism and differentiation	Clinical symptoms reduction	Ovarian failure	In association with progestins, not curative
		Clinical symptoms remission	Female hypogonadism	
	Autocrine, paracrine, and endocrine regulation	Agonists: HPO axis inhibition, anovulatory menstrual cycles	Oral contraception	In association with progestins; side effects include cholestasis and thrombosis.
	Regulation of tissue homeostasis	Reduction of bone resorption	Osteoporosis	
		Vasomotor activity	Menopausal symptoms[h]	
		Mucosal function	Mood and cognition disorders, depression	
		CNS performance		
Progesterone receptor	Cell development	Restoration of female reproductive organs	Hypogonadism	Association therapy with estrogens
	Cell differentiation and homeostasis	Control of female reproductive organs and functions	Oral contraception[i]	Association therapy with estrogens
		Blockade of estrogen undesired effects[j]	Menopausal symptoms[k]	
		Antagonists: abortion during the first trimester of pregnancy	Pharmacological interruption of pregnancy	

Novel progestins are developed which are devoid of androgenic activity and show mild antagonist activity against AR and MR, with reduced blood pressure and decreased water and salt retention

Main therapeutic target	Cellular effects	Clinical effects	Indications	Notes and choice criteria
Androgen receptor	Cell development	Hormone replacement	Hypogonadism[l]	
	Cell cycle progression	Antagonists: inhibition of cell proliferation	Prostate cancer	
			Benign prostatic hyperplasia (BPH)	
Retinoic acid receptor	Epithelial cell differentiation		Acne[m]	
			Psoriasis	
Vitamin D receptor	Inhibition of PTH synthesis and secretion in parathyroid glands (and upregulation of the calcium-sensing receptor) and renin production in kidney; stimulation of FGF23 production in bone and insulin secretion by pancreas[o]	Prevention of fractures; increased bone mineral density; improved neuromuscular function; decreased risk of falling	Osteoporosis; vitD deficiencies (rickets, osteomalacia); hypoparathyroidism and kidney failure; pathologies linked to unappropriate calcium usage	VitD supplementation (together with calcium) consists of more than 400 IU. Food supplementation with vitD. Calcium is the prevalent mineral of bone and its absorption from the intestine depends on vitamin D[n]
Thyroid hormone receptor	Energy and cholesterol metabolism	Thyroid cell development[p]	Hypothyroidism[p]	
	Cell development			
	Cell homeostasis			
Peroxisome proliferation activators receptor-alpha	Liver fatty acid utilization and ketogenesis during fasting	Reduced plasma triglycerides	Dyslipidemia	
		Increased plasma HDL	Atherosclerosis	
Peroxisome proliferation activators receptor-gamma	Adipogenesis and adipokine production in white adipose tissue	Lower plasma triglycerides	Type II diabetes[q]	
		Lipid repartitioning to fat		
		Increases insulin sensitivity		

LXR	Cholesterol efflux, catabolism and secretion[r]	
	Inhibition of intestine absorption of cholesterol[r]	
FXR	Regulation of transport of bile acids[t]	
PXR and CAR	Expression of cytochrome P450 enzymes (CYP2B, CYP3A4)	Phase I drug metabolism[u]
	Expression of sulfotransferases, glucuronosyltransferases, glutathione S-transferases	Phase II drug conjugation[u]
	Expression of drug transporters	Drug elimination[u]
	Control of energy metabolism	

PXR and CAR exhibit high binding promiscuity: some PXR ligands in humans: Rifampicin, Sulforaphane, taxol, dexamethasone; some CAR ligands: phenobarbital, 1,4-bis-[2-(3,5-dichloropyridyloxy)]benzene (TCPOBOP), androstanol, retinoic acids, clotrimazole, chlorpromazine. As xenobiotic sensors, they possess large and more flexible LBPs, providing the receptors with broad specificity. Endogenous ligands have not been identified yet.

[a] Autoimmune diseases: rheumatoid arthritis, lupus erythematosus, multiple sclerosis; COPD: chronic obstructive pulmonary disease; inflammatory bowel diseases: Crohn's syndrome, ulcerative colitis
[b] These drugs suppress the response of the body's immune system against the infective or inflammatory disease without preventing the disease itself, thus resulting in a higher susceptibility to infection.
[c] Even short-term glucocorticoid therapy tends to cause the patient to become temporarily insulin-resistant.
[d] Cushing's syndrome is characterized by a moon face (round face) and buffalo hump (accumulation of fat around the abdomen and the upper back).
[e] Resulting in reflexive vasoconstriction and hypertension.
[f] Spironolactone has antagonistic properties at the AR and agonistic properties at the PR, resulting in its main side effects; Eplerenone is a selective MR antagonist and lacks the progesterone-stimulatory and antiandrogen properties.
[g] A combination of estrogenic drugs and a SERM has been recently approved for postmenopausal therapy.
[h] Estrogenic agonists are used to prevent menopausal symptoms such as hot flashes, mucosal dryness, and mood disorders; estrogen monotherapy is restricted to hysterectomised women; adverse reactions include deep vein thrombosis, stroke and cardiovascular events both as monotherapy and as nonsequential estrogen + progestin association.
[i] The dosage is particularly lower in postmenopausal therapy than oral contraception
[j] Estrogens + progestins combination therapies are aimed at limiting estrogen's proliferative effects, particularly in uterus
[k] Adverse reactions include stroke and cardiovascular in nonsequential estrogen + progestin association
[l] Side effects include stimulatory effects on the prostate, hirsutism, hepatic toxicity
[m] All-trans retinoic acid (ATRA) provides complete responses in acute promyelocytic leukemia (APL) which develops after the translocation of RARalpha gene.
[n] Levels of 25OHD <5 ng/ml (or 12 nM) are associated with a high prevalence of rickets or osteomalacia; although there is currently no consensus for optimal levels, most experts define vitamin D deficiency as levels of 25OHD <20 ng/ml and optimal levels >30 ng/ml. Adequate sunlight exposure is the most cost-effective means of obtaining vitamin D. Current recommendations for daily vitamin D supplementation (200 IU for children and young adults, 400 IU for adults aged 51–70 years, and 600 IU for adults older than 71 years of age) are too low and do not maintain 25OHD at the desired level for many individuals. 700–800 IU appears to be the lower limit of vitamin D supplementation required to prevent fractures and falls. Unfortified food contains little vitamin D. Two forms of vitamin D exist: vitamin D3 (cholecalciferol) and vitamin D2 (ergocalciferol).
[o] Vitamin D regulates many functions in many tissues. The control of transcription requires the additional recruitment of cosuppressors or coactivators. Different tissues have varying levels of these coregulators, thus providing tissue specificity for vitamin D action.
[p] TR ligands with isoform binding selectivity are being developed. Basic research is also devoted to obtain TR antagonist for the treatment of hyperthyroidism.
[q] Side effects include weight gain, edema, and anemia.
[r] LXR isoforms are deeply involved in reverse cholesterol transport and are known as cholesterol sensors
[s] Studies are underway to identify selective LXR modulators that are devoid of lipogenic activity, while retaining the atheroprotective transcriptional activity on genes related with cholesterol homeostasis.
[t] FXR induces the expression of proteins involved in bile acids secretion and uptake from the intestine. FXR-targeted drugs did not yet reach the market, although intense studies are ongoing to identify selective FXR ligands that enhance cholesterol metabolism without effects on cholestasis and hypertriglyceridemia.
[u] PXR and CAR act as xenobiotic sensors, since they are able to recognize drugs and other exogenous molecules and activate transcriptional programs that ultimately modify the faith of these molecules inside the body. As a consequence, these receptors are also implicated in drug–drug interaction mechanisms, assuming a pivotal role in pharmacokinetics and adverse drug reactions.

in this chapter, we will focus only on the three of them that are pharmacologically relevant (Fig. 24.1a).

The receptor class defined as nuclear receptor 1 (NR1) by the official nomenclature is the largest one and includes the receptor subtypes for thyroid hormones (TR), retinoic acid (RAR), activators of peroxisomal proliferation (PPAR), oxysterols (liver X receptor (LXR)), bile acids (farnesoid X receptor (FXR)), vitamin D (VDR), steroids and xenobiotics (pregnane X receptor (PXR)), and androstane (constitutive androstane receptor (CAR)).

The nuclear receptor 2 class (NR2) includes 9-*cis* retinoic acid receptor (RXR) subtypes defined as NR2B.

Steroid hormone receptors belong to NR3 class that includes receptor subtypes for estrogen (ER), glucocorticoid (GR), mineralocorticoid (MR), progesterone (PR), and androgen (AR) steroid hormones. Orphan receptors that do not appear in Figure 24.1a belong either to the three classes listed earlier or to the NR4, NR5, and NR6 subfamilies.

Molecular Organization and Functional Domains

Receptor structural organization is highly conserved among NR1, NR2, and NR3 classes and consists of homology domains named A/B, C, D, E, and F (Fig. 24.1b).

The most conserved among all members of the family is the central C domain, which mediates receptor binding to DNA (DNA binding domain (DBD)). Through this domain, intracellular receptors recognize and bind in a highly selective manner a short nucleotide sequence, named hormone-responsive element (HRE), present in target gene promoters. The DBD consists of 66 amino acids and contains eight conserved cysteine residues. X-ray diffraction studies have led to classify this DBD as a type 2 zinc finger as it contains two zinc finger motives. A zinc finger is a DNA binding motif that contains four cysteine residues coordinating one zinc ion and that folds into two beta sheets and one alpha-helix structures. An amino acid sequence, called P-box, within the first zinc finger determines NR DNA binding specificity by forming specific hydrogen bonds with nucleotides in the HRE DNA sequence. Receptor dimerization brings two DBDs in close proximity and allows each monomer to interact with six specific nucleotides (half-site) within the HRE DNA sequence. HRE contains two half-sites separated by 1–3 nucleotides. It has been shown that a single amino acid substitution within the estrogen receptor P-box can switch its binding selectivity from estrogen-responsive element (ERE) to glucocorticoid-responsive element (GRE). The second zinc finger motif contains a domain, called D-box, which recognizes the characteristic spacing between the two half-sites of an HRE (Fig. 24.1).

A receptor dimer is able to distinguish sequence, spacing, and orientation of the half-sites, thereby discriminating between different HRE sequences. However, NRs exhibit some flexibility in DNA element recognition; selected amino acid substitutions in the helix controlling receptor–DNA interaction do not abolish DNA binding.

The E domain, named ligand-binding domain (LBD), participates in several activities including hormone binding, homo- and/or heterodimerization, interaction with heat shock proteins (HSPs), and transcriptional activation and repression. In this domain, structurally distinct components have been identified:

1. The ligand-binding pocket (LBP), in which the interaction between receptor and lipophilic molecule takes place.
2. The dimerization surface that allows association between LBD domains of two monomers forming a dimeric receptor; this region is highly conserved among different intracellular receptors.
3. An α-helix motif, named AF-2, that is mainly responsible for transcriptional activation. This region is exposed on the receptor surface upon ligand binding and recognizes LxxLL motives present in most transcriptional regulators.
4. Sites mediating receptor interaction with HSPs and other inhibitors.

X-ray diffraction studies have revealed that the E domain contains 12 α-helices numbered from 1 to 12 (H1–H12). H3, H5, and H6 contribute to delimit LBP hydrophobic cavity by exposing mainly hydrophobic amino acids.

The intracellular receptors can form homodimers, as in the case of NR2 and NR3 classes, or heterodimers with RXR as in the case of NR1 receptors.

The mechanism of activation of the AF-2 functional domain is also very similar within the receptor superfamily and involves H11 and H10 repositioning and H12 opening because of intramolecular interactions triggered by ligand binding. This affects the receptor affinity for other proteins; the high flexibility of H12 plays a key role in receptor interaction with coactivators and corepressors and thus in receptor activity.

The A/B region in N-terminus of the receptor is structurally poorly defined; its length and sequence are highly different among NRs and show very weak evolutionary conservation. This domain has a relevant role for gene transcription regulation as it contains AF-1, a region that mediates ligand-independent interaction between receptor and transcriptional machinery. Interestingly, this domain has been shown to be target of posttranslational modifications and generally contains several phosphorylation and SUMOylation consensus sites. In addition, this N-terminal region can interact with cofactors such as coactivators or other transcription factors.

Finally, other subdomains are relevant for receptor activity and regulation (Fig. 24.1). It is worth noting that, whereas the high homology in DBD and LBD sequence and structure (only LBP shows some variability) allows to easily

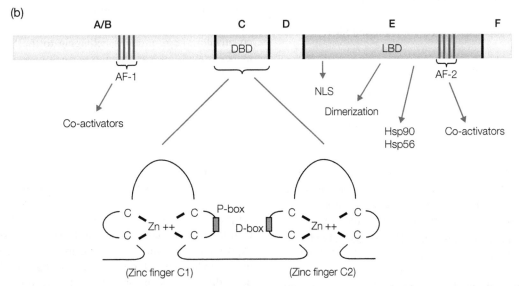

FIGURE 24.1 Intracellular receptor classification. (a) Phylogenetic analysis of genes encoding for intracellular receptors has identified different receptor classes and subfamilies. The table shows the nomenclature adopted for the classification of nuclear receptors (NR), ligands, DNA consensus sequences (hormone-responsive element (HRE)), and their orientation. The types of receptor dimers are also indicated (O, homodimers; E, heterodimers). It has been observed that receptors belonging to the same class share not only a high structural homology but also the same DNA binding and dimerization characteristics. By contrast, no relationship exists between the position of a receptor within a class and the type of ligand recognized. Based on these phylogenetic studies, an evolutionary model has been proposed suggesting that ancestral intracellular receptors were ligand independent and interacted with the DNA as homodimers. R=A or G, K=G or T; DR, direct repeat; PAL, palindrome; PI, reverse palindrome. (b) Schematic representation of intracellular receptor structure. Receptors contain several functional domains. The A/B region, highly variable both in sequence and in size, contains the AF-1 transcription activation domain (which can also operate in the absence of the ligand when isolated from the entire receptor molecule, as demonstrated in molecular pharmacology studies) and undergoes posttranslational modifications important for receptor activity. The C region includes the DNA binding domain (DBD, see text); the D region, variable in size and poorly conserved, acts as a bridge between DBD and LBD and allows the receptor to fold in different conformations without steric hindrance. The E region allows ligand binding through LBD, receptor dimerization, and transcriptional activation by exposing the AF-2 domain upon ligand binding. It contains the nuclear localization signal (NLS) and the heat shock protein (HSP) binding domain. The F region is not present in all intracellular receptors and its function is still poorly understood. A detailed representation of DBD and its schematic structure is also shown, highlighting the zinc finger structures, formed by four cysteines (C) in coordination with a zinc ion, as well as the P and D boxes.

classifying intracellular receptors on a structural basis, these receptors cannot be grouped according to their physiological or pharmacological properties; these will be discussed further later on.

INTRACELLULAR RECEPTORS AS LIGAND-REGULATED TRANSCRIPTION FACTORS

Intracellular receptors can regulate transcription through various mechanisms; in general, they can either activate or repress transcription, in a ligand-dependent or ligand-independent manner, either via direct DNA binding or via functional interference with other transcription factors. Not all family members exert all these activities, and transcriptional effects vary depending on cell type. In the absence of the ligand, nuclear receptors show distinct localization and activity: steroid receptors are associated with inhibitory proteins (HSPs) that, on one hand, allow neosynthesized receptors to fold in their appropriate tertiary structure and, on the other, prevent their dimerization and DNA binding by steric hindrance. On the contrary, RAR, TR, PPAR, and LXR receptors are associated with DNA as RXR heterodimers, and they actively repress target gene transcription through a ligand-independent mechanism involving interaction with transcriptional corepressors, such as NCoR and SMRT.

Classification of Ligands

Intracellular receptor ligands can be classified as follows:

1. Steroid hormones synthesized by endocrine glands under the stimulus of pituitary factors and regulated by feedback mechanisms. The affinity for their receptor is very high (the dissociation constant, K_d, is in the range 0.01–1 nM). In vertebrates, these hormones regulate sexual differentiation, reproduction, carbohydrate metabolism, electrolyte balance, and immunity.
2. Lipids contained in nutrients. These ligands bind with low affinity (K_d 1–10 mM) to PPAR, LXR, FXR, SXR/PXR, and CAR receptors. These receptors control lipid homeostasis; they serve as lipid sensors and modulate four major biological processes linked to lipid homeostasis, that is, lipid metabolism, storage, transport, and disposal.
3. Molecules that bind to TR, RAR, and VDR are halfway between these two classes, as they share features with both endocrine and lipid receptor ligands. In fact, they derive mainly from dietary lipids (e.g., retinoic acid) or require essential elements for their activation (e.g., ultraviolet rays for vitamin D; iodine for thyroid hormones) but regulate transcriptional programs involved in morphogenesis or development, similarly to steroid receptor ligands.

Ligand-Dependent Transcriptional Activation

Structural changes induced by ligand binding to the E domain lipophilic cavity cause a sliding of helix-12 that blocks the ligand within its binding site. In addition, these modifications cause the release of inhibitory proteins that are bound to the inactive receptors (such as HSPs), thus allowing their dimerization and interaction with DNA (Fig. 24.2a). Following HSP release, steroid receptors undergo posttranslational modifications, such as phosphorylation of the AF-1 domain and acetylation of the DBD, which affect receptor intranuclear localization and ability to interact with DNA and other signaling proteins. For NR1, conformational changes induced by the bound agonist lead to replacement of corepressors with coactivators. This event may require the activity of enzymes that promote binding of ubiquitin molecules to corepressors and their degradation by the proteasome. For these receptors, in general, HRE is a 15-nucleotide-long sequence containing two half-sites repeated in a direct, inverse, or palindromic manner and separated by 1–3 nucleotides (Fig. 24.1 and in Box 24.1). Receptor–DNA binding is required to initiate multiple interactions with coregulators and general and specific transcription factors that allow the recruitment of the preinitiation complex at target gene promoters (see chapter 22). Preinitiation complex formation is either facilitated or inhibited by coactivators and repressors, respectively. In turn, coactivators (e.g., proteins belonging to the SRC-1 family) associate with nuclear complexes, such as acetylases, which acetylate specific histone lysine residues, thereby reducing their binding affinity to DNA within nucleosomes and facilitating the recruitment of transcription preinitiation complex and RNA polymerase II to the target gene promoters (Fig. 24.3a). Conversely, corepressors recruited by inactive (not bound by agonist) NR1 receptors or by antagonist-bound intracellular receptors (see the following text) form protein complexes with deacetylases, such as HDAC3 (Fig. 24.3b), that enhance histone–DNA interactions, thus hindering transcription. Receptor–coactivator interaction is promoted by a conformational change involving helix-12 repositioning and exposure of AF-2 and AF-1 interaction domains and occurring upon binding of natural or synthetic ligands. Importantly, depending on the nature of the ligand bound, a specific conformational change occurs in the receptor that allows it to bind selected transcription coregulators or factors, thus inducing a specific transcriptional program. Indeed, several recent studies have shown that the set of genes regulated by a specific receptor in a given cell type is mostly determined by the specific ligand activating the receptor.

Ligand-Dependent Transrepression

Upon ligand binding, numerous intracellular receptors can also repress the expression of some target genes. The mechanisms involved in this ligand-dependent repression

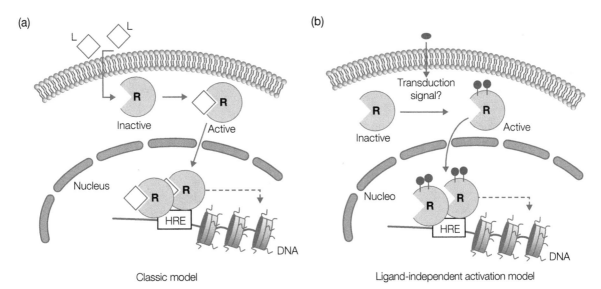

FIGURE 24.2 Mechanism of activation of intracellular receptors. (a) The classical model of ligand-dependent activation of intracellular receptors suggests that upon interaction with the ligand (L), the receptor is released from a complex with inhibitory proteins and enabled to interact with HRE regulatory sequences in target gene promoters and with other nuclear proteins. (b) The model of ligand-independent activation of intracellular receptors postulates the initial activation of protein kinases that phosphorylate the receptor (or transcription factors associated with it, thus favoring receptor binding to DNA) and alter its structural conformation, similarly to what occurs upon ligand binding; these phosphorylations promote receptor dimerization and high-affinity binding to DNA. Nucleo, nucleus.

are still under study and seem to be based on diverse molecular interactions. A well-described pathway of repression involves direct binding of intracellular receptors to other transcription regulators, such as NF-kB (a family of transcription factors regulating inflammatory response) or AP-1 (involved in cell proliferation), resulting in inhibition of their ability to stimulate expression of their target genes.

Furthermore, intracellular receptors may compete with NF-kB or AP-1 for binding to coactivators or basal transcription factors, whose availability is thus decreased. It has been hypothesized that GR, activated by dexamethasone, leads to a reduced NF-kB activity by preventing this factor to activate genes carrying NF-kB (rather than GRE) responsive elements in their promoters, thereby decreasing expression of factors that trigger and sustain inflammation (Fig. 24.3c).

Transcriptional Activation of Intracellular Receptors in the Absence of Ligand

Some members of the intracellular receptor family can be transcriptionally activated also by ligand-independent mechanisms involving posttranslational modifications such as phosphorylation, SUMOylation, and acetylation. In general, such modification are catalyzed by enzymes involved in membrane receptor signal transduction pathways (e.g., PKA or MAPK): once activated by their own ligands, several growth factor receptors activate a kinase cascade leading to phosphorylation of specific intracellular receptors, which as a result dissociate from HSPs and activate transcription of their target genes. This phenomenon has been well described for steroid hormone receptors (ER, PR, and AR) that can be stimulated by the activation of epidermal growth factor (EGF), insulin-like growth factor-1 (IGF-1), and transforming growth factor (TGF) receptors or neurotransmitter (e.g., dopamine) receptors (Fig. 24.2b). Evidence for the existence of such mechanism has been provided by studies in which receptor activation has been analyzed in animal models *in vivo* using functional imaging techniques. These methodologies have allowed observing that receptors such as AR and ER can be transcriptionally active even when circulating levels of their natural ligands are extremely low. Although the exact sequence of events leading to unbound receptor activation is still under discussion, the role of these cross-interactions between membrane receptors and intracellular receptors is generally accepted as a possible mechanism that may have developed in ancestral eukaryotic cells where intracellular receptors were serving as transcription factors sensitive to membrane receptor signaling. During evolution, this mechanism may have been favored as it allows to rapidly adjusting cell transcription program to the needs associated with changes in cell physiological state. To further support this hypothesis, it is worth mentioning that some orphan receptors that act as regulators of cell homeostasis do not show ligand-dependent activation (e.g., the NURR receptor subfamily).

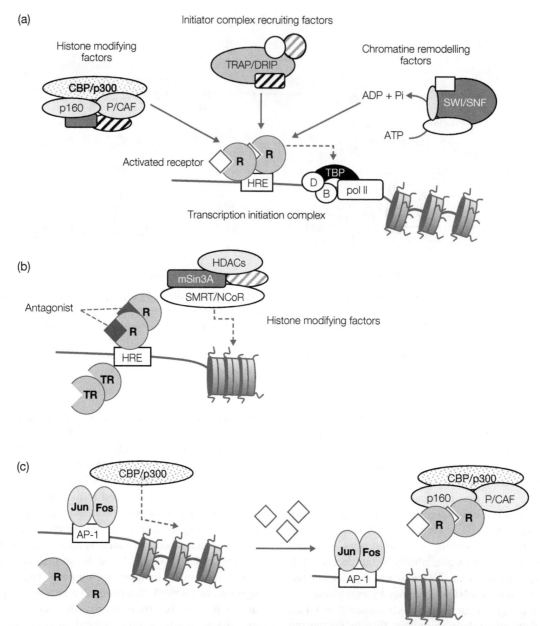

FIGURE 24.3 Nuclear activity of intracellular receptors. (a) Intracellular receptors are transcription factors and modulate gene transcription by interacting with several nuclear proteins that stabilize the initiation complex of transcription (general transcription factors, such as TFIID, and RNA pol-II). Also, other types of nuclear factor are recruited to the receptor, such as proteins called coactivators of intracellular receptors (e.g., p160 and P/CAF), that allows histone modification. These and other proteins recruited by coactivators, such as CBP/p300, are endowed with histone acetyltransferase activity: they acetylate lysine residues in chromatin histones, thereby favoring their dissociation from the DNA. Mutagenesis studies have led to the identification of the AF-2 domain as the receptor site mediating interaction between intracellular receptors and coactivators. Furthermore, activated receptors recruit chromatin-remodeling factors such as SWI/SNF, which use ATP to destabilize DNA–histone interaction promoting nucleosome repositioning; finally, receptors retrieve additional factors such as TRAP/DRIP that are able to attract RNA polymerase II to the transcription initiation site. Thus, binding of the activated nuclear receptor to target gene promoters stimulates the formation of macromolecular protein complexes of proteins that serve as potent transcription activators. (b) In the absence of their ligands, some receptors (e.g., TR and RAR) are bound to the HRE and repress gene transcription. Ligand binding to these receptor releases inhibition allowing transcription activation. The proteins (corepressors) responsible for the transcriptional inhibitory activity of TR and RAR in the absence of their ligands have been identified; of these, the most important are SMRT and NCoR. These corepressors recruit protein deacetylases on gene promoters, which deacetylate histones and contribute to maintain chromatin inactive state. It has been demonstrated that these repressors are recruited to the DNA by receptors bound to antagonists, such as tamoxifen. Accordingly, intracellular receptors can also act as potent repressors of transcription. (c) Intracellular receptors can regulate transcription independently from DNA binding, by titrating general coactivators (e.g., CBP) that therefore are no longer available for other transcription factors such as NF-kB and AP-1.

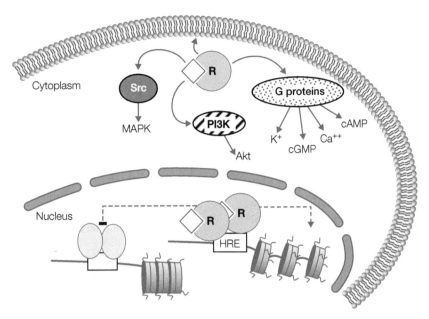

FIGURE 24.4 Estrogen receptor nuclear and cytoplasmic activity. The ER acts both in the nucleus, where it regulates gene transcription, and in the cytoplasm where, upon hormone binding, it interacts with enzymes and other proteins involved in intracellular signaling pathways, such as those regulated by Src, PI3K, and G proteins, stimulating their activity. In addition, other members of the intracellular receptor superfamily can act through this mechanism.

Temporal Oscillations of DNA Binding of Intracellular Receptors

The study of intracellular receptor transcriptional activity has been facilitated by technologies that allow genes encoding bioluminescent proteins to be fused to HRE-containing promoters (creating a reporter system) and integrated in cell genome and by the use of bioluminescent probes in protein–DNA coprecipitation assays. Through these techniques, it has been possible to demonstrate that the association of intracellular receptors with DNA and proteins of the transcriptional apparatus oscillates over time, following a cycle of assembly and disassembly processes that provide a very dynamic control of cell transcription. The time frame in which a stimulus regulates cell transcriptional response is determined by degradative processes often directed by enzymes that catalyze ubiquitination of active receptors and their coregulators. In fact, response timing and intensity is defined by the number of ubiquitin monomers that are added to a receptor at each transcriptional cycle. Initially, ubiquitin addition does not interfere with receptor activity, but as the number of transcription rounds increases, the ubiquitin load increases to a point that the receptor is recognized by the proteasome and degraded. Several other factors may contribute to interrupt receptor activity on target gene promoters (e.g., it is known that methylation inhibits intracellular receptor binding to DNA). This is functionally relevant because it allows cells to adjust the synthesis of target proteins not only to the specific hormonal stimulation but also to the entire cellular metabolism.

Extranuclear Activity of Intracellular Receptors

Pioneering electrophysiological studies carried out in the 1970s demonstrated that exposure to estradiol can induce a rapid increase in intracellular calcium mobility in neuronal cells. This event occurs within milliseconds after the stimulus is applied, too short a time interval to be mediated by protein neosynthesis. Later, it was shown that this hormone can rapidly activate protein kinase C (PKC), even in the presence of protein synthesis inhibitors. This activity can be explained by the fact that estrogen binds to ER pools outside the nucleus. In particular, a membrane-localized ER pool, which interacts with an atypical G-protein-coupled receptor lacking the seven-transmembrane structure, can rapidly stimulate cyclic nucleotide production, calcium flux, and kinase activation in response to estrogens (Fig. 24.4; see also Chapter 9). Rapid signaling can either activate or inhibit nuclear ER function and cause nongenomic effects through kinase-induced phosphorylation of substrate proteins regulating their cell localization and activity. Whether membrane-localized ER regulates transcription independently from nuclear ER, as well as the physiological relevance of this phenomenon, remains to be defined. A nongenomic mechanism of action has also been reported for AR, GR, and PR.

PHYSIOLOGICAL ACTIVITIES AND PHARMACOLOGICAL CONTROL OF INTRACELLULAR RECEPTORS

Specificity of Action of Homologous Receptors

The high degree of structural similarity shared by different members of the intracellular receptors family and their ability to recognize identical HREs raise questions on the molecular mechanisms enabling them to act so differently from each other. For example, although they recognize the same HRE sequence and have highly homologous DBD and LBD, progesterone and glucocorticoids play different physiological roles.

Three main mechanisms have been hypothesized to explain the specificity action of these receptors:

1. Difference in receptor expression levels in a given cell type. Experimental evidence supporting this hypothesis has been obtained in hepatocytes, which express GR but not PR. It has been reported that in hepatocytes experimentally engineered to express PR at similar concentrations as GR, progesterone can induce the same response observed following stimulation with glucocorticoids, indicating that if present, activated PR can bind the promoter of GR target genes and modulate their expression similarly to GR itself.
2. Use of different coregulators. Two receptors may activate transcription of distinct target genes by interacting with different coregulators and nuclear factors, thus diversifying receptor activity in a same tissue.
3. Differences in hormone metabolism. This hypothesis suggests that hormone (particularly steroid) metabolism could be tissue specific and give rise to metabolites with reduced affinity to their receptor, which therefore would be not activated.

Tissue Specificity of Nuclear Receptor Activity

The biological activity of each nuclear receptor is tissue specific. Although HREs are genomic elements identical in all cells of the organism, a given receptor can regulate different set of genes in different cell types even if they have similar receptor content. For example, genes regulated by the 17β-estradiol-ERα complex in endometrial cells are different from those modulated in hepatocytes. Several factors may contribute to define the tissue-specific activity of intracellular receptors. One is the specific cell differentiation program that causes selected portions of the genome to become inaccessible to transcription proteins. If HREs are present in those portions, they will not be accessible to activated receptors, resulting in a different transcriptional program being activated compared to cells of diverse lineage or at another differentiation stage. Another factor contributing to NR tissue-specific activity is represented by tissue-specific cytoplasmic and nuclear proteins or transcription factors participating in the regulation of a given promoter activity: the transcriptional efficiency of a gene is the result of a number of interactions taking place on its promoter and involving different regulators (besides HREs, several other response elements are present in a promoter, see Chapter 22). Thus, the contribution of intracellular receptors to the transcription of a target gene depends not only on ligand availability but also on specific coregulators expressed by the cells as well as on the transcription factors present on the gene promoter. This implies that a receptor can form different complexes in the tissues in which it is expressed, thus eliciting different transcriptional programs depending on the cell type.

Finally, another factor contributing to NR tissue-specific activity is the promoter structure: some genes may be more responsive to receptor activity because of HRE number and arrangement within their promoter. It has been shown that affinity of receptor dimers for individual response elements ($K_d = 10^{-8}$ M) is greatly enhanced by the presence of more elements in array (for two adjacent HREs, the K_d is about 10^{-11} M). For this reason, even slight variations in the number of receptors expressed in different cell types can affect the activity of a given promoter.

Receptor Agonists and Antagonists

Pharmacological treatment of pathologies in which intracellular receptors are implicated is based mainly on receptor agonists or antagonists that are extensively used in clinical practice (see Supplement E 24.1, "Ligands of Intracellular Receptors").

Receptor agonists are identified as molecules able to compete with the endogenous ligands for receptor binding and to mimic its ability to bind and regulate the transcriptional activity of the receptor AF-2 domain. Drugs acting as intracellular receptor agonists are frequently used in replacement therapies to compensate deficiencies of ligands either endogenous or deriving from the diet.

More complex is the action mechanism of antagonists. In fact, although antagonists can bind the receptor LBD causing release of inhibitory proteins and receptor dimerization on the cognate HRE, the antagonist–receptor complex has a different activity compared to the agonist–receptor complex (Fig. 24.3b). Crystallographic studies have shown that H12 rearrangement in the antagonist–receptor complex masks AF-2 and precludes its binding to coactivators. However, the antagonist-bound receptor is not inactive, as it recruits transcriptional corepressors, such as NCoR and SMRT, which prevent the access of the transcriptional machinery to the promoter (Fig. 24.5). Interestingly, most of these molecules do not behave as full antagonists, but show a mixed agonist–antagonist activity depending on the cell

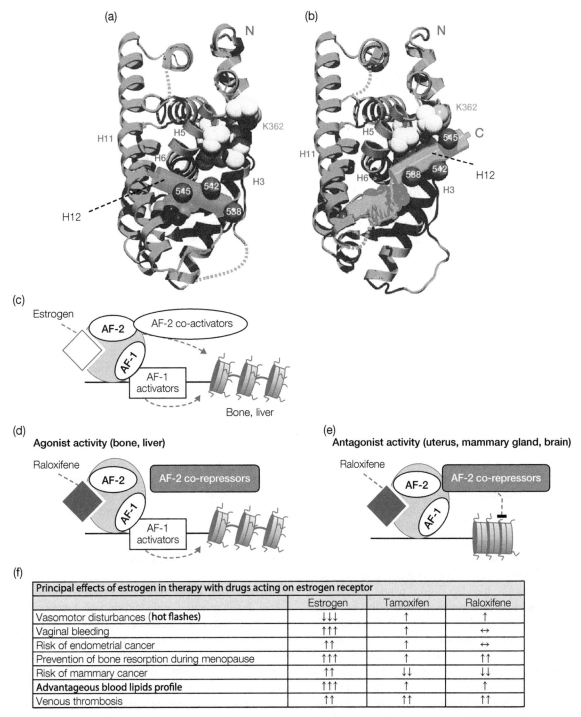

FIGURE 24.5 Helix-12 (H12) and the pharmacological activity of estrogen receptor ligands. (a) and (b) show the tertiary structure of the ERα LBD bound to estradiol (E_2) or raloxifene, respectively. H12, represented as a cylinder, is arranged in different positions when ER is bound to the agonist (a) or to raloxifene (b). The relevant hydrophobic residues and the 362 lysine residue (K362), which are necessary for E_2-dependent recruitment of specific coactivators, are highlighted in H3, H5, and H6 α-helices. The ASP538, GLU542, and ASP545 residues form a charged surface on H12 and are positioned perpendicularly to H11. This H12 position, observed in all LBDs crystallized with an agonist, is a prerequisite for transcription activation and generates an interaction surface, AF-2, which mediates the interaction with coactivators. On the contrary, in the presence of the antagonist, raloxifene (b), H12 rotates toward the N-terminal portion of LBD, thereby masking AF-2 and exposing corepressors interacting surfaces (from Brzozowski et al. Nature 1997, 389:753). (c) Involvement of AF-1 and AF-2 domains in estradiol action. (d) Agonist and (e) antagonist activity of raloxifene in different tissues. The selective, tissue-specific effect of ER drugs depends on the different availability of coregulators of AF-1 and AF-2 functions. (f) Summary of some of the principal effects of estradiol, tamoxifen, and raloxifene. The tissue-dependent selectivity of action of raloxifene, which acts as receptor agonist in the bone and liver and as receptor antagonist in other tissues such as the endometrium and mammary gland, is the rationale for the use of this SERM in osteoporosis. Similarly, tamoxifen is used in breast cancer as a powerful receptor antagonist, but it also collaterally reduces osteoporosis.

type. These molecules are named selective receptor modulators. Current hypothesis suggests that this phenomenon may be due to the different set of coregulators present in each target cells and to the ability of specific coactivators, in selected cell types, to activate the AF-1 domain or to interact with the antagonist–receptor complex.

Clinical use of intracellular receptor antagonists is primarily aimed at regulating the activity of the sex hormone-receptor subclass in hormone-responsive neoplastic diseases.

Glucocorticoid and Mineralocorticoid Receptors Cortisol, the main endogenous glucocorticoid in humans, is synthesized in response to stress by the adrenal cortex under a feedback hypothalamic control. This hormone shows a broad spectrum of physiological activities: its primary function is to regulate blood glucose levels acting on gluconeogenesis; it also regulates lipid and protein metabolism; in the immune system, it exerts anti-inflammatory and immunosuppressive actions; it also reduces bone formation and affects selected neural functions.

GR agonists (e.g., cortisone, prednisone, prednisolone, dexamethasone, and triamcinolone) are mostly used as anti-inflammatory agents in the treatment of allergic syndromes, asthma, and immune diseases (such as rheumatoid arthritis, lupus erythematosus, and multiple sclerosis). As immunosuppressants, these drugs find application in prevention of organ transplant rejection, in adjuvant chemotherapy of lymphoblastic lymphomas and leukemias, and in replacement therapy of adrenal insufficiency. However, long-term administration of these drugs is associated with a number of undesired effects: first, hypothalamic–pituitary–adrenal axis suppression, whereby abrupt discontinuation of synthetic glucocorticoid administration may result in serious morbidity and even mortality. Chronic glucocorticoid therapy is also detrimental for bone, increases plasma glucose levels, may cause general euphoria and depression, peptic ulcers, and hypogonadism. In children, these drugs may affect growth.

As explained earlier, GR activation by its ligand is associated with two main mechanisms, gene activation and gene transrepression. Intensive research is currently being carried out to identify selective GR agonists (SEGRA) that favor the interaction between GR and nuclear proteins without causing GR binding to DNA. It is believed that the transcriptional activation induced by GR may result in damaging effects, as suggested by the observation that GR binding to GREs in glucose-6-phosphatase and PEPCK gene promoters (genes encoding two enzymes involved in gluconeogenesis) underlies glucocorticoid diabetogenic effects, similarly to the activation of other genes involved in fat metabolism. Conversely, ligand-dependent GR transrepression is responsible for inhibition of genes involved in lymphocyte activation and inflammatory response, such as those encoding cytokines, chemokines, interferon-8, and hematopoietic growth factors. In fact, in transgenic mice bearing a mutation that disrupt GR dimerization and DNA binding, GR agonists can still inhibit the inflammatory process. It is still unclear what causes the onset of corticosteroid resistance that probably develops because of alterations in GR signaling pathway, such as autologous downregulation or defect in GR-interacting proteins, highlighting the complexity and diversity of glucocorticoid pharmacological effects.

Mineralocorticoids, the most important of which is aldosterone, are also synthesized in the cortical adrenal gland. By binding MR, they promote sodium-active transport in target cells. Their main site of action is the distal convoluted tubule epithelium. Synthetic MR agonists, such as fludrocortisone, are used in chronic adrenal insufficiency and hypoaldosteronism.

In case of adrenal gland hyperfunction and consequent corticosteroid overproduction, it is indicated to use drugs capable of inhibiting enzymes that catalyze steroid hormone synthesis (ketoconazole, aminoglutethimide). However, there are also antagonists that can selectively block one or the other class of receptors: spironolactone for MR and mifepristone for GR. Spironolactone belongs to the class of potassium-sparing diuretics and acts as a MR antagonist in kidney cells. In these cells, MR activation by aldosterone increases the expression of the Na^+/K^+-ATPase target gene. Therefore, by blocking this activity, spironolactone leads to a decrease in Na^+ absorption from the tubular lumen, preventing K^+ loss.

Mifepristone is a GR antagonist. However, given the high structural homology shared by GR and PR, it acts also as a PR antagonist, and it is in fact used to induce abortion in the first trimester of pregnancy.

In general, all PR and GR antagonists have poor selectivity and block both types of receptors.

Estrogen and Progesterone Receptors Estrogens coordinate developmental and reproductive functions, contribute to bone mass maintenance, regulate a variety of brain functions including verbal memory and learning, and modulate hepatic production of proteins involved in lipid homeostasis. Progesterone favors pregnancy maintenance and contributes to activity regulation. An important pharmacological application of estrogens (estradiol, ethinyl estradiol, estrone) and progestins (medroxyprogesterone acetate, desogestrel, norgestimate) is contraception. Their continued use decreases production of gonadotropins and follicle-stimulating and luteinizing hormones, through the hypothalamus–pituitary–gonads feedback system, thereby preventing ovulation. ER and PR agonists are also used in replacement therapy in case of ovarian failure as polycystic ovary syndrome or menopause. Menopause is the physiological cessation of ovarian function occurring in women in their early fifties and is frequently associated with mild or moderate symptoms (e.g., vasomotor disturbances and altered mucosal function). More severe symptoms can also occur such as osteoporosis resulting in increased risk of fractures due to reduced bone mass or alterations in mnemonic and verbal functions. Drastic reduction in estrogen synthesis has also been associated with an increased risk in cardiovascular disease. Administration of estrogens eliminates vasomotor symptoms associated with hormonal deficiency and slows down bone reabsorption, whereas its effects on the cardiovascular and nervous systems are still not entirely understood. However, it seems clear that prolonged combined therapy with estrogens and progestins increases risk of cancer to the reproductive organs (breast, ovary, uterus), whereas estrogens alone have beneficial effects when administered as soon as menopause begins and for a limited time span (5 years).

The main synthetic ligands used are tamoxifen and raloxifene, which are two selective estrogen receptor modulators (SERMs) for ER, and danazol and mifepristone, two SERMs for PR. The main therapeutic indication for tamoxifen is cancer endocrine therapy, as it blocks ER activity on expression of genes that stimulate epithelial cell proliferation in mammary gland. PR antagonists are used in endometrial cancer. Mifepristone is also used to interrupt pregnancy.

The pharmacological profile of steroid receptor drugs is complex as it varies depending on the tissue (Fig. 24.5). For example, tamoxifen is an ER antagonist in mammary cells and it is used in ER-positive breast tumors to prevent tumor recurrence. However, it acts as a partial agonist in endometrial cells, causing a slight increase in uterine cancer risk, and in bone cells, where it exerts beneficial effects analogous to those induced by circulating estrogens. Similarly, raloxifene is used as an ER agonist in preventive therapy of osteoporosis, but it acts as a receptor antagonist in the endometrium, an advantageous effect as it decreases uterine cancer risk. The use of raloxifene and tamoxifen is also associated with hot flashes, a vascular effect opposite to that of estrogens.

The different pharmacological effect of a ligand in different tissues can be explained by the following hypothesis. In cells containing a limited number of AF-1-binding coactivators, antagonist ligands behave as pure antagonists since they block AF-2 in a conformation unsuitable to interact with the transcriptional apparatus and at the same time allow the receptor to recruit corepressors. Conversely, in cells containing AF-1-binding coactivators, the same molecules behave as pure or partial agonists (Fig. 24.5). Research efforts are currently directed at the identification of selective drugs "tailored" for each tissue, called SERMs: selective ER modulators. At present, three first-generation SERMs are approved for clinical use: tamoxifen and raloxifene, already discussed, and toremifene for the treatment of advanced breast cancer.

Androgen Receptor Androgens primarily act on male reproductive system, but they also have important anabolic effects and are involved in proliferation, differentiation, apoptosis, and metabolism of several cell types. AR is mainly located in the cytoplasm; binding to androgen hormones, testosterone, and dihydrotestosterone (DHT) causes receptor activation, as in the case of other intracellular receptors. Mechanisms of ligand-independent and nongenomic activation have been described for this receptor.

Receptor agonists (methyltestosterone, oxandrolone, stanozolol, fluoxymesterone) are used in replacement therapies of testicular insufficiency and hypogonadism. Flutamide and cyproterone acetate are AR antagonists in prostate, where they block AR effect on proliferation. A glutamine repeat (poly-Q tract) encoded by CAG triplets is present in the AR N-terminal region; poly-Q tract polymorphism is associated with AR reduced expression or activity. Moreover, CAG triplet expansion over a certain number of repetitions is responsible for neuromuscular degenerative diseases.

Thyroid Hormone Receptor Thyroid hormones are synthesized from iodine and thyroglobulin in the thyroid. Through interaction with TR, these hormones regulate carbohydrate, lipid, and protein metabolism; they are also indispensable for proper somatic, sexual, and nervous development and control behavior and CNS functions. The physiologic hormone thyroxine represents the therapy of choice for hypothyroidism. However, there are also synthetic agonists, such as triiodothyronine, which are administered only in emergency therapies (myxedema coma, a hypothyroidism state). By contrast, no specific TR antagonists are available to treat hyperthyroidism; thus, this disorder is generally treated with drugs that block endogenous hormone synthesis in the thyroid gland.

Vitamin D Receptor Vitamin D (ergocalciferol, or vitamin D2, and cholecalciferol, or vitamin D3) is actually a prohormone synthesized in the skin or absorbed through the diet and plays a key role in mineral metabolism. In fact, it increases intestinal absorption of calcium and phosphates and it promotes bone calcification. The use of ergocalciferol and cholecalciferol, along with calcium, is indicated in diseases associated with hypocalcaemia and in rickets and osteomalacia (inadequate bone mineralization occurring, respectively, in infancy and adulthood), which are determined by low blood levels of vitamin D or by resistance to hormone action. Mutations in the VDR receptor-encoding gene have been identified in patients with rickets: patients carrying mutations in the DBD-specifying region of the VDR gene—either nonsense mutations blocking receptor synthesis or missense mutations disrupting its DNA binding ability—exhibit a total lack of responsiveness to vitamin D. Conversely, patients carrying missense mutations in the LBD-specifying region that reduce

VDR affinity for the ligand respond to therapy with high doses of calcitriol. Specific VDR antagonists are not available.

RAR The pharmacology of the RAR and the orphan receptor (RXR) that can dimerize with it has greatly advanced after their cloning, but it is still in a very preliminary stage. Vitamin A and its natural or synthetic analogs are called retinoids. RAR activity is involved in the control of cell differentiation. The therapeutic use of vitamin A and retinoids includes treatment of acne (isotretinoin), psoriasis (etretinate), and acute promyelocytic leukemia (APL). APL is characterized by a chromosomal translocation involving RARα gene on chromosome 17 and promyelocytic leukemia gene on chromosome 15 that gives rise to a hybrid protein with a strong inhibitory transcriptional activity that blocks blood cell differentiation. Chemotherapy with all-*trans*-retinoic acid (ATRA) in APL patients results in almost complete remission.

Peroxisomal Proliferation Activator Receptors (PPARs)
Unlike other members of the receptor superfamily, the three PPAR subtypes, α, β, and γ, contain a relatively large LBP that can accommodate a wide variety of endogenous lipids, including fatty acids and their metabolic derivatives, such as eicosanoids. These ligands must be present at high concentrations (micromolar) to activate PPARs and show poor selectivity for the individual isoforms, leaving open the question whether they truly represent physiologic ligands. In any case, PPAR activation is regulated by diet and nutritional and metabolic status, and it is involved in the control of metabolic programs governing energy homeostasis through fatty acid catabolism and lipid storage. Chemically different PPAR agonists are of considerable clinical importance.

These receptors exhibit similar tissue distribution, although each one controls specific functions in lipid homeostasis.

PPARα is expressed in metabolically active tissues, particularly in the liver, where its main function is to promote fatty acids catabolism, gluconeogenesis, and ketone body synthesis. Target genes of PPARα are involved in the coordination of metabolic reactions necessary to maintain energy. Examples of factors encoded by PPARα target genes are fatty acid-binding protein (FABP), a protein that binds intracellular fatty acids; the ABC transporter family, which carry fatty acids in peroxisomes for β-oxidation; and CYP4A liver enzymes, which catalyze ω-oxidation. Fibrates are PPARα agonist drugs that lower plasma levels of triglycerides via activation of lipid catabolism. Indeed, fibrates are used as lipid-lowering agents in the treatment of dyslipidemia (hypertriglyceridemia); drugs such as clofibrate or gemfibrozil have been used for decades to decrease triglycerides and LDL and to increase HDL, with consequent reduction of cardiovascular events in dyslipidemia patients.

PPARγ is primarily involved in the regulation of both white and brown adipose tissues. It promotes fat storage by stimulating adipocyte differentiation and expression of proteins involved in lipogenesis. Indeed, PPARγ ectopic expression in nonadipocyte cells, obtained experimentally *in vitro*, converts these cells into adipocytes, whereas PPARγ ablation in fibroblasts during embryogenesis abolishes their conversion to adipocytes. PPARγ is the receptor of drugs known as thiazolidinediones (TZDs), such as rosiglitazone and pioglitazone, which are indicated for type 2 diabetes (insulin independent) both as monotherapy and in combination with sulfonylurea, metformin, or insulin when diet, physical activity, or monotherapy fails to reduce blood glucose levels. By increasing glucose utilization in muscle and inhibiting hepatic gluconeogenesis, these drugs improve insulin action in the skeletal muscle and liver and lower blood glucose. PPARγ activation by TZD induces expression of a set of genes responsible for fatty acid transport and storage and involved in de novo adipogenesis. The increased ability of adipose tissue to uptake and store lipids causes a redistribution of fat from the muscle and liver toward adipocytes, thereby reducing negative effects induced by lipids on insulin signaling pathway in the muscle and liver. In humans, adipose expansion occurs in the subcutaneous tissue but not in the visceral sites. Furthermore, PPARγ activation in adipose tissue reduces adiponectin, resistin, and TNFα expression that alter insulin signaling pathway and thus increase insulin resistance of the skeletal muscle and liver. Finally, TZDs exert beneficial effects on cardiovascular parameters, such as lipid profile, blood pressure, inflammatory biomarkers, endothelial function, and fibrinolysis. Unfortunately, the use of TZDs is associated with weight gain and edema as secondary events to increased adiposity; these are likely a consequence of PPARγ activation in the kidney collecting duct, where it induces sodium absorption and water retention.

Since diabetic patients have increased risk for cardiovascular disease and many of them show preexisting heart disease, it is important to carefully monitor the edema that sometimes occurs during TZD administration, as it may be a sign of congestive heart failure. Therefore, it is recommended to start TDZ therapy at low doses in diabetic patients with mildly symptomatic heart disease, whereas the therapy is not recommended for patients with overt congestive heart failure.

Unlike the other two isoforms, little is still known about PPARβ. This receptor is expressed in many tissues, and it is also a regulator of fatty acid catabolism and energy homeostasis.

Basic research is currently focusing on the development of two new classes of ligands.

One class consists of tissue-selective PPARγ modulators, which induce receptor activation in certain tissues (e.g., adipose tissue) and are inactive in others (e.g., kidney). As mentioned earlier, the availability of transcriptional cofactors varies among different tissues; therefore, it is desirable to identify compounds that stimulate receptor interaction with cofactors present in adipose tissue but not in the kidney. Unfortunately, rosiglitazone for the suggested increased risk of cardiovascular events was put under selling restrictions in the United States and withdrawn from the market in Europe and is now under revision; pioglitazone is also under control in some countries for the suspected increase of bladder cancer risk.

The other class consists of dual agonists that activate both PPARγ and PPARα. These would be beneficial to simultaneously treat hyperglycemia and dyslipidemia, which are concurrent events in diabetic pathology.

LXR Receptors LXR receptors act as cholesterol sensors and respond to high concentrations of natural oxysterols by activating the transcription of genes controlling transport (e.g., the ABCA subfamily of cholesterol transporters), catabolism (such as 7a-hydroxylase CYP7A1) and elimination of cholesterol and fatty acids. LXRβ is expressed ubiquitously, while LXRα is present in those tissues where lipid metabolism occurs, such as the liver, adipose tissue, kidney, intestine, lung, adrenergic glands, and macrophages.

FXR Receptor The organic ligands of FXR are some bile acids, such as chenodeoxycholic and cholic acids and their respective conjugated metabolites. FXR is highly expressed in the enterohepatic system, where it acts as a sensor of bile acids and protects the body from high concentrations of these substances. This effect occurs via transcriptional induction of transporters that mediate bile acid efflux and secretion or translocation into the portal circulation. Moreover, FXR exerts its protective effect by increasing the expression of the small heterodimer partner (SHP) protein, a nuclear factor that represses the expression of enzymes involved in bile acid synthesis.

CAR and SXR Receptors

Detoxification and elimination of exogenous chemicals (xenobiotics) and lipids are regulated by CAR and SXR. CAR receptor, named after the initial observation of its constitutive activity, mediates the response to substances with a structure similar to phenobarbital, such as pesticides, benzene, certain androgens, and muscle relaxant zoxazolamine. It is expressed in the liver, where it regulates the expression of CYP2B, an enzyme involved in drug metabolism (see Chapter 6).

SXR receptor and its ortholog in rodents, PXR, are even more relevant for pharmacokinetics. They are activated by drugs as well as by environmental contaminants, steroids, and toxic bile acids. These receptors are mainly expressed in the liver and small intestine, where they trigger the expression of CYP3A, the enzyme responsible for the metabolism of more than 60% of all drugs, and MDR2 transporter, which is involved in drug resistance.

TAKE-HOME MESSAGE

- Intracellular receptors are transcription factors regulated by ligands.
- Activation by ligands may involve dissociation from inhibitory proteins, dimerization, translocation to the nucleus (if receptor is cytoplasmic), interaction with hormone-responsive elements present in the target gene promoters, and modulation of target gene expression.
- The DNA binding domain of intracellular receptors contains "zinc fingers" required for interaction with DNA.
- Receptor activation may occur by agonist-independent mechanisms.
- They control different biological reactions involved in the regulation of reproduction, growth and development, energy metabolism, and immune system.
- Subclasses of intracellular receptors coordinate transcription programs in different target organs with many interconnections between them.
- These receptors are targets of several hormones and related agonists and antagonists and other new potential drugs.

FURTHER READING

Bookout AL, Jeong Y, Downes M et al. (2006). Anatomical profiling of nuclear receptor expression reveals a hierarchical transcriptional network. *Cell*; 126:789–99.

Dasgupta S, Lonard DM, and O'Malley BW. (2014). Nuclear receptor coactivators: master regulators of human health and disease. *Annu Rev Med*; 65:279–92.

Deblois G and Giguère V. (2008). Nuclear receptor location analyzes in mammalian genomes: from gene regulation to regulatory networks. *Mol Endocrinol*; 22:1999–2011.

Della Torre S, Benedusi V, Fontana R, and Maggi A. (2014) Energy metabolism and fertility—a balance preserved for female health. *Nat Rev Endocrinol*; 10:13–23.

Evans RM and Mangelsdorf DJ. (2014). Nuclear receptors, RXR, and the big bang. *Cell*; 157:255–66.

IUPHAR. (2006). Compendium on the pharmacology and classification of the nuclear receptor superfamily. *Pharmacol Rev*; 58: 684–6.

Kadmiel M and Cidlowzki JA. (2013). Glucocorticoid receptor signaling in health and disease. *Trends Pharmacol Sci*; 34: 518–30.

Matsumoto T, Sakari M, Okada M et al., (2013). The androgen receptor in health and disease. *Annu Rev Physiol*; 75:201–24.

Wahli W and Michalik L. (2012). PPARs at the crossroads of lipid signaling and inflammation. *Trens Endocrin Metab*; 23:351–63.

25

RNA MOLECULE AS A DRUG: FROM RNA INTERFERENCE TO APTAMERS

VALERIO FULCI AND GIUSEPPE MACINO

By reading this chapter, you will:

- Know the mechanism of action of miRNAs.
- Understand the basic principles for the rational design of therapeutic siRNAs.
- Know the chemical modifications added to oligonucleotides and the effects on their pharmacological properties and the methods for oligonucleotide *in vivo* delivery.
- Learn the definition of aptamers and their use.
- Understand how splicing pattern can be engineered using oligonucleotides and possible applications in therapy.

In molecular biology textbooks, the mechanisms of action of drugs inhibiting specific cellular processes are often taken as typical examples to explain how drugs work. In most cases, these drugs are small molecules that exert their function by specifically binding cellular enzymes. These molecules are well represented by many antibiotics, by inhibitors of different classes of protein kinases currently used in antitumor therapy, and by more recent drugs that are able to affect the activity of histone-modifying enzymes. All these molecules interfere with the activity of specific proteins, which represent the last step of the gene expression process that from DNA takes to the synthesis of the enzymes catalyzing the biochemical reactions required for cell homeostasis.

However, great advances in molecular biology during the last decades have led to a detailed comprehension of the mechanisms controlling gene expression, paving the way to therapeutic interventions at not only the level of proteins but also at the DNA and RNA level.

Some approaches, collectively referred to as gene therapy, aim at the durable and inheritable insertion of genes into the genome of patients' somatic cells. In general, the goal of such approaches is the introduction of a functional copy of a given gene to compensate for the loss of function of the corresponding endogenous gene (see Chapters 22 and 23). Nevertheless, other techniques, such as introduction of short DNA or RNA molecules (oligonucleotides) capable of interfering with cell physiology without permanently altering the genome, are now emerging. These techniques will be described in this chapter. A glossary of relevant terms is provided in Box 25.1.

MECHANISMS OF ACTION OF RNA DRUGS

The use of DNA and RNA oligonucleotides as drugs was propelled by the discovery of the gene silencing phenomena. Over the last two decades, researchers have discovered, first in fungi and plants and later in Metazoa, cellular mechanisms allowing tiny RNA molecules to "interfere" (hence the name of RNA interference) with gene expression at posttranscriptional level. Briefly, the mechanism employs short single-stranded RNA molecules (20–24 nt long) to specifically identify, by base pairing, target messenger RNA (mRNA) molecules and trigger their degradation in the cytoplasm.

These short RNA molecules have been called small interfering RNAs (siRNAs).

In laboratory, siRNAs are routinely used in *in vitro* experiments to switch off selected genes and study their function.

General and Molecular Pharmacology: Principles of Drug Action, First Edition. Edited by Francesco Clementi and Guido Fumagalli.
© 2015 John Wiley & Sons, Inc. Published 2015 by John Wiley & Sons, Inc.

> **BOX 25.1 GLOSSARY**
>
> **Argonaute** Proteins belonging to the Argonaute family are conserved in higher eukaryotes, from fungi to Metazoa. These proteins are involved in gene silencing. Human cells contain 4 Argonaute proteins (AGO1–4) involved in posttranscriptional regulation of gene expression mediated by miRNAs. Moreover, AGO2 catalyzes endonucleolytic cleavage of RNA molecules that are fully (and not only partially) complementary to siRNAs.
>
> **MicroRNA (miRNA)** Small endogenous RNA molecules (18–24 nt long) involved in posttranscriptional regulation of gene expression. They guide Argonaute proteins to specific mRNA targets they recognize through base pairing.
>
> **Posttranscriptional gene silencing (PTGS)** It collectively refers to mechanisms that negatively regulate gene expression by inducing mRNA degradation or inhibition of mRNA translation.
>
> **RNA-induced silencing complex (RISC)** It is a ribonucleoprotein complex whose key components are a small RNA (18–24 nt long) and a protein belonging to the Argonaute family (AGO1–4 in humans). The small RNA is used as a molecular probe to recognize mRNA molecules with a partial or perfect complementarity. In the case of imperfect pairing (including a stretch of 6–8 nucleotides at 5′ end of the small RNA pairing to the target sequence), RISC inhibits translation of target mRNA and in some cases reduces its stability. On the other hand, in the case of perfect complementarity, AGO2 catalyzes endonucleolytic cleaveage of target mRNA.
>
> **Small interfering RNA (siRNA)** It is a small RNA molecule (about 22 nt long) perfectly complementary to a region of a target RNA (usually mRNA). Upon delivery into an animal cell, siRNAs trigger specific and efficient degradation of complementary RNA molecules by endonucleolytic cleavage.

However, small RNA oligonucleotides have also several possible therapeutic applications that are currently being studied.

Approaches in which antisense oligonucleotides have been used to alter cell splicing patterns are extremely promising.

A further application of RNA in pharmacology is represented by aptamers. In this case, the therapeutic effect of the RNA molecule is not due to base pairing with complementary nucleic acids in the cell, but to the three-dimensional structure of the molecule. This approach takes advantage of the striking plasticity of the RNA molecule to select oligonucleotides that in physiologic conditions bind with high specificity to target proteins inhibiting their function.

RNA Interference

RNA interference relies on a complex containing proteins and small RNAs to finely tune gene expression. The molecular complex mediating RNA interference in eukaryotic cells is known as RNA-induced silencing complex (RISC), and its catalytic engine is represented by proteins belonging to the Argonaute family, which are conserved in all eukaryotes. In human cells, Argonaute2 (AGO2) protein is associated with small endogenous RNA molecules that serve as molecular probes to identify complementary cellular mRNAs. Targeted mRNAs are either degraded or translationally inhibited.

The outlined mechanism is highly specific and efficient. In fact, in a human genome, consisting of about three billion bases, the probability to find more than once a 20 nt sequence is negligible (there are more than one thousand billion different combinations of 20 consecutive letters using a 4-letter alphabet). AGO2 and the other members of the Argonaute family can bind not only siRNAs but also other small RNAs and in particular the large family of microRNAs (miRNAs) encoded by the human genome.

Unlike siRNAs, which guide the endonucleolytic cleavage of perfectly complementary target mRNAs, miRNAs can recognize partially complementary sequences, often found in the 3′ untranslated region (3′ UTR) of target mRNAs, and, in Metazoa, elicit translation repression and, in some cases, decrease target mRNA stability. In Metazoa, miRNAs target recognition is mediated by a short sequence, called "seed," corresponding to the nucleotides 2–8 of the miRNA molecule. According to the current vision, miRNAs act as fine-tuners of gene expression rather than switches capable of turning on and off the expression of target genes.

In theory, siRNAs should work equally well regardless of the region they recognize within the target mRNA. In fact, a siRNA triggers endonucleolytic cleavage of the target mRNA giving rise to two RNA molecules, a 5′ RNA fragment lacking the poly A tail and a 3′ RNA fragment lacking the 5′ CAP, which are both expected to be readily degraded by cell RNAases. However, as mRNAs in the cytoplasm are not linear molecules, but folded in complex secondary structures, siRNAs are usually designed to recognize mRNA regions with an open conformation to favor their access to the target sequences. Several algorithms and softwares are available to select siRNAs targeting the most suitable region of the target mRNA molecule. The situation is further complicated by the fact that most human genes give rise to multiple transcripts, differing in length (because of the existence of alternative transcription start sites and alternative polyadenylation sites) and/or in base composition (because of alternative splicing events). Therefore, to obtain a complete silencing of a given gene, it is necessary to target all transcripts encoding for a functional protein. This task may be achieved either by selecting a single siRNA pairing to a sequence contained in all transcripts of the target gene or by using a mixture of different siRNAs targeting different regions of the gene.

One of the most relevant problems regarding the use of siRNAs in therapy are the "off-target effects," resulting from siRNAs acting on unintended targets. Since siRNAs are biochemically indistinguishable from miRNAs, it is reasonable to assume that cells could use them to target molecules only partially complementary, just like miRNAs do. This could lead to side effects such as mild repression of unintended target mRNAs. Several studies exist attempting to assess the actual impact of such phenomena on cell physiology, although the relevancy of "off-target" effects is still a matter of debate in the scientific community.

A further possible cause of unspecific side effects is the activation of systemic innate immunity responses by exogenous RNA molecules (see "Systemic Pharmacokinetics and Toxicology")

Inhibition of miRNA Biological Activity by Complementary Oligonucleotides

The human genome encodes for about one thousand different miRNAs involved in the regulation of key biological processes, such as cell proliferation, metabolism, differentiation, and tumorigenesis as posttranscriptional modulators of gene expression. miRNAs recognize target mRNAs molecules by Watson–Crick base pairing. Therefore, the most obvious way to inhibit their function is to introduce an excess of an RNA molecule perfectly complementary to the miRNA of interest in the target cells. The introduction of chemical modifications, and in particular the use of locked nucleic acids (LNA, which are described in detail in section "General Chemical Structure") that give rise to highly stable LNA/RNA hybrid duplexes, allows efficient and specific tethering of miRNAs. Such approach has been shown to work efficiently *in vivo* in mice. The systemic delivery by intravenous injection of a synthetic oligonucleotide complementary to miR-122 has been shown to specifically affect hepatic processes regulated by this miRNA, displaying high specificity and low toxicity.

These approaches are particularly promising, given the involvement of miRNAs in a wide range of human pathologies. Since a single miRNA has the potential to modulate even hundreds of different genes, with a dramatic impact on cell physiology, generating oligonucleotides able to specifically inhibit single miRNAs is a key task.

Antisense Oligonucleotide to Modify the Splicing Patterns

Antisense oligonucleotides have different mechanisms of action. Eukaryotic mRNAs are transcribed as precursor (pre-mRNA) molecules containing protein-encoding sequences (exons) as well as noncoding sequences (introns), which are removed by a process called splicing before mature mRNAs are exported to the cytoplasm. This important step in mRNA maturation is accomplished in the nucleus by a complex ribonucleoproteic machinery. Intron excision requires the splicing machinery to recognize evolutionary conserved consensus sequences at the intron/exon boundaries. When the splicing machinery binds to an exon/intron boundary, it removes the intron identified by the nearest intron/exon boundary toward the 3' end of the RNA molecule.

It is therefore possible to alter the splicing pattern of a specific mRNA using antisense oligonucleotides complementary to specific consensus regions to prevent their recognition (Fig. 25.1, left side). If an intron/exon boundary is masked by an antisense oligonucleotide, the splicing machinery will use the next intron/exon boundary, resulting in the removal of the exon together with the two flanking introns. Antisense oligonucleotides represent a promising therapeutic approach to inheritable diseases caused by mutations preventing proper splicing. For example, about 50% of the mutations in the ATM gene, which is responsible for ataxia telangiectasia, cause splicing errors. *In vitro* approaches have shown that antisense oligonucleotides can restore, even at very low frequencies, the correct splicing pattern, allowing the synthesis of a sufficient amount of functional protein to revert the pathologic phenotype.

This strategy may be beneficial also when diseases are caused by mutations within an exon, resulting in a truncated or nonfunctional protein. In such cases, antisense oligonucleotides can be used to mask the intron/exon boundaries of the mutated exon, allowing its excision by the splicing machinery. This technique, named **exon skipping**, yields a shorter protein that in some cases retains enough activity to decrease the severity of the disease (Fig. 25.1b, right side). This approach has been successfully tested in a murine model of Duchenne dystrophy.

Aptamers

Aptamers are small RNA molecules that, in physiologic conditions, can bind with high specificity to specific molecular targets (often proteins) in cells inhibiting their function.

The process to develop such molecules, referred to as systematic evolution of ligands by exponential enrichment (SELEX), starts from a random population of RNA oligonucleotides and consists in repeated cycles of *in vitro* selection aimed at identifying molecules capable of high-affinity

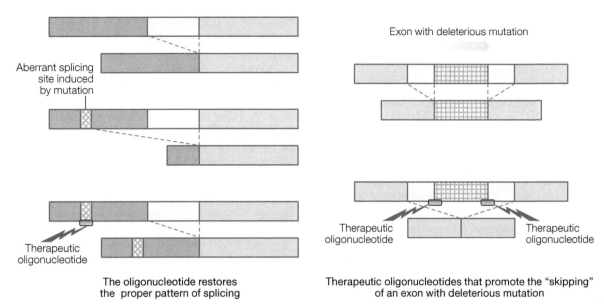

FIGURE 25.1 Mechanisms of action of RNAs designed to change splicing pattern. RNA molecules act by masking, through pairing with precursor RNA, specific consensus sequences. (Left side) Strategy to recover correct splicing pattern compromised by a mutation. (Right side) Strategy to exclude specific exons (containing a disease-causing mutation) from mature transcripts.

binding to a given target protein. After each round, binding molecules are collected, amplified, and selected again following the same procedure. After several cycles, the oligonucleotides that best interact with the target are isolated and sequenced, and their pharmacological activity is tested to identify those inhibiting target protein function. Some aptamers are already being used in clinical practice and others are being developed (Table 24.1)

DELIVERY

The main strategy employed to facilitate uptake of oligonucleotides by recipient cells consists in combining them to biologically inert compounds to enhance their solubility into biological membranes and/or stability. Given its relevance, in recent years, drug delivery has turn into a wide and autonomous research field. As for other drugs, we can distinguish delivery systems and conjugation systems. In the former case, the oligonucleotide is combined with a lipophilic compound without covalent bonding (liposomes and nanoparticles), whereas in the latter case, the oligonucleotide is covalently bound to a carrier molecule (see also Chapter 4).

Liposomes are spherical lipidic structures that can encompass large amounts of hydrophilic substances. They were first employed to deliver DNA into the cells and are still used as a delivery system in gene therapy approaches. Although packaging of oligonucleotides into liposomes is not very efficient, they are particularly interesting as they mediate direct delivery into the cells and protect nucleic acids from nucleases. For these applications, cationic lipids are employed as they easily interact with the negative charges of biological membranes. The most popular compounds are 1,2-di-O-octadecenyl-3-trimethylammonium propane (DOTMA) and 1,2-dioleoyl-sn-glycero-3-phosphoethanolamine (DOPE). A mixture of these two compounds (commercially known as Lipofectin) is widely used for delivery of oligonucleotides into cells.

Nanoparticles are colloidal delivery particles sized less than 1 μm. Currently, poly(alkyl cyanoacrylate) particles are mostly employed due to their low cost, high biodegradability, and ease of preparation. The wide use of this compound in surgical seams is a proof of its safety.

Recently, nanoparticles bearing short positively charged glucidic chains capable of binding negatively charged oligonucleotides have been developed.

Conjugation with a suitable molecule is used to improve the pharmacokinetic properties of oligonucleotides by enhancing their cellular uptake, stability, and targeted delivery to specific cell types or specific cell compartments. Oligonucleotides can be conjugated through direct covalent binding to a carrier or through electrostatic binding to a polycation that in turn can be covalently bound to a suitable molecule.

However, conjugated oligonucleotides should be considered as different molecules, depending on the function of the conjugated molecule and on the position involved in the chemical bond. For conjugated oligonucleotides to be effective, covalent binding of carrier molecule should not interfere with the oligonucleotide ability to hybridize with complementary DNA or RNA molecules and should not yield a toxic or antigenic compound.

Conjugation with cholesterol and analogs has been successfully used to increase oligonucleotide stability and to enhance their diffusion through the plasma membrane, thereby increasing their uptake by cells. Conjugation with biological ligands, such as transferrin, folic acid, and growth factors, promotes selective uptake by cell populations expressing specific receptors. In addition, methods taking advantage of transgenic bacteria (nonpathogenic *E. coli* strains) expressing siRNAs are currently being tested to reduce expression of targeted host genes specifically in intestinal epithelium.

GENERAL CHEMICAL STRUCTURE

Oligonucleotides are anionic polymers with similar molecular structure and biochemical properties; the ones that are currently studied have a molecular weight between 5000 and 20,000 Da and are synthesized automatically by cyclic reactions on solid support. Analysis and purification are performed using the same methods employed in nucleic acid research.

The molecular structure determines oligonucleotide stability in biological environments as well as its ability to cross cell membranes. New pharmacological strategies are based on molecules sharing similar biochemical properties, effects on cultured cells, and systemic pharmacotoxicology. The current task is the progressive improvement of parameters (cellular uptake, stability, biosafety) that will allow translation of promising compounds tested in experimental models into successful drugs. The main challenge is the limited uptake of oligonucleotides by the cell, which is due to their size and polyanionic nature. Indeed, being large hydrophilic polymers, oligonucleotides are unable to cross cell membranes by passive diffusion. The uptake mechanisms of naked oligonucleotides are still poorly understood but are certainly insufficient to achieve the desired therapeutic effects. One of the main tasks is to increase the intracellular concentration of oligonucleotides through modifications of the chemical structure or by delivery mediated by highly lipophilic carrier molecules. The integrity of oligonucleotides in a biological environment mainly depends on the activity of ubiquitous enzymes: exonucleases, which degrade oligonucleotides starting from either the 3′ or the 5′ end, and endonucleases, which catalyze oligonucleotide cleavage. Nucleases are ubiquitous in cells, on membranes, and in biological fluids (mainly plasma and cell culture medium). Therefore, chemical modifications of oligonucleotides aim at increasing both their cellular uptake and their stability in the biological environment without affecting their ability to hybridize with target nucleic acids. It is important to remind that any alteration of the original structure implies a change in the chemical and physical properties of the molecule, possibly affecting its biological activity. Thus far, modifications have been introduced in the oxydril groups of phosphates, in the 2′ oxydril group of ribose, and, to a lesser extent, in the bases themselves. Computer simulations of the molecular structure have been used to design new compounds that are not present in nature. Among the most popular compounds, it is worth mentioning phosphorothioates, phosphoramidates, LNAs, peptide nucleic acids (PNAs), and 2′-O-alkyl ribonucleotides (Fig. 25.2).

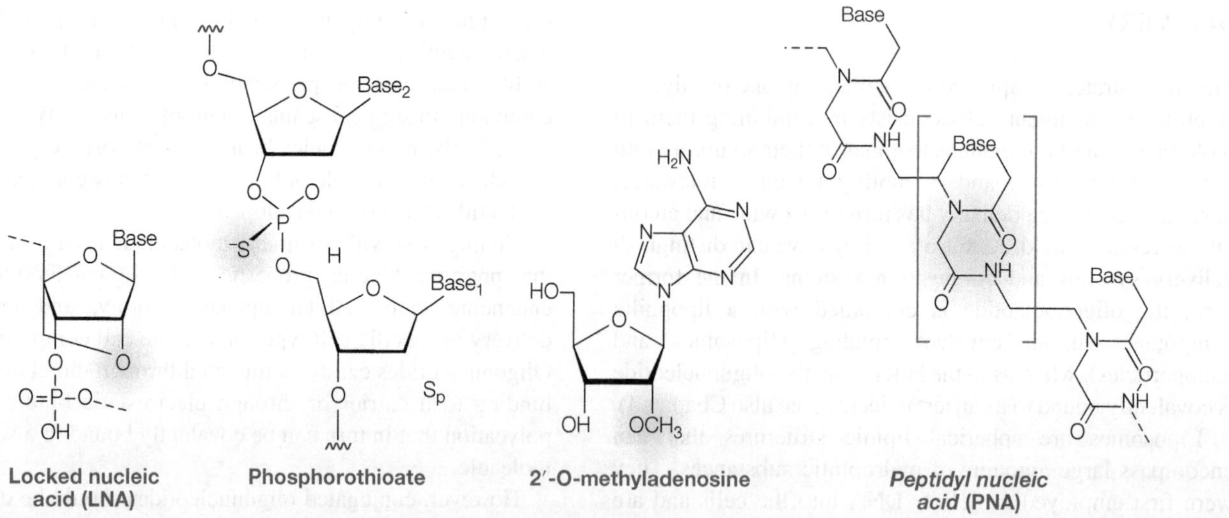

FIGURE 25.2 Some of the most common chemical modifications used to generate therapeutic oligonucleotides. The modified portion of the molecule are shaded. See text for details on single modifications.

Phosphorothioates

Phosphorothioates are analogs of natural DNA oligonucleotides in which one of the oxygen atoms of the phosphate group not involved in a phosphodiester bridge is replaced by a sulfur atom. Since phosphate is a stereocenter, phosphorothioates are diastereomers and are therefore used as diastereomer mixtures. Their success is mostly due to their high resistance to nucleases, ease of synthesis, and efficient hybridization with endogenous nucleic acids as well as to their ability to activate RNase H. However, compared to natural DNA and RNA, phosphorothioates are poorly taken up by cells.

Phosphoramidates

Phosphoramidates are analogs of natural DNA oligonucleotides in which one of the nonbridging oxygen of the phosphate group is substituted by nitrogen. Nitrogen can subsequently be replaced to introduce novel modifications. Their biological properties are quite similar to those of phosphorothioates, and they do not seem to offer any special advantage over phosphorothioates.

2′-O-Alkyl-Ribonucleotides

2′-O-Alkyl-ribonucleotides are a class of RNA derivatives in which the 2′ oxydril group is esterified with an alkyl group. Methyl ($-CH_3$) and allyl ($-CH_2-CH=CH_2$) are the alkyl groups most commonly used. Their biological stability is comparable to that of DNA, and in general, their lipophilicity and resistance to nuclease increase with the length of the alkyl chain. These molecules are employed in therapeutic strategies based on RNA oligonucleotides and to replace few bases in longer DNA oligonucleotides.

LNAs

LNAs are a class of nucleic acid analogs in which the ribose ring is locked by a methylene bridge connecting the 2′ oxygen with the 4′ carbon. Such modification drastically reduces the plasticity of the molecule locking the ribose in the most favorable conformation to form long double-stranded molecules. Oligonucleotides containing one or more LNA-modified residues form RNA/LNA and DNA/LNA hybrids that are very stable. The LNA modification is particularly suitable for the inhibition of miRNAs activity. In fact, short LNA molecules complementary to specific miRNAs can efficiently sequester them, inhibiting their activity.

PNA

PNAs have been obtained by molecular modeling. Their chemical structure is based on the substitution of the phosphodiester backbone with a peptide backbone to which purine and pyrimidine bases are linked in a way that keeps them at the same distance as in natural DNA, enabling them to recognize and pair to bases on a complementary nucleic acid. PNAs are nonionic and resistant to nucleases, proteases, and peptidases. The advantages they offer include ease of synthesis, possibility to adjust chemical and physical properties, and high stability of PNA/DNA and PNA/RNA hybrids. However, given their inability to cross cell membranes, so far, they have not been used *in vivo*, and currently, they are mainly employed in diagnostics.

Modifications of Purine and Pyrimidine Bases

Incorporation of modified bases can improve the pharmacokinetics of oligonucleotides if Watson–Crick base pairing is not prevented. Carbon atoms at positions 5 and 6 of pyrimidines are good targets for modifications that make nucleosides more lipophilic, thus enhancing their permeability through the cell membrane.

Recent findings pinpoint that 5-methyl pyrimidines confer stability to double-stranded molecules by displacing water molecules from the major groove of the duplex. Froehler has suggested taking advantage of such phenomenon by introducing a propyl group at position 5 of deoxycytidine and deoxyuridine. The consequences of such substitutions have been studied in double- and triple-helix molecules, obtaining encouraging results also in cellular systems.

Covalent conjugation with a variety of molecules has also been attempted to foster oligonucleotide entry into cells. Conjugation with cholesterol has proved extremely promising, as cholesterol does not interfere with the biologic activity of oligonucleotides but greatly increases their solubility in the plasma membrane.

PHARMACOTOXICOLOGY

A molecule with a biological activity can be profitably used in clinical practice only if it satisfies a rigorous list of requirements that can be verified only through a long and expensive process, which may ultimately give a negative outcome. Like any novel drug, also oligonucleotides need to undergo a strict evaluation of pharmacokinetics, pharmacodynamics, and toxicology before being approved as therapeutics.

Cellular Pharmacokinetics and Toxicology

Oligonucleotides are large polymers resembling the products of nucleic acid fragmentation that are normally present in cells. Their detection within living cells requires labeling with either radioactive or fluorescent markers. Given their size and polyanionic nature, cellular uptake of oligonucleotides is usually limited to 3% of the extracellular concentration. The uptake is dose and temperature dependent and can be enhanced by chemical modifications such as cholesterol conjugation. Little is known about the mechanisms whereby oligonucleotides are absorbed by cells. However, there is evidence that most oligonucleotides are taken up by pinocytosis or receptor-mediated endocytosis. The intracellular localization of oligonucleotides is still a matter of debate. It has been suggested that oligonucleotides may first localize into endosomes and only after fusion with lysosomes and partial degradation may be released into the cytoplasm. Oligonucleotide half-life is strongly affected by chemical modifications. Phosphodiester backbones are readily catabolized (within minutes), while phosphorothioates and 2'-O-alkyl derivatives are stable for hours. Efflux mechanisms have been observed having the same kinetics of uptake mechanisms.

Toxic effects emerged during the earliest studies on oligonucleotides and have been drastically reduced by improving purification processes. Toxicity may depend on interaction of oligonucleotides (or their metabolites) with cellular proteins and/or permanent genome modifications. A sequence-specific toxicity may be a consequence of the aptameric properties of oligonucleotides. Several oligonucleotides used as negative controls have been shown to specifically bind cellular proteins and to affect their functions. In particular, phosphorothioates can bind, even in a sequence-independent manner, to serum and cellular proteins mimicking heparin action.

Moreover, nucleotide residues released during oligonucleotide catabolism may imbalance endogenous metabolism. There is no evidence of any mutagenic effect of oligonucleotides.

Systemic Pharmacokinetics and Toxicology

The half-life of natural oligonucleotides delivered by intravenous injection is very short due to degradation by serum nucleases and absorption by peripheral tissues. Studies aiming at defining pharmacokinetics of more stable compounds have mainly focused on phosphorothioates and have confirmed in humans the encouraging results obtained in animal models. It has been shown that oligonucleotides are delivered to all organs with the exception of central nervous system and are mainly concentrated in the liver, kidney, and bone marrow. A strong binding of phosphorothioates to serum proteins, especially albumin and beta-2-microglobulin, has been reported, accounting for the reduced renal clearance and suggesting a role for serum as "reservoir." In summary, pharmacokinetic of oligonucleotides has proved very promising, in contrast with earlier concerns regarding their molecular size and physical and chemical properties. However, it is worth mentioning that synthetic siRNAs are efficiently recognized by Toll-like receptor 7 (TLR-7) and TLR-8, thereby triggering production of proinflammatory cytokines by innate immunity cells. Such effect is mitigated by modifications of the 2' oxygen on the ribose ring of synthetic oligonucleotides.

Nowadays, research in humans has provided a large amount of data. Phosphorothioates have been delivered by slow intravenous injection up to 9 mg/kg in a single dose or in repeated doses of 3 mg/kg. Different trials have confirmed the absence of long-term toxic effects, whereas reversible short-term effects have been described during delivery. In particular, minor hypotension, fever triggered by interleukin-1 and interleukin-6 releases, moderate activation of complement fraction 3, thrombocytopenia, and fatigue have been observed. In summary, pharmacokinetics and pharmacotoxicology studies in humans have confirmed previous observations in animal models (LD50 180–360 mg/kg in mouse and rat, 40 mg/kg in monkey) and yielded promising data, beyond expectations.

PRESENT USE AND FUTURE PERSPECTIVES

The ambitious goal of today's scientists is the identification of tools allowing simple, rational, standardized, and fast design of specific drugs targeting each single gene. So far, several oligonucleotides have reached phase I and II clinical trials. Recently, siRNAs targeting tenascin C oncogene have been used to treat patients affected by glioma during surgery, with significant increase of life expectancy. Silence therapeutics is currently studying a drug based on siRNAs targeting PKN3 gene for treatment of hepatic metastasis in colorectal cancer. A siRNA directed against SYK kinase, to be delivered by inhalation, is currently investigated for the treatment of asthma.

Moreover, strategies employing several siRNAs directed against different oncogenes at the same time are currently being studied for the treatment of glioma and breast cancer. A list of ongoing clinical trials is provided in Table 25.1 and on the CD attached to the text. The main challenges in developing oligonucleotide-based drugs include poor delivery to several cell types, short half-life in extracellular environment, and unwanted activation of immune responses, including proinflammatory cytokine production. Nevertheless, researchers are confident that chemical modifications and novel approaches to improve delivery *in vivo* will turn oligonucleotides into established therapeutics in the future.

TABLE 25.1 Oligonucleotides/siRNA in clinical trials for treatment of diseases

Company	Product	Target gene	Pathology	Delivery	Status
Alnylam Pharmaceuticals	ALN-RSV01	RSV	RSV infection	Insufflation	Phase II
ZaBeCor	Excellair	SYK kinase	Asthma	Inhalation	Phase II
Alnylam Pharmaceuticals	ALN-VSP01	VEGF e KSP	Liver cancer	Intravenous	Phase I
Quark Pharmaceuticals	RTP801i-14	RTP801	AMD	Intraocular	Phase I
Quark Pharmaceuticals	QPI-1002	p53	Acute renal failure	Intravenous	Phase I
Calando Pharmaceuticals	CALAA-01	M2 subunit of ribonucleotide reductase	Solid tumors	Intravenous	Phase I
Nucleonics	NUC B1000	HBV	HBV infection	Intravenous	Phase I
TransDerm	TD101	Single-nucleotide mutation in keratin 6a	PC	Intradermal	Phase I
Silence	Atu-027	PKN3	Solid tumors	Intravenous	Phase I
Tekmira	ApoB SNALP	ApoB	Hypercholesterolemia	Intravenous	Phase I
Sylentis	SYL-040012	ADRB2	Glaucoma	Intraocular	Phase I
Quark Pharmaceuticals	QPI-1007	Caspase-2	Glaucoma	Intraocular	Phase I

TAKE-HOME MESSAGE

- The RISC complex employs small RNAs (siRNAs, miRNAs) as molecular probes to specifically recognize mRNA molecules and induce their silencing through degradation or translation inhibition.
- Small (22 nt-long) RNA molecules may be introduced into mammalian cells to turn off gene expression with high specificity.
- Aptamers are small RNA molecules that specifically bind to and inhibit the biological activity of target proteins and are promising future drugs.
- Small RNA molecules complementary to specific regions of pre-mRNAs may be used to alter splicing patterns *in vivo* and treat genetic diseases.
- Chemical modifications are used to increase oligonucleotide stability or delivery *in vivo*.

FURTHER READING

Bennett C.F., Swayze E.E. (2010). RNA targeting therapeutics: molecular mechanisms of antisense oligonucleotides as a therapeutic platform. *Annual Review of Pharmacology and Toxicology*, 50, 259–93.

Bouchard P.R., Hutabarat R.M., Thompson K.M. (2010). Discovery and development of therapeutic aptamers. *Annual Review in Pharmacology and Toxicology*, 50, 237–57.

Chitwood D.H., Timmermans M. (2010). Small RNAs are on the move. *Nature*, 467, 415–19.

Dausse E., Da Rocha Gomes S., Toulmé J.J. (2009). Aptamers: a new class of oligonucleotides in the drug discovery pipeline? *Current Opinion in Pharmacology*, 9, 602–7.

Du L., Gatti R.A. (2009). Progress toward therapy with antisense-mediated splicing modulation. *Current Opinion in Molecular Therapeutics*, 11, 116–23.

Keefe A. Pai S., Ellington A. (2010). Aptamers as therapeutics. *Nature Reviews Drug Discovery*, 9, 537549.

Krützfeldt J., Rajewsky N., Braich R., Rajeev K.G., Tuschl T., Manoharan M., Stoffel M. (2005). Silencing of microRNAs in vivo with 'antagomirs'. *Nature*, 438, 685–9

Pereira D.M., Rodrigues P., Borralho P.M., Rodrigues C. (2013). Delivering the promise of miRNA cancer therapeutics. *Drug Discovery today*, 18, 282–9.

Pillai R.S., Bhattacharyya S.N., Filipowicz W. (2007). Repression of protein synthesis by miRNAs: how many mechanisms? *Trends in Cell Biology*, 17, 118–26.

Tiemann K., Rossi J.J. (2009). RNAi-based therapeutics-current status, challenges and prospects. *EMBO Molecular Medicine*, 1, 142–51.

Veedu R.N., Wengel J. (2009). Locked nucleic acid as a novel class of therapeutic agents. *RNA Biology*, 6, 321–3.

Wagner R.W., Matteucci M.D., Grant D., Huang T., Froehler B.C. (1996). Potent and selective inhibition of gene expression by an antisense heptanucleotide. *Nature Biotechnology*, 14, 840–4

Whitehead K.A., Langer R., Anderson D.G. (2009). Knocking down barriers: advances in siRNA delivery. *Nature Reviews Drug Discovery*, 8, 129–38.

SECTION 6

REGENERATIVE MEDICINE

26

REGENERATIVE MEDICINE AND GENE THERAPY

LUCIANO CONTI AND ELENA CATTANEO

By reading this chapter, you will:

- Learn source, origin, and functional differences among different types of stem cells and their biological properties
- Know the wide range of therapeutic opportunities arising from the discovery and use of stem cells in hematological, epithelial, cardiovascular, muscular, and neurodegenerative diseases
- Acquire the main legal issues related to cell-based drugs and their production
- Know the theoretical and practical basic concepts on gene therapy approaches to monogenic diseases and cancer

---

Recent advances in cell biology and stem cell research have led to the idea of using cells as therapeutics, in contrast to the traditional pharmacological approach employing molecules as drugs. Cell-based drugs are currently being developed as novel members of the large family of biological drugs that includes also monoclonal antibodies, ribozymes, and other biotechnological products with pharmacological properties. This has prompted a revolution in biomedical research that is increasingly focusing on potential applications of stem cell-based therapies. Significant advances have been made in this direction, but it is important to remember that to date clinical use of cellular therapies is restricted only to a small number of pathologies.

The concept of cell-based therapy was introduced in the late 1950s when the bone marrow transplant procedure was being established. The first transplant in humans was carried out by a French oncologist, Georges Mathé, who injected healthy bone marrow into six physicists whose bone marrow had been compromised by heavy radiation exposure during a nuclear reactor accident in Yugoslavia. In the following years, the transplant technique was improved, and a significant contribution came from Edward Donnall Thomas' studies demonstrating that bone marrow cells, delivered by intravenous infusion, can repopulate the bone marrow of receivers and produce new blood cells. For these studies, in 1990, Edward Donnall Thomas was awarded of the Nobel Prize for Physiology or Medicine (shared with Joseph Murray, who performed the first kidney transplant in 1954). Since then, thousands of patients affected by leukemia or immunodeficiency have been cured by bone marrow transplant. It has been estimated that currently about 70,000 bone-marrow transplants are performed worldwide every year.

During the last decade, cell-based therapies have been developed also for nonhematological disorders, with those applied to epithelial and corneal diseases being particularly successful. What diseases cell-based therapies will be able to cure in the future are hard to predict, as well as which types of stem cells will prove more effective drugs. Nevertheless, medical scientists are confident that a more in-depth understanding of biological features and handling requirements of the different stem cell types will lead to new therapeutic strategies for cardiac diseases, diabetes, muscular dystrophy, and neurological diseases. However, they are also aware that for such an ambitious goal to be reached, translation to clinic will always have to be supported by strong preclinical experimental evidence and decided in full respect of patient safety.

General and Molecular Pharmacology: Principles of Drug Action, First Edition. Edited by Francesco Clementi and Guido Fumagalli.
© 2015 John Wiley & Sons, Inc. Published 2015 by John Wiley & Sons, Inc.

PRINCIPLES OF REGENERATIVE MEDICINE

In the last decades, transplant medicine has provided tools to clinically test functional rescue after tissue and organ transplantation. The aim of traditional transplant medicine is to replace defective tissues or organs, often at the final stages of a disease, with healthy ones to restore functional homeostasis. Such approach has proved successful in curing otherwise incurable leukemias, kidney failure, cirrhosis, and cardiac conditions, but its application has always been limited by shortage of donated organs and adverse effects associated with immunosuppressive therapies. This has prompted the search for alternative solutions to restore tissue and organ functionality, opening the way to the so-called regenerative medicine. Today, the goal of regenerative medicine is to exploit regenerative potential of stem cells to permanently heal and recover damaged tissues and organs. Therefore, advancements in regenerative medicine today are tightly linked to progress in stem cell research.

Stem cells and specialized cells deriving from them, either normal or engineered, represent the "active principle" of regenerative therapies.

Stem cells are physiological constituents of the organism, providing tissues with a sort of internal repair system. They are unspecialized cells with the ability to self-renew (to a different extent depending on the stem cell type), and they are able to differentiate in specific cell types. After purification from the original tissue (in those cases in which this is possible), they can be expanded *in vitro* and "instructed" to differentiate in a specific cell type. Therefore, stem cells potentially represent an unlimited source of cells to be transplanted to heal injured tissues. Such cells could also promote recovery by releasing active molecules, such as anti-inflammatory substances or growth factors, although more knowledge is needed in support of this hypothesis. Moreover, to be better suited for the purpose, prior to transplantation, stem cells could be genetically modified to add useful genes or correct endogenous mutated genes. To date, many of these therapeutic opportunities are still being investigated and are not clinically available yet.

The aim of regenerative medicine is to permanently restore damaged tissue functionality both in congenital and acquired pathologies, as well as in aging-associated diseases. The therapeutic outcome will depend on the ability of the employed stem cells to create a niche within the pathological tissue and to promote optimal levels of tissue regeneration.

Regenerative medicine follows two main strategies. The former, the *in vivo* strategy, aims at the pharmacological stimulation of endogenous stem cells within the tissues of interest in order to trigger their regenerative potential. However, this approach is still in its infancy. The latter, the *ex vivo* strategy, involves the *in vitro* expansion and/or modification of stem cells and their subsequent transplantation in the region of interest to promote regeneration/healing. In this chapter, we will focus mainly on *ex vivo* approaches, discussing issues and examples of cell therapy in regenerative medicine and gene therapy.

DEFINITION, CLASSIFICATION, AND FEATURES OF STEM CELLS

In general, stem cells are undifferentiated cells that can be distinguished from other cells for two main characteristics:

1. Self-renewal, that is, the ability to reproduce themselves.
2. Differentiation potential, that is, the ability to differentiate in a range of different specialized cell types. The extent of such differentiation potential varies depending on the type of stem cells. According to their differentiation potential, stem cells can be ordered in a hierarchical scheme, as shown in Figure 26.1.

From a strictly functional point of view, a stem cell can be defined as a cell capable of generating all cells of the tissue in which it is found, throughout the life of an organism. Skin or blood stem cells fully exhibit this functional feature that conversely cannot be demonstrated for most stem cells from other tissues, or it is not even compatible with the stem cell role (as in the case of brain stem cells).

Stem cells can also be classified according to the timing of their appearance (Table 26.1). Stem cells do not have all the same differentiation potentials, and depending on their plasticity, they can be classified as totipotent, pluripotent, or multipotent (Fig. 26.1).

The only totipotent cell is zygote, the only cell capable of giving rise to a complete organism. It is transient and cannot be expanded as such.

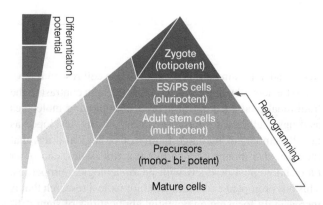

FIGURE 26.1 Hierarchical ordering of cells based on their differential potential. The zygote is placed at the top of the pyramid as it is able to generate an entire organism, including the extraembryonic tissues (totipotent). Under the zygote, there are embryonic stem (ES) cells and iPS cells (obtained by cellular reprogramming), which are able to produce all cells deriving from the three germ layers (pluripotent). Next, adult stem cells are found, which can produce only the mature cell types of the tissue they belong and their progenitors (proliferating cells that can only differentiate into one or two specific cell types). At the bottom of the pyramid, we find mature specialized cell type; even when capable of diving, these cells can only produce mature cells of their own type.

TABLE 26.1 Characteristics of the main types of stem cells

Stem cell type	iPS	ESC	Fetal SC	Blood SC	Skin SC	Brain SC	Umbilical cord SC
Source	Fetal cells, adult immature and differentiated cells	Blastocysts	Fetal tissues	Bone marrow, pheripheral blood	Skin	Brain	Umbilical cord
Differentiation potential	Pluripotent	Pluripotent	Multipotent	Multipotent	Multipotent	Multipotent	Multipotent
In vitro expansion	Yes	Yes	Depending on the tissue of origin	Limited	Yes	Yes	No
Clinical use	Phase I	Phase I	Yes	Yes	Yes	Not yet	Yes
Clinical efficacy	Not tested	Not tested	Yes	Limited	Yes	Not tested	Yes
Genetic manipulation	Yes	Yes	Limited	Limited	Limited	Limited	Limited
Cryopreservation	Yes	Yes	Yes	Yes	Yes	Yes	Yes
Tumorigenic potential	Yes	Yes	Some cases	Not seen	Not seen	Not seen	Not seen

Two cell types belong to the group of pluripotent stem cells: the embryonic stem cells (ESCs), which are transiently present in the blastocyst, and the induced pluripotent stem cells (iPSCs), which have been recently obtained by *in vitro* reprogramming of mature adult cells. In contrast to zygotes, pluripotent cells cannot differentiate into a complete organism. Nevertheless, they retain a high differentiation potential and can produce cells belonging to all three germ layers (ectoderm, mesoderm, and endoderm). This means that they can generate even more than 250 cell types making up fetal and adult organisms, which include both functionally mature specialized cell types and tissue-specific stem cells. However, in contrast to zygotes, pluripotent cells are unable to generate extraembryonic tissues (trophectoderm and placenta).

The class of multipotent stem cells includes adult tissue-specific stem cells (also called somatic stem cells) that are found in fetal and adult tissues and can differentiate only in the specific cell types of the tissue they belong. For example, a hematopoietic stem cell (HSC) can produce the eight main types of mature, differentiated blood cells. It is important to note that even if adult stem cells from the blood, skin, muscle, kidney, nervous system, etc. are grouped in a single class, each type has specific features and a different therapeutic potential and therefore is the focus of a specific research field.

Pluripotent Stem Cells

Research on pluripotent stem cells started when Leroy Stevens, a developmental biologist working at Jackson Laboratory, discovered spontaneous testicular tumors (teratocarcinoma) in the inbred 129 mouse strain. He found that cells from malignant teratocarcinoma (named embryonal carcinoma or EC cells), cultured *in vitro* and then injected into mouse blastocysts, were able to undertake multiple differentiation programs, thus contributing to the formation of the resulting chimeric mouse.

Later, in 1981, Sir Martins Evans and Matthew Kaufman succeeded in *in vitro* culturing ESCs isolated from the inner cell mass of mouse blastocysts (the preimplantation embryo), generating stable ESC lines. In the absence of stimuli, ESCs can grow and multiply *in vitro* but retain the ability to differentiate in mesodermic, endodermic, and ectodermic cells when exposed to appropriate stimuli. When ESCs are introduced into blastocysts that are then reimplanted in a pseudopregnant mouse female, they actively participate in embryo formation and give rise to chimeras, thus proving capable to differentiate in cells deriving from all three germ layers also *in vivo*. The ability to produce chimeras is used to test the "stemness" of a given cell population. It is important to note that mouse ESCs do not contribute to extraembryonic tissues, such as the trophoblastic part of the placenta, indicating that their differentiation potential is not as wide as that of zygotes. A specific feature of mouse ESCs consists in their ability to form teratomas when injected into immunodeficient mice. Teratomas are benign tumors containing a variety of cells, either highly or partially differentiated, deriving from different germ layers; this observation confirms the huge differentiation plasticity of ESCs.

In 1998, 17 years after the isolation of mouse ESCs, a paper was published in *Science* in which James Thomson and coworkers reported the derivation of ESC lines from human blastocysts produced by *in vitro* fertilization and donated after informed consent and review board approval. The result represented a significant achievement, but the procedure described by the authors involved human blastocyst dissociation. This ignited a worldwide debate on whether destroying human blastocyst could be considered ethically acceptable. The ethical controversy implies a choice between two moral duties: on one hand the duty to alleviate suffering (a goal that is pursued by the research on human embryonic stem (hES) cells) and on the other the duty to respect human life. To this respect there is discussion on whether one should consider the blastocyst a "person" just like the patients who might benefit from therapies resulting from research on hES cells. The debate is not resolved but different countries have decided to regulate research on hES cell in different ways. In the United Kingdom, Belgium, and Switzerland, it is legal to derive hES cells from blastocysts produced in excess during *in vitro* fertilization procedures. In other countries, such as Germany, only cell lines obtained before a given date (2002) were initially permitted but in 2008, as a result of pressure from scientists, the German Act was amended to move the cut-off point to 1 May 2007. In other countries, like Italy, derivation of new lines is not permitted but usage of lines obtained by others is accepted; therefore, research on hES cells is permitted but only on imported cell lines. In the United States, during the Bush administration, no public funding was allocated to research on human embryonic cell lines produced after 2001. Nevertheless, such research was permitted and went on with the support of private funding (and led to significant discoveries such as the iPSCs).

What is the reason of such an interest in hES cells? Like the murine counterpart, hES cells have the ability to differentiate in all cell types making up fetal and adult human body (except the extraembryonic tissues). hES cells can be induced to differentiate *in vitro* to generate (with different efficiency) epidermal and adrenocortical cells and keratinocytes, as well as endothelial, kidney, bone, muscle, heart, pancreas, and liver cells. Moreover, differentiation of hES cells in electrophysiologically mature cardiomyocytes and neurons has also been reported. Results obtained so far indicate that such differentiation potential of hES cells cannot be obtained with human adult stem cells. Studies in animal models have shown that transplants of cells derived from hES cells can successfully treat congenital conditions such as Parkinson's disease (PD) or diabetes. This highlights the potential of hES cells in regenerative medicine. However, given the ability of ESCs to form teratomas, the risk of tumor formation after stem cell therapies must be taken into account and is one of the highly studied topics of current research. Today, stem cell research aims at the transplant of specialized cells derived from ESCs. So far, two clinical trials of hES cell-based therapies have received FDA approval.

It is important to remind that hES cells have also other applications and can be used, for instance, in toxicological and pharmacological studies or to investigate disease mechanisms or the physiology of human development.

☞ One of the most revolutionary achievements in stem cell research is the discovery of the iPSCs (see Supplement E 26.1, "History and Development of iPS Cell Research"). iPSCs the result of mature adult cells, such as fibroblasts, that have been forced, by genetic reprogramming, to dedifferentiate and to reacquire the features of pluripotent cells. Therefore, iPSCs are very much the same as ESCs, but they have a different origin that does not involve blastocyst dissociation. iPSCs were first obtained by Shinya Yamanaka and his group, in Japan, and represent an outstanding breakthrough in stem cell research. Using the iPS technology, in the future, it could be possible to obtain pluripotent cells from any individual. This means that in the future iPSCs could be produced from patient's fibroblasts and subsequently instructed to generate the specific cell types, including germ cells, required to cure diseases in the same patient.

Multipotent (or Adult) Stem Cells

Many tissues in our body contain stem cells devoted to replacing damaged cells. Their abundance varies depending on the tissue, and in general, they tend to be more abundant in tissues with a higher ability (and need) to regenerate. For example, blood stem cells are professional stem cells; every day, they have to yield 2.5 billion erythrocytes, 2.5 billion platelets, and 1 billion leukocytes per each kg of body weight in order to replace dying cells. Skin is another tissue enriched in stem cells; every minute, 30,000 cells come off the skin superficial layer and need to be replaced for the organism to survive. In contrast, brain stem cells are very few and poorly active and have been identified only in two brain areas, hippocampus and subventricular zone. Out of the 100 billion neurons making up the brain, about 85,000 are believed to be lost every day in the subcortical area without being replaced.

Little is known about factors and mechanisms controlling the differentiation potential of adult stem cells, but there is an increasing interest in their isolation and characterization as possible therapeutic tools. Compared to ESCs, some adult stem cells are more easily accessible, and some of them retain a remarkable ability to differentiate. The HSC represent the best and most studied example of adult stem cells. HSC can be isolated in a prospective way (using specific antibodies), enriched, and employed in either autologous or allogeneic transplants to treat inherited immunodeficiencies, autoimmune diseases, and other conditions of the immune system by replenishing specific blood cell types.

From a chronological point of view, fetal stem cells are found between ESCs and adult stem cells. They are multipotent and adult, as residing in differentiated tissues, and they are believed to be similar to stem cells isolated from the corresponding adult tissues but more plastic and therefore potentially more interesting for therapeutic purposes. For this reason, much attention is being given to fetal cord blood, and there is an increasing interest in placenta and in mesenchymal stem cells (MSC). However, it is important to remember that in many cases protocols for obtaining and expanding adult stem cells have not been established yet. Moreover, with the only exceptions of HSC (often transplanted without previous *in vitro* expansion), skin stem cells (one of the very few example in which prospective isolation and *in vitro* unlimited expansion can be obtained without loss of multipotency), or bone marrow-derived MSC, isolation and *in vitro* culturing modify adult stem cell properties.

STEM CELL-BASED DRUGS

Stem cell-based drugs belong to the class of biotechnological drugs, meaning drugs resulting from a biological process. In contrast with most drugs, whose active principles consist of small molecules, in stem cell-based drugs, the active principle is represented by living cells. In the last few years, laws regulating their clinical usage have become very restrictive in order to ensure reproducibility, safety, and standardization of the protocols. Stem cell drugs can be administered in combination with biomolecules and noncellular components, such as medical devices and matrices, and can be genetically modified. The first generation of stem cell-based drugs consists of unmodified human stem cells, in their original state, such as unmodified HSC taken from the bone marrow and employed in clinics to treat blood diseases. The second generation typically consists of cells treated with growth factors, purified using tissue-specific biomarkers, or genetically modified with the aim of enhancing their therapeutic potential and reducing risks of tumor formation or immunological incompatibility. The third generation consists of stem cell with the additional property of producing and releasing soluble molecules in the transplanted tissue. The released molecules may correct a genetic defect or promote tissue regeneration or enhance stress tolerance of both endogenous and transplanted cells, thus improving the therapeutic potential of the original stem cells.

CELL THERAPY AND REGENERATIVE MEDICINE

As with many other biological products, immunotolerance is critical for the safe and effective application of stem cell-based therapies. Cellular therapies are classified as autologous, when they involve transplantation of tissues, cells, or proteins belonging to the same patient, and allogeneic (or heterologous) when a patient receives tissues, cells, or proteins from another individual of the same species.

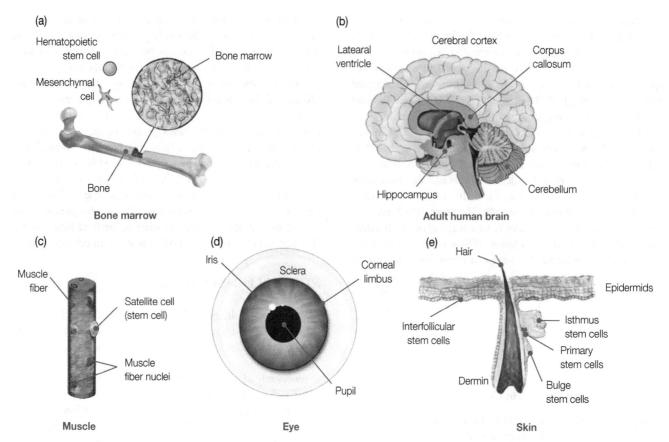

FIGURE 26.2 Examples of adult tissues in which stem cells reside. In the figure are the following: (a) hematopoietic stem cells from the bone marrow, (b) neural stem cells from the subventricular zone (SVZ) of the lateral ventricle and hippocampus, (c) muscle satellite cells, and (d) corneal stem cells from the limbus. In the skin (e), three types of "specialized" stem cells were first identified: in the dermis basal layer (interfollicular stem cells regenerating epidermis), in the follicle bulge (for hair growth), and in the isthmus (for sebaceous gland formation). Recently, a new population of skin stem cells has been identified, from which the three specialized stem cell types originate.

Cell therapies based on the allogeneic transplantation of the bone marrow or hematopoietic cells to treat different blood diseases have been used with success for about 20 years (see Supplement E 26.2, "Use of Blood Stem Cells in Hematology"). In this section, we will discuss some examples of new therapeutic approaches employing adult stem cells and in particular one procedure that has been successfully applied to treat corneal lesions (Fig. 26.2).

Regenerative Medicine Approaches to Epithelial Lesions

After a lesion, skin can regenerate a perfectly functional corneal layer in a few days, starting from different types of specialized stem cells located in the dermis basal layer (stem cells responsible for epidermal regeneration), in the follicle bulge (stem cells responsible for hair growth), and in the isthmus (stem cells involved in sebaceous glands generation). More recently, in 2010, Hans Clevers' group discovered a new population of skin stem cells that can produce all three most specialized stem cell types (Fig. 26.2). However, when endogenous skin repair is not sufficient for adequate regeneration of the lost tissue, such as in case of severe and large burns, autologous skin transplantation can be performed. New skin layers can be grown *in vitro* on collagen and matrigel matrices starting from progenitors and skin stem cells obtained from small biopsies taken from the patient. The first skin transplant using *in vitro* cultured cells was performed in 1983 by Howard Green, a pioneer in this research field, to treat three children severely burned. Since then, the same procedure has been successfully applied to treat severe skin lesions. In 1987, Yann Barrandon described a novel methodology to culture skin stem cells *in vitro* and to produce keratinocytes. However, such procedure is very expensive (it costs more than 150,000 euro to treat an adult patient with 80% of his/her body surface burned) and requires many months to generate large skin layers. Novel and cheaper methodologies to culture general skin stem cells may lead in the future to transplant of regenerated skin layers structurally and functionally identical to normal skin.

The eye corneal epithelium is another example of epithelial tissues that can be entirely regenerated *in vitro* and for which successful clinical protocols have been developed. Corneal lesions, due, for example, to chemical burns, induce the conjunctival epithelium to generate a layer, called "pannus," covering the entire eye bulb, eventually leading to blindness. This can be prevented by replacing the damaged tissue with an *in vitro* regenerated layer. Cornea can be reconstructed starting from stem cells taken from the limbus (Fig. 26.2), an area surrounding the cornea. Corneal limbus stem cells are only a few thousands and represent about 10% of the limbus and are responsible for the cornea turnover occurring every 9 months. When cultured *in vitro*, limbus stem cells can generate a layer of corneal epithelium in 3–4 weeks, which can be transplanted after removal of the damaged one. The first pioneering study on corneal regeneration from stem cells was published in 1997 by Graziella Pellegrini and Michele De Luca. In 2010, these researchers along with Paolo Rama refined the technique to allow treatment of large corneal lesions (3–4% of total lesions) for which normal stem cell transplantation is ineffective. In these cases, corneal grafting is preceded by regeneration of the limbus area using stem cells taken either from the same patient or from a compatible donor. This approach has proved very effective, leading to complete recovery in more than 75% of the treated patients. The resulting product, Holoclar, has received the green light by the European Medicines Agency on February 2015, as the very first medicinal product based on stem cells to be approved and formally registered in the Western world. Holoclar is manufactured by Holostem Advanced Therapies – a spin-off of the University of Modena and Reggio Emilia in Italy – at the Centre for Regenerative Medicine "Stefano Ferrari" (CMR) of the same University.

Regenerative Medicine to Treat Cardiac Dysfunctions

The heart has always been thought to be unable to self-regenerate, but recent studies have shown that progenitor stem cells, identified by stem cell markers such as c-kit or sca-1, are present in human myocardium. According to these studies, under normal condition, such progenitors may be able to regenerate the entire cardiomyocyte population in about 4.5 years. However, these data are controversial: other studies suggest that cardiomyocytes turnover is not as high as reported and only 1% of cardiomyocytes is actually replaced every year. Nevertheless, there is an enormous interest in developing stem cell therapies to heal cardiac tissues, and this represents one of the main research focuses in regenerative medicine. A critical question researchers need to answer is how to obtain cardiomyocytes the most similar to those damaged, as cardiomyocytes vary depending on the area in which they are found. Atrial cardiomyocytes are different from ventricular cardiomyocytes; similarly, those residing on the surface of the ventricular wall are different from those in the deep layers. Moreover, cardiac myocytes generating the electrical impulse responsible for heart beating are different from those only involved in muscle contraction. So far, functionally, cardiomyocytes have been obtained only from fetal cardiomyocytes and pluripotent cells (ESCs and iPSCs). Fetal cardiomyocytes do not represent a possible source for future clinical applications, as they are few and heterologous; on the other hand, cardiomyocytes derived from ESCs and iPSCs have only been employed in preclinical animal studies, as safety concerns still need to be solved.

In spite of the lack of convincing preclinical data, a large number of clinical trials have been performed on a variety of cardiac pathologies using different types of stem cells. However, so far, no clear evidence of significant and long-term functional rescue has been demonstrated, and no production of newly formed cardiomyocytes from transplanted cells has been observed. The first clinical studies involving patients with myocardial infarction transplanted with myoblasts taken from skeletal muscles reported ventricular arrhythmia and no significant morphofunctional recovery. The largest clinical study carried out so far included about 20 studies on more than 2000 patients with acute myocardial infarction transplanted with autologous stem cells from either bone marrow or peripheral blood. No significant clinical recovery was observed at the 3–12 months' follow-up. In some cases, a limited (3%) increase of the left ventricular ejection fraction was reported but in the absence of any improvement in the damaged area. Although unsuccessful, some would argue that this type of transplant is not dangerous for the patient and may facilitate remodeling of the infarcted tissue, probably because of the acute release of protective molecules leading to modest (and transient) enhancement of revascularization. The increased revascularization resulting from stem cell transplantation may explain the more encouraging result of clinical trials in refractory angina, which are now in phase II.

Stem Cell-Based Therapies for Skeletal Muscle Diseases

Skeletal muscle represents the largest organ of the body comprising about 40% of total body mass in humans. This tissue contains a particular cell type named satellite cells constituting a stable, self-renewing pool of stem cells (Fig. 26.2) devoted to the physiological repair of damaged skeletal muscles. In some conditions, such as Becker and Duchenne muscular dystrophies (BMD and DMD), skeletal muscles progressively degenerate and lose their intrinsic regenerative capacity.

To treat these disorders, several preclinical studies have investigated the potential of cell therapies based on the transplantation of myogenic-competent cells, including—besides satellite cells—CD133$^+$ cells (isolated from skeletal muscle or bone marrow), endothelial progenitors, and mesoangioblasts (vessel-associated stem cells). Notably, among this variety of cell populations, only satellite cells and mesoangioblasts exhibit clear myogenic competence both *in vitro* and *in vivo*. However, massive early cell death and poor proliferation and migration following transplantation represent main hurdles to be overcome for satellite cells to become a valuable therapy option. Mesoangioblasts, firstly isolated in 2003 in Giulio Cossu's laboratory from the dorsal aorta of mouse embryos, are able to differentiate into a variety of mesoderm tissues including skeletal, cardiac, and smooth muscle. Their ability to contribute to anatomical and functional muscle regeneration has been tested by arterial injections in both mouse and dog models of DMD. A phase I clinical trial based on intra-arterial delivery of donor-derived mesoangioblasts in six DMD patients has been just concluded, reporting no safety concerns related to the treatment.

Stem Cell-Based Therapies for Brain Diseases

Neurodegenerative disorders are among the most challenging and devastating illnesses in medicine. They represent a heterogeneous group of chronic and progressive diseases characterized by disparate etiologies, anatomical impairments, and symptoms. Some of these disorders, such as Huntington's disease (HD), are acquired in an entirely genetic manner. Alzheimer's disease (AD), amyotrophic lateral sclerosis (ALS), and PD mainly occur sporadically, although familiar forms caused by inheritance of gene mutations are known. On the other side, other neurodegenerative disorders, such as traumatic spinal cord injury and stroke, have no genetic (heritable) components.

By virtue of this extreme heterogeneity, different specific requirements should be envisaged when considering cell replacement as possible therapeutic strategy. We can distinguish between (i) neuronal CNS degenerative disorders caused by a prominent loss of specific neuronal populations leading to destruction of precise cerebral circuitries and (ii) nonneuronal CNS degenerative conditions characterized by loss of nonneuronal elements.

In the case of neuronal degeneration, the success of cell replacement strictly depends on the complexity and precision of the connectivity pattern that needs to be restored. In PD, the affected dopaminergic neurons in substantia nigra (SN) exert a modulator action on the target circuits (striatum) mostly through the release of diffusible molecules (dopamine). In this type of system, defined as "paracrine," even a partial pattern repair may lead to a significant functional recovery. Indeed, in PD, donor cells can be transplanted directly into the target region to circumvent the problem of long-distance neuritic growth in adult CNS. Despite the ectopic location, if grafted cells acquire a dopaminergic identity reestablishing a regulated and efficient release of dopamine, they can lead to a clinically relevant functional recovery. Differently, other diseases with selective degeneration of specific neuronal populations, such as HD, ALS, and conditions exhibiting global neuronal degeneration (trauma, stroke, and AD), require a complex pattern repair and are therefore very difficult to treat with cell-based strategies.

Differently, nonneuronal CNS degenerative syndromes, such as multiple sclerosis (MS), characterized by severe inflammation and oligodendroglial degeneration leading to axonal demyelination, represent a good target for cell replacement because of their limited requirements of pattern repair; for example, in MS, functional rescue requires grafted cells to produce oligodendrocytes capable of restoring axonal myelination.

It is important to emphasize that although pattern repair is critical to obtain permanent rescue, cells transplanted into the brain may also be beneficial via the release of molecules that may either stimulate the regenerative potential of local cells (where present) or increase the survival (and decrease disease progression) of the remaining host elements. In addition, immunomodulation activity of grafted cells could be beneficial in diseases such as MS where a prominent disease-associated inflammation contributes to establishment and progression of the disease.

In the last two decades, a number of clinical studies employing grafting of human fetal tissues have provided a proof of concept for cell-based therapies for PD and HD. Nonetheless, fetal brain tissues do not represent a practical source for large-scale therapeutic applications due to scant availability, quality concerns, and ethical considerations. Hence, the establishment of readily expandable neural stem cell (NSC) populations, maintaining their capability to generate cells belonging to the three major neural lineages and competence to regenerate the injured or diseased CNS, has represented a crucial milestone. In the last years, several NSC populations, including fetal- and adult-derived NSCs (Fig. 26.2), neural progenitors derived from human pluripotent cells (ESCs and iPSCs), have been generated, thus increasing the possibilities to ultimately uncover NSCs suitable (in terms of expandability, safety, and effectiveness) for clinical applications. A large number of studies have explored grafting behavior of several NSC typologies (and their progeny) in a variety of preclinical studies and even in some clinical trials. Nonetheless, besides the great expectations, it should be remarked that up to now an ideal NSC system is not yet available to the clinic. In the last two years, protocols to obtain authentic dopaminergic neurons from hES cells have been developed by the groups of Lorenz Studer in New York and Malin Parmar in Lund, Sweden. Upon transplantation into rodent models of PD, these donor cells exhibited important biological and functional capacities leading to behavioral recovery. A first trial in PD patients with these cells is expected to start in the next few years.

Over the last 10 years, much attention has been given to the potential use of stem cells as therapeutic agents for neurodegenerative disorders, and significant progresses have been achieved. To safely translate stem cell research into therapeutics, we need precise knowledge about molecules and signaling pathways that regulate proliferation, differentiation, and migration of NSCs *in vivo*. Several attempts have already been made to move NSC discoveries from bench to bedside. In a phase I/II clinical trial by StemCell Inc., chronic spinal cord injury patients have been transplanted with purified human adult NSCs. The same company has completed a phase I clinical trial on six children transplanted with human adult NSCs to treat Batten disease. Other examples of clinical trials in the NSC field are Neuralstem phase I clinical trials for ALS patients and ReNeuron trial of a NSC therapy for disabled stroke patients (see https://clinicaltrials.gov). The first FDA-approved clinical trial with human ESC-derived cells, oligodendrocyte progenitors for spinal cord injury, by Geron Corporation started in 2010 but was recently stopped after phase I due to shortage of financial resources for stem cell programs. These early stem cell-based clinical trials are providing scientists and clinicians with initial results on the safety of these cells for treating the diseased brain and useful data for improving the design of future clinical trials.

It is important to add that a better communication of these careful and thorough clinical trials to the public is crucial to inform patients about controlled and reliable clinical research currently ongoing and alert them against the rising number of private clinics offering unproven stem cell treatments as safe and effective therapeutic options for a wide range of different brain diseases.

GENE THERAPY

The goal of somatic gene therapy is to correct/compensate defective/abnormal gene functions responsible for disease. Its purpose is to cure a variety of both inherited and acquired diseases by introducing nucleic acids (usually genes or gene fragments) into patient's cell nuclei. Initially, gene therapy was conceived to cure monogenic hereditary diseases with a recessive phenotype (such as cystic fibrosis, hemophilia, or muscular dystrophies), by transferring a normal copy of the mutated gene into the appropriate tissue. In this context, the "drug" is represented by the encoded therapeutic protein or by the nucleic acid itself. It is important to underline that the range of nucleic acids used in gene therapy approaches includes not only those with a replacement function but also genes, DNA, or RNA fragments regulating cell activity or modulating immune system functions. Therefore, gene therapy is now extended to a vast number of acquired pathologies, such as tumors and cardiovascular diseases.

In the 1990s, the first protocol for gene therapy for the treatment of a hereditary immune system deficiency was approved in the United States. Currently, almost a thousand of protocols are in progress all over the world. In particular, most of them concern the treatment of cancers of different kind and grade. Besides these, several clinical trials on cardiovascular diseases, infective diseases (almost all about HIV), and recessive autosomal single-gene diseases, such as cystic fibrosis, are also in progress.

Overall, despite gene therapy is becoming safer for patients, its applicability remains limited, primarily because of technical issues that will be outlined in the next paragraphs.

Protocols for Gene Therapy

Two possible approaches are usually considered in gene therapy: nucleic acids transfer *ex vivo* or *in vivo* (see Fig. 26.3).

In the *ex vivo* gene therapy, patient's cells are isolated, cultivated *in vitro*, genetically modified to introduce the therapeutic gene, and then reinjected in the same patient. This approach is feasible when target cells are easily accessible and isolable from the patient and can be handled and reinjected (i.e., throughout aerosol or injection into the bloodstream). A major advantage of this approach is its reduced probability to stimulate an immune response against the vector used to transfer the gene. However, the method is laborious and expensive, as protocols have to be patient specific in order to avoid immune rejections. The number of cell types that can be maintained and expanded *ex vivo* has greatly increased in the last years. Besides lymphocytes and HSC, the list includes cells with a stem cell phenotype capable of generating different cellular subtypes, such as skin cells, vascular endothelial precursors, and muscular progenitors. The possibility of transferring genes into stem cells is expanding the therapeutic opportunities as gene therapy and cell and regenerative therapy can cooperate to yield a more efficient treatment for both genetic (i.e., muscular dystrophies) and noninherited degenerative diseases (i.e., PD and myocardial infarction).

With the *in vivo* gene therapy, the therapeutic gene is inserted directly into patient' cells or tissues. In theory, this approach is simpler than the *ex vivo*, and once optimized, it can be applied to an unlimited number of patients with the same disease. In practice, this approach raises some concerns:

1. Many tissues are difficult to reach or to transduce *in vivo* quantities that may be clinically significant. For example, it is difficult to conceive an extensive gene therapy for muscle tissue or cartilage; in these cases, the gene therapy is necessarily restricted to confined districts.
2. Off-target/side effects can be a problem due to gene transfer into cell types different from the target ones.
3. When administered *in vivo*, the vector used for gene transfer is more exposed to inactivation (retroviral vectors or specific neutralizing antibodies can be inactivated by the complement), and it can stimulate patient's immune responses.
4. Most cell types in the organism (including muscular, neural, vascular endothelial, and hepatic cells) are generally quiescent. This fact may limit the use of some type of vectors, such as retroviral vectors, that only infect cells actively dividing.

In Table 26.2, methods for gene transfer are reported with their specific characteristics, both positive and negative. We will focus on applications based on safe viral vectors (see Supplement E 26.3, "Viral Vectors for Gene Therapy in Somatic

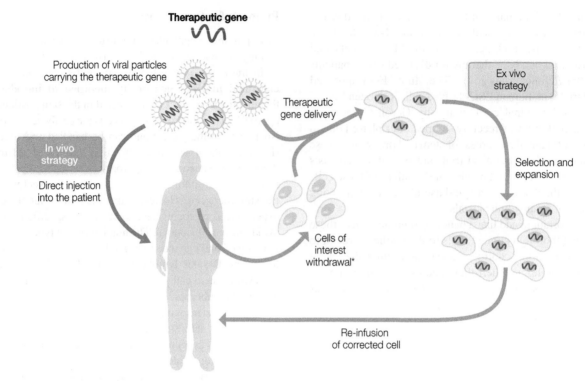

FIGURE 26.3 The two main strategies of gene therapy. There are two different strategies in gene therapy. In the *in vivo* strategy, the corrected gene is injected directly into the patient, mainly using a virus as delivery system. In the *ex vivo* strategy, cells are taken from the patient, genetically modified to introduce the curative gene, and then reinjected into the patient. *Both autologous and heterologous cells can be used; theoretically, cells of interest can be derived also from pluripotent cells (ES cells and iPS cells, the latter can be an autologous source).

Cells") that currently represent the most practiced choice (about 80% of current clinical trials employ viral vectors).

Gene Therapy for Monogenic Inherited Diseases

In 1999 took place one of the saddest events in the history of gene therapy. Jesse Gelsinger, an 18-year-old boy affected by a nonsevere form of ornithine transcarbamylase (OTC) deficiency, was the first patient ever treated with gene therapy. He died few days after the treatment because of a hyperimmune reaction caused by the excessive amount of vector injected. Indeed, he had been injected a very high dose of vector (40 billion viral particles, the highest dose ever administered in a clinical trial) carrying the normal OTC gene into his portal vein. Hundreds of gene therapy protocols have been developed since then. However, it is important to consider that about three-fourth of current clinical trials are phase I. Therefore, thorough and reliable data about their efficacy and therapeutic reproducibility are presently not available. Some of the current optimized clinical protocols for gene therapy and their relative concerns are listed in the following.

Hereditary Immunodeficiency Syndromes This group of diseases represents the gold standard for gene therapy experimentation. The first clinical trial for a hereditary immunodeficiency syndrome was developed in 1990 by three American researchers: French Anderson, Michael Blaese, and Kenneth Culver. The *ex vivo* gene therapy protocol was conceived to treat a 4-year-old girl affected by severe combined immunodeficiency disease (SCID). This pathology is caused by lack of adenosine deaminase (ADA), an essential enzyme for purine metabolism. ADA deficiency triggers selective T cell damage resulting in a severe form of immunodeficiency. The ADA-SCID patients generally suffer for recurrent severe infections, mostly concerning the airways and gastrointestinal tract, chronic dysentery, and reduced growth rate. The first symptoms onset in the neonatal period, but sometimes, they can be delayed by several months, due to a partial immune protection provided by maternal antibodies. Without early diagnosis and treatment, these patients do not survive over 24 months. Conventional treatments for ADA-SCID include hematopoietic cell transplantation and weekly ADA injection for the entire life. Gene therapy was explored as an alternative because transplantation often failed and enzymatic replacement appears useless in many patients. Another clinical trial was developed in 1991. Since no side effects emerged in the first two trials, eleven children were enrolled in a third gene therapy experimentation in 1996.

In these initial *ex vivo* protocols, the cDNA encoding the human ADA enzyme was inserted into patients' T cells by retrovirus-mediated gene transfer. The transduced T cells were then injected back into patients resulting in symptoms

TABLE 26.2 Gene delivery methods for gene therapy

Method	Vector	Advantages	Disadvantages
Direct injection	Naked DNA	Not pathogenic Easy to handle Unlimited transgene size	Low efficiency Transient transgene expression
Chemical	Liposomes	Not pathogenic Easy to handle Unlimited transgene size	Low efficiency Transient transgene expression
Physical	Electroporation	Not pathogenic Easy to handle Unlimited transgene size	Low efficiency Transient transgene expression
Viral transduction	Retrovirus	High transduction efficiency Stable integration in host genome Poor immunogenicity	Low viral titer production Possible insertional mutagenesis Targets dividing cells only
	Lentivirus	Transduction efficiency Transduction of both dividing and quiescent cells Stable integration in host genome Poor immunogenicity	Low viral titer production Possible insertional mutagenesis Targets dividing cells only
	Adenovirus	Transduction efficiency Transduction of both dividing and quiescent cells Episomal maintenance Accommodate large DNA fragments High levels of transgene expression	Transient transgene expression High immunogenicity
	Adeno-associated virus	Not pathogenic for humans Poor immunogenicity Transduction of both dividing and quiescent cells	Accommodate small DNA fragments (<5 kb) Variable transgene expression Possible insertional mutagenesis
	Herpesvirus	Accommodate large DNA fragments Transduction of both dividing and quiescent cells High tropism for nerve cells Episomal maintenance	Difficult to handle Possible immunogenicity

improvement, as long as the genetically modified cells persisted. In order to achieve a long-term solution, recent protocols have been developed by Italian researchers employing ex vivo manipulation of HSC. It is important to note that in this gene therapy protocols, patients are weekly administered their pharmacological ADA dose to avoid the ethical dilemma of blocking a potentially effective therapy (enzymatic replacement) in order to try a new treatment.

Gene therapy has also been used to treat an X-linked form of SCID (X-SCID). This lethal condition is caused by mutations in the gene encoding the common chain gamma, a protein involved in the production of different receptors in the immune system. Because of a differentiation block, this disease is characterized by deficiency of T lymphocytes and natural killer cells; lymphocyte precursors are produced, but they rapidly die by apoptosis because of the lack of gamma chain. In 2000, in France, two X-SCID pediatric patients were treated by gene therapy. Both were severely affected, manifesting pneumonia, dysentery, and skin rashes. HSC were extracted from their bone marrow, manipulated *ex vivo* using a retroviral vector carrying the healthy gene cDNA, and injected back into the patients. Notably, treated patients did not receive any other treatment. Given the encouraging initial results, additional patients were enrolled in the clinical trial. Despite the very low level of gene transfer achieved, the presence of the healthy gene had a significant beneficial effect on the transduced lymphocytes, which kept proliferating and entirely colonized the whole immune system, completely restoring the defect in a few months. At first, this trial was considered the greatest success in gene therapy history. However, two patients developed a severe acute leukemia that in one case was lethal. Subsequent studies revealed that leukemia resulted from the insertion of the retroviral vector near the proto-oncogene LMO2, which encodes a transcription factor important for hematopoiesis and whose overexpression in mice results in leukemia. Because of this, in 2003, 27 clinical trials were suspended in the United States. The year after, gene therapy for X-SCID disease was readmitted but only in patients in which bone morrow transplantation was not effective. However, a third case of leukemia in 2005 led France and the United States to suspend clinical trials again. Today, given the improved safety conditions and quality controls, experimentations have been readmitted.

Skin Genetic Diseases Several mutations in genes encoding keratins or other structural proteins have been identified as responsible of rare skin diseases, which in most severe cases lead to epidermis exfoliation from the dermis. Since the last few years, it has been possible to isolate and cultivate epidermal stem cells *in vitro* to regenerate body skin. Thus, experimentations involving autotransplantation of keratinocytes derived from genetically modified stem cells appear very interesting. In 2006, the team headed by Michele De Luca performed the first trial involving autologous transplantation of genetically modified skin stem cells in one patient affected by junctional epidermolysis bullosa (JEB), a heterogeneous group of diseases affecting about 30,000 subjects in Europe and 500,000 in the world. JEB patients present severe blisters on the face, chest, and leg skin caused by epidermis detachment from the underlying dermis. JEB can have severe complications, including infections and bleeding, that often lead to dehydration; moreover, it has been shown that 80% of JEB patients develop skin carcinoma. The disease is due to mutations in genes encoding proteins responsible for proper adhesion between skin layers, including laminin 5/6. The 36-year-old patient has shown a complete regeneration in the treated body areas, and no more lesions have occurred since then. This experimentation represents the first success of a gene therapy protocol applied to nonhematopoietic tissues.

Additional examples of gene therapy applications are reported in the Supplement material E 26.4, "Gene Therapy of Cystic Fibrosis and Tumors."

FUTURE PERSPECTIVES

Stem cells can be considered one of the most promising tools in the emerging field of regenerative medicine. Successful translation to the clinical setting will need a deeper knowledge of disease pathogenesis and stem cell biology. Also, partnership with other disciplines, such as gene therapy, transplantation immunology, and biomaterials science, will be crucial to fully translate stem cell potential in the therapeutic field. Nowadays, stem cell therapies have demonstrated therapeutic efficacy and beneficial effects in preclinical models, but applications in clinical studies are limited to few specific disorders. For this reason, stem cell therapies should be considered a pioneering experimental medicine. Although part of the public opinion would like clinical trials to start even in the absence of a comprehensive understanding of the biological aspects supporting stem cell therapies, basic science and preclinical studies are still mandatory to comprehend action mechanisms and biology of stem cell differentiation. These last few years have seen patients traveling to destinations where stem cell treatments with unproven benefits are provided (a phenomenon known as "stem cell tourism"). Indeed, clinics and companies based in countries where laws regulating cell-based treatments are lax or nonexistent and offering stem cell therapies for many conditions (e.g., spinal cord injury, MS, and cerebral palsy) are continuously increasing. This represents a serious medical and social concern that can only by counteracted by educating patients and health professionals.

To this purpose, the International Society for Stem Cell Research (ISSCR) has developed web resources (http://www.closerlookatstemcells.org) to provide patients with detailed guidelines (http://www.closerlookatstemcells.org/The_Patient_Handbook1.html) about existing legal and proven stem cell-based treatments for specific diseases. Carefully designed and ethically approved clinical trials resulting from robust preclinical pathways represent the fundamental condition to achieve a real advance in this field.

TAKE-HOME MESSAGE

- Different types of stem cells have different properties and provide different perspectives for use in regenerative medicine.
- The goal of regenerative medicine is to restore tissue function through the delivery of stem cells and/or genetic material engineered in laboratory.
- Stem cell transplantation has been used to regenerate bone marrow in neoplastic patients. Other applications of stem cell transplantation include regeneration of corneal and other epithelial, muscle, and myocardial tissues and the central nervous system. Most of these applications are under development.
- An accurate standardization and characterization of cell preparations is essential for the correct interpretation of the results of preclinical and early clinical studies.
- Wide-ranging studies are needed to enhance our confidence in the use of stem cell-based therapies. Indeed, while these therapies are commonly considered safe, extensive statistically substantiated data on the long-term effects of cell transplant are actually available only for hematopoietic stem cell clinical applications. The possibility of tumorigenicity and side effects has been raised in a number of studies.
- *Ex vivo* and *in vivo* approaches are used for gene therapy.
- Different methods are used for gene delivery.

FURTHER READING

Aiuti A., Roncarolo M.G. (2009). Ten years of gene therapy for primary immune deficiencies. *Hematology ASH Educational Program*, 2009, 682–689.

Blaese R.M., Culver K.W., Miller A.D., Carter C.S., Fleisher T., Clerici M., Shearer G., Chang L., Chiang Y., Tolstoshev P., Greenblatt J.J., Rosenberg S.A., Klein H., Berger M., Mullen C.A., Ramsey W.J., Muul L., Morgan R.A., Anderson W.F. (1995). T Lymphocyte-directed gene therapy For ADA-SCID: initial trial results after 4 Years. *Science*, 270, 475–480.

Bostrom P., Frisen J. (2013). New cells in old hearts. *New England Journal of Medicine*, 368, 1358–1360.

Gallico G.G., 3rd, O' Connor N.E., Compton C.C., Kehinde O., Green H. (1984). Permanent coverage of large burn wounds with autologous cultured human epithelium. *New England Journal of Medicine*, 311, 448–451.

Lindvall O., Kokaia Z. (2010). Stem cells in human neurodegenerative disorders-Time for clinical translation? *Journal of Clinical Investigation*, 120, 29–40.

Mavilio F., Pellegrini G., Ferrari S., Di Nunzio F., Di Iorio E., Recchia A., Maruggi G., Ferrari G., Provasi E., Bonini C., Capurro S., Conti A., Magnoni C., Giannetti A., De Luca M. (2006). Correction of junctional epidermolysis bullosa by transplantation of genetically modified epidermal stem cells. *Nature Medicine*, 12, 1397–1402.

Menasche P. (2011). Cardiac cell therapy: lessons from clinical trials. *Journal of Molecular and Cellular Cardiology*, 50, 258–265.

Nelson T.J., Martinez-Fernandez A., Terzic A. (2010). Induced pluripotent stem cells: developmental biology to regenerative medicine. *Nature Reviews Cardiology*, 7, 700–710.

Rama P., Matuska S., Paganoni G., Spinelli A., De Luca M., Pellegrini G. (2010). Limbal stem-cell therapy and long-term corneal regeneration. *New England Journal of Medicine*, 363, 147–155.

Robinton D.A., Daley G.Q., (2012). The promise of induced pluripotent stem cells in research and therapy. *Nature*, 481, 295–305.

Rossi F., Cattaneo E. (2002). Opinion: neural stem cell therapy for neurological diseases: dreams and reality. *Nature Reviews Neuroscience*, 3, 401–409.

Sheridan C. (2011). Gene therapy finds its niche. *Nature Biotechnology*, 29, 121–128.

Takahashi K., Yamanaka S. (2006). Induction of pluripotent stem cells from mouse embryonic and adult fibroblast cultures by defined factors. *Cell*, 126, 663–676.

Tedesco F.S., Dellavalle A., Diaz-Manera J., Messina G., Cossu G. (2010). Repairing skeletal muscle: regenerative potential of skeletal muscle stem cells. *Journal of Clinical Investigation*, 120, 11–19.

Thomson J.A., Itskovitz-Eldor J., Shapiro S.S., Waknitz M.A., Swiergiel J.J., Marshall V.S., Jones J.M. (1998). Embryonic stem cell lines derived from human blastocysts. *Science*, 282, 1145–1147.

Yi B.A., Wernet O., Chien K.R. (2010). Regenerative medicine: developmental paradigms in the biology of cardiovascular regeneration. *Journal of Clinical Investigation*, 120, 20–28.

SECTION 7

PHARMACOLOGICAL CONTROL OF MEMBRANE TRANSPORT

27

ION CHANNELS

MAURIZIO TAGLIALATELA AND ENZO WANKE

By reading this chapter, you will:

- Become familiar with the main principles governing function, structural organization, and classification of ion channels
- Know the role(s) played by the main classes of ion channels in different organs, tissues, and cells
- Know the clinical applications of drugs interfering with the function of each ion channel class
- Learn how functional changes resulting from drug-induced modulation of ion channels can be exploited for therapeutic purposes

---

ION CHANNELS AND TRANSPORTERS

Eukaryotic cells use about 30% of their energy to maintain the transmembrane gradients of protons (H$^+$), sodium (Na$^+$), potassium (K$^+$), chloride (Cl$^-$), and calcium (Ca^{2+}), an indication of their paramount importance for cell survival and replication.

On purely thermodynamic grounds, transmembrane transport mechanisms can be classified into active and passive. Passive processes transport ions from the side of the membrane with high electrochemical potential to the side with low electrochemical potential. Two types of proteins are responsible for passive ion transport: facilitated transporters and ion channels (Fig. 27.1), with very different transport mechanisms. Substrate binding to the transporter on one side of the membrane induces a conformational change, resulting in exposure of the substrate on the opposite side of the membrane. The substrate concentration gradient provides the energy required for the process; as the substrate movement is coupled to a conformational change of the transporter, the transfer rate is rather low. By contrast, ion channels contain aqueous pores through which permeating ions can flow at very high rates (>10^6/s, close to the diffusion rate in water), thus generating significant currents that may rapidly change the resting membrane potential (V$_{REST}$) of a cell.

Both these passive processes dissipate the energy gradient established by active transporters, which pump ions across the membrane against their concentration gradients. This process requires an energy input generally provided by ATP hydrolysis (primary active transporters). Otherwise, movement of a solute against the electrochemical gradient can be coupled to the movement of another solute down its electrochemical gradient, either in the same direction (cotransport or symport) or in the opposite (countertransport or antiporter).

Enormous progresses in the structural and functional characterization of membrane transport over the last 10–15 years have made the separation line between ion channels and transporters progressively thinner. Recent studies have shown that some toxins can convert a transporter into an ion channel and that transporters and channels can coexist within the same structural family. For instance, in the "ATP-binding cassette" (ABC) family of transporters, the cystic fibrosis transmembrane regulator (CFTR, so called as it is mutated in cystic fibrosis) is the only channel member; moreover, the Cl- channel family includes both ion channels (CLC-0, CLC-1, and CLC-2) and transporters (CLC-4 and CLC-5). Finally, within the same family, CLC-0, CLC-1, and CLC-2 are ion channels in vertebrates, whereas their bacterial counterparts behave like transporters.

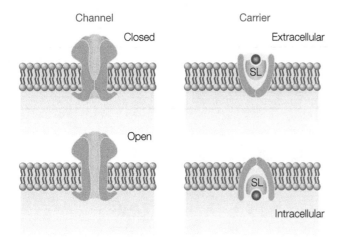

FIGURE 27.1 Ion channels and transporters. The cartoon depicts the conceptual difference between ion channels and transporters. For ion channels (left), ions diffuse through the open pore, which is thought to be controlled by one gate. For transporters (right), the transport pathway is guarded by at least two gates whose opening is coordinated so that no high conductance open pore is allowed. "SL" represents the transported substrate that binds to the binding site in the conduction pathway (Modified from Ref. [1]).

CHARACTERIZATION AND FUNCTION OF ION CHANNELS

In excitable cells like muscle cells, endocrine cells, and neurons, ion channels are responsible for generation and regulation of electrical signals required for coordinated contraction of skeletal muscle, hormonal secretion, and neurotransmitter release; furthermore, in all cells, they control cell volume and motility.

With modern electrophysiological and molecular biology techniques, several genes encoding ion channels have been identified, and the specific functional properties of their protein products have been characterized. Besides suggesting the molecular basis for the action of specific drug classes, these studies have also allowed to discover that genetically determined ion channel defects can be responsible for several human diseases (the so-called human channelopathies; see Table 27.1).

To understand the mechanism of action of drugs acting on ion channels, it is essential to recapitulate some fundamental concepts on the functional and structural properties of ion channel proteins.

Channel Classification According to Permeating Ions and Gating Mechanisms

Ion channels can be classified according to purely functional criteria. Two main properties characterize the activity of a specific ion channel: the mechanism triggering its gating and the ion species flowing through it.

In voltage-gated ion channels (VGICs), representing the third largest class of proteins involved in signal transduction, the gating trigger is represented by changes in transmembrane voltage; even few millivolts can drastically alter the opening probability of VGICs. Other gating mechanisms are represented by changes in the chemical composition of the intra- or extracellular environment (ligand-gated ion channels (LGICs); see Chapter 16), in the applied mechanical force (mechanosensitive channels), or in the environmental temperature (thermosensitive channels). Obviously, this schematic classification is an oversimplification, as several mechanisms often contribute to regulate ion channel activity in distinct pathophysiological states; for example, Ca^{2+}-dependent K^+ channels are also sensitive to changes in transmembrane voltage, some VGICs are also sensitive to changes in osmotic pressure, and some LGICs are also influenced by changes in transmembrane voltage and environmental temperature.

Permeation characteristics allow classifying ion channels based on their ion selectivity (with Na^+, K^+, Ca^{2+}, and Cl^- channels showing the greatest selectivity). Further classification criteria, such as cell- or tissue-specific expression or peculiar sensitivity to drugs and toxins, can further contribute to characterize ion channel classes.

Having introduced permeation and gating as the two main criteria for ion channel classification, now we will briefly describe these two functional properties.

Permeation and Concentration Gradients

Ion channels having the same selectivity are often discriminated based on their conductance (γ), which is the ratio between current carried (i) and electromotive force (V), the latter defined as the sum of the electrical and chemical gradient acting on the ion. In fact, each ion is subjected to both electrical forces (the membrane potential V_M is the difference between the cytoplasmic and the extracellular charges) and diffusional forces (produced by the ion concentration gradient between the intracellular and extracellular space). The equilibrium between these two forces is the Nernst potential (or reversal potential). The Nernst potential depends on the logarithm of the ion concentration ratio between the extracellular and intracellular environment, according to the equation

$$E_{Nernst} = 60 \times \log[I^+]_{out}/[I^+]_{in}$$

where $[I^+]_{out}$ and $[I^+]_{in}$ are the extracellular and intracellular concentrations, respectively, of the ion I. In most animal cells, the Nernst potentials for the various ions (E_{Na}, E_K, E_{Cl}, E_{Ca}, etc.) are +70 mV for E_{Na}, −95 mV for E_K, −30/−60 mV for E_{Cl}, and +150 mV for E_{Ca}. Thus, opening of a single ion channel species (i.e., that for K^+) will generate a current

TABLE 27.1 Main hereditary diseases caused by ion channel mutations (channelopathies)

Disease name	Acronym	Chromosome	Channel	Phenotype
Benign familial neonatal convulsions	BFNC1	20	KCNQ2	Increased neuronal excitability, epilepsia
	BFNC2	8	KCNQ3	Increased neuronal excitability, epilepsia
Autosomal congenital deafness type 2	DFNA-2	1	KCNQ4	Deafness
Long QT syndrome				
Romano–Ward (dominant)	LQTS-1	11	KCNQ1	Cardiac arrhythmias (loss-of-function mutations)
Jervell/Lange-Nielsen (recessive)	LQTS-1	11	KCNQ1	Cardiac arrhythmias and deafness (loss-of-function mutations)
	LQTS-2	7	KCNH2	Cardiac arrhythmias (loss-of-function mutations)
	LQTS-3	3	SCN5A	Cardiac arrhythmias
	LQTS-5	21	KCNE1	Cardiac arrhythmias
	LQTS-6	21	KCNE2	Cardiac arrhythmias
Andersen–Tawil syndrome	LQTS-7	17	KCNJ2	Cardiac arrhythmias (loss-of-function mutations), periodic paralysis, and dysmorphic features
Timothy syndrome	LQTS-8	21	CACNA1C	Cardiac arrhythmias (gain-of-function mutations)
	LQT-10	11	SCNA4B	Cardiac arrhythmias
Brugada syndrome	BrS	3	SCN5A	Cardiac arrhythmias
Progressive cardiac conduction disease	PCCD	3	SCN5A	Cardiac arrhythmias with conduction disorders
Sick sinus syndrome	SSS	3	SCN5A	Cardiac arrhythmias with conduction disorders
Short QT syndrome	SQTS	11	KCNQ1	Cardiac arrhythmias (gain-of-function mutations)
	SQTS	7	KCNH2	Cardiac arrhythmias (gain-of-function mutations)
	SQTS	17	KCNJ2	Cardiac arrhythmias (gain-of-function mutations)
	SQTS/BrS	21	CACNA1C	Cardiac arrhythmias (loss-of-function mutations)
Familial sinus bradycardia	FSB	15	HCN4	Sinus bradycardia
Episodic ataxia	EA-1	12	KCNA1	Ataxia, migraine, neurodegeneration
	EA-2/EA-5	19	CACNA1A	Ataxia
Spinocerebellar ataxia	SCA6	19	CACNA1A	Ataxia, migraine
Familial hemiplegic migraine	FHM1	19	CACNA1A	Migraine
	FHM3	2	SCN1A	Migraine
Kidney polycystic disease	PKD	16	PDK1	Hypertension, kidney disorders
		4	PDK2/TRPP2	Hypertension, kidney disorders
Hypokalemic periodic paralysis	HYPOK-PP	1	CaCNA1S	Low $[K^+]_o$-triggered skeletal muscle paralysis
Hyperkalemic periodic paralysis	HYPERK-PP	17	SCN4A	Low $[K^+]_o$-triggered skeletal muscle paralysis
		17	SCN4A	High $[K^+]_o$-triggered skeletal muscle paralysis
Paramyotonia congenita	PMC	17	SCN4A	Skeletal muscle hyperexcitability
Potassium-aggravated myotonia	PAM	17	SCN4A	High $[K^+]_o$-aggravated myotonia
Congenital myasthenic syndrome	CMS	17	SCN4A	Myasthenia
Hereditary erythermalgia	IEM	2	SCN9A	Pain hypersensitivity (gain-of-function mutations)
Paroxysmal extreme pain disorder	PEPD	2	SCN9A	Pain hypersensitivity (gain-of-function mutations)
Congenital pain insensitivity	CIP	2	SCN9A	Analgesia (loss-of-function mutations)
Thomsen myotonia (dominant)	TM	7	CLCN1	Myotonia

(*Continued*)

TABLE 27.1 (*Continued*)

Disease name	Acronym	Chromosome	Channel	Phenotype
Becker myotonia (recessive)	BM	7	CLCN1	Myotonia
Malignant hyperthermia	MH	19	RYR1	Drug hypersensitivity with hyperthermia attacks
		1	CACNA1S	Low $[K^+]_o$-triggered skeletal muscle paralysis
Congenital paroxysmal ventricular tachycardia	CPVT	1	RYR2	Cardiac arrhythmias
Bartter disease	IBD	11	KCNJ1	Hypokalemic alkalosis
Liddle disease	LD	12	SCNN1A (Epithelial channel)	Hypertension; pseudohyperaldosteronism (gain-of-function mutations)
Nesidioblastosis	PHHI	11	KCNJ11	Infantile hyperinsulinemia (loss-of-function mutations)
		11	SUR1	Infantile hyperinsulinemia (loss-of-function mutations)
Persistent neonatal diabetes	PNDM	11	KCNJ11	Neonatal diabetes (gain-of-function mutations); also in SUR1
Transient neonatal diabetes	TNDM	11	KCNJ11	Neonatal diabetes (gain-of-function mutations); also in SUR1
Cardiomyopathy (with ventricular arrhythmia)	CMD1O	12	SUR2A	Dilative cardiomyopathy; ventricular arrhythmias
Dent disease	DD	X	CLCN5	Proximal tubulopathy with kidney failure
Osteopetrosis	ADO, ARO	16	CLCN7	Defect in osteoclast-mediated bone reabsorption
Cystic fibrosis	CF	7	CFTR	Altered exocrine secretion of chloride
Congenital stationary night blindness type 2	CSNB2	X	CACNA1F	Changes in vision

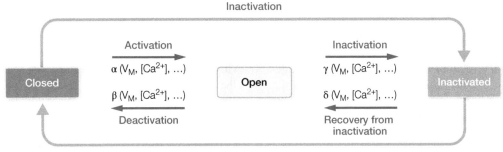

FIGURE 27.2 Kinetic states of ion channels. For each ion channel, irrespectively of the gating mechanism, kinetic models can be elaborated to account for the equilibrium among closed, open, and inactivated states by defining kinetic constants (α, β, γ, δ, etc.), which depend on variables such as V_M, $[Ca^{2+}]_i$, $[cAMP]_i$, pH_i, and others.

[according to the Ohm's law: $i = \gamma \times (V_M - E_{Nernst})$] that will drive membrane potential V_M toward the Nernst potential for the only permeating ion (K^+, E_K). Since in most cells the resting conductance for K^+ ions is larger than that for any other ion species, V_{REST} will be generally negative (between −40 and −90 mV) and close to E_K. When channels permeable to multiple ion species are open, V_{REST} reaches a value for which the algebraic sum of all inward and outward currents carried by the open channels is zero. It should be reminded that, by definition, inward currents are carried either by cations entering the cell or by anions flowing toward the extracellular space, whereas outward currents are due to cations leaving the cytoplasm or by anions entering the cell. Therefore, in most physiological conditions, inward currents cause membrane depolarization, whereas outward currents hyperpolarize the membrane.

Transmembrane Voltage Triggers Conformational Changes

The membrane potential, V_M, is the main regulator of the opening probability of VGICs. Most VGICs are activated (opened) by membrane depolarization, though few of them are activated by membrane hyperpolarization. In muscle and neuronal cells, V_{REST} is rather negative (about −60 mV or less). Therefore, activation of ion channels selective for Na^+ and Ca^{2+} ions (which are normally closed at V_{REST} and become active upon membrane depolarization) will generate a cation flux, further amplifying plasma membrane depolarization. Vice versa, activation of outward currents carried by K^+ and Cl^- channels represents the primary mechanism by which cells repolarize (or hyperpolarize) V_{REST}.

As described in Figure 27.2, in VGICs, the transition between closed and open state is defined as activation; vice versa, the term deactivation refers to the reverse transition, from open to closed. In some channels, in addition to the closed and open states, an inactivated state exists, generated by a process called inactivation. In the inactivated state, similarly to the closed state, ions cannot flow through the channel pore; however, in contrast to closed channels, inactivated channels cannot be reopened by depolarization, but need to return to the closed state from which the activation process can proceed. This process, defined as recovery from inactivation, generally requires cell repolarization to rather negative values of membrane potential. Both in the inactivation and in the recovery from inactivation processes, the channel can transit through the open state. These distinct functional states (closed, open, inactivated) correspond to different conformations of the channel protein; this has important pharmacological consequences, as most drugs acting on ion channels interact differently with each state, thus showing state-dependent actions (see the succeeding text).

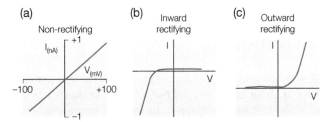

FIGURE 27.3 Current–voltage (I/V) relationships in ion channels and rectification process.

Current–Voltage Relationships and The Rectification Process

As previously illustrated, the Ohm's law allows calculating how electromotive forces influence currents carried by ion channels. In ion channels, the voltage dependence of ion channel current is experimentally measured and graphically represented by the current–voltage (I/V) relationship (Fig. 27.3). Each class of ion channel presents a "signature" I/V, whose shape depends on the specific permeation and gating properties of the channel. Panel a in Figure 27.3 shows the I/V relationship of a channel whose conductance is constant across the entire voltage range; a linear I/V indicates that the channel conductance, corresponding to the slope of the I/V relationship, is independent of voltage. Vice versa, if the conductance changes as a function of the voltage examined and the I/V is not linear across the entire voltage range examined, the channel is said to show rectification. Such divergence from the Ohm's law may take two forms: in inwardly rectifying channels, membrane conductance is larger at more negative potentials, as these channels preferentially conduct inward currents (panel b); the opposite is true in outwardly rectifying channels, which preferentially carry outward currents at depolarized potentials (panel c).

STRUCTURAL ORGANIZATION OF ION CHANNELS

VGICs are integral membrane proteins with a molecular mass of about 200–250 kDa. Na^+ and Ca^{2+} channels consist of a single large polypeptide (α-subunit) containing four homologous domains of 300–400 amino acids (domains I, II, III, and IV) (Fig. 27.4). Instead, voltage-gated K^+ channels result from the association of four smaller subunits, each corresponding to one of the domains of Na^+ and Ca^{2+} channels, to form a functional tetrameric channel. Sequence analysis suggests that voltage-gated K^+ channel subunits are phylogenetic ancestors of Na^+ and Ca^{2+} channel α-subunits (Fig. 27.4, top line), generated during evolution by gene duplication/fusion mechanisms. The higher degree of genetic heterogeneity occurring in K^+ over Ca^{2+} and, especially, Na^+ channels (Fig. 27.4, line "families") supports this hypothesis, as the longer the existence of a protein, the higher its genetic heterogeneity. Sequence homology among domains/subunits of K^+, Na^+, and Ca^{2+} channels is very high, whereas the intervening regions between domains are more divergent.

In all VGICs, each domain (or subunit) contains six segments (S1–S6) formed by mostly hydrophobic amino acids, which are likely to adopt a transmembrane topology. In the tetrameric structure, segments S5 and S6 occupy a central position, whereas the other segments are placed more radially. Each of these segments, as well as their joining regions, likely play a specific role in activation, inactivation, and selective permeability.

As already introduced, genetic and structural heterogeneity is largest in K^+ channels; in fact, voltage-independent K^+ channels of the "inward-rectifier" type consist of subunits with only two transmembrane domains (Fig. 27.4). Recently, genes encoding K^+ channel subunits characterized by two transmembrane segments repeated in tandem (4 transmembrane segments in total) have also been identified. The linker region between these two transmembrane segments is highly homologous to that between the S5 and S6 segments in classical voltage-gated K^+ channels and likely contributes to formation of the ion channel pore (see the succeeding text). On the other hand, these subunits with 2 or 4 transmembrane segments lack the S4 segment (the "voltage sensor"; see the succeeding text), explaining why channels containing these subunits are not regulated by changes in membrane voltage.

Besides the main pore-forming subunits, both voltage-dependent and voltage-independent channels contain a variable number and type of accessory subunits, contributing to correct assembly, trafficking, and plasma membrane localization, as well as to peculiar pathophysiological and pharmacological properties of the ion channel complex. Different genes belonging to the same class or alternatively spliced variants of the same gene contribute to functional heterogeneity of ion channels expressed in different tissues or cell types.

The Voltage Sensor of VGICs

In VGICs, changes in transmembrane voltage trigger pore opening. Thus, VGICs must contain a voltage sensor that "senses" the transmembrane electric field and undergoes conformational changes in response to changes in the electric field. Hodgkin and Huxley, in the 1950s, were the first to hypothesize the existence of charged particles within the ion channel region sensing the transmembrane electric field and triggering voltage-dependent gating when displaced. Hodgkin and Huxley named them "gating charges." In squid giant axon, currents generated by translocation of these gating charges within the membrane electric field were directly recorded in the 1970s in both voltage-gated Na^+ and K^+ channels ("gating currents"). In the 1980s, the primary sequences of the first VGIC genes (first, the Na^+ channel from electric fishes and rat brain, then the rabbit skeletal muscle Ca^{2+} channel, and, finally, the *Drosophila* K^+ channel named *Shaker*) were obtained. Sequence inspection revealed the presence of an amino acid stretch containing 4–8 positively charged residues (lysines and arginines) every three position and spaced by mostly hydrophobic residues in the fourth transmembrane segment (S_4) of each of the four homologous domains of the Na^+ and Ca^{2+} channel α-subunits and in each K^+ channel subunit. It was then suggested that these positive charges in S_4 could be the gating charges of VGICs and that, following changes in transmembrane voltage, their movement could represent the first conformational change leading to channel opening. Both in the resting and activated configurations of the voltage sensor, S_4 positive charges would establish distinct electrostatic contacts with the negative charges of other amino acid residues (also highly conserved among different VGICs) in the S_1, S_2, and S_3 segments of the same domain/subunit or of the membrane phospholipid head groups.

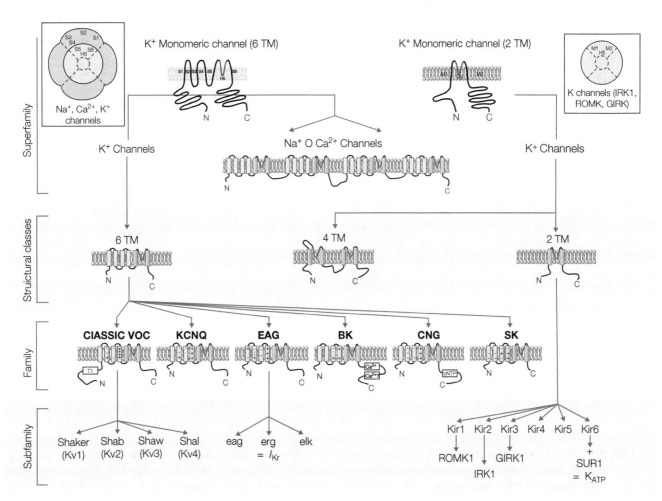

FIGURE 27.4 Structure-based classification of ion channels. Superfamilies of Na+, Ca2+, and K+ channels are outlined in the first and second lines. Voltage-gated Na+ channel (VGNC) and voltage-gated Ca2+ channel (VGCC) genes encode for proteins (α-subunits) with four highly homologous domains, each containing 6 transmembrane segments (6TM); instead, voltage-gated K+ channel (VGKC) genes encode for 6TM α-subunits analogous to one of the four domains of the VGNCs and VGCCs; VGKCs assemble as homo- or heterotetramers. In addition, a large gene family of non-VGKC genes encode for 2TM proteins, highly homologous to the 5th and 6th TM segments of the VGKCs, sometimes duplicated in tandem (4TM). The structural diversity among 6TM, 4TM, and 2TM channels is also shown in the third line. 6TM channel families (sharing amino acid sequence identity of about 25–30%) are shown in the fourth line, where they are classified on the basis of genetic and functional differences: classical VGKCs (K_v1, K_v2, K_v3, and K_v4), KCNQ (K_v7) channels, EAG channels (for *ether-a-go-go*, the name of a *Drosophila* mutant; K_v10, K_v11, and K_v12), large-conductance voltage- and Ca2+-dependent K+ channels or BK (with a cytoplasmic domain mediating their regulation by $[Ca^{2+}]_i$), CNG channels (with a C-terminal domain where cyclic nucleotides such as cAMP and cGMP bind to activate the channel), and small conductance Ca2+-dependent K+ channels or SK. The last line shows that, among these gene families, a large number of subfamilies exist (sharing an amino acid sequence identity of about 50–60%); these are mostly characterized by names of *Drosophila* mutants ("shaker," shab," "shaw," "shal"). KCNQ-type channels give rise to two important K+ currents: neuronal I_{KM} (inhibited by muscarinic receptor activation) and cardiac I_{Ks} (responsible for the slow component of the ventricular repolarizing current I_K). ERG channel subfamilies are instead responsible for I_{Kr}, the rapid component of the ventricular repolarizing current I_K, also expressed in neurons. Among K_{ir} (inwardly rectifying) channels, the best known are ROMK1 (renal outer medullary kidney, involved in K+ cycling across epithelial cells of the kidneys and other tissues), IRK1 (the classical inward rectifier of cardiac and skeletal muscle), GIRK1 (an inwardly rectifying channel activated by G-protein βγ-subunits binding to the C-terminus tail following GPCR activation), and K_{ATP} channels, regulated by $[ATP]_i$ (see Fig. 27.8).

In subsequent years, this hypothesis has been tested and confirmed by mutagenesis, fluorescence spectroscopy, and electrophysiological experiments. Over the last 15 years, crystallographic studies of bacterial and mammalian channels have provided a more detailed view of the role of S_4 positively charged residues in the activation process. However, the precise structural rearrangements occurring in the voltage sensor during activation are not fully understood yet, since structural results have been mostly obtained in the absence of transmembrane potential, thus in a "depolarized"

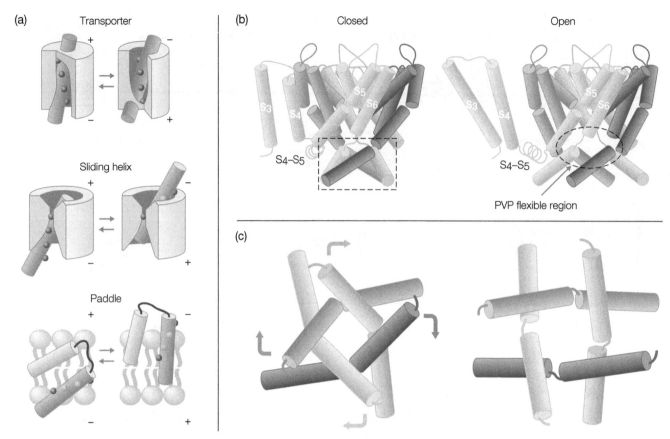

FIGURE 27.5 Models of voltage sensing in voltage-gated channels. (a) Models of voltage-sensing domain (VSD) movement during gating. Gray cylinders represent S_4 unless otherwise indicated. Protein surrounding S_4 is in light gray. The first four S_4 arginines are shown as dark spheres when they are in the foreground, as light spheres when they are behind the cylinder. To keep drawings simple, in some of the models the arginines are arranged on the same face of the S_4 helix. In each model, the S_4 resting position is shown on the left and the activated position on the right. In the helical screw model, the extent of S_4 transmembrane movement ranges from ~5 to ~13 Å, depending on the tilt of the helix (Modified from Ref. [2]). (b) VSD movements are transmitted to the pore region. The panel shows two models (closed state on the left, open state on the right) of the $K_v1.2$ channel. The boxed region has been enlarged in panel c. (c) Bottom view (from the intracellular side) of the $K_v1.2$ pore, showing the position changes of the C-terminal part of S_6 in the closed–open transition (Modified from Ref. [3]).

state of the membrane, when the voltage sensors are likely found in an activated state, whereas much less information is available on the "resting" state of the sensor.

Three main conceptual models have been proposed to describe the dynamic structural changes occurring in the voltage-sensing domain during activation (Fig. 27.5a):

- The transporter model. In this model, most S_4 arginine residues are positioned within water-accessible polar cavities, in direct contact with the intracellular or extracellular environment. Upon membrane depolarization (activation), S_4 would mostly undergo rotational movements, with an axial dislocation of about 2–4 Å. S_4 positive charges would therefore move from a cavity in direct continuity with the intracellular environment to one facing the extracellular space. The transmembrane electric field would thus be concentrated in a narrow region of the protein, only 5–10 Å wide, considerably thinner than the plasma membrane (~30 Å). This model highlights similarities between the structural changes undergoing in the voltage sensor of VGICs and those occurring in transporters during ion translocation.

- The "sliding helix" model. According to this model, in the resting state, the positively charged S_4 segment would be drawn closer to the intracellular region of the membrane by the electrostatic potential (negative inside); positive charges would interact with negative charges in the S_1, S_2, and S_3 segments. Membrane depolarization would cause a 60–180° rotation of S_4, together with a 5–15 Å axial dislocation, to allow the first three arginine residues to completely cross the membrane electric field and to form novel electrostatic contacts with neighboring protein regions.

- The "paddle" model. This model is based on the tridimensional structure of the K_vAP bacterial channel, in which S_4 and the more distal part of S_3 (S_{3b}) form a helical hairpin, or paddle, that moves across the membrane as a unit with a high degree of freedom (15–20Å). The fundamental characteristic of the paddle model is that the S_4 arginines are directly exposed to lipids in both resting and activated states, whereas in the other models, they move within the protein environment. Although this model seems to properly describe some of the functional features of K_vAP, it seems not applicable to most eukaryotic channels, whose S_{3b}–S_4 linker is longer, thus making the formation of a highly mobile "paddle" structure rather unlikely.

Finally, it must be reminded that recent studies have confirmed the role of S_4 as the voltage-sensing domain not only in depolarization-activated channels but also in channels in which the activation process is triggered by membrane hyperpolarization, such as in hyperpolarization-activated cyclic nucleotide-gated (HCN) channels (see the succeeding text).

Inactivation

As already introduced, activation is often followed by inactivation. In general, there are two mechanisms responsible for inactivation: voltage-dependent inactivation and ligand-dependent inactivation. Voltage-dependent inactivation occurs with highly variable kinetics (ranging from few milliseconds to minutes) and is likely caused by a structural rearrangement involving distinct protein regions (cytoplasmic N-terminal region of α-subunits in K^+ channels, linker region between domains III and IV in Na^+ channels or between domains I and II in Ca^{2+} channels). These intracellular hydrophobic regions may be regarded as the channel "inactivation gate," which would interrupt permeation by binding to the pore region. Ligand-dependent inactivation (such as that triggered by $[Ca^{2+}]$ rise in Ca^{2+} channels) is instead due to complex processes involving interaction of the channel with Ca^{2+}-binding proteins such as calmodulin, phosphorylation–dephosphorylation events, and interaction of the main channel subunit with accessory subunits.

Ion Selectivity

As shown by mutagenesis experiments investigating the role of specific amino acid residues in ion permeability and selectivity, as well as in potency and efficacy of drugs acting as pore blockers (such as tetraethylammonium and charybdotoxin), the linker region (P region) between the S_5 and S_6 transmembrane segments contains a stretch of about twenty amino acids that forms the walls of the central region of the channel pore. This stretch is positioned in the middle of a ring formed by the other segments and contains the molecular determinants of ion selectivity, the glycine–tyrosine–glycine (GYG) sequence, forming the so-called selectivity filter, where discrimination among ion species occurs. The role of the P region in ion selectivity was first described in K^+ channels but was later extended also to Na^+ and Ca^{2+} channels. In addition to this region, other segments contribute to the formation of the ion conduction pathways: the segment connecting S_4 and S_5 segments, the distal portion of S6, and, in 2TM channels, distinct regions in the C-terminal tail of each subunit.

In resting conditions, the pore is closed and ion flux is impeded; when the voltage sensor activates, the pore can open. During depolarization, S_4 moves outward pulling the intracellular S_4–S_5 linker, which in turn transfers these mechanical forces to the distal region of S_6 of a neighbor subunit, leading to pore opening (Figs. 27.5b and 27.5c). Recently, crystal structures of several mammalian K^+ channels have revealed that the hinges governing pore opening and closing are localized at the level of the C-terminal region of S_6. Within this region, the proline–valine–proline (PVP) sequence of each subunit (or domain) creates a bundle of four flexible helices that dislocate radially during pore opening and converge centrally during closing, thereby interrupting ion flux. In bacterial channels, lacking the characteristic PVP sequence, a similar hinge function is attributable to a glycine residue in homologous position. The different degree of movement of the distal S_6 regions might explain the differences in single-channel conductance observed between bacterial and mammalian channels; in fact, movements seem much wider in bacterial channels (having larger single-channel conductance) when compared to mammalian channels, having a ten times smaller single-channel conductance. Thus, direct contacts between the S_4–S_5 linker and the C-terminal region of S6 seem to be required for efficient electromechanical coupling between voltage sensor dislocation and pore opening during depolarization.

DRUGS AND ION CHANNELS

Characterization of the specific pharmacological properties of each class of ion channel is a major goal for modern medicine, as 5% of available pharmacological therapies for human diseases recognize ion channels as direct targets. Moreover, 70% of all drug targets include integral membrane proteins such as receptors and enzymes, whose drug-induced changes often indirectly translate into an altered functional behaviour of ion channels. Thus, understanding the pharmacological and biochemical modulation of specific ion channels is essential to unravel the mechanism of action of available drugs and to plan innovative strategies targeting ion channels.

Drugs Interacting Directly with Ion Channels

Several drug classes exert their therapeutic effects through direct interaction with specific ion channels (see Table 27.2). Some target Na^+ channels (local anesthetics, anticonvulsants,

TABLE 27.2   Clinically useful drugs acting on ion channels

Main therapeutic target	Cellular effects	Clinical effects	Indications	Drug classes (notes)
Voltage-gated sodium channels (VGSCs)[a]				
Neuronal (peripheral) (SCN9A)	Decreased neuronal excitability	Block of impulse conduction	Local anesthesia	Lidocaine and other Na1.7 channel blockers[a]
Neuronal (central) (SCN1A, 2A, etc.)	Decreased neuronal excitability	Decreased neuronal firing	Epilepsy	Carbamazepine, phenytoin, lamotrigine, etc.
Myocardial (SCN5A)	Decreased cardiac excitability	Reduced cardiac frequency	Arrhythmias, AV block	Class I antiarrhythmics
Skeletal muscle (SCN4A)	Decreased skeletal excitability	Myorelaxation	Myotonias	Mexiletine, tocainide[b]
Voltage-independent sodium channels				
Sodium channels in epithelial cells (ENaCs)	Increased sodium excretion / Decreased potassium excretion	Enhanced diuresis / Antikaliuretic effects	Hypertension	Amiloride (in association with potassium-wasting diuretics)
Voltage-gated calcium channels (VGCCs)				
Smooth muscle L type (CACNA1C)	Smooth muscle relaxation	Vasodilation → decreased blood pressure and decreased heart pre- and afterload	Hypertension	Dihydropyridines[c] (inactivation-dependent block)
Cardiac L type (CACNA1C)	Decreased contractility	Decreased heart work and oxygen demand	Angina, heart ischemia	Phenylalkylamines, benzothiazepines[c] (use-dependent block)
	Decreased excitability and conduction velocity	Delayed impulse propagation, increased atrioventricular conduction time	Heart arrhythmias	Phenylalkylamines
Cardiac T type (CACNA1G, CACNA1H)	Reduced sinoatrial excitability	Reduced heart rate	Hypertension	Mibefradil (poorly selective, withdrawn)
Neuronal N type (CACNA1B)	Impaired neurotransmitter release	Decreased pain sensation	Pain	Ziconotide
Neuronal T type (CACNA1G, CACNA1H, CACNA1I)	Reduced thalamocortical hyperactivity	Reduced neuronal activity	Epilepsy	Ethosuximide
Voltage-gated potassium channels (VGKCs)				
[d]Heart IK (KCNA5, KCNQ1, KCNH2)	Reduced repolarization	Increased atrial or ventricular AP duration, prolonged refractoriness	Arrhythmias	Amiodarone, bretylium, sotalol (also a β-blocker), class III antiarrhythmics
Neuronal I_{KM} (KCNQ2, KCNQ3)	Decreased neuronal hyperexcitability	Reduced neuronal activity	Epilepsy, pain	Retigabine (I_{KM} opener), flupirtine (I_{KM} opener)
Neuronal IK (PNS)	Increased neuronal excitability	Facilitated impulse conduction	Multiple sclerosis	4-Aminopyridine (blocker)
Voltage-independent potassium channels				
[e]KATP channels in pancreatic β-cells	β-Cell depolarization	Increased insulin release	Diabetes (type 2)	Sulfonylureas (blockers)
KATP channel in vascular smooth muscle	Reduced smooth muscle excitability	Vasodilation	Hypertension	Diazoxide, cromakalim (openers)
KATP channel in hair follicles	Follicular dermal papillae hyperpolarization	Hair growth	Hair loss	Minoxidil (opener)

TABLE 27.2 (*Continued*)

Main therapeutic target	Cellular effects	Clinical effects	Indications	Drug classes (notes)
Neuronal potassium channels	Decreased neuronal excitability	Neuronal silencing	General anesthesia	Halothane, isoflurane (openers)
Cyclic nucleotide-modulated, nonselective cation channels				
Cardiac HCN channels	Decreased cardiac excitability	Reduced cardiac frequency	Angina, ischemia	Ivabradine, zatebradine (blockers)

a All VGSC blockers can be considered "local" anesthetics; the "local" attribute only holds as long as systemic absorption is prevented. In local anesthetic preparations, these drugs always are often associated with a vasoconstrictor (e.g., adrenaline), with the exception of cocaine (a powerful vasoconstrictor due to its indirect sympathomimetic effects). Systemic absorption can lead to anti- or (rarely) proarrhythmic effects.
b Skeletal muscle Na channels are not the preferred target to produce muscle relaxation: GABA modulators or dantrolene is generally preferred.
c The selectivity for the heart or vascular muscle is partly accounted for by the mode of block (see also text):
Verapamil and diltiazem exert a use-dependent block (the drug binds to the open channel and slowly dissociates). This is particularly efficient on cardiomyocytes that undergo repetitive depolarization. The reduced heart activity is advantageous in hypertensive and ischemic patients, and even more so in arrhythmias, as the faster the heart beats, the more efficient is the channel block.
Dihydropyridines exert an inactivation-dependent block that does not affect the cardiomyocytes, which only transiently depolarize. The decrease in arterial resistance lowers arterial pressure (afterload) and lets the heart pump the same output with less work (work = volume × pressure); the slight decrease in venous reduces the preload as well: telediastolic volume (heart filling) is reduced, and contractility is similarly affected (Starling law), thereby contributing to decrease heart work and oxygen consumption (good in angina and ischemia).
d Heart voltage-dependent K⁺ channel block produces prolongation of the action potential and refractory period. This is more efficient in cells that express more channels and have a shorter action potential. This produces an efficacious means to prevent multiple reentry and ventricular fibrillation.
e KATP channels are targets for blockers and openers; drugs may specifically bind to either SUR1 or SUR2 subunits (see text):
SUR1-specific blockers are usually preferred to impair beta-pancreatic cell polarization and increase insulin release.
SUR2-specific activators are preferred to produce smooth muscle hyperpolarization and relaxation, with little effect on insulin release.

antiarrhythmics, diuretics), others Ca^{2+} channels (antihypertensive, antianginal drugs, other antiarrhythmics), and others K^+ channels (additional classes of antiarrhythmics, oral glucose-lowering drugs, vasodilators, and few anticonvulsants). These drugs will be described in detail within the following paragraphs specifically dedicated to each ion channel class. Supplement E27.1, "How to Observe Ion Channel Currents in Real Time," summarizes the general features of the electrophysiological techniques mostly used to gain information on ion channel function; Supplement E27.2, "How to Study Interactions between Drugs and Ion Channels," describes the main strategies to evaluate drug–ion channel interactions, largely employed today to identify new chemical entities acting on specific classes of ion channels.

Modulation of Ion Channel *Activity* by Drugs Acting on Receptors Functionally Coupled to Ion Channels

Several therapeutic agents act as agonists or antagonists of metabotropic receptors for hormones and neurotransmitters. Agonists of these receptors activate intracellular pathways (such as G proteins, enzymes such as cyclases or phospholipases, and second messengers like cAMP, cGMP, IP_3, diacylglycerol, arachidonic acid, and Ca^{2+}, kinases, or phosphatases) that can elicit, among other things, a robust modulation of plasma membrane ionic conductances. Given this tight functional interplay between metabotropic receptors and ion channels, drugs acting on the first might interfere with the activity of the second. Receptor–channel coupling mechanisms could be direct, involving a single intracellular biochemical step (such as activation of a G-protein subunit, which in turn binds an ion channel protein and modifies its function) or a cascade of biochemical intermediates. Ion channel regulation by hormone or neurotransmitter receptors is essential for modulation of fundamental processes such as heart control by sympathetic and parasympathetic branches of the autonomous nervous system, modulation of neurotransmitter release, muscle contraction, and hormonal secretion.

A classical example is represented by the autonomic regulation of L-type Ca^{2+} channels in cardiac myocytes. Activation of β1 receptors activates a specific G protein (G_s) that can affect L-type channels by two complementary mechanisms: a faster one, involving its direct interaction with the channel, and a slower one requiring cAMP synthesis, protein kinase A activation, and phosphorylation of the channel. Both processes increase the opening probability of L-type channels and thereby enhance Ca^{2+} influx and inotropic responses. By contrast, the negative chronotropic, dromotropic, and inotropic effects of acetylcholine are due to G_i-dependent activation of an inwardly rectifying K^+ channel in sinoatrial cells and in atrial myocytes. Upon binding to M_2 muscarinic receptors, acetylcholine released by parasympathetic postganglionic neurons can activate G_i

causing its dissociation into α and βγ subunits; the latter ones bind to a specific recognition sequence on the K+ channel, leading to its activation.

SODIUM CHANNELS

The exceptionally fast conformational change of these channels from "closed" to "open" state allows the high-speed depolarization observed in action potential (AP) in almost all excitable tissues. This mechanism is favored by the high channel density present at specific sites of the plasma membrane.

Molecular Structure and Modulation

Voltage-gated sodium channels (VGSCs) consist of a long single polypeptide chain (α-subunit) with four homologous domains each containing six transmembrane segments and accessory β-subunits that differ in number and type in various excitable tissues (Fig. 27.6). The VGSC α-subunit from the *Torpedo* electric organ was the first to be purified and cloned in the 1980s. α-Subunits contain all binding sites for the known drugs and toxins, as well as sites for post-translational modifications (such as glycosylation and phosphorylation), which are known to influence channel stability in the membrane and to regulate channel activity. In fact, phosphorylation by cAMP-dependent protein kinase A and by specific tyrosine kinases reduces ion flux both in neuronal and cardiac VGSC channels. By contrast, PKC phosphorylation, besides reducing ion flux, slows inactivation kinetics in neuronal VGSCs and reduces VGSC expression during chronic opiate exposure. Finally, direct modulation by α and βγ G-protein subunits has also been reported for VGSCs.

In mammals, each α-subunit is associated with one or two β-subunits of smaller size (35–38 kDa), containing a single-transmembrane segment, a large immunoglobulin-like extracellular domain, and a small cytoplasmic tail. Different genes encoding VGSC α-subunits are present in the human genome, each showing a distinct pattern of tissue-specific expression (Table 27.3). In general, different isoforms of VGSC α-subunits are expressed in peripheral nervous system (PNS) and central nervous system (CNS), in sensory, muscular, and cardiac tissues; each of these isoforms displays a distinct sensitivity to drugs and toxins. In particular, neuronal $Na_V1.1$, $Na_V1.2$, and $Na_V1.3$ subunits are blocked with high affinity (low nanomolar IC_{50}) by tetrodotoxin (TTX, a non-peptidic toxin isolated from ovary tissue of puffer fish from Japan), which is much less effective on $Na_V1.5$ (cardiac) and $Na_V1.8$ and $Na_V1.9$ (neuronal) VGSC α-subunits. Since TTX is known to act on VGSCs as a pore blocker, small differences in the primary sequence of the S_5–S_6 linker region among α-subunits are responsible for their different sensitivity to the toxin.

Cellular Localization of VGSCs

Besides tissue-specific expression, each VGIC isoform is often present only in a distinct cell population within a specific tissue (i.e., excitatory vs. inhibitory neurons), and within this cell population, expression may be restricted to a specific

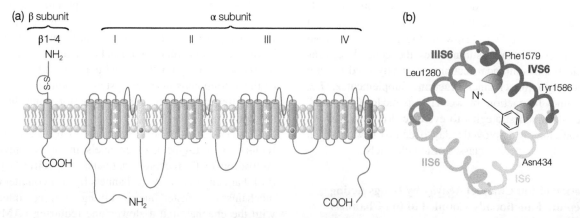

FIGURE 27.6 Membrane topology of voltage-gated Na+ channels. (a) Transmembrane segments are shown as cylinders. The main pore-forming and voltage-sensing α-subunit comprises four domains (labeled I–IV), each with six transmembrane segments. β-Subunits have a single-transmembrane segment, a short intracellular domain, and a single, extracellular immunoglobulin-like loop; $β_1$ and $β_3$ have noncovalent interactions with the α-subunit, whereas $β_2$ and $β_4$ are covalently linked to it by disulfide bridges. Site-directed mutagenesis studies have identified residues (gray dots) in transmembrane segments IS6, IIIS6, and IVS6, which are important for binding of local anesthetic and antiepileptic sodium channel blockers. (b) The typical structure of sodium channel blockers consists of a positively charged nitrogen moiety at one end and an aromatic ring at the other end. Molecular modeling of the drug binding site suggests that the positively charged amine interacts strongly with a phenylalanine in domain IV (Phe1579 in the $Na_V1.4$ channel used for modeling analysis) and, to a lesser extent, with a leucine in domain III (Leu1280 in $Na_V1.4$), whereas the aromatic group interacts with a tyrosine in domain IV (Tyr1586) and an asparagine in domain I (Asn434) (Modified from Ref. [4]).

TABLE 27.3 Classification and pharmacology of VGSC

Isoforms	Na$_v$1.1	Na$_v$1.2	Na$_v$1.3	Na$_v$1.4	Na$_v$1.5	Na$_v$1.6	Na$_v$1.7	Na$_v$1.8	Na$_v$1.9
Old names	Brain type I	Brain type II	Brain type III	μ1; skMI	h1; skMII	NaCh6;	PN4 PN1	SNS; PN3	NaN; SNS-2
Genes	*SCN1A*	*SCN2A*	*SCN3A*	*SCN4A*	*SCN5A*	*SCN8A*	*SCN9A*	*SCN10A*	*SCN11A*
Chromosome	2q23–24	2q23–24	2q23–24	17q23–25	3p21	12q13	2q24	3p21–24	3p21–24
Tissue	CNS, PNS	CNS	CNS	Skeletal muscle,	Heart, embryonic CNS (limbic system)	Cerebellum, Ranvier node, DRG	PNS, DRG, chromaffin and Schwann cells	DRG neurons (small diameter)	DRG neurons (small diameter)
Blockers (IC$_{50}$)	TTX (6 nM), local anesthetics, antiepileptics, antiarrhythmics	TTX (13 nM), local anesthetics, antiepileptics, antiarrhythmics	TTX (4 nM), local anesthetics, antiepileptics, antiarrhythmics	TTX (5 nM), lidocaine, μ-GIIIA, μPIIIA	TTX (2 μM), local anesthetics antiepileptics, antiarrhythmics	TTX (1 nM), local anesthetics, antiepileptics, antiarrhythmics	TTX (2 nM), local anesthetics, antiepileptics, antiarrhythmics	TTX (>100 μM)	TTX (40 μM)
Activators	Veratridine, β-toxins	Veratridine, β-toxins,	Veratridine, β-toxins	Veratridine, β-toxins	Veratridine, β-toxins	Veratridine, β-toxins	Veratridine, β-toxins		
Inhibitors of "inactivation"	ATXII, α-toxins	ATXII, α-toxins	ATXII, α-toxins	ATXII, α-toxins	ATXII, α-toxins	ATXII, α-toxins	ATXII, α-toxins		
Physiology	Action potential, firing	Action potential, firing, KO mice prenatally lethal	Action potential, firing, upregulated under neuronal damage	Action potential, firing	Action potential, firing	Action potential, firing	Action potential, firing	Action potential nociception	Nociception
Pharmacological role	Antiepileptic drugs	Antiepileptic drugs	Antiepileptic drugs	Antiepileptic drugs	Antiarrhythmics	Antiepileptic drugs	Local anesthetics, carbamazepine	Antiepileptic drugs	Analgesic
Channelopathies	GEFS+, FHM3	GEFS+		HPPC, PC, PAM	LQT3, BrS, PCCD, SSS	Cerebellar atrophy, mood disorders, ataxia	IEM, PEPD, CAIP		

Abbreviations: ATXII, toxin purified from sea anemone; α-toxins, neuropeptide toxin purified from scorpions; β-toxins, neuropeptide toxin purified from scorpions; BrS, Brugada syndrome; CAIP, channelopathy-associated insensitivity to pain; DRG, dorsal root sensory ganglia; FHM3, familial hemiplegic migraine type 3; GEFS+, generalized epilepsy febrile syndrome; HPPP, hyper- or hypokalemic periodic paralysis; IEM, (inherited erythromelalgia); LQT3, long QT syndrome type 3; μ-GIIIA, μ-PIIIA, toxins purified from marine snail; PAM, potassium-aggravated myotonia; PC, paramyotonia congenita; PCCD, progressive cardiac conduction defect; PEPD, paroxysmal extreme pain disorder; SSS, sick sinus syndrome; TTX, tetrodotoxin (IC$_{50}$).

subcellular site. As an example, in neurons, some channels are expressed in dendrites, others at postsynaptic or somatic regions, and others along axons or presynaptic regions. Interaction with cytoskeletal proteins or other signaling molecules govern cellular- and subcellular-specific expression of several ion channels. For example, the density of neuronal VGSCs is the highest in axon hillock, where synaptic inputs are integrated and all-or-none action potential responses are triggered. Along axons, very few channels are expressed in the internodal membrane, whereas a much higher density ($2000/\mu m^2$) is present in Ranvier nodes, causing the saltatory conduction of AP. It has been suggested that the immunoglobulin-like N-terminal domain of the β-subunit (similar to that found in many adhesion molecules) mediates interaction with adhesion molecules of glial cells, and this interaction is responsible for VGSC subcellular location. Moreover, cytoskeletal proteins such as ankyrin G are also able to interact with the VGSC α-subunits, and this interaction is determinant for their nodal localization. Proper VGSC subcellular location is also crucial for skeletal muscle functioning, where local depolarization caused by opening of ion channels associated with nicotinic receptors in the upper regions of the postsynaptic invaginations is sensed by VGSCs concentrated at the bottom of these invaginations, which integrate and amplify these signals until the AP threshold is reached.

Pharmacology of VGSCs

Some pharmacological properties of VGSCs were characterized well before the structural and functional heterogeneity of this protein class had been recognized. Indeed, the pharmacology of VGSCs is a paradigmatic example of how therapeutic drugs may help to clarify the molecular mechanisms mediating their own action. After the discovery of the potent anesthetic action of cocaine at the end of the nineteenth century, the search for synthetic derivatives devoid of the addictive properties of this alkaloid led to synthesize procaine in 1905. In the 1940s, to overcome limitations due to the short duration of procaine action, lidocaine was introduced, the prototype of a new class of local anesthetics with amidic structure. Lidocaine synthesis can be considered the divergence point between local anesthetics and antiarrhythmics. In fact, phenytoin, which exerts anticonvulsant actions with reduced sedative effects compared to barbiturates, still represents an effective antiarrhythmic. Given this historical overview, and in consideration of the role played by VGSCs in controlling excitability of a variety of cell types, it is not surprising that drugs with very different therapeutic indications exert their effects through a common mechanism of action—VGSC blockade in excitable membranes.

The most important drug classes acting on VGSCs are:

a. Local anesthetics, blocking VGSCs in peripheral nerves
b. Anticonvulsants, blocking VGSCs in central neurons
c. Class I antiarrhythmics, blocking cardiac VGSCs
d. Muscle relaxants, blocking VGSCs in skeletal muscle

In spite of their similar mechanism of action, considerations about pharmacokinetics (i.e., route of administration) and pharmacodynamics (i.e., selectivity for different isoforms) justify the classification of each molecule within a specific therapeutic class. For these drugs, the interaction mode and the molecular determinants of VGSC block are now relatively well known (Fig. 27.6b).

Frequency- and Voltage-Dependent Binding of Blockers to VGSCs As mentioned earlier, VGSCs undergo a sequential process of voltage-dependent activation, inactivation, and recovery from inactivation. A distinct pharmacological sensitivity corresponds to each of these kinetic states. A peculiar characteristic of VGSC blockade by local anesthetics is their use dependence (or phasic block); in fact, in the presence of these molecules, the Na^+ current block is proportional to the nerve stimulation frequency. This is because the drug (particularly if charged) rapidly reaches the binding site in the channel mouth when this is in the activated state and then has a high affinity for the inactivated configuration. Thus, blockade increases cumulatively and progressively during each AP elicited in the neuron; the higher the firing frequency, the faster the blockade. Since all transitions between kinetic states are voltage dependent and open and inactivated states are more represented when the membrane is depolarized, it is reasonable to hypothesize that depolarized cells are more prone to be blocked. These properties are responsible for the fact that class I antiarrhythmics preferentially act on damaged (depolarized) cardiac cells, without greatly affecting the electrical properties of normal cells, and that anticonvulsants preferentially block epileptic discharges. Moreover, local anesthetics preferentially block small size, mostly unmyelinated nervous fibers carrying sharp pain sensations as these have higher firing frequencies compared to fibers carrying other sensory or motor modalities with slower firing frequencies.

VGSC Blockers in Cardiovascular Diseases Blockade of VGSCs is one of the main mechanisms of action of antiarrhythmic drugs. Indeed, VGSC blockade can inhibit reentry arrhythmias, reduce delayed afterdepolarization-induced extrasystoles, and decrease AP duration and prevent early afterdepolarizations (EADs). VGSC blockers display negative dromotropic effects and decrease cardiac excitability. The classification of antiarrhythmic drugs is shown in Table 27.4; within VGSC blockers (class I), three subclasses—namely, IA, IB, and IC—can be identified based on drug-specific effects on AP duration. This pharmacological difference can be influenced by the state-dependent interaction of each drug with the cardiac channel molecule and by their ability to block also other cardiac ion channels. Indeed,

TABLE 27.4 Classification of antiarrhythmic drugs according to Vaughan-Williams

Class		Mechanism of action	Examples
I.		**VGSC antagonists**	
	IA	*Blockade*, medium to high; *dissociation*, medium; *AP duration increase*, medium, due also to effects on K+ channels	Quinidine, procainamide, disopyramide
	IB	*Blockade*, medium to low; *dissociation*, fast; *AP duration increase*, scarce no effects on K+ channels	Lidocaine, phenytoin, mexiletine, tocainide
	IC	*Blockade*, high; *dissociation*, slow; *AP duration increase*, scarce, but decreased AP upstroke velocity (V_{max}) and prolongation of the refractory period	Encainide, flecainide, propafenone, moricizine
II.		**β-Adrenergic antagonists**	
		Indirect block of Ca^{2+} channels due to decreased adrenergic stimulation	Propranolol, metoprolol, atenolol, esmolol, acebutolol, timolol, betaxolol, carvedilol, etc.
III.		**K+ channel antagonists**	
		Prolongation of the refractory period and reduced repolarization speed with prolongation of AP duration (long QT); poor effects on VGSC	Dofetilide, ibutilide, tedisamil, amiodarone, sotalol, azimilide, bretylium
IV.		**Ca^{2+} channel antagonists**	
		Inhibition of sinoatrial node "pacemaker cells" and atrioventricular conduction speed due to direct Ca^{2+} channel blockade	Verapamil, diltiazem, mifebradil

Abbreviations: AP, action potential; QT, cardiac action potential duration.

although all class I antiarrhythmics show a marked use dependence, IA (quinidine, procainamide, disopyramide) and IC (flecainide, encainide, propafenone) drugs have very slow dissociation rates from the channel during the diastole (>2 s); therefore, they depress Na+ currents also at normal cardiac frequency. Vice versa, the IB class molecules (lidocaine, mexiletine, tocainide, phenytoin), having a recovery speed between 0.5 and 1 s, inhibit cardiac conduction only at high frequency rates; this could explain why IB molecules show lower proarrhythmic and negative inotropic effects with respect to other group I molecules. The prolongation of the AP produced by class IA molecules is likely a consequence of their ability to block also K+ channels responsible for myocardial repolarization. Finally, it should be remembered that drugs able to activate cardiac VGSCs produce also an important positive inotropic effect that could be used in the future for the treatment of cardiac failure.

VGSC Blockers in Anesthesia VGSC blockers such as bupivacaine, lidocaine, mepivacaine, tetracaine, and many others are widely used to produce local anesthetic effects. Molecules introduced more recently display an improved efficacy and tolerability compared to these drugs, showing less neurotoxic and, especially, cardiotoxic unwanted effects (chronotropic, dromotropic, and negative inotropic effects). It has been demonstrated that the negative dromotropic effects show a high level of stereoselectivity; the S-(-) isomers of local anesthetics with prolonged effects are less effective than the corresponding racemic mixtures. Moreover, in the last years, VGSCs have gained considerable attention as targets of analgesic drugs to treat peripheral neuropathic pain caused by primary afferent fiber hyperexcitability.

Specific VGSC isoforms are expressed in dorsal root ganglia of the peripheral nervous system (Table 27.3), whose expression can be modulated by pain itself. Indeed, drugs able to selectively block these isoforms would be devoid of CNS or cardiovascular side effects.

VGSC Blockers and Neurological Disorders

Epilepsy. Loss of balance between excitatory and inhibitory inputs in specific neuronal populations represents the main pathogenetic mechanism underlying epileptic seizures; such an unbalance causes neurons in the epileptic focus to fire synchronously at rather high frequencies. None of the antiepileptic drugs currently in use are highly selective for one VGIC class or subclass; however, experimental data over the last 20 years have suggested that VGSCs are primary targets for both older (phenytoin, carbamazepine) and newer (lamotrigine, oxcarbazepine, and felbamate) anticonvulsants. VGSCs may also contribute to the anticonvulsant effects of valproic acid and topiramate. VGSC blockade by anticonvulsants has the following characteristics: (i) blockade is stronger at depolarized membrane potentials; (ii) recovery from inactivation is significantly prolonged, thereby causing a decreased availability of VGSCs during high-frequency spike trains. By these mechanisms, VGSC blockers are effective anticonvulsants but do not impair neuronal behavior under physiological conditions. The novel anticonvulsant lacosamide has been suggested to act via a distinct mechanism, namely, through a stabilization of the slow-inactivate state of neuronal VGSCs.

Muscular Disorders. Myotonia is a skeletal muscular disorder, either acquired or inherited (Table 27.1), characterized by prolonged contractions caused by abnormal and delayed relaxation of skeletal muscle fibers. Among antiarrhythmic

VGSC blockers, both mexiletine (an orally active lidocaine derivative) and tocainide have been used to alleviate myotonia symptoms. Moreover, mexiletine is very efficacious in controlling muscular paralysis attacks induced by cold in patients suffering from congenital paramyotonia. Future development of molecules highly selective for SCN4A, the VGSC isoform present in skeletal muscle, will be of great therapeutic interest for these muscular disorders.

Neuroprotection. Similarities between cellular and molecular mechanisms underlying epileptic manifestations and some neurodegenerative conditions having abnormal functioning of VGSC and voltage-dependent Ca^{2+} channels (voltage-gated calcium channels (VGCCs)) as a major pathogenetic cause have provided the rationale for using VGSC (and VGCC) blockers in acute pathological processes of the CNS such as cerebral ischemia and spinal trauma. By preventing excessive cytoplasmic Na^+ concentrations in neurons, these drugs can exert neuroprotective actions via several mechanisms, such as reduction of ATP consumption by Na^+/K^+-ATPases, inhibition of the reverse mode of operation of Na^+/Ca^{2+} exchangers and glutamate transporters (thus preventing $[Ca^{2+}]_i$ increase and excessive extracellular glutamate accumulation), and reduction of osmotic swelling. Indeed, both local anesthetics, such as lidocaine, and anticonvulsants, such as phenytoin (and its more hydrophilic derivative fosphenytoin), topiramate, valproate, and zonisamide, have shown *in vivo* neuroprotective effects in animal models of both global and focal ischemia. Riluzole, which represents a unique therapeutic option available today for the treatment of amyotrophic lateral sclerosis (a neurodegenerative disorder selectively affecting motor neurons), seems to act by blocking the neuronal persistent Na^+ current (I_{NaP}); however, this molecule also seems to act by potentiating some neuronal K^+ currents and by exerting direct antiglutamatergic effects.

CALCIUM CHANNELS

VGCCs can be subdivided according to various criteria including biophysical properties, functional roles, pharmacological sensitivity, and tissue- or cell-specific expression. Such functional heterogeneity reflects an extraordinary genetic variety, as several genes encoding for VGCCs have been discovered (Table 27.5). In biophysical terms, VGCCs can be subdivided in two main classes: low voltage activated (LVA), which open in response to membrane depolarization below −50 mV, and high voltage activated (HVA), whose activation threshold is higher than −40 mV. Besides the activation threshold, other biophysical properties distinguish LVA from HVA VGCCs; in fact, compared to HVA channels, LVA VGCCs have a marked tendency to inactivate (as they mediate T-type or "transient"-type currents); they often display slower deactivation kinetics when membrane repolarizes to negative values and are similarly permeable to both Ba^{2+} and Ca^{2+} ions (whereas HVA channels are more permeable to Ba^{2+} than Ca^{2+} ions). Moreover, among inorganic cations, Ni^{2+} is a better blocker than Cd^{2+} for LVA channels, whereas the opposite is true for HVA channels. Finally, LVA and HVA can be also distinguished for their different sensitivity to synthetic organic blockers such as dihydropyridines and to toxins such as ω-conotoxin GVIA, a blocker of certain HVA subtypes, but not of LVA (Table 27.5).

The HVA class can be further subdivided in L, N, P/Q, and R subtypes according to both functional and pharmacological criteria. L (L: long-lasting) channels are expressed in neuronal and muscular (cardiac, skeletal, and smooth) tissues and are highly sensitive to dihydropyridines like nifedipine or nimodipine. They are selectively activated by other agents like BayK8644 or SZ(+)-(s)-202-791. The N (N: neuronal) subtype is highly expressed in neurons but is insensitive to dihydropyridines, being instead blocked by ω-conotoxin GVIA, a peptide toxin from *Conus geographus*. In Purkinje cells of the cerebellum, a different subclass (P: Purkinje cells) of HVA channels is highly expressed; P-type currents are insensitive to dihydropyridines and ω-conotoxin GVIA but are selectively blocked by toxins from spiders such as ω-agatoxin IVA and by the funnel spider venom toxin FTx (see Supplement E27.3, "Natural Peptide Toxins"). Other neurons, such as cerebellar granular cells, display HVA currents that are sensitive to ω-agatoxin IVA but require much higher toxin concentrations to be blocked than the P type; it is not clear whether these currents (Q-type currents) are carried by channels encoded by the same genes encoding P-type currents. Therefore, it is common to define as P/Q-type all currents sensitive to ω-agatoxin IVA blockade. Finally, a fraction of the HVA Ca^{2+} current in cerebellar granular cells is insensitive to agents commonly used to identify L-, N-, and P/Q-type currents and have been therefore defined as R-type current (R: resistant).

Localization and Physiological Functions of VGCCs

The distinctive biophysical properties of LVA and HVA currents account for their different role in controlling cell excitability; whereas LVA-type currents are important for rhythmic firing of neurons or pacemaking in heart cells, LVA channels are responsible for the cytoplasmic increases in $[Ca^{2+}]$ that allow this ion to act as second messenger. In CNS pyramidal neurons, LVA channels are mainly expressed in apical dendrites, where they regulate cell responsiveness to synaptic inputs, whereas HVA channels are preferentially expressed in soma and basal dendrites, subcellular regions where Ca^{2+} ions control gene expression, enzyme activity, and cytoskeletal organization. N, R, and P/Q channels mainly localize at presynaptic terminals, where they regulate neurotransmitter release. In cardiac tissue, HVA channels do not display a significant site-specific expression, whereas LVA channels are mainly present in cells of the sinoatrial node

TABLE 27.5 Classification and pharmacology of VGCCs

Isoforms	$Ca_v1.1$	$Ca_v1.2$	$Ca_v1.3$	$Ca_v1.4$	$Ca_v2.1$	$Ca_v2.2$	$Ca_v2.3$	$Ca_v3.1$	$Ca_v3.2$	$Ca_v3.3$
α-Subunit	($α_{1S}$)	($α_{1C}$)	($α_{1D}$)	($α_{1F}$)	($α_{1A}$)	($α_{1B}$)	($α_{1E}$)	($α_{1G}$)	($α_{1H}$)	($α_{1I}$)
Genes	CACNA1S	CACNA1C	CACNA1D	CACNA1F	CACNA1A	CACNA1B	CACNA1E	CACNA1G	CACNA1H	CACNA1I
Tissue	Skeletal muscle	Heart, embryonic muscle, endocrine cells	CNS, pancreas, kidney, ovary, inner ear hair cells	Retina	CNS (synaptic terminals, dendrites) cochlea, hypophysis, pancreas, heart	Neurons (synaptic terminals, dendrites, cell body)	Neurons (synaptic terminals, dendrites, cell body)	Heart, neurons, placenta, ovary	Heart, neurons, placenta, ovary, kidney, liver, adrenal gland, medulla	Neurons
Blockers (IC_{50})	DHP^a, BTZ^b, and PA^c sites	DHP^a, BTZ^b, and PA^c sites	DHP^a, BTZ^b, and PA^c sites	DHP	ω-AgaIVA; ω-CtxMVIIC	ω-Ctx GVI; ω-Ctx MVIIA SNX-111 or ziconotide, ω-Ctx MVIIC	Ni^{2+} (27 μM); SNX-482 (tarantula toxin)	Ni^{2+} (250 μM); kurtoxin; mibefradil, ethosuximide	Ni^{2+} (12 μM); kurtoxin; mibefradil, ethosuximide	Ni^{2+} (216 μM); mibefradil, ethosuximide
Activators	BayK8644 FLP64176	BayK8644 FLP64176	BayK8644							
Ion current type	LVA, L type	LVA, L type	LVA, L type	LVA, L type	LVA, P/Q type	LVA, N type	LVA, R type	HVA, T type	HVA, T type	HVA, T type
Chromosome	1q31–32	12p13.3	3p14.3	Xp11.23	19p13.2	9q34	1q25–1q32	17q22	16p13.3	22q12.3–13.2
Splicing variants		$α_{1C-a}$, heart; $α_{1C-b}$, smooth muscle; $α_{1C-c}$, heart, CNS, hypophysis			$Ca_v2.1a$, $Ca_v2.1b$	$Ca_v2.2a$, $Ca_v2.2b$	$Ca_v2.3a$, $Ca_v2.3b$			
Physiology	Skeletal muscle contraction	Cardiac and smooth muscle contraction; signaling in neuroendocrine cells	Signaling in neuroendocrine cells	Neurotransmitter release in mammalian retina; cones, bipolar cell color vision	Neurotransmitter release; neuronal signaling	Neurotransmitter release; neuronal signaling	Neurotransmitter release; neuronal signaling, firing	Cardiac pacemaker; neuronal discharge; thalamic neurons oscillatory tone	Cardiac pacemaker; neuronal discharge; thalamic neurons oscillatory tone; hormone release	Cardiac pacemaker; neuronal discharge; thalamic neurons oscillatory tone
Pharmacological role		Antiarrythmics, vasodilators, antianginal			Analgesics, neuroprotectors			Antiepileptics, analgesics	Antiepileptics, analgesics	Antiepileptics, analgesics
Channelopathies	Muscular dysgenesis in mice; hypo-PP and human hyperthermic susceptibility		Deafness in KO mice, sinoatrial dysfunction, and atrioventricular deficit, poor insulin secretion	XLCSNB	EA-2; FMH; SCA-6, "tottering" mice					

Abbreviations: BTZ, benzothiazepines; DHP, dihydropyridine; EA-2, episodic ataxia type 2; FHM, familial hemiplegic migraine; hypo-PP, hypokalemic periodic paralysis; PA, phenylalkylamine; SCA-6, spinocerebellar ataxia type 6; XLCSNB, X-linked congenital stationary night blindness.

a Drugs interacting with the dihydropyridine-binding site: nifedipine, nicardipine, nitrendipine, nisoldipine, felodipine, isradipine, amlodipine, and nimodipine.
b Drugs interacting with the benzothiazepine binding site: diltiazem, clentiazem, and diclofurime.
c Drugs interacting with the phenylalkylamine binding site: verapamil, gallopamil, levemopamil, anipamil, devapamil, and tiapamil.

and of the conduction system. LVA channels are absent in the mature ventricular tissue but are transiently expressed during embryogenesis and can reappear in adult cardiac tissue during disorders like cardiac hypertrophy. In skeletal muscle, HVA channels are selectively concentrated at the triads, where they control the excitation–contraction coupling mechanism by regulating Ca^{2+} release from the sarcoplasmic reticulum (Supplement E27.4, Physiopathology and Pharmacology of Muscular Contraction). A close functional cooperativity exists between HVA channels and Ca^{2+}-dependent K^+ channels; in particular, L-type channels are coupled to SK channels in the soma, whereas N-type channels appear to be positioned in close proximity to BK channels at presynaptic terminals.

Structural Organization of VGCC

VGCCs are hetero-oligomeric complexes (Fig. 27.7a). The α_1-subunit forms the pore region and carries binding sites for most agonists and antagonists. The first α_1 ($\alpha_1 S$)-subunit was purified and cloned from the muscular triad tissue; it has the same structural organization of VGSCs, with high sequence homology in the hydrophobic, transmembrane regions. Nine different genes encoding for α_1-isoforms have been identified, often generating multiple transcripts as a result of alternative splicing. Functional characterization of each of these products in heterologous systems has led to the identification of the molecules corresponding to the previously defined VGCC subtypes: the $\alpha_1 C$ and D subunits form L channels; the $\alpha_1 B$ subunit form N channels; the $\alpha_1 A$ form P/Q channels; the $\alpha_1 E$ form R channels; and the $\alpha_1 G$, $\alpha_1 GH$, and $\alpha_1 GI$ form T-type channels. In addition to α_1, in skeletal muscles, L channels also contain the α_2 (covalently linked to the δ-subunit), β (in various spliced isoforms), and γ accessory subunits. The α_2-, δ-, and γ-subunits are membrane proteins, while the β-subunit associates intracellularly to α_1. All accessory subunits are crucial for correct membrane expression and physiological function of each VGCC. Additional functions of accessory subunits, partially or completely independent from their role in VGCCs, have also been described, such as transcriptional regulation, neuromuscular junction formation and stabilization, and control of stability of some mRNAs, including those encoding for the VGCC subunits themselves. In L and N channels in the CNS, the γ-subunit is probably substituted by a yet unknown subunit; the molecular composition of P/Q, R, and T channels has not been defined.

VGCCs exert their complex regulatory roles as part of multimolecular complexes that create "nanoenvironments" in which interactions among channels and various transduction pathways can occur. Using nanoproteomic techniques, it has been recently shown that more than 200 proteins are associated with $Ca_V 2.2$ channels: pumps, ion channels, G-protein-coupled receptors (GPCRs), kinases, phosphatases, enzymes, extracellular matrix proteins, cytoskeletal proteins, and SNARE complex proteins (necessary for neurotransmitters release; Fig. 27.7b). These complexes are normally present in cholesterol-rich lipid rafts, allowing lipid-based (cholesterol or phosphoinositides) regulatory mechanism to take place at the level of ion channel proteins.

Finally, it is important to remark that active investigations are being carried out using animal models to understand the functional role played by each VGCC subunit (Supplement E27.5, "Physiopathology of VGCCs: Genetic Studies in Animal Models and Humans").

VGCC Pharmacology

In recent years, many pharmacological tools have been used to dissect the functional roles of VGCC subtypes and to identify their tissue and cellular distribution. Among these, polyvalent cations such as Cd^{2+}, Ni^{2+}, Co^{2+}, Mn^{2+}, Zn^{2+}, La^{3+}, and Gd^{3+} are all potent blockers since they bind, in a voltage-dependent manner, to the internal pore site where they compete with Ca^{2+} itself and block ion flux. Although they block nonspecifically all VGCCs at millimolar concentrations, at micromolar concentrations, they show certain selectivity for some subclasses (Table 27.5). Moreover, peptide and nonpeptide toxins can selectively recognize different VGCC subtypes, representing therefore potent tools for pharmacological investigations (Supplement E27.3). Nevertheless, inorganic polycations and toxins purified from animal venoms both have limited therapeutic interest. By contrast, the so-called calcium antagonists are an important class of organic molecules with a large therapeutic spectrum of action, exerting antihypertensive, antiarrhythmic, and antianginal effects. Potential CNS effects have been also described. These drugs inhibit VGCCs, being active mostly on L channels, although new molecules have recently been generated with a certain degree of selectivity for N and T channels.

Drugs Acting on L-Type VGCCs The "calcium antagonist" concept was developed in the 1960s by Fleckenstein and Godfraind when they discovered that new molecules with vasodilating actions (verapamil) also had negative chronotropic and dromotropic effects that were not present in nitrate-type drugs. Indeed, in the 1970s, it was hypothesized that these molecules could interfere with the excitation–contraction coupling mechanism by blocking transmembrane Ca^{2+} inward flux, and thus, the "calcium antagonist" concept was coined. Over the following years, functional and molecular discoveries and identification of additional molecules with distinct tissue-dependent activity (dihydropyridines) confirmed the original hypothesis that the pharmacological actions of these molecules were due to their ability to inhibit Ca^{2+} flux mediated by VGCCs. The most widely used drugs belong to three chemical classes—dihydropyridines, benzothiazepines, and phenylalkylamines—and act on L-type

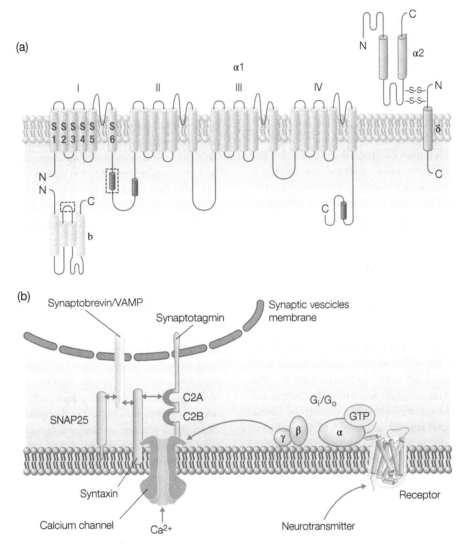

FIGURE 27.7 Membrane topology of voltage-gated Ca^{2+} channels and their role in neurotransmitter release. (a) Ca_v1 and Ca_v2 family members of VGCCs consist of a pore-forming α_1-subunit, containing four domains (I–IV), each with six transmembrane regions and a pore region in the S_5–S_6 linker. α_1-Subunits associate with intracellular β-subunits and with extracellular α_2-subunits; the latter is linked via disulfide bridges to a δ-subunit with a single-transmembrane domain. In some channels, also, a γ-subunit (not shown) may be present. Ca_v3 members may be formed by α_1-subunits alone, but their molecular composition is still uncertain. Regions contributing to G-protein subunit binding are indicated in dark (Modified from Ref. [5]). (b) Ca^{2+} entry via VGCCs triggers neurotransmitter release by promoting fusion between secretory vesicle membrane and plasma membrane, mediated by the SNARE [soluble NSF (*N*-ethylmaleimide-sensitive factor)] protein receptor complex, containing syntaxin, SNAP-25, and VAMP/synaptobrevin. SNARE protein function is regulated by several interactions with other proteins, including synaptotagmin, a Ca^{2+}-binding protein of the plasma membrane. SNARE proteins are specifically cleaved by clostridial toxins (tetanus and botulinum), which therefore inhibit neurotransmitter release. $Ca_v2.1$ and $Ca_v2.2$ VGCCs (but also other classes) interact with SNARE proteins and with synaptotagmin itself, via a cytoplasmic region called "synprint." Through these interactions, VGCCs facilitate vesicle positioning at presynaptic sites where $[Ca^{2+}]_i$ changes can efficiently translate in neurotransmitter release. Such a process is also regulated by various neurotransmitters acting on presynaptic receptors that trigger direct or indirect modulation of VGCCs by G proteins (Modified from Ref. [6]).

channels. Indeed, three different binding sites for these three drug classes are present on the α_1-subunit; these sites are allosterically linked, meaning that binding of a drug to one of them may positively or negatively influence the effects of drugs acting at different ones. In general, interaction of these drugs with VGCCs shares considerable similarity with that of local anesthetics with VGSCs, both in terms of biophysical properties (frequency and voltage dependence) and in terms of binding site location (in domains $IIIS_5$–S_6, IVS_5–S_6 linker, and IVS_6). Moreover, although each of the three drug classes interacts with L-type VGCCs, they show differential blockade of channels formed by α_1C, α_1D, or

α_1S subunits or of the proteins encoded by their alternatively spliced variants.

Cardiovascular and Noncardiovascular Indications for L-Type VGCC Blockers As previously mentioned, L-type VGCC blockade exerts antihypertensive, antiarrhythmic, and antianginal effects; however, not all three chemical classes of calcium antagonists share the same clinical indications. Dihydropyridines show a higher selectivity for L channels of vascular smooth muscles than of striated cardiac muscles. This selectivity derives from various factors such as (i) the primary sequence of the VGCC isoform (in particular the IS_6 region that is different in the α_1Cb expressed in smooth muscles compared to the α_1Ca present in cardiac muscles), (ii) the biophysical mechanism of blockade (dihydropyridines interact preferentially with the inactivated state of the channel, which is more represented in smooth muscle than in cardiac muscle), (iii) the different pharmacokinetic properties of each molecule, and (iv) the different mobilization properties of Ca^{2+} in these tissues. At vascular level, Ca^{2+} antagonists selectively block VGCCs in smooth muscle cells, without influencing (at therapeutic concentrations) $[Ca^{2+}]_i$ changes caused by depletion of intracellular stores or by activation of other Ca^{2+}-entry pathways (mostly via nonvoltage-dependent channels). At doses reducing peripheral vascular resistance, dihydropyridines are devoid of direct inhibitory cardiac effects; instead, activation of sympathetic reflexes caused by reduced vascular pressure often causes increased chronotropic and inotropic responses (particularly with "first-generation" molecules). Some dihydropyridines such as nicardipine seem to act preferentially on coronary arteries, while others, such as nimodipine, on cerebral arteries. In general, calcium antagonists are more effective on arterial than on venous smooth muscle, thereby reducing vascular resistance without influencing venous blood return.

Benzothiazepines and phenylalkylamines act on peripheral vascular resistances only at concentrations that produce blockade of cardiac VGCCs, resulting in reduction of cardiac frequency, atrioventricular conduction speed, and inotropism. These direct cardiac actions could be desirable to produce antihypertensive effects while blocking cardiostimulatory effects of cardiac sympathetic reflexes. By contrast, these two drug classes should be used with caution in patients with preexistent cardiac alterations because they may induce conduction block or cardiac failure.

Given these considerations, dihydropyridines should be considered as first choices as antihypertensive and antianginal drugs and in the prevention of cerebral posthemorrhagic vasospasm. Based on the international clinical experience over the last 40 years, Ca^{2+} antagonists may be classified in three generations: the first generation includes molecules (verapamil, diltiazem, nifedipine, felodipine, isradipine, nicardipine, nitrendipine) characterized by a relatively short duration of action, thereby requiring multiple daily administrations.

First-generation dihydropyridines cause fast-onset peripheral vasodilation, which triggers compensatory mechanisms by sympathetic and renin–angiotensin systems; therefore, their administration may cause hypotension, headache, nausea, and, rarely, pulmonary or cerebral edemas. Peripheral hypotension may decrease coronary flow, thus exposing patients to serious heart and brain ischemic episodes; meta-analysis of clinical trials suggests that, in infarct patients, first-generation dihydropyridines may increase relapse and mortality risk. This risk is lower for the other classes of calcium antagonists having a weaker vasodilation action; in these cases, association with β-blockers to block the sympathetic reflex may be advisable. By contrast, association of phenylalkylamines and β-blockers could be dangerous because of summation of their cardiac depressing effects.

Second-generation molecules often consist of the same active principles used in first-generation drugs, but in extended-release formulations; these can be administered as a single daily dose and have a lower tendency to produce reflex tachycardia. Third-generation molecules are characterized by long-term duration of action due to specific pharmacokinetic properties such as longer plasma half-life (amlodipine) or higher lipid solubility resulting in a higher partition coefficient in the plasma membrane that allows a prolonged action despite a relatively short plasma half-life (lercanidipine, lacidipine, manidipine). Clinical studies on these third-generation drugs suggest that they can reduce atherosclerosis progression; however, this effect seems to be only partially explained by their Ca^{2+}-antagonistic actions and may involve additional pharmacological properties, such as antioxidant effects, inhibition of neointimal proliferation, increase of nitric oxide production, inhibition of cytokine-induced endothelial apoptosis, and modulation of extracellular matrix deposition. On the other hand, third-generation dihydropyridines with ultrashort duration for intravenous administration (clevidipine) have been also used for controlling acute surgical hypertensive emergencies.

More recently, it has been demonstrated that some third-generation dihydropyridines, such as benidipine and efonidipine, besides blocking L channels, also behave as T-type channel blockers, paving the way for novel therapeutic applications. Indeed, T-type channels control the vascular tone of efferent glomerular arteries, and T-type selective VGCC antagonists have better efficacy, compared to other VGCC blockers, in kidney protection during hypertensive nephropathy. At cardiac level, Ca^{2+} antagonists block L-type VGCCs in the sinoatrial node, in the atrioventricular node, in conduction fibers, and in contractile myocytes. They reduce sinoatrial pacemaking rhythm, atrioventricular conduction time, and contraction force. In cardiac tissue, phenylalkylamines (verapamil, gallopamil) result much more potent than dihydropyridines, while benzothiazepines (diltiazem) have an intermediate potency; thus, phenylalkylamines are often used as antiarrhythmics (IV class; Tables 27.4 and 27.5), in particular for acute treatments of various forms of supraventricular tachyarrhythmias.

L-type VGCC blockers also have effects in other tissues. While most VGCCs in the CNS are non-L type (insensitive to classical Ca^{2+} antagonists), a fraction of L-type VGCCs are present in neuronal soma and in glial cells, and their blockade might exert neuroprotective effects.

In hormone-secreting cells such as insulin-secreting pancreatic β-cells, both L- and non-L-type channels are expressed; although no specific indication for Ca^{2+} antagonists exist at this level, Ca^{2+} antagonists might precipitate or favor clinical situations in which a decreased hormonal response plays a major pathogenetic role (i.e., decreasing insulin secretion in diabetes mellitus).

Cardiovascular Therapeutic Indications for Non-L-Type VGCC Blockers Some novel Ca^{2+} antagonists having a non-L-type profile have been evaluated in clinical trials for various therapeutic indications. Among these, mibefradil has an affinity for T-type about 10–30 times higher than for L-type VGCC channels. Indeed, this molecule seems selective for the vascular tissue, in particular for coronary arteries, without modifying cardiac inotropism; however, in 1998, one year after its commercialization as an antihypertensive and antianginal drug, it was withdrawn from the market because of various unwanted pharmacokinetic interactions with other drugs due to its extensive metabolism by CYPs 2D6 and 3A4.

Other molecules, like flunarizine and lifarizine, are under investigation as cerebral anti-ischemic drugs. Indeed, they have a large spectrum of action ranging from L- and T-type VGCCs to brain-type VGSCs, possibly explaining their neuroprotective actions. Some anticonvulsants like ethosuximide, which is particularly active in *petit mal* (absence) seizures, inhibit T-type VGCCs in the thalamus; indeed, these channels control thalamocortical synchronization, whose alteration underlies this type of epilepsy.

N- and P/Q-type VGCCs are present in nerve terminals, where they participate in the control of neurotransmitter release. A high density of N-type channels is found in dorsal root ganglia sensory neurons receiving inputs from Aδ and C nociceptive afferent fibers; their blockade interrupts pain sensations. Ziconotide is an analog of the N-type VGCC peptide blocker ω-conotoxin MVIIA, which has been approved in Europe and the United States for use in chronic pain states in patients unresponsive to other analgesic therapies. Compared to commonly used analgesics, ziconotide has a low tolerance liability, but its peptide structure requires intrathecal administration. Despite its side effects (tremors, orthostatic hypotension), ziconotide has shown neuroprotective effects in animal models of global and focal ischemia, also when administered several hours after the ischemic insult.

Also, gabapentinoids, such as gabapentin and pregabalin, are often used for neuropathic pain treatment and, more rarely, as anticonvulsants. These molecules were originally synthesized as GABA analogs with higher CNS penetration; however, their analgesic activity has been shown to be independent from GABAergic mechanisms, being rather related to their inhibition of VGCCs. Gabapentinoids bind to $α_2/δ$-subunits and affect their intracellular trafficking when bound to $Ca_V2.1$ (P/Q-type) and $Ca_V2.2$ (N-type) VGCC channels, rather than blocking channels in the plasma membrane (as other Ca^{2+} antagonists do). Moreover, their effects seem to depend on the presence of a specific type of β-subunit, β4a, in the VGCC complex.

VGCC Modulation by Indirectly Acting Drugs Several neurotransmitters acting on GPCRs can indirectly modulate VGCC activity, producing either stimulatory or inhibitory effects on current levels or activation kinetics. Agonists for some GPCRs that inhibit presynaptic VGCCs can exert analgesic or neuroprotective effects. Among these receptors are group II and III metabotropic receptors for excitatory amino acids, $GABA_B$ receptors, $α_2$-adrenergic receptors, A_1 adenosine receptors, and *k*-receptors for opioids. In most cases, they exert their inhibitory action on N-type channels, but T-, L-, and P/Q-type channels can also be modulated. This inhibition is frequently voltage dependent (being relieved at highly depolarized potentials) and can be mediated by G-protein subunits: it can be fast (regulation by βγ-subunits) or slow (regulation by intracellular second messengers). By contrast, some hormones and neurotransmitters can stimulate VGCC activity; a paradigmatic example is the regulation of L-type cardiac VGCCs by activation of $β_1$-receptors (see Chapter 11). β-adrenoceptor agonists like dobutamine can increase cardiac inotropism, an effect of therapeutic relevance during cardiogenic shock or cardiac failure caused by other acute conditions.

POTASSIUM CHANNELS

K^+ channels are the largest and the most functionally heterogeneous class of ion channels; they are expressed in all eukaryotic cells and in prokaryotes. K^+ channels mostly perform inhibitory functions by stabilizing membrane potential; their opening drives membrane potential closer to the K^+ equilibrium potential that is far away from the AP threshold in excitable cells. Moreover, activation of K^+ channels shortens duration of AP, terminates periods of intense electrical activity, reduces neuronal firing frequency, and, in general terms, decreases efficacy of cell excitatory inputs. Besides these roles, K^+ channels contribute to solute transport across epithelial membranes and in glial cells to K^+ clearance from brain interstitial spaces.

Structural Organization of Potassium Channels

More than 70 genes encoding proteins serving as K⁺ channel subunits have been identified in humans. Molecular cloning of these genes has allowed classifying K⁺ channels based on topology deduced from their primary sequences.

As illustrated in Figure 27.4, three families of subunits can form K⁺ channels:

1. the classical family with 6 transmembrane segments (6TM), which includes voltage-gated K⁺ channels (Kv channels; Kv1–Kv12). Segments S_1–S_4 form the voltage sensor domain (VSD), whereas S_5, S_6, and the intervening linker contribute to pore formation;
2. the family with 2 transmembrane segments (2TM), which are homologous to the S_5–S_6 segments of the Kv channels. Channels formed by these subunits are voltage independent as they lack the VSD. This family is formed by at least seven gene families ($K_{IR}1$–$K_{IR}7$) and includes inward-rectifier channels (both constitutively active and G protein gated);
3. the family with 4 transmembrane segments (4TM), including at least 15 different genes ($K_{2P}1$–$K_{2P}17$). While channels formed by subunits of the first two groups are tetrameric, those of the third group are dimers.

In the following paragraphs, we will describe in detail the K⁺ channel families most relevant as pharmacological targets. However, it should be reminded that structural and functional heterogeneity of K⁺ channels is not restricted to the three structural groups mentioned above. For example, some large-conductance Ca²⁺-dependent K⁺ channels (BK channels) assemble as tetramers of subunits containing seven transmembrane segments, which differ from Kv subunits for an extra transmembrane segment (S_0) at the N-terminus.

The K⁺ Channel Family with 2 Transmembrane Segments (2TM).
This family includes the so-called K_{ATP} channels. These are potassium channels largely expressed in various tissues, being particularly abundant in pancreatic ß-cells, skeletal muscle cells, cardiac myocytes, and, although at lower levels, smooth muscle cells and the brain. Their physiological role is to link electric activity to metabolic changes in cells expressing them. Indeed, these channels are open at resting membrane potential, and they close when intracellular concentrations of ATP increase because of enhanced extracellular availability and intracellular utilization of glucose. Closure of K_{ATP} channels determines cellular depolarization, resulting in VGCC activation and [Ca²⁺]$_i$ increase, thus triggering muscle contraction, as well as hormone and neurotransmitter release. K_{ATP} channels assemble as tetramers of $K_{IR}6.1$ or $K_{IR}6.2$ subunits and contain four accessory subunits belonging to the ABC (ATP-binding cassette) family, carrying two nucleotide-binding sites each. In K_{ATP} channels, these accessory proteins are called SUR, as they also represent the binding site for sulfonylureas, drugs capable of specifically blocking these channels. Each SUR subunit is arranged topologically in 17TM segments arranged in three domains (TMD0, TMD1, and TMD2), containing five, six, and six helices, respectively, being structurally similar to the CFTR chloride channel (see the succeeding text) (Fig. 27.8). The functional heterogeneity of distinct K_{ATP} channels expressed in different tissues results from the different combination of the K_{IR} (6.1 o 6.2) subunits with the SUR (SUR1, SUR2A, SUR2B) accessory subunits (see also Table 27.6). Indeed, in pancreatic β-cells, K_{ATP} channels are formed by $K_{IR}6.2$ and SUR1 subunits, in cardiac muscle by $K_{IR}6.2$ and SUR2A subunits, and in vascular and nonvascular smooth muscle cells by $K_{IR}6.1$ or $K_{IR}6.2$ and SUR2B, respectively. Finally, channels formed by almost all possible subunit combinations can be found in the brain.

K_{ATP} channels are targets for oral hypoglycemic drugs of sulfonylurea and glinide chemical classes. Sulfonylureas are orally active hypoglycemic drugs widely used to treat type II, noninsulin-dependent diabetes. They can be divided in first- and second-generation drugs: the first generation includes tolbutamide, chlorpropamide, tolazamide, and acetohexamide, whereas the second generation includes glyburide or glibenclamide, glipizide, gliciclazide, gliclazide, gliquidone, glisolamide, and glimepiride. Although members of both generations display similar glucose-lowering efficacy, second-generation molecules are generally more potent. Some new compounds (meglitinide, repaglinide, nateglinide; collectively referred to as glinides) are derivatives of benzoic acid or 3-phenylpropionic acid rather than benzylsulfonic derivatives. The mechanism of action of both sulfonylureas and glinides is based on their ability to block K_{ATP} channels in pancreatic β-cells, thus mimicking the effect of an increase in extracellular glucose or in intracellular ATP concentrations (Fig. 27.8). In β-cells, closure of K_{ATP} channels causes depolarization, activation of voltage-dependent Ca²⁺ channels, and insulin release. Therefore, these molecules are effective only if pancreas retain its insulin-secreting ability; thus, they are ineffective in insulin-dependent diabetes.

In general, SUR1-containing K_{ATP} channels are more sensitive to blockade by first-generation, short-chain sulfonylureas or by smaller-sized glinides (nateglinide), compared to SUR2-containing channels, providing a plausible explanation for the lack of extrapancreatic effects observed in clinical practice with these drugs. These differences are much less evident for long-chain, second-generation sulfonylureas such as glibenclamide or for larger glinides (repaglinide, nateglinide). In pancreatic channels, the binding site for sulfonylureas is split in two: the "A site" for the sulfonylurea part and the "B site" for the carboxamido part. Short sulfonylureas and nateglinide bind the A site, which is formed by the 14TM and 15TM segments of SUR1.

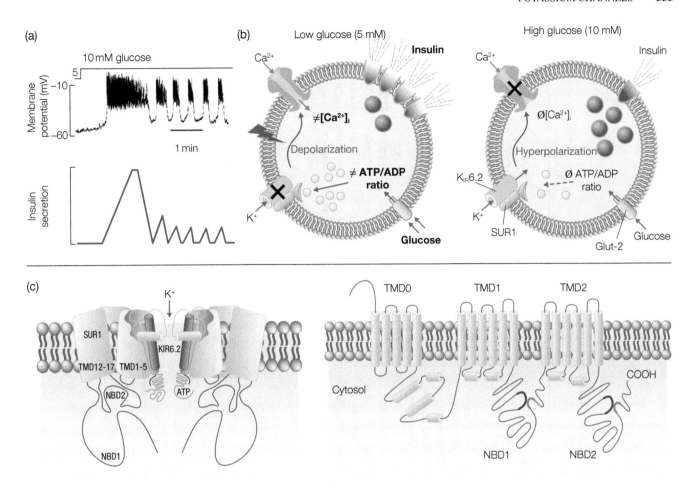

FIGURE 27.8 Physiological role, structure, and molecular determinants of K_{ATP} channels in insulin-secreting pancreatic β-cells. (a) β-Cells act as glucose sensors; when extracellular glucose concentrations increase (from 5 to 10 mM), β-cells depolarize and increase insulin secretion. (b) Role of K_{ATP} channels in glucose-dependent regulation of insulin secretion. Under low-glucose (5 mM) conditions, the ATP/ADP ratio is low, and K_{ATP} channels are open, resulting in cell hyperpolarization. Therefore, VGCCs are inactive, $[Ca^{2+}]_i$ is low, and insulin secretion is impeded. When, after a meal, glucose plasma concentrations increase (10 mM), the ATP/ADP ratio increases, causing K_{ATP} channels to close, resulting in cell depolarization, which triggers VGCC opening, $[Ca^{2+}]_i$ increase, and enhanced insulin secretion. Mutations in K_{ATP} channels (both in SUR1 and in $K_{IR}6.2$; see Table 27.1) that make them insensitive to high ATP/ADP inhibition cause chronic hyperpolarization of β cells, with consequent suppression of insulin release, even in the presence of high glucose concentration; this situation is responsible for neonatal diabetes. Vice versa, loss-of-function mutations in SUR1 or $K_{IR}6.2$ genes (such as in nesidioblastosis) lead to hypofunctional K_{ATP} channels, with chronic depolarization of β-cells and insulin hypersecretion, even in low-glucose conditions (see Table 27.1). (c) A model illustrating the hetero-octameric configuration of the pancreatic K_{ATP} channel. In the fully functional channel (left panel), four $K_{IR}6.2$ molecules are surrounded by four SUR1 proteins. The nucleotide-binding domains on SUR1, the ATP-binding site on $K_{IR}6.2$, and K^+ ions traveling through the selectivity region are highlighted. In the right panel, a topology model of SUR1 is presented with three transmembrane domains (TMD0–TMD2) and two nucleotide-binding domains (NBD1 and NBD2). NBD1 seems to preferentially bind ATP, whereas MgADP regulates K_{ATP} channels by binding preferentially NBD2 (Modified from Ref. [7]).

Glibenclamide and other long sulfonylureas, as well as glinides developed from the carboxamido part of glibenclamide, show little if any selectivity for the pancreatic channel, suggesting that the B site (formed by the N-terminus of $K_{IR}6.x$ subunits and the cytoplasmic linker between TMD0 and TMD1 in SURs) is very similar in all SURs.

In addition, differences in kinetics and modes of interaction with K_{ATP} channels might contribute to distinct pharmacological differences among insulinotropic drugs; in fact, glibenclamide and repaglinide display long-lasting effects on insulin release and reduce the ability of pancreatic β-cells to respond to physiological daily changes in glucose levels, whereas nateglinide seems to inhibit K_{ATP} channels with faster kinetics, thus displaying shorter and glucose-dependent effects. This would decrease risk of hypoglycemia associated with longer-acting molecules, providing a rationale for nateglinide use in postprandial hyperglycemias.

TABLE 27.6 Classification and functional properties of the main K⁺ channels with 2 transmembrane domains (2TM)

Isoforms	KIR1.1	KIR2.1	KIR3.1 GIRK1	KIR6.1	KIR6.2
Old names	ROMK1	IRK1	GIRK1		
Genes	KCNJ1	KCNJ2	KCNJ3	KCNJ8	KCNJ11
Tissue	Kidney cortex and medulla, spleen	CNS, heart, skeletal muscle, endothelial cells, macrophages	CNS, atrium	Widely expressed, especially in smooth muscle	Nervous system, heart, skeletal muscle, pancreas
Blockers	Ba, Cs (nonselective)	Ba, Cs (nonselective); Mg²⁺, polyamines	Ba, Cs (nonselective); Mg²⁺, polyamines	Sulfonylureas (via associated SUR receptors)	Sulfonylureas (via associated SUR receptors)
Activators			G proteins (via βγ); phosphatidylinositol 4,5-bisphosphate (PIP$_2$)	Disulfide nucleosides, KCOs	MgADP; KCOs
Chromosome	11q24	17q23.1–24.2	3q32	12p11.23	11p15.1
Physiology	Weak inward rectifier; ion transport in the kidney	Strong inward rectifier; V$_{REST}$ control; cardiac action potential repolarization	Receptor-dependent hyperpolarization of V$_{REST}$; inhibition of cardiac rate, inhibition of neuronal activity	Together with SUR2B forms the ATP-dependent K⁺ channel in vascular smooth muscle; involved in vascular tone regulation	Together with SUR1 forms the "classical" K⁺ ATP-dependent channel involved in insulin secretion acting as oxygen sensor and glucose in neurons and in cytoprotection during ischemic preconditioning
Pharmacology targets				SUR2B is the target for antihypertensive and coronary vasodilators	SUR1 is the target for sulfonylureas and KCOs
Channelopathies	Bartter syndrome	Andersen syndrome			Loss-of-function mutations, persistent hypoglycemic hyperinsulinemia of infancy (PHHI); gain-of-function mutations, persistent or transient neonatal diabetes (PNDM, TNDM)

TABLE 27.7 Classification of potassium channel openers (KCOs)

Chemical class	Drugs
1. "Openers" of ATP-sensitive K⁺ channels	
Benzopyrenes	Cromakalim, levcromakalim
Cyanoguanidines	Pinacidil, P1075
Thioformamides	Aprikalim, RP49356
Pyrimidines	Minoxidil sulfate, LP-805
Benzothiadiazines	Diazoxide
2. Drugs acting on both ATP-sensitive K⁺ channels and guanylate cyclase	
Nicotinamides	Nicorandil, KRN2391
3. Openers of large-conductance Ca²⁺-activated K⁺ channels (BK channels)	
Benzimidazoles	NS1619, NS004
Imidazopyrazines	SCA40
Terpenic derivatives	DHS-1, MaxiKdiol
Fluoro-oxindoles	BMS-204352 (also activates some members of the KCNQ family)
4. Openers of KCNQ-type voltage-gated K⁺ channels	
Derivatives of carbamic acid	Retigabine, flupirtine
Benzodiazepines	R-L3
Fluoro-oxindoles	BMS-204352
Acrylamide derivatives	S1, S2
Fenamates	Diclofenac, Meclofen
Benzamides	ICA-27243

Finally, inactivating (loss-of-function) mutations in $K_{IR}6.2$ or SUR1 are responsible for familiar hyperinsulinism of infancy (PHHI), a disease in which K_{ATP} channels are constitutively inactivated and therefore unable to exert their braking effects on insulin release; vice versa, activating (gain-of-function) mutations in the same genes have been associated with conditions of transient or persistent diabetes in neonates (see Table 27.1).

K_{ATP} channels are also targets of several potassium channel openers (KCO). The definition "potassium channel openers" includes a variety of compounds belonging to different chemical classes (Table 27.7). Among KCO having K_{ATP} as their primary target (also known as K_{ATP}COs), pinacidil, cromakalim, and nicorandil act as vasodilators, showing a clear preference for SUR2-containing K_{ATP} channels and little effects on SUR1 subunits, a pharmacological profile consistent with their lack of pancreatic effects. Diazoxide, instead, does not discriminate between SUR1- and SUR2-containing channels, acting therefore as a potent activator of both muscular and pancreatic K_{ATP} channels; in fact, intense hypoglycemic effects accompany the vasorelaxation produced by this molecule.

The most promising pharmacological action of K_{ATP}COs is their hypotensive effect, as opening of K⁺ channels hyperpolarizes smooth muscle cells, reduces Ca²⁺ entry through VGCCs, and inhibits contraction. Cromakalim and its most active isomer levcromakalim both relax airway smooth muscles, acting as bronchodilators and antiasthmatics; however, their development for these indications has been limited by the occurrence, at therapeutic doses, of migraine, possibly because of vasodilation in meninges or within the brain. Minoxidil sulfate is used as vasodilator in hypertension, usually in conjunction with diuretics and β-blockers; a characteristic indication for minoxidil in topical formulation is alopecia, since it facilitates blood supply to the scalp, favoring hair growth; direct effects on follicular dermal papillae might also contribute. Cardiac effects of K_{ATP}COs are also important; in fact, these molecules accelerate AP repolarization and shorten AP duration. Although this might be protective under certain conditions, it could also favor arrhythmias. Under condition of cardiac ischemia, these drugs would reduce myocardial cell excitability, limiting energy consumption and tissue damage. Similar molecular mechanisms might underlie the anticonvulsant and neuroprotective effects exerted by K_{ATP}COs at CNS level, in particular given the major role played by K_{ATP} channels in ischemic preconditioning. However, the low degree of tissue selectivity of these molecules has severely hampered their therapeutic development; in fact, potent vasodilation accompanies cardiac and CNS effects. Novel agents with higher selectivity toward tissue-specific K_{ATP} isoforms are needed to fully exploit the therapeutic potential of these channels. Finally, recent evidence also suggests that activation of K_{ATP} channels in the internal mitochondrial membrane (mito-K_{ATP}) might trigger cytoprotective responses (reducing free radical production, cytochrome c release, caspase activation) during ischemic preconditioning; although the molecular and pharmacological profile of mitoK_{ATP} channels is yet to be understood, it seems likely that they might mediate some of the described pharmacological actions of K_{ATP}COs.

Other 2TM K⁺ channels are regulated by intracellular factors (G proteins, Mg²⁺, polyamines, protons). The first 2TM channel cloned, renal outer medullary kidney (ROMK), belongs to $K_{IR}1$ family, a group of weakly rectifying channels that, together with the $K_{IR}4$, $K_{IR}5$, and $K_{IR}7$ families, are involved in pH-dependent K⁺ transport across epithelial and glial cells (Table 27.6). The $K_{IR}2$ family, instead, is expressed in muscular and neuronal tissues and displays strong inward rectification, conferred by a voltage-dependent block exerted by intracellular cations such as magnesium and polyamines (spermine and spermidine). $K_{IR}2$ channels, like all 2TM channels, lack a voltage-sensing region corresponding to the S_4 segment and are intrinsically voltage independent. Intracellular cations, upon binding to sites in TM2 and C-terminus, occlude the intracellular mouth of the channels preventing access of cytoplasmic K⁺ ions. Their binding is enhanced upon membrane depolarization, thus determining the characteristic inward rectification of conductance (see Fig. 27.2). These channels are responsible for the large K⁺ conductance observed in muscular and neuronal tissues at rest. The $K_{IR}3$ family (also named GIRK, for G-protein-activated inward-rectifier K channels) includes channels that are activated by binding of G-protein βγ-subunits to their cytoplasmic N- and C-terminal regions following activation of metabotropic neurotransmitter receptors. GIRK channels have strong inwardly rectifying properties and are expressed mostly in neuronal cells and the heart. Their activation following binding of acetylcholine (released by postganglionic parasympathetic neurons) to M_2 receptors mediates vagal inhibitory effects on cardiac functions. In addition to cardiac M_2 receptors, CNS receptors coupled to GIRK channels include adenosine A_1, opioid μ and δ, $GABA_B$, $α_2$-noradrenergic, D_2 dopaminergic, $5-HT_{1A}$ serotonergic, and somatostatin receptors. Functional receptor–G protein–GIRK channel complexes are often located at presynaptic nerve terminals where their activation inhibits neurotransmitter release. Such activity represents an important pharmacological target, particularly under condition of excessive excitatory neurotransmitter release following ischemic or traumatic episodes; although no drug interfering directly with these channels is currently in clinical use, their involvement in the actions of drugs activating or inhibiting previously described metabotropic receptors has to be underlined.

The K+ Channel Family with 4 Transmembrane Segments (4TM) K+ channels formed by 4TM subunits generate background voltage-independent currents. Each subunit belonging to this structural class (13 are known at the moment, denominated from $K_{2P}1$ to $Kv_{2P}17$) exhibits a characteristic topology, with four TM segments and two pore regions, resembling two 2TM subunits covalently linked in tandem. Two of these subunits form a fully functional channel. All channels formed by these subunits carry voltage-independent currents activated by various stimuli, including fatty acids and osmotic stress (TRAAK, TREK), alkalosis (TALK), or inhibited by acidosis (TASK). 4TM channels have been found in all tissues examined, and they are thought to contribute to the high K+ conductance observed at rest in most cell types. In particular, they have assumed a relevant pharmacological role following the observation that some of them can be specifically and directly activated by volatile general anesthetics (chloroform, halothane, isoflurane), providing a plausible mechanism of action for these compounds. Activation of 4TM channels at rest determines membrane hyperpolarization, neuronal firing suppression, and inhibition of synaptic transmission. Riluzole, an anti-ischemic, anticonvulsant, and sedative compound used in amyotrophic lateral sclerosis, seems to activate some members of the 4TM family, though interaction with other ion channels has also been suggested to explain the pharmacological actions of this compound.

The K+ Channels with 6 Transmembrane Segments (6TM) Although members of this family widely diverge in their biophysical, pharmacological, and tissue localization properties, they are all activated by membrane depolarization. Diverse gene subfamilies belong to this class (from K_v1 to K_v12), each comprising several members. The first four families correspond, respectively, to the K_v1 (Shaker), K_v2 (Shab), K_v3 (Shaw), and K_v4 (Shal) *Drosophila* channels, identified using genetic approaches and extensively characterized. Although each of these subfamilies serves a precise pathophysiological role, only those with clearly defined pharmacological roles will be described in the following paragraphs (Fig. 27.4; Table 27.8).

"Shaker"-type channels are important in neuronal excitability. "Shaker"-type channels (K_v1, so called as they were originally identified in the "shaker" *Drosophila* mutant) repolarize AP and modulate neuronal firing frequency. Channels containing these subunits are responsible for both the so-called "delayed rectifier" (I_{DR}) currents, lacking fast inactivation, and "A-type" (I_A) currents, showing faster activation and inactivation kinetics (Table 27.8). Though historically important (they were the first K+ channel-encoding genes to be identified in both *Drosophila* and mammals), Shaker-type channels, similarly to Shab-type (K_v2), Shal-type (K_v3), and Shaw-type (K_v4) channels, only have a limited relevance for pharmacotherapy. 4-Aminopyridine (4-AP) is a voltage- and time-dependent K+ channel blocker relatively selective for I_A currents in axons and nerve terminals of cholinergic neurons, which is currently in use (recently also in extended-release formulations) for the symptomatic treatment of multiple sclerosis. 3,4-Diaminopyridine, a more lipophilic derivative of 4-AP showing better pharmacokinetic properties, also prolongs AP duration and stimulates neurotransmitter release from nerve endings, exerting beneficial effects in myasthenic paralytic syndromes (Lambert–Eaton) and in multiple sclerosis.

$K_v1.5$ channels represent the molecular basis of the ultrarapid repolarizing current (I_{Kur}) expressed in atrial, but not ventricular, myocytes. Given the implications of I_{Kur} in arrhythmias, Kv1.5 blockers such as vernakalant and AVE-0118 are under active investigation as "atrial-selective" drugs.

EAG channels are crucial for cardiac and neuronal function. Channels of the EAG family (EAG stands for "*ether-a*-go-go," the *Drosophila* mutant where they have been first identified) have been subdivided in three gene subfamilies: eag (Kv10.1 or KCNH1), erg (Kv11.1 or KCNH2), and elk (Kv12.1 or KCNH4). These subfamilies exhibit about 50% amino acid homology among themselves and 22% homology with shaker-type channels. $K_v10.1$ has two peculiarities: it is inhibited by high $[Ca^{2+}]_i$ and displays activation kinetics strongly dependent on the preceding voltage, activating slowly when the resting voltage is very negative. Although their role is yet to be precisely defined, $K_v10.1$ channels appear to be ectopically expressed in cancer cells from diverse tissues, highlighting their potential role in cell cycle and proliferation control (this property is also shared by members of the K_v11 subfamily).

TABLE 27.8 Classification and functional properties of the main voltage-gated K⁺ channels

Subunit	$K_v1.1$	$K_v1.4$	$K_v1.5$	$K_v3.4$	$K_v4.3$	$K_v7.1$	$K_v7.2$	$K_v11.1$ (ERG1)
Gene	*KCNA1*	*KCNA4*	*KCNA5*	*KCNC4*	*KCND3*	*KCNQ1*	*KCNQ2*	*KCNH2*
Accessory subunits	β-Subunits	β-Subunits, KCNB1	β-Subunits $Kvβ_{1,2}$, $Kvβ_{1,3}$ $Kvβ_{2,1}$	KCNE3	KchIP, KCND2	KCNE1, KCNE3	$K_v7.3$ (KCNQ3), KCNE3	KCNE2
Macroscopic current	I_{DR}	I_A	I_{Kur}	I_A	I_{to}	I_{Ks}	I_M	I_{Kr}
Tissue	CNS, muscle	CNS	Heart (atrium), smooth muscle, glia	CNS, skeletal muscle	Heart, smooth muscle, CNS	Heart, inner ear, pancreas	CNS, PNS	Heart, neuroblastomas and other cancer cells
Blockers	TEA, DTX	4-AP, TEA	Vernakalant and AVE-0118	TEA, BDS-1	Bupivacaine, PaTx1	Chromanol, azimilide, linopirdine, XE-991	Linopirdine, XE-991	E-4031, astemizole, dofetilide, haloperidol (Table 27.9)
Activators						Table 27.7	Table 27.7	
Single-channel conductance	10 pS	5 pS	11 pS	14 pS	5 pS	1.8 pS	6–8 pS	2 pS
Physiology	Action potential repolarization	Action potential repolarization	Ultrarapid component of atrial repolarizing current	Action potential repolarization	Action potential repolarization	Cardiac repolarization	Neuronal excitability, spike frequency adaptation	Rapid component of ventricular repolarization, neuronal spike frequency adaptation
Channelopathies	Episodic ataxia type 1			Periodic paralysis (KCNE3)		LQTS-1, SQTS	BFNC	LQTS-2, SQTS

Channels of the K_v11 family (also called erg or *eag-related gene*, as they were identified by virtue of their sequence homology with eag channels) display a characteristic inactivation process showing much faster kinetics than the activation at depolarized potentials; thus, these channels, despite being activated by depolarization, only display small outward currents, behaving therefore as inward rectifiers. Erg channels represent the molecular basis of the fast component of the AP repolarizing current in ventricular myocytes (I_{Kr}). Vice versa, heterotetrameric complexes formed by main α-subunits encoded by KCNQ1/$K_v7.1$ gene (see the succeeding text) and by accessory proteins (KCNE1) underlie the slow component of the cardiac repolarizing current in ventricles (I_{Ks}); in neurons and endocrine cells, channels of the K_v11 family regulate AP frequency and accommodation.

$K_v11.1$ channels are targets for antiarrhythmic or proarrhythmic drugs. Class III of the antiarrhythmic drug classification originally proposed by Vaughan Williams in 1984 includes molecules increasing AP duration in cardiac myocytes (Table 27.4). AP lengthening is mostly due to blockade of K^+ channels controlling phase 3 of cardiac repolarization (I_{Kr} and I_{Ks}). Among the class III drugs are amiodarone, clofilium, dofetilide, N-acetylprocainamide, and sotalol. Sotalol and amiodarone have been the first class III drugs introduced in therapy; both have a complex pharmacological profile as, in addition to K^+ channels, they also block cardiac sympathetic stimulation (a characteristic effect of class II drugs). Subsequently, efforts were devoted to identify "pure" class III drugs: however, selective I_{Kr} blockers, such as dofetilide and d-sotalol, failed to show favorable effects on arrhythmia-associated mortality. Furthermore, in the mid of the 1990s, it was shown that mutations in genes encoding I_{Kr} and I_{Ks} channels are responsible for serious cardiac diseases (Table 27.1) and that drugs belonging to various therapeutic classes (Table 27.9) and blocking the I_{Kr}, which is carried by the K_v11 family (in particular $K_v11.1$, the only family member significantly expressed in myocardial tissue), increase duration of the QT interval on ECG and predispose patients to occurrence of a polymorphic ventricular arrhythmia known as "torsade de pointes," manifesting with syncopal episodes and cardiac arrest. Supplement E27.6, "Drug-Induced Long QT Syndrome," describes some of the main pathogenetic aspects of the long QT syndrome (LQTS), placing emphasis on the structural basis of the selective I_{Kr} blockade by "torsadogenic" drugs. Given these observations, the development of drugs selectively blocking I_{Kr} or I_{Ks} has been slowed down. However, among new class III drugs, dronedarone is an orally active antiarrhythmic having the same benzofuranic structure of amiodarone, but lacking iodine ions. Dronedarone is indicated to reduce risk of hospitalization for atrial fibrillation (AF) in patients in sinus rhythm with a history of paroxysmal or persistent AF. Additional drugs that block one or more K^+ conductances are tedisamil, affecting I_{Kr}, I_{to} (I_A), K_{ATP} and BK(Ca), and azimilide, which blocks both I_{Kr} and I_{Ks}; finally, BRL-32872 is a novel antiarrhythmic drug with both class III and class IV actions.

K_v7 channels control cardiac repolarization, neuronal excitability, and sensory transduction. All members of this gene family (KCNQ1–KCNQ5; Kv7.1–Kv5) encode for K^+ channel subunits forming depolarization-activated channels with slow activation and deactivation kinetics and lack of inactivation. Each member of this family has unique tissue distribution and pathophysiological and pharmacological relevance (Table 27.8). $K_v7.1$ subunits are expressed in cardiac cells, where, together with KCNE1 accessory subunits, underlie the slow component of the ventricular repolarizing current (I_{Ks}). Loss-of-function mutations in $K_v7.1$ are responsible for the most common form of genetically determined LQTS (LQTS-1); interestingly, a few gain-of-function mutations in both $K_v7.1$ and KCNE1 have been linked to familial atrial fibrillation (FAF) or to short QT syndrome (SQTS) (Table 27.1). Drugs acting as I_{Ks} activators (some new benzodiazepines, like

TABLE 27.9 Examples of drugs causing a prolongation of the QT interval and/or Torsade De Pointes (TdP)

Antiarrhythmics	Class Ia, quinidine, procainamide, disopyramide; class III, N-acetylprocainamide, amiodarone, sotalol
Cardiovascular (not including antiarrhythmics)	Bepridil, isoprenaline, ketanserin, mibefradil
Antipsychotics	Haloperidol, chlorpromazine, droperidol, pimozide, quetiapine, sertindole (withdrawn), thioridazine (withdrawn)
Antidepressants	Amitriptyline, desipramine, doxepin
Antimania	Lithium
Antihistamines	Astemizole (withdrawn), terfenadine (withdrawn), diphenhydramine, hydroxyzine
Macrolides	Clarithromycin, erythromycin, spiramycin
Antimalaria	Halofantrine, quinine
Antimycotics	Ketoconazole
Quinolones	Grepafloxacin (withdrawn), levofloxacin, sparfloxacin
Antiemetics	Ondansetron, granisetron
Immunosuppressants	Tacrolimus (FK-506)
Antimuscarinics	Terodiline
Various	Cisapride (withdrawn), pentamidine, pentavalent antimonial, probucol, trimethoprim/sulfamethoxazole, cocaine, arsenic, organic phosphates

R-L3) are currently under investigation as antiarrhythmics. On the other hand, $K_v7.2$, $K_v7.3$, $K_v7.4$, and $K_v7.5$ subunits are mainly expressed in both central and peripheral neurons. $K_v7.4$ seems to be mainly expressed in the inner ear and in central auditory pathways, whereas heteromeric assembly of $K_v7.2$ and $K_v7.3$ (and perhaps $K_v7.5$) subunits underlies the M current (I_{KM}, suppressed upon activation of $G_{q/11}$-linked muscarinic receptors), a crucial regulator of firing frequency in various neuronal populations. Activation of several GPCRs inhibit (and, rarely, stimulate) I_{KM}. I_{KM} plays a double pharmacological role. Blockers of this current cause neuronal depolarization and enhancement of neurotransmitter release; "cognitive enhancers," such as linopirdine or anthracenonic derivatives such as XE-991, act through this mechanism and might be of help in diseases characterized by neurotransmission deficits, such as Alzheimer's disease. On the other hand, I_{KM} enhancers hyperpolarize neuronal plasma membrane and inhibit electrical activity; retigabine is a novel KCO specifically acting on neuronal K_v7 channels that has been recently approved as an add-on for the treatment of partial seizures in adults. Given the important role of I_{KM} also in pain neurotransmission, neuropathic pain represents a potential therapeutic application for I_{KM} openers; retigabine is a structural analog of flupirtine, another neuronal K_v7 channel opener that has been approved for the treatment of various pain states (posttraumatic, dental, postsurgical, headache, dysmenorrhea).

Finally, it has been recently demonstrated that channels mainly formed by $K_v7.4$ and $K_v7.5$ subunits are expressed in both vascular and visceral smooth muscle cells; in vascular cells, selective activators of these channels might exert vasodilating actions that could be exploited in hypertensive states.

Additional Families of 6TM or 7TM Subunits: The Ca^{2+}-Dependent K^+ Channels Ca^{2+}-dependent K^+ channels (K_{Ca}) are important for spike frequency adaptation. K_{Ca} channels are expressed in several cell types and open in response to increases in $[Ca^{2+}]_i$. Depending on their single-channel conductance and pharmacological sensitivity, K_{Ca} channels can be classified as "big K_{Ca}" or BK, which have large conductance (>200 pS) and are blocked by charybdotoxin (a scorpion toxin); "intermediate K_{Ca}" or IK (10–20 pS); and "small K_{Ca}" or SK, which have a much lower conductance (<10 pS) and are blocked by apamine (a toxin from bee venom) (Table 27.10). BK channels are also activated by depolarization and contribute to AP repolarization. Both BK and SK channels are responsible for the afterhyperpolarization (AHP), which in many neurons follows a single AP or train of AP; the main role of AHP is to modulate AP frequency at high firing rates. BK channels also control vascular tone, being directly regulated by nitric oxide, an important vasoactive mediator (Chapter 38). By contrast, IK channels are expressed in several tissues, including nonexcitable tissues, and participate in the regulation of cell volume and proliferation. In K_{Ca} channels, Ca^{2+} dependence is either a consequence of their ability to directly bind this cation when its cytoplasmic concentration increases (as in the case of BK) or to associate with the ubiquitous Ca^{2+}-binding protein calmodulin, which acts as a Ca^{2+} sensor (as in the case of SK and IK); the Ca^{2+}–CaM interaction triggers a conformational change in the channel leading to pore opening.

In addition to the aforementioned KCOs acting on K_{ATP} ($K_{ATP}COs$) or K_v7 (K_v7Co_s) channels, several molecules appear selective for $K_{(Ca)}$ opening, being particularly active on BK channels (Table 27.7). Among these, there are some benzimidazoles such as NS-004 and NS-1619, which relax smooth muscles of the upper airways and could be developed as new generation antiasthmatics, and the natural terpenoid dihydrosaponine 1 (DHS-1) used to treat dysmenorrhea and asthma. Novel BK activators with fluorooxindolic structure (BMS-204352) and wide distribution in the brain have also shown neuroprotective properties in animal models of brain ischemia; however, their selectivity for BK channels over other K^+ channels has been questioned.

NONSELECTIVE CHANNELS, ANIONIC CHANNEL, AND OTHERS

This category includes channels activated by extracellular ligands (which may be selective for cations, as in the case of nicotinic receptors, or for anions, as for $GABA_A$ receptors; see Chapter 16), by intracellular molecules (cyclic nucleotides), as well as by some voltage-gated channels. We will describe here only those whose pharmacological role is well defined; we refer the reader to Supplement E27.7, "TRP Channels," for nonselective cationic "transient receptor potential" (TRP) channels; Supplement E27.8, "Nonvoltage-Dependent Na^+ Channels and Their Role as Mechanosensitive and Acid-Sensitive Transducers," for nonvoltage-gated epithelial (ENaC) and acid-sensitive (ASIC) Na^+ channels; Supplement E27.9, " Anion Channels," for anionic channels; Supplement E27.10, " Water Channels: Aquaporins," for water channels (Aquaporins); Supplement E27.11, " Voltage-Independent Ca^{2+} Channels Activated by Store Depletion," for channels refilling endoplasmic reticulum Ca^{2+} stores; and Supplement E27.12, "Pharmacological Modulation of Gap Junction Channels and Electrical Synapses," for gap junction channels and electrical synapses.

Cationic Channels Modulated by Cyclic Nucleotides

These channels are heterogeneous in their sensitivity to changes in membrane potential, and although selective for cations, they only poorly discriminate among Na^+, K^+, and Ca^{2+}. In physiological conditions, since resting

TABLE 27.10 Classification and functional properties of Ca^{2+}-activated K^+ Channels

	$K_{Ca}1.1$	$K_{Ca}2.1$	$K_{Ca}2.2$	$K_{Ca}2.3$	$K_{Ca}3.1$
Isoforms					
Old names	Maxi-K, BK, slo	SKCa1, SK1	SKCa2, SK2	SKCa3, SK3	IKCa1, SK4, Gardos channel
Genes	KCNMA1	KCNN1	KCNN2	KCNN3	KCNN4
Associated subunits	KCNMB1–KCNMB4	Calmodulin at the C-terminus	Calmodulin at the C-terminus	Calmodulin at the C-terminus	Calmodulin at the C-terminus
Tissue	Ubiquitous: brain, skeletal muscle, smooth muscle	Brain (hippocampus, thalamus), pituitary gland, glioblastomas, aorta	Ubiquitous: brain, (hippocampus, thalamus), pituitary gland, prostate, heart, etc.	Ubiquitous: brain, (hippocampus), thalamus), lymphocytes, pituitary gland, prostate, colon	Erythrocytes, lymphocytes, pancreas, smooth muscle
Blockers	TEA, charybdoTx, iberioTx, paxilline	Apamin	Apamin	Apamin	charybdoTx, mauroTx, TRAM-34
Activators	$[Ca^{2+}]_i$, benzimidazoles, terpenoids, voltage dependent	$[Ca^{2+}]_i$ (K_d 0.7 µM), voltage independent	$[Ca^{2+}]_i$ (K_d 0.6 µM), EBIO, chlorzoxazone, voltage independent	$[Ca^{2+}]_i$ (K_d 0.6 µM), EBIO, voltage independent	$[Ca^{2+}]_i$ (K_d 0.2 µM), EBIO, methylxanthines, voltage independent
Single-channel conductance	260 pS	2–3 pS	2–3 pS	2–3 pS	11 pS
Physiology	Afterhyperpolarization, repolarization of the action potential	Afterhyperpolarization,	Afterhyperpolarization,	Afterhyperpolarization,	Regulation of erythrocyte volume
Pharmacological target	"Openers" in epilepsy, asthma, hypertension, gastric hypermotility	Myotonic dystrophy, cognitive impairment	Myotonic dystrophy, cognitive impairment	Myotonic dystrophy, cognitive impairment	Blockers useful against anemia and diarrhea
Channelopathies				Schizophrenia (polyglutamines), ataxia	

TABLE 27.11 Classification and functional properties of cationic channel-modulated cyclic nucleotides

Subfamilies	CNGA1	CNGA2	CNGA3	HCN1	HCN2	HCN3	HCN4
Associated subunits	CNGB1 (the stoichiometry of the retinal rod channel is 3 CNGA1/1 CNGB1)	CNGB1, CNGA4 (the stoichiometry of the olfactory channel is 2 CNGA2/1 CNGB1/1 CGNA4)	CNGB3 (the stoichiometry of the channel in retinal cones is 3 CNGA3/1 CNGB3)				
Tissue	Retinal rods	Olfactory neurons	Retinal cones, Olfactory neurons	Central neurons (cortex, hippocampus, cerebellum, brain stem, spinal cord), retina, peripheral neurons (dorsal root ganglionic neurons) sinoatrial node	Central neurons (especially in the thalamus and brain stem nuclei), retina, heart (sinoatrial node, where it is 20% of I_f in humans, atrial and ventricular tissue)	Central neurons (especially in the olfactory bulb and some hypothalamic nuclei), retina, peripheral neurons, sinoatrial node	Central neurons (especially in the thalamus and in mitral cells of the olfactory bulb), retina, heart (sinoatrial node, where it is 80% of I_f in humans, atrioventricular node, Purkinje fibers)
Blockers	L-cis-diltiazem	Pseudechetoxin (PsTx) from the venom of the Australian king brown snake (*Pseudechis australis*) (nanomolar IC_{50})	L-cis diltiazem	Cs$^+$, ZD7288, zatebradine, ivabradine	Cs$^+$, ZD7288, zatebradine, ivabradine	Cs$^+$, ZD7288, zatebradine, ivabradine	Cs$^+$, ZD7288, zatebradine, ivabradine
Activators	cGMP>>cAMP cAMP is a partial agonist of the rod channel, with an intrinsic efficacy of 10–40% of the maximal cGMP-activated responses	cGMP~cAMP	cGMP>cAMP	cGMP<cAMP	cGMP (Ka 6 µM)<cAMP (Ka 50–500 nM)	cGMP<cAMP	cGMP<cAMP
Conductance	27 pS	35 pS	40 pS	n.d.	n.d.	n.d.	n.d.
selectivity	Ca^{2+}>K$^+$~Na$^+$	Ca^{2+}>K$^+$~Na$^+$	Ca^{2+}>K$^+$~Na$^+$	K$^+$>Na$^+$	K$^+$>Na$^+$ (PNa:PK = 1:3–1:5)	K$^+$>Na$^+$	K$^+$>Na$^+$
Physiology Channelopathies	Phototransduction CNGA1 or CNGB1 mutations cause a rare form of retinitis pigmentosa with autosomal recessive hereditary transmission	Olfactory transduction Deletion of each of the three genes alters odorant sensitivity	Daytime color vision CNGA3 or CNGB3 mutations cause complete achromatopsia	Pacemaking activity Sequence variation associated with human epilepsy	Pacemaking activity The KO mice are ataxic, epileptic, and arrhythmic. Sequence variation associated with human epilepsy	Pacemaking activity	Pacemaking activity Sinus bradycardia and/or more complex arrhythmic conditions HCN4 deletion is lethal

Modified from Refs. [10, 11].
n.d., not defined.

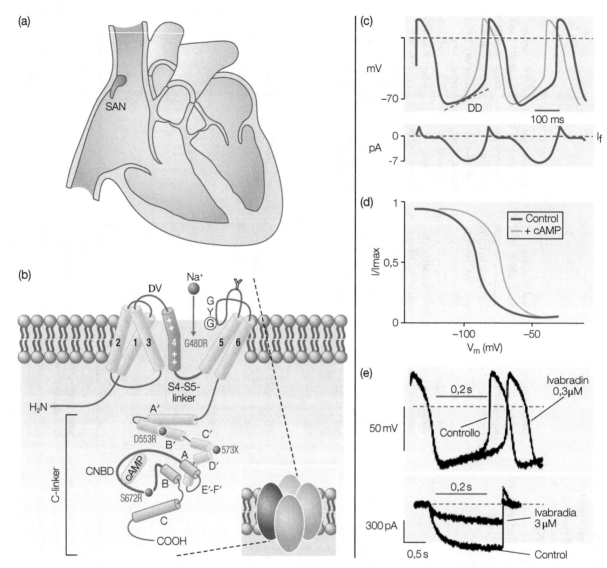

FIGURE 27.9 Physiological role, structure, and molecular determinants of cardiac I_f. (a) Localization of the sinoatrial node (SAN). (b) Structure of HCN channels. HCN channels are tetramers. One monomer is composed of six transmembrane segments including the voltage sensor domain (S1–S4) and the pore region between S5 and S6. The pore region contains the selectivity filter carrying the GYG motif. The C-terminal channel domain is composed of the C-linker (consisting of six α-helices, designated A′ to F′) and the cyclic nucleotide-binding domain (CNBD). The CNBD follows the C-linker domain and consists of α-helices A–C with a β-roll between the A- and B-helices (thick gray line). Human mutations in HCN4 involved in I_f channelopathies are indicated as gray circles. The N-glycosylation site between S5 and the pore loop is indicated as a Y. (c) Top panel: idealized pacemaker potentials in the absence (black) and presence (gray) of adrenergic stimulation. DD, diastolic depolarization. Bottom panel: proposed I_f time course during the pacemaker potential. (d) Voltage-dependent activation curve of I_f in the absence (black) and the presence (gray) of cAMP. In the presence of cAMP, the activation curve is shifted to the right, so that the current provides a higher contribution to diastolic depolarization (Modified from Ref. [8]). (e) Heart rate-reducing action of I_f inhibitors. Upper panel: ivabradine (0.3 μM) lowers the rate of spontaneous activity of isolated SAN myocytes by reducing the steepness of diastolic depolarization; this action is attributable to drug-induced I_f inhibition, as demonstrated in voltage-clamp experiments (bottom panel; ivabradine 3 μM) (modified from Ref. [9]).

membrane potential is close to the K⁺ equilibrium potential, their activation causes cell depolarization. Both cyclic nucleotide-gated (CNG) channels and HCN channels belong to this class. Similarly to voltage-dependent K⁺ channel subunits or to each domain of the pore-forming α-subunits of Na⁺ or Ca²⁺ channels, members of both classes are heterotetramers of subunits with 6TM structure and contain a cyclic nucleotide-binding domain (CNBD) in their C-terminus, which shows high affinity for cGMP or cAMP (Table 27.11).

CNG channels are poorly responsive to changes in membrane potential, are expressed in photoreceptors and olfactory epithelia, and have the following selectivity sequence: $Ca^{2+} > Na^+ \sim K^+$. Their physiological function is to transduce changes in cytosolic cGMP concentrations into electrical signals; in fact, light absorption by visual pigments or stimulation of odorant receptors trigger the G-protein (transducin)-mediated activation of cGMP phosphodiesterase, leading to cGMP hydrolysis and closure of CNG channels. CNG channels are blocked by L-cis diltiazem; mutations in retinal or olfactory CNG channels cause retinitis pigmentosa/achromatopsia or anosmia, respectively.

HCN channels are strongly activated by voltage and modulated by cAMP (Fig. 27.9). These channels are responsible for the depolarizing currents underlying cardiac and neuronal pacemaking and mediate currents commonly referred to as I_f or I_h in cardiac and neuronal cells, respectively. In contrast to most voltage-gated channels that open upon membrane depolarization, HCN channels activate when the membrane potential is hyperpolarized from rather negative values, close to the resting membrane potential.

Among monovalent cations, I_f channels are permeable to Na^+ ions, and their main role is to provide a depolarizing force at negative membrane potentials (between −80 and −55 mV), with very slow opening kinetics (500 ms). In fact, when the membrane is repolarized after an AP, activation of this current promotes entry of Na^+ ions, which slowly depolarize the membrane until the threshold of activation of the subsequent AP is reached, thereby determining the spontaneous activity of pacemaker cells. I_f channels can be blocked by cesium ions but also by organic molecules such as ZD7288, zatebradine, and ivabradine and is regulated by cAMP; higher cytoplasmic cAMP concentrations shift the I_f activation curve toward more positive values of membrane potential, making the channels more likely to open within a physiological range of membrane potential (Fig. 27.9). I_f channels are functionally upregulated by activation of the sympathetic nervous system, by β-adrenergic agonists, by VIP, by glucagon, and by phosphodiesterase blockers such as methylxanthines, whereas they are inhibited by activation of the parasympathetic branch of the autonomic nervous system and by muscarinic agonists.

I_f inhibition causes bradycardia, decreases myocardial oxygen demand, and improves blood supply to the endocardium; these actions might exert positive anti-ischemic and antianginal effects. In fact, ivabradine is an I_f blocker in clinical use as an antianginal drug. A recent large, randomized, double-blind, placebo-controlled, multicenter study, called Systolic Heart Failure Treatment with the I_f Inhibitor Ivabradine Trial (SHIFT), was conducted to evaluate the effect of heart rate reduction with ivabradine on outcomes in adults with symptomatic chronic heart failure and left ventricular systolic dysfunction. Although the study was terminated earlier than planned because of adverse events, the results confirmed the proof of concept since a clinically and statistically significant reduction of 18% in the relative risk of primary composite end point (i.e., the composite of cardiovascular death or hospital admission for worsening heart failure) occurred in the ivabradine treatment group within 3 months of the start of treatment.

TAKE-HOME MESSAGE

- Ion channels are the targets of several drug classes widely used in cardiology, neurology, endocrinology, anesthesiology, and other therapeutic areas.
- Therapeutic and toxic effects of specific drug classes can be predicted based on their pharmacological profile on specific ion channels.
- Voltage dependence, inactivation, and ion selectivity are major properties of voltage-gated ion channel.
- Voltage-gated sodium channels consist of a single polypeptide organized in multiple functional domains. Several isoforms are known, classified in up to nine different subfamilies. They are major players in the generation of action potential in excitable cells. Drugs acting on sodium channels are used in the control of cardiac arrhythmias, epileptic discharges, and local anesthesia.
- Up to 10 different subfamilies of voltage-gated calcium channels have been identified, which differ for localization, activation threshold, and rate of inactivation. Drugs acting on these channels are relevant for the treatment of several cardiovascular diseases.
- Potassium channels vary in terms of the molecular organization of their subunit(s) and of the channel complex. Potassium channels can be activated by different stimuli including voltage, Ca^{2+}, and ATP. Drugs acting on potassium channels are used in several diseases, from hypertension to diabetes and epilepsy.
- Drugs acting on membrane metabotropic receptors, by interfering with intracellular or membrane-delimited signaling pathways, can indirectly modulate ion channel function, further expanding the pharmacological role of this class of membrane proteins.

REFERENCES

1 Chen T-Y, Hwang T-C. 2008. CLC-0 and CFTR: chloride channels evolved from transporters. *Physiol Rev* 88: 351–387.
2 Tombola F, Pathak MM, Isacoff EY, et al. 2006. How does voltage open an ion channel? *Annu Rev Cell Dev Biol* 22: 23–52.

3. Webster SM, Del Camino D, Dekker JP. 2004. Intracellular gate opening in Shaker K1 channels defined by high-affinity metal bridges. *Nature* 428: 864–868.
4. Mantegazza M, Curia G, Biagini G, et al. 2010. Voltage-gated sodium channels as therapeutic targets in epilepsy and other neurological disorders. *Lancet Neurol* 9: 413–424.
5. Budde T, Meuth S, Pape HC. 2002. Calcium-dependent inactivation of neuronal calcium channels. *Nat Rev Neurosci* 3: 873–883.
6. Catterall WA. 2000. Structure and regulation of voltage-gated Ca^{2+} channels. *Annu Rev Cell Dev Biol* 6: 521–555.
7. Bennett K, James C, Hussain K. 2010. Pancreatic β-cell KATP channels: hypoglycemia and hyperglycemia. *Rev Endocr Metab Disord* 11: 157–163.
8. Biel M, Wahl-Schott C, Michalakis S, Zong X. 2009. Hyperpolarization-activated cation channels: from genes to function. *Physiol Rev* 89: 847–885.
9. Bucchi A, Barbuti A, Baruscotti M, DiFrancesco D. 2007. Heart rate reduction via selective "funny" channel blockers. *Curr Opin Pharmacol* 7: 208–213.
10. Reid CA, Phillips AM, Petrou S. 2012. HCN channelopathies: pathophysiology in genetic epilepsy and therapeutic implications. *Br J Pharmacol* 165(1): 49–56.
11. Schön C, Biel M, Michalakis S. 2013. Gene replacement therapy for retinal CNG channelopathies. *Mol Genet Genomics* 288(10): 459–467.

FURTHER READING

Ashcroft F. (2000). *Ion Channels and Disease*. London. Academic Press.

Catterall WA. (2014). Sodium channels, inherited epilepsy, and antiepileptic drugs. *Annu Rev Pharmacol Toxicol* 54: 317–338.

Harmar AJ, Hills RA, Rosser EM, et al. (2009). IUPHAR-DB: the IUPHAR database of G protein-coupled receptors and ion channels. *Nucleic Acids Res* 37 (Database issue): D680–D685 (available at http://www.iuphar-db.org/) Accessed December 13, 2014.

Hille B. (2001). *Ion Channels of Excitable Membranes*. 3rd Edition. Sunderland, MA. Sinauer Associates, Inc. Publishers.

Miceli F, Soldovieri MV, Martire M, Taglialatela M. (2008). Molecular pharmacology and therapeutic potential of neuronal Kv7-modulating drugs. *Curr Opin Pharmacol* 8: 65–74.

Roubille F, Tardif JC. (2013). New therapeutic targets in cardiology: heart failure and arrhythmia: HCN channels. *Circulation* 127(19): 1986–1996.

Waxman SG, Zamponi GW. (2014). Regulating excitability of peripheral afferents: emerging ion channel targets. *Nat Neurosci* 17: 153–163.

Wulff H, Castle NA, Pardo LA. (2009). Voltage-gated potassium channels as therapeutic targets. *Nat Rev Drug Discov* 8: 982–1001.

28

MEMBRANE TRANSPORTERS

Lucio Annunziato, Giuseppe Pignataro, and Gianfranco Di Renzo

---

By reading this chapter, you will:

- Understand the fundamental concepts on transmembrane transport
- Learn the classification of membrane pumps and transporters and the characteristics of each class
- Know the structure, function, and localization of individual membrane transporters and the clinical significance of drugs modulating their activity

---

Membrane transporters are integral membrane proteins capable of mediating transport of medium-sized ions or polar molecules, such as amino acids, sugars, or nucleotides, into the cell. Their main function is to allow molecules to pass through the plasma membrane, thus permitting cells to communicate with the extracellular environment, to fulfill their metabolic needs, and to maintain ionic homeostasis.

Although membrane transporters are numerous, they do share common characteristics including substrate selectivity and capability of undergoing conformational changes to carry out substrate translocation across the membrane. In general, membrane transporters differ in mechanism of action, cellular localization, and type of molecules transported.

Molecules that pass through the plasma membrane can be transported either individually or in conjunction with other molecules.

Uniporters mediate the transport of single substances exploiting electrochemical potential differences created by other transporters.

Cotransporters accomplish the simultaneous transport of two ionic species or other solutes and can be classified into antiporters and symporters. Antiporters mediate the simultaneous transport of two ionic species or other solutes moving in opposite directions through the membrane. Depending on the concentration gradient, one of the two substances flows down its concentration gradient, or in other words from the compartment where its concentration is high to the one where it is lower. This process provides the energy to sustain the transport of the other solute against its concentration gradient. Similarly to antiporters, symporters exploit the flow of a solute down its gradient to move another molecule against its concentration gradient, but in this case, the two solutes move in the same direction.

Finally, depending on energy consumption, two types of membrane transports can be distinguished: ATP-dependent active transports and ATP-independent passive transports.

TRANSPORTER CLASSIFICATION

ATP-Dependent Active Transporters

Active transport allows solutes to cross the plasma membrane against their concentration gradients and is mediated by membrane proteins called ATP-dependent transporters, as they require ATP for energy supply.

They are divided into ATP-binding cassette (ABC) transporters, P transporters, and F and V transporters.

ABC transporters harness the energy released by ATP hydrolysis to transport various substances (such as lipids, ions, small molecules, large polypeptides, and drugs) through cell membranes (see Table 28.1).

Mutations in genes encoding these transporters cause or contribute to a wide variety of diseases such as anemia, cystic fibrosis, neurological disorders, defects in cholesterol

---

General and Molecular Pharmacology: Principles of Drug Action, First Edition. Edited by Francesco Clementi and Guido Fumagalli.
© 2015 John Wiley & Sons, Inc. Published 2015 by John Wiley & Sons, Inc.

TABLE 28.1 ATP-dependent transporters

ABC type	Glycoprotein P (P-gp) or MDR1 (ABCB1)
	CTRF (ABCC7)
	ABC, retinal transporter
	ABCG2, drug transporter
	MIRP1, drug transporter
P type	Na^+/K^+-ATPase
	Ca^{2+}-ATPase
	H^+/K^+-ATPase
	Mg^{2+}-ATPase
	H^+-ATPase

TABLE 28.2 Examples of the ATP-independent transporters of the SLC (solute carrier) family

Ion transporters	SLC4, HCO_3^- transporter
	SLC8, Na^+/Ca^{2+} exchanger
	SLC9, Na^+/H^+ exchanger
	SLC24, Na^+/Ca^{2+} K^+ exchanger
	SLC12, Na^+/K^+ Cl^- cotransporter
Ion/neurotransmitter cotransporters	SLC2, glutamate and Na^+, K^+, OH^- cotransporter
	SLC6, monoamine (DAT, NET, SERT) and Na^+, Cl^- cotransporters

and bile transport, and abnormal drug response. Among the best-known transporters belonging to this class, we remember the cystic fibrosis transmembrane conductance regulator (CFTR) encoded by the ABCC7 gene and involved in the pathogenesis of cystic fibrosis and the multidrug resistance protein (MRP) involved in drug extrusion (see Supplement E 28.1, "Systems for Drug Extrusion").

P transporters are primarily deputed to ion transport. They consist of a single polypeptide chain that can simultaneously hydrolyze ATP and transport molecules. They are named P type for the presence of an aspartate residue serving as phosphorylation site. Examples of this class of transporters are Na^+/K^+-, H^+/K^+-, and Ca^{2+}-ATPases. Such carriers can be found on both plasma membrane and endoplasmic reticulum (ER) membranes.

F and V transporters are ATPases involved exclusively in proton transport. This class of ATPases is structurally more complex than the other two, being formed by more protein subunits that assemble on the membrane of different intracellular compartments. Their complexity is also reflected by their high molecular weight, which can exceed 500 kDa. In contrast with P-type ATPases, these transporters do not transfer any phosphate group to their amino acids. F-type ATPases are found in mitochondria, where they are responsible for the correct proton movement, which is critical for energy production. V-type ATPases, initially identified on vacuoles, are also present on lysosomes, where proton transport creates an acid environment that favors the activity of lysosomal acid hydrolases. They are of little pharmacological interest since only few molecules capable of interfering with their activity have been identified.

ATP-Independent Transporters

Unlike the active transport, passive transport takes place according to the concentration gradient and therefore does not require the biochemical energy of ATP. It may be mediated by membrane proteins called solute carrier (SLC) transporters.

The superfamily of SLC transporters includes both genes encoding for proteins mediating the so-called facilitated transport and genes encoding for proteins mediating a secondary active transport coupled to ion exchange. So far, the superfamily has been divided into 43 families corresponding to more than 300 SLC transporters. From the pharmacological point of view, the most relevant are the dopamine (DAT), norepinephrine (NET), serotonin (SERT), GABA (GAT), glutamate (EAAT) transporters, the vesicular monoamine transporters (VMAT) (see also Chapter 29), the Na^+/H^+ exchanger (NHE), the Na^+/Ca^{2+} exchanger (NCX), and the cotransporter $Na^+/K^+/Cl^-$ cotransporter (NKCC) (Table 28.2).

The pharmacological interest in these membrane transporters resides in the fact that compounds capable of modulating their activity may contribute to change the pharmacokinetic and pharmacodynamic properties of drugs. Indeed, many of these carriers represent useful targets for the control of tissue distribution as well as absorption and elimination of drugs. Other transporters, instead, represent pharmacological targets of different classes of compounds. For example, most antidepressants exert their effects by acting on NET or SERT. Some antiepileptic drugs act mainly through inhibition of GAT. Inhibition of DAT, SERT, and NET by drugs of abuse such as cocaine is the most important mechanism of their action. Digitalis glycosides exert a positive inotropic action by inhibiting Na^+/K^+-ATPase, thereby leading to the activation of the NCX. Most diuretic drugs exert their effect by inhibiting transporters expressed in the nephrons, such as the NHE and the NKCC.

Many transporters are involved in the important phenomenon of drug resistance as well as in the transport of drugs across cells forming the blood–brain barrier (BBB) (see Supplement E 28.1).

In the following sections of this chapter, we will describe the most common carriers whose pharmacological modulation is relevant to current therapeutic practice. An overview of the therapeutic significance of drugs acting on membrane transporters is provided in Box 28.1.

NA^+/K^+-ATPASE

Na^+/K^+-ATPase pump is a protein complex ubiquitously expressed in the human body, which play a relevant role in maintenance of Na^+ and K^+ homeostasis. The importance of

BOX 28.1 THERAPEUTIC OVERVIEW OF MEMBRANE TRANSPORTERS

Drugs	Target	Cellular effects	Clinical	Indications	Notes and criteria of choice
Cardiac glycosides	Na^+/K^+-ATPase	Inhibition of ion homeostasis in the heart and neurons with effects on resting/action potential	Positive inotropic effect Depression of atrioventricular nodal conduction	Heart failure Atrial fibrillation Dilated cardiomyopathy	Digoxin and other digitalis do not increase life expectancy of heart failure (HF) patient and show several side effects; their use is limited to patients with mild to moderate HF with reduced ejection fraction

Digoxin is also recommended for those HF patients in which other classes of drugs (ACE inhibitors, ARBs, β-blockers, and diuretics) have failed to reverse poorly tolerated symptoms such as dyspnea, tachycardia, and poor resistance to stress.

Adverse effects of digoxin include nausea, visual disturbances, fatigue, gynecomastia, and arrhythmias. Since digoxin binds to Na^+/K^+-ATPase by competing with K^+, conditions of hypokalemia accentuate digoxin side effects, whereas hyperkalemia reduces its therapeutic effects.

Proton pump inhibitors	H^+/K^+-ATPase	Inhibition of acidification of gastric secretion	Antiulcer effects	Peptic ulcer, gastroesophageal reflux	Omeprazole and pantoprazole are the most prescribed molecules

To be converted into their active form, proton pump inhibitors require the low pH levels present in the stomach in order to be able to bind cysteine residues in the H^+/K^+-ATPase extracellular side. Because of this mechanism of activation, there is a latency time between drug administration and response.

Loop diuretics	$Na^+/K^+/2Cl^-$ cotransporter (NKCC)	NKCC blockade. Thirty percent of Na filtered by renal glomeruli is reabsorbed via NKCC present in the ascending portion of Henle's loop	High-ceiling diuresis	Acute pulmonary edema Ascites and other forms of edema Chronic HF Nephrotic syndrome	Loop diuretics, such as furosemide, bumetanide, and ethacrynic acid, are 15–25 % of the sodium filtered by glomeruli

Following pharmacological blockade of the cotransport, a reduction in Na^+, K^+, and Cl^- reabsorption occurs. The lumen positive electrical potential that supports Ca^{2+} and Mg^{2+} reabsorption is also altered. Chronic treatment with these drugs is then associated with loss of Na^+, K^+, Ca^{2+}, and Mg^{2+} ions.

Thiazide/thiazide-like diuretics	Na^+/Cl^- cotransporter	Inhibition of transporter present in distal convoluted tubule of the nephron	Diuresis	Hypertension Prevention of stone formation in idiopathic hypercalciuria Edema by chronic heart, liver, and kidney syndromes Edema by corticosteroids	Diuretics of the distal convoluted tubule, called thiazide diuretics, include compounds such as chlorthalidone and are medium intensity in term of potency

As with other diuretics mentioned above, the reduced Na^+ reabsorption in the high parts of the nephron stimulates Na^+/K^+-ATPase activity in the collecting ducts, thus causing increased K^+ excretion. Unlike loop diuretics that inhibit Ca^{2+} reabsorption, thiazides increase Ca^{2+} reabsorption in the distal convoluted tubule with an unknown mechanism. It has been suggested that the reduction in intracellular Na^+, by stimulating the Na^+/Ca^{2+} exchanger of the basolateral membrane, may cause the increase in total Ca^{2+} reabsorption. The toxic effects of thiazides are similar to those of loop diuretics and are represented by hypokalemic metabolic alkalosis and hyperuricemia.

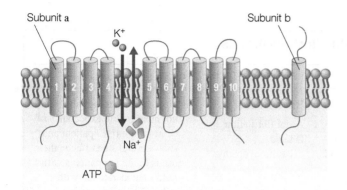

FIGURE 28.1 Structure of Na⁺/K⁺-ATPase. Na⁺/K⁺-ATPase is a heterodimer, consisting of α-subunit (left) and β-subunit (right), which may be associated with a third subunit called γ (not shown in the figure).

this transport system is highlighted by the fact that it uses 40–70% of the whole energy needed to sustain ionic transport. This energy is used to exchange intracellular sodium and extracellular potassium with a stoichiometry 3 Na⁺:2K⁺. Na⁺/K⁺-ATPase activity is also of fundamental importance for maintaining K⁺ and Na⁺ transmembrane gradients. The balance of this gradient is essential for the secondary transport of other ions such as Ca²⁺ via the NCX and protons via the NHE.

Moreover, Na⁺/K⁺-ATPase is essential for preserving cell osmolarity and resting potential. Although these functions are particularly critical for neuronal, cardiac, and renal cells, all cells of the human body ultimately depend on correct functioning of this ion pump.

Structure

Na⁺/K⁺-ATPase was the first membrane transporter to be identified more than 50 years ago, and it is still one of the most studied membrane transporters. Characterization of this transport system was carried out in 1957 by Jens Christian Skou, who in 1997 was awarded the Nobel Prize in Chemistry for this work.

Na⁺/K⁺-ATPase belongs to the P-type ATPase family, which also comprises H⁺/K⁺ pump and Ca²⁺-ATPase. Following ATP hydrolysis, members of this family undergo autophosphorylation triggering conformational changes essential for their correct functioning. Although all P-type ATPases share the same structure and consist of heterodimeric complexes, significant functional differences arise from differences in their composition. In particular, Na⁺/K⁺-ATPase consists of a heterodimeric core, composed of alpha and beta subunits, which may be associated with a third subunit called gamma (Fig. 28.1). These subunits have distinct properties and characteristics.

The alpha subunit is a 112 kDa protein containing 10 transmembrane segments, with intracellular N- and C-terminals. It contains sites for ATP binding and phosphorylation as well as for ion transport and is therefore the catalytic core of the pump.

The beta subunit is a 35 kDa glycosylated protein containing a single-transmembrane segment. It plays a key role in Na⁺/K⁺-ATPase maturation, transport, and stabilization and also presents key residues for K⁺ and ouabain binding.

The gamma subunit is a 6.5 kDa membrane protein that associates to Na⁺/K⁺-ATPase in a tissue-specific manner. Although not essential for pump expression and activity, this subunit is present in a 1:1 ratio with the alpha subunit and beta subunit, and it influences Na⁺/K⁺-ATPase function and distribution and can increase its affinity for ATP.

Biophysical Properties and Involvement in Intracellular Signaling

Na⁺/K⁺-ATPase mode of operation is commonly described according to Albers–Post's model. This model suggests that the pump alternates between two possible states, named E1 and E2, having different conformation and affinity for Na⁺ and K⁺ ions. Transition between these two stages depends on voltage and Na⁺, K⁺, and ATP concentrations and is accompanied by the consecutive export of sodium ions and import of potassium ions (Fig. 28.2).

Since Na⁺/K⁺-ATPase transports Na⁺ and K⁺ ions in a 3:2 ratio, it works as an electrogenic transporter, as each activity cycle determines a net transport of a positive charge outside the cell. Such active transport system involves both O_2 consumption and ATP hydrolysis.

Besides representing an important mechanism of ion transport, Na⁺/K⁺-ATPase is also able to activate several intracellular signal transduction pathways. In particular, activation of Na⁺/K⁺-ATPase has been associated to intracellular signals mediated by kinases of Src, Ras, MEK, and ERK families or by transcription factors such as NF-κB. The multifunctional nature of Na⁺/K⁺-ATPases is probably due to the large number of isoforms existing for each subunit that constitute this pump.

Pharmacology of Na⁺/K⁺-ATPase

A large number of drugs capable of specifically interacting with Na⁺/K⁺-ATPase pump and inhibiting its catalytic activity have been isolated from plants and animals. The largest group is represented by cardiotonic steroids, also called cardiac glycosides or cardiac steroids (see Supplement E 28.2, "Cardiac Glycosides"). These compounds and some of their derivatives are mainly used in the treatment of cardiac diseases. In particular, digitalis glycosides and derivatives are used for the treatment of heart failure and some tachycardic phenomena and dilated cardiomyopathy. For example, treatment with low doses of digoxin (0.5–0.9 ng/ml) significantly improves the general clinical conditions of heart failure patients. However, despite improving the overall clinical condition of these patients, digitalis derivatives do not increase their survival rate. From a chemical point of view, the structure of these compounds comprises

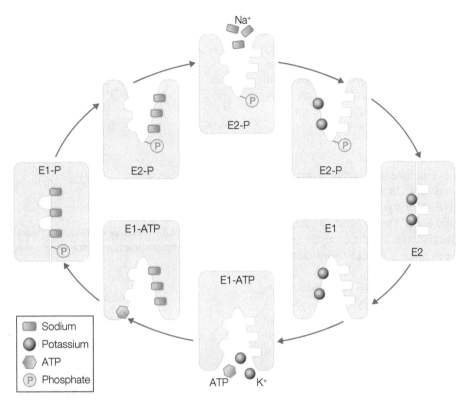

FIGURE 28.2 Na⁺/K⁺-ATPase operation model. Na⁺/K⁺-ATPase operating mode is commonly described according to the Albers–Post's model. In this model, the pump alternates between two possible states, named E1 and E2, which differ in conformation and Na⁺ and K⁺ affinity. The transition between these two states depends on voltage and Na⁺, K⁺, and ATP concentrations and is coupled to the consecutive Na⁺ export and K⁺ import.

two portions: a sugar and a steroid called aglycone. Such pharmacological compounds have been found in leaves, flowers, seeds, roots, and barks of a great variety of plants. Acid or enzymatic hydrolysis of these derivatives gives rise to steroidal aglycones, which are convulsant poisons. Among the major steroid glycosides, there are some derivatives of *Digitalis purpurea*, digoxin and digitoxin, which are currently used as cardiotonics, and derivatives of *Scilla maritima*, which were used until the thirteenth century as rat poison. The aglycone contains an R group that defines the chemical class of cardioactive glycosides and is important for their function. The sugar portion instead does not influence the mechanism of action of the molecule but is important for its binding to the pump, for modulatory activity, and for pharmacokinetics. Cardiac glycosides act by binding to extracellular portions of the α-subunits. This binding elicits a conformational change leading to dose-dependent inhibition of the pump. This inhibition subsequently leads to an increase in Na⁺ intracellular concentration that in turn induces activation of NCX in the reverse mode of operation, resulting in Na⁺ extrusion and Ca²⁺ entry. Ca²⁺ influx through this pathway activates Ca²⁺-induced Ca²⁺ receptors (CICRs) or ryanodine receptors (RyRs) located on the ER membranes, leading to an increase in intracellular Ca²⁺ concentrations (see also Chapter 12). This increase is responsible for the positive inotropic effect of cardiac glycosides. In addition, digoxin is also able to reduce sympathetic tone and to increase vagal tone, probably because of its action on baroreceptors. The increased vagal tone and, to a lesser extent, the depression that digoxin exerts directly on atrioventricular (AV) nodal conduction determine a slower ventricular response during atrial fibrillation.

Currently, digoxin is considered a second-line therapy for the treatment of heart failure, after ACE inhibitors and beta blockers. Digoxin adverse effects have been well characterized and include nausea, visual disturbances, fatigue, gynecomastia, and arrhythmias. Since digoxin binds Na⁺/K⁺-ATPase pump by competing with K⁺ ion binding, conditions of hypokalemia exacerbate its side effects. Therefore, it is important to adjust digoxin dose when administering it concomitantly with medications that interfere with K⁺ blood concentrations or kidney function.

H⁺/K⁺-ATPASE

H⁺/K⁺-ATPase belongs to the P-type ATPase family of transporters. This pump was identified approximately 50 years ago in yeast, when it was observed that, during fermentation, cells were able to secrete large quantities of acidic substances, and the pH was further lowered after glucose addition to the culture medium. This was an indication that not only yeast can secrete H⁺ ions but also that H⁺ extrusion is energy dependent. Actually, proton transport occurs in an electroneutral manner: H⁺ ions are transported from the cytoplasm to the extracellular environment, whereas K⁺ ions are transported from the extracellular space into the cytoplasm.

Structure, Distribution, and Function

H⁺/K⁺-ATPase possess heterodimeric structure comprising an alpha and a beta subunit (Fig. 28.3).

The α-subunit is encoded by the ATP4A gene and consists of about 1000 amino acids. This subunit, organized in 10 transmembrane segments, contains the catalytic sites of the enzyme and forms open pores in the cell membrane, thus allowing ionic transport.

The β-subunit is encoded by the ATP4B gene and consists of about 300 amino acids. It contains a cytoplasmic domain of 36 amino acids, a single-transmembrane domain, and a highly glycosylated extracellular domain. This subunit stabilizes the α-subunit and is required for correct functioning of the proton pump. Both alpha and beta subunits contain several signals for targeting the proton pump to the cell membrane.

H⁺/K⁺-ATPase is usually defined as a gastric proton pump. In fact, although this proton pump has been identified in many plant and animal tissues, such transporter is highly implicated in the acidification of the gastric content. In particular, the proton pump is localized in the parietal cells of the gastric mucosa (Fig. 28.4).

H⁺/K⁺ ATPase is an electroneutral transporter that uses the energy released by ATP hydrolysis to carry one proton from the cytoplasm to the extracellular space while moving one potassium ion in the opposite direction. As in the case of other P-type ATPases, a phosphate group is transferred from the ATP molecule to the carrier during the transport cycle. This phosphorylation causes a conformational change in the H⁺/K⁺-ATPase that determines the ion transfer. The transport mechanism is analogous to that of the Na⁺/K⁺-ATPase, as it proceeds through the formation of two reactive intermediates, E1 and E2, with different binding affinities for the two transported ions.

Regulation of the Gastric Proton Pump

Gastric acid secretion is a dynamic and complex process that is regulated not only by paracrine and hormonal factors, such as gastrin, histamine, ghrelin, and somatostatin, but also by neuronal factors. Furthermore, it is also modulated by mechanical stimuli, such as gastric wall distension and chemical substances such as caffeine and ethanol. In particular, the main factors capable of stimulating acid secretion are histamine, gastrin, and acetylcholine.

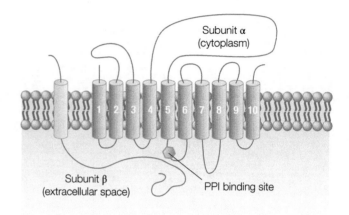

FIGURE 28.3 Structure of H⁺/K⁺-ATPase proton pump. H⁺/K⁺-ATPase is a heterodimer consisting of an α-subunit and a β-subunit.

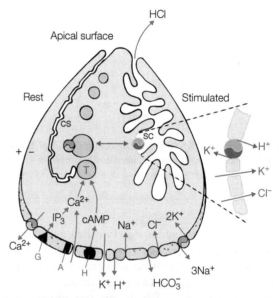

FIGURE 28.4 Gastric parietal cell. In the basolateral surface are present receptors for gastrin (G), which induces release of IP3 and then Ca²⁺; acetylcholine (A), which increases Ca²⁺ permeability; and histamine (H), which activates cAMP production. In addition, other membrane transporters such as Ca²⁺-ATPase, Na⁺/K⁺-ATPase, Na⁺/H⁺, and Cl⁻/HCO₃⁻ as well as K⁺ channels are also present. H⁺/K⁺-ATPase and K⁺ and Cl⁻ channels are found in vesicles and canaliculi membranes (enlargement). Following stimulation, the tubulovesicular intracellular organelles fuse with the plasma membrane and form the secretory canaliculi (sc). The pump is thus in contact with the extracellular fluids, which can bind the K⁺ ions present in the gastric lumen and release H⁺ in the stomach.

Histamine, released by cells known as enterochromaffin-like (ECL) cells, binds to H2 receptors that increase cAMP levels by activating adenylate cyclase. Gastrin, released by G cells, binds to CCK2 receptors, which activate phospholipase C and induce Ca²⁺ release in the cytosol. Gastrin is also able to stimulate histamine release from ECL cells.

Acetylcholine, released from intramural neurons, binds to M3 receptors resulting in increased intracellular concentrations of Ca^{2+} ions. Activation of these signal transduction mechanisms ultimately contributes to increase cAMP and Ca^{2+} intracellular levels. This in turn determines activation of intracellular kinases, thereby favoring the transfer of the proton pump from the cytoplasm to the membrane. In fact, under resting conditions, H^+/K^+-ATPase pump is contained mainly in cytoplasmic vesicles. After stimulation, these vesicles move toward and fuse with the apical plasma membrane, resulting in insertion of the proton pump in the apical cell membrane. When the secretory stimulus ceases, the proton pump is reabsorbed by endocytosis, and the tubular–vesicular compartment is reestablished. Furthermore, as discussed previously, H^+/K^+-ATPase can function properly only in the presence of K^+ ions on the extracellular side, whose main source is represented by the potassium channel KCNQ1, expressed on the plasma membrane of the same cells.

Not only the increase but also the reduction in acid secretion by the proton pump is finely regulated. Some of the factors that reduce acid secretion are leptin, glucagon-like peptide 1, and *Helicobacter pylori*. In particular, regulation of acid secretion by *H. pylori* occurs at transcriptional level, as this bacterium is capable of inhibiting the transcription of H^+/K^+-ATPase-encoding genes.

Pharmacology of the Gastric Proton Pump

The two main diseases associated with gastric hypersecretion are peptic ulcer and gastroesophageal reflux. Since gastric acid secretion is mainly mediated by H^+/K^+-ATPase, inhibition of this transport system represents nowadays the most effective way to treat any complications due to acid hypersecretion. The first generation of proton pump inhibitors (PPI) was introduced in the early 1970s when it was observed that thioacetamide, a derivative of the CMN 131 compound, had antisecretory properties. Further pharmacological developments of this class of compounds have led to the synthesis of omeprazole, which has been marketed since 1980 and today represents the treatment of choice for peptic ulcer and gastroesophageal reflux.

Binding of these substituted benzimidazoles to the extracellular portion of the proton pump requires an acidic environment and is irreversible. In fact, PPI need to be converted into their active form by the low pH present in the stomach in order to be able to bind cysteine residues present on the H^+/K^+-ATPase extracellular side. Because of this mechanism of activation, there is a latency time between drug administration and response. For this reason, novel compounds have been developed that can be activated also at higher pH values. Being independent of gastric acidity, these drugs exert a much faster therapeutic action.

Subsequently, some tertiary amines derived from SCH28080 were synthesized to improve pharmacodynamics and reduce adverse effects of the initial compounds. The new class of inhibitors can block acid secretion by competing with extracellular potassium (PCAB). More specifically, since PCABs antagonize K^+ binding to H^+/K^+-ATPase, their binding to the luminal side of the membrane is reversible. Given their mechanism of action, PCABs act faster than old pump inhibitors as they do not require prior activation. However, due to their noncovalent binding to H^+/K^+-ATPase, higher doses of these drugs are required to obtain the same blocking effect observed with omeprazole and its derivatives. Despite this slight drawback, PCABs (such as soraprazan) still represent a promising class of compounds for treating diseases associated with gastric acid hypersecretion. Other PPI include scopadulcic acid B, obtained from *Scoparia dulcis*, along with its derivatives that also exert antiviral activity, and cibenzoline, a class I antiarrhythmic agent.

Possible overdoses of these compounds can reduce proton pump activity, thereby leading to a compensatory hypergastrinemia, which, if prolonged, can generate parietal and enterochromaffin cell hyperplasia. The clinical consequence is a rebound acid secretion that can cause dyspeptic symptoms in healthy subjects and exacerbate symptoms in patients with gastroesophageal reflux or peptic ulcer.

PLASMA MEMBRANE Ca^{2+}-ATPASE

The plasma membrane Ca^{2+}-ATPase (PMCA) is an integral membrane protein belonging to the P-type family of ATPases. It is responsible for unidirectional transport of Ca^{2+} from the cytosol to the extracellular space. Proper functioning of this transporter is of vital importance for all eukaryotic cells in order to maintain cytosolic Ca^{2+} concentrations within physiological ranges (in cooperation with NCX).

PMCA is expressed in all tissues of the human body, including the brain. The energy for Ca^{2+} extrusion is supplied by ATP hydrolysis (one Ca^{2+} ion extruded for each ATP molecule hydrolyzed). This transporter has a high affinity for Ca^{2+} (Km 100–200 nM) but low V_{max}. On the contrary, NCX, the other transporter able to extrude Ca^{2+} ions, has low affinity for Ca^{2+} but high V_{max}. PMCA affinity for Ca^{2+} can be further increased (20–30 fold) by Ca^{2+}/calmodulin binding.

In the CNS, PMCA, besides being involved in intracellular Ca^{2+} regulation, is also implicated in regulation of synaptic activity and in neurotransmitter release from the synaptic vesicles.

From the topological point of view, PMCA contains 10 transmembrane segments with intracellular C- and N-terminals. A 70–200-amino-acid-long segment in the C-terminal region participates in the regulation of the pump (Fig. 28.5).

So far, four PMCA isoforms have been identified, namely, PMCA1–PMCA4, expressed in different body tissues and encoded by four different genes, denominated ATP2B1–ATP2B4. More than 20 variants of these proteins exist, resulting from alternative splicing events.

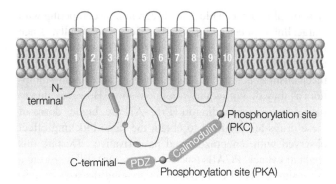

FIGURE 28.5 Structure of the plasma membrane Ca^{2+}-ATPase. The pump consists of 10 transmembrane segments and intracellular C- and N-terminals. The C-terminal region contains a 70–200-amino-acid segment having regulatory function. PKC and PKA indicate protein kinase C and protein kinase A phosphorylation sites, respectively.

PMCA1 is a ubiquitous protein and its absence is incompatible with life. PMCA2 and PMCA3 instead are mainly expressed in the muscle and CNS. PMCA4 is the other ubiquitous isoform; however, unlike PMCA1, it is not essential for life but may determine male infertility. PMCA2 and PMCA3 are activated much faster than the other two isoforms. This is important in those cells that allow the entry of high amounts of Ca^{2+} ions when excited, as is the case of the CNS and muscle cells, where these isoforms are highly expressed.

Defective PMCA activity has been described in a number of diseases such as deafness, diabetes, hypertension, and cerebral ischemia. However, it is still unclear whether such deficits in PMCA activity are the actual cause of the diseases or secondary events and whether such deficits are due to defects in PMCAs themselves or in proteins somehow related to PMCAs.

There are no selective drugs capable of inhibiting or activating PMCA. So far, only two modulators are known to inhibit PMCA: vanadate (1–5 mM) and eosin (1.5–10 μM). Both modulators act at high concentrations and in a nonselective manner and have been used only in preclinical studies.

SARCOPLASMIC/ENDOPLASMIC RETICULUM Ca^{2+}-ATPASE

ER is the largest intracellular organelle. Within the ER, a number of critical cellular processes occur, such as synthesis and transport of proteins. In addition, it represents the main site of phospholipid, phosphatidylinositol, and leukotriene synthesis. Besides these important mechanisms, the ER also takes part in reception and production of signals of crucial importance for cell activity and survival.

The ER is actually involved in all mechanisms activated by intracellular Ca^{2+} (see also Chapter 12). In particular, being the largest Ca^{2+} store, it can generate Ca^{2+} flows between the cytosol and its lumen in response to extracellular stimulation.

Ca^{2+} movements are also important for ER activities; indeed, protein synthesis and assembly is controlled by Ca^{2+} ions present in the ER lumen.

ER Ca^{2+} homeostasis is regulated by numerous molecular systems capable of mediating ion fluxes through intracellular membranes. These systems include:

1. The family of Ca^{2+}-ATPases present on the sarcoplasmic reticulum membrane called sarcoplasmic/endoplasmic reticulum calcium ATPases (SERCAs).
2. Channels controlling intracellular Ca^{2+} release into the cytoplasm, such as RyRs, inositol 1,4,5-trisphosphate receptors (IP3Rs), and probably other Ca^{2+} channels as TRPV1 receptors.
3. Ion channels that contribute to the maintenance of ER membrane potential, counterbalancing currents produced by Ca^{2+} release from ER stores. The identity of these channels is largely unknown; however, trimeric intracellular cation channels (TRICs) seem to have an important role in this process.

Among the different systems contributing to Ca^{2+} homeostasis in ER and cytoplasm, SERCA plays the most important role.

Structure, Distribution, and Regulatory Mechanisms of SERCA

SERCA structure includes four domains: a transmembrane domain (M domain) composed of 10 transmembrane segments containing Ca^{2+}-binding sites and three cytoplasmic regions called A, P, and N domains. The A (actuator) and P (phosphorylation) domains are coupled to the M domain, whereas the N (nucleotide binding) is linked to the P domain (Fig. 28.6).

Pumps belonging to the SERCA family are ubiquitously distributed in all body cells. Differences occur in the distribution of SERCA isoforms.

SERCA1a is expressed predominantly in adult skeletal muscle and SERCA1b is its alternatively spliced neonatal form.

SERCA2a is expressed in the skeletal muscle and heart and at very low levels in nonmuscle tissues. In contrast, SERCA2b is ubiquitously expressed. The role of SERCA3 has not yet been defined, and its genetic ablation does not determine clear phenotypic consequences, except for minor deficits in the relaxation mechanisms of vascular and tracheal smooth muscle. This isoform is expressed to a great

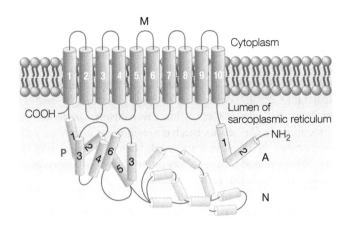

FIGURE 28.6 Structure of the sarcoplasmic/endoplasmic reticulum Ca^{2+}-ATPase (SERCA). SERCA structure consists of four domains: a transmembrane domain (M) composed of 10 transmembrane segments and containing Ca^{2+}-binding sites and three cytoplasmic domains called domains A (actuator), P (phosphorylation), and N (nucleotide binding). A and P are coupled to the main domain (M), while N domain is linked to P.

extent in the intestine, lung, and spleen, as well as in many other tissues such as skeletal muscle.

Ca^{2+} ions within the ER are the primary factors regulating SERCA activity. In particular, an increase in ER Ca^{2+} concentrations can effectively inhibit Ca^{2+} entry into the ER, whereas a decrease in ER Ca^{2+} concentrations favors its entry.

This regulatory mechanism involves the interaction between SERCA and two proteins denominated calreticulin and ERp57. At high ER Ca^{2+} concentrations, calreticulin forms a complex with ERp57 that inhibits SERCA activity. By contrast, at low Ca^{2+} concentrations, the complex calreticulin–ERp57 dissociates, leading to an increase in pump activity.

Physiological Properties and Pharmacological Modulation

The ER plays a dual role in the control of Ca^{2+} homeostasis being both a store and a source of Ca^{2+} ions for the cytoplasmic environment. SERCA pumps maintain Ca^{2+} homeostasis by promoting Ca^{+2} extrusion from the ER when concentrations are high and by allowing Ca^{2+} accumulation within the ER when Ca^{2+} concentrations are low. Therefore, SERCA is particularly important in all life processes, since Ca^{2+} acts as a critical second messenger controlling several functions. For this reason, SERCA appears to be of considerable importance in those pathologies, such as myocardial infarction and cerebral ischemia, involving changes in Ca^{2+} intracellular concentration.

Therapeutic drugs able to selectively modulate SERCA activity have not yet been described. However, compounds capable of inhibiting SERCA with high potency and selectivity, such as thapsigargin, cyclopiazonic acid, and 2,5-di-(t-butyl)-dihydroxybenzene, have recently been identified. A number of drugs (e.g., artemisinin, rapamycin, cyclosporine, coxibs, ivermectin) can also inhibit SERCA, but it is not yet established whether this inhibitory property contributes to their therapeutic effect.

Na^+/Ca^{2+} EXCHANGER

The NCX is a high-capacity and low-affinity ionic transporter that exchanges three Na^+ ions for one Ca^{2+} ion. When intracellular Ca^{2+} concentrations ($[Ca^{2+}]_i$) rise and cells need to return to resting levels, this exchanger transport mechanism couples Ca^{2+} efflux against its electrochemical gradient to Na^+ influx down its electrochemical gradient. This mode of operation, defined as forward mode or Ca^{2+} efflux or, more correctly, Ca^{2+}-exit mechanism, keeps a 10,000-fold difference in Ca^{2+} concentration between the two sides of the cell membrane. Under other physiological or pathophysiological conditions, when the intracellular Na^+ concentrations ($[Na^+]_i$) rise or membrane depolarization occurs reducing the transmembrane Na^+ electrochemical gradient, NCX reverses its mode of operation and mediates Na^+ extrusion and Ca^{2+} entry. This mode of operation is defined as reverse mode or Ca^{2+}-entry mechanism. The mode of operation of the antiporter depends on Na^+ gradient, Ca^{2+} gradient, and membrane potential.

Structure and Distribution

Three genes coding for the three different NCX proteins, NCX1, NCX2, and NCX3, have been identified in mammals. Each gene can give rise to multiple splice variants; 15 NCX1 and 6 NCX3 splice variants have been described so far. These isoforms share a sequence homology of about 70%. Despite this high sequence homology, the three proteins differ in biochemical properties as well as in cell and tissue distribution.

The Na^+/Ca^{2+} exchanger NCX1 has a molecular weight of 120 kDa and consists of 938 amino acids arranged in 9 transmembrane segments (TMS1–TMS9) subdivided into an N-terminal hydrophobic domain containing the first 5 TMSs (TMS1–TMS5) and a C-terminal hydrophobic domain containing the last 4 (TMS6–TMS9) (Fig. 28.7). NCX2 and NCX3 isoforms consist of 921 and 927 amino acids with a molecular weight of 102 and 105 kDa, respectively. The intervening region between TMS2 and TMS3 is called α1-repeat, while the region between TMS7 and TMS8 is called α2-repeat. While α1-repeat is extracellular, α2-repeat is intracellular; both participate in ionic transport. The N-terminal and the C-terminal domains are separated by a 500-amino-acid-long intracellular loop, named f loop. This loop has regulatory functions but is not involved in ionic transport. NCX activity is directly regulated by intracellular pH and

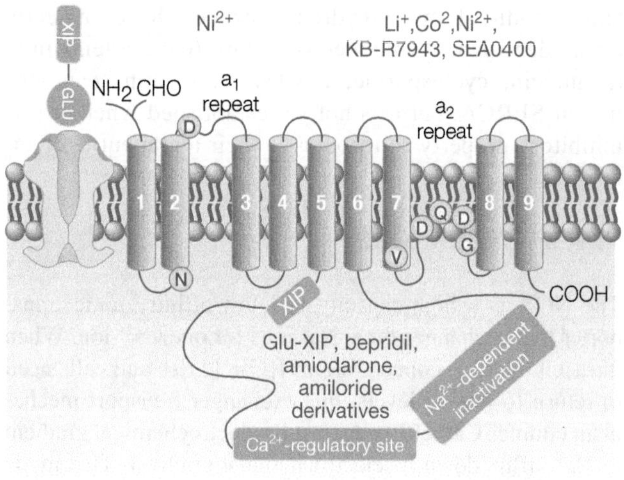

FIGURE 28.7 Structure of the Na$^+$/Ca^{2+} exchanger (NCX). The regions called α1-and α2-repeats take part in ion transport. The N-terminal domain is separated from the C-terminal domain by an intracellular f loop, which has regulatory functions and is not involved in ion transport. Binding sites for the most common pharmacological agents interacting with NCX are also shown.

numerous molecules that bind to specific sites in the f loop. These molecules include Ca^{2+} and Na$^+$ ions, nitric oxide (NO), protein kinase C, protein kinase A, ATP, and phosphatidylinositol 4,5-bisphosphate (PIP2). A very recent hypothesis suggests that the 938 aminoacids that constitute NCX are organized in 10 TMS.

The NCX is virtually present in all cell types. NCX1 is nearly ubiquitous being expressed in many tissues such as the CNS, heart, skeletal muscle, smooth muscle, kidney, eye, immune system, and blood cells. Conversely, NCX2 and NCX3 are expressed exclusively in neuronal and skeletal muscle cells.

At cellular level, recent studies have identified NCX in mitochondria and nuclei. NCX role in these intracellular organelles is currently being studied.

Biophysical Properties and Physiological Role

NCX exchanges three Na$^+$ for one Ca^{2+}. It has an affinity for Ca^{2+} 10 times lower compared to Ca^{2+}-ATPase pump. However, when activated to its maximum capacity, it shows a flow rate 10–50 times higher than that of Ca^{2+}-ATPase. Therefore, NCX is a low-affinity, high-capacity Ca^{2+} efflux system.

A density of about 400 NCX molecules for millimeter of plasma membrane generates a current of 30 pA/F. The direction depends on the amplitude of the current generated by cell membrane potential (V_m) and Na$^+$ and Ca^{2+} gradients. These gradients define the NCX reversal potential (E_{NCX}) that represents the V_m value at which NCX reverses its mode of operation.

NCX plays an important role in the repolarization of cardiac cells subsequent to cardiac contraction. In fact, muscle cell contraction that determines the physiological heart rate is strongly influenced by intracellular Ca^{2+}. Ca^{2+} can either enter the cell through voltage-gated Ca^{2+} channels (VGCCs; see also Chapter 27) or may be released from the sarcoplasmic reticulum through Ca^{2+}-activated ryanodine channels. To maintain Ca^{2+} homeostasis and ensure relaxation of cardiac muscle contraction, Ca^{2+} efflux mechanisms are required. Among these Ca^{2+} efflux mechanisms present in cardiac cell membranes, NCX plays a major role. Indeed, by operating in the "forward mode," it regulates intracellular Ca^{2+} concentrations and subsequent cardiac contractility.

In contrast, the physiological role of NCX operating in the "reverse mode" is still a matter of debate. However, it is known that such mode of operation is involved in the action mechanism of cardiac glycosides (see Supplement E 28.2, "Cardiac Glycosides").

In smooth muscle cells, NCX triggers a decline in intracellular calcium levels by working in the forward mode, thus producing muscle relaxation.

The kidney expresses high concentrations of NCX, which is involved in Ca^{2+} reabsorption in distal and proximal convoluted tubules, and contributes to maintain ionic homeostasis in the ascending limb of Henle's loop.

Equally important seems to be NCX role in pancreatic beta cells, where it facilitates insulin release.

The brain is the only organ ubiquitously expressing all three NCX isoforms. Despite the large number of studies investigating the role of each of them, it is still unclear whether they have a redundant action. However, it is well known that in the CNS, NCXs control cell depolarization by contributing to ionic homeostasis maintenance. Furthermore, these transporters are involved in the release of neurotransmitters such as dopamine, noradrenaline, and serotonin and in physiological processes related to learning as well as to short- and long-term memory.

NCX in Pathologies and Pharmacological Modulation

Given the important role played by NCXs in maintenance of intracellular Na$^+$ and Ca^{2+} homeostasis, it is easy to predict that changes in expression and activity of these transporters can contribute to pathophysiology of several human diseases.

At cardiac level, lack of NCX1 expression in adult animals improves recovery following myocardial ischemia and reduces the occurrence frequency of arrhythmias. On the other hand, overexpression of NCX1 appears to ameliorate other heart diseases such as heart failure. Furthermore, in some types of hypertension, NCX1 hyperactivity in the Ca^{2+} influx mode promotes increase in blood pressure by inducing muscle cell vasoconstriction and subsequent Ca^{2+} entry.

Recent studies have established that NCX has a strong neuroprotective role in neurodegenerative diseases such as cerebral ischemia (see Supplement E 28.4, "The Na$^+$/Ca^{2+} Exchanger as a New Molecular Target for the Development

TABLE 28.3 Inhibitors of the Na$^+$/Ca^{2+} exchanger

Amiloride derivatives	CB:DMB, DCB, DMB
Pyrrolidine derivatives	Bepridil
Isothiourea derivatives	KB-R7943
Ethoxyaniline	SEA-0400
Benzofuran containing drugs	Amiodarone
Quinazoline derivatives	SM-15811
Thiazolidine compounds	SN-6
Phenoxy pyridine derivatives	JP11092454
Nicotinamide derivatives	6-[4-[(3-fluorobenzyl)-oxy]phenoxy]nicotinamide
Piperidine derivatives	TM-252077
Acetamide derivatives	YM-270951
Peptides	XIP, Glu-XIP, FMRFa, FRCRCF
Small interference RNA	siRNA-NCX1, siRNA-NCX3
Antisense oligonucleotides	AS-NCX1, AS-NCX2, AS-NCX3
Inorganic cations	Ni^{2+}, La^{2+}, Cd^{2+}

of Drugs to Treat Cerebral Ischemia"). In fact, studies in animal models of cerebral ischemia have revealed that either pharmacological blockade of NCX or its genetic ablation worsens the outcome of cerebral ischemia. Therefore, it is conceivable that the NCX increased activity recently demonstrated with neurounina-1, a benzodiazepine-like compound, may reduce the consequences of ischemic brain damage.

In the last 40 years, many new organic and inorganic compounds that interfere with NCX activity have been identified. These include pyrrolidine derivatives (e.g., bepridil) and benzofuranic derivatives (e.g., amiodarone). However, such compounds, besides acting on NCX, are also capable of interfering with other membrane transporters and ion channels. More recently, isothiourea derivatives have been synthesized. Among these compounds, there is KB-R7943, which has been shown to selectively inhibit NCX operating in the reverse mode. However, even this compound has little selectivity for NCX. Therefore, the search for new selective compounds capable of modulating its activity has continued over time allowing the identification of new more potent and selective molecules such as those belonging to the class of ethoxyaniline (e.g., SEA0400) and thiazole derivatives (e.g., SN-6; see Table 28.3). Recently, neurounina-1, a benzodiazepine derivative capable of activating NCX, has been described.

For a systematic classification of the compounds identified so far that modulate NCX activity, it is important to take into account (i) specificity of action toward a specific isoform, since some compounds inhibit an isoform better than another; (ii) selectivity for a specific mode of action of NCX, since some compounds act better on the reverse mode; and (iii) localization of binding site on the exchanger.

In summary, there is a need to develop drugs that act selectively on each of the three isoforms and that exert their effects on NCX forward and reverse modes of operation.

Na$^+$/H$^+$ EXCHANGER

The family of NHEs is implicated in numerous pathophysiological conditions. It comprises a group of integral membrane proteins that mediate electroneutral exchange between one intracellular H$^+$ ion and one extracellular Na$^+$ ion, thereby regulating intracellular pH and cell volume.

Up to now, nine NHE isoforms have been cloned. They share an amino acid sequence homology of 25–70% and similar secondary structure. These isoforms differ in tissue and cellular distribution, kinetic properties, physiological functions, and sensitivity to inhibitors. Currently, for its numerous pathophysiological implications, NHE1 is the most studied isoform.

Structure, Distribution, and Functional Properties

The NHE1 gene is located on human chromosome 1p36.1–p35, is about 70 kb long, and is organized in 12 exons and 11 introns. Its promoter is regulated by several transcription factors, such as AP-1, AP-2, and C/EBP, and responds to reactive oxygen species (ROS).

NHE1 is a membrane protein of 815 amino acids with a molecular weight of 85 kDa. Topological analysis has allowed scientists to establish that NHE1 comprises two domains, a hydrophobic N-terminal domain of 500 amino acids and a hydrophilic cytoplasmic C-terminal domain of 315 amino acids. The N-terminal domain is composed of 12 transmembrane segments and is involved in Na$^+$ and H$^+$ ion transfer. In addition, this domain contains binding sites for most drugs acting on this protein. The C-terminal domain has important regulatory functions and contains an H$^+$ ion sensor (Fig. 28.8).

NHE1 is ubiquitously expressed on the plasma membrane of all mammal cells. In some cell types, NHE1 is not evenly distributed in the plasma membrane but accumulates preferentially in some microdomains; for example, it is located in the basolateral membrane of polarized epithelial cells, in the intercalated discs, in the transverse tubules of cardiomyocytes, and, lastly, in membrane protrusions of muscle cell fibers.

Because ion fluxes through NHE1 are generated by a Na$^+$ gradient directed inside the cell and by an H$^+$ gradient directed outside the cell, this exchanger does not require energy to work. According to the Michaelis–Menten kinetics, NHE1 depends on extracellular Na$^+$ with a K_m of 5–50 mM. In particular, extracellular Li$^+$ and H$^+$ ions compete with Na$^+$ for NHE1 binding; consequently, high extracellular H$^+$ concentrations inhibit NHE1 activity. In contrast, lowering of intracellular pH allosterically stimulates NHE1 activity with a Hill coefficient equal to 3.

NHE1 is activated by numerous stimuli including intracellular acidification, some growth factors, hormones, and cytokines. In general, it is possible to summarize mechanisms

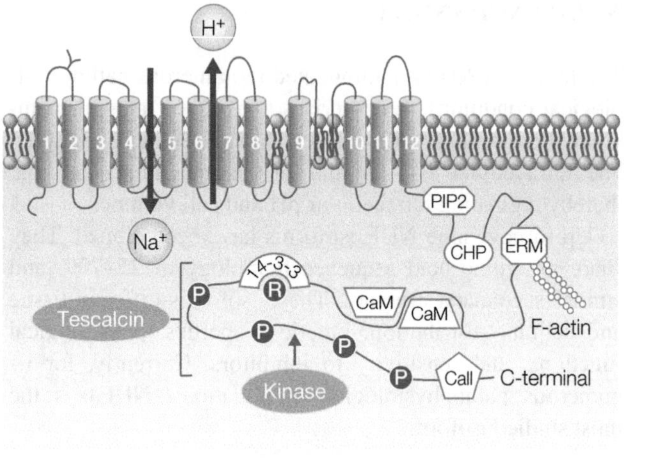

FIGURE 28.8 Structure of the Na+/H+ exchanger (NHE). NHE is constituted by two domains. The N-terminal domain is composed of 12 transmembrane segments, is involved in Na+ and H+ transfer, and contains binding site for most of the drugs acting on this protein. The C-terminal domain has important regulatory functions.

modulating NHE1 into three categories: (i) phosphorylation mechanisms, mediated by ERK, CaMK, and MAPK and affecting some C-terminal amino acid residues; (ii) interactions with regulatory proteins; and (iii) allosteric modulation affecting H+ affinity for its binding site.

Functional Significance and Pharmacological Modulation

NHE1 has several physiological roles. First, it regulates intracellular pH; indeed, NHE1 is activated by intracellular acidification and mediates H+ efflux, thus effectively restoring intracellular pH. Second, it adjusts cell volume; in fact, NHE1 activity is stimulated by cell shrinkage and mediates Na+ influx, thereby normalizing cell volume. Third, it regulates cell proliferation; in particular, NHE1-induced increases in intracellular pH promote transition from the G2 to the M phase of the cell cycle, thereby facilitating cell cycle progression. Fourth, it regulates cell growth and differentiation. Finally, it participates in apoptotic and immune processes.

Therefore, NHE1 proper functioning is implicated in all diseases involving changes in pH, ionic homeostasis, and cell volume. In particular, NHE1 activity is essential for proper body response to mechanical stress, osmotic, or metabolic disorders. Hypoxic or anoxic stimuli represent pathological conditions that, more than others, may influence NHE1 activity. Indeed, after ischemia onset, the insult causes a complex series of pathophysiological events including lowering of intracellular pH, changes in cell shape as a result of imbalances in ionic homeostasis, and kinase activation. For this reason, drugs blocking NHE1 activity have been tested in clinical trials in myocardial ischemia patients. However, very modest clinical results have emerged, and some of these studies have even been abandoned because of serious adverse effects on the cerebrovascular system.

NHE1 inhibitors are classified into three groups: (i) amiloride and benzamide analogs, (ii) benzoyl guanidine and derivatives, and (iii) bicyclic guanidines.

Amiloride, which belongs to the class of potassium-sparing diuretics, was the first NHE1 inhibitor to be synthesized. Being a partially selective drug, amiloride also acts on epithelial Na+ channels, NCX, and Na+/K+-ATPase pump.

Among the various benzoyl guanidine derivatives, HOE-694, cariporide, and eniporide are the most representative ones. These compounds, being more selective than amiloride, do not interfere with either NCX activity or Na+ channels. In addition, they are more effective in inhibiting NHE1 and NHE3 rather than NHE5.

Bicyclic derivatives, having a guanidine structure like zoniporide and BMS-284640, can also selectively inhibit NHE1.

In recent decades, NHE1 blockers have yielded promising results in animal models of ischemia/reperfusion-induced injury. To further substantiate these findings, clinical studies have been conducted to validate the effectiveness of these drug treatments in patient suffering from hypoxia/reperfusion damage. However, rather disappointing results have been obtained so far; therefore, new studies involving more selective and potent compounds are currently being carried out (see Supplement E 28.3, "Diuretic Drugs").

Na+/K+/Cl− COTRANSPORTER

The bumetanide-sensitive NKCC belongs to the superfamily of cation–chloride cotransporters (CCC). It mediates the electroneutral cotransport of 1Na+, 1K+, and 2Cl− from the extracellular to the intracellular space. This process can also occur in the opposite direction.

Two NKCC isoforms have been described: NKCC1 expressed ubiquitously in all mammalian cells and NKCC2 expressed selectively in kidney cells. NKCC1 favors the maintenance of the chloride gradient in excitable cells, thereby influencing intracellular concentrations of Cl− ions and helping to modify neuronal excitability.

Also, NKCC2 is of considerable importance from a physiological and pharmacological perspective. In fact, 30% of sodium filtered by renal glomeruli is reabsorbed in cells through the action of the NKCC2s present in the ascending portion of Henle's loop. There, the cotransporter represents the pharmacological target of loop diuretics, the most potent class of diuretics available to date.

Structure and Distribution

NKCC1 and NKCC2 share a sequence homology of about 70–90%. They are encoded by genes located on chromosome 15 and 5, respectively, and consist of about 1300 amino acids

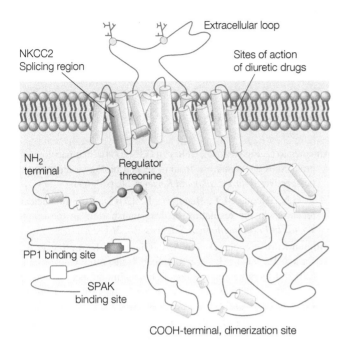

FIGURE 28.9 Structure of the Na$^+$/K$^+$/Cl$^-$ cotransporter (NKCC). NKCCC topological structure consists of 12 transmembrane segments flanked by two large cytoplasmic N- and C-terminal ends, which have regulatory functions and are subject to kinase phosphorylation. Between transmembrane segments 7 and 8, there is a large extracellular loop having regulatory functions and containing several glycosylation sites. White cylinders represent the action sites of diuretic drugs.

and have a molecular weight of 130 kDa. The topological structure includes 12 transmembrane segments flanked by two large cytoplasmic N- and C-terminal ends having regulatory functions and subjected to phosphorylation by kinases (Fig. 28.9).

Between the transmembrane segments 7 and 8, there is a large extracellular loop having regulatory functions and containing numerous glycosylation sites. The protein may be present in the form of homodimers, in which the two monomers are joined together through their C-terminal domains.

NKCC1 is ubiquitously expressed; its highest concentrations are found in epithelial cell basolateral membrane, where NKCC1 contributes to transcellular ionic transport. In the CNS, NKCC1 is expressed in neurons, glia, and endothelia. Its presence in auditory sensory neurons is particularly important. Indeed, genetically modified animals lacking NKCC1 are completely deaf.

By contrast, NKCC2 is expressed only in renal epithelial cells of the thick ascending limbs of Henle's loop.

In general, it is possible to affirm that NKCC1 is expressed to a greater extent in cells involved in secretion, whereas NKCC2 is expressed in cells involved in reabsorption.

Functional Properties

NKCC-mediated transport requires Na$^+$, K$^+$, and Cl$^-$ to be simultaneously present on the same side of the membrane. Because NKCC-mediated transport is characterized by a stoichiometry of 1Na$^+$:1K$^+$:2Cl$^-$, it is electroneutral. When high Na$^+$ concentrations and low K$^+$ concentrations persist, NKCC can also operate by carrying Na$^+$ rather than K$^+$ ions.

NKCC activity is strongly influenced by kinases that phosphorylate amino acids within its cytoplasmic N- and C-terminal domains. Important physiological and pathophysiological implications derive from phosphorylation mediated by kinases Ste20-related proline–alanine-rich kinase (SPAK) and oxidative stress response kinase (OSR1). Indeed, SPAK and OSR1 kinases are linked to numerous important cellular processes such as cytoskeletal rearrangement, cell differentiation, and cell proliferation.

Role in Cell Physiology and Pathology

In exocrine glands, NKCC1 is localized in the cell membrane portion closest to blood vessels. This localization enables NKCC to transport Na$^+$, K$^+$, and Cl$^-$ from blood into cells. Such action is carried out in cooperation with other carriers that contribute to the movement of these solutes in the same direction.

Besides its distribution in exocrine glands, NKCC1 is also found in other organs such as the inner ear, where its proper functioning is required for cochlear endolymph to be rich in K$^+$ ions. Indeed, NKCC1 inhibition with furosemide or other loop diuretics can cause deafness. Consistently, experimental data show that NKCC1 knockout animals experience hearing loss.

In addition, NKCC1 has also an important role in the CNS. Being responsible for adjusting Na$^+$, K$^+$, and Cl$^-$ intracellular homeostasis in excitable cells including neuronal cells, NKCC1 also appears to be involved in several other phenomena such as membrane hyperpolarization, a phenomenon that takes part in many CNS functions including memory, learning, and growth factor release.

Being localized in renal medulla in correspondence of Henle's loop and the juxtaglomerular apparatus, NKCC2 physiological function is restricted to processes related to kidney ionic homeostasis. In particular, NKCC2 contributes to regulation of extracellular fluid volume and osmolarity. In fact, this isoform represents the main mechanism mediating Na$^+$ reabsorption in the ascending limb of Henle's loop. NKCC2 inhibition underlies the antihypertensive effect of loop diuretics, such as furosemide and ethacrynic acid, which effectively counteract hypertension by promoting diuresis. This effect is associated with hypokalemia and hypochloremia.

Since NKCCs are involved in numerous physiological processes in the CNS and peripheral organs, changes in their function and/or expression may be involved in several pathologies. Indeed, NKCC1 mutations have been related to some diseases such as hereditary Gitelman syndrome, Bartter syndrome type I, and Andermann syndrome. In addition, changes in mechanisms regulating NKCC1 regulation appear to be related to the pathophysiology of another hereditary disease, namely, Gordon's disease.

Finally, several other experimental results indicate that these transporters are potentially involved in some of the most important polygenic disorders such as hypertension, cerebral ischemia, epilepsy, and osteoporosis.

Pharmacology

Drugs capable of inhibiting NKCC belong to the loop diuretic class (see Supplement E 28.3). Currently, there are no drugs able to act selectively on the two NKCC isoforms. However, some compounds are available that can inhibit both NKCC isoforms but with different pharmacological efficacy.

One of the most used compounds belonging to this class is bumetanide. This drug has an IC_{50} for NKCC1 and NKCC2 between 100 and 300 nM. Other derivatives are furosemide (IC_{50} 10 mM) and piretanide (IC_{50} 1 mM). More selective compounds should reduce side effects.

TAKE-HOME MESSAGE

- Transmembrane transport can be ATP dependent or independent.
- ATP-dependent transporters use energy provided by ATP hydrolysis to translocate ions, drugs, and other substrates.
- ATP-independent transporters use energy provided by electrochemical gradients of counterions to translocate ions, neurotransmitters, and other substrates.
- Several classes of medications act on membrane transporters, including diuretics, cardiac glycosides, and proton pump inhibitors.
- Cardiac glycosides inhibit Na^+/K^+-ATPase and interfere with Na^+, K^+, and Ca^{2+} intracellular concentration, thus producing depolarizing and inotropic effects.
- Inhibitors of the gastric proton pump have revolutionized the therapy of gastric ulcer.
- Different diuretics acts on ion transporters present in different segments of the nephron.

FURTHER READING

Annunziato L., Pignataro G., Di Renzo G.F. (2004). Pharmacology of brain Na^+/Ca^{2+} exchanger: from molecular biology to therapeutic perspectives. *Pharmacological Reviews*, 56, 633–654.

Chan G.N., Hoque M.T., Bendayan R. (2013). Role of nuclear receptors in the regulation of drug transporters in the brain. *Trends in Pharmacological Sciences*, 34, 361–372.

Hillgren K.M., Keppler D., Zur A.A., Giacomini K.M., Stieger B., Cass C.E., Zhang L., International Transporter Consortium. (2013). Emerging transporters of clinical importance: an update from the International Transporter Consortium. *Clinical Pharmacology and Therapeutics*, 94, 52–63.

Kell D.B., Dobson P.D., Bisland E., Oliver S.G. (2012). The promiscuous binding of pharmaceutical drugs and their transporter-mediated uptake into cell: what we (need to) know and how we can do so. *Drug Discovery Today*, 18, 218–239.

Kühlbrandt W. (2004). Biology, structure and mechanism of P-type ATPases. *Nature Reviews Molecular Cellular Biology*, 5, 282–295.

Li Y., Lu J., Paxton J.W. (2012). The role of ABC and SLC transporters in the pharmacokinetics of dietary and herbal phytochemicals and their interactions with xenobiotics. *Current Drug Metabolism*, 13, 624–639.

Molinaro P., Cuomo O., Pignataro G. et al. (2008). Targeted disruption of Na^+/Ca^{2+} exchanger 3 (NCX3) gene leads to a worsening of ischemic brain damage. *Journal of Neuroscience*, 28, 1179–1184.

Molinaro P., Cantile M., Cuomo O. et al. (2012). Neurounina-1, a novel compound that increases Na^+/Ca^{2+} exchanger activity, effectively protects against stroke damage. *Molecular Pharmacology*, 83, 142–156.

Mruk D.D., Su L., Cheng C.Y. (2011). Emerging role for drug transporters at the blood-testis barrier. *Trends in Pharmacological Sciences*, 32, 99–106.

Sun D. The "Loop" Diuretic Drug Bumetanide-Sensitive Na^+-K^+-Cl^- Cotransporter in Cerebral Ischemia. in Annunziato L. *New Strategies in Stroke Intervention*, Springer, Totowa, NJ. 2010, p. 85–108.

29

NEUROTRANSMITTER TRANSPORTERS

Gaetano Di Chiara

By reading this chapter, you will:

- Learn the general functional significance of neurotransmitter transporters (neurotransporters)
- Know the classification of the transporters and the molecular mechanism of the transport
- Understand the role of transporters in the mechanism of action of centrally acting drugs and in neurological and psychiatric diseases

---

NEUROTRANSMITTER TRANSPORTERS AND SYNAPTIC FUNCTION

Neurotransmitter transporters are localized on plasma and synaptic vesicle membranes and regulate neurotransmitter concentrations in the central nervous system (CNS) and peripheral nervous system. Two main kinds of neurotransmitter transporters have been recognized: plasma membrane transporters and vesicular transporters. Vesicular transporters regulate neurotransmitter uptake into the synaptic vesicles from the cytoplasm, where the enzymes that catalyze neurotransmitter synthesis and metabolism are localized. Plasma membrane transporters, instead, regulate neurotransmitter concentrations in the intra- and extracellular compartments and neurotransmitter clearance following its release in the synaptic cleft during synaptic transmission as well as its steady-state concentrations. Therefore, plasma membrane transporters modulate both phasic and tonic modalities of neurotransmission.

Following neurotransmitter release, the basic processes that predispose the nerve terminal to the subsequent synaptic event are (i) neurotransmitter removal from synaptic cleft and extracellular compartment, (ii) reconstitution, and (iii) refilling of synaptic vesicles. The released neurotransmitter can be removed from the extracellular milieu by enzymatic degradation and/or by transport into the same nerve terminals that has released it (reuptake) or into glial cells (see also Chapter 37).

Transport is the most common mechanism; thus, except for acetylcholine (ACh), all low molecular weight transmitters, including glutamate (GLU), glycine, gamma-aminobutyric acid (GABA), serotonin (5-HT), and catecholamines, are removed from the extracellular space by high-affinity transporters localized on the membrane of the same terminals that released them (Fig. 29.1a). Moreover, GABA, GLU, and 5-HT transporters are also found on the membrane of glial cells. By contrast, ACh is rapidly hydrolyzed by acetylcholinesterase in two inactive products: choline and acetate (see Chapters 37 and 39). Choline is taken up by a specific transporter localized on cholinergic terminals and used for de novo ACh synthesis (Fig. 29.1a). Blockade of ACh esterase markedly increases duration of ACh postsynaptic effects.

Transmitter reuptake is very efficient: about 50% of released neurotransmitter molecules are recaptured. Consistent with the importance of reuptake, its blockade by selective inhibitors strongly potentiates the effects of nerve stimulation on effectors.

Once in the cytoplasm, neurotransmitter molecules can be metabolized by intracellular enzymes, such as monoamine oxidases (MAOs) in the case of monoamines, or can be transported and accumulated in synaptic vesicles, ready for a further exocytotic release (Fig. 29.1b). Intravesicular neurotransmitter concentrations can be 10–1000 times higher than those in the cytoplasm. Transport is mediated by a specific vesicular transporter using the energy of the H^+ electrochemical gradient generated by a proton pump ATPase localized

---

General and Molecular Pharmacology: Principles of Drug Action, First Edition. Edited by Francesco Clementi and Guido Fumagalli.
© 2015 John Wiley & Sons, Inc. Published 2015 by John Wiley & Sons, Inc.

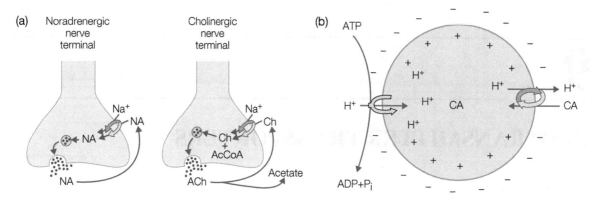

FIGURE 29.1 (a) Schematic representation of noradrenaline (NA) and choline (Ch) uptake mechanisms. (b) Schematic representation of catecholamine (CA) transport mechanism in synaptic vesicles.

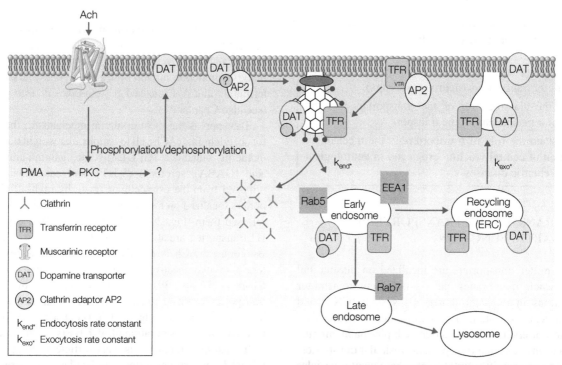

FIGURE 29.2 A model of the regulation of transporter endocytosis. Activation of protein kinase C (by phorbol esters or by stimulation of muscarinic receptors) determines phosphorylation of the transporter (DAT) or of a protein linked to it (AP2) resulting in targeting of DAT to clathrin-mediated endocytosis. As a result, DAT is internalized in the pool of early endosomes and then sent either to the lysosome pool for degradation or to recycling endosomes to be reinserted into the plasma membrane. Rab5, Rab7, EEA1, and TFR are markers of different endosomal pools (see also Chapter 38).

in the synaptic vesicle membrane (Fig. 29.1b). The K_m of different neurotransmitters for the relative transporter corresponds to their cytoplasmic concentrations. Thus, the K_m of GABA and GLU for the vesicular transporter is in the millimolar range, whereas that of catecholamines is in the micromolar range.

Monoamine transporters use the energy of the ion gradients created by ATPase pumps: the Na^+ gradient generated by Na^+/K^+-ATPase in the case of plasma membrane transporters and the H^+ gradient created by a Na^+/H^+-ATPase in the case of vesicular transporters. Thus, neurotransmitter transport against gradient is coupled to Na^+ or H^+ transport in the opposite direction (antiporter).

Regulation of Transporter Activity and Traffic

Given the importance of neurotransporters for synaptic function and for the flexibility and adaptive capability typical of chemical transmission, it is not surprising that their function is dynamically regulated in relation to the changing needs of the organism.

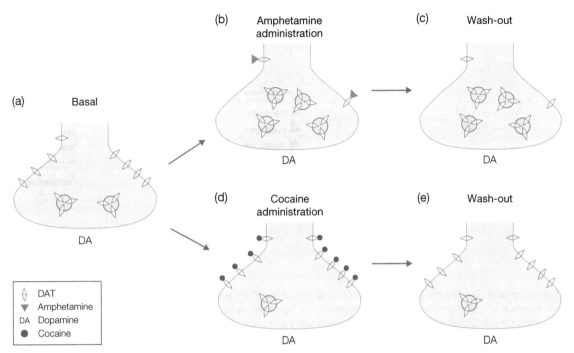

FIGURE 29.3 Regulation of dopamine transporter (DAT) activity by amphetamine and cocaine. (a) At baseline, DAT is expressed both on the plasma membrane outer surface and on intracellular membranes. (b) Amphetamine competes with dopamine for binding to DAT and reduces DAT expression on the plasma membrane outer surface. (c) DAT activity remains low for some time after removal of amphetamine, and this maintains the high extracellular dopamine concentrations. (d) Cocaine blocks DAT and increases its expression on the membrane. (e) After cocaine removal, DAT activity increases, thus reducing dopamine extracellular concentrations.

The transporter is recycled between plasma membrane and intracellular organelles, where it is sequestered in a nonfunctional form by mechanisms similar to those mediating insertion/removal of neurotransmitter receptors (see Chapter 9). Thus, if the transporter remains in the endosomes, it can be recycled back to the membrane, whereas if it is sent to the lysosomes, it will be degraded (Fig. 29.2). Various factors (e.g., substrate concentration, drugs, neurotrophins) modify the insertion of transporters into the plasma membrane, and therefore their functional activity, by affecting their turnover and intracellular compartmentalization. For example, amphetamine and cocaine have opposite effects on expression of the dopamine active transporter (DAT) in the plasma membrane of dopamine (DA) terminals (Fig. 29.3). Various stimuli affecting transporter trafficking between plasma membrane and intracellular compartments act via protein kinase C (PKC)-mediated phosphorylation of the supramolecular complexes that mediate anchoring and fusion of synaptic vesicles with the nerve terminal membrane (syntaxin, SNARE complex) (see Chapter 37).

NEUROTRANSMITTER TRANSPORTER FAMILIES

Based on molecular structure, subcellular distribution, and pharmacological properties, neurotransmitter transporters can be distinguished into three main families (Fig. 29.4): (i) Na^+/K^+-dependent plasma membrane transporters of GLU and aspartate (ASP), (ii) Na^+/Cl^--dependent plasma membrane transporters of GABA and catecholamines, and (iii) H^+-dependent vesicle membrane transporters of monoamines, ACh, and amino acids.

Na^+/K^+-Dependent Transporters for Excitatory Amino Acids

Transporters for excitatory amino acids are electrogenic transporters that can work in either direction depending on extracellular Na^+ and intracellular K^+ concentrations. The energy necessary for GLU transport is provided by the Na^+ and K^+ electrochemical gradients across the plasma membrane.

Since reuptake of a GLU molecule is coupled to the cotransport of two Na^+ and to the countertransport of one K^+ and one OH^- or HCO^{3-}, the transport is electrogenic: as, on each cycle, the countertransported anion (either OH^- or HCO^{3-}) is protonated, GLU reuptake determines a net flux of cations that depolarizes the membrane and reduces the intracellular pH (Fig. 29.4a). The transporter function is strictly dependent upon Na^+ and K^+ concentrations. Thus, an increase in K^+ or a decrease in Na^+ in the extracellular compartment (such as following prolonged high-frequency stimulation of nerve terminals or anoxia or pharmacological inhibition of Na^+/K^+-ATPase) can even result in reverse transport and substantially contribute to GLU release from nerve terminals.

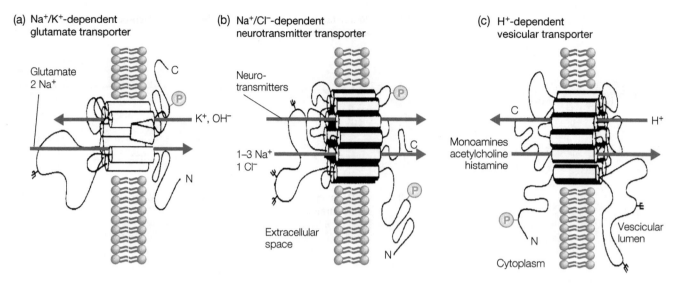

FIGURE 29.4 Structural diversity and ionic dependence of neurotransmitter transporters. High-affinity transporters localized in the membrane of nerve terminals and glial cells remove the neurotransmitter from the synaptic cleft and terminate its action on pre- and postsynaptic receptors, thus regulating timing of neuronal communication. Vesicular transporters regulate neurotransmitter storage in synaptic vesicles. Neurotransmitter transport at the plasma membrane and at the vesicular membrane is directly coupled to transmembrane ion gradients that provide the energy needed for the active transport (against the concentration gradient) of neurotransmitter. Based on molecular structure, subcellular distribution, and pharmacological properties, transporters are divided into three main groups: (a) Na$^+$/K$^+$-dependent and (b) Na$^+$/Cl$^-$-dependent plasma membrane transporters and (c) H$^+$-dependent vesicular transporters.

In contrast with exocytotic release, GLU release through the transporter is nonvesicular and calcium independent. Given the cytotoxic properties of GLU, it has been hypothesized that this mechanism might contribute to ischemic neuronal damage.

Molecular Organization and Function The mammalian genome contains five glutamate transporter genes (*EAAT1, slc1a3; EAAT2, slc1a2; EAAT3, slc1a1; EAAT4, slc1a6; EAAT5, slc1a7*) (Table 29.1). The most abundant and important for glutamate removal in the brain are *EAAT2* (GLT1) and *EAAT1* (GLAST). *EAAT3* (EAAC1) does not appear to play a role in signal transduction. EAAT4 and EAAT5 have the largest chloride conductance and, due to their high uncoupled anion conductance, act more as inhibitory glutamate receptors than as glutamate transporters. All EAATs catalyze Na$^+$- and K$^+$-coupled transport of L-glutamate as well as L- and D-aspartate, but not D-glutamate. K_m values of glutamate for EAAT2 are around 20 μM, and those for EAAT1 and EAAT3 are about twice that for EAAT2; those for EAAT4 and EAAT5 are one order of magnitude lower and higher, respectively.

The transport mediated by EAAT1–EAAT3 involves the exchange of one potassium ion with one glutamate molecule, three sodium ions, and one hydrogen ion. The coupling to three sodium ions makes these transporters less prone to reversal than the GABA transporters, which are coupled to two sodium ions. In addition to the coupled (stoichiometric) transport, there are uncoupled fluxes.

TABLE 29.1 Nomenclature of human plasma membrane glutamate transporters

Human Genome Organization (HUGO) name	Other names
Excitatory amino acid transporter 1 (EAAT1; slc1a3)	GLAST
Excitatory amino acid transporter 2 (EAAT2; slc1a2)	GLT-1
Excitatory amino acid transporter 3 (EAAT3; slc1a1)	EAAC1
Excitatory amino acid transporter 4 (EAAT4; slc1a6)	
Excitatory amino acid transporter 5 (EAAT5; slc1a7)	

Differences in rates of transporters can be measured between transporters. The cycling time of EAAT2 and EAAT3 is about 30 glutamate molecules per second at V_{max}. EAAT5 is even slower and behaves as a slow-gated anion channel with glutamate transport activity more than an order of magnitude slower than EAAT2.

Distribution EAAT2 is highly expressed in astrocytes in the brain and spinal cord, with the highest levels in the hippocampus and neocortex. Also, EAAT1 is found in astrocytes in the CNS. Immunogold labeling and electron microscopy show that both EAAT1 and EAAT2 are preferentially localized in the plasma membrane regions facing terminals rather than those facing cell bodies, or astrocytes, or pia mater. EAAT2 mRNA is also found in some neurons: pyramidal cells in CA3 hippocampus and in layer VI of parietal neocortex. In fact, EAAT2 mRNA is found in most neocortex neurons and in areas of the olfactory bulb, thalamus, and inferior olive. Finally, in normal and mature mammalian retina, EAAT2 protein is not expressed in retinal glial cells (neither in Muller cells nor in astrocytes), but in neurons (cone photoreceptors and bipolar cells). EAAT2 is responsible for about 95% of total glutamate uptake in young adult forebrain tissue. Selective deletion of the EAAT2 gene in mice confirms this conclusion, as in these animals glutamate uptake activity is reduced to 5% compared to wild-type (WT) mice. EAAT1 is the predominant glutamate transporter in the cerebellum, inner ear, circumventricular organs, and retina. Within the CNS, EAAT5 is preferentially expressed in the retina, whereas in the brain its expression is very low. EAAT3 is a neuronal transporter and is not expressed in glial cells. It appears to be expressed in most, if not all, neurons throughout the CNS but is selectively targeted to somata and dendrites, and thus, it is not present in axon terminals. Within the CNS, the highest EAAT3 concentrations are found in the hippocampus, but EAAT3 total tissue content in young adult rat brains is about 100 times lower than that of EAAT2.

EAAT4 is predominantly found in the cerebellar Purkinje cells, where it is localized on dendrites, spines in particular, but is also expressed to some extent in a subset of forebrain neurons.

Outside the CNS, EAAT2 is primarily expressed in glandular tissues, including mammary gland, lacrimal gland, and ducts and acini in salivary glands and by perivenous hepatocytes. It is not present in the heart, where EAAT1, EAAT3, EAAT4, and EAAT5 are expressed. Thus, the main role of EAAT2 is in the brain. EAAT1 is found in several nonneuronal tissues including the heart, fat cells, and taste buds, and EAAT3 is present in the kidney.

Functional Consequences of EAAT Deletion Consistently with biochemical findings, deletion of the *EAAT2* gene causes a reduction in glutamate uptake activity by about 95% and higher extracellular glutamate levels. Mice lacking EAAT2 at 3 weeks are hyperactive, epileptic, and smaller than their WT littermates. About half of the mice die from spontaneous seizures before they reach 4 weeks of age. The gradual increase in the severity of the phenotypic effects parallels the postnatal increase in EAAT2 expression in WT animals and the increase in extracellular glutamate exported via the glutamate–glutamine exchanger. Heterozygote EAAT2 knockout mice exhibit a 59% decrease in EAAT2 protein levels in the brain, but do not show any anatomical abnormality and have a similar lifespan as their WT littermates. They have moderate behavioral changes in respect to WT animals (mild sensorimotor impairment, hyperlocomotion, lower anxiety, better learning of cue-based fear conditioning but worse context-based fear conditioning) and are more sensitive to traumatic spinal cord injury.

Mice lacking EAAT1 develop normally but show symptoms of insufficient glutamate uptake in regions where EAAT1 is the major glutamate transporter. Thus, cerebellar function is affected, as indicated by reduced motor coordination and increased susceptibility to cerebellar injury, disturbance of the inner ear with exacerbation of noise-induced hearing loss, and disturbed retinal function. EAAT1 knockout mice also display poor nesting behavior, abnormal sociability, defective reward response to alcohol, and thus a reduced alcohol intake behavior. Lack of EAAT1 does not lead to spontaneous seizures like in EAAT2 knockout mice; however, when seizures are initiated, lack of EAAT1 increases seizure duration and severity.

In humans, mutations in EAAT1 are associated with episodic ataxia. Mice lacking EAAT3 develop dicarboxylic aminoaciduria and possibly a reduced spontaneous locomotor activity (open field). They do not show signs of neurodegeneration at young age and do not have epilepsy, but may age prematurely. Also, in humans, lack of EAAT3 results in dicarboxylic aminoaciduria. In addition, human EAAT3 polymorphisms have been reported to be associated with obsessive–compulsive disorders.

Pharmacology Initially, the design of GLU transporter inhibitors was based on the structure of known substrates. Thus, the first compounds were conformationally constrained analogs of dihydrokainic acid, a weak inhibitor of GLU uptake, containing additional isosteric groups, like azole-based rings, to mimic the acid group in the glutamate side chain. As a result of this approach, L-*trans*-2,4-PDC, WAY-855 (a strong inhibitor of glutamate transport in the low micromolar range with a slight selectivity for GLT1), and (+)-HIP-B (a mixed, competitive/allosteric antagonist) were developed (Fig. 29.5).

A second class of inhibitors has been derived from the aspartate analog L-threo-beta-hydroxyaspartate. A major breakthrough was the discovery of L-threo-beta-benzyloxyaspartate (L-TBOA), whose derivative (2S,3S)-3-(3-[4-(trifluoromethyl)benzoylamino]benzyloxy) aspartate (TFB-TBOA) shows an IC_{50} value in the low nanomolar range and is about 100 times more potent than L-TBOA. Another compound active in the high nanomolar range is WAY-213613, which shows a 45- and 59-fold selectivity for GLT1 over EAAC1 and EAAT1, respectively. Recently, a completely new class of inhibitors has been discovered, whose most potent analog, UCPH-101, is active in high nanomolar range ($IC_{50}=0.66\,\mu M$) and has a more than 400-fold selectivity for EAAT1 over EAAC1 and GLT1. A different kind of compounds potentially useful as therapeutic agents are those enhancing GLU transport efficiency (positive allosteric modulators). Riluzole, a drug approved for the treatment of amyotrophic lateral sclerosis (ALS),

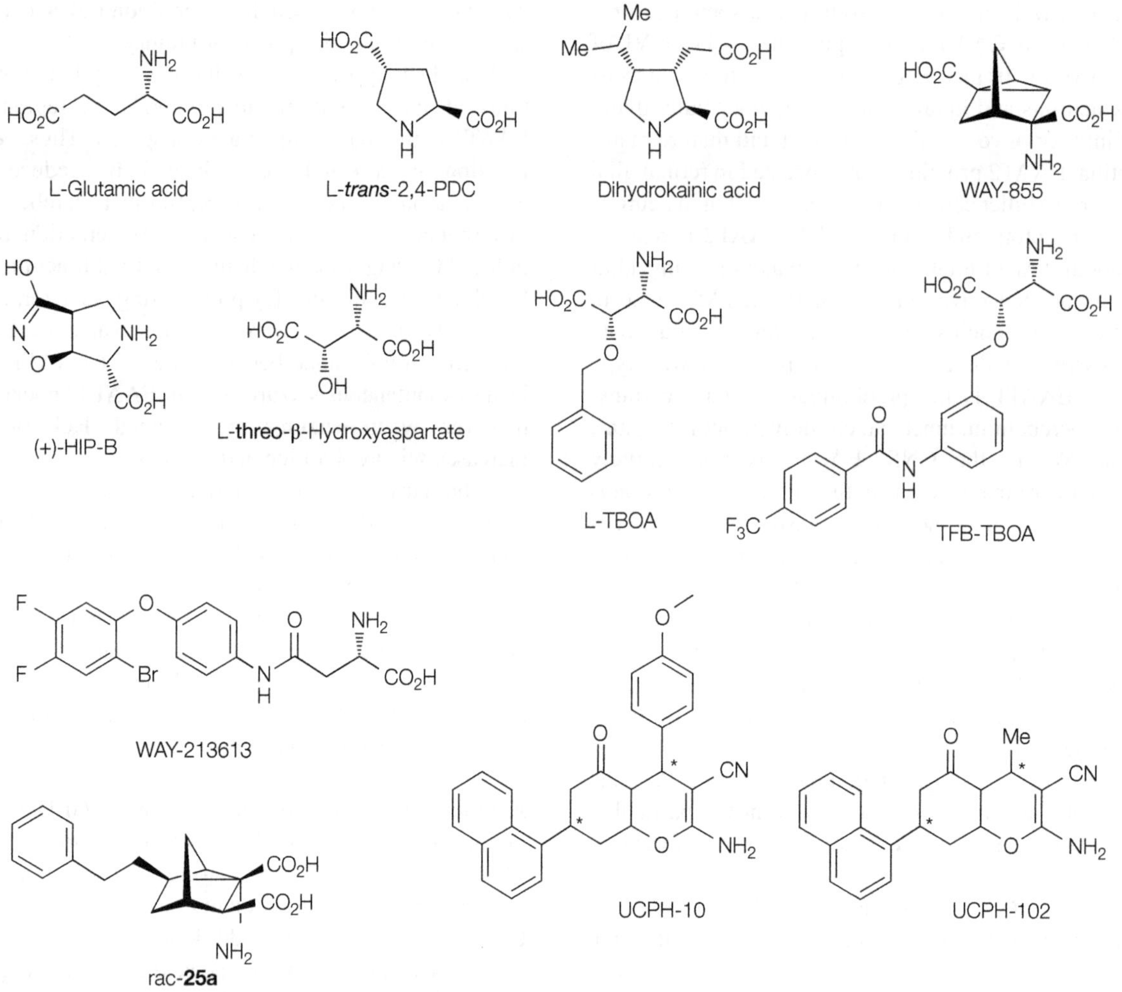

FIGURE 29.5 Structural formulas of the endogenous substrate L-Glu and of inhibitors and activators of excitatory amino acid transporters (EAATs). Their biological activities and isoform selectivities are described in the text and tables. Rac 25a is a novel selective GLT1 inhibitor.

besides other pharmacological activities, exerts neuroprotective action and increases glutamate uptake in a dose-dependent manner in cell lines expressing EAAC1, GLT1, and EAAT1 by enhancing glutamate affinity for the transporter. Another compound, Parawixin 1, extracted from the venom of *Parawixia bistriata*, a Brazilian spider, selectively enhances the activity of GLT1 versus EAAC1 and EAAT1.

NA$^+$/CL$^-$-DEPENDENT PLASMA MEMBRANE TRANSPORTERS

Molecular biology techniques have allowed the identification and characterization of a large number of Na$^+$/Cl$^-$-dependent transporters such as the transporters for GABA (Table 29.2), for monoamines (noradrenaline (NA), serotonin (5-HT), dopamine (DA); Table 29.3), and for glycine, taurine, proline, betaine, and choline. In addition, "orphan" transporters have been identified, for which physiological substrates have not been identified yet.

For all members of this family, neurotransmitter affinity is positively modulated by local concentrations of Na$^+$ and Cl$^-$ ions. They are neurotransmitter–sodium symporters (NSSs) that use the energy of the Na$^+$ gradient, produced by the Na$^+$/K$^+$-ATPase activity, for substrate translocation through the membrane. Transport efficiency is significant and can generate transmembrane concentration differences of up to 700–1800-fold. Given the ionic dependence of the transport, depending on the ionic gradient, under certain conditions, transport direction can be reversed.

Monoamine transporters (5-HT, DA, NA) constitute a subfamily characterized by high amino acid sequence homology (69–80%) and are targets of drugs such as antidepressants (tricyclics and heterocyclics), amphetamine, cocaine, and their congeners. Evolutionary studies based on the available data on the structure of these transporters in

TABLE 29.2 Na$^+$Cl$^-$-dependent plasma membrane neurotransporters

	Species			Localization			
					GABA transporters		
Rat	Mouse	Men	(neurons vs glia)	Distribution in CNS and peripheral tissues	Substrate	Inhibitors
rGAT1	mGAT1	hGAT1	N>>G	Olfactory bulb, hippocampus, cerebral cortex, thalamus, hypothalamus, subthalamus, striatum, substantia nigra, cerebellar cortex,	GABA>betaine	Nipecotic acid=L-DABA>>β-alanine
rGAT2 602 aa	mGAT3	?		Pia mater, arachnoid	GABA>betaine	β-alanine>L-DABA=nipecotic acid
rGat3 627 aa	mGAT4	hGAT3	G>N	Neurons: olfactory bulb, retina Glia: cerebral cortex, hippocampus, brainstem	GABA>betaine	β-alanine>L-DABA=nipecotic acid
rBGT	mGAT2	hBGT	N>>G	Pia mater, arachnoid, cerebellar granule cells, kidney tubule epithelial cells	Betaine>GABA	β-alanine>>nipecotic acid=L-DABA

TABLE 29.3 Na⁺/Cl⁻-dependent plasma membrane neurotransporters

Monoamine transporters	cDNA (species)	Molecular structure	Localization	Substrate	Inhibitors
1. NE transporter (NET)	Mouse, rat, men	≈600 aa, TM12	Neurons	Affinity DA > NE, V_{max} NE > DA, amphetamine	Tricyclic and heterocyclic antidepressants, *selective inhibitors (antidepressants)*: maprotiline, viloxazine, reboxetine. Cocaine, metilfenidate. D-amfetamine = L-amfetamine
2. DA transporter (DAT)	Mouse, rat, men	≈600 aa, TM12	Neurons, platelets	NE – DA, amphetamine, MDMA	*Selective inhibitors (antidepressants)*: amineptine. GBR-12935. Cocaine. D-amphetamine > L-amphetamine
3. 5-HT transporter (SERT)	Mouse, rat, men	≈600 aa, TM12	Neurons, glia, platelets, lymphocytes (see text)	5-HT, MDMA, fenfluramine, MPTP	SSRI: fluoxetine, paroxetine, citalopram, sertraline

D, dopamine; NE: noradrenaline.

different species have allowed generating a family tree in which two subfamilies can be distinguished: one including the amine carriers (5-HT, DA, NA) and the other including the GABA, taurine, creatine, and betaine transporters (Fig. 29.6).

Based on amino acid sequence analyses, a common structural model has been proposed for these transporters, characterized by 12 transmembrane segments, intercalated by 5 extracellular and 6 intracellular loops (Fig. 29.7). Both the amino and the carboxyl ends are intracellular. The long extracellular loop, located between the third and fourth transmembrane segment, contains a number of potential glycosylation sites (Fig. 29.7). This loop is the most variable region among members of this transporter family and therefore may be responsible for their substrate specificity.

Molecular Mechanism of Transport

The classic model of operation of these transporters suggests that they can assume two conformational states in which the neurotransmitter binding site faces alternately the extracellular and the intracellular environments. In normal operating conditions (transmembrane potential at rest, Na⁺ and Cl⁻ abundant in the extracellular space), the carrier binds neurotransmitter and ions while facing the extracellular space. Upon substrate binding, a conformational change occurs in the carrier molecule that leads to exposure of the neurotransmitter binding site toward the cell interior, where Na⁺ and Cl⁻ concentrations are low and the ions are released; this drastically reduces the affinity of the transporter for the neurotransmitter, determining its release into the cytoplasm. This classical model has been refined and amended by Yamashita and coworkers (2005) based on high-resolution (1.65 Å) X-ray diffraction analysis of the crystal structure of a bacterial leucine transporter (LEUT) belonging to the same family (SLC6) of monoamine and GABA transporters. Substrate and the cotransported sodium ions are located inside the transporter, in a pocket delimited by two transmembrane segments (TM1 and TM6). In this conformation, the amino acid leucine (the substrate) binds, through its carboxyl group, one of the two cotransported Na⁺ ions. In the case of amine substrates (DA, 5-HT, NA), Na⁺ binds to a carboxyl residue of aspartate. The substrate is maintained in the pocket by ionic bonds with the amino acid residues surrounding the pocket (Fig. 29.8). This configuration corresponds to an occluded state, in which substrate translocation is prevented by two molecular gates, one oriented outward and the other inward, consisting of ionic pairs formed by amino acid residues belonging to the segments that surround the pocket. In the case of LEUT, the substrate is transported with Na⁺ but not with Cl⁻ ions, and Cl⁻ does not bind to the pocket of the substrate. In the GABA transport, GABA is transported along with Na⁺ and Cl⁻, and Cl⁻ binds to the pocket together with the substrate.

The study by Yamashita and coworkers (2005) has provided a model to define, based on structure–activity relationship and bioenergetics data, the molecular mechanism underlying the activity of other transporters belonging to the same family, such as DAT, whose mechanism is shown in Figure 29.9. This model predicts that during dopamine transport, DAT shifts between four conformational states and dopamine binds two sites, called S1 and S2. According to this model, in resting conditions, the transporter has an open-outward conformation (i.e., with the internal gate closed and the outer open); upon substrate binding, DAT shifts to an occluded conformation due to substrate

NA⁺/CL⁻-DEPENDENT PLASMA MEMBRANE TRANSPORTERS 367

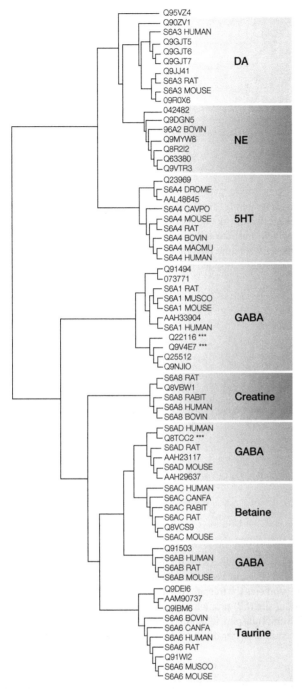

FIGURE 29.6 Evolutionary tree of Na⁺/Cl⁻-dependent transporters.

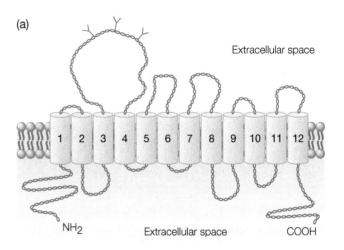

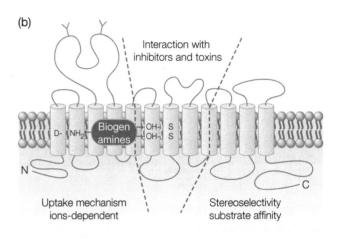

FIGURE 29.7 Topology of Na⁺/Cl⁻-dependent transporters. (a) GABA transporters. (b) Monoamine transporters. See Figures 29.8 and 29.9 for molecular mechanisms.

penetration in the transporter molecule. Subsequently, binding of a second dopamine molecule induces opening of the internal gate and transition to the open-inward conformation, which allows substrate release in the cytoplasm (Fig. 29.9).

This model can be used to explain the mechanism of DAT inhibition by drugs such as cocaine. Cocaine binds initially at the S2 site. If this binding takes place when DAT is oriented outward, cocaine prevents dopamine access to the S1 site; if it occurs in the occluded conformation, the drug prevents DAT transition to the open-inward conformation and DA translocation into the cell. Once bound to the S2 site, as a result of the conformational changes it induces, cocaine can make its way inside the transporter and bind to the primary substrate binding site (S1). This model is compatible with the existence of several binding sites for cocaine on DAT, either competitive or noncompetitive with DA.

GABA Transporters

GABA transporters can be localized on neurons and glia. GABA is the major inhibitory neurotransmitter of the brain, present in 60–75% of CNS synapses in mammals. Extracellular concentrations of GABA and the duration of its effects are regulated by Na⁺-dependent uptake by nerve endings and glial cells in the vicinity of the synaptic cleft (Table 29.2). GABA uptake in neuronal and glial cells

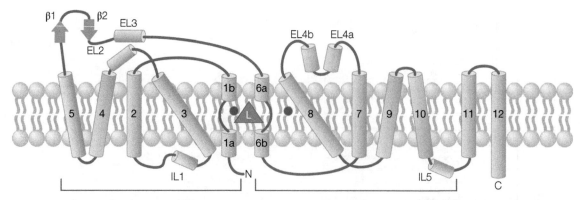

FIGURE 29.8 Topology of the leucine transporter of the monoamine transporter family. Transmembrane domains (TMs) are numbered from 1 to 12. The domains from TM1 to TM5 and from TM6 to TM10 represent two distinct but homologous groups, oriented in opposite directions in the membrane; they form the walls of a pseudochannel through which substrate and cotransported ions are translocated. In the core of the membrane, TM1 and TM6 lose the helical structure and form a pocket where the binding sites for Na⁺ ions (dots) and substrate (L-containing triangle) are located. EL, external loop; IL, internal loop; beta1 and beta2 correspond to portions with a beta structure. (Modified from Yamashita et al. (2005).)

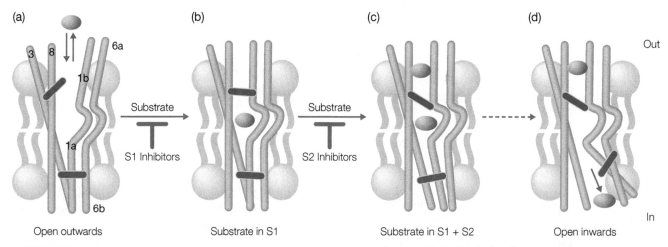

FIGURE 29.9 Model of a substrate translocation cycle for DAT. The carrier has two binding sites for the substrate (dopamine, depicted as an oval). (a) In the open-outward conformation, the transmembrane regions 1a and 6b assume a tight conformation, whereas 1b and 6a regions open up, allowing substrate binding. (b) Substrate binding to the first site (S1) promotes the transition of 1b and 6a regions to a semioccluded conformation. (c) Binding of a second substrate molecule to the second binding site (S2) promotes (d) inner gate opening, transition to the open-inward conformation, and consequent substrate release into the cytoplasm. Transport inhibitors, such as cocaine, can act by binding either S1, thus preventing substrate binding, or S2, preventing inner gate opening and translocation of the bound substrate to the S1 site. (Modified from Shan et al. (2011).)

exhibits different sensitivity of selective inhibitors. Neuronal GABA uptake is selectively inhibited by nipecotic acid and L-2,4-diaminobutyric acid (L-DABA), while glial uptake is selectively inhibited by β-alanine (Table 29.2). However, some important exceptions have been detected in the selectivity of inhibitors: for example, the GABA transporter localized in oligodendrocytes and in Müller glial cells of the retina is not inhibited by β-alanine.

Molecular Structure Four different GABA transporters have been cloned in mouse (mGAT1, mGAT2, mGAT3, and mGAT4) and three in rat (rGAT1, rGAT2, and rGAT3). The mGAT2 probably corresponds to the GABA and betaine carrier originally isolated from dog kidney (BGT1). In humans, so far, homologs of rGAT1 (hGAT1), rGAT3 (hGAT3), and rBGT (hBGT) have been cloned (Table 29.2). The nucleotide sequence of the rGAT1 (the first GABA transporter to be cloned) cDNA encodes a 67 kDa glycoprotein of 599 amino acids. rGAT2 and rGAT3 cDNAs encode proteins of 602 and 627 amino acids, respectively. These three transporters have high affinity for GABA ($K_m \cong 10\,\mu M$) and a secondary structure characterized by 12 potential transmembrane segments, with intracytoplasmic amino and carboxyl terminus (Fig. 29.7a). The intracellular loops contain phosphorylation sites for Ca^{2+}/calmodulin-dependent protein kinase. Moreover, in the domain immediately upstream of the TM9 transmembrane segment, GABA transporters have a set of amino acid residues with electric charge. MGAT1 and the rat homolog, rGAT1, are distributed in the retina, whole brain, and spinal cord; its expression is high in the olfactory bulb, neocortex, hippocampus, and cerebellar cortex. GAT1 expression is preponderant in GABAergic neurons, but this transporter is also present in postsynaptic sites of glutamatergic neurons (dendrites of pyramidal neurons) and astrocytes of rat cerebral cortex. rGAT3 (homologous of mGAT4) is particularly abundant in the retina, olfactory bulb, hypothalamus, medial thalamus, and brain stem, whereas its expression is very low in the neocortex, hippocampus, and cerebellum. While GAT1 is mainly neuronal, rGAT3 is predominantly (but not exclusively) glial. rGAT2 (rat counterpart of mGAT3) is exclusively localized in pia mater and arachnoid cells. Finally, the mRNA encoding the murine GABA and betaine transporter (mGAT2, the mouse homolog of the hBGT human transporter) is located in the epithelium of renal tubules and in leptomeninges.

Pathophysiological Role and Pharmacological Modulation GABA transporters are localized in GABAergic nerve endings as well as in glial cells and postsynaptic membranes of non-GABAergic neurons. GABA reuptake inhibitors have a clear therapeutic potential as antiepileptic agents. For example, tiagabine, an inhibitor of GABA reuptake with preferential affinity for rGAT1, possesses anticonvulsant activity in different experimental models of epilepsy; excess of extracellular GABA produced by tiagabine induces activation of inhibitory GABAB receptors on glutamatergic nerve endings, resulting in reduction of glutamate release. Both neuronal and glial transporters may be involved in the GABA extrusion (Ca^{2+} independent and nonexocytotic) that is observed when extracellular K$^+$ concentration increases dramatically during convulsions. This nonexocytotic release of GABA is an important mechanism to limit spread of epileptic activity to other brain areas, and its dysfunction may contribute to the pathogenesis of some forms of epilepsy. It has been observed that the density of GABA transporters is decreased in the hippocampus of patients with temporal lobe epilepsy. Moreover, in these patients, the increase in extracellular GABA induced by depolarizing concentrations of KCl, obtained through a microdialysis probe, is significantly lower in the epileptic focus (see Supplement E 29.1, "Therapeutic Properties of GABA Transporter Inhibitors").

The Serotonin Transporter

The serotonin transporter (SERT) plays a critical role in the termination of serotonergic neurotransmission and is the pharmacological target of some antidepressant drugs and several neurotoxic agents. For this reason, there is much interest on environmental, genetic, and pharmacological factors that might impair serotonergic function and the consequences that these changes have on brain development, synaptic plasticity, and neurodegenerative processes.

Molecular Structure The cDNAs of human and rat SERT code for a protein with 12 transmembrane domains (Fig. 29.7b). Deletion and mutagenesis studies indicate that an aspartate residue and probably a set of serine residues, respectively, located in the transmembrane domains TM1 and TM6 are involved in substrate translocation and binding of competitive inhibitors. Autoradiography studies with selective ligands, such as [3H]cyan-imipramine and [3H]citalopram, have shown a high density of 5-HT transporter in the cell bodies of the serotonergic raphe nuclei and their projection areas such as the neocortex, entorhinal cortex, hippocampal CA3 region, amygdala, substantia nigra, caudate–putamen, and hypothalamus.

Several mechanisms of signal transduction are able to influence transcription of SERT-encoding gene (Fig. 29.10). Its expression is regulated by cyclic AMP, PKC, and tyrosine kinase-dependent mechanisms. Also, SERT protein activity is finely regulated by nitric oxide and cyclic GMP, as well as by posttranslational modifications, including phosphorylation/dephosphorylation by PKC and calmodulin-dependent protein kinase A.

Human SERT gene, located on 17q11.2 chromosome, is composed of 14 exons distributed along 35 kb (Fig. 29.10). Functional mapping of the promoter has revealed numerous negative and some positive regulatory elements. In particular, key sequences are located within the 100 bp region upstream of the transcription start site; induction of promoter activity caused by cyclic AMP and PKC depends on *cis*-acting elements corresponding to potential binding sites for several

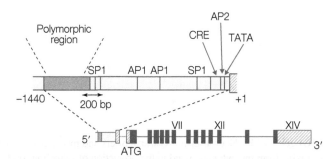

FIGURE 29.10 The human serotonin transporter (SERT) gene. Structure of the human SERT gene and its 5′ regulatory region at 17q11.2 locus. Blue boxes indicate translated regions and striped boxes indicate untranslated regions. SERT gene promoter contains a TATA-box motif and numerous putative binding sites for transcription factors AP1, AP2, SP1, and CRE. The position of the polymorphic region (5-HTTLPR) is indicated. ATG, translation start site.

transcription factors (e.g., AP1, AP2, SP1 sites, and a single CRE element) (Fig. 29.10).

Alterations of serotonin reuptake have been observed in platelets and brain of patients suffering from mood disorders. Genetic studies have shown that in most cases, functional defects result from changes in gene expression regulation. In particular, a highly polymorphic region (5-HT transporter gene-linked polymorphic region, 5-HTTLPR) (Fig. 29.10) has been identified, which seems to play an important role in the control of SERT gene expression. Recent studies have shown a high allele-dependent variability in promoter derepression by glucocorticoids and a significant association between a short variant of 5-HTTLPR (the S allele of the promoter) and anxious personality. Indeed, in a relatively large sample of subjects, it has been demonstrated that the S allele is associated with lower expression of SERT in lymphocytes and increased incidence of behaviors typical of neurotic personality.

Pharmacology of SERT SERT is a molecular target of antidepressant drugs, neurotoxins, and drugs of abuse. Tricyclic antidepressants, such as imipramine, and selective serotonin reuptake inhibitors (SSRIs), such as fluoxetine, paroxetine, citalopram, and sertraline, occupy sites pharmacologically distinct from the substrate binding site.

Some drugs including amphetamine derivatives such as 3,4-methylenedioxymethamphetamine (MDMA, "ecstasy"), fenfluramine, and the neurotoxins 1-methyl-4-phenyl-1,2,3,6-tetrahydropyridine (MPTP) and 2'-NH2-MPTP are uptaken by serotonergic neurons via this transporter (Table 29.3).

While tricyclic and heterocyclic antidepressants inhibit monoamine reuptake, numerous agents increase monoaminergic transmission (e.g., amphetamines and their analogs) by facilitating nonexocytotic serotonin release following transport reversal (i.e., rather than taking up the substrate, the transporter extrudes it). Amphetamines and their analogs, such as MDMA, are SERT and DAT substrates and are transported into nerve terminals. This produces an increase in the transporter sites available for serotonin binding on the membrane inner surface, leading to an increase in the outward serotonin transport (Fig. 29.11). This mechanism is used by MDMA and fenfluramine (an anorectic compound) to release serotonin "nonexocytotically." Released serotonin can be converted into reactive compounds (e.g., 5,7-di-hydroxytryptamine) that can induce neurodegeneration of serotonergic nerve terminals. Consistent with a role of SERT in MDMA neurotoxicity, these effects are prevented by tricyclic and heterocyclic compounds that inhibit serotonin reuptake and by SERT gene deletion. Similarly, serotonin release caused by fenfluramine is blocked by SERT inhibitors.

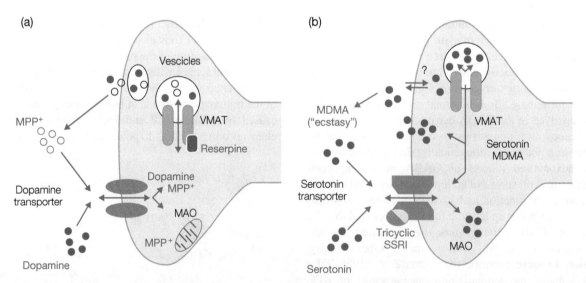

FIGURE 29.11 Neurotransmitter transporters and mechanism of action of drugs and toxins. (a) VMAT plays a key role in the elimination of toxins from the cytoplasm and in the protection of mitochondria from oxidative damage. 1-Methyl-4-phenylpyridinium (MPP+), the neurotoxic metabolite of MPTP, is imported into neurons by DAT. In pathological conditions, functional deficit of VMAT can result in increased susceptibility to toxins like MPP+. Vulnerability to neurotoxins may be further increased by reduced function of VMAT due, for example, to reduced availability of ATP, required for proton pump activity or to drug effect (reserpine). (b) Amphetamines and their analogs such as MDMA ("ecstasy") are DAT and SERT substrates. Amphetamine and MDMA may also enter nerve terminals by simple diffusion, displace biogenic amines from vesicles, and ultimately induce serotonin release via reverse transport. Although MDMA ability to induce dopamine release is probably responsible for its reinforcing properties, MDMA is also a powerful serotonin releaser, and long-term administration results in degeneration of serotonergic terminals. Serotonin release and MDMA neurotoxicity can be prevented by tricyclic and heterocyclic inhibitors of serotonin reuptake.

The Dopamine Transporter

After exocytotic release from nerve endings, also dopamine action on its receptors is terminated by a Na$^+$/Cl$^-$-dependent neuronal reuptake transporter (Table 29.3). Although mostly associated with dopaminergic neurons, the DAT is also expressed in platelets, as well as in several tumor cell lines such as neuroblastoma. After reuptake, dopamine can be transported into synaptic vesicles or metabolized by MAO associated with mitochondria (Fig. 29.11).

Molecular Structure DAT is a glycoprotein of about 80 kDa, with multiple isoforms located in different brain areas and characterized by distinct carbohydrate side chains. Human DAT cDNA encodes a protein of 620 amino acids sharing high homology with other members of the same family of transporters, such as those for 5-HT, NA, and GABA. It contains 12 hydrophobic transmembrane regions with cytoplasmic amino- and carboxyl-terminals and multiple glycosylation sites in the extracellular loops, which undergo major changes during prenatal development and aging. Moreover, several phosphorylation sites are present in the intracellular loops. An aspartic acid residue located in the first transmembrane segment is important for transport efficiency and ionic dependence, whereas a set of serine residues in the TM6, TM7, and TM8 transmembrane segments determines the affinity for the inhibitors. Finally, the carboxyl-terminal domain determines stereoselectivity and affinity for the respective substrates. The sulfhydryl groups of cysteine residues in the second extracellular loop are necessary for DAT function as they play an important role in its insertion in the plasma membrane and formation of quaternary structures. DAT protein also contains a potential leucine zipper, which may determine the formation of homotetramers.

DAT Pharmacology Cocaine and its analogs inhibit DAT activity, leading to a reduction in dopamine reuptake with consequent increase in the extracellular concentration of the neurotransmitter released from nerve endings. Thus, cocaine effect on dopaminergic transmission depends on the exocytotic dopamine release triggered by nerve impulse. Conversely, stimulation of DA transmission by amphetamine is independent of nerve impulse and is due to reversion of dopamine transport via DAT (see Supplement E 29.2, "Mechanism of Action of Amphetamine"). Indeed, transport of amphetamine into the terminal increases the number of DAT binding sites available for dopamine on the membrane internal side (see also effects on the norepinephrine transporter, NET).

Cocaine and amphetamine interaction with dopaminergic neurotransmission is the basis of their psychostimulant effects and dependence liability. In mice, selective deletion (knockout) of the DAT gene causes a dramatic increase in basal motor activity. In addition, these mice do not show any increase in motor activity in response to high doses of cocaine and amphetamine, consistent with the hypothesis that DAT is the substrate of the motor stimulant actions of cocaine and amphetamine.

However, mutant mice lacking DAT still learn to self-administer cocaine seeking its effects. Although these results appear inconsistent with the hypothesis that in normal mice cocaine reinforcement is due to DAT blockade, they may indicate that alternative mechanisms cause cocaine-reward response when DAT is deleted. In fact, brain areas belonging to the extended amygdala and its transition areas (e.g., the nucleus accumbens "shell"), where cocaine causes the most pronounced increases in extracellular dopamine concentration, show high density of noradrenergic terminals expressing the NET, which can uptake dopamine quite efficiently. Therefore, in mutant mice lacking DAT, dopamine uptake could be largely carried by NET. Given that cocaine is also an effective and powerful blocker of NET, lack of DAT does not significantly affect cocaine ability to increase extracellular dopamine concentration in the accumbens shell and to elicit reward and reinforcement.

DAT and Degenerative Diseases of the Dopaminergic System Given its selective localization in dopaminergic cells, DAT is a specific marker of these neurons. *Postmortem* studies have shown that DAT is reduced in patients with Parkinson's disease. Consistent with these data, positron emission tomography studies have shown decreases in binding of [11C]nomifensine (a selective DAT inhibitor) and in uptake of [18F]fluorodopa in the putamen of Parkinson's disease patients.

DAT plays an essential role in the degeneration of the nigrostriatal DA neurons induced by 1-methyl-4-phenylpyridinium (MPP+), the neurotoxic metabolite of MPTP. In fact, the latter is concentrated in dopaminergic nerve terminals by DAT (Fig. 29.11) (for details on the role of monoaminergic systems in mediating the effects of antidepressant drugs, see Supplement E 29.3, "Inhibitors of Amine Transporters And Antidepressant Drugs").

The Noradrenaline and Adrenaline Transporters

Nerve terminals releasing noradrenaline (NA) and adrenaline (A) express membrane transporters that take up the released neurotransmitter. However, these transporters are not selective for their relative amine but are able to recapture other amines, albeit with different efficiencies. For example, the noradrenaline transporter (NET) shows the highest efficiency (V_{max}) for NA but can also transport dopamine, which binds NET with even higher affinity than NA. The ability of NET to take up dopamine is the basis of an important interaction between dopamine transmission and drugs that block NET. Indeed, in brain areas where the density of NA terminals is higher than that of DA endings, such as the prefrontal cortex, NET substantially contribute to remove dopamine released from dopamine terminals from the extracellular compartment. Under these conditions, a NET blocker as desipramine, like any tricyclic antidepressant, can induce a more robust increase in extracellular DA in cortical areas than specific DAT blockers.

Molecular Structure and Function NET (Table 29.3) has been cloned from human (hNET), bovine (bNET), and rat (rNET) cells. The homology between the amino acid sequences of these transporters is high (91–93%), while it is less between DAT and NET (~65%). This transporter belongs to the family of the Na^+/Cl^--dependent transporters, with whom it shares several characteristics, both structural and functional. Unlike SERT, NET is not influenced by the K^+ concentration. NET has 12 transmembrane domains (TM1–TM12), intracellular amino- and carboxyl-terminal, and a long extracellular loop between TM3 and TM4, containing several potential glycosylation sites. Several phosphorylation sites for various protein kinases are also present. In the second transmembrane segment is present a leucine zipper domain, a motif that mediates the interaction with other proteins containing the same motif (Fig. 29.8a). NET (as well as DAT) has sites for selective binding of the neurotransmitter (TM9–TM12 and C-terminal tail) and of antagonists (TM5–TM8) (Fig. 29.8a). TM2 and TM3 seem to be important to determine the V_{max} of amine transport. The long extracellular loop between TM3 and TM4, which contains at least two N-glycosylation sites, is important for the subcellular localization of the transporter.

Recently, an adrenaline (A) transporter has been cloned from frog sympathetic ganglia (FET). This transporter has a high degree of homology (75%) with human NET and, like this latter, has a high affinity for tricyclic antidepressants (e.g., desipramine). V_{max} of FET is E > NE > DA, whereas that of hNET is reversed (NE > DA > E).

Noradrenaline transport is accompanied by a depolarizing current. Indeed, NET, as well as other carriers of the same family (DAT, SERT), shows a typical ion channel conductance. Thus, despite that transport stoichiometry of noradrenaline (NA) involves transfer of a single Na^+ ion for every NA, patch clamp studies demonstrate that each transported NA molecule is accompanied by a higher number of Na^+ ions. NET is permeable to Na^+ ions even in resting conditions, that is, when NA transport is not operational. In the presence of sufficient extracellular concentrations of NA, Na^+, and Cl^-, noradrenaline is bound to the carrier and translocated to the cytoplasm. For each noradrenaline molecule imported, about 300 Na^+ ions enter the cell through the transporter. Under conditions of intense transport, this Na^+ current is sufficient to depolarize the cell, with at least two consequences: (i) it modifies membrane excitability, facilitating the generation of action potentials at somatodendritic level and reducing the effect of exocytosis at the terminal level; (ii) it reduces the efficiency of the transporter. Therefore, membrane depolarization concurrent to noradrenaline transport is part of a homeostatic mechanism that keeps noradrenaline transport within predetermined limits.

NET as a Molecular Target of Antidepressant Drugs, Cocaine, and Indirect Sympathomimetics NET has several pharmacological characteristics that differentiate it from DAT, the most homologous carrier. Tricyclic antidepressants block NET with affinity at least 3 orders of magnitude higher than for DAT (Table 29.3; see Supplement E 29.3). Amphetamine isomers bind NET with the same affinity, whereas D-amphetamine has a greater affinity for DAT compared to L-amphetamine (Table 29.3).

In addition to tricyclic antidepressants, NET is also blocked by cocaine and methylphenidate. These drugs bind the carrier, but they are not substrates for its transport. This exchange mechanism is the basis of the monoamine release properties of amphetamine. It explains both the ability of cocaine and NET antagonists to block amphetamine-induced noradrenaline release and the ability of amphetamine, in contrast with cocaine and antidepressants, to increase noradrenaline extracellular concentrations even in the absence of nerve impulses and exocytotic release.

H^+-Dependent Vesicular Transporters

The electrochemical H^+ gradient generated by a vesicular proton pump provides energy for neurotransmitter transport in synaptic vesicles. The high concentration gradient generated by neurotransmitter transport into synaptic vesicles requires considerable energy consumption. In all synaptic vesicles currently known, the energy required for transport comes from the electrochemical H^+ gradient generated by a proton pump (ATPase) in the vesicle membrane (Fig. 29.1b). This pump has a remarkable sequence homology and similar drug sensitivity to the pump that generates the proton gradient in endosomes and lysosomes. It is therefore likely that secretory granules and synaptic vesicles share a common evolutionary origin.

The vesicular proton pump couples ATP hydrolysis to H^+ transport from the cytoplasm into synaptic vesicles. Depending on the ionic permeability of the vesicle membrane, proton transport can generate an H^+ gradient, a membrane potential, or both. If the synaptic vesicle is permeable to an anion (e.g., Cl^-), proton transport into the vesicle generates an inflow of negative ions, which tends to abolish the electrochemical gradient. In this case, the proton pump creates a proton gradient without altering membrane potential. Conversely, if anions cannot cross the vesicular membrane at the same rate of H^+ ions, proton transport generates a membrane potential that limits their further import.

TABLE 29.4 Na+/Cl−-dependent vesicular neurotransporters

cDNA	Molecular structure			Localization	Substrate	Inhibitors
Monoamine methamphetamine	VMAT1 (rat)	512 aa	12TM	Chromaffin cells	NE ~ DA ~ 5-HT ~ E >> MPP+	RES > TB ~
Transporter methamphetamine (VMAT)	VMAT2 (rat)	512 aa		CNS	NE ~ DA ~ 5-HT ~ E ~ MPP+	TB ~ > RES
Acetylcholine transporter VAChT	VAChT (rat)	500 aa	12TM	Cholinergic nerve endings	Acetylcholine	Vesamicol
GABA transporter (VGAT)	VGAT (rat)	≈500 aa	10TM	GABAergic nerve endings glycinergic nerve endings	GABA GABA- and glycine	Gamma-vinyl-GABA

RES, reserpine; TB, tetrabenazine.

Vesicular transporters are a family of proteins genetically and pharmacologically distinct from plasma membrane transporters. Thus, vesicular transport is inhibited by drugs such as reserpine and tetrabenazine (for amines) and vesamicol (for ACh), which are inactive on plasma membrane transporters (Table 29.4). The cDNAs encoding for three different vesicular transport systems have been isolated and characterized: vesicular monoamine transporters (VMAT) (VMAT1 and VMAT2), vesicular acetylcholine transporter (VAChT), and vesicular GABA transporter (VGAT) (Table 29.4).

The Vesicular Monoamine Transporters

The best-characterized mechanism of vesicular transport is that involved in catecholamine storage in chromaffin cell granules. In these granules, the proton pump generates an H+ concentration gradient (intravesicular pH 5.5) and a membrane potential of 60 mV (interior positive). The transporter binds catecholamines, present in the cytoplasm in nonionized form, and moves them into the vesicle where their amino group is protonated due to the acidic pH. Once in protonated form, catecholamines cannot cross the vesicle membrane to return to the cytoplasm and thus remain trapped in the granule (Fig. 29.1b). In chromaffin granules and in monoaminergic synaptic vesicles, a nucleotide transport mechanism also exists, which accumulates adenine nucleotides in the vesicles and is coupled to the electrochemical proton gradient. Formation of insoluble complexes between these adenine nucleotides (ATP) and catecholamines constitutes a fundamental mechanism for regulating the osmotic balance between intravesicular and cytoplasmic compartments.

Following neurotransmitter release, the emptied vesicles contain high concentrations of Na+ and Cl−, which move down their concentration gradient into the cytoplasm, providing part of the energy required for neurotransmitter or nucleotide import into the vesicles (for more details on synaptic vesicle dynamics, see Chapter 37).

Pharmacology of the VMAT Unlike plasma membrane transporters for DA, NA, and 5-HT that have preferential affinity for their respective substrates, the vesicular transporter has a similar affinity for different monoamines. Reserpine, a potent antihypertensive drug, prevents transport of biogenic amines into synaptic vesicles and chromaffin granules, thus causing depletion of intravesicular neurotransmitter stores. Reserpine binds irreversibly to VMAT and inhibits monoamine transport (Fig. 29.11). Other inhibitors, such as MPP+ and tetrabenazine, interact with different sites of the carrier and therefore do not antagonize reserpine binding to VMAT. However, MPP+ transport into synaptic vesicles is effectively inhibited by both reserpine and tetrabenazine (Fig. 29.11). It is important to emphasize that tetrabenazine, which has been used to treat movement disorders including tics and dystonias, causes amine depletion much more pronounced at the central than at the peripheral level. [^{11}C]Tetrabenazine has been used to study the *in vivo* distribution of VMAT in human brain by positron emission tomography.

Molecular Structure of VMAT DNA sequencing of the VMAT-encoding cDNAs from rat adrenal medulla and brain has led to the identification of two proteins, of 512 and 521 amino acids, respectively, sharing a high degree of homology. These proteins, named VMAT1 and VMAT2, have 12 potential transmembrane segments, intracytoplasmic amino- and carboxyl-terminals, and a long intraluminal loop localized between the first two transmembrane segments, probably involved in substrate translocation and binding of inhibitors. VMAT protects neurons from the toxic effects of MPP+ through intravesicular sequestration (Fig. 29.11). VMAT1 and VMAT2 have a different anatomical distribution and also differ in their functional characteristics: VMAT1 is expressed predominantly in adrenal medulla chromaffin cells and compared to VMAT2, which is present in the brain, has a lower affinity for all monoamines. In the CNS, VMAT2 is expressed by different populations of dopaminergic, noradrenergic, adrenergic, serotonergic, and histaminergic neurons. Tetrabenazine and methamphetamine inhibit VMAT2 with a much higher potency than VMAT1.

Human VMAT is encoded by a single gene and is expressed in the midbrain and platelets. Of particular interest is the role of the intraluminal loop between TM1 and TM2 in MPP+ binding: in this region, the differences in amino acid sequence between rat and primates are quite pronounced and probably responsible for the different susceptibility of primates to MPP+ neurotoxic effect, compared to rat (which is resistant). This suggests that this region is susceptible to mutations capable of determining a reduction in VMAT ability to remove cerebral neurotoxins or to regulate the concentration of dopamine transported by DAT. Congenital (or acquired) VMAT dysfunctions could increase the sensitivity to exogenous and endogenous toxins responsible for degeneration of dopaminergic neurons and thus be involved in the genesis of lesions of the aminergic systems, as in the case of Parkinson's disease. In this context, it is interesting to note that a reduction in VMAT expression accompanying the loss of dopaminergic neurons has been observed in experimental models of brain aging.

The Vesicular Acethylcholine Transporter

Cholinergic nerve terminals are enriched in VAChT, which is H+ dependent and selectively inhibited by vesamicol. This compound prevents vesicular accumulation of ACh and blocks cholinergic transmission.

The rat VAChT-encoding cDNA specifies a protein with 12 potential hydrophobic transmembrane segments and the amino- and carboxyl-terminals localized in the extravesicular space (Table 29.3). VAChT distribution in central and peripheral cholinergic neurons is similar to that of choline acetyltransferase (ChAT), the enzyme catalyzing ACh synthesis. The human VAChT-encoding cDNA is devoid of introns and is contained within the sequence of the first intron of the ChAT gene; this allows the coordinated regulation of the expression of the two proteins.

The Vesicular Transporters for Excitatory Amino Acids, GABA, and Glycine

Unlike the vesicular transport of biogenic amines and ACh, which mainly depends on pH gradient, vesicular glutamate transport is almost exclusively dependent on the electrical gradient generated by the proton pump in the vesicle membrane, whereas GABA and glycine transport depends on both the electrical and the chemical gradient. It has not yet been established whether there are two separate vesicular transporters for GABA and glycine or whether both amino acids are concentrated into vesicles by the same carrier.

The VGAT is not inhibited by glutamate and is sensitive to nipecotic acid and to γ-vinyl-GABA (Table 29.4). γ-Vinyl-GABA is an irreversible inhibitor of GABA transaminase, the enzyme that catalyzes GABA degradation, and is used in the treatment of some forms of epilepsy.

In rat, VGAT contains 10 potential hydrophobic transmembrane segments and intravesicular amino- and carboxyl-terminals. The molecular structure is very different from that of VMAT and VAChT, suggesting that the VGAT gene belongs to a different family. Besides being localized in synaptic vesicles of GABAergic nerve endings, in rat VGAT is present also in the nerve endings of neurons of the upper olive containing both GABA and glycine and, at lower density, in neurons containing only glycine. This latter finding suggests that VGAT is involved in glycine transport into synaptic vesicles of glycinergic neurons. In agreement with this hypothesis, transfection of cultured nerve cells with VGAT-encoding cDNA determines the appearance of an activity of vesicular transport for both GABA ($K_m \cong 5\,mM$) and glycine ($K_m \cong 27\,mM$). Finally, some GABAergic or glycinergic neuronal populations do not possess VGAT; these neurons may express a different vesicular transporter not yet identified.

TAKE-HOME MESSAGE

- Transporters regulate the concentration of neurotransmitters in the synaptic clefts in both central and peripheral nervous system.
- Transmitter transport against its concentration gradient is coupled to cotransport of Na+ or H+ along their concentration gradient.
- Transporters can be distinguished into three major families on the basis of their molecular structure, subcellular distribution, and pharmacological properties.
- The noradrenaline transporter can take up also dopamine, the precursor of noradrenaline.
- In brain areas, such as the prefrontal cortex, where the density of noradrenergic terminals is high compared to dopaminergic ones, NET plays a more important role than DAT in clearing dopamine from the extracellular compartment.
- Catecholamine transport is the target of cocaine and indirect sympathomimetics.
- Monoamine and serotonin transports are the targets of drugs used to treat depression.
- Membrane and vesicular transports of GABA are the targets of some antiepileptic drugs.

FURTHER READING

Seal R.P., Amara S.G. (1999). Excitatory amino acid transporters: a family in flux. *Annual Review of Pharmacology and Toxicology* 39:431–456.

Adnot S., Houssaini A., Abid S., Marcos E., Amsellem V. (2013). Serotonin transporter and serotonin receptors. *Handbook of Experimental Pharmacology* 218:365–380.

Bernstein A.I., Stout K.A., Miller G.W. (2014). The vesicular monoamine transporter 2: an underexplored pharmacological target. *Neurochemistry International* 73C:89–97.

Beuming T., Sover O.S., Goldstein R.A., Weinstein H., Javitch J. (2003). Probing conformational changes in neurotransmitter transporters: a structural context. *European Journal of Pharmacology* 479:3–12.

Buckley K.M., Melikian H.E., Provoda C.J., Waring M.T. (2000). Regulation of neuronal function by protein trafficking: a role for the endosomal pathway. *Journal of Physiology* 525(11):19.

Di Chiara G., Tanda G.L., Frau R., Carboni, E. (1992). Heterologous monoamine reuptake: lack of transmitter specificity of neuron-specific carriers. *Neurochemistry International* 20:231S–235S.

Focke P.J, Wang X., Larsson H.P. (2013). Neurotransmitter transporters: structure meets function. *Structure* 7(21):694–705.

Guillot T.S., Miller G.W. (2009). Protective actions of the vesicular monoamine transporter 2 (VMAT2) in monoaminergic neurons. *Molecular Neurobiology* 39:149–170.

Klig K.M. and Galli A. (2003). Regulation of dopamine transporter function and plasma membrane expression by dopamine, amphetamine, and cocaine. *European Journal of Pharmacology* 479:153–158.

Melandro M.S., Kilberg M.S. (1996). Molecular biology of mammalian amino acid transporters. *Annual Review of Biochemistry* 65:305–336.

Rilstone J.J., Alkhater R.A., Minassian B.A. (2013). Brain dopamine–serotonin vesicular transport disease and its treatment. *New England Journal of Medicine* 368:543–550.

Shan J., Javitch J.A., Shi L., Weinstein H. (2011). Substrate-driven transition to an inward-facing conformation in the functional mechanism of the dopamine transporter. *PloS One* 6(1): e16350.

Sulzer D. Galli A. (2003). Dopamine transport currents are promoted from curiosity to physiology. *Trends in Neurosciences* 26:173–176.

Vandenberg R.J, Ryan R.M. (2013). Mechanisms of glutamate transport. *Physiological Reviews* 93:1621–1657.

Vaughan R.A, Foster J.D. (2013). Mechanisms of dopamine transporter regulation in normal and disease states. *Trends in Pharmacological Sciences* 34(9):489–496.

Yamashita A., Singh S.K., Kawate T., Jin Y., Gouaux E. (2005). Structure of a bacterial homologue of Na^+/Cl^--dependent neurotransmitter transporters. *Nature* 437:215–223.

SECTION 8

CONTROL OF PROTEOLYSIS

30

INTRACELLULAR PROTEOLYSIS

Fabio Di Lisa and Edon Melloni

By reading this chapter, you will:

- Acquire the knowledge on the structural and functional features of the different families of proteolytic enzymes
- Learn the basic concepts on mechanisms regulating intracellular proteolytic systems
- Understand how information on protease structure–function relationships has allowed to define inhibitory mechanisms, leading to the design of new drugs
- Learn how inhibitors of intracellular proteases can be exploited for clinical use

---

GENERAL CHARACTERISTICS OF PROTEASES

Proteases are enzymes present in all tissues that catalyze the hydrolysis of peptide bonds. Besides determining protein fate, in many instances, their activity is involved in important cell functions through activation or inactivation of enzymatic proteins. Thus, although excessive protease activity underlies many pathologies as well as cell death, proteolysis is indispensable for cell life, as it contributes to maintain the balance between protein synthesis and degradation and participates in cell response to physiologic and pathologic stimuli.

Complete degradation of proteins to their constituent amino acids occurs mostly within cells and, in physiological conditions, is counterbalanced by protein synthesis. However, besides complete proteolysis, many examples of partial proteolysis exist, both intracellular and extracellular. In these cases, proteases act chiefly at specific sites, cleaving target proteins in one or more peptides. Therefore, this type of proteolysis does not counterbalance protein synthesis, but rather, it proceeds in parallel with it and can be regarded as a mechanism of posttranslational modification. However, in contrast with other covalent modifications of proteins (e.g., phosphorylation), proteolysis is irreversible: the only way to reconstitute degraded proteins is to synthesize them again. Proteolysis is therefore at the basis of irreversible processes such as coagulation, apoptosis, and production of cytokines, hormones, and neuropeptides.

Classification

In all proteases, hydrolysis of the peptide bond is based on a general acid–base reaction that starts with the attack of a nucleophilic group of the enzyme active site onto the carbon atom of the peptide carbonyl group (see Supplement E 30.1, "Protease Classification and Nomenclature"). Based on the strategy employed in the nucleophilic attack, proteases are classified into four main classes: serine protease, cysteine protease, aspartate protease, and metalloprotease (in most cases Zn^{2+}). In the last two classes, the aspartate residue or the metal ion polarizes a water molecule, making the oxygen atom nucleophilic. By contrast, in the other protease classes, the nucleophilic group is represented by the serine hydroxyl moiety in the form of oxyanion or by the cysteine sulfhydryl moiety in the form of thiolate.

Proteases are assigned to the different classes using specific inhibitors (see Table 30.1). The search for protease inhibitors is constantly ongoing. Indeed, besides their possible pharmacological applications, inhibitors are essential to distinguish and classify different proteases, to isolate and purify them through affinity chromatography, and to localize them by autoradiography when specific antibodies are not available for immunohistochemistry.

TABLE 30.1 Inhibitors used for protease characterization

Catalytic residue	Inhibitors	Limits or disadvantages
Serine	3,4-Dichloroisocoumarin (3,4-DCI)	Does not inhibit all serine proteases
	Diisopropyl fluorophosphate (DFP)	Inhibits also cysteine proteases and acetylcholinesterase
	Phenylmethanesulfonylfluoride (PMSF)	Inhibits also cysteine proteases; slow reactivity
Cysteine	E64	
	Alkylating –SH agents	React with all thiols
	Diazomethanes	Inhibit some serine proteases
	Leupeptin	Inhibits many serine proteases
Aspartate	Pepstatin	
Metals (Zn^{2+})	1,10-Phenanthroline	

TABLE 30.2 Intracellular distribution and main representatives of different protease classes

Catalytic residue	Localization	Main enzymes	Endogenous inhibitors
Serine	Lysosomes	Cathepsin A	Serpins
	Endoplasmic reticulum	Signal peptidases	
	Cytosol	Proteasome	
	Secretory vesicles	Maturation peptidases	
	Granules	Granzymes	
Cysteine	Lysosomes	Cathepsins	Cystatins
	Cytosol	Calpain	Calpastatin
		Caspases	XIAP
		Bleomycin hydrolase	
		Deubiquitinase	
Aspartate	Lysosomes	Cathepsins	
	Cytosol	Retropepsins	
	Endoplasmic reticulum and Golgi	Carboxypeptidases	
	Mitochondria	Signal peptidases	
	Plasma membrane (ectoenzymes)	Convertases	

In serine, cysteine, and aspartate proteases, the molecule is generally folded in two globular domains separated by a groove containing the active site. Protease specificity is never absolute and depends on the interaction between residues proximal to the peptide bond to be hydrolyzed and a portion of the active site, named binding or positioning site. For example, the presence of either charged or hydrophobic residues in that region of the substrate protein determines the different specificities of trypsin and chymotrypsin, respectively. Table 30.2 provides a summary of the intracellular localization of the main enzymes belonging to the different protease classes.

Serine Proteases Most serine proteases are extracellular. Intracellular serine proteases include proteases that catalyze the removal of the signal peptide (signal proteases) from proteins entering the endoplasmic reticulum (ER) (see later). In addition, one separate family is represented by the multicatalytic endopeptidase complex called proteasome (see later).

Cysteine Proteases This group includes mainly intracellular proteases. The catalytic cysteine residue forms an ionic couple with a proximal histidine residue, which is thereby essential for the catalytic activity. The ionic couple affects the ionization degree of both residues causing deprotonation of the cysteine thiol that becomes a strong nucleophilic group. In cysteine proteases, another conserved amino acid is an asparagine residue that orients the histidine imidazole ring. In this way, a catalytic triad is formed, similar to that of serine proteases.

Papain, isolated from papaya, has been the first cysteine protease to be characterized and represents the prototype of the most abundant cysteine protease family. The prototypical member of the other family is the interleukin-1β-converting enzyme (ICE).

The papain family includes mainly proteases segregated into lysosomes. In the cytoplasm, other papain-like proteins are found, but different for their higher structural complexity. Besides calpain, special examples are the deubiquitinases, which catalyze the hydrolysis of the isopeptide bond between ubiquitin and proteins, and the bleomycin hydrolase, which is involved in detoxification of this antibiotic.

ICE family includes proteases with a marked specificity for peptide bonds with aspartate in P_1. From this characteristic, shared only by granzyme B, an enzyme present in granulocytes, derives the term caspase (*c*ysteinyl *asp*artate-specific prote*ases*), which has replaced the previous nomenclature. Thirteen proteases belonging to this family have been identified and progressively numbered according to the order in which they were discovered.

Aspartate Proteases This class includes the "acidic proteases" already known at the beginning of the twentieth century, such as pepsin, secreted in the digestive tract, or the lysosomal proteases cathepsins D and E.

Resolution of their three-dimensional structure has shown two lobes connected by a short intervening sequence (<10 residues). Once bound the substrate, the active site closes up, sequestering the substrate and one water molecule essential for catalysis. Each lobe provides one of the two aspartate residues that polarize the water molecule during catalysis.

Aspartate proteases expressed in cells infected by some retroviruses, such as the human immunodeficiency virus (HIV), are included in the retropepsin family. They have bilobate morphology but act as homodimers.

Metalloproteases This is the most heterogeneous protease class and is subdivided in more than 30 families. The most represented metallic ion is Zn^{2+}. One superfamily of these proteases, specifically inhibited by phosphoramidon, includes the neprilysin family. This is constituted by ectoenzymes such as neprilysin (also called enkephalinase or *n*eutral *e*ndo*p*eptidase (NEP)), which degrades and inactivates several biologically active peptides, and the endothelin-converting enzyme (ECE), which generates the vasoactive peptide endothelin from the precursor big endothelin-1. From a structural point of view, this family includes also the Kell glycoproteins that are found on the erythrocyte membrane and contain the 24 antigenic determinants of the blood groups. Although these proteins exhibit the same structure of the neprilysin catalytic site, they do not have catalytic activity.

Other metalloproteases are the plasma membrane peptidases of the intestinal brush border and the ACE that converts angiotensin I in the prohypertensive peptide angiotensin II.

The phosphoramidon-insensitive group of metalloproteases includes endopeptidases that degrade the extracellular matrix (e.g., collagenases), intracellular enzymes such as oligopeptidases and carboxypeptidases required for maturation of biologically active peptides and hormones, and signal peptidases localized in mitochondria.

Some metalloproteases are introduced in cells by infective agents. This is the case of tetanus and botulinum toxins, activated via proteolysis in the cytoplasm of the host cell (see Chapter 37). The enzyme produced by *Leishmania*, leishmanolysin, localizes in the lysosomes and represents an example of adaptation to the environment, being one of the very few metalloproteases active at acidic pH.

CHARACTERISTICS AND REGULATION OF INTRACELLULAR PROTEOLYSIS

In contrast with the extracellular systems, the two intracellular proteolytic processes require ATP. Lysosomal protein degradation requires ATP for acidification of the intravesicular milieu through the lysosome proton pump, whereas the proteasomal degradation pathway needs ATP for polyubiquitination and subsequent demolition of the substrate. The exergonic ATP hydrolysis, coupled to proteolysis, further promotes completion of the process, which is already thermodynamically favored. Another difference from the extracellular processes is the hormonal control of some intracellular proteolytic processes (such as autophagy; see Supplement E 30.2, "Autophagy").

In general, proteolysis, and in particular intracellular proteolysis, is under tight spatiotemporal control to prevent continued and indiscriminate protein degradation that would shortly lead to tissue destruction.

Several factors contribute to regulation of intracellular proteolysis:

- Compartmentalization. It is the main measure to protect cell structures. Moreover, compartmentalization of more enzymes participating in the same metabolic process facilitates substrate interchange and significantly increases the overall speed of the process. In this respect, particularly important are posttranslational modifications of proteins, which determine their targeting and activation.
- Zymogen conversion. In general, proteases are synthesized as inactive precursors (zymogens) subsequently activated by proteolysis. Being a limiting step in proteolytic processes, this event is highly controlled. In fact, in general, the zymogen-active enzyme conversion occurs under specific conditions (e.g., acidic pH) or in response to specific intracellular or extracellular signals.
- pH. Several proteases are poorly active at neutral pH, as they are optimized to work in acidic compartments (lysosomes and endosomes). The low pH typical of these compartments determines partial denaturation of substrates, favoring proteolytic activity.
- Redox state. In cysteine proteases, the thiol group is easily oxidable. For this reason, these enzymes require a reductive environment; in the endosomes, this is maintained via cysteine accumulation.
- Inhibitors. The activity of different intracellular proteases and cytoplasmic proteases in particular is controlled by the balance with inhibitor peptides, which in general are in excess over the enzyme molecules. This is undoubtedly the most selective mechanism used by cells to regulate intracellular proteases. Several proteic

inhibitors of serine proteases are known, which are grouped in the serpin family (*serine protease inhibitors*), mainly localized in the extracellular space. However, they are also present within cells of different tissues (kidney, endocrine pancreas, some cerebral areas, small intestine) even if their targets are not known and their physiological role is still unclear. The most widespread proteic inhibitors of cysteine proteases belong to the cystatin group, which is subdivided into three families. The oligomeric proteins, which compose the family 1, are found in the cytoplasm, whereas the members of the other families are extracellular. Reversible competitive inhibition concerns mainly papain-like proteases. Calpain activity is specifically inhibited by calpastatin, whereas caspases are inhibited by both endogenous and exogenous proteins introduced by infective agents.

Besides protein catabolism, partial degradation of intracellular proteins has different biochemical and physiological roles: it may serve to convert zymogens into active enzymes but also to generate new enzymes or peptides with different and specialized functions as compared to the native proteins. This peculiar characteristic of intracellular proteolysis is well exemplified by some signal transduction processes described in detail in the next paragraphs. In general, proteases can participate in signal transduction processes in three ways:

1. By activating specific signal transduction pathways. Following increase in a second messenger (e.g., Ca^{2+}) or interaction with specific proteic effectors, the intracellular proteolytic cascade is activated, which in turn activates another cascade of enzymatic reactions leading to the cellular response.
2. By modulating receptor activity. Proteolysis of transmembrane receptor intracellular domain has functional consequences that, depending on the case, may lead to either activation or repression of specific receptor functions.
3. By regulating receptor number. Once internalized via endocytosis, membrane receptors can be recycled to the membrane or degraded in the lysosomes. The number of receptors on the cell surface depends on the balance between internalization/degradation and synthesis/membrane insertion processes (see Chapter 9).

Function and Pharmacological Modulation of the Main Intracellular Proteolytic Systems

In cells, complete protein degradation is carried out by lysosomal enzymes and proteasome. The former are mostly involved in the removal of both extracellular and intracellular proteins collected within vesicles, whereas the latter degrade part of the intracellular proteins. Since also proteasome can be considered structurally similar to an intracellular compartment (see later), sequestration of proteins to be degraded represents not only a control mechanism but also a significant difference from extracellular proteolysis.

To better define the role of different proteolytic systems and the opportunities of pharmacological intervention, we will group proteases based on their main cellular functions rather than on their catalytic mechanisms. It is worth emphasizing that poor specificity of inhibitors and difficulty to block only harmful protease functions limit the number of drugs targeting intracellular proteases available for clinical practice. As described in detail in the next paragraphs, therapeutic interventions of this type at the moment concern only the proteasome and the metalloprotease synthesized upon HIV infection. The side effects caused by inhibitors of this metalloprotease highlight how difficult it is to obtain beneficial effects in cells by inhibiting proteases responsible of pathological events.

LYSOSOMAL PROTEASES

Most intracellular proteases are removed by the proteolytic activity of the ubiquitin–proteasome system in the cytoplasm (described later) and by several proteases within lysosomes. Lysosomal proteolysis is mostly aspecific and directed against both exogenous and endogenous proteins. Proteins to be degraded in lysosomes enter these organelles via vesicles and transport systems (Fig. 30.1). These systems are collectively described with the terms heterophagy, in case of proteins coming from the extracellular environment, and autophagy for intracellular proteins.

Extracellular proteins are taken up by cells via receptor-mediated endocytosis, pinocytosis, and phagocytosis. Vesicles formed by invagination of the plasma membrane are named heterophagic vacuoles and fuse with primary lysosomes. Intracellular proteins can be taken up by lysosomes in either a nonselective or a selective way. Nonselective uptake occurs through budding of vesicles from the rough endoplasmic reticulum (RER) containing small cytoplasm portions, which may include also organelles. Degradation of the content of these vesicles, named autophagic vacuoles, occurs in lysosomes and ends up with the release of single amino acids that can be reused for protein synthesis or gluconeogenesis. This process, named macrophagy, is strictly correlated with the nutritional state, being activated by fasting (via glucagon) and suppressed by insulin. In addition, autophagy depends on ATP availability, in a still unclear way. Vesicles of smaller size are directly generated from lysosomes during the process called microautophagy. Besides these two nonselective processes, which degrade

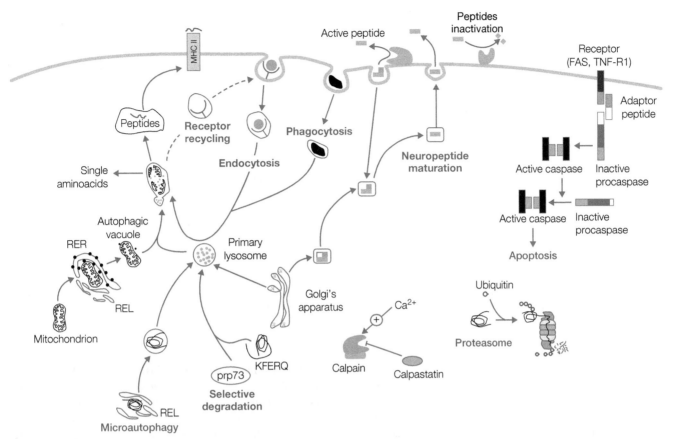

FIGURE 30.1 Schematic summary of the different intracellular proteolytic processes. On the left, the role of lysosomes in intracellular digestive processes is shown. Vesicles originated from plasma membrane or endoplasmic reticulum carry, respectively, extracellular and intracellular components to the lysosomes. Besides this type of vesicle-mediated unspecific degradation, some proteins containing the Lys-Phe-Glu-Arg-Gln (KFERQ) sequence can be specifically bound by a 73 kDa recognition protein (prp73) that transfers them to the lysosomes. Complete degradation of proteins occurring inside the lysosomes releases amino acids that become available for cellular metabolism. Conversely, partial protein degradation generates peptides that are exposed on the cell surface in combination with class II MHC proteins. On the right side, other enzymes involved in intracellular proteolysis are depicted, such as peptide maturation proteases, caspases, proteasome, and calpain, which are illustrated in more detail in the next figures. REL, smooth endoplasmic reticulum; RER, rough endoplasmic reticulum.

several proteins at the same time, autophagy includes also a mechanism mediated by some heat shock proteins (in particular HSC70 and HSP90). This process, called chaperone-mediated autophagy (CMA), is selective and involves only one protein at a time (see Supplement E 30.2).

Types and Families of Lysosomal Enzymes

Lysosomal proteases are collectively indicated as cathepsins, a term deriving from the fusion of the words pepsin and catabolism, which also refers to the intracellular localization of these enzymes. With the recent identification of new members, 11 cathepsins are now known in humans (cathepsins B, H, L, S, C, K, O, F, V, X, and W); these may be endopeptidases or exopeptidases. The most abundant are cathepsin D, an aspartate protease, and cathepsins B and L, which are both cysteine proteases. Among the exopeptidases, the most important are cathepsin A, a serine protease, and the lysosomal carboxypeptidase B, a cysteine protease. The presence of different types of proteases ensures the degradation of all sorts of proteins. Moreover, cooperation exists between endopeptidases, which generate peptides of more than eight amino acids, and exopeptidases, which complete the degradation. The function of endopeptidases is mainly to increase the number of attack sites for aminopeptidases and carboxypeptidases.

Lysosomal proteases are concentrated in lysosomes by a selection mechanism taking place in the *trans*-Golgi that recognizes a specific posttranslational modification: the presence of mannose-6-phosphate residues. A rare congenital disease, named mucolipidosis II, has been described that affects the mannose phosphorylation system and is characterized by a reduced lysosomal content in acid hydrolases.

Besides glycosylation, other posttranslational modifications are necessary to convert lysosomal proteases to their active forms. In fact, they are synthesized as inactive preproenzymes: the signal sequence, indicated by the "pre" prefix, is required to target the neosynthesized protein to the ER, where the signal sequence is cleaved off. Once in the lysosomes, proteolytic cleavage of the inactivating sequence, indicated by the "pro" prefix, activates the enzyme.

Lysosomal ATPase proton pump maintains the intravesicular pH around 5, which corresponds to the optimal pH for lysosomal enzyme activity. Cells are protected from accidental leakage of lysosomal components, since cathepsins are generally rapidly inactivated by neutral pH. However, there are exceptions to this rule, as demonstrated by recent studies on extralysosomal activity of cathepsins. One particular case is cathepsin A, which acts as carboxypeptidase, esterase, and deaminase. Moreover, by binding β-galactosidase and neuraminidase in lysosomes, it protects them from the proteolytic action of other lysosomal enzymes. Cathepsin A deficiency results in an accumulation syndrome called galactosialidosis, which is due to the uncontrolled degradative activity of the two hydrolases.

Intra- and Extracellular Effects of Cathepsins Cathepsins, and in particular cysteinic cathepsins, once believed to be involved only in nonselective lysosomal degradation of proteins, today are ascribed specific roles, both extracellular and intracellular and in both physiological and pathological conditions. Cathepsin S is the main cysteine protease involved in exposure of MHC class II-bound antigens (see Chapter 50). Cathepsin K, another cysteine protease, is particularly expressed in osteoclasts. It is the most potent elastolytic enzyme, and it is believed to play a significant role in extracellular matrix remodeling; for this reason, its inhibition has been proposed as a possible therapeutic approach to osteoporosis. Recent studies indicate the involvement of several cysteine cathepsins in apoptosis. Such proteases, released in the cytoplasm after the initial insult (in particular oxidative stress), are responsible for cleavage of the proapoptotic protein Bid, resulting in its binding to mitochondria, cytochrome c release, and sequential activation of caspase-9 and caspase-3. A role of cathepsins in programmed cell death is suggested also by experimental evidence showing an involvement of soluble proteolytic systems, such as the Ca^{2+}-dependent proteolytic system, in rupture/permeabilization of the lysosome membrane (hypothesis of cooperation between cathepsins and calpains). This protease cascade has been described in simple organisms but also in humans.

Activity, expression, and localization of different members of the cathepsin family are modified in several tumors. In particular, increase in expression and activity of cathepsins B and L invariably correlates with malignant tumor progression; in addition, also cathepsin localization may change in tumors, being no longer restricted to lysosomes but extending to plasma membrane and extracellular space. The presence of cathepsins, released from tumoral cells, in the extracellular space may favor the action of compounds poorly permeable through the plasma membrane, thus contributing to their direct action on the tumoral mass.

Furthermore, increased expression of cathepsin K has been correlated with osteoporosis, whereas hyperactivity of cathepsin S has been associated with rheumatoid arthritis.

Pharmacological Control of Cathepsins Several inhibitors of cathepsins, and in particular cathepsin K, are currently under phase I and II clinical trials. Table 30.3 summarizes the observations correlating different cathepsins with human pathologies. Given the lack of complete specificity, the variety of isoforms, and their abundance and cooperation, it is difficult to devise therapeutic strategies based on inhibition or activation of specific proteases. However, the lysosomal proteolytic activity can be modified by agents able to penetrate these organelles and to alter their internal conditions. As a result, the following effects may be observed: (i) increase in lysosomal pH caused by basic compounds and consequent inactivation of lysosomal enzymes; (ii) protease inactivation, to a different extent, followed by accumulation of undegraded proteins; (iii) lysosome rupture and enzyme release, with consequent loss of cellular integrity; and (iv) inactivation of the agent by chemical modification or entrapment. Such conditions may represent side effects of drugs or contribute to their toxicity.

TABLE 30.3 Involvement of cathepsins in human pathologies

Cathepsin	Distribution	Pathology
B	Ubiquitous	Inflammation, Alzheimer's disease, cancer
C	Ubiquitous	Inflammation
D	Ubiquitous	Inflammation, Alzheimer's disease, cancer, rheumatoid arthritis
E	Immune system	Dermatitis
F	Ubiquitous	Inflammation
G	Neutrophils, B lymphocytes	Inflammation
H	Ubiquitous	Cancer
K	Epithelia	Rheumatoid arthritis
L	Ubiquitous	Thymus pathologies, atherosclerosis, rheumatoid arthritis, cancer
S	Antigen presenting cells	Arthritis, atherosclerosis, bronchial asthma, psoriases, cancer
V	Ubiquitous	Thymus pathologies, cancer
W	T lymphocytes and NK cells	Autoimmune atrophic gastritis
X	Ubiquitous	Cancer

Given the extensive involvement of cathepsins in human pathologies, the development of new tools to control their activity has attracted the interest of pharmacologists. One of the most recently devised strategies consists in the identification of common features in tertiary structure or in substrate binding specificity among cathepsins through X-ray investigations. A second strategy is aimed at the characterization of new positive modulators of the lysosomal proteolytic system to favor elimination of protein aggregates that accompany neurodegenerative diseases such as Alzheimer's and frontotemporal dementia.

COMPARTMENTALIZED PROTEASES WITH SPECIFIC FUNCTIONS

Signal Proteases

In the ER, these enzymes have oligomeric structure, being formed by a variable number of subunits. From dog's microsomes, a five-subunit complex, called signal peptidase complex (Spc), has been purified; its subunits have a molecular weight ranging between 12 and 25 kDa. Only one of the subunits is catalytically active, whereas the others probably regulate the process of protein translocation across the membrane. The ER signal peptidases, although members of the serine protease group, exhibit two peculiar characteristics: they are inhibited by the common inhibitors used to distinguish different protease classes and their catalytic mechanism does not require histidine. The proton acceptor is probably a highly conserved aspartate residue, in analogy with the glutamate residue that exerts the same function in beta-lactamase. Both in prokaryotes and eukaryotes, signal proteases show a marked preference for small hydrophobic amino acid in P_1 and P_3. Proteolysis is favored by the presence of a lysine in P_1', whereas peptides with a proline residue in the same position act as competitive inhibitors.

In mitochondria, at least three different types of signal proteases are present. The signal peptide of proteins synthesized in the cytoplasm and transported to mitochondria is removed by the mitochondrial processing peptidase (MPP). In some proteins, an additional octapeptide is cleaved off by the mitochondrial intermediate protease (MIT), belonging to the metalloprotease family. These enzymes are metalloproteases in contrast with the mitochondrial signal peptidase, which is a serine protease and acts on proteins synthesized by mitochondria.

Maturation Proteases

Maturation proteases convert prepropeptides into bioactive peptides. Peptides are synthesized as inactive prepropeptides (Fig. 30.2). The "pre" prefix indicates the presence of a signal peptide at the peptide amino-terminus, whereas the "pro" prefix indicates the presence of an inhibitory amino acid stretch that has to be removed for the peptide to be active. Signal proteases present in RER are in charge of the first step of the maturation process; this process is not specific and similar for all soluble proteins that enter this organelle. The propeptide released in the lumen of the reticulum and subsequently transferred to the Golgi apparatus is stored within secretory vesicles in the *trans*-Golgi cisternae along with proteases that terminate the maturation process. Inside the secretory vesicles that bud from the *trans*-Golgi, the second phase of the maturation process occurs; this process is specific, requires an acidic pH, and is generally catalyzed by the sequential action of two proteases. First, the "pro" sequence is cleaved off by trypsin-like proteases,

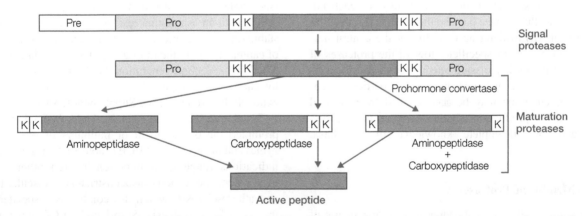

FIGURE 30.2 Proteolytic maturation of biologically active peptides. Peptides such as neuropeptides, hormones, and growth factors are synthesized as inactive prepropeptides. Signal peptidases of the endoplasmic reticulum cleave off the targeting signal (the "pre" sequence). In the secretory vesicles, the propeptide is processed by the sequential action of two proteases, releasing the active peptide. The first enzyme is a trypsin-like protease that cleaves the propeptide recognizing a couple of basic amino acids. The basic residues are then removed by either aminopeptidases or carboxypeptidases depending on whether they are attached, respectively, to the amino- or the carboxy-terminus of the peptide.

which recognize, as a signal, a couple of basic residues (Arg-Arg or Lys-Arg).

A novel group of proteases capable of specifically processing prohormones has been identified. These enzymes, named prohormone convertases, are synthesized as zymogens and require Ca^{2+} ions for their activity. The main examples of this protease class are PC1 and PC2, which have been only identified in endocrine and neuroendocrine tissues, and furin, which is almost ubiquitous and may serve as convertase of PC1 and PC2.

In case one or both basic amino acids of the couple are still present in the peptide C-terminus after the cleavage, they are removed by metalloexopeptidases, the most important being carboxypeptidase H (also called carboxypeptidase R or enkephalin convertase), which is particularly abundant in the anterior pituitary. In case the basic residues are located at the N-terminus, they are removed by aminopeptidases. Sometimes, peptide maturation is terminated extracellularly by other metalloproteases secreted and localized on the plasma membrane (see later). In other instances (e.g., endothelin convertase), isoforms are found both in the extracellular space and within the *trans*-Golgi; in this way, the peptide can be secreted either as inactive precursor or in its active form.

Often, from a single precursor molecule, multiple active peptides are produced, exerting different functions. Moreover, peptide maturation is tissue specific, meaning that from the same precursor, different mature products may originate depending on the tissue. These properties are exemplified by POMC (see Chapters 45 and 46).

Several different inhibitors of the intracellular peptide maturation pathway are currently being investigated, which will be used to characterize and distinguish different proteases involved in the process (see Table 30.4). The development of inhibitors of peptide intracellular maturation is hampered by several factors: (i) difficulty to reach not only the intracellular space but also the specific intracellular compartment, (ii) difficult accessibility of the central nervous system, (iii) poor specificity toward the proteases of interest, (iv) possibility of side effects due to tissue specificity (i.e., inhibition of the production of a particular peptide in a given tissue may be accompanied by undesired blockade of the production of other peptides in other tissues), and (v) incomplete knowledge of the basic mechanisms.

Plasma Membrane Proteases

Proteases are involved in both synthesis and degradation of neuropeptides, with the ensuing termination of their signaling. In this case, the enzymes involved are metalloproteases localized on the external plasma membrane surface (ectoenzymes). The best characterized enzyme of this group is neprilysin. Neprilysin is a homodimer, in which each subunit (90–100 kDa) contains a short cytoplasmic N-terminal domain, linked by a 22-amino-acid transmembrane stretch to a large extracellular portion, highly glycosylated and containing the catalytic site. Neprilysin is present in several tissues (ventral and peripheral nervous system, kidney, small intestine, adrenal cortex, ovary, testis, lymphocytes, and neutrophils) and can degrade, besides different opioid peptides, also substance P, bradykinin, natriuretic peptides, and the chemotactic peptide fMet-Leu-Phe. This property suggests the involvement of neprilysin in the modulation of the inflammatory response. In fact, transgenic mice lacking the neprilysin-encoding gene are 100 times more susceptible to death by endotoxic shock.

Clinical indication for thiorphan, a neprilysin inhibitor (see Table 30.4), is limited to diarrhea syndrome, given the potent effect exerted by opioid peptides on intestinal peristalsis. The limited extension of the indications is due to the lack of selectivity of the available inhibitors, which gives rise to a wide range of side effects as a result of a more or less marked inhibition of ACE and EDE (see later). The most important examples of propeptides secreted and activated by metalloproteases localized on the external cell surface are endothelin and angiotensin II convertases. The angiotensin II convertase, although not specific since it can also act as carboxypeptidase (its correct name is in fact peptidyl-dipeptidase A) on bradykinin and substance P, is the target of many drugs, called ACE inhibitors, which are extensively used in hypertension treatment (see Supplement E 30.3, "Inhibitors of Angiotensin-Converting Enzymes and Their Action in Cardiovascular Pathologies"). The first compound to be used in clinics was captopril (D-3-mercapto-2-methyl-propionyl-L-proline) in which the sulfhydryl group serves as ligand for Zn^{2+}, increasing the inhibitory effect of the alanine carbonyl group in the dipeptide Ala-Pro, which is present in the venoms of several snakes, from which the drug is derived (see Table 30.4). The proline carboxyl group forms a salt bridge with an asparagine residue in the catalytic site, thus stabilizing the enzyme–inhibitor interaction. The side effects of captopril, due to the thiol group, have been largely eliminated in the second-generation ACE inhibitors, constituted by carboxyalkyl dipeptides. For further details on these extracellular proteases, see Supplement E 30.3.

The therapeutic success obtained with ACE inhibitors has prompted the development of inhibitors of other ectoenzymes and in particular of ECE (Table 30.4). The therapeutic indications range from hypertension (endothelin is the most potent endogenous vasoconstrictor) to vascular lesions and atherosclerosis, which are conditions associated with increased ECE synthesis. Surprisingly, ECE hyperproduction accompanies also the evolution of prostatic carcinoma; for this reason, inhibition of endothelin and ECE receptors has been proposed as a potential therapy for this tumor. The main limitations in the search of effective ECE antagonists are the lack of selectivity on neprilysin and the fact that

TABLE 30.4 Experimental and clinical relevance of the main inhibitors of intracellular proteases and plasma membrane extracellular proteases

Proteases	Experimental inhibitors with possible clinical development	Possible indications	Inhibitors in clinical use	Clinical indications
Proteasome	Lactacystin, MG 132, MG115	Inflammation, oncology	Bortezomib, Carfilzomib, PR-047, NPI-0052, MLN9708	Multiple myeloma
Signal peptidases	Peptidyl CMK (furin); acetyl hexapeptides (PC1 and 2)	Neuropetide production; viral infections (furin)		
Caspases	Ac-DEVD-CHO, Z-VAD-DCB, Ac-YVAD-CHN$_2$	Apoptosis, degenerative diseases, ischemic syndromes, inflammation	Pralnacasan (ICE or caspase 1 inhibitor)	Rheumatoid arthritis
Calpain	Peptidyl halomethanes and diazomethanes, peptidyl aldehydes, α-mercapto-acrylate derivatives (PD150606)	Ischaemic damage, neurodegeneration, muscular dystrophy, cataract		
Cysteine proteases of protozoan parasites	Z-Phe-Ala-FMK	Infections by *E. histolytica*, *P. falciparum*, *T. cruzi*, etc.		
HIV protease			Indinavir, nelfinavir ritonavir, saquinavir, amprenavir, lopinavir	AIDS
Neprilysin	Phosphoramidon, acetorphan, kelatorphan	Hypertension, pain	Thiorphan	Diarrhea syndromes
ECE	FR901533, aminophosphonic acid derivatives (CGS31477), phosphoramidon	Atherosclerosis, hypertension, prostate cancer		
ACE			Captopril, enalapril	Hypertension

Proteases are listed according to the class they belong and the order of appearance in the text.

endothelin is activated inside the cells, an environment that, as we said previously, poses additional difficulties.

CYTOPLASMIC PROTEASES

The Proteasome

Many cytosolic proteins and several membrane proteins are degraded by a specific multiprotein complex called proteasome. The proteasome is located in cytoplasm and nucleus and is partly associated with ER and cytoskeleton. One of the important characteristics of this degradation pathway is that the proteins to be eliminated are tagged by multiple copies of a peptide, called ubiquitin, added to specific lysines. Both the ubiquitination signals and the enzymatic steps responsible for this important posttranslational modification are targeted by drugs whose effects influence not only the destiny of some proteins but also different and complex cellular functions. Proteasome system and ubiquitination process are described in detail in Supplement E 30.4, "Ubiquitin and Proteasome." Besides the aspecific removal of altered proteins, the specific degradation of some proteins carried out by the proteasome is necessary for several cellular functions, the most relevant being:

1. Cell cycle. Entry into the M phase depends on cyclin B degradation, which is ubiquitinated near the N-terminal residue by a specific E3 ligase. Such enzyme is activated by phosphorylation at the end of a cascade of phosphorylation events related to the activity of the cdc2 kinase (cdc stands for cell division cycle).
2. Endoplasmic reticulum-associated degradation (ERAD). The 26S proteasome participates in the quality control mechanism that takes place in the ER during post-translational maturation of proteins. Improperly folded proteins are removed and degraded by the ubiquitin–proteasome 26S system. This occurs in all cells, but it may have pathological consequences as in the case of cystic fibrosis, a condition in which a mutation in the *CFTR* gene leads to improper folding of its protein product (a chloride channel) favoring its degradation. For this reason, the ERAD system has been considered as a possible target for new therapeutic approaches to human pathologies.

3. Immune response. The peptides generated by the proteasome are transported in the ER where they bind MHC class I proteins, to be then exposed on the cell surface (see Chapter 50). Proteasome inhibition prevents exposure of antigenic peptides, indicating that these are mainly produced by this proteolytic system.

4. Signal transduction. Studies have shown that, besides the complete degradation of substrates, the proteasome can catalyze also the limited proteolysis of some proteins involved in signal transduction, such as p105, an inactive precursor of transcription factor NF-κB. The proteasome degrades only the carboxy-terminal portion of p105, releasing the mature form, called p50, which can dimerize with other cytoplasmic proteins (c-Rel, RelB, and p65), thus forming different transcription activator complexes. These heterodimers are maintained in inactive form in the cytoplasm by the binding of an inhibitory protein, named IκBα. Signals generated during inflammatory processes (e.g., stimulation by TNFα) promote ubiquitination and partial proteasomal degradation of p105 and complete degradation of IκBα. Removal of IκBα and processing of p105 to p50 allow NF-κB translocation to the nucleus, a necessary event for its transcriptional activity (see Chapter 22).

5. Protein synthesis. Some transcription factors, such as the products of *c-Fos* and *c-Jun* proto-oncogenes, are very short-lived proteins that are degraded upon ubiquitination.

6. Tumor development. Ubiquitination and degradation of p53, a protein that opposes tumor formation favoring cell apoptosis, are significantly stimulated by binding of the E6 peptide of human papillomavirus (HPV), which is correlated with the uterine cervix carcinoma.

The relationship between ubiquitinated proteins and proteasome, although prevalent, is not compulsory (see Supplement E 30.4). In fact, some proteins are degraded by the proteasome even if not bound to ubiquitin. The best example is ornithine decarboxylase (ODC), the key enzyme in polyamine synthesis.

PROTEASOME INHIBITORS AS ANTI-INFLAMMATORY AND ANTITUMORAL DRUGS

The development of proteasome inhibitors has allowed first to clarify its functions and relationships with other intracellular proteolytic system and then to set up new therapeutic approaches. Lactacystin, a peptide isolated from actinomycetes, is the most specific proteasome inhibitor commercially available (see Table 30.4) and is widely used in laboratory experiments. By reacting with the threonine catalytic residues, it inhibits by 90% intracellular proteolysis without affecting lysosomal proteases.

One of the main purposes of the clinical development of proteasome inhibitors is to contrast the formation of the NF-κB p50 subunit. Such strategy is potentially important in the treatment not only of inflammatory processes but also of neoplastic pathologies. In this case, the aim is to prevent NF-κB antiapoptotic activity, which is related to its ability to enhance expression of antiapoptotic proteins, such as Bcl-2 and inhibitor of apoptosis (IAP). Bortezomib (also known as PS-341 or Velcade) is the most clinically effective proteasome inhibitor; it is a dipeptide derivative of boronic acid. This compound reversibly inhibits chymotrypsin-like activity, and thereby it inhibits the entire catalytic process of proteasome, by interacting with the threonine catalytic residue. Its specificity is related to boron ability to interact with the threonine hydroxylic oxygen, but not with the thiol sulfur of the cysteine residues present, for example, in several lysosomal cathepsins. Therefore, the introduction of peptidic derivatives of boronic acid, previously developed as inhibitors of serine proteases, has solved the specificity problems posed by peptidic aldehydes such as MG132 (CBZ-leu-leu-leucinal) and MG115 (CBZ-leu-leu-norvalinal).

On this basis, bortezomib has been developed, which has proved clinically effective in multiple myeloma, a pathology associated with NF-κB constitutive activation. The therapeutic success has allowed its very rapid approval by the FDA. However, new inhibitors have been introduced in clinical experimentation to overcome bortezomib limitations, which are poor efficacy in solid tumors, intravenous administration, and toxic effects on the heart and nervous system. In these new inhibitors, besides the variations introduced in the peptidic structure, the reactive group of boronic acid has been replaced with epoxy ketones, aldehydes, or vinyl sulfones. Some of these compounds, such as PR-047 and MLN9078 in phase I clinical trial, can be orally administered and seem to be effective also in solid tumors.

Other Serine Proteases for Protein Quality Control

Protein quality control involves not only proteasome and autophagy but also the serine protease family A named high-temperature requirement A (HTRA). Although exhibiting a trypsin-like activity, compared to trypsin, these proteases have a more complex structure as their catalytic domain is flanked by one or more postsynaptic density of 95 kDa, discs large, and zonula occludens (PDZ) domains, which not only mediate interaction with other proteins including substrates but also allosterically modulate enzymatic catalysis. Another peculiar characteristic of HTRAs is the oligomerization that is favored by substrate binding and results in increased activity. These proteases convert from a monomeric form with high substrate binding activity to a polymeric form with enhanced catalytic activity. HTRAs are expressed also in bacteria and plants; in humans, four isoforms (HTRA 1–4) have been identified, 80% localized in the extracellular space where their action on extracellular matrix proteins suggests their probable involvement in inflammation and joint disorders.

Genetic studies have allowed clarifying HTRA role as tumor suppressors. Moreover, HTRA1 has been found to favor cell proliferation by promoting insulin growth factor (IGF) release and inactivating transforming growth factor β (TGFβ). Inside the cells, HTRA2 has been clearly shown to promote autophagic removal of altered mitochondria. Loss of HTRA activity has been related with several diseases, such as arthritis, cancer, macular degeneration, and Alzheimer's and Parkinson's diseases. However, their relationships with proteasome and autophagy in the regulation of protein degradation in these pathological processes are still unclear. Moreover, the lack of specific inhibitors has not allowed the clinical investigations that, for the proteasome, have been allowed by bortezomib and its derivatives.

Caspases: Initiators and Executors of Apoptosis

Caspases (the name is a short form for cysteinyl aspartate-specific proteases) are endopeptidases synthesized as inactive proenzymes and converted by proteolysis into tetramers formed by a couple of heterodimers, each one containing a catalytic and a regulatory subunit (Fig. 30.3). Their peptidase activity requires the presence of both subunits that exclusively catalyze reactions of limited proteolysis, such as those required for cytokine activation or involved in programmed cell death (apoptosis). Initially identified as fundamental enzymes for the proteolytic maturation of some interleukins (caspase-1 is also called interleukin-converting enzyme or ICE), members of this cysteine protease family exert an important role in several stages of apoptosis (see Chapter 32). They are classified with numbers that in general reflect the chronological order in which they were discovered. Caspase-1, caspase-4, and caspase-5 are mainly involved in interleukin maturation and therefore play a critical role in interleukin-mediated inflammatory and immune processes.

Caspase-2, caspase-3, caspase-6, caspase-7, caspase-8, caspase-9, and caspase-10 are instead involved in apoptotic processes. They are present in cells as inactive precursors and are activated by partial proteolysis, which can be catalyzed by caspases themselves. In fact, two groups of caspases can be distinguished: the proximal or initiator caspases (caspase-2, caspase-8, and caspase-10), which activate the distal or executioner caspases (caspase-3, caspase-6, caspase-7, caspase-9, and caspase-13), which execute the cell death program by catalyzing the proteolysis of a large number of target proteins. In contrast with the coagulation cascade, which is characterized by a unidirectional relationship

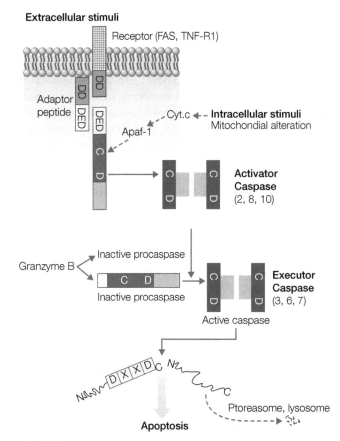

FIGURE 30.3 Caspase proteolytic cascade involved in apoptosis. Stimuli from the extracellular or the intracellular environment induce proteolytic activation of some caspases (activator caspases), which in turn can proteolytically activate zymogens of other proteases (executioner or effector caspases). Effector caspases execute the cell death program by specifically cleaving proteins carrying the Asp-Xaa-Xaa-Asp (DXXD) sequence. In their active form, caspases are tetramers formed by two heterodimers consisting of a large subunit (C), containing the cysteine catalytic residue, and a small regulatory subunit (D). Apaf, apoptosis-activating factor; Cyt. c, cytochrome c; DD, death domain; DED, death effector domain.

between proteases, the caspase activation process involves also bidirectional relationships having the function of amplifying the initial signal. However, the hierarchical order in the proteolytic cascade has been clarified only in some experimental models (such as CD95 receptor stimulation), whereas in other situations, executioner caspase activity does not seem to require preventive activation of initiator caspases.

Proximal caspases exhibit a more complex structure compared to the distal. Since their activation occurs via autoproteolysis, they are tightly controlled to ensure that the entire cascade is activated only in response to specific signals. In fact, in addition to the catalytic domain, proximal caspases are endowed with another domain (called death effector domain or DED) that mediates their connection with specific receptors. The interaction is not direct, but occurs via adaptor proteins interacting on one side with caspase DED and on the other with death domains (DD) of membrane receptor proteins such as CD95 or TNFR1. The signal generated upon interaction between these receptors and their respective ligands leads to the formation of a supramolecular complex containing adaptors and procaspases. The recruitment of an increasing number of procaspases represents the basis for their proteolytic autoactivation and for propagation of the apoptotic signal from the cell surface to the cytoplasm. The system is also endowed with control and feedback mechanisms to reduce stimulation of proximal caspases. In this subtle equilibrium, proteins such as FLICE inhibitor protein (FLIP) and apoptosis repressor with caspase recruitment domain (CARD) (ARC) act as negative regulators interposed between receptors and enzymes, inhibiting caspase activation.

Also, procaspase-9, although lacking DED, seems to be activated by autoproteolysis stimulated by aggregation with apoptosis-activating factor-1 (Apaf-1), after Apaf-1 binding to cytochrome c. This complex, called apoptosome, is particularly efficient in causing apoptosis and seems to be generated in response to stimuli capable of modifying mitochondrial function or structure. Caspase-9, which can behave also as a distal caspase, activates procaspase-3, and such process is inhibited by the antiapoptotic protein Bcl-2. Bcl-2 is mainly localized in the outer mitochondrial membrane, and its hyperproduction seems to prevent cytochrome c release in the cytoplasm. The mechanisms controlling Bcl-2 inhibition and mitochondrial contribution to apoptosis are still not completely clear.

Proteolysis of caspase zymogens can be also aspecifically catalyzed by lysosomal or viral proteases. Thus, caspase activation can become a massive and uncontrolled process. An example of this is represented by granzyme B, a serine protease introduced in target cells by cytotoxic T lymphocytes, which triggers the caspase cascade by activating some caspases (but not ICE).

Several proteins have been identified as specific targets of distal caspases. For instance, DNA fragmentation seems to be caused by an endonuclease that translocates inside the nucleus only after detachment from a cytosolic protein, called DNA fragmentation factor (DFF), that is a caspase-3 substrate. Morphological alterations and formation of apoptotic bodies seem to be due to degradation, respectively, of a cytoskeletal protein (gelsolin) and p21-activated kinase-2 (PAK-2). However, the hierarchical order of the events that, starting with caspase activation, lead to apoptotic cell death is not entirely clear in its complexity and variety.

Besides caspases, other proteases can contribute to the amplification of the initial signal. For example, calpain can induce proteolytic activation of some caspases that in turn, by processing calpastatin, can enhance calpain activity. In

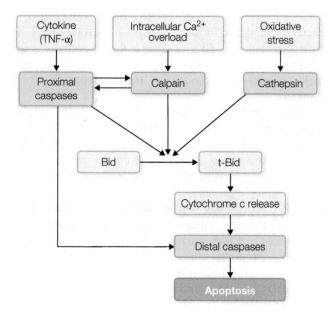

FIGURE 30.4 Relationship between different proteolytic systems involved in apoptosis. Activation of proximal caspases in response to extracellular stimuli, or activation of calpain and lysosomal proteases in response to changes in the intracellular environment, converges on mitochondria causing release of several proapoptotic proteins, among which a prominent role is played by cytochrome c. In this phase, the initial signals can be amplified by the synergistic interaction between calpain and caspases (see text). In the cytoplasm, cytochrome c causes activation of distal caspases and thereby execution of the late phase of the cell death program. A critical role is played by proteolysis of proapoptotic Bid protein, whose fragments can bind mitochondria and trigger the apoptotic response through still unclear mechanisms. This cell death pathway involving mitochondria is called intrinsic pathway. Conversely, in the extrinsic pathway, proximal caspases, in some circumstances, can directly activate distal caspases in the absence of mitochondrial alterations.

addition, activation of distal caspases (caspase-9 and caspase-3), generally occurring after mitochondrial alterations, can be the result of different events that, by activating proximal caspases (e.g., by cytokines) or releasing lysosomal cathepsins (e.g., upon oxidative stress) or activating calpain (in response to Ca^{2+} increase), lead to proteolytic cleavage of the proapoptotic protein Bid and consequent cytochrome c release (Fig. 30.4).

Pharmacological Modulation of Caspases Caspase activity is inhibited by different peptides. Mammalian cells contain a protein, X-chromosome-linked iap gene product (XIAP), member of the IAP family, that prevents proper maturation of procaspase-3 and procaspase-7. Mutations in the XIAP-encoding gene have been found associated with the onset of spinal muscular atrophy. By contrast, cytokine response modifier A (CrmA) has a viral origin, being a serpin protein encoded by the vaccinia virus. CrmA inhibits mainly caspase-1 and caspase-8, whereas all caspases are inhibited, to a different extent, by the p35 protein encoded by baculovirus.

Pharmacological inhibition can be obtained using agents that interact at the same time with the catalytic cysteine and the substrate binding site that specifically recognizes the aspartate. A tetrapeptide reproducing the P_1–P_4 sequence can work as a substrate analog, specifically recognized by the enzyme. The specificity for different caspase types is provided by the residue in P_4. In such position, caspase-3 recognizes an aspartate residue, whereas the caspase-1 subfamily recognizes a hydrophobic residue. Using as a model the YVAD cleavage sequence contained in prointerleukin-1β, a tetrapeptidic aldehyde (Ac-YVAD-CHO) has been synthesized, which inhibits ICE ($K_i < 1$ nM) with marked selectivity for caspase-3 ($K_i > 10$ μM). Conversely, the cleavage site contained in the poly-(ADP-ribose)-polymerase (PARP), a specific caspase-3 target, has been used to synthesize Ac-DEVD-CHO, which specifically inhibits caspase-3. Peptidic aldehydes, ketones, or nitriles behave as reversible inhibitors, whereas peptidyl (acyloxy)methyl ketones act as irreversible inhibitors (see Table 30.4). More recently, nonpeptidyl inhibitors have been developed, named isatins; however, *in situ*, these molecules have not shown the same effectiveness observed in *in vitro* studies.

Inhibition of apoptotic caspases is still far from being a therapeutic approach. To date, it could be used to limit damages due to ischemic events. An essential feature of an antiapoptotic chronic therapy would be the precise targeting of the inhibitor to the cells to be treated, in order to allow apoptosis to occur normally in healthy tissues. The same type of issue limits development of drugs capable of stimulating caspases and apoptotic processes, which would be very useful in oncology. By contrast, synthesis of ICE inhibitors seems to be a promising approach to develop a new class of anti-inflammatory drugs. In particular, a nonpeptidic aldehyde, pralnacasan, has been used, although with nonsignificant results, in a phase IIa clinical study for the treatment of rheumatoid arthritis. However, in November 2003, it was abandoned because of hepatic toxicity observed in experimental animals.

Calpain

Calpain is a cysteine endopeptidase activated by Ca^{2+} at neutral pH and is essential for the control of many physiological and pathological processes.

The name calpain results from the fusion of two words, "calcium" and "papain," to indicate that this protein is a cysteine protease (papain is considered one of the prototypes of this enzyme class) that totally depends on Ca^{2+} ions. Six different forms of calpain have been identified, which can be subdivided in two groups, based on their presence and distribution in tissues: ubiquitous calpains and tissue-specific calpains.

Two forms of ubiquitous calpains exist, namely, μ-calpain and m-calpain. μ-Calpain is activated by micromolar Ca^{2+} concentrations, whereas m-calpain requires millimolar concentrations; this suggests that μ-calpain is the regulated form under physiological conditions. *In vivo*, the Ca^{2+} affinity of these enzymes can be enhanced by several factors, among which are phosphatidylserine and phosphatidylinositol, suggesting that this enzyme is activated at the cell membrane.

Ubiquitous calpains are heterodimers formed by an 80 kDa catalytic subunit and a 30 kDa regulatory subunit (identical in both isoforms). The two subunits contain four EF-hand Ca^{2+} binding domains, which are structurally homologous to the Ca^{2+} binding domain of calmodulin.

Ablation of the regulatory subunit-encoding gene prevents expression of both ubiquitous forms and is embryonic lethal. Conversely, lack of the μ isoform only causes alterations in platelet aggregation; this observation is in contrast with multiple data indicating a physiological role of this protease in cell cycle control and several pathological processes.

Calpains catalyze limited proteolysis, meaning that they hydrolyze one or few peptide bonds of the target protein. An example is protein kinase C, in which cleavage of a single peptide bond results in the release of the active form of its catalytic domain. The active fragment produced by calpain exerts its maximal catalytic activity in the absence of cofactors (Ca^{2+}, phospholipids, and diacylglycerol), which in contrast are essential for activation of the kinase native form (Fig. 30.5).

Besides the two ubiquitous forms, DNA cloning and sequencing have led to the identification of other genes encoding tissue-specific forms, whose corresponding protein products have not yet been isolated though. However, some of these calpain-like genes (14 in vertebrates) have raised interest in the biomedical field. Lack of calpain 9-encoding gene seems to be a causative factor in stomach cancer; the G/A polymorphism in intron 3 of the calpain 10 gene has been associated with insulin resistance and type 2 diabetes, whereas lack of calpain 3a (also known as p94) causes shoulder girdle dystrophy.

Calpain G, a protein complex formed by calpain 8 and calpain 9, is instead essential for the defense of gastric mucosa from stress induced by alcohol, anti-inflammatory drugs, and *Helicobacter pylori*.

Calpains degrade a large number of proteins, but the physiological significance of many of these processes has not yet been defined; in fact, many of these targets have been identified only *in vitro* and may not reflect processes occurring *in vivo*.

Calpains are regulated not only via autoproteolysis but also by the presence *in vivo* of a specific intracellular and ubiquitous inhibitor: calpastatin. This protein blocks both the

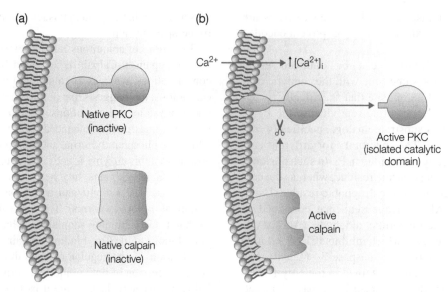

FIGURE 30.5 Example of the possible role played by calpain in transmission of extracellular signals inside the cell. (a) In basal conditions, calpain and its physiological target PKC are mainly found in the cell-soluble fraction in inactive form. (b) Upon stimulation leading to increase in Ca^{2+} concentration, both enzymes associate to the membrane and become active. Active calpain proteolytically activates membrane-bound PKC by releasing the catalytic domain that becomes active in the absence of any effector and capable of triggering a phosphorylation cascade in the cell-soluble fraction.

activation processes and calpain catalytic activity, with a still undefined competitive mechanism. In cells, calpastatin can be expressed in different forms that can undergo posttranslational modifications, generating a variety of calpastatins with different inhibitory properties and specificities.

The cytosolic localization of calpains and the mechanisms controlling their activation suggest also a pathological role for these enzymes. In the presence of excessive and/or prolonged increases in intracellular Ca^{2+}, the endogenous and exogenous systems controlling calpains could be reduced, thus allowing the proteases to act indiscriminately on cellular structure causing irreversible damages. Therefore, calpains are believed to actively participate in cellular necrosis processes. On this basis, damages due to cerebral or myocardial ischemia represent the main targets of therapies based on calpain inhibition (Table 30.4). Several inhibitors have been developed that in general interact with the cysteine thiol group in the catalytic site. Derivatives of α-mercapto acrylic acid have been synthesized, which are capable of inhibiting calpains by interacting with the Ca^{2+} binding domain. However, the specificity of these compounds is rather poor and not sufficient to discriminate between different calpains or between calpains and proteasome or other cysteine proteases.

EXOGENOUS INTRACELLULAR PROTEASES

Several viruses encode peptidic factors capable of inhibiting cell proteolytic activity, which are necessary for viral replication or cell damage. By contrast, other viruses introduce new proteases in the host cells: this is the case of HIV or the tetanus and botulinum toxins (see Chapter 37). The high pathogenicity of these agents can be ascribed to the inability of the infected cells to control or limit the activity of these proteases. Another example in which pathogenicity is related to exogenous proteases is that of some infectious protozoa. In the case of malaria, which is responsible for about 1 million deaths per year, different proteolytic activities are involved in all phases of *Plasmodium* life cycle. The development of *P. falciparum* is completely blocked by the combined inhibition of cysteine and aspartate proteases by inhibitors such as E64 and pepstatin. Thus, the development of specific protease inhibitors is a promising approach in the therapy of this disease.

HIV and Inhibitors of Viral Proteases

Therapy based on inhibition of HIV-1 proteases has represented a huge success in the treatment of HIV-1 infection. HIV-1 is a highly changeable lentivirus belonging to the retrovirus family, which are viruses having single-stranded RNA as genetic material. The RNA, injected into the cells, is used as a template by the virus-encoded inverse transcriptase to synthesize double-stranded DNA, which then becomes inserted into the host cell genome. The viral structural proteins, which include those necessary to the virus for plasma membrane binding and cell entry, are encoded by the *gag* and *env* genes, whereas the *pol* gene specifies for the transcription enzymes and for an aspartic protease (which usually corresponds to the N-terminal portion of the polyprotein encoded by the *pol* gene) belonging to the retropepsin (retroviral protease) family. This protease is necessary not only for maturation of all viral polyproteins but seems to be essential also for RNA dimerization and therefore for virus infectivity.

Also, endogenous proteases can contribute to virion maturation. In particular, the gp160 glycoprotein, which gives rise to gp120 and gp41, two proteins of the double-layered membrane that coats HIV-1, contains the REKR sequence that is recognized by furin. Inhibition of this protease represents an additional strategy to fight HIV.

Today, an effective anti-HIV strategy has as a target both the protease and the inverse transcriptase, which is inhibited by nucleoside and nucleotide analogs (such as AZT) or by nonnucleosidic compounds (Table 30.4). The combination of these two classes of transcriptase inhibitors with the protease inhibitors is used in the so-called highly active antiretroviral therapy (HAART). The 19 compounds approved to this aim can be combined in more than 3000 different therapeutic regimens.

HIV protease inhibition leads to the formation of immature noninfectious viral particles. The development of inhibitors of the HIV protease (saquinavir, indinavir, ritonavir, nelfinavir, amprenavir, and lopinavir) represents one of the most advanced examples of molecular drug design based on knowledge of tridimensional structure and mode of action of the molecular target.

The starting point has been the identification of the cleavage site. The protease cleaves peptide bonds linking a hydrophobic residue and a proline, when also the adjacent residue is hydrophobic. In particular, inhibitors have been designed based on the Phe-Pro dipeptide at positions 167 and 168 of the polyprotein *gag–pol*. Since no protease in higher metazoans is known to have cleavage site with a proline at P_1', initially, peptides were created in which the proline was replaced by a hydroxyethyl group that, by mimicking the transition state, could interact with the two catalytic aspartate residues. The further development, which has led to the compounds currently approved for clinical use, has consisted in the replacement of amino acids with groups capable of more accurately adapting to the binding sites. In saquinavir, for example, the original proline has been replaced with a quinolinic derivative that interacts with S_1' and is bound through a carboxamide bond to *t*-butyl amide residue in S_3'.

Unfortunately, the high mutation rate of HIV-1 rapidly leads to the production of proteases with preserved catalytic activity but reduced ability to interact with inhibitors. For example, one valine-to-isoleucine substitution in the S_1' site, although preserving the hydrophobicity, reduces the contact surface, whereas the opposite substitution can result in steric hindrance preventing binding.

An important limitation in the anti-HIV therapy in general and in the use of protease inhibitors in particular is the toxicity exerted on several organisms through multiple mechanisms. The main adverse effects of protease inhibitors are hypercholesterolemia, lipodystrophy, and insulin resistance. Hypercholesterolemia and lipodystrophy are due to the lack of proteolysis of the sterol regulatory element binding proteins (SREBPs). In the presence of high intracellular cholesterol, these proteins are inactive and bound to the ER. When cholesterol level decreases, SREBPs undergo proteolytic processing, first by S1P protease (or site-1 protease), an ER subtilisin inhibited by sterols, and then by S2P (site-2 protease), a metalloprotease that differs from the others for its localization in the ER membrane. Such protease is probably inhibited by HIV protease inhibitors. The SREBP proteolytic fragments translocate to the nucleus and promote the expression of the enzymes that catalyze synthesis of cholesterol and fatty acids, as well as the plasma membrane receptors required to take lipoprotein-carried cholesterol inside the peripheral cells. The absence of SREBP proteolysis alters adipocyte maturation and differentiation (lipodystrophy) because it inhibits their use of circulating lipids, thereby increasing cholesterolemia and triglyceride blood levels. In insulin resistance, reduced plasma membrane translocation of GLUT4 seems to occur; however, the mechanism responsible for the lack of translocation is still unknown.

TAKE-HOME MESSAGE

- Intracellular proteolysis is responsible for 75% of protein turnover in the organism and is carried out by four groups of proteases, acting through distinct catalytic mechanisms.
- Protease activity is finely controlled by metabolites, covalent modifications of substrate proteins, or protease segregation, in particular within lysosomes.
- Dysregulation of intracellular proteolytic activity is directly responsible for several diseases and indirectly involved in almost all pathologies through mechanisms related to cell death.
- In clinics, only a limited number of protease inhibitors are available, because of the poor specificity of the molecules currently known and their adverse effects due to the essential role played by proteases.

FURTHER READING

Adams J. (2002) Proteasome inhibition: a novel approach to cancer therapy. *Trends in Molecular Medicine*, 8, S49–S54.

Barrett A.J., Rawlings N.D., Woessner J.F. (editors) (1998) *Handbook of Proteolytic Enzymes*. London: Academic Press.

Chapman H.A., Riese J.R., Shi G.P. (1997) Emerging role for cysteine proteases in human biology. *Annual Review of Physiology*, 59, 63–68.

Choi A.M.K., Ryter S.W., Levine B. (2013) Autophagy in human health and disease. *The New England Journal of Medicine*, 368, 651–662.

Coux O., Tanaka K., Goldberg A.L. (1996) Structure and functions of the 20S and 26S proteasomes. *Annual Review of Biochemistry*, 65, 801–847.

DeLemos A.S., Chung R.T. (2014) Hepatitis C treatment: an incipient therapeutic revolution. *Trends in Molecular Medicine*, 20, 315–321.

Dick L.R., Fleming P.E. (2010) Building on bortezomib: second-generation proteasome inhibitors as anti-cancer therapy. *Drug Discovery Today*, 15, 243–249.

Goll D.E., Thompson V.F., Li H., Wei W., Cong J. (2003) The calpain system. *Physiological Reviews*, 83, 731–801.

Kroemer G., Marino G., Levine B. (2010) Autophagy and the integrated stress response. *Molecular Cell*, 40, 280–293.

Nalepa G., Rolfe M., Harper J.W. (2006) Drug discovery in the ubiquitin-proteasome system. *Nature Reviews. Drug Discovery*, 5, 596–613.

Nixon, R.A. (2013) The role of autophagy in neurodegenerative diseases. *Nature Medicine*, 19, 983–997.

Pomerantz R.J., Horn D.L. (2003) Twenty years of therapy for HIV-1 infection. *Nature Medicine*, 9, 867–873.

Pontremoli S., Melloni E. (1986) Extralysosomal protein degradation. *Annual Review of Biochemistry*, 55, 455–481.

Salvessen G.S., Dixit V.M. (1997) Caspases: intracellular signaling by proteolysis. *Cell*, 91, 443–446.

Willis M.S., Patterson C. (2013) Proteotoxicity and cardiac dysfunction. Alzheimer's disease of the heart? *The New England Journal of Medicine*, 368, 455–464.

Zhang W., Sidhu S.S. (2014) Development of inhibitors in the ubiquitination cascade. *FEBS Letters*, 588, 356–367.

31

EXTRACELLULAR PROTEOLYSIS

Francesco Blasi

> **By reading this chapter, you will:**
>
> - Understand the physiological importance of extracellular proteases in fibrinolysis, migration, differentiation, and cell growth
> - Learn the molecular structures of the components of the plasminogen/plasmin systems and matrix metalloproteases
> - Know drugs that can modulate extracellular proteolysis and the rationale of their use in cancer and cardiovascular diseases

Active extracellular proteases activate and modulate several physiological and pathological processes. Indeed, coagulation and complement activation are characterized by a prototypic proteolytic cascade that through multiple steps activates zymogens leading to the formation of one or more active proteases or multimeric opsonizing complexes in the case of complement. However, extracellular proteolysis is also important in the activation and liberation of prohormones and propeptides that interact with specific cell surface receptors leading to cellular signaling. In this chapter, we will focus in particular on two families of extracellular proteases: plasminogen activators (PAs) and matrix metalloproteases (MMPs).

THE EXTRACELLULAR MATRIX PROTEOLYTIC DEGRADATION SYSTEMS

The extracellular matrix (ECM) is a complex, not passive, structure that supports cells, composed of proteins (e.g., collagens, fibronectin, vitronectin, laminin, proteoglycans, fibrin) and complex carbohydrates. Connections between cells and ECM are mediated by specific multiprotein complexes that, besides anchoring cells to the ECM (see Chapter 20), transduce signals that modulate fundamental cell functions like differentiation, apoptosis, proliferation, and migration. A rapid, efficient, and often irreversible system to modify or modulate such interactions is represented by the degradation of ECM or cell surface proteins.

Plasminogen and pro-MMPs are zymogens activated by specific proteolytic cleavage. Plasminogen/plasmin and its activators (PAs) and MMPs are the two most involved and best studied endopeptidases participating in ECM remodeling. Both are protein complexes present in large variety and quantity and their expression/activity is tightly regulated. The final role of these complexes is to activate (and control the activation of) inactive proteases and zymogens. Plasminogen is activated to plasmin by two specific enzymes, the urokinase plasminogen activator (uPA) and the tissue plasminogen activator (tPA). MMP activation is more complex and requires the intervention of MMP themselves and, in some cases, of plasmin (see the following text). The final active enzymes (MMPs and plasmin) do not show a high degree of substrate selectivity and mainly act on ECM proteins (Table 31.1). The proteolytic cascade can affect also cell surface receptors regulating cell growth and adhesion.

In addition to proteolytic enzymes, the extracellular proteolytic complexes also contain specific proteins that bind and inhibit the enzymes. These are the plasminogen activator inhibitors (PAI-1 and PAI-2), the plasmin inhibitor α_2-antiplasmin, and the tissue inhibitors of metalloproteases (TIMPs). All inhibitors recognize two areas of the specific target: the active site and an accessory site (exosite) located outside the protease domain.

TABLE 31.1 Substrates of extracellular endoproteases

	Enzymes	Synonyms	Substrates
Matrix Metalloproteases	MMP-1	Collagenase-1	Col I, II, III, VII, X, GL, EN, LP, AG, TN, L-selectin, IGF-BP, Pro-MMP-2, -9, α2M, α1PI
	MMP-8	Collagenase-2	
	MMP-13	Collagenase-3	
	MMP-2	Gelatinase A	GL, Col I, IV, V, VII, X
	MMP-9	Gelatinase B	XI, EL, FN, LN, LP, AG, galectin-3, IGF-BP, VN, FGF receptor-1, Pro-MMP-2, Pro-MMP-9, Pro-MMP-13, MMP-3, MMP-10, AG
	MMP-7	Stromelysin-1	PG, LN, FN, GL, Col III, IV, V, IX, X, XI,LP, FB, EN, TN, VN
		Matrylisin	Pro-MMP-1, -8, -9, -13, α1PI, α2M, L-selectin
	MPP-11	Stromelysin-3	LN, FN, AG, 1PI, 2M
	MMP-12	Metalloelastase	EL, FB, FN, LN, PG, MBP, PL, α1PI
	MMP-14, 15	MT1-MMP, MT2-MMP	Col I, II, III, GL, FN, LN, VN, PG, Pro-MMP-2, -13, α1PI, a2M
Plasminogen activators and other endo-proteases	uPA	Urokinase	PLG, FN, pro-HGF, uPAR, VN
	tPA	Tissue plasminogen activator	PLG, FN, pro-HGF, uPAR
	Plasmin		FB, FN, TN, LN, AG, latent TGFb BP, PG, Pro-MMP-1, Pro-MMP-3, -9, -14, C1, C3, C5 and many other proteins
	Thrombin		FB, Pro-MMP-2, syndecan, thrombin receptor (PAR-1)

While α_2-antiplasmin function is to inhibit plasmin activity and hence its function in fibrinolysis, PAIs and TIMPs exert mainly a regulatory function, as they affect not only the proteolytic activity of their targets but also their localization and signaling function.

Inhibitor and proteolytic enzymes are generally produced by different cells. In some cases, for example, tissues (such as the trophoblast) are surrounded by a ring of cells secreting inhibitors in very specific regions in order to restrict and localize the extracellular proteolytic activity. It is interesting to note that in many physiological and pathological processes, increased expression of proteases is accompanied by a similar increase in inhibitors. This occurs, for example, during embryo implantation in the maternal decidua and in all tissue remodeling processes. This phenomenon is observed also during tumor development because of the close relationship between stromal and tumor cells. In all these cases, the presence of inhibitors (and hence the inhibition of the proteolytic activity) is not meant to block the biological process (embryo implantation, wound healing, cancer) but rather to favor its evolution and to regulate its intensity.

In addition, proteolytic enzymes often contain amino acid stretches mediating their anchoring to specific ECM or plasma membrane components. An extreme example of such spatial regulation is the urokinase receptor (uPAR) that, being anchored to the plasma membrane, allows the recruitment of the activated enzymes to specific regions of the cell surface in which interaction with the ECM is highly dynamic.

Hence, protease inhibition is an integral part of the tissue proteolytic function that allows its regulation, modulation, and localization in space.

PLASMINOGEN SYSTEM AND ITS ACTIVATORS

Plasminogen (Fig. 31.1) is a protein of about 90 kDa containing a catalytic C-terminal domain (protease) and five N-terminal kringle domains. The kringle domain, about 90 amino acids long, is a compact structure organized by three specifically located disulfide bonds. It is endowed with high affinity for the carboxy-terminal lysine residues present in fibrin and in some proteins of the cell surface and ECM. Plasminogen is proteolytically inactive (zymogen) and is activated by the hydrolysis of the peptide bond between Arg560 and Val561 and converted into plasmin, an extremely active and broad-spectrum serine protease. Plasmin physiological substrates, in addition to fibrin, are the ECM proteins fibronectin, vitronectin, laminin, and proteoglycans. Plasmin has a very short half-life (few seconds) because of the presence in blood and tissues of its inhibitor α_2-antiplasmin.

PAs as Fibrinolytic Agents

Proteolytic activation of plasminogen to plasmin depends on two serine proteases, uPA and tPA, which have one single major substrate. Also in PAs, the catalytic activity is localized in the C-terminal region and is preceded by one or more kringle domains that in tPA, but not uPA, mediate binding to fibrin.

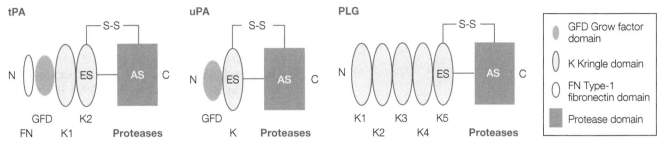

FIGURE 31.1 Molecular organization of plasminogen and its activators. Plasminogen (PLG) and its activators (PAs) contain common structural motifs. They are synthesized in the cells as single-chain polypeptides that, in the case of uPA and PLG, have no proteolytic activity (zymogens); single-chain tPA, instead, is endowed with basal enzymatic activity. Activation of all three proteins occurs through a proteolytic cleavage between a kringle domain and the protease region. In the figure, molecules are depicted from the amino-terminal (left) to the carboxy-terminal region (right). Below each scheme, the identity of each structural domain is indicated. AS, protease active site that binds substrates and inhibitors; ES, "exosite," inhibitor binding site, located outside the protease domain; FN, fibronectin type-1 domain; GFD, growth factor domain, because of its homology to the epidermal growth factor (EGF) amino acid sequence; K1, K2, etc., kringle domain, a protein motif containing three internal disulfide bridges. Protease, domain highly homologous to trypsin and other serine proteases.

TABLE 31.2 Phenotypes of genetically modified mice in one or more genes of the plasminogen activation system

Deleted gene	Fibrin deposition and thrombosis	Wound healing	Inflammatory response	Neo angiogenesis	Metastasis
tPA	Absent	Normal	n.d.	n.d.	n.d.
uPA	Rare	Normal	Deficient	Normal	Reduced
uPAR	Absent	Deficient	Deficient	n.d.	Reduced
uPA/tPA	Frequent	Deficient	n.d.	n.d.	n.d.
PLG	Frequent	Deficient	n.d.	n.d.	Reduced
uPAR/tPA	Frequent	n.d.	n.d.	n.d.	n.d.
PAI-1	Absent	Normal	n.d.	n.d.	n.d.
PAI-2	Absent	Normal	n.d.	n.d.	n.d.

n.d., no data available; PAI-1 and -2, inhibitors 1 and 2 of plasminogen; PLG, plasminogen; t-PA, tissue plasminogen activator; uPA, urokinase; uPAR, urokinase receptor.

tPA catalytic activity is quite low but strongly stimulated by fibrin binding. In this way, plasmin production is concentrated right in the area where its activity is required, that is, where a fibrin clot is present.

Unlike tPA, uPA does not bind fibrin but, through an N-terminal domain (termed growth factor domain (GFD) because of its homology to the epidermal growth factor), binds a specific receptor, uPAR (see the following text), on the plasma membrane.

These characteristics are at the basis of the use of tPA and other PAs, like streptokinase and staphylokinase, as thrombolytic drugs in cases of arterial or venal obstruction caused by fibrin deposits. Despite the lack of fibrin specificity, also, uPA is used successfully in vasal obstructions, although more frequently in venal occlusions (deep vein thrombosis). The bacterial PA streptokinase acts with a different mechanism since it does not hydrolyze plasminogen but associates with it forming a stable complex that modifies plasminogen physical–chemical properties, making its active site available to substrates.

Plasminogen activation is fundamental in fibrin deposit degradation and depends entirely on PAs, as shown by the phenotypes of knockout (KO) mice (Table 31.2). Plasminogen-deficient mice accumulate large amounts of fibrin deposits, which cause serious damages and their early death. In particular, these mice show fibrin deposits not only in the vasal system but also in the intestine, liver, lung, and brain, as well as a major delay of skin wound healing. Double KO mice lacking both PA genes (*uPA* and *tPA*) display the same phenotype as plasminogen KO mice.

On the contrary, single KO mice lacking either *uPA* or *tPA* do not have major phenotypic effect as the missing enzymatic function is taken over by the remaining PA.

Localized Activation of uPA by Interaction with a Specific Receptor

Several cell surface proteins contain C-terminal free lysines that can be bound by plasminogen and tPA kringle domains. This is fundamental to control "where" plasminogen

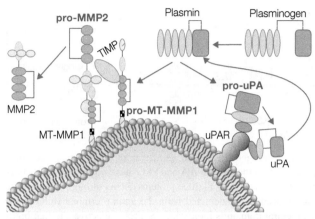

FIGURE 31.2 Pro-uPA and pro-MMP activation takes place on cell surface, with two different mechanisms. On the right, pro-uPA (single polypeptide chain) activation is shown. Upon binding to the receptor (uPAR), the uPA inactive precursor (pro-uPA) is proteolytically attacked at residue 157, in a linker region between the kringle and the protease domain. Mature urokinase (uPA) is hence made up of two polypeptide chains held together by a disulfide bridge linking the kringle to the protease domain. Activation of the MMP-2 inactive precursor pro-MMP-2 (shown on the left) requires the cleavage of the amino-terminal propeptide. The proteolytic process takes place on the plasma membrane and is executed by a membrane MMP (MT-MMP). Pro-MT-MMP expressed on the plasma membrane binds a molecule of the inhibitor TIMP-2, which in turn binds pro-MMP-2, recruiting it to the cell membrane. The recruited pro-MMP-2 is then activated by a second molecule of active MT-MMP that removes the propeptide. Propeptide removal can take place in two steps: the first step can be catalyzed by different proteases (including plasmin), whereas the second step is specifically catalyzed by a metalloprotease.

activation takes place and therefore to prevent indiscriminate plasminogen activation and active plasmin diffusion (e.g., in bloodstream).

Also, plasminogen activation by uPA takes place on the cell surface, although uPA does not bind fibrin or C-terminal lysines. This occurs because uPA can bind a specific plasma membrane receptor with high affinity. The uPA receptor, called uPAR (Fig. 31.2), is associated with the plasma membrane through a glycosylphosphatidylinositol (GPI) anchor. In addition, uPAR plays a critical role in uPA activation by binding its inactive precursor, pro-uPA or scuPA (Fig. 31.2), and promoting its conversion into the active form, uPA. uPAR induces a conformational change of the bound pro-uPA that makes it more accessible to plasmin, which activates pro-uPA by hydrolyzing a specific single peptide bond.

uPAR is constitutively expressed by few cells, but its expression is strongly increased upon induction of a physiological or pathological migratory phenotype (i.e., in monocytes, macrophages, T lymphocytes, fibroblasts, and endothelial, embryonic trophoblast, and intestinal epithelial cells as well as in keratinocytes during wound healing or in cancer cells in the neoplastic process) (see Supplement E 31.3, "uPAR, Cancer, and Stem Cells").

PAI-1 is active both on soluble uPA and on uPAR-bound uPA; when bound to the inactive complexes PAI-uPA, uPAR induces their internalization and degradation (but not when bound to soluble uPA). In this process, uPAR is recycled back to the cell surface ready to start a new round of extracellular proteolysis by binding free uPA. Through this process, cells can continuously modify uPAR localization on their surface and hence specifically direct the extracellular proteolytic activity. This mechanism, coordinated with the adhesion–deadhesion processes, allows continuous cell–ECM attachment and detachment and therefore cell migration.

Function and Pharmacological Modulation of the PA System

The PA system plays an important role in fibrinolysis, cell migration, growth, and differentiation. PAs degrade fibrin accumulated at endothelial lesions, reducing or preventing formation of thrombi. Therefore, they are used in several thrombotic pathologies (see Supplement E 31.1, "Plasminogen Activators and Cardiovascular Diseases"). The role of PAs in more complex processes, like inflammation, angiogenesis, growth, and infiltration of tumors, has been accurately defined in several studies employing KO mice (Table 31.2).

Lack of one PAs does not lead to a pathological phenotype because the activity of the missing enzyme is at least in part compensated by the residual PA. Only when both PAs are missing, the phenotype becomes markedly pathological and is almost completely identical to the phenotype due to absence of plasminogen. Outside the vascular bed, PA functions lead to degradation not only of fibrin but also of other ECM proteins, like fibronectin, laminin, and proteoglycans, regulating cell migration. Moreover, since plasmin has a role in the activation of MMPs (see the following text) and growth factors (TGF-β, basic fibroblast growth factor (FGF), etc.), lack of PAs also affects cell growth and differentiation. For example, *uPA KO mice have major phenotypes in processes requiring cell migration. Defects in the migration process affect inflammatory and immune cells with consequent hypersensitivity to infections, smooth muscle cell whose migration is required to prevent thickening of intima and restenosis following angioplasty, embryonic neurons, and cancer cells (see Supplement E 31.2, "Plasminogen Activators and Tumor Malignancy"). In particular, it is important to note that uPA and plasmin have multiple roles in cancer cell spreading as they participate not only in ECM degradation but also in activation of potentially angiogenic factors like basic FGF (basic FGF9), TGF-β, and hepatocyte growth factor (HGF). PAI-1 itself is an important angiogenic factor; indeed, development and invasivity of experimental tumors are markedly reduced in mice lacking *PAI-1*.

Another uPA substrate is in fact pro-HGF, the inactive precursor of HGF (Table 31.1). Mice lacking *uPA* or *uPAR* display a deficient activation of pro-HGF in the embryonic brain with deficient neuronal migration (Table 31.1). Precursors of other growth factors, like TGF-β and β-FGF, are indirectly activated by PAs through plasmin production. Also, these substrates are not physiologically exclusive of plasmin since their activation takes place also in mice lacking uPA, tPA, or plasminogen. These properties show the complexity of tissue proteases and how they are connected to cell growth and tissue remodeling.

In addition, angiostatin, a proteolytic product of plasminogen that contains the first few kringle domains, has a potent antiangiogenic effect (Table 31.3). This effect may be due to apoptosis of endothelial cells, but the underlying mechanisms are not very clear. Blockade of tumor neoangiogenesis could be a very important approach to tumor therapies even though doubts still exist on the relationship between angiogenesis and tumor mass growth and on the therapeutic effectiveness of tumor anoxia.

Pharmacological Inhibition of the PA System in Neoplastic Processes

Whereas uPA, tPA, and functional analogs are already important agents in clinical practice to treat cardiovascular pathologies, much less is known about the benefit of pharmacologically targeting the PA system to inhibit cancer cell migration. Despite the structural homology and functional redundancy of uPA and tPA, it is possible to identify protease-specific inhibitors, like amiloride, which specifically inhibits uPA. This molecule, well known for its diuretic potassium-saving activity (see Chapter 28), represents the starting point for the search of uPA-specific inhibitors lacking diuretic and antihypertensive effects. Table 31.3 reports some uPA-specific drugs that are being or have been studied as antimetastatic drugs.

Analogs or fragments of uPA lacking the catalytic region but still able to bind uPAR have been or are being studied as anti-invasive drugs in neoplastic processes. In particular, the GFD fragment (Fig. 31.1), the amino-terminal fragment (ATF), and the uPA kringle of uPA, produced by recombinant technology, are promising tools in this field (Table 31.3).

MATRIX METALLOPROTEASES

MMPs are proteases synthesized as zymogens that are activated by a proteolytic cleavage. They are called "metallo"-proteases because their enzymatic activity depends on a zinc atom bound to their highly conserved active site (Fig. 31.3). These enzymes were originally named collagenases or gelatinases as they are capable of hydrolyzing native or denatured (gelatin) collagen. Some MMPs are quite selective for the type of collagen they hydrolyze (like MMP-2 and MMP-9, which are specific for type IV collagen), whereas others have a broader substrate spectrum, which includes also laminin, fibronectin, cell surface selectins, receptors, and growth factors (Table 31.1).

TABLE 31.3 Pharmacological applications and effects of extracellular protease inhibitors or antagonists

Drug	Disease	Activity
uPA	Deep vein thrombosis	Thrombolytic
Pro-uPA	Myocardial infarction	Thrombolytic
tPA	Myocardial infarction	Thrombolytic
tPA variants[a]	Myocardial infarction	Thrombolytic
Streptokinase	Myocardial infarction	Thrombolytic
Stafilokinase	Myocardial infarction	Thrombolytic
GFD of uPA	Tumor metastasis	uPAR antagonist
ATF of uPA	Tumor metastasis	uPAR antagonist
Angistatin	Solid tumors	Apoptosis of endothelial cells
Amiloride and derivatives	Tumor metastasis	Inhibition of uPA
Batimastat and derivatives (withdrawn)	Tumor metastasis	Inhibition of MMP
Galardin and derivatives (withdrawn)	Tumor metastasis	Inhibition of MMP

AFT, aminoterminal domain of urokinase (aac 1-136); GFD, growth factor domain.
[a] Recombinant variants with higher affinity for fibrin and more resistant to inhibitors (PAI).

Modular Structure of MMPs

MMPs have a modular structure in which the catalytic domain is preceded and followed by specific modules mediating its interaction with inhibitors and substrates (Fig. 31.3). The hemopexin domain mediates MMP interaction with the TIMPs, some of which are known to have different specificities for the MMPs. Like PAIs, TIMPs recognize the active catalytic site of the protease. The MT-MMP subgroup presents a transmembrane domain similar to the one found in many growth factor receptors.

As we saw for PAIs, TIMPs' inhibitory activity is not only aimed at blocking but also at regulating the various biological processes in which MMPs are involved.

MMP Activation at the Plasma Membrane

Pro-MMPs are synthesized in an inactive form and are activated by two proteolytic cleavages in the N-terminal domain. The first proteolytic step may be catalyzed by a

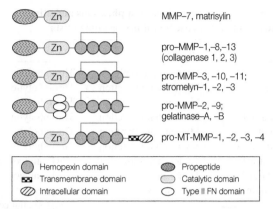

FIGURE 31.3 Modular structure of matrix metalloproteases (MMPs). Symbols and names of the structural domains are indicated at the bottom of the figure. In the case of pro-MMP-2 and pro-MMP-9 (gelatinases A and B), the catalytic domain includes also three fibronectin type II domains. All proteases are depicted in their inactive form, containing, at the amino-terminus, the propeptide that is cleaved and discarded during the process of activation (see also Fig. 31.2).

serine protease (including plasmin itself) and releases an N-terminal peptide. In the second step, the amino acid sequence uncovered by the first cut is attacked by another MMP with the consequent release of a second peptide and exposition of the catalytic site. As shown in Figure 31.2, the second step in MMP-2 activation takes place on the plasma membrane. This requires cooperation of two active MT-MMPs (metalloproteases containing a transmembrane region) with the TIMP-2 inhibitor. TIMP-2 is anchored to the plasma membrane via interaction with the active site of one active MT-MMP-1 and immobilizes a pro-MMP-2 molecule on the plasma membrane via the hemopexin domain. The membrane-anchored pro-MMP-2 is then activated by a second active MT-MMP-1 molecule on the cell surface, thus generating an active MMP-2 (gelatinase A).

TIMP-2, therefore, plays a dual function in the regulation of the enzymatic activity of its target soluble proteases: at low concentrations, it favors their cell surface localization and hence their activation by MT-MMPs; at higher concentrations, it inhibits MMP enzymatic activity. It is therefore apparent how critical the balance between proteases and inhibitors is and how the two protease families (PAs and MMPs) are functionally very similar. In fact, plasmin is one of the main enzymes involved in MT-MMP activation.

Function and Pharmacological Modulation of MMP

MMP physiological functions are still not entirely clear and their study is hindered by the large number of isoforms identified. Characterization of genetically modified mice has revealed the functional redundancy existing within the MMP family; indeed, several KO mutants do not exhibit any detectable phenotype (for instance, stromelysin 3/MMP-11 and MMP-7) or have only a minor phenotype (MMP-9). For instance, MMP-9 KO mice exhibit only reduced angiogenesis in implanted embryos and reduced length of long bones. However, genetically modified mice lacking a single MMP gene are useful to clarify the involvement of individual proteases in the neoplastic phenotypes of different organs (MMP-9 in the skin, MMP-7 in the colon, and MMP-11 in the mammary gland, Table 31.4).

The expression of MMPs and TIMPs is strongly increased during all tissue remodeling and cell migration processes, such as inflammation and neoplasias, even though given their functional redundancy the specific role of each MMP/TIMP system is often difficult to understand.

Some synthetic inhibitors have provided interesting information on MMP roles. For example, galardin, a nonselective inhibitor of collagenase activities, strongly inhibits wound healing, whereas the synthetic inhibitor batimastat is efficient in tumor therapy even though its broad spectrum of action causes undesired toxic effects. The most efficient of these inhibitors, batimastat (Table 21.3), is a derivative of hydroxamic acid (–CONHOH) that specifically recognizes the enzyme active site and reversibly binds the zinc atom. Many inhibitors recognize most MMPs but some have a moderate selectivity; batimastat, for example, is particularly efficient in inhibiting MMP-2, MMP-7 and MMP-9.

TABLE 31.4 In vivo roles of MMPs

Gene deletion	MMP-7	None
	MMP-9	Shortening of bones
	MMP-11	None
Gene overexpression	MMP-7 (colon)	Increased rate of tumor progression
	MMP-9 (skin)	Increased rate of tumor progression
	MPP-11 (mammary gland)	Increased rate of tumor progression

Effects of experimental gene deletion or overexpression in transgenic mice.

TAKE-HOME MESSAGE

- Extracellular proteases are involved in the proteolytic activation of zymogens that lead to the formation of fibrin and lytic activity of complement.
- They also alter the relationship between cells and extracellular matrix, thereby modulating cellular signals important for differentiation, proliferation, migration, and apoptosis.
- Plasminogen and metalloproteases are zymogens activated by specific activators.

- Plasminogen activators are essential for the degradation of fibrin deposits and are life-saving drugs in many diseases including thrombotic infarct, stroke, and deep thrombosis.
- Metalloproteases are activated on the cell membrane through a multimolecular process.
- Metalloproteases are involved in cell localization within tissues, cell mobilization in the organism, and metastatic processes in tumors.
- Drugs that control metalloproteases are under investigations for the treatment of tumors and wounds.

FURTHER READING

Blasi F. (1997). uPAR-uPA-PAI-1: a key intersection in proteolysis, adhesion and chemotaxis. *Immunology Today*, 18, 415–417.

Blasi F., Carmeliet P. (2002). uPAR: a versatile signaling orchestrator. *Nature Reviews. Molecular Cell Biology*, 3, 932–943.

Chapman H.A. (1997). Plasminogen activators, integrins, and the coordinated regulation of cell adhesion and migration. *Current Opinion in Cell Biology*, 9, 714–724.

Clause K.C., Barker T.H. (2013). Extracellular matrix signaling in morphogenesis and repair. *Current Opinion in Biotechnology*, 24, 830–833.

Collins R., Peto R., Baigent C., Sleight P. (1997). Aspirin, heparin and fibrinolytic therapy in suspected acute myocardial infarction. *New England Journal of Medicine*, 336, 847–849.

Frank A., David V., Aurelie T.R., Florent G., William H., Philippe B. (2012) Regulation of MMPs during melanoma progression: from genetic to epigenetic. *Anti-Cancer Agents in Medicinal Chemistry*, 12, 773–782.

Fuhrman B. (2012). The urokinase system in the pathogenesis of atherosclerosis. *Atherosclerosis*, 222, 8–14.

Hua Y., Nair S. (2014). Proteases in cardiometabolic diseases: pathophysiology, molecular mechanisms and clinical applications. *Biochimica et Biophysica Acta*, 448, 267–73.

Kelley L.C., Lohmer L.L., Hagedorn E.J., Sherwood D.R. (2014). Traversing the basement membrane in vivo: a diversity of strategies. *Journal of Cell Biology*, 204, 291–302.

Mezentsev A., Nikolaev A., Bruskin S. (2014). Matrix metalloproteinases and their role in psoriasis. *Gene*, 540(1), 1–10.

Shapiro S.D. (1997). Mighty mice: transgenic technology "knocks out" questions of matrix metalloproteinase function. *Matrix Biology*, 15, 527–533.

Skotnicki J.S., Di Grandi M.J., Levin J.I. (2003). Design strategies for the identification of MMP-13 and Tace inhibitors. *Current Opinion in Drug Discovery & Development*, 6, 742–759.

Spinale F.G., Janicki J.S., Zile M.R. (2013). Membrane-associated matrix proteolysis and heart failure. *Circulation Research*, 112, 195–208.

Tocchi A., Parks W.C. (2013). Functional interactions between matrix metalloproteinases and glycosaminoglycans. *FEBS Journal*, 280, 2332–2341.

Werb Z. (1997). ECM and cell surface proteolysis: regulating cellular ecology. *Cell*, 91, 439–442.

Wolf K., Friedl P. (2011). Extracellular matrix determinants of proteolytic and proteolytic cell migration. *Trends in Cell Biology*, 21, 736–744.

SECTION 9

CONTROL OF CELL CYCLE AND CELLULAR PROLIFERATION

32

CELL CYCLE AND CELL DEATH

Marco Corazzari and Mauro Piacentini

> **By reading this chapter, you will:**
> - Learn the molecular mechanisms regulating the cell cycle
> - Understand the differences between various modes of cell death and their relevance in human pathophysiology
> - Know the different pathways of apoptosis and the interconnection between cell cycle and apoptosis
> - Know the strategies to control cell death

CELL CYCLE

Cell cycle consists of a variety of finely regulated events resulting in DNA replication and division of a cell into two genetically identical daughter cells. This process lasts 16–24 h and involves a sequence of four phases named, in the order, G1, S, G2, and M.

During S and M, DNA replication and mitosis occur, respectively. G1 and G2 are two preparatory stages in which cells get ready for the following DNA duplication and cell division, respectively. Another phase also exists, called G0, that represents a period in which cells stop proliferating and enter a quiescent state, usually preceding the terminal differentiation process. However, G0 is a reversible condition, and under appropriate conditions, cells can reenter the cell cycle program.

Timely Regulated Expression of Cyclins and Cell Cycle Progression

Cyclins are cellular proteins responsible for the regulation of key serine/threonine kinases called cyclin-dependent kinases (Cdks; Fig. 32.1). They are characterized by two typical sequences critical for their function: the "cyclin box," essential for their interaction with specific Cdks, and the "destruction box," required for their degradation by the proteasome. The latter is also known as PEST sequence (proline (P), glutamic acid (E), serine (S), and threonine (T)).

Whereas Cdk molecules are stably expressed throughout the cell cycle, specific cyclins are rapidly synthesized and degraded during each phase.

As G1 starts, cyclin D is rapidly synthesized and interacts with either Cdk4 or Cdk6 depending on the cell type to form a stable complex that remains active throughout the G1 phase. During the late G1, cyclin E starts to be synthesized; this binds and activates Cdk2. Cyclin E level reaches a peak during the G1–S transition and decreases during the S phase. At the end of G1, also cyclin A starts to be expressed; this associates with Cdk2, forming the Cdk2/cyclin A complex that is essential for DNA replication. In contrast to cyclin E, cyclin A increases during the S phase, reaches a maximum at the beginning of mitosis (prophase), and is rapidly degraded during the subsequent metaphase. Cyclin A can also bind and activate Cdk1, whose activity is required to start and complete mitosis. However, it is believed that physiological activation of Cdk1 occurs through interaction with cyclins B, which are expressed during the late S phase and rapidly degraded once mitosis is completed (Fig. 32.1).

Cdk/cyclin complexes are activated by phosphorylation of a specific Cdk threonine residue that causes a conformational change unmasking the kinase active site. Cdk phosphorylation and activation are carried out by the Cdk-activating kinase (CAK) complex that contains Cdk7, cyclin H, and ménage à trois 1 (MAT1). Once activated, Cdk/cyclin complexes can be inhibited by a second phosphorylation event on a specific N-terminal tyrosine residue by a different kinase, the Wee kinase. Cdc25 phosphatase

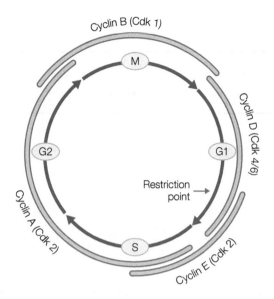

FIGURE 32.1 Cdks and cyclins regulate cell cycle in metazoans. The figure shows cell cycle phases and the Cdks and cyclins guiding their progression. The arcs outside the circle represent the periods of maximum activation of the indicated Cdk/cyclin complexes.

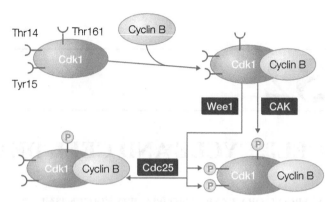

FIGURE 32.2 General mechanism of Cdk activation during cell cycle progression referred to Cdk1. Cyclin B is associated with Cdk1, which is phosphorylated on Thr61 by CAK, but is kept inactive by the simultaneous phosphorylation of residues Thr14 and Tyr15, operated by Wee; complete activation occurs after dephosphorylation of these two residues by Cdc25.

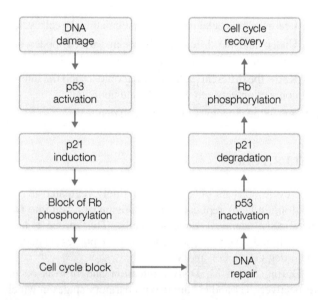

FIGURE 32.3 DNA damage-mediated cell cycle arrest. DNA damage induces activation of p53 that, in turn, increases p21 expression. p21 inhibits the activity of cyclin-dependent kinases responsible for pRb phosphorylation, thereby keeping pRb in a hypophosphorylated status and blocking G1–S transition. Once DNA has been repaired, inactivation and/or degradation of both p53 and p21 allows pRb phosphorylation and transition to the S phase.

may relieve this inhibition by removing the phosphate group (Fig. 32.2). Cdk activity is finely and promptly tuned by phosphorylation/dephosphorylation events throughout the cell cycle.

Role of Retinoblastoma Binding Protein during G1–S Transition

The ability of Cdk/cyclin complexes to regulate cell cycle progression is directly linked to their specific targets. In particular, the G1 to S phase transition is regulated by the retinoblastoma binding protein (pRb), an ubiquitous protein so called because it is initially identified for its involvement in retinoblastoma, a childhood cancer. In fact, homozygous mutations in its gene, *rb* (a tumor suppressor gene), can lead to retinoblastoma, as well as to several other types of cancer, such as leukemia, osteosarcoma, and prostate, lung, and breast cancer.

pRb controls the activity of E2F1, a transcription factor that regulates the expression of key enzymes required for DNA replication. Physiologically, pRb inhibits E2F1 activity by binding and sequestering it. This interaction is inhibited by Cdk-dependent pRb phosphorylation. In fact, during G1 phase progression, pRb is phosphorylated by cyclin D/Cdk4, cyclin D/Cdk6, and subsequently cyclin E/Cdk2, and its increasing phosphorylation results in E2F1 release. pRb persists in such hyperphosphorylated status until the late M phase, when the PP1 phosphatase progressively dephosphorylates it, restoring the inhibitory interaction between pRb and E2F1. It is important to note that cyclins A and E are also transcriptionally regulated by E2F1; therefore, when pRb is inactive or missing, E2F1 is constitutively active, leading to deregulation of cyclin expression and hence to uncontrolled cell growth and cell transformation (Fig. 32.2).

In addition, recent studies suggest that E2F sequestration is not the only mechanism whereby pRB controls G1–S transition. Indeed, unphosphorylated pRB can form a complex with E2F that inhibits transcription of specific target genes (Fig. 32.3).

Cdk Inhibitors and Cell Cycle "Checkpoints"

Cdk kinase activity can be negatively regulated by CKI (CDK inhibitor) proteins. In mammals, two main CKI subgroups have been identified: one includes p16 (INK4A), p15 (INK4B), p18 (INK4C), and p19 (INK4D), which specifically inhibit Cdk4 and Cdk6 activity, whereas the other includes p21 (CIP1/WAF1), p27 (KIP1), and p57 (KIP2), which act as general Cdk inhibitors but having Cdk4 and Cdk2 as preferred targets. CKIs inhibitory activity can be due either to Cdk sequestration, preventing cyclin binding and subsequent activation (p16), or to interaction and direct inhibition of active Cdk/cyclin complexes (p21 and p27).

CKIs play key roles during cell cycle progression, as they are the main effectors of cell cycle "checkpoints," control mechanisms that arrest cell cycle when anomalies are present. For example, it has been established that cellular stresses leading to DNA damage (e.g., radiations and chemical mutagens) activate the transcription factor p53 (also known as the "guardian" of the genome), which in turn upregulates the expression of p21, thus preventing Cdk-mediated phosphorylation of pRb and cell cycle progression (Fig. 32.3).

Besides DNA damage, p53 is also able to "sense" other cellular stresses precluding cell proliferation, like oxygen shortage (hypoxia), ribonucleotide deprivation, and mitotic apparatus dysfunctions. In response to such stressors, p53 triggers cell cycle arrest until suitable conditions for its progression are restored (see Supplement E 32.1, "p53: Regulator of Mitotic Cycle Progression and Apoptosis Inducer").

CELL DEATH

Cell death is an evolutionary conserved physiological process that is controlled by both homeostatic and morphogenetic stimuli.

Together with cell division and differentiation, cell death plays a key role in homeostasis of multicellular organisms. Indeed, in each organism, several million cells are thought to die every second, and cell proliferation and division counterbalance such cell loss. Due to its key role in embryogenesis and cell number homeostasis, cell death represents a fundamental process during the entire life of an organism. For this reason, cell death is finely regulated and its deregulation may result in pathological conditions, such as autoimmune disorders and cancer (deficient cell death), or in neurodegenerative syndromes and infections (excessive cell death). The main cell death process is called programmed cell death (also known as apoptosis) and is an active, genetically controlled mechanism regulated at both transcriptional and posttranscriptional levels. During the apoptotic process, cells gradually "degrade" themselves without releasing their contents. Their remains are recognized by macrophages or surrounding cells through specific surface antigens and removed. For this reason, apoptosis does not induce the inflammatory processes that would result from cellular lysis. Conversely, necrosis is a cell death process in which plasma membrane is disrupted and cellular components are released. Although long considered the result of an uncontrolled process, it is now emerging that necrosis might be regulated by molecular mechanisms yet to be completely elucidated.

Several other death processes have also been described that can be classified neither as apoptosis nor as necrosis, but appear as something intermediate, showing features of both apoptosis and necrosis and collectively known as necroptosis. Such evidence supports the notion that apoptosis and necrosis may represent the extremes of a "continuum" of cell death processes involving factors and mechanisms belonging either to the fully apoptotic or to the fully necrotic process (see also Supplement E 32.2, "Mitotic Catastrophe").

PROGRAMMED CELL DEATH OR APOPTOSIS

Phenomena morphologically corresponding to the apoptotic process were discovered at the end of the nineteenth century studying animal embryogenesis. Cell death was found to be a crucial event for tissue remodeling and cell selection during organism development. Indeed, it is required for removal of redundant or vestigial structures such as interdigital membranes, mesonephros, etc. It is also important during tissue development for eliminating supernumerary cells in the immune and nervous systems generated in excess during tissue expansion and/or negatively selected. Equally noteworthy is the fact that apoptosis is an active process since inhibiting transcription or translation or inhibiting ATP production results in apoptosis impairment (Table 32.1).

Apoptotic Bodies

Once the apoptotic process is completed, each cell fragments into several electron-dense particles surrounded by membranes, called apoptotic bodies, that are removed by macrophages or surrounding cells. In the apoptotic bodies, all

TABLE 32.1 Biochemical and morphological events occurring during apoptosis

- Decrease of cytoplasmic volume and cytoplasm condensation
- Increase of intracellular calcium concentration
- Activation of specific cysteine proteases (caspases)
- Alteration of mitochondria outer membrane permeability and release of cytochrome c
- DNA fragmentation
- Altered cell surface

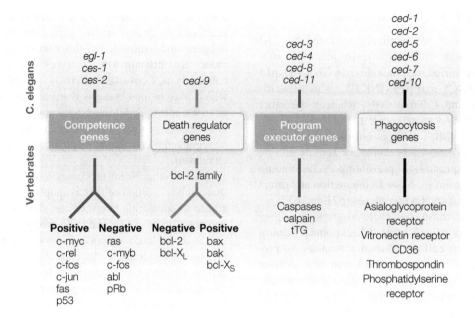

FIGURE 32.4 Genes involved in the cell death program identified in *Caenorhabditis elegans* and vertebrates.

cellular components (proteins, DNA, etc.) are cross-linked and condensed to avoid their release in the extracellular compartment and subsequent activation of the inflammatory response. The morphological changes accompanying the formation of these structures are cell shrinkage due to water loss and protein condensation; loss of membrane specialized structures, such as microvilli or cilia; loss of intercellular connections; DNA condensation (picnosis); and cell fragmentation in small membrane-bound structures. During the apoptotic process, cell organelles retain their structural integrity, and at least part of the mitochondria population remains functional to ensure energy supply.

Selective removal of the apoptotic bodies is mediated by signaling molecules called "eat me signals" that are exposed on their surface and consist of a mixture of modified glycoproteins and other molecules such as phosphatidylserine. Once identified, apoptotic bodies are engulfed by the surrounding cells or macrophages and eliminated by phagocytosis. If not removed, apoptotic bodies may become potentially harmful for the organism as they can collapse by a process called secondary necrosis, releasing their contents and inducing inflammation.

Apoptotic Biochemical Pathways

Caenorhabditis elegans has been fundamental for our understanding of the molecular and biochemical pathways regulating the apoptotic process. Apoptosis is a critical event during development of this tiny nematode, as 131 out of the 1090 cells generated in the embryo are selectively removed by genetically programmed cell death. Genetic studies of such mechanism have led to the identification of a set of genes involved in the cell death program, called *ced-9*, *ced-3*, *ced-4* (ced stands for cell death defective), and *Egl-1*. Ced-9 is required for cell survival, and its uncontrolled activation confers resistance to apoptosis, while its inactivation results in cell death induction. The product of the *Egl-1* gene can bind Ced-9, inhibiting its activity, whereas Ced-3 and Ced-4 are required for the execution of the apoptotic process. The cellular choice between survival and death depends on the balance between the competing activities of proapoptotic (ced-3 and ced-4) and antiapoptotic factors (ced-9). The mechanism of apoptosis has been conserved during evolution, and homologs of the cell death genes of *C. elegans* have been identified also in humans. The human homolog of ced-9 is B cell lymphoma 2 (Bcl-2), the first member of a family of proteins that play a pivotal role in the regulation of apoptosis. This family, called Bcl-2 family, includes both antiapoptotic and proapoptotic members. Several human homologs of ced-3 are known; indeed, ced-3 is the first identified member of a family of proteases, called **caspases**, which are sequentially activated during apoptosis and carry out proteolytic cleavage of essential cellular molecules. The human homolog of ced-4 is human apoptotic protease-activating factor-1 (hApaf-1), a protein involved in apoptotic protease activation. The high number of human factors involved in apoptosis reflects the higher complexity of the process in mammals compared to nematodes (Fig. 32.4).

Investigations into the molecular interactions between proteins involved in cell death have led to the identification of two main apoptotic pathways: the extrinsic pathway, controlled by death receptors (DR) belonging to the tumor necrosis factor (TNF) receptor family (see also Chapter 9), and the intrinsic pathway, involving mitochondria

☞ (see Supplement E 32.3, "Intrinsic Apoptotic Pathway Induced by Endoplasmic Reticulum").

The Intrinsic Pathway of Apoptosis p53 and c-myc are two of the most important factors controlling both cell cycle progression and induction of apoptosis (for a detailed
☞ description of p53 functions, see Supplement E 32.1).

c-myc is a proto-oncogene that encodes for a transcription factor regulating two sets of genes, one involved in cell proliferation and the other in programmed cell death. When human fibroblast cell lines are exposed to growth factors, c-myc expression increases and cells enter the cell cycle; conversely, growth factor deprivation results in cell cycle arrest and apoptosis induction. Therefore, the choice between these two opposite activities of c-myc—survival and cell death—strictly depends on the presence of growth factors and/or the expression of prosurvival factors. Accordingly, concomitant expression of Bcl-2 and c-myc results in cell survival and proliferation and thereby inhibition of c-myc proapoptotic activity, whereas absence of secondary prosurvival factors or impairment of growth signaling pathway(s) promotes c-myc proapoptotic activity.

Caspases: The Effectors of Apoptosis C. elegans ced-3 gene shares a high sequence homology with the mammalian *interleukin β converting enzyme* (ICE) factor that processes interleukin 1β to its active form. Other genes have been cloned as part of the ICE family, such as Inc-1/Nedd-2, ICErel II, Mch2, and CPP32/apopain/YAMA, collectively named caspases. All caspases (i) share a high sequence homology with the first member of the family, ICE/caspase-1; (ii) carry a characteristic pentapeptide (QACRG) where the cysteine (C) represents the catalytic residue (for this reason, caspases are also known as cysteine proteases); and (iii) require an aspartic acid (D) residue in the P1 position of the substrate consensus sequence -X-X-P4-P3-P2-P1-X-X that represents the typical substrate sequence recognized by all caspases. Whereas P4 and P3 residues can vary, thus conferring specificity to each caspase member, the D residue in P1 is conserved. Proteolytic cleavage takes place after the P1 residue.

Collectively, caspases share some common characteristics such as:

1. Active caspases consists of two large subunits (17–22 kDa) plus two small subunits (10–12 kDa).
2. Each enzyme is synthesized as an inactive proenzyme (procaspase), containing an N-terminal domain and two regions corresponding respectively to one large and one small subunit of the active form.
3. Activation occurs by proenzyme cleavage mediated by another active caspase, as part of a cascade of proteolytic events.

Hitherto, more than 14 caspase family members have been identified and organized into three subgroups:

- The ICE subgroup, consisting of caspase-1, caspase-2, caspase-4, caspase-5, caspase-11, and caspase-13, possibly involved in cytokine processing and inflammation
- The CED-3 subgroup, including caspase-3, caspase-6, caspase-7, caspase-8, caspase-9, caspase-10, and caspase-12, responsible for apoptosis initiation and execution
- Caspase-14, involved in keratinocyte differentiation

Caspases can also be classified as either initiator or executioner. Caspase-8 and caspase-9 are initiator caspases, as they are the first two enzymes to be activated; in turn, they activate executioner caspases, such as caspase-3 and caspase-7 (see also section "The Extrinsic Apoptotic Pathway: The CD95-R/CD95-L System").

Briefly, initiator caspases are recruited to the cytoplasmic domain of DR through a number of adaptor molecules that recognize and bind specific domains within procaspases: DED domains in procaspase-8 and procaspase-10 and *ca*spase *r*ecruitment *d*omains (CARD) in procaspase-1, procaspase-2, procaspase-4, and procaspase-9 (Figs. 32.5 and 32.6).

Once active, initiator caspases cleave and activate executioner caspases, which carry out proteolytic degradation of many, if not all, cellular proteins. Indeed, caspase-3, the main executioner caspase, is able to cleave some crucial proteins including proteins responsible for DNA stability and repair like poly-(ADP-ribose)-polymerase (PARP), DNA-dependent kinase (DNA-PK), and U1-70 kDa; lamin; cytoskeletal proteins such as actin and gelsolin; and pRb.

It is still unclear how many and which caspase substrates need to be degraded for apoptosis to be completed (for more details, see Chapter 30).

Apoptosis Regulators: The Bcl-2 Family The proto-oncogene *bcl-2* is one of the main regulators of the apoptotic pathway as it encodes a protein that can block cell death induced by a variety of stimuli (e.g., UV radiation, chemotherapeutic drugs, deregulated temperature, free radicals, TNF, and viruses).

Bcl-2 protein is localized in mitochondrial, nuclear, and endoplasmic reticulum membranes and is thought to inhibit apoptosis induction by inhibiting activation of a multiprotein complex, named apoptosome, formed by APAF-1, procaspase-9, and procaspase-3 (Fig. 32.5). This complex is activated upon cytochrome c binding, released from mitochondria. Apoptosome activation initiates the apoptotic pathway.

Although the exact molecular mechanisms underlying Bcl-2 antiapoptotic activity are yet to be fully elucidated, it has been established that it forms homodimer and

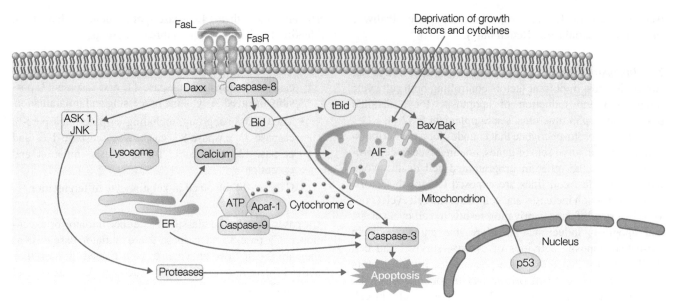

FIGURE 32.5 Schematic representation of Fas/FasL-induced apoptosis. FasL binding causes trimerization of its receptor CD95R (FasR) and assembly of a cytosolic multimeric complex containing adapter proteins recruited to the receptor via its death domain. This complex activates caspase-8 or FLICE, triggering the proteolytic cascade that leads to apoptosis.

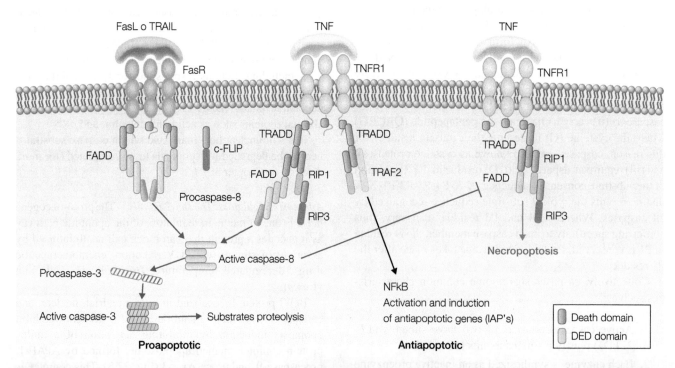

FIGURE 32.6 Schematic representation of cell death induction by TNF family receptors. FasL or TRAIL binding cause trimerization of CD95R (FasR), causing the assembly of a cytosolic complex of adapter proteins, in particular FADD, that activate caspase-8 or caspase-10. The consequent proteolytic cascade leads to apoptosis also through the intrinsic mitochondrial pathway, via recruitment and cleavage of BH3-only protein Bid. Note that the TNF receptor (TNFR) can induce different pathways depending on the adapter molecules contained in the cytoplasmic complex. In the presence of RIP-1, RIP-3, FADD, TRADD, and caspase-8, it induces apoptosis, whereas in the presence of RIP-1, TRADD, and TRAF2, it causes NF-κB activation and hence inhibition of apoptosis. TNFR may also trigger an alternative death pathway, known as "necroptosis," if RIP-3 but not caspase-8 is part of the complex bound to its death domain (DD).

heterodimers with two proapoptotic Bcl-2 family members, Bax and Bak, inhibiting their activity. When Bax and/or Bak expression exceeds that of Bcl-2—as a result of Bax/Bak overexpression or Bcl-2 downregulation—Bax and Bak can either homo- or heterodimerize to form pores in the outer mitochondrial membrane, allowing cytochrome c release and hence apoptosome activation (Fig. 32.5). Proapoptotic Bax and Bak may also be inhibited by another antiapoptotic member of the Bcl-2 family, Bcl-XL, that shares 44% sequence identity with Bcl-2. On the other hand, the antiapoptotic function of both Bcl-2 and Bcl-XL can be impaired by members of a subgroup of the Bcl-2 family, collectively named BH3-only proteins (e.g., PUMA, NOXA, HRK). These proteins are transcriptionally regulated and can bind both Bcl-2 and Bcl-XL, thus freeing Bax and Bak.

Both Bax and BH3-only proteins are also transcriptionally regulated by p53, confirming the pivotal role of this transcription factor in cell fate control.

The Intrinsic Mitochondrial Pathway Mitochondria are the main cellular site for energy production, containing the electron transport chain responsible for ATP production. However, they are also involved in apoptosis induction in response to several stimuli such as UV radiation, chemotherapeutic drugs, and growth factor withdrawal. The apoptotic pathway starts when cytochrome c, a mitochondrial factor usually involved in the electron transport chain, is released from the outer membrane into the cytosol. Here, it interacts with APAF-1, ATP, and procaspase-9 to form the apoptosome (Fig. 32.5).

As previously described, apoptosome formation determines activation of the initiator caspase-9 that, in turn, activates the executioner caspase-3.

Importantly, cytochrome c release is an early step in the intrinsic apoptotic pathway. Proapoptotic stimuli converge on the mitochondria membrane (MM), producing a temporary alteration of MM potential and permeability (MMP) resulting in the release of several factors such as cytochrome c, SMAC, and apoptosis-inducing factor (AIF). Such changes are controlled by Bax and Bak, which can homo- and heterodimerize to form pores in the outer MM (see also Chapter 34).

Bak is physiologically localized on both the outer MM and the ER membranes, whereas Bax is found also in the cytoplasm. Both Bax and Bak are bound and inhibited by Bcl-2 and Bcl-XL, but such effect is counteracted by proteins belonging to BH3-only subgroup of the Bcl-2 family, whose expression is induced by prodeath stimuli. BH3-only proteins can bind and translocate Bax to the MM but can also bind Bcl-2 and Bcl-XL, preventing their interaction with Bax and Bak.

Although Bax is crucial for cytochrome c release from mitochondria, other factors are thought to be required. The so-called permeability transition pores (PTPs) are pores crossing both the outer and inner MMs resulting from the interaction between proteins embedded in the outer membrane (voltage-dependent anionic channels (VDACs)) and protein located in the inner membrane (adenine nucleotide transferase (ANT)). PTP opening is promoted by both calcium accumulation into the mitochondrion and transmembrane potential alterations. Reactive oxygen species (ROS) are also responsible for PTP activation and mitochondrial apoptotic pathway stimulation. Indeed, antioxidant treatment inhibits ROS-stimulated apoptosis.

As mentioned previously, besides cytochrome c, other factors are released from mitochondria under prodeath stimuli, such as the proapoptotic proteins AIF and Smac. Once released, AIF is translocated to the nucleus where it activates a DNAse, whereas Smac inhibits the activity of caspase inhibitors such as IAPs. Recently, another factor has been found to be released from mitochondria under stress conditions, endonuclease G, which moves to the nucleus to degrade DNA.

The Extrinsic Apoptotic Pathway: The CD95-R/CD95-L System The extrinsic pathway is activated when a member of the transmembrane DR family (e.g., CD95-R) interacts with an extracellular soluble death protein (e.g., CD95-L). TNFR is another member of the DR family specifically binding TNF. All members of the DR family share a characteristic death domain (DD) in the intracellular portion that mediates the recruitment of cytosolic adapter proteins (Fig. 32.6). CD95-L binding induces trimerization of its receptor, CD95-R, and interaction of the cytosolic DD with the corresponding DD domain of the adapter Fas-associated protein with death domain (FADD). FADD contains also another domain, called death effector domain (DED), that recruits and binds the corresponding DED in the N-terminal tail of procaspase-8. Assembly of such multiprotein complex results in caspase-8 transactivation and apoptosis initiation. Interestingly, caspase-8 activation by the extrinsic pathway may lead to activation of the intrinsic pathway through caspase-8-dependent processing of the cytosolic protein Bid (a BH3-only protein) into a form, called tBid, which can translocate to the outer MM and mediate cytochrome c release.

Caspase-8 can be activated also by the intrinsic apoptotic pathway in a DR-independent manner.

Apoptosis and Cell Cycle Cell cycle and apoptosis are intimately interconnected since several different stimuli (e.g., cytokines and growth factors) can stimulate cell growth and at the same time inhibit apoptosis induction. Conversely, growth factor and/or cytokine shortage, ionizing radiation, and some drugs initiate the apoptotic pathway preferentially in proliferating cells, whereas terminally differentiated cells are relatively insensitive to apoptotic stimuli. This is particularly important in the context of anticancer therapies, since

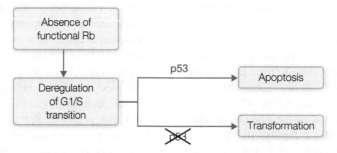

FIGURE 32.7 pRb-mediated apoptosis or cell transformation depends on p53. Inhibition of pRb activity results in dysregulation of G1–S transition, potentially triggering uncontrolled cell replication. Functional p53 prevents this phenomenon by inducing apoptosis. Conversely, when p53 is mutated or missing, cell transformation and tumor growth occur.

these agents indiscriminately kill cancer cells and cells in active proliferation such as lymphoblasts.

On the other hand, cell cycle dysregulation does not induce apoptosis but, conversely, may initiate cell transformation. However, apoptosis induction is an active process requiring specific stimuli and mechanisms usually associated with altered G1–S transition. In this context, a pivotal role is played by p53, which can actively induce apoptosis and whose inactivation results in cell transformation and tumor development because of dysregulation of G1–S transition (Fig. 32.7).

In addition, once induced, some active components of the apoptotic pathway, such as active caspase-3, may degrade key factors controlling cell cycle progression (such as Wee, which regulates Cdk1 activity, and a factor regulating cyclin A and B ubiquitination) resulting in cell proliferation arrest.

Role of Drugs in the Crosstalk between Cell Cycle and Apoptosis Apoptosis dysregulation may cause either excessive or ineffective cell elimination compared to normal conditions and is associated with several human pathologies. Increased neuronal apoptosis is associated with the progression of neurological disorder such as Parkinson's and Alzheimer's syndromes, while HIV-induced massive T lymphocytes demise results in immunodeficiency syndrome. On the other hand, deficient apoptosis due to mutations in key genes controlling the apoptotic pathway may contribute to cancer development. Therefore, pharmacological treatments capable of increasing or inhibiting apoptosis can be beneficial in the treatment of such human conditions (see also Chapter 33).

NECROSIS

Necrosis is generally described as a process of uncontrolled cell death characterized by increase in cell volume due to water influx, unspecific degradation of proteins and nucleic acids, general loss of energy, and mitochondria and plasma membrane disruption with progressive release of cellular components in the extracellular space. Thus, necrosis seems to be a passive process resulting from an acute cell stress and eventually leading to inflammation and potential tissue damage. However, accumulating data indicate that necrotic cell death is not merely the consequence of an acute stress, but similarly to apoptosis, it may be induced by excitotoxin exposure, ischemia, bacterial toxins, viruses, protozoa, and other triggers. Moreover, a physiopathological role as well as a specific genetic program controlling necrotic cell death seems to exist.

Morphological Features of Necrosis

Necrosis is typically associated with a rapid increase in cell volume that has been related to sodium uptake, initially balanced by potassium exit and subsequently followed by a further increase in intracellular sodium concentration. It is important to note that potassium (and chlorine) exit from cells is also associated with the initial decrease in cell volume occurring during early apoptotic events. Therefore, early decrease in intracellular potassium concentration (and possibly many other osmotic molecules) is a common event in both necrosis and apoptosis, but the decision to initiate one or the other of the two processes seems to depend on the availability of ATP (at least in some circumstances), required to sustain proton pump activity and other processes essential for proper ionic homeostasis. Thus, early ATP decrease would result in cytosolic sodium accumulation responsible for water influx and cellular volume expansion; on the contrary, high ATP levels, by preserving ion homeostasis, would allow a decrease in both cytosolic potassium concentration and cell volume.

However, it has also been reported that necrosis may occur even with unaffected or slightly reduced ATP levels, indicating that the ATP level is not always critical and other factors determine which death pathway cells will enter. Moreover, these observations suggest the existence of some shared mechanisms in apoptosis and necrosis, suggesting that mitochondria may represent both direct and indirect targets of the aforementioned kinases.

Role of Mitochondria in the Necrotic Process

The inner mitochondrial membrane may lose its selective permeability to ions, or other small molecules (<2 kDa), as a consequence of pore formation at the sites of contact between inner and outer mitochondrial membranes. This process, defined mitochondrial permeability transition (MPT), is regulated by several proteins, some of which belong to the Bcl-2 family. Generally, mechanisms preventing MPT (e.g., Bcl-2 overexpression, Bad translocation from mitochondria to the cytosol) or substances inhibiting MPT (e.g., cyclosporin A) are able to prevent both apoptosis and necrosis. Moreover, kinases that promote apoptosis (e.g., JNK and p38) or that inhibit it (e.g., AKT and ERK) can also play a similar role in necrosis. Interestingly, it has been reported that exposure to toxic substances may also be associated with cell survival due to activation of signaling pathways inhibiting MPT, for example, through Bad phosphorylation, which reduces Bad concentration in the mitochondria, thereby allowing Bcl-2 to exert its anti-MPT activity. Inhibition of these pathways at different stages upstream of Bad phosphorylation leads to rapid necrotic cell death, without a significant reduction of the ATP pool, as previously observed in other paradigms of toxicity leading to apoptosis.

It has been previously reported that caspases play a critical role in most apoptotic models; in the presence of endogenous or exogenous inhibitors of these enzymes or in the absence of their expression, the apoptotic process is inhibited and necrotic cell death often occurs. Thus, similarly to what occurs in conditions causing a decrease in ATP levels, in the presence of caspase inhibition, the apoptotic process is inhibited and necrosis ensues.

Interestingly, it has been recently reported that classical apoptotic inducers, such as CD95L, TNF-α, and TRAIL, are also able to stimulate a different cell death program known as necroptosis. This process may be defined as a form of necrotic cell death regulated by receptor-interacting protein-1 (RIP-1), a serine/threonine kinase. Interestingly, the inhibitor of RIP-1, necrostatin-1, prevents necroptosis with no significant effects on either apoptosis or several models of necrosis.

The discovery of necrostatin-1 has allowed understanding the role of necroptosis in tissue damage associated with cerebral ischemia, excitotoxicity, myocardial infarction, and exposure to anticancer agents.

The identification of substrates and downstream effectors of RIP-1 and similar kinases (such as RIP-3) will provide a more in-depth understanding of the necroptotic program.

In summary, data currently available suggests the existence of early events shared by apoptosis and necrosis, as well as the existence of subsequent events directing cells toward either apoptosis or necrosis. Intensity and duration of the stimulus are important parameters that can also trigger either apoptosis or necrosis independently from any effects on the cell energy state. Finally, there is clear evidence of the existence of a highly regulated necrotic death process, called necroptosis, that involves metabolic pathways different from those relevant to the apoptotic process.

Therefore, necrosis, after being long considered as an alternative cell death pathway to apoptosis, has recently acquired a significantly different role.

DRUGS AND APOPTOSIS

Proapoptotic Drugs

Most chemotherapeutic agents used in cancer treatment trigger apoptosis by causing DNA damage either directly (e.g., cis-platinum, mitomycin C, and bleomycin) or indirectly (e.g., DNA topoisomerase I and II inhibitors, such as camptothecin and teniposide) (see also Chapter 33). As we said previously, p53 is a key sensor of DNA damage, and its activation determines cell cycle arrest at the G1–S transition to allow repair of damaged DNA and, if the damage cannot be repaired, to induce apoptosis. The observation that in many tumors resistance to chemotherapy is associated with loss of p53 function has led to the idea of developing a gene therapy approach to introduce a functional copy of the p53 gene in tumor cells, using different carrier systems, such as retrovirus, adenovirus, or cationic liposomes, to deliver it. This approach, in combination with traditional therapies, has already been tested in animal models with encouraging results and has recently been attempted also in humans. Another important group of chemotherapeutic agents is that targeting cytoskeletal structures and in particular microtubules. For example, vincristine and vinblastine prevent tubulin polymerization and thereby mitotic spindle assembly, resulting in apoptosis induction. On the contrary, other compounds, such as taxol, stabilize microtubules; also, these agents can induce apoptosis by preventing chromosome separation during mitosis. However, alternative mechanisms may underlie taxol proapoptotic action and in particular its ability to promote inactivation of antiapoptotic Bcl-2 protein by phosphorylation.

Other mechanisms by which chemotherapeutic agents induce apoptosis may include production of oxygen radicals (doxorubicin) and generation of ceramide (daunorubicin). However, the mechanisms underlying the proapoptotic effect of anticancer drugs are still largely unknown. Advances in this field of research will allow developing new pharmacological strategies to selectively destroy cancer cells sparing normal healthy cells (see also Chapter 33).

Drugs That Inhibit Apoptosis

The discovery of the role of caspases in the apoptotic process suggested the possibility to develop specific inhibitors to block cell death in diseases characterized by increased apoptosis (e.g., AIDS and neurodegenerative disorders). These inhibitors usually consist of polypeptides containing the consensus sequences recognized by caspases. Although the development of such antiapoptotic drugs may have a wide clinical application, their use is nowadays feasible only in acute diseases, such as fulminant hepatitis or myocardial infarction, in which apoptosis inhibition may be crucial in preventing irreversible damage of the involved organs.

More complex is the case of chronic diseases, such as degenerative disorders, in which prolonged cell death inhibition could favor the development of tumors by inhibiting removal of damaged, precancerous cells.

In conclusion, recent studies on the molecular mechanisms regulating the genetic program of cell death have opened unexpected new avenues for the clinical use of "old" drugs in novel and innovative therapeutic regimens.

TAKE-HOME MESSAGE

- Cell cycle progression involves specific sets of cyclins.
- The transcription factor p53 controls cell cycle progression and apoptosis induction at several steps.
- In multicellular organisms, cell death is a physiological process whose dysregulation is associated with development and/or progression of several human pathologies.
- Different genes control different steps of the apoptotic pathway with caspases acting as executors of the program.
- Several signals, including mitochondria and/or plasma membrane receptors, activate apoptosis.
- Pharmacological modulation of cell death is a new frontier in the therapeutic approach to human diseases such as cancer, neurodegenerative, autoimmune, and infectious diseases.

FURTHER READING

De Falco M., De Luca A. (2010). Cell cycle as a target of antineoplastic drugs. *Current Pharmaceutical Design*, 16, 1417–1426.

Delston B.B., Harbour J.W. (2006). Rb at the interface between cell cycle and apoptotic decision. *Current Molecular Medicine*, 6, 713–718.

Green D., Kroemer G. (2004). The pathophysiology of mitochondrial cell death. *Science*, 305, 626–629.

Kroemer G. (2003). Mitochondrial control of apoptosis: an introduction. *Biochemical and Biophysical Research Communications*, 304, 433–435.

McCall K. (2010). Genetic control of necrosis—another type of programmed cell death. *Current Opinion in Cell Biology*, 22, 882–888.

Rubinsztein D.C. Codogno P., Levine B. (2012). Autophagy modulation as a potential therapeutic target for diverse diseases. *Nature Reviews. Drug Discovery*, 11, 709–729.

Suryadinata R., Sadowski M., Sacevic B. (2010). Control of cell cycle progression by phosphorylation of cyclin-dependent kinase (CDK) substrates. *Bioscience Reports*, 30, 243–255.

Tyson J.J., Novak B. (2008). Temporal organization of the cell cycle. *Current Biology*, 8, R759–R768.

Vandenabeele P., Galluzzi L., VandenBerghe T., Kroemer G. (2010). Molecular mechanisms of necroptosis: an ordered cellular explosion. *Nature Reviews. Molecular Cell Biology*, 11, 700–714.

Wyllie A.H., Kerr J.F., Currie A.R. (1980). Cell death: the significance of apoptosis. *International Review of Cytology*, 68, 251–306.

Youle R.J., Strasser A. (2008). The Bcl-2 protein family: opposing activities that mediate cell death. *Nature Reviews. Molecular Cell Biology*, 9, 47–59.

33

MECHANISMS OF ACTION OF ANTITUMOR DRUGS

Giovanni Luca Beretta, Laura Gatti, and Paola Perego

By reading this chapter, you will:

- Know the targets and the mechanisms of action of antitumor drugs
- Know the different classes of antitumor agents and distinguish them into conventional and target-specific drugs
- Learn the molecular rationale behind the clinical use of each antitumor drug group

---

Conventional antitumor therapy is based on the use of antiproliferative or cytotoxic agents that can be grouped into different classes according to their mechanism of action. Such drugs are mainly genotoxic agents that interfere with normal DNA functions in different ways. Cytotoxic agents target fundamental processes underlying cell proliferation (i.e., DNA function and cell division). The drugs that interfere with DNA functions are DNA damaging (genotoxic) compounds or antimetabolites (inhibitors of nucleic acid biosynthesis), whereas those interfering with cell division are inhibitors of mitosis acting on microtubule dynamics.

A large number of the clinically available drugs acts on targets associated with cell proliferation and are classified according to their origin, chemical structure, and action mechanism (see Box 33.1, "Therapeutic targets and drug choice criteria"). The cellular targets of cytotoxic agents in clinical use cannot be considered tumor specific, although their variable efficacy in different tumor types may suggest a partial specificity probably due to the tumor cell ability to activate a rapid process of cell death in response to some molecular modifications. Besides the high toxicity resulting from the lack of sufficient selectivity for tumor cells, another important limitation of cytotoxic agents is their reduced capability to kill drug-resistant tumor cells, especially in solid tumors. Many preclinical studies have shown that cytotoxic drugs kill tumor cells mainly by triggering their programmed cell death or apoptosis (see also Chapter 32). This process is finely regulated at genetic level and involves numerous gene products with either pro- or antiapoptotic function. Alterations affecting cell death regulation may lead to intrinsic resistance.

Our increasing knowledge on the molecular alterations present in tumor cells has led to the development of target-specific agents. The most innovative therapeutic approaches aim at interfering with factors that are critical for tumor cell survival, including components of the mitogen signal transduction pathway, which is frequently altered in tumors. These approaches differ from the classical pharmacological strategies based on genotoxic damage as they are designed to selectively hit tumors of specific histological types. Recent molecular and biochemical studies have allowed devising therapeutic strategies aimed at interfering with proteolytic processes and epigenetic changes. Some agents capable of reactivating the expression of tumor suppressor genes and proapoptotic genes are currently under clinical evaluation. Integration of information coming from biochemistry and molecular studies has prompted a renewed interest in cancer cell metabolism, which is now proposed as a potential drug target. Moreover, advances in understanding the immune response to chemotherapy, which can also be immunostimulating, are contributing to the development of strategies combining immune-based therapies with chemotherapy regimens. Despite their higher specificity toward tumor cells, the use of these new agents may still be limited by drug

---

General and Molecular Pharmacology: Principles of Drug Action, First Edition. Edited by Francesco Clementi and Guido Fumagalli.
© 2015 John Wiley & Sons, Inc. Published 2015 by John Wiley & Sons, Inc.

BOX 33.1 THERAPEUTIC TARGETS AND DRUG CHOICE CRITERIA

Drug class	Main drugs	Main target	Mechanism of action	Clinical use and notes
Alkylating agents	Dacarbazine, temozolomide	DNA	Covalent binding to nucleophilic DNA residues	Dacarbazine: melanoma, sarcoma and lymphoma Temozolomide: melanoma, glioblastoma
Platinum compounds	Carboplatin, cisplatin, oxaliplatin	DNA	Covalent binding to nucleophilic DNA residues	Carboplatin/cisplatin: ovarian carcinoma, non-small cell lung cancer; testicular cancer; head and neck carcinoma
Antifolates	Methotrexate	DNA metabolism	Binding to dihydrofolate reductase blocking DNA and RNA synthesis	Acute lymphatic leukemia and various solid tumor types in drug combination regimens
Purine analogs	6-Mercapto-purine	DNA metabolism	Purine synthesis inhibition by block of conversion of inosinic to adenylic acid	Maintenance therapy of acute lymphatic leukemia
Pyrimidine analogs	Gemcitabine	DNA metabolism	Decrease of deoxynucleotide pool and replication blockade	Pancreatic cancer, non-small cell lung cancer, breast cancer, ovarian cancer
Taxanes	Taxol	Microtubules	Stabilization of microtubules	Breast and ovarian cancer
Vinca alkaloids	Vincristine, vinblastine	Microtubules	Depolymerization of microtubules	Breast and ovarian cancer
Camptothecins	Topotecan	DNA topoisomerase I	Stabilization of the topoisomerase I-DNA complex	Topotecan: ovarian and lung cancer; irinotecan: colon cancer irinotecan
Anthracyclines	Doxorubicin, daunorubicin, epirubicin, idarubicin	DNA topoisomerase II	Stabilization of the topoisomerase II-DNA complex	Doxorubicin and epirubicin: various solid tumors and hematological malignancies; daunorubicin and idarubicin: leukemias
Anthracenediones	Mitoxantrone	DNA topoisomerase II	Stabilization of the topoisomerase II-DNA complex	Breast cancer; lymphomas
Epipodophyllotoxins	Etoposide	DNA topoisomerase II	Stabilization of the topoiso-merase II-DNA complex	Lymphoma, lung and testis cancer
Tyrosine kinase inhibitors	Imatinib, mesylate/Gleevec	BCR-Abl	ATP mimetic	Philadelphia-chromosome positive-chronic myelogenoeous leukemia; KIT-positive gastrointestinal stromal tumors
Tyrosine kinase inhibitors	Gefitinib/Iressa	EGF receptor	ATP mimetic	Advanced or metastatic non-small cell lung cancer; efficient inhibition of mutant EGF receptor
Tyrosine kinase inhibitors	Erlotinib/Tarceva	EGF receptor	ATP mimetic	Advanced or metastatic non-small cell lung cancer; pancreatic cancer in combination with gemcitabine; efficient inhibition of mutant EGF receptor

Drug class	Main drugs	Main target	Mechanism of action	Clinical use and notes
Tyrosine kinase inhibitors	PLX4032/Vemurafenib	Mutant BRAF	ATP competitive BRAF (V600E) kinase inhibitor	Melanoma with BRAF mutation
Proteasome inhibitors	Bortezomib/Velcade	20S proteasome	Selective inhibitor of the catalytic site of the 20S proteasome	Multiple myeloma and lymphomas
Monoclonal antibodies	Trastuzumab/herceptin	HER-2/neu, ERBB-2	Blockade of signal transduction	Advanced breast cancer overexpressing HER-2
Monoclonal antibodies	Cetuximab	EGF receptor	Blockade of signal transduction	Colon carcinoma, non-small cell lung cancer, head and neck cancer
Monoclonal antibodies	Bevacizumab/Avastin	VEGF	Blockade of signal transduction	Breast cancer, metastatic colon cancer

resistance and toxic effects (see also Supplement E 33.1, "Mechanisms of Drug Resistance in Cancer Chemotherapy"). Another strategy currently being pursued consists in combining different drugs capable of interfering with the multiple molecular alterations of tumor cells.

On the other hand, several approaches are emerging to decrease drug-related toxicities and overcome drug resistance, which employ nanocarriers to deliver multiple drugs and/or polymeric drug conjugates. The introduction of new technologies (e.g., comparative genomic hybridization or CGH, microarrays, proteomics) to investigate gene, mRNA, and protein functions on a genomic scale has already contributed to develop new tools to validate novel potential molecular targets. Research is also focusing on the identification of markers of response that may allow predicting patient's outcome and response to treatments, thereby leading to personalized therapy.

CONVENTIONAL ANTITUMOR DRUGS

Alkylating Agents

Alkylating agents, the first compounds used in antitumor therapy, are a heterogeneous group of drugs (Fig. 33.1) that share the capability of alkylating nucleophilic sites (in particular nitrogen, oxygen, and sulfur) of cell macromolecules. They can virtually damage all kinds of cell macromolecules, but their cytotoxic, clastogenic, and mutagenic effects are linked to the capability of damaging DNA. These agents interact with DNA either directly or through species generated by metabolic or spontaneous chemical activation (e.g., dacarbazine, temozolomide). Alkylating agents covalently bind nucleophilic DNA residues forming inter- and intramolecular cross-links. They can replace a proton belonging to another molecule with an alkyl cation. Oxygen and nitrogen atoms in DNA bases and phosphodiester bonds

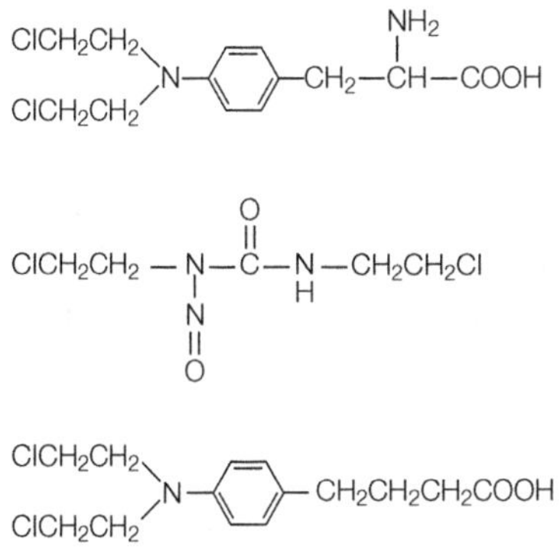

Principal alkylating agents	
Nitrogen mustard	Mechlorethamine
	Melphalan
	Chlorambucil
	Cyclophosphamide
	Iphosphamide
Nitrosureas	BCNU
	CCNU
	MeCCNU
	Streptozotocin
Alkyl alkane sulfonates	Busulfan
	Treosulfan
Aziridin	Thio-TEPA
	Mitomycin C
Compounds with N-methyl group	Hexamethyl methylamine
	Procarbazin
	Dacarbazine
	Temozolomide

FIGURE 33.1 Main alkylating agents and chemical structures.

are targets for alkylation, the most frequently damaged being N7-guanine and O6- and O3-adenine.

Alkylating agents can be grouped as mono- and bifunctional compounds based on the way they interact with DNA. In principle, the most potent and effective compounds are bifunctional agents (e.g., chlorambucil and melphalan) that mainly act by generating DNA interstrand cross-links through alkylation of N7-guanine or other nucleophilic sites on both DNA strands. This feature distinguishes bifunctional agents from tetrazines (e.g., temozolomide), which are monofunctional agents and alkylate DNA at single sites.

Alkylating agents can also be classified based on their ability to alkylate O6-guanine, forming an adduct crucial for their antitumor activity. Nitrosureas, tetrazines, and also procarbazine act through this mechanism. Currently, O6-guanine alkylating agents are mainly used in the treatment of melanoma and glioblastoma.

Platinum Compounds

Cisplatin is a Pt(II) coordination complex with a planar structure where a central Pt atom is linked to two chlorides and two aminic groups in *cis* position (Fig. 33.2). The main cellular target of this drug is DNA. Cisplatin interferes with DNA functions with a mechanism similar to that of bifunctional alkylating agents because it forms (i) intermolecular cross-links between opposite DNA strands, (ii) DNA intrastrand cross-links between bases of the same strand, and (iii) cross-links between DNA and DNA-associated proteins (DNA polymerases, DNA topoisomerases, histones; Fig. 33.2). Studies have shown that the most frequent bivalent adduct produced by cisplatin and likely responsible for its cytotoxic activity is the cross-link between adjacent purines of the same DNA strand with preference for guanines (1,2-d(GpG)). In such case, the cross-link occurs at the N7 position due to its nucleophilic nature. Indeed, when cisplatin enters the cells, because of the low intracellular chloride concentration, it loses its chlorides and becomes a potent electrophilic compound capable of interacting with nucleophilic residues of macromolecules. In addition, intramolecular links involving adenine (in particular at O6 position) and intermolecular links are formed that, although less frequent, are relevant for the drug cytotoxic effect, as they represent an obstacle to DNA replication and transcription. Cisplatin is employed in the treatment of solid tumors such as germ testicular cancer, lung cancer, osteosarcoma, and cervix and ovarian carcinoma. Another clinically available platinum compound is oxaliplatin (Fig. 33.2), which contains diaminecyclohexane (DACH) as carrier ligand and seems to generate less adducts and cross-links than cisplatin, in spite of the similar or even more marked cytotoxic effect. Thus, DNA binding is not always a reliable measurement of drug antiproliferative activity.

Antimetabolites

Antimetabolites (antifolates, purine, and pyrimidine analogs) include various compounds with chemical structure similar to the metabolites physiologically employed for DNA and RNA synthesis (Fig. 33.3). These drugs can be substrates and/or enzyme inhibitors. This means that cells exposed to these agents may erroneously incorporate them in nucleic acids giving rise to metabolically inactive compounds or may have key enzymes blocked by them. Antimetabolites that interfere with nucleic acid synthesis can be grouped into three classes: folic acid antagonists, purine analogs, and pyrimidine analogs. Figure 33.3 shows the chemical structures of some antimetabolites of clinical interest and their cellular targets.

Antifolates The antifolates are structural analogs of folic acid, an essential cellular vitamin that serves as carbon donor in many biosynthetic reactions, including amino acid (methionine, serine) biosynthesis, *de novo* purine biosynthesis, and thymidylate synthesis, as well as in methylation reactions important for regulating gene expression. Aminopterin, the first antimetabolite to be introduced in clinical use but subsequently abandoned for its severe toxic effects and replaced by methotrexate (Fig. 33.3), belongs to this class of compounds. Methotrexate is taken up by cells mainly through the reduced folate transporter (encoded by the RFC1 gene). However, if cells are exposed to high drug concentrations, it enters also by passive diffusion. Inside the cell, the drug binds dihydrofolate reductase and makes the enzyme unable to catalyze the

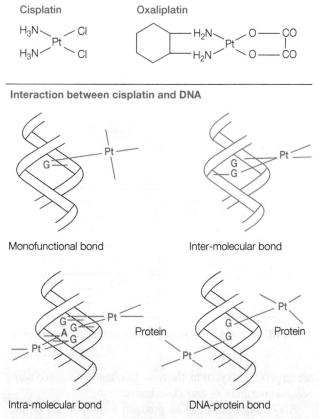

FIGURE 33.2 Chemical structures of platinum compounds and schematic representation of cisplatin–DNA interaction.

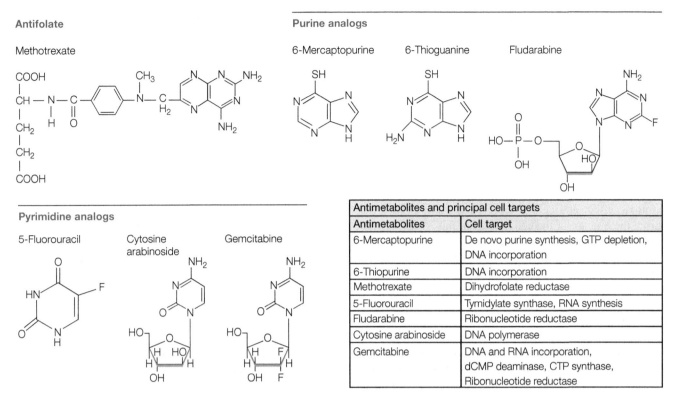

FIGURE 33.3 Chemical structures of the main antimetabolites.

reduction of dihydrofolic acid (FH2) to tetrahydrofolic acid (FH4), normally used in the synthesis of thymidylate and purines, thereby inhibiting DNA and RNA synthesis. Folate and antifolates can be metabolized to polyglutamate derivatives in a reaction catalyzed by folate polyglutamate synthetase. These derivatives are trapped in the cytoplasm and have a high affinity for specific enzymes; for example, the polyglutamate forms of methotrexate (methotrexate polyglutamates) have high affinity for thymidylate synthetase.

Purine Analogs This class of drugs includes 6-mercaptopurine and 6-thioguanine, the thiopurine analogs of hypoxanthine and guanine, respectively. Their cytotoxic action is due to inhibition of purine neosynthesis and blockade of the conversion of inosinic acid to adenylic/guanylic acid. Among the purine derivatives of therapeutic interest, the deoxyadenosine analogs, such as fludarabine (2-fluoro-ara-AMP), characterized by resistance to deoxyadenosine deaminase-mediated deamination, are of great interest. Intracellular fludarabine is converted into 5′-phosphorylated derivatives by deoxycytidine kinase; the triphosphate derivative is a potent inhibitor of DNA polymerases and ribonucleotide reductase and can be incorporated into both DNA and RNA. In addition, the triphosphate derivative inhibits also the DNA primase, thereby blocking the synthesis of RNA primer, and DNA ligase I, thus interfering with DNA synthesis. Fludarabine induces apoptosis in lymphoid and myeloid tumor cells in which it causes downregulation of the antiapoptotic Bcl-2 protein, thus promoting cell death.

Pyrimidine Analogs Among the pyrimidine analogs, fluoropyrimidines are widely used for the treatment of gastrointestinal tumors. The prototype molecule is represented by the fluorinated uracil analog 5-fluorouracil (5FU; Fig. 33.3), which is structurally similar to orotic acid and thymidine and therefore enters the same metabolic pathways of these physiological substrates. Cellular uptake of 5FU occurs both by passive diffusion and through high affinity transporters. The mechanism of action depends on the type of conversion the molecule undergoes. Indeed, to exert its activity, 5FU must be converted to:

- 5-fluorouridine-5′-triphosphate (5FUTP), which acts as RNA polymerase substrate and is incorporated into RNAs, thereby inhibiting protein synthesis.
- 5-fluoro-2′-uridine 5′-triphosphate (5FdUTP), which can be incorporated in the DNA.
- 5-fluorouridine-5′-diphosphate (5FUDP) sugars, which can interfere with protein and lipid glycosylation.
- 5-fluoro-2′-deoxyuridine-5′-monophosphate (FdUMP), which binds thymidylate synthase blocking its activity, thereby inhibiting DNA synthesis. Inhibition of thymidylate synthase by FdUMP is considered the main event in 5FU action mechanism.

The antileukemic drug cytosine arabinoside (Ara-C) is the best-known compound belonging to the cytidine analogs (Fig. 33.3). It differs from deoxycytidine for the presence of arabinose instead of ribose or deoxyribose. Ara-C enters the cell through the nucleoside triphosphate pathway. Inside the cells, Ara-C is phosphorylated by deoxycytidine kinase and, when incorporated into the DNA, inhibits DNA polymerase, thus causing arrest of DNA synthesis and repair. The main effect of Ara-C consists in the stable inhibition of DNA elongation. Several studies have shown that cellular accumulation of Ara-CTP and its incorporation into DNA are relevant determinants of Ara-C cytotoxic effect and cells are more sensitive to the drug during the S phase of the cell cycle. For this, Ara-C is considered an S phase-specific drug.

Gemcitabine (2′,2′-difluoro-2-deoxycytidine) is a pyrimidine analog structurally related to cytarabine (Fig. 33.3). The metabolism of gemcitabine is very similar to that of other nucleoside analogs, as it competes with deoxycytidine for incorporation into DNA. Although structurally similar to cytosine arabinoside, gemcitabine is pharmacologically and pharmacokinetically different, being more active on solid tumors. Gemcitabine acts through at least two mechanisms: (i) it decreases the pool of deoxynucleotides (including deoxycytidine triphosphate, dCTP), available for DNA synthesis, and competes with dCTP for incorporation into DNA (self-potentiation); (ii) it blocks replication by replacing dCTP in the nascent DNA chain (causing arrest of DNA strand elongation), a process called "masked chain termination." Therefore, gemcitabine blocks proliferating cells in G1–S phase, acting both as an inhibitor and as a suicide substrate during DNA synthesis. The mechanism of self-potentiation of gemcitabine increases drug concentration and prolongs the retention of its active nucleotides in tumor cells; this property distinguishes gemcitabine from other antimetabolites. When the drug is incorporated into the DNA strand, DNA synthesis does not stop immediately, but one or more normal nucleotides are added to the chain before DNA elongation stops. This process traps the drug in the DNA filament, preventing recognition, excision, and repair of the lesion.

Drugs Acting on Microtubules

Drugs acting on tubulin represent a class of compounds used not only as anticancer agents but also as herbicides and pesticides (Fig. 33.4). The first drug of this family used as a therapeutic agent in humans was colchicine, an alkaloid extracted from the plant *Colchicum autumnale*. Vinca alkaloids (vincristine and vinblastine) and taxanes are two additional drug groups with similar functions. The mechanism of action of these compounds is based on their ability to bind tubulin and prevent normal microtubule functions during cell division. Among all conventional chemotherapeutics, drugs acting on tubulin are peculiar in their activity because their target is the cellular mitotic apparatus rather than DNA. Antitubulin agents have a wide spectrum of effects and are used in the treatment of solid tumors (in particular breast and ovarian cancers) and hematologic malignancies. However, the modifications they produce on microtubule polymerization can cause pathological effects on axons (neuritis). Interestingly, it has been experimentally observed that taxanes, by stabilizing microtubules, contribute to axon regeneration after nervous system injury, thereby providing a potential therapeutic option for this disease.

Mechanism of Action of Antitubulin Drugs Antitubulin drugs exert their antitumor activity by altering mitotic apparatus function during cell division. They exert an effect also on interphase cells, by interfering with both cell adhesion and motility. Vinca alkaloids have a mechanism of action slightly different from colchicine. Whereas the latter binds free tubulin in the cytoplasm before its incorporation in microtubules, vinca alkaloids act directly on already organized microtubules. These drugs cause microtubule depolymerization with the formation of spiral filaments lacking microtubular structure. Taxanes bind microtubules at specific binding sites located on each tubulin dimer, and their action is tightly

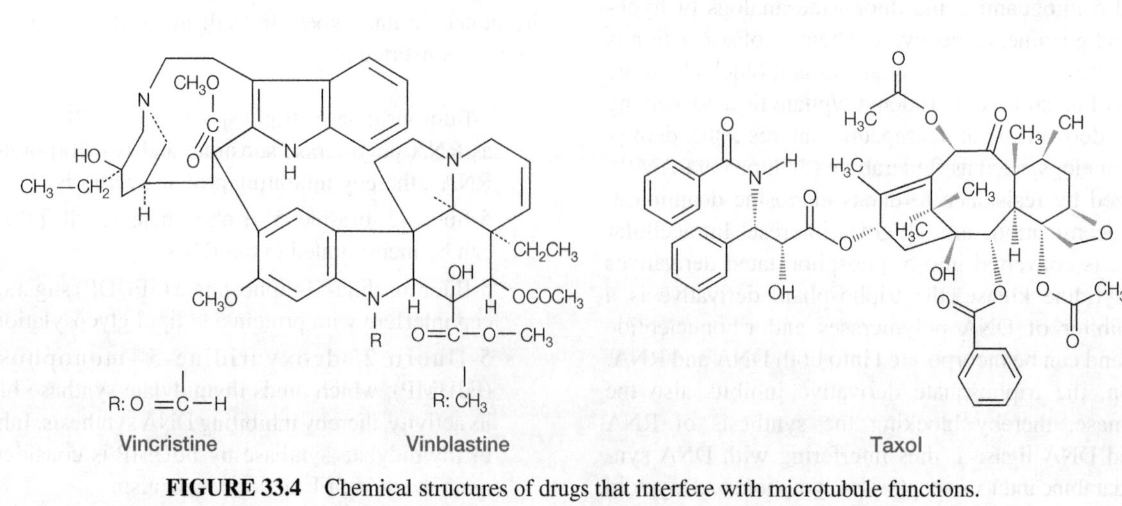

FIGURE 33.4 Chemical structures of drugs that interfere with microtubule functions.

concentration dependent. At high concentrations, they induce microtubule polymerization, whereas at low concentrations, they stabilize them by blocking addition of tubulin heterodimers at one microtubule end and dissociation at the opposite end (treadmilling). The final result of taxanes action is mitotic cell death by apoptosis. Importantly, in contrast with drugs causing DNA damage, these compounds induce apoptosis via p53-independent pathways, and for this reason, they are being used in the treatment of tumors bearing mutated p53.

DNA Topoisomerase Inhibitors

DNA topoisomerases are enzymes whose function and mechanistic properties have been extensively characterized. These proteins play an important role in many nuclear processes (DNA replication, repair, and transcription), but they are pharmacologically relevant also because they can be transformed into intracellular toxins by a group of drugs that inhibit their function. All DNA topoisomerases share two properties. The first one is the capability to cleave and then rejoin DNA phosphodiester bonds by two successive transesterification reactions. In particular, DNA topoisomerase I binds the DNA double helix and cleaves one of the two strands, whereas DNA topoisomerase II acts as a dimer and cleaves both DNA strands. In both cases, during the cleavage, a covalent intermediate between cleaved DNA and topoisomerase is formed. The second property is that once the topoisomerase–DNA intermediate is formed, the enzyme separates the broken ends and allows the passage of another DNA, single-stranded in the case of topoisomerase I and double-stranded in the case of topoisomerase II, through the break. The drugs act by stabilizing the reaction intermediate, the topoisomerase–DNA covalent complex, and promoting the formation of DNA breaks. Indeed, stabilization of the topoisomerase–DNA intermediate interferes with the activity of proteins that slide along DNA (helicase, DNA polymerase, RNA polymerase). The collision between these proteins and topoisomerase–DNA complexes produces lethal DNA strand breaks that are believed to be the basis of the antitumor activity of these drugs (Fig. 33.5).

Poisoning the DNA Topoisomerase I: Camptothecins
Camptothecin, an alkaloid extracted from *Camptotheca acuminata* that targets DNA topoisomerase I, is the prototype molecule of a class of drugs that play a primary role in the treatment of several human tumors. Following clinical studies demonstrating the poor activity and high toxicity of the natural molecule in its salt form (open lactone-ring form), structural studies have led to the synthesis of two potent camptothecin derivatives, topotecan, currently used in ovarian and lung cancers, and irinotecan (prodrug of SN-38), administered in colon cancer patients (Fig. 33.6). Other derivatives, such as ST1968 (namitecan) and ST1481 (gimatecan), are at different phases of clinical evaluation. It

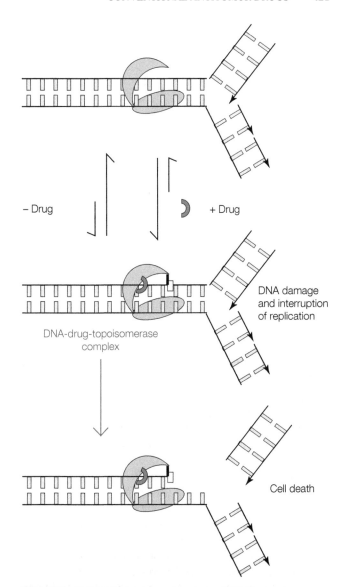

FIGURE 33.5 Mechanism of action of DNA topoisomerase inhibitors. The figure shows how DNA topoisomerase physiologically associates with DNA; enzyme inhibitors stabilize the topoisomerase–DNA complex, causing DNA replication arrest and lethal DNA breaks, thus triggering cell death.

has been established that the lactone ring and the S configuration at position 20 of camptothecins are essential for their antitumor activity. The stability of the lactone ring is a key factor, as its hydrolysis under physiological conditions produces a large amount of the inactive carboxylate derivative, which has high affinity for human serum albumin. To increase the stability of the lactone ring, new camptothecins have been synthesized, including homocamptothecins, which contain one additional methylene group between the hydroxyl and the carboxyl groups of the lactone ring. Besides this property, these compounds have the ability to stabilize the topoisomerase I–DNA complex typical of camptothecins. In particular, the 7-oxyimino methyl derivative gimatecan

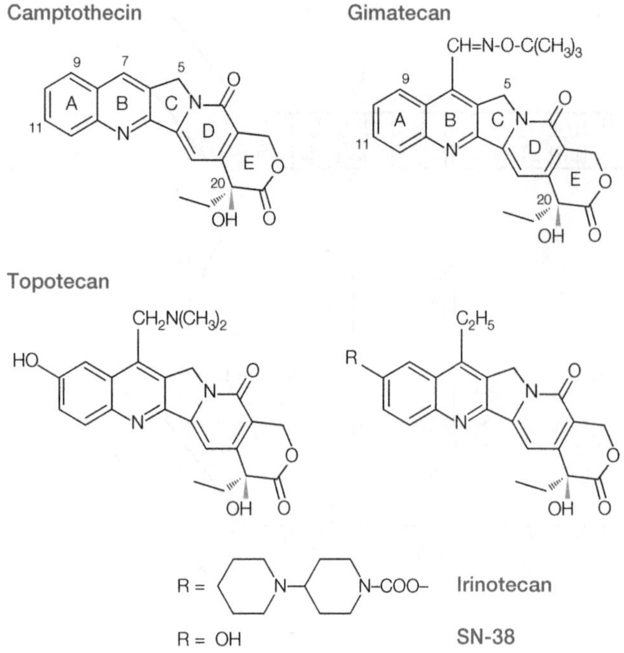

FIGURE 33.6 Chemical structures of DNA topoisomerase I inhibitors.

has shown excellent antiproliferative activity, probably depending on its greater ability to stabilize the DNA–topoisomerase I complex in comparison to other camptothecins.

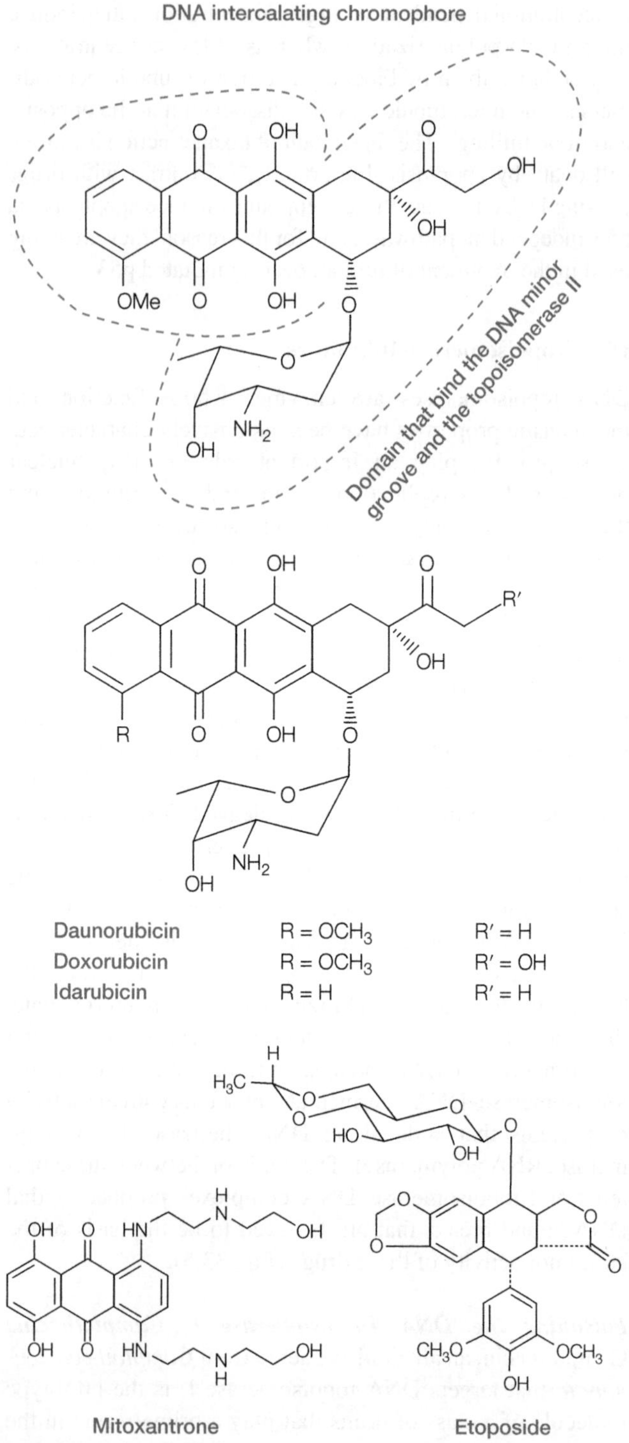

Drugs Acting on DNA Topoisomerase II: Antitumor Antibiotics

All clinically used antibiotics derive from various types of *Streptomyces* strains and include anthracyclines, actinomycin, and bleomycin. Anthracyclines, isolated from *Streptomyces peucetius* (var. *caesius*), are among the most widely used antitumor drugs. This group of antibiotics includes doxorubicin, daunorubicin, epirubicin, and idarubicin (Fig. 33.7). Their chemical structure consists of a tetracyclic ring linked to an amino sugar. These drugs are DNA intercalating agents, as their planar ring inserts between pairs of adjacent nitrogen bases, whereas the positive charge of the sugar amino group stabilizes the interaction with the negatively charged DNA phosphates (Fig. 33.7). DNA intercalating agents cause inhibition of various DNA functions. However, the mechanism relevant to the cytotoxic/antitumor activity of anthracyclines is the interference with DNA topoisomerase II functions that results in the stabilization of the topoisomerase II–DNA complex, when DNA topoisomerase II is covalently linked to the ends of the broken strand. Genotoxic stress caused by persistence of the enzyme–DNA–drug ternary complex (in particular double-strand breaks) is recognized as a fatal injury and activates cell death. Two other mechanisms, not completely elucidated, are known to play a role in anthracycline cytotoxicity. One is the ability to bind and damage cell membranes, thus altering calcium permeability. The other is the production of free radicals and reactive oxygen species

FIGURE 33.7 Chemical structures of DNA topoisomerase II inhibitors.

through the redox metabolism of the anthraquinone ring. Alteration of membrane permeability and production of radicals may be the cause of the cardiotoxicity typical of these drugs. Doxorubicin is used in the treatment of metastatic breast and ovarian cancers, leukemia and lymphomas, and various pediatric tumors, including sarcomas. Whereas doxorubicin and epirubicin are employed in various types of solid tumors and hematological malignancies (usually in combination with other cytotoxic agents), daunorubicin and idarubicin are indicated almost exclusively in leukemias.

Drugs Acting on DNA Topoisomerase II: Synthetic Drugs
The clinical success of anthracyclines has stimulated intense research in the development of more effective and less toxic analogs. These studies have led to the generation of synthetic derivatives. Among these drugs, the most studied is mitoxantrone, which belongs to the chemical class of anthracenediones. Like anthracyclines, mitoxantrone contains a planar structure typical of DNA intercalating agents, but in contrast with anthracyclines, it lacks the amino sugar. Mitoxantrone has a limited spectrum of activity with respect to doxorubicin, being used in breast cancer and lymphoma therapy. However, mitoxantrone has a better tolerability profile with reduced cardiotoxicity, which allows its use in high-dose chemotherapy. Epipodophyllotoxins are other drugs of semisynthetic origin acting on DNA topoisomerase II. The most representative members of this class are etoposide and teniposide (Fig. 33.7). Although they interact with tubulin, they have no effect on microtubule structure and function. Indeed, at pharmacological concentrations, their antiproliferative effect is dependent on topoisomerase II inhibition. Etoposide is used mainly in the treatment of lymphomas and lung and testis cancers. On the contrary, teniposide is extensively used in the treatment of pediatric leukemias.

TARGET-SPECIFIC ANTITUMOR DRUGS

Inhibitors of Survival Factors

Multiple pathways that confer a survival advantage to tumor cells over normal cells play a key role in tumor growth and progression. The best known are the PI3K–AKT–mTOR and Raf/MEK/ERK pathways that regulate cellular processes such as growth and metabolism and are aberrantly activated in human tumors (see also Chapters 13 and 18). The main causes of their alteration include inappropriate activation of receptor tyrosine kinases (RTKs) and mutations of pathway components. Given their role in cell survival, tyrosine kinases (TKs) and other enzymes belonging to survival pathways represent important therapeutic targets (Fig. 33.8).

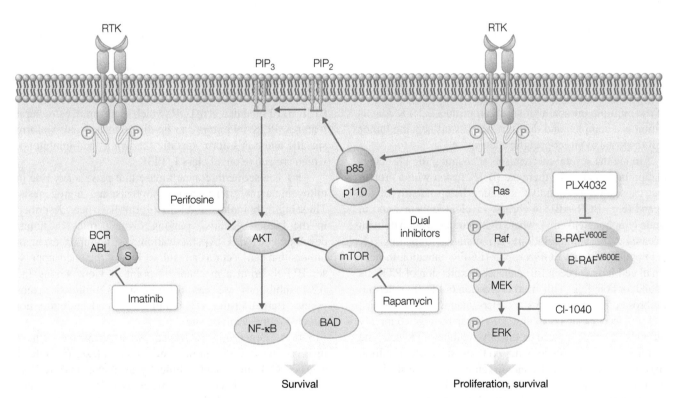

FIGURE 33.8 Survival factors as drug targets. The main survival pathways activated by receptor and nonreceptor tyrosine kinases, for example, BCR-ABL, are shown. In particular, the PI3K–AKT–mTOR and Raf/MEK/ERK axes implicated in transducing proliferative and survival signals are represented. HSP90, which acts as a chaperone for survival factors such as AKT, is also reported. The main target-specific agents and their tumor-specific targets (the BCR-ABL nonreceptor tyrosine kinase and B-RAF V600E) are represented. P, phosphate; PIP, phosphatidylinositol; RTK, receptor tyrosine kinase; S, substrate.

Most drugs developed to inhibit cell survival factors inhibit protein kinases (PKs), enzymes catalyzing the transfer of phosphate groups from high energy donor molecules (ATP) to substrates such as serine and threonine (serine/threonine kinases) and tyrosine (TKs) residues. The whole set of human kinases (kinome) has been characterized in tumor cells, revealing more than 500 kinases and the mutations affecting some of them such as B-RAF, PI3K, and c-kit.

TK Inhibitors There are two classes of TKs: receptor TKs (e.g., PDGFR) and nonreceptor TKs (e.g., ABL). All receptor TKs exhibit a similar molecular organization, consisting of a ligand-binding extracellular domain, an intracellular catalytic domain, and a transmembrane domain. In the human genome, 90 TK and 43 "TK-like" encoding genes have been found, whose products are involved in the regulation of crucial cellular functions such as proliferation and differentiation, survival and metabolism, motility, and cell cycle (see also Chapter 18). After the discovery of the oncogenic role of TKs in retrovirus-induced human tumors, several studies confirmed the deregulation of such kinases in cancer. However, TKs started to be considered good drug targets when imatinib mesylate (Gleevec) was developed for the treatment of chronic myeloid leukemia. It is now evident that the sensitivity of selected tumor types to TK inhibitors reflects their dependence on a specific factor/pathway for survival, a relationship defined as oncogene addiction. Moreover, increasing evidence supports the relevance of crosstalk among different members of the RTK family (also referred to as RTK coactivation), a process that allows tumor cells to simultaneously activate two or more RTKs because of the multiple intracellular signals transduced. RTK coactivation is a complex and dynamic process influencing tumor cell response to target-specific agents.

Among the several mechanisms accounting for TK deregulation in cancer cells, there are (i) TK fusion with a protein that causes TK constitutive activation independently of the ligand (e.g., BCR-ABL in chronic myeloid leukemia and in acute lymphoid leukemia, where BCR tetramerization domain releases ABL autoinhibition); (ii) mutation impairing TK autoregulation (e.g., EGF receptor (EGFR) mutation in nonsmall cell lung cancer); (iii) enhanced expression of RTK, its ligand, or both [e.g., enhanced expression of PDGF in dermatofibrosarcoma protuberans and Her-2/neu in breast cancer]; and (iv) decrease in TK inhibitors. TK can be targeted mainly through (i) small molecules directly inhibiting TK catalytic activity by interference with ATP or substrate binding, (ii) dimerization blockade, and (iii) antibodies against TKs or their ligands, capable of blocking signaling by ligand neutralization or by preventing binding and internalization.

TK inhibitors play a key role in the therapy of hematological tumors. Imatinib is a 2-phenylamino-pyrimidyne identified as inhibitor of BCR-ABL, a nonreceptor TK, but it inhibits also c-kit, PDGFR, and gene products related to ABL. Imatinib is an ATP-mimetic compound that in myeloid leukemia patients in chronic phase induces complete hematologic and cytogenetic remission. This drug is used also in the treatment of gastrointestinal stromal tumors (GISTs) characterized by *c-kit* mutations that make the receptor constitutively active in the absence of the ligand stem cell factor (SCF). In the search for TK inhibitors more potent and selective than imatinib, nilotinib and dasatinib have emerged, which are employed in patients not responding to imatinib. Gefitinib (Iressa) and erlotinib (Tarceva) can efficiently inhibit EGFR mutant forms.

The interest in TK inhibitors as targeted antitumor drugs is also linked to their possible use to efficiently inhibit tumor angiogenesis. In this context, the most important targets are VEGFR, FGFR, and PDGFR.

Drugs Targeting the PI3K–AKT–mTOR Axis Class I_A PI3Ks are heterodimeric enzymes implicated in tumor development consisting of regulatory subunits (p85, involved in the interaction with RTK) and catalytic subunits (p110, which phosphorylate PIP2 and generate the potent second messenger PIP3). These enzymes mediate recruitment to the plasma membrane of AKT, a crucial player of the pathway that is activated by PDK1-catalyzed phosphorylation. AKT regulates the transcriptional activity of NF-κB, which controls cell defense mechanisms and activates mTOR. The latter is involved in processes such as mRNA translation and metabolism.

Since PI3K is implicated in various processes favoring cell survival and reducing sensitivity of tumor cells to chemotherapeutic agents, several therapeutically interesting inhibitors of this kinase have been proposed. Among them, the so-called dual inhibitors can simultaneously inhibit P13K p110 subunit and mTOR, which share similar structural features. PI3K inhibitors can be distinguished into isoform specific and nonisoform specific (known as pan-inhibitors), which are active on all class I_A PI3Ks.

AKT is a serine/threonine kinase that plays a key role in mitogenic signal transduction, apoptosis, and angiogenesis. Three highly homologous AKT isoforms have been described in the human genome, playing distinct roles in tumor development. AKT hyperactivation is a common event in tumors that can occur as a result of loss of the oncosuppressor PTEN. From a mechanistic point of view, among the AKT inhibitors, we can distinguish ATP-mimetic compounds (targeting the ATP binding site) and inhibitors not directed at the catalytic site.

Drugs targeting mTOR, a serine/threonine kinase involved in two different intracellular complexes (mTORC1 e mTORC2), are already clinically available. mTORC1 is activated by AKT via TSC2, whereas mTORC2 activates AKT by phosphorylation. Besides rapamycin derivatives that inhibit only mTORC1, novel inhibitors capable of competing with ATP have been generated, which in principle can also interfere with mTORC2. The attempt to hit mTOR by interfering with mTORC1 with rapamycin and analogs in tumor cells often results in a compensatory activation of PI3K signaling. Rapamycin has been reported to be endowed with immunosuppressive effects (Chapter 50).

Drugs Targeting Raf/MEK/ERK The Raf/MEK/ERK pathway is implicated in transduction of signals generated by growth factors, mitogens, and cytokines in the extracellular microenvironment to the nucleus (see also Chapter 13). This pathway couples external stimuli to the transcription of genes controlling processes such as cell survival and death. This is achieved through regulation of multiple kinases, in particular extracellular signal regulated kinase (ERK) (ERK1 and ERK2) that phosphorylates relevant effector molecules including Bcl-2, Mcl-1, BAD, Bim, and caspase-9. ERK activity is regulated by MEK, whose inhibition has been proposed as a strategy to prevent survival of tumor cells and confirmed in early preclinical studies. The efficacy of MEK-targeting agents in inhibiting tumor cell growth or in potentiating the activity of other target-specific agents seems to depend on the genetic and molecular features of the cells. ERK1/ERK2 activation is often the result of mutations in B-RAF, the kinase most frequently mutated in human tumors. B-RAF mutations, which occur in around 10% of tumors, are extremely frequent in melanoma. Since B-RAF mutations stimulate ERK1/ERK2-dependent signaling and promote cell proliferation and transformation, mutant B-RAF has emerged as a potential tumor-specific therapeutic target. Some B-RAF inhibitors have already reached the clinical development phase or have been approved for clinical use. Among them, PLX4032 (vemurafenib), one of the first compounds to be developed, has shown, besides the inhibitory effect on RAF kinase activity, also an effect on RAF dimerization.

HSP90 Inhibitors Heat shock protein 90 (HSP90) is considered a cancer therapeutic target because it plays a key role in regulating stability and function of more than 200 "client" proteins, many of which have oncogenic function. Since HSP90 client proteins (e.g., AKT) are implicated in cell survival and proliferation, in principle, HSP90 inhibition can simultaneously interfere with multiple pathways. For this reason, HSP90 is recognized as a critical factor for tumor cell survival, representing a further example of "oncogene addiction."

At present, several HSP90 inhibitors are undergoing clinical evaluation, and important progress has been achieved to identify strategies to combine such inhibitors with other antitumor agents.

Proteasome Inhibitors

The proteasome regulates multiple cellular functions being involved in stress response, intracellular protein turnover, and cell homeostasis maintenance (see also Chapter 30). By taking part to the degradation of damaged or misfolded proteins, the proteasome contributes to the regulation of crucial processes such as cell cycle, cell death, and DNA repair (Fig. 33.9). The proteasome system comprises enzymes that activates (E1), conjugates (E2), and binds (E3) ubiquitin (Ub) to substrates that need to be degraded and deubiquitinating enzymes that remove Ub.

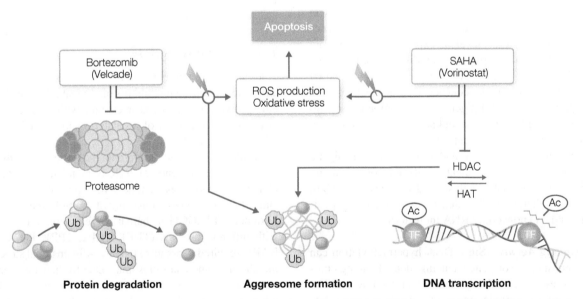

FIGURE 33.9 Mechanism of action of proteasome and histone deacetylase (HDAC) inhibitors and their potential synergistic activity. Proteasome inhibition leads to a defense response consisting in aggresome formation. HDAC inhibitors interfere with such process, contributing to the final apoptotic effect. ; Ac, acetylation; bortezomib, proteasome inhibitor; HAT, histone acetyltransferase; ROS, reactive oxygen species; SAHA, histone deacetylase inhibitor; TF, transcription factor; Ub, ubiquitin.

The validation of the Ub–proteasome pathway as antitumor drug target was first obtained with bortezomib (PS341 or Velcade; Fig. 33.9), employed in the treatment of multiple myeloma and lymphoma, diseases for which this drug represent the standard therapy. Bortezomib is a boronic acid dipeptide and acts as selective inhibitor of the 20S proteasome catalytic site, targeting a threonine residue of the chymotrypsin-like subunit. Recent studies have documented the efficacy of proteasome inhibition as monotherapy, as chemosensitizing agent to overcome tumor drug resistance. In addition, synergistic effects have been obtained by combining bortezomib and conventional cytotoxic agents.

Second-generation proteasome inhibitors comprise NPI-0052 (salinosporamide A), PR-171 (carfilzomib), epoxomicin, and lactacystin that are not structurally related to bortezomib and therefore may be active in patients refractory to bortezomib.

One of the possible mechanisms of action of proteasome inhibitors consists in the induction of cell cycle arrest through interference with degradation of cyclins and other proteins crucial for cell cycle regulation. Other possible mechanisms underlying the effects of proteasome inhibitors are aggresome formation, NF-κB inhibition, endoplasmic reticulum stress, and the so-called unfolded protein response (UPR).

Tumor cells are more dependent on proteasome activity as compared to normal cells, because they accumulate a higher amount of damaged, mutant, and misfolded proteins, which are tagged for degradation by the proteasome.

Epigenetic Factors as Drug Targets

Epigenetic mechanisms play a crucial role in tumor development and progression and may therefore represent therapeutic targets. In fact, deregulation of specific genes resulting in their altered level of expression and aberrant function is often due to epigenetic modifications, which are inheritable changes in gene expression not accompanied by modification of the DNA sequence (see also Chapter 22 and Supplement E 22.1). Epigenetic inactivation (silencing) of genes is fundamental in eukaryotic processes such as differentiation, development, and imprinting. The best-known epigenetic mechanisms underlying gene silencing are (i) DNA methylation; (ii) covalent modification of chromatin, in particular of core histones (histone deacetylase (HDAC)); (iii) physical alteration of nucleosomes (nucleosome positioning); and (iv) noncoding RNA molecules.

Demethylating Agents Since DNA hypermethylation can lead to oncosuppressor gene silencing, demethylating compounds represent a promising class of antitumor agents. The most common demethylating agents are all deoxycytidine derivatives. Azacytidine (Vidaza), approved for the treatment of myelodysplastic syndromes, has shown cytotoxic activity toward abnormal bone marrow hematopoietic cells. This effect has been related to hypomethylation of DNA in the absence of a marked inhibition of its synthesis.

HDAC Inhibitors HDACs comprise a family of enzymes that catalyze deacetylation of histones and other macromolecules (Fig. 33.9). Histone acetylation induces chromatin opening and allows the transcription machinery to interact with gene promoters. Thus, chromatin acetylation correlates with transcriptional activity, whereas deacetylation with gene silencing. In mammals, 18 HDACs have been described, which are grouped in four classes: class I (HDAC1–HDAC3, HDAC8), class II (HDAC4–HDAC7, HDAC9–HDAC10), class III (Sirtuins 1–7), and class IV (HDAC11).

HDAC inhibitors represent a heterogeneous class of compounds including hydroxamic acid derivatives, benzamides, electrophilic ketones, and cyclic peptides (Fig. 33.9). The majority of such compounds are pan-inhibitors, that is, compounds inhibiting class I, II, and IV HDACs. However, some inhibitors preferentially target class I enzymes. HDAC inhibitors are under clinical evaluation also in combination with conventional cytotoxic agents (topoisomerase inhibitors, platinum compounds, taxanes) and with target-specific agents (trastuzumab, erlotinib, gefitinib, bortezomib, ATRA, tamoxifen, imatinib, sorafenib).

Therapeutic Potential of Noncoding RNA Molecules Recent advances in molecular biology have brought the attention of the scientific community to the potential therapeutic advantage of noncoding RNA molecules. In this context, the mechanism of specific gene silencing based on double-stranded RNAs (known as "RNA interference") could be exploited to target cancer-associated genes. However, at present, such approaches are mainly limited to a preclinical setting (see also Chapter 25).

Telomeres and Telomerase

Telomeres and telomerases represent potential targets in antitumor therapy, but inhibitors of clinical relevance have not emerged yet. Telomerase is an enzyme with reverse transcriptase activity composed of a catalytic subunit (hTERT, with RNA-dependent DNA polymerase activity) and an RNA molecule (hTR), which represents the template for the DNA synthesis. The presence of hTR is mandatory for telomere elongation. Telomeres are repetitive sequences at the terminal ends of chromosomes, consisting of multiple species-specific hexamers (TTAGGG in humans). Telomeres are associated with different proteins (TRF1, TRF2, TPP1, POT1, TIN2, RAP1) required to organize the three-dimensional structure of the chromosome end (telosoma). Due to their position, telomeres are difficult to replicate, and each time a cell duplicates, it loses part of its telomeric sequences. This event is important as it regulates the lifespan of a cell. When telomeric sequences become short, cells activate a series of biochemical pathways that lead to cell death. Telomerase counteracts telomere shortening by synthesizing new telomeric sequences using hTR as a template. Telomerase is expressed in germline cells and is generally inactive in somatic cells. On the contrary, telomerase is expressed in about 90% of tumors. The fact that telomerase is almost absent in most healthy somatic human cells has made telomerase a very attractive and selective target for cancer therapy. Three strategies can be followed to inhibit telomerase respectively aiming at:

1. The catalytic subunit, hTERT.
2. The RNA subunit, hTR.
3. The telomerase substrates, the telomeres. In this case, the goal is to inhibit the protective function of telomeres or to stabilize telomeres to prevent their telomerase-mediated elongation.

Nucleoside analogs (such as AZT) act by inhibiting polymerization of the nucleotide chain. Among the compounds with nonnucleoside structure that inhibits hTERT, epigallocatechin (contained in green tea) and its synthetic derivatives are the most important.

The most promising strategy in the field of hTR inhibitors involves the use of a chimeric oligonucleotide called 2-5A antisense that causes selective degradation of hTR. This chimeric oligonucleotide consists of two components, a 2′,5′-linked tetra–adenylate group (2-5A) linked to an antisense DNA oligonucleotide complementary to a portion of hTR. GRN163 and the corresponding lipid derivative GRN163L are an example of these hTR inhibitors.

An opposite strategy is based on telomere stabilization. Guanine-rich DNA sequences (G4) can form secondary DNA structures known as G-quadruplex. These structures, which possess important biological functions, are composed of four guanines linked to each other by eight hydrogen bridges and a coordinating cation. These structures are located at both telomeric and internal chromosomal regions. Stabilization of these DNA secondary structures interferes with telomere replication and gene expression. New compounds have been developed that stabilize G4, causing head-to-head fusion of chromosomes with induction of cellular senescence also in telomerase-negative tumor cell lines. These compounds show some selectivity toward tumor cells and an excellent ability to inhibit growth of tumors intrinsically resistant to conventional chemotherapeutics.

MONOCLONAL ANTIBODIES IN CLINICAL USE

Mechanism of Action

The development of the methodology to obtain unlimited amounts of highly specific monoclonal antibodies has been an important discovery for oncologists (see also Chapters 20 and 21). However, the first monoclonal antibodies introduced in clinical trials did not provide positive results, mainly because, being of murine origin, they were recognized as foreign by patients' immune system. To avoid this problem, chimeric antibodies, containing murine variable regions conferring antigen specificity fused to human constant regions to reduce immunogenicity, were produced (Fig. 33.10). Subsequently, it was discovered that it is not necessary to include the entire variable region to determine the antigen specificity of the antibody, but only a shorter region, called complementary determining region (CDR), is required. This has allowed the generation of humanized antibodies, whose sequence is 95% of human origin. An example of humanized antibody is trastuzumab (Herceptin).

Monoclonal antibodies can exert their action through three main mechanisms: (i) delivery of cytotoxic substances, (ii) stimulation of cell-mediated (antibody-dependent cellular cytotoxicity, ADCC) or complement-mediated cytotoxicity (CDC), and (iii) blockade of signal transduction from surface receptors, such as RTK, or blockade of antigens important for cancer cell survival (e.g., CD20, the target of rituximab).

ADCC consists in cell lysis mediated by binding of the monoclonal antibody to effector cells of the immune system, such as macrophages and natural killer (NK) cells. This phenomenon has been observed in the case of trastuzumab and rituximab. Conversely, CDC is due to binding of the complement system to the antibody. This event leads to cell membrane lysis mediated by activation of the complement pathway and to the recruitment of phagocytes, resulting in death and phagocytosis of the cancer cell. Another key mechanism underlying monoclonal antibody activity consists in blockade of RTK signal transduction. Antibodies may interfere with the receptor extracellular domain and compete with growth factors for receptor binding or interact with growth factors preventing their binding to the receptor. This latter is the mechanism of action of bevacizumab (Avastin), which binds vascular endothelial growth factor (VEGF), an important factor involved in tumor angiogenesis. In other cases, the antibody (e.g., cetuximab/Erbitux) prevents ligand binding in an indirect manner by stabilizing the receptor in a conformation unable to bind the growth factor. Other antibodies prevent the interaction between receptor monomers, such as pertuzumab, which inhibits heterodimerization of HER-2/neu with other members of the EGFR family. Finally, many monoclonal antibodies are able to promote receptor internalization, thereby reducing the number of receptors on the cell surface. For example, it has been shown that trastuzumab stimulates internalization and subsequent intracellular proteasome degradation of HER-2/neu.

Trastuzumab

The target of the monoclonal antibody trastuzumab is HER-2 (HER-2/neu, ERBB-2), a member of the EGFR family that also includes HER-1 (ERBB-1, EGFR), HER-3 (ERBB-3),

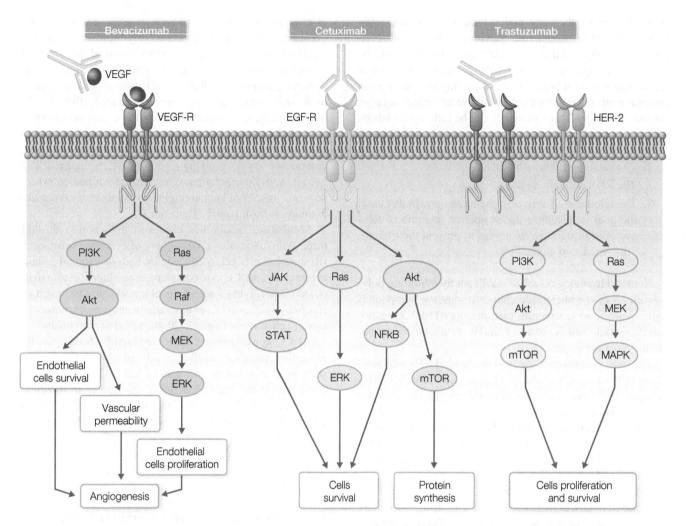

FIGURE 33.10 Cellular targets of clinically available monoclonal antibodies. Bevacizumab, cetuximab, and trastuzumab, which target respectively vascular endothelial growth factor (VEGF), EGF receptor, and HER-2, are shown. The figure depicts the PI3K–AKT–mTOR and Raf/MEK/ERK pathways indicating the main cellular functions they control.

and HER-4 (ERBB-4). All these receptors possess an extracellular domain and an intracellular domain bearing TK activity. Upon ligand (epidermal growth factors, EGFs) binding, the receptors form homo- and heterodimers and activate their TK domain by cross-phosphorylation. Subsequently, additional proteins (PI3K/AKT and MAPK) are recruited that allow signal transduction inside the cell and to the nucleus. HER-2 plays a key role in normal cell growth and differentiation, and many studies have revealed an association between its overexpression and neoplastic transformation. About 15–20% of breast tumors overexpress this protein, and HER-2-positive tumors are very aggressive, resulting in a less favorable clinical prognosis than tumors not overexpressing HER-2. The interaction of trastuzumab with HER-2 causes (i) inhibition of the molecular events downstream of the receptor (PI3K/Akt and MAPK cascades), (ii) downregulation of the cell surface receptor, and (iii) effects on tumor vasculature by reducing the expression of proangiogenic factors (VEGF, angiopoietin 1, and TGF-α) and increasing that of the antiangiogenic factors (thrombospondin 1). Trastuzumab is approved as monotherapy in second-/third-line therapy and in first-line treatment in combination with paclitaxel/docetaxel for the treatment of advanced breast cancer overexpressing HER-2.

Cetuximab

Colorectal cancer, non-small cell lung cancer, and head and neck carcinomas are tumors overexpressing EGFR. Increased EGFR expression is associated with tumor aggressiveness, worse prognosis, and lower sensitivity to chemotherapy. Cetuximab, a chimeric monoclonal antibody, binds EGFR with a 5–10-fold greater affinity than the endogenous ligands (EGF, TGF-α). Besides ADCC, the antibody inhibits the receptor activity by blocking its downstream signaling (Ras/Raf/MAPK, PI3K/AKT) and inducing its downregulation from the cell surface. The resulting effects in tumor cells are cell cycle arrest, apoptosis, and inhibition of angiogenesis, as well as of cellular invasiveness and metastasis. The efficacy of cetuximab seems to depend on the molecular background of cancer cells. Mutation in K-RAS gene (which occurs in about 40% of patients with colorectal carcinoma) constitutively activates the protein making tumor cells resistant to the effect of the monoclonal antibody. The efficacy, safety, and tolerability of cetuximab have been evaluated through numerous clinical studies, particularly in combination with other chemotherapeutic drugs and radiotherapy.

Bevacizumab

Angiogenesis, which is the generation of new blood vessels from the existing vasculature, is an important aspect of cancer pathogenesis. New endothelial cells may originate from differentiated endothelial cells preexisting in mature vessels (local angiogenesis or sprouting) or from endothelial cell progenitors deriving from bone marrow and mobilized in the peripheral circulation. Altering tumor vasculature through inhibition of the aforementioned processes seems a promising antitumor strategy. VEGF has been identified as a crucial pro-angiogenic factor and ideal target for the development of new anticancer molecules. Upon ligand binding, VEGF receptor activates downstream signaling cascades (Raf/MEK/ERK, p38MAPK, PI3K/AKT) that induce gene transcription and stimulate endothelial cell proliferation and migration, as well as vascular permeability. The interaction of bevacizumab with VEGF prevents the activation of such signals, reducing vessel formation. Bevacizumab is used in combination with standard chemotherapy. Changes in vascular permeability induced by its administration cause a reduction in intratumoral pressure that in turn enhances the uptake of chemotherapy drugs. Bevacizumab is approved in combination with irinotecan, 5FU in bolus, and folinic acid for the first-line treatment of metastatic colorectal cancer patients and for second-line therapy in combination with FOLFOX and XELOX (capecitabine and oxaliplatin). Bevacizumab is administered as monotherapy or in combination with FOLFOX in adjuvant chemotherapy. Angiogenesis plays an important role also in breast tumors overexpressing HER-2 and producing high levels of VEGF.

TAKE-HOME MESSAGE

- The field of anticancer agents is complex and under continuous and fast evolution.
- Knowledge of the molecular alterations of cancer cells provides the basis for making therapeutic decisions and for identifying new compounds.
- In general, conventional antitumor agents target DNA replication and mitosis: they may bind either DNA preventing its synthesis or proteins involved in DNA duplication or in microtubule dynamics.
- New antitumor drugs act on receptors and/or components of signal transduction pathways involved in cell proliferation.
- New potential targets for antitumor drugs include proteasome, chromatin-modifying enzymes, and noncoding RNA molecules.
- Although the new target-specific drugs offer new therapeutic options, they do not fully replace conventional cytotoxic agents.

FURTHER READING

Bellacosa A., Kumar C.C., Di Cristofano A., Testa J.R. (2005). Activation of AKT kinases in cancer: implications for therapeutic targeting. *Advances in Cancer Research*, 94, 29–86.

Chames P., Van Regenmortel M., Weiss E., Baty D. (2009). Therapeutic antibodies: successes, limitations and hopes for the future. *British Journal of Pharmacology*, 157, 220–233.

Chen G., Emens L.A. (2013). Chemoimmunotherapy: reengineering tumor immunity. *Cancer Immunology, Immunotherapy: CII*, 62, 203–216.

Courtney K.D., Corcoran R.B., Engelman J.A. (2010). The PI3K pathway as drug target in human cancer. *Journal of Clinical Oncology*, 28, 1075–1083.

Galluzzi L., Senovilla L., Zitvogel L., Kroemer G. (2012). The secret ally: immunostimulation by anticancer drugs. *Nature Reviews Drug Discovery*, 11, 215–233.

Garzon R., Marcucci G., Croce C.M. (2010). Targeting microRNAs in cancer: rationale, strategies and challenges. *Nature Reviews Cancer*, 9, 775–789.

Green D., Walczak H. (2013). Apoptosis therapy: driving cancers down the road to ruin. *Nature Medicine*, 19, 131–133.

Harley C.B. (2008). Telomerase and cancer therapeutics. *Nature Reviews Cancer*, 8, 167–179.

Hurley L.H. (2002). DNA and its associated processes as targets for cancer therapy. *Nature Reviews Cancer*, 2, 188–200.

Kelly T.K., De Carvalho D.D., Jones P.A. (2010). Epigenetic modifications as therapeutic targets. *Nature Biotechnology*, 28, 1069–1078.

Krause D.S., Van Etten R.A. (2005). Tyrosine kinases as targets for cancer therapy. *New England Journal of Medicine*, 353, 172–187.

Ledford H. (2013). Sizing up a slow assault on cancer. *Nature*, 496, 14–15.

McCubrey J.A., Steelman L.S., Abrams S.L., Chappell W.H., Russo S., Ove R., Milella M., Tafuri A., Lunghi P., Bonati A., Stivala F., Nicoletti F., Libra M., Martelli A.M., Montalto G., Cervello M. (2010). Emerging MEK inhibitors. *Expert Opinion on Emerging Drugs*, 15, 27–47.

Orlowski R.Z., Kuhn D.J. (2008). Proteasome inhibitors in cancer therapy: lessons from the first decade. *Clinical Cancer Research*, 14, 1649–1657.

Powers M.V., Workman P. (2006). Targeting of multiple signalling pathways by heat shock protein 90 molecular chaperone inhibitors. *Endocrine-Related Cancer*, 13, S125–S135.

Reichert J.M., Dhimolea E. (2012). The future of antibodies as cancer drugs. *Drug Discovery Today*, 17, 954–963.

Sheperd C., Puzanov I., Sosman J.A. (2010). B-RAF inhibitors: an evolving role in the therapy of malignant melanoma. *Current Oncology Reports*, 12, 146–152.

Stegmeier F., Warmuth M., Sellers W.R., Dorsch M. (2010). Targeted cancer therapies in the twenty-first century: lessons from imatinib. *Clinical Pharmacology and Therapeutics*, 87, 543–552.

Tennant D.A., Duran R.V., Gottlieb E. (2010). Targeting metabolic transformation for cancer therapy. *Nature Reviews Cancer*, 10, 267–277.

Thorley A.J., Tetley T.D. (2013). New perspectives in nanomedicine. *Pharmacology and Therapeutics*, 140, 176–185.

Yao S., Zhu Y., Chen L. (2013). Advances in targeting cell surface signalling molecules for immune modulation. *Nature Reviews Drug Discovery*, 12, 130–146.

SECTION 10

CONTROL OF CELLULAR METABOLISM

34

MITOCHONDRIA, OXIDATIVE STRESS, AND CELL DAMAGE: PHARMACOLOGICAL PERSPECTIVES

CLARA DE PALMA, ORAZIO CANTONI, AND FABIO DI LISA

By reading this chapter, you will:

- Learn about the mechanisms of cellular damage caused by mitochondrial dysfunctions that result in production of reactive oxygen species (ROS) and Ca^{2+} dysregulation
- Understand the most important consequences of increased mitochondrial ROS generation on cell biology and the pharmacological strategies aimed at preserving mitochondria and preventing oxidative stress
- Learn about possible new mitochondrial targets for drug therapy directed in particular to the mitochondrial permeability transition pore (PTP)

REACTIVE OXYGEN SPECIES (ROS)

The oxygen radicals, such as superoxide ($O_2^{\bullet-}$) and hydroxyl radical ($OH^{\bullet}$), are reactive oxygen species (ROS) displaying a high rate of chemical reactivity due to the presence of unpaired electrons. They react with a wide variety of molecules, including lipids, nucleic acids, and proteins. Their formation in mitochondria or other intracellular and extracellular sites causes extensive damages to biomolecules, explaining the crucial role of these radicals in a large number of pathologies. Over the last two decades, research on ROS has focused mainly on the characterization of the damages they cause, in the attempt to associate them with specific pathological traits. This has led to the misconception that oxygen radicals are merely toxic species and to the widespread administration of antioxidant molecules as a sort of cure-all method for most pathologies, especially of the degenerative type. However, it is now well established that ROS are also involved in the regulation of physiological functions, in particular because of their role in intracellular signaling.

This does not change the pathological role of ROS, but allows a better definition of their action mechanisms and tasks in specific pathologies.

It is important to remember that, while oxygen radicals can be classified as ROS, other molecules, such as the hydrogen peroxide (H_2O_2), are not radicals and their definition as ROS is improper.

ROS have various biological properties:

- $O_2^{\bullet-}$, the product of a one-electron reduction of O_2, is a sort of "primary" ROS, easily generated, which reacts with other molecules to produce secondary ROS. $O_2^{\bullet-}$ is stable in most organic solvents, but has a short life in aqueous solution where it undergoes dismutation to H_2O_2. $O_2^{\bullet-}$ production occurs mostly inside mitochondria during oxidative phosphorylation.
- $^{\bullet}OH$ is a very reactive ROS, whose generation depends on the Fenton reaction. It oxidizes many organic molecules, and for this reason, it is considered highly toxic. It has a very short half-life *in vivo* and does not diffuse, thus leading to reactions very close to its generation site.
- H_2O_2 is a poor and stable oxidant that seldom reacts, mainly with iron–sulfur cluster and weakly bound metals. H_2O_2 is not a proper free radical, and its biological relevance resides mostly in its ability to cross membranes and its role in the Fenton and Haber–Weiss reactions. H_2O_2 can react with methionines, free cysteines, and thiols.

General and Molecular Pharmacology: Principles of Drug Action, First Edition. Edited by Francesco Clementi and Guido Fumagalli.
© 2015 John Wiley & Sons, Inc. Published 2015 by John Wiley & Sons, Inc.

In normal conditions, ROS production is kept under control and balanced by several antioxidant systems, establishing a "redox tone" that is essential for several physiological responses. Such physiological balance between ROS formation and removal is lost in many pathological conditions. Indeed, ROS accumulation, which underlies oxidative stress, can result from abnormal ROS production, or impairment of the chemical or enzymatic detoxification systems, or a combination of both.

Mitochondrial Formation of ROS

At present, a wide consensus exists that the most relevant fraction of intracellular ROS is produced by mitochondria. In these organelles, various reactions lead to the partial reduction of oxygen. Under physiological conditions, these reactions are balanced by cellular systems for ROS scavenging and removal.

The best characterized systems for mitochondrial ROS formation are the electron transport chain (also called respiratory chain), the monoamine oxidases (MAOs), and the adaptor protein p66^{Shc} (Fig. 34.1). A brief overview of the structure, function, and dynamics of mitochondria is provided in the Supplement E 34.1, "Structure, Organization, and Dynamics of Mitochondria."

Mitochondria are also targets of ROS. Sites particularly susceptible to oxidative stress are the respiratory chain complex I in the inner mitochondrial membrane (IMM) and the aconitase and lipoic acid-containing dehydrogenases, which are matrix enzymes playing a key role in the Krebs cycle. These enzymes are inhibited under oxidative conditions. Consequently, cellular respiration and oxidative phosphorylation enzymatic cascades are altered, leading to a general impairment of the oxidative metabolism. Furthermore, peroxidation-induced degradation of cardiolipin alters IMM integrity. This event contributes to the release of the cytochrome c, which is linked to the IMM external face via ionic bonds with cardiolipin negative charges.

The relationships among mitochondrial sources of ROS are not entirely understood. Matters of debate are also the mechanisms leading to their activation under pathological conditions and the functional links with other intracellular systems, in particular NADPH oxidases.

Physiological Role of ROS

Cells respond to specific external stimuli through a complicated process of signal transduction across the plasma membrane leading to specific biological responses. In general, signaling pathways involve receptors, signal transducing proteins, second messengers, and effectors. Many signaling pathways involve protein kinases as critical mediators, which phosphorylate target substrates affecting their activity and/or function. However, a relevant role in signal transduction is now ascribed also to covalent changes of proteins due to oxidation/reduction (i.e., redox) reactions and in particular to mitochondria-generated ROS. Several ROS types can trigger these processes, although mitochondria-generated radical species can hardly propagate the signal outside of these organelles.

$O_2^{\cdot-}$ is transformed by either spontaneous or enzymatic dismutation into the more diffusible H_2O_2, a less reactive molecule that can reach distal targets. Its activity is modulated by catalase and glutathione peroxidase, which inactivate it. An interesting aspect of H_2O_2 signaling is its organization in microdomains; in fact, it has been observed that high levels of H_2O_2 are generated within specific subcellular microdomains containing important H_2O_2 targets, such as protein tyrosine phosphatases, and relative low levels of H_2O_2-inactivating enzymes (a similar organization has been described also for $[Ca^{2+}]_i$: microdomains have been found in which Ca^{+2} concentration is high enough to elicit selective Ca^{2+}-dependent responses).

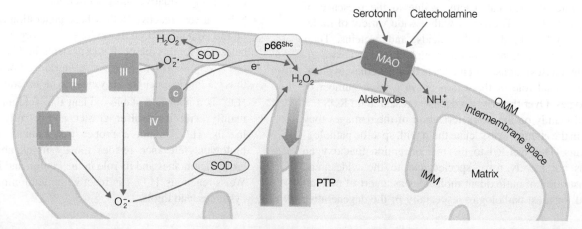

FIGURE 34.1 Mitochondrial ROS formation. Mitochondria are the most important source of intracellular ROS, the major systems involved in their production are the respiratory chain, in particular complexes I and II, the monoamine oxidases (MAOs), and the adaptor protein p66^{Shc}.

The best characterized effect of H_2O_2 is the oxidation of the thiol group in cysteinyl residues that leads to changes in protein–protein interactions or enzyme activities. For examples, protein tyrosine phosphatases, such as type 1B, PTEN, and MAPK phosphatase, are inhibited by H_2O_2 that conversely activates some protein tyrosine kinases.

These examples highlight the importance of the so-called redox signaling in the integration of cellular responses mediated by protein (de)phosphorylation. Additional indications of the importance of redox signaling come from the observation of the mutual positive feedback that often links ROS and Ca^{2+}. In fact, Ca^{2+} promotes mitochondrial formation of ROS, which in turn mobilize Ca^{2+} from its binding sites. In other words, ROS can start and propagate signals in a feed-forward way also via Ca^{2+} release, resulting in a precise spatiotemporal amplification of the redox signaling.

Pathophysiological Role of ROS in Mitochondrial Ca^{2+} Homeostasis

The pathological role played by ROS accumulation is enhanced by concurrent increase in $[Ca^{2+}]$; indeed, the combination of increased $[Ca^{2+}]$ and ROS generation may lower the injury threshold. Also in this case, mitochondria play a crucial role. These organelles have a great ability to accumulate Ca^{2+} via a Ca^{2+} uniport that uses the mitochondrial membrane potential (negative inside) as a driving force. This implies that Ca^{2+} uptake into the mitochondria is in competition with ATP synthesis, since both mechanisms use the electrochemical gradient established by mitochondrial respiration. In accordance with the Nernst equation, in theory, this gradient could allow mitochondria matrix $[Ca^{2+}]$ close to 0.1 M to be attained, that is, six orders of magnitude greater than the cytoplasm $[Ca^{2+}]$ (around 100 nM). This condition, which would significantly alter mitochondrial function and structure, is prevented by the parallel Ca^{2+} outflow. In fact, the electrophoretic Na^+/Ca^{2+} exchanger couples the entry of 3 Na^+ into the matrix with the release of 1 Ca^{2+} into the intermembrane space from which Ca^{2+} diffuses to the cytosol. Then, the Na^+/H^+ exchanger mediates the release of 1 Na^+ in exchange with 1 proton entry. Therefore, mitochondrial Ca^{2+} uptake is driven by the electrical component of the electrochemical potential (i.e., $\Delta\Psi_m$), whereas Ca^{2+} release depends on its chemical component (i.e., ΔpH).

Under physiological conditions, the amount of Ca^{2+} inside mitochondria matches the higher energy demand dictated by its increase in the cytoplasm with a greater ATP supply. In fact, inside the matrix, Ca^{2+} stimulates the oxidative decarboxylation of pyruvic, isocitric, and α-ketoglutaric acids, thereby increasing the availability of reducing equivalents in the form of $NADH(H^+)$ and $FADH_2$. Oxidation of these coenzymes by the respiratory chain establishes a feed-forward mechanism whereby a higher rate of electron flow in the respiratory chain eventually results in an increased ATP synthesis. The K_m for Ca^{2+} of the three matrix dehydrogenases is around 500 nM. This value seems to represent a threshold: indeed, at concentration below 500 nM, Ca^{2+} plays a physiological role, whereas at higher concentrations, it causes functional and structural alterations of mitochondria and the cell. Actually, Ca^{2+} activates several hydrolases, such as phospholipases and proteases, which catalyze degradation of phospholipids and proteins. The damaging action of Ca^{2+} is contributed also by the opening of the mitochondrial permeability transition pore (PTP, discussed in the following text).

It is worth pointing out that mitochondrial Ca^{2+} overload is a consequence of its abnormal increase in the cytosol. In fact, mitochondrial Ca^{2+} accumulation is a cellular system designed to reduce cytoplasmic Ca^{2+} concentration and its detrimental effects. This protective mechanism occurs as long as mitochondrial Ca^{2+} uptake does not limit ATP synthesis, and it does not cause PTP opening; indeed, when this occurs, it turns into a lethal mechanism.

Besides mitochondrial Ca^{2+} overload, especially in case of damages to the respiratory chain, mitochondria try to maintain their transmembrane potential through the reversal activity of the F_0F_1 ATP synthetase. In these conditions, hydrolysis of the ATP produced by glycolysis allows protons to be pumped into the intermembrane space, thus reestablishing the negative potential of the matrix. In this way, however, not only mitochondria stop synthesizing ATP, but they also become its main site of consumption. Therefore, this inverse operation greatly contributes to ATP depletion paving the way to cell death.

Although increase in mitochondrial $[Ca^{2+}]$ depends on cytosolic Ca^{2+} concentration, mitochondrial $[Ca^{2+}]$ changes actually depend on the close spatial relationship with the endoplasmic reticulum (ER, or sarcoplasmic reticulum (SR) in muscle cells). In response to membrane receptor stimulation, Ca^{2+} released from ER/SR is selectively taken up by mitochondria due to the juxtaposition between these organelles. Microdomains exist where, upon ER/SR release, the increase in $[Ca^{2+}]$ is large enough to activate the Ca^{2+} uniporter resulting in an effective mitochondrial Ca^{2+} uptake.

Various experimental models have documented a direct relationship between ROS accumulation and Ca^{2+} overload possibly due to a vicious cycle amplifying the initial injury. In fact, Ca^{2+} can promote $O_2^{\cdot-}$ formation by increasing the intermediate QH concentration in the Q cycle (the series of oxidation and reduction reactions of ubiquinone resulting in the net pumping of protons across a lipid bilayer, the inner membrane in case of mitochondria). In other words, stimulation of oxidative phosphorylation, and thereby increase in oxygen consumption, entails an increase in the electron flow along the respiratory complexes that might favor their escape, allowing the partial reduction of oxygen. In addition, Ca^{2+} also stimulates the activity of constitutive NO synthases (see Chapter 48); in turn, NO might stimulate $O_2^{\cdot-}$ production in the Q cycle through inhibition of cytochrome c oxidase activity. Finally, both Ca²⁺ and ROS contribute to PTP opening, which amplifies the oxidative stress by causing complex III inhibition and loss of reduced glutathione from mitochondria.

THE MITOCHONDRIAL PTP

Permeability transition (PT) is a functional term that describes the increase in IMM permeability due to opening of the PTP, which is a voltage-dependent channel. Energy conversion through oxidative phosphorylation requires the maintenance of an electrochemical proton gradient ($\Delta\mu_H$) or protomotive force (Δp). As already mentioned, mitochondrial carriers catalyze the selective exchange of ions and metabolites. Their activity is aimed at maintaining electrical and chemical gradients across the IMM as well as the osmotic equilibrium. PT occurrence makes IMM suddenly permeable to molecules with a mass up to 1.5 kDa due to PTP opening (Fig. 34.2).

The molecular identity of PTP is still uncertain. Genetic studies have shown that the voltage-dependent anion channels, ANT and VDAC, located in the outer mitochondrial membrane (OMM) do not constitute the PTP, but they contribute to its modulation. Recent studies suggest an important role for F_0F_1 ATP synthase, which, in a dimeric form, might constitute the PTP itself. Among the proteins involved in PTP regulation, cyclophilin D (CypD), a cyclophilin family member (see Chapter 43) located in the mitochondrial matrix, is particularly important. Indeed, deletion of CypD encoding gene is associated with a remarkable reduction in sensitivity to PTP opening.

CypD is also an important pharmacological target. Similarly to cyclophilin A (CypA), CypD binds cyclosporin A (CsA), and the interaction results in a reduced probability of PTP opening, comparable to that obtained by CypD genetic ablation. CsA effect is generally defined as inhibitor of PTP opening and not blocker since it simply shifts the Ca²⁺ requirement for PTP opening. Notably, protective effects of CsA depend also on mechanism other than CypD–PTP interaction. Recent studies have shown that mitochondrial protection exerted by CsA can be due to inhibition of calcineurin, a Ca²⁺-regulated serine/threonine phosphatase. Calcineurin catalyzes Drp1 dephosphorylation in the cytosol. Dephosphorylated Drp1 translocates to mitochondria, where it promotes mitochondrial fragmentation leading to cell death. The central role of CypD in PTP modulation is highlighted by the fact that most compounds reported to antagonize PTP opening (such as Nim 811, Debio 025, and sanglifehrin A) act by interacting with CypD.

The balance between PTP opening and closure is tightly controlled by a large array of compounds and processes that act physiologically as agonists or antagonists. PTP opening is enhanced not only by increases in intramitochondrial [Ca²⁺]

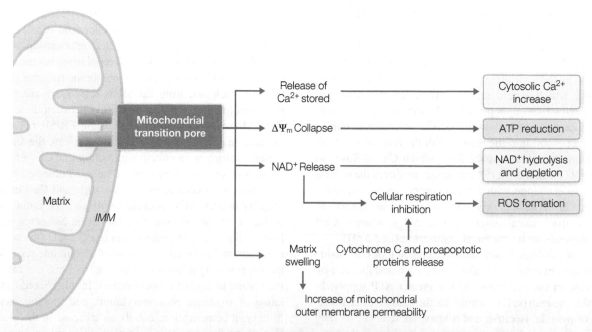

FIGURE 34.2 Schematic representation of the permeability transition pore (PTP) opening effects. Permeability transition (PT) is a functional term to describe the increase of IMM permeability due to PTP opening. PTP is a voltage-dependent channel whose opening causes immediate collapse of $\Delta\Psi_m$ followed by Ca²⁺ release from cellular stores and ATP depletion. In addition, an osmotic imbalance is generated and causes water entry inside mitochondria leading to matrix swelling and cristae reorganization. These events account for the release of cytochrome c and trigger apoptotic cell death.

but also by ROS and reduced mitochondrial membrane potential ($\Delta\Psi_m$), whereas it is physiologically antagonized by decreases in pH, Mg^{2+}, and high $\Delta\Psi_m$ values. Although its physiological role is still to be defined, prolonged PTP opening is believed to create conditions hardly compatible with cell viability. At mitochondrial level, PTP opening causes an immediate collapse of $\Delta\Psi_m$, followed by dissipation of ionic gradients and ATP depletion. Moreover, PTP opening allows efflux of the metabolites mostly compartmentalized within the matrix, such as pyridine nucleotides and coenzyme A, eventually hampering oxidation of nutrients. In addition, it generates an osmotic imbalance that draws water inside the mitochondria. Swelling of the mitochondrial matrix and reorganization of the cristae may lead to rupture of the OMM, causing cytochrome c release and apoptotic cell death.

Therefore, PTP opening may represent an interesting target for the treatment of specific diseases. Indeed, its stimulation might favor cell death in tumors, whereas several studies have shown the effectiveness of PTP inhibition in models of liver damage, neurodegeneration, muscular dystrophy, and ischemic damage of the heart and brain. Recently, the therapeutic value of the pharmacological control of PTP has been clinically demonstrated in studies showing the effectiveness of CsA in reducing both necrosis in myocardial infarction and myocyte apoptosis in dystrophy caused by collagen VI alterations.

Alongside the pharmacological approaches directly targeting PTP, other strategies exist to reduce the probability of pore opening without directly interfering with PTP. These include derivatives of coenzyme Q, which presumably act at the IMM, and compounds such as the TRO40303, which interact with the mitochondrial translocator protein (TSPO), also known as peripheral benzodiazepine receptor (PBR) at the OMM.

DRUGS AND MITOCHONDRIA

Based on the functional characteristics of mitochondria described earlier and their role in the control of ROS production and Ca^{2+} homeostasis, pharmacological approaches have been devised to prevent or stimulate mitochondrial dysfunction. For instance, inhibition of ROS production may be helpful in cardiovascular and degenerative diseases, whereas stimulation may help killing tumor cells. Here, we provide a few examples of the approaches currently under development in this rapidly advancing field. In the near future, therapeutic strategies aimed at modulating mitochondrial function are expected to gain great relevance in clinical practice.

Antioxidants Directed to Mitochondria

The mitochondrial membrane potential allows the selective accumulation of lipophilic cations within the matrix. Conjugation products combining these cations to antioxidants and bioactive molecules have been developed to selectively antagonize mitochondrial oxidative stress. Experimental evidence has shown that this strategy could be useful to treat cardiovascular diseases (e.g., myocardial infarction and stroke), neurodegenerative diseases (e.g., Parkinson's disease and Alzheimer's disease), and chronic inflammatory diseases (e.g., rheumatoid arthritis, metabolic diseases, diabetes, and obesity).

Drugs Acting on Mitochondrial Channels to Prevent Mitochondrial Dysfunction

Given the crucial role of PTP in mitochondrial dysfunction and cell death, its modulation represents the most important target of pharmacological approaches to mitochondria dysfunction. Besides the previously described PTP modulating drugs, PTP can be modulated also indirectly, by interfering with signaling pathways that either enhance (e.g., via Bad dephosphorylation) or inhibit (e.g., via Bad phosphorylation) the anti-PTP activity of Bcl-2 or Bcl-XL.

In addition, molecules have been characterized that either favor (e.g., diazoxide) or inhibit (e.g., glibenclamide) opening of the mitochondrial ATP-sensitive potassium channels.

Drugs Promoting Mitochondrial Dysfunction for Possible Antineoplastic Treatments

Given the central role of mitochondria in the intrinsic apoptotic pathway, pharmacological approaches have been devised to induce mitochondrial dysfunction that may be beneficial in antitumor therapy. Among the identified drugs, there are lonidamine, which is an inhibitor of hexokinase (a component of the PTP) and an ANT ligand, and bisphosphonates, which may act through a similar mechanism. Another trigger of mitochondrial dysfunction is As_2O_3, which interacts with thiol groups whose redox state appears to modulate the probability of PTP opening, being favored by thiol–disulfide transition. Mitochondrial dysfunction may be stimulated also by other drugs as a collateral effect of their main mechanism of action. This undesired mitochondrial dysfunction is responsible for relevant adverse effects, such as cardiotoxicity in the case of the antitumor drug anthracyclines.

Drugs Acting on the Mitochondrial Metabolism

Several drugs, long used in clinical practice, have been discovered to produce adverse effects partly due to their ability to interfere with specific mitochondrial functions. This is the case of some nonsteroidal anti-inflammatory drugs, which inhibit fatty acid β-oxidation and cause uncoupling of oxidative phosphorylation, partially acting as PTP agonists, and several local anesthetics, which uncouple oxidative phosphorylation and inhibit the F_0F_1 ATPase.

A strategy that is considered to be important in ischemic syndromes is the reduction of fatty acid oxidation to favor glucose oxidation. This goal can be achieved, also clinically, by inhibiting carnitine palmitoyltransferase I via perhexiline or by stimulating pyruvic dehydrogenase with either carnitine or dichloroacetate.

Also, drugs used in the therapy of the metabolic syndrome (diabetes, obesity, and atherosclerosis) exert their action in part by interfering with mitochondrial activity. The best-known example is provided by fibrates that, by interacting with the peroxisome proliferator-activated receptor alpha (PPARα), stimulate mitochondrial biogenesis and expression of several enzymes involved in fatty acid oxidation. Likewise, mitochondrial biogenesis and lipid oxidation are stimulated by thiazolidinediones via interaction with PPARγ, especially in the liver. In addition, the antidiabetic drug metformin has been shown to inhibit mitochondrial respiratory chain complex I. This effect appears to play an important role in the therapeutic efficacy of metformin against type 2 diabetes.

TAKE-HOME MESSAGE

- Mitochondria play a crucial role in the mechanisms of ROS production, involving different systems often working together in synergic way.
- By enhancing the opening of permeability transition pore (PTP), high ROS generation is one of the most important events that induce apoptosis.
- Pharmacological treatments preventing PTP opening and oxidative stress are currently under investigation in different models of human pathologies.

FURTHER READING

Balaban R., Nemoto S., Finkel T. (2005). Mitochondria, oxidants, and aging. *Cell*, 120, 483–495.

Bernardi P. (1999). Mitochondrial transport of cations: channels, exchangers, and permeability transition. *Physiological Reviews*, 79, 1127–1155.

Bernardi P., Krauskopf A., Basso E., Petronilli V., Blachly-Dyson E., Di Lisa F., Forte M.A. (2006). The mitochondrial permeability transition from in vitro artifact to disease target. *FEBS Journal*, 273, 2077–2099.

Di Lisa F., Canton M., Carpi A., Kaludercic N., Menabò R., Menazza S., Semenzato M. (2011). Mitochondrial injury and protection in ischemic pre- and postconditioning. *Antioxidants & Redox Signaling*, 14, 881–891.

Droge W. (2002). Free radicals in the physiological control of cell function. *Physiological Reviews*, 82, 47–95.

Li X., Fang P., Mai J., Choi E., Wang H., Yang X. (2013). Targeting mitochondrial reactive oxygen species as novel therapy for inflammatory diseases and cancers. *Journal of Hematology & Oncology*, 6, 19.

Manji H., Kato T., Di Prospero N.A., Ness S., Beal M., Krams M., Chen G. (2012). Impaired mitochondrial function in psychiatric disorders. *Nature Reviews. Neuroscience*, 13, 293–307.

Murphy M.P., Smith R.A. (2007). Targeting antioxidants to mitochondria by conjugation to lipophilic cations. *Annual Review of Pharmacology and Toxicology*, 47, 629–656.

Smith R.A., Hartley R.C., Cochemé H.M., Murphy M.P. (2012). Mitochondrial pharmacology. *Trends in Pharmacological Sciences*, 33, 341–352.

Wasilewski M., Scorrano L. (2009). The changing shape of mitochondrial apoptosis. *Trends in Endocrinology and Metabolism*, 20, 287–294.

35

PHARMACOLOGICAL CONTROL OF LIPID SYNTHESIS

Lorenzo Arnaboldi, Alberto Corsini, and Nicola Ferri

> **By reading this chapter, you will:**
>
> - Acquire basic notions on the importance, function, and metabolism of lipids and correlated pathologies (e.g., dyslipidemias, metabolic syndrome, tumors, and central nervous system pathologies)
> - Identify the cellular and biochemical targets of drugs and natural compounds used to control lipid metabolism
> - Know the pharmacological approaches used to control biosynthesis of cholesterol, fatty acids, and triglycerides (statins, bisphosphonates, DGAT inhibitors) and the transcription of genes involved in lipid metabolism (LXR ligands, PPAR agonists) and lipid transfer (ACAT, CETP, MTP)

Lipids are structurally and metabolically essential molecules for living organisms. Cholesterol is the basic constituent of plasma membranes and the precursor of bile acids and steroid hormones, whereas triglycerides (TG) represent a source of energy that is released by oxidation of their constituent fatty acids. On the other hand, phospholipids are essential components of cell membranes and lipoproteins and regulate exocytosis/endocytosis processes, synaptic fusion, and nervous transmission. Finally, specific lipids serve as mediators and intracellular antioxidants, such as inositides, eicosanoids, platelet-activating factor (PAF), lysophospholipids, sphingolipids, and plasmalogens.

Due to their essential functions, the metabolism of lipids is finely regulated and conserved in the evolutionary scale, through the control of key enzymes. Alterations and deregulations of lipid metabolism are involved in atherosclerosis, tumors, obesity, metabolic syndrome, and nonalcoholic hepatic steatosis, pathologies that are increasingly spreading in the Western world.

Since pharmacological modulation of lipid biosynthesis is a cornerstone in the treatment of the aforementioned diseases, in this chapter, we will illustrate natural, synthetic, and biotechnological inhibitors (in clinical stage and under development) of the enzymes involved in the synthesis of cholesterol, TG, and fatty acids (Fig. 35.1).

For details, see additional information on Supplements E 35.1–E 35.6. Inhibitors of protein posttranslational modifications, eicosanoid biosynthesis, lipoxins, resolvins, and protectins, will not be described in this chapter. The main acronyms used in the chapter are indicated in Box 35.1.

CHOLESTEROL BIOSYNTHESIS

The cholesterol pool in the body derives both from diet (intestinal absorption) and hepatic biosynthesis. Cholesterol is the major end product of the mevalonate (MVA) pathway, wherein the 3-hydroxy-3-methyl-glutaryl-CoA (HMG-CoA) reductase represents the most finely regulated key enzyme (Fig. 35.2). The human enzyme is a glycoprotein constituted of three domains, functionally active in a tetrameric form. Its activity is regulated at transcriptional level as well as by phosphorylation/dephosphorylation and modulation of its half-life time on the basis of the intracellular cholesterol content (see Supplement E 35.1, "Pharmacology of the Mevalonate Pathway"). Since two-thirds of human cholesterol is derived from hepatic synthesis, HMG-CoA reductase represents the ideal target for the treatment of hypercholesterolemia.

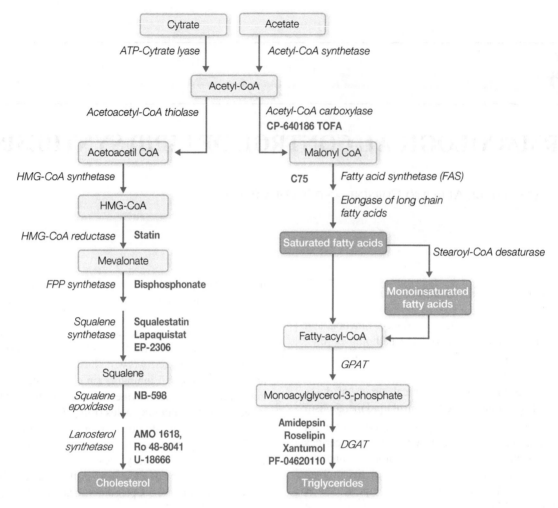

FIGURE 35.1 Schematic representation of cholesterol, fatty acid, and triglyceride biosynthetic pathways. The key enzymes involved in the biosynthesis of cholesterol, fatty acids, and triglycerides, together with their inhibitors, are described in this picture, in this chapter, and in the online supplements.

BOX 35.1 THERAPEUTIC TARGETS AND DRUG CHOICE CRITERIA

Main therapeutic target	Cellular effects	Clinical effects	Indications	Notes and choice criteria
HMG-CoA reductase	Inhibition of HMG-CoA reductase and therefore cholesterol biosynthesis	Upregulation of LDL receptor and increased uptake of LDL-C from bloodstream; decrease in LDL-C levels	Hypercholesterolemia, coronary heart disease (inactivation-dependent blockade)	Statins: lovastatin, pravastatin, fluvastatin, simvastatin, atorvastatin, rosuvastatin, pitavastatin
PPARα agonists	Activation of PPAR alpha and induction of lipoprotein lipase	Decrease in triglyceride levels, increase in HDL-C levels	Hypertriglyceridemia	Fibrates: clofibrate, gemfibrozil, etofibrate, bezafibrate, fenofibrate
PPARγ agonists	Activation of PPAR gamma	Reduction of glucose levels; reduction of TG, total cholesterol, and LDL-C; increase in HDL-C	Type 2 diabetes and associated dyslipidemia	Thiazolidinediones: rosiglitazone, pioglitazone, troglitazone

CHOLESTEROL BIOSYNTHESIS

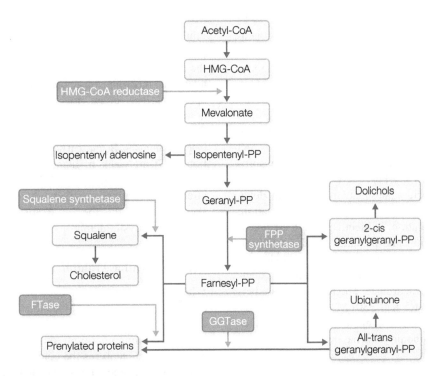

FIGURE 35.2 The mevalonate pathway. The key enzymes and intermediates that lead to the formation of cholesterol and isoprenoid molecules essential for cellular homeostasis (farnesol, geranylgeraniol, ubiquinone, dolichols) are illustrated (From Arnaboldi L., Baetta R., Ferri N., et al. (2008). Inhibition of smooth muscle cell migration and proliferation by statins. *Immunology, Endocrine & Metabolic Agents in Medicinal Chemistry*, 8, 122–140, modified.)

Statins: Inhibitors of the HMG-CoA Reductase

The most effective hypocholesterolemic therapy is based on inhibition of the HMG-CoA reductase (Fig. 35.2) by statins, the pharmacological agents most commonly prescribed to reduce low-density lipoprotein cholesterol (LDL-C) and associated cardiovascular diseases. These molecules have an HMG-like moiety (also in the inactive lactone form, which *in vivo* is enzymatically hydrolyzed to active hydroxy acid) and a more rigid portion constituted by hydrophobic groups linked to the HMG-like portion (Fig. 35.3). The founder, mevastatin or compactin, isolated in 1979 from cultured *Penicillium*, although not employed in clinics, has allowed the development of statins that are currently used for the treatment of dyslipidemias (Fig. 35.3). The reduction in hepatic cholesterol biosynthesis induced by statins is followed by an increased expression of the LDL receptor (LDLR) and increased uptake of circulating LDL (with consequent reduction of plasma LDL-C) (for a more in-depth description of action mechanism of statins, see Supplement E 35.1). The standard therapy with statins (e.g., 20–40 mg/*die* of simvastatin) reduces LDL-C by approximately one-third, although it is possible to observe up to 50% reduction with higher doses or with more potent statins (40–80 mg/*die* of atorvastatin or 10–40 mg of rosuvastatin). Moreover, statins reduce TG by 10–20% and increase HDL cholesterol (HDL-C) by about 5–10%. A meta-analysis, conducted in

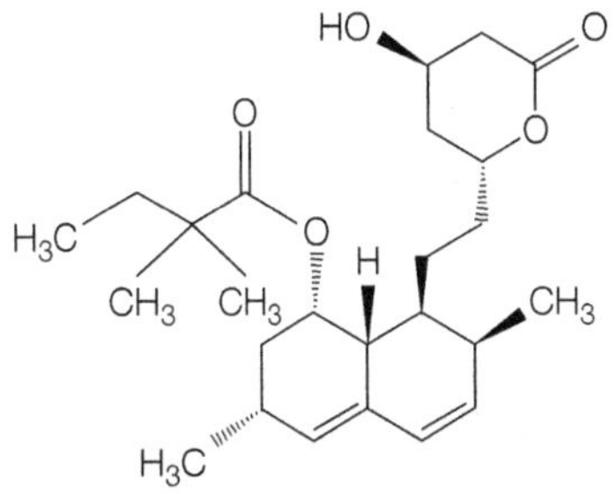

FIGURE 35.3 Chemical structure of a statin.

2005 on randomized clinical trials, has demonstrated that a reduction of 1 mmol/l in LDL-C reduces by about one-fifth the incidence of serious cardiovascular events, revascularizations, and ischemic strokes at 5 years. More recent data from a meta-analysis conducted on 26 clinical trials (170,000 patients) demonstrate that a further reduction in LDL-C (1–2 mmol/l) affects even more significantly the incidence of cardiovascular events.

Farnesyl Pyrophosphate Synthase Inhibitors (Nitrogen-Containing Bisphosphonates)

The farnesyl pyrophosphate synthase (FPPS) is a key enzyme of the MVA pathway that catalyzes the condensation of dimethylallyl pyrophosphate with two units of 3-isopentenyl pyrophosphate to generate farnesyl pyrophosphate (FPP) (Fig. 35.2). Its main pharmacological inhibitors are nitrogen-containing bisphosphonates, such as risedronate, pamidronate, zoledronate, alendronate, and ibandronate (Fig. 35.4), and stable analogs of the inorganic pyrophosphates present in nature. Their biological degradation is hampered by substitution of the oxygen between the phosphate groups with a carbon atom. Bisphosphonates represent the class of drugs more often used in the treatment of osteoporosis and other pathologies characterized by increased bone reabsorption, because they can avidly bind the bone, inhibit the de novo precipitation of calcium phosphate from solution, delay the amorphous-to-crystalline hydroxyapatite transformation, and inhibit crystal dissolution. Subsequently to their use, it has been discovered that bisphosphonates are able to inhibit FPPS with the following potencies: etidronate = clodronate (extremely weak) <<<< pamidrate < alendronate < ibandronate < risedronate < zoledronate. The IC_{50} values are between 4.1 nM (zoledronate) and 330 nM (alendronate). By inhibiting FPPS, the nitrogen-containing bisphosphonates reduce prenylation of bone proteins having GTPase activity that are essential for osteoclast homeostasis. In addition, this inhibition leads to accumulation of the precursor isopentenyl pyrophosphate (IPP) and to production of a new metabolite, the ApppI, an ATP analog formed by condensation between adenosine monophosphate and IPP, which induces osteoclast apoptosis. This effect, together with the accumulation of nonprenylated small GTPase proteins, is directly responsible for the pharmacological effect of nitrogen-containing bisphosphonates. Because it has been observed that administration of 70 mg/week of alendronate does not significantly reduce total cholesterol and TG in postmenopausal women affected by osteoporosis (due to alendronate tropism for the skeletal tissue), this therapy is not considered valid for the treatment of dyslipidemia.

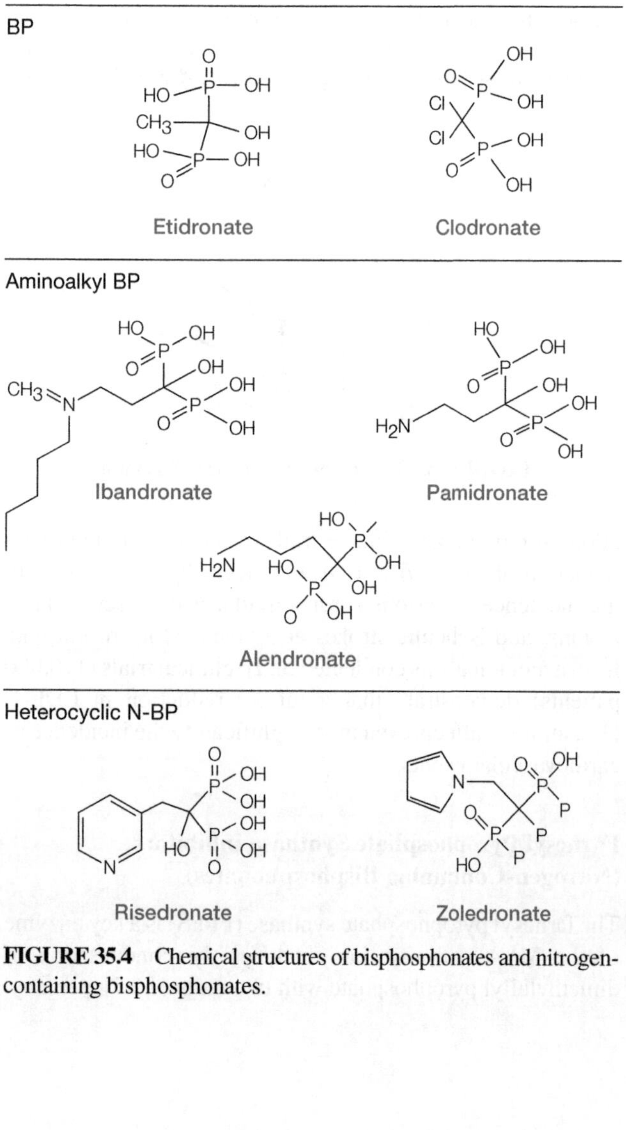

Squalene Synthase Inhibitors

Whereas early and final stages of cholesterol biosynthesis take place in peroxisomes, the metabolism of MVA occurs in the endoplasmic reticulum. Squalene synthase (SQS) is a microsomal monomeric protein present in animals, plants, and fungi that catalyzes the conversion of two molecules of FPP to presqualene diphosphate and then converted to squalene (a C30 isoprenoid) in a two-stage reaction. Since SQS is the first enzyme of the sterol biosynthesis branch of the MVA pathway, it represents a tantalizing target for antihypercholesterolemic therapy (Fig. 35.2). In theory, inhibition of SQS should lead to a reduction in cholesterol biosynthesis without affecting the MVA pathway branch devoted to the synthesis of geranyl pyrophosphate derivatives (dolichols, ubiquinone) or altering protein prenylation, thus reducing the common side effects caused by statins. SQS is not influenced by posttranslational modifications or positive feedbacks, but its activity is regulated by intracellular cholesterol concentrations. As a consequence of SQS inhibition, LDLR are upregulated (see Supplement E 35.1).

Since the 1970s, several SQS inhibitors have been discovered as selective hypocholesterolemic drugs and undergone early-stage development (Fig. 35.5). Besides squalestatin (zaragozic acid), the most studied are 1,1-bisphosphonates (substrate analogs), quinuclidines, 2-diphenylmorpholines, propylamines, dicarboxylic acid derivatives, and 4,1-benzoxazepines. Squalestatin, despite its proved hypocholesterolemic, antitumoral, and antifungal properties, has been withdrawn due to its side effects on the retina and nervous system, whereas 1,1-bisphosphonates and quinuclidine have been abandoned due to hepatonephrotoxicity and cinchonism. Different SQS inhibitors (EP-2306, BMS-187745, BMS-188494, YM-53601, and ER-27856) have demonstrated lipid-lowering (LDL-C, non-HDL-C, TG) and anti-atherosclerotic properties *in vitro* and in several animal models, apparently with no toxicity. In animals, SQS inhibitors are potent hypotriglyceridemic molecules, via a reduced hepatic TG biosynthesis and/or an increased clearance of

FIGURE 35.4 Chemical structures of bisphosphonates and nitrogen-containing bisphosphonates.

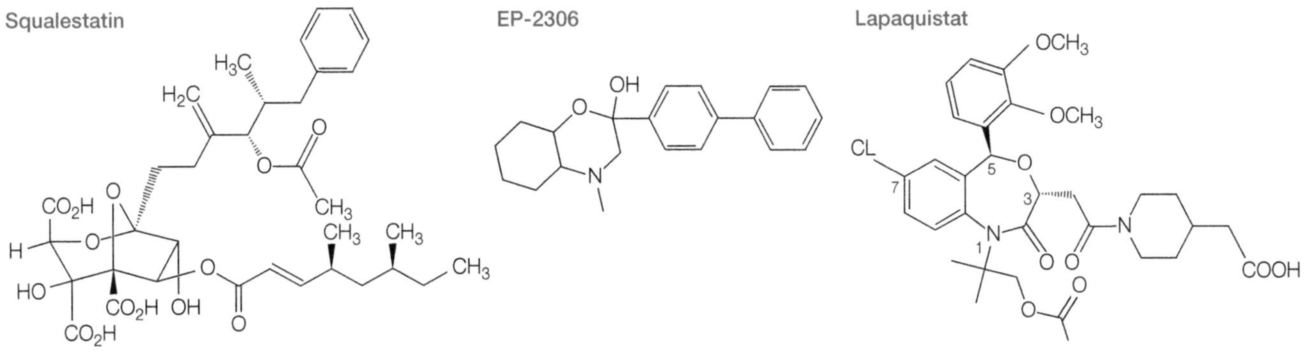

FIGURE 35.5 Chemical structure of SQS inhibitors: squalestatin, EP-2306, and lapaquistat.

plasma TG, possibly mediated by farnesoid X-activated receptor and PPAR alpha. Lapaquistat acetate (a 4,1-benzoxazepine) decreases cholesterol, TG biosynthesis, LDL-C, and TG in various animal models, by upregulating LDLR and lowering apoB100 production and statin-induced myotoxicity. Lapaquistat is the only SQS inhibitor that has entered phase II/III clinical trials.

Data from 12 randomized, double-blind, placebo- or active-controlled trials with lapaquistat alone or coadministered with other lipid-modifying drugs in dyslipidemic patients (e.g., severe homo- and heterozygous familial hypercholesterolemia), including a large ($n=2121$) 96-week safety study, have been pooled. Compared with placebo, lapaquistat 50 and 100 mg decreased LDL-C by 18 and 23%, respectively, and, coadministered with statins, reduced LDL-C by an additional 14 and 19%, respectively. Non-HDL-C, total cholesterol, VLDL-C, apolipoprotein B, TG, and hs-CRP were also reduced versus placebo or coadministration with statins.

Unfortunately, signs of hepatotoxicity halted the late-stage clinical development of lapaquistat 100 mg. The incidence of elevated transaminases was 2–3% compared with <0.3% of placebo or statin monotherapy. In two patients, the elevation of ALT and total bilirubin fulfilled the FDA Hy's law for the likelihood of progression to hepatic failure. These effects were not present at 50 mg dose, but this regimen was not commercially viable when compared with bile acid sequestrant and cholesterol absorption inhibitors that could lower LDL-C to a similar extent. Therefore, in 2007, the clinical development of lapaquistat for the treatment of hypercholesterolemia was stopped, based on the fact that its LDL-C-lowering efficacy was quite modest, the incidence of hepatotoxicity exceeded that of any statin at its highest dose, and no significant reductions in the incidence of muscle-related side effects could be seen. SQS inhibitors are currently being investigated because of their ability to inhibit sterol-related molecules (ergosterols) essential for the survival of *Leishmania* and *Trypanosoma cruzi*, microorganisms affecting approximately 18 million people. New chemotherapeutics against these parasitic protozoa are strongly needed, as current therapies are expensive and not satisfying, due to rapidly growing resistance, with high mortality and morbidity rates. SQS inhibitors may also function as antifungal agents (*Candida*) and antibiotics; they render *Staphylococcus* susceptible to normal host immune clearance, by blocking the synthesis of staphyloxanthin (whose initial biosynthesis steps are similar to those of cholesterol biosynthesis), the pigment that protects bacteria from oxidative stress. Other drugs active at subsequent steps of the cholesterol biosynthetic branch of the MVA pathway are described in Supplement E 35.1.

BIOSYNTHESIS OF FATTY ACIDS

Obesity and metabolic syndrome are characterized by alterations in the biosynthetic and catabolic pathways of fatty acids, which are reciprocally controlled by malonyl-CoA, whose elevated concentrations stimulate lipogenesis and inhibit beta-oxidation. Biosynthesis and oxidation act in opposite directions and their balance depends on the metabolic state. Lipogenesis is mainly active in the liver cytoplasm and adipose tissue, while oxidation takes place in the liver and muscles. Some hepatic enzymes involved in the metabolism of fatty acids and TG are also regulated by PPAR alpha, X receptors, nuclear hepatic factor-4alpha, and SREBP-1c. The first and most important regulatory step in fatty acid biosynthesis is the formation of malonyl-CoA from bicarbonate and acetyl-CoA, by acetyl-CoA carboxylase (ACC, two isoforms). Malonyl-CoA affects hepatic ketogenic response after starvation or diabetes and the choice of the energy substrate (fats or carbohydrates) to be used in the muscle. Finally, it stimulates insulin secretion by pancreatic beta cells and interferes with hypothalamic control of appetite.

Fatty Acid Synthase and Its Inhibitors

Fatty acid synthase (FAS) is the key enzyme in long-chain fatty acid synthesis. This enzyme synthesizes the long-chain

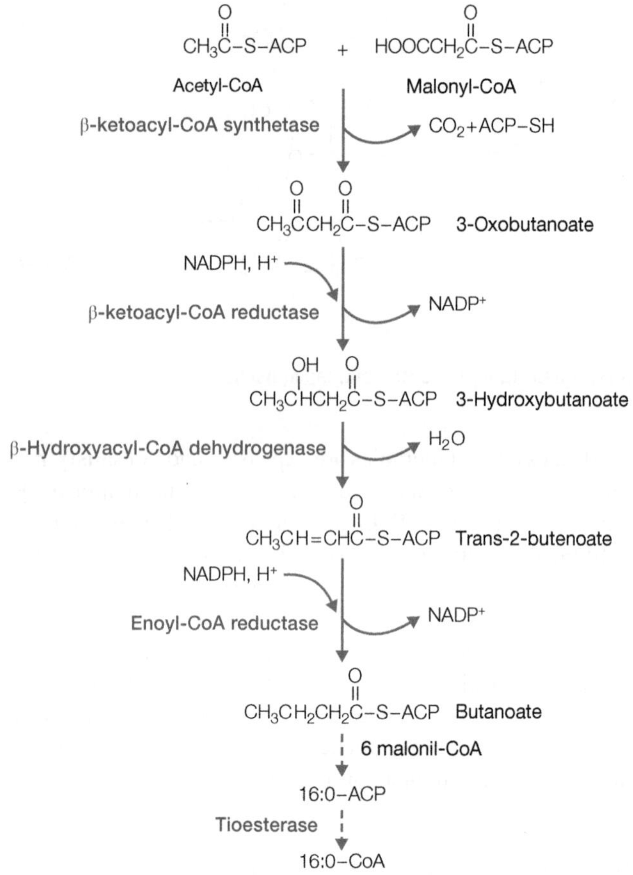

FIGURE 35.6 Metabolic pathway of FAS-mediated fatty acid synthesis. The multienzymatic complex FAS synthesizes palmitic acid (C16:0) from two simple molecules (malonyl-CoA and acetyl-CoA), using 7 molecules of NADPH and 14 of ATP.

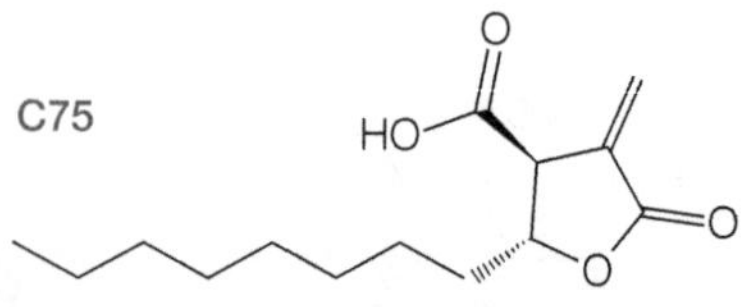

FIGURE 35.7 Chemical structure of the FAS inhibitor C75.

Fatty Acid Desaturases and Their Inhibitors

Desaturation and elongation are key processes in the synthesis of polyunsaturated long-chain fatty acids. Among the three different families of fatty acid desaturases (identified by the position of the desaturation they introduce), the most studied is the stearoyl-CoA desaturase (SCD), which is the rate-limiting enzyme in the synthesis of palmitoleate and oleate from palmitate and stearate, respectively (Fig. 35.8), and is controlled by chREBP. In humans, two isoforms exist: SCD-1 is particularly expressed in the liver and adipose tissue, while SCD-5 is mainly present in the brain and pancreas (for details on desaturases and elongases, see Supplement E 35.2).

SCD-1 inhibition could represent a novel target in the treatment of diabetes, obesity, metabolic syndrome, and nonalcoholic steatohepatitis. However, the phenotype of SCD-1 knockout (KO) mice, which is characterized by alopecia, epidermal dysfunctions (in water barrier function and thermoregulation), increased aortic atherosclerosis, intestinal inflammation, macrophage activation, and toxicity caused by saturated fatty acid accumulation in pancreatic beta cells, makes this approach questionable. Since coadministration of SCD-1 inhibitors and ω-3 fatty acids significantly reduces inflammation and atherosclerosis, many specific and potent small molecules are currently being developed (pyridazines, arylbenzimidazoles, thiazoles, benzylpiperidines). These molecules are also studied as antiproliferative drugs, since SCD-1 activity is abnormally increased in tumors, where it provides energy for the aberrant growth.

TRIGLYCERIDE BIOSYNTHESIS

Hypertriglyceridemia is an important risk factor for cardiovascular events. Elevated TG concentrations are particularly negative when associated with insulin resistance, metabolic syndrome, combined hyperlipidemia, dysbetalipoproteinemia, and diabetes. Hypertriglyceridemia may derive from:

- Excessive fat assumption
- Decreased efficiency in their catabolism
- Increased hepatic biosynthesis

When plasma TG are too elevated, they are transferred at a high rate from VLDL to HDL and LDL by cholesteryl ester transfer protein (CETP) (see later) and become preferred substrates for hepatic lipase that produces the atherogenic small-dense LDL and reduces HDL.

(C16:0) palmitic acid from malonyl-CoA and acetyl-CoA, using 7 molecules of NADPH and 14 of ATP. FAS, active in the liver, adipose tissue, and mammary gland during lactation, is essential during embryonic development to provide fatty acids and in adults to regulate the conversion of carbohydrates in fats (Figs. 35.1 and 35.6) (see Supplement E 35.2, "Pharmacology of the Biosynthesis of Fatty Acids"). The best-known FAS inhibitor is C75 (Fig. 35.7), whose antiobesity effect in mice is also due to central effects. Besides omega-3 fatty acids (see Supplement E 35.3, "Pharmacology of the Biosynthesis of Triglycerides"), which exert antiobesity activity also by inhibiting FAS and SCD-1, FAS-selective molecules are being developed, such as soy and orlistat derivatives, green tea polyphenols, and flavonoids (luteolin, quercetin, kaempferol), which are also under study as anticancer drugs. Moreover, since fatty acid synthesis in plants, bacteria, and microorganisms is dependent on a different system (FAS II), inhibition of FAS II may be useful to treat pathologies (e.g., malaria), without affecting the host.

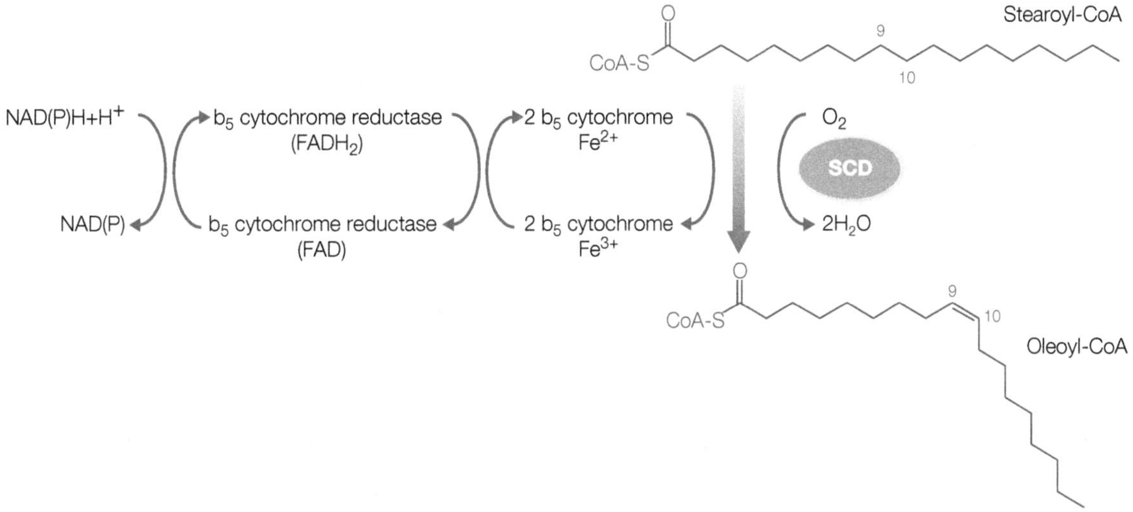

FIGURE 35.8 Metabolic pathway of stearoyl-CoA desaturase (SCD). This is the enzyme that synthesizes palmitoleic (C16:1) and oleic acids (C18:1), respectively, from palmitate (C16:0) and stearate (C18:0).

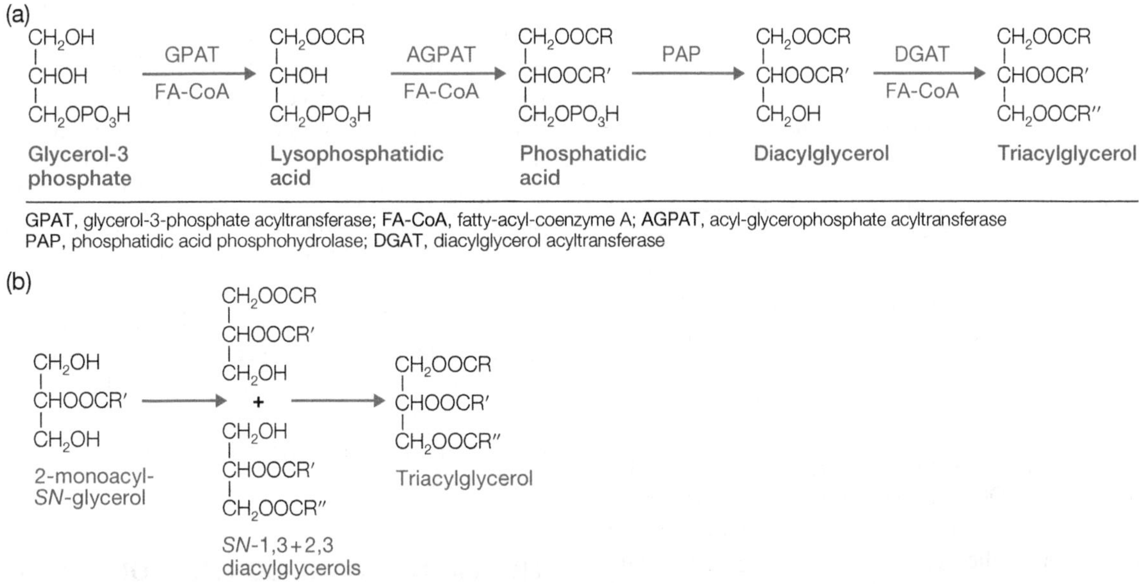

FIGURE 35.9 (a) TG biosynthetic pathway from glycerol-3-phosphate (hepatic). (b) TG biosynthetic pathway from monoacylglycerol (intestinal).

Inhibitors of Diacylglycerol Acyltransferase

All cells, but particularly intestinal and hepatic cells, synthesize TG that, besides being fuel and essential fatty acid reservoirs, remove the excess of potentially toxic bioactive lipids (FFA, diacylglycerol). They are synthesized by two different pathways (Fig. 35.9). While the glycerol-3-phosphate pathway (Fig. 35.9a) is mainly active in the liver and adipose tissue, the monoacylglycerol (Fig. 35.9b) pathway is responsible for 75% of enterocyte-produced TG. 2-Monoacyl-sn-glycerols and FFA released from dietary TG by pancreatic lipase are taken up and acylated by acyl-coenzyme A–monoacylglycerol acyltransferase to generate 1,2-diacylglycerols and, to a lesser extent, 2,3-diacylglycerols. The last rate-limiting step, common to both pathways, is regulated by diacylglycerol acyltransferase (DGAT), the only enzyme dedicated to synthesis of TG, which are later incorporated in VLDL. These characteristics suggest a role for the pharmacological inhibition of DGAT in atherogenic dyslipidemia, metabolic syndrome, and hepatic steatosis. In fact, in case of obesity, hepatic TG biosynthesis rises from 10% up to 25%, and postprandial or fasting hypertriglyceridemia represents an independent risk factor for cardiovascular events. The TG in excess are accumulated in adipose tissue and, in case of insulin resistance, also in nonadipose tissues. DGAT exists in

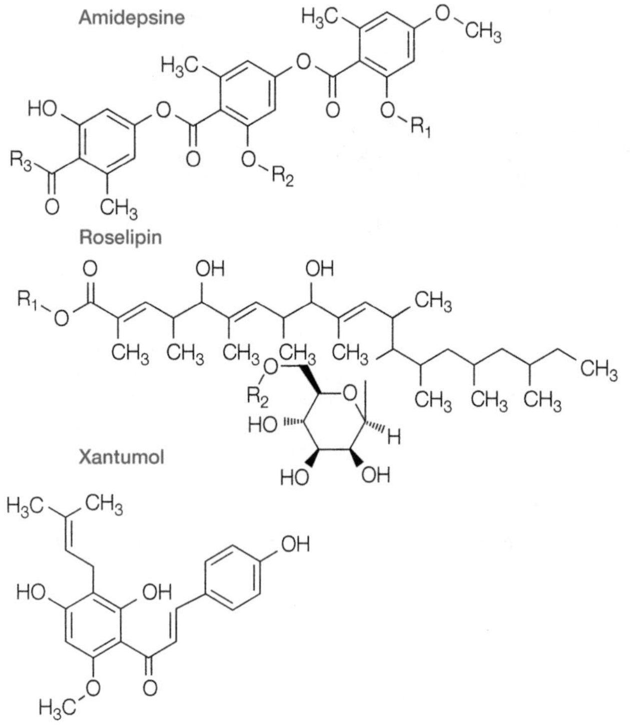

FIGURE 35.10 Chemical structure of some DGAT inhibitors.

two isoforms (1 and 2), with different functions and topology (see Supplement E 35.3). Studies suggest that the inhibition of human DGAT1 may be useful in reducing obesity, metabolic syndrome, hepatic steatosis, and acne. DGAT inhibitors have been discovered using assays on isolated enzymes and intact cells. Fungal and vegetal derivatives such as roselipins, xanthohumol, prenylflavonoids, sage, ginseng, liquorice derivatives, and ω-3 EPA inhibit DGAT *in vitro* and in animal models, but their effect is not totally specific (Fig. 35.10). Inhibition of either DGAT1, with small molecules, or of DGAT2, with antisense oligonucleotides, has been shown to significantly improve the lipid profile of dietary and genetic animal models of dyslipidemia. It is worth noting that part of the lipid-lowering effects of niacin is due to a noncompetitive inhibition of DGAT2 (see Supplement E 35.3). Intensive research is ongoing aimed at increasing specificity, selectivity, and potency of these molecules and at developing antisense oligonucleotides for the treatment of hepatic fibrosis (DGAT1/2 inhibitors). The final goal is to prevent cardiovascular and metabolic risk by acting on a unique enzyme. Preclinical studies on the effects of DGAT1 overexpression or deletion suggest that in humans DGAT1 inhibition may:

- Increase insulin sensitivity, with consequent reduction of plasma glucose in patients affected by type 2 diabetes mellitus
- Reduce body weight and induce resistance to diet-induced weight gain
- Improve plasma lipid profile, in particular reducing postprandial lipid blood levels
- Reduce hepatic steatosis
- Modify the secretion of intestinal polypeptides and diminish food consumption

Among DGAT1 inhibitors in clinical evaluation, pradigastat (LCQ908) is currently in phase III development for the treatment of familial chylomicronemia syndrome (FCS) and also in phase II clinical trials for type 2 diabetes and severe hypertriglyceridemia. FCS, known as lipoprotein lipase (LPL) deficiency or type I hyperlipoproteinemia, is a rare autosomal recessive disorder usually present in childhood. Since LPL hydrolyzes the TG component of chylomicrons, VLDL, and other TG-rich lipoproteins, reductions in its activity cause accumulation of these particles in the bloodstream. FCS is defined as chylomicronemia accompanied by one or more of eruptive xanthoma, lipemia retinalis, abdominal pain, acute pancreatitis, and/or hepatosplenomegaly. Abnormal circulating TG cause blood hyperviscosity and may result in symptoms similar to transient ischemic attacks. In adults, chylomicronemia syndrome is primarily caused by familial hyperlipoproteinemia; however, excessive alcohol intake, uncontrolled diabetes mellitus, pregnancy, estrogens, and retinoids can also trigger acute hypertriglyceridemic pancreatitis.

The formerly promising compound PF-04620110, after preliminary evaluation in normal and obese subjects, was discontinued in 2011. Safety, tolerability, pharmacokinetics, and pharmacodynamics of the selective DGAT1 inhibitor AZD7687 have been recently studied in healthy male subjects. Despite that the drug attenuates postprandial TG excursion (DGAT1 inhibition), dose- and diet-related gastrointestinal side effects (nausea, vomiting, and diarrhea) may compromise its further development (as well as that of other DGAT1 inhibitors).

TRANSCRIPTIONAL CONTROL OF GENES INVOLVED IN LIPID METABOLISM

Liver X Receptors and Liver X Receptor Synthetic Ligands

The liver X receptors (LXRs) are ligand-activated transcription factors belonging to the nuclear receptor superfamily and playing a key role in the control of lipid metabolism by acting as regulators of fatty acids and cholesterol (see also Chapters 22 and 24). LXRs exert their effects on cholesterol metabolism at different levels:

1. Intestinal absorption
2. Biosynthesis and liver uptake
3. Elimination via conversion into bile acids and secretion with bile

4. Reverse transport (see Supplement E 35.4, "Role of Liver X Receptors in Lipid Metabolism and Their Pharmacological Role")

Besides the endogenous LXR ligands (i.e., oxysterols, 22(R)-hydroxycholesterol, 20(S)-hydroxycholesterol, 24(S), 25-epoxycholesterol), a series of synthetic ligands has been developed, including double agonists that activate both LXRα and LXRβ, with favorable effects on cholesterol metabolism but unfavorable on fatty acid metabolism. Among these, the most studied are T0901317 and GW3965, compounds with steroid structure that are structurally distinct and exhibit different pharmacological action. The discovery of the key role of LXR on cholesterol metabolism has led to a further development of new chemical entities active on the LXR for the treatment of atherosclerosis. Structure–activity studies have allowed the synthesis of different LXR agonists, including AT-829 and the DMHCA, two compounds with steroid structure currently under clinical development, which are capable of selectively activating the expression of genes only in some tissues without affecting genes involved in lipogenesis. In addition, it is important to mention the compound riccardin C, a natural nonsteroidal product with agonist properties on LXRα and antagonist effect on LXRβ.

Peroxisome Proliferator-Activated Receptors and Their Pharmacological Modulation

The peroxisome proliferator-activated receptors (PPARs) are nuclear receptors that control lipid metabolism and regulate homeostasis of lipids and glucose. Three isoforms have been identified: PPARα, expressed in the liver, kidney, skeletal muscle, and adipose tissue; PPARβ/δ (here referred to as PPARβ), expressed in the brain, adipose tissue, and skin; and PPARγ, expressed almost ubiquitously in all tissues. Peroxisome proliferators and fatty acids are able to activate PPARα and PPARβ, which induce peroxisomal enzymes that catalyze β-oxidation of fatty acids in the liver, skeletal muscle, and adipose tissue. On the other hand, PPARγ are important regulators of hepatic and adipose tissue lipogenesis as well as of the response to insulin in the regulation of plasma glucose levels. Therefore, the pharmacological modulation of different tissue-specific receptor subtypes can lead to new therapeutic approaches to the control of dyslipidemia in diabetes (see Supplement E 35.4).

PPARα Agonists: The Fibrates PPARα and PPARγ are the molecular targets of several drugs, including fibrates (PPARα activators) and thiazolidinediones (PPARγ activators). Fibrates (Fig. 35.11) are a class of lipid-lowering drugs mainly used to control hypertriglyceridemia; in addition, they also exert anti-inflammatory and antithrombotic activities, which contribute to the reduction of cardiovascular

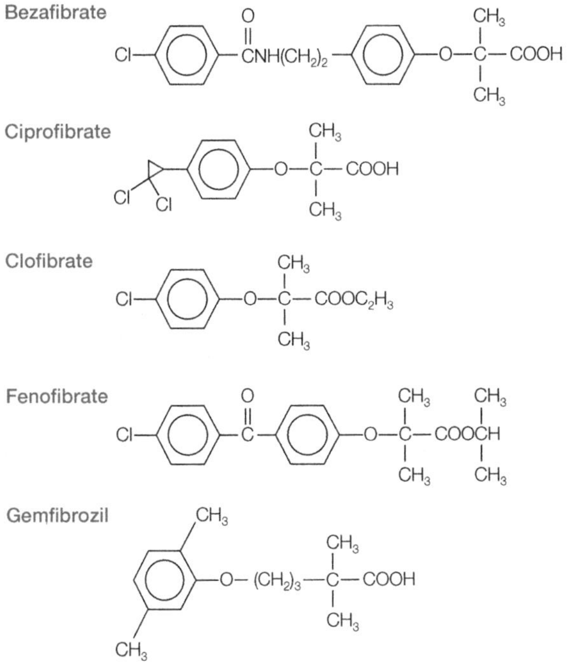

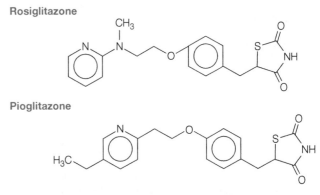

FIGURE 35.11 Chemical structure of fibrates.

FIGURE 35.12 Chemical structure of thiazolidinediones.

events. A meta-analysis conducted on placebo-controlled clinical trials with fibrates has revealed their positive effect on the reduction of total cholesterol (−8%) and TG (−30%) together with an increase of HDL-C (+9%). The modest reduction in LDL-C (−9%) varies from a 10% increase in mixed hyperlipidemia and hypertriglyceridemia up to a 20% reduction in pure hypercholesterolemic patients. A meta-analysis conducted in 2010 on clinical trials from 1950 onward has also revealed a beneficial effect of fibrates on serious cardiovascular events (prevention of coronary diseases).

PPARγ Agonists: Thiazolidinediones Approved in 1990 for adjuvant therapy in type 2 diabetes mellitus and associated diseases, rosiglitazone, pioglitazone, and troglitazone have entered clinical use (Fig. 35.12). Besides reducing plasma glucose, the therapy with thiazolidinediones ameliorates dyslipidemia, since it selectively activates nuclear receptor PPARγ and modulates the transcription of insulin-sensitive genes involved in the control of glucose and lipid metabolism. Indeed, pioglitazone and rosiglitazone have raised interest in their use for the therapy of diabetic patients at high cardiovascular risk. However, the EMA, after a review of the clinical results, found that treatment with rosiglitazone increased the cardiovascular risks in treated patients and that the therapeutic versus toxic effects were not positives and, therefore, retired it from the market. Pioglitazone seems to have a better profile; in the presence or absence of concomitant antidiabetic therapy, it reduces TG, total cholesterol, and LDL-C while increasing HDL-C in dyslipidemic patients. Both as monotherapy or combination therapy, pioglitazone seems to have a higher beneficial effect on lipid profile compared to rosiglitazone. Moreover, the results of the "PROactive" clinical trial have shown an encouraging 16% reduction in the combined events of death, nonfatal myocardial infarction, and nonfatal stroke, after treatment with pioglitazone, compared to placebo. This effect has prompted the development of dual PPARα/γ agonists. Nevertheless, these beneficial effects of thiazolidinediones are not sufficient to consider this drug class as a first-choice therapy for dyslipidemia treatment. Troglitazone has been dropped from the market due to an increased incidence of hepatitis.

Dual PPARα/γ Agonists As expected, these drugs show high effectiveness, when compared to specific PPARγ agonists, improving both glucose and lipid homeostases. However, their clinical development has been blocked because of the observed adverse effects, that is, increase in creatinine plasma levels, reduction in glomerular filtration (tesaglitazar), and increase in cardiovascular events (muraglitazar). However, since the adverse effects appear to be specific for each type of compound rather than being associated with the therapeutic class, it might still be possible to develop new agonists with a better safety profile. For example, aleglitazar is a dual PPARα/γ agonist designed to optimize the positive effects on lipid and glucose levels while reducing the action on increased body weight and edema (Fig. 35.13). Preclinical trials have demonstrated a favorable effect of aleglitazar on sugar levels, insulin response, and dyslipidemia, compared to rosiglitazone, with a reassuring general profile in terms of toxicity. The phase II study SYNCHRONY, conducted versus placebo and pioglitazone, has demonstrated that aleglitazar has a similar effectiveness on glycemia compared to pioglitazone but with a better lipid profile (in particular a significant reduction in LDL-C).

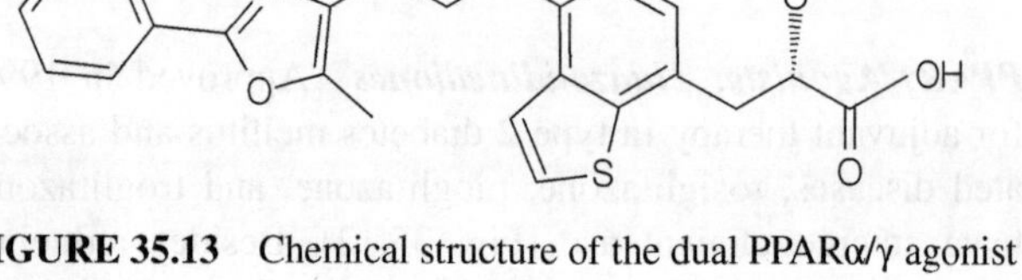

LIPID TRANSFER PROTEINS

Plasmatic transport of TG and cholesterol is essential not only to supply tissues with these lipids but also to remove them from the body, when in excess. Since these lipids are highly lipophilic, their plasma transport requires the presence of macromolecular complexes called lipoproteins (see Supplement E 35.5, "The Lipoproteins"), characterized by a lipid core (constituted by free and esterified cholesterol, phospholipids, and TG) and a protein moiety (apolipoprotein). To accommodate these components, lipoproteins are spherical shaped and covered by a phospholipid layer where the hydrophilic heads face the aqueous environment and the hydrophobic tails the lipid core. Apolipoproteins are able to interact with phospholipids on one side and with the aqueous phase on the other. While free cholesterol is on the surface, the most hydrophobic TG and CE are mainly found in the core of the structure. This peculiar structure confers stability to lipoproteins in the plasma and allows transport of nonpolar molecules. Therefore, the increase in plasma lipids correlates with an increase in circulating lipoproteins. Lipid transport is mediated by the interaction between TG-rich lipoproteins and lipolytic enzymes of vessel endothelium and transport proteins but also by interaction with specific receptors (scavenger receptors for LDL and LDL receptor-related protein (LRP) for remnants, beta VLDL, and HDL) that promote uptake of some lipoproteins (or their components), therefore modulating their concentrations.

Some lipid transfer proteins will be discussed here due to their actual pharmacological interest: acyl-CoA–cholesterol acyltransferase (ACAT), CETP, and microsomal TG transfer protein (MTP) (for other details and inhibitors, see Supplement E 35.6, "Lipid Transfer Proteins").

ACAT and Its Inhibitors

The ACAT is an integral membrane enzyme that catalyzes the synthesis of CE from free cholesterol and activated long-chain fatty acids. ACAT is essential for the formation of apoB-containing lipoproteins and contributes to the transformation of macrophages and smooth muscle cells in foam cells, thus increasing the risk of atherosclerosis but reducing the accumulation of potentially toxic free cholesterol. Chylomicrons, VLDL remnants, and LDL carry CE synthesized by hepatic and intestinal ACAT; the latter promotes cholesterol absorption by a diffusion gradient through the enterocyte membrane. In mammalians, ACAT exists in two isoforms (1 and 2), regulated by translational and posttranslational mechanisms. While ACAT1 is ubiquitous, ACAT2 is

FIGURE 35.13 Chemical structure of the dual PPARα/γ agonist aleglitazar.

mostly found in the liver and intestine, where it plays important roles, respectively, in lipoprotein formation and cholesterol absorption (see Supplement E 35.5, "The Lipoproteins," and 35.6, "Lipid Transfer Proteins").

Between 1980 and 1995, various nonselective ACAT inhibitors from different chemical classes were developed to block cholesterol esterification in macrophages, thus decreasing hepatic, intestinal CE, and finally apoB-containing lipoproteins. Among these compounds, GW447C88, glibenclamide, CI-976, and eflucimibe have been extensively tested. CI-976 and even more eflucimibe decrease plasma cholesterol in hypercholesterolemic animals and reduce aorta macrophage accumulation and lesions in rabbits; clinical data are not yet available. Urea derivatives and, among these, avasimibe and pactimibe demonstrated beneficial antiatherosclerotic effects *in vitro* and in *in vivo* models. However, since clinical trials did not confirm these properties revealing even a worsening of the clinical scenario (pactimibe was associated with increased incidence of major cardiovascular events), these molecules have been withdrawn. This failure is partially due to the low dosage employed and the different roles played by ACAT in animal models and humans. However, the main reason is that unspecific inhibition of ACAT decreases ABCA1-mediated reverse cholesterol transport and increases free cholesterol, apoptosis, and proinflammatory genes. Interestingly, avasimibe has shown antitumoral properties, by inhibiting ACAT1 expression, CE synthesis, and proliferation and inducing caspase-3 and caspase-8 in glioblastoma cells.

Selective ACAT2 inhibitors have shown beneficial antiatherosclerotic effects in animal models and may lead to antiatherosclerotic and hepatoprotective molecules in humans provided they can reach pharmacologically meaningful concentrations and good safety profile. On the other hand, since natural compounds such as beauveriolides (preferential ACAT1 inhibitors) have been shown to reduce atherosclerotic lesions in LDLR KO and apoE KO mice, without apparent side effects, modifications in their structure are being studied to create highly selective ACAT1 or ACAT2 inhibitors (Fig. 35.14). The only ACAT1-selective inhibitor still under development is K-604. Recent advances in the search for new ACAT inhibitors include the study of pyrazolines, polypropylene derivatives (ACAT2 selective), and tetrahydroisoquinoline and sulfamide derivatives.

CETP and Its Inhibitors

CETP is predominantly secreted in the liver and circulates in plasma bound to HDL. It facilitates lipid transfer among lipoproteins by:

- Homoexchange: two-way lipid exchange between lipoproteins.

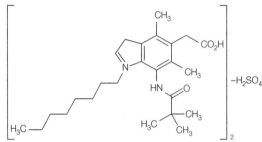

FIGURE 35.14 Chemical structure of the ACAT inhibitors avasimibe and pactimibe.

- Heteroexchange: net CE transfer from HDL to VLDL and LDL (with consequent formation of atherogenic small-dense LDL) and TG transfer from VLDL to LDL and HDL (creation of short-lived HDL but more efficient in reverse cholesterol transfer; see Supplement E 35.5 and E 35.6). CETP acquires neutral lipids from a donor particle, transports them in the aqueous phase, and vehiculates them to the acceptor lipoprotein by (i) its binding with lipoproteins, (ii) lipid exchange at the interphase CETP–lipoproteins, (iii) phospholipid and neutral lipid transfer, and (iv) release of assembled lipoproteins. CETP is expressed only in those animal species (rabbit, turkey, chicken, humans) susceptible to diet-induced atherosclerosis. The elevated activity of CETP in patients affected by dyslipidemia or metabolic syndrome and hypertriglyceridemia is associated with an increased risk of coronary heart disease and correlates with decreased HDL, increased LDL, atherosclerosis progression, and reduction of LDL and HDL size.

Pharmacological inhibition of CETP causes a significant increase in HDL-C, together with a concomitant reduction in LDL-C, non-HDL-C (−31%), Lp(a) (−35%), and apoB (−20%). Antisense oligonucleotides targeted to the liver or vaccinations demonstrated benefits in animals overexpressing CETP, and their safety and immunogenicity have already been tested in humans. However, they have not been developed further because of an inconsistent effect on HDL-C. Torcetrapib, dalcetrapib, anacetrapib, and evacetrapib are the only molecules that have reached the clinical phase; they act by complexing CETP with HDL, blocking the enzymatic activity (Fig. 35.15) (see Supplement E 35.6). The first CETP inhibitor, torcetrapib, after nine phase III studies has been withdrawn (2006), since the results of the ILLUMINATE trial showed a significant increase in blood pressure, cardiovascular events, and total mortality due to CETP-unrelated effects. The less potent dalcetrapib is the only CETP inhibitor that binds the enzyme in a covalent and irreversible fashion. Dalcetrapib decreases CETP activity in humans, rabbits, macaques, marmosets, and hamsters and, when administered per os in normolipidemic animals, increases HDL-C while decreasing the non-HDL/HDL-C ratio. Its efficacy and safety have been tested in clinical studies. Dalcetrapib shows no side effect, but its clinical development was interrupted in 2012 after the phase IIb dal-VESSEL and phase III dal-OUTCOMES trials revealed a lack of efficacy in reducing cardiovascular events. Anacetrapib causes a significant increase in HDL and in their functionality. Associated with simvastatin, anacetrapib is safe and exerts additional effects on LDL-C (see Supplement E 35.6). In the phase III DEFINE study, conducted in patients with or at risk of coronary events already on statin therapy, anacetrapib decreased LDL-C (−39.8%) and non-HDL-C and increased HDL-C (138.1%) and apoAI (Fig. 35.15), with no apparent side effects. The clinical effectiveness of anacetrapib in combination with atorvastatin treatment is now being studied in 30,000 patients affected by cardio- and cerebrovascular diseases (REVEAL study). The last CETP inhibitor, evacetrapib, shows similar efficacy compared to anacetrapib in monotherapy or in combination with statin and is currently evaluated in phase III clinical trials.

The MTP and Its Inhibitors

Assembly and secretion of apoB-containing lipoproteins, including VLDL in the liver and chylomicrons in the intestine, are mediated by MTP, which is found in the endoplasmic reticulum of hepatocytes and enterocytes. Since MTP mediates CE and TG transport toward nascent apoB, its inhibition decreases VLDL-C, LDL-C, and TG. MTP is a heterodimer (two subunits of 55 and 97 kDa, respectively) encoded by the *mttp* gene. Its functional significance is highlighted by the observation that abetalipoproteinemic patients carry loss-of-function mutations in the *mttp* gene and *mttp* missense mutations are one of the causes of heterozygous familial hypobetalipoproteinemia. Several MTP inhibitors have been developed, and they all proved beneficial in reducing LDL-C and TG (among these are CP-346086, implitapide, lomitapide, dirlotapide and analogs, JTT-130, SLx-4090, and antisense oligonucleotides; Fig. 35.16). A very recent study conducted in LDLR KO mice fed a Western diet for 16 weeks and then switched to chow diet containing the MTP inhibitor BMS 212122 for 2 weeks documented a reversal of hyperlipidemia accompanied by beneficial changes in plaque composition and inflammatory state. On the other hand, MTP inhibition increases hepatic cholesterol content, generates fatty liver, and elevates transaminases. Consequently, nonabsorbable, enterocyte-selective inhibitors have been developed, which have been demonstrated to reduce intestinal MTP mRNA *in vitro* and *in vivo*. Target populations for these drugs are patients affected by metabolic syndrome, type 2 diabetes mellitus, familial combined hyperlipidemia, and homo- and heterozygous hypercholesterolemia. The small-molecule lomitapide (at incremental doses from 5 to 60 mg/*die* per os) reduces LDL-C by 50% and in hypercholesterolemic patients (incremental doses from 5 to 10 mg/*die*) reduces LDL-C by 20–30% and decreases apoB, LDL-C, non-HDL-C, and Lp(a). When administered in combination with other blood lipid-lowering treatments, the effect on LDL-C is additive. Gastrointestinal side effects and hepatic fat accumulation (that remains stable after the initial increase) are documented, but no hepatic toxicity has been reported.

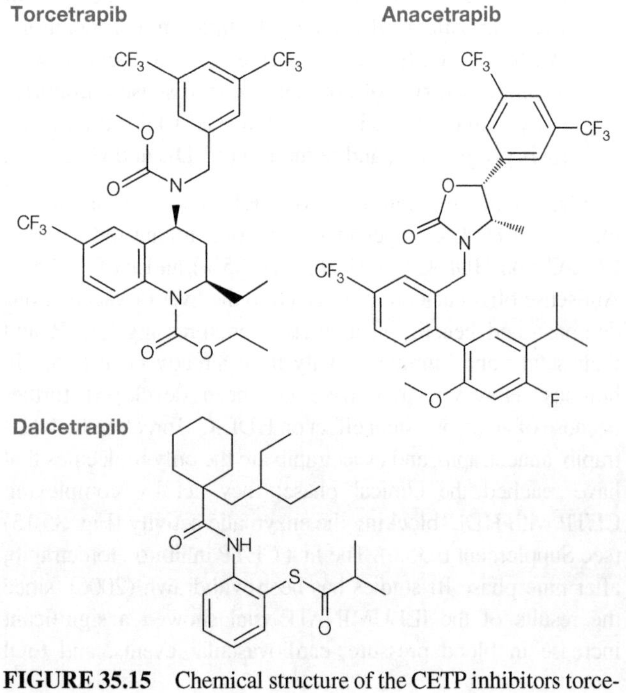

FIGURE 35.15 Chemical structure of the CETP inhibitors torcetrapib, anacetrapib, and dalcetrapib.

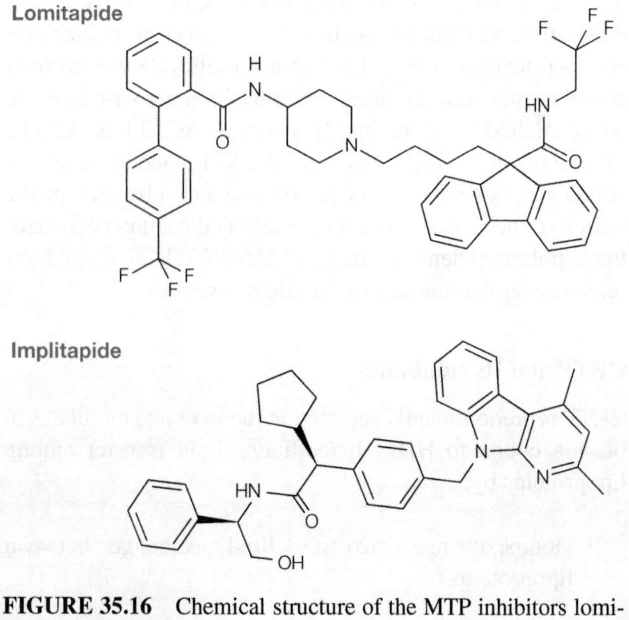

FIGURE 35.16 Chemical structure of the MTP inhibitors lomitapide and implitapide.

However, long-term evaluation will be required to exclude the development of liver pathologies. Lomitapide has been recently approved by the FDA and EMA as an adjuvant to a lipid-poor diet in combination with other hypolipidemic treatments to reduce LDL-C in homozygous familial hypercholesterolemia, with significant results. Dirlotapide is approved by the FDA for the management of obesity in dogs.

TAKE-HOME MESSAGE

- Hyperlipoproteinemias are relevant risks for several and common cardiovascular diseases.
- Knowledge in biochemistry and pharmacology of lipid metabolism opened the stage for identification of new targets and novel therapeutic approaches.
- Among several pharmacological classes, statins are the therapeutic cornerstone for hyperlipoproteinemias, since they are effective and safe in the treatment of hypercholesterolemia, mainly in reducing LDL-C and, consequently, CV risk.
- The pharmacological control of the synthesis of fatty acids and triglycerides and of the transcription of genes involved in lipid metabolism and of lipid transfer proteins offers new valid approaches for the control of dyslipidemias, hepatic steatosis, and metabolic syndrome.

FURTHER READING

Arnaboldi L., Baetta R., Ferri N., et al. (2008). Inhibition of smooth muscle cells migration and proliferation by statins. *Immunology, Endocrine & Metabolic Agents in Medicinal Chemistry*, 8, 122–140.

Birch A.M., Buckett L.K., and Turnbull A.V. (2010). DGAT1 inhibitors as anti-obesity and anti-diabetic agents. *Current Opinion in Drug Discovery & Development*, 13, 489–496.

Brown J.M. and Rudel L.L. (2010). Stearoyl-coenzyme A desaturase 1 inhibition and the metabolic syndrome: considerations for future drug discovery. *Current Opinion in Lipidology*, 21, 192–197.

Chen H.C. and Farese R.V. Jr. (2005). Inhibition of triglyceride synthesis as a treatment strategy for obesity: lessons from DGAT1-deficient mice. *Arteriosclerosis, Thrombosis, and Vascular Biology*, 25, 482–486.

Cuchel M., Bloedon L.T., Szapary P.O., et al. (2007). Inhibition of microsomal triglyceride transfer protein in familial hypercholesterolemia. *New England Journal of Medicine*, 356, 148–156.

Cuchel M., Meagher E.A., du Toit Theron H., et al. (2013). Efficacy and safety of a microsomal triglyceride transfer protein inhibitor in patients with homozygous familial hypercholesterolaemia: a single-arm, open-label, phase 3 study. *Lancet*, 381, 40–46.

Davidson M.H. (2009). Novel nonstatin strategies to lower low-density lipoprotein cholesterol. *Current Atherosclerosis Reports*, 11, 67–70.

Elsayed R.K. and Evans J.D. (2008). Emerging lipid-lowering drugs: squalene synthase inhibitors. *Expert Opinion on Emerging Drugs*, 13, 309–322.

Ginsberg H.N. and Reyes-Soffer G. (2013). Niacin: a long history, but a questionable future. *Current Opinion in Lipidology*, 24, 475–479.

Gotto A.M. Jr. and Moon J.E. (2012). Recent clinical studies of the effects of lipid-modifying therapies. *American Journal of Cardiology*, 110, 15A–26A.

Harris W.S. and Bulchandani D. (2006). Why do omega-3 fatty acids lower serum triglycerides? *Current Opinion in Lipidology*, 17, 387–393.

Hewing B., Parathath S., Mai C.K., et al. (2013). Rapid regression of atherosclerosis with MTP inhibitor treatment. *Atherosclerosis*, 227, 125–129.

http://lipidlibrary.aocs.org. Accessed November 7, 2014.

Julius U. and Fischer S. (2013). Nicotinic acid as a lipid-modifying drug—a review. *Atherosclerosis. Supplements*, 14, 7–13.

King A.J., Judd A.S., and Souers A.J. (2010). Inhibitors of diacylglycerol acyltransferase: a review of 2008 patents. *Expert Opinion on Therapeutic Patents*, 20, 19–29.

Kourounakis A.P., Katselou M.G., Matralis A.N., et al. (2011). Squalene synthase inhibitors: an update on the search for new antihyperlipidemic and antiatherosclerotic agents. *Current Medicinal Chemistry*, 18, 4418–4439.

Lalloyer F. and Staels B. (2010). Fibrates, glitazones, and peroxisome proliferator-activated receptors. *Arteriosclerosis, Thrombosis, and Vascular Biology*, 30, 894–899.

Matsuda D. and Tomoda H. (2007). DGAT inhibitors for obesity. *Current Opinion in Investigational Drugs*, 8, 836–841.

Meuwese M.C., Franssen R., Stroes E.S., et al. (2006). And then there were acyl coenzyme A: cholesterol acyl transferase inhibitors. *Current Opinion in Lipidology*, 17, 426–430.

Quintao E.C. and Cazita P.M. (2010). Lipid transfer proteins: past, present and perspectives. *Atherosclerosis*, 209, 1–9.

Ratni H. and Wright M.B. (2010). Recent progress in liver X receptor-selective modulators. *Current Opinion in Drug Discovery & Development*, 13, 403–413.

Rozman D. and Monostory K. (2010). Perspectives of the non-statin hypolipidemic agents. *Pharmacology and Therapeutics*, 127, 19–40.

Saggerson D. (2008). Malonyl-CoA, a key signaling molecule in mammalian cells. *Annual Review of Nutrition*, 28, 253–272.

Stein E.A., Bays H., O'Brien D., et al. (2011). Lapaquistat acetate: development of a squalene synthase inhibitor for the treatment of hypercholesterolemia. *Circulation*, 123, 1974–1985.

Toutouzas K., Drakopoulou M., Skoumas I., et al. (2010). Advancing therapy for hypercholesterolemia. *Expert Opinion on Pharmacotherapy*, 11, 1659–1672.

Weber O., Bischoff H., Schmeck C., et al. (2010). Cholesteryl ester transfer protein and its inhibition. *Cellular and Molecular Life Sciences*, 67, 3139–3149.

Weintraub H. (2013). Update on marine omega-3 fatty acids: management of dyslipidemia and current omega-3 treatment options. *Atherosclerosis*, 230, 381–389.

Wierzbicki A.S., Hardman T.C., and Viljoen A. (2012). New lipid-lowering drugs: an update. *International Journal of Clinical Practice*, 66, 270–280.

36

GLUCOSE TRANSPORT AND PHARMACOLOGICAL CONTROL OF GLUCOSE METABOLISM

Paolo Moghetti and Giacomo Zoppini

> **By reading this chapter, you will:**
>
> - Learn the main mechanisms underlying the homeostatic control of glycemia
> - Know the different mechanisms regulating insulin release and its signal transduction in different cell types
> - Have an overview of the pharmacological approaches to control glucose homeostasis
> - Know how drugs may interfere with insulin secretion and/or signaling and the significance of the integrated use of drugs for the management of diabetes

Glucose is an energetic substrate that plays an essential role, especially in those tissues that are vital for the functioning of whole organism, such as the nervous system, red blood cells, and leukocytes. The key metabolic role of glucose derives from its ability to supply energy rapidly and, under certain conditions, even in the absence of oxygen. Both these properties make glucose unique as a metabolic substrate, even though it produces less energy per molecule than other substrates, and it can be stored (as glycogen) only in limited amounts. It is noteworthy that there is a hierarchy in glucose utilization in the body. In vital tissues, glucose uptake is not subject to hormonal regulation and depends exclusively on blood levels, whereas in other tissues, such as muscle and adipose tissues, glucose uptake is finely regulated by specific hormones. These characteristics may explain why there is a tight physiological control of blood glucose levels.

MECHANISMS OF GLYCEMIC CONTROL

Glycemia is finely regulated, and many factors contribute to maintain it constant throughout the day, thus ensuring a continuous glucose supply to the brain and other vital tissues. Insulin plays a central role in this regulation, and its secretion is mainly controlled by blood glucose levels via feedback mechanisms. Insulin secretion increases when blood glucose level increases, and it decreases when blood glucose level decreases even modestly. The hypoglycemic action of insulin is counteracted by several hyperglycemic hormones, generally defined counterregulatory hormones: adrenaline, glucagon, cortisol, and growth hormone (GH). Both counterregulatory hormones and insulin control many other biological actions besides glycemia, often showing opposite effects. The ventromedial nucleus of the hypothalamus is regarded as the main area controlling the response to hypoglycemia, although other areas of the central nervous system are probably involved.

The hypoglycemic effect of insulin results from the simultaneous stimulation of glucose uptake in insulin-dependent tissues, mainly muscles, and inhibition of glucose production, mainly in the liver. Under insulin stimulation, glucose can be immediately taken up by cells either to supply energy or to be stored as glycogen. Other fundamental metabolic actions of insulin are inhibition of lipolysis and ketogenesis and regulation of protein metabolism.

The apparent redundancy of the counterregulatory system highlights the importance of mechanisms capable of counteracting any drop in blood sugar level. This redundancy allows one hormone to at least partially replace another if

---

General and Molecular Pharmacology: Principles of Drug Action, First Edition. Edited by Francesco Clementi and Guido Fumagalli.
© 2015 John Wiley & Sons, Inc. Published 2015 by John Wiley & Sons, Inc.

TABLE 36.1 Activation sequence, biological significance, and relevance of endocrine and behavioral changes induced by acute hypoglycemia

Phenomenon	Glycemic threshold (mg/dl)	Main mechanism of adaptation to hypoglycemia	Relevance for adaptation to hypoglycemia
Insulin reduction	80–85	Increased glucose production	Primary (the earliest phenomenon)
Glucagon increase	65–70	Increased glucose production	Primary
Epinephrine increase	60–70	Increased production + reduced glucose utilization	Secondary (critical in case of glucagon deficit)
Growth hormone increase	65–70	Increased production + reduced glucose utilization	Secondary
Cortisol increase	60–65	Increased production + reduced glucose utilization	Secondary
Symptoms	50–55	Food ingestion	Variable

defective. The mechanisms underlying counterregulatory actions are hormone specific. Glucagon exerts its action in the liver, where it increases endogenous glucose production through both glycogenolysis and gluconeogenesis. Glucagon action is crucial in the response to acute hypoglycemia (Table 36.1). Catecholamines act at various levels, including the regulation of secretion of all other hormones participating in the glucose regulation system. Their effects are rapid and include both reduction of insulin sensitivity in peripheral tissues and a strong stimulation of lipolysis in adipose tissue. The metabolic effects of catecholamines are interconnected with their hemodynamic effects and fit with the fundamental role of these hormones in organism adaptation to any homeostasis perturbation. Cortisol acts mainly on muscles by stimulating protein catabolism. This effect leads to release of amino acids that can be used as gluconeogenic substrates in the liver. Cortisol also stimulates lipolysis and reduces insulin sensitivity. GH hyperglycemic effect is mostly due to reduction of glucose uptake in hormone-dependent tissues and stimulation of lipolysis. Moreover, GH strongly stimulates protein synthesis and cellular proliferation.

The lipolytic effect of many counterregulatory hormones not only provides an alternative fuel for cell metabolism but has also important consequences for glucose metabolism. In fact, there is a competition for oxidation between glucose and fatty acids, and excess of the latter reduces both glucose utilization and tissue sensitivity to insulin action. These mechanisms seem to play an important role in conditions characterized by insulin resistance, such as type 2 diabetes.

For further details on the counterregulatory hormone system, see Supplement E 36.1, "The Controinsular Hormones."

In order to be utilized by cells, glucose has to cross the plasma membrane. In some tissues, as in the apical pole of intestinal or renal cells, glucose is actively moved across the plasma membrane against its concentration gradient by specific molecules. At least six of such molecules, called sodium–glucose linked transporters (SGLTs), have been

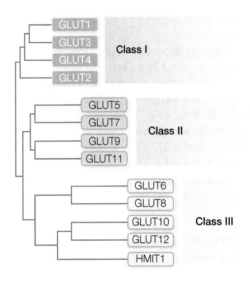

FIGURE 36.1 Glucose transporters. The transporters are grouped in several classes, based on their molecular characteristics.

described and they are all sodium dependent. Conversely, the bidirectional movement of glucose for metabolic purposes occurs through an energy-independent passive mechanism, mediated by a family of membrane proteins carrying glucose and other polyols down their concentration gradients. At present, 14 members of this family have been identified, which are referred to as GLUT1–14 and HMIT1 (H^+/myo-inositol cotransporter, or GLUT13). These proteins are grouped in three classes, based on their molecular characteristics (Fig. 36.1) and belong to the major facilitator superfamily (MFS) of membrane transporters. GLUT proteins are about 500 amino acid long and contain 12 transmembrane-spanning α-helices with amino- and the carboxy-terminal domains located in the cytoplasm and a single oligosaccharide chain on the extracellular side (Fig. 36.2). Members of this family differ in substrate specificity (not all of them transport glucose), kinetic properties, and tissue distribution (Table 36.1).

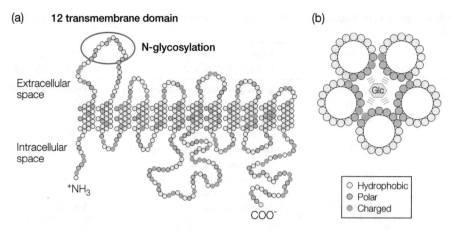

FIGURE 36.2 Structure of GLUTs. (a) Proteins of the GLUT class have 12 transmembrane domains and an extracellular oligosaccharide chain. (b) Spatial configuration of a GLUT transporter, showing the association of 5 amphipathic helices. The polar part faces the internal cavity. Glucose interacts with the protein through hydrogen bridge bonds.

GLUT1 is mainly expressed in red blood cells and in the endothelium of the blood–brain barrier. GLUT3 is the major neuronal glucose transporter, present in both dendrites and axons. GLUT3 is also expressed in monocytes/macrophages and platelets. In these cells, it is localized inside vesicles that, upon cellular stimulation, move to the plasma membrane. GLUT1 and GLUT3 are essential in mediating glucose transport in vital tissues.

GLUT2 is found in the pancreatic β-cells and liver and is characterized by a uniquely high K_m (around 17 mM), which allows the equilibrium between extra- and intracellular glucose in an ample range of glycemia, in both normal and diabetic conditions. In these tissues, the limiting step in glucose metabolism is glucose phosphorylation by hexokinase, rather than its transport. Whereas in general intracellular glucose concentrations are low and, consequently, glucose is taken up by cells, in the liver and kidney, where gluconeogenesis takes place, under specific conditions, glucose can be moved toward the extracellular space. The pancreatic GLUT2 is part of the mechanism mediating regulation of insulin secretion by glucose. Due to its high K_m, this transporter ensures a proportionate relation between glycemia and secretion of insulin up to glycemic values as high as 10–12 mM.

GLUT4 is the glucose transporter modulated by insulin and muscular contraction.

Glucose-regulated insulin secretion is a complex process involving several steps: phosphorylation and metabolism of glucose inside β-cells, ATP synthesis with consequent increase in ATP/ADP ratio, closure of ATP-sensitive potassium channels, and, finally, depolarization of the plasma membrane. Depolarization of the plasma membrane triggers the opening of voltage-dependent Ca^{2+} channels with subsequent Ca^{2+} influx stimulating exocytosis of insulin-containing vesicles and thus insulin secretion (Fig. 36.3). In this process, the limiting step is glucose phosphorylation by glucokinase that should be considered as the true glucose sensor. Following insulin release from secretory granules, *ex novo* synthesis of the hormone starts. High-frequency pulsatility is a characteristic of insulin secretion, with peaks occurring every 10–15 min. Secretory stimuli increase amplitude of each single peak. The mechanisms underlying this secretory pattern are not completely understood, but the integrity of this system is a condition *sine qua non* for biological efficacy of insulin. Insulin degradation is quite rapid, with a $t_{1/2}$ of about 4 min, and takes place mainly in the liver. This process starts with endocytosis of the hormone–receptor complexes.

Insulin exocytosis is enhanced by increase in cyclic AMP occurring in β-cells upon stimulation by glucagon, through a paracrine signaling, and by other gastrointestinal hormones. Such effect may explain the larger insulin secretion elicited by oral glucose ingestion compared to intravenous infusion of an isoglycemic amount of glucose, the so-called incretin effect (Fig. 36.4). There are two main hormones in the incretin system: glucose-dependent insulinotropic peptide (GIP) and glucagon-like peptide-1 (GLP-1).

GIP is synthesized by K cells in the first part of the intestine. The alanine residue at position 2 of its amino acid sequence is important for the catalytic actions of dipeptidyl peptidase-4 (DPP-4), an enzyme present both on cellular membranes and in body fluids that catalyzes rapid ($t_{1/2}$ of about 7 min) degradation of GIP. Kidneys play a crucial role in GIP degradation, as the brush borders of their tubular cells contain high quantity of DPP-4. This explains why subjects with kidney failure show elevated blood levels of GIP. Interestingly, insulin secretion stimulated by GIP is strictly glucose dependent. Unlike other secretagogues, GIP also stimulates β-cell proliferation, mainly through an antiapoptotic action. This phenomenon could have important clinical implications as reduction of β-cell mass is central in the pathogenesis of diabetes. GIP exerts its activity by binding to a specific receptor that exists in two isoforms of 466 and 493 amino acids, respectively.

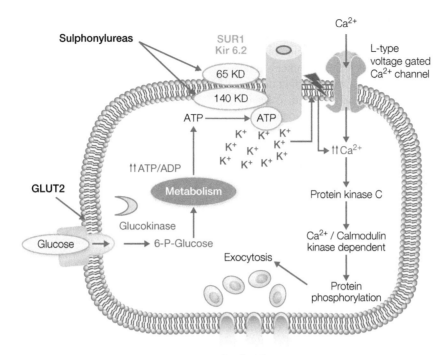

FIGURE 36.3 Regulation of insulin secretion in pancreatic β-cells. The metabolic pathway underlying glucose-dependent insulin release and the mechanism of action of sulfonylureas and glinides are shown. Sulfonylureas stimulate insulin release by binding to a membrane structure, called sulfonylurea receptor (SUR1/ABCC8), corresponding to a subunit of the ATP-dependent potassium channels. These drugs mimic the effects of an increased ATP/ADP ratio induced by glucose metabolization in β-cells. Glinides act with a similar mechanism, although through a different binding site on SUR1.

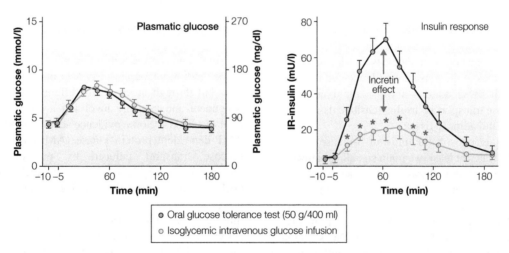

FIGURE 36.4 Incretin effect. Insulin release induced by oral glucose ingestion is greater than that triggered by an isoglycemic intravenous injection.

GLP-1 is generated by posttranslational enzymatic cleavage of the proglucagon molecule in specific cells localized in the distal portion of the intestine. The cleavage is catalyzed by the PC1/3 enzyme, which belongs to the convertase family. The biologically active peptide GLP-1 (1–37) is present in the bloodstream in two equipotent isoforms: GLP-1 (7–37) and GLP-1 (7–36). The GLP-1 (7–36), corresponding to the COOH-terminal, is the most abundant active isoform in humans (at least 80% of circulating GLP-1). The physiologic stimulus for GLP-1 release is ingestion of a mixed meal. The secretion profile of this hormone is typically biphasic, with a peak about 15 min after the beginning of the meal followed by a prolonged increase 30–60 min later. GLP-1 secretion is modulated also by other stimuli from the endocrine or nervous system, in particular, from the vagus nerve through M1 muscarinic receptors. Like GIP, GLP-1 has an alanine residue in position 2, which explains its sensitivity to the catalytic action of DPP-4. GLP-1 binds a specific receptor, called GLP-1R, on target cells. GLP-1R is a G-protein-coupled receptor of 463 amino acids, which is expressed mainly, but not uniquely, in pancreatic β-cells. Upon binding to its receptor, GLP-1 activates an intracellular enzymatic cascade leading to insulin secretion. Blockade of ATP-dependent potassium channels involved in regulation of plasma membrane electrical activity contributes to this process. GLP-1 also induces proinsulin gene expression by activating the PDX-1 transcription factor, one of the most important factors controlling insulin synthesis and secretion. GLP-1 has both trophic and antiapoptotic effects on β-cells. Another metabolic activity of GLP-1 is inhibition of glucagon secretion, mainly exerted during postprandial hyperglycemia. GLP-1 receptors have also been found in several cerebral structures—hypothalamus, thalamus, amygdala, and cortex—where GLP-1 modulates food intake and satiety.

Acute hyperglycemia strongly increases insulin secretion. Conversely, chronic hyperglycemia determines reduction in insulin secretion, a phenomenon called "glucose toxicity." The molecular basis of this phenomenon is not clearly defined, although alterations in the expression of several genes controlling insulin secretion are presumably involved.

Modulation of GLUT4 Transporter Function by Insulin and Physical Activity

Insulin and physical activity modulate GLUT4 transporter through different mechanisms that may act synergistically. GLUT4 is the main glucose transporter in insulin target tissues, and the hormone both stimulates its translocation from intracellular vesicles to the plasma membrane and increases its activity. Both these actions allow a rapid and large increase in glucose transport in insulin-sensitive tissues, particularly muscle and adipose tissue.

Insulin activates phosphatidylinositol 3-kinase (PI3K), which in turn activates at least two main signaling pathways: the AKT → AS160 → Rab pathway and the Rac → actin → α-actinin-4 pathway. Their activation leads to release of GLUT4 from its anchoring proteins and translocation to the plasma membrane (Fig. 36.5). The AKT pathway involves AKT-dependent phosphorylation of 160 kDa substrate, which in turn inhibits Rab-GTPase activity resulting in stabilization of Rab target proteins in the GPT-bound form that is presumed to stimulate vesicular traffic inside cells. The second pathway revolves around the α-actinin-4 protein, which is able to bind GLUT4 and actinin filaments. Insulin activation of this pathway seems to promote fusion of GLUT4 containing vesicles with the plasma membrane, a process mediated by VAMP2 proteins associated to the SNAP receptor (SNARE), as well as by sintaxin4 and SNAP-23 along with their regulatory partners MUNC18C and SYNIP (see also Chapter 37).

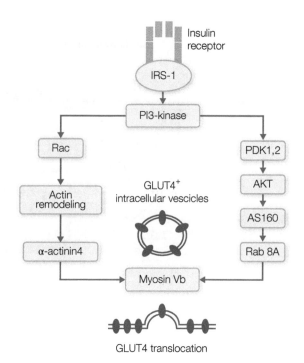

FIGURE 36.5 Insulin effect on GLUT4. Insulin induces GLUT4 translocation from intracellular sites to the plasma membrane, with a concurrent increase of transporter activity. Downstream of phosphatidylinositol-4, 5-bisphosphate 3-kinase (PI3K), the signaling process follows two distinct pathways, one involving the AKT → AS160 → Rab system and the other involving the Rac → actin → α-actinin-4 system. These two pathways ultimately converge, releasing GLUT4 from its anchoring proteins and causing fusion of GLUT4-containing vesicles with the plasma membrane.

Muscular contraction, membrane depolarization, and mitochondrial uncoupling can also induce GLUT4 endocytosis but through mechanisms different from those insulin dependent and possibly involving a different intracellular pool of GLUT4. Some evidence seems to indicate a role of AMP-dependent protein kinase (AMPK) in the increased glucose transport induced by muscular contraction. Differences in regulation and/or intracellular location of GLUT4 may explain the additive effect that insulin and physical activity have on glucose transport in the skeletal muscle. This interaction between insulin and physical activity has great clinical importance for the adaptation of therapy with insulin or insulin secretagogue drugs during and after exercise.

Insulin Receptors and Signal Transduction Pathway

Insulin exerts its specific cellular actions by activating a membrane receptor, the insulin receptor (IR), which is structurally very similar to the insulin-like growth factor-1 (IGF-1) receptor. Both receptors belong to the group of tyrosine kinase receptors (see Chapter 18). IR occurs in two isoforms, IR-A and IR-B, resulting from alternative splicing of exon 11. The A isoform (without exon 11) prevails during embryonic development and in tumor cells, has mainly mitogenic and growth activities, and displays a strong affinity for both insulin and insulin-like growth factor-2 (IGF-2). Conversely, the B isoform is expressed in postnatal life and mainly induces metabolic responses. It has a very high affinity for insulin only.

Despite having a similar affinity for insulin and IGF-2, the A isoform of the receptor elicits biological responses that are somehow different when stimulated by the two hormones. Indeed, the A isoform mediates an attenuated metabolic response to insulin and a dysfunction of the splicing mechanisms controlling the balance between the two receptor isoforms that may cause reduction in insulin sensitivity or even diabetes. In myotonic dystrophy, a pathology characterized by severe insulin resistance, an increased expression of CUG-BP, a protein that can affect IR splicing, has been reported. However, the physiological significance of this regulatory mechanism is still not entirely clear. For example, in β-cells, insulin binding to IR-A induces insulin gene expression, whereas binding to IR-B induces glucokinase gene expression, a key element in the secretory response to metabolic stimuli. Chronic hyperglycemia reduces expression of IR-A in β-cells and increases IR-B relative amount. Both these effects are associated with reduced insulin secretion.

Substrates of the IR Tyrosine Kinase Insulin binding induces receptor dimerization and autophosphorylation. Autophosphorylation occurs on tyrosine residues of the β-subunits and activates its intrinsic tyrosine kinase activity. Phosphotyrosine residues in the β-subunit recruit different substrate adaptors, such as the insulin receptor substrate (IRS) family of proteins. This family is composed of at least 6 proteins (IRS1/IRS6), three of which are present in humans. Phosphorylation of IRS proteins is required for activation of a cascade of reactions, mediating different biological responses to insulin. IRS proteins do not have enzymatic activity but contain several binding sites for signaling molecules (docking proteins) around the tyrosine residues activated by phosphorylation. The docking proteins generally bind to specific domains of IRSs called src-homology-2 (SH2).

Conversely, phosphorylation at serine residues of IRS proteins caused by metabolic alterations (e.g., chronic hyperglycemia or chronic hyperlipidemia) and several adipokines (e.g., TNFα) inhibits insulin signaling. IRS serine phosphorylation seems to serve as a short-term inhibitory mechanism. By contrast, long-term inhibition seems to be induced by IRS proteasome degradation. Chronic hyperinsulinemia, occurring as a compensatory response to insulin resistance, is one of the factors increasing IRS degradation. Accordingly, the ubiquitin–proteasome system shows increased activity in type 2 diabetes, and in animal models, its inhibition can improve insulin sensitivity.

Insulin and IGF-1 have many similar effects and share some intracellular signaling pathways. Even though IRSs mediate the activity of different hormones, they play an import role in integrating their responses and in determining the specificity of action of each of them. In particular, IRS-1 and IRS-2 have a key role in the control of glucose metabolism. Transgenic mice with a selective defect in IRS-2 expression develop diabetes and show hepatic and peripheral insulin resistance, as well as alterations of insulin secretion that are very similar to those characterizing type 2 diabetes in humans.

Activation of PI3K is central to many metabolic effects of insulin, and this molecule is also a common ligand for both IRS-1 and IRS-2. PI3K products activate specific serine kinase cascades, such as the AKT and the atypical PKC isoform pathways that ultimately stimulate glucose transport (Fig. 36.5). These cascades are altered in obese subjects. However, weight loss relieves such alterations, suggesting that they are mostly acquired defects.

AKT also regulates activity of glycogen synthase, a key enzyme in nonoxidative glucose metabolism, and synthesis of several enzymes involved in the energetic metabolism by interacting with transcription factors, such as FKHRL1 (belonging to the family of forkhead transcription factors) that control expression of enzymes involved in gluconeogenesis.

It is now widely accepted that the reduction of β-cell secretory capacity that accompanies the transition from compensated insulin resistance to diabetes does not result from a functional exhaustion of these cells due to overactivity, as previously hypothesized, but from alterations of the proliferation/apoptosis processes, triggered by different causes, either genetic or acquired.

It is noteworthy that insulin effects exhibit different sensitivity to hormonal regulation. For example, the insulin concentration required to inhibit both lipolysis and ketogenesis is lower than that capable of stimulating glucose utilization. This hierarchy in insulin actions explains why in type 2 diabetes, in which the insulin defect is partial, even severe hyperglycemia is usually not accompanied by ketosis. Moreover, mediators of insulin responses have different tissue distributions: IRS-1 is the main transducer of insulin effects in muscle and adipose tissue, while IRS-2 is mainly acting in the liver and β-cell. These differences may contribute to determining the tissue-specific effects of insulin.

PHARMACOLOGY OF GLYCEMIC CONTROL

Several drugs are available for management of type 2 diabetes, targeting different aspects of the physiopathology of this condition (see Box 36.1). In this part of the chapter, insulin therapy will not be discussed as its mechanisms of action have already been described.

BOX 36.1 THERAPEUTIC TARGETS AND DRUG CHOICE CRITERIA

Main therapeutic target	Cellular effects	Clinical effects	Indications	Notes and choice criteria
Mitochondrial complex I AMPK; glucagon activity	Reduced hepatic gluconeogenesis	Reduced glycemia	Type 2 diabetes	Metformin
	Reduced lipid synthesis	Reduced lipid levels Increased insulin sensitivity		Insulin sensitization (consider renal function). First-choice drug
PPR-γ	Differentiation of adipocytes	Reduced glycemia	Type 2 diabetes	Pioglitazone
	Increased glucose utilization	Reduced lipid levels		Insulin sensitization (consider heart function)
	Reduced FFA levels	Decreased visceral fat Increased subcutaneous fat		
α-glucosidase	Reduced/delayed digestion of carbohydrates	Reduced post-prandial hyperglycemia	Type 2 diabetes	Acarbose
Pancreatic β-cell, K-ATP channel	SUR 1 blockade	Increased insulin secretion Reduced glycemia	Type 2 diabetes	Sulfonylureas and glinides Insulin secretagogues (moderate hypoglycemia risk)
GLP-1 receptor	Increased [Ca^{2+}]	Increased insulin secretion	Type 2 diabetes	Incretins (GLP-1 analogs and DPP-4 inhibitors)
		Reduced glucagon secretion		Insulin secretagogues (low hypoglycemia risk)
		Reduced glycemia		
Renal glucose transporter SGLT2	SGLT2 blockade	Reduced glycemia	Type 2 diabetes	Gliflozins
		Increased glycosuria		
Dopamine receptor	Increased dopaminergic tone	Increased insulin sensitivity and secretion Reduced glycemia	Type 2 diabetes	QR-bromocriptine
Inflammation	NFκB inhibition Increased tyrosine phosphorylation/ reduced serine phosphorylation of IRSs	Reduced glycemia Increased insulin sensitivity	Type 2 diabetes	Salsalate
All insulin-dependent targets	Insulin action	Reduced glycemia	Type 1 and type 2 diabetes	Insulin
				Insulin deficiency (different formulations)

AMPK, adenosine monophosphate kinase; DPP-4, dipeptidyl peptidase-4; GLP-1, glucagon-like peptide-1; IRSs, insulin receptor substrates; NFκB, nuclear factor κB; PPR-γ, peroxisome proliferator receptor gamma; SGLP2: sodium–glucose transporter 2; SUR1, sulfonylurea receptor 1.

Pharmacological Stimulation of β-Cell Activity

Among the drugs stimulating insulin secretion, sulfonylureas are those that have been used for longer, more than 40 years. This class includes several molecules, which differ mainly in their pharmacokinetic characteristics. More recently, glinides, another class of medications able to increase insulin secretion, have been introduced into clinical use. This class comprises several molecules, such as repaglinide, a derivative product of meglitinide, and nateglinide, a derivative product of phenylalanine, both chemically different from sulfonylurea, although similar in the mechanisms of action. In fact, both classes of insulin secretagogue drugs bind the same receptor on β-cell surface, which is referred to as sulfonylurea receptor (SUR1/ABCC8) (Fig. 36.3) and consists in a subunit of the ATP-sensitive potassium channels (see also Chapter 27). Sulfonylureas and glinides act on β-cells mimicking the increased ATP/ADP ratio that follows metabolic stimuli. However, in SUR1, the binding site for sulfonylurea is distinct from that for glinides. Moreover, there are differences in the modalities of interaction between these two drugs and SUR1 receptor, the glinide binding being more short lasting than that of sulfonylurea. In addition, sulfonylureas possess other biological activities, which are documented *in vitro* but not well *in vivo*; they increase GLUT4-dependent glucose transport in the skeletal muscle, apparently via PI3K→IRS-1 pathway. Interestingly, loss-of-function mutations of either SUR1 or KIR6.2 (the other subunit of the potassium channels) cause unrestricted insulin secretion, leading to severe hypoglycemic familiar syndromes, whereas gain-of-function mutations lead to neonatal diabetes. Sulfonylureas can also bind to the cardiac SUR2 receptor. This interaction has potential unfavorable consequences, such as arrhythmias and impairment of myocardial ischemic preconditioning, although the clinical significance of this phenomenon remains unclear. The interaction with SUR2 may vary depending on the sulfonylurea molecule and its blood concentration. Glinides show lower affinity for SUR2.

For many years, it has been known that insulin secretion can be inhibited by diazoxide, and for this reason, this drug is used for the treatment of hypoglycemic syndromes caused by excessive insulin secretion. This drug acts by activating ATP-sensitive potassium channels through SUR1 receptors, thus inducing β-cell hyperpolarization and thereby inhibition of β-cell secretion.

As mentioned earlier, food ingestion causes release of multiple intestinal hormones that regulate many activities including pancreatic endocrine secretion. The glucose-dependent insulinotropic action of some of these hormones is of particular interest and has been recently exploited in management of type 2 diabetes (Table 36.2). Two different strategies have been developed to approach the pharmacology of the incretin system:

1. Generation of GLP-1 analogs with a longer half-life compared to the natural hormone, which are given by subcutaneous injection.
2. Synthesis of oral inhibitors of the catabolic enzyme DPP-4, which increase endogenous incretin half-life by slowing down their degradation. Several aspects of this novel pharmacological approach to the treatment of type 2 diabetes are particularly interesting: insulin stimulus is glucose dependent, greatly reducing risk of hypoglycemia, inhibition of glucagon secretion, potential anti-apoptotic effect on β-cells, and reduction of body weight (this effect is restricted to GLP-1 analogs) (Table 36.3).

Modulation of Insulin Signaling

Many substances can modulate IR signaling, either potentiating or inhibiting it. The agonist effect is particularly relevant for the treatment of insulin resistance, which is a major mechanism in the pathogenesis of hyperglycemia. However, the mechanisms of action of these drugs are still partially unknown. Drugs belonging to the biguanide class, phenformin and metformin, have been known for long time to be able to enhance insulin action and to induce an adaptive reduction of insulin secretion. While phenformin has been almost completely abandoned because of risk of lactic acidosis, metformin has been used for many years in Europe to treat diabetes, and recently, its use has also been approved in the United States. This drug is the only one belonging to the biguanide class available today. Evidence of multiple beneficial effects of metformin on cardiovascular risk factors as well as on cardiovascular and all-cause mortality in type 2 diabetes patients has led this old drug to be regarded as the first-line therapy for this disease.

The molecular mechanism of action of metformin has been unknown for a long time. Recently, it has been documented that this substance can stimulate AMPK, and this effect is hypothesized to be the major mechanism underlying metformin metabolic activities. AMPK represents the final element of a protein kinase cascade activated by increase in AMP/ATP ratio. This increase signals a reduction in intracellular energy due to either decreased ATP production, such as during hypoxia, or reduction in substrate availability. Physical activity, which causes ATP consumption, can also increase AMP/ATP ratio, thus activating AMPK activity. AMPK activation increases oxidation of fatty acids and energy production and inhibits anabolic processes. Esterification of fatty acids decreases along with reduction of intracellular adipose content. AMPK can stimulate

TABLE 36.2 GIP and GLP-1 effects and potential role in type 2 diabetes (DT2) therapy

GIP	Alterations in DT2	GLP-1
Ineffective in DT2	→ Insulin secretion	← Insulin secretion ↑
Glucagon secretion	→ Hyperglucagonemia	← Glucagon secretion ↓
β-cell apoptosis	→ β-cell apoptosis	← β-cell apoptosis ↓
β-cell replication	β-cell mass	β-cell replication ↑
Fat deposition	→ Obesity	← Food ingestion ↓, body weight ↓
No effect	→ Gastric emptying ↑=↓	← Gastric emptying
No effect	→ Hyperlipidemia	← Postprandial triglycerides ↓, postprandial free fatty acids ↓
No effect	→ Insulin resistance	← No rapid changes (insulin sensitivity ↑)

TABLE 36.3 Incretin system

Action	GLP-1 agonist	DPP-4 inhibitor
Administration route	Subcutaneous	Oral
Insulin secretion	Increased ↑↑	Increased ↑
Glucagon secretion	Suppressed ↓↓	Suppressed ↓
Postprandial hyperglycemia	Lowered ↓↓	Lowered ↓
Gastric emptying rate	Markedly lowered	No effect
Appetite	Suppressed	No effects
Satiety	Induced	No effects
Body weight	Lowered	Neutral
β-cell function	Preserved, improved proinsulin/insulin ratio	Preserved, improved proinsulin/insulin ratio
Gastrointestinal adverse effects	Frequent	Rare

Differences in the pharmacological actions of GLP-1 agonists and DPP-4 inhibitors.

glucose transport by insulin-independent mechanisms. This enzyme seems to play a central role in regulation of glucose and lipid metabolism on the basis of the cellular energy status. Thus, the insulin-like effect that follows AMPK activation can be at least partially indirect. The effect of metformin on AMPK appears to be independent from changes in AMP/ATP ratio and could be mediated by phosphorylation at threonine-172, a key regulatory site in AMPK catalytic subunit.

However, recent studies have showed that metformin possesses other major, AMPK-independent, metabolic effects. It can inhibit mitochondrial function by acting on complex I and antagonize glucagon activity on adenylate cyclase, thereby reducing hepatic gluconeogenesis.

Other drugs can stimulate the AMPK system. Among these, pioglitazone, an insulin sensitizer belonging to the thiazolidinedione class, is used in clinical practice. This drug is an agonist of peroxisome proliferator-activated receptor-gamma (PPAR-γ), a nuclear receptor whose natural ligands are fatty acids and other lipidic substrates. Activation of this receptor triggers transcription of genes controlling adipocyte differentiation and directly or indirectly modulating glucose and lipid metabolism and thereby insulin secretion. The molecular mechanisms underlying pioglitazone effect on AMPK are still unclear, but they seem to be different from those of metformin. The improvement of insulin sensitivity induced by this molecule is attributed to indirect mechanisms, the most important being the reduction of serum free fatty acids. At hepatic and muscular level, reduction of free fatty acids leads to increased glucose utilization. Other important effects are increased production of adiponectin and reduced secretion of inflammatory adipokines. In this context, it is worth mentioning that pioglitazone induces redistribution of fat from the omentum, liver, and muscles toward subcutaneous adipocytes, possibly explaining the beneficial metabolic effects of the drugs. However, the effect of thiazolidinediones on renal tubules, where they promote water and saline retention thereby increasing risk of heart failure, should be considered in clinical practice. Moreover, recent studies suggest that these drugs can increase risk of bone fractures.

Modulation of Intestinal Glucose Absorption

Acarbose is an old drug with a unique mechanism of action. It binds with high-affinity and high-specificity α-glucosidase, an enzyme located in small intestine brush border. This enzyme hydrolyzes both disaccharides and more complex carbohydrates to glucose and other monosaccharides. As a competitive inhibitor of α-glucosidase, Acarbose slows digestion and absorption of carbohydrates, leading to reduction in postprandial glycemia. Acarbose has been shown to safely and effectively prevent diabetes and to control hyperglycemia in diabetic patients, with a major activity on postprandial glycemia.

In normal subjects, glucose filtered by the kidneys is completely reabsorbed by sodium–glucose cotransporters (SGLTs). In particular, SGLT2 is a high-capacity, low-affinity glucose transporter accounting for about 90% of reabsorbed glucose. It is located in the luminal surface of the epithelial cells that line mainly the proximal tubule S1 segment. In diabetes, the capacity for glucose reabsorption is increased, and SGLT2 expression seems to be upregulated. Pharmacological inhibition of SGLT2 has been recently investigated as an option to reduce hyperglycemia in diabetic patients. Nowadays, many highly selective SGLT2 inhibitors have been developed or are currently being developed to decrease hyperglycemia in an insulin-independent manner. Among these molecules, canagliflozin has been approved by the US Food and Drug Administration and dapagliflozin by the European Medicines Agency for the treatment of type 2 diabetes. Available data suggest that SGLT2 inhibitors could be used either in monotherapy, to treat early diabetes, or in combination with insulin, in more advanced stages of the disease. Because the effectiveness of SGLT2 inhibitors depends on the amount of glucose filtered through the glomeruli, medication efficacy wanes as the glomerular filtration rate declines as a result of renal impairment. During treatment with these drugs, an increase in urinary infections has been reported. In addition, concerns have been raised about their cardiovascular safety, in particular regarding a possible higher risk of stroke associated to canagliflozin usage. At present, a cardiovascular outcome trial is ongoing.

INSULIN RESISTANCE AND NEW THERAPEUTIC PERSPECTIVES

In type 2 diabetic subjects, hypothalamic dopaminergic tone declines early in the morning. Activation of hypothalamic–pituitary–adrenal axis resulting from this decline has been postulated to increase insulin resistance, thus contributing to hyperglycemia. Bromocriptine is a dopaminergic agonist, which has been used for a long time in the treatment of prolactinoma. More recently, the once-daily quick-release bromocriptine (QR-bromocriptine) formulation has been proposed to address the central nervous system defect observed in subjects with insulin resistance and diabetes. Its early-morning administration may normalize circadian clock dysfunctions in these patients. In accordance with this hypothesis, some RCTs carried out in nondiabetic obese and type 2 diabetic subjects have reported improved insulin sensitivity, insulin secretion, and glycated hemoglobin levels after treatment with QR-bromocriptine. The timing of administration is critical, as this preparation must be given within 2 h after waking. Currently, a single 0.8 mg dose is available; the drug should be titrated once weekly by increasing the dose, up to a maximum of 4.8 mg or until the desired glucose effect is achieved or intolerance occurs.

In recent years, evidence has pinpointed a close relationship between chronic inflammation and insulin resistance, which is partially explained by several cytokines activating serine kinases. Accordingly, in animal models, salicylates have proved able to attenuate hyperinsulinemia and its associated metabolic alterations, with concurrent reduction of serine phosphorylation and increased tyrosine phosphorylation of IRSs. This activity of salicylates has a promising application in the treatment of type 2 diabetes. The potential efficacy of salicylates in the cure of diabetes was first suggested 100 years ago, when reduction of glycosuria in diabetic patients assuming high doses of salicylates was reported. Such observation has been recently confirmed by the targeting inflammation using salsalate in type 2 diabetes (TINSAL-T2D) study. In this RCT, treatment with salsalate led to a 0.5% reduction in glycated hemoglobin, which is comparable to the results obtained by other drugs currently approved for the treatment of type 2 diabetes. In addition, the treatment was reported to decrease triglycerides and increase adiponectin, an adipokine with antidiabetic and anti-inflammatory activities. Salsalate is a prodrug for salicylate that has been used for a long time in the treatment of rheumatic pain and shows an acceptable safety profile. Nonacetylated salicylates, such as salsalate, do not inhibit platelet aggregation and are insoluble in the acidic gastric environment. These properties reduce risk of hemorrhages and gastric upset. Salsalate, unlike other nonsteroidal anti-inflammatory drugs, inhibits nuclear factor kappa-light-chain-enhancer of activated B cells (NFκB) but not cyclooxygenase. NFκB transcription factor plays a critical role in the expression of many molecules implicated in inflammation, such as TNFα, IL-6, IL-1β, resistin, and others. It also stimulates recruitment of macrophages in adipose tissue. Although it is still unclear whether inflammation is either a cause or a consequence of insulin resistance, the TINSAL-T2D study suggests that targeting inflammation may represent a promising strategy to cure metabolic diseases.

TAKE-HOME MESSAGE

- Cells uptake glucose through specific membrane transporters.
- Deficiency in insulin secretion or intracellular signaling induces type 2 diabetes.
- Insulin binds a transmembrane tyrosine kinase receptor, triggering phosphorylation and activation of the insulin receptor substrates, which mediate the assembly of supramolecular complexes transducing insulin signal.
- Pharmacological agents employed in management of type 2 diabetes include drugs stimulating insulin secretion or its intracellular signaling and agents interfering with glucose kinetics.
- Metformin is the first-line drug in the treatment of type 2 diabetes.
- Medications targeting either the incretin system or renal glucose reabsorption are novel and effective tools for diabetes management.

FURTHER READING

Belfiore A., Frasca F., Pandini G., Sciacca L., Vigneri R. (2009). Insulin receptor isoforms and insulin receptor/insulin-like growth factor receptor hybrids in physiology and disease. *Endocrine Reviews*, 30, 586–623.

Doyle M.E., Egan J.M. (2003). Pharmacological agents that directly modulate insulin secretion. *Pharmacological Reviews*, 55, 105–131.

Hanefeld M., Schaper F., Koehler C. (2008). Effect of acarbose on vascular disease in patients with abnormal glucose tolerance. *Cardiovascular Drugs and Therapy*, 22, 225–231.

McClenaghan N.H. (2007). Physiological regulation of the pancreatic β-cell: functional insights for understanding and therapy of diabetes. *Experimental Physiology*, 92, 481–496.

McCrimmon R.J. (2012). Update in the CNS response to hypoglycemia. *Journal of Clinical Endocrinology and Metabolism*, 97, 1–8.

Morino K., Petersen, K.F., Shulman G.I. (2006). Molecular mechanisms of insulin resistance in humans and their potential links with mitochondrial dysfunction. *Diabetes*, 55 (Suppl 2), S9–S15.

Quinn C.E., Hamilton P.K., Lockhart C.J., McVeigh G.E. (2008). Thiazolidinediones: effects on insulin resistance and the cardiovascular system. *British Journal of Pharmacology*, 153, 636–645.

Rena G., Pearson E.R., Sakamoto K. (2013). Molecular mechanism of action of metformin: old or new insights? *Diabetologia*, 56, 1898–1906.

Riser Taylor S., Harris K.B. (2013). The clinical efficacy and safety of sodium glucose cotransporter-2 inhibitors in adults with type 2 diabetes mellitus. *Pharmacotherapy*, 33, 984–999.

Russell-Jones D., Gough S. (2012). Recent advances in incretin-based therapies. *Clinical Endocrinology*, 77, 489–499.

Taborsky G.J. Jr., Mundinger, T.O. (2012). Minireview: the role of the autonomic nervous system in mediating the glucagon response to hypoglycemia. *Endocrinology*, 153, 1055–1062.

Thorens B., Mueckler, M. (2010). Glucose transporters in the twenty-first century. *American Journal of Physiology. Endocrinology and Metabolism*, 298, E141–E145.

SECTION 11

INTERCELLULAR COMMUNICATION

37

PHARMACOLOGICAL REGULATION OF SYNAPTIC FUNCTION

MICHELA MATTEOLI, ELISABETTA MENNA, COSTANZA CAPUANO, AND CLAUDIA VERDERIO

> **By reading this chapter, you will:**
> - Understand the molecular mechanisms underlying synapse formation and function
> - Learn the different targets of drug action at synaptic level
> - Understand how alterations of synaptic function may lead to neurological and psychiatric disorders (synaptopathies)

THE SYNAPSE

A fundamental property of neurons is their ability to communicate with each other rapidly, over long distances and in an exceptionally precise manner. The sites of interneuronal communication are called synapses, a term coined by Sherrington in 1987. All functions of an organism, including aspects of awareness such as attention, perception, learning, decision-making processes, mood, and affection, are regulated by information transmission at synaptic level, also known as synaptic transmission. The number of synapses present in the human brain is estimated to be 10^{14}. From a structural perspective, synapses are cell-to-cell contact points (synaptic contacts, synaptic junctions) that allow a nerve cell to transmit an electrical or chemical signal to a target cell (either another nerve cell or a different cell type). Aside from a few exceptions, synapses form between the terminal part of a presynaptic axon and a postsynaptic cell. Two main types of synaptic junctions exist in nature: electrical synapses and chemical synapses. The former allow direct contact between the cytoplasm of the cells involved and therefore direct transmission of electrical impulses without chemical intermediates. These types of synapse are frequently found in invertebrates and are much less common in more complex organisms.

In electrical synapses, called gap junctions, the membranes of the presynaptic and postsynaptic cells are separated by a 3–4 nm gap, typically spanned by protein channels. These are regions of close apposition that provide direct cytoplasmic continuity between cells. Gap junctions are formed by two juxtaposed hemichannels, or connexons, composed of smaller subunits called connexins. They allow the bidirectional movement of ions, metabolites, and intracellular messengers, hence conferring electrical and metabolic "continuity" between presynaptic and postsynaptic cells (see also Supplement E 27.12).

However, the majority of synapses found in the central nervous system (CNS) are chemical. In terms of function, these junctions can be defined as specific sites where information transmission occurs in the absence of direct contact between the presynaptic neuron, which transmits the signal through chemical messengers, and the postsynaptic cell, be it a nerve cell or a different cell type. Signal transmission occurs in one direction and is quantal, meaning that it is formed by signals of fixed intensities. At a morphological level, chemical synapses are highly asymmetric, micron-sized structures. Their structure consists of a presynaptic terminal, characterized by the presence of vesicles containing neurotransmitters (synaptic vesicles (SVs)) and an electron-dense region, or active zone (AZ), and a postsynaptic terminal, specialized in receiving and responding to neurotransmitters. The two synaptic junction components are separated by a 20–25 nm gap, or synaptic cleft, where neurotransmitters are released and across which they diffuse.

General and Molecular Pharmacology: Principles of Drug Action, First Edition. Edited by Francesco Clementi and Guido Fumagalli.
© 2015 John Wiley & Sons, Inc. Published 2015 by John Wiley & Sons, Inc.

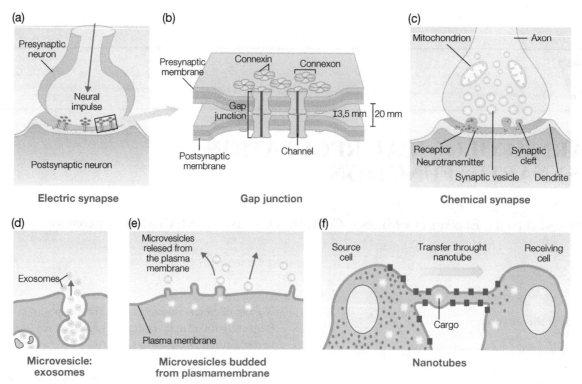

FIGURE 37.1 Mechanisms of intercellular communication. (a) and (b) Electrical synapse and gap junction. The cells involved are in direct contact, allowing electrical signal exchange without chemical intermediates. Presynaptic and postsynaptic cell membranes are separated by a 3–4 nm gap, spanned by protein channels called connexons. These junctions, termed gap junctions, provide cytoplasmic continuity between the two cells. Connexons are formed by two juxtaposed hemichannels composed of smaller subunits called connexins. This allows the bidirectional movement of ions, metabolites, and intracellular messengers. (c) Chemical synapse. These junctions are specialized sites that allow information transmission in the absence of direct contact between the presynaptic neuron, transmitting the signal (chemical messenger), and the postsynaptic cell (a nerve cell or a different cell type), receiving it. (d) and (e) **Microvesicles** are derived from either the plasma membrane (shedding vesicles) or intracellular compartments (exosomes). These are released into the extracellular space and mediate transfer of proteins, lipids, and genetic material from the cellular source to the target cell. They play a crucial role in communication between neurons and glial cells. (f) Nanotubes are cytoplasmic bridges that allow transfer of cytosolic components or even whole organelles between cells that are not in direct contact.

The presynaptic and postsynaptic compartments are connected and held in position by various cell adhesion molecules, or transsynaptic cell adhesion molecules, and by extracellular matrix proteins (Fig. 37.1).

Neurotransmitter release into the synaptic cleft represents the main mechanism whereby neurons communicate within the nervous system. However, alternative mechanisms that commonly operate among cells outside the nervous system are also used by neurons to communicate with each other and other cell types. These processes include nanotube formation and extracellular microvesicle release. Nanotubes are cytoplasmic bridges that allow the transfer of cytosolic components, or even whole organelles, between cells that are not in direct contact. Similarly, extracellular membrane microvesicles, derived from either the plasma membrane or intracellular compartments, mediate the transfer of proteins, lipids, and genetic material from the cellular source to the target cell. Due to their small size (50 nm to 1 μ), microvesicles can diffuse from their sites of release and are thought to play an important role as vectors of pathogenic and inflammatory agents (Fig. 37.1).

Synaptic Organization Complexity and Synaptopathies

The Tripartite Synapse Besides the neural components, the role of glial cells, in particular astrocytes, is fundamental to synapse development and function. Astrocytes are the most abundant type of glial cells in the brain, exceeding the number of neurons in both rodent and mammal nervous tissue (see Supplement E 37.1, "Role of Astrocytes in Synaptic Transmission"). The 3D reconstruction of astrocytes has shed light on their structure, showing numerous projections extending and occupying a defined space in the brain, without overlapping the area controlled by adjacent astrocytes. A single astrocyte comes into contact with hundreds of neurites and neuronal cell bodies, and its projections interact with tens of thousands of synapses. The term "tripartite synapse," which suggests a close morphological resemblance between neurons and astrocytes at synaptic level, indicates a structure composed of the astrocyte wrapped around the synapse, presynaptic and postsynaptic terminals. This concept recognizes the importance of glial cells as active components of synaptic transmission. Although astrocytes act on neural transmission at the level of single synapses, these cells are also capable of activating compensatory homeostatic mechanisms. This astrocyte-mediated compensatory response, or homeostatic plasticity, is essentially coordinated by the release of the pro-inflammatory cytokine TNFα by astrocytes, which regulates the strength of synaptic connections. Furthermore, astrocytes are able to sense overall alterations in neural circuit activity, for example, in the integrated rhythmic cerebral activity associated with sleep rhythm generation (Fig. 37.2).

Synaptic Networks If we consider that a neuron receives and forms an average of 1000 synaptic connections and that there are approximately 10^{11} neurons in the human brain, we can estimate the number of synapses in the human brain to be approximately 10^{14}. This multitude of connections displays different functional characteristics that depend not only on the neurotransmitter involved but also on the high variability of the molecular components present in synapses using the same neurotransmitter. This concept supports the now widely accepted view according to which synapse classification based upon the type of transmitter fails to adequately represent the broad diversity of synapses present in the nervous system. In addition, we must consider that synapses are not stable structures, but on the contrary, they maintain remarkable structural and functional plasticity, which persists throughout the life of an individual (see Section "Synapse Formation, Maintenance, and Plasticity").

However, molecular, cellular, and functional heterogeneities are not the only factors contributing to overall synapse complexity. Synapses that operate in a coordinate manner within neural networks, which in turn coordinate functions like attention, perception, learning, decision-making, mood,

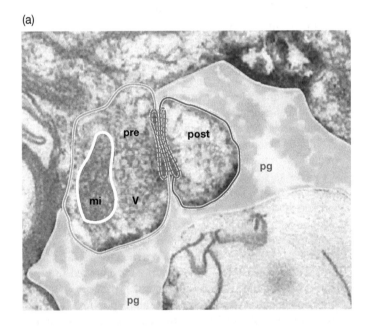
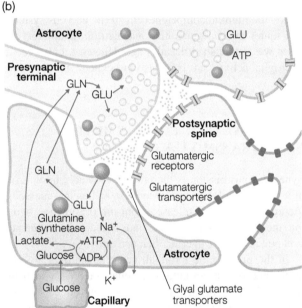

FIGURE 37.2 Interactions between synapse and glial cells: the tripartite synapse. (a) Electron microscope image of a cortical synapse. The light grey line identifies the presynaptic compartment (pre) and the black line the postsynaptic compartment (post). Glial cell processes (pg, white line) are marked to reveal the presence of the enzyme glutamine synthetase. The dotted lines encircle the pre and postsynaptic densities. V, synaptic vesicles; mi, mitochondrion. The bar represents 0.5 μm (From: Xu-Friedman MA *et al.*, 2001. Copyright permission 2006 from the Society of Neuroscience). (b) Schematic representation of the cellular mechanisms by which astrocytes can control synaptic function. Glutamate, released at synaptic level, is taken up by astrocytes through specific transporters and converted into glutamine by the enzyme glutamine synthetase. Astrocytes provide neurons with glutamine which they reconvert into glutamate. Glutamate is co-transported along with sodium ions, causing an increase in intracellular sodium concentration that leads to activation of the Na+/K+ATPase pump. This stimulates glycolysis in astrocytes, resulting in production of lactate which is then released and used by neurons as energy source. Astrocytes can also release numerous neurotransmitters, such as glutamate and ATP. These are released in response to stimuli that induce increase of intracellular calcium. Neurotransmitter release from astrocytes occurs via a regulated exocytotic mechanism similar to neuronal exocytosis. In fact, astrocytes express proteins of the SNARE complex: synaptobrevin 2, syntaxin 1, and SNAP 23. Astrocytic glutamate release is blocked by tetanus toxin which cleaves the protein synaptobrevin 2. EAAC1, neuronal glutamate transporters; GLN, glutamine; GLU, glutamate (Modified from: Bacci A., Verderio C., Pravettoni E., Matteoli M. (1999) The role of glial cells in synaptic function. *Philosophical Transactions of the Royal Society London B: Biological Sciences* 354, 403–409).

and other cognitive processes of an organism, also need to be considered.

Synaptopathies It is not surprising that synaptic alterations may induce pathological states. In fact, mutations in genes encoding synaptic proteins are often present in various neurological and psychiatric disorders or in neurodegenerative diseases. The term "synaptopathy" describes disruptions in synaptic structure and function underlying pathological states. The fact that many psychiatric and neurological disorders are now recognized as "synaptopathies" has led to significant progresses in the fields of psychiatry and neurology. As a result, synapses are now targeted for the pharmacological treatment of such pathologies. It has now become clear that glial cell dysfunctions may cause or contribute toward the manifestation of neurological conditions. A large amount of experimental evidence suggests that reactive astrogliosis, characterized by astrocyte hypertrophy and release of inflammatory mediators and cytokines, favors neuronal hyperexcitability and onset of epileptic impulses.

The link between functional alterations of microglial cells—the immunocompetent cells of the nervous system—and major neurodegenerative and neuroinflammatory diseases was established decades ago. However, only recently microglial activity has been also associated with psychiatric disorders.

THE PRESYNAPTIC COMPARTMENT: NEUROTRANSMITTER RELEASE

Synaptic transmission is the process in which, upon arrival of a nerve impulse (action potential), a chemical messenger (neurotransmitter) is released from the presynaptic neuron into the synaptic cleft where it binds to receptors on the postsynaptic neuron, triggering their activation.

In the presynaptic terminal, neurotransmitters are compartmentalized in vesicular organelles, or SVs, which then fuse with the presynaptic membrane at thickened areas termed AZ (see Supplement E 37.2, "Dynamic Organization of Synaptic Vesicle Pools"). Voltage-gated Ca^{2+} channels are expressed at high concentrations in AZ. The role of these channels is to translate electrical signals into an elevated Ca^{2+} influx, which hits the vesicles docked at AZ, reaching millimolar concentrations, and triggers exocytosis and release of vesicular contents into the synaptic cleft. Neurotransmitter release is said to be quantal, to indicate the relatively constant number of neurotransmitter molecules stored in SVs and subsequently released via SV fusion. A quantum is the minimum amount of neurotransmitter that is released and evokes a postsynaptic effect (miniature potential) with replicable amplitude. Thickenings of the postsynaptic membrane (or postsynaptic densities) are present on postsynaptic neurons, opposite to AZ. In the case of excitatory glutamatergic synapses (formed on excitatory neurons), these postsynaptic densities are localized on specialized dendritic structures termed spines. Although excitatory and inhibitory synapses differ in the structure of their postsynaptic apparatus, both have postsynaptic densities containing high concentrations of neurotransmitter receptors, signal transducers (e.g., heterotrimeric G proteins), enzymes, and ion channels, mediating the conversion of the chemical, neurotransmitter-mediated signal into an electrical (changes in postsynaptic membrane electric potential) or a metabolic signal (Figs. 37.3 and 16.1). This type of synaptic transmission, restricted to neural circuits interconnected by synapses, is also termed wiring transmission (WT). Neuronal transmission takes also place through volume transmission (VT), which consists in the release of chemical messengers into extracellular spaces in the brain and cerebrospinal fluid, which can therefore diffuse in larger volumes and over a longer time period, thus transferring the signal to a higher number of distant receptor cells.

Chemical Messengers: Neurotransmitters and Neuropeptides

Neurons are extremely polarized cells, with some axons extending up to one meter from their cell bodies as is the case of the sciatic nerve motor neurons in humans. Neurons use small molecules such as amino acids and amines (known as classical neurotransmitters) as chemical messengers. The advantage is that such small molecules can be synthesized locally by cytoplasmic enzymes present at the terminals and reuptaken into the terminals from the synaptic cleft after exocytosis.

Furthermore, the organelles responsible for neurotransmitter storage and release (SVs) can be recycled via an economic and efficient exo-endocytotic pathway (see Supplement E 37.3, "Endocytotic Process of Synaptic Vesicles"). Because of these properties, classical neurotransmitter release is a unique process compared to other forms of regulated secretion in other tissue types. The functional autonomy of the cell body provides neurons with remarkable resistance to secretory exhaustion (or fatigue), which is only reached in conditions of extreme and extended stimulation. Additionally, SV exocytosis is an extremely rapid process that takes only a few hundred milliseconds to reach completion and is followed by an extremely precise and efficient endocytotic reuptake process. Lastly, despite the constant amplitude of the action potential, neurotransmitter release is a plastic and regulated process. Neurons may secrete different quantities of neurotransmitter in response to a single action potential depending on their previous activity and the specific biochemical environment surrounding presynaptic terminals. These properties are of fundamental importance for the information integration occurring at synaptic level and in synaptic plasticity processes.

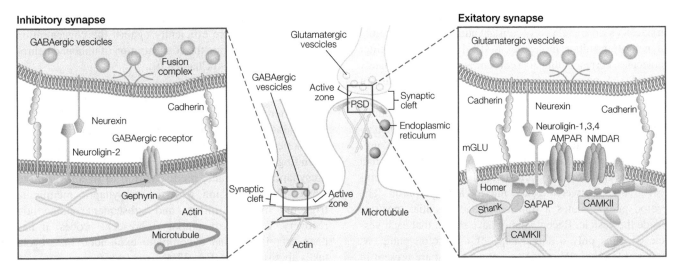

FIGURE 37.3 Inhibitory and excitatory synapses. CNS inhibitory (left) and excitatory (right) synapses express different arrays of specific proteins. Synaptic vesicles, containing GABA or glutamate, fuse with the plasma membrane at the active zone. The presynaptic terminal is separated from the postsynaptic compartment by a gap (synaptic cleft). Numerous adhesion proteins span the synaptic cleft, holding close pre- and postsynaptic membranes and in fixed positions. Neurotransmitter receptors are located on the postsynaptic membrane, opposite the fusion sites of synaptic vesicles, where they arrange in macromolecular complexes that also encompass a large number of structural (scaffold, gephyrin, homer, PSD-95, and shank) and signaling (e.g., CaMK II) proteins. These components form the postsynaptic density, which has the main function of receiving and transmitting signals. In the CNS, glutamatergic excitatory synapses typically form on dendritic spines, while inhibitory GABAergic synapses can be found on dendritic branch or soma. AMPAR, AMPA receptors; mGLU, metabotropic glutamate receptors; NMDAR, NMDA receptors; PSD, postsynaptic density (From McAllister AK. (2007). Dynamic aspects of CNS synapse formation. *Annual Review of Neuroscience* 30: 425–450.)

Neuropeptides need a separate discussion. These protein-based secretion products play a crucial role in a wide array of biological processes, such as extracellular matrix degradation, neuronal survival, synaptic transmission, and learning processes. In this case, neurons use a different system; peptides are synthesized in the cell body, stored in large morphologically distinct vesicles and transported to the peripheral area.

This system implies a different use of these neurotransmitters (see Supplement E 37.4, "Neuropeptide Secretion").

Secretory Granules and SVs

Both classical neurotransmitters and neuropeptides are stored in vesicles from which they are released into extracellular spaces following an increase in cytosolic Ca^{2+} concentration. However, the organelles storing these neurotransmitters have different morphology, cell cycle, and secretion mode. Classical neurotransmitters (glutamic acid, glycine, GABA, acetylcholine, and biogenic amines) are stored in small SVs with surprisingly homogeneous diameters (40–50 nm). These accumulate in proximity of AZ, where neurotransmitter release occurs. In contrast, neuropeptides are stored, either alone or with classical neurotransmitters, in larger vesicles of more heterogeneous sizes (100–300 nm in diameter). These secretory vesicles have an electron-dense core due to a high concentration of proteins (large synaptic dense-core vesicles) (Fig. 37.2). Unlike small SVs (hereinafter referred to as SVs or SV), large dense-core vesicles (hereafter referred to as secretory granules) are produced in the cell body and can dock and fuse with the plasma membrane at various sites within the neuron, including cell body, dendrites, and axon. Interestingly, while there is a close correspondence between classical neurotransmitters (at presynaptic level) and their receptors (at postsynaptic level), there is often a lack of anatomical correspondence between the localization of neuropeptides and that of their receptors at cerebral level ("mismatch"). This is a further proof of the long-distance action of this type of neurotransmitters, which is typical of endocrine/paracrine transmission. Therefore, we can say that in general SVs are responsible for WT, a spatially well-defined signaling mode (with the aforementioned exceptions), whereas secretory granules are involved in VT, a nonspatially defined transmission mode.

At presynaptic terminal, SVs are organized in different functional groups: some are anchored to the dense cytoskeletal matrix of the terminal, which mainly consists of actin and spectrin (reserve pool); the remaining vesicles are mobile and available for release (recycling pool). A small percentage of the recycling pool vesicles are already docked at the AZ (readily releasable pool) (see Supplement E 37.2). Secretory granules show a similar organization (see Supplement E 37.4).

Neurotransmitter Storage Being remarkably homogeneous in size, SVs store a relatively constant number (or "quantum") of neurotransmitter molecules (~10^3–10^4 neurotransmitter molecules, corresponding to a concentration of approximately 100 mM). A quantum represents the minimum quantity of neurotransmitter that is released and evokes a postsynaptic effect, termed miniature potential, with replicable amplitude. Loading of SVs is an active process during which neurotransmitters are taken up against their concentration gradient. An electrogenic proton pump (an ATPase that transports H^+ into the vesicles) establishes an electrochemical gradient between vesicle interior and cytoplasm. The proton pump creates a transmembrane potential, positive inside the vesicle. Recent studies have shown that each vesicle contains only a single copy of the proton pump, as opposed to other SV membrane proteins that are present in multiple copies. The electrochemical gradient generated by the pump is then used for the active transport of neurotransmitters by transporter proteins specific for biogenic amines, acetylcholine, glutamic acid, and GABA/glycine.

Neurotransmitters are stored in vesicles together with Ca^{2+}, ATP, and—especially in the case of peptide neurotransmitters—proteins and peptidoglycans. These components appear to reduce osmolarity of secretory vesicle contents (therefore preventing osmotic swelling) and to allow transporters to work more efficiently against the concentration gradient (by decreasing the concentration of unbound neurotransmitters) (see also Chapter 29).

The Exocytotic Process The complex process of neurotransmitter release consists of a sequence of mechanisms: a preparatory phase, during which SVs detach from the cytoskeleton, followed by a series of "mandatory" reactions including targeting and docking of vesicles to AZ on the presynaptic membrane, predisposition to fusion (priming or hemifusion), fusion with the presynaptic membrane triggered by Ca^{2+} influx, and subsequent endocytotic recovery. Using extracellular fluorescent probes, it has been established that a complete exo-endocytotic cycle takes about 1–2 min (Fig. 37.4).

The processes of SV docking, hemifusion, and fusion with the plasma membrane require specific vesicle membrane proteins (synaptobrevin, or v-SNARE) and target membrane proteins (syntaxin and SNAP-25, or t-SNAREs). SNARE complex assembly is a highly regulated process that requires SM (Sec1/Munc18-like) proteins, which spatially and temporally organize trans-SNARE complexes, and complexins, which stabilize the complex in the metastable state

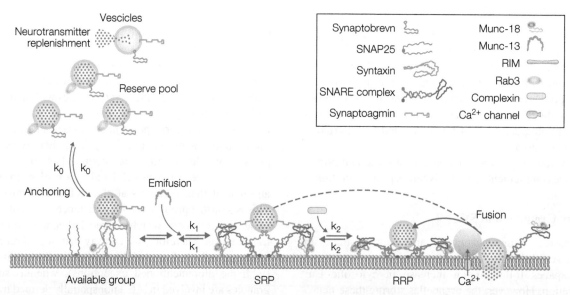

FIGURE 37.4 Neurosecretion and reuptake of synaptic vesicle components. During calcium-dependent exocytosis, synaptic vesicles interact with the presynaptic membrane. The diagram shows the major synaptic vesicle and plasma membrane proteins involved in the fusion process. A chain of molecular mechanisms controls the events that ultimately lead to synaptic vesicle exocytosis, when intracellular calcium concentration rises. The docking process consists in the transfer of vesicles from the reserve pool to the available pool (unprimed pool (UPP)) and their attachment to the active zone. The priming process consists in the transfer of vesicles from the available group to the release group (slowly releasable and readily releasable) and in the assembly of the SNARE complex, which brings the vesicle in close contact with the presynaptic membrane (hemifusion). Munc-18 and Munc-13 proteins regulate these processes, facilitating docking and priming, respectively. Complexins favor hemifusion. Vesicle fusion with the membrane is triggered by calcium binding to its sensor, synaptotagmin. The main proteins in synaptic vesicle and presynaptic membrane are schematically represented (see insert in figure). For further details, see text. RRP, readily releasable pool; SRP, slowly releasable pool (From Becherer U, Rettig J. (2006). Vesicle pools, docking, priming, and release. *Cell and Tissue Research* 326: 393–407.)

of hemifusion until Ca^{2+} influx within the terminal triggers full-blown fusion. Ca^{2+} binds to synaptotagmin, an SV integral protein that acts as a cytosolic Ca^{2+} sensor. Synaptotagmin is able to interact with both the SNARE complex and the acidic phospholipids in the cytosolic layer of the presynaptic membrane in a Ca^{2+}-dependent manner. This interaction destabilizes the hemifusion complex, triggering the complete fusion of the SV with the membrane, and hence neurotransmitter release. Vesicles of the reserve pool interact with the cytoskeleton through synapsin-mediated mechanisms. Synapsins are a family of SV-associated proteins that interact with actin and undergo dissociation following phosphorylation by protein kinases. Therefore, the phosphorylation state of synapsins (which is regulated by intraterminal Ca^{2+} levels) contributes to regulate the availability of SVs for fusion. (Details of the SV exocytotic cycle can be found in the Supplements E 37.2, E 37.3, and E 37.5).

It is important to note that, despite research has mainly focused on the role of synaptic proteins in vesicle fusion and endocytosis, it is now clear that the lipid-based environment in which proteins operate plays a crucial role in neural transmission modulation. Lipids regulate synaptic function by binding to regions on synaptic proteins that regulate prefusion phases (e.g., synaptotagmin). Furthermore, posttranslational modification of synaptic proteins, such as palmitoylation or conjugation with other fatty acids, represents an important mechanism in the regulation of exocytosis (see Supplement E 37.5, "The Role of Lipids in the Exo-endocytotic Cycle of Synaptic Vesicles").

The Endocytotic Process Proteins and lipids that make up SVs need to be recovered from the plasma membrane following exocytosis and recycled to form new vesicles to be loaded with more neurotransmitters. This occurs via a selective mechanism of membrane recovery, termed exocytosis (Fig. 37.4). In 1973, Heuser and Reese demonstrated for the first time the SV cycle using an extracellular peroxidase solution. The recent introduction of optical tracers and subsequent use of pH-sensitive protein probes (pHluorin) have been particularly helpful in the study of exocytosis, also at the level of small synaptic terminals. These techniques allow specific monitoring of SV fate and visualization of endocytosis even after a single action potential.

It is now widely accepted that vesicle recovery is mediated by clathrin, a protein composed of three heavy chains and three light chains, assembled in structures termed triskelions. These wrap around the vesicle being recovered from the plasma membrane, forming a basketlike structure. This results in the formation of the so-called clathrin-coated pits. From an anatomical–structural point of view, clathrin-coated pits are found at the lateral sides of AZ and therefore distant from presynaptic membrane regions dedicated to SV fusion. These terminal regions, termed sites of endocytosis, are characterized by high actin concentration. Fission of clathrin-coated vesicles from the plasma membrane involves a number of proteins, including GTPase dynamin and polyphosphoinositide phosphatase synaptojanin (see Supplement E 37.3). Following membrane fission, clathrin-coated vesicles can either fuse with endosomes, from which new vesicles are then generated, or lose their coating without fusing with the endosome; in both cases, the recovered vesicle is acidified by the proton pump and subsequently loaded with newly synthesized neurotransmitters.

After observing that clathrin-coated pits are extremely rare in synaptic terminals at rest, whereas they are numerous in terminals that have undergone massive exocytosis, Bruno Ceccarelli hypothesized the existence of an alternative exocytosis mechanism. This process, referred to as "kiss and run," may be predominant in conditions of physiological stimulation. According to this mechanism, vesicles release their contents through a fusion pore, but they do not fuse completely with the presynaptic membrane. In this way, they can be rapidly recovered by sealing of the fusion pore and dissociation of the two membranes. Thus, the "kiss and run" is an efficient time- and energy-saving mechanism (Fig. 37.4). Another mechanism may also exist, termed "kiss and stay," in which vesicles never move away from the membrane and are reloaded with neurotransmitter *in loco*. This mechanism has not been well defined yet.

Neurotransmitter Fate

Neurotransmitters released in the synaptic cleft can undergo one of the following processes, which are all aimed at decreasing their concentration (hence promoting their detachment from receptors and inhibition of postsynaptic activity) while preserving their molecules (saving biosynthetic energy):

- Neurotransmitters (e.g., monoamines or GABA) can be transferred into nerve terminals by transporters present in the plasma membrane. This is the most important mechanism to stop neurotransmitter action. Once the neurotransmitter has been internalized in the cytoplasm, it can be metabolized or stored back into SVs by specific vesicular transporters.

- Neurotransmitters can be taken up and subsequently metabolized by glial cells. Neurotransmitters can be metabolized in the synaptic cleft. This represents the main mechanism inhibiting acetylcholine action.

Many drugs developed to treat psychiatric, neurologic, or neurodegenerative disorders specifically act on mechanisms of neurotransmitter uptake and metabolism. These include antidepressants inhibiting serotonin and/or noradrenaline reuptake or degradation; drugs for the treatment of Alzheimer's disease that inhibit acetylcholine degradation, thereby increasing its availability in the synaptic cleft; and certain antiepileptic drugs that inhibit GABA reuptake or degradation by neurons or glial cells.

Released neuropeptides undergo a very different fate compared to classical neurotransmitters as their release generally occurs outside the synaptic cleft (parasynaptic release), directly into the extracellular space, and there are no transporters mediating their reuptake into the terminal. For these reasons, neuropeptides have two possible fates: extrasynaptic diffusion and extraneuronal metabolism due to peptidase action. Both processes have important functional implications. Peptidase action can either biologically inactivate the peptide, or it can produce fragments with a different biological activity compared to the original peptide. Extrasynaptic diffusion allows neuropeptides to act over long distances, in a paracrine modality, on high-affinity extrasynaptic receptors (VT).

THE POSTSYNAPTIC COMPARTMENT: SIGNAL RECEPTION

In a synapse, the postsynaptic compartment is the structure specialized in receiving and processing the signal. The action exerted by a neurotransmitter on the postsynaptic compartment does not depend on its chemical features, but on the properties of the receptor it binds. For example, acetylcholine can stimulate some neurons and inhibit others, and in some cases, it can also have both effects on the same neuron. Therefore, it is the receptor to determine whether a cholinergic synapse is excitatory or inhibitory and whether an ion channel is directly activated by the transmitter (electrical response) or indirectly via secondary messengers (metabolic or metabotropic response).

All neurotransmitter receptors share two biochemical characteristics: (i) they are integral membrane proteins in which the extracellular domain recognizes and binds the neurotransmitter; (ii) they have a biological effect on postsynaptic cells; typically, they have a direct or indirect influence on the opening and closing of ion channels, or they induce complex metabolic modifications (see Chapter 9).

In the brain, one neuron is connected with hundreds of other neurons through either excitatory or inhibitory connections. Therefore, a single neuron is continuously exposed to multiple signals mediated by different neurotransmitters affecting the activity of different ion channels and has to integrate such incoming signals into a coordinated response. CNS synapses are moderately efficient, meaning that stimuli produced by a single presynaptic neuron are often insufficient to trigger an action potential in the postsynaptic neuron. The postsynaptic response depends on the combination of stimuli the neuron receives from other neurons through synaptic contacts. However, it is also influenced by various factors such as localization, size and shape of synapses, as well as proximity and relative strength of nearby synapses.

The process by which postsynaptic neurons integrate synaptic signals is termed neuronal integration.

All neuronal compartments—axon, terminals, cell body, and dendrites—can act as presynaptic and postsynaptic sites. However, most contacts are axoaxonal, axosomatic and axodendritic. Axodendritic contacts can be established between an axon terminal and a dendritic branch or spine. The proximity of a synapse to the neuronal region where the action potential is triggered (axon hillock) has a greater influence on the postsynaptic response compared to other synapses, which are found on the distal part of the dendritic tree. Synapses found on the cell body are in general inhibitory, whereas synapses found on dendritic spines are often excitatory.

As previously stated, receptors are the most important postsynaptic molecules and exert two main functions: specific neurotransmitter recognition and effector activation (see also Chapter 9). In particular, ionotropic receptors mediate rapid (few milliseconds) synaptic transmission, whereas G-protein-coupled receptors (metabotropic receptors) and tyrosine kinase receptors mediate long-lasting neurotransmitter effects. The cascade of events triggered by secondary messengers may also culminate in gene expression changes, and its effects may last for days or even weeks.

SYNAPSE FORMATION, MAINTENANCE, AND PLASTICITY

The ability of the nervous system to propagate information depends on correct formation of neural connections. Axonal growth and synaptogenesis are two distinct yet closely related processes and are essential for the formation of functional connections during brain development. The trajectory marked by the axonal growth cone to the target area can be divided into different segments, which are defined by the contact between the axon and the guide cells (glial cells) releasing various chemoattracting or chemorepellent molecules. These molecules respectively favor and inhibit axonal growth, guiding the axon through heterogeneous cellular elements, until the appropriate neuronal target is reached. Once there, axon terminals start to branch out into multiple, fine structures, initiating synaptogenesis (Fig. 37.5). In mammalian CNS, immature postsynaptic apparatus initially protrudes from the dendrite surface as small, actin-rich filopodia. Through cytoskeletal remodeling and recruitment of postsynaptic components, filopodia evolve into mature, functional spines. This process occurs in parallel with the elaboration of presynaptic axon terminals accompanying the stabilization of intercellular connections in mature synapses. Recent studies have suggested that molecules involved in directing axonal growth also contribute to synaptogenesis by regulating cytoskeletal dynamics and promoting recruitment of pre- and postsynaptic components of the neurotransmission

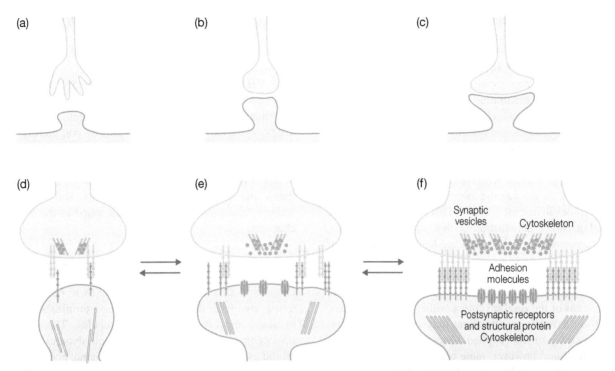

FIGURE 37.5 Schematic representation of synapse formation (a–c) and maturation (d–f). (a) Election of synaptic site. (b) Formation of synaptic contact. (c) Synapse maturation. (d–f) Role of cell adhesion molecules in synaptogenesis: N-cadherin and catenins regulate morphology and strength of dendritic spines. In the early stages of development, dendritic spines display an elongated morphology and are highly motile. (d) Synaptic contacts are then stabilized by recruitment of cell adhesion molecules. Interactions between pre- and postsynaptic adhesion molecules activate intracellular signaling cascades and specific secondary messengers that can alter the organization of receptors and synaptic cytoskeleton. (e) This chain of events regulates synaptic maturation and is also fundamental to synaptic plasticity. (f) See text for further details (From Giagtzoglou N, Ly CV, Bellen HJ. (2009). Cell adhesion, the backbone of the synapse: "Vertebrate" and "Invertebrate." *Cold Spring Harb Perspect Biol* 1: a003079.)

machinery. Other molecules, activated after synaptic contact formation, then contribute to synapse stabilization, expansion, and maturation. These molecules include transmembrane semaphorins, tyrosine kinase receptors that bind the membrane-associated protein ephrin, and various classes of adhesion proteins (e.g., neuroligin–neurexin complex, leucine-rich repeat (LRR) proteins, members of the Ig family, members of the SynCAM family (synaptic adhesion proteins), cadherins).

The interaction between presynaptic neuroligins and postsynaptic neurexins is one of the best-studied examples of interaction between adhesion proteins involved in synaptogenesis. When expressed in nonneuronal cells, neuroligins and neurexins induce the differentiation of the presynaptic and postsynaptic compartments, for example, in hippocampal neurons cocultured with cells expressing the adhesion proteins. The interaction between neuroligin and the postsynaptic protein PSD-95 mediates coupling of the adhesion protein to postsynaptic components, allowing the transsynaptically activated signal to propagate in the dendritic spine (Fig. 37.3).

Correct development of synaptic contacts and appropriate connectivity models are crucial for the assembly of functional neuronal circuits. Alterations in synaptic contact formation may result in neuropsychiatric disorders such as mental retardation and autism and are also likely to play a role in neurodegenerative disorders such as dementia and Alzheimer's. To date, mutations in neurexin 1 and neuroligins 3 and 4 have been associated with autism and intellectual disability. Furthermore, SHANK3, a protein that interacts with neuroligin and has a structural role in postsynaptic compartment maintenance, has been associated with autism. In addition, more than 140 different mutations in L1CAM have been associated with neurological disorders, including intellectual disability. Therefore, the study of proteins involved in synaptogenesis can be useful in defining the pathological mechanisms underlying neurological disorders.

During postnatal CNS development, synaptic maturation is followed by a regulated removal of several synaptic contacts (synaptic pruning). During this phase, functionally important synapses are strengthened, and less relevant synapses are weakened and subsequently eliminated. Removal of redundant synapses represents the terminal phase of neural network formation. Alterations of this stage may underlie psychiatric disorders such as adolescent-onset schizophrenia. It has recently been shown that a genetic alteration resulting in defective synaptic pruning and persistence of hyperconnectivity in the adult brain causes excessive synchronous firing between groups of neurons and epileptic activity. Synaptic contacts remain plastic even after completion of their maturation in the adult brain. Synaptic plasticity and structural changes of neural circuits are believed to underlie memory, learning, and brain function. This has recently been demonstrated using new cell imaging tools (two-photon microscopy) and transgenic mouse lines expressing fluorescent proteins (such as the green fluorescent protein (GFP)) to analyze the morphology of axonal terminals and spines in animal brains *in vivo*. These studies have provided a direct demonstration that synaptic structures undergo a remodeling process during learning processes or following sensory experiences, showing that the processes of learning a motor task or exposure to enriched environments or visual stimuli induce formation of new spines in specific areas of the cerebral cortex within hours. The duration of these synaptic modifications may vary depending on the stimulation protocols applied. In some cases, new spines may last for long; however, more often, the net number of spines returns to control values in the 2 weeks following the stimulus as a result of enhanced pruning speed.

Although spine plasticity has not been entirely proved to be essential for acquisition of new skills and memory retention, a strong body of evidence suggests its fundamental role in these phenomena.

THE PHARMACOLOGY OF NEUROSECRETION

As synapses represent the preferred targets of CNS-active drugs, synaptic pharmacology almost completely corresponds with neuropharmacology. In this chapter, we will focus on drugs acting on neurotransmitter release mechanisms (Fig. 37.6). Drugs affecting neurotransmitter synthesis and metabolism, their receptors, and intracellular transduction processes will be discussed in the following chapters.

Drugs Interfering with Secretory Vesicle Transport

Granules and secretory vesicles (or their precursors) are transported from the cell body to nerve terminals via a form of active movement along microtubules. Therefore, substances that interfere with microtubule dynamics will also inhibit exocytosis; these include colchicine, nocodazole, and vinca alkaloids (vincristine). Brefeldin A is a fungal metabolite that severely affects secretory mechanisms. In fact, this drug inhibits secretion and considerably alters the endomembrane structure, forming elongated structures that destroy

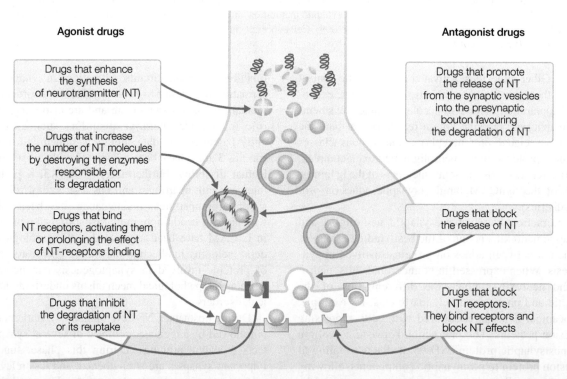

FIGURE 37.6 Mechanisms of drug action at synaptic level. Among the main drug categories currently used for the treatment of neurological and psychiatric pathologies, there are agonists or stimulants that increase synthesis of neurotransmitters or inhibit their degradation, or drugs that activate neurotransmitter receptors, and drugs with antagonistic or inhibitory effect that promote degradation of neurotransmitters at presynaptic level or inhibit their release, or drugs that inhibit neurotransmitter receptors.

the Golgi apparatus. Brefeldin A also interferes with endocytosis by inhibiting clathrin coat assembly. Because of its strong effects on secretion in eukaryotic cells, brefeldin A, initially designed for antibiotic therapy, is no longer used for patient treatment. However, it has proved useful to study the mechanisms involved in secretory organelle biogenesis and transport.

Drugs Interfering with Neurotransmitter Loading into Vesicles

Since vesicle loading with neurotransmitters requires vesicle acidification, substances that interfere with the proton pump activity inhibit this refilling process. These substances include bafilomycin, a macrolide antibiotic derived from *Streptomyces griseus* and mainly used for research purposes that specifically inhibits vacuolar ATPase function and therefore vesicle acidification. By dissipating acid intravesicular pH, bafilomycin inhibits dissociation and transfer of tetanus and botulinum toxin catalytic subunits to the cytosol, hence preventing proteolysis of SNARE proteins (further details ahead).

Among the substances directly inhibiting vesicular neurotransmitter transporters, there are reserpine and tetrabenazine. These drugs act by blocking monoamine vesicular transporters, which transfer noradrenaline, serotonin, and dopamine into SVs. Because of this property, reserpine has been used for the pharmacological treatment of hypertension and certain types of psychotic disorders in the past. However, the design of more efficient compounds with fewer side effects has made this drug obsolete, resulting in its withdrawal from the market in various countries.

Vesamicol is a substance that interferes with choline transport into SVs.

Drugs and Toxins Interfering with Late Steps of Neuroexocytosis

Many toxins interfere with exocytosis. Among these, there are botulinum and tetanus toxins, α-latrotoxin, and certain snake venom phospholipase toxins.

The seven serotypes of botulinum toxins (A–G) and the tetanus toxin are produced by *Clostridium botulinum* and *Clostridium tetani*, respectively. These two toxins have very similar mechanisms of action. Differences in the neurological syndromes they cause—flaccid paralysis in the case of botulism and spastic paralysis in the case of tetanus—are attributed to the fact that, although both these toxins inhibit exocytosis, they act on different synapses. Botulinum toxins act at nerve terminals of neuromuscular junctions by inhibiting acetylcholine release and hence skeletal muscle activation, causing flaccid paralysis. Tetanus toxin, which also enters the nervous system at the neuromuscular junction, migrates to motor neuron cell bodies in the spinal cord anterior horn by retrograde axonal transport along motor nerves. Here, the toxin reaches transsynaptically the inhibitory interneuron terminals, possibly via exosome release, and inhibits glycine release. This causes motor neuron overexcitation and leads to simultaneous and uncoordinated contraction of agonist and antagonist muscles, causing spastic paralysis.

Clostridial toxins are composed of two subunits linked by a disulfide bridge: the heavy subunit (100 kDa) allows the toxin to bind specific protein/lipid receptors present on the target neuron membrane, favoring their internalization; the light subunit (50 kDa) is a Zn protease that, following translocation to the cytoplasm, causes SNARE protein proteolysis, inhibiting exocytosis (Fig. 37.5). Botulinum toxins A, C, and E cause SNAP-25 proteolysis; toxins B, D, F, and G and tetanus toxin cleave VAMP-2; and serotype C also causes syntaxin 1A proteolysis. Clostridial toxins cause neuron intoxication through a multistage mechanism that includes binding of toxins to the neuronal surface, internalization mostly via SV endocytosis, dissociation of heavy and light subunits, light subunit translocation to the cytosol, and subsequent SNARE protein proteolysis. The currently accepted model for clostridial toxin binding to peripheral nerves involves a double interaction: initially, the toxin binds to gangliosides on the terminal membrane, and subsequently, it interacts with a protein receptor composed of luminal regions of the synaptotagmin I/II or SV2 vesicular proteins. Both protein and lipid receptors of botulinum toxins are widely distributed throughout the central and peripheral nervous systems. It is still to be determined whether the preferential action of such toxins on cholinergic synapses is due to the abundance of isoform-specific receptors and gangliosides at cholinergic junctions, as it has been suggested.

The identification of SNARE proteins as molecular targets of clostridial toxins has played an essential role in understanding secretory mechanisms. Indeed, it has highlighted the essential role of SNARE proteins in exocytosis by showing that single-site cleavage of any of these proteins (synaptobrevin, syntaxin, or SNAP-25) is sufficient to cause irreversible inhibition of neurotransmitter release. Additionally, since the 1980s, botulinum toxin A (and to some extent also B) has been successfully used for the treatment of disorders caused by hyperactivity of cholinergic nerves innervating different muscles (strabismus, blepharospasm, clubfoot, neck ache, occupational cramps, and esophageal achalasia) and glands (hyperhidrosis). They are administered via intramuscular injection, and despite their extensive use, so far, no major long-term side effects have been reported. In the past years, studies have shown that various serotypes of botulinum toxins inhibit the release of algogen mediators, including CGRP and substance P, by sensory neurons. These results open up new possibilities for the clinical application of botulinum toxins to treat neuropathic pain and migraine in patients who are unresponsive to current treatments.

α-latrotoxin represents the main toxic component of the black widow spider venom. It is a high-molecular-mass protein that acts by binding to specific membrane receptors, called latrophilins, exclusively present on neuron presynaptic membrane. α-latrophilins are a new family of adhesion G-protein-coupled proteins. Binding of the toxin to these receptors causes calcium increase in the presynaptic terminal and a large asynchronous neurotransmitter release. In the absence of extracellular Ca^{2+}, activation of exocytosis is accompanied by inhibition of endocytosis, leading, within several minutes, to a decrease in SV content and permanent incorporation of vesicle membrane into the presynaptic plasma membrane. Some snake venom toxins (notexin, β-bungarotoxin, taipoxin, and textilotoxin) stimulate neuroexocytosis, ultimately causing depletion of SVs, probably via simultaneous inhibition of endocytosis. It has been shown that these toxins have phospholipase activity and can hydrolyze phospholipids in the external layer of the presynaptic membrane. Subsequent formation of fatty acids and lysophospholipids alters membrane curvature and permeability, promoting fusion of SVs followed by inhibition of synaptic transmission.

TAKE-HOME MESSAGE

- Neuron synapses are complex subcellular structures that mediate information transfer between nerve cells.
- Synapse functions are based upon interactions among hundreds of proteins present in the presynaptic and postsynaptic compartments that regulate synapse formation, protein and organelle trafficking involved in neurotransmitter release, signal reception at postsynaptic level, and synaptic plasticity.
- Several drugs currently used for the treatment of brain disorders target synaptic functions, including neurotransmitter uptake and degradation and receptor function.
- The concept of synaptopathy has been a major breakthrough in both psychiatry and neurology opening new avenues for the development of drugs specifically acting on the synaptic components that are altered in such pathologies.

FURTHER READING

Alabi AA, Tsien RW. (2012). Synaptic vesicle pools and dynamics. *Cold Spring Harbor Perspectives in Biology* 4:a013680.

Caroni P, Donato F, Muller D. (2012). Structural plasticity upon learning: regulation and functions. *Nature Reviews. Neuroscience* 13:478–490.

Cesca F, Baldelli P, Valtorta F, Benfenati F. (2010). The synapsins: key actors of synapse function and plasticity. *Progress in Neurobiology* 91:313–348.

Dittman J, Ryan TA. (2009). Molecular circuitry of endocytosis at nerve terminals. *Annual Review of Cell and Developmental Biology* 25:133–160.

Gundelfinger ED, Fejtova A. (2012). Molecular organization and plasticity of the cytomatrix at the active zone. *Current Opinion in Neurobiology* 22:423–430.

Halassa MM, Haydon PG. (2010). Integrated brain circuits: astrocytic networks modulate neuronal activity and behavior. *Annual Review of Physiology* 72:335–355.

Melom JE, Littleton JT. (2011). Synapse development in health and disease. *Current Opinion in Genetics and Development* 2:256–261.

Nishimune H. (2012). Active zones of mammalian neuromuscular junctions: formation, density, and aging. *Annals of the New York Academy of Sciences* 1274:24–32.

Panatier A, Vallée J, Haber M, Murai KK, Lacaille JC, Robitaille R. (2011). Astrocytes are endogenous regulators of basal transmission at central synapses. *Cell* 146:785–798.

Pérez-Alvarez A, Araque A. (2013). Astrocyte-neuron interaction at tripartite synapses. *Current Drug Targets* 14:1220–1224.

Rizzoli SO. (2014). Synaptic vesicle recycling: steps and principles. *The EMBO Journal* 33:788–822.

Sala C, Segal M. (2014). Dendritic spines: the locus of structural and functional plasticity. *Physiological Reviews* 94:141–188.

Südhof TC. (2013). Neurotransmitter release: the last millisecond in the life of a synaptic vesicle. *Neuron* 80:675–690.

Xu-Friedman MA, Harris KM, Regehr WG. (2001). Three-dimensional comparison of ultrastructural characteristics at depressing and facilitating synapses onto cerebellar Purkinje cells. *The Journal of Neuroscience* 21:6666–6672.

38

CATECHOLAMINERGIC TRANSMISSION

PIER FRANCO SPANO, MAURIZIO MEMO, M. CRISTINA MISSALE, MARINA PIZZI, AND SANDRA SIGALA

By reading this chapter, you will:

- Acquire the basis for understanding the anatomical and functional distribution of the catecholaminergic system in the central and peripheral nervous systems
- Know synthesis, metabolism, and receptors of catecholamines
- Understand the most important functions of the central and peripheral catecholaminergic systems
- Learn the rational basis for the therapeutic use of drugs active on the catecholaminergic system

Catecholamines play a fundamental role in the control of autonomic, motor, and psychic functions. They are the main neurotransmitter released from the terminals of post-ganglionic fibers in the peripheral sympathetic system and from the adrenal medulla and are therefore fundamental in controlling the activity of the cardiovascular system. In the central nervous system (CNS), catecholaminergic neurons represent a minor population among neuronal cells, but their importance is documented by the clinical utility of drugs acting on this system in the treatment of mental disorders (antidepressants and antipsychotics), by the psychological effects of drugs of abuse (e.g., amphetamines), and by the role of dopaminergic neuron degeneration in the genesis of Parkinson's disease.

THE CATECHOLAMINERGIC SYSTEM IN THE AUTONOMIC NERVOUS SYSTEM

Blood circulation, respiration, digestion, water and salt balance, metabolic balance, and several other visceral functions are controlled by two peripheral nervous systems that generally function in opposition to each other: the (ortho)sympathetic and the parasympathetic systems. These systems form the autonomic nervous system that modulates the function of cells and organs in response to environmental and emotional stimuli and generates two typical symptomatic patterns: the sympathetic system is responsible for the "fight and flight" response, whereas the parasympathetic is responsible for a series of responses that can be described as the effects of a substantial meal.

The "fight and flight" response is characterized by a series of visceral activities that prepare the subject to deal with physical challenges requiring strength, endurance, and prompt reaction. The cardiovascular system is intensely stimulated (increase in strength and frequency of heartbeats and cardiac output). At the same time, the blood supply of tissues essential for the "fight and flight" response (e.g., skeletal muscles) increases and decreases in other tissues that are either not involved in the reaction (e.g., gastrointestinal apparatus) or potentially exposed to the risk of damage and bleeding (e.g., skin with fear-induced facial pallor). In addition, the capsular smooth muscle of the spleen contracts to squeeze blood in the circulation and increase its volume. Respiration is facilitated by bronchial dilation and inhibition of bronchial secretions, and the force of contraction of the skeletal muscles is increased. The activity of the gastrointestinal system and associated glands is inhibited, but fatty acid and glucose mobilization from storage is increased, to provide the brain and muscles with the energy required to face the emergency. Mydriasis (pupil dilation) occurs, which may represent an advantage for animals that hunt and fight mostly at night. Piloerection causes an apparent increase in size of the animal that may help frightening enemies; increased production of sweat may help to slip out of a grab or to release repellent substances (especially in amphibians).

The sympathetic activity of the autonomic nervous system is mediated by the catecholaminergic neurotransmitters noradrenaline (NA), adrenaline (A), and dopamine (DA). NA is the principal neurotransmitter released from the synaptic terminals of sympathetic nerve fibers, whereas A is mostly produced by the adrenal medulla and is released directly into the bloodstream. Since A and NA exhibit different potency on the various adrenergic receptors, sympathetic nerves and adrenal medulla stimulation may produce different effects (e.g., in the splanchnic district, NA induces vasoconstriction, whereas A produces increased blood flow by reducing peripheral resistances). In general, compared to sympathetic nerves, activation of the adrenal medulla is induced by stressful stimuli of greater intensity.

The effects of the sympathetic system are generally counteracted by the cholinergic parasympathetic neurons. Indeed, the activities of the parasympathetic system are more closely related to digestive functions. The inotropic and chronotropic effects of the parasympathetic system reduce cardiac output, which is associated with vasodilation in extensive areas such as the skin (warm skin and redness) and splanchnic regions. The resulting reduction in blood pressure and blood supply to the brain are responsible for the typical postprandial somnolence. The increased secretion of salivary glands and other glands of the digestive tract and the increased intestinal mobility support digestion. Bronchoconstriction and increased bronchial secretion as well as miosis (pupil constriction) are also induced by parasympathetic stimulation.

The activity of the parasympathetic nervous system is mediated by the neurotransmitter acetylcholine, which acts on peripheral muscarinic receptors (see Chapter 39). Part of the fibers of the autonomic nervous system does not release either catecholamines or acetylcholine: these are called nonadrenergic noncholinergic (NANC) fibers, and most of them use nitric oxide as transmitter (see Chapters 44, 45, and 48). Both catecholaminergic and cholinergic neurons can release neuropeptides, which in general act synergistically with the neurotransmitter.

Anatomical Organization of the Sympathetic System

Preganglionic neurons of the sympathetic system, which synthesize and release acetylcholine, extend from the first thoracic spinal segment to the lower lumbar segments and are located in the intermediolateral gray matter of the spinal cord (Fig. 38.1). Their axons emerge from the spinal cord through the ventral roots, pass in the spinal nerves, and project through the white rami communicantes to the paravertebral and prevertebral ganglia where they form nicotinic synapses with postganglionic sympathetic neurons. These distribute to the peripheral organs together with blood vessels and innervate and regulate the activity of effector organs by synthesizing and releasing catecholamines (Fig. 38.1). On the effector organs/sites (and also in the CNS), adrenergic fibers have the microscopic appearance of chains of a rosary wherein each enlargement represents an area of accumulation of synaptic vesicles; interestingly, the catecholaminergic terminals are usually not faced by postsynaptic densities enriched of receptors on the recipient cells.

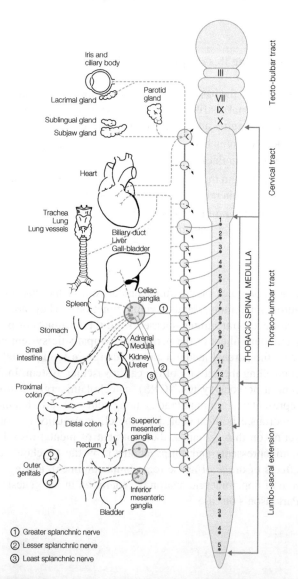

FIGURE 38.1 Anatomical distribution of the catecholaminergic sympathetic system. Postganglionic sympathetic fibers are depicted as dashed lines.

Indeed, in the innervated organs, the neurotransmitter can rapidly diffuse from the site of release and stimulate many surrounding cells at the same time (volume transmission). For this reason, activation of catecholaminergic nerves produces a more widespread stimulation than the cholinergic.

The autonomic sympathetic system also includes the adrenal glands; these are suprarenal endocrine glands consisting of a cortical (outer) part that produces steroid hormones and an inner part or medulla. The adrenal medulla cells (chromaffin cells) derive from the ectoderm of the neural crest and produce A and, to a lesser extent, NA (A:NA=4:1). The adrenal medulla is innervated by preganglionic cholinergic fibers running within the splanchnic nerve and is considered a kind of sympathetic ganglion secreting A in the bloodstream. A generalized and sustained activation of the sympathetic system is accompanied by an increased hormone secretion from the adrenal medulla, which enhances the effects of the locally released NA.

Cardiovascular Effects of the Sympathetic System

Myocardium, sinoatrial (SA), and atrioventricular (AV) node cells and conduction tissue express both α_1 and β_2 receptors (the classification of adrenergic receptors is discussed later in this chapter), although show a prevalence of the β_1 receptor subtype. Activation of β_1 receptors in the SA and AV nodes and conduction tissue induces, respectively, an increase in heart rate (positive chronotropic effect), an increase in the speed of AV conduction (dromotropic-positive effect), and abnormal rhythms (bathmotropic-positive effect). In particular, catecholamines potentiate the I_f current responsible of the slow depolarization of the SA node cells during diastole (phase 4 of the action potential) thus increasing heart rate. Catecholamines accelerate diastolic depolarization also in Purkinje cells and can induce activation of latent pacemakers (induction of automation). By a direct effect on the myocardium, catecholamines increase the contraction force (positive inotropic effect), cardiac output, cardiac work, and oxygen consumption.

The effects of catecholamines released by the sympathetic system on smooth muscles of the vasculature vary depending on the type of receptor expressed and tissue distribution. Vascular territories where β receptors predominate will undergo vasodilation and increased blood flow (e.g., skeletal muscles); at the level of the skin, mucous membranes, and kidney, α receptors are more abundant and vasoconstriction predominates. Together with activation of the renin–angiotensin–aldosterone system (an effect induced by stimulation of β receptors), the consequence of generalized sympathetic system stimulation is an increase in blood pressure.

Catecholamine effects are mostly evident at the level of arterioles and precapillary sphincters, but it is present also at the level of veins and large arteries. Since A and NA have different potency spectrum on the various receptors (e.g., NA is more active on α than on β receptors), activation of the sympathetic system may have different effects on peripheral resistance, depending on the degree of activation.

Receptors for the catecholamine DA have been shown at different levels in the cardiovascular system. It is not yet clear whether the peripheral DA comes from specific neurons or is coreleased in adrenergic neurons. Immunohistochemical studies have demonstrated the existence of dopaminergic interneurons at ganglionic level, controlled by preganglionic cholinergic neurons and with terminations on postganglionic noradrenergic neurons.

In the cardiovascular system, DA has mainly vasodilatory effects, which can be either direct, because of the presence of specific receptors on smooth muscles of some vascular beds (in particular renal and mesenteric), or indirect, via inhibition of catecholamine release. Furthermore, DA inhibits the release of aldosterone induced by angiotensin II and has inhibitory activity on the ganglionic transmission. Stimulation of the dopaminergic system involves peripheral effects, such as increase in glomerular filtration rate and in renal blood flow and natriuresis, which are of considerable therapeutic interest. As discussed later, DA agonists may be useful in the control of cardiogenic or hypovolemic shock, in which the increase in sympathetic activity may impair kidney function.

Other Noncardiovascular Effects of Catecholamines

Catecholamines have multiple effects on the smooth musculature in different tissues depending on the type of receptor expressed. They relax gastrointestinal muscles through activation of β_2 receptors; in particular, they decrease motility and tone of stomach and intestine muscles. By contrast, they induce contraction of the sphincter muscle, which expresses α_1 receptors. NA has also an indirect myorelaxant effect mediated by α_{2A} receptors localized on the intramural parasympathetic ganglia of the stomach, whose stimulation causes a decrease in acetylcholine release. The α_{2A} receptor stimulation also causes inhibition of gastric acid secretion induced by stimulation of the vagus nerve.

The bronchial muscles express β_2 receptors and are relaxed by catecholamines: thus, sympathetic activation improves the ventilatory capacity, an effect of fundamental importance in emergency situations (for a detailed description of the mechanisms of adrenergic control of muscle contraction/relaxation, see Supplement E 27.4).

At various levels of the urinary tract, both α and β receptor subtypes are present. Catecholamines influence the activity of these receptors with a major effect on urinary incontinence. α Receptors are the targets of the α_1 antagonists, which are used to treat disorders of the lower urinary tract (e.g., benign prostatic hypertrophy; see later in this chapter). β Receptors, and in particular the β_2 and β_3 subtypes, are expressed throughout the urinary tract and, upon NA binding, induce smooth muscles relaxation; they are also expressed in urothelium, although their functional significance is not clear.

As part of the fight and flight reaction, catecholamines have a number of effects on metabolism. The sympathetic system increases glucose and free fatty acid blood levels. In the liver, catecholamines interact with β_2 receptors whose stimulation cause activation of glycogen phosphorylase and inhibition of glycogen synthase. The resulting decrease in glycogen synthesis and increase in glucose production leads to a rise in glucose blood level (see also Chapter 36).

Adipose tissue expresses β_3 receptors whose stimulation induces triglyceride lipase activity thus causing lipolysis and increasing the plasma level of free fatty acids (see also Chapter 35).

THE CATECHOLAMINERGIC SYSTEMS IN THE CNS

In the CNS, NA, A, and DA have specific distributions that are often associated with distinct functions. The dopaminergic system plays an important role in the control of extrapyramidal movement: alterations in some dopaminergic pathways are at the basis of Parkinson's disease. Dopaminergic and noradrenergic systems are also involved in the control of other high brain activities, and functional abnormalities in the central catecholaminergic systems are believed to be responsible for several aspects of psychosis and depression.

Distribution and Functions of the Adrenergic and Noradrenergic Systems

Two main groups of noradrenergic neurons have been identified in the CNS, namely, locus coeruleus and lateral tegmental area, whose axonal extensions innervate many brain areas.

The locus coeruleus contains 43% of the brain noradrenergic neurons, which project diffusely to the entire cerebral cortex, as well as to the hippocampus, thalamus, hypothalamus, and spinal cord (mainly to the ventral horns) and to the sensory nuclei of the brain stem and cerebellum (Fig. 38.2a).

Three main groups of neurons have been identified in the lateral tegmental system: the dorsal motor nucleus of the vagus nerve, the solitary tract nucleus, and the lateral tegmental nucleus. The neurons of the lateral tegmental nucleus innervate the hypothalamus, olfactory bulb, and amygdala. The solitary tract nucleus receives input from more rostral centers of the vegetative system and from the centers of the bulbopontine reticular formation. Descendant fibers originate from the solitary tract nucleus and exert a tonic and phasic control on preganglionic neurons of the spinal cord and sympathetic side chain.

The system of the lateral tegmental nuclei, and in particular the solitary tract nucleus, which receives fibers from the aortic and carotid baroreceptors and chemoreceptors, controls the activity of the autonomic nervous system. A reduction in blood pressure causes a decrease in baroreceptor firing and therefore a deinhibition of this noradrenergic nucleus. The resulting activation of the sympathetic tone causes an increase of peripheral vascular resistance and cardiac output and restores normal blood pressure.

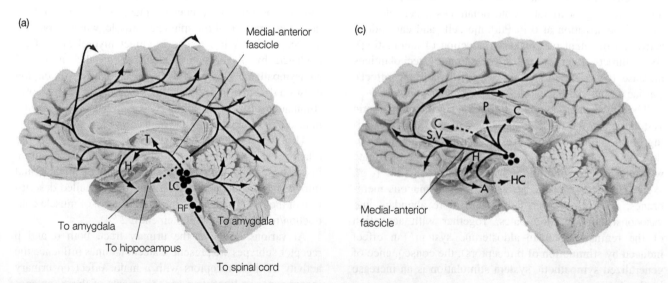

FIGURE 38.2 (a) Catecholaminergic and (b) dopaminergic pathways in the human central nervous system. A, amygdala; C, caudate nucleus; H, hypothalamus; HC, hippocampus; LC, locus coeruleus; RF, reticular formation; P, putamen; S, septal nuclei; V, ventral striatum (From Bear MF, Connors BW, Paradiso MA. (2007). *Neuroscience: exploring the brain*. 3rd ed. Baltimore, MD: Lippincott Williams & Wilkins, modified).

Stimulation of postsynaptic α_2 receptors localized in the solitary tract nucleus and locus coeruleus decreases the activity of the efferent noradrenergic neurons, thereby reducing peripheral sympathetic tone and blood pressure. Agonists at these receptors (e.g., clonidine and α-methyl-DOPA, discussed later in the chapter) can be used in arterial hypertension therapy.

The Noradrenergic Systems in Depression, Wakefulness, and Feeding Behavior The noradrenergic systems are involved in several functions of the brain and in the genesis of functional disorders. In the case of depression, several lines of evidences have demonstrated that this disorder is associated with an altered activity of the α_2 receptors that control NA secretion and an increased number of β_2-adrenergic receptors in the frontal cortex and limbic system. These alterations may account for the presence of anxiety, agitation, insomnia, anorexia, and disorders of the autonomic nervous system in depressed patients. The therapy with antidepressant drugs causes, in the long term, desensitization of β_1 receptors in the cortex and hippocampus. Certain classes of antidepressant drugs that act on NA reuptake mechanisms normalize noradrenergic transmission and β_2 receptors responsiveness.

Sleep–wake functions depend on the integrated of activity of diffuse projection systems: the noradrenergic, serotonergic, cholinergic, and histaminergic systems located in the brain stem. In particular, the waking state is controlled by the activation of the cholinergic and catecholaminergic systems, while the "slow sleep" state is controlled by the serotonergic system. The diffuse neuronal network originating from locus coeruleus controls vigilance and attention; the activity of adrenergic neurons in this nucleus is increased in the waking state and reduced during sleep. The NA effect on the sleep–wake function is mediated by α_1 receptors. In fact, pharmacological stimulation of α_1 receptors in the locus coeruleus projection areas promotes wakefulness, whereas stimulation of α_2 receptors on neuron cell bodies and dendrites of locus coeruleus produces sedation.

In the CNS, catecholamines regulate feeding behavior through a facilitatory effect mediated by α_2 receptors in the hypothalamus medial part and an inhibitory effect mediated by β_2 receptors in the lateral hypothalamus. It is important to underline that the perifornical nucleus receives also dopaminergic fibers whose stimulation causes a significant reduction in food consumption, mediated by D_2 receptors.

The Adrenergic and Noradrenergic Control of Pituitary Function Adrenergic and noradrenergic systems exert both a stimulatory and an inhibitory control on adrenocorticotropic hormone (ACTH) secretion by acting on both the hypothalamus and hypophysis. The stimulatory action is mediated by α_1 receptors, whereas the inhibitory effect is mediated by β-type receptors.

The noradrenergic system modulates growth hormone (GH) secretion via α-type receptors in the hypothalamus. In particular, α receptor stimulation inhibits both the phasic and the tonic release of GH, while α_{2A} receptor stimulation elicits the opposite response. Besides these effects, it should be noted that β receptors mediate inhibition of GH secretion, probably due to an increase in somatostatinergic tone.

The noradrenergic system also controls thyroid-stimulating hormone (TSH) secretion, having a facilitatory effect mediated by α receptors. This effect is due to a direct action at hypothalamic level where NA stimulates the release of thyrotropin-releasing hormone (TRH).

The gonadotropins, follicle-stimulating hormone (FSH) and luteinizing hormone (LH), are synthesized and secreted from the anterior pituitary, under hypothalamic control mediated by the luteinizing hormone-releasing hormone (LHRH) peptide. NA is the main neurotransmitter responsible for the pulsatile LH release, through its stimulatory effects mediated by receptors on LHRH-producing hypothalamic neurons.

Distribution of the Dopaminergic System

The first hypotheses on a physiological role of DA beyond its function as NA and A precursor were developed in the late 1950s. Around that time, specific neurons containing DA as primary neurotransmitter were identified in the CNS, in the extrapyramidal, limbic, and hypothalamic–pituitary systems (Fig. 38.2b). Today, strong evidence indicates that DA is hierarchically very important in the CNS, being involved in the modulation of psychic and motor functions, mood, and secretion of some pituitary hormones and possibly in the regulation of some components of cognitive processes.

DA neurons form systems that have major functional significance and have been extensively characterized. These include the nigrostriatal dopaminergic system, the mesolimbic and mesolimbocortical systems, and the mesothalamic and the tuberoinfundibular and tuberopituitary dopaminergic systems.

Due to its involvement in movement control and Parkinson's disease, the nigrostriatal dopaminergic system has been extensively characterized. Dopaminergic neurons of the ventral midbrain are distributed into three groups. A10 neurons are localized in the ventral tegmental area (VTA) and midbrain, A9 neurons in the substantia nigra (SN), and A8 neurons in the caudal retrorubral field. The dorsal component of the mesostriatal system takes origin in A9 neurons in the SN pars compacta, which in humans are highly pigmented. The dendrites of this neuronal population innervate the SN pars reticulata where DA regulates the activity of afferent endings originating in the basal ganglia. The axons of A9 neurons project to the caudate nucleus and putamen, which altogether form the corpus striatum. In the basal ganglia, two types of dopaminergic endings have been identified: one gives rise to a dense and diffuse innervation, and the other form highly innervated areas (striatosomes) in the striatum. These two different endings arise from different populations of A9 neurons and exhibit different basal activity and responsiveness to DA agonist and antagonist drugs, suggesting the existence of a highly complex control of basal ganglia activity. Besides the SN, the dorsal striatum is innervated by the sensory and motor cortex, whereas the ventral part is innervated by regions of the limbic cortex, such as the piriform cortex, amygdala, and entorhinal cortex. The striatonigral system originates from the basal ganglia and consists of GABAergic neurons located mostly in the caudate nucleus and in the pallidum, which innervates the SN pars reticulata where dendrites of A9 dopaminergic neurons branch out.

The ventral component of the mesostriatal dopaminergic system originates mainly in the VTA A10 area and in the medial part of the SN and innervates the nucleus accumbens, olfactory tubercle, and interstitial nucleus of the stria terminalis (mesolimbic system). Other fibers, originating from A10, innervate the septum (especially the lateral septum nucleus), hippocampus, amygdala, entorhinal cortex, prefrontal cortex, perirhinal cortex, and piriform cortex (mesolimbocortical system). More recent data indicate that dopaminergic fibers innervate also regions such as the visual and association areas of the neocortex.

Mesencephalic dopaminergic neurons have been shown to be heterogeneous as they contain different cotransmitters: some neurons contain only DA, others contain cholecystokinin or neurotensin, and others contain DA, cholecystokinin, and neurotensin.

A10 neurons also innervate structures of the pons, diencephalon, and telencephalon. The mesothalamic pathway originates in the VTA and ends in the habenula, especially in its lateral and medial sides. The lateral habenula is an important station of the extrapyramidal system because the ventral pallidum and dorsal pallidum send afferents to it that, in turn, project to the reticular formation and SN.

Finally, the tuberoinfundibular and tuberopituitary dopaminergic systems originate from dopaminergic cell bodies of the arcuate and periarcuate nuclei of the hypothalamus (called A12 area). The tuberopituitary system originates in the front part of the A12 area and innervates the intermediate and posterior pituitary, where it inhibits secretion of melanocyte-stimulating hormone (α-MSH) and β-endorphin (pituitary intermediate) and of oxytocin and vasopressin hormones (posterior pituitary). The tuberoinfundibular system neurons innervate the outer layer of the median eminence, where they are in close contact with the capillaries of the hypophyseal portal system. DA is released into the pituitary portal circulation and reaches the anterior pituitary where it inhibits prolactin secretion. In addition, DA nerve terminals in the median eminence are in close contact with LHRH- and TRH-producing neurons and regulate their activity.

Functions of the Dopaminergic Systems in the CNS

The Dopaminergic Nigrostriatal System and Movement Control The nigrostriatal dopaminergic system is part of the extrapyramidal system and controls muscle tone and motor coordination. DA released from the striatum can interact with specific receptors on dopaminergic terminals (autoreceptors), cortical glutamatergic terminals, and postsynaptic intrastriatal cholinergic and GABAergic neurons projecting to the globus pallidum and/or SN pars reticulata. Since DA is an inhibitory neurotransmitter, its activity in the striatum results in reduced release of DA (presynaptic autoregulation), glutamate, and acetylcholine and, at pallidum–nigra level, of GABA. This constant inhibitory tone is reduced in at least two conditions: (i) degeneration of SN dopaminergic neurons in Parkinson's disease and (ii) chronic treatment with receptor antagonists (neuroleptics). Similar events occur in both situations, such as fine resting tremor, rigidity, and akinesia.

The discovery that degeneration of dopaminergic nigrostriatal neurons causes Parkinson's disease symptoms and in-depth studies on the physiological role of other neurotransmitter systems in the control of basal nuclei activity have led to the development of a rational pharmacological therapy for extrapyramidal diseases employing drugs acting on the dopaminergic and other neurotransmitter systems (e.g., cholinergic and purinergic) (see for more details Supplement E 38.1, "Dopamine and Parkinson's Disease").

DA and Behavior Disorders The importance of central dopaminergic systems in behavioral disorders is demonstrated by the linear correlation existing between efficacy of "typical" antipsychotic drugs (neuroleptics) used in schizophrenia therapy and their ability to block DA receptors, especially D_2 subtype (see classification of DA receptors later in this chapter). The hypothesis that alterations in dopaminergic transmission are responsible for the pathogenesis of schizophrenia was based on the observation that antipsychotic drugs often induce Parkinson's-like symptoms (iatrogenic parkinsonism) as side effects. Consistently, drugs that increase DA levels (e.g., L-DOPA, cocaine, amphetamines) cause a psychosis similar to the paranoid form of schizophrenia. More recent *in vivo* data, obtained using the positron emission tomography (PET) technique, have revealed a considerable increase in D_2 receptors in caudate and accumbens nuclei and olfactory tubercle of schizophrenic patients not yet exposed to pharmacological treatment. However, it should be keep in mind that these changes may not be the primary cause of psychosis but may result from functional alterations in other neuronal systems.

The aforementioned dopaminergic systems seem to play partially different functions. The mesolimbic system seems to be involved in memory and emotional processes: changes in perception and intellectual abilities, similar to those observed in schizophrenic patients, appear in certain types of psychomotor epilepsy. Lesions of the VTA cause dementia and onset of psychotic episodes. The mesocortical DA system projects to the prefrontal cortex, which is involved in important higher mental functions such as attention, motivation, planning, temporal organization of behavior, and socialization. Activation of this system is believed to play a critical role in addiction to drugs and substances of abuse, such as heroin, cocaine, nicotine, and alcohol. It is important to remember that Parkinson's disease patients, which have a generalized reduction of dopaminergic neurons, exhibit extrapyramidal motor symptoms as well as a reduced capacity of affection and motivation and lack of spontaneity.

Anatomical, functional, pathological, and pharmacological findings suggest that two different alterations in the dopaminergic systems coexist in psychotic patients with schizophrenia. One is due to an increased activity of the mesolimbic system, probably mediated by D_2, D_3, and D_4 receptors; this is responsible for the "positive" symptoms of schizophrenia and responds quite well to treatment with typical antipsychotics. The other, characterized by a reduced dopaminergic activity in the prefrontal cortex, is responsible for the "negative" symptoms and is more responsive to "atypical" rather than to "typical" antipsychotic drugs.

The proposed unifying hypothesis suggests that the primary defect in schizophrenia consists in a reduced feedback activity of mesocortical dopaminergic projections to the prefrontal cortex, which, under normal conditions, exerts an inhibitory effect. The lack of prefrontal inhibition would lead to the dopaminergic mesolimbic pathway hyperactivity typical of schizophrenia. However, it is important to underline that the dopaminergic hypothesis, although attractive, is only one of the several hypotheses and it is not immune to criticism. For example, several antipsychotic drugs lack selectivity and are also active on 5-HT$_2$ serotonin receptors; moreover, symptoms often recede only after several days or weeks of treatment and not immediately as it would be expected for a simple receptor blockade. Therefore, probably, these dysfunctions are due to alterations of complex regulatory circuits involving also other neurotransmitter systems.

A further relevant behavior function where DA plays an important role is reward. The mesolimbic DA pathways are the main anatomical structures responsible for reward and pleasure that are involved in addiction in relation to drug abuse, sexual activity, gambling, binge eating, and even aesthetic enjoyment. Several neurotransmitters are involved in the reward system, but almost all of them act through modulation of the DA mesolimbic pathways. Therefore, the pharmacological control of this behavior can be achieved by several drugs acting on different targets (see also Chapters 10, 29, 40, and 47 and Supplement E 38.2, "Brain Reward Circuits").

Effects of DA on the Chemoreceptor Trigger Zone, Hypothalamus, and Pituitary DA receptors are present in the chemoreceptor trigger zone (CTZ) that controls the vomiting center. CTZ is located in the area postrema in the fourth ventricle floor of the medulla oblongata and contains also histaminergic and cholinergic receptors. Activation of DA receptors (mainly D_2) induces vomiting, whereas DA receptor antagonists, muscarinic and histamine H$_1$ receptor antagonist, and serotonin 5-HT$_3$ receptor agonists have antiemetic activity. It is worth noting that DA receptors are also present in the gastric wall where they seem to mediate the inhibition of gastric motility that occurs in the course of nausea and vomiting. Although antidopaminergic drugs may be useful to control vomiting (e.g., metoclopramide), the risk of chronic effects (parkinsonism) and the extent of sedation they produce favor the prescription of antihistamines instead and restrict the use of antidopaminergic drugs to severe situations such as the treatment of vomiting induced by anticancer drugs or vomiting during pregnancy. It is important to remember that the CTZ is located in a CNS area where the efficiency of the blood–brain barrier is particularly reduced. Administration of drugs with poor lipid solubility (and therefore relatively unable to cross the blood–brain barrier), such as the D_2 antagonist domperidone, reduces the extent of the central effects and allows their use, for example, in the control of vomiting induced by drugs in the course of the dopaminergic therapy of Parkinson's disease.

DA neurons and receptors have also been identified in specific hypothalamic nuclei. In particular, activation of D_2 DA receptors in the perifornical nucleus produces a decrease in food consumption, whereas administration of antipsychotic drugs with antidopaminergic activity induces increase in body weight. Body temperature is also, in part, regulated by DA. In fact, hypothalamic stimulation of D_2 receptors induces hypothermia in experimental animals; in addition, psychiatric patients treated with antipsychotic drugs may exhibit alterations of the thermoregulatory systems with marked hyperthermia (neuroleptic malignant syndrome).

DA is the major inhibitor of prolactin secretion. Neurons mediating this effect originate in the hypothalamus arcuate nucleus and are part of the tuberoinfundibular system. DA exerts its inhibitory effect on prolactin release through D_2-subtype pituitary receptors, whose blockade by neuroleptics causes marked hyperprolactinemia with amenorrhea and galactorrhea in women and impotence and gynecomastia in men. Conversely, D_2 agonists (e.g., bromocriptine, lisuride, pergolide, cabergoline) are used to inhibit prolactin secretion in the treatment of prolactinomas.

DA is also involved in the regulation of pulsatile GH secretion and the inhibitory control of MSH.

SYNTHESIS OF CATECHOLAMINES

Catecholamine synthesis starts with phenylalanine and tyrosine, two essential amino acids that are derived from proteins in the diet (Fig. 38.3).

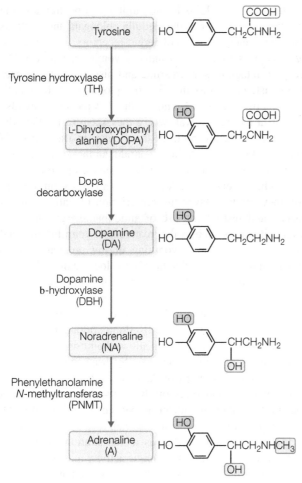

FIGURE 38.3 Catecholamine biosynthesis pathway.

Tyrosine intake from the diet is generally sufficient for catecholamine synthesis. However, phenylalanine can be converted to tyrosine by the enzyme phenylalanine hydroxylase. This enzyme is absent in the congenital disease called phenylketonuria.

Tyrosine hydroxylase is the enzyme that catalyzes the hydroxylation of the three position of the tyrosine phenol, with formation of dihydroxyphenylalanine (DOPA). The reaction speed, rather slow, represents the limiting step in catecholamine synthesis and is tightly controlled by neuronal activity. In fact, the rate of synthesis increases under conditions of high neuronal activity (neural firing), mainly due to an increase in the number of enzyme molecules present in the neuron. By contrast, an increase in intracellular catecholamine concentration has an inhibitory effect on the enzyme activity. Tyrosine hydroxylase is associated with the endoplasmic reticulum of catecholaminergic neurons and requires tetrahydropteridine and Fe^{2+} ions for its activation. The drug α-methyltyrosine inhibits tyrosine hydroxylase activity by competing with its natural substrate and reduces L-DOPA synthesis. L-DOPA is not accumulated in catecholaminergic nerves since it is very slowly produced and rapidly converted to DA by the DOPA decarboxylase. This enzyme, also called aromatic amino acid decarboxylase because of its lack of substrate specificity, requires pyridoxal phosphate as a coenzyme and has a widespread localization (it is found also in the kidney and liver). The enzyme acts only on levogyre isomers. L-DOPA conversion to DA is competitively inhibited by a number of drugs, such as methyldopa, carbidopa, and benserazide. Given their inability to cross the blood–brain barrier, carbidopa and benserazide are used in combination with L-DOPA in the treatment of Parkinson's disease; in this way, peripheral conversion of L-DOPA to DA is inhibited, and possible side effects due to stimulation of peripheral DA receptors (e.g., vomiting, hypotension) are reduced.

Chromaffin cells of the adrenal gland and noradrenergic and adrenergic neurons are characterized by the ability to express the DA β-hydroxylase, an enzyme that converts DA to NA. The enzyme contains copper as a prosthetic group and requires ascorbic acid as a cofactor. In chromaffin cells and adrenergic neurons, NA is methylated to A by the enzyme phenylethanolamine-*N*-methyltransferase.

VESICULAR STORAGE AND RELEASE OF CATECHOLAMINES

Catecholamines are found in nerve terminal either free in the cytoplasm or stored within synaptic vesicles where they can be either free or bound to macromolecular complexes. Equilibrium exists between the vesicular and the cytoplasmic forms. NA uptake into the vesicles involves an active transport mechanism that uses, as energy source, the pH gradient generated across the vesicle membrane by an ATP-dependent proton pump (see Chapter 29). In catecholaminergic terminals and chromaffin cells, synaptic vesicles or secretory granules contain also ATP (one ATP molecule every four catecholamine molecules); high amounts of acidic water-soluble proteins, called chromogranins; and the enzyme DA-β-hydroxylase.

Acidic chromogranins form a macromolecular complex with ATP and divalent ions (Ca^{2+} and Mg^{2+}) that binds catecholamines (or other amines) inside the vesicles, thereby reducing the intravesicular concentration of free neurotransmitter and favoring the vesicular transporter activity.

Drugs that compete with catecholaminergic neurotransmitters for granin binding (indirect sympathetic drugs) can induce a rapid increase in the intravesicular concentration of free amines causing their release into the cytoplasm.

Many phenylethylamine derivatives can compete with catecholamines for chromogranin binding thereby shifting the equilibrium toward the intravesicular free form of the neurotransmitter, which rapidly equilibrates with the cytoplasmic free form. As a result, the free cytoplasmic form markedly increases, and the catecholamine is released from the nerve terminal via a nonvesicular mechanism. Such nonvesicular release is mediated, at least in part, by the membrane transporter that normally reuptakes the released catecholamines; under these conditions, the transporter works in the opposite direction (see also Chapter 29). Drugs having this mechanism of action are called indirect sympathomimetic as their effects are not due to receptor activation but to the neurotransmitter released from the nerve terminal. Some drugs of abuse (e.g., amphetamine) have indirect sympathetic activity (Table 38.1).

TABLE 38.1 Indirect sympathomimetic drugs

Drug	Mechanism of action	Treatment
Ephedrine	Mixed	Nasal congestion
Pseudoephedrine	Mixed	Nasal congestion
Phenylpropanolamine	Mixed	Nasal congestion
Amphetamine	Indirect	—
Methamphetamine	Indirect	—
Methylphenidate	Indirect	Attention-deficit/hyperactivity disorder (ADHD)
Pemoline	Indirect	ADHD

CATABOLISM AND REUPTAKE OF CATECHOLAMINES

The main routes of catecholamine degradation are oxidative deamination catalyzed by the monoamine oxidase (MAO) and O-methylation catalyzed by the catechol-O-methyltransferase (COMT). Metabolic pathways and formation of the major catecholamine urinary metabolites are shown in Figure 38.4. The 3-methoxy compounds formed by COMT from catecholamines serve as MAO substrates; catechols with alcoholic or acidic side chains formed by MAO from catecholamines serve as COMT substrates. Normally in humans, the following amounts of endogenous catecholamine metabolites are excreted over 24 h of urinary excretion: 2–4 mg of methoxy-hydroxy mandelic acid, 0.1 mg of normetanephrine, and 0.1–0.2 mg of metanephrine. Normally, about 2–5 μg of not metabolized A is excreted. The plasma levels of catecholamines and their metabolites may increase up to more than a hundred times in the presence of neoplastic degeneration of chromaffin cells (pheochromocytoma). Often, spontaneous catecholamine release from pheochromocytomas clinically manifests with a substantial

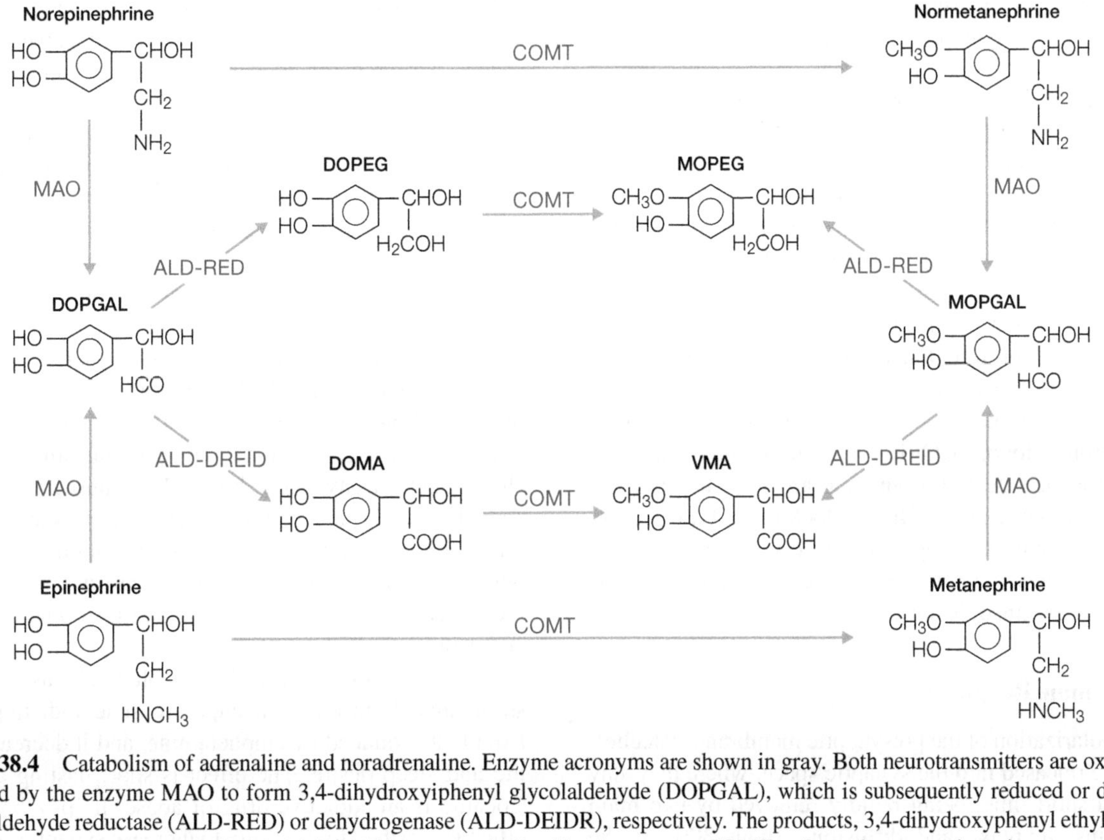

FIGURE 38.4 Catabolism of adrenaline and noradrenaline. Enzyme acronyms are shown in gray. Both neurotransmitters are oxidized and deaminated by the enzyme MAO to form 3,4-dihydroxyiphenyl glycolaldehyde (DOPGAL), which is subsequently reduced or dehydrogenated by aldehyde reductase (ALD-RED) or dehydrogenase (ALD-DEIDR), respectively. The products, 3,4-dihydroxyphenyl ethylene glycol (DOPEG) and 3,4-dihydroxymandelic acid (DOMA), are metabolized by the enzyme COMT to form the two major urine-excreted metabolites: 3-methoxy-4-hydroxyphenyl ethylene glycol (MOPEG) and 3-methoxy-4-hydroxymandelic acid (VMA). An alternative catabolic pathway of the two neurotransmitters involves the initial methylation by COMT, followed by the oxidative deamination by MAO. The common intermediate product, 3-methoxy-4-hydroxyphenylglycolaldehyde (MOPGAL), is then reduced or dehydrogenated to MOPEG or VMA, respectively.

rise in blood pressure, frequently unresponsive to drug therapy. Differential diagnosis with other forms of hypertension is based on the evaluation of catecholamine (or their metabolites) plasma levels. Neuronal MAOs are mainly located on the outer membrane of mitochondria present in the catecholaminergic nerve terminal. The product of the oxidative deamination is different depending on the enzyme substrate: 3,4-dihydroxyphenylglycol aldehyde (DHPGAL) is formed when the substrate is NA, whereas 3,4-dihydroxyphenylacetic acid (DOPAC) is formed when the substrate is DA. There are at least two subtypes of neuronal MAOs, MAO-A and MAO-B, encoded by two different genes. MAO-B is predominantly neuronal, whereas MAO-A is ubiquitous. MAO-A is specialized in the deamination of the ethylamine side chain of indole derivatives present in the nerve endings of adrenergic and serotonergic neurons in the central and peripheral nervous systems. MAO-B is specialized in the deamination of aromatic phenylethylamine nuclei. Excellent substrates for MAO-B are DA, bisphenylethanolamine, and tyramine. Pharmacologically, the two forms of MAO exhibit a different sensitivity to inhibitory drugs. Iproniazid, an irreversible MAO inhibitor, was the first drug used for the treatment of depression. However, irreversible MAO inhibitors (phenelzine and tranylcypromine) increase the availability of tyramine contained in many foods (e.g., wine, cheese, salami), causing severe hypertensive crisis. Drugs showing prevalent and reversible inhibition of MAO-B (selegiline) have been developed, which increase the availability of synaptic DA. Selegiline is used, either alone or in combination with L-DOPA, in the treatment of Parkinson's disease. Because of its specific action on MAO-B, selegiline has poor antidepressant activity and does not enhance the cardiovascular effects induced by tyramine.

COMTs are preferentially localized at postsynaptic sites, mostly in the cytoplasm and, to a lesser extent, in the plasma membrane. They use S-adenosylmethionine as methyl group donor. COMTs are ubiquitous enzymes, selective for the catechol group, and are activated by Mg^{2+} and other divalent cations. Drugs blocking COMT activity are currently being developed to improve the pharmacokinetic and pharmacodynamic properties of L-DOPA in the treatment of Parkinson's disease.

Catecholamine Reuptake

Upon depolarization of the presynaptic membrane, catecholamines are released into the synaptic space, where they stay only for a short time, being rapidly removed by two main mechanisms: reuptake and diffusion. About 80% of the released catecholamines is reuptaken into the presynaptic terminal by an active, saturable, Na^+-dependent mechanism mediated by specific membrane transporters (see Chapters 29 and 37).

Transporters for catecholamines are not selective as they can also carry serotonin. However, they exhibit different binding affinities for their substrates; based on this feature, three transporters can be distinguished: one for A/NA, one for serotonin, and one for DA. The selectivity of action depends on their cell-specific pattern of expression: central and peripheral noradrenergic neurons selectively express the NA transporter, DA neurons express the DA transporter, and serotonin neurons express the serotonin transporter.

Many drugs are capable of inhibiting catecholamine reuptake at synaptic level. Their action enhances and prolongs the functional effects of catecholamines, both centrally and peripherally. Among these drugs, it is worth mentioning some classes of tricyclic antidepressants, cocaine, and phenoxybenzamine.

Tricyclic antidepressants exert their effect by blocking monoamine reuptake systems. They can be subdivided into two categories: those preferentially acting on the serotonin reuptake system (e.g., amitriptyline) and those preferentially acting on the NA reuptake system (e.g., imipramine). Antidepressants that selectively inhibit the reuptake of serotonin (fluoxetine), NA (reboxetine), or both (duloxetine, venlafaxine) have also been developed (see Chapter 40 and E 29.2). "Atypical" antidepressants are a large group of drugs with different chemical structures and mechanisms of action, such as bupropion, trazodone, mirtazapine, and atomoxetine.

Regardless of their primary mechanism of action, all antidepressants produce a decrease in cortical β-adrenergic receptor density after continued treatment for at least 2 weeks. The delayed effect induced by antidepressants correlates well with clinical data indicating that a treatment of at least 2 weeks is needed to attain the therapeutic effects.

The lack of a single mechanism of action responsible for all pharmacological effects is very common in pharmacology. As we said, the antidepressant action of tricyclic derivatives can be ascribed to the blockade of monoamine reuptake; however, it is important to remember that these drugs have also a certain degree of antihistaminic and anticholinergic activity, due to blockade of muscarinic receptors and H_1 histamine receptors. Therefore, a variety of side effects limit dosages and choice of the patients to be treated with these drugs. Such side effects are not present in the most recent antidepressants, such as fluoxetine, reboxetine, and venlafaxine.

Cocaine belongs to the pharmacological class of psychostimulants. It produces a euphoric state indistinguishable from that produced by amphetamine, and it decreases appetite and sleep desire. The effect is short, lasting about 1 h. Cocaine is an addictive drug of abuse. Its psychostimulant effect is ascribed to its ability to increase synaptic concentrations of DA by inhibiting its reuptake in dopaminergic neurons of the mesolimbic cortex system.

However, cocaine produces several cardiovascular effects, independent from blockade of DA reuptake. Coronary vasoconstriction and myocardium sensitization, both due to inhibition of NA reuptake at peripheral level, are the main factors responsible for brain and heart ischemia and arrhythmias, which often occur in cocaine users. The central effects of psychostimulants undergo peripheral tolerance more quickly than the cardiovascular, and for this reason, the cardiovascular system is often damaged in cocaine users. It is important to remember that cocaine has also a strong local anesthetic action that probably contributes significantly to the onset of its peripheral and central effects.

Phenoxybenzamine, another well-known psychostimulant, is a drug with multiple pharmacological activities and sites of action; in fact, it irreversibly blocks α_1 receptors, is a noncompetitive antagonist on DA receptors, and inhibits NA reuptake.

TABLE 38.2 Adrenergic receptor classification

Family	Subtype	Transduction mechanism
α_1	α_{1A}	↑ IP_3/DAG
	α_{1B}	↑ IP_3/DAG
	α_{1D}	↑ IP_3/DAG
α_2	α_{2A}	↓ cAMP
		↓ K^+ permeability
		↓ Ca^{2+} permeability
	α_{2B}	↓ cAMP
		↓ Ca^{2+} permeability
	α_{2C}	↓ cAMP
β	β_1	↑ cAMP
	β_2	↑ cAMP
	β_3	↑ cAMP

CATECHOLAMINE RECEPTORS

All receptors of the catecholaminergic system belong to the G-protein-coupled receptor superfamily and consist of a single polypeptide that crosses the plasma membrane seven times. They are divided into adrenergic (A and NA) receptors and DA receptors. Based on sensitivity to different drugs, sequence homologies, and transduction mechanisms, both the adrenergic and the DA receptors can be further subdivided into families each including various members.

Adrenergic Receptors

The existence of different populations of adrenergic receptors was first proposed by Ahlquist in 1948. While studying the order of potency of some sympathomimetic amines in different effector systems, he observed that NA was very potent in causing contraction of smooth musculature and very weak in determining relaxation. Conversely, he found that isoproterenol caused only relaxation, whereas A was able to produce both contraction and relaxation, with the same potency. Based on these findings, Ahlquist proposed the existence of two types of adrenergic receptors, called α and β.

The subsequent development of radioreceptor binding techniques and the introduction of molecular methods for receptor studies have unequivocally demonstrated the existence of at least nine adrenergic receptors, of which three are similar to the β receptor and six to the α receptor. Thus, the paradigm originally suggested by Ahlquist has been modified, and the existence of two receptor families (identified as α and β) rather than two receptor types has been proposed (Table 38.2).

The α-Adrenergic Receptors In the mid-1970s, several experimental observations revealed the existence of several α receptor subtypes. Receptors sensitive to the antagonist phenoxybenzamine, and present in the vascular smooth muscle where they mediate vasoconstriction, were defined as α_1. Receptors insensitive to phenoxybenzamine but sensitive to the agonist clonidine were defined as α_2; these are located on noradrenergic terminals where they inhibit NA release and on the intramural parasympathetic ganglia of the stomach where they inhibit acetylcholine release. Subsequent studies have shown that α_2 receptors are not only presynaptic but are also present on vascular smooth muscle, and catecholamine-induced vasoconstriction is mediated by a mixed population of postjunctional α_1 and α_2 receptors.

Molecular biology studies have revealed further heterogeneity among α_1 and α_2 receptors. In fact, three types of α_1 receptors, identified as α_{1A}, α_{1B}, and α_{1D}, and three types of α_2 receptors, named α_{2A}, α_{2B}, and α_{2C}, have been cloned (Table 38.2). Sequence analysis of α receptor subtypes indicates that they all have seven transmembrane regions and therefore belong to the G-protein-coupled receptor family.

Functional studies have shown that α_{1A}, α_{1B}, and α_{1D} receptors preferentially stimulate the $G_{q/11}$ protein, with activation of the phosphatidylinositol cascade by phospholipase C. Therefore, their activation induces the release of inositol triphosphate (IP_3) and diacylglycerol (DAG) from membrane phospholipids, with consequent increase in intracellular calcium and activation of protein kinase C (Fig. 38.5).

α_1 Receptors mediate most contractile effects exerted by the sympathetic system on smooth muscles. They are mainly located on vessels, smooth muscles of the gastrointestinal system, and the genitourinary tract and in the heart (Table 38.3). At central level, both α_{1A} and α_{1B} receptors are present in the areas innervated by noradrenergic systems, such as cortex, hippocampus and hypothalamus.

α_2 Receptors mediate many of the metabolic effects of the peripheral sympathetic nervous system, facilitate platelet aggregation, and participate in the vasoconstrictor response induced by A. α_2 Receptors play an important role in the control of blood pressure by inhibiting catecholamine release

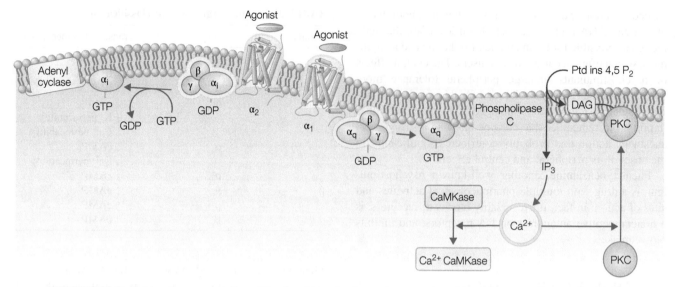

FIGURE 38.5 Schematic representation of the signal transduction pathways activated by α receptors. See text for details.

from synaptic terminals of the sympathetic system (acting as inhibitory presynaptic autoreceptors) and the activity of the central vasomotor centers (Table 38.3).

α_{2A} Receptors interact with a G_i protein and inhibit cAMP production (Fig. 38.5 and Table 38.2). They can also modulate ion channel activity by promoting K^+ efflux and inhibiting Ca^{2+} influx.

α_{2B} Receptors directly inhibit the activity of voltage-dependent Ca^{2+} channels. This seems to be responsible for the α_2 receptor-mediated inhibition of acetylcholine release at the intramural parasympathetic ganglia of the stomach and NA release from sympathetic endings.

α_{2A} Receptors are expressed centrally, in the brain stem, cerebral cortex, hippocampus, cerebellum, pituitary, and spinal cord, and locally, in the kidney, skeletal muscles, bronchial muscles, and blood vessels.

α_{2B} Receptors have been identified only in brain areas of the limbic system and are not detectable in peripheral tissues.

α_{2C} Receptors have been found both peripherally, only in the liver and kidney, and centrally, in the basal ganglia and cerebellum.

The β-Adrenergic Receptors Three β receptors, named β_1, β_2, and β_3, have been identified. All three β receptors are coupled to a G_s protein and activate adenylate cyclase and cAMP formation (Table 38.2, Fig. 38.6).

β_1 Receptors are located in the heart, kidney juxtaglomerular apparatus, and CNS. NA and A are almost equipotent in stimulating this receptor. In the heart, β_1 receptors mediate the inotropic-, chronotropic-, dromotropic-, and bathmotropic-positive effects induced by catecholamines. In the kidney juxtaglomerular apparatus, β_1 receptors mediate the stimulatory effect of NA on renin secretion (Table 38.3).

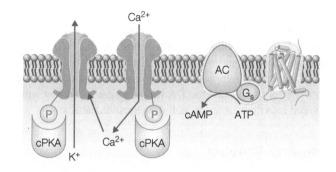

FIGURE 38.6 Schematic representation of the signal transduction pathways activated by β receptors. See text for details.

β_2 Receptors are located in the smooth musculature, where they induce relaxation; skeletal muscles, where they increase the contraction force; liver; and CNS (Table 38.3).

β_3 Receptors are found in adipose tissue, where they are involved in the regulation of thermogenesis and lipolysis, and in the gastrointestinal and genitourinary smooth muscles (Table 38.3). Selective β_3 agonists have shown a good anti-obesity and antidiabetes effect in experimental animals. β_3 agonist drugs are currently at advanced stage of development for the treatment of overactive bladder (mirabegron) and irritable bowel syndrome (solabegron).

Polymorphic variants of the β_3 receptor have been identified; a Trp-Arg substitution at position 64 decreases the receptor transduction efficiency resulting in defective activation of the lipoprotein lipase and reduced lipolysis. Although this variant receptor is normally present in the population, its probability of expression is much higher in obese individuals and has been associated with early onset of insulin-resistant diabetes.

TABLE 38.3 Adrenergic receptor localization and function

	Localization	Function
α$_1$-*Adrenergic receptors*		
Eye	Radial muscle of iris	Contraction
Vascular smooth muscle	Arterioles and venules	Contraction
Smooth muscle tissues	Stomach	Relaxation
	Bowel wall	Relaxation
	Bowel sphincter	Contraction
	Genitourinary tract	Contraction
Liver		Glycogenolysis
CNS		Sleep–awake regulation
		Regulation of ACTH and LH secretion
Heart		Increase of myocardial contractility
		Arrhythmia
α$_2$-*Adrenergic receptors*		
Noradrenergic terminals	Presynaptic receptors	Inhibition of NA release
Stomach	Intramural ganglia	Inhibition of ACh release; relaxation
Vascular smooth muscle	Arterioles	Contraction
Kidney	Proximal tubule	Decrease of Na$^+$ excretion
Pancreas	Langerhans islets	Decrease of insulin secretion
Platelets		Aggregation
CNS		Sedation
		Food intake increase
		Sympathetic tone reduction
β$_1$-*Adrenergic receptors*		
Heart	SA node	Heart rate increase
	Atrium	Contractility increase
	AV node	Increase of conduction and automaticity
	His–Purkinje bundle	Increase of conduction and automaticity
	Ventricle	Increased contractility
		Increased conduction and automaticity
Kidney	Juxtaglomerular apparatus	Renin secretion increase
β$_2$-*Adrenergic receptors*		
Vascular arterioles	Coronary	Relaxation
	Skeletal muscles	
	Lung	
	Kidney	
Smooth muscle	Lung	Relaxation
	Stomach	
	Bowel	
	Genitourinary tract	
	Uterus	
Liver		Glycogenolysis
CNS		Food intake inhibition
β$_3$-*Adrenergic receptors*		
Fat		Lipolysis
Smooth muscle	Bowel	Relaxation
	Genitourinary tract	Detrusor relaxation

DA Receptors

Based on the pharmacological results obtained since the 1970s, DA receptors are classified into two categories, named D$_1$ and D$_2$. D$_1$ receptors have been identified as sensitive to DA, apomorphine, and other DA agonists; insensitive to bromocriptine; and blocked by neuroleptics. D$_2$ receptors are selectively stimulated by bromocriptine and lisuride and selectively blocked by sulpiride.

The advent of new recombinant DNA technologies has allowed identifying several DA receptor subtypes that, according to both sequence homology and pharmacological profile, can still be grouped into the two classic families D$_1$ and D$_2$. D$_{1A}$ and D$_5$ receptors belong to the D$_1$ family,

TABLE 38.4 Classification of dopamine receptor subtypes

Family	Subtype	Signal transduction	Selective agonists	Selective antagonists
D_1	D_1	↑ cAMP ↑ PLC	Fenoldopam	SCH-23390
	D_5	↑ cAMP ↑ PLC	The pharmacology of D_5 is superimposable to D_1	The pharmacology of D_5 is superimposable to D_1
D_2	$D_2{}^a$	↓ cAMP ↑ K$^+$ permeability ↓ Ca^{2+} permeability	Bromocriptine	Sulpiride
	D_3	?	Quinpirole	
	D_4	↓ cAMP		Clozapine

?, mechanism undetermined.

^{a} The D_2 receptor gene encodes two different messenger RNA, which differ by the presence (long, D_{2l}) or absence (short, D_{2s}) of a sequence of 39 amino acids, located in the third intracellular region. The two subtypes have a superimposable pharmacological spectrum.

whereas D_2, D_3, and D_4 receptors belong to the D_2 family (Table 38.4). Sequencing of the D_2 receptor-encoding cDNAs have revealed the existence of two isoforms, called D_{1s} and D_{2l} (short and long), of this receptor originating from alternative splicing.

The two families, D_1 and D_2, differ also in the signal transduction systems they use. D_1 receptors stimulate cAMP production and increase phospholipase C activity, whereas D_2 receptors inhibit adenylate cyclase, activate K$^+$ efflux, and inhibit Ca^{2+} influx.

D_1 and D_2 receptors are also differentially distributed both in brain neurons and endocrine cells. For example, neurons of caudate-putamen that receive projections from the dopaminergic SN (pars compacta) express both D_1 and D_2 receptors, whereas only D_2 receptors are expressed in lactotroph cells and only D_1 receptors are expressed in parathyroid glands.

The identification of different populations of DA receptors has had important implications in drug research, leading to the development of many drugs acting preferentially or even selectively on D_1 or D_2 receptors and to a more in-depth knowledge of the pathophysiological significance of these two receptors. *In vitro* studies and *in vivo* investigations in humans have shown that DA receptors are subject to deep functional modifications in Parkinson's disease and schizophrenia, the two diseases in which the dopaminergic system is significantly involved. In particular, *postmortem* analyses of schizophrenic patient brains have revealed that in this condition both D_1 and D_2 receptors become supersensitive. It is still debated whether such supersensitivity is due to treatment with neuroleptic drugs or is an integral part of the disease etiopathology. However, convincing evidence demonstrates the important role of the dopaminergic systems (and receptors) in major psychoses.

D_1 and D_5 receptors have overlapping pharmacological properties (Table 38.4). D_1 is more widely expressed than D_5, which is predominantly expressed in the hippocampus and thalamus parafascicularis nucleus.

The receptors of the D_2 family, D_{2s}, D_{2l}, D_3, and D_4, have similar but not identical pharmacological properties. Compared to D_2 and D_4, D_3 receptors have a higher affinity for quinpirole and DA, whereas D_4 receptors, unlike the other family members, are extremely sensitive to clozapine (Table 38.4). The receptor subtypes differ for localization and level of expression. The most common and most expressed are the D_2 receptors, which are present in all areas of dopaminergic innervation, where they represent 85–95% of all D_2 receptors.

The physiological significance of the D_{2s} and D_{2l} isoforms has not yet been clarified, and they are often expressed in the same areas although at different level. For example, the D_{2l}/D_{2s} ratio is 20:1 in the pituitary, 8:1 in the striatum, and 1:1 in the cortex.

D_3 receptors are mainly located in limbic areas, such as the olfactory tubercle, hippocampus, and pars compacta of the SN. D_4 receptors are found in the cortex, hippocampus, and hypothalamus. It is interesting to note that the D_4 receptor mRNA content in the atrium cells is about 20-fold greater than in the brain. Moreover, a further modulation of DA receptor activity is achieved also by interactions with cell proteins (see Supplement E 38.3, "Modulation of the Functional Properties of Dopamine Receptors by Interaction with Other Proteins").

PRINCIPLES OF DRUG ACTION ON CATECHOLAMINERGIC RECEPTORS

Drugs Acting on α-Adrenergic Receptors

$α_1$ Receptor Agonists Sympathomimetic drugs are a group of compounds capable of activating, in a relatively selective manner, $α_1$ receptors in vascular smooth muscles, where they induce vasoconstriction leading to increased peripheral vascular resistance and increased blood pressure. The increase in blood pressure is associated with sinus bradycardia caused

by activation of vagal reflexes and is blocked by atropine (a muscarinic cholinergic receptor antagonist; see Chapter 39). The clinical use of these drugs is limited to the treatment of some hypotension or shock patients. They are also employed as vasoconstrictors and decongestants for topical use, at the level of the nasal mucosa and eye. These drugs may also be associated with local anesthetics to increase duration of anesthesia, since they reduce the diffusion rate of the local anesthetic into systemic circulation.

α_2 Receptor Agonists The α_2 receptor-selective agonists (Table 38.5) are used in the treatment of hypertension. Following the development of other drugs with greater efficacy and fewer side effects, today, α_2 agonists are no longer considered the first choice in the antihypertensive treatment, although they still offer a valuable therapeutic aid in case of hypertensive forms particularly resistant to therapy. Their effectiveness is mediated by the α_2 receptors of the brain stem (solitary tract nucleus). Activation of these receptors decreases the sympathetic tone and increases the parasympathetic tone, resulting in decreased blood pressure. Part of the antihypertensive effect is also due to inhibition of NA release from terminals of the sympathetic system (presynaptic α_2 receptor inhibitors). Administered intravenously, these drugs may initially cause transient increases in blood pressure, secondary to the activation of α_2 receptors present in the arterioles.

Activation of central α_2 receptors is also responsible for the sedative properties of clonidine and other centrally acting selective α_2 agonists, such as azepexol and dexmedetomidine, which can be used in premedication and neuroleptoanesthesia. These drugs significantly reduce the doses of anesthetics or opioids necessary to induce anesthesia.

α Receptor Antagonists α Receptors antagonists are divided into selective and nonselective based on their ability to discriminate between α_1 and α_2 receptors (Table 38.5). Some of the selective agents (e.g., prazosin and derivatives) are more active on α_1 than α_2 receptors, whereas others (e.g., yohimbine) have an opposite potency. Nonselective antagonists are active on both receptor subtypes. The nonselective antagonist phenoxybenzamine also binds to the α receptors, in an irreversible manner.

α_1 Receptor antagonists, such as prazosin and its derivatives, cause vasodilation and decrease in blood pressure, but do not change the heart rate because, at therapeutic doses, they do not block inhibitory α_2 autoreceptors and therefore do not induce NA release from sympathetic endings. For these features, these drugs are used in the treatment of hypertension.

Yohimbine, the only relatively selective antagonist at α_2 receptors, has opposite effects to clonidine and has long been used to treat male impotence. Yohimbine and other antagonists, such as ergot alkaloids, indoramin, ketanserin, and several other neuroleptics, inhibit also nonadrenergic receptors.

Drugs Acting on β-Adrenergic Receptors

β_2 Receptor Agonists β_2 Agonist drugs are important therapeutic tools in the treatment of bronchoconstriction in patients with asthma or obstructive pulmonary disease (COPD) and in the treatment of preterm labor. In all these situations, β_2 agonists have a relaxant effect on smooth muscles.

A is more active on the β receptors compared to NA; such difference is due to the presence of a methyl moiety on the amino group of the β-phenylethylamine (β-phenylethylamine represents the basic chemical structure of catecholamines and many of their agonists) in the A but not in the NA molecule. Indeed, it has been established that increasing the size of the alkyl substituents at this position increases the activity of the β agonist. For example, isoproterenol, the first β-selective drug developed in the 1940s, has a $CH(CH_3)_2$ group at this position. However, this drug, which can be used as a cardiac stimulant in emergency situations (e.g., cardiac arrest), does not discriminate between β_1 and β_2 receptors. Drugs with greater selectivity for the β_2 receptors have subsequently been developed, which have proved effective in the control of symptoms of asthma and other COPD (Table 38.5).

It is important to note that β_1 receptor activation has stimulatory effects on the heart and blood pressure (due to increased cardiac output and renin release), whereas β_2 receptor activation has more variable and different effects, reflecting the wider distribution of these receptors. β_2 Receptor activation involves pressure changes due to peripheral vasodilation, resulting in reflex adaptation phenomena that may lead to the onset of reflex tachycardia. Stimulation of receptors on striated muscle fibers can cause tremors and β-blockers are active in essential tremors. Systemic administration of high doses of β_2 agonist drugs determines an increase in glucose, lactic acid, and free fatty acid plasma concentration and a decrease in K^+ plasma concentration. The occurrence of these side effects has led to the identification of operational strategies to obtain a long-lasting stimulation of β_2 receptors in bronchial muscles with minimal effects on other tissues.

Pharmaceutical research has tried to find molecules not only more selective for the β_2 receptors but also more resistant to enzymatic degradation and orally absorbable. A higher resistance to degradative enzymes allows either to prolong the time interval between administrations or to reduce the single dose, whereas an oral administration allows attaining lower peaks of plasma drug concentrations compared to those obtained with the same dose administered intravenously but with more constant concentrations in time. Methods and tools have also been developed to achieve a controlled administration of these drugs by inhalation to treat bronchial asthma and COPD, in order to obtain effective concentrations at bronchial level without significant systemic effects. All of these properties (administration by inhalation, low peak plasma levels, prolonged action

TABLE 38.5 Drugs acting on adrenergic receptors and their therapeutic use

α Receptors

Agonists

$α_1$ *Preferring drugs*

Adrenaline[a]	Cardiac arrest
	Anaphylaxis
	Allergic reactions
Ephedrine[a]	Nasal congestion
Ethylephrine	Postural hypotension
Phenylephrine	Postural hypotension
Phenylpropanolamine	Nasal and eye congestion
Midodrine	Nasal congestion
Naphazoline	Postural hypotension
	Nasal and eye congestion
Noradrenaline[a]	Cardiac arrest
	Shock
Oxymetazoline	Nasal and eye congestion
Pseudoephedrine	Nasal congestion
Tetryzoline	Eye congestion
Tramazoline	Nasal congestion
Xylometazoline	Nasal congestion
Dipivefrine[a]	Glaucoma

Selective $α_2$

Clonidine	Primary hypertension
Methyldopa	Primary hypertension
Apraclonidine (Iopidine)	Glaucoma
Guanfacine	Primary hypertension
Guanabenz	Primary hypertension

Antagonists

Not selective

Phenoxybenzamine	Pheochromocytoma
Phentolamine	Pheochromocytoma
Tolazoline	Pheochromocytoma

$α_1$ *Selective*

Alfuzosin	Benign prostatic hypertrophy
Doxazosin	Primary hypertension
	Benign prostatic hypertrophy
Tamsulosin	Benign prostatic hypertrophy
Terazosin	Primary hypertension
	Benign prostatic hypertrophy
Urapidil	Hypertensive crises

$α_2$ *Selective*

Yohimbine	Sexual dysfunctions

β Receptors

Agonists

$β_1$ *Selective*

Dobutamine	Cardiogenic shock
Dopamine[b]	Cardiogenic shock
Isoproterenol	Cardiogenic shock
	Cardiac arrhythmias

$β_2$ *Selective*

Clenbuterol	COPD; asthma
Fenoterol	COPD; asthma
Formoterol	COPD; asthma
Indacaterol	COPD; asthma
Ritodrine	Premature labor
Salbutamol	COPD; asthma
Salmeterol	COPD
Terbutaline	COPD; asthma

$β_3$ *Selective*

Mirabegron	Overactive bladder

Antagonists (β-blockers)

$β_1$ *Selective*

Acebutolol	Primary hypertension
	Angina pectoris
	Cardiac arrhythmias
Atenolol	Primary hypertension
	Angina pectoris
	Cardiac arrhythmias
Betaxolol	Glaucoma
Bisoprolol	Primary hypertension
	Angina pectoris
	Chronic heart failure
Celiprolol[c]	Primary hypertension
Esmolol	Cardiac arrhythmias
Metoprolol	Angina pectoris
	Cardiac arrhythmias
	Migraine prophylaxis
Nebivolol[c]	Primary hypertension
	Chronic heart failure

Not selective

Carteolol[d]	Glaucoma
Carvedilol[e]	Primary hypertension
	Chronic heart failure
	Angina pectoris
Labetalol[e]	Primary hypertension
	Hypertensive crises
Levobunolol	Glaucoma
Metipranolol	Glaucoma
Oxprenolol	Primary hypertension
Pindolol	Primary hypertension
	Angina pectoris
Propranolol	Primary hypertension
	Angina pectoris
	Cardiac arrhythmias
	Portal hypertension
	Hyperthyroidism
	Migraine prophylaxis
Sotalol	Cardiac arrhythmias
Timolol	Glaucoma
	Primary hypertension
	Angina pectoris
	Migraine prophylaxis

[a] Also β agonists. [b] Dose-dependent effect. [c] Also NO donor. [d] Also $β_2$ agonists. [e] Also $α_1$ antagonists.

duration) make current β_2 agonist drugs more effective, with less frequent and smaller side effects.

β Receptor Drug Antagonists The identification of β receptors, their detection in the cardiovascular system, and the characterization of their functions have led to the development of a class of drugs, called β-blockers, which have an enormous importance in the treatment of many cardiovascular diseases, such as hypertension, arrhythmias, myocardial ischemia, and heart failure. β-Blockers are a family of β receptor competitive antagonists; their prototype is propranolol, developed in the 1960s and defined as nonselective β-blocker as it can bind to both β_1 and β_2 receptors.

Drugs developed subsequently are characterized by the following features: (i) ability to discriminate between receptor subtypes (e.g., metoprolol, atenolol, and, more recently, nebivolol are selective for the β_1 receptors); (ii) a certain degree of sympathetic–mimetic intrinsic activity (e.g., pindolol and acebutolol, which act as partial agonists); and (iii) associated vasodilating properties, as in the case of celiprolol and carvedilol (Table 38.5).

The most important therapeutic effects and side effects of β-blockers can be derived from Table 38.3, where the β receptor-mediated effects in various organs and systems are reported.

The therapeutically most important effects of β-blockers are cardiovascular. By interacting with cardiac β_1 receptors, β-blockers prevent their interaction with catecholamines, causing a decrease in force of contraction, cardiac output, and heart rate. Therefore, they exhibit inotropic-, chronotropic-, bathmotropic-, and dromotropic-negative effects, which are particularly evident when the sympathetic system is activated, such as during exercise or stress conditions.

The therapeutic indications of β-blockers include numerous cardiovascular diseases, as mentioned earlier. Their cardiovascular effects make β-blockers particularly effective in the treatment of angina, a condition characterized by a discrepancy between demand and supply of oxygen to the heart. Administration of β-blockers reduces the oxygen demand by the heart, thereby improving exercise tolerance.

β-Blockers have important effects on cardiac rhythm and automatism. They reduce the sinus node frequency, decrease depolarization rate of ectopic pacemakers, slow atrial conduction and AV nodal conduction, and increase the refractory period of the AV node. They are therefore an essential class of antiarrhythmic drugs, mainly used in the treatment of supraventricular arrhythmias.

The decrease in cardiac output and frequency and, above all, the decreased renin production and reduced activation of the renin–angiotensin–aldosterone system account for the indication of these drugs in the hypertension treatment, either alone or in combination with other hypotensive drugs.

Another important observation that emerged in recent years is the therapeutic benefit of administering low doses of β-blockers in the treatment of early-stage congestive heart failure. The use of β-blockers in this condition allows breaking the vicious cycle of reflex sympathetic activation induced by the cardiovascular system to compensate for the reduced cardiac output at the basis of the pathophysiology of heart failure. The β-blockers reduce the reflex sympathetic activation, the tachycardia and vasoconstriction, and the activation of the renin–aldosterone–angiotensin system, which is detrimental to a failing heart. The issue with this therapy is the negative inotropic effect; for this reason, the therapy is limited only to patients with a cardiac output not severely compromised. In addition, the drug dose should be low and "titrated" for each patient.

β Antagonists are also used to reduce anxiety symptoms in players or speakers and nonparkinsonian essential tremor; the mechanism of action could be due to a combination of the central and peripheral effects of β-blockers.

Besides the therapeutic effects, β-blockers have side effects related to their mechanism of action that limit their use in some clinical situations. In particular, it is important to remember that nonselective β-blockers, by interacting with β_2 vascular receptors, cause vasoconstriction, which may be significant in patients with obstructive arteriopathy; in addition, their interaction with β_2 receptors in bronchial muscles causes bronchoconstriction in bronchial asthma or COPD patients. Their interaction with liver β_2 receptors cause decreased glucose production, thus enhancing insulin hypoglycemia in diabetics.

Drugs Acting on DA Receptors

As previously discussed, drugs active on DA receptors are important in the pharmacological treatment of Parkinson's disease and psychoses. Dopaminergic stimulation has also become part of the therapeutic strategies of acute heart failure. The influence of DA on emesis and hormone secretion has opened the way to the symptomatic treatment of vomiting and of dysfunctions dependent on the prolactin system.

DA Agonists in Cardiogenic Shock Shock is a clinical syndrome characterized by an insufficient tissue perfusion associated with hypotension. Causes of shock can be hypovolemia secondary to dehydration or hemorrhage and heart failure secondary to myocardial infarction. The fall in blood pressure occurring during shock generally induces a marked activation of the sympathetic system that, in turn, produces peripheral vasoconstriction and increases heart rate and force of cardiac contraction. At early stages, these mechanisms may help to maintain blood pressure and cerebral blood flow, but later they decrease renal blood flow, causing decreased urine production and development of metabolic acidosis.

Low DA doses mainly activate D_1 DA receptors present in renal, splanchnic, and coronary vessels. The increase in adenylate cyclase activity causes dilation of renal arteries with consequent increase in renal blood flow, glomerular filtration rate, and Na^+ excretion. These effects occurring only at kidney level are particularly useful in the cardiogenic and the hypovolemic shock. Higher DA doses also activate heart β_1 receptors, causing tachycardia, in part via direct activation of cardiac receptors and in part via stimulation of neurotransmitter release from catecholamine nerve terminals (presynaptic receptors stimulators). At high concentrations, DA activates also α_1 receptors and produces a generalized vasoconstriction.

DA Agonists in Disorders of the CNS In addition to L-DOPA, administered to increase the amount of neurotransmitter in the remaining dopaminergic neurons of the nigrostriatal system, a number of ergot-derived dopaminergic agonists (bromocriptine, pergolide, lisuride, dihydroergocryptine, cabergoline) are used in the therapy of Parkinson's disease.

Bromocriptine and lisuride act as agonists at D_2 receptors and as antagonists at D_1 receptors; lisuride is also endowed with serotonergic activity. By contrast, pergolide acts as an agonist at both DA receptor subtypes. In addition to ergot derivatives, also apomorphine and its derivatives are endowed with dopaminergic activity (both at D_1 and D_2 receptors). Other DA agonist drugs, which are not ergot derivatives, such as ropinirole, pramipexole and piribedil (D_2 and D_3 receptor agonists), and rotigotine (D_1, D_2, and D_3 agonist) are also available. The use of DA agonist drugs is secondary to that of L-DOPA; bromocriptine is used in combination with levodopa in patients showing inadequate symptomatic remission with the primary treatment alone or in patients with too frequent "on–off" episodes. The use of these drugs is mainly limited by the occurrence of side effects due to peripheral dopaminergic activity, such as nausea, vomiting, and postural hypotension.

DA Antagonists Neuroleptics or antipsychotics are a chemically heterogeneous class of drugs capable of controlling the core symptoms of schizophrenia, such as psychomotor agitation, and sensory–perceptual disorders. They are used both to control acute episodes and to prevent recurrence. They are divided into two categories: classic and atypical. The first category includes phenothiazine derivatives (chlorpromazine, thioridazine, fluphenazine), thioxanthene derivatives (chlorprothixene, flupenthixol), butyrophenones (haloperidol, droperidol), and piperidine derivatives (pimozide).

A common feature of antipsychotic drugs (neuroleptics) is their ability to block DA D_2 receptors (which is the basis of the dopaminergic theory of schizophrenia). However, many of these drugs can also block, to variable extent, D_1 receptors (butyrophenones, such as haloperidol, and benzamides, such as sulpiride, are those with a higher selectivity for D_2 receptors) and are often characterized also by antagonistic activity at adrenergic receptors; in particular, the antipsychotic clozapine is a potent antagonist of α_1 receptors and seems to exhibit minimal antidopaminergic activity. In addition, many of these drugs possess antimuscarinic and antihistaminic activity.

Treatment with antipsychotic drugs induces different neurological syndromes: some appear acutely in association with the administration of the drug, and others are delayed and may appear years after the beginning of the treatment. Of these syndromes, the so-called iatrogenic parkinsonism is due to the competition at DA receptors. Its incidence varies depending on the drug and in some patients, it may not occur at all. Gynecomastia and galactorrhea appear also at doses lower than those producing the antipsychotic effects (probably because of the absence of the blood–brain barrier at the pituitary level) and do not give rise to tolerance. Endocrine abnormalities also affect gonadotropin, estrogen, and progesterone secretion, which are probably involved in the possible development of amenorrhea in women treated with chlorpromazine. The same drug can lead to reduced secretion of ACTH, GH, ADH, and insulin.

Other side effects involve the cardiovascular system, the most common being the appearance of hypotension, mostly of orthostatic type.

Risperidone is one of the molecules recently introduced in antipsychotic therapy. This drug is a benzisoxazole derivative and exhibits a marked antagonistic activity at $5-HT_2$ receptors, in addition to that at D_2 receptors. The unique receptor profile of risperidone may explain the reduced incidence of side effects and the broader therapeutic spectrum of this drug.

Antidopaminergic drugs are also used as antiemetic and in anesthesiology. As previously discussed, the dopaminergic system plays an important role in the control of CTZ and gastrointestinal motility. For this reason, the antiemetic therapy takes advantage of antidopaminergic drugs devoid of antipsychotic effects (e.g., metoclopramide; note that these drugs may induce extrapyramidal syndromes and galactorrhea because they can cross the blood–brain barrier) or capable of crossing the blood–brain barrier but in limited amount, such as domperidone. The antiemetic activity is also due to the prokinetic effect on gastric emptying.

Some neuroleptics such as droperidol, a butyrophenone derivative, produce a deep sedation state characterized by reduction of anxiety and motor activity; it also increases the sensitivity to other drugs that depress CNS activity.

The concomitant administration of an opioid analgesic turns sedation into neuroleptanalgesia; the use of nitrous oxide allows converting analgesia into neuroleptoanesthesia.

TAKE-HOME MESSAGE

- The autonomous nervous system regulates visceral functions through the sympathetic and parasympathetic nervous systems. These normally act in opposition to each other.
- The catecholamines adrenaline, noradrenaline, and dopamine mediate neurotransmission in the autonomic sympathetic system and in the brain.
- In the brain, the adrenergic system has a discrete distribution and is involved in the control of basic motor and behavioral functions. Its alteration leads to severe brain diseases such as Parkinson's disease, schizophrenia, and depression.
- Catecholamines are synthesized from tyrosine and metabolized by MAO and COMT.
- Catecholamines are concentrated in synaptic vesicles by a vesicular monoamine transporter inhibited by reserpine, secreted into synaptic cleft, and reuptaken into synapses by transporters inhibited by cocaine.
- Catecholaminergic receptors are GPCRs and can be grouped in several types and subtypes.
- Drugs interfering with catecholamine synthesis, catabolism, secretion, and reuptake or interacting with their receptors are widely used in different clinical situations, in particular in depression, schizophrenia, and Parkinson's disease and in the control of blood pressure and cardiac rhythm and contraction and of smooth muscle contraction in cardiogenic shock and vomiting.

FURTHER READING

Bear M.F., Connors B.W., Paradise M.A. (2007). *Neuroscience: exploring the brain*. 3rd ed. Baltimore, MD: Lippincott Williams & Wilkins.

Beaulieu J.M., Gainetdinov R.R. (2011). The physiology, signaling, and pharmacology of dopamine receptors. *Pharmacological Reviews*, 63, 182–217.

Cote L., Crutcher M.D. (1985). Motor function of the basal ganglia and transmitter diseases of metabolism. In: Kandel E.R., Schwartz J.H. (eds.), *Principles of neural science*. 2nd ed. New York: Elsevier Press, pp. 523–534.

Emery A.C. (2013). Catecholamine receptors: prototypes for GPCR-based drug discovery. *Advances in Pharmacology*, 68, 335–356.

Frishman W.H. (2008). Beta-adrenergic blockers: a 50-year historical perspective. *American Journal of Therapeutics*, 15, 565–576.

Goldstein D.S., Kopin I.J., Sharabi Y. (2014). Catecholamine autotoxicity. Implications for pharmacology and therapeutics of Parkinson disease and related disorders. *Pharmacology & Therapeutics*, 144, 268–282.

Hawrylyshyn K.A., Michelotti G.A., Coge F., Guénin S.P., Schwinn D.A. (2004). Update on human alpha1-adrenoceptor subtype signaling and genomic organization. *Trends in Pharmacological Sciences*, 25, 449–455.

O'Connell T.D., Jensen B.C., Baker A.J., Simpson P.C. (2013). Cardiac alpha1-adrenergic receptors: novel aspects of expression, signaling mechanisms, physiologic function, and clinical importance. *Pharmacological Reviews*, 66, 308–333.

Shen M.J., Zipes D.P. (2014). Role of the autonomic nervous system in modulating cardiac arrhythmias. *Circulation Research*, 114, 1004–1021.

Weis W.I., Kobilka B.K. (2008). Structural insights into G-protein-coupled receptor activation. *Current Opinion in Structural Biology*, 18, 734–740.

39

CHOLINERGIC TRANSMISSION

GIANCARLO PEPEU

By reading this chapter, you will:

- Know the anatomy of the cholinergic system in CNS and PNS and its role in motor activity, neurovegetative regulation, and cognitive processes
- Know the molecular mechanisms regulating acetylcholine synthesis, release from nerve endings, and inactivation
- Learn the classification, molecular structure, and signal transduction of membrane receptors activated by acetylcholine
- Know the classification, properties, effects, and therapeutic and toxic significance of drugs acting on different components of the cholinergic system

Acetylcholine (ACh) is the chemical agent that mediates cholinergic transmission (Fig. 39.1). It was synthesized by Bayer in 1866, long before its detection in living organisms. ACh was identified in the ergot (*Claviceps purpurea*) by Ewins in 1914, in horse's adrenomedulla by Dale and Dudley in 1929, and in frog's brain by Chang and Gaddum in 1933. A century ago, in 1913, by comparing the effects of vagal stimulation with those of ACh administration, Dale hypothesized that ACh might act as a neurotransmitter. Otto Loewi in 1921 demonstrated in frog's heart that vagal stimulation induces the release of a substance, which he called "vagusstoff," responsible for bradycardia and decrease in contraction strength, two actions similar to those exerted by ACh. In the 1930s, using bioassays, it was demonstrated that electrical stimulation of splanchnic nerves, vagus, preganglionic sympathetic fibers, and motor nerves induces the release of an ACh-like substance. The release of an ACh-like substance from the spinal cord was demonstrated in 1941 and from the cerebral cortex by Elliott in 1950. The ultimate identification of the "vagusstoff" with ACh was obtained in 1970 with the introduction of mass spectrometry.

DISTRIBUTION AND FUNCTION OF THE CHOLINERGIC SYSTEMS

Cholinergic Transmission in the Peripheral Nervous System

ACh is the neurotransmitter responsible for nervous impulse transmission in the following peripheral synapses:

1. Synapses between parasympathetic nerve endings and glandular cells, cardiac pacemaker, and smooth muscle cells.
2. Synapses between postganglionic sympathetic fibers and sweat glands. Nerve fibers regulating sweat secretion belong anatomically to the sympathetic system, but most of them release ACh (cholinergic sympathetic system).
3. Synapses between preganglionic fibers and ganglionic neurons in the sympathetic and parasympathetic systems.
4. Synapses between the sympathetic fibers (splanchnic nerves) that innervate the adrenal medulla and regulate catecholamine release.

General and Molecular Pharmacology: Principles of Drug Action, First Edition. Edited by Francesco Clementi and Guido Fumagalli.
© 2015 John Wiley & Sons, Inc. Published 2015 by John Wiley & Sons, Inc.

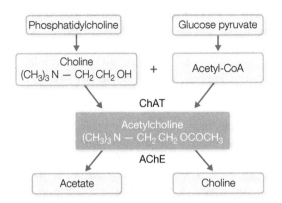

FIGURE 39.1 Synthesis and metabolism of ACh.

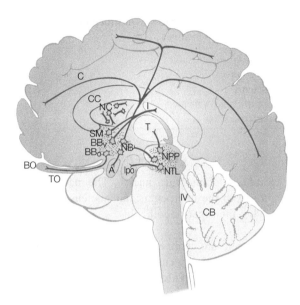

FIGURE 39.2 Schematic drawing of cerebral cholinergic pathways. A, habenula; BBh, Broca diagonal band, horizontal limb; BBv, Broca diagonal band, vertical limb; BO, olfactory bulb; C, cortex; CB, cerebellum; CC, corpus callosum; DTN, dorsolateral tegmental nucleus; H, hippocampus; Hypo, hypothalamus; IV, fourth ventricle; MS, medial septum; NB, nucleus basalis magnocellularis; NC, caudate nucleus and putamen; OT, olfactory tract; PTN, pontine tegmental nucleus; T, thalamus.

5. Junctions between motor neuron nerve endings and striate muscle cells (neuromuscular plaques). ACh induces muscle contraction by depolarizing the terminal plaque of the neuromuscular junction.

In many synapses, cholinergic transmission may be modulated by the corelease of peptide cotransmitters, such as cholecystokinin, substance P, and vasoactive intestinal peptide (VIP). An example, in cats, is the potentiation of ACh stimulation of salivary secretion by VIP coreleased from the parasympathetic fibers innervating the salivary glands. ACh released from vagal nerve endings also modulates the inflammatory response by inhibiting TNF release from macrophages.

The Nonneuronal Cholinergic System In humans and other animal species, ACh and choline acetyltransferase (ChAT), its synthesizing enzyme, are present in many nonnervous tissues such as erythrocytes, immune cells, and endothelial and epithelial cells, including the respiratory airways, and in the placenta. Small amounts of ACh can be detected also in the blood. ACh has been found in bacteria, seaweeds, protozoa, and several plants. Both muscarinic and nicotinic ACh receptors have been detected in many peripheral cells. The role of the nonneuronal cholinergic system is not yet fully understood and is considered of autocrine type. It contributes to the regulation of blood circulation since ACh induces a generalized vasodilation by stimulating nitric oxide (NO) release from vascular endothelia and erythrocytes (see also Chapter 48). NO relaxes the vessel smooth muscle fibers and modifies hemorheological properties. ACh released from bronchial mucosal cells regulates mucociliary activity and mucosal repair. It is likely that nonneuronal ACh contributes to the cholinergic control of inflammation, together with ACh released from vagus nerves.

The Cholinergic Transmission in the Central Nervous System

The cholinergic neuron distribution in human brain is schematically depicted in Figure 39.2. Moving rostrocaudally, we find the following groups of neurons:

1. The forebrain magnocellular cholinergic neurons that form a series of nuclei without precise correspondence with anatomical structures. They were classified by Mesulam using the abbreviations Ch1 for those in the medial septum, Ch2 for those in the ascending limb, Ch3 for those in the horizontal limb of the Broca's band, and Ch4 for those in the nucleus basalis of Meynert. The cholinergic neurons represent 50–75% of the cells of these nuclei and their projections form the cholinergic input to the hippocampus, cerebral cortex, olfactory bulb, and amygdala.

2. The cholinergic interneurons of caudate nucleus and putamen.

3. The cholinergic nuclei of the brain stem, namely, the pontine tegmental nucleus (Ch5), the dorsolateral tegmental nucleus (Ch6), the medial habenular nucleus (Ch7), and the parabigeminal nucleus (Ch8). Cholinergic neurons in Ch5 and Ch6 project to the thalamus, hypothalamus, nucleus pallidum, and

TABLE 39.1 Cognitive processes in which the forebrain cholinergic system is involved

Attention, sustained attention
Working memory
Spatial memory
Explicit or declarative memory
Information encoding

forebrain cholinergic nuclei. The neurons in C7 project to the interpeduncular nucleus and those in Ch8 to the superior colliculus.

Finally, there are cholinergic neurons in the ventricular part of the arcuate nucleus in the hypothalamus, which project to the medial eminence.

Role of Brain Cholinergic System in Learning, Memory, and Movement

Animal and human experiments have demonstrated the involvement of the forebrain cholinergic system in the cognitive processes listed in Table 39.1 (see also Supplement E 39.1, "Methods for Investigating the Role of the Brain Cholinergic System in Learning and Memory").

Though not involved in long-term memory formation, the cholinergic system has such a pervasive role in cognitive processes that some authors have raised the question of whether it underlies all human behaviors, the conscience, and the self. This hypothesis is supported by the impairment of attention and the amnesia caused by administration of atropine, scopolamine, and other anticholinergic drugs. Moreover, degeneration of the forebrain cholinergic neurons is a pathogenetic feature of Alzheimer's disease and alcoholic and pugilistic dementias, conditions in which loss of attention and memory deficits represent the first and most important symptoms.

The projections from brain stem Ch5 and Ch6 cholinergic nuclei to the thalamus, mesencephalic dopaminergic nuclei, and forebrain cholinergic nuclei modulate sleep–wakefulness cycle and REM sleep, and their activation underlies arousal. The cholinergic fibers originating from the forebrain nuclei regulate blood flow in the cerebral cortex and hippocampus since they stimulate NO production.

The intrinsic cholinergic neurons in the caudate nucleus and putamen are involved in the extrapyramidal regulation of movement. In Parkinson's disease and parkinsonisms, degeneration or dysfunction of the dopaminergic neurons of the substantia nigra disrupt the dopaminergic control of the cholinergic interneurons. The hyperactivity of the latter is a pathogenetic component of rigidity and tremors, which are the main symptoms of these pathological conditions. For this reason, anticholinergic drugs can be prescribed for the therapy of Parkinson's disease and for reducing the extrapyramidal side effects of antipsychotic agents, drugs that block dopaminergic receptors.

The striatal cholinergic neurons play a role in the formation of implicit, nondeclarative memory, such as the procedures needed to drive a car or solve a problem.

The brain stem cholinergic system is involved in cardiovascular regulation as demonstrated by the increase in blood pressure in laboratory animals and humans induced by the administration of cholinesterase (ChE) inhibitors and muscarinic agonists. The cholinergic neurons in the arcuate nucleus participate in the regulation of growth hormone secretion through the modulation of the hypothalamic secretion of somatostatin.

The mechanism through which ACh affects cognitive processes involves activation of nicotinic and muscarinic receptors (see following text) and early genes and facilitation of the responses to sensory input through a long-term potentiation-like mechanism. This mechanism improves the signal-to-background noise ratio and facilitates neuronal plasticity leading to memory formation.

ACh SYNTHESIS AND METABOLISM

ACh is formed from choline and acetyl coenzyme A by the enzyme ChAT (Fig. 39.1). Under physiological conditions, the amount of ACh in cholinergic nerve endings remains at a constant level because ACh synthesis keeps up with the release. When the depolarization-induced release increases, ACh synthesis is stimulated in order to restore its intraneuronal level. Precursor availability, concentration of the end product, and enzyme activity are the factors regulating ACh synthesis. ACh released from nerve endings into the synaptic cleft is hydrolyzed to choline and acetic acid by ChEs, mainly by acetylcholinesterase (AChE) but also by butyrylcholinesterase (BChE). It is also removed by diffusion.

ChAT is a 60–70 kDa enzyme codified by a single gene located in the cholinergic locus of the human chromosome 10 at 10q11.2-10. This gene also contains the coding sequence for the vesicular ACh transporter (vesicular ACh transporter 1 (VACHT1)). ChAT is expressed in cholinergic neurons and represents one of their molecular markers. Low ChAT concentrations can be found also in nonneuronal cells like erythrocytes, lymphocytes, epithelial cells, and placenta and in bacteria and plants, as previously mentioned. In cholinergic neurons, ChAT is in excess and is not a limiting factor for ACh synthesis as it is not saturated by the choline concentrations present in cells. The rate-limiting factor for ACh synthesis is choline transport. Thyroid hormones, estrogens, and several neurotrophins, the most important being nerve growth factor (NGF), increase neuronal ChAT expression and regulate its long-term activity. The short-term regulation of its catalytic activity depends on changes in ionic environment in nerve endings, level of phosphorylation controlled by kinases, and enzyme subcellular distribution. In nerve endings, ChAT exists in soluble and membrane-bound forms characterized by different physicochemical and biochemical properties. The soluble form is hydrophilic, whereas the membrane-bound form is amphiphilic. This form is considered physiologically the most important as it is coupled to the choline transport system. The soluble form is considered the precursor of the bound form. Specific ChAT inhibitors have been synthesized for experimental purposes; of them, the most used is N-methyl-4-(1-naphtylvinyl) pyridinium.

ACh Precursors

The choline needed for ACh synthesis comes from plasma free choline and phospholipids, mainly phosphatidylcholine (lecithin). The latter is provided to the organism by food. Between 35 and 50% of the choline deriving from ACh hydrolysis in the synaptic cleft is reutilized for ACh synthesis. An efficient homeostatic system keeps a steady level of choline in plasma (~10 µM) and extracellular brain fluids (~3 µM) and under normal conditions prevents the increase in ACh synthesis following choline administration. Choline in the body is mostly used for the synthesis of cell membrane phospholipids and is transported through the membrane by a low-affinity uptake mechanism present in all cells. Conversely, choline for ACh synthesis is supplied by a high-affinity, Na^+-dependent uptake mechanism called choline transporter 1 (CHT1). CHT1 is a 60 kDa protein containing 13 transmembrane domains and is homologous to members of the glucose Na^+-dependent transporter family. CHT1 distribution has been investigated using specific antibodies, showing that it is only expressed in cholinergic neurons and their nerve endings. Choline availability is the rate-limiting factor for ACh synthesis, and the uptake system activity increases proportionally with ACh release and a higher need in neurotransmitter synthesis. For example, in rat frontal cortex, ACh synthesis increases during behaviors requiring cholinergic activation, such as attention. Therefore, changes in the V_{max} of choline transporter reflect cholinergic neuronal activity. CHT1 is inhibited by hemicholinium (HC3). HC3 administration blocks ACh synthesis causing a progressive decrease in its release and blockade of cholinergic transmission.

The acetyl-CoA necessary for ACh synthesis originates from glucose and pyruvate and is formed in mitochondria. Acetyl-CoA is transported through the mitochondrial membrane by a transporter (ACATN) belonging to the solute carrier family (see also Chapter 29). It has been shown that under conditions of intense stimulation leading to large ACh release, it is possible to increase ACh synthesis by administering glucose and choline.

ACh Hydrolysis by ChEs

In the brain and peripheral tissues, there are two types of ChEs: AChE and BChE (see Supplement E 39.2, "Cholinesterases and Cholinesterase Inhibitors"). AChE and BChE belong to the serine hydrolase family and are expressed by two different genes sharing 65% amino acid sequence homology. AChE preferentially hydrolyzes ACh, whereas BChE hydrolyzes butyrylcholine, a synthetic compound used as a substrate in order to distinguish between the two enzymes. Both enzymes are present in the CNS. However, in human brain, AChE activity is from 1.5 to 60 times higher than that of BChE (with regional differences), and both enzymes are present in neurons and glia. AChE is also found in erythrocytes and BChE in plasma and liver; therefore, they may also exert functions unrelated to cholinergic transmission.

In cholinergic neurons, AChE is synthesized in the cell body and is then transported to dendrites and nerve endings where it is inserted in the membranes and acts as an ectoenzyme. AChE, whose function is to rapidly inactivate ACh released in the synaptic cleft, is a highly efficient enzyme as it hydrolyzes 5000 ACh molecule per second per enzymatic site.

Both ChEs exist in different molecular forms deriving from the same gene by alternative splicing. They show different quaternary structures, solubility, and binding to the synaptic membranes but exert the same catalytic activity. Of these enzymes, we know tridimensional structures; localization of the active site, formed by an anionic site that attracts the ACh-positive charge and an esteratic site; and catalytic mechanism. In the past, it was believed that, given its distribution and rapid action, AChE was the only enzyme responsible for the hydrolysis of ACh released from nerve endings. BChE, which is also present in plasma, was not believed to be involved in nervous transmission but in the hydrolysis of different exogenous esters. However, it has recently been demonstrated that administration of N1-phenylethylnorcymserine, a selective BChE inhibitor, causes an increase in extracellular ACh levels. Moreover, AChE gene knockout mice survive, although with many abnormalities, because BChE can partially compensate for AChE loss.

Genetic alterations of BChE expression are known. Affected subjects show reduced enzyme activity and abnormal responses to drugs normally hydrolyzed by ChEs, such as succinylcholine.

Many cholinesterase inhibitors (ChEIs) have been synthesized. They are classified as reversible or irreversible and selective for AChE and BChE or nonselective (Fig. 39.3). ChEIs are alternative substrates that upon binding to the enzyme are slowly hydrolyzed, leading to carbamylated or phosphorylated ChE. Among the reversible inhibitors, we list edrophonium (with short-lasting action as it only binds to the anionic site of the enzyme); carbamates (physostigmine

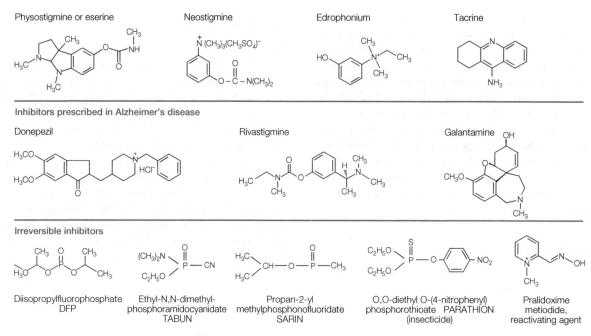

FIGURE 39.3 Cholinesterase inhibitors.

or eserine, neostigmine, pyridostigmine, and rivastigmine); ambenonium, a diethylammonium derivative; tacrine (tetrahydroacridin-9-amine); donepezil; and galantamine. Irreversible ChEIs are organic phosphor esters (alkylphosphates) such as DFP, tabun, sarin, parathion, ecothiopate, and metrifonate. The hydrolysis of the complex formed by enzyme and carbamate inhibitors requires from several minutes (for physostigmine) to some hours (for rivastigmine). During this time, ACh hydrolysis is prevented. Reversible inhibitors are used as therapeutic agents in all situations in which it may be beneficial to potentiate cholinergic transmission (e.g., myasthenia gravis, glaucoma, postoperative paralytic ileum, anesthesia, and Alzheimer's disease). Edrophonium, due to its short-lasting action, is used for diagnosis *ex adjuvantibus* of myasthenia gravis. Conversely, the binding between alkylphosphates and the enzyme is stable, and complex hydrolysis takes longer than synthesis and turnover of new enzyme. Therefore, the inhibition can be considered irreversible. The most potent and volatile organophosphorous compounds are used as chemical weapons (nerve agents), and the less potent as insecticides. Most ChEIs show higher affinity for BChE than for AChE. Few compounds selectively inhibit AChE; among them, we mention ambenonium and donepezil used in myasthenia gravis and Alzheimer's disease, respectively. N1-phenylethyl-norcymserine is the most active BChE-selective inhibitor. During the first phases of the reaction between enzyme and organic phosphorous compounds, it is possible to reactivate the enzyme by administering highly nucleophilic compounds, which remove the phosphorous residue from the enzyme. The best-known nucleophilic cholinesterase reactivator is 2-pyridine aldoxime methyl iodide (pralidoxime). Its prompt administration allows blocking or attenuating ChEI poisoning. After a given time, this becomes no longer possible as the bond between enzyme and inhibitor becomes permanent (aging). The aging process is caused by hydroxylation of the phosphorous atom bound to the enzyme, which cannot be dephosphorylated anymore. Peripheral symptoms, caused by parasympathetic hyperfunction, and impairment of neuromuscular transmission, leading to muscle paralysis, are the main consequences of ChEI poisoning.

ACh INTRACELLULAR STORAGE AND RELEASE

In nerve endings, ACh is stored within synaptic vesicles at a concentration 100 times higher than in the cytoplasm, corresponding to about 2000 molecules per vesicle. The vesicle loading is similar to that of other classical neurotransmitters and involves at least two macromolecules in the vesicular membrane. One is an ATPase, which pumps proton into the vesicle, acidifying the intravesicular medium. The other is the VACHT1, which exchanges protons with cytoplasmic ACh molecules. VACHT1 is a glycoprotein formed by 500–600 amino acids with 75 kDa molecular weight (see Chapter 29). The gene encoding this transporter is contiguous to the ChAT encoding gene within the so-called cholinergic operon. VACHT1 is selectively blocked by vesamicol [2-(4-phenyl-1-piperidyl)cyclohexan-1-ol]. ACh is stored in the vesicles along with ATP, in a ratio ranging from 5 to 10 ACh molecules for one ATP molecule. ATP is coreleased with ACh upon nerve ending stimulation. The physiological role of this corelease is unknown. However, ATP may act as a neurotransmitter by itself (see Chapter 44). Moreover, vesicles contain Ca^{2+}, Mg^{2+}, Na^+, K^+, and negatively charged proteoglycan. Their role could be to maintain the osmotic and electrical equilibrium in the vesicles.

ACh Release

ACh release following depolarization induced by nervous stimuli occurs through a Ca^{2+}-dependent exocytosis mechanism involving several proteins. Among them, the SNARE complex is responsible for the fusion between the vesicle membrane and the plasma membrane. Among the SNARE proteins, synaptotagmin (a Ca^{2+} ion sensor) and complexin regulate ACh release through the formation of pores, but the exact nature of the process is still debated (see Chapter 37).

ACh exocytosis is blocked by toxins produced by the anaerobic bacterium *Clostridium botulinum*. Botulinum toxins are metalloproteins; 7 serologically different types of them exist. After selectively binding to the cholinergic nerve endings, botulinum toxins are internalized and degrade SNARE proteins blocking synaptic vesicle exocytosis and causing long-lasting impairment of peripheral cholinergic transmission. Botulinum toxins may cause severe foodborne poisoning. Botulinum toxins A and B are used as therapeutic agents by local injection. Their use spans from blepharospasm and strabismus to several neurological and medical conditions caused by muscular hyperactivity such as cervical dystonia, tics, tremors, and spastic situations involving smooth muscles. Finally, botulinum toxins A and B are widely used in cosmetic medicine for the reduction of facial wrinkles.

ACh release from nerve endings is quantal. However, at neuromuscular junctions, together with the spontaneous quantal release (generating miniature end-plate potential, MEPP) and the evoked release (originating the excitatory postsynaptic potential, EPSP), a continuous nonquantal ACh release occurs whose physiological function is still unknown. The amount of the ACh released depends on the firing rate of the cholinergic neurons and is modulated by presynaptic muscarinic and nicotinic autoreceptors, particularly in CNS, and by heteroreceptors. There are regional differences in presynaptic modulation. Activation of muscarinic autoreceptors results in a 30–40% inhibition of ACh release, which, presumably, helps in preventing ACh accumulation in the synaptic cleft. Nicotinic autoreceptors, detected in cholinergic neurons of the nucleus basalis and nerve endings of motor nerves, stimulate ACh release. The physiological conditions in which they are activated are still unknown. Presynaptic heteroreceptors are inhibitory and are represented by 5-HT receptor subtypes and by $GABA_A$, noradrenaline α_2, and adenosine A_1 receptors. Firing of cholinergic neurons is regulated by several neurotransmitters released from fibers establishing synaptic contacts on them. Some important regional differences exist. For example, in striatum, cholinergic interneurons are inhibited by D_2 dopaminergic receptors. Cholinergic neurons of the septum and nucleus basalis are modulated by GABAergic interneurons, which in turn are regulated by glutamatergic fibers. On the same cholinergic neurons, there are D_1 dopaminergic and NMDA glutamatergic receptors. Firing of cholinergic neurons is inhibited by enkephalins, galanin, and histamine and stimulated by TRH, substance P, and cholecystokinin, with regional differences. The elaborate pre- and postsynaptic modulation of central cholinergic neurons indicates that they represent an important common final pathway responding to the input from many neuronal systems.

CHOLINERGIC RECEPTORS

The subdivision of cholinergic receptors into nicotinic and muscarinic proposed by Henry Dale in 1913 based on the responses to nicotine and muscarine (alkaloid extracted from the mushroom *Amanita muscaria*) is supported by the differences in their molecular structure and transduction mechanisms: the nicotinic receptors are ligand-gated ion channels, whereas muscarinic receptors are G-protein-coupled receptors.

Nicotinic Receptors

ACh acts on nicotinic receptors (nAChRs) (Table 39.2) located at the neuromuscular junctions, ganglia, and chromaffin cells and in the CNS. nAChRs belong to the receptor family of the cation-specific ligand-gated ion channels (see Chapter 16). Their activation by ACh or nicotine leads to the opening of an ion channel followed by depolarization and effector response. nAChRs exist in different functional conformations: at rest, open, and desensitized. Channel opening induced by agonists is followed by a desensitization, which may be rapid or slow according to agonist concentrations. Receptor desensitization with synaptic transmission blockade may result from persistent high concentrations of ACh in the synaptic cleft, as during ChEI poisoning. Receptor desensitization is also caused by nicotine and depolarizing neuromuscular blockers, such as succinylcholine. Antagonists are defined as ligands that stabilize nAChRs in close conformation such as the competitive, nondepolarizing curare (see Supplement E 39.3, "Neuromuscular Blocking Agents"). It is known since long time that the nAChRs of the neuromuscular junction are pharmacologically different from those of the chromaffin cells, ganglia, and CNS. The former are called neuromuscular, and the latter neuronal nAChRs. For instance, low concentrations of α-bungarotoxin block muscle nAChRs but do not affect the majority of the neuronal nAChRs. Conversely, mecamylamine blocks ganglionic nAChRs but has no effect on muscle receptors. Therefore, on the basis of their effects, the peripheral nAChR antagonists

TABLE 39.2 The Nicotinic Receptors

Type	Muscle	Neuronal (main subtypes)			
Subunity composition	$\alpha_1\beta_1\delta\gamma/\epsilon$	α_7	$\alpha_4\beta_2$	$\alpha_3\beta_4$	$\alpha_4\alpha_5\beta_2$	$\alpha_4\alpha_6\beta_2$
Localization	Neuro-muscular junctions	CNS, ganglia, pre- and post-synaptic	CNS presynaptic	Hippocampus, pineal gland, locus coeruleus, retina	Striatum, nucleus accumbens	Nucleus accumbens, VTA-SN
Ionic permeability	Na$^+$	Ca^{2+}	Na$^+$	Na$^+$	Na$^+$	Na$^+$
Agonists	ACh, succinyl-choline	Choline, epibatidine	Epibatidine, nicotine	Cytisine, nicotine	T-2559	—
Selective antagonists	α-Bungarotoxin, α-conotoxin, pancuronium	α-Bungaro-toxin, methyllyeacaconitine	DHβE,	α-Cntx AuIB, mecamylamine	—	α-CntxMII

α-Cntx MII, α-conotoxin MII; α-Cntx AuIB, α-conotoxin AuIB; DHβE, dihydro-β-erythroidine; TC-2559, 4-(5-ethoxy-3-pyridinyl)-N-methyl-(3E)-3-buten-1-amine difumarate.

are classified in gangioplegics, blocking ganglionic transmission, and curares, blocking neuromuscular junctions.

The Muscle nAChR The molecular structure of the muscle nAChR is well known and is described in Chapter 16. The receptor is located on the crest of the postsynaptic membrane folds and is formed by 5 subunits arranged symmetrically, in a rose-cut shape of 9 nm diameter, around a central pore of 1.5–2 nm size. Its molecular weight is 250 kDa, and the 5 subunits, named α, β, γ, δ, and ε, show this stoichiometry: $\alpha:2\beta:\delta:\varepsilon/\gamma$, $2:1:1:1$. The agonist binding site is formed by a hydrophobic pocket at the interface between two subunits. The largest part of the pocket is formed by an α-subunit loop. The different subunits show 80% of amino acid sequence homology. The region responsible for ion passage shows an enlarged extracellular portion whose negative charges contribute to concentrate cations in its inside. The entrance narrows forming a transmembrane channel with a bottleneck of 0.7–0.8 nm diameter. Agonist binding to the receptor determines channel opening in a few microseconds through rearrangement of the hydrogen bonds between the amino acids adjacent to the binding site.

The Neuronal nAChR The neuronal nAChRs are also pentamers and are subdivided in homomeric and heteromeric. The former are constituted by only one type of subunit, usually $\alpha 7$, and the latter by two types of subunits, α and β, in the stoichiometric ratio $2\alpha:3\beta$. Eight types of α-subunit and 3 types of β-subunit have been cloned sharing 50–70% homology. They show a nonuniform distribution in the brain. The most frequent subunits are $\alpha 4$, $\alpha 7$, and $\beta 2$, and 90% of receptors are represented by the $\alpha 4\beta 2$ form. In the hippocampus, homomeric $\alpha 7$ represents 70% of the total nAChRs. In peripheral ganglia, nAChRs are mostly constituted by $\alpha 3$- and $\beta 4$-subunits. The role of each subunit in determining the physiological properties of nAChRs has been studied by injecting cDNAs encoding the different subunits (cloned in appropriate vectors) in heterologous cell systems. The receptors made of $\alpha 3$- or $\alpha 4$- and $\beta 2$- and $\beta 4$-subunits form channels with high Na^+ conductance. These are opened by nicotine and blocked by β-erythroidine and n-bungarotoxin, but not by α-bungarotoxin. Since the ACh binding site is located at the interface between an α-subunit and an adjacent subunit, it appears that the type of β-subunit is important in determining the pharmacological properties of neuronal nAChRs. For instance, substitution of $\beta 2$ for $\beta 4$ results in high sensitivity to cytisine and low to nicotine. The $\alpha 7$- and $\alpha 8$-subunits bind α-bungarotoxin and form a channel permeable to Ca^{2+} ions, even if no β-units are present.

Generation of knockout or knock-in mice, in which one or more subunit encoding genes have been deleted or mutated, has widened our understanding of the physiological role of the different nAChR subtypes. For example, in $\beta 2$ knockout mice, nicotine cannot induce dependence nor facilitate learning.

Nicotinic Receptors in Nonmuscle Nonnervous Cells Nicotinic receptors are present also in keratinocytes, macrophages, microglia, and endothelial and epithelial cells where they modulate important functions such as adhesion, motility, differentiation, and proliferation. Some of the toxic effects of tobacco smoking may be due to interaction of nicotine or nicotinic agents contained in tobacco with these receptors.

Muscarinic Receptors

Muscarinic receptors are G-protein-coupled receptors. Muscarinic receptors (mAChRs) mediate many cholinergic actions in the peripheral and central nervous systems and are target of many drugs.

Five genes (m1–m5) have been identified that encode five mAChR subtypes, subdivided in two subclasses: M1, including M1, M3, and M5 receptors, and M2, including M2 and M4 receptors. Their characteristics and properties are summarized in Table 39.3. mAChRs belong to the G-protein-coupled receptor superfamily described in Chapter 17. They show a high degree of amino acid sequence homology, mainly in the transmembrane domain, and marked differences in the intracytoplasmic loops. The orthosteric agonist binding site is located in a pocket in the middle of a ring formed by the 7 transmembrane domains. Agonist binding triggers conformational changes leading to coupling with a G protein. mAChRs exert different functions depending on the G proteins they are coupled with, and the associated transduction mechanisms. M1, M3, and M5 receptors, by coupling with the pertussis toxin-insensitive $G_{q/11}$ and G_p proteins, activate phospholipase C, which produces IP3 and DAG second messengers, and inhibit K^+ currents. M3 receptors activate adenylcyclase and protein kinase A via a G_s protein. M2 and M4 receptors, by coupling with members of the pertussis toxin-sensitive G_i/G_o families, inhibit adenylcyclase as a primary transduction mechanism. In addition, by coupling with proteins of the G_s and G_q/G_{11} families, M2 receptor activates adenylcyclase and phospholipase C, respectively, as a secondary transduction mechanism. Moreover, M2 and M4 receptors open inwardly rectifying K^+ conductances and inhibit Ca^{2+} conductances. In cardiac cells of the sinoatrial node, M2 receptors are coupled to a G_o protein that opens K^+ channels causing an outward K^+ current, which slows down the heart rate.

The generation and characterization of knockout mice for each subtypes as well as the use of muscarinic toxins (MT) have significantly contributed to clarify the specific physiological roles of mAChR subtypes. MT are small peptides of 65 amino acids, deriving from the venom of African snakes that bind with high affinity and selectivity for different mAChR subtypes, exerting agonistic and antagonistic effects. For instance, MT7 binds to the M1 receptors with pmolar affinity and to the other mAChR subtypes with nanomolar affinities.

TABLE 39.3 Muscarinic Receptors

Subtype	M_1	M_2	M_3	M_4	M_5
G protein	$G_{q/11}$	$G_{i/o}$	$G_{q/11}$	$G_{i/o}$	$G_{q/11}$
Signal transduction	↑ IP_3, DAG	↓ cAMP	↑ IP_3, DAG	↓ cAMP	↑ IP_3, DAG
	↑ $[Ca^{2+}]i$	↑ K^+ channel	↑ $[Ca^{2+}]i$	↑ K^+ channel	↑ $[Ca^{2+}]i$
	↑ cAMP	↓ Ca^{2+} currents	↑ cAMP	↓ Ca^{2+} currents	↑ cAMP
Agonists	ACh, CCh	ACh, CCh	ACh, CCh	ACh	ACh
Selective antagonists	Pirenzepine MT7, 4-DAMP	Tripitramine AFDX384	4-DAMP	MT3 darifenacin	4-DAMP

CCh, carbachol; DAG, diacylglycerol; IP_3, inositol phosphate; ↑, increase; ↓, decrease MT3 and MT7, green mamba snake toxins; 4-DAMP, (1,1-dimethylpiperidin-1-ium-4-yl) 2,2-di(phenyl)acetate; AFDX384, N-(2-[(2R)-2-[(dipropylamino)methyl]piperidin-1-yl]ethyl)-6-oxo-5H-pyrido[2,3-b] [1,4]benzodiazepine-11-carboxamide.

DRUGS ACTING ON CHOLINERGIC RECEPTORS

Drugs Active on Nicotinic Receptors

Nicotinic Agonists Nicotinic agonists have few clinical uses. Besides ACh, nicotine is the principal agonist of nicotinic receptors. Nicotine is an alkaloid extracted from *Nicotiana tabacum*, a plant of the Solanaceae family native to Central America. Nicotine, absorbed mostly through cigarette smoke, exerts complex stimulatory and inhibitory actions on the central and autonomous nervous systems and is strongly addictive. Nicotine is administered through gums, transdermal patches, and electronic cigarettes for treating nicotine dependence. Its use has also been proposed for cognitive deficits in Alzheimer's dementia, in Gilles de la Tourette syndrome, and in Crohn's ileitis, but the clinical results are limited. Other natural agonists of nAChRs are cytisine, a partial agonist of β2-containing receptors; epibatidine, a potent but not very selective agonist of α4β2 and α3β4 receptors; and choline, a weak agonist of α7 receptors. Many agonists have been also synthesized. Few are therapeutic agents: carbachol, a choline ester acting on both nAChRs and mAChRs, and varenicline (7,8,9,10-tetrahydro-6,10-methano-6H-pyrazino[2,3-h] [3]benzazepine), a cytosine derivative that is a partial agonist of α4β2 receptors and is used to treat nicotine addiction. Some new α7 agonists are in advanced clinical experimentation for the treatment of cognitive deficits in schizophrenia and Alzheimer's dementia.

Nicotinic Antagonists Nicotinic antagonists are classified in ganglioplegic and neuromuscular blocking agents or curares. Among the ganglioplegic, only trimetaphan and mecamylamine are still prescribed. Trimetaphan is used in severe hypertensive crisis and for inducing controlled hypotension to reduce bleeding in neurosurgery. Mecamylamine is a secondary amine that crosses the blood–brain barrier and blocks the α3β4 nAChR. It was initially prescribed as an antihypertensive agent; it is now proposed for the treatment of nicotine addiction.

The nondepolarizing, competitive antagonists D-tubocurarine, pancuronium, and atracurium and the agonist succinylcholine (a depolarizing curare) are used as adjuvant in anesthesia in order to obtain muscle relaxation during surgery (see Box 39.1).

Drugs Active on Muscarinic Receptors

Muscarinic Agonists There are several natural muscarinic agonists, besides ACh (Fig. 39.4); among them, muscarine, arecoline, and pilocarpine are alkaloids extracted from the fly agaric (*A. muscaria*), the betel nut (*Areca catechu*), and the leaves of *Pilocarpus jaborandi*, respectively. Many mAChR agonists have been obtained by synthesis. The first were choline esters including bethanechol, carbachol, and methacholine. Carbachol is more active on nAChRs than mAChRs. Pilocarpine is used as a miotic and to stimulate salivation in Sjögren syndrome patients. Cevimeline, a more recent M3 agonist, is also prescribed for the latter indication. Pilocarpine, carbachol, and aceclidine are used as miotics and in the narrow-angle glaucoma. Bethanechol may be used to treat urinary retention, whereas methacholine is administered in some forms of tachycardia. The central effects of arecoline are widely exploited by South Asian populations who chew both betel nuts and leaves in order to obtain light euphoria associated with increased salivation. Clinical trials on arecoline and the synthetic muscarinic agonists xanomeline and cevimeline as symptomatic treatments for cognitive deficits in Alzheimer's dementia have failed. All muscarinic agonists induce parasympathomimetic side effects whose severity depends on the dose.

Muscarinic Antagonists Muscarinic antagonists are also called parasympatholytics since they inhibit the effect of parasympathetic stimulation by reversibly occupying ACh binding sites. They are also called anticholinergics. The two main parasympatholytic drugs are atropine, an alkaloid extracted from *Atropa belladonna*, and scopolamine, which

BOX 39.1 THERAPEUTIC TARGETS AND DRUG CHOICE CRITERIA

Drug class	Main therapeutic target	Cellular effects	Clinical effects	Indications	Notes and choice criteria
AChE inhibitors	AChE	AChE and BChE inhibition; increased [ACh] in synaptic cleft	Activation of peripheral and central cholinergic transmission; miosis, stimulation of smooth muscle motility, bradycardia, facilitation of neuromuscular transmission	Glaucoma, anti-cholinergic poisoning, myasthenia gravis, paralytic ileum, urinary bladder atony, Alzheimer's disease	Physostigmine, edrophonium, neostigmine, pyridostigmine, ambenonium, donepezil, rivastigmine, galantamine

Physostigmine under the form of eye drops is used for the symptomatic therapy of narrow-angle glaucoma; by injection, it is used as an antidote to poisonings by atropine and other muscarinic receptor blockers. **Edrophonium** is a short-acting cholinesterase inhibitor used as a tool for the diagnose of myasthenia gravis (*ex adjuvantibus*). **Neostigmine, pyridostigmine**, and **ambenonium** do not cross the blood–brain barrier and are administered orally for the symptomatic therapy of myasthenia gravis. **Neostigmine**, injected or orally, is used in cases of paralytic ileum to reactivate peristaltic activity and in urinary bladder atony. Parenteral neostigmine is used to shorten curarization induced by competitive, nondepolarizing neuromuscular blockers. **Donepezil, rivastigmine**, and **galantamine** are indicated for the symptomatic treatment of cognitive deficits in Alzheimer's disease. **Donepezil** is a selective AChE inhibitor, **rivastigmine** inhibits both AChE and BChE, and **galantamine** is an AChEI endowed with some agonistic activity at nicotinic receptors. In spite of these pharmacological differences, the clinical efficacy is similar, usually short lasting and mostly confined to a delay of disease progression.

Potassium channel blockers	Neuronal K+ channels	Action potential prolongation and increased ACh release	Facilitation of synaptic transmission at cholinergic synapses	Multiple sclerosis	Fampridine (4-aminopyridine)
Muscle relaxant	SNARE proteins in cholinergic nerve endings	Inhibition of ACh exocytosis from synaptic vesicles	Long-lasting impairment of peripheral cholinergic transmission	Blepharospasm, strabismus, spasticity, hyperhidrosis, neurogenic detrusor overactivity, cosmetic medicine (facial rejuvenation)	Onabotulinumtoxin A, abobotulinumtoxin A, botulinum toxin B

Fampridine stimulates ACh release from cholinergic nerve endings and is prescribed for the symptomatic treatment of the motor disturbances of multiple sclerosis. The different **botulinum toxins** are injected locally in a number of medical situations in which it is beneficial to inhibit ACh release. In cosmetic medicine, it is used to reduce facial wrinkles. Botulinum toxins are highly toxic and cause severe food-borne poisonings and side effects due to diffusion from the injection sites.

CNS drugs	Neuronal nicotinic receptors	Activation/block of receptors (partial agonism)	To mimic central nicotine action	Nicotine addiction	Varenicline
Muscle relaxant	Muscle nicotinic receptors	Activation of receptors, depolarization of end plaque	Fasciculation followed by short-lasting muscle relaxation	Muscle relaxation (curarization) of ultrashort duration during general anesthesia	Succinylcholine
Muscle relaxant	Muscle nicotinic receptors	Receptor blockade (competitive antagonism)	Muscle relaxation	Muscle relaxation during general anesthesia	D-tubocurarine, pancuronium, atracurium, vecuronium, rocuronium
Ganglionic blocker	Neuronal nicotinic receptors of ganglion type	Receptor blockade (competitive antagonism)	Blockade of sympathetic and para-sympathetic ganglionic transmission (gangioplegic action)	Hypertensive crisis, controlled hypotension to reduce bleeding in neurosurgery, nicotine addiction	Trimetaphan, mecamylamine

Varenicline is a recently registered partial agonist of $\alpha_4\beta_2$-subtype nicotinic receptors. It reduces nicotine craving and pleasurable effects and is prescribed to help in quitting smoking. **Succinylcholine** is a depolarizing curare. It is a dicholine ester hydrolyzed by BChE, the reason of its short duration of action. Of the nondepolarizing, competitive neuromuscular blockers, D-**tubocurarine** and **pancuronium** induce a long-lasting curarization, and **atracurium**, **vecuronium**, and **rocuronium** a curarization of intermediate duration. The curarization can be interrupted by administering neostigmine. Rocuronium shows a rapid onset of action, and the action can be interrupted by **sugammadex** administration, a cyclodextrin that removes the curare from the receptors. **Trimetaphan camsilate** is the only ganglioplegic still available for controlling hypertensive crisis and inducing a controlled hypotension in neurosurgery. **Mecamylamine**, initially used as an antihypertensive drug, is now prescribed to control nicotine addiction since it crosses the blood–brain barrier and blocks neuronal nicotine receptors.

Parasympatho-mimetics	Muscarinic receptors	Nonselective and selective activation of muscarinic receptors	Activation of parasympathetic system and cholinergic sympathetic system: miosis, salivation, bradycardia, increased gastric secretion, increased peristalsis, contraction of bladder detrusor muscle, bronchoconstriction, increased bronchial secretions	Narrow-angle glaucoma, postoperative abdominal distension, gastric atony, postoperative urinary retention, chronic neurogenic bladder, xerostomia post head and neck radiation and in Sjögren syndrome, tachycardia	Pilocarpine, bethanechol, carbachol, methacholine, xanomeline, cevimeline
Parasympatholytics	Muscarinic receptors	Selective and non-selective blockade of peripheral and central muscarinic receptors inhibition of parasympathetic system and cholinergic sympathetic system and central muscarinic receptors	Mydriasis, cycloplegia, reduced salivation (dry mouth); tachycardia, bronchodilation and reduced bronchial secretions, reduced gastric secretion, reduced gastrointestinal tone and motility, decreased urinary bladder and ureter tone and motility, antispastic action on gallbladder and sphincters, reduced sweating; drowsiness, amnesia, fatigue	Iridocyclitis, keratitis, examination of the eye fundus; excessive salivation; peptic ulcer, intestinal hypermotility and spasms; biliary colics; bradycardia of various causes; over-active urinary bladderdisease; acute rhinitis, bronchial asthma, chronic obstructive pulmonary disease; motion sickness, Parkinson's disease, drug-induced parkinsonism; poisonings by muscarine and cholinesterase inhibitors	Atropine, homatropine; scopolamine, atropine methyl nitrate, butylscopolamine, pirenzepine; tolterodine, solifenacin; ipratropium and tiotropium, benztropine, biperiden, procyclidine, diphenhydramine

Muscarinic agonists. **Pilocarpine**, eye drops, is used in open-angle glaucoma; **bethanechol** given orally or subcutaneously is used in postoperative abdominal distension, gastric atony, postoperative urinary retention, and chronic neurogenic bladder; **pilocarpine** and **cevimeline** (M3 agonist) are prescribed for increasing salivation in xerostomia including Sjögren syndrome; **methacholine** is now rarely used as peripheral vasodilator and in tachycardia because duration and intensity of its action are unpredictable.

Muscarinic antagonists. Eye drops of **atropine** and **homatropine** are used to induce mydriasis and cycloplegia for the examination of the eye fundus and the treatment of iridoclyclitis and keratitis; **pirenzepine** can be used in gastric ulcer, but more effective therapy is available; **atropine methylnitrate** and **butylscopolamine**, which do not cross the blood–brain barrier, are used as spasmolytics in biliary and urinary cholics and to reduce gastrointestinal hypermotility; for the treatment of overactive urinary bladder disease, the M3-selective **tolterodine** and **solifenacin** are preferred; **ipratropium** and **tiotropium** by inhalation are used for bronchial asthma and chronic obstructive pulmonary disease. **Scopolamine** patches are applied to prevent motion sickness; **benztropine**, **biperiden**, **procyclidine**, and **diphenhydramine** are the anticholinergic used for treating Parkinson's disease when other therapies cannot be used because of side effects and for antipsychotic-induced extrapyramidal movements. **Atropine** and **scopolamine** by parenteral administration are used as adjuvants in anesthesia to inhibit secretions and vagal reflexes.

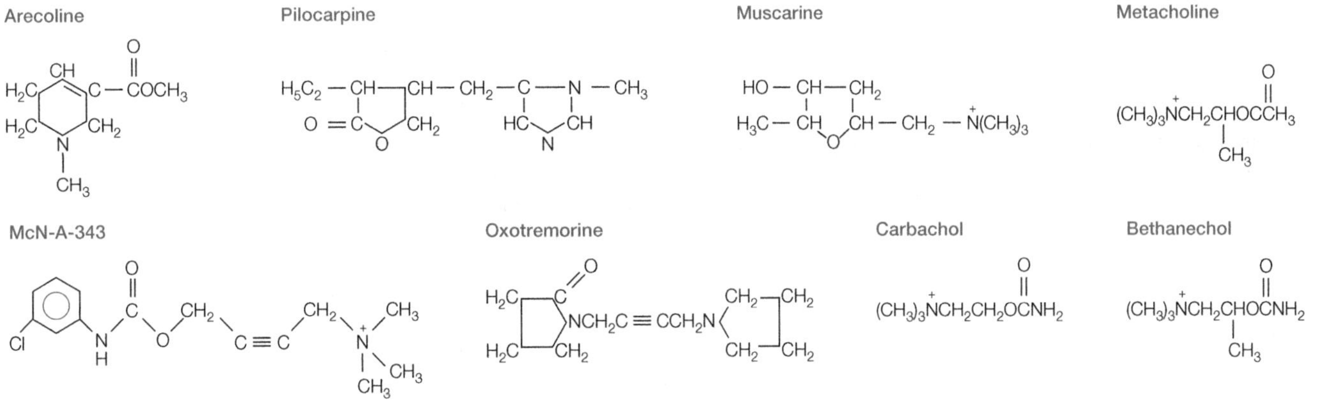

FIGURE 39.4 Muscarinic receptor agonists.

is found in *Hyoscyamus niger*, *Scopolia carniolica*, and *Datura stramonium*, plants of the Solanaceae family. There are no substantial differences between the actions of the two drugs, although scopolamine is somewhat more active on the CNS. The quaternary derivatives of alkaloids, atropine methylnitrate and methyl- and butylscopolamine, do not cross the blood–brain barrier and exert peripheral actions only. Atropine, scopolamine, and their derivatives are nonselective antagonists of mAChRs and block all their subtypes. Synthetic muscarinic antagonists showing selectivity for some receptor subtypes have been obtained. Among them, pirenzepine is an M1 antagonist used to treat gastrointestinal pathologies, and tiotropium is a bronchoselective M1 and M3 antagonist. Muscarinic antagonists are used in gastroenterology, cardiology, urology, neurology, pneumology, and ophthalmology (see Box 39.1). Their main side effects are dry mouth, mydriasis with blurred vision, urinary retention, reduced intestinal motility, increased heart rate, amnesia, and, at high doses, hallucinations and delirium. Side effects are more frequent in aging subjects. It should be remembered that many histamine H1 receptor antagonists, tricyclic antidepressants, and phenothiazines also block muscarinic receptors and may induce atropine-like side effects.

TAKE-HOME MESSAGE

- Acetylcholine is synthesized in nerve endings and stored in vesicles; after release, it is rapidly degraded by the highly widespread enzyme acetylcholinesterase (AChE).
- Acetylcholine receptors are classified as nicotinic (nAChRs) and muscarinic (mAChRs). nAChRs are ligand-gated ion channels, and mAChRs are G-protein-coupled receptors.
- nAChRs are present in the postsynaptic membrane of the neuromuscular junction (muscle AChR) and in synaptic and extrasynaptic sites of PNS and CNS neurons. Muscle and neuronal AChRs can also be distinguished for their specific sensitivities to selective drugs.

- Five mAChRs (named M1–M5) have been identified and cloned; M1, M3, and M5 are mostly (but not exclusively) coupled to G_q and thus activate PLC and IP3/DAG production; M2 and M4 are mostly coupled to $G_{i/o}$ and thus inhibit cAMP production and activate K^+ currents.
- Drugs activating the cholinergic system include cholinomimetic drugs, which can be used to increase salivary secretion and intestinal motility and to improve cognitive impairment in some brain pathologies (Alzheimer's disease), and AChE inhibitors, which can be subdivided in reversible and nonreversible, the latter being frequently used as pesticides with potential toxicity for humans.
- Drugs inhibiting cholinergic transmission are classified as parasympatholytic, which block muscarinic receptors of the parasympathetic system (atropine and analogs), and neuromuscular blockers (curare) and ganglioplegics, which block nicotinic receptors at neuromuscular junctions and ganglia, respectively. Drugs inhibiting cholinergic transmission have widespread indications in gastroenterology, cardiology, pneumology, neurology, and anesthesiology.

FURTHER READING

Albuquerque E. X., Pereira E. F. R., Alkondon M., Rogers S. W. (2009). Mammalian nicotinic acetylcholine receptors: from structure to function. *Physiological Reviews*, 89, 73–120.

Bennaroch E. E. (2013). Acetylcholine in the cerebral cortex: effects and clinical implications. *Neurology*, 75, 659–665.

Fadel J. R. (2011). Regulation of cortical acetylcholine release: insights from in vivo microdialysis studies. *Behavioural Brain Research*, 221, 527–536.

Giacobini E., Pepeu G. *The brain cholinergic system in health and disease*, Informa Healthcare, Abingdon, 2006.

Goldstein D. S. (2013). Differential responses of components of the autonomic nervous system. *Handbook of Clinical Neurology*, 117, 13–22.

Gotti C., Clementi F., Fornari A., Gaimarri A., Guiducci S., Manfredi I., Moretti M., Pedrazzi P., Pucci L., Zoli M. (2009). Structural and functional diversity of native brain neuronal nicotinic receptors. *Biochemical Pharmacology*, 78, 703–711.

Picciotto M. R., Higley M. J., Mineur Y. S. (2012). Acetylcholine as a neuromodulator: cholinergic signaling shapes nervous system function and behavior. *Neuron*, 76, 116–129.

Quik M., Zhang D., Perez X. A., Bordia T. (2014). Role for the nicotinic holinergic ystem in movement disorders; therapeutic implications. *Pharmacology & Therapeutics*, 144, 50–59.

Rosas-Ballina M., Tracey K. J. (2009). Cholinergic control of inflammation. *Journal of Internal Medicine*, 265, 663–679.

Sarter M., Parikh V., Howe W. M. (2009). Phasic acetylcholine release and the volume transmission hypothesis: time to move on. *Nature Reviews. Neuroscience*, 10, 383–390.

Servent D., Fruchart-Gaillard C. (2009). Muscarinic toxins: tools for the study of the pharmacological and functional properties of muscarinic receptors. *Journal of Neurochemistry*, 109, 1193–1202.

Wess J., Eglen R. M., Gautam D. (2007). Muscarinic acetylcholine receptors: mutant mice provide new insights for drug development. *Nature Reviews. Drug Discovery*, 6, 721–733.

Wessler I., Kirkpatrick C. J. (2008). Acetylcholine beyond neurons: the non-neuronal cholinergic system in humans. *British Journal of Pharmacology*, 154, 1558–1571.

40

THE SEROTONERGIC TRANSMISSION

MAURIZIO POPOLI, LAURA MUSAZZI, AND GIORGIO RACAGNI

By reading this chapter, you will:

- Know the neuroanatomy of the serotonergic system in CNS and PNS and its role in cerebral, cardiovascular, gastrointestinal, genitourinary, endocrine, and nociceptive functions
- Know the molecular mechanisms that regulate serotonin synthesis, vesicular accumulation and release, catabolism, and signal turnoff mechanisms
- Learn classification, molecular structure, and signal transduction of membrane receptors activated by serotonin and their localization and physiological functions
- Know the role of serotonin in brain pathologies and the effects of pharmacological compounds modulating serotonergic transmission

---

The discovery of 5-hydroxytryptamine (5-HT) is attributed to Italian pharmacologist Vittorio Erspamer who, in the 1930s, observed that chromaffin cells of the intestinal mucosa contained significant amounts of a substance of unknown structure but showing a range of biological activities. He called this substance enteramine. In subsequent studies, the group headed by I.H. Page characterized a serum factor with vasoconstrictor activity that was called serotonin. Afterward, it was discovered that enteramine and serotonin are the same molecule, 3-(β-aminoethyl)-5-hydroxyindole, more commonly called 5-HT.

FUNCTIONS AND DISTRIBUTION OF THE SEROTONERGIC SYSTEM IN THE BODY

About 95% of the approximately 10 mg of 5-HT present in the body is synthesized in enterochromaffin cells of the intestinal tract and released in blood, where it is captured by platelets. Platelets accumulate 5-HT in secretory granules and release it during aggregation. Furthermore, 5-HT is produced by neurons in the central nervous system (CNS) and peripheral nervous system (PNS). For many years, 5-HT has been mainly studied for its role in the modulation of behavior and in the pathophysiology of neuropsychiatric disorders. However, it is now clear that it is involved in the modulation of a wide range of physiological and pathophysiological processes throughout the body (Fig. 40.1), including regulation of platelet aggregation, functions of cardiovascular and respiratory systems, peristalsis of the gastrointestinal system, blood sugar and energy balance, bone metabolism, sexual function, sleep–wake cycle, control of pain, and nociceptive system.

5-HT in the Nervous System

Modulation of Behavior and Neuropsychiatric Diseases Although neurons producing and releasing 5-HT are no more than 300,000 throughout the CNS (out of a total of about 100 billions), this neurotransmitter virtually modulates all processes related to psychobiological behavior, including mood, aggression, perception, attention, gratification, memory, appetite, and sexuality. The cell bodies of serotonergic neurons are primarily located in the brain stem at the level of the bulb, pons, and midbrain, where they are particularly concentrated in raphe nuclei (Fig. 40.2). The projections of serotonergic neurons from raphe nuclei diffusely innervate different areas of the telencephalon, diencephalon, spinal cord, and cerebellum. The ventrolateral medulla and raphe magnus project to the spinal cord and innervate especially laminae I and II of the dorsal horn; they also innervate the main sensory nucleus and spinal trigeminal nucleus where serotonergic pathways

510 THE SEROTONERGIC TRANSMISSION

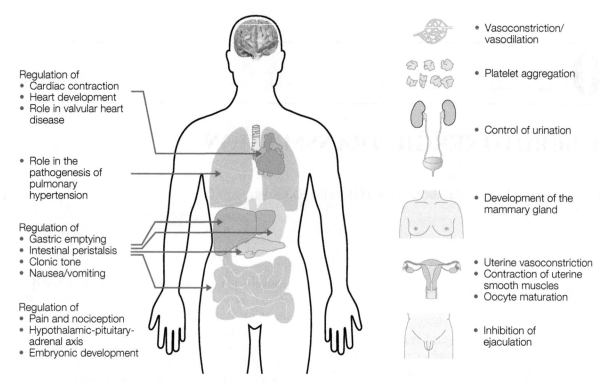

FIGURE 40.1 Role of serotonergic system in the modulation of physiologic and pathophysiological processes.

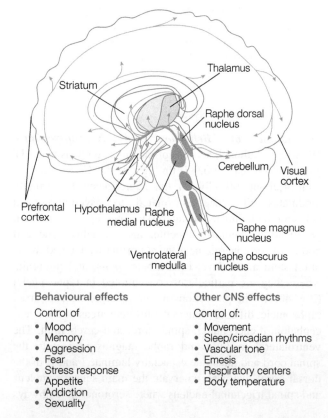

FIGURE 40.2 Main serotonergic pathways in human central nervous system and their effects.

are involved in the control of nociceptive pathways. In addition, the nucleus raphe magnus projects to the periaqueductal gray nucleus, the locus coeruleus, and other noradrenergic nuclei involved in supraspinal modulation of pain perception. The innervation of the cerebellar nuclei and inferior olivary nucleus by serotonergic neurons of the medial and dorsal raphe nuclei explains the role of this mediator in timing and coordination of motor sequences. All raphe nuclei project to various areas of the cortex, thalamus, amygdala, basal ganglia, nucleus pallidus, and nucleus accumbens. Serotonergic innervation is also present in the hippocampus, hypothalamus, and neurohypophysis. Serotonergic neurons often contain other neurotransmitters, such as galanin, substance P, and enkephalins, and also express the enzymes that synthesize nitrogen monoxide. This makes the framework of serotonergic transmission highly complex. Furthermore, the serotonergic system is closely linked, both anatomically and functionally, with other neurotransmitter systems; thus, the effects of drugs active on a system are generally accompanied by changes in the functional state of another. The variety and importance of psychobiological processes modulated by 5-HT, the relevance of drugs active on the serotonergic system in psychiatry (in particular, the 5-HT reuptake inhibitors), and, finally, the fact that different substances of abuse interfere with the function of this system have led to the conclusion that there is no disturbance of the psychic sphere in which the serotonergic system is not involved. It is widely recognized that 5-HT is involved in the pathophysiology of psychiatric disorders, particularly depression, anxiety, obsessive–compulsive disorder, psychosis, drug addiction, and eating disorders. In close connection with this role of 5-HT, many drugs acting on serotonergic system are clinically used for the treatment of psychiatric disorders, particularly depression and anxiety (see Supplement E 40.1, "Drugs Acting on Serotonergic System"). Seven families of 5-HT receptors, with at least 14 different subtypes, have been cloned and pharmacologically characterized (see the following). For many of them, the involvement in pathophysiology has been investigated; for example, anxious behavior seems to be regulated primarily by receptors 5-HT_{1A} and 5-HT_{2C}. 5-HT_{2C} receptor has also been implicated in the regulation of metabolism and in addiction. Since different behaviors are regulated by several serotonin receptors, drugs with affinity for one specific receptor act on different behavioral processes.

Lesions of the serotonergic system in experimental animals produce aggressive and uninhibited behaviors. For example, KO mice for 5-HT_{1B} receptor develop aggressive behavior (Table 40.1), which is observed also in mice and men carrying a mutation in the gene encoding monoamine oxidase type A (MAO-A), the main enzyme responsible for 5-HT catabolism. These observations have suggested that the serotonergic system has the important function to suppress aggressive behavior or other activities that involve psychomotor adverse consequences. It is widely recognized that 5-HT deficiency is associated with aggressive behavior in both rodents and primates. Clinical studies on suicidal patients showed a labile serotonergic system, accompanied by reduced levels of serotonin transporters in CNS, increased expression of 5-HT_{1A} and 5-HT_{2A} receptors, and lower levels of 5-HT and 5-hydroxy-3-indoleacetic acid (5-HIAA, a metabolite of 5-HT) in the prefrontal cortex and cerebrospinal fluid. Moreover, 5-HT has inhibitory effects on sexual behavior (mainly due to modulation of sex hormones production) and modulates sensitivity to pain and appetite, body temperature, mood, and vomiting control. 5-HT is also an important regulator of the sleep–wake system, controlling both the mechanisms of awakening and those of sleep, depending on the receptors activated in selected neural structures. Thus, drugs interfering with the activity of serotonergic pathways have important effects on sleep, which may be positive (reduction of insomnia in depressed patients) or negative (induction of insomnia during treatment with selective serotonin reuptake inhibitors (SSRIs)) (see Supplement E 40.3, "Pineal Gland, Melatonin, Serotonin, and Regulation of Circadian Rhythms").

Pain Control and Anesthesia The serotonergic system controls pain perception at various levels in both CNS and PNS. Inflammation causes local release of 5-HT, which sensitizes peripheral nerve fibers carrying nociceptive information to the CNS. In addition, midbrain raphe nuclei send descending projections to the spinal cord reducing input of nociceptive information. Finally, raphe nuclei send ascending projections modulating pain perception to cortical and limbic regions of the brain. It has been demonstrated that patients suffering from mood disorders show increased perception of pain, associated with alterations in serotonergic modulation of pain perception. The complex involvement of the serotonergic system in modulation of pain explains the effectiveness of serotonergic drugs in the treatment of pain. Some tricyclic antidepressants are used for the treatment of pain, and triptans relieve headache by activating thalamic receptors (see Supplement E 40.2, "Serotonergic System and Modulation of Pain Perception").

5-HT and the Cardiovascular System

The effects of 5-HT on the cardiovascular system are multiple and complex (Fig. 40.1), mediated by different receptors with considerable differences between species. In humans, 5-HT controls hemostasis and platelet function, blood pressure, and cardiac function. Platelets lack the enzymes responsible for 5-HT synthesis, but they have a transporter capable of picking up the neurotransmitter present in blood. Indeed, almost all 5-HT present in blood is stored in platelets. When platelets are activated by specific signals, such as those produced during hemorrhage (vascular endothelium damage, ischemia), they release various factors, including 5-HT, which promote hemostasis. 5-HT promotes aggregation and thrombus formation by binding 5-HT_{2A} receptor on platelets. Moreover, 5-HT mainly exerts a vasoconstrictor effect on blood vessels. The effects on vascular tone are complex and dependent on different mechanisms such as (i) activation of 5-HT_{1A} receptors localized on neurons of vasomotor nuclei, mediating hypotension; (ii) inhibition of noradrenaline release from orthosympathetic terminations; and (iii) activation of nitric oxide (NO) production by endothelial cells (probably through activation of 5-HT_2 receptors). Various receptors are involved in the effects on vascular tone (5-HT_{1B}, 5-HT_{2A}, 5-HT_{2B}, 5-HT_4, 5-HT_7), but the specific role of each receptor is still under study. It should be noted that, in the coronary district, 5-HT induces constriction of large- and medium-caliber vessels and dilation of small arteries. Selective 5-HT reuptake inhibitors (SSRI; see Supplement E 40.1), the class of antidepressants most widely used in clinic, can reduce platelet aggregation and increase bleeding duration because they inhibit 5-HT reuptake and accumulation by platelets. Conversely, numerous observations suggest that treatment with SSRIs may reduce risk of myocardial infarction.

In addition, 5-HT exerts its effects on cardiac function through the raphe nuclei neurons. By activating sympathetic and parasympathetic mechanisms, 5-HT exerts chronotropic

TABLE 40.1 Serotonin receptors: main effectors and phenotypes of corresponding knockout mouse (when available)

Name	Effector	Distribution	Function	Knockout mouse (phenotype)
$5\text{-}HT_{1A}$	$G_{i/o}$; ↓ AC; ↑ PLC; ↑ K^+ channel; ↓ Ca^{2+} channel	Hippocampus, entorhinal and cingulate cortex, septum, thalamus, amygdala, olfactory bulb, raphe, hypothalamus, bone, kidney, gastro-intestinal tract (GT)	Autoreceptors in raphe serotonergic neurons; postsynaptic receptors in limbic areas; ↑ acetylcholine release in GT	↑ Responsiveness to five 5-HT; ↑ anxious phenotype; ↑ response to stress; ↑ sleep; ↓ learning and long-term memory
$5\text{-}HT_{1B}$	$G_{i/o}$; ↓ AC; ↑ PLC	Basal ganglia (substantia nigra, globus pallidus, caudate, putamen), frontal cortex, raphe, hippocampus; cerebral arteries	Autoreceptors and heteroreceptors on GABA, acetylcholine and glutamate neurons	↑ Aggression; ↓ short-term and long-term memory; ↑ response to cocaine
$5\text{-}HT_{1D}$	$G_{i/o}$; ↓ AC; ↑ PLC	Basal ganglia (substantia nigra, globus pallidus, caudate, putamen), raphe, spinal cord	Autoreceptors and heteroreceptors on GABAergic, acetylcholine and glutamate neurons	
$5\text{-}HT_{1E}$	$G_{i/o}$; ↓ AC	Olfactory cortex, caudate, putamen, hippocampus, raphe, frontal cortex and cingulate, claustrum, amygdala, hypothalamus, thalamus	Autoreceptors (?)	
$5\text{-}HT_{2A}$	$G_{q/11}$; ↑ PLCβ	Cerebellum, septum, hypothalamus, amygdala	Heteroreceptors	Lethal
$5\text{-}HT_{2C}$	$G_{q/11}$; ↑ PLCβ	Choroid plexus, hippocampus, amygdala, piriform and cingulate cortex, olfactory bulb, thalamic and sub-thalamic nuclei, substantia nigra	Somatodendritic heteroreceptors on dopamine, GABA, acetylcholine and glutamate neurons	Obesity; spontaneous seizures; ↓ learning and memory; ↑ response to cocaine
$5\text{-}HT_3$	Cation channel	Dorsal vagal complex, hippocampus, amygdala, cerebral cortex, PNS, GT, cardiovascular system	Heteroreceptors on GABA, acetylcholine and glutamate neurons	↓ Nociception ($5\text{-}HT_{3A}$)
$5\text{-}HT_4$	G_s; ↑ AC; ↓ K^+ channel	Septum, hippocampus, amygdala, striatum, substantia nigra, GT, cardiovascular system	Somatodendritic heteroreceptors on GABA, acetylcholine and glutamate neurons	↓ Stress response; ↑ response to convulsant agents
$5\text{-}HT_5$	$G_{i/o}$; ↓ AC	Hippocampus, hypothalamus, olfactory bulb, cerebral cortex, thalamus, striatum, pons, cerebellum	Heteroreceptors on GABAergic neurons (?)	↑ Exploration activity; ↓ response to LSD
$5\text{-}HT_6$	G_s; ↑ AC	Striatum, amygdala, nucleus accumbens, hippocampus, olfactory bulb, cerebral cortex	Heteroreceptors on GABA, acetylcholine and glutamate neurons	Altered alcohol response
$5\text{-}HT_7$	G_s; ↑ AC	Cardiovascular system, intestinal smooth muscle, thalamus, hippocampus, cerebral cortex, amygdala, suprachiasmatic nucleus	Heteroreceptors on glutamate and GABA neurons	"Antidepressant-like" phenotype

and inotropic effects on the heart. Activation of $5\text{-}HT_{1A}$ receptors induces sympathetic inhibition and bradycardia, while activation of $5\text{-}HT_{2A}$ induces sympathetic activation, with increase of pressure and tachycardia. The $5\text{-}HT_3$ receptor, expressed on vagal afferent endings, mediates an effect known as "von Bezold–Jarisch reflex," which represents one of the possible causes of syncope (neurocardiogenic syncope), characterized by bradycardia and hypopnea. The reflex can be triggered by vasodilation of peripheral veins with a compensatory increase of cardiac contractility and consequent activation of a paradoxical vagal response, which produces a rapid and transient bradycardia, followed by a hypotensive response and tachycardia. In particular, tachycardia seems to be due to activation of cardiac 5-HT$_4$ receptors, mediating positive inotropic and chronotropic effects in the myocardium.

The effects of stimulation of 5-HT$_4$ receptors on the myocardium have been suggested also by detection of atrial fibrillation in patients with tumors inducing overproduction of 5-HT and by increased expression of 5-HT$_4$ receptors in heart failure. 5-HT also plays a role in the pathogenesis of valvular heart disease; this has been discovered using a combined treatment with fenfluramine and fenfluramine/phentermine (feh/phen), an anorexigenic drug cocktail withdrawn from the market because of its dangerous side effects.

5-HT in Gastrointestinal and Genitourinary Systems

The functions of the serotonergic system in the gastrointestinal tract (GT) are complex, nonhomogeneous, and very different from species to species. In humans, 5-HT regulates GT functions at various levels, including gastric emptying, secretions, and intestinal peristalsis (Fig. 40.1). When food enters the GT, it is pushed along the intestine by peristaltic waves modulated by 5-HT released from enterochromaffin cells in response to various signals (acetylcholine, sympathetic stimulation, increased intraluminal pressure, lowering of the pH). 5-HT activates receptors both on intrinsic (located within the thickness of GT) and extrinsic neurons (outside the GT). The first induce peristaltic contractions and secretory reflexes, while the latter allow communication between sympathetic and parasympathetic systems and GT. These connections are involved in nausea, vomiting, pain in GT, as well as in side effects of drugs that increase 5-HT availability (SSRIs; see Supplement E 40.1). 5-HT also controls nausea and vomiting at cerebral level, in particular through 5-HT$_3$ receptors (see the following). Moreover, 5-HT released in GT regulates secretion of pancreatic enzymes according to state and content of GT.

The serotonergic system plays an important role also in the control of genitourinary functions. 5-HT increases ejaculation latency and delays orgasm through stimulation of 5-HT$_{2C}$ receptors and 5-HT$_{1B}$ while exerting an opposite effect via 5-HT$_{1A}$ receptor. The net effect is a delay in ejaculation and orgasm. 5-HT modulates urination in a similar way to ejaculation. The spinal circuits that control bladder activity are intensely innervated by serotonergic fibers. Activation of these fibers leads to reduction of parasympathetic tone and activation of the sympathetic system, with consequent increase of bladder capacity due to a decrease of its emptying.

5-HT in Metabolism and Endocrine System

The serotonergic system is also important in central control of energy balance and in modulation of hypothalamic–pituitary–adrenal (HPA) axis (Fig. 40.1). It is believed that hypothalamic 5-HT$_{2C}$ receptors have a central role in regulation of glucose homeostasis and energy balance. For this reason, 5-HT$_{2C}$ agonists have been proposed for the treatment of obesity and diabetes. Some evidence has suggested that 5-HT has a hypoglycemic effect, mediated by 5-HT$_{2A}$ receptors expressed in skeletal muscle, due to increased levels of glucose transporter in the plasma membrane. The levels of 5-HT are low in subjects suffering from type 1 diabetes, and it has been suggested that this 5-HT deficiency may contribute to the increased incidence of psychiatric disorders associated with the disease.

Moreover, the serotonergic system regulates secretion of steroid hormones and prolactin, regulates HPA axis at various levels, and exerts complex effects on stress response. There is also evidence of a serotonergic regulation of bone metabolism. Indeed, it has been demonstrated that osteoblasts and osteocytes express 5-HT transporters and receptors and that SSRIs, by increasing 5-HT availability, decrease bone mass.

SYNTHESIS AND METABOLISM OF SEROTONIN

The 5-HT precursor is the amino acid L-tryptophan, which is provided from the diet and transported into cells by the neutral amino acid transporter. Since neutral amino acids compete for the transporter, L-tryptophan bioavailability depends on the diet; carbohydrate diets promote L-tryptophan synthesis. Also other amino acids may modulate cerebral concentration of tryptophan, competing for the transport at the blood–brain barrier. 5-HT synthesis begins with the enzymatic hydroxylation of the precursor and the formation of hydroxytryptophan (5-OH-tryptophan) by tryptophan hydroxylase (Fig. 40.3). This enzyme is not inhibited by its reaction product, and its activity is positively modulated by stimulation of serotonergic neurons. Moreover, tryptophan hydroxylase catalyzes the first step in the synthesis of melatonin, the neurotransmitter produced in the pineal gland, which plays a key role in the sleep–wake cycle regulation (see Supplement E 40.3). Many isoforms of this enzyme have been described, some of which potentially associated with predisposition to suicide. 5-OH-tryptophan can be administered as exogenous compound and converted into 5-HT. The efficiency of 5-OH-tryptophan is higher compared to L-tryptophan, probably because it is not used in other metabolic processes or for protein synthesis. 5-OH-tryptophan is then decarboxylated to 5-HT by an aromatic amino acid decarboxylase (Fig. 40.3). Tryptophan hydroxylase and synthesis of 5-HT can be permanently blocked by parachlorophenylalanine, which is exclusively used as experimental drug. Typically, drugs that interfere with 5-HT synthesis are useful in experimental models, but have no clinical interest.

The main catabolic pathway of 5-HT starts with its oxidation to 5-hydroxy-3-indolacetaldehyde by MAO-A, localized in mitochondria within nerve endings (Figs. 40.3 and 40.4, step 11); this aldehyde is immediately transformed

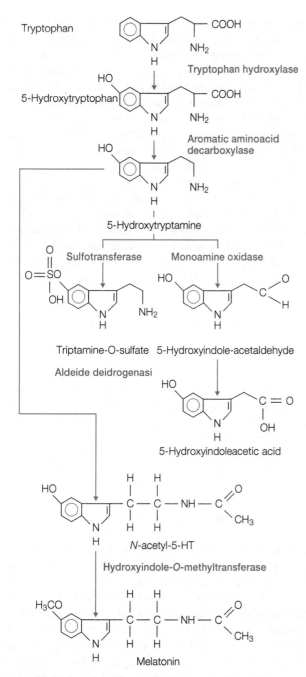

FIGURE 40.3 Synthesis of 5-hydroxytryptamine.

into 5-HIAA by aldehyde dehydrogenase. Accumulation of 5-HIAA in cerebrospinal fluid is indicative of 5-HT turnover. About one-third of 5-HT urinary metabolites come from its transformation in O-sulfate by a sulfotransferase.

VESICULAR STORAGE, RELEASE, AND EXTRACELLULAR CLEARANCE OF SEROTONIN

In neurons, newly synthesized (or reuptaken by membrane transporter) 5-HT is concentrated in vesicles at synaptic terminals (step 5 in Fig. 40.4) by a transporter similar to the one

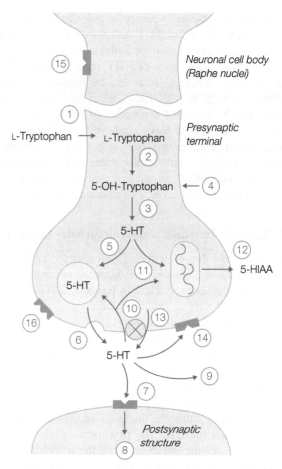

FIGURE 40.4 A serotonergic neuron and its pharmacological targets. 1, Uptake of L-tryptophan; 2, hydroxylation of L-tryptophan to 5-OH-tryptophan (inhibited by parachlorophenylalanine); 3, decarboxylation of 5-OH-tryptophan to 5-HT; 4, uptake of 5-OH-tryptophan; 5, accumulation of 5-HT in synaptic vesicles (inhibited by reserpine and tetrabenazine); 6, release of 5-HT by exocytosis; 7, interaction with postsynaptic receptors; 8, intracellular signaling pathways; 9, detachment from the receptor and passage into blood vessels; 10, reuptake in the synaptic terminal mediated by the selective transporter (inhibited by fluoxetine, chlorimipramine, fluvoxamine, etc.); 11, oxidative deamination by MAO (inhibited by moclobemide); 12, elimination of the deaminated metabolite 5-HIAA; 13, release of 5-HT mediated by the transporter (blocked by reuptake inhibitors; see step 10); 14, presynaptic autoreceptor; 15, somatodendritic autoreceptor (activated by 5-HT$_{1A}$ agonists such as gespirone, buspirone, and ipsapirone); and 16, presynaptic heteroreceptors for different neurotransmitters.

used for norepinephrine and dopamine vesicular accumulation. The natural alkaloids reserpine and tetrabenazine inactivate these vesicular transporters and cause depletion of 5-HT, noradrenaline, and dopamine in their synaptic terminals. The action of reserpine is longlasting because of its irreversible binding to the transporter. Reserpine has been used as antipsychotic and antihypertensive drug, but its side effects, sometimes very serious and mostly due to its lack of selectivity against biogenic amines, have made reserpine a drug with an illustrious past but presently used only as a research tool.

The physiological release of 5-HT from nerve terminals occurs by calcium-dependent exocytosis from presynaptic vesicles. The exocytotic release of 5-HT can be positively or negatively modulated by other neurotransmitters, which activate presynaptic nonserotonergic receptors (presynaptic heteroreceptors; Fig. 40.4, step 16) located on 5-HT-releasing synaptic terminals. Muscarinic cholinergic heteroreceptors and GABAergic α_2-adrenergic receptors have been found on serotonergic synaptic terminals. Also presynaptic serotonergic receptors (autoreceptors) modulate 5-HT release (Fig. 40.4, step 16): 5-HT$_{1D}$ receptor inhibits 5-HT release in the substantia nigra and basal ganglia, while 5-HT$_{1A}$ somatodendritic autoreceptor decreases the firing of serotonergic neurons in the raphe. These presynaptic receptors are potential targets for new drugs able to modulate serotonergic transmission.

The 5-HT Reuptake System

The concentration of 5-HT released in the synaptic space is reduced by two processes: (i) dilution in the extracellular spaces and passage into blood vessels and (ii) selective reuptake in nerve terminals by a specific membrane transporter. The transporter belongs to the same gene superfamily of aminergic transporters (see Chapter 29). Its binding to 5-HT occurs in the thickness of the plasma membrane, in a sort of pocket formed by hydrophobic residues. The affinity of the binding site is modulated by Na$^+$ and Cl$^-$, which are cotransported within the cell. The V_{max} of the transporter can be modulated by protein kinase A phosphorylation. Following activation, the transporter can be internalized, allowing the cell to regulate the number of transporters on its surface. The transporter is able to both reuptake and release 5-HT (steps 10–13 of Fig. 40.4), a phenomenon underlying the mechanism of action of indirect agonist drugs. Many recent genetic studies have shown that a variant of the 5-HT transporter encoding gene is a vulnerability factor for some psychiatric diseases and regulates the response to treatment with SSRIs.

5-HT Indirect Agonists: Effects on Appetite and Drug Addiction Some drugs, such as amphetamine and similar substances of abuse, as 3,4-methylenedioxy-N-ethylamphetamine (MDEA) or 3,4-methylenedioxy-N-methylamphetamine (MDMA, known as "ecstasy"), cause 5-HT release from synaptic terminals. The release occurs via the transporter located in the plasma membrane of terminals, which is normally used for 5-HT reuptake but can also work in the opposite direction (see Chapter 29). Ecstasy causes a massive release of dopamine and 5-HT. In rodents, the drug causes a delayed and impressive neurotoxic effect on serotonergic neurons (which can appear months after the administration of the drug), an effect that can be prevented by administration of blockers of 5-HT reuptake. Recent experiments have shown that administration of MDMA in primates (monkeys) for only four days induces degeneration and death of serotonergic fibers in the brain cortex and limbic areas, suggesting deep damage of serotonergic neurons. These lesions are not completely recovered up to 7 years after administration.

5-HT Reuptake Inhibition and Antidepressant Drugs Tricyclic and nontricyclic antidepressants are able to inhibit 5-HT reuptake by reversibly binding a site on the extracellular side of the transporter. Actually, many tricyclic antidepressants (amitriptyline, imipramine, nortriptyline) can also inhibit the activity of transporters for other monoamines. SSRIs (fluoxetine, fluvoxamine, sertraline, paroxetine, citalopram, and escitalopram) are nontricyclic antidepressants with different selectivity for the serotonin transporter. All these drugs are used in the treatment of mood and anxiety disorders and other psychiatric illnesses (see Supplement E 40.1).

CLASSIFICATION OF SEROTONIN RECEPTORS

The first demonstration of the existence of at least two types of 5-HT receptors was obtained by Gaddum and Picarelli in 1957, although the first work hypothesizing the existence of 5-HT receptor subtypes had been published by these authors in 1954. Since then, numerous studies have led to the identification of seven receptor families with at least 14 different subtypes, characterized by specific structural, biochemical, or pharmacological features (Table 40.1).

The 5-HT1 receptor family includes 5-HT$_{1A}$, 5-HT$_{1B}$, 5-HT$_{1D}$, 5-HT$_{1E}$, and 5-HT$_{1F}$. Their sequence homology is 40–60%. They are G-protein-coupled receptors, mainly coupled with $G_{i/o}$, and their activation leads to reduction of cAMP intracellular levels. The receptor subtypes 5-HT$_{1B}$ and 5-HT$_{1D}$ are homologous and selectively expressed in rodents and humans, respectively, but with different pharmacological properties. Their activation limits release of 5-HT and other transmitters from synaptic terminals.

The 5-HT$_2$ receptor family consists of three subtypes: 5-HT$_{2A}$, 5-HT$_{2B}$, and 5-HT$_{2C}$. The sequence homology among these receptors is 45–50%, and they are $G_{q/11}$-protein-coupled receptors. Their activation induces increased levels of IP$_3$ and release of intracellular calcium ion (see Chapter 12).

5-HT$_3$ receptors are pentameric ligand-gated channels, structurally similar to nicotinic acetylcholine receptors. This family includes the isoforms 5-HT$_{3A}$, 5-HT$_{3B}$, 5-HT$_{3C}$, 5-HT$_{3D}$, 5-HT$_{3E}$, and 5-HT$_{3Ea}$. Furthermore, the 5-HT$_{3A}$ subunit, through alternative splicing, can give rise to at least four different isoforms (5-HT$_{3A(a)}$, 5-HT$_{3A(b)}$, 5-HT$_{3AT}$, 5-HT$_{3AL}$). 5-HT$_3$ receptors are located on central and peripheral neurons, where they induce rapid depolarization by influx of sodium and calcium ions and efflux of potassium.

Experimental evidence has shown that $5\text{-}HT_3$ functional receptors are heteromeric, all containing one $5\text{-}HT_{3A}$ subunit ($5\text{-}HT_{3A}/5\text{-}HT_{3B}$, $5\text{-}HT_{3A}/5\text{-}HT_{3C}$, $5\text{-}HT_{3A}/5\text{-}HT_{3D}$, $5\text{-}HT_{3A}/5\text{-}HT_{3E}$, $5\text{-}HT_{3A}/5\text{-}HT_{3Ea}$).

The $5\text{-}HT_4$ receptor family consists of eight isoforms: $5\text{-}HT_{4(a)}$, $5\text{-}HT_{4(b)}$, $5\text{-}HT_{4(c)}$, $5\text{-}HT_{4(d)}$, $5\text{-}HT_{4(f)}$, $5\text{-}HT_{4(g)}$, $5\text{-}HT_{4(n)}$, and $5\text{-}HT_{4(hb)}$. They are G_s protein-coupled receptors that positively modulate adenylate cyclase leading to increased cAMP levels. The $5\text{-}HT_{5A}$ and $5\text{-}HT_{5B}$ isoforms of the $5\text{-}HT_5$ receptor family have a sequence homology of 70–80% and are coupled to $G_{i/o}$ proteins. Their activation determines a reduction in cAMP intracellular levels.

Receptors belonging to the $5\text{-}HT_6$ and $5\text{-}HT_7$ families ($5\text{-}HT_{7A}$, $5\text{-}HT_{7B}$, $5\text{-}HT_{7C}$, $5\text{-}HT_{7D}$) are coupled to G_s proteins and increase intracellular levels of cAMP via activation of adenylate cyclase.

Because of the continuous evolution of this topic, continuous update is necessary.

DRUGS ACTING ON SEROTONIN RECEPTORS

The pharmacology of serotonergic receptors is constantly evolving. For some receptors, isoform-selective agonists and antagonists have been developed (Table 40.2), some of which showing clinically relevant pharmacological properties. However, the majority of drugs active on serotonergic system have a mixed receptor profile (Table 40.3). Although 5-HT is particularly important in the regulation of cardiovascular function, to date, no serotonergic drugs have been approved for the treatment of cardiovascular diseases, except for ketanserin (Table 40.3). Main reasons for this are the complexity of the serotonergic regulation, involving at least 14 different receptor subtypes not yet fully characterized, and the absence of highly selective ligands for some 5-HT receptors.

Pharmacology of the $5\text{-}HT_1$ Receptors

Partial Agonists of $5\text{-}HT_{1A}$ Receptors have Anxiolytic Properties $5\text{-}HT_{1A}$ receptors act as somatodendritic autoreceptors in raphe nuclei, from which the majority of serotonergic CNS pathways originates. The activation of autoreceptors causes a reduction of the discharge frequency of raphe nuclei. The consequent inhibition of 5-HT seems to be responsible for the anxiolytic effects of some partial agonists of $5\text{-}HT_{1A}$ receptors (buspirone, gepirone, ipsapirone) and their side effects. Indeed, stimulation of $5\text{-}HT_{1A}$ receptors causes hyperphagia and hypothermia. In addition, these receptors play an important role in modulation of cognitive function and negative symptoms of schizophrenia: some atypical antipsychotics (aripiprazole, clozapine, quetiapine, olanzapine) are $5\text{-}HT_{1A}$ partial agonists. Finally, receptor agonists are currently being developed for the treatment of cognitive disorders associated with Alzheimer's disease and mood disorders.

TABLE 40.2 Selective receptor agonists and antagonists

Name	Selective agonists	Selective antagonists
$5\text{-}HT_{1A}$	8-OH-DPAT, 8-OH-PIPAT, pirons (partial agonists)	p-MPPF, p-MPPI, WAY 100635, WAY 101405
$5\text{-}HT_{1B}$	CP 93129, CP 94253, GR 46611	SB 216641, SB 224289, GR 55562 ($5\text{-}HT_{1B}$ and $5\text{-}HT_{1D}$ potent agonist)
$5\text{-}HT_{1D}$	PNU 109291, PNU 142633	LY 310762, BRL 15572
$5\text{-}HT_{1E}$	BRL 54443 ($5\text{-}HT_{1E}$ and $5\text{-}HT_{1F}$ agonist)	
$5\text{-}HT_{1F}$	LY 334370, LY 344864, BRL 54443 ($5\text{-}HT_{1E}$ and $5\text{-}HT_{1F}$ agonist)	
$5\text{-}HT_{2A}$	DOI, DOB, DOM	4F 4PP, M 100907, R 96544, sarpogrelate
$5\text{-}HT_{2B}$	α-Me-5-HT, BW 723C86	LY 272015, RS 127445, SB 204741
$5\text{-}HT_{2C}$	MK 212, Ro 60-0175, CP 809191, WAY 163909	RS 102221, SB 242084, lorcaserin
$5\text{-}HT_3$	SR 57227, MD-354, RS 56812	Bemesetron, dolasetron, granisetron, itasetron, ondansetron, zotasetron
$5\text{-}HT_4$	Benzamides (saprides), benzimidazolones	GR 113808, GR 125487, RS 67532
$5\text{-}HT_5$		SB 699551
$5\text{-}HT_6$	E-6801, WAY 181187, WAY 208466	Ro 04-6790, SB 258585, SB 399885
$5\text{-}HT_7$	AS 19, LP 12, LP 44	SB 258719, SB 2699710

Agonists of $5\text{-}HT_{1B}$, $5\text{-}HT_{1D}$, $5\text{-}HT_{1E}$, and $5\text{-}HT_{1F}$: Use in Migraine Triptans (rizatriptan, sumatriptan, zolmitriptan) are $5\text{-}HT_{1B}$, $5\text{-}HT_{1D}$, $5\text{-}HT_{1E}$, and $5\text{-}HT_{1F}$ receptor agonists used in the treatment of acute migraine. Their therapeutic efficacy is believed to be due to activation of these receptors at the level of sensory trigeminal fibers associated with large meningeal vessels. This would inhibit release of the vasodilator substances (neuropeptides, such as CGRP, or NO), which are held responsible for the onset of migraine attacks (see Supplement E 40.2).

Pharmacology of the $5\text{-}HT_2$ Receptors

$5\text{-}HT_{2A}$ receptors are believed to modulate the biochemical and behavioral responses to drugs of abuse. For example, LSD, a powerful hallucinogen, although acting on a large

TABLE 40.3 Drugs acting on the serotonergic system

Drugs	Pathology	Pharmacological properties
Buspirone, gepirone, ipsapirone	Anxiety	5-HT$_{1A}$ partial agonists
Sumatriptan, zolmitriptan, rizatriptan, naratriptan, almotriptan, eletriptan, frovatriptan	Migraine	5-HT$_{1B}$, 5-HT$_{1D}$, 5-HT$_{1E}$, 5-HT$_{1F}$ agonists
Ritanserin	Extrapyramidal disorders	5-HT$_{2A}$ and HT$_{2C}$ antagonist
Metisergide	Migraine	5-HT$_2$ antagonist
Sarpogrelate	Atherosclerosis	5-HT$_{2A}$ antagonist
Ketanserin	Hypertension	5-HT$_{2A}$ e HT$_{2C}$, α1-adrenergic, histaminergic H1 antagonist
Dolasetron, granisetron, ondansetron, palonosetron, tropisetron, zotasetron	Chemotherapy-induced emesis	5-HT$_3$ antagonists
Prucalopride	Constipation	5-HT$_4$ agonist
Atypical antipsychotics (chlorpromazine, haloperidol)	Schizophrenia	D$_2$, D$_4$, 5-HT$_{2A}$, HT$_{1A}$ antagonist
Atypical antipsychotics (aripiprazole, clozapine, quetiapine, olanzapine, risperidone, sertindole, ziprasidone)	Schizophrenia	Mixed receptor profile (dopaminergic, serotonergic, histaminergic, adrenergic)
Mirtazapine	Depression	α1-adrenergic, 5-HT$_2$, 5-HT$_3$, 5-HT$_7$ antagonist
Fluoxetine, sertraline, citalopram, escitalopram, fluvoxamine, paroxetine	Depression	Selective 5-HT reuptake inhibitors
Agomelatine	Depression	Melatonergic MT1 and MT2 agonist, 5-HT$_{2C}$ antagonist
Clomipramine	Depression	5-HT, noradrenaline, dopamine reuptake inhibitor; 5-HT$_{2A}$, 5-HT$_{2C}$, 5-HT$_3$, 5-HT$_6$, 5-HT$_7$, dopaminergic, α1-adrenergic and histaminergic H1 antagonist
Amitriptyline	Depression	5-HT and noradrenaline reuptake inhibitor; 5-HT$_{2A}$, 5-HT$_{2C}$, 5-HT$_6$, 5-HT$_7$, α1-adrenergic and histaminergic H1 antagonist
Nortriptyline	Depression	Noradrenaline and 5-HT reuptake inhibitor; Histaminergic H1, 5-HT$_2$ and α1-adrenergic antagonist
Amisulpride	Depression	Dopaminergic and HT$_7$ antagonist

number of receptors (serotonergic, dopaminergic, adrenergic), exerts its effect mainly through agonism at 5-HT$_{2A}$ receptors. In addition, several typical (chlorpromazine, haloperidol) and atypical antipsychotics (risperidone, sertindole, clozapine, olanzapine, quetiapine) and some antidepressants (mirtazapine, fluoxetine, agomelatine) have an antagonistic action on 5-HT$_2$ receptors (see Supplement E 40.1). Ritanserin, an antagonist at 5-HT$_{2A}$ and 5-HT$_{2C}$ receptors, is used in combination therapy with typical antipsychotics to reduce extrapyramidal effects. Moreover, also methysergide, used for the treatment of migraine, blocks 5-HT$_2$ receptors at therapeutic doses.

5-HT$_{2A}$ receptors are widely distributed in the periphery, where they induce platelet aggregation and increased capillary permeability. The 5-HT$_{2A}$ selective antagonist sarpogrelate is clinically used as antiplatelet agent for the treatment of arteriosclerosis. The antihypertensive activity of ketanserin has been thought for years to result from antagonism at 5-HT$_{2A}$ receptors present in bulbar vasomotor nuclei. Actually, ketanserin acts as an antagonist of both 5-HT$_{2A}$ and 5-HT$_{2C}$ and shows high affinity also for α$_1$-adrenergic receptors and histamine H$_1$ receptors.

Pharmacology of the 5-HT$_3$ Receptor

Antagonists of 5-HT$_3$ receptor are antiemetic drugs devoid of extrapyramidal side effects and potential antipsychotics. 5-HT$_3$ receptors are particularly concentrated in brain areas involved in the pharyngeal or gag reflex (nucleus of the solitary tract and area postrema). Setrons (ondansetron, granisetron) are selective 5-HT$_3$ receptor antagonists used in the treatment of pernicious vomiting caused by some chemotherapy drugs. Setrons are not antagonists of dopamine receptors. Moreover, 5-HT$_3$ receptors are not expressed in the nigrostriatal system. Therefore, unlike classic antidopaminergic antiemetics (such as metoclopramide), these drugs do not induce extrapyramidal disorders. Finally, activation of 5-HT_3 receptors facilitates release of several neurotransmitters, including dopamine. Selective antagonists of these receptors are thus potential neuroleptics without extrapyramidal effects.

Pharmacology of the 5-HT_4 Receptors

5-HT_4 receptors are located in the gastrointestinal system and heart, where they regulate peristaltic activity and force of contraction of the cardiac atrial fibers, respectively. In the past, some full agonists (benzamides, cisapride, mosapride) and partial agonists (tegaserod) were used for the treatment of gastrointestinal diseases. However, they have been withdrawn from the market because of their side effects, especially in the cardiovascular system. The only benzamide currently used for the therapy of constipation is prucalopride because of its minor side effects.

5-HT_4 receptors are widely distributed also at central level, where they seem to play a role in learning and memory processes and in control of mood and appetite. These properties make 5-HT_4 receptor a potential target for the development of agonists for the treatment of Alzheimer's disease and depression and antagonists for the treatment of anorexia nervosa.

Pharmacology of the 5-HT_6

Although 5-HT_6 receptors are not located on cholinergic neurons, their selective antagonists increase cholinergic transmission and positively regulate learning and memory processes. For this reasons, they are currently under investigation for the treatment of cognitive dysfunction in Alzheimer's disease and have also been proposed as potential antipsychotics and antidepressants. On the other hand, some antipsychotics (clozapine, olanzapine) and antidepressants (clomipramine, amitriptyline, nortriptyline) show antagonistic activity on 5-HT_6 receptors.

Finally, a 5-HT_6 antagonist is studied for the treatment of obesity.

Pharmacology of the 5-HT_7 Receptors

5-HT_7 receptors, identified only in the early 1990s, have been shown to be involved in regulation of circadian rhythms, thermoregulation, learning, and memory processes and possibly also in psychiatric and neurological disorders, particularly mood disorders. Indeed, many antidepressants (both tricyclics and SSRI) play an antagonistic action on 5-HT_7 receptors and inhibition of this receptor potentiates the effect of antidepressants. In addition, the antidepressant efficacy of amisulpride has been shown to be mainly due to its antagonistic action on 5-HT_7 receptors.

TAKE-HOME MESSAGE

- Serotonin modulates important psychobiological processes including mood, aggression, perception, attention, reward, memory, appetite, and sexuality.
- Peripheral roles of serotonin include modulation of cardiovascular system, gastrointestinal tract, genitourinary system, metabolism, endocrine system, and nociception.
- Serotonin is synthesized from the amino acid L-tryptophan and is stored in vesicles and released from nerve endings by calcium-dependent exocytosis.
- Synaptic levels of serotonin are mainly regulated by diffusion and reuptake by a specific membrane transporter that is the target of some antidepressant drugs.
- The metabolism of serotonin involves its oxidation by mitochondrial MAO-A.
- Seven families of serotonin receptors have been identified: six are GPCRs and one (5-HT_3) is a cationic LGIC.
- Drugs acting on 5-HT receptors have important clinical activities including antimigraine, anxiolytic, antipsychotic, antidepressant, and antiemetic effects.

FURTHER READING

Berger M., Gray J.A., Roth B.L. (2009). The expanded biology of serotonin. *Annual Review of Medicine*, 60, 355–366.

Caspi A., Hariri A.R., Holmes A., Uher R., Moffitt T.E. (2010). Genetic sensitivity to the environment: the case of the serotonin transporter gene and its implications for studying complex diseases and traits. *American Journal of Psychiatry*, 167, 509–527.

Davies P.A. (2011). Allosteric modulation of the 5-HT_3 receptor. *Current Opinion in Pharmacology*, 11, 75–80.

Fakhfouri G., Rahimian R., Ghia J., Khan W., Dehpour A. (2012). Impact of 5-HT_3 receptor antagonists on peripheral and central diseases. *Drug Discovery Today*, 17, 741–747.

Filip M., Bader M. (2009). Overview on 5-HT receptors and their role in physiology and pathology of the central nervous system. *Pharmacological Reports*, 61, 761–777.

Hannon J., Hoyer D. (2008). Molecular biology of 5-HT receptors. *Behavioural Brain Research*, 195, 198–213.

Harmar A.J., Hills R.A., Rosser E.M., Jones M., Buneman O.P., Dunbar D.R., Greenhill S.D., Hale V.A., Sharman J.L., Bonner T.I., Catterall W.A., Davenport A.P., Delagrange P., Dollery C.T., Foord S.M., Gutman G.A., Laudet V., Neubig R.R., Ohlstein E.H., Olsen R.W., Peters J., Pin J.P., Ruffolo R.R., Searls D.B., Wright M.W., Spedding M. (2009). IUPHAR-DB: the IUPHAR database of G protein-coupled receptors and ion channels. *Nucleic Acids Research*, 37(Database issue), D680–D685.

Jonnakuty C., Gragnoli C. (2008). What do we know about serotonin? *Journal of Cellular Physiology*, 217, 301–306.

Millan M., Marin P., Bockaert J., Mannoury la Cour C. (2008). Signaling at G-protein-coupled serotonin receptors: recent advances and future research directions. *Trends in Pharmacological Sciences*, 29, 454–464.

Müller C.P., Homberg J.R. (2015). The role of serotonin in drug use and addiction. *Behavioural Brain Research*, 277, 146–192.

Racagni G., Popoli M. (2010). The pharmacological properties of antidepressants. *International Clinical Psychopharmacology*, 25, 117–131.

Selvaraj S., Arnone D., Cappai A., Howes O. (2014). Alterations in the serotonin system in schizophrenia: a systematic review and meta-analysis of postmortem and molecular imaging studies. *Neuroscience and Biobehavioral Reviews*, 45, 233–245.

Villalon C.M., Centurion D. (2007). Cardiovascular responses produced by 5-hydroxytriptamine: a pharmacological update on the receptor/mechanisms involved and therapeutic implications. *Naunyn-Schmiedeberg's Archives of Pharmacology*, 376, 45–63.

41

HISTAMINERGIC TRANSMISSION

Emanuela Masini and Laura Lucarini

> **By reading this chapter, you will:**
> - Know the tissue distribution of histamine and its synthesis, metabolism, release, and role in physiological processes, including inflammatory reaction, and the nervous, cardiovascular, and gastrointestinal systems
> - Learn the structure and the functional significance of the different histamine receptors
> - Understand how the histaminergic system is the target of drugs with potential therapeutic implications

Histamine, as a chemical curiosity, was synthesized via histidine decarboxylation by Windaus and Vogt in 1907, before its biological significance had been recognized. In 1910, Barger and Dale isolated histamine from rye extract. Subsequently, they found it also in intestinal mucosa and in many other animal tissues, discovering that it stimulated contraction of intestinal smooth muscles and induced vasodilation. Ten years later, its effect on gastric acid secretion was described.

The term "histamine" (from the Greek word *histos* (ιστοσ) that means "tissue") appeared for the first time in 1912 in research articles authored by Fuhner, Frohlich, and Pick. In 1940, Feldberg found that histamine released from mast cells causes bronchial constriction during anaphylactic shock. In 1941, Kwiatkowski detected histamine in the brain, and in 1959, White demonstrated its synthesis and catabolism.

Antihistamines (synthesized by Bovet and Staub in 1937) have been used for allergic diseases for the past 70 years, and their sedative side effects led Monnier in 1967 to define histamine as a "waking substance." Major advances in the biochemistry of the histaminergic system in the brain came in 1970 from the group headed by Jean Charles Schwartz, who located histaminergic neurons and their projections through lesion experiments. Electrophysiological recordings on central neurons as well as behavioral studies upon central infusion of histamine supported the role of histamine as a neurotransmitter. In 1972, Black and coworkers identified the second histamine receptor and revolutionized the treatment of peptic ulcers. Histamine H_2 receptor (H_2R) is also abundant in the central nervous system (CNS), and its central actions were soon detected. The groups of Watanabe and Panula in 1984 brought a breakthrough in research on histaminergic system with the immunohistochemical localization of histaminergic neurons in the tuberomammillary nucleus (TMN) of the posterior hypothalamus and its wide projections using antibodies against histidine decarboxylase (HDC) or histamine. Schwartz's group identified the H_3 autoreceptors that control activity of histaminergic neurons, histamine synthesis and release, and its electrophysiological action. The most recent addition is H_4 receptor (H_4R), whose presence and function in the nervous system has still to be completely defined.

Histamine (2-(imidazol-4-yl)ethylamine) is an important hydrophilic chemical messenger regulating a wide variety of physiological responses. It is produced via histidine decarboxylation by the L-HDC enzyme. The ethylamine chain is also present in other neurotransmitters (dopamine, norepinephrine, and serotonin), whereas the imidazole core, absent in other neurotransmitters, confers distinct chemical properties to histamine, such as tautomerism, a property whereby histamine exists in two different forms. The tautomeric properties of histamine are important for activation of different histamine receptors. At physiological pH, histamine is prevalently in cationic form, with a positively charged amino-terminal group (Fig. 41.1).

General and Molecular Pharmacology: Principles of Drug Action, First Edition. Edited by Francesco Clementi and Guido Fumagalli.
© 2015 John Wiley & Sons, Inc. Published 2015 by John Wiley & Sons, Inc.

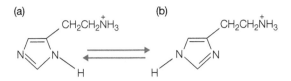

FIGURE 41.1 Chemical structure of histamine in its two tautomeric forms. The imidazole ring has two N atoms available for protonation. The chemical form with a single proton can exist in two tautomeric forms: (a) Nπ-histamine and (b) Nτ-histamine.

Chemical modifications of the imidazole ring and ethylamine structure change histamine receptor affinity and cause different biological effects. The different biological activities of histamine are due to its interaction with four different receptors, named H_1, H_2, H_3, and H_4, all belonging to the G-protein-coupled receptor superfamily (G_q, G_s, and $G_{i/o}$).

DISTRIBUTION AND FUNCTION OF THE HISTAMINERGIC SYSTEM

Histamine is widely (but unevenly) distributed throughout the animal kingdom, and it is present in many venoms, bacteria, and plants. Almost all mammalian tissues contain histamine in amounts ranging from less than 1 to more than 100 μg/g.

Concentrations in plasma and other body fluids are very low, except in human cerebrospinal fluid (CSF). Mast cells are the most important storage sites for histamine (10–20 pg/cell); in particular, the skin, bronchial tree, and intestinal mucosa contain high concentration of histamine, because they contain a large number of mast cells. In blood, histamine is stored in basophils. Mast cells and basophils synthesize histamine and store it in secretory granules. In the gastric mucosa, histamine release from enterochromaffin-like cells (ECL) stimulates gastric secretion by parietal cells, while histamine release upon mast cell degranulation leads to several of the known symptoms of allergic conditions in skin and airway preparations.

At pH 5.5, as found in secretory granules, histamine is positively charged and complexed with negatively charged acidic groups on other constituents of the granules, primarily proteases and heparin or chondroitin sulfate proteoglycans. The turnover rate of histamine in secretory granules is slow, and when tissues rich in mast cells are depleted of their histamine stores, it may take weeks before histamine concentrations return to normal levels.

Other sites of histamine formation or storage include epidermis, gastric mucosa, neurons within the CNS, and cells in regenerating or rapidly growing tissues. Turnover is rapid at these non-mast cell sites because histamine is synthesized and released "on demand" rather than stored. Non-mast cell sites of histamine production contribute significantly to daily excretion of histamine metabolites in urine. Histamine, ingested or produced by gastrointestinal bacteria, is rapidly metabolized and its metabolites are eliminated in urine.

Histaminergic Neurons in the CNS

In the brain, histamine has a tissue concentration of 0.5 nmol/g but is unevenly distributed. In humans, the highest histamine concentration is found in the hypothalamus. In the CNS, this amine is synthesized by a restricted population of neurons located in the TMN of the posterior hypothalamus. These neurons project diffusely to most cerebral areas and have been implicated in various brain functions in mammals.

There are two major pathways of histamine metabolism in humans. The most important involves ring methylation to form N-methylhistamine, catalyzed by histamine-N-methyltransferase, which is widely distributed. Most N-methylhistamine formed is then converted to N-methylimidazoleacetic acid by monoamine oxidase (MAO); this reaction can be blocked by MAO inhibitors. Histamine may undergo oxidative deamination catalyzed by the enzyme diamine oxidase (DAO), yielding imidazole acetic acid, which is then converted to imidazoleacetic acid riboside. These metabolites have little or no activity and are excreted in urine.

Specific enzymatic systems allow synthesis and catabolism of cerebral histamine, and their distribution is similar to that of endogenous histamine. Histamine has to be synthesized locally because it hardly crosses the blood–brain barrier. For this, cerebral histamine is localized in synaptic sites, together with the enzyme L-HDC. Experimental studies with α-fluoromethyl-histidine, an HDC inhibitor, have highlighted the presence of two different histamine pools: a neuronal pool rapidly synthesized and released and a nonneuronal pool, in vascular system and mast cells, with a longer half-life. Cerebral mast cells are located along vessels and regulate blood flux, vessel permeability, and immunological reactivity. In the dura mater, mast cells are localized near sensory nerve fibers and regulate release of inflammatory mediators. Cerebral mast cells seem to be involved in neurodegenerative diseases, such as multiple sclerosis, Alzheimer's disease, or Wernicke encephalopathy. In the past, the role of cerebral histamine was inferred only from the sedative and orexigenic effects of classic antihistamine H_1 drugs and typical and atypical antipsychotic drugs and antidepressants. Those side effects were due to blockade of histamine H_1 receptors (H_1R), which play a pivotal role in the regulation of sleep–wake cycle and food assumption.

Despite this evidence, many H_1-antihistamine drugs normally used in clinics are not employed in sleep therapy, because of their long half-life and side effects. Histamine H_1 agonists are not used to treat insomnia and eating disorders, because they are not sufficiently permeable through the blood–brain barrier and devoid of peripheral side effects.

The pathophysiological role of cerebral histamine was recognized at the beginning of the 1980s, when histamine H_3

receptor (H_3R)—an autoreceptor controlling the activity of histaminergic neurons—was discovered and antibodies for histamine and HDC became available.

The presence of histaminergic neurons, diffusely projecting to widely divergent cerebral areas and arising from the TMN of the posterior hypothalamus, has been highlighted in rodents and other animal species using electrophysiological and immunochemical techniques. Histaminergic neurons have been found also in human fragments from autopsy in TMN of the posterior hypothalamus, sending projections to the septal area and to the medial preoptic cortex.

In the CNS, histamine is considered a modulating neurotransmitter (Table 41.1). In all species studied so far, bipolar or multipolar cell bodies with a diameter of 20–30 μm are found only in TMN, where the presence of HDC mRNA has been demonstrated. Many histamine pathways have been identified in large cerebral areas using fluorescent antibodies against HDC and retrograde fluorescent labeling methods. The highest density has been found in the hypothalamus, medial septal area, and ventral tegmental area. An intermediate density has been observed in the cortex, striatum, hippocampus, amygdala, and substantia nigra. A low concentration of histaminergic fibers has been observed in the thalamus, cranial nerve nuclei, and ventral septal area (Fig. 41.2).

Functions of the Histaminergic System

Figure 41.3 shows the most important physiological roles of histamine.

Histamine in the CNS

Histamine is involved in the regulation of homeostatic mechanisms of the hypothalamus: histaminergic neurons in TMN of the posterior hypothalamus and their projections to the cortex are involved in the regulation of sleep–wake mechanism via H_1R. H_1R knockout mice show a deficit in locomotor system and in explorative behavior. Histamine is involved in emotional control and cognitive process through modulation of cholinergic transmission; significant low histamine levels have been found in the hypothalamus, hippocampus, and temporal cortex of deceased Alzheimer's disease patients compared to subjects deceased because of other pathologies. Histamine is involved in regulation of body temperature and neuroendocrine functions; indeed, it modulates secretion of vasopressin, oxytocin, prolactin, adrenocorticotropic hormone (ACTH), and β-endorphin via H_1R and H_2R. Reduction of cerebral histamine content, via inhibition of its synthesis with α-fluoromethyl-histidine, causes inhibition of ACTH release in surrenalectomized rat.

The involvement of histamine in autoimmune diseases and neuroinflammation has been recently reported. The recent discover of histamine H_4R in dendritic cells, lymphocytes, and mast cells seems to be important for the involvement of this receptor in pathophysiological interactions between the nervous and immune systems.

Histamine regulates food and water consumption and body weight control. A high concentration of histamine receptors is present in the ventromedial nucleus of the hypothalamus, where the satiety center is supposed to be. Moreover, leptin, a physiological inhibitor of appetite, acts through activation of histamine H_1R. Indeed, cyproheptadine, an antiserotonergic drug with antihistaminic actions, is used to stimulate appetite.

Modifications in the number and morphology of histaminergic neurons have been highlighted in neurological pathologies, such as multiple sclerosis, Alzheimer's disease, Down syndrome, and Wernicke encephalopathy. Histamine released from neurons and mast cells could be involved in these pathologies by modifying vascular functions, blood–brain barrier, and immune system. Cell culture studies on

TABLE 41.1 Effects of histamine on other neurotransmitters

Neurotransmitter	Effect	Receptor involved
Ach	↓ Release	H_3
Ach	↑ Release	H_2
CGRP	↓ Release	H_3
DA	↓ Release	H_3
GABA	↓ Turnover	H_3
NE	↓ Release	H_3
NE	↑ Release	H_1
5-HT	↓ Release	H_3

5-HT, 5-hydroxytryptamine; ↑, increase; ↓, decrease; Ach, acetylcholine; CGRP, calcitonin gene-related peptide; DA, dopamine; GABA, γ-aminobutyric acid; NE, norepinephrine.

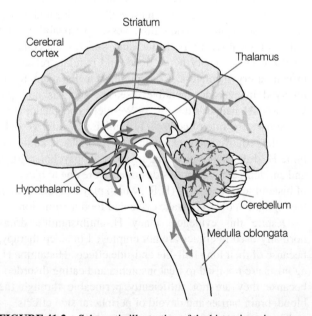

FIGURE 41.2 Schematic illustration of the histaminergic system in the human brain. The arrows indicate fibers projecting from cell bodies.

Histamine in the Cardiovascular System

Intradermal injection of histamine causes the so-called triple response of Lewis, which consists of erythema, local edema, and peripheral hyperemia. These effects involve three different cell types: smooth muscles of microcirculation, capillary and venular endothelium, and sensitive nervous fibers. Similar local effects can be induced by injecting histamine-releasing substances in the derma or by applying an antigen on the skin of sensitized subjects. H_1R and H_2R are critical for histamine vascular effects. Histamine increases the expression of adhesion molecules (P-selectin) in the endothelium, stimulating migration of polymorphonucleated cells in tissues. These actions occur via activation of H_1R and H_4R and are important for the pathogenesis of tissue inflammation.

In the heart, histamine determines increase of frequency, contractibility, and coronary flow as well as decrease of atrioventricular conduction velocity. The decrease in vessel tone and the increase in coronary flow determined by histamine are due to nitric oxide (NO) production by endothelial and smooth muscle cells via H_1R and H_2R activation. Histamine effects at cardiac level are multifactorial. The positive chronotropic and inotropic effects of histamine seem to be related to activation of histamine H_2R. The negative dromotropic effect on atrioventricular conduction is due to activation of H_1R, whereas both H_1R and H_2R are involved in coronary flow control.

Effects of Histamine on Platelet Aggregation and Tissue Proliferation Histamine is an intracellular mediator in platelet aggregation and tissue proliferation. Histamine amount increases after platelet exposure to aggregation stimuli and in rapidly growing tissues: this effect is prevented by α-fluoromethyl-histidine, an HDC inhibitor, and by N,N,diethyl-2-[4-(phenylmethyl)phenoxy] ethanamine (DPPE) that blocks at microsomal level the receptor site for estrogens (tamoxifen). A histamine H_4R has been identified in hematopoietic progenitor cells, and recent *in vitro* data show that this receptor controls cell growth and proliferation.

Histaminergic System in the Stomach

Histamine secreted from ECL cells reaches parietal cells and activates H_2R, which is coupled to adenylate cyclase, stimulates intracellular production of adenosine 3′,5′-cyclic monophosphate (cAMP) and subsequent transfer of H^+/K^+ ATPase pump from intracellular tubular–vesicular structures to the plasma membrane. As a result, H^+ ions are released in the gastric lumen, and phosphorylation by cAMP-dependent protein kinase (PKA) increases proton pump activity. Specific receptors for gastrin and cholinergic drugs are identified in parietal cells. Gastrin, secreted from antral G cells, stimulates gastric secretion and proliferation

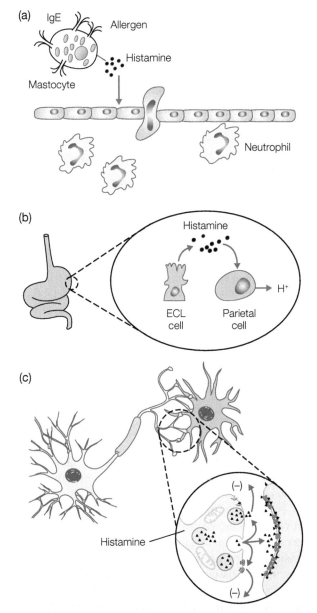

FIGURE 41.3 (a) Histamine is a mediator of inflammation; (b) it is secreted in the stomach by ECL cells and acts on parietal cells stimulating acid secretion; (c) it is stored in neurons and acts as a neurotransmitter at central and peripheral level.

hippocampus cells and *Xenopus* oocytes expressing NR2B and NR1 of *N*-methyl-D-aspartate (NMDA) have shown that histamine is able to enhance the current induced by NMDA receptor activation, and this may explain some of its neurotoxic effects. Nevertheless, in some cerebral ischemia model, histamine seems to exert a neuroprotective action and to stimulate learning processes. For what concern the electrophysiological actions, histamine induces hyperpolarization with slow onset and long lifetime. These entire data highlight the complexity of histamine action in the brain, whose effects vary depending on cerebral pathways, areas, and receptors involved as well as the species.

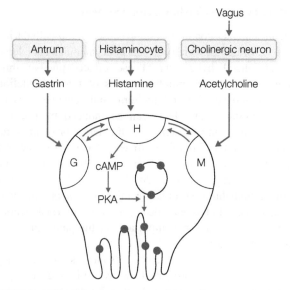

FIGURE 41.4 Schematic representation of gastric secretion. G, gastrin receptor; H, histamine H_2 receptor; large dots, proton pump; M, muscarinic receptor. See Chapter 28 for more information on the role of proton pump in H^+ ion release.

of mucous cells. Its action is mediated by the G-coupled cholecystokinin-B receptor (CCKB or gastrin receptor), which is localized on ECL cells and upon stimulation causes histamine secretion. Secreted acetylcholine from postganglionic neurons directly activates parietal cells through cholinergic M_3 receptor and indirectly through activation of cholinergic M_2 and M_4 receptors, which inhibit somatostatin secretion (Fig. 41.4).

Histamine not only acts as secretagogue but also as mucous membrane growth factor. HDC knockout mice present alterations in cellular differentiation, an effect that is counterbalanced by an increase in gastrin synthesis induced by the decrease in gastric acidity. By contrast, H_2R KO mice do not have alterations in cellular differentiation, despite the presence of hypergastrinemia, suggesting that histamine also acts via histamine H_3R.

Histamine Effects on Smooth Muscles

Histamine causes contraction of intestinal smooth muscle through stimulation of histamine H_1R. Indeed, contraction of the guinea pig ileum is a classic biological test for the amine. H_2R control the release of several neurotransmitters, whereas H_3R have an inhibiting role, modulating neurotransmission at presynaptic level. Histamine causes bronchoconstriction in guinea pig and humans. In guinea pig, bronchoconstriction causes death for histamine toxicity. In humans, in normal conditions, bronchoconstriction evoked by physio–logical concentration of histamine is not pronounced; however, patients affected by bronchial asthma are extremely sensitive to histamine. Contraction of bronchial smooth muscles is mediated by H_1R, but a negative regulation via H_2R is also present: selective antagonists of H_2R do not cause bronchoconstriction but potentiate bronchoconstriction in asthmatic patient. Activation of histamine H_3R on cholinergic nerves and parasympathetic ganglia inhibits substance P release and neurogenic inflammation and modulates bronchoconstriction.

Histamine and the Immune Response

Histamine participates in the complex crosstalk between cytokines and cells involved in differentiation of Th1 and Th2 lymphocytes. The latest discovery of histamine H_4R in mast cells and other inflammatory cells has allowed investigating the therapeutic role of H_4R-selective antagonist in allergic asthma and in acute and chronic inflammatory diseases. H_4R knockout mice show a minor pulmonary inflammation after immunological stimulation and a reduction in Th2-dependent cytokine production. H_4R has been detected in synovial tissue of patients affected by rheumatoid arthritis and osteoarthritis and in ocular tissue of patients affected by uveitis. Differences in H_4R expression correlate with variability in disease duration and severity.

SYNTHESIS AND METABOLISM OF HISTAMINE

Although exogenous histamine could be absorbed by different tissues, histamine present in food or synthesized by intestinal bacteria does not contribute significantly to the endogenous amine pool. Histamine ingested with food is inactivated and turned into N-acetyl-histamine by intestinal bacteria; only a small amount of it is absorbed and rapidly metabolized in hepatic and extrahepatic sites.

Mammalian tissues contain two different enzymes that decarboxylate histidine: an aromatic L-amino acid decarboxylase (AADC), which decarboxylates also other aromatic amino acids and has low affinity for histidine, and an HDC specific for L-histidine, whose activity in many tissues is 100 times lower than that of AADC.

HDC uses pyridoxal phosphate (vitamin B_6) as a cofactor. The purified enzyme is composed of two 55 kDa subunits whose amino acidic sequence is partly similar to that of rat DOPA decarboxylase. Biochemical, biophysical, and immunological studies have revealed the presence of HDC isoenzymes resulting from posttranslational changes and allelic variants.

HDC is particularly abundant in the stomach, mast cells, and histaminergic neurons. In neurons, histamine is accumulated into synaptic vesicles by the vesicular monoamine transporter. HDC activity can be modulated through phosphorylation by cAMP-dependent PKA, but its increase depends on ex novo synthesis, as it is blocked by protein synthesis inhibitors. Some compounds (α-chloromethylhistidine, α-trifluoromethylhistamine) can inhibit HDC; in particular, α-fluoromethyl-histidine (α-FMH) acts as a "suicide" substrate activating an irreversible HDC inhibitor and may have a therapeutic role in some conditions such as mastocytosis.

Histamine Metabolism

In humans, histamine metabolism occurs mainly via N-demethylation of the aromatic ring by imidazole-N-methyltransferase that forms *tele*-methylhistamine and S-adenosyl-L-homocysteine, an inhibitor of this reaction, by transferring the methyl group of S-adenosyl-L-methionine.

Overall, 1–2% of histamine is eliminated as unmodified amine, 8% as *tele*-methylhistamine, 60% as *tele*-methylimidazole acetic acid, and 25% as imidazole acetic acid partly conjugated to ribose (Fig. 41.5).

In the CNS, histamine metabolism is rapid, and neuronal histamine is mostly inactivated by histamine methyltransferase, an enzyme selective for histamine, and by monoamine oxidases B (MAO-B). Cerebral histamine turnover is rapid, having its maximal activity in the hypothalamus (2–10 times faster than in other cerebral regions), and it seems to diminish in stress conditions. Turnover in mast cells and vascular tissues is slow.

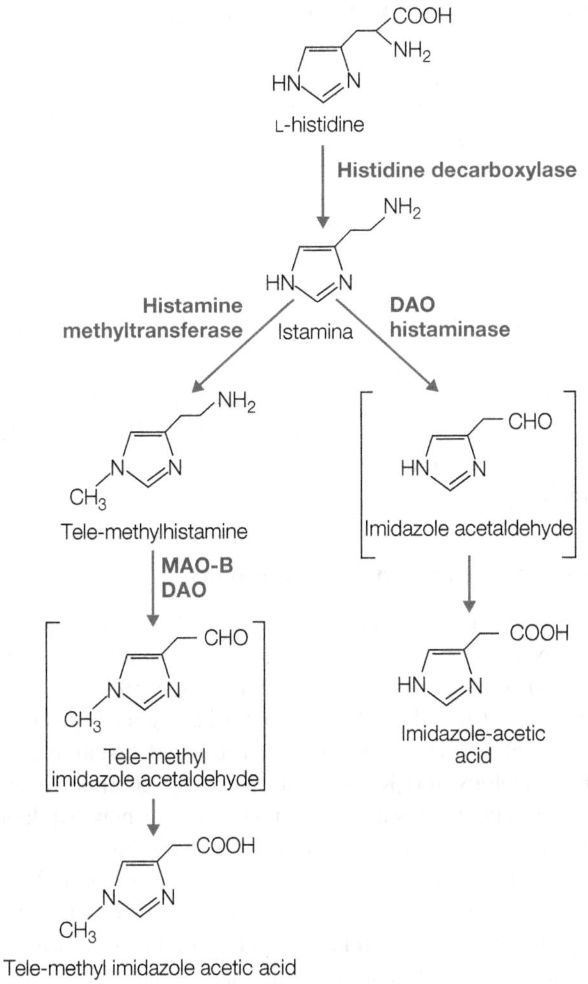

Histamine methyltransferase is found in tuberomammillary neurons along with histamine but also in glia and CSF. The enzyme is competitively inhibited by antimalarial drugs (such as quinacrine and amodiaquine) and pyrimidine analogs (such as metoprine and etoprine) that are not in clinical use, but are important research tools to investigate cerebral histamine.

Most *tele*-methylhistamine is deaminated by MAO-B enzyme to *tele*-methylimidazole acetic acid, the principal urinary metabolite of histamine. Thus, *tele*-methylhistamine is the principal histamine metabolite in the brain.

The second metabolic way of histamine consists in oxidative deamination by DAO (histaminase) to imidazole acetic acid, which is excreted in urine as ribosyl-imidazole acetic acid.

Imidazole acetic acid and its ribosylated metabolite have been detected in some rat brain areas; the former can act as agonist at GABA receptors, while the latter acts on the benzodiazepine site, reducing its activity.

Polyamines, as putrescine, are substrates for histaminase. In humans, high levels of histaminase are present in the gastrointestinal tract, liver, kidney, and placenta. MAO-B enzymes are inhibited by deprenyl, an effective antidepressant, whereas aminoguanidine and semicarbazide are DAO inhibitors but not in clinical use.

STORAGE AND RELEASE OF HISTAMINE

Neuronal and extraneuronal pools of histamine are present in the CNS. In the neuronal pool, histamine is stored within synaptic vesicles and is released upon depolarizing stimuli. Its release depends on extracellular calcium entering the cell through a voltage-dependent Ca^{2+} channel. In the CNS, there is no evidence of high affinity uptake of histamine; this represents important criteria to distinguish histamine from other biogenic amines (serotonin, noradrenalin, dopamine).

In other tissues, most histamine is stored in mast cells and circulating basophils within metachromatic granules along with other active substances, such as heparin, peptides, proteases, and proteoglycans.

Histamine can be released by mast cell granules via two different mechanisms. Endogenous and exogenous compounds (such as compound 48/80, dextran, polymyxin, d-curarine, ionophores as A23187 and X527A, lectins as concanavalin A, neurotensin, substance P) determine histamine secretion by sequential exocytosis. This process involves fusion of perigranular membranes with the plasma membrane and subsequent exposure of granules to the

FIGURE 41.5 Structure, synthesis, and catabolism of histamine.

extracellular environment, where histamine is secreted. Otherwise, histamine can be released through a nonexocytotic process involving lysis of mast cell membrane.

An important pathophysiological mechanism for histamine release is the immunological mechanism that results from interaction of specific antigens with the corresponding IgE exposed on the plasma membrane of tissue mast cells and circulating basophils. Histamine release can occur during immunological IgG- or IgM-mediated reactions after complement activation (Fig. 41.3).

Histamine release can be inhibited by β_2-adrenergic agonists, while it can be enhanced by muscarinic and α-adrenergic agonists.

Moreover, histamine release is inhibited by prostaglandin E that activates adenylate cyclase and by histamine itself via H_2R. Some compounds (sodium cromoglycate and nedocromil) can prevent histamine release; for this reason, they are used in the prophylaxis of allergic disorders, such as asthma.

Pharmacological Modulation of Histamine Metabolism and Release

Many drugs currently in clinical use interfere with histaminergic transmission (Fig. 41.6).

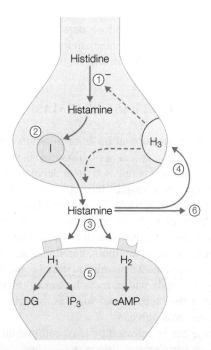

FIGURE 41.6 Histaminergic neuron and its drug targets. (1) Histidine decarboxylation to histamine (inhibited by α-fluoromethyl-histidine); (2) storage of histamine in granules; (3) histamine interaction with postsynaptic H_1R, H_2R, and H_3R; (4) histamine interaction with presynaptic H_3R (inhibition of histamine synthesis and release); (5) transduction systems; and (6) histamine removal from synaptic cleft and metabolic inactivation.

Synthesis. In experimental studies, L-histidine has been shown to increase synthesis of cerebral histamine, but this is not useful in therapy. α-Fluoromethyl-histidine is an irreversible inhibitor for HDC; also, this compound has no clinical use.

Release. The release of neuronal histamine is stimulated by reserpine; many other drugs (especially alkaline drugs, such as d-tubocurarine, morphine, codeine, dextran) trigger histamine release from mast cells; this action can be inhibited by sodium cromoglycate and nedocromil, which are used as antiallergy drugs.

Reuptake. No molecules capable of interfering with histamine reuptake are known.

Catabolism. Antimalarial drugs (quinacrine, amodiaquine), analogs of pyrimidine (metoprine and etoprine), and an acetylcholinesterase inhibitor (tacrine) inhibit histamine-N-methyltransferase. Deprenyl, a MAO-B inhibitor, is an effective antidepressant; aminoguanidine and semicarbazide are DAO inhibitors, but without clinical use.

HISTAMINE RECEPTORS AND THEIR PHARMACOLOGICAL MODULATION

Four different receptors for histamine have been identified: H_1, H_2, H_3, and H_4 (See also Supplement E 41.1 "The Four Histamine Receptors and Their Signal Transduction Pathways"). All of them are G-protein-coupled (metabotropic) receptors. The antagonist [^{3}H]-mepyramine was the first selective H_1R ligand identified. The antagonist [^{3}H]-tiotidine is used as H_2R ligand. The agonist [^{3}H]-R-(a)-methylhistamine and the antagonist [^{3}H]-thioperamide were the first H_3R ligands identified. Later, other radioligands for H_3R were described: [^{125}I]-iodofenpropit and [^{125}I]-iodoprossifan. The first generation of imidazolic ligands for H_3R shows high affinity also for H_4R. The antagonist [^{3}H]-JNJ7777120 and [^{3}H]-histamine are used as radioligands for H_4R (Table 41.2).

Drugs Active on Histamine Receptors

Drugs active on histamine receptors have an important therapeutic role. In 1937, the first antihistamine drug (H_1R antagonist) was discovered. Later, in the 1970s, the discovery of H_2R and the development of antagonists prompted a new interest in the physiological roles of histamine. The pharmacology of H_3R began more recently; compounds with good selectivity toward these receptors are now available, and some of them are currently under clinical trial for cognitive disorders. In addition, antagonists for H_4R are promising candidates for the treatment of inflammatory pathologies, allergic rhinitis, asthma, rheumatoid arthritis, pruritus, and neuropathic pain, and some have already entered clinical trials for allergic diseases.

TABLE 41.2 Pharmacological classification of histamine receptors

R	Location/function	Effector	Agonists	Antagonists	Radioligands
H_1	Smooth muscles of airways, intestine, blood vessels (contraction); CNS (sleep–wake, food intake, thermoregulation); heart (inotropic and chronotropic positive effects)	↑ IP_3/DAG	2-Thiazol ethylamine, 2-phenyl histamine	Mepyramine, triprolidine	[^{3}H]-mepyramine
H_2	Heart (inotropic and chronotropic positive effects); stomach (acid secretion); uterus (contraction); SNC (neuroendocrine processes)	↑ cAMP	Dimaprit, impromidine	Cimetidine, ranitidine, tiotidine	[^{3}H]-tiotidine
H_3	CNS (inhibition of histamine synthesis and release); lung (inhibition of histamine synthesis, inhibition of neurogenic contraction); intestine (inhibition of neurotransmitter release); stomach (inhibition of acid secretion)	↓ cAMP ↓Ca^{2+} influx	R-(α)-methyl-histamine, imetit	Thioperamide, clobenpropit	[^{125}I]-iodofenpropit
H_4	CNS (hyperpolarization of neurons); bone marrow; dendritic cells; spleen; eosinophils, neutrophils, monocytes, macrophages; mast cells; $CD4^+$ cells; $CD8^+$ T cells; fibroblasts	↓cAMP	Imetit, immepip	JNJ7777120, JNJ10191584	[^{3}H]-histamine JNJ7777120

Pharmacology of the H_1R Many H_1 antagonists have a therapeutic use: for example, diphenhydramine mepyramine, antazoline, terfenadine, and loratadine are used as antiallergy drugs. Cyclizine and meclizine are used to treat kinetosis and Ménière syndrome. Promethazine is used as an antiemetic drug and to control stiffness in Parkinson's patients and for its anticholinergic properties. Cyproheptadine stimulates appetite.

H_1R agonists are abundant. Most H_1R antagonists are substituted ethylamine, and their formula is shown below:

$$\begin{array}{c} Ar' \\ \diagdown \\ X-C-C-N \\ Ar \diagup \diagdown \end{array} \begin{array}{c} R \\ R \end{array}$$

R are usually methyl groups; Ar and Ar′ are arylic or heteroarylic rings, conferring high liposolubility to the molecule. X determines if the molecule is an ethanolamine ($X=O$, e.g., diphenhydramine), ethylenediamine ($X=N$, e.g., mepyramine, antazoline, tripelennamine), and alkylamine ($X=C$, e.g., chlorpheniramine, triprolidine). Cyclizine and meclizine are piperazine antihistamines; terfenadine is a piperidine antihistamine; and promethazine is a phenothiazinic antihistamine. Anti-H_1 antihistamine drugs (such as chlorpheniramine, mepyramine, and promethazine) are mostly used for prevention and therapy of allergic reactions.

The most recently developed H_1 antagonists with low sedative effects (fexofenadine, loratadine) are largely used to treat allergic rhinitis and chronic urticaria. Cyclizine and meclizine are used for the treatment of kinetosis and Ménière syndrome.

Sedative effects are due to H_1R blockade. H_1 antagonists, less lipophilic and less capable of crossing the blood–brain barrier and probably with high affinity for peripheral H_1R, have recently been marketed (terfenadine, loratadine). Many of these compounds have pharmacological actions other than antihistamine. They are anticholinergic with atropine-like activity, especially ethanolamine and ethylenediamine derivatives. This action may cause urinary retention and blurred vision; moreover, they have significant effects on Parkinson-like syndromes accompanying the use of antipsychotic drugs.

Weak α-blocker effects have been demonstrated for some antihistamine belonging to the subgroup of phenothiazines; such effects are responsible for episodes of orthostatic hypotension. Some H_1 antagonists also have local anesthetic properties (dimethindene, mepyramine) and quinidine-like activity. Tricyclic (particularly doxepin) and nontricyclic (mianserin) antidepressants and some antidopaminergic neuroleptics also have antihistamine H_1 actions; this may explain their sedative effects and contribute to their specific therapeutic effects. H_1-antihistamines are well absorbed after oral administration; their effect begins within 15–30 min, reaches its maximum after 1 h, and lasts for 3–6 h. At therapeutic doses, they cause a number of side effects; sedation is the most common effect caused by all anti-H_1 drugs although the extent varies depending on the drug. The incidence of gastrointestinal disorders is high. Allergic reactions, even anaphylactic type, may occur after their administration; the histamine-liberating ability of some drugs may explain the occurrence of these phenomena.

Pharmacology of the H_2R: The Treatment of Peptic Ulcer
Agonists of H_2R (betazole and dimaprit) are used to assess gastric acid secretion. Cimetidine, ranitidine, tiotidine, nizatidine, and famotidine are H_2R-selective antagonists and are used to treat peptic ulcer disease and Zollinger–Ellison syndrome. The antidepressant mianserin acts as antagonist at H_2R.

The availability of agonists (dimaprit, impromidine) and antagonists (cimetidine, ranitidine, famotidine, nizatidine) has led to the characterization of the H_2R, which has a primary role in the control of gastric chloropeptic secretion. The most important action of H_2 antagonists consists in inhibition of acid secretion, stimulated by histamine, gastrin, cholinergic drugs, and vagal stimulation, by decreasing both H^+ concentration and amount of secretion. These molecules are widely used in peptic ulcer, gastroesophageal reflux disease, and Zollinger–Ellison syndrome. In peptic ulcer disease, H_2R antagonists relieve symptoms and promote lesion healing. In Zollinger–Ellison syndrome, acid hypersecretion is due to a tumor secreting gastrin; H_2-blockers are effective in controlling the symptoms linked to acid hypersecretion, and they are used both as primary treatment and as presurgery therapy.

A pharmacological antagonism at central H_2R may occur in particular situations, causing restlessness, disorientation, and sometimes hallucinations; hyperprolactinemia has also been observed.

These drugs are well tolerated and only 1–2% of patients experience side effects. Cimetidine can cause gynecomastia in men, galactorrhea in women, and granulocytopenia. Cimetidine inhibits oxidative metabolism reactions catalyzed by CYP3A4 cytochrome giving rise to numerous drug interactions.

Pharmacology of the H_3R and H_4R R(α)-methylhistamine, imetit, and immepip are H_3 agonists, and thioperamide and clobenpropit are H_3 antagonists; at present, they have no clinical use. Pitolisant, an inverse agonist, is in phase III clinical study for cognitive disorders.

Imetit and immepip are H_4 agonists. JNJ777120, UR65318, and UR63325 are H_4 antagonists. JNJ39758979, PF3893787, and UR63325, all H_4R antagonists, are in clinical study for allergic diseases.

The discovery of histamine receptors is fairly recent, and to date, only preliminary indications for their therapeutic application are available. H_3R is present in the cerebral cortex, hippocampus, amygdala, striatum, basal ganglia, and hypothalamus. Its localization is mainly presynaptic; as an autoreceptor, it modulates histamine secretion, whereas as a heteroreceptor, it primarily regulates release of glutamate, acetylcholine, noradrenaline, dopamine, GABA, and serotonin. In the CNS, H_3R are involved in sleep–wake mechanism and in cognitive and memory processes. H_3 inverse agonists have been proposed for the treatment of attention deficit/hyperactivity disorder (ADHD), memory disorders, narcolepsy, and neuropathic pain. Drugs active on H_4R are in experimental and clinical studies with the purpose to reduce pruritus, neuropathic pain, neurogenic inflammation, and immunological diseases.

TAKE-HOME MESSAGE

- The histaminergic system has a wide variety of effects on the CNS, microcirculation, cardiac function, regulation of gastric juice composition, and immune response.
- Histamine is synthesized by decarboxylation from L-histidine and degraded by histamine-N-methyltransferase and MAO-B.
- Histamine receptors are GPCRs; four subtypes have been identified, which are coupled to different G proteins and have specific tissue distributions.
- H_1 receptor antagonists are mainly used in the control of allergies; by contrast, since the H_2 receptor controls HCl gastric secretion, its antagonists are used for the control of gastric ulcer.
- The discovery of new receptors and active drugs has suggested new potential applications for the pharmacological modulation of the histaminergic transmission in the control of the CNS and immune system function.

FURTHER READING

Panula P., Nuutinen S. (2013). The histaminergic network in the brain: basic organization and role in disease. *Nature Review Neuroscience*, 14, 472–487.

Stark H. (2013). *Histamine H_4 receptor: a novel target in immunoregulation and inflammation*. Berlin: Versita.

Thurmond R.L., Gelfand E.W., Dunford P.J. (2008). The role of histamine H_1 and H_4 receptors in allergic inflammation: the search for new antihistamines. *Nature Reviews. Drug Discovery*, 7, 41–53.

Tiligada E., Zampeli E., Sander K., Stark H. (2009). Histamine H_3 and H_4 receptors as novel drug targets. *Expert Opinion on Investigational Drugs*, 18, 1519–1531.

Tiligada E., Kyriakidis K., Chazot P.L., Passani M.B. (2011). Histamine pharmacology and new CNS drug targets. *CNS Neuroscience and Therapeutics*, 17, 620–628.

Vohora D. (2009). *The third histamine receptor: selective ligands as potential therapeutic agents in CNS disorders*. Boca Raton, FL: CRC Press.

42

GABAergic TRANSMISSION

Mariangela Serra, Enrico Sanna, and Giovanni Biggio

By reading this chapter, you will:

- Know the function of GABAergic neurotransmission and the mechanisms of GABA synthesis, secretion, reuptake, and catabolism
- Learn the classification, molecular structure, and signal transduction mechanisms of $GABA_A$ and $GABA_B$ receptors
- Understand the concept of "phasic" and "tonic" inhibition
- Know the major classes of drugs and substances that act on different GABA receptors and their mechanism

---

γ-aminobutyric acid (GABA) is the major neurotransmitter with inhibitory function in the central nervous system (CNS) of mammals. GABA was first identified in 1950 by E. Roberts and S. Frankel in brain extracts from different animal species. Further studies allowed the classification of GABA as an amino acid serving as inhibitory neurotransmitter (i.e., a substance produced and released by nerve cells in the synaptic cleft to convey an inhibitory message to other nerve cells) in mammal CNS. GABA is the most abundant inhibitory neurotransmitter in the mammalian brain (about 35–40% of the synapses are GABAergic) and has an important role in the control of various brain functions and therefore in the pathophysiology of numerous mental and neurological diseases. The understanding of the molecular events involved in GABA activity at synapses has allowed the identification of some of the biological mechanisms involved in the control of emotions and neuronal excitability as well as the discovery of new therapeutic agents to treat anxiety disorders and epileptic pathology.

GABA DISTRIBUTION, SYNTHESIS, AND METABOLISM

GABA and glutamate decarboxylase (GAD), the enzyme responsible for its synthesis, are not evenly distributed in the CNS. GABA is the most common inhibitory neurotransmitter, and maximal concentrations are found in the substantia nigra, globus pallidus, hypothalamus, quadrigeminal bodies, cerebral cortex, cerebellum, and hippocampus. Lower concentrations are present in the pons, medulla oblongata, and white matter.

In the striatum, GABAergic neurons form the striatonigral GABAergic pathway, which originates in the caudate nucleus and globus pallidus and projects to the substantia nigra pars reticulata. From the substantia nigra originate the nigrocollicular and nigrothalamic pathways, which from the pars reticulata project to the deep layers of the superior colliculus and to the thalamic nuclei (ventral–medial, middorsal, and intralaminar), and the nigrotegmental pathway, which projects to the pontomesencephalic tegmentum. GAD-positive neurons are also found in arched nucleus and periventricular nuclei of the mediobasal hypothalamus. In the hippocampus, specific GABAergic interneurons form synaptic contacts with hippocampal pyramidal cells.

In the cerebellum, GABA has the highest concentration in the Purkinje cell layer and at the level of the deep nuclei, where GABAergic axons of Purkinje cells project, representing the only efferent pathway of the cerebellum. High GABA concentrations are also present in basket and stellate cells, with which Purkinje cells make synaptic connections, as well as in Golgi type II neurons that are in synaptic contact with granule cell dendrites. GABA is also present in glial cells, which contribute to its removal from the synaptic cleft through a specific system of uptake and catabolism (Fig. 42.1).

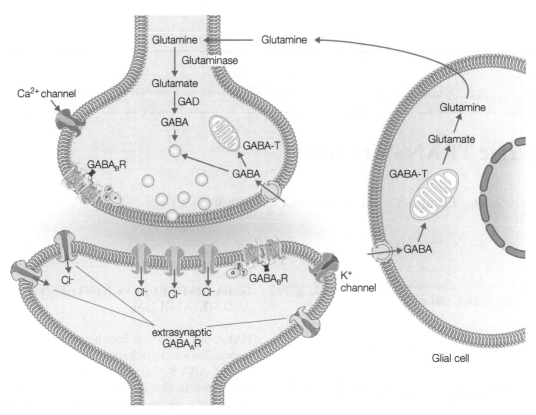

FIGURE 42.1 Schematic representation of the GABAergic synapse. The presynaptic ending releases GABA that acts on postsynaptic receptors ($GABA_A$ and $GABA_B$). GABA synthesis and catabolism occur both in the neuron and in the glial cell. On the postsynaptic membrane, synaptic and extrasynaptic $GABA_A$ receptors are shown.

GABA is produced by decarboxylation of glutamic acid by GAD, a highly specific enzyme that requires pyridoxal phosphate as a cofactor and is inhibited by several antagonists of pyridoxal phosphate, such as hydrazinic compounds (isoniazid and thiosemicarbazide). GAD distribution in the CNS reflects that of GABA. GABA, synthesized in the cytoplasm, is stored in synaptic vesicles present in axon terminals (Fig. 42.1). Vesicular accumulation of GABA is carried out by a specific transporter using the energy stored in electrical and pH gradients generated across the vesicular membrane by the vesicular H^+-ATPase (proton pump). GABA is degraded by the GABA-α-ketoglutarate transaminase enzyme (GABA-T), which deaminates it to succinate semialdehyde; this is oxidized by a NAD-dependent succinate semialdehyde dehydrogenase to form succinic acid, which enters the Krebs cycle. The amino group is transferred from GABA-T to a molecule of α-ketoglutarate to form glutamic acid, which is then reused for the synthesis of new GABA. GABA-T and succinate semialdehyde are always bound to mitochondria not only in the axon terminal but also in postsynaptic structures and glial cells; in the latter, glutamic acid is converted into glutamine by glutamine synthetase. Glutamine is released and then recaptured by the GABAergic terminal where it is first converted into glutamic acid by a glutaminase and then into GABA. Pharmacological inhibition of GABA-T (by drugs such as ethanolamine-O-sulfate, γ-vinyl-GABA, and valproic acid) results in an increase in neurotransmitter concentration and potentiation of GABAergic synapses. Some of the drugs that can inhibit GABA-T activity are widely used in the treatment of some forms of epilepsy (see Supplement E 42.1, "Antiepileptic Drugs with a GABAergic Mechanism of Action").

GABA RELEASE AND REUPTAKE

In vitro studies have shown that GABA is released either spontaneously or upon nerve stimulation. GABA release induced by depolarization is a Ca^{2+}-dependent phenomenon, whereas the spontaneous release is not dependent on depolarization of the terminal.

Multiple carrier proteins (GAT-1, GAT-2, GAT-3, and BGT-1; see Chapter 29) are present in GABAergic endings, which rapidly remove GABA from the synaptic cleft, thus terminating its action. These carriers cotransport GABA, Cl^-, and Na^+ ions and are present both at nerve ending and on glial cells. They are pharmacologically distinguishable based on their different sensitivity to carboxylic acid *cis*-1,3-aminocyclohexanol (ACHC) (higher affinity for the neuronal transporter) and to β-alanine (higher affinity for the glial transporter). It is believed that preservation of low GABA concentrations in the extracellular space is mainly due to glial transporter activity. Drugs inhibiting reuptake systems, such as 2,4-diaminobutyric acid (DABA), nipecotic acid, and β-alanine, have antiepileptic activity. Tiagabine, an analog of nipecotic acid with similar properties, is currently in advanced stage of clinical trials for the treatment of partial epilepsy that is refractory to other treatments (see Supplement E 42.1).

GABA RECEPTOR CLASSIFICATION

Electrophysiological and biochemical studies have demonstrated the existence of two distinct GABA binding sites, conventionally referred to as $GABA_A$ and $GABA_B$ receptors, which differ in molecular structure, signal transduction mechanism, distribution, function, and pharmacological profile.

$GABA_A$ receptors are ligand-gated channels permeable to Cl⁻ ions, characterized by high sensitivity to bicuculline and muscimol (selective and high-affinity antagonist and agonist of GABA, respectively). These receptors contain specific binding sites for benzodiazepines and barbiturates, which modulate their function.

$GABA_B$ receptors are coupled to inhibitory G proteins and are selectively activated by the GABA derivative, β-p-chlorophenyl GABA (baclofen); unlike $GABA_A$ receptors, they are insensitive to bicuculline and muscimol.

$GABA_A$ RECEPTORS

$GABA_A$ receptors are ligand-gated Cl⁻ channels and are the main mediators of GABA "fast" (few milliseconds) inhibitory signal to the postsynaptic neuron. Initially characterized for their sensitivity to the selective agonist muscimol, as well as bicuculline (a competitive antagonist) and picrotoxin (Cl⁻ channel blocker), $GABA_A$ receptors play an important role in virtually all neuronal functions and also represent the molecular target of multiple drug classes of clinical relevance.

Given that Cl⁻ is the main ion permeating through the $GABA_A$ receptor-associated channel, activation of these receptors keeps membrane potential close to the Cl⁻ equilibrium potential, which normally is around −70 mV. Thus, by increasing membrane conductance (shunting effect) or by causing membrane hyperpolarization, activation of these receptors reduces the probability that excitatory postsynaptic potentials may reach the threshold value required for generating an action potential. However, in some areas, such as presynaptic endings in the spinal cord and dendrites in hippocampal and cortical neurons, the Cl⁻ potential equilibrium seems to be shifted to less electronegative values (probably due to lack of Cl⁻ extrusion systems), and activation of the receptor causes a loss of negative charges (Cl⁻ ions exit the cell); the resulting depolarization can generate an action potential as a consequence of the activation of voltage-dependent Na⁺ channels. A similar mechanism seems to take place in many types of neurons during the developmental phase.

The $GABA_A$ receptor is made of multiple subunits. The $GABA_A$ receptor is a heteromeric complex consisting of five glycoprotein subunits belonging to different classes. 19 different genes encoding for different subunits have been identified, and based on their sequence, they have been classified as α(1–6)-, β(1–3)-, γ(1–3)-, δ-, ε-, θ-, π-, ρ(1–3)-subunits (Fig. 42.2). Amino acid sequence analysis shows a high degree of homology between all $GABA_A$ receptor subunits, as well as between these and those forming other members of the "cys-loop" ionotropic receptors superfamily, which includes nicotinic cholinergic receptor, glycine receptor, and $5-HT_3$ receptor for serotonin. The distribution of individual subunits in the CNS is extremely varied: some subunits, such as $α_1$ and $β_2$, are expressed in a diffuse manner in most brain regions, whereas others are found in more discrete areas, such as the $α_6$-subunit, which is exclusively expressed in cerebellar granule cells.

The most common form of these receptors is represented by a pentamer composed of 2 α-subunits, 2 β-subunits, and 1 $γ_2$-subunit, which is the structure required for the formation of the high-affinity binding site for benzodiazepines. In a minor amount of receptors, the $γ_2$-subunit can be replaced by a δ- or ε- or π-subunit, while the β-subunit can be replaced by a θ-subunit.

In 1998, members of the committee responsible for the nomenclature of $GABA_A$ receptors of the International Union of Pharmacology (IUPHAR) proposed a classification, based on receptor subunit composition, which is summarized in Table 42.1.

Drug Binding Sites on $GABA_A$ Receptors

In the macromolecular $GABA_A$ receptor complex, several binding sites for specific molecules have been identified (Fig. 42.2).

The binding site for GABA, GABA-mimetic drugs (muscimol), and GABA antagonists (bicuculline) is located at the interface between α- and β-subunits, and it is well conserved among different isoforms of the two subunits. The interaction with two GABA molecules results in a conformational change of the receptor complex that determines opening of the Cl⁻ channel. Bicuculline and the compound SR95331 (gabazine) block GABA-receptor interaction with a competitive mechanism.

The binding site for benzodiazepines and benzodiazepine-like compounds is characteristic of receptors formed by α-, β-, and γ-subunits, and it is localized at the interface between the α- and γ-subunits. The sensitivity of the

TABLE 42.1 GABA$_A$ receptor subtypes

Subunit composition	Pharmacological profile
$\alpha_1\beta_n\gamma_2$	High affinity and efficacy for benzodiazepines, imidazopyridines (zolpidem), pyrazolopyrimidine (zaleplon), 2-oxoquazepam, and β-carbolines
$\alpha_1\beta_n\gamma_3$	Like $\alpha_1\beta_n\gamma_2$, but 400 times less sensitive to zolpidem and zaleplon
$\alpha_1\beta_n\gamma_1$	Like $\alpha_1\beta\gamma_3$, but flumazenil and Ro 15-4513 have low affinity and act as inverse agonists such as β-carbolines
$\alpha_2\beta_n\gamma_2$	High affinity for benzodiazepines and β-carbolines
$\alpha_2\beta_n\gamma_1$	Agonists have an affinity 2–20 times lower than for $\alpha_2\beta_n\gamma_2$ The affinity of zolpidem is five times greater, but its efficacy is much lower Insensitive to the antagonist flumazenil
$\alpha_3\beta_n\gamma_2$	Affinity for benzodiazepines and β-carbolines similar to $\alpha_1\beta_n\gamma_2$ receptors Intermediate action (10-fold lower affinity) for zolpidem and zaleplon
$\alpha_4\beta_n\gamma_2$	Insensitive to classical benzodiazepines, zolpidem, and zaleplon and many other agonists Intermediate affinity for β-carbolines (inverse agonists) Flumazenil is agonist. Propofol and pentobarbital have no direct action
$\alpha_5\beta_{1/3}\gamma_2$	High affinity for the classic benzodiazepines. Insensitive to imidazopyridines
$\alpha_5\beta_3\gamma_3$	Affinity similar to $\alpha_5\beta_{1/3}\gamma_2$ except triazolam and β-carbolines (30 times lower affinity)
$\alpha_6\beta_1\gamma_2$	Insensitive to all ligands of benzodiazepine receptor Flumazenil and Ro 15-4513 act as partial agonists, and some β-carboline inverse agonists act as antagonists
$\alpha_6\beta_{2/3}\gamma_2$	Same properties of $\alpha_6\beta_1\gamma_2$. It is selectively antagonized by furosemide
$\alpha_6\alpha_1\beta_{2/3}\gamma_2$	It combines the properties of $\alpha_1\beta_n\gamma_2$ and $\alpha_6\beta_{2/3}\gamma_2$

different GABA$_A$ receptor subtypes to this drug class is mainly determined by the specific isoforms of the α- and γ-subunits present in the receptor. Interaction with benzodiazepines or benzodiazepine-like ligands (agonists) increases GABA ability to activate the Cl$^-$ channel. This effect, *in vivo*, results in a reduction of anxiety, sedation, hypnosis, and muscle relaxation. The same site is also recognized by ligands that act as inverse agonists (β-carbolines). These molecules can reduce the interaction of GABA with its recognition site and induce opposite effects (anxiety and convulsions) to those of benzodiazepines. The benzodiazepine binding site is also recognized by drugs devoid of intrinsic activity (competitive antagonists such as flumazenil) but able to antagonize the action of both agonists and inverse agonists. This site, now called central benzodiazepine receptor, is one of the few examples of receptor capable of mediating opposite effects (anxiolytic–anxiogenic, anticonvulsant–convulsant, and hypnotic–somnolytic) when activated by agonists and inverse agonists, respectively.

The allosteric binding sites for barbiturates and their antagonists (e.g., picrotoxin) and for some organophosphoric derivatives such as t-butyl-bicycle-phosphorothionate (TBPS) lie within or close to the Cl$^-$ channel. This localization is of fundamental functional importance: unlike benzodiazepines, barbiturates can enhance Cl$^-$ conductance even in the absence of GABA (GABA-mimetic effect). Picrotoxin and TBPS (direct and allosteric barbiturate antagonists, respectively) block the function of the Cl$^-$ channel and induce opposite pharmacological effects (anxiety and convulsions).

Finally, recent studies have identified two different sites of interaction for neuroactive steroids, such as allopregnanolone and THDOC, at the level of the GABA$_A$ receptor. One site is located in the first transmembrane domain (M1) of α-subunit and mediates the effect of positive modulation exerted by GABA. The other site is located at the interface between α- and β-subunits and mediates the direct GABA-mimetic action of neuroactive steroids. The δ-subunit, whose presence increases the action of neuroactive steroids at GABA$_A$ receptors, does not appear to be directly involved in the formation of these binding sites.

Although distinct, the binding sites for GABA, benzodiazepines, barbiturates, and steroids are functionally linked to each other: activation or inhibition of one site by its specific agonist or antagonist alters the ability of others to interact with their specific ligands. This may result in a positive (facilitation) or negative (inhibition) modulation of the function of the receptor channel complex (Fig. 42.2).

It is important to emphasize that activation of the benzodiazepine binding site favors interaction of both GABA and barbiturates with their respective binding site. Similarly, the presence of a barbiturate molecule at the binding site facilitates binding of both GABA and benzodiazepines at their respective recognition site. On the other hand, both benzodiazepines and barbiturates reduce the capability of negative modulators, such as picrotoxin and TBPS, to interact with the GABA$_A$ receptor. These effects may explain at molecular level the pharmacological synergism between benzodiazepines, barbiturates, and steroids.

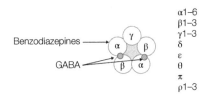
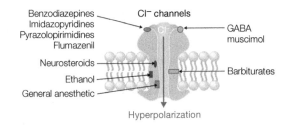
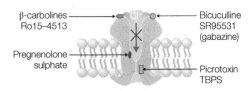

FIGURE 42.2 Schematic representation of GABA$_A$ receptor molecular structure. (a) The three different subunits, α, β, and γ, necessary to assemble a functional receptor sensitive to both barbiturates (α–β) and benzodiazepines (αβγ) are shown. Subunit stoichiometry (in the figure, 2α:2β:1γ) may vary in different CNS areas. Two hypothetical functional states of the receptor Cl$^-$ channel are shown: (b) activated and (c) inhibited. Binding sites of different (b) positive and (c) negative modulators are shown.

Subunit Composition of Native Receptors

As shown in Table 42.1, different GABA$_A$ receptor subtypes markedly differ in their pharmacological profile. Among the various subtypes, the most abundant in the CNS (about 45%) is $\alpha_1\beta_2\gamma_2$, which is present in most brain areas. $\alpha_2\beta_n\gamma_2$ and $\alpha_3\beta_n\gamma_2$ receptors are also relatively abundant: the former is present in spinal motor neurons and hippocampal pyramidal cells, the latter on cholinergic and monoaminergic neurons where it seems to play a role in controlling the release of neurotransmitters involved in the physiology and pharmacology of emotional states and mood. Other receptor subtypes are less abundant.

A particular subpopulation of GABA$_A$ receptors that is insensitive to bicuculline, positive modulators (benzodiazepines, barbiturates, and steroids) as well as baclofen (GABA$_B$ agonist) is expressed in neurons of the retina. These receptors have an intrinsic Cl$^-$ channel and consist of homomeric or heteromeric complexes of ρ_1-, ρ_2-, and ρ_3-subunits. While the ρ_1-subunit is probably expressed only in the retina, the ρ_2-subunit is also present in various CNS regions. Even though ρ-subunits show a 30–38% amino acid homology with other GABA$_A$ receptor subunits, they do not assemble with α-, β-, γ-, ε-, and δ-subunits and also differ in the molecular organization of the drug binding sites.

The identification of the GABA$_A$ receptor subtypes that are actually expressed *in vivo* in the CNS (native receptors) represents a major problem. The IUPHAR Committee has recently reviewed all experimental data available, and by applying strict criteria, it has compiled the following list of native GABA$_A$ receptor subtypes for which the expression in the CNS is certain:

$\alpha_1\beta_2\gamma_2$; $\alpha_2\beta\gamma_2$; $\alpha_3\beta\gamma_2$; $\alpha_4\beta\gamma_2$; $\alpha_4\beta_2\delta$; $\alpha_4\beta_3\delta$; $\alpha_5\beta\gamma_2$; $\alpha_6\beta\gamma_2$; $\alpha_6\beta_2\delta$; $\alpha_6\beta_3\delta$; and ρ

Instead, the following subtypes have been placed on a list of "probable" native receptors:

$\alpha_1\beta_3\gamma_2$; $\alpha_1\beta\delta$; $\alpha_5\beta_3\gamma_2$; $\alpha\beta_1\gamma/\alpha\beta_1\delta$; $\alpha\beta$; and $\alpha_1\alpha_6\beta\gamma/\alpha_1\alpha_6\beta\delta$

Other subtypes, whose existence is not supported by sufficient experimental data, have not been included in either of the two lists.

Extrasynaptic GABA$_A$ Receptors: Phasic and Tonic Inhibition

One of the most recent and interesting discovery in the field of GABA$_A$ receptors is the finding that the "fast" inhibitory GABAergic transmission can be mediated by two different mechanisms: a phasic component and a tonic component. The "phasic" component is mediated by "classic" GABA$_A$ receptors clustered in the postsynaptic membrane and activated by GABA released from the presynaptic terminal. The "tonic" component is mediated by receptors located at the edge or outside of the synapse and therefore called peri- or extrasynaptic receptors. The tonic component of the GABAergic inhibition was initially described using electrophysiological techniques, but the existence of extrasynaptic GABA$_A$ receptors in neurons of different brain areas (e.g., cerebellum, thalamus, hippocampus, and cerebral cortex) was subsequently demonstrated by electron microscopy and immunofluorescence techniques. Extrasynaptic GABA$_A$ receptors differ from the synaptic ones both for subunit composition and for some critical functional properties. The most abundant extrasynaptic receptors are formed by α_4-, $\beta_{2/3}$-, and δ-subunits and are expressed in granule cells of the dentate gyrus, thalamus, and cerebral cortex, while cerebellar granule cells express $\alpha_6\beta_{2/3}\delta$ receptors. It has also been proposed that $\alpha_5\beta\gamma_2$ and $\alpha_1\beta\delta$ receptors may further contribute to the generation of a tonic conductance in the hippocampus. Extrasynaptic receptors have a higher affinity (~0.5 μM, roughly corresponding to GABA extracellular concentration) for GABA and show a highly reduced rate of desensitization in comparison with synaptic receptors. Activation of these receptors by GABA, diffused outside the synapse (spill over), generates a continuous (tonic) hyperpolarizing current that does not undergo inactivation. It has been calculated that the average amount of transferred charge (Cl⁻) per unit of time through the extrasynaptic receptor channels is more than three times higher than that transferred through synaptic receptors, suggesting that tonic inhibition represents a highly efficient mechanism for the control of neuronal excitability. Extrasynaptic receptors are also particularly sensitive to the modulatory action of neurosteroids. Since these hormones are synthesized and secreted also by CNS cells, they may also further contribute, through their regulation of tonic current, to the fine-tuning of neuronal excitability. Lastly, extrasynaptic receptors can be activated by the preferential agonist THIP (4,5,6,7-tetrahydroisoxazol-[5,4-c]-pyridine-3-ol, gaboxadol) and inhibited by high concentrations of picrotoxin or bicuculline.

Pharmacology of GABA$_A$ Receptors

The evidence that drugs of choice in the treatment of anxiety disorders, sleep disorders, and certain seizure pathologies, such as benzodiazepines and barbiturates, recognize specific binding sites at the GABA$_A$ receptor channel complex and the subsequent discovery that general anesthetics, steroids, and ethanol potentiate the synaptic action of GABA have clarified the pharmacological role that GABA$_A$ receptors play in mammalian CNS.

Benzodiazepines Benzodiazepines have binding sites on the GABA$_A$ receptor that are distinct from the GABA binding site. Activation of the benzodiazepine binding site facilitates the interaction of GABA with its receptor site, resulting in an increased frequency of channel opening in the presence of GABA. Therefore, benzodiazepines, although interacting with binding sites different from that of GABA, can positively modulate the GABAergic transmission (allosteric modulation). The IUPHAR suggested a new nomenclature for GABA$_A$ receptor subtypes and benzodiazepine binding sites [1] (Table 42.1). Taking into account GABA$_A$ receptor heterogeneity and that the α-subunit contributes to the formation of the benzodiazepine binding sites (to which also nonbenzodiazepine molecules such as imidazopyridines, pyrazolopyrimidine, cyclopyrrolones, β-carbolines, etc. can bind), a nomenclature of central receptors has been proposed based on the affinity of various ligands for binding sites located on receptors containing different α-subunit isoforms. These receptors mediate the classic pharmacological effects of benzodiazepines.

GABA$_A$ Receptor Subtypes Mediating the Effects of Benzodiazepines The gene knock-in strategy has recently allowed clarifying the role of certain GABA$_A$ receptor subtypes (Fig. 42.3). Using this experimental approach, a point mutation was introduced in the benzodiazepine binding sites of different "diazepam-sensitive" isoforms (α$_1$, α$_2$, α$_3$, and α$_5$) of the α-subunit to make them diazepam insensitive.

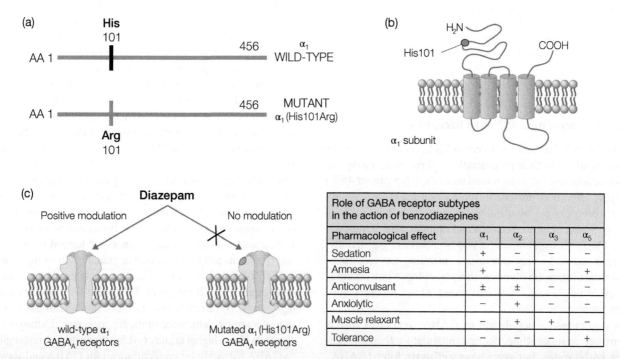

FIGURE 42.3 Site-specific mutation introduced in the α$_1$-subunit of the GABA$_A$ receptor to generate a specific knock-in mouse line. (a) The point mutation replaces the histidine 101 (His101) of the α$_1$-subunit with an arginine (Arg101). As illustrated in (b), His101 is localized in the extracellular N-terminal domain of the α$_1$-subunit and is crucial for benzodiazepine binding. (c) GABA$_A$ receptors containing the mutated α$_1$ (His101Arg) subunit do not bind diazepam and are therefore insensitive to this benzodiazepine. The same mutation has also been introduced in the α$_2$, α$_3$, and α$_5$ isoforms, thus generating different knock-in mouse lines. The table summarizes the pharmacological effects of diazepam mediated by specific GABA$_A$ receptor subtypes containing different α-subunits.

This mutation replaces a histidine residue (H) in position 101 (numbering refers to the α_1-subunit) with an arginine residue (R) [α_1 (H101R), α_2 (H101R), α_3 (H126R), and α_5 (H105R)]. In this way, different knock-in mice were generated, each one expressing the mutated form of a single α-subunit, so that all receptors containing that subunit were "diazepam insensitive." Mice were then analyzed to establish which behavioral effects of benzodiazepines were canceled or reduced in the presence of a specific diazepam-insensitive receptor subtype. These studies have shown that α_1 (H101R) mice are resistant to the sedative effect of diazepam, suggesting that this effect is mediated by receptors containing the α_1-subunit. However, in these mice diazepam administration induces EEG alteration similar to that observed in wild-type mice, thus allowing distinguishing between sedative effect and changes in the physiological sleep pattern induced by benzodiazepines. In addition, the α_1 (H101R) mutation partially reduces the anticonvulsant effect of diazepam in the pentylenetetrazol test, suggesting that this $GABA_A$ receptor subtype may also, at least in part, mediate anticonvulsant effect. The same mutation (H101R) introduced in the α_2-subunit has allowed to discover that α_2-containing receptors mediate the anxiolytic and muscle relaxant effects of benzodiazepines. Moreover, with the same gene knock-in strategy, it has been shown that α_5-containing $GABA_A$ receptors are important for the development of tolerance to benzodiazepine sedative properties and also for the impairment of cognitive processes and learning induced by these drugs (Fig. 42.3).

Translocator Protein or Peripheral Benzodiazepine Receptor The benzodiazepine binding site located on glial cells and other cell types in various organs (e.g., liver, kidney, testes, lung, ovary, and adrenal gland), initially called peripheral benzodiazepine receptor and recently renamed translocator protein (TSPO) 18 kDa, has no functional relationship with the $GABA_A$ receptor subtypes present in the CNS (Table 42.2). The physiological and pharmacological role of TSPO has not been fully defined yet. TSPO is mainly found in the mitochondrial membrane of cells that synthesize steroids, including those in CNS and PNS, and plays a key role in cholesterol translocation within mitochondria, in concert with the steroidogenic acute regulatory protein (StAR protein) (Table 42.2).

Drugs Acting on the $GABA_A$ Receptor Cl^- Channel Barbiturates potentiate GABA action by interacting with specific receptor sites on the luminal surface of the $GABA_A$ receptor Cl^- channel: occupation of these sites has the main effect of increasing the channel mean open time, also independently of the presence of the neurotransmitter (Fig. 42.4). Barbiturate administration results in a functional activation of all central GABAergic synapses with generalized, dose-dependent depression of the CNS with blockade of respiratory centers and death. It is worth remembering that benzodiazepines modulate receptor channel function by facilitating GABA action and are completely ineffective in the absence of GABA. The CNS depressant effects of benzodiazepines are self-limiting (if the synapse is so inhibited that it does not release GABA, benzodiazepine effect

TABLE 42.2 Translocator protein 18 kDa (TSPO)

	CNS	Peripheral organs
Distribution	Glial cells	Adrenal gland, testis, ovary, lung, kidney, liver, heart, skeletal muscle, platelets, lymphocytes, red blood cells
Subcellular localization	Mostly mitochondrial	
Molecular structure	Heteromeric complex composed of three different subunits: a subunit of 18 kDa, which contains the binding site, associated with two proteins having a molecular weight of 32 kDa (voltage-dependent anion channel, VDAC) and 30 kDa (nucleotides translocator, ANT)	
Synthetic ligands	Benzodiazepines: diazepam, flunitrazepam, 4-clorodiazepam Imidazopyridine derivatives: alpidem, zolpidem, CB series (34, 50, and 54) Isoquinoline derivatives: PK 11195 Arylindole derivatives: 1-X FGIN series	
Endogenous ligands	Cholesterol, porphyrins, antraline Diazepam binding inhibitor (DBI), peptide of 104 amino acids; its biological activity is due to the fragment (TTN) between amino acids 17–50	
Physiological role	The peripheral benzodiazepine receptors (PBRs) are not functionally linked to $GABA_{A-B}$ receptors and do not mediate the central effects of benzodiazepines. PBR activation results in steroidogenesis induction through cholesterol transport into mitochondria and conversion into pregnenolone that translocates to the endoplasmic reticulum where it is converted into progesterone. Two progesterone metabolites with high efficacy modulate $GABA_A$ receptor function PBRs also play a role in the molecular mechanisms involved in tumor growth, mitochondrial respiration, gliosis, inflammatory response, nerve regeneration, insulin secretion, Ca^{2+} influx, anionic transport, and apoptosis	

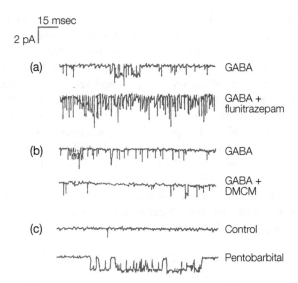

FIGURE 42.4 Cl⁻ currents mediated by $GABA_A$ receptor. Single-channel Cl⁻ currents activated by GABA and various $GABA_A$ receptor modulators were recorded in rat cortical (a and b) or spinal (c) neurons. (a) Currents activated by GABA are enhanced by the benzodiazepine flunitrazepam via increase in channel opening frequency. (b) Current activated by GABA is inhibited by the β-carboline DMCM via decrease in channel opening frequency. (c) Currents activated by pentobarbital in the absence of GABA. Barbiturates can promote receptor Cl⁻ channel opening even in the absence of the agonist. The drug prolongs the mean opening time of the channel.

disappears) and variable in extent between brain areas (depending on the receptor subtype expressed), and generally, they do not lead to death by respiratory depression.

Steroids and $GABA_A$ Receptors The evidence that steroid hormones have biological activity at CNS level was provided by Selye in 1941 with the demonstration that some of these compounds exert sedative and anesthetic effects. Most recent research has shown that some steroid hormones also possess anxiolytic and anticonvulsant effects. The rapidity with which these effects occur suggests that they are mediated by a direct action of these hormones on specific membrane receptors, thus excluding a genomic mechanism. In fact, two metabolites of progesterone, allopregnanolone and allotetrahydrodeoxycorticosterone, are among the most potent and effective positive modulators of Cl⁻ currents mediated by $GABA_A$ receptors, both *in vitro* and *in vivo*. Electrophysiological studies on native and recombinant receptors have shown that these compounds facilitate GABAergic transmission by increasing frequency and duration of $GABA_A$ receptor channel opening.

Receptor subunit composition influences the effectiveness of some steroids. In general, recombinant receptors containing α_1-, α_3-, or α_6-subunits in combination with β_1- or β_2-subunits are more sensitive to allopregnanolone action compared to recombinant receptors containing α_2-, α_4-, or α_5-subunits. On the other hand, isoforms of the β-subunit have no effect on the sensitivity of steroid receptors, whereas the enhancement of GABA-stimulated Cl⁻ currents by allopregnanolone is greater in receptor combinations containing the γ_1-subunit. In mutant mice lacking the δ-subunit, the behavioral effects of the anesthetics alfaxalone and ganaxolone (a synthetic analog of allopregnanolone) are markedly reduced, demonstrating the importance of this subunit for the activity of these compounds. Finally, receptors containing the ε-subunit and homomeric receptors composed only of the ρ-subunit are relatively insensitive to steroid action.

Steroids are synthesized by glial cells and neurons in the CNS and PNS from cholesterol or steroidal precursors from peripheral sources and are called neurosteroids; steroids synthesized by peripheral organs, such as ovary and adrenal, are called neuroactive steroids. In contrast with progesterone metabolites, some sulfate steroid derivatives (pregnenolone sulfate and dehydroepiandrosterone sulfate) negatively modulate $GABA_A$ receptor function by inhibiting the frequency of receptor Cl⁻ channel opening. In fact, administration of these derivatives to mice and rats induces anxiogenic and proconvulsant effects. This discovery is relevant as it demonstrates the existence of endogenous allosteric modulators capable of activating or inhibiting $GABA_A$ receptor function. These hormones appear to represent a key component of the physiological mechanisms controlling the activity of brain areas involved in emotion regulation and more generally neuronal excitability.

It is important to remember that progesterone and its metabolites undergo dramatic physiological fluctuations during menstrual cycle, pregnancy, and menopause. Recent experimental evidence has shown that these hormones have a crucial role in regulating gene expression and $GABA_A$ receptor synthesis under these conditions. Therefore, alterations in the mechanisms regulating the peripheral or central secretion of these hormones can lead to changes in the expression pattern of $GABA_A$ receptors and thus in neuronal excitability threshold and effectiveness of drugs acting on these receptors.

General Anesthetics and $GABA_A$ Receptors General anesthetics form a class of drugs with highly heterogeneous chemical structure and whose precise mechanism of action remains unknown. Although, in the past, evidence suggested that general anesthetics exert a nonspecific action on the lipid component of neuronal membranes (anesthetic potency highly correlates with their oil–water partition coefficient, $P_{o/w}$), in recent years, numerous studies have convincingly shown that some membrane proteins (receptors and ion channels) are primary targets for the pharmacological action of these compounds. Among these membrane proteins, $GABA_A$ receptor is particularly sensitive to the action of anesthetics, which, at clinically effective concentrations, increase its function. Their effect in general consists in a marked facilitatory action, much more pronounced than that of anxiolytic and hypnotic benzodiazepines. However, in contrast with benzodiazepines, many general anesthetics (in particular the intravenous ones) at high concentrations have also the ability to directly activate the receptor Cl⁻ channel even in absence of GABA (GABA-mimetic action).

The action of certain anesthetics (e.g., propofol and etomidate) is heavily influenced by the specific subunits that form the receptor complex. Mutagenesis studies on cDNA coding for the different $GABA_A$ receptor subunits have allowed the identification of two amino acids (Ser270 in the M2 domain and Ala291 in M3 domain) in the α_1-subunit and one amino acid (Ser265 in the M2 domain) in the β-subunit, which are critical for the effects of the anesthetics isoflurane and enflurane but also for ethanol. Therefore, it is conceivable that these amino acids participate in the formation of a hydrophobic site or "pocket," between the M2 and M3 transmembrane segments, capable of hosting these molecules and mediating their effects on receptor function. In addition, the modulatory action of etomidate and propofol is markedly higher on recombinant receptors containing the β_2- or β_3-subunit, compared to that measured on β_1-containing receptors. The β_2- and β_3-subunits contain an asparagine residue in position 265, whereas the β_1-subunit contains a serine at the same position. Studies with β_2 (N265S) knock-in mice demonstrated that β_2-containing receptors mediate only the sedative effects induced by subanesthetic concentrations of these compounds, as the anesthetic effect was still present in these animals. On the other hand, the anesthetic effect of etomidate and propofol was markedly reduced in β_3 (N265M) knock-in mice, suggesting that this effect is primarily mediated by $GABA_A$ receptors containing the β_3-subunit. Studies in thalamic nuclei have also shown that GABAergic tonic currents are robustly enhanced by isoflurane at clinically relevant concentrations. Since thalamic nuclei are important in sleep regulation and the GABAergic system in this area plays a key role in the control of sleep–wake states, it can be assumed that the increase in isoflurane-induced tonic inhibition in this area can contribute to its general anesthetic effect.

Ethanol Potentiates GABAergic Transmission Ethyl alcohol, or ethanol, potentiates $GABA_A$ receptor function and can produce a spectrum of pharmacological effects partially overlapping that of benzodiazepines and barbiturates. The effects of ethanol are enhanced by GABAergic agonists, such as muscimol, and reduced by antagonists such as bicuculline. Numerous experimental studies show that anxiolytic–sedative concentrations (5–20 mM) of ethanol facilitate $GABA_A$ receptor function in nerve cells of different brain areas although with different efficacy. This finding could be at least partially related to the specific $GABA_A$ receptor subtypes expressed in various brain regions or in different neuronal subpopulations. In this regard, it has recently been shown that recombinant $\alpha_4\beta_3\delta$ and $\alpha_6\beta_3\delta$ receptors, which are exclusively extrasynaptic in neurons, are positively modulated by extremely low ethanol concentrations (3–10 mM) compared to those required (>50 mM) to enhance $\alpha_1\beta_2\gamma_2$ recombinant receptors. This result has been confirmed in experiments on rodent brain slices showing that tonic currents, mediated by extrasynaptic receptors in granule cells of the dentate gyrus (which express $\alpha_4\beta_{2/3}\delta$ receptors), are enhanced by low (30 mM) ethanol concentrations. These results suggest that extrasynaptic $GABA_A$ receptors may be responsible for ethanol anxiolytic effect, while synaptic receptors, which are less sensitive to ethanol, may mediate more marked depressant effects, such as sedation and motor incoordination, as well as toxic effects.

Ethanol, at pharmacologically relevant concentrations (25–60 mM), increases the probability of GABA release from presynaptic terminals, resulting in a greater stimulation of postsynaptic $GABA_A$ receptors and thus in an inhibitory signal. This presynaptic action of ethanol may involve an increase in Ca^{2+} release from the terminal intracellular compartment, and requires activation of adenylate cyclase/PKA- and PLC/IP_3/PKC-dependent transduction pathways.

Ethanol can exert a modulatory action on $GABA_A$ receptors also through an alternative indirect mechanism, involving neuroactive steroids. In fact, systemic administration of ethanol in rats induces a significant increase in plasma and brain concentrations of these hormones, which enhance $GABA_A$ receptor function. The demonstration that finasteride, an inhibitor of neurosteroid biosynthesis, prevents such increase as well as other pharmacological effects of ethanol supports the hypothesis that neuroactive steroids may mediate some ethanol effects at central level.

$GABA_B$ RECEPTORS

In 1980, while studying the mechanism of GABAergic control of neurotransmitter release in different rat brain preparations, Bowery and coworkers observed that GABA inhibitory effect on norepinephrine release evoked by K⁺ was mediated by receptors with pharmacological properties totally different from classical GABA receptors. Indeed, the GABA effect was not blocked by bicuculline, a competitive antagonist of classical GABA receptors; it was stereoselectively mimicked by baclofen (β-para-chlorophenyl-GABA, a clinically used spasmolytic); and it was independent from Cl⁻ ions. Subsequent receptor binding studies provided direct evidence of the existence of distinct recognition sites for baclofen on central neuronal membranes. The term $GABA_B$ receptor was then introduced to distinguish this site from the bicuculline-sensitive site, which was therefore designated as $GABA_A$.

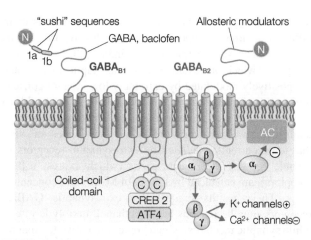

FIGURE 42.5 Structure of the $GABA_B$ receptor. The $GABA_B$ receptor consists of a heterodimer formed by the $GABA_{B1}$ and $GABA_{B2}$ subunits interacting with each other via a coiled-coil region in their C-terminal domains. GABA and baclofen bind to the $GABA_{B1}$ N-terminal region, whereas $GABA_{B2}$ is responsible for coupling with the G protein and binding of allosteric modulators.

In the CNS, $GABA_B$ receptors are located on both neurons and glial cells. At the periphery, they are located on smooth muscle cells, liver, and other tissues. In the brain, $GABA_B$ receptors have a distribution very similar to that of $GABA_A$ receptors. At subcellular level, these receptors are expressed on presynaptic membranes, both on GABAergic (autoreceptors) and non-GABAergic terminals (heteroreceptors), as well as on postsynaptic membranes, dendrites, and cell bodies. The main functional roles of $GABA_B$ receptors include the inhibitory control of the release of several neurotransmitters and, at postsynaptic level, the generation of the slow component of the inhibitory postsynaptic potential.

Molecular cloning of the $GABA_B$ receptor, in 1997, revealed that it has the molecular structure typical of all G-protein-coupled receptors, characterized by seven transmembrane regions, and a long extracellular N-terminal domain similar to that of metabotropic glutamate receptors (Fig. 42.5). Two closely related receptor subunits, $GABA_{B1}$ and $GABA_{B2}$, have subsequently been identified, which consist of approximately 950 amino acids and share about 35% identity and about 50% similarity. In addition, two isoforms of the $GABA_{B1}$ subunit are present in mammalian brain, $GABA_{B1a}$ and $GABA_{B1b}$, encoded by the same gene and originated by alternative splicing. In fact, $GABA_{B1a}$ and $GABA_{B1b}$ differ for the presence of two "sushi" regions in the extracellular N-terminal domain of $GABA_{B1a}$ protein, which are missing in the $GABA_{B1b}$ protein.

The functional $GABA_B$ receptor consists of a $GABA_{B1}$–$GABA_{B2}$ heterodimer, the first example of multimeric complex among the metabotropic receptors to be discovered. The two subunits of the $GABA_B$ receptor interact through their intracellular C-terminal domains via parallel α-helical regions called coiled-coil. In the heterodimer, $GABA_{B1}$ subunit forms the binding site for GABA and agonists/antagonists, while $GABA_{B2}$ subunit mediates coupling to the G protein (Fig. 42.5). The interaction between the two subunits is also crucial for the final localization of the receptor complex on the cell membrane. Indeed, in $GABA_{B2}$ knockout mice, the $GABA_{B1}$ subunit fails to reach the cell membrane and remains confined in the endoplasmic reticulum. This occurs because $GABA_{B1}$ subunit contains a sequence of recognition and interaction (RSRR) for special coat proteins (COPI), which favor its retention in the endoplasmic reticulum. In the heterodimer, the RSRR sequence is masked by $GABA_{B2}$ thus allowing transport of the receptor to the cell membrane.

Interaction of GABA, or baclofen, with the heterodimeric $GABA_B$ receptor activates a G protein belonging to the family of pertussis toxin-sensitive $G\alpha_i/G\alpha_o$ proteins. Effector systems coupled to $GABA_B$ receptors include adenylate cyclase, Ca^{2+} and K^+ channels. In particular, $GABA_B$ receptor activation results in adenylate cyclase inhibition and thereby reduction in cAMP formation and PKA-dependent phosphorylation processes. At presynaptic level, $GABA_B$ receptor activation causes a decrease in conductance of voltage-dependent Ca^{2+} channels of N ($Ca_v2.2$) or P/Q ($Ca_v2.1$) type, via the G protein βγ dimer (Gβγ). In turn, reduction of Ca^{2+} influx determines inhibition of neurotransmitter release. The positive modulation of K^+ conductance mediated by Gβγ through K^+ channels of the Kir-3-type inward rectifier is mainly associated with $GABA_B$ receptors localized at postsynaptic level. The increased membrane permeability to K^+, and the resulting K^+ efflux from the cell, determines hyperpolarization as well as a shunting effect thus increasing the threshold of nerve cell excitability. The role of $GABA_{B1a}$ and $GABA_{B1b}$ proteins has been investigated using genetically modified mice that selectively express only one of the two isoforms. Morphological and electrophysiological studies suggest that the two isoforms may carry out separate functions, at least in certain brain areas, as a result of their differential localization at the subcellular level. In fact, in pyramidal cells of the hippocampus and neocortex, at presynaptic level, the $GABA_{B1a}$ isoform is expressed selectively on glutamatergic terminals, while both isoforms are found on terminals of GABAergic interneurons. At postsynaptic level, the $GABA_{B1b}$ isoform appears to be selectively localized on the membrane of dendritic spines facing glutamatergic excitatory terminals; by contrast, both isoforms appear to be expressed in the membrane of the dendritic trunk, probably as extrasynaptic $GABA_B$ receptors. It is believed that the two "sushi" domains present in $GABA_{B1a}$, but not in $GABA_{B1b}$, may contribute to regulate the subcellular localization of these proteins. Since these "sushi" regions are well-conserved elements involved in protein–protein interactions, they may mediate binding of $GABA_{B1a}$ isoform to extracellular adhesion proteins required for localization at axonal and/or dendritic level.

Several laboratories have observed that the $GABA_{B1}$ subunit can interact with some transcription factors, such as CREB2 and ATF4, through the coiled-coil region present in its intracellular C-terminal domain. It has been reported that following receptor activation these transcription factors translocate to the cell nucleus and the expression of CREB2/ATF4-sensitive reporter genes increases. These data demonstrate a direct interaction between metabotropic receptors and transcription factors and suggest a further mechanism by which these receptors can activate long-term responses involving regulation of the expression of specific genes.

Pharmacology of $GABA_B$ Receptors

The prototype agonist drug of $GABA_B$ receptors is baclofen, which was introduced in clinical practice as a spasmolytic agent in the 1970s, well before the discovery of $GABA_B$ receptors. Table 42.3 summarizes the pharmacological effects of baclofen. Its muscle relaxant effect is mediated by a central action. Given its marked efficacy, baclofen is the drug of choice for the treatment of spasticity, including that associated with multiple sclerosis. However, the appearance of side effects often associated with high dosages can sometimes limit its use. Recently, a system has been developed to administer baclofen at the level of the spinal subarachnoid space (intrathecal baclofen therapy), showing clinical efficacy at low doses, thereby reducing also possible development of tolerance. Baclofen effectiveness has also been tested in the treatment of addiction from drugs such as cocaine, ethanol, nicotine, and morphine-like substances.

In addition to baclofen, $GABA_B$ receptor agonists include 3-aminopropylphosphonic (3-APPA) and its methylated analog that has a 10–100 times higher affinity for the receptor. Conversely, two analogs of baclofen, faclofen and saclofen, as well as the SCH50911 compound, have an antagonistic action on the receptor. Some of these antagonists have been tested to improve cognitive processes and exert a positive effect on epileptic seizures.

More recently, a new class of compounds has been proposed, the positive allosteric modulators, which have a different mechanism of action. Two examples are the CGP7930 and GS39783 compounds, which are able to potentiate the effects of GABA or baclofen by binding to a site different from that of the agonist, probably localized on the $GABA_{B2}$ subunit of the receptor complex (Fig. 42.5). These positive allosteric modulators have been examined in animal models of cocaine addiction, and both have proven effective in reducing self-administration of the substance without the need to coadminister baclofen. These results suggest that allosteric modulation of $GABA_B$ receptors can be an effective strategy that may reduce occurrence of side effects associated with agonists such as baclofen.

In this context, it is worth remembering that $GABA_B$ receptors mediate many of the pharmacological actions of γ-hydroxybutyric acid (GHB), a drug of abuse with depressant effects on brain function.

SCHEMATIC OVERVIEW OF THE MAIN PHARMACOLOGICAL INTERVENTIONS ON GABAERGIC SYNAPSES

i. Synthesis, Uptake, and Catabolism. With the exception of vigabatrin, an antiepileptic drug that inhibits GABA transaminase, no drugs are known that can alter these functions and have a clinical application. In contrast, some GAD inhibitors (isoniazid and amino-oxyacetic acid) that reduce GABA synthesis and induce seizures are used for experimental purposes.

ii. $GABA_A$ Receptor. Receptor agonists, such as muscimol, isoguvacine, THIP, and imidazole acetic acid, capable of activating synaptic and extrasynaptic $GABA_A$ receptors, have not found therapeutic application. Receptor antagonists, such as bicuculline, have convulsant effects.

iii. Benzodiazepine Binding Site. Drugs with different chemical structures can bind to this site and induce either positive (agonists) or negative (inverse agonists) effects. The agonists include benzodiazepines (diazepam, flunitrazepam, clonazepam, chlordiazepoxide, alprazolam, etc.), cyclopyrrolones (zopiclone and suriclone), imidazopyridines (zolpidem), and pyrazolopyrimidines (zaleplon). These drugs have anxiolytic, anticonvulsant, sedative, and hypnotic activity. Inverse agonists, such as β-carbolines (DMCM, FG-7142), induce anxiogenic and convulsant effects. The effects of both agonists and inverse agonists can be blocked by the antagonist flumazenil, which in itself is devoid of pharmacological effects.

TABLE 42.3 Pharmacological effects of baclofen mediated by $GABA_B$ receptor

In vivo actions	Side effects
Decrease in cognitive functions	Drowsiness
Antinociception	Dizziness
Muscle relaxation	Nausea
Antipanic	Hypotension
Suppression of craving for drugs of abuse	Seizures
Increased food intake	Hallucinations
Epileptogenesis	Confusion
Bronchial relaxation	
Antihiccups	
Aggravation of absence seizure	
Reduction of intestinal peristalsis	
Contraction of uterine musculature	

iv. Binding Site for Barbiturates. Barbiturates induce a hypnotic and anticonvulsant effect. The binding site for barbiturates, located at the receptor Cl⁻ channel, is also recognized by convulsant drugs such as picrotoxin and organophosphate compounds (e.g., TBPS).
v. Binding Site for Neurosteroids. Allopregnanolone and allotetrahydrodeoxycorticosterone are among the most potent and effective positive modulators of $GABA_A$ receptors both *in vitro* and *in vivo*. They have anxiolytic, anticonvulsant, and sedative activity *in vivo*. Two different binding sites have been identified.
vi. General Anesthetics and Ethanol. Although the existence of specific binding sites for these compounds in $GABA_A$ receptor has only been hypothesized, these molecules potentiate GABAergic transmission. General anesthetics include halothane, enflurane, methoxyflurane, isoflurane, propofol, and etomidate. Ethanol strongly enhances the CNS depressant actions of benzodiazepines and barbiturates.

TAKE-HOME MESSAGE

- γ-aminobutyric acid (GABA) is the major inhibitory neurotransmitter in mammalian CNS.
- GABA is synthesized in the presynaptic terminals via glutamic acid decarboxylation catalyzed by GAD and is degraded by GABA transaminase. Following secretion, GABA is taken up by both the nerve endings and the surrounding glial cells.
- GABA acts on two classes of receptors: $GABA_A$ and $GABA_B$.
- $GABA_A$ are pentameric, cys-loop anionic LGICs composed of different subunits. Two GABA binding sites are present in $GABA_A$ at the interface between α- and the β-subunits.
- Benzodiazepines bind to $GABA_A$ α-subunits and exert a positive allosteric modulation on GABA at its receptor. Benzodiazepines also bind to a different (non-GABA receptor) binding site expressed by glial cells.
- $GABA_A$ are also targets of barbiturates, alcohol, some anesthetic drugs, and neurosteroids.
- $GABA_B$ are GPCRs and are active as dimers and target of baclofen.
- Drugs acting on GABA system are useful for the treatment of anxiety, sleep disorders, anesthesia, and epilepsy.

REFERENCE

Barnard E.A., Skolnick P., Olsen R.W. et al. (1998). Subtypes of gamma-aminobutyric acid receptors: classification on the basis of subunit structure and receptor function. *Pharmacological Reviews*, 50, 291–313.

FURTHER READING

Bettler B., Tiao J.Y. (2006). Molecular diversity, trafficking and subcellular localization of $GABA_B$ receptors. *Pharmacology and Therapeutics*, 110, 533–543.

Biggio G., Sanna E., Serra M., et al. (eds.) *Advances in Biochemical Psychopharmacology*. Vol. 48. Raven Press, New York, 1995.

Biggio G., Purdy R.H. (eds.) *Neurosteroids and Brain Function. International Review of Neurobiology*. Vol. 46. Academic Press, Waltham (MA), 2001.

Concas A., Mostallino M.C., Porcu P. et al. (1998). Role of brain allopregnanolone in the plasticity of γ-aminobutyric acid type A receptor in rat brain during pregnancy and after deliver. *Proceedings of the National Academy of Sciences of the United States of America*, 95, 13284–13289.

Hosie A.M., Wilkins M.E., da Silva H.M., et al. (2006). Endogenous neurosteroids regulate $GABA_A$ receptors through two discrete transmembrane sites. *Nature*, 444, 486–489.

Luscher B., Fuchs T., Kilpatrick C. (2011). $GABA_A$ receptor trafficking-mediated plasticity of inhibitory synapses. *Neuron*, 12, 385–409.

Olsen R.W., Sieghart W. (2008). International Union of Pharmacology. LXX. Subtypes of gamma-aminobutyric acid receptors: classification of the basis of subunit composition, pharmacology, and function. Update. *Pharmacological Reviews*, 60, 243–260.

Rupprecht R., Papadopoulos V., Rammes G., et al. (2010). Translocator protein (18 kDa) (TSPO) as a therapeutic target for neurological and psychiatric disorders. *Nature Reviews. Drug Discovery*, 9, 971–988.

Sanna E., Mostallino M.C., Busonero F., et al. (2003). Changes in GABA(A) receptor gene expression associated with selective alterations in receptor function and pharmacology after ethanol withdrawal. *Journal of Neuroscience*, 23, 11711–11724.

Smith S.S. (ed.) *Neurosteroids Effects in the Central Nervous System. The Role of the $GABA_A$ Receptor*. CRC Press, New York, 2004.

Ulrich D., Bettler B. (2008). $GABA_B$ receptors: synaptic functions and mechanisms of diversity. *Current Opinion in Neurobiology*, 17, 298–303.

Wulff P., Goetz T., Leppä E., et al. (2007). From synapse to behaviour: rapid modulation of defined neuronal types with engineered $GABA_A$ receptors. *Nature Neuroscience*, 10, 923–929.

43

GLUTAMATE-MEDIATED NEUROTRANSMISSION

FLAVIO MORONI

> **By reading this chapter, you will:**
>
> - Learn the process of glutamate synthesis, storage, release, and inactivation in the excitatory synapses
> - Learn classification, molecular structure, and signal transduction mechanisms of ionotropic and metabotropic glutamate receptors
> - Know the role of different glutamate receptors in excitatory transmission with special reference to synaptic plasticity and excitotoxicity
> - Understand the therapeutic potentials and limits of glutamate receptor pharmacological modulation

Glutamate, a common dicarboxylic amino acid, was deeply studied in the first 60 years of the last century for its role in the metabolism of brain nitrogen and the metabolic fate of cerebral ammonia and in the synthesis of important cofactors (folic acid, glutathione) and obviously of proteins. From 1960 to 1970, the role of glutamate and glutamate decarboxylase (GAD) were mostly studied in the framework of GABA synthesis and brain inhibitory neurotransmission. In the early 1970s, after the demonstration that glutamate application to neuronal cells could increase their firing rate, it was difficult for the scientific community to accept the idea that a common amino acid such as glutamate could itself be a neurotransmitter.

It is now well demonstrated that glutamate is present at high concentration (up to 100 mM) in synaptic vesicles and is released from neuronal terminals in a Ca^{2+}-dependent manner and, once released, it depolarizes the postsynaptic membranes by activating specific receptors. Transporters capable of removing the glutamate released in the synapse have also been identified in neuronal and glial membranes, and it is now considered fully demonstrated that glutamate excitation has a key role in numerous brain functions (ordered perception of sensations and pain, learning, memory, control of mood, and motor function). Pathological changes in excitatory synaptic function have been demonstrated in key neurological conditions such as seizures, neuropathic pain, postischemic brain damage, and neurodegeneration and in common psychiatric diseases such as depression and schizophrenia.

Unfortunately, the clinical results so far obtained with therapeutic agents acting on glutamate synapses have not met the expectations raised by the exciting scientific breakthroughs obtained in this field in the last 30 years.

GLUTAMATE SYNTHESIS AND METABOLISM

Glutamate is the most abundant free amino acid in the central nervous system (CNS) where it is present in both neurons and glial cells. Since the blood–brain barrier is almost impermeable to glutamate, the amino acid is locally synthesized. Glucose is considered the main glutamate precursor: glycolytic pathway and Krebs cycle lead to α-ketoglutarate, which is then transaminated to glutamate (see Fig. 43.1). Another important pathway of glutamate neosynthesis is mediated by glutaminase, which metabolizes glutamine to glutamate (see Fig. 43.1).

Glutamate dehydrogenase, an enzyme catalyzing the conversion of glutamate to α-ketoglutarate with the production of ammonia, is the main enzyme responsible of glutamate

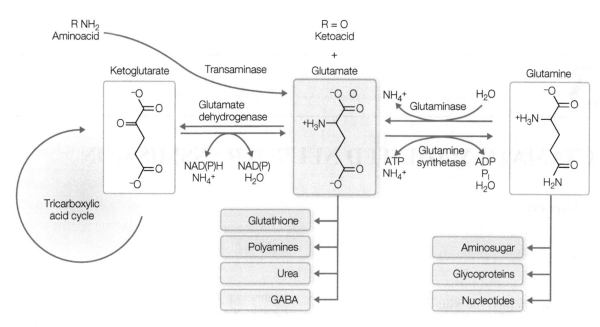

FIGURE 43.1 Glutamate synthesis and metabolism. The two key enzymes in glutamate synthesis are α-ketoglutarate transaminase and glutaminase. The metabolism is catalyzed by glutamate dehydrogenase and glutamine synthetase. Most brain glutamate is formed by transamination between an amino acid and α-ketoglutarate, a keto acid produced in the tricarboxylic acid cycle. Glutamine is formed by incorporation of one ammonia molecule into glutamate. Glutamate and glutamine have key roles in numerous metabolic processes, some of which are listed in the bottom part of the figure.

catabolism. A second catabolic pathway is mediated by the enzyme glutamine synthetase, which converts glutamate to glutamine and is particularly active in glial cells. Glutamine can then be eliminated (transported outside the blood–CSF barrier) or used by neurons for the synthesis of the neurotransmitter glutamate pool (see Fig. 43.1). Another enzymatic activity involved in glutamate metabolism is GAD, which is mostly present in GABA neurons.

GLUTAMATE TRANSPORTERS, VESICULAR ACCUMULATION, AND SIGNAL INACTIVATION

Glutamate concentration in the nervous tissue is quite high (up to 10 mM). However, in the extracellular spaces of the brain and in the cerebrospinal fluid, glutamate concentration is 10,000 times lower (<1 μM). This is due to the presence of efficient transporters that transfer glutamate from outside to inside the cells, in neuronal and glial membranes (Fig. 43.2). Therefore, the transporters contribute to the inactivation of the excitatory signal and to maintain extracellular glutamate concentrations within physiological levels.

Membrane Transporters

At least five cDNAs encoding proteins capable of transporting glutamate inside the cell have been described. In humans, these transporters are named excitatory amino acid transporter (EAAT)1–5. EAAT1, EAAT2, and EAAT3 are similar to the transporters previously identified in rat and, respectively, called glutamate and aspartate transporter (GLAST), glutamate transporter 1 (GLT1), and excitatory amino acid carrier (EAAC). EAAT1 and EAAT2 are predominantly expressed in astrocytes, EAAT3 is expressed in neurons, EAAT4 is expressed in cerebellar Purkinje cells, and EAAT5 is expressed in photoreceptors and bipolar cells of the retina. Glutamate transport from the extracellular to the intracellular space is carried out using the energy stored in ion gradients generated by the Na^+-/K^+-ATPase across the plasma membrane. Three Na^+ ions and one proton (H^+) are cotransported with a glutamate molecule, while one K^+ ion is transferred outside the cell and the ionic balance is maintained by a Cl^- current.

Changes in the expression or function of glutamate transporters have been associated with a high number of pathological conditions such as amyotrophic lateral sclerosis (ALS), epilepsy, postischemic and posttraumatic brain damage, Huntington's chorea, AIDS, dementias, drug addiction, etc. Therefore, glutamate transporters could become important targets for drugs. It should be noted that a common antibiotic, ceftriaxone, is able to increase the expression of EAAT2 (GLT1) and to improve survival in a mouse model of ALS. The molecular mechanisms underlying this effect remain to be clarified.

The role of membrane transporters in physiology and pathology is extensively described in Chapter 29.

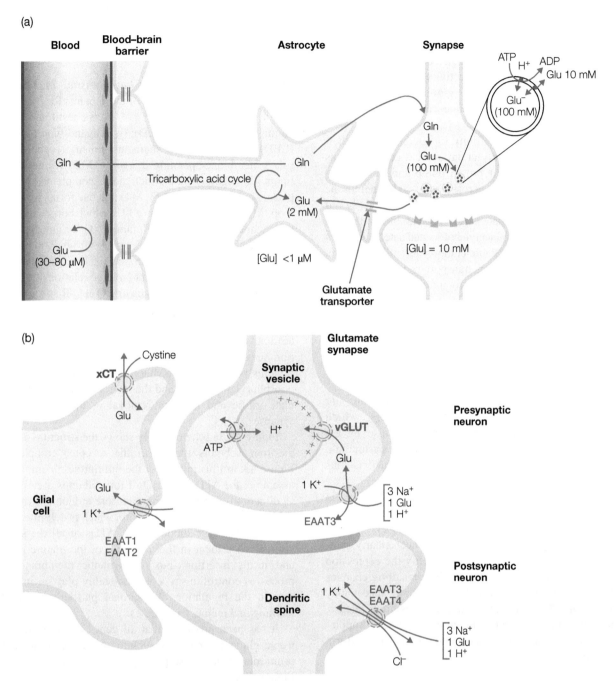

FIGURE 43.2 Compartmentalization of brain glutamate. (a) Glutamate is accumulated in synaptic vesicles by a transporter that uses the proton gradient across the vesicle membrane as energy source. Glutamate released during synaptic activity is taken up either by neurons or by glial cells that synthesize glutamine. Glutamine can be recycled in the glutamate neurotransmitter pool or transported outside the blood–brain barrier, thus allowing eliminating ammonia from the brain. Glial cells play a key role in keeping glutamate concentrations in brain extracellular spaces at rather low levels (~1 µM). Only in pathological conditions glutamate extracellular levels may increase up to toxic concentrations (excitotoxicity). (b) Location and function of membrane transporters and cystine–glutamate exchanger. Glial cells express the glutamate transporters GLT1 and GLAST (EAAT1 and EAAT2) and participate in shutting down the synaptic signals. Neurons express EAAT3 (EAAC), whereas cerebellar Purkinje cells express mostly EAAT4. The cystine–glutamate exchanger participates in the regulation of excitatory transmission by controlling glutamate concentrations in the perisynaptic areas where metabotropic receptors (mGluR2/3) are located. Activation of these receptors inhibits neurotransmitter release.

Vesicular Transporters

Once transported inside the cells, glutamate can be stored in special vesicles at particularly high concentrations (up to 100 mM). This is due to the activity of other specific transporters, called vesicular glutamate transporters (vGLUTs), belonging to the solute carrier (SLC) family described in Chapter 29.

The vGLUTs have low affinity for glutamate and their action become effective only at amino acid concentrations above 1 mM. The vesicular proton pump and the electrochemical gradient existing across the vesicle membrane provide the energy for the active transport. Once stored inside the vesicles, glutamate is easily releasable. In fact, its storage differs from that of acetylcholine and catecholamines that are anchored to semistable components of the vesicle. This may explain why, when transporters in the synaptic vesicle membrane are unable to maintain an adequate electrochemical gradient (under hypoxia or hypoglycemia conditions), glutamate diffuses and may reach the extracellular spaces thus causing excitotoxicity and neuronal death.

The Cystine–Glutamate Exchanger

In the CNS, extracellular glutamate concentrations are regulated not only by EAAT1–5 transporters, which contribute to terminate synaptic signals, but also by the cystine–glutamate exchanger. This latter carries cystine into the cell by exchanging it with glutamate, which is thus transferred to the extracellular space.

This exchanger, present in glial cells, has a key role in providing sufficient amount of cystine/cysteine for the synthesis of glutathione, a tripeptide that plays a key role in the defense against oxidative stress. It has been clearly demonstrated that cocaine addiction can modify the expression of the cystine–glutamate exchanger. The resulting changes in glutamate concentrations in the perisynaptic spaces may stimulate type 2 mGluR, thus reducing transmitter release, synaptic transmission, and neuronal plasticity (see Supplement E 43.1, "Long-Term Potentiation and Long-Term Depression of Excitatory Synaptic Transmission").

GLUTAMATE RECEPTORS

Glutamate may activate both ion channel (ionotropic) and G-protein-coupled (metabotropic) receptors (GPCRs).

Glutamate Ionotropic Receptors

Ionotropic receptors are polymeric complexes consisting of four subunits assembled to form ion channels opened upon glutamate binding. They have different ionic conductance and kinetic of activation/inactivation or desensitization and are classified into three categories based on their selective agonists: (i) AMPA receptors, (ii) kainate receptors, and (iii) NMDA receptors.

Molecular and Structural Characteristics of Ionotropic Glutamate Receptors Each of the subunits forming the ion channel receptor complex (Fig. 43.3) is constituted by (i) an amino-terminal domain (ATD); (ii) a ligand-binding domain (LBD), formed by two segments called S1 and S2; and (iii) a membrane-spanning domain (TMD) formed by four lipophilic segments, of which three completely cross the plasma membrane (M1, M3, M4) and one forms a re-entrant loop (M2). In a precise area of M2, which constitutes one of the channel walls, either a glutamine (Q) or an arginine (R) residue can be present. Receptors with an R at this site, called Q/R site, are characterized by low permeability to calcium ions, and their current–voltage relationship curve is linear. By contrast, receptors having Q at Q/R site are permeable to calcium ions, and their current–voltage relationship has a plateau phase. Finally, each subunit has a carboxyl-terminal domain (CTD, Fig. 43.3a) containing phosphorylation sites and sequences capable of interacting with other proteins present in postsynaptic densities (PSD). Both the phosphorylation status and the interacting proteins may regulate traffic and insertion/removal of the subunits in the synapses.

Figure 43.3a schematically shows the structure of one of the four subunits that form the receptor complex, and Figure 43.3b illustrates how the multiprotein complex may assemble: the ATD and LBD of four subunits are organized as dimers of dimers (having therefore a double symmetry). Each dimer forms half of the pore, which is delimited by the TMDs of the four subunits. The ATDs of all the subunits seem to be critical in the assembly of the mature receptor and in its insertion into the synaptic membranes. The processes controlling receptor assembly play an important role in the regulation of neuronal plasticity, as well as learning and memory processes.

It has been proposed that all genes encoding the ionotropic glutamate receptor subunits originate from a single primordial gene whose phylogenetic evolution is shown in Figure 43.4.

AMPA Receptors AMPA receptors are responsible for the fast excitatory (depolarizing) responses typical of glutamatergic synapses. The receptor ion channel complex is localized in the postsynaptic membranes and is formed by four subunits (GluA1–GluA4). As previously mentioned, their kinetics of activation/inactivation and desensitization are very fast (1–2 ms). Since most AMPA receptors contain the GluA2 subunit, which is the most abundant and contains an arginine (R) at the Q/R site, they are permeable to sodium but virtually impermeable to calcium ions. Cells that do not express GluA2, such as the cerebellar Bergmann

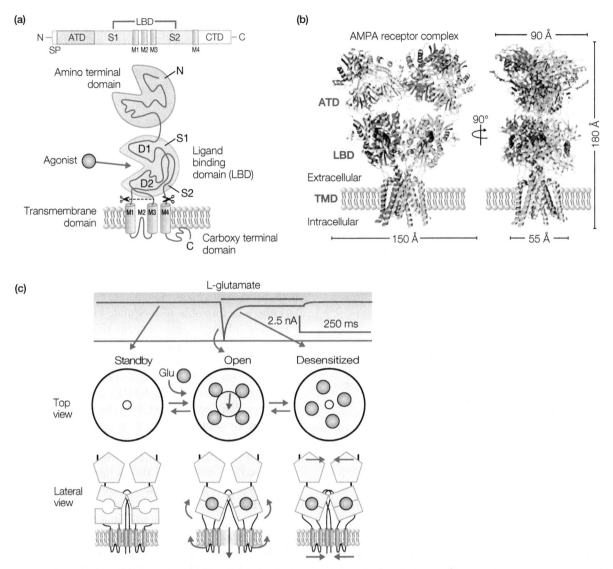

FIGURE 43.3 Structure, function, and molecular organization of ionotropic glutamate receptors. (a) Schematic representation of an ionotropic receptor subunit. The amino-terminal portion (ATD) is extracellular and is followed by the S1 domain (a portion of the ligand-binding domain, LBD) and by three hydrophobic regions (M1, M2, M3). M1 and M3 cross the membrane, whereas M2 forms a loop in the membrane. Next, there is the S2 domain (a second portion of the LBD), followed by another transmembrane hydrophobic region (M4), which is part of the transmembrane domain (TMD), and a cytoplasmic C-terminal portion containing regions of interaction with PSD proteins. (b) Three-dimensional images of the receptor channel complex. The tetrameric complex is formed by the assembly of two dimers. (c) Schematic representation (top and a lateral view) of the receptor complex under basal conditions (channel closed), after glutamate application (channel open), and in desensitized state (glutamate is present, but the channel remains closed). The electrophysiological responses to glutamate application are designed in the upper part of the figure.

glial cells, respond to glutamatergic stimulation with a strong calcium influx.

It is interesting to note that, at the Q/R site, the *GluA2* gene contains a codon specifying for glutamine (Q), which in the mRNA is converted into an arginine (R) codon by a process called "mRNA editing." The physiological properties of cells expressing AMPA receptors are thus governed not only by mechanisms controlling the activity of the subunit-encoding genes but also by the mRNA editing enzymes.

Kainate can stimulate AMPA receptors causing their prolonged depolarization. This may lead to important changes in cellular homeostasis and may eventually lead to excitotoxic cell death.

Kainate Receptors Functional kainate receptors may be obtained in oocytes injected with GluK1–GluK5 encoding mRNAs (see Fig. 43.4). Since messengers for these receptor subunits are widely expressed in the CNS, it is

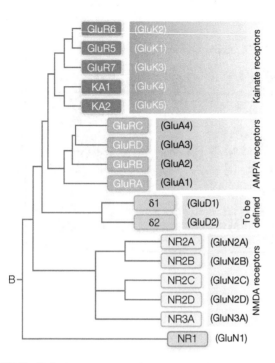

FIGURE 43.4 Dendrogram of ionotropic glutamate receptor subunits. The original and the IUPHAR names of the subunits are indicated (see Traynelis S.F., Wollmuth L.P., McBain C.J., Menniti F.S., Vance K.M., Ogden K.K., Hansen K.B., Yuan H., Myers S.J., Dingledine R. (2010). Glutamate receptor ion channels: structure, regulation, and function. *Pharmacological Reviews*, 62, 405–496).

usually assumed that kainate receptors are important for the physiological functioning of neural networks. When studied in brain preparations, the electrophysiological responses mediated by specific kainate receptors are masked by kainate depolarizing effects on AMPA receptors (see earlier). It has been demonstrated that kainate receptors are particularly abundant in presynaptic membranes and may modulate neurotransmitter release and strength of synaptic connections.

In oocytes, responses to applied kainate can be observed after injections of mRNAs encoding GluK1–GluK3 but not GluK4–GluK5. However, coexpression of GluK4 or GluK5 drastically changes the electrophysiological responses of oocytes injected with GluK1–GluK3, which are otherwise similar to those observed in neurons. This suggests that physiological receptors are formed by a combination of several subunits. Both GluK1 and GluK2 may have R or Q at the Q/R site. Furthermore, GluK2 has two other sites in which mRNA editing may occur, resulting in functional differences. Thus, overall, the *GluK2* gene can give rise to eight different proteins, with different electrophysiological properties. This implies that the properties of a neuron are determined not only by gene expression mechanisms but also by editing processes and events regulating these processes.

NMDA Receptors NMDA activates a particular type of receptor channel complex having a high permeability to Ca^{2+} and significantly slower (hundreds milliseconds) kinetics compared to AMPA channels. The complex is formed by the GluN1 subunit, at least one of either GluN2A or GluN2D subunits, and in many cases also by either GluN3A or GluN3B. In order to open the channel, the simultaneous presence of both glutamate and glycine (or D-serine) is required. Glutamate interacts with the GluN2 subunits, while glycine (or D-serine) interacts with the dimers formed by the GluN1 or GluN3 subunits. At resting membrane potential, NMDA receptors are inactive. This is due to voltage-dependent blockade of the channel pore by Mg^{2+}. However, when the cell is depolarized (e.g., during sustained activation of AMPA channels), Mg^{2+} blockade is removed and the channel becomes permeable to ions. All the subunits forming the receptor complex have an asparagine residue (N) at the Q/R site in M2: this amino acid is critical in determining both the Mg^{2+} blockade and the high Ca^{2+} permeability.

Others factors able to modulate channel openings are (i) intracellular polyamine content, (ii) extracellular proton concentration, (iii) oxidation/reduction state of the -SH groups present in the receptor complex proteins, and (iv) activation of metabotropic glutamate receptors (see the following text). Thus, NMDA receptor activation is rather peculiar and different from that of all other receptor channels. To summarize, the pore opens only under depolarization conditions (required to remove Mg^{2+} blockade), activation requires two neurotransmitters (glutamate and glycine/serine), and the ion flux is finely tuned by the state of the synapses. A possible reason for such a complex modulation of the NMDA-dependent Ca^{2+} influx in neurons is the elevated permeability of the channel pore that in the absence of a tight control could lead to important biological effects also on cell survival (see excitotoxicity). Other important biological effects mediated by NMDA channel activation, especially during brain development, are neural network formation and regulation of synaptic efficiency (see long-term potentiating, LTP, and long-term depression, LTD).

Metabotropic Receptors

The metabotropic glutamate receptors (mGluRs) are a group of eight different GPCRs classified into three subgroups based on sequence similarity, intracellular signaling mechanisms, and pharmacology:

1. The first subgroup includes mGluR1 and mGluR5 and is coupled to G_q/G_{11} proteins and activation of PLC and inositol cascade. DHPG is the most used agonist for both mGluR1 and mGluR5.
2. The second subgroup is composed of mGluR2 and mGluR3 and is coupled to G_i proteins and adenylyl

cyclase inhibition. DCG-IV is a selective agonist for these receptors.

3. The third subgroup includes mGluR4, mGluR6, mGluR7, and mGluR8. All receptors of this group are associated with G_i proteins and adenylyl cyclase inhibition. L-AP4 is the most used agonist for these receptors.

The mGluRs, together with the $GABA_B$-, the pheromone-, the vomeronasal- and the taste-receptors, belong to the C family of GPCRs (see also Chapter 17). Other members of this GPCR family are the Ca^{2+} sensor receptors (CaSR) expressed in cells controlling calcium homeostasis (parathyroid, kidney, or bone).

As shown in Figure 43.5, the molecular structure of mGluRs includes (i) a long N-terminal domain consisting of two globular portions (LB1 and LB2) that enclose the glutamate recognition site, (ii) a cysteine-rich domain, (iii) seven hydrophobic membrane-spanning domains (M1–M7) connected by three cytoplasmic loops, and (iv) a CTD containing phosphorylation sites.

The eight mGluRs are differentially distributed in the CNS, and multiple subtypes can coexist in the same neuronal population and in the same synapses. The final effects of their activation can be both excitatory and inhibitory. For example, activation of presynaptic mGluRs capable of regulating neuron transmitter release mechanisms may lead to interesting modulation of neuronal network: stimulation of mGluR2, mGluR4, or mGluR7 can reduce Ca^{2+} entry in nerve terminals by directly inhibiting N-type voltage-dependent calcium channels, thereby inhibiting synaptic transmitter release. As a result, inhibition of GABA release may increase the system excitability, while inhibition of glutamate release decreases synaptic excitation.

In the cerebellum, for example, stimulation of mGluR1 leads to activation of Ca^{2+}-dependent potassium channels with consequent hyperpolarization; conversely, in the hippocampus, activation of mGluR1 increases neuronal excitability possibly because these receptors can modulate different types of ion flux.

Of particular interest is the modulation of hippocampal neuronal circuits mediated by mGluRs. Activation of mGluR4 or mGluR7 reduces glutamate release, whereas mGluR5 activation amplifies ionotropic NMDA and AMPA responses and the overall excitability of the circuit. Stimuli of low intensity are therefore blocked, while higher intensity stimuli, able to overcome presynaptic inhibition, are amplified. In this way, the strategic location of different mGluRs leads to the formation of filtering systems capable of increasing the signal/noise ratio, thereby improving the network efficiency.

In a typical excitatory synapse, ionotropic and metabotropic receptors coexist in both pre- and postsynaptic membranes. The molecular mechanisms regulating the efficacy of synaptic transmission and synaptic plasticity are actively investigated in many neuroscience laboratories. In recent years, the discovery of allosteric sites in the seven transmembrane spanning domains of each mGluR has allowed to obtain allosteric modulators having either positive (positive allosteric modulators (PAM)) or negative (negative allosteric modulators (NAM)) actions on each of the different mGluRs. Hopefully, these modulators will be helpful not only to study the physiology of excitatory transmission but also to improve the symptoms of several neurodegenerative disorders, such as Parkinson's disease and Huntington's chorea. Modulation of mGluR function has also been proposed as a possible approach to the treatment of seizures, pain, anxiety, mood disorders, and postischemic brain damage (see Table 43.1).

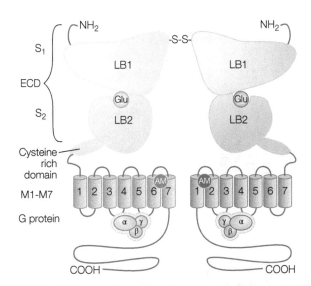

FIGURE 43.5 Metabotropic glutamate receptor (mGluRs) structure. The mGluRs have a large N-terminal extracellular domain (ECD) formed by two globular regions (S_1 and S_2) containing the glutamate binding site (LB1 and LB2) and a cysteine-rich domain that connects the extracellular part with the seven transmembrane domains. The intracellular loops contain interaction sites for different G proteins. In the TM6–TM7 region, there is a binding site for allosteric modulators (AMs). The receptor is depicted in a dimeric form.

GLUTAMATE NEUROTRANSMISSION IN PHYSIOLOGY AND PATHOLOGY

Glutamate and Excitotoxicity

An increased glutamate concentration in brain extracellular spaces is toxic as it can activate a particular type of neuronal damage commonly named "excitotoxicity." Excitotoxic brain damage seems to play an important role in stroke, brain trauma, cardiac arrest, and seizures. Other disorders in

TABLE 43.1 Drugs acting on metabotropic glutamate receptors

	Group I		Group II		Group III			
	mGlu1	mGlu5	mGlu2	mGlu5	mGlu4	mGlu6	mGlu7	mGlu8
Transduction		Phospholipase C Adenylyl cyclase Phospholipase D Jun kinase mTOR/p70 kinase	PI3 kinase Adenylyl cyclase		Adenylyl cyclase PI3 kinase MAP-kinase			
Agonists	DHPG	DHPG CHPG	DCGIV	DCGIV	L-AP4	L-AP4	L-AP4	L-AP4
Competitive antagonists	AIDA LY367385 3-Matida							
NAM	CPCCOEt	MTEP Fenobam					MMPIP	
PAM	Ro67-4853	CPPHA ADX47273	LY487379		PHCC		AMN82	
Proposed therapeutic uses								
Antagonists and NAMs	Neuro- protection Pain	Addiction Obesity X fragile syndrome Schizophrenia	Schizophrenia Anxiety	Neuro- protection	Parkinson's disease Neuro- blastoma		Epilepsy	
Agonists and PAMs	Nootropics Ataxia					Retinitis pigmentosa	Anxiety Parkinson's disease	Addiction

which excitotoxicity has been involved are Alzheimer's and vascular dementias, Huntington's chorea, Parkinson's disease, HIV, or other viral meningoencephalitis.

Neuronal death due to excessive glutamate accumulation in the extracellular space may have the characteristics of either necrosis or programmed cell death depending on the area involved and the cerebral metabolic status. A key role in excitotoxic neuronal loss is played by excessive calcium accumulation in the cytosol resulting in activation of several enzymes such as peptidases (caspases, cathepsins), lipases, endonucleases, nitric oxide synthase, and xanthine oxidase. The increased activity of some of these enzymes can lead to free radical formation and DNA damage that in turn can activate poly(ADP-ribose) polymerases with depletion of cellular ATP storage and activation of death processes.

Excessive stimulation of glutamate receptors may occur as a result of excessive neurotransmitter release, reduced transporter function, or impaired function of the blood–brain barrier. It is also possible that molecules active on glutamate receptors are ingested with food or formed during intermediary metabolism. In 1987, a significant number of inhabitants of Prince Edwards Island, Canada, presented abdominal pain and diarrhea, which was followed by amnesia and dementia. In the few cases that reached autopsy, a diffuse loss of hippocampal pyramidal cells with hippocampal gliosis was detected. It was subsequently demonstrated that both the symptoms and the neuropathology were due to domoic acid ingestion. Domoic acid, an agonist of kainate receptors, was found in abnormally high concentrations in mussels eaten by the affected individuals. The toxic compound is synthesized by a type of algae that flourished in an extremely abundant manner in 1987 and had been accumulated in those mussels.

Serious neurological diseases may also be caused by excessive ingestion of foods such as the flour prepared from palm tree *Cica circinalis* seeds, which are rich in β-methyl amino aniline (BMAA) and are widely used in the island of Guam (Comoro). A high incidence of degenerative diseases such as ALS, Parkinson's disease, and dementia has been demonstrated in the Guam population.

Another neurological disease possibly due to ingestion of excitotoxic substances is lathyrism, a condition characterized by spastic paraplegia, which affects population eating abundant amount of *Lathyrus sativus* (grass pea). These peas contain β-oxalylamino-alanine (BOAA), another amino acid possibly active on glutamate receptors. The disease is common in East Africa and South Asia where the grass pea is widely used.

Glutamate and Depression

In the last few years, it has been repeatedly reported that ketamine, a general anesthetic capable of antagonizing NMDA receptors, rapidly improves mood tone when administered to depressed patients even those unresponsive to the available treatments (SSRIs, SNRIs). This finding is consistent with the common observation that large alcohol doses may attenuate "melancholy" (alcoholics are often depressed) and that alcohol (at relatively elevated concentrations) inhibits NMDA receptors.

Depression symptoms improve few hours after low doses of intravenous drug administration and the therapeutic action last 3–4 days. Since common antidepressants need a prolonged treatment (3–6 weeks) prior their results can be appreciated, ketamine actions appear particularly interesting. Unfortunately, since ketamine may cause hallucinations and drug dependence, it is difficult to propose its use in depressed patients.

Ketamine is a noncompetitive antagonist of most NMDA receptor channel subtypes. Since selective antagonists of receptors formed exclusively by GluN1 and GluN2B subunits (e.g., Ro 25-6981) are now available, clinical trials with these selective antagonists have been proposed and are in progress. The results emerging from these studies should allow assessing risks and benefits of selective NMDA antagonists in the numerous SSRI-resistant depressed patients. It remains to be clarified which is the action mechanism of NMDA receptor antagonists in depression and how ketamine, which has a half-life of about 2 h, can reduce depressive symptoms for 3–4 days. A clear answer to these questions is mandatory before clinical use of these molecules outside carefully controlled clinical trials could be permitted. However, NMDA antagonists are a therapeutic option for unresponsive, depressed patients, and it would be important to not underestimate the relevance of their therapeutic use, especially in the presence of high suicide risk.

DRUGS AND EXCITATORY NEUROTRANSMISSION

Despite the initial enthusiasm raised by the promising results obtained with glutamate receptor antagonists in cellular or animal models of stroke or head trauma, the results of several controlled clinical trials that have shown that many molecular entities proposed for therapeutic use have an unacceptable risk/benefit ratio. For example, antagonists of AMPA and kainate receptors (e.g., CNQX or DNQX) were reported to cause damage to renal tubules, even at low doses. NMDA receptor antagonists were abandoned because they caused hallucinations, and finally, selective antagonists of each of the eight metabotropic glutamate receptors were abandoned because of pharmacokinetic issues.

Other therapeutic agents able to inhibit ionotropic glutamate receptors are barbiturates and most volatile anesthetics. However, their specificity is rather poor.

The only commercially available therapeutic agents selectively interacting with glutamate-mediated neurotransmission are the three molecules reported in Table 43.2. The first one is ketamine, a general anesthetic (defined also dissociative anesthetic because of its modest effects on the autonomic nervous system), which is a potent, noncompetitive antagonist of the NMDA receptor channel complex; ketamine is still used not only in veterinary anesthesiology but also in humans possibly because it can also be administered intramuscularly and this can be an advantage in emergency situations. Unfortunately, ketamine may cause hallucinations and even neuronal death, and when repeatedly administered, it causes addiction. The second is memantine, which is also active on the NMDA receptor Ca^{2+} channel complex. By virtue of its low affinity for the target, it seems to have a favorable risk/benefit ratio, and it is the only NMDA receptor antagonist authorized for the treatment of vascular dementia and moderate or severe Alzheimer's disease. Memantine seems capable of limiting glutamate excitotoxic effects, thus slowing the disease progression.

Finally, the third therapeutic agent is riluzole, a molecule capable of reducing glutamate release. Chronic use of riluzole has been proposed to reduce progression of ALS. Controlled clinical studies have shown that long-term use of this molecule may increase survival of ALS patients by few months.

In spite of the relative failure of glutamatergic drugs as therapeutics, a number of compounds are available for experimental investigation of the glutamatergic system (Fig. 43.6).

TABLE 43.2 Drugs and excitatory neurotransmission

Drug	Mechanism of action	Clinical use
Ketamine	High-affinity NMDA channel blocker	General anesthetic
Memantine	Low-affinity NMDA channel blocker	Dementia
Riluzole	Glutamate release inhibitor	Compassionate use in amiotrophic lateral sclerosis

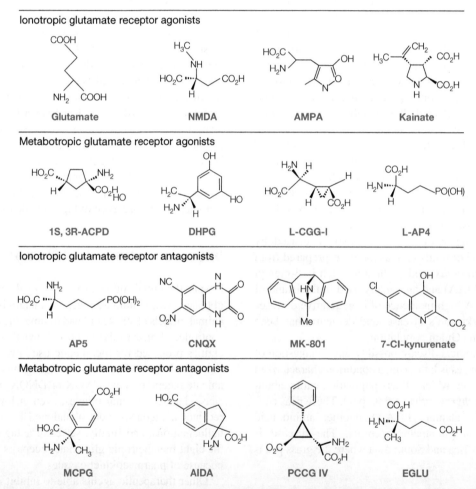

FIGURE 43.6 Agonists, antagonists, and modulators of ionotropic and metabotropic glutamate receptors used for experimental purposes. NMDA, AMPA, and kainate are the prototype agonists of ionotropic receptors. 1S, 3R-ACPD is a nonspecific agonist of metabotropic receptors, and DHPG is an agonist of the first mGluR group; L-CGG-1 stimulates the second mGluR group; L-AP4 stimulates the third mGluR group. AP5 is a competitive antagonist of NMDA receptors, CNQX is a competitive antagonist of AMPA receptors, and MK-801 is a noncompetitive antagonist of NMDA receptors; 7Cl-kynurenate is an antagonist of the glycine site on NMDA receptors; MCPG is a nonspecific antagonist of mGlu receptors, and AIDA is an mGluR1 antagonist; PCGIV is an mGluR2/mGluR3 antagonist; EGLU is an mGluR3 antagonist.

It remains to be noted that given the importance of metabotropic receptors in synaptic function, the therapeutic interest has now moved from antagonists at ionotropic receptors to molecules capable of modulating mGluR receptor responses. PAM and NAM are currently being evaluated in a large variety of psychiatric or neurological conditions (anxiety, depression, schizophrenia, addictions, chronic pain, some forms of epilepsy, learning disorders, stroke, and other neurodegenerative diseases).

GLUTAMATE-MEDIATED NEUROTRANSMISSION IN BRIEF

1. Glutamate is the most abundant excitatory neurotransmitter in the CNS. It is synthesized from glucose and glutamine, and it is stored at high concentrations in the synaptic vesicles.
2. Glutamate can act on both ionotropic and GPCR (metabotropic receptors). Ionotropic glutamate receptors are subdivided into NMDA, AMPA, and kainate receptors. NMDA receptors are permeable to calcium ions, while AMPA and kainate receptors are mostly permeable to sodium. Metabotropic receptors (mGluRs) can be distinguished in (i) receptors associated with phospholipase C activation (group 1) and (ii) receptors associated with adenylyl cyclase inhibition (groups 2 and 3).
3. AMPA and kainate receptors mediate fast synaptic transmission, while NMDA receptors, with their peculiar requirement of two transmitters (glycine and glutamate) and their sensitivity to membrane voltage and Mg^{2+} ions, are involved in synaptic plasticity, learning, and memory process. Metabotropic receptors contribute to the control of synaptic efficiency and neuronal network functioning.
4. An excessive concentration of glutamate in the extracellular spaces of the CNS may cause excitotoxic neuronal death. This type of neuronal loss plays a role in brain damage occurring after stroke, trauma, hypoglycemia, or prolonged seizures.
5. NMDA receptor activation can be prevented by blocking the channel (ketamine, MK801, phencyclidine) or by competing with either glutamate (AP5) or with glycine (7Cl-kynurenate). AMPA and kainate receptors are antagonized in a competitive manner by quinoxalines (CNQX or DNQX) and by noncompetitive antagonists such as GYKI-52466.
6. NMDA receptor antagonists are effective in preventing seizures and in reducing excitotoxic cell death. A noncompetitive NMDA receptor antagonist, memantine, has been approved for the treatment of vascular dementia. Another, ketamine, has been used for decades as an anesthetic agent and a drug of abuse. Its use has been proposed in depressed patients but requires further investigation.
7. The use of numerous mGluR agonists, antagonists, and selective modulators has been suggested in a large number of neurological and psychiatric diseases. Clinical trials with these agents are currently in progress.

TAKE-HOME MESSAGE

- Glutamate is the major excitatory neurotransmitter in CNS and is taken up by specific transporters present in neurons and glial cells. Intracellular glutamate is concentrated in presynaptic vesicles by a vesicular glutamate transporter ($vGLUT_{1-3}$).
- Three classes of tetrameric ionotropic glutamate receptors (AMPA, kainate, and NMDA receptors) exist differing for subunit composition and pharmaco- and electrophysiological properties.
- AMPA receptors are cationic LGICs, highly conductive for Na^+ and made up of GluA subunits. Presence of arginine (R) at the Q/R editing site of the $GluA_2$ subunit makes most AMPA receptors impermeable to Ca^{2+}.
- Kainate receptors are widely expressed cationic channels composed of GluK subunits. They are heteromeric and both pre- and postsynaptic receptors. Q/R editing sites are present in $GluK_1$ and $GluK_2$ mRNAs.
- NMDA receptors contain a $GluN_1$ and at least one $GluN_2$ subunit and are highly permeable to Ca^{2+}. Receptor activation requires glutamate and glycine binding and membrane depolarization. Calcium entered through these channel receptors acts as second messenger and induces specific responses ranging from synaptic memory to cell death.
- Eight genes coding for metabotropic glutamate receptors (mGluRs) have been identified. mGluRs are divided in three classes: group I (mGluR1 and mGluR5) is mostly coupled to G_q, whereas groups II (mGluR2 and mGluR3) and III (mGluR4, mGluR6, mGluR7, mGluR8) are mostly coupled to G_i. Direct agonists/antagonists and allosteric modulators of mGluRs are in clinical trials for the therapy of neurological and psychiatric disorders.

FURTHER READING

Dolgin E. (2013). Rapid antidepressant effects of ketamine ignite drug discovery. *Nature Medicine*, 19, 8.

Kew J.N., Kemp J.A. (2005). Ionotropic and metabotropic glutamate receptor structure and pharmacology. *Psychopharmacology*, 179, 4–29.

Nicoletti F., Bockaert J., Collingridge G.L., Conn P.J., Ferraguti F., Schoepp D.D., Wroblewski J.T., Pin J.P. (2010). Metabotropic glutamate receptors: from the workbench to the bedside. *Neuropharmacology*, 60, 1017–1041.

Niswender C.M., Conn P.J. (2010). Metabotropic glutamate receptors: physiology, pharmacology, and disease. *Annual Review of Pharmacology and Toxicology*, 50, 295–322.

Paoletti P., Bellone C., Zhou Q. (2013). NMDA receptor subunit diversity: impact on receptor properties, synaptic plasticity and disease. *Nature Reviews. Neuroscience*, 14, 383–400.

Sheldon A.L., Robinson M.B. (2007). The role of glutamate transporters in neurodegenerative diseases and potential opportunities for intervention. *Neurochemistry International*, 51, 333–355.

Traynelis S.F., Wollmuth L.P., McBain C.J., Menniti F.S., Vance K.M., Ogden K.K., Hansen K.B., Yuan H., Myers S.J., Dingledine R. (2010). Glutamate receptor ion channels: structure, regulation, and function. *Pharmacological Reviews*, 62, 405–496.

Urwyler S. (2011). Allosteric modulation of family C G-protein-coupled receptors: from molecular insights to therapeutic perspectives. *Pharmacological Reviews*, 63, 59–126.

Watkins J.C., Jane D.E. (2006). The glutamate story. *British Journal of Pharmacology*, 147(Suppl. 1):S100–S108.

44

PURINERGIC TRANSMISSION

Stefania Ceruti, Flaminio Cattabeni, and Maria Pia Abbracchio

> **By reading this chapter, you will:**
>
> - Learn about the release, metabolic consequences and catabolism of purines and pyrimidines, their role as neurotransmitters and neuromodulators and their receptor-mediated and receptor-independent actions in various organs and systems.
> - Understand the molecular structure, pharmacology, signalling mechanisms and effects mediated by P1 and P2 receptors for nucleosides and nucleotides in the nervous, cardiovascular, respiratory and immune systems;
> - Acquire the background to understand the mechanism of action of anti-platelet drugs already in use in therapy for many years (namely, dipyridamole, ticlopidine, clopidogrel and, more recently, ticagrelor) and of natural compounds (like caffeine, theophylline and theobromine), and to discuss the reasons why specific P1 and P2 receptor subtypes represent new molecular targets to develop innovative drugs for the therapy of human diseases, like thrombosis, cystic fibrosis, Parkinson's disease, neuropathic pain and acute and chronic neurodegenerative disorders.

It has long been known that purines (adenine nucleosides and nucleotides, such as adenosine and adenosine triphosphate (ATP)) have a role in energy metabolism and in nucleic acid synthesis, but their role in intercellular signaling has been demonstrated only recently. The first indication that purines could also have a specific activity dates back to 1929, when Drury and Szent-Györgyi described their potent vasodilator and bradycardic effects. However, it was only in 1963, when production of adenosine was observed in heart hypoxia, that a role of purines as regulators of coronary blood flow was hypothesized. The demonstration that the adenosine effects were blocked by methylxanthine caffeine suggested the existence of specific membrane receptors, antagonized by methylxanthines. At the same time, the hypothesis that ATP could have a role in neurotransmission was being corroborated by studies performed by Geoffrey Burnstock, who, in 1972, proposed that ATP served as a neurotransmitter in "noncholinergic, nonadrenergic" (NANC) synapses in the autonomic nervous system. Initially, this "purinergic theory" was not well accepted, as it was difficult to believe that such ubiquitous molecules as nucleosides and nucleotides, at that time merely considered metabolic precursors, could take part in such a specific activity as synaptic transmission. In the following years, mounting evidence confirmed that adenosine and ATP are important regulators of homeostasis in the central nervous system (CNS) and peripheral nervous system, as well as in the cardiovascular, respiratory, gastrointestinal, renal, and immune systems. Additional studies have shown the role of purines in emergency situations such as cerebral and myocardial infarct, epileptic seizures, and infection where these molecules serve as "danger signals." Moreover, it has been ascertained that alterations of the purinergic system have a role in the pathogenesis of many human diseases, such as Alzheimer's disease, multiple sclerosis, immune deficits, and tumors. In addition, characterization of purinergic receptors has led to the demonstration that pyrimidine derivatives such as UTP and UDP (and more recently pyrimidines bound to sugars, such as UDP-glucose and UDP-galactose) (Fig. 44.1) can also activate these receptors and that these molecules can be considered as potential neurotransmitters as well. Today, the concept of "purinergic transmission" is accepted worldwide, and it refers to the intercellular signaling sustained by purine and pyrimidine derivatives.

General and Molecular Pharmacology: Principles of Drug Action, First Edition. Edited by Francesco Clementi and Guido Fumagalli.
© 2015 John Wiley & Sons, Inc. Published 2015 by John Wiley & Sons, Inc.

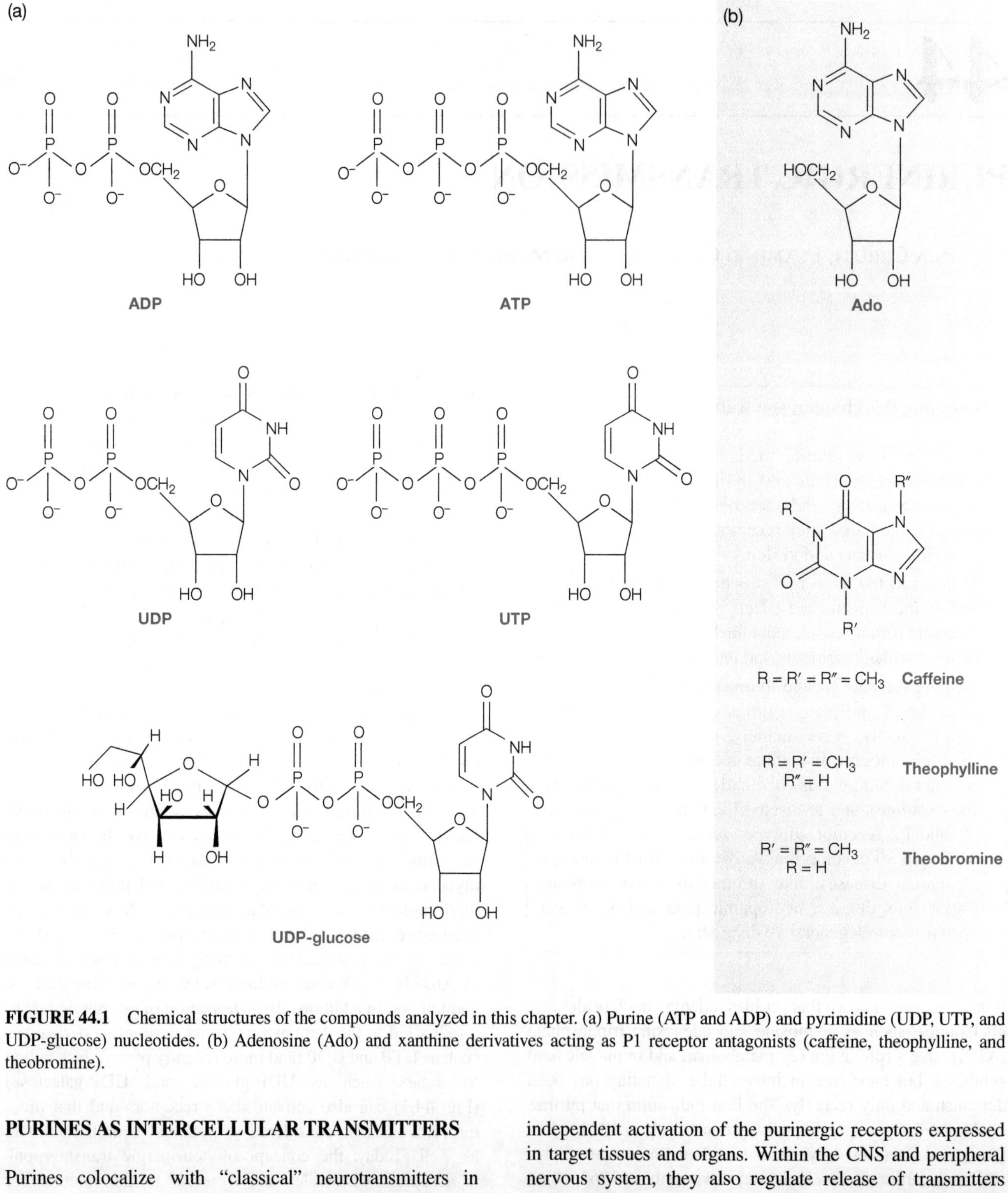

FIGURE 44.1 Chemical structures of the compounds analyzed in this chapter. (a) Purine (ATP and ADP) and pyrimidine (UDP, UTP, and UDP-glucose) nucleotides. (b) Adenosine (Ado) and xanthine derivatives acting as P1 receptor antagonists (caffeine, theophylline, and theobromine).

PURINES AS INTERCELLULAR TRANSMITTERS

Purines colocalize with "classical" neurotransmitters in various synapses of the CNS and peripheral nervous system. They also colocalize with some releasable mediators in blood circulating cells, such as platelets. Their effects are partially due to modulation of the effects induced by the transmitters they are coreleased with, and partially to the independent activation of the purinergic receptors expressed in target tissues and organs. Within the CNS and peripheral nervous system, they also regulate release of transmitters through presynaptic receptors. Further complexity to the purinergic system is added by the following evidence: (i) the neurotransmitter is not a single type of molecule, but is indeed a family of compounds, all biologically active (ATP, ADP, and adenosine), resulting from stepwise hydrolysis of ATP

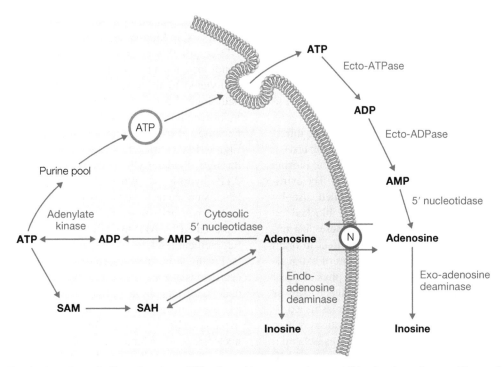

FIGURE 44.2 Synthesis and metabolism of purines. ATP, released by exocytosis or possibly also through a specific membrane transporter, is metabolized in the extracellular space by ectoenzymes to ADP, AMP, and adenosine. Adenosine inactivation may occur by deamination to inosine thanks to the exo-adenosine deaminase enzyme or by intracellular reuptake through the bidirectional nucleoside transporter (N), followed by phosphorylation and recycle in the intracellular purine pool. Apart from ATP hydrolysis, adenosine can also derive from the intracellular metabolism of S-adenosylmethionine (SAM) to S-adenosylhomocysteine (SAH).

(Fig. 44.2); (ii) these neurotransmitters are released not only by neurons and platelets but also by several other tissues, especially in metabolic emergencies (see the following text). These factors, together with the difficulty to clearly distinguish the "transmitter" pool of purines from that involved in metabolic and cellular processes (energy production, nucleic acid synthesis, enzyme catalysis, etc.), have made the identification of truly purinergic transmission very difficult.

Source, Metabolism, and Release of Purines

In this section, we will describe how formation, metabolism, and release of purines involved in neurotransmission are regulated. We will not describe the processes of biosynthesis, catabolism, and conversion of purine nucleotides in cellular metabolism (better described in text of classical biochemistry) except for the role of the metabolic pool of purines in conditions of emergency, such as ischemia and hypoxia (see the following text and Supplement E 44.1, "Purines and Ischemia").

In platelets, adenine nucleotides are stored as dense aggregates and are released by exocytosis upon platelet activation. In the presynaptic terminals of the nervous system, ATP is stored in synaptic vesicles along with "classical" neurotransmitters (i.e., noradrenaline, acetylcholine, dopamine, serotonin, excitatory amino acids, and peptides). In presynaptic terminals of the peripheral nervous system, catecholamines and ATP are stored at a 4:1 molar ratio. Whether adenosine is accumulated in synaptic vesicles is still a matter of debate. In physiological conditions, only a calcium-dependent purine release from the presynaptic vesicular pool is observed. In emergency conditions (e.g., hypoxia), where alterations of plasma membrane permeability (and eventually membrane rupture) occur, purines are also released from their metabolic pool, not only by nervous cells, but also by endothelial, muscle, and red blood cells and by any other cell types damaged by the cytotoxic insult. Thus, huge increases in the local purine and pyrimidine concentrations are detected in ischemic/hypoxic or traumatic areas, also due to degradation of nucleic acid released from dead cells (see Supplement E 44.1).

Generation of Active Metabolites by ATP Hydrolysis

After release, ATP is a substrate of enzymes called ecto-ATPases (ectonucleotidases), which degrade it sequentially to ADP, AMP, and adenosine (Fig. 44.2). All these compounds are able to activate, with different efficacy, various purinergic receptors (see the following text). Recent findings have shown pyrimidine di- or triphosphates (UDP and UTP) as well as UDP-glucose and UDP-galactose (i) are also substrates of ecto-ATPases and (ii) can activate the same G-protein-coupled receptors (GPCRs) activated by ATP and ADP (see the following text), suggesting that also these molecules function as intercellular mediators, in addition to their well-known role in cellular metabolism. To further support their biological role, UTP (which is particularly concentrated in circulating erythrocytes) has been shown to be significantly released in blood of patients undergoing heart failure. Although release and catabolism of uridine nucleotides are far less known, they seem to occur in a way similar to ATP. For UTP, the final product is uridine, which instead does not seem to have any biological activity. Both UTP and UDP are active on P2 purinergic receptors, even if with a response profile different from that of adenine nucleotides (see the following text).

Adenosine is degraded to the inactive metabolite inosine by the enzyme exo-adenosine deaminase (ADA). The extracellular concentration of adenosine is controlled also by reuptake via specific transporters, which are blocked by dilazep and dipyridamole. It is assumed that these pharmacological compounds exert their therapeutic effect in brain and cardiac ischemia by increasing the amount of extracellular adenosine (see also Supplement E 44.1). In physiological conditions, reuptake is the most relevant mechanism to lower extracellular concentrations of adenosine, whereas its extracellular deamination is more relevant in situations where adenosine concentrations are much higher, such as ischemia.

Following reuptake, adenosine is phosphorylated to AMP by an intracellular adenosine kinase, and it is therefore recycled in the purine cellular pool; intracellular adenosine can also be deaminated to inosine by an endo-ADA. K_m values of adenosine kinases are much lower than those of endo-ADAs; for this reason, it is assumed that under normal conditions, the majority of the reuptaken nucleoside is phosphorylated to form again ATP. On the other hand, under pathological conditions, characterized by massive increases in intracellular concentrations of adenosine, deamination becomes the most relevant mechanism. As shown in Figure 44.2, intracellular adenosine may also be produced through S-adenosylhomocysteine (SAH) hydrolysis and ATP hydrolysis. Whereas in physiological conditions adenosine seems to be mostly produced via SAH hydrolysis, in hypoxia and ischemia, the main metabolic pathway is represented by ATP hydrolysis (see the following text and Supplement E 44.1).

Finally, it has been recently demonstrated that tissue concentrations of ectonucleotidase enzymes increase substantially under particular physiological conditions (i.e., in development-associated neurogenesis) or during damage repair (i.e., after hypoxia or brain trauma). This suggests that nucleotides and nucleosides play a fundamental role in migration of embryonic stem cells, as well as in their terminal differentiation and recruitment of adult stem cells following traumatic events (see the following text).

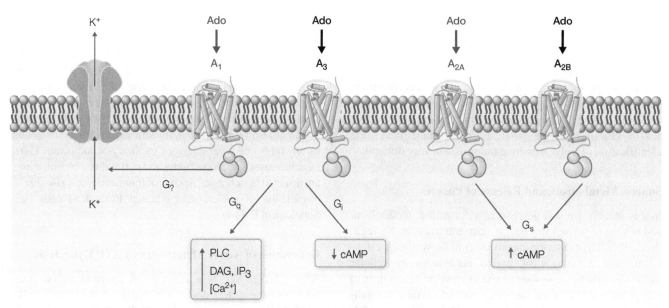

FIGURE 44.3 P1 adenosine receptors. Four receptor subtypes activated by adenosine are currently recognized (viz., the A_1, A_{2A}, A_{2B}, and A_3 subtypes). The A_1 and A_{2A} subtypes are activated by low (nanomolar) concentrations of the endogenous nucleoside, as detected in tissues under physiological conditions. Conversely, the A_{2B} and A_3 subtypes are preferentially activated by micromolar adenosine concentrations and are therefore recruited mostly under pathological conditions, when a marked increase in the extracellular purine levels is observed. A_1 and A_3 receptor subtypes are coupled both to adenylyl cyclase inhibition, leading to a reduction in cAMP levels, and to activation of phospholipase C, which in turn promotes IP_3 production and calcium release from the intracellular stores. Moreover, the A_1 receptor subtype can also promote potassium ion outflow from the cytosol, thanks to its coupling to a yet-to-be-identified G protein, leading to plasma membrane hyperpolarization. Conversely, the A_{2A} and A_{2B} receptor subtypes are generally coupled to Gs members of the G-protein family, which positively regulate adenylyl cyclase and cAMP production.

RECEPTORS FOR PURINES

In 1978, Geoffrey Burnstock proposed the existence of two types of receptors for purines that he named P1 (activated by adenosine) and P2 (activated by ATP) receptors. P1 receptors are competitively inhibited by xanthines, such as caffeine, theophylline, and theobromine (Fig. 44.1), which are completely inactive on P2 receptors. This subdivision is maintained today and has served as basis for current classification and nomenclature of purinergic receptors (see also Supplement E 44.2, "Drugs Acting on Purinergic System").

Each of the two families is composed of a number of receptor subtypes with peculiar molecular structure, transduction mechanisms, and pharmacological profile.

P1 Adenosine Receptors

Four different P1 receptors have been cloned and characterized: A_1, A_{2A}, A_{2B}, and A_3. All these receptors are coupled to G proteins. Like all GPCRs, P1 receptors are made of a single polypeptide spanning sevenfold the membrane, with extracellular N-terminus and intracellular C-terminus. A_1 and A_3 receptors inhibit adenylyl cyclase, whereas A_2 (both A_{2A} and A_{2B}) subtypes stimulate it through Gi and Gs proteins, respectively (Fig. 44.3). In some tissues, A_1 and A_3 receptors also stimulate phospholipase C. A_1 receptor can also interact with various other types of G_i, G_o, and G_k proteins leading to inhibition of Ca^{2+} and stimulation of K^+ conductance, respectively.

A_1, A_2, and A_3 receptors differ significantly for their tissue distribution, the effects they evoke (Table 44.1), and the pharmacological profile they show when activated with nonhydrolyzable analogs of adenosine. The chemical structure of adenosine is shown in Figure 44.1; by adding various substituents both on purine ring and on ribose moiety, molecules with different affinity and selectivity for various receptor subtypes have been synthesized. Similarly, first antagonists have been obtained by modifying the structure of natural xanthines, such as theophylline (Fig. 44.1). More recently, molecules with different chemical structures but having potent antagonist activity have been also synthesized.

P2 Receptors for ATP

On the basis of their membrane topology and transduction mechanism, two major families of P2 receptors are known: ion channel receptors (P2X) and GPCRs (P2Y). Each of them comprises numerous subtypes.

P2X Ion Channel Receptors

P2X receptors mediate fast and transient responses to ATP and contribute to depolarizing responses in excitable tissues. They are channels permeable to cations (Na^+, Ca^{2+}, and K^+) and are inhibited by the trypanocide suramin and, in some instances, by the nonselective compound pyridoxalphosphate-6-azophenyl-2′,4′-disulphonic acid (PPADS). Cloning of all 7 P2X receptors (P2X1-7) has led to the discovery of a new structure of ion channel receptors. Indeed, these receptors were supposed to have a topology similar to that of other already known ligand-operated receptors, such nicotinic cholinergic or glutamatergic receptors (see also Chapter 16). P2X receptors, instead, show a molecular organization more similar to amiloride-sensitive epithelial channels permeable to Na^+ and to some K^+ channels: they have only two transmembrane domains, both amino- and carboxy-terminals facing the cytoplasmic domain and a large extracellular loop (Fig. 44.4). This suggested the fascinating hypothesis that ATP could have been the first true chemical transmitter to be used by cells to communicate with the surrounding environment. To transmit its own message to distal cells, cells would have used a mediator already present in the cytoplasm and involved in

TABLE 44.1 Biological effects induced by adenosine

Effects	Receptor subtype
Central nervous system	
Inhibition of neurotransmitter release[a]	A_1
Sedation[a]	A_1
Reduction of motor activity[a]	A_{2A}
Anticonvulsant activity[a]	A_1
Chemoreceptor stimulation	A_{2A}
Cardiovascular system	
Vasodilation[a]	A_{2A}, A_{2B}, A_3
Vasoconstriction	A_1
Bradycardia	A_1
Negative inotropism[a]	A_1
Platelet inhibition[a]	A_{2A}
Renal system	
Reduction in glomerular filtration	A_1
Inhibition of diuresis[a]	A_1
Inhibition of renin release	A_1
Respiratory system	
Bronchodilation	A_2
Bronchoconstriction[a]	A_1, A_{2B}, A_3
Control of chloride ion secretion	A_1, A_{2B}
Immune system	
Immunosuppression	A_{2A}
Neutrophil chemotaxis[a]	A_1
Superoxide ion generation	A_{2A}
Mastocyte degranulation[a]	A_{2B}, A_3
Gastrointestinal system	
Inhibition of acid secretion	A_1
Metabolism	
Inhibition of lipolysis[a]	A_1
Increased sensitivity to insulin	A_1
Stimulation of gluconeogenesis	A_2

[a]Most important.

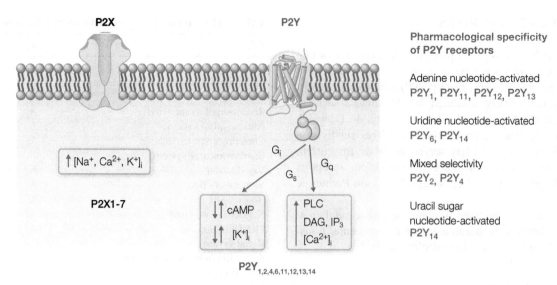

FIGURE 44.4 P2 nucleotide receptors. P2X receptors are ligand-operated ion channels permeable to mono- and divalent cation ions (Na^+, K^+ and Ca^{2+}). They therefore mediate rapid and excitatory cellular responses. Seven P2X receptor subtypes (from 1 to 7) have been cloned and characterized, which can homo- or heterodimerize in the plasma membrane (see text). P2Y receptors are seven transmembrane domain receptors coupled to G protein. To date, 8 P2Y receptor subtypes are officially recognized in mammals (viz., $P2Y_1$, $P2Y_2$, $P2Y_4$, $P2Y_6$, $P2Y_{11}$, $P2Y_{12}$, $P2Y_{13}$, and $P2Y_{14}$), which can be further subdivided based on their transduction mechanism or on their sensitivity to endogenous ligands. As for transduction mechanisms, group I receptors (i.e., $P2Y_1$, $P2Y_2$, $P2Y_4$, $P2Y_6$, $P2Y_{11}$) are generally coupled to Gq proteins, leading to PLC activation, IP_3 production, and calcium release from the intracellular stores of the endoplasmic reticulum. The $P2Y_{11}$ receptor subtype (whose rodent ortholog has not been identified so far) can also couple to adenylyl cyclase activation, through Gs. P2Y receptors belonging to group II ($P2Y_{12}$, $P2Y_{13}$, and $P2Y_{14}$) are instead coupled to Gi, with consequent adenylyl cyclase inhibition and reduction of the intracellular cAMP levels.

principal metabolic events, such as ATP. Upon release in the extracellular space, ATP would have bound to "adapted" ion voltage-gated channels that were widely expressed in very primitive cells.

The first direct evidence that ATP takes part in excitatory neurotransmission in the CNS was reported in 1992, and it is well known today that this is mediated by a fast depolarization induced by P2X receptors (Table 44.2). This response is also responsible for many other effects induced by ATP (see also the following text), such as nociception, induction of long-term potentiation (LTP) in the hippocampus (a brain area involved in learning and memory), contraction of vascular and intestinal smooth muscles, cardiac positive inotropism and chronotropism, and many other effects still to be characterized, as suggested by the extensive distribution of P2X receptors in the organism (Table 44.2). Fully functional P2X receptors result from the assembly of three subunits to form the ion channel, and it is now evident that these trimeric receptors might be formed by different P2X subunits (e.g., P2X2 and P2X3 with variable stoichiometry). These heteromers display pharmacological and functional activities that are different from the receptor made of three identical subunits, adding further complexity to the system.

P2X3 receptors are particularly interesting: their selective expression in sensory neurons in the dorsal roots in spinal horns (where they mediate nociceptive signaling often as heteromers with P2X2 subtype) makes them an interesting pharmacological target to develop analgesic compounds with a new mechanism of action and possibly devoid of side effects. Also, the P2X7 receptor (previously known as P2z) is interesting as it mediates apoptotic cell death induced by ATP derivatives (see the following text). Indeed, at physiological ATP concentrations, this receptor behaves like a normal ligand-operated receptor. However, when activated repeatedly or by high ATP concentrations, it becomes a large and relatively nonselective membrane pore. Its activation leads to a massive increase in cellular Na^+ and/or Ca^{2+}, thus leading to necrosis (cellular lysis) or apoptosis (delayed death due to increased intracellular Ca^{2+} concentrations). This receptor is typical of mastocytes and macrophages, and it has been recently identified also on microglial cells. Since the discovery that this receptor is also expressed on tumor cells, an important role has also been proposed for the control of neoplastic transformation. In fact, it has been hypothesized that some immune cells might induce P2X7 expression in transformed cells and, once activated, trigger their death by engaging tumor cells in an "immunological" synapse and massively releasing ATP. Recently, ligands with agonistic/antagonistic activities for various P2X receptors have been synthesized.

TABLE 44.2 Biological effects induced by ATP and other extracellular nucleotides

Effects	Receptor subtype
Central nervous system	
Inhibition of neurotransmitter release[a]	$P2Y_{2,4}$
Neuron activation	P2X
Astrocyte activation	P2Y
Nociception[a]	$P2X3, P2Y_2$
Peripheral nervous system	
Modulation of neurotransmission in sensory and intrinsic[a] ganglia in various organs	$P2Y_2$
Pre- and postjunctional modulation of neurotransmission in neuroeffector junctions	P2Y
Cardiovascular system	
Vasodilation[a]	$P2Y_{1,2,4}$
Positive inotropism	$P2X, P2Y_{2,6,11}$
Platelet activation[a]	$P2Y_1, P2Y_{12}$
Smooth muscles	
Contraction	P2X
Relaxation (intestine, vessels)	P2Y
Respiratory system	
Secretion of mucin and surfactant	$P2Y_2 (P2Y_6)$
Pulmonary vasoconstriction	P2X
Immune system	
Lymphocyte activation	P2X7
Secretion of histamine from mastocytes[a]	P2X7
Regulation of cell differentiation	$P2Y_{14}, P2Y_{11}$
Endocrine system	
Induction of pancreatic secretion of insulin	P2Y
Gastroenteric system	
Gastric acid secretion	P2Y
Inhibition of intestinal mobility	P2Y
Salt and fluid transport in duodenal epithelium[a]	$P2Y_4$
Renal system	
Vasoconstriction of afferent arterioles	P2X
Induction of renin secretion	$P2Y_{11}$
Musculoskeletal system	
Increased activity of osteoblasts and apoptosis of osteoclasts[a]	P2X7
Reduction of bone mineralization	$P2Y_2$
Metabolism	
Modulation of cholesterol transport	$P2Y_{13}$

[a] Most relevant.

P2Y GPCRs

P2Y receptors are ubiquitous and particularly abundant in the brain, vasculature, kidney, endocrine pancreas, and liver. To date, eight different subtypes are known in mammals: $P2Y_1$, $P2Y_2$, $P2Y_4$, $P2Y_6$, $P2Y_{11}$, $P2Y_{12}$, $P2Y_{13}$, and $P2Y_{14}$. They are characterized by relatively low sequence homology (40–60%) and by different response profiles to adenine (ATP and ADP) and uridine (UTP and UDP) nucleotides, to diadenosine polyphosphates (APxA), and to sugar-bound nucleotide (UDP-glucose and UDP-galactose) (Fig. 44.4; see also the following text). On the basis of their phylogenetic and structural features, P2Y receptors have been further divided in two subfamilies: one encompassing $P2Y_1$, $P2Y_2$, $P2Y_4$, $P2Y_6$, and $P2Y_{11}$, which are primarily associated to Gq proteins (and phospholipase C activation), and one encompassing $P2Y_{12}$, $P2Y_{13}$, and $P2Y_{14}$, which are coupled to Gi proteins (and adenylyl cyclase inhibition). Under some circumstances, some P2Y receptors can also modulate phospholipase A2 or, in the case of $P2Y_{11}$, couple to Gs and adenylyl cyclase stimulation. Receptors that are missing in the numerical sequence correspond to nonmammalian cloned receptors (whose mammalian orthologs are not yet known), to yet uncharacterized receptors, or to receptors that have been previously and erroneously assigned to the P2Y family.

There is evidence that more P2Y receptors may exist. Indeed, it has been reported that some responses induced by nucleotides and mediated by GPCR display a pharmacological profile different from already cloned receptors. Moreover, the completed sequence of the human genome has revealed that the total number of GPCR is between 700 and 1000 and several of these, whose endogenous ligand is not yet known (the so-called "orphan" receptors), are structurally similar to the known P2Y (see also the following text). One of these receptors, GPR17, has been demonstrated to respond to both uracil nucleotides and the arachidonic acid-derived cysteinyl leukotrienes, suggesting a role in inflammatory neurodegenerative diseases.

Even if several P2Y receptors are antagonized by suramin and PPADS, agonists or antagonists selective for the various subtypes do not exist, except for $P2Y_1$, $P2Y_2$, and $P2Y_{12}$ (see Supplement E 44.2). At present, efforts are being made to obtain new selective derivatives resistant to ecto-ATPases.

Among the characterized receptors, $P2Y_2$ receptor has recently raised a lot of enthusiasm. $P2Y_2$ is abundantly expressed in the epithelial respiratory system (where it regulates fluid and salt secretion) and is activated by both ATP and UTP, the only compounds able to normalize secretion of respiratory epithelia in cystic fibrosis patients. Cystic fibrosis is a lethal genetic disease characterized by an abnormal function of epithelia due to reduction in chloride ion flux. This in turn leads to dehydration of secretions and severe obstructive complications especially in the lungs. Activation of $P2Y_2$ receptors normalizes the flux of chloride ions, probably through phosphorylation of the anion channel. Agonists for this receptor have been proposed for the treatment of cystic fibrosis and for normalizing hydration in syndromes of ocular dryness and corneal lesions (for the latter application, the $P2Y_2$ agonist diquafosol is currently on the market in Japan; see Supplement E 44.2).

Unfortunately, in 2012, the ongoing clinical trial with the P2Y$_2$ agonist denufosol for cystic fibrosis has been interrupted due to lack of efficacy. In 2003, a similar activity has been demonstrated for the P2Y$_4$ receptor in the regulation of salts and fluid at intestinal level, suggesting that alterations in the function of this receptor may have a role in obstructive intestinal pathologies and that P2Y$_4$-selective agonists may represent new pharmacological tools for the treatment of these diseases.

P2Y receptors present on endocrine pancreatic cells have also attracted a lot of interest, since their activation stimulates insulin release improving glucose tolerance in diabetic patients.

The P2Y receptor present on platelets (previously known as P2$_t$, now called **P2Y$_{12}$**) has been cloned in 2001; this receptor is associated to adenylyl cyclase inhibition and is responsible for the potent proaggregant effects induced by ADP. It is interesting to note that ATP can antagonize the activation induced by ADP (see also Supplement E 44.1) and that antiplatelet drugs already in use (ticlopidine and clopidogrel) and others in development (i.e., ticagrelor; see also Supplement E 44.2) exert their antithrombotic activity by antagonizing this receptor.

The receptor responsive to UDP-glucose and UDP-galactose has been recognized as a new family member (P2Y$_{14}$). This receptor is activated by immune stimuli, and its expression (and capability of inducing chemotaxis) has been recently demonstrated in progenitor cells able to differentiate in hematopoietic cells. This finding might have important implications for the use of stem cells in human diseases. It is interesting to note that the nucleotide sequence of this receptor has been known for many years, but until recently, it was considered an "orphan" receptor; the identification of its natural ligands represents an excellent example of how "deorphanization" of receptors might contribute to identify new pharmacological targets for human diseases (see also Chapter 17).

P2Y$_{13}$ receptor seems to have important roles in the immune system. Recently, characterization of knockout (KO) animals lacking this receptor has shown that it has an important role in the inverse transport of cholesterol. This effect could be exploited therapeutically to promote cholesterol elimination and therefore improve lipid profile in hypercholesterolemic subjects.

P$_{2D}$ receptor owes its name to diadenosine polyphosphates (tetra-, penta-, and esaphosphates indicated by the common acronym APxA) that seem to preferentially activate it. However, this receptor has not been cloned and it is still unclear if it really exists. The effects of diadenosine polyphosphates could be mediated by other already known members of the family.

Additional information on P2 receptor distribution and their effects are summarized in Table 44.2 and in the following paragraphs.

BIOLOGICAL ROLES OF PURINES

In some biological systems, ATP and adenosine induce similar effects; in others, they induce opposing effects. This functional antagonism is particularly interesting, since in many cases adenosine derives from hydrolysis of ATP released from nerve terminals. ATP effects are rapid, transient, and normally associated with excitatory events; these are followed by a slower and generally inhibitory/modulatory effect of adenosine. Adenosine, therefore, mediates an inhibitory feedback mechanism on the excitatory effects induced by ATP and by classical neurotransmitter coreleased with ATP from nerve terminals. Therefore, the higher the initial activation (and hence the ATP release), the higher the inhibitory effects induced by adenosine (as higher amounts of adenosine will be formed from ATP). A schematic overview of the potential clinical use of drugs acting on purinergic receptors is provided in the Box 44.1.

Effects on the Cardiovascular System

In heart, ATP induces positive inotropism, which adds to that produced by β-adrenergic agonists. This is due to activation of P2X receptors and stimulation of calcium conductance. The negative inotropic and chronotropic effects of adenosine are much better characterized from a therapeutic point of view; the receptor involved is the A$_1$ subtype that functionally antagonizes the positive inotropism and chronotropism produced by catecholamines. In the atrium, the effects of adenosine are direct and are due to activation of K$^+$ conductance; in the ventricles, however, the effects are indirect and due to decrease in Ca^{2+} influx through L channels induced by β$_1$-adrenoceptors. This activity is therapeutically exploited in the control of supraventricular arrhythmia (i.v. Adenocard®; see Supplement E 44.2).

Recently, a role for adenosine has been described in ischemic heart preconditioning, a short (1–2 min) ischemia protecting myocardium from a subsequent and prolonged ischemia. This activity is due to both A$_1$ and A$_3$ receptors; at molecular level, the mechanism of cellular protection involves activation (and migration to the membrane) of protein kinase C and phosphorylation of a cellular substrate that protects from ischemic damage. The contribution of A$_3$ receptors is also indirect and due to massive degranulation of residential cardiac immune cells during the short-lasting ischemia and, as a consequence, to a substantial decrease of releasable cytotoxic compounds during the second prolonged ischemia. Recent data suggest that adenosine, via A$_{2A}$ receptors, has an important anti-inflammatory role by reducing production of cytokines like TNFα and IL-6, which are present at high levels in patients with chronic cardiovascular diseases, like heart failure. Activation of A$_{2A}$ receptors could contribute to delay disease progression and would offer new pharmacological tools for this disease.

BOX 44.1 THERAPEUTIC TARGETS AND CURRENT OR POTENTIAL USES

Main therapeutic target	Cellular effects	Clinical effects	Indications	Name of drug (if already available) and notes
Myocardial A_1 adenosine receptor	Inotropic and chronotropic effect	Decreased heart work and oxygen demand, decreased heart rate and delayed impulse propagation	Supraventricular arrhythmia	
Platelet and vessel (smooth muscle) A_{2A} receptors, bronchial A_{2B} receptors	Antiaggregant effect, vasodilation, reduced bronchoconstriction	Antithrombotic effect, antihypertensive effects, antiasthmatic effect	Hypertension, thrombosis, cardiac ischemia, asthma, bronchoconstriction	Dipyridamole[a], bamifylline, theophylline
Platelet $P2Y_{12}$ receptor	Inhibition of platelet aggregation	Antithrombotic effect	Thrombosis, prevention of secondary myocardial infarction and stroke, infarction and stroke	Ticlopidine, clopidogrel, prasugrel, ticagrelor
$P2Y_2$ receptors on bronchial epithelial cells	Increased water content in bronchial secretions	Fluidification of bronchial secretions	Cystic fibrosis	$P2Y_2$ agonists[b]
$P2Y_2$ receptors on eye secretory glands	Increased water content in tears	Increased eye hydration and tear production	Hypolacrimia, dry eye syndrome	Diquafosol[c]
Striatal A_{2A} receptors	Blockade of adenosine receptors mediating functional antagonism of dopamine in basal ganglia	Amelioration of muscle tone and movement control	Parkinson's disease patients	SYN115, preladenant, KW-6356[d]

[a] Dipyridamole does not target the A_{2A} adenosine receptors directly, but rather increases the extracellular concentrations of adenosine, which, in turn, activates A_{2A} receptors on platelets and smooth muscle cells.
[b] In 2011, results from the phase III study TIGER 2 showed that the $P2Y_2$ agonist denufosol tetrasodium was not superior to placebo in cystic fibrosis patients, leading to study discontinuation.
[c] This drug is currently only marketed in Japan, where only safety studies are required for registration.
[d] In 2013, an initial review of data from three separate phase III trials on the adenosine A_{2A} receptor antagonist preladenant for the treatment of Parkinson's disease did not provide evidence of efficacy compared to placebo SYN115, preladenant, KW-6356****.

At vascular level, both ATP and adenosine are potent vasodilators. The effect of adenosine is direct and mediated by vascular receptors of the A_2 subtypes (both A_{2A} and A_{2B}) on smooth muscle cells that stimulate intracellular production of cyclic AMP (cAMP). On the other hand, the effect of ATP is indirect, as a consequence of the NO production resulting from activation of endothelial $P2Y_{2,4}$ receptors. As outlined in detail in the Supplement E 44.1, purines play a fundamental role in vascular tone in ischemia/hypoxia. Adenosine is a potent platelet inhibitor, whereas ADP is a potent trigger of aggregation. The effect of adenosine is mediated by the A_{2A} receptors that stimulate cAMP production, whereas the effect of ADP is due to activation of $P2Y_{12}$ receptors (see the preceding text). Two classes of drugs already exploit these activities therapeutically. Dipyridamole, by inhibiting adenosine reuptake, increases its extracellular concentrations and potentiates its activity on vascular and platelet A_{2A} receptors, with consequent increase of local vasodilation and reduction of platelet aggregation. Ticlopidine, clopidogrel, and other antithrombotics of new generation, such as ticagrelor, remarkably reduce platelet aggregation by acting as selective inhibitors of $P2Y_{12}$ receptor and represent drugs of first choice in thrombosis (see Supplement E 44.2).

Effects on the CNS and Peripheral Nervous System

As already mentioned, purines colocalize with other neurotransmitters in the CNS. After release from presynaptic terminals, ATP generally induces rapid and transient excitatory responses mediated by P2X receptors. Through these receptors, ATP has an important role as excitatory transmitter and has been involved in phenomena like hippocampal LTP, which plays a key role in learning and memory. P2X receptors, especially P2X2/P2X3 types, are present on peripheral sensitive fibers where they allow transmission of pain stimuli to the spinal cord when activated by ATP released by the surrounding cells as a consequence of trauma or inflammation. Sensitive fibers projecting to the spinal cord, in turn, release various excitatory neurotransmitters, such as glutamate, substance P, and again ATP, which contributes to central pain transmission via P2X2/P2X3 receptors expressed by nociceptive neurons. Selective antagonists of these receptors are currently at advanced phases of development as analgesics for inflammatory pain and some forms of otherwise untreatable neuropathic pain. On the other hand, adenosine is one of the most potent inhibitor of excitatory neurotransmission and neurotransmitter exocytosis (viz., of catecholamines, acetylcholine, and excitatory amino acids) via the presynaptic A_1 receptor subtype coupled to both stimulation of K^+ and reduction of Ca^{2+} conductance. Indeed, most of the sedative–hypnotic, anticonvulsant, neuroprotective, and antianxiety properties of adenosine are due to this activity. These effects are also enhanced by inhibitory postsynaptic A_1 receptors that are present in all brain areas (particularly abundant in the hippocampus and cortex) and antagonize the excitatory effects induced by other transmitters like glutamate. Interestingly, during epileptic seizures, large amounts of purines are released because of repeated depolarizations. This leads to formation of adenosine that reduces seizure spreading by inhibiting neurotransmitter release. There are now efforts to exploit this endogenous neuroprotective mechanism to reduce neurodegenerative damage following epileptic seizures; so far, this effort has been hampered by the difficulty to "separate" the desired activation of A_1 cerebral receptors from the effects mediated by A_1 and A_{2A} receptors in the cardiovascular system. It is important to stress that antagonism of A_1 cerebral receptors is responsible for the stimulatory properties (i.e., increase of attention, concentration, and learning and, more generally, of intellectual performances) of methylxanthines such as caffeine, theophylline, and theobromine, which are present in drinks or food (coffee, tea, cola, and chocolate).

A_{2A} receptors are almost exclusively present in basal ganglia, where they colocalize with D_2 dopaminergic receptors on striatal interneurons containing GABA and enkephalins. On these neurons, which are important components of the extrapyramidal system, A_{2A} and D2 receptors exert opposing effects. Namely, adenosine inhibits whereas dopamine stimulates locomotor activity. In Parkinson's disease, there is a progressive degeneration of dopaminergic neurons, and symptoms consist in bradykinesia and tremor; in rodent models of Parkinson's disease, antagonists of A_{2A} receptors significantly alleviate symptoms by reducing adenosine inhibitory effect on motor activity. This mechanism also underlies the locomotor stimulation induced by methylxanthines at the same doses that induce stimulatory effects on the CNS. This role of A_{2A} receptors has been also confirmed by analyses of A_{2A} receptor KO mice. These mice exhibit a spontaneous rise in blood pressure (in line with the vasodilator effect mediated by A_{2A} receptors; see the preceding text) and do not show any stimulation of locomotor activity after caffeine administration. All this evidence has been the rationale for recent clinical trials of A_{2A}-selective antagonists as new therapeutic agents potentially useful in Parkinson's disease (see Supplement E 44.2). Unfortunately, in 2013, an initial review of data from three separate phase III trials on the adenosine A_{2A} receptor antagonist preladenant for the treatment of Parkinson's disease did not provide evidence of efficacy compared with placebo.

Moreover, A_{2A} receptor KO animals show aggressiveness and hypoalgesia, suggesting that A_{2A} receptors have a role in the control of behavior and in mediating nociceptive signals. An excessive activation of brain A_{2A} receptors, especially those present on microglia, seems to induce cell death and may contribute to neurodegeneration. This appears to be in contrast with the anti-inflammatory activity mediated by A_{2A} receptors on circulating cells and with the neuroprotective activity mediated by A_1 receptors. Recent experimental evidence also suggests a role for the A_{2A} receptor in neurodegenerative damage associated with ischemia and other chronic neurodegenerative diseases of basal ganglia, such as Huntington's chorea.

Even if present at very low levels in the brain, the A_3 receptor has been implicated in regulation of cell survival. Depending on the pathophysiological conditions, activation of this receptor in neurons and astroglial cells induces either trophic and differentiating effects (and in turn cellular protection) or necrosis and/or apoptosis, with possible implications in acute (i.e., ischemia) and chronic aging-associated diseases. The cytoprotection mediated by this receptor has interesting analogies with the myocardial protection previously discussed and could be therapeutically exploited to enhance brain "resistance" to neurodegeneration after trauma or ischemia. The role of ATP in modulating neurodegeneration after ischemia/hypoxia is still controversial (see also Supplement E 44.1).

Effects on the Respiratory System

In asthmatic and rhinitis patients, adenosine induces bronchoconstriction. These effects are antagonized by methylxanthines such as theophylline, which has been used for a long time in asthma therapy (see Supplement E 44.1). However, it is still unclear whether this antiasthmatic activity is due to antagonism of adenosine or, rather, to inhibition of bronchial phosphodiesterases with consequent increase of intracellular cAMP in smooth muscle cells, mediating dilation. More recently, it has been shown that at bronchial level, adenosine also potentiates release of histamine and proinflammatory and procontracting mediators from mastocytes and basophils by activating the A_{2B} and A_3 receptor subtypes. Selective A_{2B} and A_3 antagonists could therefore be useful in preventing (or reducing) bronchoconstriction caused by endogenous adenosine. This hypothesis has also been confirmed by the finding that the xanthine derivative bamifylline, an antiasthmatic agent already used in clinical practice, acts as a selective A_{2B} antagonist, although with a relatively weak activity. Adenosine also inhibits acetylcholine release from vagal nerve terminals. Indeed, it has been reported that aerosol administration of an antisense oligonucleotide for A_1 receptors in an animal model of allergic asthma considerably reduces bronchoconstriction induced by subsequent administration of adenosine or allergens. This suggests a specific role of A_1 receptors and opens up new avenues for antisense therapy in asthma and related diseases.

As previously described, nucleotide derivatives (especially UTP) have important effects on fluid and salt excretion in respiratory epithelia via $P2Y_2$ receptors. Aerosol administration of these compounds was initially found to normalize secretion of epithelia in patients affected by cystic fibrosis, suggesting new therapeutic perspectives for this serious disease. This activity, however, has not been confirmed in a recent clinical study (for details, see Supplement E 44.2).

Effects on Other Systems

P1 and P2 receptors have been found in all organs and systems, even if their role, in some cases, is still unclear (Tables 44.1 and 44.2).

On bladder smooth muscle, adenosine acts as a potent relaxant via A_2 receptors, whereas ATP induces rapid and transient contraction; this biphasic activity of purines has suggested that they may have a role in bladder emptying, with possible implications in the treatment of urinary incontinence.

In the gastrointestinal system, ATP controls hydrochloric acid secretion from gastric epithelial cells and exerts an inhibitory effect on intestinal motility; moreover, together with adenosine, it stimulates glycogenolysis and pancreatic secretion of insulin and glucagon.

Adipocytes express A_1 and A_2 receptors that modulate lipolysis and glucose oxidation; in line with this findings, high doses of methylxanthines induced weight loss. A number of preparations containing very high concentrations of caffeine have recently become available, with no prescription needed, as products to induce weight loss. Many of these products are potentially dangerous as they induce tachycardia (see Section "Effects on Cardiovascular System").

P2X7 receptor subtype is highly expressed on both osteoclasts and osteoblasts, playing a fundamental role in bone remodeling and deposition. Selective ligands for this receptor subtype are currently under evaluation as new pharmacological approaches to osteoporosis. Purinoceptors are also expressed on spermatozoa, amniotic cells, chromaffin cells, and acinar cells in parotids. This suggests that purines might be involved also in exocrine and endocrine regulations and in reproduction.

Drugs targeting the purinergic system are currently under development for dry eye syndrome, a serious pathology characterized by lack of appropriate eye hydration that dramatically affects patients' everyday life (due to continuous pain, blurred vision, and other symptoms) and eventually leads to severe corneal damage. Two classes of drugs targeting different purinergic receptors have been proposed: agonists at the A_3 adenosine receptor subtype, which exert a general anti-inflammatory activity, and agonists at $P2Y_2$ receptors, which directly promote fluid secretion and teardrop production (see Supplement E 44.2).

Adenosine Effects Independent of Receptor Activation

The "ADA deficiency syndrome" is characterized by lack of ADA, the enzyme that catabolizes adenosine. This is a genetic disease characterized by a buildup of intracellular adenosine and deoxyadenosine, particularly in immune cells. The high increase in concentration of these nucleosides disrupts cellular energy equilibrium and activates death pathways, leading to lymphocyte apoptosis and serious immunodeficiency and to patients' demise within few years. Today, this pathology can be treated using an efficient gene therapy approach (developed by an Italian team) that allows patients to lead a normal life without being totally isolated from the outside world to avoid lethal infections. Characterization of this disease has led to the discovery that some of adenosine effects, in particular those on cell survival, are directly induced by intracellular adenosine. This has suggested the idea of synthesizing purine analogs capable of accumulating in the cytoplasm and inducing apoptosis that could be useful in antitumor therapies (see also the following text). Even if not mediated by membrane receptors, these actions are considered highly specific, since they involve well-defined intracellular pathways. Indeed, through these biochemical pathways, adenosine plays an important role in embryonic development, contributing to tissue remodeling by eliminating unnecessary structures (e.g., interdigital membranes and supernumerary neurons) and to the final configuration of the immune system by recognizing (and eliminating by apoptosis) potentially autoreactive clones. Furthermore, *in vitro* studies have shown that adenosine has cytotoxic effects on cellular precursors of muscle fibers, suggesting that alterations of this system might have a role in the pathogenesis in some forms of muscular dystrophy. Therefore, it has become quite clear that studying basic intracellular mechanisms underlying adenosine effects could provide new insights into the pathogenesis of some human diseases and new ideas for the development of novel drugs.

Purines and Antitumor Therapy

Many molecules used as antitumor drugs are purine or pyrimidine derivatives chemically modified to replace natural molecules in nucleic acid synthesis and induce metabolic damages that are not compatible with cell survival (see Chapter 33). Indeed, recent studies have shown that these molecules can also activate intracellular death pathways, such as caspase activation, shedding light on the mechanisms underlying their specific actions and therapeutic effects. New therapeutic application in anticancer treatment may arise from studies on A_3 receptors: these molecules exert toxic effects on tumor cells and at the same time stimulate immune cell proliferation, thus increasing natural host defense. Some of these molecules are in an advanced phase of clinical trials in Israel, even if it is not clear whether their effect is solely due to A_3 receptor stimulation or also to nonreceptor-mediated mechanisms. These molecules, besides being tested for various forms of carcinoma, are also in advanced clinical trials as anti-inflammatory drugs in some autoimmune diseases, such as rheumatoid arthritis or psoriasis. The rationale for their use is based upon their ability to induce death of the autoreactive immune clones responsible for the development of these diseases (see Supplement E 44.2).

Studies performed years ago suggested ATP could exert antitumor effects by activating the cytotoxic receptor P2X7 (see the preceding text), but there has been no clinical follow-up. On the contrary, ATP beneficial activity in cachexia has been confirmed by i.v. administration in patients undergoing chemotherapy resulting in great improvement of their quality of life.

TAKE-HOME MESSAGE

- ATP is stored in neurotransmitter-containing vesicles and released; ecto-ATPases dephosphorylate ATP to adenosine, which is then degraded to inosine; adenosine is also taken up by a specific membrane transporter and used for intracellular synthesis of new ATP molecules.
- Purinergic receptors are classified as P1 and P2.
- P1 receptors are named "adenosine receptors"; they are GPCRs and are encoded by four distinct genes (A_1, A_{2A}, A_{2B}, and A_3). A_1 and A_3 activate PLC and inhibit adenyl cyclase, and A_{2a} and A_{2b} inhibit cAMP generation. Signaling via A receptors is relevant in the CNS and PNS; gastrointestinal, cardiovascular, immune, and metabolic systems; kidney; and lungs.
- P2 receptors are named "ATP receptors." P2Xs are trimeric cationic LGICs, whereas P2Y are GPCRs. P2Xs are involved in synaptic transmission and in control of several functions; when strongly stimulated, P2X7 receptors form a large cationic pore leading to cell death.
- By activating P1 and P2 receptors, purine and pyrimidine nucleotides and nucleosides exert specific functions in neurotransmission and communication between different cell types in the nervous, cardiovascular, respiratory, and immune systems.
- Defects and dysregulation of purinergic transmission underlie several human diseases, such as brain ischemia, neurodegeneration, tumors, and immune pathologies.
- Some drugs targeting the purinergic system are already used as antithrombotic, antiarrhythmic, antiasthmatic, and antitumor agents.

FURTHER READING

Abbracchio M.P., Burnstock G., Boeynaems J.M., Barnard E.A. et al. (2006). International Union of Pharmacology LVIII: update on the P2Y G protein-coupled nucleotide receptors: from molecular mechanisms and pathophysiology to therapy. *Pharmacological Reviews*, 58, 281–341.

Burnstock G. (2008). Purinergic signalling and disorders of the central nervous system. *Nature Reviews. Drug Discovery*, 7, 575–590.

Burnstock G. (2009). Purinergic receptors and pain. *Current Pharmaceutical Design*, 15, 1717–1735.

Chen J., Eltzschig H., Fredholm B. (2013). Adenosine receptors as drug targets-what are the challenges? *Nature Reviews. Drug Discovery*, 12, 265–286.

Surprenant A., North R.A. (2009). Signalling at purinergic P2X receptors. *Annual Review of Physiology*, 71, 333–359.

Wilson C.N., Mustafa S.J. *Adenosine Receptors in Health and Disease*. Handbook of Experimental Pharmacology, Vol. 193, Springer–Verlag, Berlin–Heidelberg, 2009.

45

NEUROPEPTIDES

LUCIA NEGRI AND ROBERTA LATTANZI

> **By reading this chapter, you will:**
> - Know the mechanisms of synthesis and secretion of peptides acting as neurotransmitters
> - Know the functional differences between peptidergic transmission and classical neurotransmitter signaling

Neuropeptides are linear chains of amino acids linked by peptide bonds. They form families that we can find today in living species derived from common ancestral genes originated many million years ago, which over time have undergone changes as a result of duplication and mutation processes.

In 1931, von Euler and Gaddum demonstrated for the first time the presence of an active proteic ingredient in mammalian tissues by isolating substance P (SP), whose amino acid sequence was determined only 40 years later by Chang and Leeman. In the meantime, Vittorio Erspamer in Italy (see Supplement E 45.1, "Characteristics of Some Neuropeptides") had already extracted and identified, from invertebrates and amphibian, other peptides (e.g., eledoisin, physalemin, and kassinin) with similar biological properties to SP. He identified their sequence and classified them as members of the tachykinin family. When the structure of SP was uncovered, it appeared similar to that of the already known tachykinin, as expected based on their similar biological properties. Therefore, SP was included in the tachykinin peptide family as well. Afterward, Erspamer isolated many other peptides from frog skin, and analogous peptides were subsequently isolated from mammalian brain and gut; for example, caerulein, sauvagine, and bombesin were found to be the amphibian counterparts, respectively, of cholecystokinin, corticotrophin-releasing factor (CRF), and gastrin-releasing hormone (GRH).

In the 1950s, the hypothalamic neurohormone oxytocin had already been chemically identified as a nonapeptide (du Vigneaud was awarded the Nobel Prize in 1953 for its isolation, structural identification, and synthesis). At the end of the 1950s, Viktor Mutt identified new peptides in the gastrointestinal tract, which he subsequently found also in the brain. In 1975, Hughes and Kosterlitz isolated two brain pentapeptides, the enkephalins, capable of reproducing many of the pharmacological effects of morphine. The presence of peptides in the nervous system and the development of immunohistochemistry demonstrate that neurons can synthesize and release peptides (for more details, see Supplement E 45.1).

Neuropeptides are produced by both central and peripheral neurons and released via the regulated secretory pathway. They modulate and mediate neuronal functions acting on neuronal receptors, which in general are seven-transmembrane spanning G-protein-coupled receptors (GPCRs). More than 70 neuropeptide-encoding genes have been identified, and more than 100 neuropeptide transmitters are known to play diverse roles in the nervous system, including regulation of sleep and arousal, emotion, reward, feeding and energy balance, pain and analgesia, and learning and memory.

Other peptides secreted by cells (it is still debated whether they should be included among neuropeptides) modulate functions of the nervous system; these are mainly chemokines and growth factors produced by neurons and glial cells (especially microglia). There is increasing evidence for expression of chemokines and chemokine receptors and for their electrophysiological effects in neurons. Chemokines are also implicated in neuronal migration and neuroinflammation. Some growth factors belonging to the neurotrophin

---

General and Molecular Pharmacology: Principles of Drug Action, First Edition. Edited by Francesco Clementi and Guido Fumagalli.
© 2015 John Wiley & Sons, Inc. Published 2015 by John Wiley & Sons, Inc.

family, such as BDNF, show characteristics of neuropeptides. The BDNF gene is widely expressed in neurons, and its primary product is pre-BDNF, a precursor molecule containing signal peptide and cleavage sites for typical prohormone convertases. BDNF is present in dense-core vesicles and undergoes stimulated release, two characteristics typical of neuropeptides.

Moreover, there are peptidergic hormones of nonneuronal, peripheral origin, as leptin and insulin, which send signals to the brain. From 1975 onward, a plethora of new peptides has been identified through new experimental approaches based on molecular biology techniques. The human genome sequencing, the detection of transcripts encoding for potentially active peptide, and the cDNA library construction have permitted to identify orphan receptors but also orphan peptides whose function(s) still needs to be defined.

NEUROPEPTIDE SYNTHESIS

In contrast with small-molecule neurotransmitters, neuropeptide synthesis, like protein synthesis, requires DNA transcription into messenger RNA (mRNA) and mRNA translation into protein. A striking aspect of neuropeptide synthesis is that it consists in multiple regulated steps so that a single gene can give rise to several different neuropeptides in a tissue-specific manner (Fig. 45.1).

Diversity can be generated by alternative splicing of primary RNA transcripts resulting in the production of different mRNAs. For example, calcitonin and calcitonin gene-related peptide (CGRP) are generated in a tissue-specific fashion by alternative splicing of a single primary transcript to yield peptides with very different biological actions. Calcitonin is produced in parafollicular C cells of the thyroid gland and plays a central role in Ca^{2+} and phosphorus metabolism. CGRP is produced in neurons and, among its other actions, is a potent vasodilator; in fact, its receptors are potential target for treating migraine. The tachykinin family (whose members share the same C-terminal sequence) represents another example of alternative splicing of neuropeptide-encoding mRNAs. In humans, two genes encode tachykinins, preprotachykinin (PPT) A and B. PPT-A transcripts undergo alternative splicing, and as a result, three distinct prepropeptides are produced that in turn give rise to multiple peptides, including SP and neurokinin A. PPT-B gives rise to neurokinin B (see Supplement E 45.1) Following splicing, the mature mRNA is exported to the cytoplasm for translation. The initial protein resulting from translation is not an active signaling molecule, but a precursor polypeptide (a prepropeptide) that undergoes processing and is converted into one or more neuropeptides. Prepropeptides contain an N-terminal sequence (or "pre" sequence) that directs the newly synthesized protein into the lumen of the rough endoplasmic reticulum (ER) and thus into the proper,

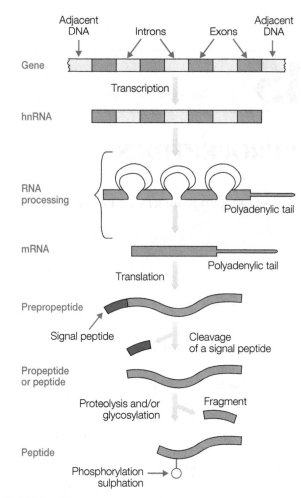

FIGURE 45.1 Different stages in the expression of a neuropeptide-encoding gene. The whole DNA sequence is transcripted into an hnRNA that, by intron removal and addition of a polyadenylic tail, is turned into an mRNA that is translated into a prepropeptide. This, after removal of a signal sequence and through reactions of hydrolysis and/or glycosylation and/or phosphorylation or sulfation, is transformed into a mature peptide.

regulated secretory pathways. The signal sequence is cleaved by a signal peptidase even before translation is completed, a mechanism described as cotranslational processing. Removal of the signal sequence generates a propeptide that is released from the ribosome after translation is completed, transferred to the Golgi complex, and subsequently packaged within large dense-core vesicles (LDCVs). Inside Golgi complex and LDCVs, the propeptide undergoes additional post-translational processing, involving additional cleavages and covalent modifications.

Propeptide Processing

Frequently, different types of peptides are contained in the same precursor molecule, as in the case of opioid peptides. Sometimes, the precursor contains several copies of the same peptide as the TRH precursor that contains five copies of the TRH tripeptide. In general, in the prohormone sequence, the active peptide sequence is delimited by pairs of basic amino acid residues (basic motifs, Lys-Arg or Lys-Lys, Arg-Arg or Arg-Lys), which represent the recognition sites for the prohormone convertase, a little family of cell-specific endoproteases (prohormone convertases 1 and 2, PC1 and PC2) that cleaves the propeptide at distinct positions generating different peptides from the same precursor.

For example, ACTH and β-lipotropin are the main hormones derived from proopiomelanocortin (POMC) in the periphery, while α-MSH and β-endorphins are the main POMC-derived neuropeptides in the brain. This difference depends on the different processing pathways existing in anterior pituitary and hypothalamic neurons. Otherwise, peptides of different lengths may be generated from the same primary sequence: for example, procholecystokinin (pro-CCK) contains the sequence of at least five CCK-like peptides that differ in length (from 4 to 58 amino acids). CCK (33 amino acids) is the main peptide produced in the gut, while the brain produces mainly CCK-8 (see Supplement E 45.1).

In the propeptide sequence, the regions corresponding to the active peptides are separated by long sequences whose function is unknown. In some occasions, unknown peptides (e.g., CGRP) have been discovered by observing the distribution of the cleavage sites in the propeptide molecule. Nevertheless, there are many peptidic sequences whose function is still unknown.

Prohormone convertase-generated peptides can be subject to further modification: C-terminal amidation, glycosylation, acetylation, sulfation, or phosphorylation of lateral chains. C-terminal amidation is catalyzed by peptidyl-aminotransferase (PAM) that uses a C-terminal glycine as amide donor for the preceding amino acid. This amide is present in many neuropeptides and is essential for their biological activity. Modifications can have different effects on neuropeptide activity: for example, N-terminal acetylation significantly increases the biological activity of α-MSH, while it decreases that of β-endorphin.

Some endopeptidases are targets of drugs, such as the inhibitors of angiotensin-converting enzyme, captopril, enalapril, and lisinopril, widely used to treat hypertension.

STORAGE AND SECRETION OF NEUROPEPTIDES

In contrast with "fast-acting neurotransmitters" (e.g., excitatory amino acids and catecholamine) that in central neurons are packaged into small clear synaptic vesicles (SSVs) localized at the synaptic terminals, neuropeptides are stored in LDCVs, assembled in the Golgi apparatus, and transported to the neuronal cell body. LDCVs can be detected near the plasma membrane or concentrated in axonal varicosities or spread in the cytoplasm not only in presynaptic terminals but also in soma and in dendrites of neuronal cells (see also Chapter 37). In response to membrane depolarization, secretory granules fuse with the cell membrane and release the peptides into the synaptic cleft (regulated secretion, controlled by a bioelectric signal) but also outside the synaptic space (volume transmission). This phenomenon is Ca^{2+} dependent, similarly to the release of classical neurotransmitters from SSVs.

Peptide release occurs in response to high-frequency stimulation of neurons.

Indeed, while a single action potential can cause SSVs to fuse with the cell membrane, a rapid train of action potentials is required to trigger neuropeptide release from LDCVs. Thus, specific patterns of electrical activity in neurons may lead to preferential release of either a neuropeptide or a small-molecule neurotransmitter or may prompt the release of both. Because neuropeptides tend to be released under conditions of sustained activity, they may regulate strongly stimulated synapses by providing positive or negative feedback.

In the same neuron, LDCVs (containing different peptides) commonly coexist with SSVs (containing small-molecule fast neurotransmitters), and they may be simultaneously released (cotransmission) (Table 45.1). An example is the corelease of CGRP and SP (at a 1:1 stoichiometric ratio) that occurs at both central and peripheral endings of DRG neurons. Some neurons contain more than one peptide, such as some neurons of the hypothalamic supraoptic nucleus containing oxytocin, enkephalin, cholecystokinin, and cocaine- and amphetamine-regulated transcript (CART).

Calcium-dependent activator protein for secretion (CAPS) is an LDCV-associated Ca^{2+} sensor that makes

TABLE 45.1 Examples of colocalization of neuropeptides and neurotransmitters in the same neuron terminal

Neuropeptides	Classical transmitters	Colocalization sites
Neuropeptide Y (NPY)	Noradrenaline	Neurons of locus coeruleus; sympathetic preganglionic neurons
VIP	Acetylcholine	Parasympathetic preganglionic neurons
CGRP	Acetylcholine	Motor spinal neurons
Neurotensin, cholecystokinin	Dopamine	Neurons of substantia nigra
TRH, substance P, enkephalin	5-HT	Neurons of raphe nuclei
Enkephalin	GABA	Striatal neurons projecting to globus pallidus
Dynorphin, substance P	GABA	Striatal neurons projecting to pars reticulata of substantia nigra

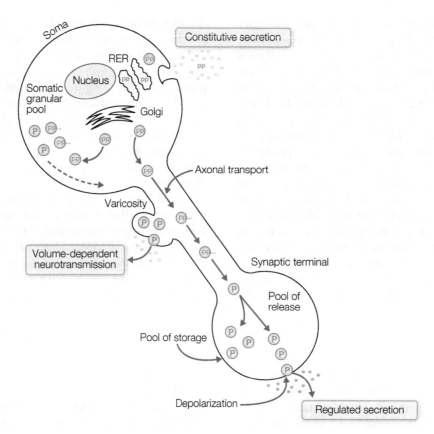

FIGURE 45.2 The peptidergic neuron. Schematic representation of the processes of synthesis, accumulation, transport, and release of a neuropeptide and a proneuropeptide. pp, propeptide; P, mature peptide; RER, rough endoplasmic reticulum.

LDCVs highly sensitive to the small increases in cytoplasmic Ca^{2+} concentration occurring far from the synaptic terminal, thus allowing extrasynaptic exocytosis. Part of the propeptide may be released by constitutive secretion, a mechanism that does not involve concentration of peptides in LDCVs or their molecular remodeling by proteolytic processes and does not show preferential sites of exocytosis. In general, this type of secretion occurs at the level of cell body. The big molecule of secreted precursor does not act as neurotransmitter but, in some cases, as growth factor (Fig. 45.2). This is particularly evident in some neuronal populations during embryonic development, and it probably plays an important role in organization and development of the neuronal network.

PEPTIDERGIC TRANSMISSION

Peptidergic transmission shows both similarities and differences with the "conventional" transmission mediated by nonpeptide neurotransmitters, as amines and amino acids. Intravesicular packaging and Ca^{2+}-dependent release are common aspects to peptidergic and conventional neurotransmissions. The effects of peptides on target cells involve specific receptor-mediated mechanisms. They can be excitatory or inhibitory, can be pre- or postsynaptic, and can be exerted near or at significant distance from the release sites.

Other than synthesis, the main differences concern receptor interaction and signal extinction (see also Chapter 37).

Once released, generally outside the synaptic space (parasynaptic release), peptides can reach receptors located at significant distance from the release sites. This type of endocrine transmission is called volume-dependent neurotransmission or volume neurotransmission (VNT) or extrasynaptic transmission.

Neuropeptide Receptors

Neuropeptide receptors (there are exceptions such as insulin) belong to the superfamily of G-protein-coupled receptors that produce slower responses than ligand-gated ion channels. Moreover, many of the actions mediated by G proteins and second messengers alter the response properties of neurons resulting in modulation rather than simple excitation or inhibition.

Neuropeptides have a molecular weight 50-fold higher than classical neurotransmitters; they have more recognition sites and show greater affinity (nM vs. mM) and selectivity for their receptors compared to classical neurotransmitters.

In most cases, a given neuropeptide can bind several receptor subtypes differently distributed in the nervous system, with different affinities. Often, related peptides share the same receptors. This is the case of the corticotrophin-releasing factor (CRF) receptor. The CRF_1 receptor shows high affinity for both CRF and the related peptide urocortin, whereas the CRF_2 receptor binds preferentially urocortin.

Each of the melanocortin receptor family members, named MC_{1-5}, is activated by different POMC-derived peptides, with ACTH, α-MSH, and γ-MSH displaying varying degrees of potency. MC_4 receptors can also be antagonized by a distinct peptide called agouti-related peptide (ARP); indeed, MC_4 was the first receptor identified having both an endogenous agonist and an endogenous antagonist. These MC_4 ligands are involved in the regulation of feeding behavior.

Neuropeptide receptors are localized not only in synapses but also in the plasma membrane of axons, cell bodies, and dendrites. Neuropeptides are degraded by endo- and exopeptidases (different from those involved in synthesis) that reside in the plasma membrane. Their concentration and activity are such that peptides can diffuse relatively large distances in the nervous system. In the extracellular space, the half-life of a neuropeptide is very long: for example, half-lives of oxytocin and vasopressin are 20 min in the brain and only 2 min in the blood. Neuropeptides are not reuptaken from synaptic terminals. The only exception is cholecystokinin for which a high-affinity membrane transporter has been described. So, in contrast with classical neurotransmitters that can be recycled through mechanism of reuptake, neuropeptides need to be resynthesized and transported to the terminals through axonal transport. Neuropeptide receptors undergo internalization after sustained binding to a ligand; subsequently, the internalized receptors are either recycled to the plasma membrane or degraded. The neurokinin 1 receptor is internalized 5 min after agonist binding and reinserted in the plasma membrane within 30 min. For several neuropeptide receptors, such as neurotensin receptors, internalization may lead to transport from synapse to cell body. Interestingly, neuropeptide–receptor complexes have been detected near the cell nucleus. In the same nucleus, colocalization and possible corelease (cotransmission) of neuropeptides and classical neurotransmitters permit fast (2–5 ms) and slow (100–500 ms) synaptic communication.

NEUROPEPTIDE FUNCTIONS AND THERAPEUTIC POTENTIAL

Neuropeptides, as well as classical neurotransmitters, not only act as messengers but also contribute to formation of structural elements during the ontogenetic development of the central nervous system (CNS) and maintenance of neuronal plasticity in the adult. Some neuropeptides, expressed during perinatal development, are downregulated after birth but may reappear during adulthood in particular functional and pathological conditions (e.g., VIP, NPY, PK2 in sensitive neurons after nerve lesion; galanin in sensitive neurons in β-amyloid plaques; CGRP in motoneurons). The expression levels of neuropeptides vary according to the diverse conditions, being subject to circadian variations and experimental manipulations. Neuropeptides exert their functions especially when the brain is under stress or pathological conditions. According to Tomas Hökfelt, this suggests that peptidergic communication represents a physiological language used by the brain when damaged and is therefore an interesting target for the development of new drugs (see Supplements E 45.1; E 45.2, "Hypothalamus and Neuropeptides"; and E 45.3, "Neuropeptides and Nociception").

The lack of agonists and antagonists selective for different receptor subtypes has represented a major issue preventing in-depth understanding of the peptidergic system functions. Studies of neuropeptide function have relied mainly on the generation of genetically engineered mice, either under- or overexpressing genes encoding for peptides and their receptors.

Research on neuropeptides, begun in the 1960s, has generated and continues to produce a wealth of data: a dozen of great peptide families and many other little families have been identified (Table 45.2 and E 45.1). Nevertheless, pharmacological manipulation of peptidergic transmission pathway is still lagging behind. Efforts have been made to identify nonpeptidergic molecules, since peptides are subject to enzymatic hydrolysis and cannot be orally administered. Moreover, in the case of neuropeptides acting on the CNS, the compounds need to be able to cross the blood–brain barrier. Actually, only few peptides, generally synthetic analogs of endogenous peptides, are clinically used and administered by injection or by nasal spray. Design of "peptoids" and randomized screening of nonpeptidic molecule libraries have allowed to identify selective antagonists for the main neuropeptide receptors. Nevertheless, the expectations rose by experimental results, obtained both *in vitro* and in animal models, have been often disappointed by clinical experimentation. So, only few peptide antagonists are nowadays used in spite of their therapeutic potential.

TAKE-HOME MESSAGE

- Neuropeptides are small proteins or polypeptides that serve as neurotransmitters in the nervous system, usually by interacting with GPCRs.
- In general, neuropeptides act as neuromodulators and are often released in association with other peptides or nonpeptidic transmitters (cotransmission).

TABLE 45.2 Neuropeptides grouped according to their chemicopharmacological properties

Peptide	R	Ant	Ag	Use
Opioids				See Chapter 46
Enkephalins	δ			
β-Endorphin	μ			
Dynorphin A, dynorphin B	κ			
Nociceptin (orphanin FQ)	NOP			
POMC derivatives				
Melanocortins				
Melanocyte-stimulating hormones				
α-MSH	MC3, MC4		+	Obesity, metabolic disorders
β-MSH				
γ-MSH				
ACTH			+	Renal failure; diagnosis
Vasopressin/oxytocin				
Vasopressin (VP)	V_{1A}	+		Anxiolytic
	V_{1B} (V_3)	+		(SSR149415) Anxiolytic
	V_2		+	(Lypressin) Diabetes insipidus
Oxytocin (OT)	OT		+	(WAY267464) Anxiolytic
Cholecystokinins				
Gastrin-34, gastrin-17, gastrin-4	CCK_2	+		(Lorglumide/LY225910) To reduce gastric secretion
CCK-8, CCK-33, CCK-58	CCK_1	+		(Proglumide) Pain, coadjuvant in obesity
	CCK_2		+	
Somatostatin				
SS-12, SS-14, SS-28, antrin	SST_{1-5}		+	(Octreotide) Acromegaly, intestinal tumors
NPY				
NPY	Y1		+	Anxiolytic, analgesic
	Y2	+		Anxiolytic, antidepressant
	Y5	+		Obesity
	Y4			
PPY				
PYY, Pyy-(3–36)				
Calcitonin				
Calcitonin	CLR		+	Paget's disease
α-CGRP	CLR/RAMP1	+		Olcegepant/BIBN4096BS, e.v; telcagepant/MK-0974, per os; BMS-694153, nasal spray; migraine
β-CGRP	CLR/RAMP1			
Adrenomedullin	CLR/RAMP2/3			
Bombesin				
Gastrin-releasing peptide (GRP)	BB3		+	(MK5041) Obesity in dogs and rodents
Neuromedin C	BB2	+		Anxiolytic
Neuromedin B	BB1	+		Anxiolytic
VIP–glucagon				
VIP, PHM-27/PHI-27, PHV-42	VPAC1		+	(IK312532, RO-251553) Diabetes, asthma
	VPAC2		+	Inflammation, obesity
PACAP-38, PACAP-27, PRP-48	PAC1		+	Neurodegenerative diseases, Parkinson's disease
CRH				
CRH or CRF	CRF1	+		Depression, pain, epilepsy, Alzheimer's disease, cognitive, and alimentary disturbances
	CRF2			
Urotensin	UT			
Sauvagin	CRF1			
Kinins and tensins				
Substance P	NK1	+		(Aprepitant/MK-869) Depression, anxiety, emesis
Neurokinin A	NK2	+		Anxiety, depression
Neurokinin B or neuromedin K	NK3	+		(Osanetant) Schizophrenia
Bradykinin	B1	+		Inflammatory pain

TABLE 45.2 (*Continued*)

Peptide	R	Ant	Ag	Use
	B2			
Angiotensin I, angiotensin II, angiotensin (I–VII)	AT1, AT2			(Sartans) Antihypertensive
Neurotensin (NT), neuromedin N	NT1	+		Psychosis, drug dependence
	NT2			
	NT3			
Galanin				
Galanin, galanin message-associated peptide (GMAP)	GAL1	+		Depression
	GAL2		+	Depression, neuropathic pain, neuroprotection
	GAL3	+		(SNAP37889) Anxiety, depression
Galanin-like peptide (GALP)				
Other neuropeptides				
Melanin-concentrating hormone (MCH)	MCH1	+		(GW803430) Anxiety, depression, obesity, steatosis
	MCH2			
Orexin A (hypocretin-1)	OX1	+		Alimentary disturbances
Orexin B (hypocretin-2)	OX2		+	Sleep disorders, narcolepsy, depression
CART-(1–39), CART-(42–89)				
AGRP				
Prokineticin-1 (PK1), endocrine gland-derived VEGF (EG-VEGF)	PKR1	+		Inflammatory pain
Prokineticin-2 (PK2)	PKR2			

R, receptors; Ag, agonists; Ant, antagonists.
Receptors are indicated with their official IUPHAR names.
Possible or future clinical applications of agonist and antagonist ligands are listed.

- Neuropeptides are encoded by genes. A single gene can generate multiple peptides as a result of alternative RNA splicing or posttranslational processing.
- Synthesis of neuropeptides requires DNA transcription, RNA translation, posttranslational maturation, and its transport to the site of release.
- Neuropeptides are synthesized as prepropeptides that undergo extensive posttranslational processing. The prefix "pre" refers to the N-terminal signal sequence that targets newly synthesized peptides to the endoplasmic reticulum. When the presequence is cleaved off, the remaining propeptide undergoes further cleavage to generate the mature peptide(s).
- Neuropeptides are secreted at synaptic and/or extrasynaptic level. They can diffuse in the extracellular space (volume-dependent secretion) and bind also distant receptors.

FURTHER READING

von Euler U.S., Gaddum J.H. (1931). An unidentified depressor substance in certain tissue extracts. *The Journal of Physiology (London)*, 72, 74–87.

Harmar A.J., Fahrenkrug J., Gozes I., Laburthe M., May V., Pisegna J.R., Vaudry D., Vaudry H., Waschek J.A., Said S.I. (2012). Pharmacology and functions of receptors for vasoactive intestinal peptide and pituitary adenylate cyclase-activating polypeptide: IUPHAR. *British Journal of Pharmacology*, 166, 4–17.

Hökfelt T., Broderger C., Xu Z.D., et al. (2000). Neuropeptides—an overview. *Neuropharmacology*, 39, 1337–1356.

Katz P.S., Lillvis J.L. (2014). Reconciling the deep homology of neuromodulation with the evolution of behavior. *Current Opinion in Neurobiology*, 29C, 39–47.

Nestler E.J., Hyman S.E., Malenka R.C. (2008). *Molecular neuropharmacology: a foundation for clinical neuroscience*. New York: McGraw-Hill Professional.

Ogren S.O., Kuteeva E., Elvander-Tottie E., Hökfelt T. (2010). Neuropeptides in learning and memory processes with focus on galanin. *European Journal of Pharmacology*, 626(1), 9–17.

Salio C., Lossi L., Ferrini F., Meringhi A. (2006). Neuropeptides as synaptic transmitters. *Cell and Tissue Research*, 325, 583–598.

van den Pol A.N. (2012). Neuropeptide transmission in brain circuits. *Neuron*, 76, 98–115.

Zhang C., Truong K.K., Zhou Q.Y. (2009). Efferent projections of prokineticin 2 expressing neurons in the mouse suprachiasmatic nucleus. *PLoS One*, 4, 1–12.

46

THE OPIOID SYSTEM

PATRIZIA ROMUALDI AND SANZIO CANDELETTI

By reading this chapter, you will:

- Know distribution and functions of the opioid system
- Know synthesis and metabolism of endogenous opioids and structure and distribution of opioid receptors
- Learn the role of opioid system in nociception and other CNS and visceral organ functions
- Understand the mechanisms responsible for opioid tolerance
- Know the drugs active on opioid receptors and their therapeutic and toxicological significance

---

The opioid system represents one of the endogenous systems involved in neurotransmission and/or neuromodulation and consists of specific peptidergic ligands and their corresponding receptors. It is present in the CNS and periphery and modulates different functions. Besides its well-known modulation of nociceptive transmission, this system is also involved in the regulation of gastrointestinal (GI), endocrine, and autonomic functions, in reward and dependence mechanisms, as well as in memory and learning processes.

The existence of the opioid system explains the pharmacological actions of the alkaloid morphine and its derivatives, the opiates (see Box 46.1 and Supplement E 46.1, "Opiate Drugs").

The word "opiate" refers to both natural and synthetic substances related to morphine, which is the prototype molecule of narcotic analgesics, a term now obsolete, usually replaced by opiates and more recently, but improperly, by "opioids." In fact, a distinction between "opiates" (exogenous drugs) and "opioids" (the endogenous substances that are natural ligands of opioid receptors) was made in the 1980s. For this reason, in this chapter, the two terms will be used to distinguish between endogenous and exogenous substances.

ENDOGENOUS OPIOID PEPTIDES

Opioid neuropeptides are short amino acid sequences serving as natural endogenous ligands for opioid receptors. All classical opioid peptides may be classified into three families: enkephalins, endorphins, and dynorphins. Each family derives from a distinct precursor: proenkephalin, pro-opiomelanocortin (POMC), and prodynorphin, respectively (Fig. 46.1).

Each precursor, encoded by its corresponding gene, undergoes enzymatic hydrolysis (processing), producing shorter amino acid sequences, biologically active, able to interact with opioid receptors. POMC processing produces the opioid peptide β-endorphin but also α-MSH and ACTH, suggesting the existence of close neuroendocrine correlations between opioid system and responses to stress conditions (hypothalamic–hypophysis–adrenal axis). Proenkephalin gives rise to few copies of met-enkephalin and one of leu-enkephalin, whereas prodynorphin produces three main opioid peptides—dynorphin A, dynorphin B, and α-neoendorphin—all having the same five amino acid leu-enkephalin sequence at their N-terminal end. Another precursor, named pronociceptin, has been cloned and isolated; its processing produces a peptide called nociceptin or orphanin FQ (F stands for phenylalanine and Q for glutamine, which are its first and last amino acids). Pronociceptin also contains nociceptin 2 and another peptide named nocistatin.

General and Molecular Pharmacology: Principles of Drug Action, First Edition. Edited by Francesco Clementi and Guido Fumagalli.
© 2015 John Wiley & Sons, Inc. Published 2015 by John Wiley & Sons, Inc.

BOX 46.1 THE MAIN OPIATE DRUGS

Name	Indications	Contraindications	Side effects
Morphine	Acute pain and severe chronic pain	Severe asthma	Severe constipation, respiratory depression, itching, nausea, vomiting, and urinary retention
	Management of dyspnea caused by pulmonary edema and left ventricular failure	Paralytic ileus	
		Respiratory depression/hypoventilation	
		Upper airway obstruction	
Codeine	Moderate pain	During delivery of a premature infant	Same as morphine; in addition, seizure with high dose
	Suppression of the cough reflex	Premature infants	
Hydromorphone	Acute and chronic pain from moderate to severe	Same as morphine	Less side effects (itching, sedation, nausea, vomiting)
Oxycodone	Pain, from moderate to severe	Same as morphine	Same as morphine
Tramadol	Light–moderate pain	Respiratory depression	Myocardial infarction, pancreatitis, anaphylaxis, dyspnea; less severe side effects compared to morphine (nausea, vomiting, dizziness, headache, seizure, and constipation); risk of serotoninergic syndrome
Levorphanol	Acute and chronic pain	Hypersensitivity to levorphanol	Same as morphine; less nausea and vomiting compared to morphine
Dextromethorphan	No analgesic effects	Coadministration of monoamine oxidase inhibitor (MAOI)	Dizziness, somnolence, fatigue
	Cough suppressor		
Pentazocine	Scarcely used for acute pain treatment	Cardiovascular diseases	Dysphoric and psychotomimetic effects; high doses produce severe respiratory depression, increase of blood pressure, and tachycardia
Nalbuphine and butorphanol	Pain, moderate to severe	Hypersensitivity to either drugs	Hypotension, palpitations, tinnitus, respiratory depression, upper respiratory infection
			Dizziness, sedation, insomnia, nasal congestion with long-term intranasal administration
Meperidine	Not recommended in chronic pain therapy	Recent or concomitant MAOI	Same as morphine; less constipation and urinary retention compared to morphine; excitatory symptoms, hallucinations, convulsions, and mydriasis; serotoninergic syndrome
	It is preferred to morphine during labor since it causes less respiratory depression in neonates		
Diphenoxylate, loperamide	Diarrhea		
Fentanyl	Acute pain and severe chronic pain	Same as morphine	Same as morphine; marked muscular rigidity; less cardiovascular effects compared to morphine
	Adjuvant in anesthesia		
Sufentanil, remifentanil, alfentanil	Potent analgesics in short surgery practices or in severe chronic pain therapy by infusion	Same as morphine	Same as morphine
Methadone	Treatment of opiate dependence	Hypersensitivity to methadone	Same as morphine; additionally, delayed respiratory depression, prolonged QT interval, torsades de pointes
	Severe chronic pain therapy		

Name	Indications	Contraindications	Side effects
Dextropropoxyphene	Light to moderate pain	Hypersensitivity to propoxyphene	Same as morphine
Buprenorphine	Acute and severe chronic pain	Hypersensitivity to buprenorphine	Hypotension, palpitations, tinnitus, respiratory depression, upper respiratory infection
	Opiate dependence treatment		Dizziness, sedation, insomnia, nasal congestion with long-term intranasal administration
Tapentadol	Acute and chronic pain from moderate to severe	Same as morphine, coadministration of MAOI	Less side effects compared to morphine
Naloxone and naltrexone	Acute intoxication by opiate agonists	Acute hepatitis or liver failure (naltrexone)	Cardiac arrhythmia, hypertension, hypotension, hepatotoxicity, pulmonary edema, opioid withdrawal
	Small doses to control the side effects caused by the intravenous or epidural administrations of opiates		Deep vein thrombosis, pulmonary embolism (naltrexone)
	Opiate psychic dependence therapy		
	Treatment of alcoholism		

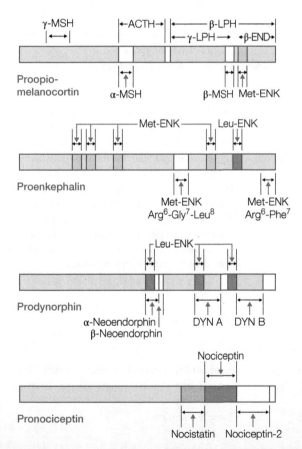

FIGURE 46.1 Opioid peptide precursors. The four opioid peptide precursors are shown: pro-opiomelanocortin, proenkephalin, prodynorphin, and pronociceptin. Pairs of basic amino acids, lysine and arginine, are recognized by enzymes that process the precursors into active peptides.

Nociceptin displays different characteristics from the other opioid peptides. Although in structural and phylogenetic terms this system belongs to the endogenous opioids, biologically and physiologically, it represents an independent system, which can also functionally antagonize the classical opioid system.

In addition to these well-characterized peptides, other opioid receptor-interacting peptides have been isolated: deltorphins and dermorphins, isolated from amphibian skin, and endomorphins, tetrapeptides isolated from rodent brains that bind with high affinity and selectivity the μ opioid receptor (Table 46.1).

Opioid System Distribution

In the CNS, β-endorphin is present in the arcuate nucleus, solitary tract nucleus, dorsal parvocellular area of the hypothalamic paraventricular nucleus, ventral septum, nucleus accumbens (NAc), medial thalamus, medial amygdala, periaqueductal gray (PAG) matter, in locus coeruleus, and bulbopontine reticular area. Besides the CNS and hypophysis, β-endorphin is also produced in some peripheral tissues such as the pancreas, gastric antrum mucosa, placenta, testis, and adrenal medulla.

Enkephalins are widely distributed in the CNS, peripheral nerve plexuses, and adrenal medulla; in the CNS, they are present in short interneurons in areas involved in the modulation of nociceptive transmission (laminae I and II of the spinal cord, spinal trigeminal nucleus, PAG matter), in the control of affective behavior and memory (NAc, amygdala, hippocampus, locus coeruleus, anterior olfactory nucleus, cerebral cortex), in the control of motor activity (substantia

TABLE 46.1 Endogenous opioid peptides, receptor selectivity, and amino acid sequences (one-letter code)[a]

	Receptors	Amino acid sequence of peptides
Endogenous opioid		
Met-enkephalin	δ μ	Y-G-G-F-M
Leu-enkephalin	δ μ	Y-G-G-F-L
β-Endorphin	μ δ	Y-G-G-F-M-Y-S-E-K-S-Q-T-P-L-V-T-L-F-K-N-A-I-I-K-N-A-Y-K-K-G-E
Dynorphin A	κ μ	Y-G-G-F-L-R-R-I-R-P-K-L-K-W-D-N-Q
Dynorphin B	κ	Y-G-G-F-L-R-R-Q-F-K-V-V-T
α-Neoendorphin	κ	Y-G-G-F-L-R-K-Y-P-K
β-Neoendorphin	κ	Y-G-G-F-L-R-K-Y-P
New opioid-related endogenous peptides		
Nociceptin/orphanin FQ	NOP	F-G-G-F-T-G-A-R-K-S-A-R-K-L-A-N-Q
Endomorphin 1	μ	Y-P-W-F-NH_2
Endomorphin 2	μ	Y-P-F-F-NH_2

[a] The structural requirement for binding to opioid receptor is the tyrosine (Y) at position 1, followed, at the critical steric distance (position 3 or 4), by the phenylalanine (F). Nociceptin lacks Y at position 1, and it does not bind to the classic opioid receptors.

nigra, caudate), and in the regulation of the autonomic nervous system (medulla oblongata) and neuroendocrine functions (hypothalamus).

Although often present in distinct neuronal populations, in the CNS, dynorphins show a distribution quite similar to enkephalins, being localized in lamina II of the spinal cord and in the anterior hypothalamic nucleus whose axons project to the posterior hypophysis, reticular formation, caudate, hippocampus, and different regions of the cerebral cortex.

Nociceptin is present in the CNS and peripheral tissues (see Supplement E 46.2, "Nociceptin").

The presence of endomorphins has been demonstrated (only by immunochemistry) in the outer layers of the spinal cord dorsal horns, spinal trigeminal nucleus, nucleus ambiguous, NAc, septum, thalamic nuclei, hypothalamus, amygdala, locus coeruleus, and PAG.

OPIOID RECEPTORS

All opioid receptors are G-protein-coupled receptors (GPCRs) interacting with different G proteins.

There are three types of opioid receptors that, according to the most common nomenclature, are named μ, δ, and κ. Over the last 15 years, they have also been named as OP3, MOP, or MOR (for μ receptor); OP1, DOP, or DOR (for δ); and OP2, KOP, or KOR (for κ).

In addition, another receptor, initially named ORL-1 or OP4 and currently named NOP, has been identified as the binding site for the endogenous ligand nociceptin. Despite its high structural homology with classical opioid receptors, NOP exhibits a different pharmacology. In fact, its activation causes different effects from those classically ascribed to the opioid system, at least at the supraspinal level. Current experimental evidence suggests that the nociceptin/NOP system might functionally antagonize the classical opioid system.

The three classical opioid receptors and NOP are the only receptors cloned so far. Hence, despite *in vitro* and *in vivo* pharmacological studies seem to suggest the existence of different subtypes for each of the three receptors, so far this has not been confirmed by molecular biology data. Such discrepancy between the pharmacological properties of natural and synthetic ligands on one side and the molecular evidence on the other may be due to several phenomena such as heterodimerization between different types of opioid receptors or between these and other GPCRs (see also Chapters 9 and 17) or alternative splicing of their mRNAs.

Finally, a met-enkephalin-binding receptor, named opioid growth factor receptor (OGFr), structurally different from classical opioid receptors and endowed of inhibitory properties on cell proliferation and tissue organization, has been also identified.

Signal Transduction

Generally, μ, δ, and κ opioid receptors are coupled to pertussis toxin (PTX)-sensitive $G_{\alpha o}$ and $G_{\alpha i2}$ proteins, showing differences related to the type of tissue or cells in which they are expressed. For example, the μ receptors expressed in DRG sensory neurons transduce through $G_{\alpha o}$ subunit, while in brain areas it is preferentially coupled to $G_{\alpha i1-3}$ proteins. κ receptors preferentially activate the $G_{\alpha o}$ subunit, while δ receptors activate $G_{\alpha i1}$ proteins. Stimulation of presynaptic μ receptors decreases depolarization-dependent neurotransmitter release by inhibiting N-type Ca^{2+} channels, whereas stimulation of postsynaptic μ receptors produces hyperpolarization by activating K^+ channels and by inhibiting L-type Ca^{2+} channels. Therefore, opioids tend to inhibit neuronal transmission, such as the transmission of impulses produced by noxious stimuli.

More recent evidences indicate that μ, δ, and κ opioid receptors interact with five different isoforms of $G_{\alpha i/o}$ (G_{i1-3} and G_{oA-B}) and thereby regulate signal transduction through different effectors, such as adenylate cyclase (AC) (AC1, AC5, AC6, AC8), ion channels, and mitogen-activated protein kinases (MAPK).

In addition, all three opioid receptors can also transduce inhibitory signals through PTX-insensitive G proteins, like $G_{\alpha z}$ proteins, the only $G_{\alpha i}$ protein to be insensitive to PTX because of a missing cysteine residue in the carboxy-terminal portion (target of the ADP-ribosylation catalyzed by PTX). $G_{\alpha z}$ is expressed in nervous tissue and colocalizes with opioid receptors in neuronal cell lines; it is coupled with μ receptor in PAG, where it mediates supraspinal analgesia. Besides inhibiting AC activity, $G_{\alpha z}$ is able to regulate MAPK and Ca^{2+} and K^+ channel activity and to interact with several recently identified effectors, belonging to the regulator of G-protein signaling (RGS) family, such as G-protein-regulated inducers of neurite outgrowth (GRIN) and Rap1-specific GTPase-activating protein (Rap1GAP).

Finally, opioid receptors can couple to other PTX-insensitive G proteins, such as $G_{\alpha 14}$ and $G_{\alpha 16}$, which activate a signal transduction cascade mediated by phospholipase C (PLC) stimulation (PLCβ) (usually activated by $G_{\alpha q}$ protein) with activation of c-Jun N-terminal kinase (JNK) and effects on cellular growth. The distribution of these G proteins, which is mainly restricted to hematopoietic cells and peripheral tissues, suggests a role in the immunomodulatory activity of opioids.

It has been also suggested that opioids exert some of their actions, such as analgesia, tolerance, and dependence, through the $G_{\alpha s}$ protein, and this has been recently confirmed by the observation that in CHO cells μ receptors transduce the signal for caveolae formation through a $G_{\alpha s}$ protein.

To add to the complexity of the opioid receptor signaling, it has been shown that μ receptor localization and signaling in primary cultures of trigeminal sensory neurons can be modulated by integrins that may change the type of G protein coupled to the receptor (e.g., from a $G_{\alpha i}$ to a $G_{\alpha s}$).

As in the case of other GPCRs, the $G_{\beta\gamma}$ complex, after dissociation from α-subunit, plays a very important role in the diversification of the opioid-activated signal transduction. It has been demonstrated that a high number of effectors and proteins involved in signal transduction may functionally interact with $G_{\beta\gamma}$ subunits, determining a wide range of biological responses (Fig. 46.2), also opposite to those triggered by the $G_{\alpha s}$ subunit: for example, the AC2-induced cAMP production might be regulated by $G_{\beta\gamma}$ and $G_{\alpha i}$ in opposite ways.

The G-protein-coupled inwardly rectifying K^+ channels (GIRK channels) of type 3 are activated by direct interaction with the βγ-subunits after GPCR stimulation; this phenomenon is considered part of the mechanism of opioid inhibition of nociceptive transmission.

Binding of distinct agonists exhibiting different intrinsic activities may cause the same receptor to undergo different conformational changes, leading to formation and stabilization of specific receptor/G-protein complexes and thereby to the induction of specific signal transduction cascades. In recent years, it has also been proposed that dimerization could occur between different types of opioid receptors, generating homo- and heterodimers, also with other GPCRs. Dimerization can cause changes in ligand affinity, receptor transduction mechanisms, and cellular trafficking; nevertheless, the functional significance of this phenomenon is still uncertain.

OPIOID RECEPTOR DISTRIBUTION AND EFFECTS

Opioid receptors are widely distributed in the CNS and also in the periphery (see Supplement E 46.3, "Opioid Receptor Distribution"). Their localization in cerebrospinal areas modulating specific functions suggests the physiological roles played by the endogenous opioid system, and it explains the pharmacological effects of opiates. The μ opioid receptors are the most widespread and abundant and mediate many pharmacological effects of opiate analgesics.

The functions modulated by the endogenous opioid system can be pharmacologically defined using either exogenous ligands, such as morphine-like opium alkaloids and their derivatives, or synthetic molecules acting as antagonists at opioid receptors.

The opioid peptides derived from POMC, proenkephalin, and prodynorphin precursors exhibit a moderate selectivity toward their receptors, whereas endomorphins 1 and 2 are highly selective for the μ receptor and dermorphin and deltorphin for the δ.

Modulation of Nociceptive Transmission

The ascending spinothalamic–cortical and the descending brain–spinal systems represent the major CNS neuronal pathways involved in transmission, modulation, and control of nociception. The strategic localization of opioid receptors, as well as their transduction mechanisms, assigns to the endogenous opioid system a primary role in the modulation of nociceptive input and analgesia (Fig. 46.3).

By activating specific structures (nociceptors) located at the end of small-diameter primary afferent fibers (C and A_δ), high-intensity stimuli (noxious stimuli) of different types (thermal, mechanical, and chemical) induce action potentials that reach the spinal cord dorsal horn.

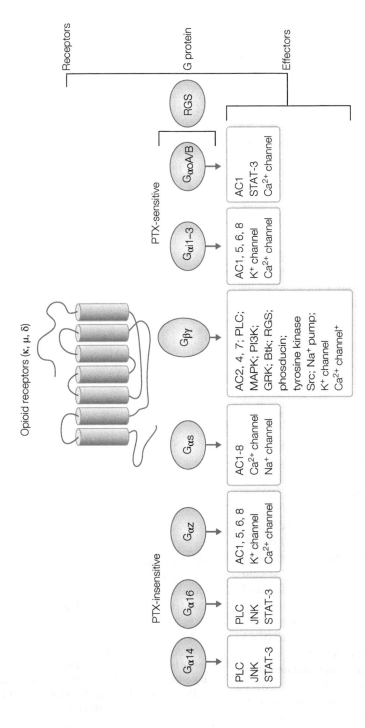

FIGURE 46.2 Opioid receptor signal transduction. Opioid receptors are coupled to pertussis toxin (PTX)-sensitive or PTX-insensitive G protein. The $G_{\alpha i}$, $G_{\alpha o}$, and $G_{\alpha z}$ GTPase activity is modulated by regulators of G-protein signaling (RGS) proteins. AC, adenylate cyclase; Btk, Bruton's tyrosine kinase; GRK, G-protein-coupled protein kinase; JNK, c-Jun N-terminal kinase; MAPK, mitogen-activated protein kinase; PI3K, phosphoinositide-3-kinase; PLC, phospholipase C; STAT3, signal transducer and activator of transcription 3.

Laminae II and III of the dorsal horn (substantia gelatinosa (SG)) contain small interneurons that produce and release opioid peptides (enkephalins, dynorphins). Opioids are able to modulate/inhibit transmission between primary afferent fibers and spinothalamic neurons through opioid receptors located presynaptically on the spinal end of the primary neuron and/or postsynaptically on the second-order (spinothalamic) neuron.

The descending inhibitory system originates from neurons in the PAG that receive impulses from the cortex, hypothalamus, and thalamus in particular.

PAG neurons project to some nuclei of the medulla oblongata, such as the nucleus raphe magnus (NRM), nucleus reticularis magnocellularis (NRMC), and nucleus reticularis paragigantocellularis (NRPG). From these nuclei originate descending aminergic fibers that run along the dorsolateral funiculus of the spinal cord and terminate in the SG (Fig. 46.3), where they modulate afferent nociceptive transmission, either directly or through activation of opioid interneurons (gate control theory).

The analgesic effects of opioids and opiates are due to their ability to directly inhibit the ascending transmission of nociceptive impulses from the spinal cord dorsal horns and to activate the descending pain control circuits from the thalamus to the dorsal horn.

At spinal level, most of μ, δ, and κ receptors are localized presynaptically on the dorsal horn terminals of primary afferent fibers. In the descending inhibitory pathways, μ and κ receptors are always coexpressed, whereas the δ receptors are mainly present in the PAG, NRM, and thalamus. At supraspinal level, μ and κ receptors are mainly localized in the locus coeruleus (Fig. 46.3).

Activation of μ receptors always causes analgesia, for example, through removal of GABAergic inhibition on PAG neurons projecting, downstream to the spinal cord, in the descending system. On the other hand, activation of κ receptors can produce either analgesia or hyperalgesia, for example, by inhibiting neurons descending from the PAG.

It should be taken into account that the overall opioid-dependent analgesia results from the sum of the effects on nociceptive input transmission and the effects on the emotional reactions to the inputs themselves (see Supplement E 46.4, "Neurobiological Basis of Acute and Chronic Pain").

The presence of opioid receptors on the peripheral terminals of sensitive neurons suggests the involvement of opioids in nociceptive modulation also at the periphery.

Current knowledge indicates that during inflammation, an increase of μ and κ receptors occurs on the terminals of primary afferents, along with a migration of immune cells from vessels to inflamed tissue. Here, leukocytes release opioid peptides that interact with the opioid receptors upregulated by inflammation and induce analgesia by decreasing sensory ending excitability and/or release of proinflammatory neuropeptides (Fig. 46.4).

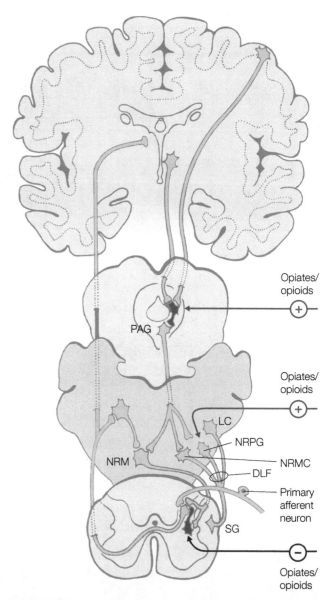

FIGURE 46.3 The descending inhibitory pathway controlling nociceptive inputs. Periaqueductal gray (PAG) neurons project to the nucleus raphe magnus (NRM) and nucleus reticularis magnocellularis (NRMC) of medulla. Descending fibers, arising from these nuclei, reach the spinal substantia gelatinosa (SG) through the dorsolateral funiculus (DLF). Other spinal projections of this pain descending system originate from the reticular paragigantocellular nucleus (NRPG), controlled by PAG, and from pontine–medullary noradrenergic cells (locus coeruleus (LC)). NRM receives activating signals from the adjacent NRPG that in turn receives inputs from the spinothalamic ascending fibers. The descending system exerts an inhibitory control on spino-thalamic ascending neurons either directly or indirectly via opioid inhibitory SG interneurons. Right side, inhibitory descending system; left side, spino-thalamic ascending neurons; in dark and pointed by arrows, opioid peptide-containing neurons.

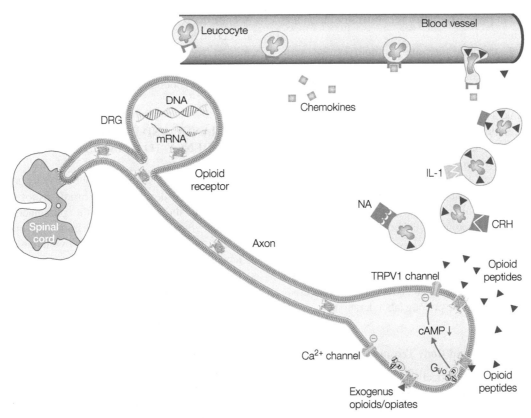

FIGURE 46.4 Opioid-mediated peripheral analgesia. Opioid peptide-containing leucocytes migrate from the vessel to the inflamed tissue, where they release peptides following stimulation by various factors, such as chemokines, CRH, and noradrenaline. Opioid receptors are synthesized in the dorsal root ganglion and then transported to the dorsal horn of the spinal cord and the peripheral nerve ending of the primary afferent neuron. During inflammation, the expression of opioid receptors increase, and they are therefore available to bind the released opioid peptides and the exogenous molecules applied in the periphery. (from Busch-Dienstfertig M., Stein C. (2010). Opioid receptors and opioid-producing leukocytes in inflammatory pain—basic and therapeutic aspects. *Brain, Behavior, and Immunity, 24*, 683–694, modified.)

During inflammation, peripherally administered opiates induce less tolerance to their analgesic effects due to a higher availability of opioid receptors in nerve peripheral terminals, possibly caused by increased receptor endocytosis and recycle induced by leukocyte-released endogenous opioids.

Respiratory Depression

Opioid peptides and opiate drugs depress respiration through a direct effect on brain stem respiratory centers. In fact, activation of μ opioid receptors causes a marked inhibition of neurons of the medulla oblongata respiratory centers that, in physiological conditions, trigger the respiratory act in response to the increase in blood CO_2 partial pressure produced by energy metabolism. Therapeutic doses of morphine in humans reduce respiratory frequency and volumes and can also determine irregular and periodic breathing. This effect is dose dependent and can lead to death for respiratory arrest; however, clinically significant respiratory depression rarely occurs at therapeutic doses of morphine.

Cardiovascular Effects

Experimental data show that opioid receptor activation causes contrasting effects on blood pressure and heart rate depending on dose of agonist, route of administration, and site of injection, as well as on conditions of anesthesia or stress.

In supine patients, therapeutic doses of opiates have no major effects on blood pressure or cardiac rate and rhythm. Nevertheless, orthostatic hypotension and fainting may occur since opiates can produce peripheral vasodilation, reduced peripheral resistance, and baroreceptor reflex inhibition. The peripheral arteriolar and venous dilation produced by morphine involves several mechanisms, including histamine release.

Effects on the GI Tract and Other Smooth Muscles

Activation of μ opioid receptors causes a delay in gastric emptying, thus increasing the possibility of esophageal reflux, along with a relevant increase in muscle tone and a marked decrease in propulsive peristaltic waves in the small and large intestine.

GI effects of morphine are due to both central and peripheral drug actions.

Stimulation of μ, δ, and κ receptors of the intramural nerve plexuses causes hyperpolarization (due to increased potassium conductance) and blocks acetylcholine release. However, intracerebroventricular or intrathecal injection of μ agonists can inhibit GI propulsive activity, provided that the extrinsic innervation to the bowel is intact. The tolerance to GI morphine effects is low, and constipation is a major side effect after chronic use, especially with oral formulations. Administration of analgesic doses of morphine increases pressure in the biliary tract because of constriction of the sphincter of Oddi.

Stimulation of the brain and spinal cord μ and δ receptors inhibits emptying reflex of the urinary bladder and increases vesical sphincter tone. κ receptor stimulation promotes diuresis. Therapeutic doses of morphine prolong labor duration.

Effects on Food Intake and Body Temperature

Opioid peptides and opiates affect food and liquid intake. In fact, it is well known that opioid agonists stimulate food intake, while antagonists inhibit it. Recent data suggest that the opioid system may play a role not only in the control of the amount of ingested food but also in its palatability, or in other words its choice based on appreciation.

This suggests an involvement of the opioid system in rewarding circuits activated in the control of food intake, as well as by substances able to induce abuse behavior. Opioid peptides affect the hypothalamic mechanisms that regulate body temperature with variable effects also depending upon external temperature. μ receptor agonists increase body temperature in subjects exposed to external temperatures exceeding 30°C, whereas μ and δ agonists decrease body temperature in subjects exposed to cold (4°C). It is possible that body temperature regulation could be closely related to eating behavior modulation.

Effects on the Immune System

Different cells of the immune system produce peptides deriving from POMC, proenkephalin, and also pronociceptin. β-Endorphin increases the cytostatic activity of human monocytes and recruitment of killer cell precursors; it also exerts a powerful chemotactic effect. Several opioid peptides exert naloxone-sensitive effects on macrophage and monocyte functions, while morphine produces mainly inhibition of lymphocyte rosette formation.

In experimental animals, morphine suppresses NK cell cytostatic activity and promotes growth of implanted tumors.

The effects of opioids on the immune system are complex and derive part from direct actions on immune system cells and part from centrally mediated neuronal mechanisms.

The overall effect is immunosuppressive with differences, also marked, among various opiate drugs.

TOLERANCE AND PHYSICAL DEPENDENCE TO OPIATES

Tolerance and physical dependence are two pharmacological phenomena that develop after chronic exposure to opiates. Tolerance is the decrease of the pharmacological effect occurring after repeated administration of opioid receptor agonists, which causes the need to increase the dose to achieve the same effect.

Once doses much higher than the starting ones are reached, the body loses its homeostasis, and physical dependence takes place; in these conditions, alterations in its equilibrium determine the onset of abstinence crisis after abrupt discontinuation of agonist administration.

Tolerance and drug dependence are therefore related to each other and independent of psychic dependence (see also page 581, Chapter 10 and Supplement E 46.5, "Opiate Addiction").

It has been thought for many years that tolerance to opiates could be the result of cellular adaptation processes, such as acute desensitization and receptor downregulation, leading to a reduction in the number of functional receptors present on cellular membrane.

It is now believed that neuronal adaptation to opiates' chronic effects involves a complex series of molecular and cellular events, acting in opposite directions, where receptor modulation represents only the first step.

The most studied cellular adaptation downstream of receptor signaling consists in the superactivation of the cAMP pathway, with increase in cyclic AMP response element DNA-binding protein (CREB) and Fos proteins, that represents the molecular mechanism underlying the homeostatic response to long-term inhibition of AC due to chronic exposure to opiate agonists.

Molecular Mechanisms of Cellular Adaptation to Chronic Exposure to Opiates

Tolerance can be observed either when receptor exposure to the agonist occurs for a short time (acute tolerance) or when it is prolonged (chronic tolerance); in both cases, it leads to a decreased downstream effect. Recent evidence has modified the classical hypotheses on tolerance by adding important information, particularly related to cellular mechanisms underlying the earliest steps of this complex phenomenon.

Similarly to other receptors belonging to the GPCR family, it has been demonstrated that agonist binding to opioid receptors, besides activating signal transduction, leads to short-term events of desensitization that may lead to tolerance and that involves receptor phosphorylation by PKC and other kinases, such as PKA and β-ARK.

Following desensitization, internalization occurs through endocytosis, and depending on receptor type, different pathways of cellular trafficking may be activated. It has been suggested that different ligands can determine activation of distinct intracellular pathways, accounting for the different efficacy of various agonists, either endogenous or opiate drugs, and possibly also for the different drug abuse liability.

Opioid receptor internalization is partially mediated by G-protein-coupled receptor kinases (GRKs) that phosphorylate the agonist-bound receptor favoring its interaction with β-arrestin, which uncouple the receptor from the G protein and promote its internalization. This phenomenon does not occur with morphine, which is not able to translocate β-arrestin, and therefore, it does not determine receptor internalization. This observation confirms that signal transduction activation and receptor internalization are two distinct mechanisms.

AC Superactivation At molecular level, chronic tolerance has been related to hyperactivation of AC (with increase in cAMP levels) and to adaptations of $G_{\alpha i/o}$ subunits and βγ complex.

Opioid-induced AC hyperactivation is isoenzyme specific: AC2 and AC4 seem to be directly activated by the $G_{\beta\gamma}$ complex, whereas AC1, AC5, AC6, and AC8 seem to be indirectly activated through effectors regulated by $G_{\beta\gamma}$, such as CREB, MAPK (Ras-dependently regulated), or JNK (Fig. 46.5).

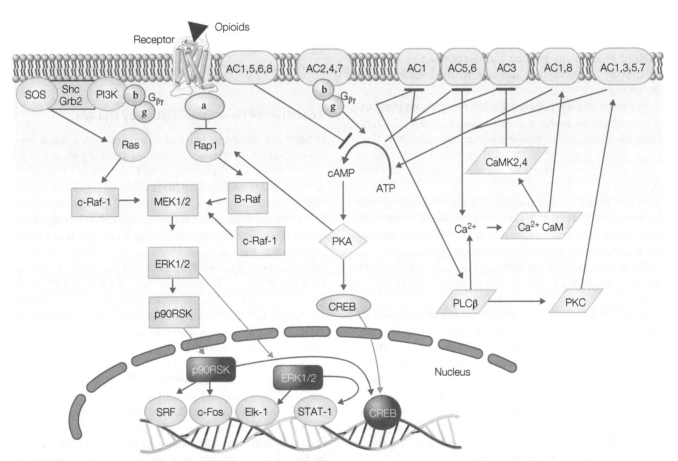

FIGURE 46.5 Opioid receptor signal transduction and trafficking. Opioid receptor signal transduction and trafficking involve adenylate cyclase (AC), cAMP response element-binding protein (CREB), mitogen-activated protein kinase (MAPK), and different transcription factors. The $G_{\alpha i/o}$ proteins prevent activation of cAMP-dependent protein kinase A (PKA) by inhibiting cAMP production by AC1, AC5, AC6, and AC8. The $G_{\beta\gamma}$ complex directly stimulates AC2, AC4, and AC7 but inhibits AC1. Moreover, $G_{\beta\gamma}$ inhibits AC1, AC3, AC5, and AC6 via Ca^{2+} mobilization induced by phospholipase Cβ (PLCβ) and activation of calmodulin kinase 2 and 4 (CaMK2 and CaMK24). However, at the same time, it stimulates AC1 and AC8 via Ca^{2+}/calmodulin (activated by PLCβ) and AC1–AC3, AC5, and AC7 via PKC. CREB activity is regulated by the MAPK cascade through β-Raf or c-Raf-1. Opioid receptors can modulate the MAPK cascade and other transcriptional events not only through the $G_{\alpha i/o}$ proteins but also via $G_{\beta\gamma}$ following recruitment of phosphatidylinositol-3-kinase (PI3K), Shc, Grb2, and SOS.

In addition, many other proteins belonging to the signal transduction system are upregulated; these include RGS proteins of which numerous and different isoforms have been identified.

Relationship between Receptor Stimulation and Their Endocytosis Recent experimental evidence has modified the classical hypothesis on the development of opioid tolerance. Different opioid agonists, although all activating the signal transduction of μ opioid receptors, determine widely diversified biochemical responses: for example, morphine and other low-efficacy agonists do not promote μ receptor internalization, whereas other agonists do it very quickly. This means that the μ receptor desensitization and downregulation are agonist-dependent phenomena involving different receptor configurations determined by stereochemically distinct ligands and produce different downstream cellular events. As for other GPCRs, endocytosis or internalization could lead either to receptor degradation or to its dephosphorylation and recycling to the cell membrane. This means that receptor internalization can lead to signal reduction as a result of receptor inactivation and degradation or to signal increase by recycling the receptor to the membrane for further use.

In conclusion, it is believed that tolerance to opioid effects is not related to receptor desensitization, but to its lack. Thus, morphine, which does not induce internalization, does not promote recovery of recycled receptors and leads to major adaptations downstream, increasing chronic tolerance to its effects.

In this regard, a new parameter, named relative agonist versus endocytosis (RAVE) has been proposed, which consists in the ratio between signaling of the opioid agonist (measured as the ability to activate K$^+$ currents) and its ability to induce endocytosis (i.e., to cause internalization); this may represent a modern and effective tool to predict the potential development of tolerance to each specific agonist.

Indeed, recent *in vitro* and *in vivo* studies have shown that morphine, which has a high RAVE ratio (as it does not induce internalization), develops tolerance very rapidly, whereas other μ opioid agonists, such as fentanyl or DAMGO, having a low RAVE ratio (as they induce a rapid internalization) develop low tolerance. In particular, it has been observed that addition of low doses of fentanyl or DAMGO to morphine causes a delayed development of tolerance to the analgesic effect of the alkaloid, as they activate recycling and therefore receptor reuse.

Even if the RAVE theory has been recently criticized, these interesting observations will hopefully improve our understanding of the differences between endogenous modulation of the opioid system and its activation by drugs, enabling us to implement therapeutic strategies and to develop analgesic drugs with fewer side effects, such as tolerance and physical dependence.

Molecular Mechanisms of Withdrawal

Discontinuation of a chronic treatment with opiates causes a severe withdrawal syndrome characterized by drug seeking, agitation, nausea, vomiting, muscle pain, spasms, dysphoria, insomnia, diarrhea, piloerection, and sweating.

Variety and severity of these symptoms contribute to relapse. Among the molecular mechanisms proposed to explain this, it has been observed that chronic treatment with opiates determines cAMP upregulation in several brain nuclei. In the NAc, this compensatory upregulation of cAMP–PKA pathway causes an increase in CREB-mediated prodynorphin transcription, leading to a marked stimulation of presynaptic κ opioid receptors, with inhibition of dopamine release, therefore contributing to the dysphoria typical of the opiate abstinence syndrome.

Since the activity of locus coeruleus, the cerebral nucleus most enriched in adrenergic neurons, is associated with somatic manifestations of the withdrawal syndrome and chronic administration of opiates in this nucleus produces increased activation of the cAMP–PKA–CREB pathway, this mechanism also contributes to manifestations of dysphoria and somatic symptoms of the withdrawal syndrome.

Regulation of Gene Transcription by Opioids

It has been demonstrated that treatment with opiates produces changes in gene expression of several proteins (e.g., neurotransmitters, mediators, and structural molecules involved in synapse formation) in different tissues and cells. These alterations of the transcriptional pattern may contribute to the long-term molecular and cellular adaptations occurring during tolerance and dependence. The best-known phenomena are increased expression of some molecular components of the cAMP transduction pathway, of specific subunits of the AMPA receptor, and of CRF and NO synthase genes and expression of new isoforms of the Fos and Jun family of transcription factors. Furthermore, alterations in the expression of opioid precursor and receptor-encoding genes have been observed.

Opioids are also able to modulate different transcription factors, such as CREB, DARPP32, NF-κB, and members of the MAPK cascade. CREB is phosphorylated by PKA on Ser133, and, once activated, it is able to translocate to the nucleus where it regulates early gene transcription. It has been shown that the upregulation of AC1 and AC8 occurring in the locus coeruleus upon opioid chronic treatment is induced by CREB; CREB phosphorylation is increased in the dorsal horn of the spinal cord of mice tolerant to the analgesic effect of morphine; CREB phosphorylation is also increased in the NAc of morphine-addicted animals. CREB depends on different regulatory factors, such as PKA, MAPK cascade, DARPP32, Ca^{2+}/calmodulin, and PKC (Fig. 46.5). Opioid receptors are also able to stimulate all MAPK pathways. Activation of μ, δ, and κ receptors, through the $G_{\beta\gamma}$ complex and the Ras-dependent pathway, increases ERK1/ERK2 phosphorylation, both in cellular systems and *in vivo* (ventral tegmental area (VTA) and locus coeruleus), leading to activation of transcription factors such as Elk-1. In cell lines, activation of the classical μ, δ, and κ opioid receptors produces only a weak stimulation of JNK and regulates p38 MAPK activation by PKA and PKC.

It is worth mentioning that opioid system gene expression may be altered in several pathological conditions, such as Parkinson's disease, epilepsy, or upon exposure to different classes of addictive drugs.

Epigenetic Mechanisms Recently, the development of advanced techniques to study chromatin remodeling has allowed investigating the epigenetic mechanisms underlying the alterations in gene expression occurring after chronic exposure to opioid receptor agonists. It has been observed that μ, κ, and δ opioid receptor genes undergo different kinds of epigenetic regulation, suggesting that this level of modulation of gene expression may contribute to their different cellular specificity, in particular in response to physiological, environmental, and nutritional conditions. In fact, it has been reported that chronic treatment with morphine induces changes in epigenetic mechanisms, such as DNA methylation or histone acetylation. Furthermore, μ receptor gene activation has been associated with histone modifications (trimethylation and acetylation) in its promoter. These occur, for example, in T lymphocytes where IL4 induces expression of the μ opioid receptor; the initial activation is mediated by phosphorylated STAT6 transcription factor and occurs in a few minutes, but after several hours, numerous epigenetic changes take place, shaping gene architecture and affecting its expression.

ADDICTION TO OPIOIDS

It is well known that individuals treated with opiates may develop psychic dependence, which can be distinguished from tolerance and physical dependence.

Psychic dependence is a chronic recurrent disorder characterized by a compulsive behavior, which consists in loss of control over drug seeking and drug taking, regardless of the negative consequences for the subject and others. The English word "addiction" derives from the Latin "addicere" or enslave and indicates a compulsive, uncontrolled, search behavior, looking for a situation or a condition of reward. The reinforcing effects of all substances of abuse are due to their actions on the mesocorticolimbic system, a circuit mainly consisting of dopaminergic neurons that project from the VTA to the shell of the NAc, amygdala, and prefrontal cortex (PFC). This circuit also includes glutamatergic projections from PFC and amygdala to NAc and GABAergic projections from NAc to VTA. Opioid interneurons are present in NAc, amygdala, and VTA (Fig. 46.6). The positive reinforcement properties of all substances of abuse result from the activation of the mesolimbic dopaminergic projections. Endogenous opioids and opiates facilitate dopamine release directly by activating μ and δ receptors in NAc and indirectly by activating μ receptors on GABAergic neurons of the VTA; at VTA level, inhibition of GABAergic neurotransmission increases firing of dopaminergic neurons.

In contrast, activation of κ receptors, induced by opioids and opiates, inhibits dopaminergic transmission in the mesocorticolimbic circuit: κ receptors are localized on cell bodies of the dopaminergic neurons in VTA and on terminations in NAc.

Dopamine release in NAc is associated with both the effects of the substance of abuse and the environmental circumstances in which its administration occurs. Thus, the attention to environmental conditions is enhanced, and their recurrence facilitates their recognition as a warning or a sign anticipating drug effects.

Although in physiological conditions rewarding stimuli often activate dopamine release from this neuronal circuit, it seems that this neurotransmitter itself does not produce rewarding or otherwise pleasant effects, as previously believed, but rather facilitates learning of experiences associated with its release, regardless of their affective content. In this context, dopamine may also play a role in duration of short-term memory. This hypothesis could explain why opiates and other drugs of abuse produce reinforcement for both positive affective (pleasure, gratification) and negative affective (aversion) experiences and why they may function as discriminative stimuli in operational decisions (the subject chooses what was previously associated with drug administration). This ability of opioids to function as

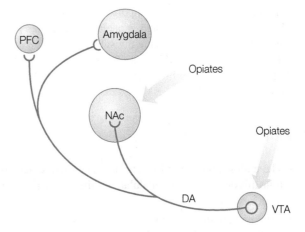

FIGURE 46.6 The mesocorticolimbic system. Dopaminergic neurons project from the ventral tegmental area (VTA) to the nucleus accumbens (NAc), prefrontal cortex (PFC), and amygdala.

reinforcing and discriminative stimuli probably represents the main incentive to the abuse of these drugs.

The opioid system has also been involved in the molecular mechanisms underlying the effects of numerous substances of abuse such as cocaine, amphetamines, and alcohol, also as a result of its close interaction with other neurotransmitter systems involved in these dynamics, such as the cannabinoid system.

At least three different types of factors contribute to the susceptibility to develop addiction: factors related to the effects of the substance, as well as genetic and environmental factors (see Supplements E 46.1 and E 46.2).

Pharmacogenetic studies and identification of polymorphisms in genes encoding opioid system components have provided new perspectives to understand psychic dependence. It has been proposed that at least 25–60% of factors determining vulnerability to develop psychic dependence are genetic, that is, it is due to genetic polymorphisms, for example, in the μ opioid receptor gene.

A crucial role has been attributed to environmental conditioning factors. It has been proposed that development of psychic dependence and vulnerability to relapse after deprivation are the result of CNS neuroadaptive processes that oppose the action of reinforcement of drugs of abuse. The long-term effectiveness of drug-associated stimuli in causing compulsive seeking behavior is observed in animal models of relapse and in humans and reflects the continuing responsiveness to conditioned stimuli that favor the drug-seeking behavior. This confirms a significant role of learning and conditioning factors in the persisting dependence on addictive drugs.

A therapeutically appropriate use of opiates for the treatment of chronic pain has been hindered to date by the incorrect belief that their use will inevitably lead to addiction. The actual prevailing hypothesis suggests that the therapeutic use of opiates is not associated with the conditional environmental stimuli that determine the positive reinforcement responsible for compulsive use. The condition in which the drug is taken and especially the underlying painful pathology do not provide the substrate and the context for development of addiction; clinical findings in the field of pain medicine confirm that the phenomenon of abuse is observed very rarely.

TAKE-HOME MESSAGE

- The endogenous opioid system is a peptidergic neurotransmission system modulating the activities of neurons and other cells.
- Opioid peptides belong to three classes of small peptides—enkephalin, endorphin, and dynorphin—that originate from selective processing and cleavage of three larger precursors, proenkephalin, pro-opiomelanocortin, and prodynorphin.
- Opioid receptors are GPCRs. They are named μ, δ, κ, and NOP and mostly coupled to G_i and G_o.
- Opioid drugs are analgesic; they also depress respiration and intestine motility and interfere with several endocrine and immunological functions.
- Chronic use of opioid drugs is associated with tolerance, physical dependence and may evolve into addiction.

FURTHER READING

Alexander S.P. (2009). Guide to receptors and channels, 4th edition. *British Journal of Pharmacology*, 158(Suppl. 1), S78–S79.

Alvarez V., Arttamangkul S., Williams J.T. (2001). A RAVE about opioid withdrawal. *Neuron*, 32, 761–763.

Bali A., Randhawa P.K., Jaggi A.S. (2014). Interplay between RAS and opioids: opening the Pandora of complexities. *Neuropeptides*, 48, 249–256.

Busch-Dienstfertig M., Stein C. (2010). Opioid receptors and opioid-producing leukocytes in inflammatory pain—basic and therapeutic aspects. *Brain, Behavior, and Immunity*, 24, 683–694.

Chu Sin Chung P., Kieffer B.L. (2013). Delta opioid receptors in brain function and diseases. *Pharmacology & Therapeutics*, 140, 112–120.

Civelli O. (2008). The orphanin FQ/nociceptin (OFQ/N) system. *Results and Problems in Cell Differentiation*, 46, 1–25.

Gutstein H.B., Akil H. (2006). Opioid analgesics. In: Brunton L.L., Lazo J.S., Parker K.L. (eds.), *Goodman and Gilman's the pharmacological basis of therapeutics*. 11th ed. New York: McGraw-Hill, pp. 547–590.

Koch T., Hollt V. (2008). Role of receptor internalization in opioid tolerance and dependence. *Pharmacology & Therapeutics*, 11, 199–206.

Pan H.L., Wu Z.Z., Zhou H.Y., et al. (2008). Modulation of pain transmission by G protein-coupled receptors. *Pharmacology and Therapeutics*, 117, 141–161.

Pasternak G.W., Pan Y.X. (2013). Mu opioids and their receptors: evolution of a concept. *Pharmacological Reviews*, 65, 1257–1317.

Trigo J.M., Martin-Garcia E., Berrendero F., et al. (2010). The endogenous opioid system: a common substrate in drug addiction. *Drug and Alcohol Dependence*, 10, 183–194.

Van Rijn R.M., Whistler J.L., Waldhoer M. (2010). Opioid-receptor-heteromer-specific trafficking and pharmacology. *Current Opinion in Pharmacology*, 10, 73–79.

Van't Veer A., Carlezon W.A. Jr. (2013). Role of kappa-opioid receptors in stress and anxiety-related behavior. *Psychopharmacology*, 229(3), 435–452.

Wei L.-N., Loh H.H. (2011). Transcriptional and epigenetic regulation of opioid receptor genes—present and future. *Annual Review of Pharmacology and Toxicology*, 51, 75–97.

Williams J.T., Ingram S.L., Henderson G., Chavkin C., von Zastrow M., Schulz S., Koch T., Evans C.J., Christie M.J. (2013). Regulation of μ-opioid receptors: desensitization, phosphorylation, internalization, and tolerance. *Pharmacological Reviews*, 65, 223–254.

47

THE ENDOCANNABINOID SYSTEM

DANIELA PAROLARO AND TIZIANA RUBINO

By reading this chapter, you will:

- Learn the distribution, metabolism, and function of the endogenous cannabinoid system
- Know the structure and function of endocannabinoids and their pharmacological modulation
- Learn the potential use and toxicity of drugs acting on the endocannabinoid system

---

The discovery in 1990 that a 473-amino-acid G-protein-coupled receptor (CB1) encoded by a rat brain cDNA clone mediated the effects of Δ9-tetrahydrocannabinol (THC), the main psychoactive component present in *Cannabis sativa*, represented the starting point for the identification and characterization of the now known as endocannabinoid system (ECS). Indeed, the cloning of CB1 receptor (CB1R) was followed by the identification of a second G-protein-coupled receptor called CB2 and the discovery that mammalian tissues can both synthesize and release endogenous agonists of these receptors called endocannabinoids (ECs). Based on the intense research carried out in the last decades, we now know that the ECS is implicated in the modulation of numerous physiological functions and pathological conditions not only in the central nervous system (CNS) but in the entire organism.

So far, the ECS comprises two main receptors, CB1 and CB2; ECs (the best known are anandamide (AEA) and 2-arachidonoylglycerol (2-AG)); and the enzymes responsible for their synthesis and degradation.

CANNABINOID RECEPTORS

To date, the International Union of Basic and Clinical Pharmacology recognizes two types of cannabinoid receptors, called CB1 and CB2, both belonging to the G-protein-coupled receptor superfamily. The possible inclusion in this list of a third receptor for cannabinoids, formerly known as orphan GPR55 receptor, is still debated. Moreover, different experimental data demonstrate that other classes of receptors, in particular the vanilloid TRPV1 receptor and the nuclear peroxisome proliferator-activated receptors (PPAR)α and PPARγ, are involved in the pharmacological effects of ECs.

CB1R

CB1Rs are found in many mammals, including humans, and are particularly abundant in the CNS. High CB1R densities are present in the basal ganglia (substantia nigra, globus pallidus, entopeduncular nucleus, lateral caudate–putamen), cerebellum, hippocampus, and cortex. Lower CB1R densities have been detected in limbic areas such as the nucleus accumbens and amygdala and in the hypothalamus, midbrain, medulla, and spinal cord (Fig. 47.1). Moderate receptor density is also found in peripheral tissues such as the adipose tissue, liver, reproductive organs, heart, intestine, vascular tissue, and bone.

CB1Rs are coupled to $G_{i/o}$ proteins that inhibit adenylyl cyclase and N-type and P/Q-type calcium channels and vice versa stimulate MAP kinases, NO production, and A-type and inwardly rectifying potassium channels.

In the CNS, CB1Rs are mainly located in presynaptic membranes, and their activation inhibits neurotransmitter release from axon terminals. It is widely demonstrated that

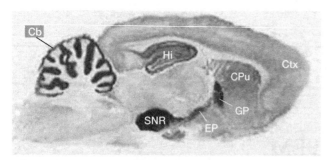

FIGURE 47.1 Autoradiographic localization of CB1 receptors in rat brain. High-density areas are the basal ganglia (substantia nigra, globus pallidus, entopeduncular nucleus, lateral caudate–putamen), cerebellum, hippocampus, and cortex. Cb, cerebellum; CPu, caudate–putamen; Ctx, cerebral cortex; EP, entopeduncular nucleus; GP, globus pallidus; Hi, hippocampus; SNR, substantia nigra reticulate.

most effects of cannabinoids in the CNS are mediated by CB1R activation.

CB2 Receptors

CB2 receptors (CB2Rs) were cloned a few years after CB1R, and initially, their localization was thought to be restricted to peripheral tissues and in particular to immune tissues such as the spleen and liver. We now know that CB2Rs are located also in the lung, gastrointestinal tract, and adipocytes as well as in cardiac and bone tissue. In the brain, CB2Rs are expressed by microglia during inflammatory and neurodegenerative processes, by blood vessels, and by some neurons. However, the role of neuronal CB2Rs is currently under debate.

CB2Rs share 48% protein identity with CB1Rs and are coupled to G_i but most likely not to G_o proteins. They inhibit adenylyl cyclase, activate MAP kinase and NO release, but have no effect on ion channels.

GPR55 Receptors

Recent evidence shows that ECs can bind the "orphan" G-protein-coupled receptor called GPR55, which is conservatively defined as "atypical" receptor for cannabinoids. GPR55 is present in many organs and tissues, including the brain, and shares less than 20% sequence homology with CB1R. Unlike CB1R and CB2R, it activates phospholipase C and members of the small GTPase family such as RhoA through G_q and $G_{12/13}$ proteins. The classic CB1/CB2 antagonists behave as agonists at GPR55, while the synthetic cannabinoid agonist CP-55,940 acts as an antagonist/partial agonist.

TRPV1 Receptors

The vanilloid receptor (VR1), recently defined TRPV1 because of its similarity with the transient receptor potential (TRP) family of ion channels, is a nonselective cation channel highly expressed in neurons of the primary afferent fibers where it mediates inflammatory pain. Besides, TRPV1 is also expressed in different brain areas and sometimes colocalizes with CB1R. It may be activated by a wide variety of exogenous and endogenous physical and chemical stimuli, such as heat, protons, and vanilloid compounds like capsaicin, the pungent compound in hot chili peppers. AEA, but not 2-AG, binds to a cytosolic site of the receptor increasing intracellular calcium concentration. AEA affinity for TRPV1 receptor (TRPV1R) is lower than that for CB1Rs and CB2Rs.

PPARs

PPARs could represent a third nuclear branch of the cannabinoid receptor family. Since PPARs are sensors of fatty acid levels and ECs are fatty acid derivatives, it is not surprising that an increasing body of evidence suggests that ECs activate PPARs and that this action could at least partly mediate some of the biological effects of cannabinoids, such as regulation of glucose and lipid metabolism as well as inflammatory responses.

ENDOCANNABINOIDS

Cloning of cannabinoid receptors opened the way to the identification of their endogenous ligands, called ECs. The first identified endogenous ligands are AEA and 2-AG. Both are arachidonic acid derivatives and are produced from phospholipid precursors through activity-dependent activation of specific phospholipases. Other ECs have also been proposed during the last years, including 2-arachidonyl-glycerol ether (noladin ether), N-arachidonoyl-dopamine (NADA), and virodhamine, but their pharmacological activity and metabolism have not yet been thoroughly investigated.

The endogenous ligands do not share the same metabolic or biosynthetic pathways; indeed, distinct regulatory mechanisms for AEA and 2-AG have been demonstrated.

ECS acts differently from most neurotransmitter systems. More specifically, ECs are released "on demand" from the postsynaptic cell, cross the synapse as retrograde messengers, and bind to CB1Rs expressed presynaptically, thus inhibiting neurotransmitter release.

On this basis, ECS may be considered as an important modulator of neurotransmission and participates in the regulation of synaptic plasticity. In the next paragraphs, the degradative and biosynthetic pathways of the two best characterized ECs, AEA and 2-AG, are analyzed.

Anandamide

Anandamide (AEA) originates from a phospholipid precursor, N-arachidonoyl-phosphatidylethanolamine (NArPE), that is formed from the N-arachidoylation of phosphatidylethanolamine via both Ca^{2+}-sensitive and Ca^{2+}-insensitive N-acyltransferases (NATs). NArPE is then transformed into AEA by four possible alternative pathways, the most direct of which (direct conversion) is catalyzed by an *N*-acyl-phosphatidylethanolamine-selective phosphodiesterase (NAPE-PLD) (Fig. 47.2).

Several alternative enzymes for AEA biosynthesis from NArPE have been recently proposed, such as the formation of phospho-AEA from hydrolysis of NArPE catalyzed by phospholipase C enzyme(s), followed by its conversion into AEA by protein tyrosine phosphatase N22. Finally, the biosynthesis of AEA might also occur via conversion of NArPE into 2-lyso-NArPE by a soluble form of phospholipase A2, followed by the action of a lysophospholipase D (Fig. 47.2).

After cellular reuptake, AEA is metabolized by the fatty acid amide hydrolase (FAAH), a membrane-bound enzyme, into ethanolamine and arachidonic acid (Fig. 47.2). Moreover, AEA (as well as 2-AG) may also be metabolized by cyclooxygenases, lipoxygenases, and cytochrome P450, leading to formation of bioactive metabolites that may activate CBR-independent mechanisms. It is important to note that FAAH is also responsible for the degradation of numerous potentially bioactive lipids. Thus, the biological consequences of the inhibition of this enzyme are not necessarily a result of enhanced AEA levels. AEA acts as an agonist at CB1R, CB2R, TRPV1R, and PPAR, showing the highest affinity for CB1R.

2-Arachidonoylglycerol

Like AEA, 2-AG is found in both brain and periphery, although in the brain at concentrations approximately 150 times that of AEA.

Although several potential pathways exist for the formation of the *sn*-1-acyl-2-arachidonoylglycerols, the direct biosynthetic precursors of 2-AG, these are mostly produced by phospholipase C beta (PLC-β) acting on membrane phosphatidylinositols and then converted to 2-AG by the action of either of two isoforms of the same enzyme, the *sn*-1-diacylglycerol lipases α and β (Fig. 47.2). In neurons, 2-AG biosynthesis appears to be calcium sensitive. A second pathway leading to 2-AG features a 2-arachidonoyl-lysophosphatidylinositol (lyso-PI) intermediate; this pathway involves the sequential actions of a phosphatidylinositol-preferring phospholipase A1, producing the lyso-PI intermediate, and a lysophosphatidylinositol-selective phospholipase C

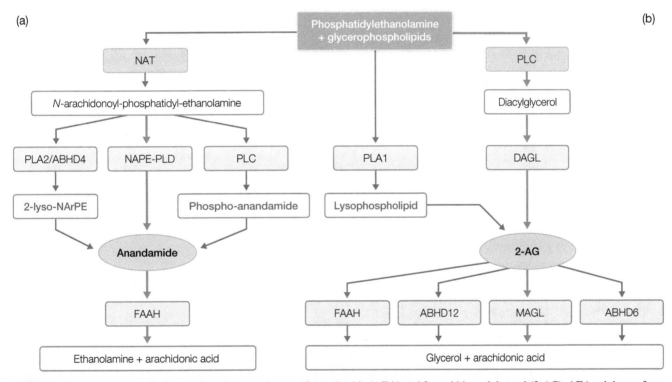

FIGURE 47.2 Main biosynthetic and degradative pathways of anandamide (AEA) and 2-arachidonoylglycerol (2-AG). AEA originates from a phospholipid precursor, *N*-arachidonoyl-phosphatidylethanolamine, through four alternative pathways. The most direct of them (direct conversion, see bold line) is catalyzed by NAPE-PLD. 2-AG is mainly produced via diacylglycerol hydrolysis by diacylglycerol lipases (see bold line). After their cellular reuptake, AEA is metabolized by fatty acid amide hydrolase (FAAH) and 2-AG by monoacylglycerol lipase (MAGL). 2-AG can also be metabolized by other recently identified lipases, the αβ-hydrolases 6 (ABH6) and 12 (ABH12), as well as FAAH. ABHD4, ABHD6, and ABHD12, alpha/beta-hydrolases 4, 6, and 12; DAG, diacylglycerol; DAGL, diacylglycerol lipase; FAAH, fatty acid amide hydrolase; MAGL, monoacylglycerol lipase; NAPE-PLD, *N*-acyl-phosphatidylethanolamine-selective phosphodiesterase; NAT, N-acyltransferase; PLA1, phospholipase A1; PLA2, phospholipase A2; PLC, phospholipase C; 2-lyso-NArPE, 2-lyso-*N*-arachidonoyl-phosphatidylethanolamine.

(lyso-PLC) producing 2-arachidonoylglycerol (Fig. 47.2). After its cellular reuptake, 2-AG is metabolized via monoacylglycerol lipase (MAGL) and to some extent by other recently identified lipases, the alpha/beta-hydrolases 6 (ABHD6) and 12 (ABHD12), as well as FAAH (Fig. 47.2). Like AEA, 2-AG may also be metabolized by cyclooxygenases, lipoxygenases, and cytochrome P450 to give rise to the corresponding hydroperoxy derivatives. 2-AG is an agonist at CB1R and CB2R, with a greater efficacy than AEA, whereas it does not interact with TRPV1R but binds PPAR γ.

The Transporter

ECs are produced on demand and released in the extracellular space, where they can bind and activate CB1Rs and CB2Rs. However, their hydrolysis is an intracellular event. Although ECs, lipophilic in nature, can freely cross cell membranes, evidence suggests the existence of mechanisms facilitating EC internalization such as specific transporters. For example, despite that a specific AEA membrane transporter (AMT) has not been cloned so far and its very existence is a matter of debate, the saturable, temperature-dependent, and specific nature of AEA uptake argues against passive diffusion through the membrane. In addition, specific inhibitors of AEA uptake but not hydrolysis bolster the claims for the existence of AMT. Recently, intracellular transporters (fatty acid-binding proteins 5 and 7) have been identified that facilitate the delivery of AEA to FAAH through the hydrophilic cytosol. This finding may help reconcile the data supporting both passive membrane diffusion and carrier-mediated kinetics for AEA uptake.

BIOLOGICAL FUNCTIONS OF ENDOCANNABINOID SYSTEM

There is now strong evidence that CB1R and CB2R localization is not restricted to the CNS and periphery, respectively, but they are widely distributed throughout the body (Fig. 47.3) and both can modulate central and peripheral functions (i.e., neuronal development, synaptic transmission, inflammation, hormone activity and release, bone production, as well as

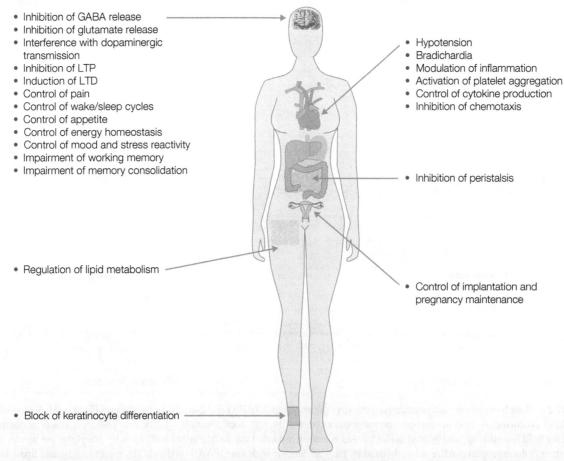

FIGURE 47.3 Biological functions mediated by the endocannabinoid system in the central nervous system and periphery. (Modified from Maccarrone M., Gasperi V., Catani M.V., Diep T.A., Dainese E., Hansen H.S., Avigliano L. (2010). The endocannabinoid system and its relevance for nutrition. *Annual Review of Nutrition*, 30, 423–440.)

cellular functions such as cellular proliferation, motility adhesion, and apoptosis). Alterations in the ECS (receptors and/or ligands) can be involved in the physiopathological regulation of different organs and tissues.

EC-Mediated Synaptic Plasticity

The notion that the EC signaling may have a general role in the regulation of synaptic transmission has been discussed for a long time. Since ECs are synthesized during periods of intense neuronal activity, the presynaptic localization of CB1Rs suggests that they might participate in a form of neurotransmitter release from the axon terminal. Indeed, this appears to occur in a number of synapses throughout the CNS, from the spinal cord to the cortex. This phenomenon is referred to as "EC-mediated plasticity." This mechanism serves to either attenuate or enhance excitability, depending on whether the neurotransmitter whose release is inhibited is excitatory or inhibitory (e.g., glutamate or GABA). Different forms of EC-mediated plasticity exist such as depolarization-induced suppression of inhibition (DSI), depolarization-induced suppression of excitation (DSE), and long-term depression/long-term potentiation (LTD/LTP). Accumulating evidence suggests that 2-AG may be a more suitable candidate, rather than AEA, for mediating synaptic plasticity, at least at central synapses. This is supported by the different anatomical localization of synthetic and degradative enzymes for AEA (preferentially postsynaptic) and 2-AG (postsynaptic biosynthesis but presynaptic degradation), which suggests a prevalent action of 2-AG at presynaptic sites where it is then metabolized by MAGL (Fig. 47.4). Biosynthesis, action, and degradation of ECs are triggered on demand and are normally restricted in time and space, also thanks to the lipophilic nature of these compounds, clearly suggesting a paracrine function. However, in chronic condition, time and space restriction might be lost, leading to a more prolonged action also at different target sites, favoring progression of pathologies.

Other Biological Functions

Many biological functions are modulated by the ECS, and a thorough analysis of each of them is beyond the scope of this

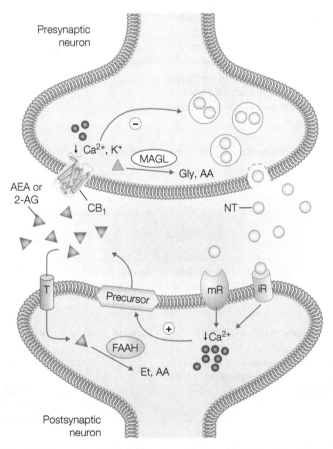

FIGURE 47.4 Endocannabinoid-mediated synaptic plasticity. Postsynaptically released endocannabinoids travel retrogradely through the synaptic cleft to bind presynaptic CB1 cannabinoid receptors. This leads to a short-term suppression of neurotransmitter (e.g., GABA and glutamate) release by inhibiting voltage-gated calcium channels and by activating potassium channels. Endocannabinoid neuromodulator signaling is terminated by an unidentified membrane-transport system followed by intracellular degradation mediated by enzymes such as FAAH and MAGL. (Modified from Guzman M. (2003). Cannabinoids: potential anticancer agents. *Nature Reviews Cancer*, 3, 745–755.)

chapter. In the following sections, the most relevant pathologies in which ECS modulation could represent a new therapeutic approach are described (Box 47.1). The issue of *Cannabis* abuse is examined in the supplementary material (E 47.1, "Chemical Structure of Drugs Acting on the Endocannabinoid System," and E 47.2, "Cannabinoids and Drug Dependence").

Eating Disorders Although the stimulating effect of *Cannabis* on food intake (mainly palatable one) has been known from centuries, only recently evidence has been accumulated for a role of ECS in regulating food intake and energy balance, both centrally and peripherally. In particular, ECs act through CB1Rs, as demonstrated by the observation that, in animal models, CB1R agonists are able to increase food intake, whereas antagonists exert opposite effects. In addition, CB1R knockout mice eat less than wild-type littermates and are resistant to diet-induced obesity. In the CNS, ECs modulate food intake acting both at hypothalamic and

---

BOX 47.1 MODULATION OF ENDOCANNABINOID SYSTEM COULD REPRESENT A NEW THERAPEUTIC APPROACH

Disorder or symptom	Proposed drug	Clinical effects	Note
Acute or postoperative pain, persistent inflammatory pain, neuropathic pain, cancer pain including bone cancer pain	CB1 receptor agonists (inhibition of neurotransmission), CB2 receptor agonists (attenuation of inflammatory pain via unknown mechanisms), brain-impenetrant FAAH inhibitors (peripherally restricted increase in endocannabinoids), dual COX-FAAH inhibitors, CBD, THC + CBD preparation	Analgesia and attenuation of inflammatory pain	Sativex (oromucosal THC + CBD) is the only approved drug in several European countries and Canada for neuropathic pain and pain associated with advanced cancer
Obesity	CB1 receptor antagonists/inverse agonists that do not cross the blood–brain barrier, CB1 neutral antagonists, inhibitors of endocannabinoid production	Weight loss	Rimonabant (CB1 receptor antagonist/inverse agonist) already authorized in some European countries for obesity was withdrawn from the market due to centrally mediated side effects (depression and anxiety)
Nausea and vomiting	CB1 receptor agonists	Antiemetic	THC (dronabinol, Marinol) and nabilone (Cesamet) have been approved by the FDA for chemotherapy-induced nausea and vomiting
Wasting syndrome	CB1 receptor agonists	Appetite stimulation	THC (dronabinol, Marinol) and nabilone (Cesamet) have been approved by the FDA for stimulating appetite in wasting disorder (AIDS, tumor cachexia)
Anxiety and depression	CB1 receptor agonists, FAAH inhibitors, dual FAAH–TRPV1 inhibitors, CBD	Anxiolytic and antidepressive effects	So far, no approved drug is present on the market. URB597 reached phase II clinical trial but was stopped for pharmacokinetic problems
Neurodegenerative/neuroinflammatory disorders (multiple sclerosis (MS) and Alzheimer's, Parkinson's, and Huntington's disease)	CB1 receptor agonists (attenuation of excitotoxicity) CB2 receptor agonists (attenuation of inflammation and immunosuppression), CBD, Sativex	Relief of some symptoms	Sativex licensed for MS in several European countries and Canada as a treatment for neuropathic pain, spasticity, and disturbed sleep. Dronabinol and nabilone have given similar results but are less effective in improving spasticity and motor functions
Cancer (various tumors)	CB1/CB2 receptor agonists, CBD: context-dependent attenuation of tumor growth (apoptosis, angiogenesis, proliferation)	Antineoplastic effect	So far, no drug has been approved for marketing

limbic sites. In the hypothalamus, ECs (particularly 2-AG) act as local pro-orexigenic mediators. Indeed, EC levels increase during fasting, whereas they are reduced in satiety; accordingly, direct injection of CB1R agonists into the hypothalamus exerts hyperphagic effects, whereas CB1R antagonists lead to a reduction in appetite. ECs are also involved in motivational processes linked to appetite regulation. Reward mediated by food or food memories can be modulated by EC–dopamine interactions at the level of the intrinsic projecting neurons of the nucleus accumbens. In the periphery, CB1Rs are widely distributed in the adipose tissue, liver, pancreas, and skeletal muscle, thus controlling body weight and increasing adipogenesis and insulin and glucagon secretion. This supports its role in energy expenditure, regulation of appetite, and lipid metabolism. In humans, blood levels of ECs are higher in obese female subjects with a binge-eating disorder. This observation has been exploited for the development of CB1 antagonists (e.g., rimonabant and taranabant) as new antiobesity drugs. Rimonabant reached phase III clinical trials and was even authorized in some European countries for dyslipidemia, obesity, and diabetes. However, the appearance of important centrally mediated side effects such as anxiety and depression prompted its withdrawal from the market. To date, new CB1 antagonists/inverse agonists that do not cross the blood–brain barrier, CB1 neutral antagonists, CB1 partial agonists, as well as drugs inhibiting EC production are under investigation. The use of THC (dronabinol, Marinol) and its synthetic analog nabilone (Cesamet) has been approved by the FDA for the treatment of chemotherapy-induced nausea and vomiting and for stimulating appetite in wasting disorders (e.g., AIDS, tumor cachexia, etc.). Rather unexpectedly, high blood levels of ECs are detectable also in anorexic female subjects. Although this apparently conflicts with data on obesity, but it can be hypothesized that in both cases ECs escape from the tonic inhibitory action of leptin. Thus, changes in EC levels seem to represent either an adaptive response to induce intake of food (or to cope with the lack of it) or a disrupted orexigenic mechanism that participates in hyperphagia, fat accumulation, and obesity.

Pain and Inflammation Pain perception can be effectively controlled by neurotransmitters that operate in the CNS but also at terminals of afferent nerve fibers outside the CNS. It has become clear that cannabinoid analgesia is predominantly mediated by peripheral CB1Rs in nociceptors. This has provided the rationale for selectively targeting peripheral CB1Rs by peripherally restricted (brain impermeable) agonists, thereby eliminating the undesirable CNS consequences of CB1 stimulation induced by THC-based drugs. However, phase II clinical studies with novel, peripherally restricted, orally bioavailable CB1/CB2 agonists (AZD1940 and AZD1704) have not given conclusive results. Moreover, clinical conditions associated with neuropathic pain or inflammation are accompanied by peripheral elevations of AEA levels, probably representing adaptive reactions aimed at reducing pain and inflammation. Accordingly, URB937, a potent FAAH inhibitor that does not cross the blood–brain barrier and that causes marked antinociceptive effects in rodent models of acute and persistent pain, paved the way for a possible therapeutic application of this class of compounds. However, a phase II clinical study with PF-04457845 (an impenetrant irreversible FAAH inhibitor) failed to show any analgesic efficacy in patients with pain due to osteoarthritis of the knee. A promising alternative is the development of molecules with more than one target, such as dual COX-FAAH or COX-MAGL inhibitors. Inhibition of EC inactivation produces gastroprotection against nonsteroidal anti-inflammatory drug-induced gastric hemorrhages in animal models, thus raising the possibility that dual inhibitors of COX and EC hydrolytic enzymes might be not only more efficacious but also safer than the corresponding drugs with only one target. Another interesting approach is the combination of THC with opioid drugs that enhances their efficacy in pain and limits their side effects. Sativex, an oromucosal spray, containing THC and the nonpsychoactive plant cannabinoid cannabidiol (CBD), has recently been approved in Canada, the United Kingdom, and several other European countries for the symptomatic relief of neuropathic pain and spasticity associated with multiple sclerosis (MS) and as an adjunctive analgesic treatment for adults with advanced cancer. Surprisingly, some experimental data have shown that CB1 and CB2R antagonists can exert analgesic and anti-inflammatory actions, possibly because ECs can behave as both pro- and anti-inflammatory mediators.

Anxiety and Depression A possible association between ECS and depression is rooted in the fact that consumption of *Cannabis sativa* in humans has profound effects on mood. ECS is widely distributed throughout the brain corticolimbic circuits implicated in the etiology and treatment of depressive illness. Impairments in EC signaling produce behavioral disturbances reminiscent of the symptom clusters associated with depression in rodent models. Accordingly, clinical populations diagnosed with depression are found to have reduced levels of circulating ECs. Collectively, these data suggest that a deficiency in EC signaling could be involved in the generation or maintenance of a depressive episode. Moreover, augmentation of CB1R signaling, either through direct agonists at CB1R or through agents that inhibit EC metabolism, such as URB597, is sufficient to produce a series of behavioral and biochemical effects reminiscent to those induced by conventional antidepressant treatments (i.e., increase in monoaminergic neurotransmission, reduction in HPA axis activity, increase in neurotrophin content, neurogenic processes, and cellular resilience within the hippocampus). The use of inhibitors of EC degradation should be preferred for several reasons: CB1R agonists could mimic the psychoactive effect of THC and would have abuse liability, whereas the inhibitors would not. However, very high doses of FAAH inhibitors can increase anxiety levels, an effect that is believed to be mediated by the promiscuous activation of TRPV1Rs by AEA. Given these potential problems, preclinical and clinical studies are now focused on the possible use of multitarget drugs (such as dual FAAH–TRPV1R blockers) and nonpsychotropic cannabinoids (such as CBD).

Immune System ECS is a potent regulator of immune responses, with CB2R being the key component due to its high expression by all immune cell subtypes. CB2R has been shown to regulate immunity by a number of mechanisms including development, migration, proliferation, and effector functions. In addition, CB2R has been shown to modulate the function of all immune cell types examined to date, but its action shows high complexity. Both *in vitro* and *in vivo* studies indicate that CB2R is capable of suppressing immune responses, which raises the possibility that CB2-selective agonists could potentially be used as anti-inflammatory therapeutics. However, CB2R in certain models has been shown to induce or aggravate inflammatory responses and CB2-selective antagonists/inverse agonists alleviate inflammation. Thus, factors and molecular mechanisms responsible for these contrasting outcomes still need to be clarified. It is possible that various cellular components of the immune system are differentially regulated by CB2Rs and that different inflammatory microenvironments shape CB2-mediated responses in distinct ways. Since both endogenous and synthetic cannabinoids bind to multiple cannabinoid receptors with different affinities, their biological outcomes cannot be accurately predicted. Thus, more careful and comprehensive studies are needed to understand which immune cells are positively or negatively regulated by CB2R in order to direct CB2-mediated effects in a specific direction. A number of *in vivo* disease studies indicate that CB2R expressed by immune cells is a good therapeutic target for the treatment of autoimmune diseases where suppression of immune responses is the goal (i.e., MS, rheumatoid arthritis, type 1 diabetes, Crohn's disease, and psoriasis). Conversely, in infectious disease and cancer, immune suppression could be detrimental. Finally, cannabinoids can modulate immune reactions also in the brain, influence T cell subset balance and cytokine expression, and play a role in the balance between neuroinflammation and neurodegeneration via the immune system or independent pathways. Therefore, CB2R in the CNS is an attractive target to develop drugs specifically acting on neuroinflammation and neurodegeneration.

Neurodegenerative Diseases There is anecdotal and scientific evidence that cannabinoids can provide symptomatic relief in diverse neurodegenerative disorders. These include MS, Huntington's disease (HD), Parkinson's disease (PD), Alzheimer's disease (AD), and amyotrophic lateral sclerosis (ALS). These findings imply that dysregulation of ECS may be responsible for some of the symptomatology of these diseases. In HD and AD, as well as in ALS, pathologic changes in EC levels and CB2 expression are induced by the inflammatory environment. Activation of CB2R by upregulated ECs goes some way toward halting microglial activation; however, this innate compensation is insufficient to prevent the subsequent inflammatory damage to neurons, which may also suffer from the loss of protection conferred by the downregulated CB1R in HD and AD. Studies on the potential therapeutic utility of cannabinoids CB1 and CB2 agonists and antagonists as well as uptake inhibitors in PD and HD have produced conflicting results. Interestingly, preclinical positive results were obtained with the use of CBD alone for both pathologies or even better in combination with THC (Sativex) for HD. Regarding AD, THC (dronabinol) has been shown to alleviate behavioral disturbances and weight loss and nighttime agitation symptoms in severe dementia. ECS is involved in many pathogenic mechanisms of MS and its experimental models. Cannabinoids exert their regulatory action at several levels, largely due to CB2-mediated modulation of inflammatory response and CB1-mediated neuroprotection. Experimental evidence shows that treatment with cannabinoids (CB1 and CB2 agonists) suppresses inflammatory responses in the CNS of animal models and improves their neurological symptoms. Sativex is licensed for use in MS in several countries, following demonstrations of its efficacy as a treatment for symptoms of neuropathic pain and disturbed sleep. It is generally well tolerated even upon long-term use. Similar findings have emerged from trials delivering THCs (dronabinol or nabilone), but they are less effective in improving spasticity and motor function.

Cancer Both CB1 and CB2 agonists were found to counteract carcinogenesis and growth and invasiveness of several types of cancers (i.e., breast, prostate, glioma, colon, lung, etc.) by modulating processes ranging from mitosis and apoptosis to angiogenesis and cancer cell migration and metastasis. It is evident that both endogenous and (endo) cannabinoid-like molecules, acting through cannabinoid receptor-dependent and cannabinoid receptor-independent mechanisms, target key signaling pathways affecting all cancer hallmarks. However, examples of proproliferative effects of ECs have also been reported, although in most cases they do not seem to be mediated by CBRs. Interestingly, the CB1 antagonist rimonabant has also been shown to inhibit cancer cell proliferation by as yet unidentified molecular targets. Finally, accumulating evidence suggests that CBD is a potent inhibitor of both cancer growth and spread. The efficacy of CBD is linked to its ability to target multiple cellular pathways, and its low toxicity even after long-term treatment suggests that CBD might be worthy of clinical consideration for cancer therapy.

DRUGS AFFECTING THE ECS

Drugs acting on the ECS can be classified in two main categories: one includes direct agonists/antagonists of CB1/CB2Rs, whereas the other includes drugs that elevate the endogenous tone through inhibition of AEA and 2-AG degradative enzymes (the so-called indirect agonists), as well as drugs that reduce the endogenous tone by affecting the biosynthetic enzymes (Fig. 47.5).

Direct Agonists

CB1/CB2 Agonists Compounds most commonly used in the laboratory as CB1/CB2R agonists fall essentially into one of four chemical groups: classical cannabinoids, nonclassical cannabinoids (CP-55,940), aminoalkylindoles (WIN 55,212-2), and eicosanoids (AEA and 2-AG) (Fig. 47.6). Many widely used CB1/CB2R agonists contain chiral centers and generally exhibit signs of marked stereoselectivity in pharmacological assays in which the measured response is CB1 or CB2R mediated.

CB1-Selective Agonists The starting point for the development of the first CB1-selective agonist was the AEA molecule. The first compounds synthesized were O-585, containing a fluorine atom inserted on the terminal 2′ carbon, and (R)-(+)-methanandamide, in which a hydrogen atom on the 1′ or 2′ carbon is replaced by a methyl group. These modifications enhance the affinity for CB1R but also confer greater resistance to the hydrolytic action of FAAH. Other potent CB1-selective agonists so far developed include arachidonyl-2′-chloroethylamide (ACEA) and arachidonyl-cyclopropylamide (ACPA), both exhibiting reasonably high CB1 efficacy but, unlike methanandamide, no sign of resistance to enzymatic hydrolysis. Another arachidonic acid derivative worth mentioning as a CB1-selective agonist is 2-arachidonylglyceryl ether (noladin ether).

CB2-Selective Agonists The CB2-selective agonists most widely used are JWH133 and JWH015, aminoalkylindole compounds developed by Dr. John Huffman. Other notable CB2-selective agonists include the GlaxoSmithKline compound GW 405833, which behaves as a potent partial agonist at CB2R, and HU308 and AM1241. Interestingly, AM1241 may be a "protean agonist" as it has been reported to behave as an agonist in tissues in which CB2Rs are naturally expressed but not in tissues in which they have been overexpressed.

Antagonists

Selective CB1R Antagonists/Inverse Agonists The first of these to be developed was the diarylpyrazole, SR141716A, a highly potent and selective CB1R ligand that prevents or

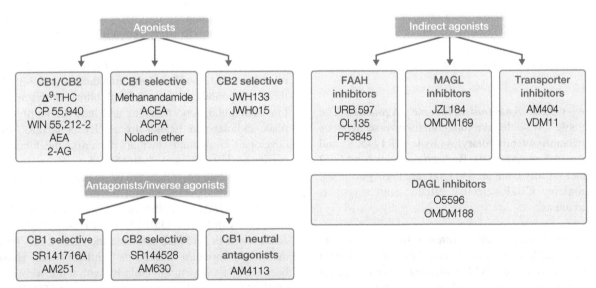

FIGURE 47.5 Most frequently used cannabinoid ligands. CB1/CB2 receptor ligands include either endogenous agonists AEA and 2-AG or exogenous agonists such as Δ9-tetrahydrocannabinol (Δ⁹-THC), the main psychoactive compound present in *Cannabis sativa*, as well as synthetic derivatives such as CP-55,940 and WIN 55,212-2. The diarylpyrazole family includes both CB1-selective (SR141716A, AM251) and CB2-selective (SR144528, AM630) antagonists/inverse agonists. AM4113 is instead a neutral CB1 antagonist. The endocannabinoid system can be also modulated via inhibition of the AEA and 2-AG degradative enzymes through URB597 and JZL184, respectively, or via inhibition of the cellular reuptake (AM404). Recently, drugs inhibiting 2-AG synthesis have been developed (O5596 and OMDM188).

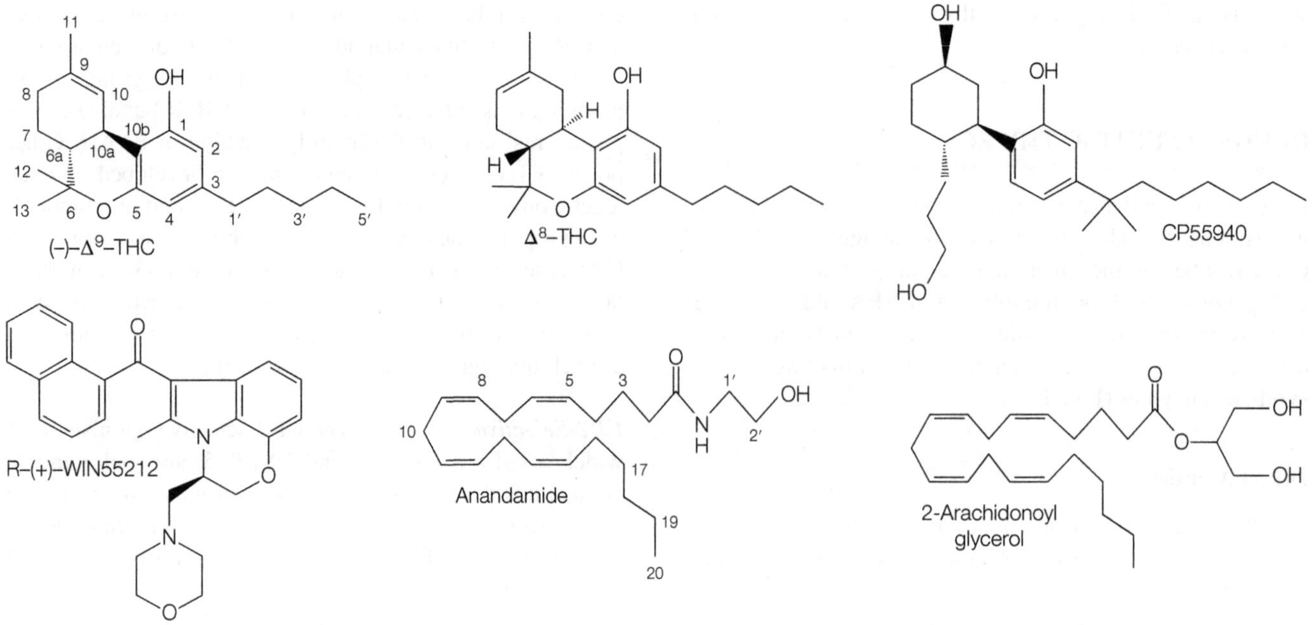

FIGURE 47.6 Chemical structure of CB1/CB2 receptor ligands. CB1/CB2 receptor ligands can be classified as classical cannabinoids (Δ9-THC and Δ8-THC), nonclassical cannabinoid (CP-55940), aminoalkylindoles (WIN 55,212-2), and eicosanoids (AEA and 2-AG).

reverses CB1-mediated effects both *in vitro* and *in vivo*. Other CB1-selective antagonists are AM251 and AM281. There is now plenty of evidence that SR141716A, AM251, and AM281 are not "neutral" antagonists but can by themselves induce responses in some CB1R-containing tissues that are opposite in direction from those elicited by CB1R agonists. These "inverse cannabimimetic effects" may in some cases be attributable to a direct antagonism of responses evoked at CB1R by released ECs, but this is not always the underlying mechanism and they are in fact classified as "inverse agonists" reducing the constitutive activity of CB1Rs.

Selective CB2R Antagonists/Inverse Agonists The most notable CB2-selective antagonists/inverse agonists are the Sanofi-Aventis diarylpyrazole SR144528 and 6-iodopravadoline (AM630). Both compounds bind CB2 with higher affinity compared to CB1, are highly effective at antagonizing CB2R-selective effects, and behave as inverse agonists.

Neutral Cannabinoid Receptor Antagonists There is currently considerable interest in the possibility of developing potent neutral CB1 and CB2R antagonists, that is, ligands for either CB1 or CB2Rs devoid of significant agonist or inverse agonist efficacy. A neutral antagonist could be used to distinguish between tonic cannabimimetic activity arising from stimulation of CB1Rs or CB2Rs upon EC release (which would be opposed by a neutral antagonist) and tonic activity due to constitutively active CB1Rs or CB2Rs (which would not be affected by a neutral antagonist). The most promising compound is AM4113.

Indirect Agonists The identification of the metabolic pathways involved in EC degradation has led to the development of indirect agonists, compounds able to selectively inhibit EC degradation or uptake.

Among inhibitors of the enzymatic hydrolysis of ECs are compounds that selectively inhibit FAAH activity. Numerous FAAH inhibitors have been described over the years, the most used as research tool being URB597. Other selective inhibitors are OL-135 and the urea derivative PF-3845. All these compounds produce their inhibitory effects *in vitro* at low nanomolar concentrations and lack the ability to displace radiolabeled ligands from CB1 or CB2Rs. Recent developed compounds that do not cross the blood–brain barrier are URB937 and PF-04457845.

Considerable progresses are being made toward the characterization of potent and selective inhibitors of 2-AG hydrolysis (MAGL inhibitors). For example, the MAGL competitive inhibitor OMDM169 increases 2-AG but not AEA levels in N18TG2 cells and increases 2-AG levels at the site of formalin-induced paw inflammation. However, because of its structural similarity with tetrahydrolipstatin, OMDM169 also inhibits pancreatic lipase and DAGL-α. The carbamate-based derivative JZL184 is a highly potent and selective MAGL inhibitor. Administration of JZL184 increases 2-AG but not AEA levels in the CNS and peripheral tissues and results in strong cannabinoid-related effects. Because JZL184 inhibits MAGL without affecting ABHD6 or ABHD12 activity, it is a powerful tool to investigate MAGL role in regulating 2-AG levels without the confounding effect of the additional 2-AG-hydrolyzing activities.

Among inhibitors of AEA uptake, the first to be developed was *N*-(4-hydroxyphenyl) arachidonylamide (AM404). However, this compound is not particularly selective as it also inhibits FAAH, binds to CB1Rs, and activates TRPV1Rs at the same (or lower) concentrations at which it inhibits AEA uptake. Other inhibitors such as VDM11, UCM707, OMDM1, and OMDM2 are now available, but they also behave as CB1R and CB2R ligands and as TRPV1R agonists. It is currently unclear whether the compounds so far found to inhibit AEA cellular uptake act by targeting an AEA transport protein or by attenuating FAAH-mediated metabolism of AEA, thus causing an intracellular accumulation of this fatty acid amide that prevent AEA entry into the cell by diffusion.

Finally, the interest has been recently focused on the identification of EC biosynthesis inhibitors. Such inhibitors should be useful tools because reducing EC production in those tissues in which it is pathologically overactive might be an interesting alternative to CBR antagonists. Although very attractive, this approach still represents a challenge because NAEs and 2-AG are produced through a multitude of synthetic pathways and they share these pathways with numerous other bioactive lipids. Thus, inhibition of a single enzyme might not be sufficient to actually decrease 2-AG or NAE production, but it may result in unexpected changes in cell membrane composition or bioactive lipid production. The few EC synthesis inhibitors reported to date target 2-AG biosynthesis. Compounds such as tetrahydrolipstatin and RHC80267 have been extensively used to inhibit DAGL-mediated 2-AG production in neuronal cell cultures, without affecting MAGL. Second-generation inhibitors such as the fluorophosphonate-based O-5596 and the tetrahydrolipstatin analog OMDM188 might represent interesting starting points for more selective drugs, because they inhibits DAGL in a nanomolar range without any effect on NAPE-PLD, FAAH, or MAGL.

TAKE-HOME MESSAGE

- The endocannabinoid system is a homeostatic cell communication system with pleiotropic functions, present in and outside the CNS.
- Endogenous ligands are lipids; the best known are anandamide and 2-arachidonoylglycerol; both are arachidonic acid derivatives and are produced from phospholipid precursors through different metabolic pathways.
- Endocannabinoid receptors CB1 and CB2 are GPCRs.
- In the CNS, endocannabinoids are released on demand and act as retrograde messengers, modulating neuronal plasticity.
- Alterations in the endocannabinoid system are present in several pathologies, such as anxiety and depression, pain, inflammation, neurodegenerative and eating disorders, and cancer.
- Drugs acting on the endocannabinoid system are classified as direct agonists (natural and synthetic, with different selectivity toward CB1 and CB2 receptors); CB1-/CB2-selective antagonists (often with properties of inverse agonists); indirect agonists (able to increase endocannabinoid levels through inhibition of their degradation or reuptake); and inhibitors of endocannabinoid synthesis.
- Use of cannabinoids is associated with addiction. The modulation of endocannabinoid system has potential therapeutic applications.

FURTHER READING

Castillo P.E., Younts T.J., Chávez A.E., Hashimotodani Y. (2012). Endocannabinoid signaling and synaptic function. *Neuron*, 76, 70–81.

Fernández-Ruiz J., Moreno-Martet M., Rodríguez-Cueto C., Palomo-Garo C., Gómez-Cañas M., Valdeolivas S., Guaza C., Romero J., Guzmán M., Mechoulam R., Ramos J.A. (2011). Prospects for cannabinoid therapies in basal ganglia disorders. *British Journal of Pharmacology*, 163, 1365–1378.

Guindon J., Hohmann A.G. (2011). The endocannabinoid system and cancer: therapeutic implication. *British Journal of Pharmacology*, 163, 1447–1463.

Katona I., Freund T.F. (2012). Multiple functions of endocannabinoid signaling in the brain. *Annual Review of Neuroscience*, 35, 529–558.

Marco E.M., Romero-Zerbo S.Y., Viveros M.P., Bermudez-Silva F.J. (2012). The role of the endocannabinoid system in eating disorders: pharmacological implications. *Behavioural Pharmacology*, 23, 526–536.

Micale V., Di Marzo V., Sulcova A., Wotjak C.T., Drago F. (2013). Endocannabinoid system and mood disorders: priming a target for new therapies. *Pharmacology and Therapeutics*, 138, 18–37.

Pacher P., Kunos G. (2013). Modulating the endocannabinoid system in human health and disease—successes and failures. *FEBS Journal*, 280, 1918–1943.

Pertwee R.G., Howlett A.C., Abood M.E., Alexander S.P., Di Marzo V., Elphick M.R., Greasley P.J., Hansen H.S., Kunos G., Mackie K., Mechoulam R., Ross R.A. (2010). International Union of Basic and Clinical Pharmacology. LXXIX. Cannabinoid receptors and their ligands: beyond CB1 and CB2. *Pharmacological Reviews*, 62, 588–631.

Pryce G., Baker D. (2012). Potential control of multiple sclerosis by cannabis and the endocannabinoid system. *CNS & Neurological Disorders Drug Targets*, 11, 624–641.

Zogopoulos P., Vasileiou I., Patsouris E., Theocharis S.E. (2013). The role of endocannabinoids in pain modulation. *Fundamental and Clinical Pharmacology*, 27, 64–80.

48

PHARMACOLOGY OF NITRIC OXIDE

EMILIO CLEMENTI

By reading this chapter, you will:

- Know the molecular basis of nitric oxide (NO) production and the significance of NO as a diffusible messenger
- Understand how NO interaction with its cellular receptors modulates the cellular response to different physiological and pathological stimuli
- Know the responses to NO generation in different organs
- Know the pharmacodynamics and pharmacokinetics of NO-donating drugs currently used in clinical practice

Nitric oxide (NO, or nitrogen oxide) is a nitrogen radical originally studied for its toxic properties as an environmental pollutant and as a component of cigarette smoke. However, more than 20 years ago, it was discovered that it is also an important physiological mediator, playing a pivotal role in the regulation of vascular tone. It is also involved in the regulation of relevant biological functions outside the cardiovascular system, especially in the immune system, in nervous transmission, and in metabolism and function of skeletal muscles. Being a gas, it is particularly exploitable by biological systems for intercellular communication.

In this chapter, we will describe the chemical properties of NO that account for its biological actions, the mechanisms controlling its generation, and its effects at cellular level. These concepts will provide the basis for a subsequent discussion on NO systemic and organ effects. Although NO pharmacology is still bound up with standard drugs used in cardiovascular diseases (mainly nitroprussiate sodium and organic nitrates), we will briefly describe new pharmacologic potentials that are opening up today thanks to innovative NO-donating drugs.

CHEMISTRY AND BIOSYNTHESIS OF NO

NO is a free nitrogen radical present under normal conditions as an atmospheric pollutant and in cigarette smoke. Free radicals are chemical species in which electrons of the most external energy level are unpaired; they tend to respond actively in biological tissues causing variations in the chemical–physical state of the molecules they get in contact with. NO is in an intermediate state of oxidation, and for this reason, it is capable of both oxidizing and reducing other compounds.

NO can interact in various ways with living matter, turning into more stable compounds that can be both active at a biological level and useful as biochemical indices of NO production. Molecules that can react with NO are summarized in Table 48.1. Among the various possible NO reactions, those with guanylate cyclase and cytochrome c oxidase, two heme iron-containing proteins, are the most biologically relevant as they mediate the vast majority of NO physiological effects. Further significant reactions at the border between pathology and physiology are the interactions with thiols. Interactions with amines, tyrosines, superoxide anion, and sulfured groups mediate mainly pathological effects. The interactions are described in detail in Supplement E 48.1, "Chemistry of Nitric Oxide."

General and Molecular Pharmacology: Principles of Drug Action, First Edition. Edited by Francesco Clementi and Guido Fumagalli.

BIOSYNTHESIS OF NO

NO biosynthesis in endothelial cells and other organism districts occurs primarily through transformation of L-arginine into L-citrulline catalyzed by enzymes called NO synthase (NOS) (Fig. 48.1). NO can also be generated through nonenzymatic mechanisms, by either disproportionation or reduction of nitrites. Typically, an acid pH is required for these chemical to occur. Therefore, nonenzymatic NO generation can be relevant upon tissue damage, when the pH value decreases (ischemia). Analogs of L-arginine or ornithine, such as N^{ω}-monomethyl-L-arginine (L-NMMA), L-nitroarginine (L-NA), L-nitroarginine methylester (L-NAME), or L-iminoethyl ornithine (L-NIO), act as false substrates of NOS and are selective inhibitors of NO production (see Table 48.2). Such NOS inhibitors significantly reduce production of both NO and L-citrulline in cells and organisms. The action of L-NMMA can be antagonized by L-arginine but not by D-arginine.

NO biosynthesis is a stepwise process catalyzed by NOS that leads to oxidation of the L-arginine guanidinic group. The process uses five electrons; requires NADPH, FAD, and FMN as cofactors; and results in the stoichiometric formation of NO and of L-citrulline (Fig. 48.2).

NO Synthases

Three isoforms of NOS have been identified: neuronal (nNOS or NOS I), inducible (iNOS or NOS II), and endothelial (eNOS or NOS III). Their structure is represented

TABLE 48.1 Groups or chemical compounds that interact with NO

Oxygen
Superoxide anion
Ozone
Oxyhemoglobin and other hemoproteins
Sulfurated compounds
Amines
Tyrosines
Thiol groups

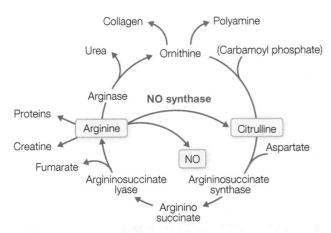

FIGURE 48.1 Importance of endothelium in modulating vasodilation induced by acetylcholine. Administration of noradrenaline (NA) to rabbit aortic rings kept *in vitro* produces vasoconstriction. Addition of acetylcholine (ACh) induces vasodilation, which reaches its maximum at a 100 nM (10^{-7}) concentration. In the absence of endothelium, ACh is totally ineffective even at higher concentrations, and vessel relaxation can be obtained only by washing the preparation to remove NA (w).

TABLE 48.2 Drugs active on the metabolic conversion L-arginine–NO

Drugs enhancing nitrergic transmission	
1. Precursors	L-Arginine, L-ornithine
2. Drugs causing NO release	Through intracellular Ca^{2+} increase: A. Agonists of receptors coupled to polyphosphoinositide hydrolysis and IP_3 generation (G-coupled and tyrosine kinase receptors) B. Calcium channel receptors (NMDA) C. Calcium ionophores (ionomycin) Through NOS II stimulation: LPS, IL-1β, TNFα, IFN-γ, SDF-1, other cytokines and chemokines
3. Drugs releasing NO	Spontaneously in salt solution: Sodium nitroprusside, DETA-NONOate, GSNO, S-nitroso-N-acetylpenicillamine (SNAP) Through enzymatic degradation: Organic nitrates (nitroglycerin, isosorbide, mono- or dinitrate, molsidomine, and other nitrovasodilators) NO donors associated with other drugs, for example, naproxcinod
4. Drugs active on cGMP-dependent pathway	Phosphodiesterase inhibitors: sildenafil, tadalafil, vardenafil, some xanthines including theophylline (extend cGMP half-life) Activators of guanylate cyclase: YC-1, cinaciguat, ataciguat, riociguat (bind the enzyme at a different site from NO)
Drugs inhibiting nitrergic transmission	
1. Drugs inhibiting NOS	Nonselective compounds: drugs based on methylated arginine (L-NMMA, L-NA, L-NAME), aminoguanidine, L-NIO, 7-nitroindazole, diphenylene iodonium Selective compounds: ARL17477 selective for NOS I versus NOS III (but only five times more selective for NOS I vs. NOS II); 1400W, GW273629, GW274150, and AR-C102222, selective for NOS II
2. Drugs inhibiting guanylate cyclase activity	Inhibitors of the enzyme: methylene blue (nonspecific), H-[1,2,4]oxadiazolo[4,3-α]quinoxalin-1-one (ODQ, more selective)
3. Drugs inhibiting G kinase	Modified analogs cGMP (RP compounds), KT5823

schematically in Figure 48.3. Like many other cellular enzymes, NOSes function as oligomers: in particular, they only work as homodimers.

NOS I is present in the CNS and PNS neurons, astrocytes, neurosecretory cells, skeletal muscles, pancreatic islets, gastric and lung epithelium, and cells of the kidney macula densa. NOS II is typically present in cells of the monocyte–macrophage lineage wherever located, including microglia, and in astrocytes. However, NOS II expression can also be induced in cells that are activated by cytokines and by bacterial products, including endothelium, smooth and heart muscle, keratinocytes, hepatocytes, and mast cells.

NOS III is expressed in endothelium, nervous system and kidney epithelium, cardiac and skeletal fibers, T and B lymphocytes, and hepatocytes and in many other cell types.

NOSes I and III differ from NOS II for being Ca^{2+} dependent. NOSes I and III are regulated by variations in cytosolic Ca^{2+} concentration, whereas NO generation by NOS II is independent from it. Actually, all three NOSes bind the Ca^{2+}/calmodulin complex and can only function when this complex is present. But in the case of NOSes I and III, formation of the complex with Ca^{2+}/calmodulin is directly dependent on $[Ca^{2+}]_i$ increase; when $[Ca^{2+}]_i$ returns to basal values, the complex dissociates and the enzyme returns to its inactive status. By contrast, small differences in NOS II binding site for Ca^{2+}/calmodulin make the NOS II–Ca^{2+}/calmodulin complex virtually inseparable even at low cation concentrations. This difference in NOS modulation is very important: NOSes I and III generate NO in a regulated, not continuous, manner and at nanomolar concentrations, while NOS II generates NO continuously and at concentrations often reaching the micromolar range.

Besides Ca^{2+}-dependent regulation, other mechanisms have been discovered regulating NOS enzymatic activity by phosphorylations and protein–protein interactions. These regulations are described in detail in the Supplement E 48.2, "Catalytic Activity, Molecular Features, and Regulation of NO Synthase." Here, we highlight only the essential information to understand how molecular differences between NOSes account for their different biological roles.

NOS I is able to stably interact with two membrane proteins, dystrophin in skeletal muscle and PSD95 in postsynaptic densities. Interaction with these two proteins allows NOS I to be held in strategic locations under the plasma membrane of neurons and myofibers. In the nervous tissue, localization in postsynapses allows NOS I to sense changes in Ca^{2+} entry at the plasma membrane and to modulate synaptic transmission

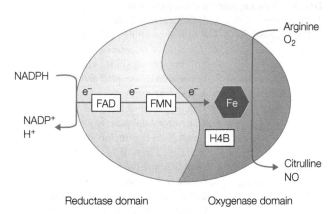

FIGURE 48.2 Chemical events occurring during NO generation by NOSes. Electrons donated by NADPH catalyze the oxidation of L-arginine by the oxygen, initially generating a reaction intermediate, N^ω-hydroxy-L-arginine, which is extremely labile and devoid of biological effects, and ultimately NO and L-citrulline. The sites of action of L-NMMA, an NOS inhibitor, are shown (from Alderton, W.K., Cooper, C.E. and Knowles, R. Biochem. J. 357, 593–616, 2001, with permission).

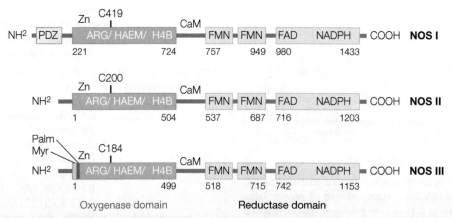

FIGURE 48.3 Structure of the three NOS isoforms. All NOSes share two principal domains: one containing the heme prosthetic group (HAEM) and endowed with oxygenase activity and the other endowed with reductase activity. All NOS isoforms contain binding sites specific for NADPH, FAD, FMN, calmodulin (CaM), tetrahydrobiopterin (H4B), and L-arginine (ARG). Specific of NOS III are myristoylation (Myr) and palmitoylation (Palm) sites that allow its anchoring to cellular membranes. Specific of NOS I is the presence of an N-terminal PDZ domain mediating protein–protein interactions with membrane proteins (from Alderton, W.K., Cooper, C.E. and Knowles, R. Biochem. J. 357, 593–616, 2001, with permission).

(see further on). Similarly, localization in the plasma membrane of myofibers allows NOS I to generate NO in response to muscle contractile activity, thus facilitating vasodilation and expression of GLUT4 glucose transporter and allowing contracting muscles to receive a greater energy supply.

NOS III is characterized by the presence of consensus sequences for myristoylation and palmitoylation, which promote its dynamic and reversible association with cell membranes, thus mediating its localization not only at the plasma membrane but also at the Golgi complex and possibly at the mitochondria. Another feature of NOS III is that it is activated not only by cytosolic Ca^{2+} increases but also by phosphorylations mediated by protein kinase B (also termed Akt) whose activation depends on phosphatidylinositol 3 kinase (PI3K) stimulation. This activation pathway is triggered by stimuli such as growth factors, cytokines, and chemokines and accounts for the important role NOS III plays in generating trophic effects in tissues expressing it, first of all cardiovascular and muscular tissues.

An important feature distinguishing NOS I and III isoforms from NOS II isoform is the regulation of expression. NOSes I and III are expressed at basal levels in cells; conversely, NOS II is normally absent and its transcription is induced by specific stimuli, consisting of cytokines, such as interferon γ (IFN-γ), tumor necrosis factor α (TNFα), interleukin-1β (IL-1β), and other cytokines acting in the presence of bacterial degradation products, including lipoarabinomannan and lipopolysaccharide (LPS). This explains why NOS II is expressed electively during immune responses in immunocompetent cells and in cells expressing cytokine receptors and products of bacterial degradation. For this reason, NOSes I and III are defined as "constitutive" and NOS II as "inducible." However, NOS I and III expression can also be modulated, although in a more restricted way (see Supplement E 48.2 for further details).

NO Control of NOS Activity The signal transduction system triggered by NO has evolved to sense intracellular concentrations of NO and to check closely NO generation. In the case of NOS II, the control by NO is exerted on the synthesis of the enzyme. NO is able to bind I-κB, the cytosolic factor that inhibits N-κB transcription factor, stabilizing the I-κB/NF-κB complex. This event prevents NF-κB nuclear translocation and subsequent stimulation of NOS II gene transcription. NF-κB regulates many genes; therefore, the inhibitory effect of NO does not affect only NOS II synthesis, but it can have a wider impact on the modulation of gene transcription.

NO can also bind three NOSes directly inhibiting their enzymatic activity. It is not yet clear whether inhibition results from NO binding directly to NOS heme group or also by its binding to critical thiols.

Regulation of NO Synthesis by L-Arginine The availability of L-arginine is one of the factors regulating NO generation both in a positive and a negative way. The two constitutive NOSes are relatively insensitive to this regulation, since they generate NO at low concentrations and after stimulation through second messengers. On the contrary, prolonged production of high concentrations of NO by NOS II is sensitive to regulation by L-arginine. Many cytokines regulating NOS II expression also regulate L-arginine availability. LPS and IFN-γ, for example, can stimulate the expression of cationic amino acid carriers specific for this amino acid (CAT1, CAT2A, CAT2B, and CAT3). In addition, these stimuli can increase the activity of argininosuccinate synthetase (Fig. 48.1), thus facilitating regeneration of L-arginine intracellular pool. However, L-arginine availability is also subject to negative regulation. Indeed, many cytokines can also increase the expression of the constitutive arginase and induce the expression of an inducible isoform of this enzyme (arginase II). The prevalence of one or the other regulatory system (positive or negative) depends on the specific biological conditions in which activation of NO synthesis system occurs.

BIOCHEMISTRY OF NO

As mentioned earlier, NO generation by NOSes takes place essentially in two modes: constitutive NOSes generate NO at nanomolar concentrations and for limited time spans (minutes); NOS II can instead generate NO at micromolar concentrations and for prolonged periods. At nanomolar concentrations, NO acts with certainty on two intracellular targets: it activates cytosolic cyclase guanylate, and it inhibits cytochrome c oxidase. NO induces S-nitrosylation of proteins usually at higher concentrations (Table 48.3). Several studies have demonstrated that NO ultimate biological effect often results from a combination of events mediated by concomitant activation of guanylate cyclase, inhibition of cytochrome c oxidase and S-nitrosylation. A clear example of this is provided by the control NO exerts on programmed cell death (see Supplement E 48.3, "Nitric Oxide and Control of Cell Death").

The other chemical reactions involving NO, namely, deamination of DNA and nitrotyrosine formation, do not play an important role in physiological responses, as they tend to occur under extreme conditions (strong cellular oxidation, presence of high concentrations of oxygen, and NO reactive species). Instead, they are an index of nonspecific and terminal cellular damage.

NO and Activation of Guanylate Cyclase

Guanylate cyclase, an enzyme located both in the cytosol and on the inner face of the plasma membrane, is a heterodimer composed of an α- and a β-subunit. There are two isoforms of each subunit generating enzymes with different activity. 3′,5′-Cyclic-GMP (cGMP) production leads to the activation of a number of cGMP-dependent proteins including kinases, phosphodiesterases, and some membrane channels. Stimulation of cGMP-dependent protein kinases (G kinase) is important in the control of Ca^{2+} homeostasis; G kinase, in fact, inhibits

TABLE 48.3 Protein S-nitrosylation: Targets and diseases

Arginase	Asthma, endothelial dysfunction
SNC5a cardiac sodium channel	Prolonged QT syndrome
Dynamin-related protein 1 (Drp-1)	Alzheimer's disease
Glutathione	Asthma, cystic fibrosis
Glyceraldehyde 3-phosphate dehydrogenase	Parkinson's disease
Hemoglobin	Type 1 diabetes, pulmonary hypertension
HIF 1α	Angiogenesis, pulmonary hypertension, sensitivity to chemotherapeutic agents
Insulin receptor β	Type 2 diabetes
Insulin receptor substrate 1 (IRS-1)	Type 2 diabetes
Matrix metalloproteinase 9	Cerebral ischemia
Parkin	Alzheimer's disease, Parkinson's disease
Peroxiredoxin 2	Alzheimer's disease, Parkinson's disease
Protein disulfide isomerase (PDI)	Alzheimer's disease, Parkinson's disease
Protein kinase B/Akt	Type 2 diabetes
Ras	Adaptive immunity
Ryanodine type 1 receptor	Duchenne muscular dystrophy, malignant hyperthermia, sudden cardiac death
Ryanodine type 2 receptor	Duchenne muscular dystrophy
Serum albumin	Preeclampsia
Surfactant protein D	Pulmonary inflammation
X-linked inhibitor of apoptosis	Alzheimer's disease, Parkinson's disease
Class I histone deacetylase 2	Duchenne muscular dystrophy

(from Foster, M.W., Hess, D.T. and Stamler, J.S. Trend Mol. Med. 15, 391–404, 2009, with permission)

phospholipase C β and γ activity, phosphorylates and inhibits the receptor for inositol 1,4,5-trisphosphate (IP_3), inhibits Ca^{2+} entry through second messenger-operated channel (SMOC) and voltage-operated Ca^{2+} channel (VOCC), and stimulates Ca^{2+} extrusion from the cytosol by activating the Na^+/Ca^{2+} exchanger and Ca^{2+}-ATPases of the reticulum and plasma membrane. As its role is to negatively modulate $[Ca^{2+}]_i$, G kinase mediates a negative "feedback" control between the NO and the Ca^{2+} systems; indeed, NO generation, stimulated by an increase in $[Ca^{2+}]_i$, in turn limits Ca^{2+} increase.

NO/Ca^{2+} interaction is very important for fine regulation of vascular tone (Fig. 48.4): increase in $[Ca^{2+}]_i$ stimulates activation of myosin light chain kinase, eliciting contraction of vascular smooth muscle. This stimulation system is juxtaposed in parallel by the one triggered by NO: increase in $[Ca^{2+}]_i$ induced by vasoconstrictors agonists stimulates NO production by NOS III in endothelial cells. NO diffuses to the underlying muscle smooth cells where it activates guanylate cyclase and then G kinase. The consequent decrease in $[Ca^{2+}]_i$ leads to inhibition of smooth muscle contraction and then to vasodilation. The vasodilation effect is also partly due to actions of G kinase on targets not directly involved in the regulation of calcium homeostasis. Among these actions, there are stimulation of Ca^{2+}-activated K^+ channels and activating phosphorylation of myosin light chain phosphatase.

Via cGMP, NO controls also the metabolism of sphingolipids. Hydrolysis of sphingomyelin by sphingomyelinase leads to generation of ceramide that is then metabolized into sphingosine 1 phosphate. This latter, acting on its own specific G-protein-coupled S1P receptors ($S1P_1$–$S1P_5$), induces both increases in $[Ca^{2+}]_i$ and activation of PI3K/Akt, with stimulation of constitutive NOSes I and III (see the preceding text). The resulting NO inhibits sphingomyelinase activity in a G kinase-dependent mode, generating a negative feedback control loop, which has an important role in maturation of antigen-presenting cells, angiogenesis, migration, and chemotaxis.

The effects produced by the NO/G kinase system on other intracellular pathways are multiple, and it is impossible to describe them all. For example, the NO/G kinase system modulates the function of some structural proteins of the cytoskeleton, such as vimentin, desmin, connexin, and calponin, through the activation of phosphoproteins and protein phosphatase 2A. In addition, the NO/G kinase system inhibits the activity of Na^+ channels sensitive to amiloride and can induce gene transcription by increasing the expression of c-fos and PGC1-α, by activating the AP-1 complex and the so-called cyclic GMP-responsive elements present on specific gene promoters. Among the genes regulated by cGMP via G kinase, there are enzymes involved in cell cycle regulation including cyclins and proteins inhibiting their kinases. Finally, NO can promote mitochondrial biogenesis by increasing the expression of PGC1-α, which in turn promotes the expression of transcription factors controlling the level of mitochondrial proteins encoded by both nuclear and mitochondrial genes.

cGMP-dependent signaling takes place also independently of G kinase. cGMP directly stimulates the hydrolytic activity of some phosphodiesterases (2, 5, 6, and 10) catalyzing the hydrolysis of both cGMP itself and cAMP. This mechanism acts as a negative feedback system controlling excessive expansion of the NO/cGMP signal. Finally, cGMP increases the opening probability of some specific membrane channels permeable to Na^+ and Ca^{2+}.

NO and Inhibition of Cytochrome c Oxidase

Cellular respiration is a regulated process that is critical for cell metabolic fate. Cellular respiration is controlled at the level of mitochondrial electron transport chain in which oxygen is consumed to build the transmembrane proton gradient necessary for ATP synthesis. In hypoxic conditions, cells react by reducing oxygen consumption and therefore ATP synthesis; this slows metabolic processes and reduces

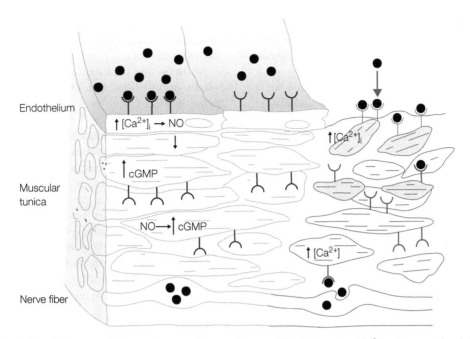

FIGURE 48.4 Endothelium in the control of vascular tone. A transmitter capable of triggering [Ca^{2+}]$_i$ raise may stimulate either vasodilation or vasoconstriction depending on where it is released and on the presence/absence of endothelial cells. An example is bradykinin: bradykinin (filled circles) in bloodstream can activate its receptors on endothelial cell membranes (left portion of the figure) inducing an increase in [Ca^{2+}]$_i$ that activates NOS to produce NO. NO diffuses in nearby smooth muscle cells of the vessel wall where it enhances cGMP generation leading to vasodilation. By contrast, bradykinin released from intramural nerve endings (bottom in the figure) or in the absence of endothelium (such as in atherosclerotic plaques) activates its receptors on smooth muscle cells inducing an increase in [Ca^{2+}]$_i$ that leads to vasoconstriction.

energy consumption. This internal control occurs at the level of cytochrome c oxidase, the terminal complex of the mitochondrial transport chain that binds oxygen "consumed" during respiration. NO is the intracellular messenger responsible for regulation of cytochrome c oxidase activity and represents, along with cytochrome c oxidase, an intracellular sensor for acute hypoxia. NO exerts this control by reversibly binding cytochrome c oxidase at the same site where oxygen binds, and with similar affinity, resulting in a competition between the two gases. In conditions of tissue normoxia (30 μM of O$_2$), NO binding is disfavored, and respiration proceeds normally. When oxygen concentration decreases, NO binding to cytochrome c oxidase is favored, reducing enzyme activity and consequently cellular respiration. Thus, mitochondrial activity is gradually and progressively reduced, and cells have to switch to glycolytic pathway to produce energy in anaerobic conditions. In this regard, it is important to emphasize that nanomolar concentrations of NO inhibit also stabilization of HIF-1 factor, a cellular factor that serves as an alert signal under hypoxia helping cells to adapt to long-term oxygen deprivation.

Nitrosylation of Thiols

Nitrosylation of thiols (i.e., conversion of SH into S-NO groups) is a pathophysiological mechanism of action of NO. To understand its pathophysiological role, it is necessary to remember that the cellular environment is usually extremely reducing, containing high concentrations (in the order of 3–10 mM depending on cell type) of mercaptans, such as reduced glutathione (GSH). It has been demonstrated that GSH SH groups are quickly accessible to NO and its reactive derivatives ($^-$OONO) that convert GSH into S-nitrosoglutathione (GSNO). GSH SH groups compete very effectively for NO/$^-$OONO with the SH groups present on proteins, thus acting as a buffer system preventing protein S-nitrosylation by NO. High concentrations of NO are needed to S-nitrosylate proteins in a normal cellular redox state. Under conditions in which NOS II is expressed, such as inflammation, S-nitrosylation can thus easily take place and effectively contribute to generate pathophysiologic signals. Generation of NO and its derivatives by NOS II can produce concentrations capable of significantly decreasing the concentration of free SH groups on GSH and other mercaptans, thus allowing protein S-nitrosylation to occur. Table 48.3 lists key proteins that can be subject to S-nitrosylation and pathologies in which they have been identified.

S-Nitrosylation may occur also under normal redox conditions, when constitutive NOSes are in physical contact or colocalize with target proteins; a transient decrease in GSH in the environment surrounding the target proteins, facilitated by proximity to the NO source, can originate reversible phenomena of S-nitrosylation without changes in the overall cell redox. To this category belongs S-nitrosylation of HIF-1 (the hypoxia sensor), β-adrenergic receptor, and type 2 ryanodine receptor, which seems to justify, at least in part, the cardioprotective effects of organic nitrates in cardiac ischemia (see the following text).

In general, S-nitrosylation has an inhibitory effect, but in some cases (e.g., ryanodine receptors), it has a stimulatory function.

NO and MicroRNAs

MicroRNAs (miRNAs) are small noncoding RNA molecules that exert a regulatory function on gene expression. So far, more than 460 miRNAs have been identified, and it has been proved that alterations in their expression are involved in various pathological conditions, including coronary artery diseases, neurodegenerative diseases, Duchenne muscular dystrophy, different autoimmune diseases, and tumors, in which miRNAs act either as promoters or oncosuppressors. Recent studies have shown that some miRNAs can regulate NO generation and in turn NO can regulate the expression of some of them. For example, in Duchenne muscular dystrophy, it has been observed that dysregulation of NOS I activity (see the succeeding text) is accompanied by altered expression of many miRNAs involved in the regulation of muscular energy metabolism and reparative myogenesis (probably resulting in their inhibition). In the cardiovascular system, miR-21 miRNA expression has been shown to increase in endothelial cells in response to stimuli, such as shear stress, mimicking the friction between blood and blood vessels. In turn, miR-21 increases Akt-dependent NOS III phosphorylation and hence NO generation, which determines vessel vasodilation thus counteracting the potentially pathologic stimulus. NOS inhibition with the endogenous NOS inhibitor asymmetric dimethylarginine (ADMA) reduces miR-21 expression and causes functional deficits in stem cells that form blood vessels. Finally, miR199a-5p, which is expressed in patients affected by severe acute heart failure, increases ADMA expression resulting in reduced NO production that may contribute to the pathogenesis of this cardiac dysfunction. miR-155 and miR-661 can, respectively, increase and decrease NOS II expression; however, the biological significance of this event is not clear.

SYSTEMIC AND ORGAN EFFECTS OF NO

Effects in the Cardiovascular System

NO relaxes vascular smooth muscles and contributes to regulation of blood pressure. Vascular smooth muscles represent one of the principal targets of NO biological action. Compounds capable of releasing NO directly or indirectly (nitro derivatives such as nitroglycerin and sodium nitroprusside; see Table 48.2) are powerful vasodilators that have been long used in the treatment of some cardiovascular diseases, such as angina pectoris, heart failure, and hypertensive emergencies.

An important contribution to our understanding of the role in vasodilation of endogenous NO released by endothelial cells has been provided by studies with the NOS inhibitors L-NMMA and L-NAME. These drugs not only inhibit vasodilation induced by agents such as acetylcholine, bradykinin, and substance P, but they also provide some important vasoconstrictor effects. This implies that NO regulates basal vascular tone. Imbalance in NO regulation is involved in the pathogenesis of some important human diseases such as hypertension and atherosclerosis. Regulation of the vascular tone by NO is also important in the physiology of skeletal muscular contraction (see the following text). The effects of NO on vessels are shown schematically in Figure 48.4.

NO is important in the regulation of coronary circulation. Coronary circulation is a vascular district that strictly depends on NO release. Vessel calibers and peripheral resistances in this district are adjusted at each cardiac cycle through the release of short-term action endothelial mediators, including NO. A reduced biosynthesis of NO by circulating blood cells and platelets can have harmful effects on coronary circulation and be at the basis of some forms of ischemic heart disease, such as unstable angina, which are characterized by association of coronary vasoconstriction and increased platelet aggregation, a phenomenon inhibited by NO in a cGMP-dependent way.

The importance of NO emerges clearly in myocardial ischemia. It is known that when perfusion of ischemic tissue resumes, oxygen free radicals, such as O_2^-, are generated locally. These compounds can react with NO, forming ^-OONO, which contributes to the harmful effects on cell membranes and on the function of numerous proteins. In addition, the resulting NO consumption has a further negative effect on the regulation of intramural flows that are largely dependent on NO release. These pieces of evidence explain why nitrovasodilators (see the following text) still represent one of the most effective drug classes in the treatment of heart perfusion deficits.

The Effects of NO in Hypertension The discovery of the existence of a stable vasodilating tone maintained by a constant production of NO keeping blood pressure within physiological limits has totally changed our understanding of the regulation of arterial blood pressure and pathogenesis of hypertension and has stimulated the use of NO donors and supplements of L-arginine in hypertension therapy (see the succeeding text).

The L-arginine–NO metabolic pathway is particularly important in kidneys, and its alteration can lead to renal and renovascular hypertension. NO positively controls glomerular function through vasodilation of the afferent arteriole leading to a net increase of the glomerular filtration rate. In addition, NO produced by renal artery endothelial cells or by nephron cells regulates natriuresis. In this context, it is interesting to note that angiotensin II, the main positive regulator of glomerular filtration pressure, exerts inhibitory effects on NOS expression at the level of macula densa cells, the component of juxtaglomerular apparatus releasing renin. The increase in vascular resistance induced in the kidney by angiotensin II alongside the inhibition of NO biosynthesis in the juxtaglomerular apparatus can amplify the damage observed in hypertension forms characterized by overstimulation of the angiotensin system.

The Role of NO in Atherosclerosis Proper functioning of endothelial cells is necessary to counteract atherogenic events including platelet activation, aggregation, and degranulation. Platelet degranulation is a process that releases substances with vasoconstrictor and proatherogenic activities as well as mitogens that contribute to the transformation of subendothelial smooth muscle cells into "foam cells."

Among the important properties of NO, there are inhibition of platelet aggregation and inhibition of leukocyte adhesion to endothelium (see the following text). In the early stages of atherosclerosis, the alteration of adhesive properties of the endothelium that promotes atherosclerotic plaque occurrence temporally correlates with reduction in NO production by NOS III. In several experimental models of hypercholesterolemia, including genetic models, addition of L-arginine supplements to the diet can restore normal expression of adhesion molecules, reduce negative effects on endothelial cells caused by prolonged exposure to platelet aggregates, enhance antiatherogenic activities of endothelial cells, and reduce xanthomatotic lesions. Thus, NO generated by endothelial cells contributes significantly to protect the subendothelium from metabolic atherogenic factors.

NO and the Respiratory System

NOS III is present in endothelial cells of cerebral and pulmonary vessels, where NO liberation represents a key event in the regulation of circulation function. It should be noted that in both humans and test animals, exhaled air contains measurable concentrations of NO. In the lungs, NOS is also present in epithelial cells, in nonadrenergic noncholinergic nerve endings, macrophages, mast cells, and neutrophils. The clinical use of NO, administered by inhalation, is proposed today in some countries in the treatment of chronic pulmonary hypertension and neonatal respiratory insufficiency, clinical situations in which NO significantly increases oxygenation of arterial circulation. In addition, stimulation of intracellular cGMP production, both through inhibition of phosphodiesterase 5 and activation of guanylate cyclase, is effective in the treatment of pulmonary hypertension.

NO and Metabolic Diseases

The nitrergic system seems to be involved in the pathogenesis of the most prevalent metabolic diseases in Western-style societies, such as insulin resistance, type 2 diabetes mellitus, and the metabolic syndrome (i.e., the combination of visceral fat accumulation, arterial hypertension, insulin resistance, and hypertriglyceridemia).

There is a potentially relevant correlation between insulin resistance and plasma concentrations of ADMA, the endogenous NOS inhibitor, suggesting that high plasma levels of ADMA may contribute to endothelial dysfunction observed in patients with insulin resistance. This observation is reinforced by the fact that administration of drugs that improve insulin sensitivity, such as thiazolidinediones (in particular rosiglitazone, used in therapy of type 2 diabetes) significantly reduces ADMA plasma levels.

In addition, high circulating levels of ADMA have been found in essential arterial hypertension, dyslipidemia, hyperglycemia, hyperhomocysteinemia, and renal insufficiency. High ADMA plasma levels have also been associated with increased risk of cardiovascular diseases.

NO in Central and Peripheral Nervous Systems

NOS I is present in many CNS areas, in particular the cerebellum, olfactory bulb, and hippocampus. In the cerebellum, the enzyme is localized in granule cells and interneurons (such as basket cells), while it is not present in Purkinje cells. In the hippocampus, NOS I is present in dentate gyrus granule cells and interneurons, but not in pyramidal neurons. NO can also be generated by glial cells. Astrocytes are endowed with NOS III and, in hypoxic conditions, can express also NOS II.

The extremely complex relationships existing between endothelial components of microcirculation vessels, neurons, and glial cells capable of releasing NO make the role played by NO in the nervous system particularly complicated. In particular, astrocytes, being located between cerebral microcirculation vessels and nerve cells, are anatomically and functionally important interconnecting elements between different CNS cellular systems. Given their close contiguity with neurons in the synaptic cleft, astrocytes can regulate neurotransmission at the adjoining synapses by releasing NO that modulates neurotransmitter release (see Fig. 48.5).

NO, Synaptic Plasticity, and Memory Plasticity phenomena are important in shaping synaptic connections and are at the basis of many aspects of learning and memory processes. Among these are the long-term potentiation (LTP) and the long-term depression (LTD). LTP and LTD can be induced in the hippocampus by specific frequency stimulation of some neuronal circuits. They consist, respectively, in

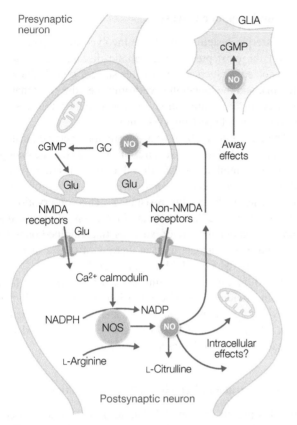

FIGURE 48.5 Functional interplay between excitatory neurotransmission, NO synthesis, and synaptic communication. Glutamate (GLU), released from nerve ending into synaptic cleft following depolarization at presynaptic terminal, activates non-NMDA (mainly AMPA) receptors on postsynaptic membrane, leading to Na^+ influx and depolarization. This allows Ca^{2+} influx in postsynaptic neuron via GLU-activated NMDA receptors. $[Ca^{2+}]_i$ increase activates NOS I and NOS III, if present, which are both Ca^{2+}/calmodulin dependent. NO thus formed triggers intracellular signaling in postsynaptic neuron and, most importantly, diffuses to presynaptic terminal where it activates guanylate cyclase (GC) in both presynaptic nerve ending and nearby glial cells. cGMP production facilitates fusion of neurotransmitter-containing vesicles leading to GLU release. These events are believed to explain, at least in part, NO role in sustaining LTP.

enhancement and reduction of the synaptic response and are considered as electrophysiological correlates of learning and memory processes. Here, NO action is favored by NOS I localization in postsynaptic density close to ionotropic AMPA and NMDA glutamate receptors. Such localization is mediated by interaction between NOS I and PSD95 protein (see Supplement E 48.2 for further details). In LTP, glutamate released from the presynaptic side generates NO in the postsynaptic side, through a pathway that involves Ca^{2+} influx via AMPA and NMDA receptors (Fig. 48.5). The NO thus generated diffuses in the presynaptic side and activates guanylate cyclase. The increase in cGMP levels and consequent activation of specific kinases increase neurotransmitter secretion, producing modulatory effects on synaptic activity. A correlation between receptor activation and NO production has also been described for non-NMDA glutamate receptors and muscarinic receptors. However, it is important to underline that NO does not act ubiquitously in LTP and LTD phenomena, but only under certain circumstances and in defined neuronal circuits.

NO Neurotransmission in the Nonadrenergic Noncholinergic Vegetative System Nitrergic pathways use NO as a neurotransmitter and are important components of the vegetative system where acetylcholine and noradrenaline do not act as natural neurotransmitters (known as nonadrenergic and noncholinergic (NANC) fibers). Nitrergic pathways are responsible for autonomous innervation of gastrointestinal smooth muscles, pelvic organs, and respiratory and genitourinary systems. They regulate gastric emptying and peristaltic wave progression, stimulate gastrointestinal absorption of electrolytes, and increase gastroduodenal secretion of mucus and bicarbonate. Finally, their activity prevents excessive acid gastric secretion, thus exerting a protective and eupeptic role. Therefore, it is not surprising that some diseases of the gastroesophageal tract, for example, esophagus achalasia, depend on degeneration of nitrergic terminations of the myenteric plexus (in the specific case, in the distal one-third of the esophagus). NO is also important in nerve transmission at penile level. Stimulation of nitrergic fibers in penile cavernous tissue and anus-coccyx muscles is accompanied by NO release produced by NOS I, which is extremely abundant in these nerve fibers. Consequent cGMP increase and dilation of postcapillary vessels increase turgor of the corpora cavernosa, thus contributing significantly to penile erection.

NO and Skeletal Muscle

In muscles, NO is produced by a specific isoform of NOS I called NOS Iμ (for further details, see Supplement E 48.2) and by NOS III. NO controls muscle physiology and regulates muscle-repairing processes. Therefore, it is particularly important when muscles get damaged not only because of traumas or degenerative diseases but also during their regular contractile activity. The role of NO in skeletal muscle repair mechanisms has been studied in depth and this has suggested new opportunities of pharmacological intervention in the treatment of muscle degenerative diseases such as muscular dystrophies (see the succeeding text). The role of NOS Iμ as an NO generator in skeletal muscles appears to be particularly important as this enzyme is localized at the dystrophin–glycoproteins complex (Fig. 48.6). The dystrophin–glycoprotein complex is an aggregate of integral membrane proteins and peripheral glycoproteins grouped into at least three major subassemblies: dystroglycan, sarcoglycan,

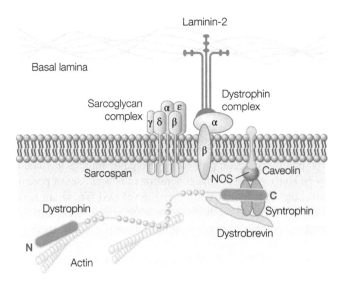

FIGURE 48.6 Molecular organization of dystrophin–glycoprotein complex in skeletal muscle. NOS Iμ binds the complex via interaction of its N-terminal PDZ domain with the α-syntrophin PDZ domain. Various components involved in dystrophin–glycoprotein complex are shown.

and syntrophins. In muscles, the dystrophin–glycoprotein complex acts as a bridge across the plasma membrane linking cytoskeleton actin to extracellular matrix. In this way, not only it mechanically strengthens myofibers, but it also mediates signal transduction and serves as sensor of muscle contraction. Within the dystrophin–glycoprotein complex, α-syntrophin has a PDZ domain that interacts with the N-terminal PDZ domain of NOS Iμ mediating its anchorage to the complex (see Supplement E 48.2 for further details). This locates NOS Iμ close to voltage-dependent Ca^{2+} channels, allowing it to sense muscular activity. Given this functional connection with muscular activity, NO controls excitation–contraction coupling so as to minimize muscle damage during contraction itself.

The second relevant parameter regulated by NO in muscles is energy balance, since NO contributes significantly to couple increased energy demand in contracting muscle with increased energy availability. This occurs through multiple mechanisms: induction of vasodilation resulting in increased blood flow and supply of oxygen and nutrients (see the succeeding text), stimulation of GLUT4 (glucose transporter; see Chapter 36) expression, induction of mitochondria biogenesis, regulation of mitochondrial respiration, and regulation of enzymes critical for energy balance such as glyceraldehyde 3-phosphate dehydrogenase, aconitase, and creatine kinase. Such pleiotropic effect of NO in muscles is mediated by all its biochemical actions, that is, cGMP increase, cytochrome c oxidase inhibition, and S-nitrosylation.

NO plays an important role also in myogenesis, a relevant process not only in embryonic and fetal life but also in adult life because it allows damaged muscles to be repaired. This action is performed by myogenic stem cells whose morphofunctional characterization is currently being carried out. Among these, the most important and characterized are the satellite cells, which are involved in muscle regeneration during pathological degeneration or after mechanical damage. They are a small population of undifferentiated cells residing in a quiescent status in adult skeletal muscles, between the sarcolemma and the basal lamina. They are activated upon muscle damage, proliferate, and finally fuse with the damaged myofibers or with one another to repair the damage. NO, acting through cGMP-dependent mechanisms, promotes activation of these cells, facilitates their fusion, and blocks the activity of myostatin, a negative regulator of skeletal muscle mass proliferation. Finally, NO promotes maintenance of the pool of myogenic progenitor cells preventing its depletion also in case of repetitive and/or severe damage.

NO and the Immune System

The role of NO in inflammatory processes has been long thought to be based on its ability to induce cytostasis and direct antimicrobial and antitumor effects. However, over the last ten years, some studies have led to the discovery of many other roles of NO that is now regarded as an essential mediator in fine regulation of the immune response.

Localization of NOS in Immunocompetent Cells Many cells of the immune system express NOS III or NOS II (the latter usually after cellular activation). NOS III has been found in T and B lymphocytes, macrophages, and natural killer cells. Expression of NOS III in these cells is generally nonconstitutive but induced by activation stimuli specific for each cell type. Expression of NOS II can be observed in activated macrophages, activated B cells, and natural killer cells; in some cases, even dendritic cells, once in lymphatic organs, can express this enzyme. Expression of NOS II can be stimulated by cytokines, including IL-2, IL-18, IL-12, IFN-γ, and chemokines, such as SDF-1, either acting alone or in different combinations. Moreover, NOS II expression can increase following bacterial or viral infections.

Role of NO in Infections Generation of large amounts of NO by NOS II exerts a lytic action. In experimental models of *Leishmania* or *Schistosoma* infections, pretreatment with L-NMMA reduces or completely abolishes macrophage cytotoxic power toward these protozoal species. Furthermore, in NOS II knockout mice, poor resistance to intracellular bacterial infection has been detected. The importance of NO as a defense agent against infections is demonstrated by the efficacy of ointments containing an NO donor (SNAP, see Table 48.2) in treating human skin lesions induced by *Leishmania*.

The mechanism underlying NO bactericidal action has not yet been clarified univocally. NO, which is produced in large quantities, is capable of generating ⁻OONO and blocking the activity of bacterial dehydrogenases, including those responsible for microorganism respiration, by depriving Fe–S clusters of iron. In addition, NO induces apoptotic death of infected macrophages, thus facilitating the elimination of the infection. In addition, NO can also generate cytotoxic species such as OH and H_2O_2. Finally, many pathogens, for example, *T. cruzi*, *G. lamblia*, and *S. mansoni*, require L-arginine for polyamide synthesis and for their proliferation. The expression of NOS II and the effect of cytokines on arginases (see the succeeding text) help reduce the bioavailability of this amino acid. The bactericidal action of NO is also exerted via γδ T lymphocytes acting as a first barrier to bacteria. Antigenic stimulation induced by the presence of bacteria causes a marked expression of NOS III in T lymphocytes that persists as long as the antigenic stimulus is present. The NO thus produced facilitates T lymphocyte proliferation allowing them to provide a more effective defense against intracellular pathogens. NO can also facilitate antigen presentation by dendritic cells. In particular, it increases their ability to phagocytize antigens at the site of infection and to activate their specific T lymphocyte clones.

In addition, experimental and clinical evidence indicates an involvement of NO in chronic inflammation, autoimmune diseases, chemotaxis, leukocytes adhesion, and tumors. These more specialist aspects are discussed in the Supplement E 48.4, "Role of NO in Inflammation and in Tumor Pathology."

PHARMACOLOGY OF NO

Drugs that interfere with NO generation or with its proximal transduction pathways are summarized in Table 48.2. Some of these drugs have been used in clinical practice for long, even before the discovery of NO. In the following paragraphs, we will discuss the impact in therapy of the most important of these drug classes.

Nitrate Vasodilators

The discovery of the antiangina properties of amyl nitrite and nitroglycerin dates back to the second half of the nineteenth century. Later, other drugs were identified, such as isosorbide mono- and dinitrate and molsidomine, and used to treat angina, myocardial infarction, and cardiac failure. Another drug, sodium nitroprusside, has been used in therapy to resolve acute hypertensive crises. The discovery of the biological role of NO has provided the basis to understand how these drugs work. Despite that a discussion of these drugs is beyond the scope of this chapter, one important aspect concerning the use of nitroglycerin, amyl nitrite, and isosorbide dinitrate has to be pointed out, that is, the tolerance to their pharmacological effects that typically occurs in the absence of a daily interruption of the treatment. Several mechanisms have been proposed to explain this phenomenon. Type 2 mitochondrial aldehyde dehydrogenase (ALDH-2) is the major responsible for this effect, as well as for the development of cross-tolerance among organic nitrates.

Nitric Esters of Known Drugs

The pharmacological properties of organic nitrates mediated by NO have prompted the generation of new classes of potential drugs that are currently being developed. These are new molecules with multiple action mechanisms that are built by binding organic nitrates to drugs already used in therapy through chemical structures called "spacers." Representative examples are the nonsteroidal anti-inflammatory (NSAIDs) NO donors, also called cyclooxygenase-inhibiting nitric oxide donors (CINODs). CINODs are hydrolyzed by cell and tissue enzymes to yield NSAIDs initially and then a slow NO release. It is known that NSAIDs produce side effects such as gastric mucosa and intestinal injuries and increase systolic blood pressure with consequent increase of adverse cardiovascular events. The addition of NO donors to NSAID molecules attenuates harmful gastrointestinal effects by activating local microcirculation and inhibiting inflammatory processes such as leukocyte adhesion. In addition, NO prevents the blood pressure rise induced by NSAIDs. Therefore, CINODs display a better tolerability and safety compared to NSAIDs. An example that deserves attention is naproxcinod, an NO donor derived from naproxen that is currently being developed for the treatment of Duchenne muscular dystrophy.

Other potential drugs based on this molecular approach are being clinically studied for the treatment of ocular pathologies. For example, an NO donor derivative of latanoprost is being clinically assessed for patients with high ocular hypertension, the main risk factor for glaucoma. Finally, NO-donating steroids show promise for the treatment of diabetic retinopathy because NO prevents the glaucoma risk induced by steroids.

Stimulators of cGMP Action

Among the modulators of the NO pathway, inhibitors of cGMP-selective phosphodiesterase 5 have been successful in clinical practice. These drugs inhibit cGMP hydrolysis thus increasing intracellular cGMP concentrations and activating several pharmacological activities typically associated with NO/cGMP signaling. The progenitor of the class is sildenafil (Viagra). This drug is effective in the treatment of male impotence, as it increases dilation of postcapillary vessels, thus increasing turgor of the corpora cavernosa, and consequently penile erection. The success of sildenafil has stimulated the search for similar drugs, such as vardenafil and tadalafil. The therapeutic application of phosphodiesterase 5 inhibitors has been extended; indeed, sildenafil and tadalafil have proven effective and have been approved for the treatment of pulmonary hypertension. Recent studies have also revealed a possible use in heart disease associated with Duchenne muscular dystrophy.

The search for mechanisms capable of increasing intracellular cGMP concentrations has also led to the development of guanylate cyclase activators. Some compounds, such as riociguat, cinaciguat, and ataciguat, are being studied for the treatment of congestive heart failure and pulmonary hypertension.

NOS Inhibitors

Whereas strategies designed to enhance NO/cGMP pathway have shown a remarkable therapeutic validity, so far, this has not occurred for strategies aimed at inhibiting NOS activity. In spite of the extensive research carried out over the years, no molecules of clinical interest have being identified yet.

TAKE-HOME MESSAGE

- NO is a short half-life messenger acting on different targets depending on the mode of its endogenous generation or exogenous administration.
- NO is an ubiquitous endogenous messenger acting in all regions of the organism, generating complex coordinated responses.
- At NO nanomolar concentrations, its effectors are cytochrome c oxidase and guanylate cyclase leading to physiological responses that regulate signal transduction and cellular metabolism. By contrast, higher NO concentrations, which mainly occur under pathophysiological conditions such as in inflammation and cancer, lead to S-nitrosylation.
- NO-based drugs are currently in use in the therapy of cardiovascular diseases. New molecules and new approaches with NO-based drugs are under clinical development; they may open new therapeutic perspectives in several diseases, including degenerative diseases of skeletal muscle.

FURTHER READING

Historical Readings

Beckman J.S., Koppenol W.H. (1996). Nitric oxide, superoxide, and peroxynitrite: the good, the bad, and the ugly. *American Journal of Physiology*, 271, C1424–C1437.

Furchgott R.F., Zawadzki J.V. (1980). The obligatory role of endothelial cells in the relaxation of arterial smooth muscle by acetylcholine. *Nature*, 288, 373–376.

Gryglewski R.J., Palmer R.M., Moncada S. (1980). Superoxide anion is involved in the breakdown of endothelium-derived vascular relaxing factor. *Nature*, 320, 454–456.

Moncada S., Palmer R.M., Higgs E.A. (1991). Nitric oxide: physiology, pathophysiology, and pharmacology. *Pharmacological Reviews*, 43, 109–142.

Palmer R.M.J., Ferrige A.G., Moncada S. (1987). Nitric oxide accounts for the biological activity of endothelium-derived relaxing factor. *Nature*, 327, 524–526.

Palmer R.M., Ashton D.S., Moncada S. (1988). Vascular endothelial cells synthesize nitric oxide from L-arginine. *Nature*, 333, 664–666.

Reviews

Alderton W.K., Cooper K., Knowles R. (2001). Nitric oxide synthases: structure, function and inhibition. *Biochemical Journal*, 357, 593–615.

Bailey J.C., Feelisch M., Horowitz J.D., Frenneaux M.P., Madhani M. (2014). Pharmacology and therapeutic role of inorganic nitrate and nitrite in vasodilatation. *Pharmacology & Therapeutics*, 144, 303–320. .

Brown G.C. (1999). Nitric oxide and mitochondrial respiration. *Biochimica et Biophysica Acta*, 1411, 351–369.

Farrugia G., Szurszewski J.H. (2014). Carbon monoxide, hydrogen sulfide, and nitric oxide as signaling molecules in the gastrointestinal tract. *Gastroenterology*, 147, 303–313.

Foster M.W., Hess D.T., Stamler J.S. (2009). Protein S-nitrosylation in health and disease: a current perspective. *Trends in Molecular Medicine*, 15, 391–404.

Gladwin M., Kim-Shapiro D. (2012). Nitric oxide caught in traffic. *Nature*, 491, 344–345.

Gross S.S., Wolin M.S. (1995). Nitric oxide: pathophysiological mechanisms. *Annual Review of Physiology*, 57, 737–769.

Holscher C. (1997). Nitric oxide, the enigmatic neuronal messenger: its role in synaptic plasticity. *Trends in Neurosciences*, 20, 298–303.

Ignarro L.J. (2000). *Nitric oxide, biology and pathobiology*. San Diego, CA: Academic Press.

Lucas K.A., Pitari G.M., Kaserounian S., Ruiz-Stewart I., Park J., Schulz S., Chepenik K.P., Waldman S.A. (2000). Guanylyl cyclases and signaling by cyclic GMP. *Pharmacological Reviews*, 52, 376–413.

Nisoli E., Clementi E., Carruba M.O., Moncada S. (2007). Defective mitochondrial biogenesis: a hallmark of the high cardiovascular risk in the metabolic syndrome? *Circulation Research*, 100, 795–806.

Paul B.D., Snyder S. (2012). H_2S signalling through protein sulfhydration and beyond. *Nature*, 13, 499–507.

Rastaldo R., Pagliaro P., Cappello S., Penna C., Mancardi D., Westerhof N., Losano G. (2007). Nitric oxide and cardiac function. *Life Sciences*, 81, 779–793.

Taylor C.T., Moncada S. (2010). Nitric oxide, cytochrome C oxidase, and the cellular response to hypoxia. *Arteriosclerosis, Thrombosis, and Vascular Biology*, 30, 643–647.

Vincent S.R. (2010) Nitric oxide neurons and neurotransmission. *Progress in Neurobiology*, 90, 246–255.

49

ARACHIDONIC ACID METABOLISM

CARLO PATRONO AND PAOLA PATRIGNANI

By reading this chapter, you will:

- Learn the molecular basis of the biosynthesis of arachidonic acid-derived mediators and the key enzymes involved in eicosanoid synthesis
- Learn the eicosanoid receptor function and the mechanisms that translate their activation into specific functional responses
- Know the pharmacodynamics of the main drug classes used to inhibit eicosanoid synthesis or to antagonize their receptors

---

Eicosanoids belong to the family of autacoids, local mediators that induce functional responses by interacting with specific receptors on the same cells producing them and/or other cells nearby. Therefore, eicosanoids mediate short-distance intercellular communication. Their short-distance effect is related to their chemical instability and the presence of enzymatic pathways rapidly metabolizing and inactivating them. In contrast with neurotransmitters, which are synthesized continuously and stored within synaptic vesicles, eicosanoids are produced only in response to specific stimuli and immediately released in the extracellular space. Eicosanoids are synthesized from arachidonic acid, an essential fatty acid containing four double bonds (5,8,11,14-eicosatetraenoic acid), which is mainly found, esterified to the second position of the glycerol backbone, in three main membrane phospholipids: phosphatidylinositol, phosphatidylcholine, and phosphatidylethanolamine. Since the concentration of free arachidonic acid in cells is very low, eicosanoid biosynthesis depends primarily on cell ability to release it from membrane phospholipids in response to various stimuli. Arachidonic acid can be transformed into a number of biologically active substances through oxidative reactions catalyzed by specific enzymes or oxygen radicals. Four groups of eicosanoids have been identified (Fig. 49.1): prostanoids, which are products of prostaglandin H synthases (PGHS); hydroxyeicosatetraenoic acids (HETE) and leukotrienes (LT), which are produced through the action of various lipoxygenases (5-, 12-, and 15-lipoxygenase); epoxyeicosatrienoic acids and hydroxyeicosatetraenoic acids, whose formation is catalyzed by cytochrome P450-containing enzymes; and isoeicosanoids, a series of eicosanoid isomers (isoprostanes, isothromboxanes, isoleukotrienes, and isoHETE), which are formed via lipid peroxidation catalyzed by oxygen radicals. Eicosanoids produced and released *in vivo* undergo extensive enzymatic degradation leading to biologically inactive metabolites that are excreted in the urine.

Oxidative modifications of polyunsaturated fatty acids, such as arachidonic acid, are at the center of local regulatory mechanisms (Table 49.1) affecting immune and inflammatory processes, hemostasis and thrombosis, bronchial responsiveness, renal function, female reproduction, gastric cytoprotection, intestinal epithelial cell apoptosis, and vascular function. Therefore, pharmacologic or dietary modulation of these reactions may have complex clinical consequences—favorable or unfavorable—which are not always predictable in the individual patient.

General and Molecular Pharmacology: Principles of Drug Action, First Edition. Edited by Francesco Clementi and Guido Fumagalli.
© 2015 John Wiley & Sons, Inc. Published 2015 by John Wiley & Sons, Inc.

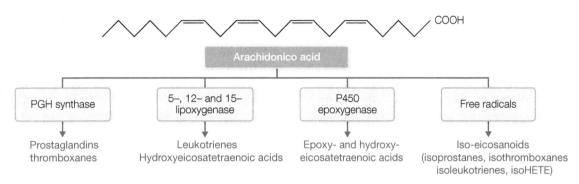

FIGURE 49.1 Main pathways of arachidonic acid metabolism. Arachidonic acid is metabolized through at least four main pathways: (i) prostaglandin H synthase (PGH-synthase-1 and PGH-synthase-2) pathway leading to prostanoid formation, (ii) lipoxygenase (5-, 12-, and 15-lipoxygenase) pathway leading to hydroxyeicosatetraenoic acid and leukotriene production, (iii) cytochrome P450 enzymes leading to epoxyeicosatrienoic acid and hydroxyeicosatetraenoic acid production, and (iv) nonenzymatic lipid peroxidation catalyzed by oxygen radicals leading to isoeicosanoid formation. The three enzymatic pathways operate on free arachidonic acid, while nonenzymatic peroxidation occurs on esterified arachidonic acid.

TABLE 49.1 Biological activities of the main eicosanoids and isoprostanes

Eicosanoid	Production site	Biological response
TXA_2	Platelets, lung, renal cortex	↑ Platelet aggregation and degranulation; ↑ vascular and bronchial smooth muscle tone; ↓ renal blood flow and glomerular filtration rate
PGE_2	Monocytes, renal medulla, hypothalamus, thymus	↓ Vascular smooth muscle tone; ↓ gastric secretion neurotransmitter release in autonomic nervous system; ↓ lipolysis in adipocytes; ↓ pain threshold; ↑ diuresis, natriuresis; ↑ body temperature; ↑ regulation of lymphocyte differentiation
$PGF_{2\alpha}$	Uterus, renal medulla, platelets	↑ Vascular smooth muscle tone; ↑ luteolysis
PGD_2	Mast cells, platelets, brain	↓ Vascular smooth muscle tone; ↑ bronchial smooth muscle tone; ↓ platelet aggregation; ↑ sleep
PGI_2	Blood vessels, renal cortex	↓ Vascular smooth muscle tone; ↓ platelet aggregation; ↑ renal blood flow and glomerular filtration rate
LTC_4	Monocytes, eosinophils	↑ Vascular and bronchial smooth muscle tone; ↑ vascular permeability
LTB_4	Monocytes, neutrophils	↑ Aggregation, degranulation, and neutrophil chemotaxis
8-Iso-$PGF_{2\alpha}$	Cell membranes, LDL	↑ Platelet aggregation; ↑ vascular smooth muscle tone; ↑ proliferation of vascular smooth muscle cells

ARACHIDONIC ACID RELEASE FROM MEMBRANE LIPIDS

Arachidonic acid esterified to membrane phospholipids can be released in response to different stimuli reaching the plasma membrane: physiologic stimuli (histamine, bradykinin, vasopressin, angiotensin II, interleukin (IL-1), growth factors, proteases such as thrombin); physical stimuli (shear stress, ischemia); and pharmacologic agents (calcium ionophore A23187, phorbol esters such as PMA). Physiologic stimuli interact with specific receptors, usually coupled to G proteins, present on the cell membrane. These G proteins in turn activate phospholipases (A_2, C, and D) capable of releasing arachidonic acid from membrane phospholipids. Moreover, some cell types use arachidonic acid esterified in low-density lipoproteins (LDLs) to synthesize eicosanoids or isoeicosanoids. Phospholipase A_2 catalyzes hydrolysis of the sn-2 ester bond of phosphatidylcholine releasing arachidonic acid with the formation of 1-acyl-phosphatidylcholine. It is important to note that the anti-inflammatory action of glucocorticoids appears to be due, at least in part, to their ability to interfere with this step of arachidonic acid metabolism; these hormones induce expression of lipocortin 1, an endogenous inhibitor of phospholipase A_2 (see Supplement E 49.1, "Mechanism of Action of Glucocorticoids as Anti-inflammatory Drugs").

Two different calcium-dependent phospholipases A_2 involved in arachidonic acid release from membrane phospholipids have been characterized: a 14 kDa secretory phospholipase (sPLA_2), which is secreted in the extracellular environment and causes release of fatty acids (including arachidonic acid) esterified to the second position of membrane phospholipids, and an 85 kDa cytosolic phospholipase (cPLA_2) specific for arachidonic acid (see Supplement E 49.1).

ENZYMATIC METABOLISM OF ARACHIDONIC ACID

As shown in Figure 49.1, enzymatic transformations of arachidonic acid involve PGH-synthases, lipoxygenases, and cytochrome P450-containing enzymes. In the following sections, we will discuss in detail the two main metabolic pathways of pharmacologic interest, the PGH-synthase and 5-lipoxygenase pathways.

PGH-Synthase Pathway

The second step in prostanoid biosynthesis is the conversion of arachidonic acid to PGH_2 (Fig. 49.2) catalyzed by PGH-synthase, an enzyme with two distinct catalytic activities: a cyclooxygenase (COX) activity that leads to PGG_2 formation and a peroxidase activity, which reduces PGG_2 to PGH_2.

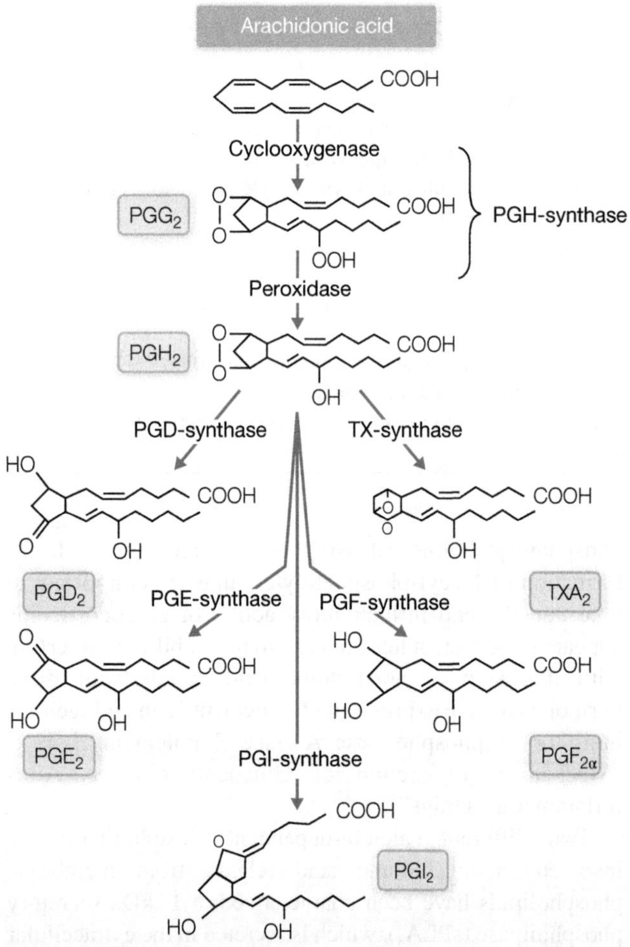

FIGURE 49.2 PGH-synthases and prostanoid biosynthesis. The two enzymatic activities of PGH-synthases catalyze arachidonic acid conversion into PGH_2. PGH_2 is metabolized to thromboxane (TXA_2), prostacyclin (PGI_2), and prostaglandins (PGD_2, PGE_2, $PGF_{2\alpha}$) by specific synthases.

Two isoenzymes named PGH-synthase-1 (PGHS-1 or COX-1) and PGH-synthase-2 (PGHS-2 or COX-2) have been characterized.

The COX-1 gene (approximately 22 kb long) is a "housekeeping" gene, meaning that it is constitutively expressed by virtually all cells of the human body. However, COX-1 gene can be regulated in some conditions, such as monocytic cell differentiation, thus acting as a "delayed-response gene."

COX-1 plays an important role in physiological biosynthesis of prostanoids involved in intercellular communication as well as in amplification or local modulation of diverse homeostatic functions such as gastrointestinal, platelet, and renal functions (Table 49.1).

COX-1 is a homodimeric integral membrane protein of 72 kDa with a globular structure and characterized by three components: (i) an epidermal growth factor (EGF)-homologous region, (ii) a membrane-binding motif anchoring the enzyme to the inner leaflet of the phospholipid bilayer, and (iii) a region containing COX and peroxidase catalytic sites. These two sites are structurally distinct as you can see in Figure 49.3 showing the three-dimensional structure of the sheep enzyme obtained by X-ray crystallography. The COX catalytic site consists of a long and narrow hydrophobic channel, whose entrance is found near the enzyme membrane-binding motif, a strategic position to capture released arachidonic acid. Immunohistochemical techniques have shown that COX-1 is localized in the endoplasmic reticulum (ER) **membranes**, facing the ER lumen, and at perinuclear level.

During its COX activity, COX-1 undergoes a rapid inactivation through a mechanism not fully elucidated yet. Such regulatory "suicidal" strategy limits COX-1 activity induced by arachidonic acid release but at the same time stimulates its de novo synthesis, thus responding to cellular needs.

Although constitutive expression of a second type of enzyme, named PGHS-2 (or COX-2), has been shown in some tissues, such as the brain, testes, prostate, kidney, and vascular endothelium, COX-2 is mainly an inducible enzyme. This protein is encoded by an 8 kb long gene belonging to the class of "immediate early genes," which are rapidly expressed in response to appropriate stimuli and characterized by highly unstable mRNAs due to the presence of AUUUA sequences in their 3' untranslated region. COX-2 shares 60% homology with COX-1, with all the amino acid residues essential for COX-1 catalytic activities being conserved in COX-2. Under physiologic conditions, COX-2 mRNA is detectable only in the aforementioned tissues.

Proinflammatory stimuli (e.g., IL-1, lipopolysaccharide (LPS), and growth factors including platelet-derived growth factor (PDGF)) can cause a rapid and intense induction of COX-2 expression in different cell types such as monocytes/macrophages, fibroblasts, and vascular endothelial cells. Prostanoid production contributes to the characteristic signs

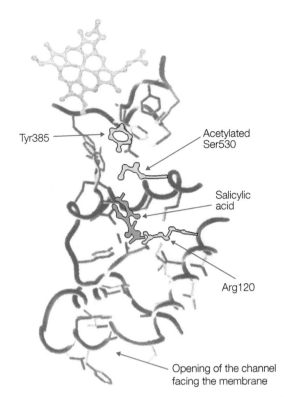

FIGURE 49.3 Tridimensional structure of the ovine COX-1 channel crystallized with acetylsalicylic acid. The carboxyl group of aspirin interacts reversibly with COX-1 Arg120 residue, a common binding site for all traditional NSAIDs. This produces a local pool of acetylating residues just below Ser530 (529 in the human enzyme), which explains the selective acetylation of this amino acid by aspirin. Acetylated Ser530 is strategically located within the COX-1 channel, close to Tyr385, a critical residue for initiation of cyclooxygenase catalytic cycle. Thus, any arachidonic acid entering the channel cannot interact with Tyr385 (kindly provided by Dr. P.L. Loll, Department of Pharmacology, University of Pennsylvania School of Medicine, Philadelphia, PA).

and symptoms of the inflammatory response (i.e., vasodilation, edema, and hyperalgesia) and mainly results from local induction of COX-2 expression in inflammatory cells. In addition to their role in inflammation, prostanoids produced by COX-2 are involved in the control of some pathophysiological events affecting female reproductive system, renal function, and intestinal carcinogenesis.

Inhibition of COX-1 and COX-2 by Nonsteroidal Anti-inflammatory Drugs Nonsteroidal anti-inflammatory drugs (NSAIDs) (Table 49.2) compete with arachidonic acid for a binding site within the COX channel, thus preventing its oxidation to PGG_2. By contrast, the peroxidase activity is not affected by NSAIDs. NSAID therapeutic activities (i.e., analgesic, antipyretic, and anti-inflammatory effects) as well as many of their adverse effects (e.g., gastrointestinal bleeding complications) can be ascribed to the same mechanism of action, that is, inhibition of COX-2 and COX-1 (Table 49.3).

NSAIDs can be divided into three classes (Table 49.4) based on their mechanism of COX-isozyme inhibition: class I molecules act via a simple competitive mechanism (e.g., ibuprofen); class II via a time-dependent competitive, slowly reversible mechanism (e.g., indomethacin and selective inhibitors of COX-2); and class III via a time-dependent competitive, irreversible mechanism (e.g., acetylsalicylic acid).

Class I NSAIDs reversibly compete with arachidonic acid for a common binding site (Arg120) within the COX channel of PGHS isozymes, rapidly forming an easily reversible enzyme–inhibitor (EI) complex. This leads to an equipotent inhibition of COX-1 and COX-2.

Class II includes some traditional NSAIDs, such as indomethacin, flurbiprofen and diclofenac, and coxibs (rofecoxib, celecoxib, valdecoxib, etoricoxib, and lumiracoxib), a new class of selective COX-2 inhibitors, some of which have been withdrawn from the market because of cardiovascular (rofecoxib), skin (valdecoxib), or hepatic (lumiracoxib) toxicity. Class II inhibitors rapidly and reversibly bind to both isoenzymes, forming an EI complex. However, if retained in the enzyme binding site for long enough, they cause a conformational change of the two isozymes slowly producing a stable EI* complex. Inhibitors dissociate from the active site very slowly. Coxibs are time-dependent reversible inhibitors of COX-2, while they are simple competitive inhibitors of COX-1.

Therapeutic effects of traditional NSAIDs and coxibs, as well as their cardiorenal toxicity, depend on COX-2 inhibition, whereas their gastrointestinal toxicity depends mainly, but not exclusively, on COX-1 inhibition (Table 49.3). In fact, large randomized trials have established that some highly selective coxibs (e.g., rofecoxib and lumiracoxib) have a better gastrointestinal safety as compared to traditional NSAIDs (e.g., naproxen and ibuprofen). It is important to note that the selectivity for COX-2 is a variable, with some first-generation coxibs (e.g., celecoxib) partially overlapping with some traditional NSAIDs (e.g., nimesulide and diclofenac) (Fig. 49.4).

Acetylsalicylic acid (aspirin) is the only NSAID belonging to class III that causes irreversible inactivation of COX isozymes through acetylation of strategically located serine residues. Aspirin binds COX-1 and COX-2 through a weak-affinity interaction with their Arg120 residue and covalently modifies them by subsequent acetylation of Ser529 and Ser516 in human COX-1 and COX-2, respectively (Fig. 49.3). Although salicylate is not an efficient inhibitor of COX-isozyme activity, when administered *in vivo*, it exerts an anti-inflammatory effect through an unclear mechanism. *In vitro* studies have shown that millimolar concentrations of the drug inhibit activation of proteins regulating gene transcription involved in inflammatory response, such as NF-κB.

Irreversible inhibition of platelet COX-1 by low-dose aspirin has clinical relevance (see Supplement E 49.2, "Mechanism of Action of Aspirin as Antithrombotic Drug").

TABLE 49.2 Mechanism of action and therapeutic uses of the main chemical classes of nonsteroidal anti-inflammatory drugs

Chemical class	Generic name	Main therapeutic uses	Mechanism of action
Salicylates	Salicylic acid, diflunisal, acetylsalicylic acid (ASA or aspirin)	*Analgesic*: inflammatory diseases of the musculoskeletal system, cervical and spinal pain syndromes, neuralgia, sciatic nerve pain, radiculitis, trigeminal and cervicobrachial neuralgia, headaches and migraines, toothaches, neoplastic diseases, posttraumatic and postoperative painful syndromes, dysmenorrhea, cold and flu symptoms *Anti-inflammatory*: acute articular rheumatism, rheumatoid arthritis *Antiplatelet* (ASA but not salicylic acid)	Salicylic acid: weak inhibitor of COX-1 and COX-2; high concentrations *in vitro* inhibit activation of transcription factor NF-κB. ASA: irreversible inhibitor of COX-1 and COX-2. Relatively selective COX-1 inhibitor at low dose
Pyrazole derivative	Phenylbutazone, aminopyrine, sulfinpyrazone	*Anti-inflammatory*: rheumatoid arthritis and osteoarthritis *Analgesic–antipyretic*: symptoms of cold and flu syndrome *Antiplatelet, uricosuric* (sulfinpyrazone)	COX-1 and COX-2 weak inhibitors
Para-aminophenol derivatives	*Acetaminophen* or paracetamol	*Antipyretic, analgesic, with very weak anti-inflammatory activity* (it is used as an alternative to ASA or in combination with it)	COX-1 and COX-2 weak inhibitor
Propionic acid derivatives	Fenoprofen, flurbiprofen, ibuprofen, ketoprofen, naproxen, oxaprozin	*Anti-inflammatory*: antirheumatic for rheumatoid arthritis, osteoarthritis periarthritis, lumbago, sciatic nerve pain and radiculoneuritis, fibrosis, myositis, tenosynovitis *Analgesic*: in different pain syndromes (including dysmenorrhea and headache) *Antipyretic*	COX-1 and COX-2 inhibitors
Acetic acid derivatives	Diclofenac sodium, etodolac, indomethacin, ketorolac, nabumetone (*prodrug, active metabolite: 6-methoxy-2-naphthylacetic acid*), sulindac (*prodrug, active metabolite: sulindac sulfide*), tolmetin	*Anti-inflammatory*: rheumatoid arthritis and osteoarthritis *Analgesic–antipyretic*	COX-1 and COX-2 inhibitors. Diclofenac and etodolac: moderate selectivity for COX-2
Fenamate (anthranilic acids)	Sodium meclofenamate, mefenamic acid, flufenamic acid	*Anti-inflammatory* *Analgesic*: rheumatic diseases, musculoskeletal disorders, and dysmenorrhea *Antipyretic*	COX-1 and COX-2 inhibitors
Oxicam derivatives	Piroxicam, cinnoxicam, droxicam, tenoxicam, meloxicam	*Anti-inflammatory*: rheumatoid arthritis and osteoarthritis *Analgesic–antipyretic*	COX-1 and COX-2 inhibitors. Meloxicam: moderate selectivity for COX-2
Butyric acid derivatives	Indobufen	*Antiplatelet*	COX-1 inhibitor
Sulfonanilide derivatives	Nimesulide	*Anti-inflammatory*: rheumatoid arthritis and osteoarthritis *Analgesic–antipyretic*	Moderate selectivity for COX-2
Coxib	Celecoxib, rofecoxib, valdecoxib, etoricoxib, lumiracoxib	*Anti-inflammatory*: rheumatoid arthritis and osteoarthritis *Analgesic–antipyretic*	Variable selectivity for COX-2, from moderate (celecoxib) to high (etoricoxib, lumiracoxib)

Site-specific mutagenesis and crystallographic studies of COX-1 have shown that Ser529 is not important for the catalytic activity but its acetylation obstructs the COX channel in its narrowest point, thereby preventing access of the substrate to Tyr385 (a residue critical for COX catalysis) at the apex of the channel (Fig. 49.3).

Duration of aspirin inhibitory effect is not dependent on its pharmacokinetics (15–20 min half-life) but on the rate of

TABLE 49.3 Side effects of COX-1 and COX-2 inhibition

Effects	COX-1	COX-2
Gastrointestinal toxicity	+	?
Nephrotoxicity	+	+
Hemorrhagic complications	+	
Pharmacodynamic interactions with antihypertensive drugs	+	+
Pharmacodynamic interactions with aspirin	+	?

TABLE 49.4 Nonsteroidal anti-inflammatory drug classification according to the mechanism of PGH-synthase inhibition

Class I: simple competitive mechanism	$E + I \rightleftarrows EI$
Ibuprofen	
Piroxicam	
Sulindac sulfide	
Naproxen	
6-Methoxy-2-naphthylacetic acid (6-MNA)	
Flufenamic acid	
Mefenamic acid	
Class II: slowly reversible time-dependent competitive mechanism	$E + I \rightleftarrows EI \rightleftarrows EI^*$
Indomethacin	
Flurbiprofen	
Meclofenamic acid	
Diclofenac	
Rofecoxib	
Celecoxib	
Valdecoxib	
Etoricoxib	
Lumiracoxib	
Class III: irreversible time-dependent competitive mechanism	$E + I \rightleftarrows EI \rightarrow EI^*$
Acetylsalicylic acid	

E, enzyme; EI, rapidly reversible enzyme–inhibitor complex; EI*, slowly reversible or irreversible enzyme–inhibitor complex; I, inhibitor.
Modified from Smith W.L., DeWitt D.L. (1995). Biochemistry of prostaglandin endoperoxide H synthase-1 and synthase-2 and their differential susceptibility to nonsteroidal anti-inflammatory drugs. *Seminars in Nephrology*, 15, 179–194.

COX-isozyme synthesis (a few hours in nucleated cells). This explains the variable duration of different pharmacologic effects of aspirin: 4–6 h for analgesic, antipyretic, and anti-inflammatory effects (dependent on de novo protein synthesis), in contrast with a minimum of 24 h for its antiplatelet effect (dependent on platelet turnover).

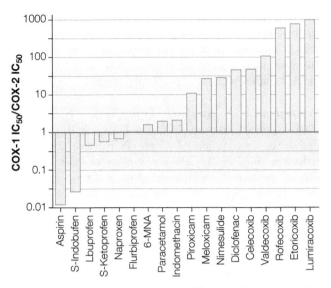

FIGURE 49.4 The selectivity of traditional NSAIDs and coxibs for COX-2 is a continuous variable. The ratio of IC_{50} values for inhibition of platelet COX-1 and monocytic COX-2 is a measure of the selectivity of traditional NSAIDs and coxibs for COX-2 and behaves as a continuous variable.

Cyclic Endoperoxide Metabolism

Thromboxane Synthase Thromboxane synthase (TX-synthase) is a membrane-bound hemoprotein in which, as in other P450 monooxygenases, the heme group is bound to the thiol group of a cysteine residue. TX-synthase exhibits only 16% homology to other P450 monooxygenases and belongs to the CYP5 subfamily. TX-synthase catalyzes the conversion of PGH_2 into thromboxane A_2 (TXA_2; Fig. 49.2). TX-synthase also forms 12(S)-hydroxyheptadeca-5(Z)-8(E)-10(E)-trienoic acid (HHT) and malondialdehyde (MDA). HHT does not have biological activity. Human TX-synthase consists of 533 amino acids and has a molecular weight of 60.5 kDa. TX-synthase mRNA is widely distributed in human tissues and is very abundant in platelets, leukocytes, spleen, lung, and liver. TX-synthase expression can be induced by phorbol ester and dexamethasone.

After catalysis, suicidal inactivation of TX-synthase restrains TXA_2 biosynthesis. The loss of activity is proportional to the amount of PGH_2 converted into TXA_2. This phenomenon seems to be very important for the control of TXA_2 biosynthesis.

Imidazole and pyridine derivatives or PGH_2 analogs cause selective inhibition of TX-synthase. TX-synthase inhibition promotes PGH_2 accumulation, increasing biosynthesis of PGE_2, prostaglandin D_2 (PGD_2), and prostaglandin $F_{2\alpha}$ ($PGF_{2\alpha}$) in platelets and PGI_2 in the vascular tissue. In activated platelets, PGH_2 accumulation causes occupation and activation of platelet TXA_2 receptor (TP), thus explaining the poor inhibitory effects on platelet function associated with administration of TX-synthase inhibitors in humans.

These inhibitors reduce deterioration of renal function in some animal models of progressive glomerular disease and reduce mortality in a murine model of lupus nephritis. However, their clinical development has not gone beyond phase 2 studies because of conflicting results.

Prostaglandin F Synthase Prostaglandin F synthase (PGF-synthase) converts PGH_2 to $PGF_{2\alpha}$. $PGF_{2\alpha}$ and its derivatives are produced from PGH_2 via 2-electron reduction (Fig. 49.2). Cells can synthesize $PGF_{2\alpha}$ and its derivatives through multiple metabolic pathways including (i) PGH_2 reduction to $PGF_{2\alpha}$ by an endoperoxide reductase (PGF-synthase), (ii) PGD_2 reduction to $9_\alpha,11_\beta\text{-}PGF_{2\alpha}$ by a 9-cheto-reductase, and (iii) PGE_2 reduction to $PGF_{2\alpha}$ by a 9-cheto-reductase.

$PGF_{2\alpha}$ plays an important role in the pathophysiology of uterine contraction, and its increased production at the end of pregnancy contributes to labor onset and maintenance. Structural analogs of $PGF_{2\alpha}$ metabolically stable (e.g., 15-methyl-$PGF_{2\alpha}$ or carboprost) are used as oxytocic drugs.

Prostaglandin I Synthase Prostaglandin I synthase (PGI-synthase) converts PGH_2 to prostacyclin. Prostacyclin (PGI_2), a platelet inhibitor and vasodilator substance mainly produced by vasculature, is synthesized through the activity of PGI-synthase (Fig. 49.2). Like TX-synthase, PGI-synthase is a member of the P450-monooxygenase family. PGI-synthase belongs to a new subfamily, named CYP8. The human enzyme is composed of 500 amino acids and has a molecular weight of 57 kDa. The enzyme is associated with nuclear, ER, and plasma membranes. PGI-synthase is inactivated during catalysis.

PGI_2 produced in the vascular endothelium contributes, along with other endothelial mediators, to thromboresistance of the intimal surface and to regulation of vascular tone. Furthermore, PGI_2 increases activity of enzymes metabolizing cholesterol esters in vascular smooth muscle cells, inhibits accumulation of cholesterol esters in macrophages, and prevents release of certain growth factors. Although PGI_2 biological actions are compatible with a protective role of this eicosanoid in atherothrombosis, lack of inhibitors selective for PGI-synthase has not allowed to experimentally testing this hypothesis. However, deletion of PGI_2 receptor (IP) encoding gene increases mice susceptibility to arterial thrombosis. Recent experimental and clinical data have shown that PGI_2, produced in the vasculature mainly by COX-2, is an important atheroprotective and antithrombotic mediator. Its inhibition by high-dose regimens of traditional NSAIDs or coxibs is associated with a moderate increase in risk of major cardiovascular events (mainly coronary).

PGI_2 is rapidly metabolized *in vivo* (plasma half-life in humans is 3–5 min). Analogs, chemically and metabolically more stable than the natural compound, have been developed. Some of these (e.g., iloprost) have been used in clinical trials and have shown efficacy in the treatment of peripheral arterial disease.

Prostaglandin D Synthases Prostaglandin D synthases (PGD-synthase) convert PGH_2 into PGD_2. PGD_2 is formed through nonoxidative rearrangement of PGH_2 (Fig. 49.2). Although the physiologic role of PGD_2 has not been defined yet, it has been shown that it acts as an inhibitor of platelet function, induces vascular smooth muscle relaxation, causes bronchoconstriction, and seems to be involved in dream induction (Table 49.1). Two different proteins with PGD-synthase activity have been characterized: lipocalin-type PGD-synthase (L-PGD-synthase) and hematopoietic PGD-synthase (H-PGD-synthase). These proteins are not homologous and carry out different functions. L-PGD-synthase is expressed in endothelial cells in response to laminar shear stress and could play an atheroprotective role. H-PGD-synthase, a glutathione-requiring enzyme, is mainly expressed in antigen-presenting cells, mast cells, and megakaryoblasts, and its role remains elusive. Moreover, albumin seems to be able to catalyze nonenzymatic conversion of PGH_2 into PGD_2.

Prostaglandin E Synthases Rearrangement of PGH_2 into PGE_2 (Fig. 49.2) may occur both nonenzymatically and enzymatically through the action of prostaglandin E synthases (PGE-synthases). PGE-synthases have been found both in the microsomal and cytosolic compartments and therefore named mPGE-synthase and cPGE-synthase, respectively. Two different types of mPGE-synthase have been characterized, mPGE-synthase-1 and mPGE-synthase-2.

mPGE-synthase-1 is a 17.5 kDa enzyme with glutathione-dependent activity. It belongs to the Membrane-Associated Proteins in Eicosanoid and Glutathione metabolism (MAPEG) superfamily and is induced in response to proinflammatory stimuli. It is also expressed constitutively in some tissues such as the kidney, ovary, and bladder. mPGE-synthase-1 and COX-2 coordinate expression has been demonstrated in different cell types, such as macrophages and pulmonary tumor cell lines.

mPGE-synthase-2, a 33 kDa protein with glutathione-dependent activity, is expressed constitutively in different tissues and produces PGE_2 from PGH_2 originated by both COX-1 and COX-2. This enzyme is also expressed in human colorectal cancers.

cPGE-synthase is a constitutive and ubiquitous protein identical to p23, a protein associated with numerous proteins including heat shock protein (Hsp) 70 and Hsp 90. cPGE-synthase seems to promote rapid PGE_2 synthesis mediated by COX-1 and to represent the primary source of PGE_2 production *in vivo*. PGE_2 serves a variety of functions in different tissues: it enhances PGH_2 platelet proaggregant effect, promotes vasodilation and cell growth, and may cause hyperalgesia and edema if produced in excess. PGE_2 (dinoprostone) and structural analogs of PGE_1 (e.g., misoprostol) are used for gastric cytoprotection and for abortion induction.

Selective mPGES-1 inhibitors are currently in preclinical development. The aim is to obtain drugs with analgesic and anti-inflammatory actions (through inhibition of PGE_2 synthesis) with an improved cardiovascular safety profile as compared to traditional NSAIDs and coxibs. Indeed, these compounds should not inhibit the vascular synthesis of prostacyclin and may actually increase its production by redirecting PGH_2 metabolism.

The Lipoxygenase Pathway

Synthesis of Leukotrienes 5-Lipoxygenase transforms arachidonic acid into 5-hydroperoxyeicosatetraenoic acid (5-HPETE), which is unstable and is converted into 5-hydroxyeicosatetraenoic acid (5-HETE) either enzymatically, by a peroxidase, or nonenzymatically. 5-Lipoxygenase has also a second catalytic activity as it catalyzes 5-HPETE conversion into leukotriene (LT) A_4, an unstable epoxide (Fig. 49.5). LTA_4 is then processed enzymatically to produce two types of biologically active substances, LTB_4 and sulfidopeptide leukotrienes (LTC_4, LTD_4, and LTE_4). LTB_4 is a powerful chemotactic substance for neutrophils. Sulfidopeptide leukotrienes are constricting substances for vascular and bronchial smooth musculature. Furthermore, they modulate mucus production, vascular permeability, and bronchial hyperreactivity. Based on these biological activities, sulfidopeptide leukotrienes may play an important role in pathophysiology of bronchial asthma.

5-Lipoxygenase is a 78 kDa protein expressed in white blood cells and other cell types. In nonstimulated cells, this enzyme can be found associated with membranes or in the cytoplasm. Upon cell activation, $[Ca^{2+}]_i$ increases and arachidonic acid—released mainly by cytosolic phospholipase activity—is transferred to 5-lipoxygenase by a 28 kDa membrane protein named 5-lipoxygenase-activating protein (FLAP) (Fig. 49.6) that binds arachidonic acid. In certain cell types, such as blood neutrophils and peritoneal macrophages, cell activation leads to perinuclear localization of both 5-lipoxygenase and FLAP. While FLAP is also localized in the nuclear membrane of nonactivated cells, 5-lipoxygenase shuttles from the cytosol to the nucleus only in response to cell stimulation. The catalytic activity of 5-lipoxygenase requires the presence of calcium and

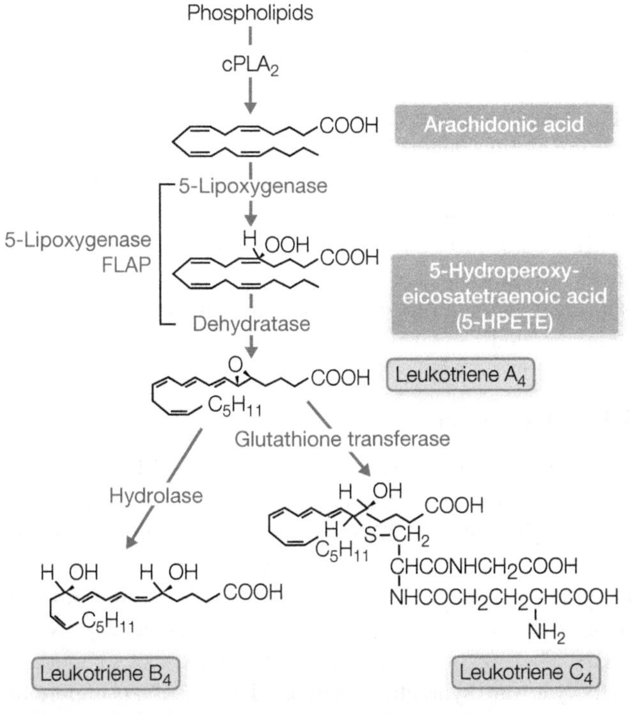

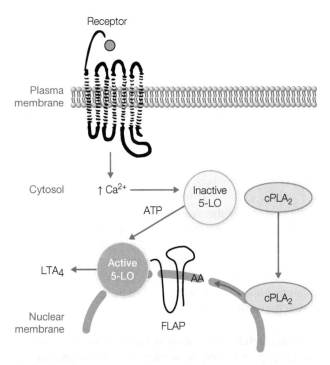

FIGURE 49.5 5-Lipoxygenase pathway and leukotriene formation. 5-Lipoxygenase has two catalytic activities: an oxygenase activity that leads to 5-hydroperoxyeicosatetraenoic acid (5-HPETE) formation from arachidonic acid and a dehydratase activity (LTA_4-synthase) that transforms 5-HPETE into LTA_4, an unstable epoxide. LTA_4 can be converted to LTB_4 (5,12-hydroxyeicosatetraenoic acid) by the LTA_4-hydrolase enzyme or it can be conjugated to glutathione, forming LTC_4.

FIGURE 49.6 5-Lipoxygenase activation. Upon cell activation, due, for example, to interaction of a stimulus with specific membrane receptors, an increase in cytosolic free calcium occurs causing cytosolic phospholipase A_2 ($cPLA_2$, 85 kDa) and 5-lipoxygenase (5-LO) to move from the cytosol to the nuclear membrane. 5-Lipoxygenase-activating protein (FLAP) is localized in the nuclear membrane, where it binds arachidonic acid (AA) released from membrane phospholipids by the activated $cPLA_2$ and transfers it to 5-LO. Then, 5-LO converts AA into leukotriene A_4 (LTA_4).

ATP. Similarly to other enzymes involved in arachidonic acid metabolism, 5-lipoxygenase undergoes suicide inactivation during catalysis.

5-Lipoxygenase Inhibitors The catalytic activity of 5-lipoxygenase can be inhibited by different types of compounds:

1. Compounds named "redox" (phenidone and its analogs, like BW755c) that seem to act by reducing the enzyme (probably reducing the ferric group in the active site), hence generating intermediate radicals. The possibility that these compounds may interact with other cellular redox systems leading to methemoglobin formation has limited their development for clinical use.
2. Compounds capable of interacting with the iron in the enzyme active site (hydroxamic acid derivatives, e.g., zileuton).
3. Nonredox inhibitors that specifically interact with the enzyme active site (methoxytetrahydropyran derivatives, e.g., ICI 211965 and ZD2138).
4. FLAP inhibitors. These compounds do not inhibit directly the enzyme activity. Nevertheless, by binding FLAP, they interfere with arachidonic acid transfer to 5-lipoxygenase (Fig. 49.6) (e.g., MK-0591 and BayX1005). Characterization of FLAP tridimensional structure has allowed the development of novel FLAP inhibitors, such as AM-103 and GSK-2190915 (currently in phase II for the treatment of inflammatory diseases, including asthma).

The 5-lipoxygenase inhibitor zileuton has been launched on the market. Although evidence of greater efficacy of this inhibitor has not yet been provided, it may show advantages over inhibitors of individual leukotriene receptors as it blocks completely the synthesis of products deriving from lipoxygenases. It is used for asthma treatment, and it could be useful for other inflammatory conditions. However, the use of this drug is associated with sleep disorders and behavioral alterations, which can be severe. Other 5-lipoxygenase inhibitors are currently being studied for possible applications in the cardiovascular field.

Arachidonic Acid Metabolism by 12- and 15-Lipoxygenases The 12- and 15-lipoxygenase enzymes convert arachidonic acid to 12- and 15-HPETE, respectively; hydroperoxides are then reduced nonenzymatically to the respective hydroxyl acids.

12-Lipoxygenase products have chemotactic activity toward neutrophils and can induce migration of vascular smooth muscle cells.

Not much is known about the role of direct products of 15-lipoxygenase, such as eoxins, which are primarily generated in eosinophils and bronchi, particularly in asthmatic individuals. However, it is interesting that coordinated action of 5- and 15-lipoxygenases leads to formation of lipoxins LXA_4 (i.e., 5,6,15-trihydroxy-7,9,11,13-eicosatetraenoic acid) and LXB_4 (i.e., 5,14,15-trihydroxy-6,8,10,12-eicosatetraenoic acid). Lipoxins possess biological actions on various organ systems; for example, they activate leukocytes, contract bronchial smooth muscle, and relax vascular smooth muscle. The pathophysiological importance of these eicosanoids remains to be defined.

Acetylation by aspirin of human COX-2 Ser516 residue does not completely inhibit the enzyme COX activity, but modifies its catalytic activity resulting in production of 15R-hydroxyeicosatetraenoic acid (15R-HETE) instead of PGG_2. The biological properties of 15R-HETE have not yet been characterized; however, it can be metabolized by 5-lipoxygenase to form 15-epi-lipoxins (15-epi-LXA_4 and 15-epi-LXB_4), a new series of biologically active eicosanoids, which are potent inhibitors of cell adhesion and proliferation.

Lipoxins and 15-epi-lipoxins are lipid mediators that may exert anti-inflammatory actions by inducing mechanisms responsible for resolution of the inflammatory response. Recently, new lipid mediators have been identified, namely, resolvins and protectins, which are generated from omega-3 fatty acids (eicosapentaenoic acid and docosahexaenoic acid) by enzymatic pathways activated during the inflammation resolution phase. Resolvins possess anti-inflammatory and immunoregulatory activities; in particular, they control infiltration of polymorphonuclear neutrophil leukocytes (PMN) and promote inflammation resolution. Whether these interesting compounds are formed *in vivo* in sufficient amounts to exert these activities remains to be demonstrated.

LTA_4 Metabolism

Leukotriene A_4 Hydrolase In contrast with 5-lipoxygenase that is selectively distributed in myeloid lineage cells, leukotriene A_4 hydrolase (LTA_4-hydrolase) is present in different cell types (leukocytes, erythrocytes, renal cells). This enzyme catalyzes LTA_4 hydrolysis to LTB_4, a dihydroxylic leukotriene (Fig. 49.5) with potent chemotactic activity for PMN. The enzyme is a zinc protein and possesses peptidase activity toward synthetic substrates. The zinc ion is important for both hydrolase and peptidase activities. This enzyme undergoes suicide inactivation during catalysis, losing both enzymatic activities.

A few specific inhibitors of LTA_4-hydrolase have been synthesized and characterized. Both LTA_4 analogs (LTA_3 and LTA_5) and eicosapentaenoic acid are inhibitors of LTA4 to

LTB$_4$ conversion. Bestatin is an inhibitor of aminopeptidases and also inhibits the enzyme hydrolase activity.

Leukotriene C$_4$ Synthase Leukotriene C$_4$ synthase (LTC$_4$-synthase) is a microsomal enzyme that transfers glutathione to LTA$_4$ leading to formation of LTC$_4$ (Fig. 49.5), a powerful smooth muscle constrictor substance. LTC$_4$-synthase is part of the glutathione-S-transferase family and is widely distributed in various tissues, including endothelium and vascular smooth muscle. Some inhibitors of the hepatic glutathione-S-transferase system (ethacrynic acid, caffeic acid and its phenethylester) can inhibit LTC$_4$-synthase activity *in vitro*. A limitation to the development of LTC$_4$-synthase selective inhibitors is represented by the possible accumulation of LTA$_4$ that could be used by LTA$_4$-hydrolase to form LTB$_4$.

Transcellular Metabolism of PGH$_2$ and LTA$_4$

Although eicosanoid biosynthesis may depend on the presence of all necessary enzymes in the same cell, eicosanoid production can also occur through cell cooperation. In fact, the products of PGH-synthase and 5-lipoxygenase activity, PGH$_2$ and LTA$_4$, respectively, are partly released in the extracellular environment and can be used by other cells to form biologically active eicosanoids (Fig. 49.7).

Prostacyclin synthesis can result from cooperation between platelets and endothelial cells and lymphocytes. Recent studies have shown that human platelets having COX-1 activity suppressed by low-dose aspirin can recover their ability to synthesize TXA$_2$ in the presence of endothelial cells expressing COX-2 (poorly sensitive to low concentrations of aspirin). This occurs because PGH$_2$ produced by endothelial COX-2 activity can be used by TX-synthase of aspirinated platelets to form TXA$_2$. This transcellular biosynthesis of TXA$_2$ may contribute to aspirin-resistant TXA$_2$ biosynthesis *in vivo*, despite complete blockade of platelet COX-1 activity. The LTA$_4$ produced by PMN can be used by erythrocytes possessing LTA$_4$-hydrolase activity to form LTB$_4$. In the same way, endothelial cells and platelets not possessing 5-lipoxygenase activity can use PMN-released LTA$_4$ and turn it into LTC$_4$ through the action of glutathione transferase.

NONENZYMATIC METABOLISM OF ARACHIDONIC ACID

Isoeicosanoids represent a family of arachidonic acid metabolites produced by nonenzymatic mechanism of lipid peroxidation catalyzed by oxygen radicals (Fig. 49.1). These include prostaglandin isomers called

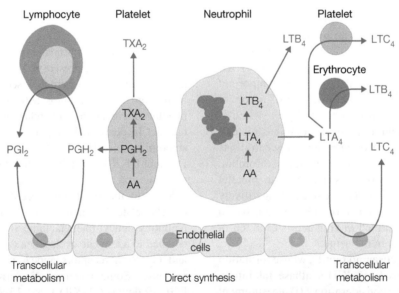

FIGURE 49.7 Interactions between blood cells and vessel wall in prostanoid and leukotriene formation. Platelets and neutrophils (generating cells), once activated, have the enzymatic systems to synthesize PGH$_2$ and LTA$_4$ that they release into the extracellular environment. Here, these may be used by other cells (acceptor cells) that do not need to be activated, but possess enzymes such as PGI-synthase (endothelial cells and lymphocytes), LTA$_4$-hydrolase (red blood cells), and LTC$_4$-synthase (platelets and endothelial cells) to produce PGI$_2$, LTB$_4$, and LTC$_4$, respectively. Through this process of transcellular biosynthesis, lymphocytes, which possess low PGH-synthase activity, can produce PGI$_2$, while red blood cells, platelets, and endothelial cells, which do not possess 5-lipoxygenase, can produce LTB$_4$ and LTC$_4$.

TABLE 49.5 Areas of pharmacological intervention mediated by changes in eicosanoid biosynthesis or action

Drugs	Mechanism(s)	Effects
Glucocorticoids	Induction of anti-PLA$_2$ proteins; inhibition of PGHS-2 expression	Anti-inflammatory effect
NSAIDs	PGHS-1 and PGHS-2 inhibition	Analgesic, antipyretic, anti-inflammatory; gastrotoxicity, nephrotoxicity, cardiovascular toxicity
Low-dose aspirin	Platelet PGHS-1 inhibition	Antithrombotic effect; increased bleeding
Coxibs	PGHS-2 inhibition	Analgesic, anti-inflammatory; nephrotoxicity; cardiovascular toxicity
PGH$_2$/TXA$_2$ antagonists	Antagonism at platelet and vascular TP receptors	Antithrombotic effect; increased bleeding
LTC$_4$/D$_4$ antagonists	Antagonism at bronchial receptors	Antiasthmatic effect
n-3 fatty acids	Competition with arachidonic acid?	Multiple effects, potentially relevant to inflammation and atherosclerosis

isoprostanes and thromboxane analogs called isothromboxanes. Similarly, leukotriene and hydroxyeicosatetraenoic acid (HETE) analogs have been denominated isoleukotrienes and isoHETE, respectively. One of the F-series isoprostanes, 8-iso-PGF$_{2\alpha}$, has important biological activities: it is a vasoconstrictor of coronary and renal vessels; it has mitogenic activity on vascular smooth muscle cells; and it induces complete platelet aggregation in the presence of very low concentrations of platelet agonists (such as ADP, collagen, thrombin, and arachidonic acid). Vasoconstriction and platelet aggregation induced by 8-iso-PGF$_{2\alpha}$ can be competitively antagonized by PGH$_2$/TXA$_2$ receptor (TP) antagonists. In humans, 8-iso-PGF$_{2\alpha}$ is formed *in vivo* and is excreted in easily detectable amounts in the urine.

The interest for this eicosanoid family, whose formation is not substantially altered by PGH-synthase inhibitors, derives from the following considerations: (i) measurement of urinary isoprostanes, such as 8-iso-PGF$_{2\alpha}$, represents a noninvasive index of lipid peroxidation *in vivo* and can provide a useful biochemical end point to assess the effects of antioxidants in humans; (ii) increased F$_2$-isoprostane formation in association with several cardiovascular risk factors (cigarette smoking, hypercholesterolemia, diabetes mellitus, obesity) may represent a common mechanism induced by oxidative stress to trigger specific forms of cell activation, such as platelet activation and vascular smooth muscle cell proliferation, that characterize atherosclerosis development; and (iii) some pharmacological effects of antioxidants may be explained by a reduced formation of biologically active isoeicosanoids (Table 49.5).

EICOSANOID RECEPTORS

To date, all eicosanoid receptors characterized and cloned are G-protein-coupled receptors. Although sharing a common molecular organization with seven-transmembrane domains, they differ in terms of specificity, signal transduction mechanism(s), and tissue distribution.

Prostanoid Receptors

Prostanoid receptors can be classified into five types based on their affinity for five natural prostanoids: PGD$_2$, PGE$_2$, PGF$_{2\alpha}$, PGI$_2$, and TXA$_2$ (Table 49.6). These receptors are called P-receptors, and each one is designated by a letter indicating the natural prostanoid it has the highest affinity for (i.e., DP, EP, FP, IP, and TP, respectively). Furthermore, EP receptors are divided into four subtypes (EP1, EP2, EP3, and EP4) based on their affinity for a number of agonists and antagonists. These receptors have been cloned and their relative signal transduction mechanisms characterized (Table 49.6).

PGD$_2$ Receptors Pharmacological studies have shown that PGD$_2$ receptors are present in the vascular smooth muscle, platelets, and brain, where they seem to be involved in sleep regulation. PGD$_2$ receptor activation causes inhibition of platelet aggregation and relaxation of vascular smooth muscle. In contrast, PGD$_2$ contracts tracheal and bronchial smooth muscle. Two PGD$_2$ receptors have been identified: DP$_1$ and DP$_2$/CHRT$_2$ (chemoattractant receptor-homologous molecule receptor expressed on Th2 cells). While DP$_1$ belongs to the prostanoid receptor family, DP$_2$/CHRT$_2$ is closely related to chemotactic receptors for the *N*-formyl-methionyl-leucyl-phenylalanine (FMLP) peptide. They are coupled to adenylate cyclase via two different G proteins, G$_i$ and G$_s$. Both receptors appear to be involved in allergic responses. Some nonenzymatic derivatives of PGD$_2$, such as PGJ$_2$ (9-deoxy-Δ9-PGD$_2$) and 15-deoxy-Δ12,14-PGJ$_2$, can stimulate peroxisome proliferator-activated receptor-γ (PPAR-γ), a nuclear receptor belonging to the family of transcription factors. PPAR-γ activation in monocytes/macrophages is associated with inhibition of proinflammatory gene expression. However, the pathophysiologic relevance of these PGD$_2$ derivates has been questioned.

TABLE 49.6 Classification of prostanoid receptors

Type	Subtype	Agonist	Antagonist	Mechanism of signal transduction
DP		BW245C ZK110841; RS93520, SQ27986	BWA868C; SH6809	↑ cAMP
EP	EP1	17-Ph-ω-PGE$_2$ iloprost[a], sulprostone[b]	SC19220, AH6809, ZD-6416	↑ IP$_3$/DAG/Ca^{2+}
	EP2	Butaprost; AH13205; misoprostol; AY23626[c]	AH6809[d]	↑ cAMP
	EP3	Enprostil; GR63799; sulprostone, misoprostol, AY23626	DG-041	↓ cAMP; ↑ IP$_3$/DAG/Ca^{2+}
	EP4	NA	AH22921; AH23848[e]	↑ cAMP
FP		Fluprostenol; cloprostenol; latanoprost	NA	↑ IP$_3$/DAG/Ca^{2+}
IP		Cicaprost; octimibate; iloprost	NA	↑ cAMP
TP		U46619; U44069; STA$_2$; IBOP; AGN192093	GR32191; sulotroban; ifetroban; daltroban; teruroban	↑ IP$_3$/DAG/Ca^{2+}

[a] Iloprost is a partial agonist at EP1 receptor but a full agonist at IP receptor.
[b] Sulprostone is a more potent agonist at EP3 receptor.
[c] Misoprostol and AY23626 are also agonists at EP3 receptor.
[d] AH6809 is also an antagonist at EP1 and DP receptors.
[e] AH22921 and AH23848 are also TP receptor antagonists.
cAMP, cyclic AMP; DAG, diacyl*glycerol*; IP$_3$, *inositol* trisphosphate; NA, not available.

PGE$_2$ Receptors Binding sites for PGE$_2$ have been identified in different tissues and cell types, such as adipocytes, uterus, kidney, gastric mucosa, platelets, and macrophages. Four subtypes of PGE$_2$ receptors have been cloned. Molecular signaling events triggered by the natural agonist PGE$_2$ and different synthetic agonists are detailed in Table 49.6. The receptor subtypes EP1 and EP2 are widely distributed on smooth muscle cells, where they mediate contraction and relaxation responses, respectively. The EP3 subtype is widely distributed in different cell types; via adenylate cyclase inhibition, it causes inhibition of neurotransmitter release in the autonomic nervous system, gastric acid secretion, and adipocyte lipolysis. This receptor subtype can also affect phosphatidylinositol turnover resulting in increased intracellular Ca^{2+} and can induce Ca^{2+} influx from the extracellular space, thus inducing smooth muscle contraction and platelet shape change. Several EP3 receptor isoforms have been identified that result from alternative splicing in the C-terminal-encoding region of the mRNA and are coupled to different G proteins. In the kidney, and in particular in tubular cells of the collecting duct and in the thick ascending limb of Henle's loop, PGE$_2$ plays a physiologic role in controlling sodium and water reabsorption via interaction with two receptor subtypes, EP1 and EP3. PGE$_2$–EP4 interaction regulates intestinal homeostasis, inflammatory responses, and bone formation by increasing cAMP levels.

Numerous EP1 antagonists have been developed; in particular, the ZD-6416 compound has been successfully tested in animal models of inflammation and analgesia. In healthy volunteers, this compound reduces esophageal pain caused by electrical stimulation. Among the numerous EP3 receptor antagonists, the DG-041 compound is in phase I clinical trial and represents a promising drug for the treatment of peripheral arterial disease. Studies on EP4 antagonists carried out *in vitro* and in animal models have demonstrated that PGE$_2$–EP4 interaction is involved in colon tumorigenesis.

The FP Receptor for PGF$_{2\alpha}$ FP receptor is present in the uterus, vasculature, and bronchus. PGF$_{2\alpha}$–FP interaction stimulates phospholipase C activity, thus inducing [Ca^{2+}]$_i$ increase and smooth muscle contraction in these tissues. FP receptor activation in the corpus luteum is responsible for luteolysis. Furthermore, PGF$_{2\alpha}$–FP interaction lowers intraocular pressure.

The IP Receptor for PGI$_2$ PGI$_2$ is the most abundant product of arachidonic acid metabolism in the vascular tissue. It inhibits platelet aggregation and induces vasodilation by interacting with IP receptor that in turn activates adenylate cyclase through a G$_s$ protein. IP receptors are widely distributed in arterial smooth muscle cells, platelets, and central nervous system. Several potent IP agonists (cicaprost, octimibate, iloprost) are available, but many of these also recognize other prostanoid receptors, especially EP receptors.

TP Receptors for TXA$_2$ Due to chemical instability of TXA$_2$ and its precursor PGH$_2$ (30 s and 5 min *in vitro* half-life, respectively), the biological actions of these prostanoids have been studied using chemically stable analogs (e.g., U46619, IBOP; see Table 49.6). Binding sites for these compounds have been identified on platelets, bronchial and vascular smooth muscle cells, and mesangial cells of renal glomerulus. In platelets, TXA$_2$ and PGH$_2$ act on a common receptor with similar affinity. TP receptor activation causes an increase in [Ca^{2+}]$_i$ by opening of membrane channels and by phospholipase C activation. Platelet TP activates phospholipase C through a pertussis toxin-insensitive G protein. There are at least two platelet TP isoforms: one characterized by low affinity for the IBOP agonist and regulating granule secretion and platelet aggregation, and another with high affinity for IBOP and controlling platelet shape change. Molecular signaling associated with TP receptor activation in the vascular tissue has not been extensively studied. However, it seems that an increase in cytoplasmic free calcium plays an important role in TXA_2-induced vascular and mesangial contraction. Although pharmacological studies suggest the existence of receptor heterogeneity, only one TP receptor-encoding gene has been cloned from a placenta cDNA library. A receptor variant resulting from alternative splicing in the C-terminal-encoding sequence has been cloned from a cDNA library of human umbilical endothelial cells. The two receptor isoforms have been designated TPα and TPβ. Only TPα has been identified in human platelets, whereas both isoforms are present in endothelial and smooth muscle cells. The activation of the receptor causes its desensitization.

Activation of TP receptors can be blocked by specific antagonists (sulotroban, ifetroban, daltroban, GR32191, terutroban). Although it was pharmacologically plausible that TP antagonists would produce beneficial effects against atherothrombosis not shared by low-dose aspirin (e.g., antagonism of aspirin-insensitive TP agonists, such as COX-2-derived TXA_2 or F_2-isoprostanes), terutroban—the only TP antagonist that has completed phase 3 clinical development—did not prove more effective than aspirin in preventing major vascular complications in a clinical trial (PERFORM) of approximately 19,000 patients with a recent ischemic stroke.

Leukotriene Receptors

Two classes of leukotriene receptors have been characterized: receptors for sulfidopeptide leukotrienes LTC_4, LTD_4, and LTE_4 (abbreviation: CysLT) and LTB_4 receptors (abbreviation: BLT).

Receptors for Sulfidopeptide Leukotrienes Sulfidopeptide leukotrienes exert several proinflammatory effects and are important mediators of asthma, allergic rhinitis, and other chronic inflammatory conditions.

Two receptors for sulfidopeptide leukotrienes have been characterized: $CysLT_1$ and $CysLT_2$. $CysLT_1$ (previously called LTD_4 receptor) is blocked by several antagonists with different structures, named "classical antagonists," and by BAYu9773. By contrast, $CysLT_2$ (previously called LTC_4 receptor) is blocked by BAYu9773 but not by classical antagonists. $CysLT_1$ activation results in intracellular free calcium increase and stimulation of phosphatidylinositol turnover. Studies suggest that this receptor may be associated with more than one G protein, with only one sensitive to pertussis toxin. $CysLT_1$ undergoes homologous desensitization. Several $CysLT_1$ receptor antagonists have been developed as antiasthmatic drugs and are used for the treatment of asthma and allergic rhinitis. Among these, zafirlukast, montelukast, and pranlukast are now commercially available in many countries. Because of their anti-inflammatory action, antileukotrienes (5-lipoxygenase inhibitors and $CysLT_1$ receptor antagonists) are appropriate in maintenance therapy of asthma patients when bronchodilators are not sufficient.

Moreover, sulfidopeptide leukotrienes appear to play an important role in asthma induced by aspirin and exercise; therefore, antileukotriene drugs may represent the treatment of choice for this condition.

$CysLT_2$ has been identified in pulmonary macrophages, bronchial smooth muscle cells, cardiac Purkinje cells, vascular endothelium, adrenal medulla cells, peripheral blood leukocytes, and brain.

LTB_4 Receptors Two LTB_4 receptors are known: BLT_1 that is mainly expressed in leukocytes and is associated with chemotactic response and BLT_2 that is ubiquitously expressed but has a lower affinity for LTB_4 compared to BLT_1.

TAKE-HOME MESSAGE

- Eicosanoids are a family of biologically active compounds generated from arachidonic acid through both enzymatic and nonenzymatic metabolic pathways.
- In contrast with circulating hormones, eicosanoids are autacoids acting near the site of production through specific receptors mainly belonging to the GPCR family.
- Biological activities of prostanoids and leukotrienes can be modulated by drugs that inhibit their synthesis or block their action on receptors.
- NSAIDs inhibit PGH-synthase (COX) activity and exert analgesic, antipyretic, and anti-inflammatory effects but can cause serious adverse effects on gastrointestinal and cardiovascular systems.
- Different types of COX have been identified; this has allowed the generation of COX-2-selective NSAIDs devoid of adverse gastrointestinal effects.
- Low-dose aspirin is used as antithrombotic agent because of its relatively selective inactivation of platelet COX-1.
- CysLT1 receptor antagonists are used in asthma treatment.
- Isoeicosanoids derive from nonenzymatic lipid peroxidation catalyzed by oxygen radicals and include biologically active isomers of the major classes of eicosanoids.
- F_2-isoprostanes are $PGF_{2\alpha}$ isomers used as oxidative stress markers in humans.

FURTHER READING

Bell-Parikh L.C., Ide T., Lawson J.A., McNamara P., Reilly M., FitzGerald G.A. (2003). Biosynthesis of 15-deoxy-Δ12,14-PGJ2 and the ligation of PPARγ. *Journal of Clinical Investigation*, 112, 945–955.

Boutaud O., Oates J.A. (2010). Study of inhibitors of the PGH synthases for which potency is regulated by the redox state of the enzymes. *Methods in Molecular Biology*, 644, 67–90.

Evans J.F. (2002). Cysteinyl leukotriene receptors. *Prostaglandins and Other Lipid Mediators*, 68–69, 587–597.

FitzGerald G.A., Patrono C. (2001). The coxibs, selective inhibitors of cyclooxygenase-2. *New England Journal of Medicine*, 345, 433–442.

Funk C.D. (2001). Prostaglandins and leukotrienes: advances in eicosanoid biology. *Science*, 294, 1871–1875.

Jones R.J., Giembycz M.A., Woodward D.F. (2009). Prostanoid receptor antagonists: development strategies and therapeutic applications. *British Journal of Pharmacology*, 158, 104–145.

Maclouf J., Folco G., Patrono C. (1998). Eicosanoids and iso-eicosanoids: constitutive, inducible and transcellular biosynthesis in vascular disease. *Thrombosis and Haemostasis*, 79, 691–705.

Mayer R.J., Marshall L.A. (1993). New insights on mammalian phospholipase A2(s); comparison of arachidonoyl-selective and -nonselective enzymes. *FASEB Journal*, 7, 339–348.

Narumiya S., FitzGerald G.A. (2001). Genetic and pharmacological analysis of prostanoid receptor function. *Journal of Clinical Investigation*, 108, 25–30.

Patrono C., FitzGerald G.A. (1997). Isoprostanes: potential markers of oxidant stress in atherothrombotic disease. *Arteriosclerosis, Thrombosis, and Vascular Biology*, 17, 2309–2315.

Patrono C., García Rodríguez L.A., Landolfi R., Baigent C. (2005). Low-dose aspirin for the prevention of atherothrombosis. *New England Journal of Medicine*, 353, 2373–2383.

Patrono C., Baigent C., Hirsh J., Roth G. (2008). Antiplatelet drugs: American College of Chest Physicians evidence-based clinical practice guidelines. 8th ed. *Chest*, 133, 199S–233S.

Picot D., Loll P.J., Garavito M.R. (1994). The X-ray crystal structure of the membrane protein prostaglandin H2 synthase-1. *Nature*, 367, 243–246.

Rodríguez M., Domingo E., Municio C., Alvarez Y., Hugo E., Fernández N., Sánchez Crespo M.(2014). Polarization of the innate immune response by prostaglandin E2: a puzzle of receptors and signals. *Molecular Pharmacology*, 85, 187–197.

Samson A., Holgate S. (1998). Leukotriene modifiers in the treatment of asthma. *British Medical Journal*, 316, 1257–1258.

Smith W.L., Garavito R.M., Dewitt D.L. (1996). Prostaglandin endoperoxide H synthases (cyclooxygenases)-1 and -2. *Journal of Biological Chemistry*, 271, 33157–33160.

Yazid S., Norling L.V., Flower R.J. (2012). Anti-inflammatory drugs, eicosanoids and the annexin A1/FPR2 anti-inflammatory system. *Prostaglandins & Other Lipid Mediators*, 98, 94–100.

Yokoyama U., Iwatsubo K., Umemura M., Fujita T., Ishikawa Y. (2013). The prostanoid EP4 receptor and its signaling pathway. *Pharmacological Reviews*, 65, 1010–1052.

SECTION 12

PHARMACOLOGY OF DEFENCE PROCESSES

50

PHARMACOLOGICAL MODULATION OF THE IMMUNE SYSTEM

CARLO RICCARDI AND GRAZIELLA MIGLIORATI

By reading this chapter you will:

- Understand the pharmacological immunomodulatory strategies used to treat inflammatory/autoimmune diseases and transplant rejection
- Know molecular action mechanisms, therapeutic applications, and clinically significant adverse effects of immunomodulatory drugs

---

Immunopharmacology is a research and medical field that focuses on pharmacological agents affecting the immune system. Some therapeutic approaches, such as transplants, require suppression of the immune response through drugs called immunosuppressants (or immunosuppressive drugs). By contrast, others require activation of the immune response and employ immunostimulant drugs. Some authors identify all the agents (including drugs, bacteria, and biological products) affecting the immune response, either inhibiting or enhancing it, as biological response modifiers (BRMs).

THE IMMUNE RESPONSE

The immune system is a complex, highly autoregulated system that allows the organism to recognize and tolerate its own healthy constituents, called "self" and to detect and inactivate all the "nonself" agents such as pathogens, exogenous, or transformed cells. This system is critical for organism survival, homeostatic control of development, and tissue and organ integrity. Protection from infections and diseases is provided by two main constituents of the immune system: the innate immunity and the adaptive (or acquired) immunity. The innate immunity represents the first line of defense against exogenous antigens, and its main effectors are the complement, the lysozyme, and several cell types such as macrophages, granulocytes, mast cells, and natural killer (NK) cells. In contrast to the innate immunity, which provides a nonspecific defense, the adaptive immunity provides a highly specific response and generates a long-lasting immunological memory of nonself agents encountered. Main effectors of the acquired immunity are B lymphocytes, which produce antibodies mediating humoral immunity, and T lymphocytes (cytotoxic, helper, and suppressor), which have different functions and are responsible for cell-mediated immunity. Immunoglobulins expressed on B lymphocytes external membrane act as receptors for soluble antigens (protein, carbohydrate, and lipid antigens) which they identify through epitopes exposed on the molecular surface. By contrast, T lymphocytes identify antigens presented on the surface of specific cells, called antigen-presenting cells (APCs), like macrophages, dendritic cells, and B lymphocytes. APCs play a critical role in the immune response as they expose antigens in association with proteins of the class I and II major histocompatibility complex (MHC). Many different T lymphocyte subpopulations can be distinguished for their antigen receptor composition (either TCR $\alpha\beta$ or $\gamma\delta$), specificity and effector role. T lymphocytes carrying TCR $\alpha\beta$ respond to antigens presented by class I or II MHC molecules, whereas those expressing TCR $\gamma\delta$ are usually independent from MHC and involved in tissue stress response, both in the presence and in the absence of infection. T- helper CD4+ lymphocyte activation depends on the interaction between the TCR and the peptide-class II MHC complex on the APCs. T cytotoxic CD8+ lymphocyte activation requires the TCR to identify the peptide-class I MHC complex on the APCs. Both T CD4+ and CD8+ activation involves co-stimulatory signals provided by molecules on the APCs. Cytokines are soluble proteins that play a fundamental role in the complex network of interaction within the immune system, acting both as regulators and direct effectors. Acquired immunity is involved not only in normal immune response to pathogens and tumoral cells but also in autoimmune diseases and transplant rejection.

The Concept of Autoimmunity

Autoimmunity occurs when the immune system fails to distinguish among self and nonself and attacks normal and healthy tissues of the organism. Despite the complex mechanisms controlling development and maintenance of self-tolerance, many potentially autoreactive T and B lymphocytes are present in the body. Stimulation of autoreactive lymphocytes may be provoked by different factors such as an altered concentration and/or sequence of a potentially immunogenic peptide presented by the APCs (due to defects in the processing mechanism). Whereas presentation of a potentially immunogenic peptide in appropriate concentrations guarantees maintenance of the tolerance, variations in its concentration or sequence may trigger an immune response. Myasthenia gravis, multiple sclerosis, rheumatoid arthritis, diabetes mellitus type I, and lupus erythematosus are among the main autoimmune disorders; in all these pathologies, defective epitope recognition causes epitopes belonging to healthy tissues to be attacked by the immune system as they are nonself. Major attention is currently being given to specific therapeutic procedures which, for example, aim at inducing tolerance toward specific peptides; nevertheless, so far the most effective and more often adopted therapeutic approach relies on immunosuppressive drugs. Moreover, it is important to remember that immunosuppressants are not only used to treat autoimmune diseases, such as those mentioned earlier but are also often employed in conditions that are more suitably defined as inflammatory (e.g., ulcerative rectocolitis and Crohn's disease) in which the autoimmune component plays a relevant role in determining onset and duration of the disease (see Box 50.1).

IMMUNOSUPPRESSIVE DRUGS

Immunosuppressive drugs are used in organ transplantation to inhibit either graft rejection by the host or graft versus host disease (GVHD) in incompatible bone marrow transplants, in autoimmune diseases and, often, in inflammatory diseases with a strong autoimmune component.

Anticancer Chemotherapeutic Agents

Cyclophosphamide Cyclophosphamide is a nitrogen mustard containing a cyclophosphamide group. It is used to treat several types of cancer and is often employed, either alone or in combination mainly with cyclosporine and corticosteroids, in immunosuppressive therapies.

Cyclophosphamide action is directly linked to its ability to add an alkyl group to the N7 atom of the DNA guanine base, causing the following alterations of DNA structure and function:

1. Modification of the genetic code, as the alkyl group promotes guanine–thymine pairing rather than guanine–cytosine
2. Breakage of guanine imidazole ring
3. DNA depurination, as a consequence of guanine base loss, resulting in DNA strand break
4. Crosslinking of guanine bases on opposite strands, thus preventing DNA replication

As in the case of other alkylating agents not cyclo-specific, cyclophosphamide immunosuppressive activity results from its ability to inhibit cell proliferation and to induce lymphocytes death. Although cyclophosphamide can bind DNA even in nonproliferating cells (G_0 cells), its action is most powerful in S phase proliferating cells that then become arrested in premitotic G_2 phase and die (mitotic death). Cyclophosphamide is active on both B and T lymphocytes and is very effective in suppressing the humoral immune response. For this reason, it is particularly useful in those pathologies, like lupus erythematosus, where antibodies play a pathogenic role.

Cyclophosphamide is a prodrug that is orally well absorbed and is activated by metabolic conversion, mostly in the liver. Metabolic activation is carried out by microsomal enzymes of the P450 cytochrome system and consists in ionization of the chloride atoms and formation of cyclic ethyl ammonium ion. Its plasma half-life is about 7 h.

In some clinical applications, cyclophosphamide has been partially replaced by cyclosporin A (CsA), but it is still the drug of choice to treat polyarteritis nodosa in combination with prednisone. It is used to treat most cases of glomerulonephritis (in combination with glucocorticoids (GCs) and azathioprine), Wegener's granulomatosis and other forms of systemic vasculitis (as the drug of choice but also in combination with prednisone), and pulmonary fibrosis, when the GC therapy fails. It is also used to treat patients affected by dermatomyositis–polymyositis, autoimmune hemolytic anemia, nephrotic syndrome, thrombocytopenic purpura unresponsive to steroid therapy, and systemic lupus erythematosus, in particular when kidneys and central nervous system are implicated. It may be effective in treating myasthenia gravis. Moreover, cyclophosphamide is used in most conditioning protocols prior to bone marrow transplantation.

BOX 50.1 DRUGS USED IN IMMUNOPATHOLOGIES

Drug	Properties
Cyclophosphamide	
Target (effects)	DNA (alkylation, interstrand link formation; alteration of the genetic code)
Cellular effects	Inhibition of lymphocyte proliferation; induction of apoptosis
Clinical effects	Antitumor, immunosuppressive (slowdown of disease progression)
Indications	Lymphatic and solid tumors, nodular polyarthritis, autoimmune vasculitis, rheumatoid arthritis, and nephrotic syndrome
Azathioprine	
Target (effects)	DNA and RNA (inhibition of *de novo* synthesis of purines; inhibition of DNA, RNA, and protein synthesis)
Cellular effects	Inhibition of proliferation and cytotoxic effect on T and B lymphocytes
Clinical effects	Prevention/treatment of transplant rejection, immunosuppression (with slowdown of disease progression)
Indications	Organ transplantation, rheumatoid arthritis, intestinal inflammatory disease (in association with GCs), and multiple sclerosis
Methotrexate	
Target (effects)	Dihydrofolate reductase (inhibition of DNA and RNA synthesis)
Cellular effects	Inhibition of T and B lymphocyte proliferation
Clinical effects	Prevention/treatment of GVHD and immunosuppression (with slowdown of disease progression)
Indications	GVHD, rheumatoid arthritis, psoriasis, and systemic lupus erythematosus
Mycophenolate mofetil	
Target (effects)	Inosine monophosphate dehydrogenase (inhibition of *de novo* synthesis of purines)
Cellular effects	Inhibition of lymphocyte proliferation and activity
Clinical effects	Prevention/treatment of transplant rejection
Indications	Organ transplantation (in combination with GC or calcineurine inhibitors), psoriasis, and myasthenia gravis
Thalidomide and lenalidomide	
Target (effects)	Not defined (inhibition of IL-6 and TNFα synthesis and release; enhancement of IL-2 and IFN γ production)
Cellular effects	Antiproliferative, proapoptotic, and antiangiogenic effects
Clinical effects	Anticancer, immunosuppression (with slowdown of disease progression)
Indications	Multiple myeloma and rheumatoid arthritis
Cyclosporine-A and FK506 (tacrolimus)	
Target (effects)	Calcineurine (inhibition of IL-2 and IL-4 synthesis; inhibition of TCR signal transduction)
Cellular effects	Inhibition of T lymphocyte activation and proliferation
Clinical effects	Prevention/treatment of transplant rejection, immunosuppression (with slowdown of disease progression)
Indications	Organ transplantation, rheumatoid arthritis, and psoriasis
Rapamycin (sirolimus)	
Target (effects)	mTOR (target of rapamycin)-kinase (inhibition of IL-2 and other cytokine signaling)
Cellular effects	Inhibition of T lymhpocyte activation and proliferation
Clinical effects	Prevention/treatment of transplant rejection
Indications	Organ transplantation (in combination with GC-o with calcineurine inhibitor)
Glucocorticoids	
Target (effects)	Promoters of genes encoding for many proteins and transcription factors such as AP-1 and NF-κB (inhibition of transcription of many cytokines, chemokynes, and their receptors; inhibition of synthesis and/or release of arachidonic acid and its metabolites; transcription induction of cytokines, transcription factors and modulators with anti-inflammatory function, lymphocytotoxicity)
Cellular effects	Inhibition of humoral and cell-mediated immunity; inhibition of inflammatory response
Clinical effects	Prevention/treatment of transplant rejection, immunosuppression, and anti-inflammatory
Indications	Organ transplantation; most autoimmune and inflammatory diseases
Muromonab CD3 (mouse monoclonal antibody)	
Target (effects)	CD3 (induction of cell-mediated antibody-dependent toxicity and CD3 positive cell complement-dependent cytotoxicity of CD3 positive cell)
Cellular effects	Depletion of T lymphocyte in bloodstream and lymphatic organs
Clinical effects	Prevention/treatment of transplant rejection
Indications	Organ transplantation

BOX 50.1 (Continued)

Drug	Properties
Infliximab, adalimumab, etanercept, certolizumab, and golimumab	
Target (effects)	TNFα (inhibition of TNF receptor activation)
Cellular effects	Inhibition of inflammatory cytokines and adhesion molecules production; inhibition of lymphocyte activity
Clinical effects	Prevention/treatment of transplant rejection, immunosuppression (with slowdown of disease progression)
Indications	Crohn's disease, rheumatoid arthritis, psoriatic arthritis, and ankylosing spondylitis
Basiliximab and daclizumab	
Target (effects)	CD25 (inhibition of CD25 receptor and IL-2 receptor activation, and reduction of α subunit expression)
Cellular effects	Inhibition of T lymphocyte activation
Clinical effects	Prevention/treatment of transplant rejection
Indications	Organ transplantation
Rituximab, epratuzumab, and alemtuzumab	
Target (effects)	CD20, CD22, and CD52, respectively (several effects on lymphocytes through modulation of receptor function, complement-dependent lysis, and cell-mediated death)
Cellular effects	Apoptosis of B lymphocytes
Clinical effects	Antitumor, immunosuppressive
Indications	Non-Hodgkin lymphomas, chronic lymphatic leukemia, rheumatoid arthritis, and organ transplantation
Abatacept, belatacept, and alefacept	
Target (effects)	CD80/86, CD80/86, and CD2, respectively (blockade of T lymphocyte co-stimulatory signal)
Cellular effects	Inhibition of T-cell-mediated response (mainly memory cells) and induction of T lymphocyte apoptosis
Clinical effects	Prevention/treatment of transplant rejection, immunosuppression
Indications	Kidney transplant, rheumatoid arthritis, and psoriasis
Efalizumab	
Target (effects)	CD11a subunit of LFA-1 (blockade of interaction between LFA-1 and adhesion molecules)
Cellular effects	Inhibition of T cell activation, migration, and activity
Clinical effects	Immunosuppression
Indications	Psoriasis
Natalizumab and vedolizumab	
Target (effects)	Integrine α4 (inhibition of T lymphocyte-antigen-presenting cells interaction)
Cellular effects	Inhibition of lymphocyte activation and migration
Clinical effects	Immunosuppression with reduction of disease reactivation
Indications	Multiple sclerosis, Crohn's disease
Omalizumab	
Target (effects)	IgE (blockade of IgE binding to high- and low-affinity receptors)
Cellular effects	Blockade of IgE-mediated activation of basophils, mast cells, T and B lymphocytes
Clinical effects	Anti-inflammatory, prevention of asthma attacks
Indications	Asthma
Anakinra, tocilizumab, and ustekinumab	
Target (effects)	IL-1, IL-6, IL-12/IL-23, respectively (inhibition of the cytokine activity)
Cellular effects	Inhibition of inflammatory response
Clinical effects	Immunosuppression
Indications	Psoriasis and rheumatoid arthritis

Some of the main cyclophosphamide adverse effects are due to its cytotoxic activity on proliferating cells and consist in bone marrow depression with leukopenia and thrombocytopenia, increased susceptibility to infection, alopecia, and damage of germinal tissues in gonads. It may also cause hemorrhagic cystitis, secondary malignant tumors, interstitial pulmonary fibrosis, nausea, vomiting, and anorexia.

Azathioprine Azathioprine is a phase-specific purine antimetabolite, which has been used for many years, in association with prednisone, in the immunosuppressive treatment of transplants. Its immunosuppressive activity is mainly linked to the antiproliferative effect it exerts on lymphocytes.

The 6-mercaptopurine (thionosinic acid) resulting from azathioprine metabolic transformation can be incorporated in DNA and RNA molecules, inhibiting their synthesis. As a consequence, protein synthesis is also inhibited. Azathioprine exerts its action on both T and B lymphocytes and is mainly active on proliferative cells, particularly during the S phase. Cell cycle inhibition results in cell death.

Azathioprine is well absorbed in the gastrointestinal tract and is mainly metabolized to 6-mercaptopurine that is converted by the xanthine oxidase, a critical enzyme in purine catabolism, in the principal inactive metabolite, the 6-thiouric acid. Drugs inhibiting xantine oxidase (i.e., allopurinol) can enhance azathioprine toxicity. Plasma half-life is about 10 min for azathioprine and 1 h for 6-mercaptopurine.

Azathioprine is used in immunosuppressive therapy in organ transplantation, and sometimes in combination with cyclosporine and prednisone (e.g., in heart transplant). It is also used to treat some forms of rheumatoid arthritis that are resistant to other drugs, autoimmune hemolytic anemia and in patients affected by dermatomyositis–polymyositis that are refractory to corticosteroid treatment. It may be used in combination with corticosteroid in ulcerative colitis and Crohn's disease, and to treat chronic GVHD.

Severe adverse effects limit the use of azathioprine. It causes bone marrow depression with leukopenia and thrombocytopenia, gastrointestinal and hepatic toxicity, and nausea and vomiting. It may also increase the susceptibility to infection. However, the main reason for azathioprine-limited application is its mutagenic and cancerogenic activity, resulting in an increased incidence of secondary cancer development after long-term treatments.

Methotrexate Methotrexate is a phase-specific antineoplastic drug inhibiting dihydrofolate reductase. It binds dihydrofolate reductase in a noncovalent way and prevents conversion of dihydrofolic acid to tetrahydrofolic acid. Tetrahydrofolic acid is in turn converted into N^5, N^{10}-methylenetetrahydrofolate a cofactor that is essential for thymidylate, purine, methionine, and glycine synthesis. Lack of dihydrofolic acid production inhibits thymidylate synthase. Therefore, conversion of deoxyuridylic acid in thymidylic acid is prevented and DNA and RNA synthesis cannot proceed.

Methotrexate immunosuppressive action is due to inhibition of T and B lymphocyte proliferation following the arrest of DNA synthesis.

Methotrexate is rapidly absorbed from the gastrointestinal tract, but bioavailability is low; usually more than 25 mg/m^2 are administered intravenously.

In humans, methotrexate metabolism is in general negligible, but accumulation of nephrotoxic metabolites may occur at high dosage. Methotrexate is eliminated mainly by renal excretion.

Main Clinical Application Methotrexate is used as immunosuppressant in combination with cyclosporin A in GVHD prophylaxis of patients undergoing bone marrow transplantation. Today, GVHD does not represent a serious problem as in the past since new procedures are available to remove T lymphocytes from donor bone marrow prior to transplantation. In addition, methotrexate is used to treat rheumatoid arthritis, even in low-dose monotherapy, as well as polymyalgia rheumatica and psoriasis. It can be used also to treat lupus erythematosus and dermatomyositis, as an alternative to GC therapy.

Major adverse effects resulting from low-dose long-term immunosuppressive therapy with methotrexate are different from those observed following antineoplastic treatment (mucositis, nausea, vomiting, diarrhea, anaphylaxis, hepatic necrosis, and kidney failure) and consist in hepatic and pulmonary fibrosis, cirrhosis, and aseptic chronic pneumonia.

Immunosuppressive Drugs with Higher Specificity

In order to obtain more selective and less toxic immunosuppressive drugs, several compounds have been tested. Although they cannot be defined as specific immunosuppressants, they exhibit a lower risk–benefit ratio compared to the aforementioned anticancer agents.

Mycophenolate Mofetil Mycophenolate mofetil (MFM) is the prototype of a class of compounds, such as mizoribine, merimepodib (VX-497), and others still being tested, which act by inhibiting inosine monophosphate dehydrogenase. This enzyme is currently regarded as the most important target of immunosuppressive agents that are more selective and less toxic compared to cyclophosphamide, azathioprine, and others. Some of these compounds exert antiviral activity when combined with IFNα to treat chronic hepatitis C.

MFM is the 2-morpholinoethyl ester of mycophenolic acid (MA); this is an active metabolite that is a reversible and noncompetitive inhibitor of IMPDH and purine *de novo* synthesis. The drug acts on T and B lymphocytes in a moderately selective way as they depend on this metabolic pathway for purine biosynthesis because their hypoxanthine-guanine phosphoribosyltransferase (HGPRT) pathway is poorly active. By contrast, other bone marrow cells use more easily the IGFT pathway and therefore are not significantly sensitive to MFM treatment. Thus, MFM inhibits lymphocyte proliferation and several other lymphocyte activities, like antibody production and leukocyte migration. In particular, the effect on leukocyte migration seems to be due to inhibition of glycosylation of cell surface proteins involved in endothelial cell adhesion.

MFM is orally well absorbed and rapidly hydrolyzed to MA, whose plasma half-life is about 16 h. In liver, MA is then converted in a pharmacologically inactive glucuronide derivative. Such derivative is excreted in the bile and in the intestine can be converted again into MA by bacterial glucuronidases; this produces high concentration of free MA and an enterohepatic circle responsible for the intestinal toxicity of the drug.

MFM in combination with CsA, prednisone or both, is currently used in kidney transplantation in place of the more toxic azathioprine. Its possible application in liver, heart, pancreas, and kidney/pancreas transplantations is also being tested. Moreover, studies are currently underway to assess the effectiveness of MFM either in monotherapy or in association with steroids only, or with steroids and low-dose CsA; such therapeutic schemes may allow to avoid or reduce CsA nephrotoxicity. MFM is also effective in the treatment of rheumatoid arthritis, psoriasis, myasthenia gravis, and other inflammatory-autoimmune conditions.

Given its partially selective action on lymphocytes, the drug has less adverse effects compared to cyclophosphamide and azathioprine, in particular on bone marrow, and therapy interruption is only seldom required. The main effects are gastrointestinal intolerance, thrombocytopenia, leukopenia, and infections.

Thalidomide

Thalidomide is an orally administered anxiolytic–hypnotic drug that was withdrawn from the market because of its teratogenic effects when taken during pregnancy. However, thalidomide has a significant antitumor activity and is an effective immunomodulating agent with a wide range of potential clinical application. Its mechanism of action is not completely understood, but it has been proven to suppress TNFα production and stimulate IL-2 and IFNγ production. Moreover, it exerts antiproliferative, proapoptotic, and antiangiogenic effects. Lenalidomide, a thalidomide analogue, seems to be a more potent immunomodulatory compound with less antiangiogenic effects. Thalidomide has been approved to treat multiple myeloma, and is currently being tested for chronic lymphocytic leukemia. Leflunomide is prescribed to treat rheumatoid arthritis and psoriatic arthritis and as a second-line therapy in multiple myeloma. The antiproliferative effects of lenalidomide are enhanced by concurrent administration of dexamethasone. The adverse effects include, besides teratogenicity, neurotoxicity, hypothyroidism, and increased risk of thrombosis.

Calcineurin Inhibitors

The calcineurin (Cn) antagonists CsA and FK-506 (or tacrolimus) are among the most potent immunosuppressive agents currently available for clinical use; their introduction in therapy represented a major turning point in transplant medicine as it led to a marked increase in the survival rate of transplanted patients as well as to a reduction of graft rejections and infections leading to an improved lifestyle following transplantation. CsA and FK-506, along with rapamycin (discussed later), are also defined as selective immunosuppressive drugs; indeed, they block T lymphocyte and inhibit their activation and proliferation.

Cyclosporin A and FK-506 From a functional point of view, CsA represents the prototype of a number of molecules, either chemically correlated (like the well-studied cyclosporine-A, -B, -C, -D, -E, -G, and -H) and or not chemically correlated (the macrolides FK-506 and rapamycin), exerting a similar effect on T lymphocytes. In particular, CsA and FK-506 share the same mechanism of action and have very similar therapeutic uses and adverse effects, even if FK-506 seems to be 10–100 times more potent than CsA. Rapamycin exerts its effect on T lymphocyte through a different mechanism that will be discussed later.

The immunosuppressive activity of CsA and FK-506 is mainly due to inhibition of T lymphocytes proliferation through blockade of Ca^{+2} dependent regulation of *IL-2* gene transcription, which results in blockade of G_0–G_1 transition. They interact with cytosolic proteins of different molecular weight called immunophilins; in particular, CsA interacts with a group of proteins, called cyclophilins. One of them, cyclophilins A (CyPA), is considered the most important CsA receptor. It is a ubiquitous intracellular protein of 165 amino acids (18 kDa) with peptidilpropil cis–trans isomerase activity that apparently is not involved in CsA action mechanism. Drug–receptor interaction leads to the formation of a macromolecular complex containing two overlapping CyPA pentamers.

In a similar way, FK-506 macrolide binds proteins of different molecular weight (e.g., FK-binding protein, FKBP 12, 25, and 56–59 kDa), and in particular a 12 kDa cytoplasmatic receptor (FKBP12) with peptidylprolyl isomerase activity, like CyPA.

The CsA–CyPA and FK-506–FKBP12 complexes can interact with another cytosolic protein, called Cn, a Ca^{+2}-calmodulin-dependent serine–threonine phosphatase which is essential for TCR signal transduction. Such interaction results in inhibition of Cn phosphatase activity and dephosphorylation of members of the NF-AT (nuclear factor of activated T lymphocytes) family of transcription factors. Dephosphorylation is required for NF-AT translocation from the cytoplasm to the nucleus. There, in combination with other transcription factors like AP-1 and NF-κB, it activates transcription of several genes involved in T lymphocyte activation process, such as IL-2, IL-4, and CD40L encoding genes (Fig. 50.1).

The mechanisms described explain, at least in part, how CsA and FK-506 inhibit T lymphocyte activation and proliferation. However, other mechanisms may be involved, such as inhibition of transcription of other cytokine encoding genes besides IL-2 and IL-4, and inhibition of IL-4 receptor and IL-2 receptor alpha chain synthesis. Recently, it has been shown that CsA and FK-506 are capable of inhibiting the signal transduction pathway mediated by c-Jun N-terminal kinase (JNK) and p38, proteins belonging to the mitogen-activated protein (MAP) kinase superfamily (see

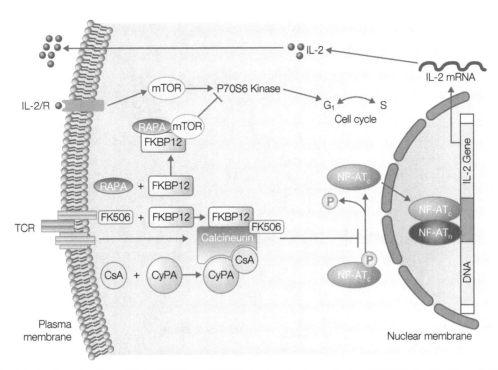

FIGURE 50.1 Mechanism of action of CsA, FK506, and RAPA. In the cytoplasm, CsA and FK506 bind CyPA and FKBP12, respectively. The CsA/CyPA and FK506/FKBP12 complexes inhibit calcineurin (a Ca^{+2}-calmodulin dependent phosphatase) and prevent nuclear translocation of a cytoplasmic subunit of NF-AT required for the expression of IL-2 and other cytokines. RAPA binds FKBP12 too, and the complex RAPA/FKBP12 interacts with and inhibits the mTOR protein kinase, blocking the activation of p70S6 kinase and other kinases controlling cell cycle G_1–S transition. Other mechanisms, not shown in the figure, can contribute to the immunosuppressive effect of these drugs (see text).

Chapter 13) and involved in T cells response to antigen stimulation. It is still unclear whether this mechanism is a consequence of Cn inhibition or not. It is important to remember that FK-506 interacts also with the P56 (56 kDa) immunophylin, which is a component of the macromolecular complex of the glucocorticoid receptor (GR) (Fig. 50.2). FK-506–P56 interaction results in increased GR nuclear translocation, and therefore enhances the transcriptional effects mediated by GCs; this mechanism may contribute to FK-506 immunosuppressive action and provides an explanation for some of its side effects such as hirsutism. All these mechanisms explain how CsA and FK-506 can inhibit both T lymphocyte and T cell-dependent B lymphocyte responses. New therapeutic applications, besides immunosuppression, are currently being investigated for these drugs. In experimental models, it has been demonstrated that CsA and FK-506 have neurotrophic and neuroprotective effects. More recently, CsA and FK-506 have been shown to inhibit myocardial growth in mouse models with cardiac hypertrophy induced by different stimuli; Cn/NF-AT pathway seems to have a crucial role in such pathological conditions. These studies may pave the way to the identification of new therapeutic applications, different from immunosuppression, for CsA and FK-506.

Pharmacokinetics CsA is a liposoluble peptide of 11 amino acids, that can be administered either intravenously in ethanol solution or orally in oil suspension, but whose absorption may vary. The drug reaches its maximum pick in 3–4 h and has a half-life of 10–27 h. It is actively metabolized in the liver to produce at least 18 inactive metabolites, and it is mainly excreted in the bile. FK-506 is available in preparation that can be either injected or orally administered, even if its gastrointestinal absorption is variable and incomplete. Its plasma half-life is about 12 h. It is metabolized in the liver by P450A cytochrome to release at least nine metabolites, some of which active; it is excreted mainly in the bile.

Clinical Use CsA is used in patients undergoing bone marrow, kidney, heart, liver, and lung transplantation, often in association with azathioprine, GCs and, more recently, MFM or rapamycin. It is used also in acute and chronic GVHD, where it is preferred to other immunosuppressants as it does not inhibit proliferation and regeneration of the transplanted bone marrow. Other therapeutic applications of CsA are rheumatoid arthritis (in particular, the severe forms unresponsive to methotrexate) and psoriasis. Other possible applications, when classical therapies fail, are myasthenia

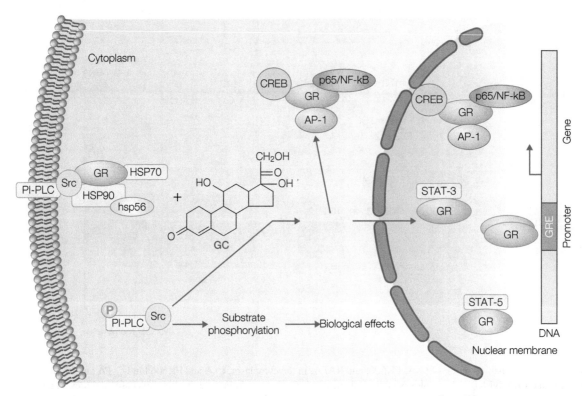

FIGURE 50.2 Mechanism of action of glucocorticoids (GCs). GCs bind a cytoplasmic receptor (GR) causing its release from a macromolecular complex (containing HSP70, HSP90, and immunophilin 56) and its nuclear translocation. In the nucleus, GR interacts as a dimer with the DNA, activating or inhibiting genes whose promoters contain either glucocorticoid response element (GRE) or nGRE consensus sequences respectively. GR exerts its action also by interacting with other transcription factors, either in a negative (AP-1, CREB, and NF-κB) or positive (STAT-5) way. In lymphocytes, inhibition of NF-κB may be caused also by induction of the I-κB inhibitory factor that is responsible for NF-κB inactivation and cytoplasmic localization. Upon glucocorticoid binding, kinase proteins (PI-PLC and Src) may also dissociate from the receptor complex and mediate rapid, biological, not transcriptional effects. Other molecules, such as SRC-1, GRIP-1, CBP/p300, and SWI (not shown in the picture), act as GR transcriptional coactivators.

gravis, atopic dermatitis, intestinal inflammatory conditions, and nephrotic syndromes.

Therapeutic applications of FK-506 in organ transplantation are similar to those described for CsA; it is used to control rejection when CsA is not effective. Preliminary clinical studies seem to indicate that topical application of FK-506 is highly effective in psoriasis, possibly due to its ability to negatively modulate the IL-8/IL-8R system and positively regulate p53 expression, which is involved in the etiopathogenesis of many diseases characterized by inflammation and cutaneous hyperplasia.

Adverse Effects Given the relative specificity of their action mechanism and the absence of effects on rapidly dividing bone marrow cells, CsA and FK-506 do not facilitate development of infections, an event that is often observed using other immunosuppressants. The most relevant side effect, which may limit its therapeutic use, is kidney toxicity that results in vasoconstriction, reduced glomerular filtration rate, interstitial fibrosis, and tubular damage. Some of these effects are a consequence of Cn inhibition, and increased endothelin-1 production, which determine vasoconstriction and induce apoptosis of tubular cells; others, such as interstitial fibrosis, are due to increased TGFβ1 production. Other adverse effects that may occur are neurotoxicity, hypertension, hyperkalemia, hyperlipidemia, hirsutism, gingival hyperplasia, gastrointestinal problems, hyperglycemia, and diabetes (mainly in combination with GCs). One additional issue is the absence of a clear correlation between plasma concentration and adverse effects. Finally, it is important to bear in mind that clinically relevant interactions may occur between CsA or FK-506 and a large number of drugs at different pharmacokinetic stages, and often it is beneficial to monitor the plasmatic level of these drugs, to avoid toxicity due to their accumulation (see Supplement E 3.1).

Rapamycin Rapamycin (RAPA) or sirolimus is a lipophilic macrolide produced by *Streptomyces hygroscopicus*, structurally analogous to FK-506. Like the Cn inhibitors, it exerts a potent immunosuppressive effect by binding immunophilins.

Rapamycin binds the same cytoplasmatic receptor as FK-506, but the molecular target of the RAPA–FKBP12 complex is not calcineurin, but the kinase protein TOR (target of rapamycin) or FRAP (FKBP and rapamycin target). Mammalian TOR (mTOR) belongs to the phosphatidylinositol kinase-kinase (PIKK) family and is involved in the regulation of many cellular functions controlling the balance between protein synthesis and degradation; in this context, mTOR controls the activity of kinases involved in the mitogenic response and cell cycle regulation, such as p70S6 and cyclin-dependent kinases p34cdc2 and p34cdk2, acting downstream in cell cycle regulation. Therefore, inhibition of mTOR kinase activity interferes with signal transduction from cytokine receptors and blocks cell cycle G_1–S transition. RAPA is particularly effective in inhibiting T cell proliferation in response to IL-2. It is important to stress that the action mechanism described previously is the more widely accepted to date, but is not fully demonstrated. Some researchers claim that RAPA effect is not due to direct inhibition of mTOR kinase activity but rather due to the ability of the RAPA–FKBP12 complex to inhibit proliferation of vascular smooth cells; such property makes RAPA particularly effective in preventing artheriopathy coupled to chronic transplant rejection.

Pharmacokinetics RAPA is rapidly absorbed orally and has a plasma half-life of 62 h. It is extensively metabolized by the CYP3A4 enzyme family to produce several metabolites, some of which are active. It is excreted mainly in the bile.

Clinical Use RAPA is used to prevent transplant rejection, often in combination with Cn inhibitors and steroids, or steroids and MFM to avoid CsA and FK-506 nephrotoxicity. It is used in psoriasis, uveoretinitis, and systemic lupus erythematosus. It is also used in stents to be implanted in cardioplastic surgery to prevent stent occlusion by intimal cells. In such application, the drug is adsorbed to the stent using different procedures to obtain a slow and long-lasting release, and restrict the antiproliferative effect of the drug to the intimal muscle cells. Studies are currently being undertaken to assess RAPA effectiveness in cardiovascular diseases, tumors, and diabetes. The RAPA-derivative everolimus is orally administered and is employed in rejection prophylaxis of kidney–heart transplantation. Everolimus and temsirolimus are being evaluated for the treatment of kidney carcinoma.

The most common adverse effect of RAPA is hyperlipidemia (abnormal levels of triglycerides and cholesterol in the bloodstream) that may require pharmacological intervention. It is not nephrotoxic *per se*, but clinical studies have shown that it may increase CsA nephrotoxicity in combination therapy. Other relevant adverse effects are anemia, leukopenia, and thrombocytopenia (possibly due to inhibition of IL-1 signal transduction). Like other immunosuppressants, RAPA can increase the risk of neoplasia (in particular, lymphomas) and susceptibility to infections.

Corticosteroids Corticosteroids with GC activity have been used for many years to treat autoimmune conditions and in immunosuppressive therapies for transplanted patients. Unfortunately, their potent therapeutic effect is associated to a wide range of adverse effects, limiting their clinical use.

Therapeutic activity of GCs relies on their ability to interfere with gene regulation mediated by their intracellular receptors, which act as transcription factors. Several GC receptor (GR) isoforms have been identified, differentially expressed depending on organ and individual. Alternative splicing of exon 9 gives rise to two receptor isoforms, α and β; β can function as a dominant negative. In addition, eight methionines in the initial portion of the translated region can produce 8α and 8β isoforms. Some of these isoforms are mainly cytoplasmatic, whereas others are found in the nucleus. All these isoforms are differentially expressed in different tissues and their expression may vary in the population.

The cytoplasmatic isoforms, and in particular GRα, are found as macromolecular complexes containing several chaperones and other proteins (see Fig. 50.2 and Chapter 24). GR activation is often favored by drug binding. However, it is still unclear what role these drugs play in the regulation of different receptor isotypes. The number of receptor isoforms and their variable expression in different organs and individuals represent a problem for the therapeutic use of GCs because the response to the treatment may vary. Further studies are needed to optimize and personalize the therapeutic approach. Moreover, several polymorphisms have also been identified in the GRs, and some of them can predispose to specific diseases.

Upon steroid binding, GRα undergoes a conformational change resulting in dissociation from the macromolecular complex and exposure of its nuclear localization signal. This allows GRα translocation to the nucleus where it binds, as a dimer-specific DNA sequences called GRE, stimulating the transcription (transactivation) of specific target genes. The dimeric receptor can also bind to specific nGRE sequences and inhibit transcription. Besides these mechanisms of direct *cis* regulation of transcription, indirect mechanisms resulting from interaction between GR and other transcription factors (i.e., NF-κB and AP-1) may also contribute to GC effects (transrepression).

It has also been shown that GCs can have rapid, nontranscriptional effects, which are dependent on GR binding but are partially mediated by activation of Src kinase associated to the receptor and by other kinases included in the macromolecular receptor complex (Fig. 50.2). The real contribution of such nontranscriptional effects to the anti-inflammatory and immunosuppressive activity of these drugs is still unclear.

Selective GR agonists (SEGRAs) are a new set of molecules that have been designed to obtain the GC therapeutic effects without their side effects. These compounds are currently being tested and results are still preliminary. Another interesting development in this area concerns delayed-release formulations that can provide a constant low-level of the drug to minimize adverse reactions.

GCs exert significant regulatory effects on glucose metabolism, salt homeostasis, and modulation of inflammatory and immune responses. In particular, the most relevant therapeutic effects of GCs on inflammatory and immune processes are as follows:

1. Inhibition, by cells in the inflammatory infiltrate, of synthesis of chemotactic compounds and factors mediating the increase in capillary permeability, resulting in inhibition of leukocyte and monocyte recruitment in the areas where inflammatory process occurs. A few hours (4–5 h) after the administration of a single dose of the drug, a rapid and transient reduction in the number of circulating lymphocytes, eosinophils, basophils, and monocytes occurs, with a corresponding increase in the bone marrow. By contrast, following GC administration, the number of circulating neutrophils increases; however, some fundamental functions of these cells are inhibited, such as chemotaxis, phagocytosis, and superoxide anion production. The altered distribution of lymphocytes and monocytes is critical for GC clinical effectiveness in transplant rejection, as rejection depends mainly on T lymphocyte infiltrate formation. Corticosteroids reduce the number of mast cells and inhibit granulation; moreover, GCs inhibit many, if not all, functions of tissue macrophages by inhibiting the release of soluble mediators and synthesis of adhesion molecules. Inhibition of adhesion molecule expression occurs also in endothelial cells.
2. Inhibition of synthesis and/or release of
   a. arachidonic acid (AA) and its metabolites (PG and leukotrienes, LTs, see also Chapter 49), due to both inhibition of PLA_2 synthesis and induction of annexin-1 (also named lipocortin, inhibits PLA_2) synthesis. Inhibition of PG synthesis results also from inhibition of the inducible form of cyclooxigenase (COX-2).
   b. platelet-activating factor (PAF); macrophage-inhibiting factor (MIF), which are capable of provoking macrophage accumulation at the inflammation site; plasminogen activator, that facilitates leukocyte migration to the inflammation site by converting plasminogen into plasmin (see also Chapter 31).
   c. several cytokines (i.e., IL-1, -2, -3, -5, -6, -11, -12, -17, TNFα, and GM-CSF) and their receptors (i.e., those for IL-2, -3, -4, and -12). This action mainly results from inhibition of their transcription but in some cases also from enhanced degradation of their mRNAs.
   d. some cytokines, like IL-8, regulated upon activation T-cell expressed and secreted (RANTES), monocyte chemoattractant protein (MCP)-1, -3, and -4, whose production is increased in patients with inflammatory and autoimmune diseases.
3. Increased TGFβ production, and increased $IL-1_B$ receptor synthesis and release in a free form. The $IL-1_B$ receptor is a "decoy target" that is not capable of transducing the signal but probably acts by sequestering IL-1, thus preventing its binding to the $IL-1_A$ receptor which in contrast can transduce the signal.
4. Induction of lymphocyte apoptosis, which results in thymic aplasia induction (mostly in rodents) because of the massive decrease in thymocyte number occurring upon GC administration *in vivo*.

The cytotoxic effect exerted on lymphocytes is observed on both thymic and peripheral T lymphocytes and is believed to significantly contribute to the immunosuppressive effect; moreover, it represents the rationale for the clinical use of these agents as antitumor drugs to treat leukemia and lymphomas. In general, the effects on T lymphocytes and cellular immunity are much stronger than those on B lymphocytes, and are those mainly responsible for GC immunosuppressive activity.

Pharmacokinetics Being highly lipophilic, GCs are efficiently absorbed orally; by contrast, their hydrophilic esters can be parenterally administered. Prolonged topical application may result in systemic absorption of significant amount of the drug. GCs are extensively metabolized, mainly in the liver, initially through reactions (reductions) producing inactive metabolites, and subsequently by sulfate or glucuronic acid conjugation (see also Chapter 6). Conjugates are mainly excreted by kidneys. Hydrocortisone has a plasma half-life of about 90 min, whereas in general the other compounds have a longer half-life that varies depending on their structure as this affects the conversion reaction rate. It must be noted that, as for most of the drugs acting through transcriptional mechanisms, GC plasma half-life is different (and much shorter) than their biological half-life.

Main Clinical Uses Corticosteroids are used not only in substitution therapies and in leukemia–lymphoma treatments but also as immunosuppressive agents in most autoimmune disease and in transplants. They represent the therapy of choice and can increase the survival rate of patients affected by chronic nonB or nonA and nonB hepatitis. At a high dosage, they represent the treatment of choice (prednisone) for dermatomyositis–polymyositis. They are effective to treat autoimmune hemolytic anemia (in responsive patients) as they inhibit both antibody synthesis and clearance of red cells carrying immunoglobulins attached on their surface. They can temporarily reduce the severity of symptoms in multiple sclerosis and they are effective (prednisone alone or in association with azathioprine or CsA) in the treatment of myasthenia gravis. They are indicated together with sulfasalazine to treat ulcerative colitis and Crohn's disease, and they are administered orally (prednisone) or topically (hydrocortisone microclisma) or parenterally (intravenous hydrocortisone) in most severe cases. They can be used in scleroderma in patients with myositis and pericarditis, in rheumatoid arthritis and systemic lupus erythematosus. Either alone or in combination with cyclophosphamide, GCs are important therapeutic agents for most vasculitis syndromes. They can also be used to limit allergic reactions that may occur after administration of antilymphocyte sera or monoclonal antibodies.

Adverse Effects Corticosteroid toxicity is directly correlated to dosage and duration of the treatment. The main adverse effects observed during immunosuppressive therapy with GCs are increased susceptibility to infections, hyperglycemia, hypertension, edema, glaucoma, hydroelectric imbalance, osteoporosis, body fat redistribution, capillary fragility, gastric ulcer, behavioral alterations, irritability, insomnia, psychosis, and endocrine alterations such as hirsutism, acne, and impotence, which are all symptoms of Cushing's syndrome. In children with posterior subcapsular cataract and growth inhibition or arrest, the most severe side effects are observed during chronic treatment (as in most cases immunosuppressive therapy is).

Further complications are observed upon treatment interruption after a prolonged therapy. For this reason, GC therapy interruption need to be planned over a long-time interval (in some cases, even 1 year), during which the dosage is progressively reduced and in the final stages the drug is administered at alternate days, in order to minimize effects associated with pituitary–adrenal axis suppression.

Dosage and treatment plans for different GCs are in most cases empirically determined; recent studies suggest the possibility to use lower dosage of these agents to reduce risk and severity of adverse effects.

Antibodies as Selective Immunosuppresants

Antibodies against cytokines, cytokine receptors or antigen expressed on immune cells are used as selective immunosuppressive agents. Polyclonal antibodies against lymphocyte antigens have been used for many years to prevent and treat transplant rejection. This has become possible with the advent of the hybridoma technology that has increased purity, specificity, and immunogenicity of the antibodies to be used as immunosuppressive drugs. A further improvement has been the development of chimeric or humanized antibodies; humanization in general consists in the replacement of the mouse Fc region and portion of the Fab segment not involved in ligand binding with the corresponding human regions, while preserving antibody specificity. However, it must be noted that humanized antibodies, although less immunogenic, can still induce the production of neutralizing antibodies. A more promising approach is represented by fully human antibodies obtained with technologies such as selection of antibody fragments from phage expression libraries or transgenic mice expressing the human Ig locus. Following immunization, these mice can produce human-against-human antibodies having very low immunogenicity.

Polyclonal Antibodies Polyclonal sera against lymphocytes (antilymphocyte globulin, ALG) or thymocytes (antithymocyte globulin, ATG) are prepared by repeated inoculation of animals such as horses, rabbits, and goats, with human lymphocytes or thymocytes followed by purification of the immunoglobulin fraction. These preparations contain cytotoxic antibodies that bind CD2, CD3, CD4, CD8, CD11a, CD25, CD44, CD45, and HLA molecules of class I and II on T lymphocyte surface, and reduce lymphocyte activity and number. ALG and ATG are used to treat transplanted patients both to induce immunosuppression, at the beginning of the therapy, and to treat acute rejection in association with other immunosuppressants. The adverse effects of polyclonal antibodies, besides those generally resulting from antibody usage (see later) are extremely variable, as well as their efficacy.

Monoclonal Antibodies Clinical studies have shown that antibodies raised against the CD3 T lymphocyte surface antigen exert a potent immunosuppressive effect. In particular, muromonab-CD3, the first mouse monoclonal antibody introduced in clinical usage, has proved extremely effective in the treatment of kidney transplant rejection, and it has been used for many years. This antibody works by binding the CD3 TCR component, thus inducing internalization of the receptor complex. Its administration results in T cell depletion both in bloodstream and in organs such as spleen and lymph nodes; this effect is due to cell death (following induction of apoptosis and complement activation) as well as to redirection of lymphocytes to nonlymphoid organs such as lungs. Moreover, it reduces the activity of the remaining T lymphocytes, and inhibits the production of the main cytokines, including IL-2. The antibody is currently being used to treat acute transplant rejection. Recently, it has been suggested that it may also be capable of inducing immunological tolerance (see later), opening to new therapeutic applications.

TNFα Antagonists TNFα is a proinflammatory cytokine whose synthesis is increased in several inflammatory/autoimmune pathologies, such as rheumatoid arthritis, psoriasis, and intestinal inflammatory diseases. Its antagonists play a relevant role in clinical practice. Infliximab is a chimeric monoclonal antibody containing the human constant region (Fc) and the mouse variable regions. Infliximab binds TNFα, preventing the production of inflammatory cytokines, including IL-1 and IL-6, and the expression of adhesion molecules involved in lymphocyte activation. Infliximab has been approved for the treatment of Crohn's disease, rheumatoid arthritis, ankylosing spondylitis, and psoriatic arthritis. Adalimumab, a human monoclonal antibody against human TNFα is a new agent approved for rheumatoid arthritis treatment. It is the first fully human monoclonal antibody used in therapy and its immunogenicity is much lower compared to any other antibody in use. We describe here also another TNFα antagonist called etanercept, even if it is not a monoclonal antibody. Etanercept is a chimeric protein that consists of the human Ig constant regions fused to the TNFα receptor; it binds TNFα preventing its interaction with the endogenous receptor. This drug is approved for the treatment of rheumatoid arthritis, psoriatic arthritis, and some forms of juvenile arthritis. certolizumab is a novel humanized monoclonal antibody capable of neutralizing both soluble and membrane TNFα. It is conjugated to polyethylene glycol, and it is indicated for Crohn's disease and, in association with methotrexate, for rheumatoid arthritis. Golimumab, a human monoclonal antibody, has the same action mechanism as certolizumab; it is used in rheumatoid arthritis, psoriatic arthritis, and ankylosing spondylitis.

TNFα antagonists display variable plasma half-life and in general are subcutaneously injected one or two times a week at different dosages. TNFα antagonists are characterized by several severe adverse effects arising in response to infusion or injection, including fever, dyspnea, and hypotension, severe infections, mainly of the airways (i.e., reactivation of tuberculosis and opportunistic infections), lymphoproliferative diseases, iatrogenic lupus, and worsening of cardiac failure.

Other Biological Drugs with Immunomodulatory Activity

Daclizumab (humanized antibody) and basiliximab (chimeric antibody) are antibodies against the IL-2 receptor (CD25). They bind with high affinity to the IL-2 receptor α subunit expressed on T lymphocyte surface, thus blocking all IL-2-mediated effects. Daclizumab has been withdrawn from the market because of its serious toxicity; basiliximab is used in transplant rejection prophylaxis in combination with GC and calcineurin inhibitors, with or without azathioprine or MFM.

Besides those mentioned previously, many other monoclonal antibodies are available as therapeutics; some are currently at late stages of clinical trial, and others have already been approved. It is worth mentioning, for example, rituximab, a mouse–human chimeric antibody that binds CD20 on B lymphocytes and it is used in some forms of lymphomas. Epratuzumab binds B lymphocytes too, but it is directed against CD22 and it is used in non-Hodgkin lymphomas.

Alemtuzumab is a humanized antibody that binds CD52 on mature lymphocytes triggering death by antibody-mediated cytotoxicity, and it is used in chronic lymphatic leukemia.

Some CTLA-4Ig fusion proteins, such as abatacept and belatacept, are indicated in rheumatoid arthritis and prevention of transplant rejection, respectively.

Alefacept is a recombinant fusion protein that blocks CD2 co-stimulatory signal and is approved for the treatment of psoriasis. Efalizumab is a humanized monoclonal antibody against the LFA-1 (CD11a) α subunit; it prevents interaction between LFA-1 and ICAM-1, interfering with T cell activation and cytotoxic T cell migration. It is approved for the treatment of severe forms of psoriasis in adult patients; tocilizumab, a human monoclonal antibody against the IL-6 receptor, has also been recently approved for the same application. A humanized monoclonal antibody, natalizumab, is available for multiple sclerosis and Crohn's disease. It prevents lymphocyte adhesion and migration to inflammation sites by binding to the α4 integrin. Its use is affected by severe side effects, such as risk of progressive multifocal leukoencephalopathy, which seems to result from its effect on T cells in the nervous system. In terms of safety, vedolizumab may represent a good alternative to natalizumab for the treatment of intestinal inflammatory diseases. It is a humanized antibody selectively acting on intestinal T cells with the same mechanism of natalizumab.

Ustekinumab is a human monoclonal antibody blocking IL-12 and IL-23 functions that has been approved for the treatment of psoriasis; it is also under clinical trial for intestinal inflammatory diseases; and for this purpose it may represent a safer alternative to TNFα antagonists.

Omalizumab, a humanized monoclonal anti-IgE antibody is approved for the therapy of allergic asthma in patients unresponsive to inhaled corticosteroids.

Among the agents available for rheumatoid arthritis, it must be mentioned anakinra, an antagonist at IL-1 receptor blocking the cytokine effect.

The fundamental role played by cytokines in the etiopathogenesis of autoimmune diseases suggests that factors involved in transduction of cytokine signaling may represent pharmacological targets for the treatment of such disease. In particular, facitinib, an inhibitor of the transduction pathway mediated by Janus kinases (JAKs) and signal transducer and activators of transcription (STAT) proteins has been approved for the treatment of rheumatoid arthritis and is currently under clinical trial for psoriasis and intestinal inflammatory diseases.

Issues Associated with the Use of Antibodies as Pharmacological Agents

The most frequent adverse effects resulting from antibody use are rhinitis, fever, breathing problems, hypotension, gastrointestinal problems, tremor, lymphoproliferative syndrome, and lymphoma. In the case of anti-CD3 antibodies, a CNS syndrome characterized by hemicranias, photophobia, and fever has been reported, that may be due to secretion of many cytokines. The *in vivo* injection of monoclonal antibodies is further complicated by their immunogenicity and the consequent production of neutralizing antibodies by the patient which may suppress their therapeutic activity. As mentioned before, to reduce or avoid the antigenicity of these agents, it is necessary to employ humanized (human/animal hybrids) or fully human antibodies.

IMMUNOSTIMULANT DRUGS

Immunostimulation is beneficial in all conditions characterized by immunodeficiency and in the treatment of viral and bacterial infections. In addition, immunostimulant agents are used, both in animal models and in clinical practice, in anticancer therapy. It must be noted that the real benefits of immunostimulants in antitumor therapy are still controversial and the mechanisms underlying their activity are not understood. However, the rationale for using immunostimulants in cancer therapy is based on two observations: (i) tumors induce immunodepression and (ii) peptides, produced by specific mutations and capable of triggering immune tolerance (anergia) rather than the immune response, can be found associated with the MHC molecules on tumoral cells. The anergy is at least partially due to the inability of tumor cells to stimulate CD4$^+$ T helper cells and as a consequence to the lack of cytokines.

Although among the therapeutic agents currently available, besides cytokines, there are also microbial products, thyme extracts and chemically defined compounds, clinical observations suggest that their activity as immunostimulants is poor. For this reason, since part of their effects are due to cytokine induction, these agents tend to be replaced by the recombinant cytokines now available for therapeutic use, administered either individually or in combination.

Cytokines

Cytokines are soluble glycoproteins controlling the physiological development of the immune system and the immune response. In these past few years, they have received major attention as clinically relevant immunostimulant drugs. A large number of interleukins (e.g., IL-1 and IL-2) are known, as well as three different families of interferons or IFNs (α, β, and γ), several growth factors (e.g., PDGF, EGF, and TGF), colony-stimulating factors (e.g., GM-CSF, G-CSF, and M-CSF) and chemotactic factors. In theory, given their role in immune response regulation, all cytokines can be considered potential immunostimulatory (or immunosuppressive) drugs. However, more studies are needed to understand whether it is feasible to adopt cytokines as immunostimulants in clinical practice. Here we briefly describe the recombinant cytokines that are currently used as therapeutics (see also Chapter 19).

Interleukin 2 Interleukin 2 (IL-2) is a 15.4 kDa glycoprotein, 133 amino acid long, that can activate and stimulate proliferation by interacting with specific receptors. IL-2 receptor system (IL-2R) consists of three membrane proteins, α, β, and γ (Fig. 50.3). The α chain (55 kDa), typical of the IL-2R, is expressed only on activated T lymphocytes; the β chain (70 kDa), found also in the IL-15 receptor, is constitutively expressed on CD8$^+$ T lymphocytes and its expression is increased upon activation. The γ chain, which is found also in other cytokine (IL-4, IL-7, IL-9, and IL-15) receptors, is constitutively expressed in lymphoid cells. The combination of these three chains, either as dimers or trimers, gives rise to a number of receptors with different affinity for the ligand, with the trimeric receptor (αβγ) having the highest affinity. The β chain is responsible for the transduction of the proliferation signal but only in the presence of the γ chain. It is believed that interaction of the β chain with IL-2 stimulates expression of the α chain and

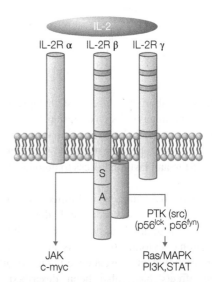

FIGURE 50.3 IL-2 receptor (IL-2R). The three receptor chains α, β, and γ, can be expressed in different combinations to generate dimeric or trimeric receptors with different affinity for IL-2. The trimeric receptor is the one with the highest affinity. β and γ chains are involved in proliferative signaling. The β chain is essential for signal transduction and contains a serine-rich region (S), important for myc induction and JAK activation, and an acidic region (A), important for tyrosine kinase activation and signal transduction.

induces the assembly of the three chains in the trimeric high affinity receptor. The IL-2 receptor does not possess an intrinsic kinase activity; however, it can rapidly induce tyrosine phosphorylation of several proteins, including the β chain itself. This occurs because, upon ligand binding, the cytoplasmic (acidic) segment of the β chain can interact and activate tyrosine kinases belonging to the *src* family ($p56^{lck}$ and $p59^{fyn}$), whose activation is correlated to the activation of several transduction pathways, such as the JNK–STAT pathway that includes several transcription factor, the Ras-MAP kinase and the PI3 kinase pathways.

IL-2 has significant effects on:

1. proliferation, as it promotes entry into the G1 phase and activation of gene products required for progression to the S phase. In this way, it stimulates proliferation of T and B lymphocytes and NK cells;
2. activation, by enhancing IL-2R expression, B lymphocyte antibody production, tumoricidal activity of T lymphocytes and NK cells (increase of LAK activity), IFNγ production which in turn is responsible for macrophage and granulocytes activation;
3. on survival, by inhibiting and/or enhancing lymphocyte apoptosis.

Clinical Use A human IL-2 recombinant product (adesleukin) is used in clinical practice. Adesleukin mediates the same biological effects as the native protein but differs from it in that it contains some amino acid substitutions and it is not glycosylated. It is used in therapy as immunostimulant, via a systemic, or regional or local route of administration, to treat patients affected by neoplasia or immune deficiency. In cancer therapy, IL-2 is currently used only individually or in association with IFNα.

Because of its short half-life, most therapeutic schemes involve either continuous infusion or multiple intermittent dosing. Protocols are currently being tested in which IL-2 is either included in liposomes or conjugated to polyethylene glycol to extend its half-life and favor its access to immune cells.

Adverse Effects IL-2 therapy is limited by a number of toxic effects mainly observed at high dosage. The most important is the "capillary leak syndrome" which is probably due to induction of cytokine (e.g., IFNs and TNFα) secretion and whose most evident consequence is pulmonary edema. Other side effects have also been described, such as hypotension, cardiovascular, renal and hematological toxicity, neurotoxicity, fever, nausea and vomiting, diarrhea, fatigue, and discomfort and cutaneous rash.

Interferons IFNs are cytokines with antiviral, immunomodulatory, and antiproliferative effects. IFNα is mainly produced by leukocytes, IFNβ by fibroblasts and epithelial cells, whereas IFNγ is produced by T lymphocytes and NK cells in response to antimitogenic stimuli, mitogens and cytokines. IFNs exhibit different biological activity depending on the class they belong: IFNα has potent antiviral and antitumor effects, whereas IFNγ has a stronger immunomodulatory activity as it triggers macrophage and monocyte activation, induces class II MHC expression, and enhances the cytotoxic activity of NK cells and T lymphocytes.

Among the IFN effects, the antitumor and antiviral are the most exploited in therapy. The mechanisms underlying IFN antitumor activity are not completely understood, but some events appear to be critical for it, such as (i) inhibition of the expression of proto-oncogenes like c-myc, c-fos, and c-H-ras; (ii) reduced availability of intracellular nutrients like tryptophan; and (iii) enhanced expression of tumor cell antigens favoring their identification and disposal.

Clinical Use IFNα therapy is effective in some tumors of the immune system such as hairy cell leukemia, T cell cutaneous lymphoma, chronic lymphocytic leukemia at early stages, and low and intermediate grade malignancy non-Hodgkin' lymphomas. It is also used in colon carcinoma, in bladder cancer and genital cancers induced by papilloma virus. It can be employed, but with poorer therapeutic results, also in multiple myeloma, metastatic melanoma, and Kaposi's sarcoma associated to AIDS. IFNγ is used to treat chronic granulomatous disease, metastatic forms of kidney and ovary carcinoma, and melanoma and T lymphomas. Both IFNα and INFγ are active on chronic myeloid leukemia.

The effects we have just described seem to be due to an antiproliferative action exerted directly on tumor cells rather than to the immunostimulant activity.

IFNβ has been recently introduced in therapy to treat multiple sclerosis (MS) with some positive results. Although the mechanism underlying such therapeutic effect is not completely understood yet, it seems to be due to the antiviral or antiproliferative effects IFNβ exerts on autoreactive lymphocytes, as well as to modulation of cytokine production which is altered in MS patients, or to effects on transcription factors (e.g., NF-κB) involved in lymphocyte activation.

Adverse Effects The most common adverse effects at the beginning of the treatment consists in flu symptoms, like fever, muscle aches, headaches, cold, and gastrointestinal problems; these are generally treated with antipyretics (like paracetamol) and gradually disappear. The most serious side effect, limiting systemic high-dosage therapies is bone marrow depression with thrombocytopenia and granulocytopenia. Neurotoxicity, cardiotoxicity, and kidney dysfunction may also arise.

G-CSF, GM-CSF, and IL-3 In the context of immunostimulant therapies, hematopoietic growth factors, like G-CSF, GM-CSF, and IL-3, are used to reduce leukocytopenia induced by anticancer chemotherapies to accelerate myeloid regeneration and reduce risk of infections. This type of therapy may also allow increasing the dosage of the chemotherapeutic agents.

G-CSF has been approved for the prophylaxis of neutropenia induced by chemotherapy as not only it can enhance granulocyte proliferation and differentiation, but it can also improve mature cell functionality (chemotaxis and antibody-dependent cytotoxicity). GM-CSF is used in autologous bone marrow transplantation and after intensive chemotherapy to promote myeloid recovery. A conjugation product, GM-CSF-polyethylene glycol, is also available; this is more resistant and has a longer half-life, but it is probably less effective on immature bone marrow cells.

Some recently developed molecules seem to open up new therapeutic perspectives for growth factors in bone marrow transplantation and chemotherapy; among them, we remember: thrombopoietin, a human recombinant cytokine stimulating megakaryocytopoiesis, and an IL-3/GM-CSF fusion product that increases circulating elements including platelets.

Cytokines as Immunoadjuvants

The administration of cytokines as adjuvants represents another possible therapeutic application to enhance the immune response. When administered to healthy voluntaries, IFNα and IFNγ have demonstrated an adjuvant activity in vaccination against *Plasmodium falciparum*. IL-2, IFNα, and IFNγ have shown a clear adjuvant effect when coupled to vaccines against hepatitis B virus, as demonstrated by an increased antibody titer compared to the controls inoculated with the vaccine only. This is just one of the possible applications of cytokines in the immunostimulant therapies that are currently being studied and might be available in the future.

New Molecular Targets: Phosphodiesterase 4 and p38 MAP Kinases Inhibitors

Starting from the assumption that inhibiting the production of inflammatory cytokines, and in particular TNFα, could be beneficial in all conditions characterized by cytokine accumulation, attempts have been made to generate selective compounds capable of blocking TNFα synthesis. Several classes of drugs have been obtained, which block cytokine production through different mechanisms. One group exerts its action by inhibiting phosphodiesterase 4 (PDE4) production. These enzymes are responsible for cAMP conversion into AMP, and their inhibition results in intracellular accumulation of cAMP and increased protein kinase A (PKA) activity.

The PDE4 enzymatic isoform is the most expressed in inflammatory cells, where its inhibition and the consequent cAMP increase block activation of transcription factors, including NF-κB, and production of mediators of inflammation, such as TNFα. cilomilast, and ruflomilast are selective PDE4 inhibitors that can be effective in inflammatory airways diseases. In addition, it is known that the signal transduction pathway leading to TNFα production is also controlled by proteins of the MAP kinase family, in particular p38 MAP kinases (see also Chapters 13 and 32). They belong to a group of proteins that phosphorylate serine and threonine residues, such as extracellular signal-regulated kinase(ERK) and c-Jun N-terminal kinase (JNK). Drugs capable of inhibiting p38 MAP kinases cause inhibition of TNFα, IL-1, and nitric oxide synthesis. The most promising compounds belonging to this class (SCIO 469-talmapimod, VX-702, and SB-203580) are currently being tested in rheumatoid arthritis and intestinal inflammatory disease. It must be noted that all these drugs are small molecules orally administered, highly diffusible, nonimmunogenic, with much lower production costs compared to other TNFα antagonists, such as monoclonal antibodies or soluble peptides.

TAKE-HOME MESSAGE

- Some immunosuppressive drugs act by inhibiting cell proliferation; new drugs selectively inhibit B and/or T lymphocyte proliferation.
- GCs, calcineurin inhibitors, and rapamycin act by interfering with signals generated by antigens and cytokines, and by inhibiting the production of cytokines/interleukins, especially IL-2.
- Mono- or polyclonal antibodies against various CDs, HLA molecules, cytokines or their receptors are used for immunosuppressive therapies or in the control of neoplastic disorders of the immune system.
- Substances obtained from bacteria or extracted from thymus have immunostimulating activity.

FURTHER READING

Kadmiel M., Cidlowski J.A. (2013). Glucocorticoid receptor signaling in health and disease. *Trends in Pharmacological Sciences*, 34 (9), 518–530.

Keogh B., Parker A.E. (2011). Toll-like receptors as targets for immune disorders. *Trends in Pharmacological Sciences*, 32, 435–442.

Kopf M., Bachmann M.F., Marsland B.J. (2010). Averting inflammation by targeting the cytokine environment. *Nature Reviews. Drug Discovery*, 9, 703–718.

Lichtenstein G.R. (2013). Comprehensive review: antitumor necrosis factor agents in inflammatory bowel disease and factors implicated in treatment response. *Therapeutic Advances in Gastroenterology*, 6 (4), 269–293.

Musson R.E., Smith N.P. (2011). Regulatory mechanisms of calcineurin phosphatase activity. *Current Medicinal Chemistry*, 18, 301–315.

Sathish J.G. Sethu S., Bielsky M.C., De Haan L., French N.S., Govindappa K., Green J., Griffiths C.E.M., Holgate S., Jones D., Kimber I, Moggs J., Naisbitt D.J., Pirmohamed M., Reichmann G., Sims J., Subramanyam M., Todd M.D., Van Der Laan J.W., Weaver R.J., Park K. (2013). Challenges and approaches for the development of safer immunomodulatory biologics. *Nature Reviews. Drug Discovery*, 12, 306–324.

Shea J.J., Holland S.M., Staudt L.M. (2013). JAKs and STATs in immunity, immunodeficiency, and cancer. *New England Journal of Medicine*, 368, 161–170.

Steinman L., Merrill J.T., McInnese I.B., Peakman M. (2012). Optimization of current and future therapy for autoimmune diseases. *Nature Medicine*, 18 (1), 59–65.

Tracey D., Klareskog L., Sasso E.H., Salfeld J.G., Tak P.P. (2008). Tumor necrosis factor antagonist mechanisms of action: a comprehensive review. *Pharmacology & Therapeutics*, 117, 244–279.

51

MECHANISM OF ACTION OF ANTI-INFECTIVE DRUGS

FRANCESCO SCAGLIONE

> **By reading this chapter you will:**
>
> - Learn the molecular mechanisms of action of antibacterial, antimycotic, and antiviral drugs
> - Learn the molecular mechanisms underlying the resistance to antibacterial, antimycotic, and antiviral drugs
> - Learn the notions to classify anti-infective drugs according to target and action mechanism

The twentieth century saw the birth of the "antibiotic era," with the beginning of the antimicrobial therapy. The identification of almost all pathogens and their pathogenetic mechanisms, and the introduction of antimicrobial agents have led to the possibility to control infectious diseases, something that was unthinkable until the nineteenth century. However, the use of antimicrobials has also raised new pharmacological issues such as antibiotic resistance, drug interactions, and toxicity. These issues are related both to the improper use and to the abuse of these drugs; it has been estimated that 100,000 tons of antibiotics are used every year. Although antimicrobials are a large group of medicines including agents with different structure and mechanism of action against bacteria, viruses, fungi, and parasites, some general and important considerations on their use can be done. In this chapter, we will describe the general classes of antimicrobial drugs, their mechanisms of action and resistance. We will briefly discuss the principles for the correct use of antibiotics.

Clinically relevant microorganisms belong to three categories: bacteria, viruses, and fungi. Thus we will start by classifying antimicrobial drugs based on the pathogens they act on (i) antibacterials, (ii) antufungals, and (iii) antivirals.

Within each of these large categories, drugs can be distinguished in different families according to their chemical features and biochemical properties. The molecules of antimicrobial drugs should be considered as ligands whose receptors are microbial proteins that are inactivated by ligand binding, leading to death of the microorganism. The main characteristic of antimicrobials is their selectivity for the microbial rather than the organism cells.

ANTIBACTERIAL DRUGS AND THEIR MECHANISMS OF ACTION

To exert its action, an antibiotic needs to be able to

- enter the bacterial cell and reach its target site,
- bind to the molecular target, which is involved in a process essential for bacterial viability,
- inhibit the process in which the target molecule plays a fundamental role,
- evade modifications by the bacterial cell.

Bacterial drugs are distinguished in bacteriostatic and bactericidal depending on whether they can simply prevent bacterial proliferation or can cause death of the microorganism. However, the difference is somehow arbitrary since this property depends on (i) phase of bacterial growth (e.g., penicillin is bactericidal only on actively replicating cells), (ii) antibiotic concentration, and (iii) bacterial species the molecule acts on.

Also in clinics, often no differences are observed between bacteriostatic and bactericidal agents. Nevertheless, bactericidal drugs are believed to offer some additional advantages

---

General and Molecular Pharmacology: Principles of Drug Action, First Edition. Edited by Francesco Clementi and Guido Fumagalli.
© 2015 John Wiley & Sons, Inc. Published 2015 by John Wiley & Sons, Inc.

TABLE 51.1 Mechanism of action of antibiotics in clinical use

Antibiotics	Mechanism of action	Examples
Compounds acting on cell wall and membrane integrity		
β-lactam		
Penicillins	Inhibition of cell wall synthesis at the level of transpeptidases (PBPs)	Penicillin G, Ampicillin, Ticarcillin, Piperacillin, Amoxicillin
Cephalosporins	As above	Cefotaxime, Cephalexin, Cefaclor, Ceftriaxone, Ceftazidime
Cephamycins	As above	Cefoxitin, Cefbuperazone
Carbapenems	As above	Imipenem, Meropenem
Monobactams	As above	Aztreonam
Clavam	β-lactamase inhibition	Clavulanic acid (+amoxicillin), Sulbactam (+ampicillin), Tazobactam (+piperacillin)
Peptidic		
Glycopeptides	Inhibition of cell wall synthesis via binding to the d-Ala-d-Ala terminal peptide of the peptidoglycan precursor	Vancomycin, Teicoplanin, Telavancin
Polypeptides	Inhibition of dephosphorylation and regeneration of lipid carrier required for peptidoglycan synthesis	Bacitracin
Cyclic polypeptides	Alteration of cell membrane	Polymyxin, Colistin
Lipopeptides	Depolarization of bacterial membrane and inhibition of protein synthesis	Daptomycin
Phosphomycin	Inhibition of MurA during the first step of peptidoglycan synthesis	Phosphomycin
Cycloserine	Inhibition of D-alanine racemase and D-alanine-D-alanine synthetase	Cycloserine
Compounds acting on nucleic acid synthesis		
Quinolones	Inhibition of DNA replication via interaction with DNA gyrase and topoisomerase IV	Nalidixic acid, Norfloxacin, Pefloxacin, Ciprofloxacin, Levofloxacin, Moxifloxacin
Coumarin	Inhibition of DNA synthesis via binding to DNA gyrase B subunit	Novobiocin
Dihydrofolate reductase inhibitors	Inhibition of folic acid metabolism	Trimethoprim
Sulfonamides	Inhibition or conversion of para-amino-benzoic acid (PABA) to dihydrofolate	Sulfamethoxazole, Sulfadiazine, Dapsone, para-aminosalicylic acid (PAS)
Rifamycins	Inhibition of RNA synthesis via formation of DNA–RNA complex	Rifampicin, Rifabutin, Rifaximin
Inhibitors of protein synthesis		
Tetracyclines	Binding to ribosome 30S subunit and blockade of peptide elongation	Tetracycline, Doxycycline, Minocycline
Macrolides		
14- and 15-membered rings	Binding to ribosome 50S subunit and blockade of peptide translocation	Erythromycin, Clarithromycin, Azithromycin, Roxithromycin
16-membered rings	As above	Josamycin, Miokamycin, Spiramycin Rokitamycin
Lincosamides	As above	Lincomycin, Clindamycin
Streptogramins (A and B)	Binding to ribosome 50S subunit and inhibition of peptide translocation and peptide bond formation	Pristinamycin, Quinupristin-dalfopristin
Aminoglycosides	Irreversible binding to ribosome 30S subunit, blockade of mRNA translation and translation mistakes	Streptomycin, Neomycin, Kanamycin, Tobramycin, Gentamicin, Amikacin, Netilmycin, Isepamycin
Fusidanes	Inhibition of elongation factor G (EF-G). No peptide translocation	Fusidic acid
Pseudomonic acid	Inhibition of isoleucine-tRNA synthetase	Mupirocin
Chloramphenicol	Binding to ribosome peptidyl transferase center and blockade of peptide translocation	Chloramphenicol
Oxazolidinones	Binding to ribosome 50S subunit with inhibition of initiation complex formation and blockade of mRNA translation	Linezolid, Posizolid, Tedizolid, Radezolid, Cycloserine

Table 51.1 (*Continued*)

Antibiotics	Mechanism of action	Examples
Antibiotics causing macromolecular damages		
Nitrofurans	Damages to proteins and nucleic acids via multiple mechanisms including hydroxyl radical production and DNA-adduct formation	Nitrofurantoin
Nitroimidazoles	DNA damages via free radical production	Metronidazole

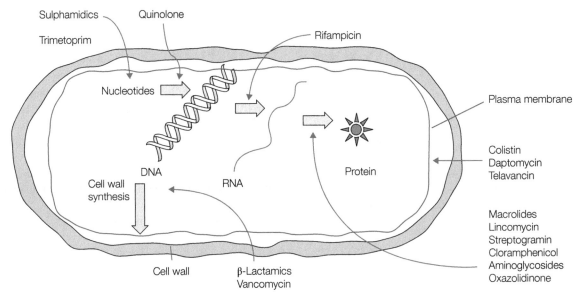

FIGURE 51.1 Mechanism of action of antibacterial drugs. The antibacterials currently available can be grouped according to the biological processes they interfere with. Thus, we can distinguish antibacterials acting on: cell wall synthesis, replication, transcription and translation of nucleic acids, cytoplasmic membrane functionality, and cell metabolism.

in case of neutropenic patients or in situations of poor tissue penetration of the antibiotic, such as in the case of endocarditis.

The bacterial functions inhibited by the antibiotics currently available for therapeutic use can be schematically subdivided as follows (see Table 51.1 and Fig. 51.1):

- Synthesis of cell wall peptidoglycan
- DNA transcription and RNA translation
- DNA synthesis and replication
- Cytoplasmic membrane functions
- Cell metabolism

Inhibitors of Peptidoglycan Synthesis

Bacterial cell wall is essential for bacterial growth and development. Peptidoglycan is a heteropolymeric component of the rigid cell wall which provides mechanical stability through its highly reticulated structure. It comprises glycan chains, formed from linear filaments of two alternating amino sugars, N-acetylglucosamine (GlcNAc), and N-acetylmuramic acid (MurNAc), cross-linked to peptide chains whose composition varies depending on the bacterial strain (e.g., in *Escherichia coli* the composition is L-ala+D-glu+m-DAP+D-ala-D-ala). The peptidoglycan unit thus comprises a heteropolymer GlcNAc–MurNac to which short peptides are linked (to MurNAc). During cell wall formation, two peptide chains of adjacent units are covalently linked to form overlapping layers that can be up to one hundred in Gram-positive bacteria, whereas they are only one or two in Gram-negative (Fig. 51.2). During cell division and growth, the existing peptidoglycan is cleaved and new material is inserted. This process is extremely delicate and requires a constant dynamic balance. When such balance is altered by the action of an antibiotic, the cell wall can no longer control the internal cell pressure. In a hypotonic environment, this leads to cell swelling and lysis.

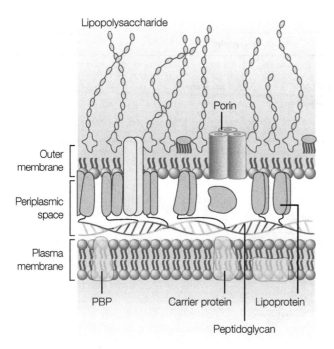

FIGURE 51.2 Structure of Gram-negative bacteria cell wall and plasma membrane. Gram-positive bacteria do not have an external membrane and have multilayer peptidoglycan. PBP, penicillin-binding protein.

Peptidoglycan synthesis occurs through tens of enzymatic reactions taking place in distinct cellular compartments (Fig. 51.3).

The first reaction occurs in the cytoplasm and consists in the transfer of phosphoenolpyruvate (PEP) to uridine diphosphate-*N*-acetylglucosamine (UDP-GlcNAc), catalyzed by the enzyme N-acetylglucosamine-3-*O*-enolpyruvyl transferase (MurA) that can be inhibited by phosphomycin. The next reaction is a reduction that leads to uridine diphosphate-*N*-acetyl-muramic (UDP-MurNAc) and requires NADPH. Some amino acids (5 in *E. coli*, 10 nonlinear in *Staphylococcus aureus*) are added to UDP-MurNAc, of which the last two are always D-alanines. L-alanine racemization and formation of the dimer that is then added to the other amino acids is catalyzed by the enzyme D-alanyl-D-alanine synthetase, that can be inhibited by cycloserine, an antibiotic structurally analogous to D-alanine. The UDP-MurNAc-peptide binds to the lipid transporter bactoprenol present in the cytoplasmic membrane. Here, another GlcNAc sugar is added to the UDP-MurNAc-peptide and the peptide disaccharide (MurNAc-GlcNAc-peptide) is formed, which represents the peptidoglycan monomer. This is then translocated by the lipid transporter to the cytoplasmic membrane external side. The lipid carrier (lipid-PP, Fig. 51.3) is then reused to transport other disaccharide units.

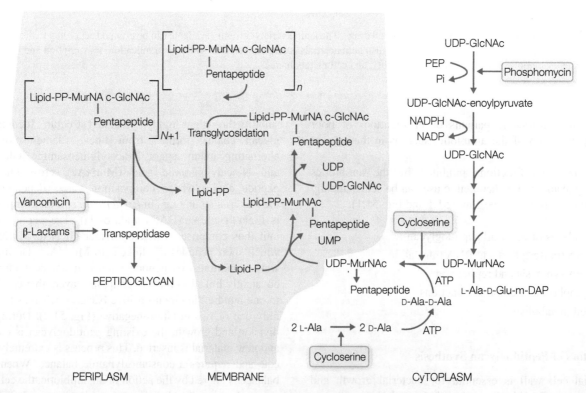

FIGURE 51.3 Diagram of peptidoglycan biosynthesis and site of action of some antibiotics that interfere with cell wall deposition. The gray shadow indicates the plasma membrane (internal membrane in Gram-negative bacteria).

Phosphomycin Phosphomycin is a broad-spectrum bactericidal antibiotic and is the only member of this class to be used in clinics. Phosphomycin is actively transported inside the bacterial cell by the hexose 6-phosphate and L-γ-glycerophosphate transport systems, and can reach very high intracellular concentrations.

Phosphomycin acts as a phosphoenolpyruvate analog, and covalently and irreversibly binds the catalytic site of MurA, the enzyme that transfers phosphoenolpyruvate to N-acetylglucosamine.

Cycloserine D-cycloserine acts as a competitive inhibitor of two enzymes that work sequentially in D-alanyl-D-alanine synthesis: the L-alanine racemase and the D-alanyl-D-alanine synthetase. D-cycloserine is a structural analog of D-alanine and its efficacy is inhibited by high concentrations of the amino acid.

Inhibitors of Peptidoglycan Polymerization

The extracellular part of peptidoglycan synthesis involves transglycosylation reactions, which allow monomer elongation, and transpeptidase reactions, which produce the peptidoglycan polymer.

The transpeptidase reaction involves always the second-last of the two D-alanines and another amino acid of an adjacent peptide chain (m-DAP in *E. coli*). The terminal D-alanine is released. This transpeptidase reaction is inhibited by β-lactams and glycopeptides, with different mechanism.

Glycopeptides Glycopeptides (vancomycin and teicoplanin) were discovered in the 1950s as secondary metabolites of two soil *Streptomyces*. They consist of a large eptapeptide core and several sugars or aminosugars. The antimicrobial spectrum of glycopeptides is limited to Gram-positive bacterial species. In fact, being large polar molecules, they cannot easily penetrate through the lipid external membrane typical of Gram-negative bacteria and therefore they cannot reach their specific targets on peptidoglycan.

These antibiotics inhibit peptidoglycan polymer synthesis by forming a high affinity molecular complex between their glycopeptide target and the terminal D-alanine dimer of the glycan pentapeptide; as a result, steric hindrance prevents the transglycosylase activity required to link the precursors to the nascent peptidoglycan. Similarly also the transpeptidase activity that terminate bacterial cell wall synthesis is prevented (Fig. 51.3).

β-Lactams This is the largest class of antibiotics characterized by the presence of a four-atom azetidinone ring (β-lactam nucleus) whose integrity is essential for antibacterial activity: metabolic transformations or chemical alterations of this portion of the molecule abolish its biological activity.

Over the past 40 years the class of β-lactam antibiotics (penicillins, cephalosporins, and monobactams) has undergone an exceptional development: more than 100 molecules belonging to this class have entered clinical practice (Table 51.1).

The isolation of the penamic nucleus and the cephemic nucleus has lead to an expansion of this drug class, since changes introduced in the nucleus and addition of lateral chains at various positions have allowed to obtain semisynthetic derivatives endowed with antibacterial activity and pharmacokinetic behavior completely different from those of the original molecule.

The cellular targets of β-lactams are transpeptidases, enzymes that catalyze peptide bond formation between amino acid chains of adjacent peptidoglycan polymers. These enzymes, commonly named penicillin-binding proteins (PBPs, Fig. 51.2), are associated to the plasma membrane and have the catalytic site facing the outside. PBPs also include bacterial proteins that are bound by β-lactam antibiotics but do not have transpeptidase function and are involved in multiple functions such as bacterial shape maintenance and proliferation (carboxypeptidase and endopeptidase).

Penicillins, like other β-lactam antibiotics, are structural analogs of the D-alanyl-D-alanine dipeptide; they acylate serines in the PBP active site (with β-lactam ring opening at the level of the CO–N amide bond) forming inactive derivatives (Fig. 51.4).

Inhibition of the transpeptidase reaction leads to weakening of the cell wall and ensuing bacterial lysis, probably with the contribution of autolysins (also called murein hydrolases) which are normally expressed during cell division processes.

The action of β-lactams (like that of other cell wall inhibitors) strongly depends on the bacterial growth rate: rapidly growing cells are killed more rapidly than those slowly multiplying, and the bacteriolytic effect is absent in

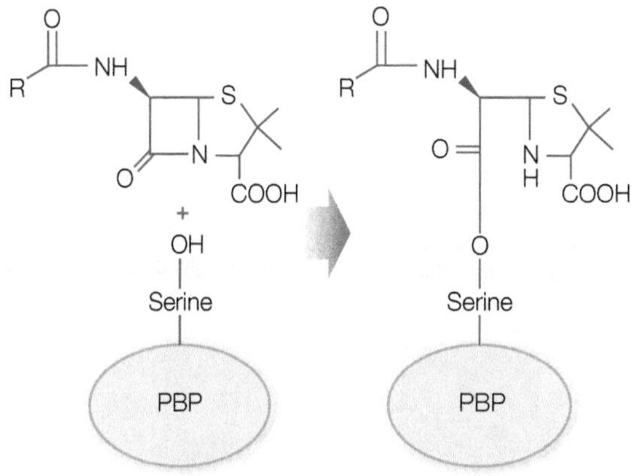

FIGURE 51.4 Inhibition of penicillin-binding proteins (PBPs) via acylation of the serine-hydroxyl group by a β-lactam antibiotic.

MECHANISM OF ACTION OF ANTI-INFECTIVE DRUGS

quiescent cells. Inhibition of other PBPs, different from transpeptidases, is probably responsible for the other alterations of bacterial biology observed, such as appearance of filamentous cells.

Transcription Inhibitors

In bacteria, transcription and translation are coupled. The molecular differences existing between the prokaryote and eukaryote processes are sufficient to guarantee the marked antibacterial selectivity of the drugs belonging to this class. The molecular differences existing between the prokaryote and eukaryote protein synthesis processes are summarized in Table 51.2.

Rifamycins Rifamycins, also called ansamycins, are a family of antibiotics discovered in Italy and produced by the bacterial strain *Streptomyces mediterranei*. They contain a naphtohydroquinone chromophore, with a long aliphatic chain which bridges two positions of an aromatic nucleus. They are active against *Mycobacterium tuberculosis* and other bacteria. Rifaximin is different from the others in that it is not absorbed in the gastroenteric tract and therefore is used only to treat gastrointestinal infections.

Rifamycin blocks transcription in bacterial cells by interfering with RNA polymerase, a four subunit complex (2α, β, and β'). It has been demonstrated that the specific site of action is the β subunit, although rifamycins can stably bind only to the holoenzyme or to a partially reconstituted complex. Therefore, they block transcription initiation, preventing the interaction between enzyme and first nucleotide, but they do not interfere with RNA synthesis once it has started; in this way, they prevent synthesis of new RNA molecules, and ultimately of new proteins. This mechanism of action has been demonstrated in different bacterial species. Rifamycins do not inhibit mammalian RNA polymerase but, at very high concentrations not reached with normal dosage, they can inhibit mitochondrial RNA synthesis.

Rifamycins can also inhibit RNA polymerase of some DNA viruses, but at relatively high concentrations (in the order of 100 mg/l), and also inverse transcriptase.

Translation Inhibitors

Protein synthesis is a complex process in which three phases can be distinguished: tRNA activation, elongation, and termination of the peptide chain (Fig. 51.5). The first phase involves 30S ribosomal subunit, a macromolecular complex formed by 21 proteins and one molecule of the 16R ribosomal RNA (rRNA); activated tRNA; mRNA; and initiation factors IF1, IF2, and IF3. The 50S ribosomal subunit participates in the translocation of the first activated tRNA to the ribosome P site (where peptide synthesis starts). The second phase (elongation phase) requires delivery of a new aminoacyl-tRNA to the ribosome A site (acceptor site), with the contribution of the elongation factors EF-Tu and EF-Ts.

TABLE 51.2 Differences in protein synthesis between eukaryotic and prokaryotic cells

	Eukaryotes	Prokaryotes
Ribosome size	80S	70S
rRNAs	5S, 29S, 18S	5S, 23S, 16S
First aminoacyl-tRNA in translation initiation	Met-tRNA	F-met-tRNA
Initiation factors	eIF1, eIF2, eIF3, eIF4	IF1, IF2, IF3 (different from eukaryotic)
Elongation factors	EF-1α, EF-1β, EF2	EF-Tu, EF-Ts, EF-G (different from eukaryotic)
Termination factors	eRF	RF1, RF2, RF3

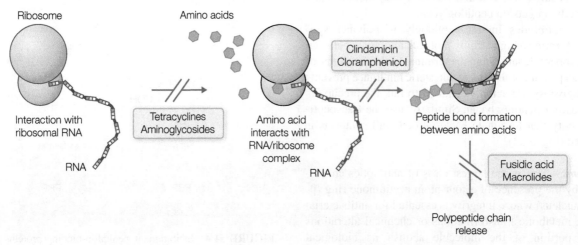

FIGURE 51.5 Main steps in protein synthesis and their inhibition by antibiotics.

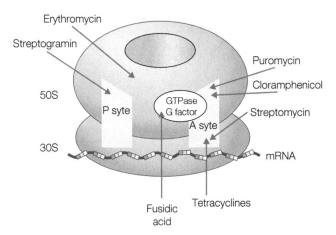

FIGURE 51.6 Bacterial ribosome and binding sites for antibacterial drugs. Site A: the acceptor site, which is generally occupied by the aminoacyl tRNA. Site P: the puromycin-sensitive site, which is generally occupied by the peptidyl-tRNA.

The peptidyltransferase activity of the 50S subunit catalyzes the transfer of the new amino acid to the amino group of the last amino acid of the growing peptide chain on the P site. The EF-G factor allows the ribosome to move one codon down the mRNA. The unloaded tRNA is then released. In the third phase (termination phase), stop codons signal the end of the translation process, causing the recruitment of the release factors: RF1, RF2, and RF3.

Several antibiotics inhibit bacterial translation by binding to specific ribosome components. Figure 51.6 schematically depicts a bacterial ribosome, showing binding sites and mechanisms of action of several antibiotics. The bacteriostatic and bactericidal effects of these antibiotics are consequence of their inhibition and/or alteration of bacterial synthesis.

Tetracyclines Tetracyclines are broad-spectrum, bacteriostatic antibiotics that inhibit bacterial protein synthesis by binding the 30S ribosome subunit and blocking tRNA access to the A site. Among the first broad-spectrum antibiotics to be developed, their clinical use is now significantly reduced because of the spreading of bacterial resistance resulting from their large use in infectious disease and as additives in food for stock animals. They are still useful in the treatment of rickettsia, chlamydia, and mycoplasma infections and in airway infections.

Aminoglycosides This group includes gentamicin, tobramycin, amikacin, netilmicin, isepamycin, kanamycin, streptomycin, paromomycin, and neomycin. These drugs are mainly used in the treatment of Gram-negative aerobic bacterial infections; streptomycin is an important agent for tuberculosis treatment and paromomycin is used, orally, for intestinal amoebiasis and in the management of hepatic coma. In contrast with most inhibitors of bacterial protein synthesis, which are bacteriostatic, aminoglycosides are bactericidal. They are polycations and their polarity is responsible for their poor oral absorption and scarce distribution to the cerebrospinal fluid. Although widely employed, the toxicity limits their use: all members of this group have the same toxicity spectrum, in particular nephrotoxicity and ototoxicity.

These drugs contain amino sugars linked to a ring, called aminocyclitol, by glycosidic bonds. The sugar moiety may be a streptidine (in streptomycin) or a 2 deoxystreptamine (in the other aminoglycosides). In general, also spectinomycin is included among aminoglycosides because it contains the aminocyclitol ring, although it lacks the aminosugar.

The antimicrobial spectrum is broad and includes *Staphylococci*, *Enterobacteriaceae*, and other Gram-negative bacteria, such as *Pseudomonas aeruginosa*.

Aminoglycosides inhibit bacterial protein synthesis by binding to a complex site involving at least three different proteins of the 30S subunit (in the case of streptomycin) and also other proteins of the 50S subunit (for the other aminoglycosides). Antibiotic binding results in distortion of the A site, which normally binds the first amino acid of the peptide chain that in prokaryotic protein synthesis is a formylmethionine (Fig. 51.6). Consequently, the entire ribosome remains stuck at the initiation codon and protein synthesis cannot proceed (this phenomenon is called "accumulation of streptomycin monosomes"). In addition, besides blocking translation with different effectiveness depending on the compound (probably because of the different binding site recognized on the ribosome), aminoglycosides cause mistakes (misreading) during amino acid incorporation in the nascent protein chain, leading to mutated or truncated protein products.

The extent of the bactericidal effect depends on the different ability of antibiotics to enter the bacterial cell, via porin channels in the periplasmic space of Gram-negative bacteria or via oxygen-dependent active transport across the internal or cytoplasmic membrane.

In the process of diffusion within the bacterial cell, two phases can be distinguished: an energy-dependent phase I (EDP1) and an EDP II (EDP2). After the EDP1, which can be blocked or inhibited by divalent cations (Ca^{2+} and Mg^{2+}), anaerobiosis, hyperosmolarity, and reduced pH, the aminoglycoside irreversibly bind the ribosome. This results in translation inhibition, and possible production of aberrant proteins that, once inserted in the bacterial membrane, may alter its permeability, thus further enhancing drug uptake (EDP2 phase) and ultimately leading to membrane rupture.

Macrolides, Lincosamin, Streptogramin, and Chloramphenicol Despite their different chemical structure, macrolides, lincosamin, streptogramin, and chlorampenicol are described together because they exert the same action on the ribosomal 50S subunit. They all have a bacteriostatic effect on most sensitive species. Macrolides, lincosamins, and streptogramins compete for the same binding site on the ribosome that seems to be near the peptidyl-transferase P-site and to involve 23S rRNA and some proteins (Figs. 51.5 and 51.6). In fact, *in vitro* all these molecules inhibit the peptidyl-transferase reaction. In addition, bacteria containing ribosomes in which the 23S rRNA adenine 2058 is demethylated are resistant to macrolides, lincosamin, and streptogramin (this type of resistance is called MLS), but not to chloramphenicol. These antibiotics inhibit protein synthesis through slightly different mechanisms. Chloramphenicol seems to inhibit the transfer of the aminoacyl-tRNA to the acceptor site on the 50S subunit. As a result, the peptidyl-transferase activity, which normally catalyzes the addition of the new amino acid to the nascent polypeptide chain, cannot find its substrate and protein synthesis stops. The stage targeted by the other antibiotics seems to be the translocation of the nascent chain from the A- to the P-site. From a functional point of view, also in this case the peptidyl-transferase activity is inhibited.

Oxazolidinones These drugs represent a new class of protein synthesis inhibitors discovered during a random screening on a large collection of synthetic compounds. They are bacteriostatic antibiotics acting on Gram-positive bacteria and can be administered both parenterally and orally.

Linezolid was the first molecule of this class to enter clinical practice for the treatment of severe nosocomial infection by Gram-positive (*Staphylococci*, *Streptococcus pneumoniae*, and *Enterococci*) bacteria resistant to glycopeptides.

Structurally, these molecules are characterized by the N-aryl-oxazolidinone group, which gives the name to the class. To this group, which is essential for drug activity, are linked an acyl-aminomethyl group and a morpholine group, in *para* position.

The mechanism of action of oxazolidinones is unique, as they block a very early stage of protein synthesis—the initiation complex formation—by reversibly binding the 23S rRNA of the 50S ribosomal subunit. Antibiotic binding alters the 50S P-site that normally binds the peptidyl-tRNA, thus preventing adaptation of the fMet-tRNA to the site and assembly of the two subunits during the initiation complex formation (Fig. 51.7). In particular, they bind the P-site when already occupied, and localize, apparently without interfering with the peptide chain, in the space normally occupied by the aminoacyl group of the aminoacyl-tRNA bound to the A-site. In this way, they prevent binding or proper positioning of the aminoacyl-tRNA in the peptidyl transferase active site.

Blockade of protein synthesis at such an early stage allows avoiding symptoms due to different virulence factors (e.g., coagulases, hemolysins, and protein A) of *Staphylococci*

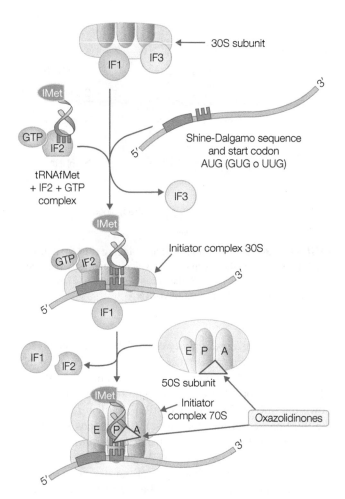

FIGURE 51.7 Mechanism of action of oxazolidinones. The antibiotic molecule binds to the 50S ribosome subunit preventing its assembly with the 30S subunit. In this way, the functional 70S initiation complex is not formed and protein synthesis is blocked. E,P,A, tRNA-binding sits.

and *Streptococci*. In addition, it prevents cross-resistance with other antibacterial drugs such as chloramphenicol, macrolides, lincosamids, streptogramins, and tetracyclines that inhibit later stages of protein synthesis, blocking peptide chain elongation. However, the possibility of cross-resistance with chloramphenicol and quinupristin-dalfopristin has recently been reported.

Fusidic Acid Fusidic acid interferes with the elongation phase of bacterial protein synthesis, acting on the EF-G elongation factor. As a result, it stabilizes the ribosome–EFG-GDP ternary complex and inhibits ribosomal translocation. Fusidic acid is one of the few protein synthesis inhibitors with bactericidal activity. It is active on some Gram-positive, but it is mainly used to treat *Staphylococcus aureus* infections.

Mupirocin Mupirocin is an antibiotic for topical use (when administered by other routes it is completely metabolized and converted to inactive products) that prevents aminoacyl-tRNA formation, by competitively inhibiting the isoleucyl-tRNA synthetase enzyme. Mupirocin is active only on Gram-positive bacteria as it is unable to cross the external membrane of the Gram-negative bacteria.

Inhibitors of DNA Synthesis and Replication

DNA replication involves a number of different enzymatic reactions, from the synthesis of nucleotide precursors to the polymerization of the nucleic acid chain.

Quinolones First-generation quinolones, such as nalidixic acid, were synthesized between 1962 and 1974 and mostly employed as urinary antiseptics. However, second-generation quinolones (ciprofloxacin, ofloxacin, and pefloxacin) have acquired a role as systemic drugs, given their broad spectrum of action, which includes aerobial Gram-negative bacterial species (e.g., *pseudomonaceae*) and Gram-positive cocci, and their high bioavailability and diffusion in the extravascular compartment. The third-generation quinolones (levofloxacin and moxifloxacin) include molecules metabolically more stable compared to the second generation, characterized by a broader antimicrobial activity which include *Streptococci*, *Haemophilus influenzae*, *Legionella pneumophila*, and in some cases also anaerobic species. Their molecular mechanisms of action are complex and still not entirely clarified. These antibiotics exert their main action on bacterial DNA gyrase or topoisomerase II (subunit A) and topoisomerase IV, which are critical enzymes for DNA replication and repair, recombination, and transcription of some operons, and moreover indispensable to maintain the supercoiled structure of bacterial DNA.

DNA supercoiling represents a form of storage of potential energy to be used during DNA replication and transcription, and is therefore essential for cell duplication and survival. DNA gyrase comprises two protein subunits, A and B of 97 and 90 kDa, respectively. The enzyme is active as a tetramer made up of two A and two B subunits. The energy required for the supercoiling process is supplied by ATP hydrolysis mediated by the B subunit. The A subunit is responsible for all the other processes leading to DNA supercoiling, except the DNA cleavage which is catalyzed by topoisomerase IV, a gyrase-homologous enzyme with high decatenating activity.

The primary target of fluoroquinolones is the DNA gyrase, which they inhibit at lower concentrations compared to topoisomerase IV. Moreover, DNA gyrase inhibition is more lethal for bacterial cells, probably because, compared to topoisomerase IV, it causes DNA damages that are less efficiently repaired. However, it is still unclear whether such model can be applied to all bacteria and in particular to Gram-positive species. Moreover, distinct quinolones exhibit different affinity and intrinsic activity toward the two enzymes. It could be possible to hypothesize that, in a given bacterial species, the primary target may vary depending on structural differences among members of this drug group.

Sulfonamides and Trimethoprim Sulfonamides, p-benzensulfonamide derivatives, have been the first chemotherapeutics to be effective in fighting systemic bacterial infections in humans. These bacteriostatic antibiotics are structural analogs and competitive inhibitors of the p-aminobenzoic acid (PABA), a fundamental intermediate in the synthesis of folic acid, an essential factor for bacterial growth. The enzymatic target of sulfonamides is the dihydropteroate synthetase (DHPS). Inhibition of this enzyme results in blockade of folic acid synthesis (Fig. 51.8).

Trimethoprim (Tp), like sulfonamides, is a synthetic drug that inhibits dihydrofolate reductase (DHFR), an enzyme that catalyzes the reaction following the one catalyzed by dihydropteroate synthetase (sulfonamide target). Trimethoprim is bacteriostatic when used alone, but is bactericidal when used in combination with sulfonamides. Trimethoprim and sulfonamides do not act on eukaryotic cells, since these do not contain sulfonamide target, and their DHFR enzyme is sufficiently different from the prokaryotic one.

Inhibitors of Cytoplasmic Membrane Functions

Bacterial cytoplasmic membrane has a different composition compared to other cells; it does not contain the lipids typical of fungi, such sterols, or the cholesterol, highly enriched in eukaryotic plasma membrane. This feature is exploited by

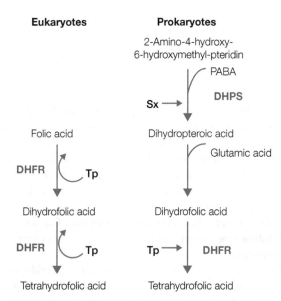

FIGURE 51.8 Diagram of tetrahydrofolic acid biosynthesis in eukaryotes and prokaryotes, and targets of sulfonamides (Sx) and trimethoprim (Tp). DHFR, Dihydrofolate reductase; PABA, 4-aminobenzoic acid.

some compounds capable of binding specific membrane phospholipids. In contrast with other antibiotics, the selectivity of these drugs towards prokaryotes is very modest; therefore their toxicity is high and the use is limited to local applications.

Polymyxins Polymyxins include several antibiotics, the best known being polymyxin B and colistin (colimycin). Polymyxin B consists of an eptapeptide ring linked to a lipophilic chain that penetrates by passive diffusion in the cellular membrane and disorganizes its structure by interacting with phospholipids. Bacterial sensitivity depends on the phospholipid content of their membrane. Disorganization of the lipid layer causes the formation of artificial pores through which ions, nucleotides, or other small molecules can leave the bacterial cell. Polymyxin B binds and inactivates the endotoxin of the external membrane of Gram-negative bacteria thus reducing its pathogenicity. Colistin, particularly active on *Pseudomonas aeruginosa*, once was scarcely used in clinics but has been recently rediscovered mainly because of the emergence of bacterial resistance to other drugs. Today, it is not unusual to find *Pseudomonas* strains sensitive only to colistin.

Daptomycin Daptomycin is a natural cyclic lipopeptide product active only against Gram-positive bacteria. The drug is not absorbed in the gastrointestinal tract and therefore is only administered intravenously. Its mechanism of action is related to its ability to bind (in the presence of Ca^{2+}) to bacterial membranes both during the proliferative and the stationary phases, and to favor K^+ release thus causing depolarization and rapid inhibition of protein, DNA, and RNA synthesis. This action leads to bacterial cell death with negligible cell wall rupture. The drug is bound by the pulmonary alveolar surfactant which inactivates it; for this reason, it cannot be used in bronchopulmonary infections.

Telavancin It is a new lipoglycopeptide antibiotic which exerts a rapid broad-spectrum bactericidal action on Gram-positive pathogens. Telavancin is a semisynthetic vancomycin derivative characterized by a hydrophobic lateral chain (decyl-aminoethyl) linked to vancosamine and by a hydrophilic phosphomethylamine group linked to an amino acid of the peptide chain.

This antibiotic has a double mechanism of action, as it both inhibits peptidoglycan synthesis and causes membrane depolarization.

Like vancomycin, telavancin inhibition of peptidoglycan synthesis is substrate-dependent; moreover, through the lipophilic lateral chain, it increases membrane permeability causing loss of ATP and K^+ with mechanisms similar to that of daptomycin. Telavancin is constantly active against *Staphylococcus aureus*, including strains resistant to methicillin, vancomycin intermediates, and linezolid.

Mechanisms of Resistance to Antibacterial Drugs

Antimicrobial drugs, when introduced in clinical practice, were initially welcomed like almost miraculous therapies. However, soon after the discovery of penicillin, it became evident that drug resistance can rapidly develop among bacteria, with the risk of making these drugs useless. This very serious phenomenon is always present, with each new antimicrobial agent, and may lead to the end of the antimicrobial era. Today, every main class of antibiotics is associated with the emergence of resistance. Two main factors are responsible for this: microorganism evolution and clinical-environmental practice. When exposed to a selective pressure menacing its survival, due, for example, to chemicals or other agents, a bacterial species undergoes changes and develops mechanisms to survive in the unfavorable conditions. Such evolution is largely promoted by an improper therapeutic use of antibiotics in humans and by an indiscriminate use of them in agriculture and breeding. Figure 51.9 schematically shows the main mechanisms of resistance.

Bacteria can become resistant either through mutational events changing their own genes or by acquiring resistance genes from other bacteria. Once a new mutation responsible for antibiotic resistance has occurred, it can spread to other microorganisms in several ways depending on whether it is determined by chromosomal or extra-chromosomal genes. In case of chromosomal genes, the transfer among bacterial cells can occur by:

1. Transduction: a process mediated by bacterial viruses (bacteriophages) that can transfer DNA fragment encoding antibiotic resistance genes
2. Bacterial conjugation: a process whereby genetic material is transferred from a bacterial cell to another via a cytoplasmic bridge

Extra-chromosomal antibiotic resistance genes may be contained in plasmids, R-factors, or transposones. These extra-chromosomal DNA elements can carry more than one gene, and simultaneously transfer the resistance to several antibiotics, even belonging to different classes (Fig. 51.9).

In all the mechanisms mentioned previously, once inside the bacterial cell, the antibiotic resistance genes either integrate in the bacterial chromosome or remain as plasmid DNA independently replicating. The resistance genes carried by independently replicating plasmids spread more rapidly within bacterial populations, compared to chromosomal resistance. For a more in-depth analysis of the mechanisms of resistance to antibacterial drugs, see Supplement E 51.1 "Mechanisms of Antibiotic Resistance in Bacteria."

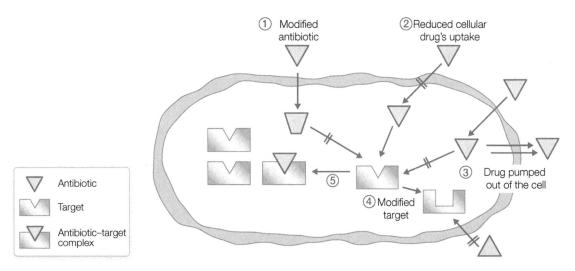

FIGURE 51.9 Main mechanism of resistance to antibiotic. Antibiotic resistance may be due to (i) antibiotic inactivation resulting from alteration of its chemical structure; (ii) reduced antibiotic uptake, via changes in cell wall and/or membrane properties; (iii) increased antibiotic extrusion; (iv) changes in the antibiotic target properties, due to mutations or post-translational modifications; and (v) increased expression of the antibiotic target.

ANTIFUNGAL DRUGS

The incidence of fungal infections, both systemic and dermal, has significantly increased over the past decades; such phenomenon is related to several factors, and in particular to an increased use of cytotoxic drugs (which alter the immune response) and an increased number of immune compromised subjects, as a result of viral infections (e.g., HIV) or immunosuppressive treatments.

Until the 1980s, the treatment of systemic fungal infection was based on amphotericin B. Only in the past few years, new drugs have been developed, thanks to investments in new technologies that have allowed:

1. To identify and characterize new selective targets,
2. To evaluate more rapidly and less expensively both natural and synthetic compounds,
3. To design, *in silico*, molecules capable of inhibiting molecular targets or biochemical processes essential for fungal cell survival.

Systemic fungal infections are more problematic to cure compared to the dermal ones, and for some of them no effective pharmacological therapy exists.

Among the antifungal drugs currently available, we can distinguish four main classes: allylamines, azoles, polyenes, and echinocandins. Another group includes flucytosine, griseofulvin, and morpholines. Although to a lesser extent compared to antibacterials, also for antifungal drugs the phenomenon of resistance is the reason of concern, in particular for azoles and flucytosine.

Figure 51.10 shows the different sites of intervention of antifungal drugs.

Drugs Acting on the Cell Wall

The main function of the cell wall is to control the internal turgor pressure; thus, compounds that interfere with this structure lead to cell lysis and death. In most fungal species, the main macromolecular components of the cell wall are chitin and β or α glucans that are chiefly present in the most internal layer and determine cell wall shape and resistance. In the internal part of the cell wall, mannoprotein derivatives are present, which regulate cell porosity, adhesivity, and antigenic characteristics. The main drugs altering the cell wall are echinocandins, pramicidins, and nikkomycin.

Echinocandins They are a family of cyclic hexapeptide antibiotics. Those in clinical use are caspofungin, anidulafungin, and micafungin. Echinocandins are highly effective against both *Candida albicans* and *non-albicans* on which they exert fungicidal activity. They have also a significant fungistatic activity on *Aspergillus*. The mechanism of action consists in the inhibition of the β(1,3)-D-glucan synthase. The resulting reduction in glycan incorporation in the fungal cell wall causes instability and reduced resistance to osmotic pressure with ensuing cell lysis.

Pramicidins These are quinone derivatives that act on the cell wall by forming mannoprotein complexes.

Nikkomycin It is a competitive inhibitor of chitin synthase.

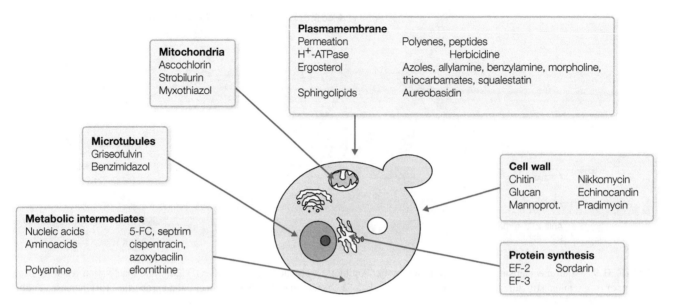

FIGURE 51.10 Mechanism of action of the main antifungal drugs.

The last two are still in preclinical phase and represent the starting point for developing new antifungal drugs with novel targets different from the current ones.

Drugs Acting on the Cytoplasmic Membrane

The chemical composition of the cytoplasmic membrane of fungal cells differs from that of bacteria and eukaryotic cells as it includes different lipid types, represented by phospholipids and mostly sterols (the most abundant being ergosterol).

Allylamines, thiocarbamates, azoles, morpholines, and polyenes interfere with ergosterol synthesis.

Allylamines and Thiocarbamates Only two allylamines, naftifine and terbinafine, and one thiocarbamate (tolfanate) are used (both topically and systemically) in clinics for the treatment of dermatomycosis and some forms of candidiasis. Allylamines and thiocarbamates are reversible, noncompetitive inhibitors of squalene 2,3-epoxidase, the enzyme that, along with 2,3 oxydosqualene cyclase, is responsible for squalene cyclization to lanosterol (Fig. 51.11). The cytotoxic effects are due to squalene accumulation and reduced ergosterol membrane concentration, with functional damages in particular on exchange processes and nutrient accumulation.

Azoles These are synthetic drugs subdivided in two classes: imidazoles and triazoles. Azoles are mainly used in the topical treatment of dermatophytosis and the surface treatment of candidiasis; only some of them can be systemically administered. They damage the cytoplasmic membrane by inhibiting the 14-α-demethylase, a P-450-dependent microsomal enzyme involved in ergosterol biosynthesis. The consequences of such inhibition are a reduced ergosterol concentration and accumulation of 14-α-methyl sterols that are responsible for functional alterations of several membrane enzymatic systems, such as the electron transport systems. At high concentrations, generally attained by topical application, some azoles (e.g., clotrimazole) can exert fungicidal action, directly damaging cell membrane (increased cell permeability).

Morpholines With the exception of amorolfine, morpholines are synthetic phenylmorpholine derivatives that interfere with ergosterol synthesis by inhibiting δ-8(14)-reductase and δ-24-methyltransferase, two enzymes involved in lanosterol conversion to ergosterol. Inhibition of these enzymes results in accumulation of 24-methylene-ignosterol in the cytoplasmic membrane, with severe changes in its physiological properties. Their use is only topical.

Polyenes More than 100 polyenes have been described, but only few molecules, among which amphotericin B and nystatin, are used in humans. Amphoterin B has a very broad spectrum which includes not only *Candida* and *Aspergillus* but also *Zygomycetes* and *Fusarium*. It was the first systemic antimycotic employed in invasive mycosis and it is still largely used (Fig. 51.12). Polyenes are cyclic amphiphilic macrolides, characterized by a cyclic ester with conjugated double bonds. Their antifungal activity depends, at least in part, on the formation of stable bonds with ergosterol and other membrane sterols. At low concentrations, this leads to the formation of pores or channels which allow low-molecular-weight molecules (glucose, amino acid, and ions) to leave the cell. At higher doses, the pore size increase and even cytosolic proteins can be lost.

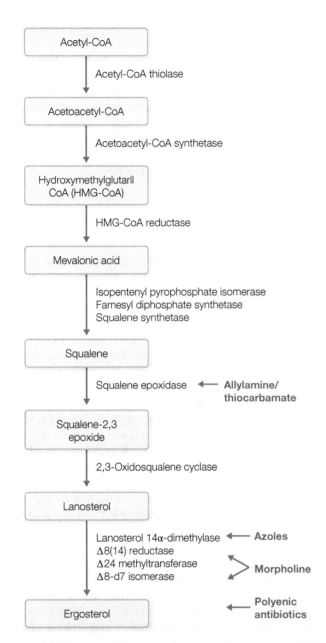

FIGURE 51.11 Steps in ergosterol biosynthesis and molecular targets of antifungal drugs in clinical use.

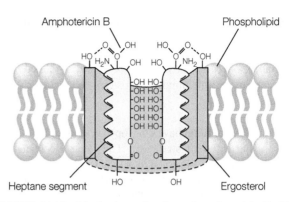

FIGURE 51.12 Mechanism of action of amphotericin B. The molecule inserts in the membrane causing the formation of pores and loss of selective permeability.

Inhibitors of DNA and Protein Synthesis

The pyrimidine derivative 5-fluorocytosine (5-FC) is transported inside the cell by a cytosine permease. Once inside, it is deaminated to 5-fluorouracil (5-FU), a powerful antimetabolite (for more details see Chapter 33). It selectively acts on fungal cells because mammalian cells are unable to convert 5-fluorocytosine to 5-FU.

Sordarins These are antibiotics of natural origin currently under clinical development, which highly selectively act on protein synthesis by interfering with the EF-2 elongation factor activity.

Inhibitors of Enzymatic Metabolic Pathways

Amino Acids Analogs
Several analogs of natural amino acids have shown antifungal activity. RI-331 is a selective inhibitor of homoserine dehydrogenase, the enzyme involved in the biosynthesis of methionine, isoleucine, and threonine. Azoxybacillin and its ester derivative exhibit broad-spectrum antifungal activity and inhibit the biosynthesis of sulfur-containing amino acids because they interfere with both expression and enzymatic activity of sulfite reductase.

Polyamines Difluoromethylornithine, difluoromethylarginine, and monofluoromethyldihydroornithine are suicidal inhibitors (as they irreversibly bind the target) of the ornithine carboxylase, an enzyme involved in polyamine synthesis.

Mitotic Inhibitors

Griseofulvin Similarly to colchicine and vinca alkaloids, this compound binds microtubules, but at a different binding site on tubulin. Moreover, it also binds some microtubule associated proteins (MAPs). The selectivity toward fungal cells is based on differences in binding affinity which in mammalian cells is low even at high doses. Griseofulvin action is evident also at morphological level with the production of multinucleate cells, as a result of blockade of mitotic spindle activity.

Mechanisms of Resistance to Antifungal Drugs

The phenomenon of resistance to antifungal drugs is becoming increasingly important, also considering the increasing number of immune-compromised patients (because of pathological or pharmacological reasons). Here in the following text, the most frequent mechanisms of resistance to each drug have been described.

Amphotericin B Although the drug has been used for over 60 years, the resistance is still low, probably because of the relative lack of specificity of its mechanism of action. The resistance, observed mostly in *Aspergillus*, is related to the replacement of membrane ergosterol with other sterol precursors.

Azoles Several fungi have developed resistance to this drug class, using different strategies. The main mechanism of resistance to azoles, observed mostly in *C. albicans*, is related to the presence of mutations in *ERG11*, the 14-α-sterol demethylase-encoding gene. The mutation prevents azoles from binding their site on the enzyme. Mutations usually confer resistance to all drugs of this class. Hyper-expression of ABC and MFS membrane transporters determines increased drug efflux from fungal cells, resulting in antibiotic resistance. This mechanism has been detected in *C. albicans* and *Candida glabrata*. Other, less frequent, mechanisms are the increased production of 14-α-sterol demethylase and mutations in *ERG3*, the sterol-reductase encoding gene.

Flucytosine The resistance to this drug is observed in about 30% of patients affected by cryptococcosis treated with flucytosine. This high incidence of resistance is one of the reasons that limit the use of this drug in therapy. The onset of resistance can be due to mutations in any of the genes encoding enzymes involved in flucytosine conversion and integration into RNA, but more often it involves the uracil phosphoribosyltransferase. Resistant strains of *C. glabrata* (but not *C. albicans*) have been isolated, carrying mutations in the cytosine permease and cytosine deaminase genes.

Echinocandines Resistance to this class of antimycotics is still rare, and it can appear during very prolonged treatments. The observed resistance is due to mutation of the *FKS1* gene that encodes essential proteins of the 1,3-β-D-glucan synthase complex.

ANTIVIRAL DRUGS

Viruses are obligate cellular parasites that need to exploit cell biosynthetic machinery for their multiplication cycle. Thus, it is not surprising that drugs exerting antiviral activity may be poorly selective and damage both virus and host cell. Many compounds exist with an adequate *in vitro* activity that cannot be used *in vivo* for their high toxicity, often due also to difficulty to reach suitable concentrations in the target organ. Another limitation of the antiviral drugs currently available is that they are effective only on actively replicating viruses (and therefore unable to inactivate latent or nonreplicating viruses). Moreover, it is important to remember that complete eradication of viral infections requires an active participation of the host immune system.

In the past years, remarkable progresses have been done in the development of new drugs with antiviral activity, in particular against HIV and hepatitis viruses. However, research on antivirals presents intrinsic difficulties related to lack of experimental models, both laboratory animals for *in vivo* studies and isolated cells for *in vitro* testing, that may faithfully reproduce the human disease. Moreover, viral infections are often tissue specific; thus, an agent effective against infections of epithelial tissues may be completely ineffective on other tissues or cellular types.

Mechanisms of Action of Antiviral Drugs

Figure 51.13 and Table 51.3 summarize the virus multiplication cycle and possible inhibitors.

Viruses infect eukaryotic cells by binding to one or more receptors usually present on the plasma membrane of the infected cells (e.g., CD4 for HIV). Once tightly bound to specific membrane sites, the virus can fuse with the plasma membrane, leaving outside the capsid and injecting its nucleic acids in the host cytoplasm. Otherwise, the virus–receptor's complex is internalized by endocytosis and sent to the endosomal compartment (receptor-mediated endocytosis, see Chapter 9 and Supplement E 9.3). The acidic pH of this compartment favors the fusion of the viral capsid with the intracellular membranes; in this way, the virus releases its nucleic acids in the host cytoplasm, which will form new viral particle by exploiting the host biosynthetic apparatus. In the case of retroviruses, which are RNA viruses (e.g., HIV), a viral reverse transcriptase catalyzes the retrotranscription of the viral RNA into a double-stranded DNA which then integrates into the host genome. For other RNA viruses with positive polarity (e.g., picornaviruses and coronaviruses), the virus RNA itself serves as mRNA and is translated into a giant polypeptide, that is subsequently cleaved to generate several distinct proteins. Virus production requires the synthetic machinery of the host cell, but viral DNA amplification may involve virus-encoded DNA polymerases. The proteins, synthesized on either free or ER-bound ribosomes, may be post-translationally modified before being assembled in the viral capsid. Viruses, containing their nucleic acids, can be released in the extracellular space via an active mechanism or upon cellular lysis. In theory, it is possible to interfere with each step of the virus multiplication cycle.

Inhibitors of Viral DNA Replication

Most inhibitors of viral DNA replication consist of derivatives of purine (e.g., analogs of 6-aminopurine and 6-oxypurine) and pyrimidine (e.g., uracil, cytosine and isocytosine derivatives, thiopyrimidines, and 6-azapyrimidines) nucleosides.

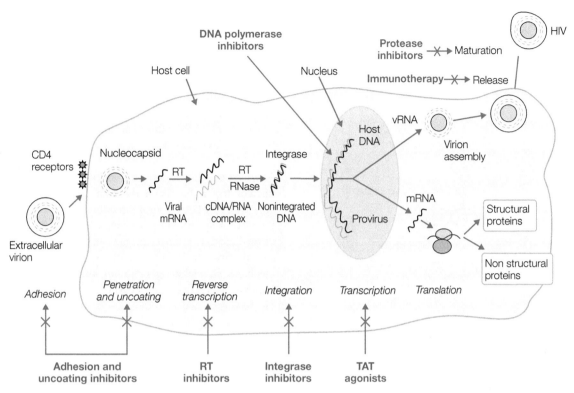

FIGURE 51.13 Multiplication cycle of HIV-1, as an example of retrovirus, with indicated antiviral drug targets. RT, reverse transcriptase; cDNA, complementary DNA; mRNA, messenger RNA; TAT, protein regulating viral transcription and replication rate; vRNA, viral RNA.

TABLE 51.3 Antiretroviral drugs approved for clinical use

Nucleoside reverse transcriptase inhibitors	Zidovudine	ZDV (AZT)
	Didanosine	ddI
	Stavudine	d4t
	Zalcitabine	DDC
	Lamivudine	3TC
	Abacavir	ABC
	Tenofovir disoproxil	TDF (PMPA)
	Emtricitabine	FTC
Non-nucleoside reverse transcriptase inhibitors	Nevirapine	NVP
	Efavirenz	EFV
	Delavirdine	DLV
	Etravirine	ETV
Protease inhibitors	Saquinavir	SQV
	Indinavir	IDV
	Ritonavir	RTV
	Nelfinavir	NFV
	Amprenavir	APV
	Lopinavir	LPV/r
	Atazanavir	ATV
	Fosamprenavir	FPV
	Tipranavir	TPV
	Darunavir	DRV
Inhibitors of virus entry	Enfurtide	T-20
	Maraviroc	MVC
Integrase inhibitors	Raltegravir	RAL

Acyclovir, Ganciclovir, Famciclovir, and Penciclovir
Prototype of this class of antiherpes (Herpes simplex virus, HSV) drugs is acyclovir. Once taken up by cells, it is phosphorylated to acyclovir monophosphate (acyclo-GMP) by the thymidine kinase. Acyclovir and its analogs (ganciclovir, famciclovir, and penciclovir) are significantly more efficiently phosphorylated by the thymidine kinase of HSV-1 and -2 (or by a cytomegalovirus phosphotransferase, in the case of ganciclovir), than by the host cell enzyme. Acyclo-GMP is then phosphorylated by cellular enzymes to di- and tri-phosphorylated compounds (acyclo-GDP and acyclo-GTP) which are highly hydrophilic and therefore cannot diffuse outside the cell (Fig. 51.14). As a consequence, concentrations of the phosphorylated drug can be up to 200 times higher in infected cells than in healthy ones.

Acyclo-GTP (and the triphosphate derivatives of this drug class) competes with deoxyguanosine triphosphate for incorporation into the DNA; also in this case, the drug is a much better substrate for the viral polymerase than for the cellular enzyme. Acyclo-GTP binding competitively inhibits the polymerase. Nevertheless, acyclo-GTP is incorporated in the viral DNA but, since it lacks the 3-OH group, it does not allow further incorporation of nucleotides, thus serving as a sort of DNA synthesis terminator. The nascent DNA chain containing acyclovir binds the polymerase and irreversibly inhibits it.

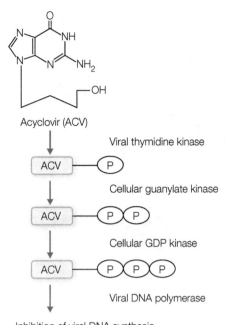

FIGURE 51.14 Acyclovir metabolic activation.

HSV resistance to acyclovir can occur with one of the following mechanisms: (i) reduced production of viral thymidine kinase; (ii) altered substrate specificity of thymidine kinase (resulting, e.g., in phosphorylation of thymidine but not of acyclovir); and (iii) alterations of the viral DNA polymerase. Alterations of viral enzymes are caused by point mutations, insertions, or deletions in the corresponding genes. The most common resistance mechanism observed in clinical isolates of HSV consists in the absence or the defective activity of the viral thymidine kinase; mutant viruses encoding altered DNA polymerase are rare.

Iodoxuridine This drug is a nucleoside thymidine analog that upon phosphorylation interferes with several enzymes of the HSV family involved in replication. The described toxic effects are due to the poor selectivity of the drug because the iodoxuridine triphosphate derivative is incorporated in both the viral and the cell DNA. Such DNA is more easily subject to transcription interruption, and therefore leads to the synthesis of altered viral proteins. The emergence of iodoxuridine resistance is highly frequent.

Vidarabine This drug is a nucleoside adenosine analog that selectively inhibits viral DNA synthesis. Production of the active triphosphate form of vidarabine is catalyzed by a cellular kinase that is more efficient than the viral enzymes. Vidarabine action is believed to be based on inhibition of herpes simplex-like virus DNA polymerase.

Foscarnet Chemically, it is the phosphonoformic acid. It noncompetitively inhibits the RNA polymerase of type-A influenza virus and the DNA polymerases of HSV-1, HSV-2, cytomegalovirus (CMV), Epstein–Barr virus (EBV), varicella zoster virus (VZV), and hepatitis B virus (HBV), at concentrations that do not interfere with the host polymerase. Foscarnet reversibly binds to the pyrophosphate binding site of the viral polymerase; in this way it noncompetitively inhibits removal of the pyrophosphate group from the deoxynucleotide triphosphate. This inhibition arrests DNA chain growth. Foscarnet resistance does not easily develop in humans.

Fomivirsen It is a phosphorothioate oligonucleotide of 21 nucleotides; it has been the first antisense therapy to be approved by FDA for viral infections. It is complementary to the mRNA specified by the cytomegalovirus (CMV) major immediate-early transcriptional unit; it inhibits CMV replication through both sequence-specific and nonspecific mechanisms, including inhibition of virus binding to cells. Fomivirsen is active against CMV strains resistant to ganciclovir, foscarnet, and cidofovir. Fomivirsen is administered via intravitreal injection in the treatment of CMV retinitis in patients intolerant or unresponsive to other therapies.

Drugs Active against Influenza Virus

Influenza, or flu, is a ubiquitous disease, which has been spreading in the human population for centuries; first descriptions of influenza epidemic date back to the fifth century BC in Greece and over the centuries multiple epidemic or pandemic of different severity have occurred, with an increase in morbidity and mortality.

Three types of influenza virus have been identified: type-A and type-B, responsible for the typical flu symptoms, and type-C which is scarcely relevant in clinics as it generally causes asymptomatic infection, or cold-like symptoms. Influenza epidemiology is explained by the high changeability of influenza viruses; in other words, they tend to undergo changes in their surface proteins which allow them to escape the immune protection due to previous infections. Anti-influenza vaccination represents the most effective medical intervention for preventing the disease. At pharmacological level, four compounds are available: two adamantanes, oseltamivir, and zanamivir.

Adamantanes These are drugs capable of interfering with the type-A influenza virus uncoating process (removal of the external coat). Although the best known compound is amantadine, rimantadine is the one with the best pharmacological-therapeutic profile. Amantadine inhibits the uncoating, which is the initial phase of the viral multiplication process. Amantadine target is probably M2, a viral membrane protein that serves as a channel and is necessary to acidify the viral internal space. This event is essential for the dissociation of the ribonucleoproteins from the viral membrane and the subsequent transfer of the viral nucleic acids to the host cytoplasm. The drug interferes neither with the binding of type-A influenza virus to the host plasma membrane nor with the subsequent endocytosis, but it accumulates in cell endosomes and lysosomes, increasing their intravesicular pH. This probably hampers the fusion of the viral with the host membrane. Moreover, the altered activity of the M2 protein and the increased pH of the cytoplasmic organelles are probably responsible also for the alterations occurring during synthesis and maturation of the neosynthesized viral hemagglutinin and viral capsid assembly. The final effect of this drug class is the blockade of viral replication.

Several examples of strains resistant to adamantanes have been described. It is a cross-resistance to all drugs of this class and can be transmitted among strains; it is associated to mutations affecting the amino acids at position 27, 30, and 31 of the M2 protein.

Oseltamivir and Zanamivir These are inhibitors of viral neuraminidase. Influenza virus neuraminidase binds the terminal sialic acid residues and destroys the site of attack recognized by the viral hemagglutinin on cell surface and in respiratory secretions. This enzymatic action is essential for the release of virus particles from the infected cells. The interaction between oseltamivir or zanamivir and neuraminidase causes a conformational change in the enzyme active site inhibiting its activity. This leads to virus aggregation at the cell surface and reduced diffusion in the respiratory tract. Virus variants carrying mutations in the hemagglutinin and/or neuraminidase genes exhibit drug resistance.

Drugs Mainly Active on Hepatitis C Viruses (HCV)

Hepatitis C is one of the most common chronic infections in developed countries and is associated with significant morbidity and mortality. If not treated, it can lead to progressive hepatocellular damage with fibrosis and cirrhosis; moreover, it can be an important risk factor for the development of hepatocellular carcinoma. HCV is a RNA virus; its genome does not integrate into the chromosomal DNA of the host cell and therefore the virus does not become latent. This implies that, in theory, the disease could be cured in all individuals. The current standard cure consists in a combination treatment with interferon α and ribavirin, with a high frequency of healing in infections by selected viral genotypes. The recent development of highly effective oral drugs, such as the HCV protease inhibitors (PIs) and the HCV polymerase inhibitors, has raised hopes for a radical change in the therapy of this infection in the next decade, with highly positive results.

Interferons (IFNs) These are cytokines endowed with potent antiviral, immunomodulatory, and antiproliferative activities. These proteins are synthesized by host cells in response to various stimuli, and in turn cause biochemical changes leading to antiviral response. Currently, three main classes of interferons are known to humans, with significant antiviral activity. The clinical treatment of HCV infections is based on recombinant IFNs, consisting of unglycosylated proteins of 19.5 kDa. The PEGylated forms are the most used in clinical practice.

The mechanism of action is very complex. Via binding to specific cellular receptors, IFNs activate the JAK–STAT transduction pathway leading to nuclear translocation of a protein complex that specifically recognize and bind IFN-stimulated response elements (ISREs) in the promoter of certain genes. This results in the expression of many proteins involved in resistance to viruses during virus penetration.

Inhibition of protein synthesis is the most effective antiviral effect, for many viruses. The IFN-induced proteins include 2′-5′-oligoadenylate (2-5(A)) synthetase and an IFN-induced kinase, both capable of inhibiting protein synthesis in the presence of a double-stranded RNA. The 2-5(A) synthetase produces RNA species that activate a latent endoribonuclease (RNase L) which cleaves both the cellular and the viral single-stranded RNA. The IFN-induced kinase selectively phosphorylates and inactivates the initiation factor 2 (eIF2) involved in protein synthesis; in addition, it can also act as an important apoptosis effector. IFN induces also a phosphodiesterase that inactivates a portion of tRNAs, thus preventing translation elongation.

A virus can be blocked at different stages of its multiplication cycle, and the main inhibitory effect may vary among different virus families. Some viruses can counteract IFN effects by inhibiting production or activity of the IFN-induced proteins. For example, one of the mechanisms of HCV resistance to IFNs is based on inhibition of the IFN-induced kinase.

Ribavirin This drug is a synthetic guanosine analog with broad-spectrum antiviral activity. It is active against both DNA and RNA viruses. Ribavirin is phosphorylated by cellular enzymes and converted to monophosphate, diphosphate, and triphosphate forms. This drug is believed to exert antiviral activity because of its ability to reduce the intracellular nucleotide concentration, or to inhibit mRNA capping, or viral RNA polymerase. Studies have demonstrated that the relevance of the inhibited enzymatic systems is different among different virus groups.

In most viruses, resistance to ribavirin is not common, but it has been described in HCV. The mechanism seems to be related to the absence of ribavirin phosphorylation to its active form. Ribavirin is also highly toxic to embryos and fetuses.

Protease Inhibitors Several PIs are currently being studied both in monotherapy and in combination therapy with classical anti-HCV drugs. The two drugs closest to clinical use are telaprevir and boceprevir, which have proved effective in reducing the viral charge in several clinical studies. The mechanism of action is directed to a specific protein of the protease complex involved in virion formation. HCV RNA encodes a single polyprotein precursor that is enzymatically processed by both host and viral proteases. The N-terminal portion of the polypeptide is cleaved by host proteases generating three structural proteins (core, E1, and E2 glycoproteins) and P7 (a membrane protein whose function has not yet been clarified). The remaining part of the polypeptide contains six nonstructural (NS) proteins required for virus maturation and replication, including NS3, a multifunctional enzyme containing an N-terminal serine protease domain and a C-terminal RNA helicase/NTPase domain. The proteolytic process requires also NS4A, an essential cofactor for peptidase activity. Eventually, a heterodimeric protease is formed, called NS3-4A that is specifically inhibited by boceprevir and telaprevir.

New HCV drugs

Even if PEG–RBV combination were the standard of care until 2013, associated with a PI for patients with genotype 1 HCV, the drawbacks of INF-based regimes are well know (a long therapeutic course, low barrier to resistance, IFN toxicity, and low efficacy in cirrhotic patients).

Intelligent drug design has made it possible to specifically target practically all the proteins synthesized by HCV and new directly acting antivirals (DAAs) are virus or even genotype specific.

There are actually three major viral proteins effectively targeted by news drugs: NS3/4A, NS5B, and NS5A.

NS3/4A inhibitors Two of the three already approved drugs targeting viral protease were described previously (see Boceprevir and Telaprevir). These PIs are highly selective for genotype 1 and have a low barrier to resistance. The new PIs partially overcome the problem associated with IFN therapy and drug–drug interaction (DDI) observed with first-generation PIs.

Simeprevir is a more tolerated PI and with a favorable pharmacokinetics. It has a narrow genotype specificity (genotype 1) and low barrier to viral resistance. Antiviral combinations may overcome these problems. It seems that there are no differences in virological response when RBV is added to simeprevir regime. Nevertheless more studies are needed to prove the real risk of DDIs, since SIM is a mild inhibitor of CYP1A2 and metabolized by CYP3A4, furthermore *in vitro* study demonstrate its inhibitory activity on OATP1B1 and MRP2 at a concentration reached during HCV therapy at the daily dose of 150 mg.

NS5B inhibitors Sofosbuvir is a nucleotide analogue inhibitor of NS5B viral RNA-dependent polymerase effective against all HCV genotypes. It is now approved by the US FDA to treat genotypes 1 and 4 infection in combination with PEG–RBV and in an all-oral regime for genotypes 2 and 3, associated with RBV. There are no differences in efficacy between IFN-based regimes and IFN-free regimes in the treatment of genotypes 2 and 3. Recently published data on VALENCE study showed that also HCV genotype 3 infections reached good response with a prolonged Sofosbuvir-RBV regime of 24 weeks (85% SVR). The response rate was 68% in the cirrhotic group, well above the historical cohort treated with IFN.

NS5A inhibitors Ledispavir is an HCV NS5A inhibitor in phase III trials active in virus with S282T mutation, find in some sofosbuvir resistant genotypes. Extraordinary results were also obtained with ledipasvir in patients with HCV genotype 1, naive and null responders, in a regime containing Sofosbuvir with or without RBV. The most common adverse events in all these trials were fatigue, nausea, headache, and insomnia.

Sofosbuvir combined with daclatasvir, another NS5A inhibitor, led to SVR12 rates of 86–100% in patients with genotypes 1, 2, or 3 both previously treated and untreated. This combination was well tolerated both with and without RBV, reporting few grade 3 or 4 laboratory abnormalities like low phosphorous and hyperglycaemia, atypical of other regime.

Future clinical studies will better clarify the proper doses and the most useful associations with other PIs.

Other Drugs Active on HCV

Viramidine (RBV ProDrug) Many methods to predict RBV-induced anaemia and improve treatment compliance were tested with different results. There is no approved laboratory test or adjuvant drug to personalize RBV therapy based on inter individual variability of haemolysis or pharmacokinetics data.

Ribavirin prodrug viramidine (VRD or taribavirin) is less toxic but its efficacy in comparison with RBV is still a matter of debate.

Miravirsen Miravirsen is a nucleic acid-modified antisense oligonucleotides (2′-O-methyl or 2′-O-methoxyethyl) that target with high affinity the 5′ region of mature miR-122 involved in HCV-RNA maturation and highly expressed in the liver. In phase-1 trial it showed efficacy in reducing HCV-RNA by 2–3 log at a weekly subcutaneous dose of 3–7 mg/kg. Miravirsen seems virus specific, well tolerated, without apparent DDI and can be a new approach to reduce viral resistance.

Drugs Active against Hepatitis B virus (HBV)

In contrast with HCV, HBV is a DNA virus whose genome can be integrated in the host cell genome, establishing a chronic infection in about 10% of patients. Subjects with chronic HBV can develop active hepatitis which may lead to fibrosis and cirrhosis, with also increased incidence of hepatocellular carcinoma. Interferon in monotherapy or combined therapy with interferon and ribavirin can cure patients with chronic infection; however, this therapy is associated with a high number of side effects, which often lead to treatment suspension. Several nucleoside antiretrovirals or nucleotide analog polymerase inhibitors, such as lamivudine, telbivudine, and tenofovir, exert a potent anti-HBV activity and have provided an alternative oral treatment, either as monotherapy or in combination. These regimens are much better tolerated compared to combination therapies with IFNs.

Adefovir This is an acyclic nucleoside phosphonate analog of adenosine monophosphate. In therapy, adefovir dipivoxil is used, which is a diester prodrug that, once entered the cells, is de-esterified to adefovir. *In vitro* it exerts inhibitory action on a variety of DNA and RNA viruses, but its clinical use is limited to HBV infections. Adefovir is converted, by cellular enzymes, to diphosphate which acts as a competitive inhibitor of the viral DNA polymerase and reverse transcriptase and also as chain terminator in viral DNA synthesis. Its selectivity is due to its higher affinity for HBV polymerase than for the cellular one. Resistance to adefovir has been observed in a small percentage (4%) of patients with chronic HBV infection during 3-year treatment. The resistance is caused by point mutations in the HBV polymerase gene, and do not interfere with virus sensitivity to lamivudine.

Entecavir, Lamivudine, and Telbivudine These are nucleoside analogs of guanosine (entecavir), cytidine (lamivudine), and thymidine (telbivudine) with selective activity against HBV-DNA polymerase. Once in the cells, these drugs are phosphorylated to their triphoshate forms that compete with the respective endogenous triphosphate nucleosides, inhibiting all activities of the HBV polymerase. The resistance is mainly due to point mutations in the viral polymerase gene or due to excessive production of viral DNA polymerase.

Drugs Active against Immunodeficiency virus (HIV)

The human immunodeficiency virus (HIV) is believed to have originated in sub Saharan Africa from a mutated animal (probably monkey) retrovirus which in the twentieth century started to spread among the human populations leading to a global epidemic. UNAIDS and WHO have estimated that since the discovery of the syndrome, 25 million deaths have occurred, meaning one of the most devastating epidemic in human history. Globally, it has been estimated that about 33 million people are affected by HIV (UNAIDS estimate, 2007). Only in 2005, 4.3–6.6 million people became infected and 2.8–3.6 million of people died by AIDS, 570,000 of them being children: the highest figures since 1981.

HIV is a typical retrovirus with a small RNA genome (9300 bases). Two copies of the genome are contained in the viral capsid which is enveloped by a lipid layer deriving from the host plasma membrane (Fig. 51.13). The viral genome contains three main coding sequences: *gag* encodes a polyprotein that is processed to generate the main structural proteins of the virus; *pol*, partially overlapping *gag*, encodes three important enzymes (an RNA-dependent DNA polymerase or reverse transcriptase with RNase activity, a protease and a viral integrase); *env* encodes a large transmembrane protein which is responsible for virus entry into the cell. In addition, the genome contains several small genes (*tata*, *rev*, *nef*, and *vpr*) encoding regulatory proteins that increase virion production or antagonize host defense response.

HIV infection is currently treated with the so-called highly active antiretroviral therapy (HAART) involving appropriate combinations of antiretroviral drugs. Its use, started in 1995, has allowed to drastically reducing morbidity and mortality in infected subjects. HAART can prevent progression of the HIV disease. The pharmacological therapy of HIV infection is a rapidly evolving field, and many antiretroviral drugs are currently available for clinical use (Table 51.3). Thanks to these drugs, AIDs has now turned from a lethal to a chronic disease. The strategy in the HAART approach is to employ combinations of different drug classes to decrease the probability of emergence of viral resistance, to use the minimal number of drugs in order to have alternatives in case of resistance, and to start the therapy during the initial phases of infection. The final goal, that is, the complete eradication of the virus from the intracellular stores to eradicate the infection, has not yet been reached.

Reverse Transcriptase Inhibitors

The HIV RNA-dependent DNA polymerase, also called reverse transcriptase, catalyzes the synthesis of a proviral double-stranded DNA using viral RNA as a template. The proviral DNA then integrates into the host genome. The available reverse transcriptase inhibitors are nucleoside/nucleotide analogs or non-nucleoside inhibitors.

Nucleoside and nucleotide transcriptase inhibitors Like all the antiretroviral drugs available, the nucleoside and nucleotide reverse transcriptase inhibitors (NRTI) prevent infection of sensitive cells, but do not eliminate the virus from cells already carrying HIV proviral DNA integrated in their genome. Nucleoside and nucleotide analogs have to enter the cells where they are phosphorylated and terminate nascent proviral DNA elongation by competition or incorporation. With only one exception, all drugs of this class are nucleosides that need to be triphosphorylated in order to be active. Only exception is tenofovir which requires the addition of two phosphate groups to become fully active. These compounds inhibit both HIV-1 and HIV-2 and some have broad-spectrum activity against other human and animal retroviruses; emtricitabine, lamivudine, and tenofovir are active also against HBV, and tenofovir is also active against herpes viruses.

Non-nucleoside Reverse Transcriptase Inhibitors Non-nucleoside reverse transcriptase inhibitors (NNRTIs) include a variety of compounds that bind the hydrophobic pocket of the HIV-1 reverse transcriptase p66 subunit. Although this pocket is not essential for enzyme activity and is distant from the active site, these compounds induce a conformational change in the enzyme tridimensional structure that strongly reduces its activity. Thus, these drugs act as noncompetitive inhibitors. In contrast with nucleoside and nucleotide reverse transcriptase inhibitors, NNRTIs do not require intracellular phosphorylation to become active. Since the NNRTI binding site is strain-specific, these agents are active against HIV-1 but not against HIV-2 or other retroviruses. In addition, these compounds are not active on the host DNA polymerase. The two agents of this class most frequently used are efavirenz and nevirapine.

HIV Protease Inhibitors

The HIV PIs are peptide-like substances that competitively inhibit the activity of the HIV aspartate protease. This is a homodimeric protein, in which each 99 amino acid monomer contains an aspartic residue essential for catalysis. The preferential cleavage site of this enzyme is the N-terminal side of proline residues, in particular between phenylalanine and proline. Human aspartate proteases (renin, pepsin, gastrin, and cathepsins D and E) contain only one polypeptide chain and are not significantly inhibited by the HIV PIs.

These drugs prevent proteolytic processing of the *gag*-encoded polypeptide, of precursors of structural polypetides (P17, P24, P9, and P7) and enzymes (inverse transcriptase, protease, and integrase) as well as of viral components encoded by the *pol* gene. Therefore, these drugs prevent maturation of the HIV-infective particles. Development of resistance to these drugs requires the accumulation of multiple resistance mutations. The primary mutation conferring resistance to saquinavir is a leucine to methionine substitution at codon 90 of the HIV protease. Another important mutation observed is a glycine to valine substitution at codon 48. Secondary resistance mutations affect codons 36, 46, 82, 84, and others. Usually cross-resistance occurs with most HIV PIs.

Inhibitors of Virus Entry into the Host Cells

Two drugs are available in this class, enfuvirtide and maraviroc, which have different mechanism of action. Enfuvirtide is a 36 amino acid synthetic peptide whose sequence derives from the HIV-1 gp41 transmembrane region, which is involved in the fusion between the lipid double layer of the virus membrane and the host plasma membrane. Its pharmacological action is aimed at inhibiting virus binding to the host cell. Enfuvirtide is not active on HIV-2. Maraviroc blocks binding of the virus coat protein gp120 to the CCR5 chemokine receptor. In this way, it prevents the virus from entering the cell. Another antagonist at the CCR5 receptor, vicriviroc, is at an advanced stage of clinical development.

Resistance may occur in mutant viruses that do not use CCR5 to enter the host cell.

Integrase Inhibitors

Chromosome integration is a distinctive feature of the retrovirus life cycle and allows viral DNA to remain inactive or latent in the host nucleus for a long time. Since human genomic DNA is not known to undergo excision and reintegration events, the integration process represents an interesting target for antiviral approaches. The first HIV integrase inhibitor, raltegravir, was approved in 2007. It prevents the formation of covalent bonds between viral and host DNAs, by blocking the catalytic activity of the HIV integrase, thereby blocking virus DNA integration in the host chromosome. Given its unique mechanism of action, raltegravir maintains its activity against viruses that have become resistant to antiviral agents of other classes. The two main mechanisms of resistance to raltegravir observed involve mutations in the integrase encoding gene.

TAKE-HOME MESSAGE

- The advent of antimicrobial drugs has allowed controlling infectious diseases. However, the emergence of resistance to these drugs may drastically reduce their beneficial effects.
- Antibacterial drugs are distinguished into bacteriostatic and bactericidal depending on whether they block bacterial proliferation or cause death of bacterial cells.
- Bacterial functions inhibited by antibiotics can be schematically subdivided as follows: (i) synthesis of cell wall peptidoglycan; (ii) RNA and protein synthesis; (iii) DNA synthesis and replication; (iv) cytoplasmic membrane functions; and (v) cellular metabolism.
- Antibiotic resistance may result from (i) reduced antibiotic uptake by bacterial cells, (ii) increased antibiotic extrusion via efflux pumps; (iii) release of microbial enzyme inactivating the antibiotic, (iv) alterations in the enzymes that convert prodrugs to active molecules, (v) alteration of target proteins, and (vi) development of alternative metabolic pathways to avoid those inhibited by the antibiotic.
- The emergence of antibiotic resistance has been and is promoted by improper therapeutic use and indiscriminate use in agriculture and breeding.

- Treatment of systemic fungal infections is based on amphotericin B and inhibitors of cell wall and cytoplasmic membrane components, inhibitors of metabolic enzymes, and mitotic inhibitors.
- Resistance to antifungal drugs occurs less frequently compared to antibacterials and is due to mutations in genes encoding antifungal drug targets.
- Antiviral drugs act through one or more of the following mechanisms: (i) virus inactivation before recognition or penetration into the host cell, (ii) inhibition of the interaction between virus and specific host plasma membrane receptors, (iii) inhibition of viral uncoating, (iv) inhibition of viral DNA integration in host genome, (v) inhibition of viral RNA and protein synthesis, and (vi) alterations of steps in capsid formation and virus release processes.

FURTHER READING

Anglemyer A., Rutherford G.W., Egger M., Siegfried N. (2011). Antiretroviral therapy for prevention of HIV transmission in HIV-discordant couples. *Cochrane Database of Systematic Reviews*, 5, CD009153.

Arts E.J., Hazuda D.J. (2012). HIV-1 antiretroviral drug therapy. *Cold Spring Harbor Perspectives in Medicine*, 2, a007161.

Barabote R.D., Thekkiniath J., Strauss R.E., et al. (2011). Xenobiotic efflux in bacteria and fungi: a genomics update. *Advances in Enzymology and Related Areas of Molecular Biology*, 77, 237–306.

Bartenschlager R., Lohmann V., Penin F. (2013). The molecular and structural basis of advanced antiviral therapy for Hepatitis C virus infection. *Nature Reviews. Microbiology*, 11, 482–496.

Fernández L., Breidenstein E.B., Hancock R.E. (2011). Creeping baselines and adaptive resistance to antibiotics. *Drug Resistance Updates*, 14, 1–21.

Hurdle J.G., O'Neill A.J., Chopra I., Lee R.E. (2010). Targeting bacterial membrane function: an underexploited mechanism for treating persistent infections. *Nature Reviews. Microbiology*, 9, 62–75.

Kohanski M.A., Dwyer D.J., Collins J.J. (2010). How antibiotics kill bacteria: from targets to networks. *Nature Reviews. Microbiology*, 8, 423–435.

Koul A., Arnoult E., Lounis N., et al. (2011). The challenge of new drug discovery for tuberculosis. *Nature*, 469, 483–490.

Medina R.A., García-Sastre A. (2011). Influenza A viruses: new research developments. *Nature Reviews. Microbiology*, 9, 590–603.

Ostrosky-Zeichner L., Casadevall A., Galgiani J.N., et al. (2010). An insight into the antifungal pipeline: selected new molecules and beyond. *Nature Reviews. Drug Discovery*, 9, 719–727.

Perno C.F. (2011). The discovery and development of HIV therapy: the new challenges. *Annali dell'Istituto Superiore di Sanità*, 47, 41–43.

Theuretzbacher U., Mouton J.W. (2011). Update on antibacterial and antifungal drugs—can we master the resistant crisis? *Current Opinion in Pharmacology*, 11, 429–432

Turel O. (2011). Newer antifungal agents. *Expert Review of Anti-Infective Therapy*, 9, 325–338.

SECTION 13

TOXICOLOGY AND DRUG INTERACTIONS

52

INTRODUCTION TO TOXICOLOGY

HELMUT GREIM

> **By reading this chapter, you will:**
> - Know the basic concepts for hazard identification and risk assessment
> - Understand the principles to derive acceptable exposure levels
> - Know the different *in vitro* and *in vivo* methods to determine toxicity

The discipline of toxicology is concerned with health risks of human exposure to chemicals or radiation. According to Paracelsus (1493–1541) "All things are poisons, for there is nothing without poisonous qualities. It is only the dose which makes a thing poison". Since it depends only upon the dose whether a substance is poison, toxicology is interested in describing the adverse effects of chemicals quantitatively by evaluating the dose–response and the exposure responsible for a given adverse effect. Overall toxicology describes the intrinsic dangerous properties of an agent (hazard identification) and determines the amount of the agent that produces adverse effects (risk characterization). Based on this information, tolerable levels of exposure are determined to define safe and acceptable concentrations of the agent in food, drinking water, consumer articles, environment, and workplace.

Humans may be exposed to chemicals through air, water, food, or skin. The external dose at which a chemical exerts its toxic effects is a measure of its potency; thus, a highly potent chemical produces its effects at low doses. Ultimately, the response to the chemical depends upon duration and route of exposure, toxicokinetics of the chemical, dose–response relationship, and individual's susceptibility. To characterize the risk, we need to identify the adverse effects and to evaluate the dose–response in order to define the exposure level that causes them. Therefore, risk characterization comprises three elements:

- Hazard identification, that is, description of the agent's toxic potential
- Evaluation of the dose–response, and in particular the concentration, above which the agent induces toxic effects, that allows the establishment of the no observed adverse effect level (NOAEL)
- Exposure assessment to identify the concentration of the agent in the relevant medium, time, and routes of human exposure

Toxicological evaluations may take different forms whether they are applied to a new or an existing chemical. In the case of newly developed drugs, pesticides, or new chemicals, a stepwise procedure is used starting from structure–activity considerations and simple *in vitro* and *in vivo* short-term tests. Depending on the hazard potential of the agent, studies are extended to evaluate long-term effects by repeated dose studies, investigating toxicokinetics, and toxic mode of action. For existing chemicals, the available information is collected and a risk assessment is performed based on exposure, dose–response relationship for the critical effects, and mode of action.

In general, a stepwise procedure provides information on the reactivity of the tested compound, on its absorption and distribution in the organism, and on the critical targets. This allows deciding whether further testing is required by repeated dose studies in animals for 28 and 90 days. Depending on the results and the intended use of the chemical, a 6-month or lifetime study is performed to evaluate potential effects upon long-term exposure including carcinogenicity.

Information on toxicokinetics, mechanisms, or mode of action allows evaluating the relevance of the findings to humans and appropriate risk assessment for a given or potential human exposure. For regulatory purposes, acceptable exposure limits can be defined; in case of non-threshold mechanisms (e.g., genotoxic carcinogens), the risk at a given exposure is determined.

The sensitivity of analytical procedures is so advanced that it is possible to identify chemicals at any concentrations in any product in human use and even in humans themselves. Several international institutions publish documentations on chemicals (Table 52.1).

To evaluate the risk of such low exposure, it is essential to understand that it is not the presence of a toxic chemical per se, which is of concern, but rather its concentration, which defines whether it is likely to cause an adverse effect.

TABLE 52.1 International institutions that publish documentations on chemicals

American Conference of Governmental Industrial Hygienists (ACGIH): www.acgih.org/TLV/
The Agency for Toxic Substances and Disease Registry (ATSDR): www.atsdr.cdc.gov/
BUA—Advisory Committee on Existing Chemicals (of the GDCh, the German Chemical Society): www.gdch.de/taetigkeiten/bua/berichte__e.htm
The Canadian Centre for Occupational Health and Safety: http://www.ccohs.ca/
Dutch Expert Committee on Occupational Standards (DECOS): www.gr.nl/adviezen.php?Jaar=2012
Environmental Protection Agency (EPA): www.epa.gov
European Centre for Ecotoxicology and Toxicology of Chemicals: www.ecetoc.org/
UK Health and Safety Executive (HSE): www.hse.gov.uk/
International Agency for the Research of Cancer (IARC): http://monographs.iarc.fr/ENG/Classification/index.php
International Programme on Chemical Safety: www.inchem.org/
The Japanese Association of Industrial Health: http://joh.med.uoeh-u.ac.jp
MAK Commission (German Research Foundation): http://www3.interscience.wiley.com/cgi-bin/mrwhome/104554790/HOME
NIOSH: www.cdc.gov/niosh/homepage.html
The Nordic Expert Group: www.nordicexpertgroup.org//
OSHA: www.osha.gov/
SCOEL–EC Scientific Committee on Occupational Exposure Limits–European Commission: http://ec.europa.eu/employment_social/health_safety/docs_en.htm
US EPA Assessing Health Risks from Pesticides (Pesticides: Topical and Chemical Fact Sheets): www.epa.gov/pesticides/factsheets/riskassess.htm

COMPONENTS OF RISK ASSESSMENT

Hazard Identification

Chemicals induce local and/or systemic effects (e.g., embryotoxicity, hepatotoxicity, neurotoxicity) after absorption from the gastrointestinal tract, skin, or lungs. Reactivity, solubility, and metabolism of the chemical, its metabolites, and their distribution within the organism determine the organ that is the target of the critical effects. Irritation or corrosion may occur when the chemical comes in contact with the skin, mucous membranes of the eye, gastrointestinal tract, or respiratory system. Distribution and metabolism of the chemical can result in various systemic effects in the critical target organ (e.g., liver, kidney, central and peripheral nervous systems). Histopathological and biochemical changes have been the major parameters used to detect organ toxicity. Increasing availability of sensitive methods in analytical chemistry and molecular-biological approaches including toxicokinetics and the various "omics" has significantly improved our knowledge on toxicity mechanisms whereby cellular and subcellular functions are impaired, cell responses to toxic insults, differences among species, and consequences of exposure to high and low concentrations over times.

Acute intoxication usually occurs in response to large doses. Chronic effects are seen after repeated exposure during which the chemical is present at critical concentrations at the target organ, leading to persistent and possibly cumulative damage. Some chemicals, such as 2,3,7,8-tetrachlorodibenzo-p-dioxin (TCDD), accumulate in tissues, and in particular in body fat, because they are lipophilic and not well metabolized. In humans, the half-life of TCDD excretion is about 8 years. In laboratory animals and humans, TCDD induces tumors in various organs. Since TCDD does not induce DNA damage or mutations, the carcinogenic effect is believed to have a threshold, meaning that there is a dose below which no adverse effects are observed.

Sensitization and allergic responses to sensitizing agents are considered to also require a threshold dose to be reached, although at very low doses. The NOAELs of these effects are rarely known. When establishing acceptable exposure standards, thresholds are not considered acceptable for genotoxic carcinogens because any genotoxic event is considered persistent.

Dose–Response and Toxic Potency

The Paracelsian admonition teaches us that occurrence and intensity of toxic effects are dose dependent (see also Chapters 1, 8, and 9). His paradigm addresses the concept of threshold effect, which implies knowledge of the dose–response relationship. Animal or human exposure is usually expressed as dose (i.e., mg of the chemical/kg body weight/day) resulting from oral, inhalation, or dermal exposure or a sum

COMPONENTS OF RISK ASSESSMENT 667

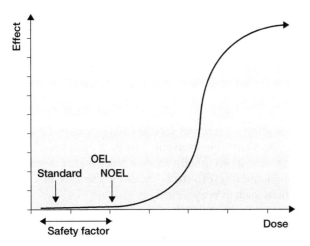

FIGURE 52.1 Dose–response curve and regulatory values. The log dose is on the *x*-axis and % response (effect) is on the *y*-axis. The figure illustrates the location of regulatory values, such as NOEL, occupational exposure levels (OELs), or environmental standards such as acceptable daily intake (ADI). Note that a doubling of dose in the lower or upper part of the S-shaped curve results in small increases in the effects, whereas in the steep part, the increases are much more prominent.

of them. The external dose leads to a specific internal dose, which depends on the amount absorbed via the different routes. Absorption rates via the different routes can vary significantly, although inhalation exposure usually leads to the highest internal doses compared to oral or dermal exposure. For example, about 50% of cadmium in inhaled air (e.g., in tobacco smoke) is absorbed in the lung, whereas cadmium absorption rate in the gastrointestinal tract is about 10%. Ultimately, it is the dose reaching the cellular target over a given time period that defines the toxicologically relevant response. The dose that defines the toxic potency of a chemical is the product of the interrelated external, internal, and target doses. No toxic effects will be seen if the dose is below the NOAEL, whereas effects will increase with increasing exposure beyond this dose. The dose–response curve may be expressed using a variety of mathematical formulas. Using the semilogarithmic form of the dose–response relationship, the curve is sigmoid in shape and varies in slope from chemical to chemical. Thus, if the curve is shallow, doubling of the dose results in a small increase in the effects, whereas if the slope is steep, doubling of the dose results in a severalfold increase in the effects (see Fig. 52.1). The log of the dose is plotted on the x-axis and increases toward the right. The location of the curve on the x-axis is a measure of the potency of the chemical.

Exposure Assessment

Exposure defines the amount of a chemical an individual or a population comes in contact with via inhalation, oral, and dermal routes. According to the general principle of toxicology, the consequences of human or environmental exposures depend on the amount of the chemical an individual (or a population) is exposed to and the duration of the exposure. Thus, exposure assessment or prediction of exposure is an ultimate requirement to assess the risk and to decide on the need for regulations. Since occupational exposure is regular and repetitive, it can be easily measured, for example, in the air of the workplace, using personal monitoring equipment or biomonitoring. Exposure of the general population is more difficult to assess. In general, it depends on the simultaneous presence of the chemical in indoor/outdoor air, drinking water, food, and other products (aggregate exposure). Moreover, frequency, duration, and site of exposure and concentration and weight of substance in the products need to be considered. Children represent a special case, as they may be exposed to chemicals released from items, such as toys, during mouthing, via skin contact, or ingestion of contaminated dust or soil. Exposure can be modeled based on data such as frequency of mouthing, release rate of the specific compound from the toy during mouthing, and absorption rates from oral cavity and gastrointestinal tract.

The rate of absorption through the skin also determines the internal exposure (body burden) to the chemical. However, external exposure does not always correlate to internal exposure; thus, using the aforementioned parameters to predict exposure often leads to overestimation of the latter. Biomonitoring of the compound or its reaction products in the exposed individuals provides the most reliable estimate of internal exposure. However, dose–response curves usually provide a correlation between external dose and effects. Therefore, risk assessment of an internal exposure either requires knowledge of the dose–response of internal exposure versus adverse effects or information on the extent of correlation between external and internal doses. Risk assessment is even more complicated when mixtures of chemicals are the source of exposure.

Ultimately, it is the dose that reaches the cellular target over a given time period that results in the toxicological response. Thus, the toxic potency of a chemical is the product of the interrelated external, internal, and target doses, which results from the multiple pathways and routes of exposure to a single chemical (aggregate exposure). In case of existing chemicals, an appropriately designed program to measure the chemical in the different media will provide the necessary information.

The measurement of the external dose is either done on collected samples (e.g., food) or by direct measurement (e.g., in the ambient air). In case collected samples are used, representative sampling and appropriate storage conditions as well as accurate and reproducible measurement techniques are essential. This also applies to biomonitoring programs.

In the case of new chemicals, such data are not available; therefore, modeling of exposure is the only option.

In the EU Technical Guidance Document on Risk Assessment Part I [1], the following core principles for human exposure assessment for new and existing chemicals and biocides are listed:

- Exposure assessments should be based upon sound scientific methodologies. The basis for conclusions and assumptions should be made clear and be supportable and any arguments developed in a transparent manner.
- The exposure assessment should describe the exposure scenarios of key populations undertaking defined activities. Such scenarios that are representative of the exposure of a particular (sub)population should, where possible, be described using both reasonable worst-case and typical exposures. The reasonable worst-case prediction should also consider upper estimates of the extreme use and reasonably foreseeable other uses. However, the exposure estimate should not be grossly exaggerated as a result of using maximum values that are correlated with each other. Exposure as a result of accidents or from abuse shall not be addressed.
- Actual exposure measurements, provided they are reliable and representative for the scenario under scrutiny, are preferred to estimates of exposure derived from either analogous data or from the use of exposure models.
- Exposure estimates should be developed by collecting all necessary information (including that obtained from analogous situations or from models), evaluating the information (in terms of its quality, reliability, etc.), and thus enabling reasoned estimates of exposure to be derived. These estimates should preferably be supported by a description of any uncertainties relevant to the estimate.
- In carrying out the exposure assessment, the risk reduction/control measures that are already in place should be taken into account. Consideration should be given to the possibility that, for one or more of the defined populations, risk reduction/control measures that are required or appropriate in one use scenario may not be required or appropriate in another (i.e., there might be subpopulations legitimately using different patterns of control that could lead to different exposure levels).

Biomonitoring of exposure is the best tool to determine actual individual exposure through measurements of the concentration of the chemical or its metabolites in blood, critical organs, urine, or exhaled air. It allows us to measure:

- The amount of a chemical taken into the organism by all routes (aggregate exposure)
- The metabolic fate of the chemical, its persistence in the organism, its rate of elimination, and the total body burden at the time of measurement
- The amount of the chemical and/or metabolites that reach the target organs

This procedure is also helpful to evaluate whether an environmental exposure, such as increased indoor air concentrations or contaminated dust or soil that might be ingested by children, actually leads to an increased body burden affecting macromolecules like proteins or DNA. The latter does not automatically mean a genotoxic effect, because mutations usually occur at much higher doses.

Biomonitoring of effects measures changes in cellular functions such as enzyme activity.

Risk Characterization

During the process of risk assessment, it is necessary to distinguish between reversible and irreversible effects. The dose–response curves for chemicals that induce reversible effects display a region below which no effect can be observed. The highest dose at which no effect is observed is the NOAEL; the point at which effects become detectable is the lowest observed adverse effect level (LOAEL). A threshold is not the equivalent of a NOAEL, since it describes concentration or exposure where the slope of a dose–response curve changes.

If damage resulting from exposure is not repaired, the effect persists and accumulates upon repeated exposure. In such cases, a NOAEL cannot be determined and every exposure is related to a defined risk. Reversibility depends on the regenerative and repair capacity of cells, subcellular structures, and macromolecules during and after exposure. For example, epithelial cells of intestinal tract or liver have a high regenerating capacity and rapidly replace damaged cells by increasing cell replication; by contrast, the highly specialized cells of the nervous system have lost this capacity during natal and postnatal development. Consequently, damaged neurons, in general, are not replaced, at least in the adult, as only in very few brain areas some neuroregeneration can take place.

For chemicals that induce reversible effects, the NOAEL of the most sensitive end point is determined and compared with the human exposure to describe the margin of exposure (MOE) (or margin of safety (MOS)), which describes the difference between NOAEL and actual human exposure. If the NOEL is derived from animal experiments, a MOE of 100 is desirable, which takes into account intra- and interspecies differences, each of them contributing by a factor of 10. Thus, a MOE of 10 is sufficient if the NOEL is derived from human data (see Fig. 52.2). The acceptable MOE can be further reduced when there is evidence that no individual differences in metabolism and sensitivity exist.

Covalent binding of genotoxic mutagens and carcinogens to DNA has been considered as an irreversible event. Since the dose–response of mutations parallels that of DNA adducts and is measured at higher exposures, DNA adducts,

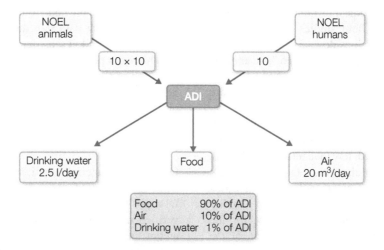

FIGURE 52.2 Acceptable daily intake (ADI) and maximal acceptable concentrations in air, food, and drinking water.

rather than mutations, are taken as indicators of exposure. Moreover, increasing knowledge about DNA repair mechanisms, role of tumor suppressor genes, apoptosis, and background mutation rates suggests that the assumption that genotoxic effects also exhibit a threshold becomes increasingly plausible (for further reading, see [2]). However, to date, the general agreement remains that the potency of genotoxic carcinogens increases with increasing dose and that a NOAEL cannot be identified. As a consequence, any exposure is associated with a certain risk and the risk at a given exposure needs to be estimated by linear extrapolation from the dose–response data obtained from experimental studies in animals or from data obtained from humans.

TOXICOLOGICAL EVALUATION OF NEW AND EXISTING CHEMICALS

The various toxic effects that chemicals may exert and the different applications for which chemicals are designed require in-depth understanding of the cause and effect relationship, that is, knowledge of the specific organs targeted by a given chemical. As a result, toxicologists tend to focus on specific organs, specific applications (e.g., pesticides or drugs), specific compounds (e.g., metals or solvents), or specific effects (e.g., carcinogenicity).

General Requirements for Hazard Identification and Risk Assessment

Sufficient information on the hazardous properties of a chemical requires investigation of:

- Acute, subchronic, and chronic toxicity (oral, inhalation, dermal)
- Irritation (skin, mucous membranes, eye) and phototoxicity
- Sensitization and photosensitization
- Genotoxicity (*in vitro* and *in vivo* methods)
- Carcinogenicity (lifetime studies)
- Reproductive toxicity
- Toxicokinetics
- Mode and mechanism of action

In all studies, it is essential to identify the slope, possible thresholds, NOAEL, and LOAEL of the dose–response curves. The maximal tolerated dose (MTD) is required to show that the test system is sensitive in measuring the toxicity. In animal studies, a 10% reduction in body weight or any other effect is the preferred parameter. The problem with this approach is that usually MTDs are high as compared to human exposure. Therefore, they may lead to disturbances of metabolism, damage repair systems, etc. that do not occur at the lower concentrations to which humans are exposed. For reproducibility of the tests and for acceptance by regulatory agencies, standardized study protocols for each test have been developed (see "Test Guidelines").

Acute Toxicity, Subchronic Toxicity, and Chronic Toxicity

Acute toxicity studies describe toxic effects assessed after a single administration of the chemical to rodents and are primarily aimed at establishing a dose range in which the chemical is likely to produce lethality. After dosing, the animals are observed over a period of 1–2 weeks to determine immediate or delayed effects. It is possible to include additional investigations to examine specific toxic effects and/or mechanisms. Having established the lethal dose range, the chemical may be examined for effects produced upon repeated administration. Common practice in such repeated dose studies is to treat animals each day for a few weeks

(28 or 90 days: subchronic studies) or months (6 months or lifetime: chronic studies). These studies usually include rodents, but larger species such as dogs and, in the case of new drugs, monkeys or apes may be employed. The animals must be observed for effects on both general and specific organ toxicities. At the end of these studies, the animals are examined for gross and microscopic pathology.

The intent of these studies is to examine the likelihood of the development of adverse effects after long-term exposure. Usually, three doses are applied: low to identify the NOAEL, mid to identify the LOAEL of the critical targets, and high, which should be the MTD. Ideally, mid and high doses should lead to clear toxic effects without producing lethality because death of animals precludes observations for the entire intended time period and impairs evaluation of the results. To assure that the entire effective dose range is adequately encompassed, a chronic study is always preceded by a short-term dose-finding study, whose results are used to select appropriate doses for the chronic study. Lifetime studies are specifically focused on cancer.

Irritation and Phototoxicity

Dermal and eye irritation resulting from exposure to compounds is evaluated by studies in animals (and humans) prior to testing for sensitization. For eye irritation, appropriate *in vitro* alternatives are available. Dermal studies in animals and humans are usually performed by using a single occluded patch under the same conditions applied when testing skin sensitization. Phototoxicity and photoallergic reactions have to be expected when compounds show significant absorption in the ultraviolet range (290–400 nm). Using the test strategy for irritation, an additional patch site is irradiated immediately after application of the test substance or after patch removal. Phototoxicity can also be tested by validated *in vitro* tests, such as uptake of Neutral Red by 3T3 cells. If such test is negative, further *in vivo* testing may not be necessary.

Sensitization and Photosensitization

To detect the sensitizing potential of products, the choice of a relevant animal is crucial. In many cases, animal models may be inappropriate for detection of a sensitizing potential; thus, most dermatologists prefer studies in humans. An acceptable alternative may be studies in nonhuman primate species, like cynomolgus or rhesus monkeys. Generally, the Buehler guinea pig test and the popliteal lymph node assay (PLNA) in mice are used in the preclinical testing program. PLNA has received great attention because it is the only reliable test for screening compounds that cause sensitization via routes other than skin. So far, the test has been successfully applied to determine relative potencies of contact allergens and has been reported to closely correlate with NOELs established from human repeat patch testing. When the animal data indicate a weak contact sensitizing potential, human skin sensitizing testing is conducted usually by a human repeated insult patch test (HRIPT). In any case, detection of antibodies in the serum during the studies may be appropriate, using specific ELISA methods or other bioassays to measure antibodies.

Genotoxicity

Generally, a bacterial mutation assay and an *in vitro* cytogenetic assay are performed. Because they are afflicted with false-positive and false-negative effects, they should be verified using the mouse bone marrow micronucleus test and/or the chromosomal aberration test, which are reliable and widely used test systems and detect aneugens as well as clastogens. Chemicals that yield positive responses to these tests frequently do not undergo further development. Conversely, those that appear not genotoxic may be carried forward and evaluated in lifetime carcinogenicity studies in rodents, when a long-term exposure to humans is expected.

Carcinogenicity

The design of carcinogenicity studies *in vivo* is similar to that of chronic studies. At least three adequately spaced doses are tested, the highest dose being the MTD. Usually, relatively large doses/concentrations are used to maximize the chance of finding a possible increase in tumor incidence in the relatively few animals (50 per sex/dose). If necessary, additional animals are included for investigations at 12 and/ or 18 months. Rats and mice are used because of their relatively short lifespan (about 2 years) and the availability of information on their susceptibility to tumor induction, physiology, and pathology. A large historical database on tumor incidence in most strains and tissues exists, which is important given the large variability in tumor incidence existing among untreated animals and among different strains. The incidence of spontaneous and substance-induced tumors increases in older animals. Therefore, it is necessary to terminate the study after a defined period to avoid the impact of different lifespans on data interpretation.

Since MTD is difficult to predict, severe toxicity may occur at this dose. This requires specific consideration when interpreting the results because in such cases, the metabolism of the animal may be overwhelmed and/or detoxifying mechanisms may not be operative. It should also be remembered that a number of tumors are species specific and, therefore, not relevant to man. One example is the induction of renal tumors mediated by the male rat specific urinary protein $2\alpha u$-globulin. Moreover, compared to humans, rodents are more sensitive to compounds that disturb thyroid hormone metabolism and are also much more sensitive to compounds that induce peroxisome proliferation in the liver.

Tumors resulting from these types of mechanisms are poorly relevant to humans. In any case, incidence and type of tumors found have to be evaluated by experts, and the underlying mechanisms as well as possible high-dose effects, such as overwhelmed metabolism of the test compound, have to be taken into account when judging whether a substance should be considered carcinogenic to humans.

Toxicity for Reproduction

Reproductive toxicity tests investigate the effects on fertility and development. They may only be needed if there is indication that the analyzed chemical, or its critical metabolites, can reach the gonads of the parent animals and/or the embryo or fetus. Testing includes male and female fertility and teratological or developmental effects induced prenatally in utero or after birth in the offspring during lactation. Tests such as the reproduction/developmental toxicity screening test (OECD 421), the combined repeated dose toxicity study with the reproduction/developmental toxicity screening test (OECD 422), or the appropriate standard tests to evaluate the effects on reproduction (One-generation reproduction toxicity, OECD 415) and prenatal developmental study (OECD 414) may be performed (for OECD test guidelines, see [3]).

Toxicokinetics

A chemical may enter the body via food, air, or skin contact. The amount absorbed depends on its concentration in the different media and on physical–chemical parameters such as solubility in water and fat, stability, and route of exposure. Toxicokinetics describe absorption, distribution, metabolism, and elimination (ADME) of a chemical in experimental animals and humans. Of specific importance for interpretation of animal studies and evaluation of intraspecies differences is the comparative information on the external exposure and the actual dose that reaches the critical target organ, cellular or subcellular targets (Fig. 52.3).

Upon inhalation or skin penetration, the compound directly enters the circulation and distributes to the organs. When absorbed from the gastrointestinal tract, the chemical enters the liver via the portal vein. The epithelial cells of the gut wall and the liver demonstrate a large capacity for metabolizing chemicals; therefore, a compound may be extensively metabolized by this "first-pass effect" before entering the (cardiovascular) systemic circulation. Larger metabolites such as the glucuronosyl conjugates can be excreted via the biliary system into the duodenum where the conjugates may be hydrolyzed; in this way, the original compound is reabsorbed and reenters the liver. This process is defined as enterohepatic circulation. Inhalation or dermal exposure to a chemical and intravenous or intraperitoneal injection may result in different effects compared to oral exposure because of the "first-pass effect" (see also Chapters 4 and 6).

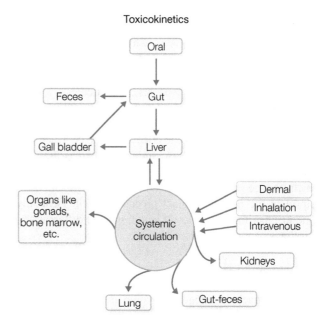

FIGURE 52.3 Routes of exposure and systemic distribution of a compound within the organism. After oral ingestion, the compound reaches the liver, where it can be extensively metabolized. Upon inhalation or dermal exposure and intravenous application, the compounds reach the circulation without major metabolism.

After entering the cardiovascular system, the chemical or its metabolites distribute to the organs where they can accumulate in tissues such as fat or bones or be further metabolized. Reactive metabolites may interact with tissue components and induce cellular damage (see also Chapters 3, 4, 5, and 7). This "tissue dose," that is, the concentration of a chemical or its metabolite in the critical target over a given time, is an important factor that helps to understand the correlation between internal and external (environmental) exposures in relation to toxicity. Comparing tissue doses in different species at similar exposures helps us to understand species differences in sensitivity to chemicals as well as interindividual variations. The chemical or its more water-soluble metabolites are primarily excreted via the kidneys or biliary system. Volatile compounds may be exhaled. The great variety of processes observed during absorption, metabolism, distribution, and excretion cannot be predicted by modeling or by *in vitro* experiments without confirmatory data from animals and humans.

The array of enzymes involved in the metabolic activation (phase I reactions) and inactivation (phase II reactions) of chemicals including drugs is described in Chapter 6. Phase I reactions introduce groups (e.g., —OH) into molecules that make them more polar and more reactive to phase II metabolism. Phase II reactions, such as conjugation with more polar moieties like sulfate, glucuronic acid, or glutathione, usually result in sufficiently water-soluble complexes for excretion (see Chapter 6 for further details).

Mode and/or Mechanism of Action

There are many mechanisms by which chemicals or other stressors, like heat or radiation, can lead to toxicity. Knowing modes or mechanisms underlying chemical toxicity is essential to understand species specificities, species differences, and sensitive populations and to interpret data regarding threshold or nonthreshold effects. They also help evaluate the relevance of toxic effects to humans when data are derived from experimental animals. Whereas the toxic mechanism is often not known in detail, modes of action, which can be described in a less restrictive manner, are helpful in the risk assessment process as well. Generally, toxic mechanisms or modes of action may be differentiated as follows:

1. Physiological changes are modifications to the physiology and/or response of cells, tissues, and organs. These include mitogenesis, compensatory cell division, escape from apoptosis and/or senescence, inflammation, hyperplasia, metaplasia and/or preneoplasia, angiogenesis, alterations in cellular adhesion, changes in steroidal estrogens and/or androgens, and changes in immune surveillance, which however may be adaptive and do not necessarily lead to adverse effects.
2. Functional changes include alterations in cellular signaling pathways that manage critical cellular processes. These include dose-dependent alterations in phase I and phase II enzyme activities, depletion of cofactors and their regenerative capacity, alterations in the expression of genes that regulate key functions of the cell (e.g., DNA repair, cell cycle progression), posttranslational modifications of proteins, regulatory factors that determine the rate of apoptosis, secretion of factors related to stimulation of DNA replication and transcription, or gap junction-mediated intercellular communication.
3. Molecular changes include reversible or irreversible changes in cellular structures at molecular level, including genotoxicity. These include formation of DNA adducts and DNA strand breaks, mutations in genes, chromosomal aberrations, aneuploidy, and changes in DNA methylation patterns.

Data derived from gene expression microarrays or from high-throughput testing of agents will become increasingly available. As long as the information is not related to functional or morphological changes, their applicability is poor and there is the possibility to overinterpret such data, although they might be useful in assessing mechanisms.

Mechanistic information is most relevant for the evaluation and classification of carcinogens. If the carcinogenic effect is induced by a specific mechanism that does not involve direct genotoxicity, such as hormonal deregulation, immune suppression, and cytotoxicity, the detailed search for the underlying mode of action may allow identification of a NOEL. This can also be considered for materials, such as poorly soluble fibers, dusts, and particles, which induce persistent inflammatory reactions as a result of their long-term physical presence that ultimately may lead to cancer.

TEST GUIDELINES

For reproducibility and acceptance by regulatory agencies, standardized study protocols for each test have been developed. Moreover, only qualified laboratories are accredited to perform tests for regulatory purposes, and they must adhere to good laboratory practice (GLP) guidelines. These guidelines describe how to report and archive laboratory data and records. The GLP guidelines also require standard operating procedures (SOPs), statistical procedures for data evaluation, instrumentation validation, material certification, personnel qualification, proper animal care, and independent quality assurance (QA). GLP regulations have been developed by the US Food and Drug Administration (FDA) (http://www.fda.gov/ora/compliance_ref/bimo/glp/78fr-glpfinalrule.pdf) and by the Organization for Economic Cooperation and Development (OECD) (http://www.oecd.org/document/63/0,2340,en_2649_201185_2346175_1_1_1_1,00.html).

The regulation on Registration, Evaluation, Authorisation and Restriction of Chemicals (REACH) issued by the European Commission (EC) to ensure a high level of protection of human health and environment from risks that can be posed by chemicals requires use of test methods as described in the Commission Regulation (EC) No 440/2008 or any methods based on internationally recognized scientific principles (Table 52.2). Use of the data for classification and labeling (C&L) of chemicals according to the Globally Harmonized System (GHS) is described in the European Chemicals Agency (ECHA) Guidance on the Application of the CLP Criteria, Version 4.0, November 2013. For pharmaceuticals, the International Conference on Harmonisation of Technical Requirements for Registration of Pharmaceuticals for Human Use (ICH) has produced a comprehensive set of safety guidelines to uncover potential risks (www.ich.org; Safety Guidelines). These include repeated dose toxicity testing, toxicokinetics and pharmacokinetics, genotoxicity, carcinogenicity, and immunotoxicology studies. They refer in part to OECD guidelines for test descriptions and include guidance for result interpretation. Guidelines prepared by the US FDA or the European Medicine Agency (EMA) are generally in agreement with the ICH guidelines.

Alternatives to Animal Experiments

In recent years, the refinement, reduction, and replacement (3R) strategy for substituting animal experiments has

TABLE 52.2 Use (tons of annual production plus import)-dependent data requirement of the European REACH regulation for protection of human health

Annual production (years)	1–10 t/a	10–100 t/a (2014–2017)	<100 t/a (2011–2013)	>1000 t/a (2008–2010)
Toxicokinetics		Assess	Assess	Assess
Acute toxicity		Oral, dermal, inhalation	Oral, dermal, inhalation	Oral, dermal, inhalation
Irritation	Skin/eye *in vitro*	Skin/eye *in vivo*	Skin/eye *in vivo*	Skin/eye *in vivo*
Sensitization	LLNA	Immuntox	Immuntox	Immuntox
Repeated dose		28 days	90 days	≥12 months
Genotoxicity carcinogenicity	Bacterial test *in vivo*	Cytogenetic gene mutant *in vitro*	*In vitro, in vivo* mutagenicity	Mutagenicity carcinogenicity
Reproduction		Reproduction/development screen test, toxic development	Development 2 generation	Development 2 generation

LLNA, local lymph node assay; t/a, tons of annual production plus import.

received increasing attention. The aim of this strategy is to preferentially use *in vitro* studies and consideration of structure–activity relationships and mathematical models to waive and reduce studies in laboratory animals. Most success in the development of alternative methods has been obtained in local toxicity and acute toxicity testing, when effects of the chemical, but not of its possible metabolites, are determined. The standardized *in vitro* mutagenicity/genotoxicity tests also include an evaluation of drug metabolism. However, since these tests are frequently afflicted by false-positive results, it is necessary to verify their outcome by appropriate *in vivo* testing. *In vitro* studies allow identification of hazardous properties of substances but only those detectable by the specific test system. Even when the test system includes an evaluation of metabolism of chemicals, its appropriateness must be verified in intact organisms. Consequently, identification of all relevant end points, their dose–response, thresholds, and NOELs can only be determined in the intact animal by repeated dose studies. In the absence of such information, hazard identification is incomplete, and without information of the dose–response relationship obtained from these studies, there is no basis for appropriate assessment of the risk of human exposure. Thus, validated alternatives to the methodologies for long-term testing for systemic and reproductive toxicity and carcinogenicity, which require the highest number of animals, are lacking. In spite of this, Regulation (EC) No 1223/2009 requires validated alternative methods, in particular *in vitro* replacement methods for the safety evaluation of cosmetic substances and products.

GENERAL APPROACH FOR HAZARD IDENTIFICATION AND RISK ASSESSMENT

For safety evaluation of substances, all available scientific data are considered, including the physical and chemical properties of the compound under investigation, results obtained from quantitative structure–activity relationship (QSAR) calculations, chemical categories, grouping, read-across, physiologically based pharmacokinetics (PBPK)/toxicokinetics (PBTK) modeling, *in vitro* experiments, and data obtained from animal studies. Also, clinical data, epidemiological studies, information derived from accidents, and any other human data are taken into consideration.

The stepwise procedure usually starts with the determination of the LD_{50} and the evaluation of genotoxicity by an *in vitro* bacterial test system (Ames test) and cytogenicity in mammalian cells. In case of positive results, the results are verified *in vivo* usually through the mouse bone marrow genotoxicity test. For further evaluation, additional tests including studies on toxicokinetics or potential genotoxic mechanisms are performed (see Fig. 52.4).

This procedure provides information on the reactivity of the test compound and its absorption and distribution in the organism and possibly to the critical target. Based on this information, it is possible to decide whether further testing is required, by repeated dose studies in animals for 28 and 90 days. Depending on their outcome and the intended use of the chemical, a 6-month or lifetime study may be also performed to evaluate potential effects, including carcinogenicity, resulting from long-term exposure. When the detailed toxicological evaluation excludes genotoxic and carcinogenic effects, the NOAEL estimated in long-term studies is the starting point to set the acceptable daily intake (ADI), which is usually 100-fold below the NOAEL (Fig. 52.2). This factor considers the existence of a 10-fold difference between the sensitivity of the experimental animals and humans and another 10-fold due to possible interindividual differences among the human populations. Thus, if the NOAEL is derived from studies in humans, the ADI is set 10-fold below the NOAEL, since only possible interindividual differences have to be taken into account. These factors can be reduced if specific information is available demonstrating that the species–species or intraspecies differences are less than 10-fold. Based on the ADI,

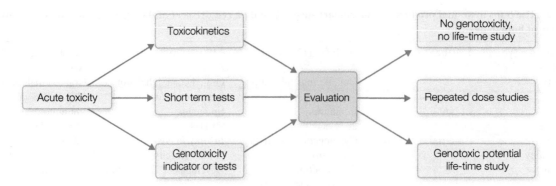

FIGURE 52.4 Stepwise procedure to evaluate the toxic potential of a chemical.

permissible concentrations in food, drinking water, consumer products, indoor and outdoor air, and other environmental compartments are established.

Detailed information on the general principles for the risk assessment of new substances, existing substances, biocidal active substances, or substances of concern is described in the guidance on the information requirements and chemical safety assessment [4].

Evaluation of Mixtures

Humans and their environments are exposed to a wide variety of substances. The potentially adverse effects of substances when present simultaneously have been analyzed in several reviews and documentations. Most recently, the available scientific literature has been analyzed by [5]. The general conclusions are that chemicals with common modes of action will act jointly to produce combination effects that are larger than the effects of each component applied singly (dose/concentration addition). However, effects only occur when the concentrations of the individual compounds are near or above their zero effect levels. For chemicals with different modes of action (independently acting), no robust evidence is available that exposure to a mixture of such substances is of health or environmental concern if the individual chemicals are present at or below their zero effect levels. If no mode of action information is available, the dose/concentration addition method should be preferred. Prediction of possible interaction requires expert judgment and hence needs to be considered on a case-by-case basis.

Evaluation of Uncertainties

Derivation of ADI, derived no-effects level (DNEL), or determination of carcinogenic risk includes uncertainties. For example, the NOAEL may not be a real NOAEL for statistical reasons in that too few animals have been used in the specific experiment. Or the NOAEL may be rather conservative because the next higher dose that determines the LOAEL of a weak adverse effect is 10-fold higher. Usually, this uncertainty is covered by applying assessment factors that build in a margin of error so as to be protective of the population from risks. In case of ADIs or DNELs, the uncertainty factor of 100 covers the uncertainties of inter- and intraindividual differences unless toxicodynamic and/or toxicokinetic information allows its reduction. Whereas the experts who have performed the risk assessment are usually aware of uncertainties, the risk managers tend to use the numbers as such, with the consequence that any exposure even slightly higher than the ADI or DNEL is not considered to be acceptable. It is possible to acknowledge these uncertainties by using statistical approaches to characterize and weight the different assumptions from various components (including dose–response, emissions, concentrations, exposure, valuation). This will help understand how the risk may vary in a population, thereby allowing better mean estimates of risk and magnitude of risk for different individuals. For uncertainty analysis, see [4, 6].

TOXICOLOGICAL ISSUES RELATED TO SPECIFIC CHEMICAL CLASSES

Jurisdictions and regulatory agencies around the world have established a variety of guidelines for risk assessment and permissible exposure standards for chemicals at workplace, at home, and in the general environment. Health risks of chemicals designed for specific applications (e.g., consumer products, drugs, or pesticides) must be assessed when people are exposed in many types of environments. Therefore, all elements of risk assessment—hazard identification, dose–response, exposure, and risk—have to be thoroughly evaluated.

Data requirements for both new and existing chemicals usually depend on annual production rate and extent of human exposure. When there is considerable exposure, regulatory requirements demand an extensive toxicological evaluation of the potential adverse effects of the specific chemical and the likelihood of their expression under the conditions of use or exposure, and the definition of the MOE or the health risk under defined conditions of exposure.

For drugs, special emphasis must be placed on efficacy, therapeutic index, potential side effects, and effects of overdosage.

For pesticides, the relative impact of the chemical on the target versus on the people is a critical requirement. Thus, because of the possibility of contamination of food and other consumer products with the pesticide, an ADI and the MOS need to be established.

To regulate chemicals at the workplace, various governmental and non-governmental institutions are involved in setting occupational exposure standards. Since institutions publish the complete toxicologically relevant information and the justification for the proposed limit values, these documentations are valuable sources for the toxicological database of the compounds. Institutions that publish these documents are listed in Table 52.1.

In 1992, the European Commission estimated that about 100,000 existing chemicals are in use and produced in quantities ranging from less than a ton to several million tons per year. Except drugs and pesticides, data requirement for existing or new chemicals had not been regulated. Since there have been high-volume products with a relatively small database, several programs have been launched to obtain knowledge at least for compounds with high annual production rates. In the United States, EPA has initiated a High Production Volume (HPV) program. In an international cooperation, the OECD has launched the ICCA program, which evaluates and documents the available information on environmental and human health hazards and risks for about 1000 chemicals. In Europe, Risk Assessment Reports under the Existing Chemical Program for about 150 compounds have been produced. Since the REACH regulation became effective in 2008, producers or downstream users have to provide the necessary information to the ECHA. The extent of information to be submitted depends on the amount produced or imported annually, and requirements are highest for compounds of >1000 t/a, less for <100 t/a, and lowest for 10–100 t/a chemicals (Table 52.2; t/a is tons of annual production + import).

As long as there is no specific risk, chemicals will be registered for the intended use. Special attention is paid to carcinogens, mutagens, and reproductive toxicants (CMR compounds) and to chemicals that show bioaccumulation, persistence, and toxicity (BPT compounds) in the environment. Member states can propose classification and labelling of chemicals and restrictions. The consequence of a CMR classification in category 1A or 1B is a ban for consumer exposure. In such cases, industry can apply for authorization for a specific use by providing evidence that there are no alternatives and the risk of consumer exposure is low. Proposals for restrictions also need to demonstrate that there are no less toxic and economically acceptable alternatives and, in case the chemical is further used, the risk of consumers is not tolerable. Both the proposals for authorization and restriction are evaluated by the RAC and the Socio-Economic Committee (SEAC) of ECHA. The latter committee performs a cost/benefit analysis for the restriction or authorization of the chemical and its possible alternatives.

CLASSIFICATION AND LABELLING (C&L) OF CHEMICALS

Criteria for C&L of chemicals have been developed by several national and international agencies. For the most recent guidance, see [7]. It is based on the GHS of classification, labeling, and packaging of chemicals, which has been developed for worldwide harmonization in the evaluation of chemicals. In the Health Hazard part, the ECHA Guidance addresses all end points of toxicological relevance as well as the presence of such chemicals in mixtures. Since the CLP criteria only consider the toxicological potential, not the potency, consideration of high-dose effects that are irrelevant to humans may lead to unnecessary C&L. For example, the hazard concept for classification of carcinogens is based on qualitative criteria and reflects the weight of evidence available from animal studies and epidemiology. Mode of action and potency of a compound are either not taken into account or at best are used as supporting arguments. The advancing knowledge of reaction mechanisms and the different potencies of carcinogens at least have triggered a discussion for a reevaluation of the traditional concept.

The proposals for restriction and authorization require estimation of cancer risk at a given exposure, which is estimated by a linear or sublinear extrapolation from the high-dose effects observed in animals to the usually lower human exposure. However, the European Food and Safety Authority (EFSA) has recommended avoiding this extrapolation because of the inherent uncertainties. Instead, the MOE between a benchmark dose and the T25 calculated from a carcinogenicity study in animals and human exposure should be determined (T25 is the dose giving a 25% incidence of cancer in an appropriately designed animal experiment). A MOE of 10,000 and more is considered to be of minor concern. The advantage is that neither a debatable extrapolation from high to low doses is performed nor are hypothetical cancer cases calculated.

THE THRESHOLD OF TOXICOLOGICAL CONCERN CONCEPT

The threshold of toxicological concern (TTC) is a concept to establish a level of exposure for chemicals, regardless of their chemical-specific toxicity data, below which there is no appreciable risk to humans. The concept is based on knowledge of the chemical structure for evaluating structural alerts, the amount of a specific chemical in a product, and the daily human exposure. So far, the TTC concept has been applied to chemicals in food. It is defined as a nominal oral dose that poses no or negligible risk to human health after a daily lifetime exposure. At a mean dietary intake below the TTC level, toxicology safety testing is not necessary or warranted. In this way, the TTC concept can contribute to a reduction in the use of animals for safety tests. The TTC concept may also represent an appropriate tool to evaluate or prioritize the need for toxicological testing. There is ongoing discussion on its general applicability for safety assessment of substances that are present at low levels in consumer products, such as cosmetics or impurities or degradation products. For a recent evaluation of the general applicability of the TTC concept by the nonfood Scientific Committees of the European Commission, see SCCS/SCHER/SCENIHR [8].

THE PRECAUTIONARY PRINCIPLE

The precautionary principle is a measure to enable rapid response in the case of a possible danger to human, animal, or plant health or to protect the environment. In particular, where scientific data do not permit a complete evaluation of the risk, this principle may be used, for example, to stop distribution or order withdrawal from the market of products likely to be hazardous. Since this description allows various interpretations, a more precise definition is given in the Communication from the Commission of February 2, 2000. This document states that the precautionary principle may be invoked when a phenomenon, product, or process may have a dangerous effect, identified by a scientific and objective evaluation, if this evaluation does not allow the risk to be determined with sufficient certainty. The Commission specifically stresses that the precautionary principle may only be invoked in the event of a potential risk and that it can never justify arbitrary decisions.

Other states use slightly different definitions. For example, the Canada definition is as follows: "The precautionary principle is an approach to risk management that has been developed in circumstances of scientific uncertainty, reflecting the need to take prudent action in the face of potentially serious risk without having to await the completion of further scientific research."

All definitions of the precautionary principle refer to the Rio Conference on Environment and Development (Principle #15 of the June 1992 Declaration).

CONCLUSIONS

Toxicology is charged with describing the adverse effects of chemicals in a qualitative sense and with evaluating them quantitatively by determining how much of a chemical is required to produce a given response. This requires sufficient information on the hazardous properties of a compound, their relevance to humans, and human and environmental exposure, which is a prerequisite for appropriate risk assessment.

TAKE-HOME MESSAGE

- Toxicology is concerned with the health risks of human exposure to chemicals. It describes the adverse effects of chemicals in a qualitative sense and, by determining the no observed adverse effect level (NOAEL), determines whether a chemical at a given exposure is a health risk to humans.
- Since toxic effects are dose dependent, knowledge of the extent and duration of exposure is an integral part of the risk assessment process. Exposure defines the amount of a chemical to which a population or individuals are exposed via inhalation, oral, and dermal routes. Animal or human exposure is commonly defined by mg of the chemical/kg body weight/day.
- The sensitivity of analytical procedures allows identification of chemicals at any concentration in any product for human use and even in humans.
- It is essential to understand that not the mere presence of a toxic chemical is a risk.

REFERENCES

1. European Commission. Technical Guidance Document on Risk Assessment Part I. European Communities, Bruxelles, 2003.
2. Greim H., Albertini R. (eds.) *The Cellular Response to the Genotoxic Insult. The Question of Threshold for Genotoxic Carcinogens*. RCS Publishing, London, 2013.
3. Spielmann H., Vogel R. (2006). REACH testing requirements must not be driven by reproductive toxicity testing in animals. *ATLA*, 34, 365–366.
4. ECHA (2010). Guidance on the information requirements and chemical safety assessment, European Chemicals Agency, 2010. http://echa.europa.eu/. Accessed on December 18, 2014.
5. SCCS/SCHER/SCENIHR (2011). Toxicity and Assessment of Chemical Mixtures. European Union, Brussels.
6. EFSA (2006). Guidance of the Scientific Committee on a request from EFSA related to uncertainties in dietary exposure assessment. *The EFSA Journal*, 438, 1–54.
7. ECHA (2013) Guidance on the application of the CLP criteria, Version 4.0, November 2013. http://echa.europa.eu/. Accessed on December 18, 2014.
8. SCCS/SCHER/SCENIHR (2012). Use of the Threshold of Toxicological Concern (TTC) approach for human safety assessment of chemical substances with focus on cosmetics and consumer products (June 08, 2012). European Union, Brussels.

FURTHER READING

EFSA (European Safety Authority). Opinion of the Scientific Committee on a request from EFSA related to a harmonized approach for risk assessment of substances which are both genotoxic and carcinogenic. http://www.efsa.europa.eu/en/science/sc_commitee/sc_opinions/1201.html. Accessed on December 18, 2014.

Greim H., Snyder R. (eds.) *Toxicology and Risk Assessment. A Comprehensive Introduction.* Wiley, Hoboken, NJ, 2008.

IARC (International Agency for the Research of Cancer). *Monographs on the Evaluation of Carcinogenic Risks to Humans.* WHO, International Agency for Research on Cancer, Lyon, France, 2004.

Opinions of Scientific Committees of DG SANCO. http://ec.europa.eu/health/scientific_committees/environmental_risks/opinions/index_en.htm#id9. Accessed on December 18, 2014.

Paracelsus. Selected Writings, J. Jacobi (ed.), translated by N. Guterman, Princeton University Press, Princeton, NJ, 1951, p. 95.

The Globally Harmonized System for Hazard Communication (2009). In: *Hazard Communication. Occupational Safety & Health Administration*, United States Department of Labor, (June 10, 2009).

53

DRUG INTERACTIONS

Achille P. Caputi, Giuseppina Fava, and Angela De Sarro

By reading this chapter, you will:

- Understand the mechanisms of drug–drug interactions (DDIs) when multiple drugs (including herbal remedies and/or food supplements) are concomitantly administered
- Learn how DDIs can modify drug effects and duration of action

---

In clinical practice, the concomitant use of more than one drug (polypharmacotherapy) is often necessary to reach the desired therapeutic end point in many disease states, such as hypertension, AIDS, heart failure, cancer, etc. However, polypharmacotherapy can lead to interactions of various kinds. The larger the number of drugs taken by a patient, the greater the likelihood of a drug–drug interaction (DDI). Given the large interpatient variability, the precise effects of DDIs are not easily quantifiable. DDIs can give rise to physiological responses to drugs that are exaggerated or insufficient with respect to therapeutic goals (Table 53.1). Some DDIs have beneficial effects that manifest themselves in patients' clinical profile and are therefore clinically relevant (Table 53.2). In some cases, the effects of some DDIs are not clinically significant.

The risk of adverse drug reactions (ADRs) due to DDI increases greatly when the drugs involved have clinically significant toxic effects and narrow therapeutic index, in particular when administered to certain categories of patients such as those with chronic pathologies on long-term medication, patients in intensive care units, and organ transplant patients. The large list of drugs often implicated in clinically relevant DDIs includes anticoagulant, antiarrhythmic, anticonvulsant, antidepressant, antibacterial, antifungal, and antiviral compounds and various antineoplastics and immunosuppressants.

To evaluate the effects of an interaction, several factors should be considered:

- Patient characteristics, for example, geriatric patients are at a higher risk of DDI because they are often under polytherapy for comorbidities.
- Pharmacological anamnesis, for example, the risk of DDI is correlated to the number of different physicians prescribed drugs or to arbitrary self-administered medications, such as over-the-counter (OTC) medicines or herbal/dietary supplements.

In general, DDIs are classified on the basis of their mechanism of action, that is, based on whether they exert a pharmacokinetic or pharmacodynamic action (Fig. 53.1).

PHARMACOKINETIC DRUG INTERACTIONS

Drugs can interact at any pharmacokinetic stage during the processes of absorption (e.g., calcium and other cations interfere with the absorption of other drugs), distribution (e.g., drug displacement from plasma protein binding due to another drug having higher affinity for plasma proteins), biotransformation (e.g., induction or inhibition of drug metabolism by another drug), and excretion (e.g., competition between drugs for tubular secretion) (Table 53.3).

TABLE 53.1 Examples of drug interactions exploited for therapeutic purposes

Drugs	Clinical indication
β-Lactams and aminoglycosides	Bacterial infections
Transcriptase inhibitors and protease inhibitors	AIDS
NSAIDs and codeine	Pain
Diuretics and β-blockers, ACE inhibitors, calcium antagonists	Arterial hypertension
$β_2$ agonists and corticosteroids, antimuscarinic agents, leukotriene antagonists	Bronchial asthma
Corticosteroids and tacrolimus, mycophenolate, azathioprine	Autoimmune diseases and transplants
Antimetabolites and alkylating agents, anthracyclines	Cancer
L-DOPA and DOPA decarboxylase inhibitors, MAO-B inhibitors, antimuscarinic agents	Parkinson's disease
Carbamazepine and phenytoin, valproate	Epilepsy

TABLE 53.2 Examples of drug combinations that can produce exaggerated or inadequate drug responses

Antacids and fluoroquinolones	Reduction of antibiotic absorption from the small intestine	Reduced antibacterial effect
Antacids and ketoconazole	Reduced absorption from the small intestine	Reduced antifungal effect
Ampicillin and estrogens	Interruption of the enterohepatic circulation of estrogens	Risk of pregnancy
NSAIDs and glucocorticoids	Increased risk of gastric mucosal lesions	Gastric hemorrhage
NSAIDs and methotrexate	Increase in plasma levels of methotrexate	Possible increase in methotrexate toxicity
NSAIDs and β-blockers	NSAIDs inhibit prostaglandin synthesis in the kidney	Reduced antihypertensive effect of β-blockers and ACE inhibitors
Erythromycin and warfarin	Inhibition of biotransformation	Risk of hemorrhage
Erythromycin and cyclosporine	Inhibition of biotransformation	Risk of nephrotoxicity
Ketoconazole and warfarin	Inhibition of biotransformation	Risk of hemorrhage
Ketoconazole and cyclosporine	Inhibition of biotransformation	Risk of nephrotoxicity
Rifampicin and warfarin	Inhibition of biotransformation	Reduced anticoagulant effect
Rifampicin and oral contraceptives	Inhibition of biotransformation	Risk of pregnancy
Aspirin and warfarin	Potentiation of warfarin effects	Risk of hemorrhage
Statins and fibrates	Potentiation of muscular toxicity	Risk of rhabdomyolysis

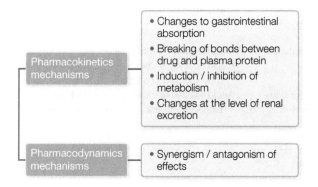

FIGURE 53.1 Classification of drug interactions.

Interactions during Drug Oral Absorption

There are several ways in which a drug may interfere with the absorption of another drug along the entire gastrointestinal tract (GIT). In addition, the bioavailability of several orally administered drugs may be influenced by the absence or presence of food in the GIT (Table 53.4).

Changes in Gastric pH Antacids and inhibitors of gastric secretions (e.g., H_2 receptor antagonists and proton pump inhibitors), by increasing the pH of the gastric milieu, can alter ionization and lipophilicity of other drug species, thereby affecting their absorption through lipid membranes. Acidic drugs such as salicylates, some antibiotics (e.g., fluoroquinolones), and antifungal agents (e.g., ketoconazole) are absorbed in smaller quantities if the gastric pH is elevated, resulting in a reduction of their therapeutic effect (Table 53.2). On the other hand, elevated pH favors absorption of basic drugs such as some antidiabetic agents (e.g., tolbutamide), resulting in enhancement of their therapeutic effect.

Neutralization Neutralization (see also section "Pharmaceutical Interactions and Incompatibility") is a physicochemical reaction between two drugs that occurs at any point along the GIT before absorption and results in the formation of chelates or otherwise nonabsorbable drug complexes. Antacids, such as magnesium hydroxide, aluminum hydroxide, or iron salts, form chelate complexes with tetracyclines, thus reducing their absorption. Similarly, cholestyramine, a resin with ionic exchange potential, adsorbs various drugs such as warfarin,

TABLE 53.3 Examples of pharmacokinetic interactions

Drug A		Drug B	Effect on drug B
Absorption phase			
Cholestyramine	+	Digitalis	Reduced bioavailability
Calcium salts	+	Bisphosphonates	Reduced bioavailability
Distribution phase			
Phenylbutazone	+	Warfarin	Potentiation (displacement)
Quinidine	+	Digoxin	Potentiation (displacement)
Metabolic phase			
Cimetidine	+	Warfarin	Potentiation (enzymatic inhibition)
Ketoconazole	+	Cyclosporine	Potentiation (enzymatic inhibition)
Phenobarbital	+	Phenytoin	Reduction (enzymatic induction)
Rifampicin	+	Tacrolimus	Reduction (enzymatic induction)
Elimination phase			
Probenecid	+	Penicillin	Potentiation
Salicylates	+	Methotrexate	Potentiation
Thiazide diuretics	+	Lithium carbonate	Potentiation

TABLE 53.4 Interactions between drugs and dietary components

Drug	Mechanism of interaction	Outcome of interaction or conditions for relevance	Dietary recommendation
ACE inhibitors	Pharmacodynamic interaction	Risk of hyperkalemia	Avoid excessive intake of potassium (banana, spinach)
Alendronic acid	Chelation	Risk of therapeutic failure	Avoid concomitant intake of food or milk
MAO inhibitors	Inhibition of deamination of amines	Risk of hypertensive crisis	Diet should be low in tyramine-rich foods such as present in diet cheese (mozzarella, Parmigiano), yoghurt, sour cream, processed meats (sausage, hams), liver, dried fish, caviar, avocado, banana, yeast extract, raisins, cabbage, soy sauce, beans, red wine, some beers, and certain caffeine products
Itraconazole capsule	Acid-dependent solubility	Improved therapeutic response	Ingest with food
Itraconazole solution	Food increases the first-pass metabolism of the drug	Advantage in anorexic patients	Ingest without food
Norfloxacin	Chelation	Risk of therapeutic failure	Avoid calcium-containing foods
Spironolactone	Pharmacodynamic interaction	Risk of hyperkalemia	Avoid excessive intake of potassium-containing food (spinach, banana)
Warfarin	Direct antagonism by dietary vitamin	Relevant in the case of daily intake of potassium-rich foods	Avoid excessive intake of vitamin K (broccoli, Brussels sprouts, spinach, and cabbage)
Tetracyclines	Chelation	High risk of therapeutic failure	Avoid calcium-containing foods

Adapted from Schmidt L.E., Dalhoff K. (2002). Food–drug interactions. *Drugs, 62,* 1481–1502.

digitalis, and thyroxin, inhibiting their absorption. Bisphosphonates and calcium salts, often used in the management of osteoporosis, can each reduce the absorption of the other, potentially resulting in therapeutic failure (Table 53.3).

Changes in Gastrointestinal Motility and Mucosa Changes of gastrointestinal motility caused by some drugs can affect the degree or speed of absorption of other drugs (absolute bioavailability). A reduced speed of gastric emptying or a reduced intestinal motility caused by opioid agonists (e.g., morphine and loperamide) or by drugs with anticholinergic activity (e.g., atropine, antihistamines, or phenothiazines) can slow down the absorption of other drugs such as macrolides. On the other hand, an increased gastric motility as induced by prokinetic drugs (e.g., metoclopramide, domperidone) can facilitate the absorption of drugs that are primarily taken up by the upper GIT.

Some drugs that can potentially cause gastrointestinal toxicity, such as antineoplastic agents, can induce states of malabsorption, thereby reducing the bioavailability of other drugs.

Modification of Intestinal Commensal Bacteria The disruption of the intestinal commensal bacteria is yet another mechanism whereby some drugs can affect absorption of other drugs. Oral administration of antibiotics, in particular broad-spectrum antibiotics, can result in a reduced number of intestinal commensal bacteria. This can indirectly potentiate the effect of drugs such as digoxin and levodopa that are largely metabolized by intestinal bacteria. On the other hand, antibiotics can reduce the hydrolytic activity exerted by intestinal bacteria on the metabolites of certain drugs (e.g., ethinyl estradiol) that are secreted with bile, in a conjugated form, in the intestine. This results in reduced intestinal reabsorption of the drug active form through the enterohepatic circulation and consequently in reduced therapeutic effect. For example, in the case of ethinyl estradiol, this results in a reduced contraceptive effect (Table 53.2).

Interactions during Drug Distribution

Drug Displacement from Plasma Protein Binding Sites Many drugs are reversibly bound to plasma proteins. Thus, at steady state, equilibrium exists between bound and free drug concentration. Depending on drug concentration and affinity for the plasma protein binding sites, a drug can successfully compete with another drug, displacing it from such binding sites and rapidly increasing its concentration in the aqueous portion of plasma. For example, if plasma protein binding of a drug is reduced from 99 to 95%, the concentration of the free form increases from 1 to 4%, that is, four times the original concentration. Drug displacement results in a significant increase in free and active drug levels only when most of the drug is found in plasma rather than tissues. Therefore, only drugs with a low volume of distribution can be subject to such phenomenon. Examples of such drugs include sulfonylureas, such as tolbutamide (96% bound, $V_D = 10 l$); anticoagulants, such as warfarin (99% bound, $V_D = 9 l$); and phenytoin (90% bound, $V_D = 35 l$).

Nevertheless, even if many drugs can interact with each other, only few examples are clinically relevant. In many cases, this phenomenon is transient and is clinically relevant only for drugs that are usually more than 90–95% bound to plasma protein and for drugs with a low therapeutic index. In these cases, the increase in circulating free drug can determine the onset of adverse, sometimes quite severe, reactions. A typical example of this is the increased likelihood of hemorrhage during the course of warfarin therapy in patients concomitantly receiving nonsteroidal anti-inflammatory drugs (NSAIDs).

In some conditions, drugs capable of altering plasma protein binding of other drugs, besides displacing them, can inhibit their biotransformation or elimination. For example, phenylbutazone displaces warfarin from the plasma protein albumin and at the same time selectively inhibits the metabolism of its pharmacologically active S-isomer, prolonging prothrombin time and causing an increased propensity to bleeding. Salicylates displace methotrexate from its albumin binding sites, also reducing its renal secretion by competing with methotrexate for tubular transport (see also Chapter 28 and Supplement E 28.1 and Tables 53.2 and 53.3). Quinidine and several other antiarrhythmic agents, such as verapamil and amiodarone, displace digoxin from its tissue binding sites and at the same time reduce its renal secretion, thereby favoring the development of digoxin-induced arrhythmias.

Interactions during Drug Biotransformation

Drug interactions during the process of drug biotransformation are not uncommon, and some are clinically very important (see also Chapter 6).

Enzymatic Induction and Inhibition Drug interactions occurring during the biotransformation stage can increase or decrease active drug concentration through the induction or inhibition of drug-metabolizing enzymes (Fig. 53.2 and Table 53.3). Even though several tissues have drug-metabolizing properties, the liver is mainly responsible for drug metabolism. Hepatic metabolism is mediated by cytochrome P450 (CYP450) and its several isoenzymes (more than a hundred isoforms), all encoded by different genes. In mammals, at least 14 such genes have been identified; those playing a key role in drug metabolism are *CYP1*, *CYP2*, and *CYP3*. CYP450 family is expressed not only in hepatocytes but also in other tissues such as the small intestine enterocytes, kidneys, lungs, and brain. CYP3A4 is the subfamily primarily expressed in the liver (~60% of the total P450

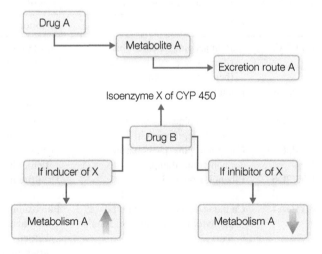

FIGURE 53.2 Drug interactions occurring during biotransformation. Induction or inhibition of cytochrome P450 activity by drug B can either accelerate or slow down the metabolism of drug A.

amount) and small intestine (~70%). Other isoenzymes that play a major role in drug metabolism include 2C9, 2C19, 2D6, 2E1, 1A1, and 1A2 (Fig. 53.3). These isoforms differ from each other in amino acid sequence, specificity of the reactions they catalyze, and sensitivity to specific inhibitors or inducers (Fig. 53.3 and Table 53.5). On the other side, enzyme inhibitors exhibit different selectivity for various enzyme isoforms and can be distinguished for the mechanism of action (competitive or noncompetitive).

Two or more drugs that are substrates of the same isoenzyme will compete for its binding site; as a result, the drug with lower affinity for the isoenzyme will be less metabolized than the other. Some drugs can compete for the active site even if they are not substrates of the enzyme themselves. For example, quinidine, a potent competitive CYP2D6 isoenzyme inhibitor, is not a substrate of the isoenzyme. For example, one noncompetitive inhibitor is ketoconazole, which binds Fe^{3+} in the CYP3A4 heme group and reversibly inhibits the isoenzyme. Erythromycin behaves in a similar way and can inhibit both CYP3A4 and CYP1A2 (two isoforms responsible for the inactivation of several drugs), resulting in a prolonged, intensified, or exaggerated response to the drug with possible development of ADRs. From a clinical point of view, very important are the interactions between erythromycin and cisapride or terfenadine (CYP3A4 substrate) and between erythromycin and triazolam (CYP1A2 substrate). In the case of cisapride and terfenadine, the interaction with erythromycin can result in severe form of ventricular tachycardia. The interaction between triazolam and erythromycin can result in marked sedation (Table 53.5). In recent years, terfenadine and cisapride have been withdrawn from the market after report of serious cardiotoxicity events, some even fatal, in patients concomitantly administered with erythromycin and azole antifungal agents.

Cimetidine is another drug that substantially reduces the activity of hepatic microsomal enzymes, thereby potentiating the effects of other drugs including benzodiazepines, carbamazepine, warfarin, and theophylline. On the other hand, some drugs and xenobiotics (e.g., phenobarbital, carbamazepine, phenytoin, rifampicin, ethanol, and tobacco smoke) induce hepatic as well as extrahepatic CYP450. Alcohol induces isoenzymes CYP1A2 and CYP2E1; thus, chronic alcohol consumers are at a higher risk of hepatotoxicity due to toxic metabolites of paracetamol, which is a CYP1A2 substrate. Similarly, a component of cigarette smoke, benzopyrene, can induce the metabolism of CYP1A2 substrates such as clozapine, fluvoxamine, theophylline, or caffeine (Table 53.5).

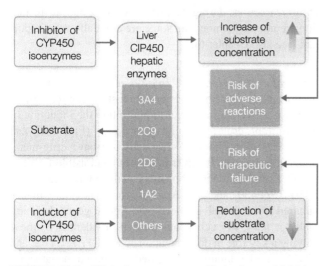

FIGURE 53.3 Clinical consequences of induction or inhibition of cytochrome P450s.

TABLE 53.5 Some inducers, inhibitors, and substrates of the main CYP450 isoforms involved in drug metabolism

Isoenzymes	Inducers	Inhibitors	Substrates
CYP1A2	Carbamazepine, phenobarbital, rifampicin, cigarette smoke (benzopyrene), ethanol	Fluvoxamine, erythromycin, clarithromycin	Paracetamol, theophylline, tacrine, clozapine, amitriptyline, fluvoxamine
CYP2C8	Rifampin	Montelukast, quercetin, Trimethoprin, Gemfibrozil	Paclitaxel
CYP2C9	Carbamazepine, phenobarbital, phenytoin, rifampicin	Amiodarone, fluconazole, fluvoxamine, valproic acid	ibuprofen, diclofenac, celecoxib, phenytoin, tolbutamide, warfarin, losartan
CYP2C19	Carbamazepine, phenobarbital, phenytoin, rifampicin	Felbamate, fluvoxamine, omeprazole, ticlopidine	Omeprazole, imipramine, citalopram, phenytoin, diazepam, warfarin
CYP2D6	Ethanol, isoniazid	Disulfiram	Ethanol, halothane, isoniazid, felbamate, phenobarbital, methoxyflurane
CYP3A4	Carbamazepine, phenobarbital, phenytoin, rifampicin, St. John's wort	Fluvoxamine, erythromycin, ritonavir, indinavir, itraconazole, ketoconazole, terfenadine, erythromycin, grapefruit juice	Carbamazepine, midazolam, imipramine, clozapine, nifedipine, diltiazem, verapamil, cyclosporine, tacrolimus, cisapride, clarithromycin, tamoxifen, amiodarone, quinidine, itraconazole, ketoconazole, ritonavir, indinavir

TABLE 53.6 Examples of substrates, inhibitors, and inducers of P-glycoprotein

Substrates	Inducers	Inhibitors
Digoxin, diltiazem, nifedipine, warfarin, indinavir, nelfinavir, morphine, cyclosporine, oral contraceptives	Rifampicin, phenobarbital, midazolam, nifedipine, St. John's wort	Quinidine, verapamil, cyclosporine, amiodarone, ketoconazole, itraconazole, erythromycin, clarithromycin, azithromycin, grapefruit juice

Other drugs, such as the protease inhibitors, sometimes behave as enzyme inducers and sometimes as enzyme inhibitors. In particular, ritonavir is the protease inhibitor that interacts with the majority of enzymes, inducing the 1A2 isoform and strongly inhibiting the 3A4 isoform and moderately inhibiting the 2D6, 2C9, 2C10, and 2C19 isoforms; therefore, potentially, it represents the drug with the greatest number of interactions. Another interesting case is represented by the COX-2 (cyclooxygenase-2) inhibitors celecoxib and rofecoxib. These are differently metabolized in the liver: rofecoxib is primarily metabolized by cytosolic enzymes, whereas celecoxib by CYP2C9. Therefore, celecoxib may be sensitive to the effects of CYP2C9 inducers and inhibitors. In addition, celecoxib inhibits CYP2D6 and may therefore influence the concentrations of drugs that are CYP2D6 substrates, such as antidepressants and antipsychotics (Table 53.5).

To summarize, the most important consequences of the induction of CYP450 enzymes by drug therapy include reduction of the pharmacological effects following increased drug metabolism, reduction of drug toxicity due to a more rapid drug biotransformation, or increase in drug toxicity as a result of an increase in the concentration of active drug metabolites, as is the case of paracetamol or some cyclic hydrocarbons (Fig. 53.3).

Unlike enzyme inhibition, which results in an almost immediate response, enzyme induction is a slower process that can reduce drug plasma concentration, thus compromising its efficacy in a time-dependent manner. Enzyme induction implies an increase in gene transcription and in the synthesis of microsomal enzymes as an adaptive response aimed at protecting cells. This occurs, for example, in the presence of toxic xenobiotics such as benzopyrene to increase the detoxifying activities in the organisms. It is difficult to predict the time needed for enzyme induction. Various factors, including drug half-life and enzyme turnover, can influence whether or not and how quickly enzyme induction occurs. For example, St. John's wort and grapefruit juice (Tables 53.5 and 53.6) are two substances found in the diet or in plants that are capable of modulating CYP450 and P-glycoprotein (P-gp; see Supplements E 6.2 and E 53.1, "Interactions between Drugs and Grapefruit Juice"). St. John's wort is an OTC product that is used for self-medication of depressive symptoms. This drug is known to reduce plasma levels of several drugs such as warfarin, digoxin, theophylline, oral contraceptives, indinavir, and cyclosporine by inducing hepatic as well as intestinal CYP3A4 and P-gp to a clinically significant degree (Table 53.6). Grapefruit juice interacts with calcium antagonists, statins, and cyclosporine, but unlike St. John's wort, they increase drug plasma levels. Even in this case, experimental studies have shown an inhibitory interaction between CYP3A4 and intestinal P-gp (Tables 53.5 and 53.6; see Supplement E 53.1).

Another metabolic interaction having toxicological effects is that occurring between allopurinol (a xanthine oxidase inhibitor used in gout treatment) and 6-mercaptopurine. In fact, as 6-mercaptopurine is metabolized by xanthine oxidase, allopurinol prolongs and potentiates its effects, increasing the risk of toxicity.

However, some drug interactions are deliberately exploited for therapeutic purposes. For example, in Parkinson's disease therapy, levodopa is almost always combined with benserazide, which inhibits peripheral DOPA decarboxylase and prevents the peripheral metabolism of levodopa. This allows the active form of levodopa to reach the central nervous system in larger concentrations and also allows reducing the dose and thereby the peripheral unwanted drug effects (Table 53.1).

Changes in Hepatic Blood Flow A number of very lipid-soluble drugs, such as lidocaine, undergo substantial biotransformation in the intestine and liver (first-pass effect). During this process, some drugs may influence the first-pass metabolism of others. An example is the increase in lidocaine plasma levels due to changes in hepatic blood flow induced by β-blockers used in cardiovascular pathologies.

Interaction during Drug Excretion

A drug can significantly influence renal excretion of other drugs through:

- Changes in protein binding and therefore changes in degree of filtration
- Inhibition of renal secretion
- Changes in renal blood flow, urinary pH, and electrolyte balance

Drugs that use the same transport system in renal tubules can compete for renal excretion (see Supplement E 53.1). The most significant and representative example of this is probenecid, which inhibits tubular secretion of penicillins and other drugs, thereby reducing their urinary concentrations and increasing their plasma concentrations. Similarly, NSAIDs can interfere with renal excretion of methotrexate, resulting in an increased risk of toxicity (Table 53.2). The increase in digoxin serum concentrations caused by azithromycin (which unlike erythromycin does not modulate hepatic CYP450 in a clinically significant manner) also seems to be due to reduced tubular secretion of digoxin due to the P-gp in the kidney (Table 53.6).

Changes in urinary pH can influence the ionization state of a drug, influencing its reabsorption from the renal tubule. Therefore, changes in urine pH that increase the amount of ionized drug (urine alkalinization in the case of acidic drugs and acidification in the case of basic drugs) increase drug excretion. In this way, dissociation of weak acids such as barbiturates is favored in the presence of alkalizing drugs such as sodium bicarbonate, while dissociation of weak bases such as amphetamines is favored by acidifying drugs such as ammonium chloride. Diuretic drugs tend to increase the excretion of various drugs, but this is rarely clinically significant. Thiazide and loop diuretics indirectly increase Li^+ absorption in the proximal convoluted tubule, and this may therefore translate into an increased risk of toxicity in patients on lithium carbonate pharmacotherapy. By reducing plasma K^+ concentrations, potassium-sparing diuretics can increase myocardial sensitivity to cardiac glycosides (e.g., digoxin), which may result in drug toxicity.

PHARMACODYNAMIC DRUG INTERACTIONS

Pharmacodynamic interactions occur via several main mechanisms; based on their effects, they can be additive, synergistic, or antagonistic. Drugs can compete for receptor binding sites (direct interaction) or for different receptor or physiological systems (indirect interaction), giving rise to similar or opposing effects (Table 53.7). Two drugs that produce similar effects when concomitantly administered cause additive effects if the final result corresponds to the sum of their individual effects. This can occur when the drugs exert their action on the same receptor system but also if they act on different systems. For example, the concomitant use of drugs that act on different receptor systems (anxiolytics, hypnotics and H_1 antihistamines, alcohol) can result in harmful additive effects on the central nervous system, such as increased sedation and somnolence, and reduced psychomotor abilities, which can interfere, for example, with car driving.

TABLE 53.7 Examples of pharmacodynamic interactions

Drugs	Effects
Anxiolytics and hypnotics	Increased sedation
Paracetamol and codeine	Increased analgesic effect
Aspirin and warfarin	Increased anticoagulant effect
Naloxone and morphine	Antagonism
Flumazenil and benzodiazepines	Antagonism

The antiplatelet effect of aspirin can potentiate the effect of oral anticoagulants, increasing the risk of hemorrhage. The concurrent use of statins and fibrates is associated with an increased risk of myotoxicity (rhabdomyolysis) (Table 53.2).

On the other hand, when drugs having the same effect give rise to a response that is greater than the sum of their individual effects, the interaction is termed synergistic. One example of synergistic interaction often exploited in clinical practice involves trimethoprim and sulfamethoxazole, which inhibit bacterial folate production at different stages of the bacterial metabolic pathway. Another example is the analgesic combination of paracetamol and codeine, which exert their effects on two different mediators.

Conversely, antagonism implies that the response of a drug is reduced by the concomitant use of another drug. Antagonism can be receptor based, when the antagonist drug binds the receptor binding site, preventing agonist binding. Otherwise, it can be functional (physiological), when two drugs exert opposite effects on the same functional parameter. Typical examples of receptor mediated antagonisms are naloxone, which antagonizes the effect of morphine; flumazenil, which antagonizes the effect of benzodiazepines; and β-blockers, which antagonize the bronchodilating effect of salbutamol or orciprenaline. Examples of functional antagonists include drugs such as insulin and glucocorticoids, which are both diabetogenic but with a different molecular mechanism.

CHEMICAL ANTAGONISM

Antidotes

Antagonism refers to a chemical reaction between two substances in which the effect of an active drug or a toxin is neutralized. If the reaction neutralizes a toxin, it is more appropriate to use the term "antidotism" (Table 53.8). Antidotes can modify the chemical characteristics of a toxin, making it less harmful or reducing its absorption. For example, chelating agents, such as dimercaprol, are used in the treatment of heavy metal (e.g., lead, zinc, etc.) poisoning: they reduce the toxicity of heavy metals by binding them, forming inactive complexes.

Pharmaceutical Interactions and Incompatibility

When chemical antagonism occurs *in vitro*, one or both drugs become inactive. When two drugs are present in the same infusion liquid, various physicochemical interactions (neutralization, salt precipitation) can occur, resulting in the formation of an inactive product or a toxic compound. This effect is more properly referred to as incompatibility, and occurs, for example, between imipenem in solution and heparin, chloramphenicol and calcium gluconate, methicillin and gentamicin, and adrenaline and sodium bicarbonate (Table 53.8). Therefore, it is good practice to avoid combining drugs in the same package unless their compatibility is confirmed.

INTERACTIONS BETWEEN HERBAL REMEDIES AND DRUGS

In recent years, the increasing number of herbal remedies has prompted a great interest in their potential interactions with medicinal drugs. Some herbs contain substances that can potentially interact with many drugs, causing changes in their therapeutic response, sometimes nullifying it or even inducing noxious responses (Table 53.9).

Various studies show that St. John's wort (*Hypericum perforatum*), used in the treatment of mild to moderate depression, can modulate plasma levels of various drugs, including indinavir, cyclosporine, and digoxin. Another example is the interaction between myrtle-based products (*Vaccinium myrtillus*), mainly used in the ophthalmic settings and to treat lower urinary tract infections, and warfarin, resulting in an increased international normalized ratio (INR).

The concomitant use of herbal remedies and medicinal drugs is problematic because, being natural products, herbal remedies are considered innocuous. In the United States, a study on herbal remedies found that 40% of people interviewed thought that concomitant use of herbal medicines and a medicinal drug improved the efficacy. Actually, this is not the case; indeed, as a large body of literature suggests, some of the resulting interactions can seriously endanger patient's lives. This represents a significant public health issue. Thus, providing correct information in this area is essential. The principles of safety and efficacy as used in conventional medicine and in good clinical practice should also apply to herbal remedies (See also Supplements E 1.1, "Alternative or Non-Conventional Therapies and Supplement" and E6.3, "Modulation of Efficacy due to Interactions Between Synthetic and Herbal Drugs").

INTERACTIONS BETWEEN DIETARY SUPPLEMENTS AND DRUGS

Some dietary supplements or even components of a person's diet can give rise to interactions with drugs. This issue should not be ignored (Table 53.4). Such interactions can

TABLE 53.8 Examples of chemical antagonism

Antidotes	Dimercaprol use in lead poisoning
	Ammonia in formaldehyde poisoning
Incompatibility (in solution)	Penicillin G + heparin
	Penicillin G + vitamin B or C complex
	Imipenem + lactate
	Imipenem + other antibiotics
	Gentamycin + ampicillin
	Gentamycin + heparin
	Gentamycin + vitamin B or C complex

TABLE 53.9 Herbal medicine–drug interactions

Source	Drugs	Clinical interaction	Possible mechanism
Garlic	Chlorpropamide	Hypoglycemia	Additive effect on glycemia
	Saquinavir	Reduction of AUC and C_{max} of saquinavir	Not known
	Paracetamol	Changes in paracetamol pharmacokinetics	Not known
Ginkgo	Alprazolam	Mild reduction of AUC of alprazolam	Weak inhibition of CYP3A4
	Thiazide diuretics	Increase in blood pressure	Not known
	Omeprazole	Decrease in omeprazole plasma concentrations	Induction of CYP2C19
	Risperidone	Priapism	Not known
St. John's wort	Atorvastatin	Reduction of efficacy atorvastatin	Induction of CYP3A4 and/or induction of P-glycoprotein
	Cyclosporine	Reduction of cyclosporine plasma levels	Induction of CYP3A4 and/or induction of P-glycoprotein
	Digoxin	Reduction of digoxin plasma levels	Induction of P-glycoprotein
	Erythromycin	Increase in erythromycin metabolism	Induction of CYP3A4

AUC, area under the curve/time; C_{max}, maximum plasma concentration.
Adapted from Izzo A.I., Ernst E. (2009). Interactions between herbal medicines and prescribed drugs. *Drugs*, 69, 1777–1798.

give rise to an increase or decrease in drug effect with resulting loss of therapeutic response or appearance of ADRs. This can lead to an increased risk of morbidity, additional need of other drugs, and even hospitalization. These interactions can occur through various mechanisms, most commonly pharmacokinetic mechanisms. This means that the dietary supplement can interact with the medicinal drug during absorption, distribution, metabolism, and elimination.

The clinical significance of such interactions is poorly characterized. The main clinically relevant interactions are those that involve drugs with a narrow therapeutic index such as theophylline, digoxin, and warfarin.

A well-known example of a dietary supplement/component interacting with medicinal drugs is grapefruit juice, which inhibits the CYP3A4 isoforms of CYP450. Some studies have shown that grapefruit juice can significantly increase the bioavailability of drugs whose bioavailability is subject to systemic metabolism by CYP3A4, which is highly expressed by enterocytes in the intestinal mucosa, mainly in the small intestine. It has been shown that furanocoumarins, substances found in grapefruit juice, are primarily responsible for such an effect (see Supplement E 54.2). Another example is the interaction between foods rich in vitamin K (cauliflower, broccoli, and spinach) and oral anticoagulants, with subsequent reduction of therapeutic effect. Another interaction is that occurring between foods rich in potassium (banana and spinach) and potassium-sparing diuretics, with resulting hyperkalemia.

Finally, other factors can contribute to the occurrence of interactions between dietary components or supplements and drugs, such as age, gender, physical and nutritional state, and number of drugs concomitantly used.

TAKE-HOME MESSAGE

- DDIs can occur at any level during absorption. For oral administration, drugs affecting gastric pH, intestinal motility, and composition of the intestinal flora may alter bioavailability of other drugs.
- Changes in plasma protein binding, as a result of DDI, can significantly modify drug distribution and plasma concentration.
- Several DDIs occur at metabolic and excretory levels; several drugs and nutrients have been shown to induce or inhibit CYP enzymes.
- DDIs may occur when drugs share mechanisms of action or targets.
- Several interactions have been shown to occur between drugs and dietary components or herbal products.

FURTHER READING

Fugh Berman A., Ernst E. (2001). Herb–drug interactions: review and assessment of report reliability. *British Journal of Clinical Pharmacology*, 52, 587–595.

Genser D. (2008). Food and drug interaction: consequences for the nutrition/health status. *Annals of Nutrition and Metabolism*, 52 (Suppl 1), 29–32.

Izzo A.A., Ernst E. (2009). Interactions between herbal medicines and prescribed drugs. *Drugs*, 69, 1777–1798.

Kane G.C., Lipsky J.J. (2000). Drug–grapefruit juice interactions. *Mayo Clinic Proceedings*, 75, 933–942.

Lin J.H., Lu A.J.H. (1998). Inhibition and induction of cytochrome P450 and the clinical implications. *Clinical Pharmacokinetics*, 35, 361–390.

Michalets E.L. (1998). Update: clinically significant cytochrome P-450 drug interactions. *Pharmacotherapy*, 18, 84–112.

Pirmohamed M.,(2013). Drug–grapefruit juice interactions. *British Medical Journal*, 346, 9.

Schmidt L.E., Dalhoff K. (2002). Food–drug interactions. *Drugs*, 62, 1481–1502.

Singh B.N. (1999). Effects of food on clinical pharmacokinetics. *Clinical Pharmacokinetics*, 37, 213–253.

Tornio A., Niemi N., Neuvonen P.J. Backman J.T., (2012). Drug interactions with oral antidiabetic agents: pharmacokinetic mechanisms and clinical implications. *Trends in Pharmacological Sciences* 33, 312–322.

VandenBrink B.M., Isoherranen N. (2010) The role of metabolites in predicting drug–drug interactions: focus on irreversible cytochrome P450 inhibition. *Current Opinion in Drug Discovery & Development*, 13, 66–77.

Yu D.K. (1999). The contribution of P-glycoprotein to pharmacokinetic drug–drug interactions. *Journal of Clinical Pharmacology*, 39, 1203–1211.

SECTION 14

DRUG DEVELOPMENT

54

PRECLINICAL RESEARCH AND DEVELOPMENT OF NEW DRUGS

ENNIO ONGINI

By reading this chapter, you will:

- Learn the contribution of scientific and technological advancement in the research and development of past, present, and future drugs
- Understand the key factors in the selection of therapeutic fields and projects in industrial drug development
- Realize the importance of knowing the molecular mechanisms of diseases, the contribution given by new technologies, and the need for a multidisciplinary approach to drug discovery
- Understand the critical issues in the identification and validation of active compounds and the importance of patents, pharmacokinetics, and preclinical development

---

Drugs have deeply changed the impact of diseases on patients: a real revolution has occurred in particular during the last 60 years. Already existing drugs have been improved (e.g., antibiotics and analgesics), and new drug classes have been discovered to treat psychiatric disorders, high blood pressure and ulcer or to reduce cholesterol levels, immunosuppressants, and anticancer agents. Besides medications obtained from natural sources or produced by pharmaceutical chemistry, over the last 30 years, a new type of drugs has been developed, the so-called "biological" drugs, such as therapeutic proteins and monoclonal antibodies. And there is the prospect of further discoveries. The discovery of new drugs requires great technological and scientific progress as well as huge investments in research field by pharmaceutical industries (for further details, see Supplement E 54.1, "The Significant Contribution Given by Small Research Groups (Start-Ups and Spin-Offs)"). The identification and subsequent development of new molecules require considerable resources (in both human and economical terms) and time (10–15 years on average). Moreover, the risk of incurring failure is high: after years of efforts, the investigated molecule may not have the expected pharmacological activity or may show unacceptable side effects, or it may become outdated because of new better molecules developed by other research groups in the meantime. The drug "cemetery" is very large and includes all the "submerged" molecules, those effective in disease models but useless in humans because of their poor bioavailability or too high toxicity as well as those abandoned because they are obsolete or because of changes in the strategies or financial possibilities of pharmaceutical industries.

TECHNOLOGICAL INNOVATION AND SCIENTIFIC KNOWLEDGE IN CURRENT PHARMACEUTICAL RESEARCH

Advances in scientific knowledge and available technologies have deeply changed research strategies (see Fig. 54.1). Some important steps on this path can be identified, and they coincide with critical milestones in biomedical scientific knowledge. During the 1950s and 1960s, great efforts were made to investigate natural compounds, coming both from

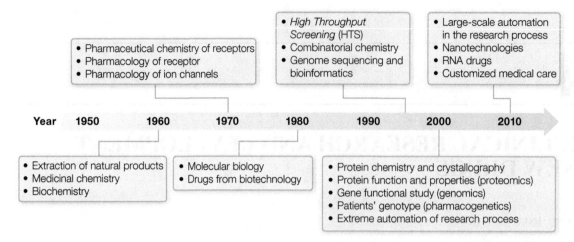

FIGURE 54.1 Evolution of drug discovery. The rapid expansion of scientific knowledge and the availability of innovative technologies have deeply changed the way drug research and development is conducted.

animal and plant kingdoms. Thanks to the ever-improving chemical know-how and the availability of sophisticated scientific equipment, compounds with a biological activity were isolated using extraction techniques. In the following years, starting from these active prototypical molecules, a great number of derivatives were synthesized, whose biological activity was assessed through screenings either in *in vitro* (e.g., isolated organs) or *in vivo* (e.g., laboratory animals) models with the aim of selecting molecules for further development. Over the years, this has allowed new medicines to become available for patients. In the 1960s, the tragedy of thalidomide (a sedative drug used to alleviate sickness during pregnancy that turned out to have severe side effects, being responsible for thousands of phocomelia-affected babies) led many countries to implement restrictive rules for the approval of new drugs. Toxicology studies became a crucial phase, clinical pharmacology became increasingly important, and the steps needed to study new drugs in humans were set more accurately (e.g., study on volunteers, use of placebo, subdivision of clinical research into well-separated phases). As a consequence, a remarkable rise in research and development (R&D) costs occurred, and the time span needed for new molecules to be approved expanded considerably.

In the early 1980s, technologies and knowledge coming from molecular biology allowed researchers to formulate new hypotheses on the mechanisms underlying diseases and to identify new molecular targets for drug development.

In the 1990s, a crucial event changed the course of research: the Human Genome Project, an international scheme that led to the sequencing not only of the entire human genome but also of simpler model organisms, such as prokaryote (e.g., *Bacillus subtilis*, *Haemophilus influenzae*) and eukaryote microorganisms (e.g., *Saccharomyces cerevisiae*), the nematode *Caenorhabditis elegans*, and mouse. The quick progress in information technology and the availability of more and more powerful computers have provided researchers with a remarkable support at multiple stages of the drug discovery process, from the identification of the biological target to the selection of the active compounds. One of the technologies developed in the 1990s that has become very popular in all the laboratories worldwide is the high-throughput screening (HTS), a series of highly automated systems that enable a rapid evaluation of thousands of chemical compounds in biologic assays based on several screening methods. Among all the screened compounds, only those showing the desired activity are selected to enter the drug discovery phase, whereas the others are abandoned (for further details, refer to Supplement E 54.2, "High-Throughput Screening"). Concurrently, some methods have been developed to synthesize a large variety of chemical compounds (e.g., combinatorial chemistry).

However, in the last decade, despite the huge investments in terms of technological innovation and financial resources, the yearly number of new drugs reaching the market has not increased with respect to the past (Fig. 54.2). This has recently led to the development of more complete and integrated approaches, in which rather than focusing on a biological target isolated from its context (e.g., a receptor or an enzyme), the whole pathological process is studied and the effects are evaluated on cells and tissues. To this aim, HTS has been modified to directly study the entire cell rather than a single isolated biochemical step, and more sophisticated and sensitive systems for image analysis have been devised to measure intracellular signals. Using this multidisciplinary approach, increasing amount of knowledge is being gathered on the complex molecular mechanisms underlying diseases, and research projects are being designed and conducted to find out the drugs of the future.

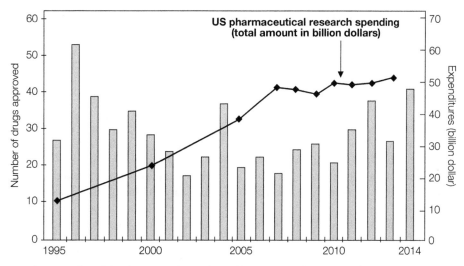

FIGURE 54.2 Investment in research and development of new drugs. The figures refer to the entire spending in the United States (line and diamond). Note the increase of expenditures over the last 10 years in contrast with the stable number of new drugs approved by the FDA (bars).

THE RESEARCH STRATEGIES

Starting a research program to discover a new drug is a great event and a big challenge. It is always the result of an accurate evaluation and discussion of different hypotheses and proposals in which several elements have to be taken into consideration: medical needs, skills of the research team, state of the art of knowledge, accurate study on the progression of diseases, and investigation on what research is being carried out in other laboratories and in particular in competing pharmaceutical groups (refer to Fig. 54.3). It is a challenge: for a few years, the efforts of a lot of researchers will be focused on the selected research program, and they will try, with dedication and professional competence, to turn the initial hypothesis into molecules able to modify the pathological process, using the most representative laboratory models of the disease. All research programs alternate between enthusiasm and frustration, but if conditions are good, a molecule will be found that deserves to proceed to development: this is already a big success, even though just a preliminary one, on the long route to obtain a new drug (for further details, refer to Supplement E 54.3, "The Birth of a Project on a New Drug").

The therapeutic area to invest in is selected based on different factors. For example, medical areas in which the drugs available have reached a good level of efficacy and tolerability in general are ignored. The regulation system is strict, and policies are made increasingly restrictive by European and American authorities: the European Medicines Agency (EMA) and Food and Drug Administration (FDA) are the regulatory authorities appointed to authorize the marketing of new drugs. For new drugs to be approved, especially for widespread pathologies, totally flawless results are required,

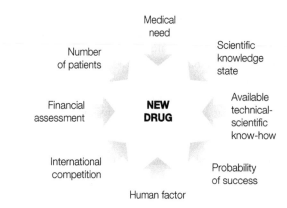

FIGURE 54.3 Items affecting investment decision. The decision of investing in a research project depends on a number of both scientific and financial factors.

obtained through hard and very expensive studies, conducted on a large number of patients treated for long periods.

The growing diffusion of the so-called "generic" drugs has significantly affected this scenario. Because of expiry of patent protection for many drugs (e.g., anti-inflammatory, antidepressant, and antihypertensive drugs, some statins, antibiotics, and many more), the big pharmaceutical groups that have invested enormous resources to develop them have lost their monopole. This has opened up the market to new companies producing and distributing "equivalent" drugs (for further details, refer to Supplement E 54.4, "The Patent"). In this context, a new trend has emerged, as pharmaceutical companies have started to direct research toward pathologies that are more complex but imply lower financial efforts and, in theory, shorter time to obtain results. Indeed, pharmaceutical companies are now increasingly investing in rare diseases (orphan diseases) that had been ignored until recently (for further details, refer to Supplement E 54.3).

Patent as a Driving Force for Innovation

Once the molecules of interest have been identified, pharmaceutical and biotechnological companies invest huge resources for their development only if granted to be the owner of the final product: the drug. The authorities concerned are thus requested to issue the patent for the drug. A document is deposited at the Patent Office describing all the details of the invention. Some of the key requirements in a patent application are:

1. The novelty (the invention has not been known before)
2. The inventive activity (the invention shall not be obviously derivable from the current knowledge status)
3. The reproducibility (in the application, all information required to reproduce the invention has to be provided)
4. The industrial feasibility

As a general rule, patents are a booster for research. In fact, by protecting and publicizing inventions, they stimulate competition, encourage inventors, and foster the continuous development of new solutions and products. This is why patents play a major role in the progress of the medical and pharmaceutical sector. Patents are valid for a limited time, 20 years since patent deposit date (a few year extension can be obtained in some cases). After patent expiry, drugs can be produced by other pharmaceutical companies (in the form of generic or biosimilar drugs) (for further details, refer to E 54.4).

THE RESEARCH STAGES

Along the drug discovery path, two well distinct phases can be distinguished: Research & Development (R&D). The research phase consists in the search of new molecules and the progressive definition of their profile, up to the identification of the "lead" compound. New ideas come from innovative results, exploiting the latest technologies, and they may develop toward hardly foreseeable directions where nothing can be taken for granted. For this reason, today laboratory research is conducted according to very different approaches from those followed not later than 10 years ago (see Fig. 54.1). In the next paragraphs, we will describe the research path currently followed by pharmaceutical and biotechnology companies. These are the approaches that will lead to the drugs of the future.

Start-Up of a New Project and Identification of the Lead Compound

A new project originates from a rational analysis, from biological hypotheses supported by the international scientific literature, or from the results of internal research (e.g., bioinformatics analysis of genes and gene products involved in a specific disease). The most important step in the research path is the identification of a biological target (e.g., an enzyme or a receptor) that plays a critical role in the molecular mechanisms of the pathology. This is a major step on the research path and shall not be undervalued: a mistaken target means wasting a lot of time and money. The hypothesis needs the support of strong evidence demonstrating that (i) the target is involved in the pathogenesis and progression of the disease and (ii) all actions taken on it significantly modify the mechanisms of the disease. Once the target has been identified, it is necessary to identify the most appropriate biological assay, either cellular or biochemical, to be used for the HTS. The automatic analysis of numerous chemical compounds starts here. Each pharmaceutical team explores its own molecule collection (from tens of thousands to millions) or employs chemical libraries of external groups—after formal joint work agreements—thus increasing the probability to find out active compounds (the so-called hits). These compounds represent the basic structures on which pharmaceutical chemists and biologists will focus their efforts with the aim of identifying the prototype molecule, namely, the "lead" compound. Thanks to increasingly advanced equipment and technologies (e.g., molecular modeling) in a relatively short time, it is possible to select molecules with an interesting pharmacologic profile, to compare them with other similar structures described in the scientific literature, and to design the synthesis of new molecules with an improved profile (see Fig. 54.4).

Pharmaceutical Chemistry: From the Mainstream Approach to the Molecular Modeling

The pharmaceutical chemist has always played a major role in the project of new drugs. A great number of today's drugs have been obtained through the "rational synthesis," starting from knowledge of the selected biological target. This is particularly true for the drugs designed on the model of the natural mediator for either a receptor or an enzyme. For example, several agonists and antagonists of the catecholamine, histamine, and acetylcholine receptor systems, as well as various inhibitors of angiotensin II-converting enzyme and acetylcholinesterase, have been obtained in this way.

This traditional approach has been subsequently integrated with molecular modeling, which relies on the availability of sophisticated computers and software packages (for instance, computer-aided drug design, CADD). The information collected on geometry, possible structures, and chemical–physical properties (e.g., energy, molecular orbitals, electronic density, etc.) of the selected molecule allows anticipating how the drug will interact with the targeted binding site. Some molecular modeling software programs enable researchers to determine the three-dimensional characteristics of active molecules. Based on all these

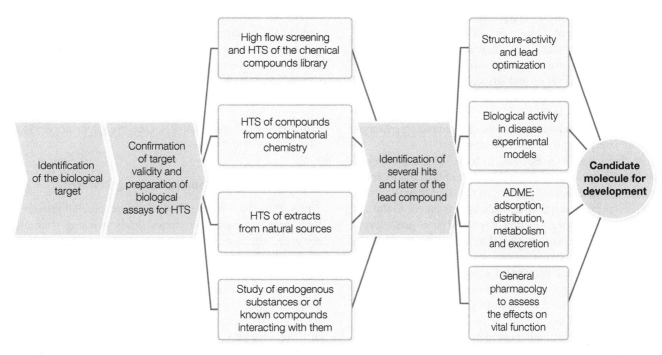

FIGURE 54.4 Crucial steps in the path to drug discovery. The pathway depicted is particularly valid for investigations into low molecular weight molecules with a decisive contribution given by chemical synthesis. The discovery of therapeutic proteins, vaccines, gene therapy, and antisense oligonucleotides follows a different path.

analyses, it is possible to make hypotheses on the biologically active structure, which is the one capable of interacting with the receptor or with the enzyme active site, and to define the relationship between chemical–physical properties (e.g., quantitative structure–activity relationships, QSAR) and the pharmacological activity of the molecule.

The development of combinatorial chemistry, in the mid-1980s, represented an important breakthrough. Combinatorial chemistry is a revolutionary approach to increase the number of molecules available to researchers for testing the biological activity, thus increasing the probability of finding out active molecules. This approach has allowed producing millions of structurally related molecules (see Figs. 54.1 and 54.4).

A drug project can also start from the knowledge gathered on the targeted binding site and therefore about macromolecules such as proteins. This is a fast progressing sector as it runs concurrently with the growing knowledge in molecular biology and with the continuous identification of proteins and their functions. Powerful analysis equipments are also available for the resolution of complex structures (e.g., X-ray crystallography, nuclear magnetic resonance spectrometry). Using these techniques, the structures of important enzymes have been resolved (e.g., HIV protease, active sites of cytochrome P450 isoforms), thus boosting the drug project development. The number of proteins whose structure has been deposited in data banks has rapidly increased over the last few years, ranging from roughly some hundreds in the 1990s to over 70,000 in 2013. An important approach for a more focused choice of the new compounds is based on "virtual screening," which consists in using computational models (through a computer-aided or in silico research) to identify the most suitable chemical structures. In the in silico selection of the compounds, the main chemical–physical properties of the biological target are considered, the molecules having potentially toxic functional groups are left out, and much attention is given to the properties considered as critical features of an effective drug (e.g., solubility, molecular weight, lipophilicity, absorption, and bioavailability). Through powerful computer models, comparisons with chemical structures with known pharmacological activity are also carried out, and the probability of interactions with cytochrome P450 isoforms is evaluated alongside the possibility of side effects, such as mutagenesis. The final result is the selection of a high number of compounds having the highest probability to become "hits" and, later, "lead" compounds to be developed further.

Drug Selection: *In Vitro* Assays and Experimental Models

Regardless of the source, being it a collection of synthetic compounds or natural products, a basic step is the screening of the molecules or the products featuring significant pharmacological properties or in other words capable of acting on the established target according to fixed modes (e.g., receptor agonist or antagonist, enzyme inhibitor, etc.). During this crucial phase, the availability of a quick, sensitive, and reproducible biological assay, relevant to the pathology under examination, is fundamental. Once the biological assay is set up and adapted to the HTS robotized system, the screening process of thousands of molecules can start. The final objective is to obtain a potent and highly selective compound that acts on the desired biological target without interfering with other systems.

At this stage, it is essential to have valid experimental models that can reproduce *in vitro* and *in vivo* the morphological and biochemical features of the disease under study. In this context, transgenic animals are extremely useful not only to investigate the function of specific genes and relevant products but also as experimental models to be employed in the search for new drugs.

Many molecules featuring interesting pharmacological actions can induce adverse effects on organs and systems different from those considered in the therapeutic activity. Thus, the study of the possible side effects on the cardiovascular, gastrointestinal, and central nervous systems, as well as on renal excretion, is fundamental (for further details on the topic, refer to Supplement E 54.7, "Safety Pharmacology: How to Evaluate Whether a New Molecule Will Cause Side Effects on Vital Functions").

Biological Drugs

The drugs obtained from chemical synthesis have played a primary role in medical treatments, and they still do. Anyway, in the late 1970s, with new knowledge emerging in the molecular biology field and with recombinant DNA technology available and the discovery of monoclonal antibodies, the foundations have been laid for the development of the "biotechnological drugs." After more than 30 years, over 200 highly innovative drugs are now available, and many more are being developed or are awaiting approval. Examples of biotechnological drugs are insulin, interferons α and β, interleukin-2, erythropoietin, the colony-stimulating factors G-CSF and GM-CSF, activator of tissue plasminogen (alteplase or tPA), platelet glycoprotein IIb/IIIa inhibitors, monoclonal antibodies, and many others used in different therapies and in particular in cancer and immunologically based diseases. It is worth mentioning the increasing number of human antibodies available for therapy, which represents a significant scientific progress with respect to both murine and humanized monoclonal antibodies (see Table 54.1) (for further details, refer to Supplement E 54.5, "Biotech Drugs").

The main drawback of biotechnological drugs is that they cannot be administered orally, but they require parenteral administration; this makes them hard to handle, increases costs, and limits the development of such drugs for the treatment of widespread pathologies.

The recent introduction of the so-called biosimilars, namely, the generic version of the original biological drug, is raising new issues. In fact, compared to synthesized drug, more controls are needed to guarantee that a biosimilar drug ("copy" protein) retain the same therapeutic properties and safety (e.g., lack of immunogenicity) of the source drugs. The guidelines of the different law systems are rapidly evolving (for further details, see Supplement E 54.6, "Toxicology Tests").

Impact of Genomics Studies on Drug Research

Since 1992, the Genome Project and the parallel growth of bioinformatics have been considerably boosting genomics research, attracting huge investments in it. The prospect of the thorough human genome sequencing raised great expectation about the possibility to rapidly discover genes responsible for diseases and therefore to design innovative drugs targeting the specific causes of diseases. However, so far, such expectations have been only partially met because of several reasons. In particular, the time needed to unravel the role played by each single gene and protein both in physiological and pathological conditions is longer than expected. Thus, the identification of proteins with a potential medical use or low molecular weight compounds capable of interacting with new targets identified by genomics studies requires years of intensive research. However, innovative drugs will definitely be generated by this kind of knowledge in the future. If we consider that all drugs currently available for therapies act on less than 500 biological targets, the study of the human genome is clearly expected to promote the discovery of new relevant targets for the development of more selective drugs.

Furthermore, genomic technologies allow to investigate the genetic differences underlying the different sensitivity and response to drugs. This is already having, and will increasingly have, a significant impact on therapy by allowing to decide the most appropriate pharmacological treatment for each patient based on his/her genotype (see also Chapter 23). A mutual relationship between genotype and patients' pharmacological response can be established using the available technologies to stratify patients according to their gene profile. Some pharmaceutical groups are investing in the research of diagnostics (theranostics) to identify the most appropriate patients for a specific therapy, in order to achieve the highest efficacy with the least side effects—a sort of à la carte pharmacology: therapeutically speaking, a real revolution.

Pharmacokinetic Studies

A pill a day, this is the way a good drug should be. It should be effective when orally administered, and its action duration should permit a single daily administration. However,

TABLE 54.1 Some examples of biotechnological drugs available in therapy

Product	Disease	Year
Therapeutic proteins		
Insulin	Diabetes type I	1982
Somatotropin	Growth disorders	1984
Interferons α and β	Antiviral, hepatitis C, multiple sclerosis	1986–1996
Alteplase (tPA)	Thrombosis	1987
Erythropoietin	Erythropoiesis	1988
Filgrastim (G-CSF)	Leukopenia	1991
Factors VIII and IX	Hemophilia	1993–1997
Reteplase	Thrombosis	1995
Follitropins α and β	Female infertility	1996
Etanercept	Rheumatoid arthritis	1998
Anakinra	Rheumatoid arthritis	2002
Therapeutic monoclonal antibodies		
Muromonab (OKT3)	Organ transplant	1986
Abciximab	Thrombosis	1995
Daclizumab	Organ transplantation	1997
Infliximab	Crohn's disease Rheumatoid arthritis	1998–2000
Basiliximab	Organ transplantation	1998
Palivizumab	Infectious diseases	1998
Rituximab	Non-Hodgkin lymphoma	1997
Trastuzumab	Breast cancer	1998
Gemtuzumab	Acute myeloid leukemia	2000
Alemtuzumab	Chronic lymphocytic leukemia	2001
Adalimumab	Rheumatoid arthritis	2002
Bevacizumab	Rectal cancer	2004
Natalizumab	Multiple sclerosis and Crohn's disease	2006
Panitumumab	Rectal cancer	2006
Ranibizumab	Macular degeneration	2006
Golimumab	Rheumatoid arthritis, ulcerative colitis	2009–2013
Ustekinumab	Psoriasis	2009
Denosumab	Bone metastasis	2010–2011

compounds that show interesting biological activity in *in vitro* experimental models often reveal a low oral bioavailability, meaning that the fraction of drug reaching the blood circulation is too small or they are not active *in vivo* (see Chapters 3, 4, 5, 6, 7, and 55).

Up to few years ago, pharmacokinetic studies (absorption, distribution, metabolism, and excretion (ADME)) were conducted at a late stage of the drug development process together with toxicology studies. However, the observation that over a third of the molecules under development were abandoned because of ADME problems (e.g., insufficient absorption, too fast liver metabolism, toxic metabolites, enzyme induction, too short or too long excretion half-life, etc.) led to the decision to carry out pharmacokinetic tests during the first stages of research. According to the latest criteria, the ADME profile is assessed already during the phase of identification of active molecules (hits and then lead compounds) both in silico, through computer simulations, and *in vitro* (see Fig. 54.4). Through computation approaches, the chemicophysical properties are compared with specific ADME features either desired or not. Some *in vitro* assays are later performed to evaluate intestinal absorption (e.g., using Caco-2 cells), metabolic stability in human and rodent hepatic microsomes or in the S9 microsomal fraction, interactions with cytochrome P450 isoenzymes, and binding to plasma proteins (see Chapter 6).

The collection of such information during the early phases of the drug discovery process allows carrying out a stricter screening. Next step, the *in vivo* profile of the selected compounds is assessed. The analytical methods now available (e.g., mass spectrometry combined with high-pressure liquid chromatography or high-resolution nuclear magnetic resonance) permit a quick analysis of plasma levels and concentrations in target organs of the compounds administered to small laboratory animals. We know that several animal species differ from one another for their hepatic microsomal enzymes, and thereby, they can affect drugs in a different way. Thus, it is crucial to select the most suitable animal

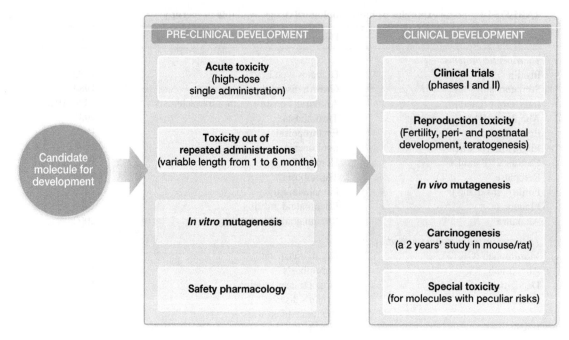

FIGURE 54.5 Toxicology studies to test the safety of new drugs. Note that the candidate molecule for development is evaluated in acute toxicity tests, in toxicity tests for repeated administrations, and in *in vitro* mutagenesis tests. If no significant effects are detected, it is possible to move forward to phase I and II clinical trial. In parallel, a more complete toxicology scheme is carried out.

species for toxicology studies, so that accurate and reliable information can be gathered on the safety of drugs destined to humans. More in-depth pharmacokinetic studies are then conducted during drug development in parallel with toxicology studies, both to better define the profile of the selected molecules and to meet the requirements of regulatory authorities (see Fig. 54.5).

THE DEVELOPMENTAL STAGES

After identification of a molecule of interest, the development process begins, which represents the longest and most expensive stage of the drug discovery process (see Fig. 54.6). Thus, during this phase, the two decisive factors are timing and costs. This implies that sometimes it is necessary to choose among different molecules; indeed, a large number of interesting compounds may be identified during the research phases, and it may not be possible, in terms of money and resources, to develop them all further. Thus, before carrying on with the development, the project has to be accurately revised. The key steps in drug development are the following: (i) production of large amounts of the drug candidate with a high degree of purity, (ii) formulation of the compound, (iii) pharmacokinetics and toxicology studies, and (iv) preparation of the documentation required by regulatory authorities for clinical studies in humans to be approved. This stage is called preclinical development to distinguish it from the subsequent stage called clinical development.

The Role of National and Supranational Legislation

The entire development phase is regulated by the pharmaceutical laws that dictate very clear rules to guarantee efficacy and safety of new drugs. These laws regulate two crucial moments in the interaction between pharmaceutical companies and national regulatory authorities (FDA in the United States, EMA in Europe): the authorization to start studies in humans, being them healthy volunteers or patients, and the formal approval to use and market the drug.

During drug development, the pharmaceutical laws establish strict criteria of quality, integrity, traceability, and reproducibility of data and outputs. To meet these requirements, guidelines are available, first produced in the United States and later adopted by other countries. More specifically, these are the regulations to comply with:

1. Good Laboratory Practices (GLP). They establish the principles to comply with during toxicology and pharmacokinetic studies.
2. Good Clinical Practices (GCP). Set of rules to guarantee quality and reproducibility of data during experimentation on humans.

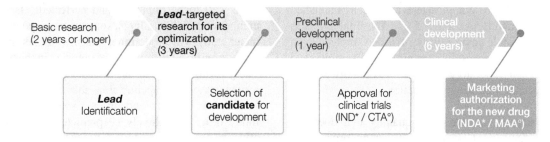

FIGURE 54.6 The complete path of research and development (R&D) of new drugs. Duration of the R&D path has significantly increased since the 1960s. The average duration, from the first stage of research to the final approval, is currently more than 10 years.
(*): Application for authorization to the USA regulatory authority, Food and Drug Administration (FDA)
(°): Application for authorization to the European regulatory authority, European Medicinal Agency (EMA)

3. Good Manufacturing Practices (GMP). Guidelines for drug production and control in order to guarantee their quality over the whole term of their validity.

Scaling Up from Laboratory to Industrial Drug Preparation

During the first research stages, the molecules to be tested are prepared in chemical or biotechnological laboratories, as only very small amounts of each compound are needed (e.g., milligrams up to few grams). When toxicology studies first and later clinical experimentation start, larger quantities of compound are requested: for drugs of a chemical origin, kilograms of active principle are needed. During this scale-up process, both chemical synthesis and biotechnological production need to be adapted and improved in order to meet rapidity, cost-effectiveness, and safety criteria. For chemical synthesis, an efficient and cheap purification process is generally desired, too dangerous solvents or reactions are eliminated, and reactions producing toxic waste are avoided. Indeed, an additional criterion recently adopted concerns the preferred use of synthetic process that does not introduce pollutants in the environment (green chemistry). In case of proteins obtained from genetically modified microorganisms or cell lines, it is fundamental to obtain stable processes to guarantee the final properties of the product (e.g., proper three-dimensional structure, no contamination from endotoxins or viruses). The legislative framework also requires a tight control of impurities, which must not exceed a prefixed limit and must not involve a mutagenic risk.

Choice of the Pharmaceutical Form

Administration route and pharmaceutical form are fundamental features of a drug. The correct use is only possible if the drug is easy to take; in particular, for the treatment of chronic pathologies, oral route and minimal number of daily administrations (e.g., once a day) are preferred. In case of acute diseases or treatments in hospitals, other administration routes are acceptable, sometimes even necessary, such as the parenteral route (e.g., intravenous route), which permit to quickly attain the therapeutic plasmatic concentrations.

Nevertheless, obtaining a pharmaceutical form that is simple and capable of providing the best patient compliance is not a simple task. The biggest issues depend on the physicochemical properties of the active molecule. Some compounds have low oral bioavailability, but they are developed anyway because of their significant therapeutic effectiveness. An example is provided by those biological drugs, such as peptides and proteins, which are administered by parenteral route only. This represents a considerable drawback for biological drugs, and studies are ongoing (e.g., insulin) to obtain their oral absorption, although with no success so far. Important administration routes are also the inhalation and the transmucosal routes, that is, through the mouth, nose, eye, and rectum (see also Chapter 4). The transdermal route (e.g., patches) is also very interesting. Studies are in progress to improve drug delivery directly to the diseased tissues. To this purpose, the rapid evolution of nanotechnologies is opening up new perspectives in particular for the treatment of cancer, by creating nanovectors to carry the drug specifically to the affected tissue (see Chapter 4).

Toxicology Studies

Over the last 50 years, there has been a raise in the number of biological tests available to reduce the risk that new compounds could have toxic effects in humans once converted into drugs. If we consider the very nature of a drug, this may seem an ambitious target, as a drug is a compound that alters the biological mechanisms, often even deeply, and is therefore bound to cause also toxic effects.

The compulsory tests are established by the regulations in force, modified over the years in different ways in different countries. A harmonization process of all the requirements from different countries has only recently begun, and internationally valid toxicological tests have been established (see Chapter 52 and E 54.6 and E 54.7). The final aim is to evaluate the possible toxic effects of a new molecule on organs and systems, using in *vitro* biological samples and *in vivo* tests on laboratory animals. Toxicology studies should provide the following information:

1. Definition of the maximum quantity that does not induce any direct or indirect effect on organs and systems
2. Definition of the quantity inducing toxic effects and the type of alterations it causes
3. Definition of the relationship between therapeutic and toxic quantities
4. Identification of the target of the toxic effect (cell structure, organ, or system) of both the original compound and its metabolites
5. Establishing whether the effects are reversible or irreversible

A number of studies demonstrate that most toxic effects induced by chemical compounds in humans can be replicated in laboratory animals: hence the validity of the toxicological tests required by the international regulatory institutions. High drug quantities, in general multiples of the therapeutic dose, are employed to emphasize those effects that could appear with low frequency in a large population of patients. The use of animals for toxicology studies has always stirred strong critics; however, no alternative models approved by the official regulations are available to date, even if encouraging results have been obtained in toxicity studies carried out in cell lines. Despite all efforts made so far, only the *in vitro* mutagenesis tests have been included among the compulsory tests required for toxicology studies. The whole set of experimental toxicology protocols has been undergoing critical revision for a few years now to further improve prediction value and experimental conditions (e.g., reduction in the number of animals employed) of the tests to be conducted on new molecules.

Although data obtained using different laboratory models are accurate in terms of efficacy and safety, new potential drugs have to be studied also in humans, through a regulated protocol of clinical studies (see Fig. 54.6). The international law defines stringent criteria for the investigation in humans.

THE DRUGS IN 2020

In the year 2020, although it seems so far away from today, in the research world, the drugs that will be possibly available on that date are already on the planning schedule. They are now at the beginning of a long route, in their embryonic stage, still to be improved and perfected. Despite a great number of possible failures, some new compounds currently under study will be definitely successful. As in the last 60 years, significant results will be obtained in the therapy field. The growing focus on rare diseases paves the way to good results in those therapeutic areas that have been neglected so far (orphan diseases and drugs). Some of the features of future drugs can already be predicted: they will be more effective, easier to use, and safer. They will be more effective because only drugs showing a clear pharmacological activity and able to improve the already existing therapy will be developed further. In addition, the research is increasingly focusing on diseases that the current therapy can hardly or inadequately cure. Future drugs will be easier to use because big progresses are being made in drug delivery and administration route. They will be safer because the experimental models available today allow accurate investigations of possible side effects, so that only drugs with a high therapeutic ratio and a very low risk of adverse reaction will succeed in the drug R&D process.

In addition, pharmacogenetics is progressing fast, offering the possibility to choose the most appropriated therapy based on the genetic and metabolic profile of the patient to guarantee a safer and more effective and therapeutic intervention. However, some limits to à la carte pharmacology will also have to be considered: ease of handling, costs, and patient's privacy protection. All new technologies and therapeutic innovations have always to take into account that, beyond the scientific value, patients are at the center of the therapy, with their need for remedies that are not only effective but also cheap and simple.

TAKE-HOME MESSAGE

- Human and financial commitment to drug research is high and needs ever-increasing investments and resources.
- Procedures for identifying new compounds have dramatically changed in recent years due to the introduction of high-throughput screening methodologies.
- End results of the discovery process are achieved after several years of intense and expensive research efforts; thus, appropriate and efficient strategies must be set up to monitor advancements during the drug discovery and development processes. The most advanced and promising approach is multidisciplinary.
- Good Laboratory Practices, Good Clinical Practices, and Good Manufacturing Practices are required at all stages of the drug development process.
- Preclinical studies are required for both target and lead identification and for defining pharmacological and toxicological profiles of drug candidates.

- International health authorities have established stringent criteria for the assessment of safety and efficacy of new drugs.

FURTHER READING

Abraham D.J., Rotella D.P. (2010). *Burger's Medical Chemistry and Drug Discovery. Vol. 1: Methods in Drug Discovery, Vol. 2: Discovery Lead Molecules, Vol. 3: Drug Development*. 7th Edition. New York: John Wiley & Sons, Inc.

Black J.A. (1996). Personal view of pharmacology. *Annual Review of Pharmacology and Toxicology*, 36, 1–33.

Drews J. (2006). Case histories, magic bullets and the state of drug discovery. *Nature Reviews. Drug Discovery*, 5, 635–640.

Haffner M.E. (1991). Orphan products: origins, progress, and prospects. *Annual Review of Pharmacology and Toxicology*, 31, 603–620.

Kaitiu K.I., Di Masi J.A. (2011). Pharmaceutical innovation in the 21st century: new drug approvals in the first decade 2000–2009. *Clinical Pharmacology & Therapeutics*, 89, 183–188.

Kueller R. (2010). The importance of new companies for drug discovery: origins of a decade of new drugs. *Nature Reviews. Drug Discovery*, 9, 867–882.

Nelson A.J., Dhimolea E., Reichert J.M. (2010). Development trends for human monoclonal therapeutics. *Nature Reviews. Drug Discovery*, 9, 767–774.

Page C.P., Schaffhausen J., Shankley N.P. (2011). The scientific legacy of Sir James W. Black. *Trends in Pharmacological Sciences*, 32, 183–188.

Spilker B. (2009). *Guide to Drug Development: A Comprehensive Review and Assessment*. New York: Lippincott Williams & Wilkins.

Stanton M.G., Colletti S.L. (2010). Medicinal chemistry of siRNA delivery. *Journal of Medicinal Chemistry*, 53, 7887–7901.

WEB SITES

www.efpia.eu. Accessed December 19, 2014.
www.epris.org. Accessed November 13, 2014.
www.phrma.org. Accessed November 13, 2014.

55

ROLE OF DRUG METABOLISM AND PHARMACOKINETICS IN DRUG DEVELOPMENT

SIMONE BRAGGIO AND MARIO PELLEGATTI

By reading this chapter, you will:

- Understand the role of pharmacokinetics in the process of research and development of new drugs
- Understand how pharmacokinetics contribute to the definition of drug clinical profile

---

In the complex and multidisciplinary process of drug discovery and development, drug metabolism and pharmacokinetics (DMPK) plays a delicate and important role in interfacing with the various disciplines involved (e.g., medicinal chemistry, *in vitro* and *in vivo* pharmacology, preclinical development, safety assessment, clinical development, and regulatory affairs).

Ideally, a drug should be absorbed, reach the site of action, and remain there for a time sufficient to carry out its effect and then be eliminated without producing toxic effects. In addition, each therapeutic class requires specific pharmacokinetic (PK) characteristics; an anesthetic molecule should be eliminated from the body fairly quickly, whereas on the contrary an anti-inflammatory drug should be eliminated much more slowly; an antibacterial or antiviral drug must reach concentrations that exceed levels that inhibit microorganisms growth; drugs for the central nervous system must be able to cross the blood–brain barrier; drugs for respiratory disorders must reach the lung, whereas anticancer drugs should be able to reach and penetrate the tumor mass.

As a consequence, DMPK has emerged as an applied science, arising from the need to understand the biological and physical processes associated with the administration of therapeutic chemical agents to animals and humans. Accordingly, in modern strategies, DMPK is involved from the very beginning of the discovery process throughout all the development and clinical stages, as summarized in Figure 55.1.

DMPK IN DRUG DISCOVERY

Recent data indicate that the discovery and development of a new small-molecule drug cost close to one billion dollars and it may take approximately 10 years before the drug reaches the market. Considering these impressive figures, it is critical that efforts are made to reduce attrition of drug candidates during the various stages of drug discovery and development. A source of attrition can be inappropriate drug disposition characteristics, and old data suggested that about 40% of all lead candidates fail in their development due to poor PK. More recent data indicate that attrition due to PK reasons has been reduced to 10% and this can be ascribed largely to integration of absorption, distribution, metabolism, and excretion (ADME) in drug discovery, resulting in an improved predictability of human PK based on *in vitro* and *in vivo* experiments. Currently, one of the most significant reasons for attrition is represented by limited or no efficacy in phase II trials. In some cases, the targeted mechanism does not appear affected as the preclinical pharmacodynamic observations seem to indicate, but sometimes, the limited or absence of efficacy may be due to lower than expected systemic exposure. The latter situation is extremely unfortunate, because the data do not allow assessing the relevance of mechanistic intervention and, therefore, the project team does not know whether a backup compound is appropriate or not.

General and Molecular Pharmacology: Principles of Drug Action, First Edition. Edited by Francesco Clementi and Guido Fumagalli.
© 2015 John Wiley & Sons, Inc. Published 2015 by John Wiley & Sons, Inc.

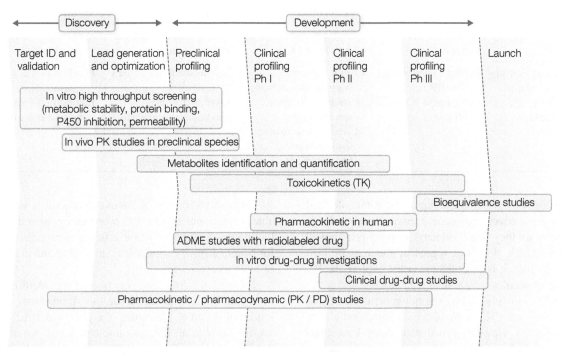

FIGURE 55.1 Role of drug metabolism and pharmacokinetics in the research and development of a new drug. The role of drug metabolism and pharmacokinetics in the long and complex process that leads to the development of a drug is an issue of extreme importance both in research phase, where the goal is to select the drug candidates with the best pharmacokinetic characteristics, and in development stages, where the predictions based on *in vitro* or preclinical studies are tested in the clinic.

Evolution of DMPK in Drug Discovery

The previous scenario suggests that continued vigilance is required to ensure that candidate drugs selected for further development have the desired PK properties.

Although elements and concepts of modern DMPK were already available and understood in the 1980s, DMPK studies in pharmaceutical industry were mainly descriptive and not aimed at providing mechanistic insight into the fate of candidate drugs in biological systems. DMPK studies were carried out only at advanced stage of drug development, mainly because it is required by regulatory authorities for registration. Their main goal was to gather information in order to reduce risks in clinical use (by anticipating the risk of interactions with other drugs and optimizing the interval between doses) or to study new formulations. In practice, the molecule to be developed was selected based on its *in vitro* affinity for the biological target and its pharmacological efficacy as evaluated in animal models.

The realization that failure of drug candidates in clinical development was primarily caused by inappropriate human PK led to a progressive involvement of DMPK as a key component of the overall drug discovery process. This effort to embed DMPK within the drug discovery phase paved the way to higher-throughput *in vitro* and *in vivo* studies. Indeed, the change in strategy implied the rapid execution of high amounts (hundreds per week) of PK and toxicity experiments (known by the acronym ADMET: absorption, distribution, metabolism, excretion, and toxicity), having the same high-quality standard of previous experiments in which a relatively limited number of compounds were analyzed. The timely emergence of techniques, such as liquid chromatography–mass spectrometry (LC–MS) and liquid chromatography–tandem mass spectrometry (LC–MS/MS), played a crucial role in this process, by speeding up method development and sample analysis.

Most major pharmaceutical companies and many biotech companies have made investments to implement early *in vitro* DMPK determinations, and today, many *in vitro* ADMET assays are available, which are predictive, to different extents, of *in vivo* outcome (Table 55.1). For example, to assess the risk of clinical drug–drug interactions (DDI), assays are available for measuring the ability of a new chemical entity to inhibit the turnover of a specific probe substrate in human liver microsomes. Models such as parallel artificial membrane permeability assay (PAMPA) and MDCK and Caco-2 transwell assays can be used to predict drug permeability. Finally, turnover of a new chemical entity by liver microsomes or hepatocytes *in vitro* can be used to predict *in vivo* data. In typical drug discovery projects, large numbers of compounds are screened using a battery of *in vitro* assays performed in multiwell plates, and most promising drug structures are selected for further investigations.

TABLE 55.1 Main *in vitro* assays for the characterization of ADMET properties in the early stages of drug discovery

Physicochemical properties	Permeability	Metabolic stability	Inhibition of cytochrome P450 enzymes	Protein binding
In silico physicochemical properties calculation Measurements by chromatographic systems	Artificial membranes (PAMPA) Cell cultures (Caco-2, MDCK)	Liver microsomes (various species/man) Hepatocytes (various species/man) Microsomes from other tissues (intestine, lung, brain, muscle)	Recombinant isoenzymes (CYP450) Liver microsomes (various species/man)	Equilibrium dialysis (plasma, blood) Equilibrium dialysis (brain, muscle, other tissues)

Although *in vitro* systems use human or human-derived materials, drug discovery programs cannot entirely rely on *in vitro* assays as they are conducted in nonphysiological environment, which does not represent the real *in vivo* scenario. Therefore, it is essential to further explore the PK of promising molecules *in vivo* in animal models.

In the early discovery phase, it is not generally essential to obtain a complete PK profile of a compound and a rank order is sufficient. However, once candidates have been selected using the aforementioned *in vitro* assays, the pace of the drug discovery program depends on the rapidity at which weak compounds are identified and discarded. Indeed, the prompt identification and elimination of the candidates that exhibit undesirable *in vivo* PK profile allow scientists to focus on fewer potential and more promising lead compounds. Relatively high-throughput *in vivo* screens, known as "cassette dosing" or "N-in-One dosing," have since long been used to rank molecules according to their oral plasma concentrations or systemic clearance after an intravenous dose. These approaches consist in the administration of a cocktail (or cassette) containing a large number of compounds to the same animal. The advantage of these screens is that they allow testing a large number of molecules using a limited number of animals, thus reducing interanimal variations and providing a single consolidated report for several compounds. On the other hand, with these approaches, the possibility of false-positive and false-negative DDI cannot be ruled out. A false-positive result implies that a compound exhibits acceptable PK when administered in a combination with several compounds, but as a single compound, its PK is unfavorable. On the contrary, a false-negative result is obtained when a compound with favorable PK exhibits unacceptable PK in a cassette. False-negative results are more serious as compared to false positives because good compounds may be rejected. If required, the PK profile of selected compounds can then be assessed by administering a single dose of a compound to reconfirm the results obtained with *in vivo* high-throughput screening and obtain reliable and accurate PK parameters. This also facilitates the identification of *in vivo* metabolites, which are quite difficult to identify in the chromatogram of a cassette containing several compounds.

Early identification of chemical structures with acceptable *in vitro* and *in vivo* PK properties is also desirable to facilitate determination of the efficacy in preclinical models, for better PK/pharmacodynamic modeling and identification of biomarkers.

The goal of a drug discovery project is to identify, through the synthesis and characterization of hundreds (or thousands) of different molecules, the chemical structure (drug candidate) with all the desirable characteristics an optimal drug should have. Each company and even each project has a different or tailored lead optimization screening strategy (Fig. 55.2). It is anyway important to define, as early as possible, a candidate drug target profile by identifying the desirable efficacy, safety, and pharmaceutical properties to be searched and also the target population. In simple terms, the aim of structure design in a drug discovery project is to establish the area of "chemical overlap" between acceptable activity–selectivity–toxicity (safety) and acceptable DMPK. DMPK properties are a central component of this candidate drug target profile and are dependent on the therapeutic hypothesis (e.g., how long and to what extent the target needs to be modulated).

DMPK IN THE REGULATORY PHASE OF DRUG DEVELOPMENT

Once a certain molecule has been selected according to the drug candidate target profile, additional properties need to be evaluated about its safety and effectiveness, as required by regulatory agencies. At this stage, PK takes on a more traditional role of characterization and in-depth description of the PK and metabolic characteristics of the drug. In other words, PK information collected at this stage serves mainly as "indicators" of the therapeutic or unwanted effects of the drug in the organism. Nowadays, given the different studies conducted during the discovery phase, a reasonably good knowledge of the PK of the compound is already available at the beginning of the development process. This allows reducing the investigations required by regulatory agencies to progress new molecules toward clinical testing and ultimately patients. During the development stage, DMPK

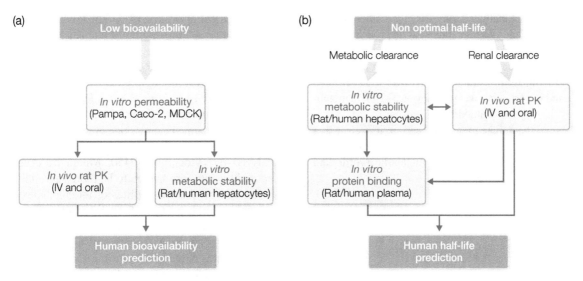

FIGURE 55.2 Examples of screening strategies focused on fixing specific liabilities. Strategies and ADMET assays during the discovery steps vary depending on the pharmacokinetic parameters that need to be optimized.

studies almost invariably include excretion balance, metabolic profile in plasma and excreta, plasma levels of drug-related material (DRM) and parent compound in toxicological species, and a tissue distribution study, normally in rat, performed with the whole-body autoradiography (WBA) technique. These studies are carried out administering the compound by the same route used for toxicological investigations. For oral drugs, sometimes, also intravenous route is investigated to evaluate the bioavailability and role of first-pass effect on metabolic profile and excretion pattern.

Starting from the 1980s, also toxicokinetic studies, involving the determination of plasma levels of the compound in animals treated during toxicology studies, have become common and required. A few other preclinical *in vivo* ADME studies, such as placental transfer in reproductive toxicology (reprotox) species and milk transfer, may be required.

Among the regulatory DMPK studies that are generally conducted during the development phase, it is worth mentioning those involving the use of radioactive compounds. Radiolabeled drugs (and related metabolites) can be followed quite easily and allow to analyze tissue distribution, even of the entire sections of an animal (WBA; see Fig. 55.3), to determine plasma levels and concentration of radioactivity in the excreta, and the excretion balance, that is, the percentage of the dose excreted in urine and feces. All these data can help to understand time of drug permanence in the body, storage organs, drug absorption after oral administration, and so on.

Safety of Metabolites

As discussed in Chapter 6, the process of biotransformation or metabolism consists in the conversion of the parent drug in one or more metabolites (pharmacologically active or

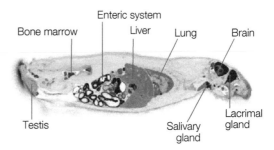

FIGURE 55.3 Whole-body autoradiography (WBA) in rat after oral administration of a radioactive drug. The WBA is a technique based on the properties of the radiation to impress a photographic film. If the film is placed in contact with a section of the whole body of a rat previously administered with a radioactive drug, an image is obtained (autoradiogram) that reproduces the localization and concentration of the radioactivity present in the section.

inactive) with higher polarity, therefore more water soluble and easily removable with urine or bile. This process involves the transformation of the functional groups of the molecule (phase I reactions, e.g., oxidation, reduction, and hydrolysis) and subsequent conjugation with endogenous substances (phase II reactions, e.g., glucuronidation and sulfation). The resulting metabolites may produce therapeutic effects and/or side effects, which may be similar, different, or completely new compared to the starting molecule (Fig. 55.4). Therefore, the evaluation of the activity and safety of a drug depends also on the effects produced by its metabolites.

For this reason, the study of metabolites has become of fundamental importance, and in recent years, new strategies of investigation have been discussed and implemented. In particular, in 2002, a guideline sponsored by the "Pharmaceutical Research and Manufacturers of America" (PhRMA) was published offering a number of considerations related to the study and evaluation of metabolites during toxicity studies of drug candidates, suggesting possible practical approaches. This guideline, which coined the term metabolites in safety testing (MIST), stirred a debate between pharmaceutical industry and the Food and Drug Administration (FDA) over the suggested approaches and in particular over the relevance given to metabolites formed in very small amounts in humans.

Regardless of the debate mentioned earlier, it is clear that an increasing amount of information and assessments will be required about the drug candidate metabolic profile and particular attention will be paid to metabolites circulating in the blood after administration to humans, even if present in small quantities.

Another important aspect is the role of the so-called reactive metabolites. Although the metabolism of exogenous substances and drugs in most cases leads to stable chemical entities, sometimes, it produces highly reactive intermediates. This can happen at any stage of the drug biotransformation process but seems to concern mainly phase I reactions catalyzed by cytochrome P450. Moreover, drugs having electrophilic sites are known to be more susceptible to bioactivation. Unstable metabolites or intermediates can react with nucleophilic groups of vital constituents of the cell (e.g., hydroxyl, sulfhydryl, and amine groups of functional proteins or nucleic acids), compromising their functionality or causing irreversible inhibition, as in the case of cytochrome P450 metabolizing enzymes. Moreover, covalent bonding between these unstable compounds and proteins may give rise to macromolecules recognized as exogenous by the immune system, causing idiosyncratic reactions (Fig. 55.5).

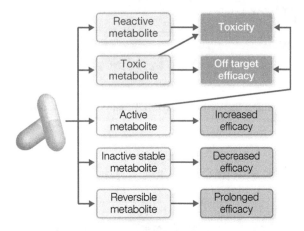

FIGURE 55.4 End results of drug metabolism. The metabolism of drugs includes different types of chemical reactions that occur in the body to change or degrade the drug itself. A drug, after being absorbed, is transformed into different molecules that may be inert, toxic, or pharmacologically useful.

PK IN CLINICAL TRIALS

The fundamental value of clinical PK as a tool to guide and understand clinical outcomes has been repeatedly pointed out. Clinical studies are initiated on the basis of preclinical results, with DMPK contributing to all phases of drug development in humans to optimize safe and efficacious use of drugs.

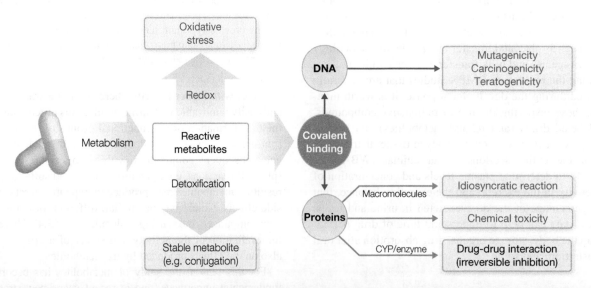

FIGURE 55.5 Possible effects of the bioactivation process. Bioactivation is a metabolic process that increases the toxicity of the substance through the formation of unstable intermediates or metabolites with highly reactive chemical bonds capable of reacting with organic macromolecules. In particular, binding to DNA can lead to carcinogenic effects. The covalent bond to the same cytochrome P450 enzymes may lead to an inactivation of the enzymes themselves (irreversible inhibition), while binding to other macromolecules may trigger the phenomena of idiosyncratic reaction to the drug, a major cause of adverse drug reaction.

TABLE 55.2 Clinical pharmacokinetic studies

First time in human study at single and increasing doses
Repeated dose study
Evaluation of plasma exposure linearity with the dose
Determination of PK differences in target populations:
    Male/female
    Fasten/fed
    Young/elderly
    Renal-impaired subjects
    Liver-impaired subjects
ADME study in healthy volunteers
Drug–drug interaction evaluation
PK/PD studies in healthy volunteers and patients
Bioequivalence studies of different formulations

The first phase of clinical trials is designed to gather information on the pharmacokinetics, safety, and pharmacodynamic of the new substance.

Dose–plasma concentration and plasma concentration–effect (both therapeutic and toxic) relationships need also to be defined as early as possible in clinical studies. The plasma concentration–effect relationships describe changes in pharmacological effect resulting from different levels of exposure (or doses) to a given drug during different dose regimens in humans. All these data are essential to establish the safest and most effective therapeutic dose schedules for initiating and adjusting therapy in general patient populations, as well as in special subsets of patients. They are particularly relevant for drugs with a narrow therapeutic range and when a close relationship between plasma concentration and therapeutic and/or toxic effect can be demonstrated or expected. Table 55.2 summarizes the typical clinical studies involving PK evaluations. The first studies conducted in healthy volunteers are generally defined as "tolerability and PK," to indicate that drug PK characterization is a main end point of the study. Data gathered during this study include PK parameters such as area under the curve (AUC, exposure), C_{max} (maximum concentration), T_{max} (time to C_{max}), half-life, clearance, volume of distribution, bioavailability, steady-state plasma concentrations, accumulation ratio, linear or nonlinear PK, time-dependent PK (autoinduction), plasma protein binding, and metabolite identity and their PK. It is worth pointing out that at this stage drug PK parameters in humans are compared with those estimated in the animal species used in preclinical studies to evaluate suitability of the animal models used for toxicology assessments. Clinical PK studies are carried out until the drug is registered and also afterward. In particular, studies are aimed at identifying correlations between PK parameters and either therapeutic efficacy or toxicity, at assessing potential PK differences in patients compared to healthy volunteers or within selected populations (e.g., elderly, children), or at devising new formulations.

Two aspects of particular interest during drug clinical development that may represent critical issues and in which PK plays a significant role are interindividual variability and DDI.

Interindividual Variability

In general, the drug dosage in humans is chosen to provide a convenient dose for the general patient population. However, the standard dosage regimen of a drug may prove therapeutically effective in most patients, ineffective in some patients, and toxic in others. Many drugs show high, sometimes very high, PK variability. At the same dose, very different (up to 10 times) blood levels can be observed between individuals and, sometimes, even in the same individual on different occasions. The reasons of such inter- and intraindividual variability can be several and complex, and most of them can be analyzed and evaluated in the context of metabolic and PK studies providing information about specific metabolizing enzyme(s) or preferred clearance/elimination route(s). If a drug is predominantly excreted in urine as such, it is unlikely that reductions in liver function alter its PK, whereas it is probable that plasma exposure is increased if renal functions are reduced. Conversely, drug exposure may be affected for drug metabolically cleared by liver enzymes and when hepatic functions are altered. This information allows to assess risks and suggest to conduct studies in individuals with reduced hepatic and/or renal function to check how these conditions alter plasma levels (see also Chapter 7).

Moreover, if a drug is metabolized by polymorphic enzymes, such as CYP2D6, it can be predicted that groups of individuals will show reduced clearance and therefore higher blood levels. All this information is used to anticipate the risk of variability and understand the causes.

Variability in drug response becomes an important problem for drugs that have a narrow therapeutic window. At the same dose, we could have patients that do not respond and others that show side effects. Warfarin is a good example; the range of daily dose (<2 to >11 mg) of warfarin needed to produce a similar prothrombin time is very wide.

DDI

The more drugs are administered to the same patient, the higher the risk of DDI that may lead to serious side effects. There are two types of drug interactions: pharmacodynamic and PK. PK interactions are relatively straightforward and are relatively predictable. Coadministration of different drugs may cause significant changes in plasma exposure in one or more of the drugs in the cocktail. Most frequent interactions involve metabolizing enzymes and drug transporters. This aspect is extensively discussed in Chapters 6 and 53. Here, we would like to highlight the importance of PK studies designed to understand, define, and predict these types of interactions. The interactions a new compound may give rise to have to be assessed in detail, during preclinical *in vitro* and *in vivo* studies and candidate selection and throughout the preclinical and clinical development. Since formal *in vivo* studies of all possible drug interactions are neither practicable nor conclusive, a careful selection of a limited number of drug combinations to be investigated *in vivo* during the development phase is indicated. Based on PK and biopharmaceutical data, the combination of *in vitro* and *in vivo* interaction studies allows establishing the clinically relevant PK DDI a new drug may cause and the proper dose recommendations.

PK IN DRUG FORMULATION OF GENERIC EQUIVALENT DRUGS

In recent years, there has been a rise in prescribing and dispensing of generic drugs, as many industrialized countries are trying to reduce health costs. Using high-quality, cheaper generic drugs can contribute to reduce sanitary costs while providing equally effective treatments as the original drugs (see also Chapter 1). According to the FDA designation of therapeutic equivalence, a generic formulation is bioequivalent to the brand formulation and is expected to have "equivalent clinical effects and no difference in their potential for adverse effects."

In the drug approval procedure, interchangeability between a generic drug and its corresponding brand-name drug is established based on the criterion of "essential similarity," meaning that the generic drug needs to have the same amount and type of active principle, administration route, and therapeutic effectiveness as the original drug, as demonstrated by bioequivalence studies. Two preparations of a drug are defined bioequivalent when they are pharmaceutically equivalent; this implies that their bioavailabilities in the same molar dose and their effects, in terms of both efficacy and safety, can be expected to be the same. Pharmaceutical equivalence indicates the same amount of the same active substance(s), given in the same dosage form and via the same route of administration and having the same or comparable PK standards.

Bioequivalence, or pharmacological equivalence to the brand formulation, has to be assessed for all generic formulations, and this is usually generally done at the registration stage. Bioequivalence is the major requirement to register a generic drug as it establishes whether two pharmaceutically equivalent drugs are identical.

The main tests to demonstrate drug bioequivalence are the evaluation of bioaccessibility and PK features in human volunteers and comparative clinical tests. The choice of the testing method depends on form, physical, chemical, and pharmacological characteristics of the drug and indications to its use. The standard bioequivalence (PK) study is conducted using a two-treatment crossover study design including a limited number of subjects. According to the regulatory authorities, two products are considered to be bioequivalent if 90% clearance (CI) of the relative mean C_{max} and AUC of the generic drug to the brand-name drug is within 80–125%.

NEW PERSPECTIVES

In recent years, DMPK evolution as a discipline has certainly contributed to reduce the number of potential drugs discarded at various stages of the development process for reasons closely related to the PK of the compound (poor bioavailability, variability, short half-life, drug interactions). However, even if the number of compounds discarded for poor PK properties is now drastically reduced, the percentage of molecules that successfully complete the development phase and become available to patients is not increased. Today, the major causes of failure are often unsatisfactory efficacy or unacceptable toxicity.

So in what direction should the discipline of PK and metabolism evolve as a science to make the research and development process more productive and efficient?

Although at first glance current reasons for failure appear to reduce the possibilities of intervention, efficacy, toxicity, and PK are closely interrelated and cannot be discussed and analyzed separately. PK has to move toward a greater integration of these three aspects that ultimately determine the therapeutic efficacy of a new drug. The leading role of PK in drug research and development is legitimized by the fact that any effect, whether therapeutic or toxic, is determined by kinetics and concentrations of the drug and its metabolites in blood and various tissues. A closer association between drug concentration and effect (or toxicity) can be found for example by estimating more systematically the actual concentrations at the site of action (biophase), which in most cases are significantly different from the plasma concentrations used to describe drug PK. Compounds that are very effective in *in vitro* tests and exhibit excellent PK (e.g., bioavailability, half-life) often are unable to reach their biological target (receptor, protein, DNA) in sufficient concentration and therefore lack the desired effects. Concepts such as "drug efficiency," which indicates the actual drug concentration available to interact with the biological target (concentration in the biophase), start to enter into the optimization process of lead compounds and certainly will contribute to the design of better drugs.

Integration of pharmacogenetics, pharmacogenomics, and metabonomics into the more traditional aspects of PK and metabolism will surely help to better understand and anticipate aspects such as DDI, metabolism and elimination pathways, mechanisms of bioactivation, and toxicity, ultimately leading to better medicines.

TAKE-HOME MESSAGE

- Traditionally, the study of drug metabolism and pharmacokinetics (DMPK) is aimed at describing the classical aspects of absorption, distribution, metabolism, and excretion (ADME). Over the last two decades,

DMPK has become a critical tool for increasing the success rate of the drug discovery process, being involved in predictive studies during design, definition, and selection of new drug candidates.

- *In vitro* DMPK studies define chemical parameters relevant for pharmacokinetic properties, including cell permeability, metabolic stability, and binding to plasma and tissue proteins.
- *In vivo* DMPK studies have identified new strategies for increasing oral bioavailability and/or improving pharmacokinetic properties.
- The increased interest in DMPK has produced a better understanding of the role of drug metabolites in their pharmacological activity and toxicity.

FURTHER READING

Baillie T.A. (2008). Metabolism and toxicity of drugs. Two decades of progress in industrial drug metabolism. *Chemical Research in Toxicology*, 21, 129–137.

Ballard P., Brassil P., Bui K.H., et al. (2012). The right compound in the right assay at the right time: an integrated discovery DMPK strategy. *Drug Metabolism Reviews*, 44, 224–252.

Braggio S., Montanari D., Rossi T., Ratti E. (2010). Drug efficiency: a new concept to guide lead optimization programs towards the selection of better clinical candidates. *Expert Opinion on Drug Discovery*, 5, 609–618.

Hop C.E., Cole M.J., Davidson R.E., et al. (2008). High throughput ADME screening: practical considerations, impact on the portfolio and enabler of in silico ADME models. *Current Drug Metabolism*, 9, 847–853.

Pellegatti M. (2012). Preclinical in vivo ADME studies in drug development: a critical review. *Expert Opinion on Drug Metabolism & Toxicology*, 8, 161–172.

Singh S.S. (2006). Pre-clinical pharmacokinetics: an approach towards safer and efficacious drugs. *Current Drug Metabolism*, 7, 165–182.

Summerfield S., Jeffrey P. (2009). Discovery DMPK: changing paradigms in the eighties, nineties and noughties. *Expert Opinion on Drug Discovery*, 4, 207–218.

Yengi L.G., Leung L., Kao J. (2007). The evolving role of drug metabolism in drug discovery and development. *Pharmaceutical Research*, 24, 842–858.

56

CLINICAL DEVELOPMENT OF A NEW DRUG AND METHODOLOGY OF DRUG TRIALS

CARLO PATRONO

By reading this chapter, you will:

- Learn the principles and rules of clinical investigation of new drugs
- Know the objectives, end points, dimension, duration, and cost of the phases eventually leading to marketing authorization of a new drug
- Understand the differences in methodology among the clinical studies and trials
- Evaluate the possible future modifications in clinical studies to improve the safety and speed of the transfer of experimental findings into clinical practice

---

Drug therapy as a science is based on the careful examination of individual patients and the critical evaluation of the evidence of efficacy and safety of a particular drug treatment. The evidence providing the most solid basis for therapeutic decisions is represented by the results of randomized, controlled clinical trials of high methodological quality, namely, the so-called level "A" evidence in most treatment guidelines. High-quality evidence may also come from properly conducted observational studies yielding very large effects. However, guidelines for evidence-based approaches to treatment recommend to take into account also other factors, besides evidence, such as patient values and preferences as well as resource expenditure. Although practicing physicians tend to rely mainly on disease-oriented treatment guidelines issued by international organizations and professional societies, it is important for them to become familiar with the principles, methodology, and terminology of randomized controlled trials (RCTs), in the context of clinical development of new drugs or old drugs for new indications.

CLINICAL DEVELOPMENT OF A NEW DRUG

Principles and Rules of Clinical Investigation

The preclinical path described in the previous chapter can yield a candidate molecule (or a series of structurally related molecules) that interacts with a disease-related target (e.g., an enzyme or receptor involved in the synthesis or action of a disease-related mediator) and exhibits beneficial effects on the experimental disease by altering target function. If no toxicity related to the mechanism of action of the candidate drug is observed at therapeutic doses and no serious off-target effects appear during the preclinical toxicology program, the candidate drug can move forward to phase 1 clinical studies. The whole process of clinical development is a highly regulated path, governed by a fundamental set of ethical principles (the World Medical Association's Declaration of Helsinki), regulatory guidelines issued by the Food and Drug Administration (FDA) and European Medicines Agency (EMA), and a detailed set of procedures, the Good Clinical Practices (GCP). These regulations are intended to guarantee integrity and safety of the subjects enrolled in clinical trials of experimental drugs, as well as to ensure that any potential hazard of a new drug is justified by its potential benefits. Thus, about 50 years ago, the Declaration of Helsinki introduced the ethical principles of research in humans, including the requirement for informed consent of participating subjects. At about the same time, the US Congress passed the Harris–Kefauver

Amendments to the Food, Drug and Cosmetic Act establishing the requirement for proof of efficacy and relative safety in terms of assessment of the drug risk/benefit ratio in relation to the specific disease to be treated. As a consequence of these amendments, the FDA has introduced rigorous protocols for controlled clinical trials in which the effects of the experimental treatment in a group of patients are compared to the effects of a placebo (a pharmacologically inert preparation with the same appearance as the experimental drug but devoid of its active moiety; see also Chapter 1, Supplement E 1.2) in a comparable group of patients. GCP provide guidelines for the preparation of study protocols (a detailed account of study procedures and responsibilities that should be made publicly available prior to initiation of the trial) and operational aspects, including approval of protocol and informed consent form(s) by an independent Ethics Committee at each participating center, and monitoring of trial conduct and documentation. Strict adherence to GCP is meant to protect the rights and safety of participating subjects as well as to ensure the integrity and validity of study results that will eventually form the basis for drug approval. An institutional Ethics Committee is an independent, multidisciplinary group of experts (both from health professions and lay community) who review the protocol and the informed consent form to protect patients' rights and safety. A proper evaluation of the scientific merit of the proposal represents the basis for an ethical judgment of the clinical trial. This should include rationale, target population, clinical relevance of the primary end point, plausibility of the hypotheses underlying sample size calculation, and risk minimization strategies. Compliance with GCP is often monitored by a contract research organization (CRO) hired by the sponsor (the pharmaceutical company developing the new drug or an institution conducting independent research) to oversee study conduct at each participating center, as well as by regulatory authorities. In the case of a multinational study, the FDA has the authority to inspect non-US centers participating in the trial and contributing to the overall study results, because its primary responsibility is to protect public health by assuring safety and efficacy of human drugs.

Before initiating clinical trials of an experimental drug in the United States, the sponsor must file an Investigational New Drug (IND) application with the FDA, in order to allow a 30-day evaluation of rationale, preclinical evidence for efficacy, and preliminary assessment of safety (besides chemistry, manufacturing, etc.) by the agency. In Europe, the same documentation is submitted to national regulatory authorities that are mainly governed by common guidelines issued by the EMA.

The Path of Clinical Development of a New Drug

The path eventually leading to marketing authorization of a new drug is typically subdivided into three phases having quite different objectives, end points, dimension, duration,

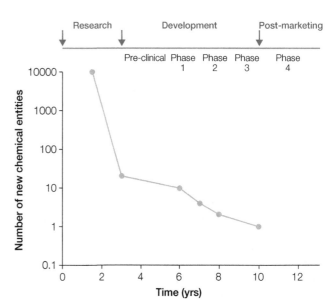

FIGURE 56.1 The phases, time frame, and attrition that characterize research and development of new drugs.

and cost. The overall process may take up to 8 years, with an overall failure rate over 90% (Fig. 56.1).

The primary objective of phase 1 is to assess tolerability and pharmacokinetics. Usually, a limited number of healthy volunteers are enrolled in open-label studies characterized by single administration of escalating doses to establish the maximal tolerated dose of the drug and provide guidance for the choice of doses to be explored in phase 2. Moreover, an important feature of phase 1 studies is the timed collection of blood and urine samples following oral, parenteral, or intravenous dosing, to measure systemic concentrations of the parent drug and its (active and/or inactive) metabolites. The emergence of unexpected side effects (e.g., an increase in liver enzymes) or unacceptable pharmacokinetic features (e.g., poor oral bioavailability) may kill the new drug after a first-in-man study. In the case of some anticancer or anti-HIV drugs, for which cellular toxicity is mechanism based and expected, phase 1 studies are carried out in patients with the disease targeted by the drug, because it would be unethical to expose healthy volunteers to undue toxicity.

Phase 2 aims at the characterization of the expected pharmacodynamic effect (e.g., lipid lowering, blood pressure lowering, analgesic) in the target patient population and to assess its dose dependence in order to provide guidance for the choice of the dose(s) to be used in pivotal phase 3 trials. Phase 2 studies are usually randomized, placebo controlled, and often blinded (single blinded, when only the patient is unaware of the assigned treatment, or double blinded, when both patient and treating physician are unaware) and involve a much larger sample size (a few hundreds to a few thousands) and repeated daily dosing (for a few weeks to several months, depending on the nature of the experimental treatment and its target disease). The primary end point of phase 2 studies is typically represented by pharmacodynamic measurements that may provide a "surrogate" indication of efficacy, but are not necessarily an accurate predictor of clinical efficacy. Thus, a novel antiplatelet agent may well inhibit platelet aggregation in phase 2 studies of patients at high cardiovascular risk, but this may not translate into a clinically relevant benefit in phase 3 trials assessing whether it reduces atherothrombotic vascular complications (e.g., myocardial infarction or ischemic stroke). Although such vascular events, and their safety counterparts (i.e., bleeding complications, in the case of an antiplatelet agent), are often listed as secondary end points, phase 2 studies do not have the statistical power (because of inadequate sample size) to reliably assess efficacy and safety, which are properly evaluated only in phase 3 trials. Although unavoidable for obvious reasons (e.g., pleasing investors), overemphasizing positive results of phase 2 studies on secondary end points related to efficacy and/or safety is potentially misleading in overestimating the former and underestimating the latter for the design of phase 3 trials.

The primary objective of phase 3 is to reliably assess efficacy (usually, the primary end point) and to provide some preliminary assessment of safety (usually, a secondary end point). The sample size (a few thousands to a few tens of thousands of patients) and duration (up to a few years) of these "pivotal" trials (usually, two with a similar design or a very large one) are primarily meant to provide reliable statistical assessment of the drug efficacy on "hard" (i.e., clinically relevant) end point(s). By contrast, adequate evaluation of the safety of a new drug cannot be provided by phase 3 trials, because the incidence of serious adverse effects is often so low to escape detection during clinical development. Therefore, safety assessment must continue after approval and marketing of the new product through phase 4 studies that include observational studies as well as new RTCs.

An additional tool for assessing safety during phase 4 is represented by meta-analyses of all the available randomized trials of a new drug (or class of new drugs) versus placebo and/or comparator drug(s). This approach may provide more reliable and less statistically uncertain estimates of specific risk(s) associated with a new agent than those obtained from individual trials. Ideally, these meta-analyses should be based on individual participant data that may allow characterizing the baseline determinants of risk and its dose and time dependence. Not infrequently, approved drugs are withdrawn from the market because of safety concerns that did not clearly emerge during phase 3. An example is represented by the withdrawal of three out of five marketed coxibs, that is, rofecoxib, valdecoxib, and lumiracoxib, because of cardiovascular, skin, and liver toxicity issues, respectively. Moreover, serious safety issues of old drugs (e.g., traditional nonsteroidal anti-inflammatory drugs (NSAIDs); see also Chapter 49) may not become apparent for decades after their approval because of the relatively small sample size and short duration of the original phase 3 trials and lack of adequate phase 4 studies. Thus, it was widely assumed that traditional NSAIDs were safe for the heart and could actually exert aspirin-like cardioprotective effects, until the much larger and longer coxib trials demonstrated that high-dose regimens of widely used NSAIDs (e.g., diclofenac) carried the same vascular risks as coxibs.

OBSERVATIONAL STUDIES AND RANDOMIZED CLINICAL TRIALS

Observational studies allow to establish a potential association between exposure to a particular drug (e.g., aspirin) or drug class (e.g., NSAIDs) and the occurrence of untoward events (e.g., upper gastrointestinal bleeding or myocardial infarction) responsible for hospitalization of the exposed subjects. By comparing risks of these events among exposed and nonexposed subjects, it could be possible to suggest that aspirin increases gastrointestinal bleeds but reduces coronary events. However, this observation is subject to multiple confounding factors because exposed and nonexposed subjects are not necessarily comparable and may actually differ for a number of variables (both known and unknown) that may influence the measured outcomes, independently of drug treatment. Therefore, observational studies can only point out an association between drug exposure and increased or decreased risk of a certain outcome, but cannot establish a causal relation between the two. Moreover, the accuracy of recorded information on drug utilization (including daily dose and duration) derived from electronic databases can vary substantially among studies. Despite these limitations, a consistent finding of a relative risk greater than or equal to 2.0 for a serious complication must be taken into account in the critical assessment of the benefit/risk profile of the drug.

Observational studies continue to play an important role in detecting relatively rare side effects (those occurring in 0.1–1.0% of exposed subjects). However, assessment of drug efficacy relies exclusively on randomized clinical trials. Thus, the numerous observational studies suggesting a chemopreventive effect of aspirin and other NSAIDs against colorectal cancer are not sufficient for regulatory approval of a new indication for these widely used agents nor for issuing treatment guidelines advocating their prescription for chemoprevention. In fact, we cannot exclude that a third variable (either unknown or unmeasured) may be responsible for both NSAID use and the reduced risk of colorectal cancer observed.

Under particular circumstances in which placebo-controlled, randomized trials cannot be performed (e.g., oral contraceptives), cohort observational studies represent the only source of information on drug safety. The long-standing debate on the potential association between use of hormonal contraceptives and increased (or decreased) incidence of some cancers of the female reproductive tract reflects the uncertainty in interpreting observational data.

The randomized, controlled clinical trial represents the fundamental tool to assess the efficacy of a new drug and evaluate its safety in a preliminary fashion. Randomized means that patients (selected based on inclusion and exclusion criteria defined by the protocol) are randomly assigned to receive the experimental or the control treatment (placebo or active comparator). Randomization ensures that each patient has the same probability of receiving treatment A or treatment B. The key role of randomization is to ensure that all variables that may influence the primary outcome(s) are equally distributed among the two treatment groups, thereby making them quite comparable in terms of measured variables (i.e., virtually identical baseline characteristics) and, presumably, also of the unmeasured ones.

The control group is important as it provides a description of the natural history of the disease and its complications, under the standard conditions of the time and socioeconomic context in which the trial is being conducted. Therefore, the control group is represented by patients treated with a placebo (of course, added to all treatments of proven efficacy for a given disease) or with an active comparator belonging to the same therapeutic class as the experimental drug (e.g., clopidogrel as a comparator for a novel $P2Y_{12}$ antagonist).

THE KEY ROLE OF THE PRIMARY HYPOTHESIS

The primary hypothesis underlying an interventional trial has two components related to (i) the pathophysiology, that is, the mediator, enzyme, or receptor, that is thought to be relevant to the occurrence of a specific outcome and (ii) the expected effect of the intervention in both qualitative (i.e., the parameter that is expected to be modified) and quantitative (i.e., the extent of the modification as compared to control) terms. For example, we expect a novel drug for cardiovascular prevention to modify the natural history of the atherosclerotic process or its serious complications to a clinically relevant extent as compared to the standard of care. The design of phase 3 trials to assess the efficacy of a new drug can use one of the three main options: (i) to test superiority of the new drug over the gold standard of the same therapeutic class (e.g., by comparing a novel antiplatelet agent with aspirin), (ii) to assess additive effects of the new drug when combined with the gold standard (e.g., by comparing a novel antiplatelet agent plus aspirin versus aspirin alone), and (iii) to test noninferiority of the new drug when compared to the gold standard of the same therapeutic class (e.g., by comparing a novel oral anticoagulant with warfarin). Noninferiority trials have been criticized on ethical grounds, although such a design may be acceptable from a regulatory perspective. Noninferiority has been defined as a "kind of similarity" within a limit. The limit is the variable extent of what is considered "acceptable inferiority" of the new drug compared with the standard treatment. An experimental drug is considered "noninferior" when the point estimate and 95% CI of its effect do not fall outside the preset, largely arbitrary noninferiority margin. This means that a "noninferior" experimental drug could actually be less effective than the comparator, but not enough to be recognized as such. Thus, noninferiority trials expose patients to clinical experiments without any assurance that the experimental drug is not worse than the standard treatment and without really exploring whether it is better.

Verifying the validity of the primary hypothesis may be jeopardized by several factors. Among these, we should consider the following: (i) overestimation of the importance of a particular mechanism of disease and (ii) occurrence of off-target effects masking potential beneficial effects of the experimental drug. An example of the former is the failure of a potent thromboxane receptor (TP) antagonist, terutroban, in showing superiority versus low-dose aspirin in a phase 3 trial of approximately 19,000 patients with cerebral ischemic events. Most likely, the pathophysiological importance of aspirin-insensitive TP agonists (e.g., F_2-isoprostanes) in ischemic stroke has been overemphasized. One example of the latter is the failure of torcetrapib, a cholesteryl ester transfer protein (CEPT) inhibitor, in reducing major vascular events in a phase 3 trial of approximately 15,000 high-risk patients. An off-target effect of torcetrapib on blood pressure is likely to have contributed to its failure.

THE CHOICE OF THE PRIMARY END POINT

The primary end point of a clinical trial represents the biochemical, functional, or clinical variable chosen by the investigators as the primary measure of the effect of the randomized intervention. In phase 1 studies, the primary end point consists of biochemical and pharmacokinetic parameters. Phase 2 studies usually measure pharmacodynamic end points. In phase 3 studies, the primary end point is represented by clinically relevant events ("hard" end points, up to mortality).

The main result of a clinical trial must be judged on the basis of the primary end point in the overall population of randomized patients. If the difference between the two treatment groups is not statistically significant, we should conclude that the study failed to validate its primary hypothesis, regardless of any statistically significant differences in secondary end points or in particular subgroups of patients. Because the sample size calculation is performed on the primary end point of the study, analyses of secondary end points have merely descriptive value. Similarly, subgroup analyses are only hypothesis generating because they lack adequate statistical power and compare subgroups of patients that are not necessarily comparable. Moreover, when performing numerous subgroup analyses, one or more may turn out to be statistically significant just by chance, as illustrated by the analysis of the International Study of Infarct Survival 2 (ISIS-2) results based on the astrological sign of birth of the randomized patients. In this trial of approximately 17,000 patients with a suspected acute myocardial infarction, the survival benefit of low-dose aspirin (a one-quarter reduction in vascular mortality at 5 weeks) was not apparent in those born under the signs of Gemini and Libra. A similar (unpublished) analysis of the Physicians' Health Study suggested that the benefit of aspirin in preventing a first myocardial infarction in healthy male physicians was not apparent in those born under the sign of Scorpio, but was highly statistically significant in all the other astrological signs including Gemini and Libra. Although we tend to laugh at these analyses, the prevailing attitude is to take the results of subgroup analyses very seriously, particularly when there is a plausible explanation for the apparent heterogeneity in treatment effects or when the results meet the pathophysiological expectations. In fact, results of subgroup analyses should be taken with great caution, regardless of whether they are based on astrological signs of birth or sophisticated biochemical and genetic biomarkers. The most reliable estimate of the treatment effect in a particular subgroup of patients is that provided by the overall study result as applied proportionally to the higher or lower risk of that subgroup.

SAMPLE SIZE CALCULATION

Two main hypotheses concur to sample size calculation: (i) the expected frequency of the primary end point in the control group and (ii) the expected difference in the frequency of the primary end point between the two treatment groups. The lower these numbers, the higher the sample size needed to test these hypotheses with adequate statistical power (e.g., 80–90% probability of demonstrating as statistically significant the expected difference). The choice of the sample size may represent a compromise between the biological plausibility of the expected treatment effect and cost considerations by the sponsor. Thus, if the sample size calculation is based on the hypothesis that a novel antiplatelet agent may reduce the risk of major vascular events by 50% as compared to aspirin, then a relatively small number of patients (a few thousands) will be required to test this hypothesis. However, given the multifactorial nature of atherothrombosis, this hypothesis is not biologically plausible. More realistically, a novel antiplatelet agent may reduce atherothrombotic risk by 10–15% as compared to standard treatment. Testing this more conservative and biologically plausible hypothesis may require a sample size of approximately 20,000 patients, a much more expensive effort. Clearly, the interests of the medical–scientific community do not necessarily coincide with those of the pharmaceutical industry when it comes to integrating cost considerations into scientific reasoning. Ethics Committees have a specific responsibility in evaluating whether the agreed compromise is acceptable in the interest of patients.

The high failure rate in clinical drug development depicted in Figure 56.1 is, at least in part, related to inadequate plausibility of the primary hypothesis and/or unrealistic estimate of the expected difference in efficacy in phase 3 trials. The serious adverse events suffered by patients during failed clinical trials should not be considered as a natural catastrophe, but rather thought of as avoidable medical errors.

A LOOK AT THE FUTURE

Historically, phase 1 studies have been largely carried out in healthy volunteers (with the exceptions noted earlier). However, it is anticipated that new approaches to drug development will progressively replace studies in healthy volunteers with an in silico transition phase between preclinical and clinical assessments and earlier involvement of patients in first-in-man studies. The latter offer some distinct advantages as they allow pharmacodynamic assessments that may lead to (i) characterization of early biomarkers of drug response, (ii) faster transfer of novel therapeutic strategies from experimental models to human disease (translational medicine), and (iii) exploring novel mechanisms of action through proof-of-concept studies.

Moreover, several strategies are currently being explored to make the assessment of efficacy and safety of new drugs a more efficient process through the development of molecular and genetic biomarkers that should guide the identification of subgroups of patients particularly susceptible to the beneficial (or untoward) effects of the experimental treatment.

Some members of the oncological community have argued for the need of a faster approval procedure for innovative anticancer drugs, on the basis of phase 1 studies with unusually positive results in subgroup of patients selected for being carriers of molecular/genetic abnormalities that a highly targeted drug may be able to modify.

Finally, several initiatives are currently being implemented to help the pharmaceutical industry integrate pharmacogenomics within the process of clinical development of a new drug (see also Chapter 23). The success of "personalized medicine" or "precision medicine" depends on the availability of accurate diagnostic tests to identify patients who may benefit from a specific targeted therapy. Approximately 10% of approved drug labels contain pharmacogenomic information that may guide a rational selection of the patients eligible for the treatment. An example is provided by the development of diagnostic tests to determine which breast tumors overexpress human epidermal growth factor receptor type 2 (HER-2), a finding associated with worse prognosis but predicting a more favorable response to trastuzumab. It is realistic to expect that progress in this field will lead to more efficient clinical trials based on a better understanding of the genetic basis of diseases. Also, it has been suggested that drugs previously discarded during clinical development because of lack of efficacy or withdrawn from the market because of safety issues may actually prove effective and safe in particular subgroups of patients with particular genetic signatures.

TAKE-HOME MESSAGE

- Principles and rules of clinical investigation are governed by a fundamental set of ethical principles, by national and international regulatory guidelines, and a detailed set of procedures, the Good Clinical Practices (GCP). The regulations are intended to guarantee integrity and safety of the subjects enrolled in clinical trials and to ensure that any potential hazard of a new drug is justified by its potential benefits.
- The path eventually leading to marketing authorization of a new drug is typically subdivided into three phases having quite different objectives, end points, dimension, duration, and cost.
- Randomized, controlled clinical trial represents the fundamental tool to assess the efficacy of a new drug and evaluate its safety in a preliminary fashion.
- In phase 1 studies, the primary end point consists of biochemical and pharmacokinetic parameters; phase 2 studies usually measure pharmacodynamic end points; and in phase 3 studies, the primary end point is represented by clinically relevant events.
- Safety assessment must continue after approval and marketing of the new product through phase 4 studies that include observational studies as well as new randomized controlled trials.

FURTHER READING

Barter P.J., Caulfield M., Eriksson M., Grundy S.M., Kastelein J.J., Komajda M., Lopez-Sendon J., Mosca L., Tardif J.C., Waters D.D., Shear C.L., Revkin J.H., Buhr K.A., Fisher M.R., Tall A.R., Brewer B., ILLUMINATE Investigators. (2007). Effects of torcetrapib in patients at high risk for coronary events. *New England Journal of Medicine*, 357, 2109–2122.

Bousser M.G., Amarenco P., Chamorro A., Fisher M., Ford I., Fox K.M., Hennerici M.G., Mattle H.P., Rothwell P.M., de Cordoüe A., Fratacci M.D., PERFORM Study Investigators. (2011). Terutroban versus aspirin in patients with cerebral ischaemic events (PERFORM): a randomised, double-blind, parallel-group trial. *Lancet*, 377, 2013–2022.

Chabner B.A. (2011). Early accelerated approval for highly target cancer drugs. *New England Journal of Medicine*, 364, 1087–1089.

Collins F.S., Varmus H. (2015). A new initiative on precision medicine. *New England Journal of Medicine,* 372, 793–795.

Coxib and traditional NSAID trialists' (CNT) Collaboration, Bhala N., Emberson J., Merhi A., Abramson S., Arber N., Baron J.A., Bombardier C., Cannon C., Farkouh M.E., FitzGerald G.A., Goss P., Halls H., Hawk E., Hawkey C., Hennekens C., Hochberg M., Holland L.E., Kearney P.M., Laine L., Lanas A., Lance P., Laupacis A., Oates J.A., Patrono C., Schnitzer T.J., Solomon S., Tugwell P., Wilson K., Wittes J., Baigent C. (2013). Vascular and upper gastrointestinal effects of non-steroidal anti-inflammatory drugs: meta-analyses of individual participant data from randomised trials. *Lancet*, 382, 769–779.

Davì G., Patrono C. (2007). Platelet activation and atherothrombosis. *New England Journal of Medicine*, 357, 2482–2494.

Garattini S., Bertelè V. (2007). Non-inferiority trials are unethical because they disregard patients' interests. *Lancet*, 370, 1875–1877.

Hamburg M.A., Collins F. (2010). The path to personalized medicine. *New England Journal of Medicine*, 363, 301–304.

ISIS-2 (Second International Study of Infarct Survival) Collaborative Group. (1988). Randomised trial of intravenous streptokinase, oral aspirin, both, or neither among 17, 187 cases of suspected acute myocardial infarction: ISIS-2. *Lancet*, 2, 349–360.

Schünemann H.J., Cook D., Guyatt G. (2008). Methodology for antithrombotic and thrombolytic therapy guideline development. American College of Chest Physicians evidence-based clinical practice guidelines (8th Edition). *Chest*, 133, 113S–122S.

INDEX

Note: Page numbers in **bold** refer to a text discussion of the entry

abatacept, **636,** 695
abciximab (ReoPro), 233
ABC transporters, 263
    polymorphisms, **263**
    substrates, 263
absorption, 21, **31–44**
    rate, **39, 74**
    time constant, **74**
absorption distribution metabolism and elimination (ADME), 671, 695, 700–707
absorption distribution metabolism excretion and toxicity (ADMET), 701
abuse *see* dependence
acamprosate, **128–9**
acarbose, 458, 460
acceptable daily intake (ADI), **669**
ACE *see* angiotensin-converting enzyme (ACE)
acebutol, 325, 492–3
aceclidine, 504
acetylation, 62, 67, 135
    histone acetyltransferase, 135
    histone deacetylase, 135
acetylcholine (ACh), 190, 194, **496–501,** 512, 513, 515
    binding protein, 116, 179
    ganglionic neurotransmission, 479, 503, 506
    precursors, 499
    release, 118, 321, 351, 475, 479, 487, 488, 501, 512, 580
    release cotransmitters, 497
    synthesis and metabolism, **498–501**
    vesicles, 500
acetylcholine receptors, **501–4**
    muscarinic (mAChR), 189, 321, 339, 360, 478, 486, 498, **503,** 524, 604
    muscarinic agonists (parasympathomimetics), 504, 506–7
    muscarinic antagonists (parasympatholytics), 338, 498, 504, 506–7
    muscarinic autoreceptors, 501
    muscle-type receptor, 503
    nicotinic, 112, 126, 140, 179, 180, 182, 184, 186–7, **504–6**
    nicotinic (nAChR), **501–3**
    nicotinic agonists, 504
    nicotinic antagonists, 504
    nicotinic autoreceptors, 501
acetylcholinesterase (AChE), 65, 137, 359, 497–8, 507, 692
    inhibitors (AChEIs), **499–500,** 505, 526
acetylcholine transporters, 360
    vesicular ACh transporter 1 (VAChT1), 498, 500
acetyl-CoA, 67, 440, 441, 497, 499, 653
    acetyl-CoA transporter (ACATN), 499
acetylprocainamide, 338
acidification, 56, 161, 169, 355–6, 381, 475, 684

acquired immuno deficiency syndrome (AIDS), 3, 387, 414, 542, 590–591, 638, 659, 678–9
    HIV infection, 382, 659
acrylamide derivatives, 335
action potential, 468
activator protein 1 (AP-1), **630**
acyclovir, 15, 655–6
Acyl-CoA-cholesterol-acyltransferase (ACAT), 448
    inhibitors, 448–9
adalimumab, 239, 627, **636,** 695
adamantanes, **656–7**
adaptation, 42, **118, 121–9,** 580, 582, 648
adaptor protein, **112, 113,** 118, 218, 219, 231, 434
addiction, 7, **122–9,** 583, 584
adefovir, 659
adenine nucleotide translocase (ANT), 411, 436
adenosine, 157, 190, 192, 336, 501, **553–64,** 656
adenosine deaminase (ADA), 304, 419, 555–6
    ADA deficiency syndrome, 563
adenosine diphosphate (ADP), **554–64**
adenosine kinase, 556
adenosine monophosphate (AMP), **555–64**
adenosine triphosphate (ATP), 190, **553–64**
adenylate kinase, 555
adenylyl cyclase, **195–6**
adesleukin, **638**

General and Molecular Pharmacology: Principles of Drug Action, First Edition. Edited by Francesco Clementi and Guido Fumagalli.
© 2015 John Wiley & Sons, Inc. Published 2015 by John Wiley & Sons, Inc.

adherent junctions *see* junctions
adhesion molecules, 219, 220, 222, 227, 229, 233, 324, 466, 473, 523, 628, 634, 636
adhesive receptors, **225–6**
   drug targeting, **233–4**
adipocytes, 282, 393, 458, 460, 563, 586, 609, 619
adrenocorticotropic hormone (ACTH), 190, 221, 481, 489, 494, 523, 567, 570, 574
administration routes, **31–43**
   dermal route, **37–8**
   drip injection, 35
   drug delivery, **42–3**
   drug targeting, 43
   enteral route, **33**
   inhalation route, **36–7**
   intradermal route, 36
   intramuscular route (i.m.), **36**
   intravascular route (i.v.), **35**
   mucosal routes, 38
   oral route, **33–5**
   parenteral routes, **35–6**
   rectal route, 35
   subcutaneous route, 36
   sublingual route, 35
   systemic administration, 31
   topical/regional routes, 31, 37
   transcutaneous route, **37–8**
adrenal cortex, 280, 386
adrenaline, 321, 371–2, 452, 478, 484–5, 492, 685
adrenal medulla, 168, 373, 478, 479, 496, 574, 620
adrenergic and noradrenergic systems, 480–481
   control of pituitary functions, 481
   depression, wakefulness and feeding behavior, 481
   distribution, 480
   functions, 480
adrenergic receptors, 487
   α-adrenergic receptors, 190, 191, 194, 487
      α1 agonists, 490
      α2 agonists, 491
      α antagonists, 491
   β-adrenergic receptors, 190–192, 194, 196, 198–200, 488
      β1, 264
      β2, 265
      β2 agonists, 491
      β antagonists, 15, 251, 263, 264, **325**, 349, 493
      β1 polymorphisms, **264**
      β2 polymorphisms, **265**
   drugs and therapeutic use, 490–493
   functions, 489
   localization, 489

adrenocorticotrope hormone (ACTH), 190, 221, 481, 489, 494, 523, 567, 570, 574
adverse reactions, 7, 36, 259, 262, 290, 291, 682, 698
affinity, **95**, 96, 99, 101, 103, 104, 107, 108
agomelatine, 517
agonist, **102**
agonists
   inverse, **107–8**
   partial, **104–5**
Agouti-related peptide (ARP), 569
AIDA, 550
AIDS *see* acquired immuno deficiency syndrome (AIDS)
A-kinase-anchoring protein (AKAP), 196, 198
Akt, 169, 170, 413
alanine (β-), 365
Albright's hereditary osteodystrophy, 194
albumin, 21, 50, 55, 59, 243, 290, 421, 614, 681
alcohol dehydrogenase, 65, 574
alcoholism, 574
aldehyde dehydrogenase, 514
aldosterone, 167, 273, 280, 479, 493
alefacept, 628, **636**
aleglitazar, **448**
alemtuzumab, 239, 628, 636, 695
alfentanil, 573
alfuzosin, 492
alkalinization of urine, 56, 169, 684
alkylamines, 527, 651
alkylating agents, 417
allopurinol, 629, 683
allosteric modulators, 102, **103**
   agonist, 103
   antagonists, 103
   potentiators, 103
almotriptan, 517
alprazolam, 71, 157, 539, 685
alteplase, 694–5
alternative splicing, 115, 167, 203, 237, 286, 351, 456, 499, 566, 619
alternatives to animal experiments, **672**
Alzheimer's disease, 169, 505
amanita muscaria, 501
amantadine, 656
ambenonium, 500, 505
amidepsine, 440, **446**
amikacin, 642, 647
amiloride, 49, 320, 354–6, 399, 557, 600
amineptine, 366
amine transporters, 360
aminoethyl-hydroxyindole, 509
aminoglycosides, 642–3, 647, 679
aminopyridine, 147, 320, 336, 505
aminopyrine, 612
amiodarone, 49, 71, 320, 325, 338, 354–5, 681–3

amisulpride, 517–18
amitriptyline, 35, 49, 71, 338, 486, 515, 517–18, 682
amoxicillin, 34, 642
AMPA receptors *see* glutamate receptors
amphetamines, 124, 126, 361, 366, 485, 515
   3,4-methylenedioxy-*N*-ethylamphetamine (MDEA), 515
   3,4-methylenedioxy-*N*-methylamphetamine (MDMA), 122, 124, 366, 370, 515
amphotericin B, 651–4, 661
ampicillin, 14, 34, 642, 679, 685
amprenavir, 387, 393
anacetrapib, 449–**50**
anakinra, 628, 636, 695
analgesia, 576–9
anandamide (AEA), 112, 585–7, 587, 593–5
anaphylaxis, 492, 573, 629
androgen receptor, 272, **281**
   poly-Q tract polymorphism, 281
androgens, 272, 281, 283, 672
anesthetics, 12, 21, 31, 33, 37, 94, 319, 321, 323–6, 336, 437, 491, 534, 540, 549, 567
   general, 336, 534, 536–7, 540
   local, 3, 21, 33, 37, 94, 319, 323–6, 329, 437, 491
angina, 492–3, 602
angiotensin-converting enzyme (ACE), 264, 381, **386**, 567
   inhibitors, 15, 42, 264, 347, 349, **386**, 679–80
angiotensin II, 15, 190, 199–200, 381, 386, 479, 571, 603, 609
animal models, 15, 46, 163, 254, 275, 290, 298, 326, 328, 413, 442, 443, 449, 461, 539, 569, 590, 592, 619, 637, 702
animal testing *see* toxicity
antagonists, **102**
   competitive, 102
   insurmountable, 102–3
   surmountable, 102
antazoline, 527
anthracenediones, 416, 423
anthracyclines, 416, 422–3, 437, 679
antiarrhythmic drugs, 320, 338
   class II: β-adrenergic antagonists, 15, 251, 263, 264, **325**, 349, **493**
   class III: VGKC blockers, 320, **325**
   class IV antiarrhythmic: VGCC blockers, 320, **325**
   class I: VGSC blockers, 323, **325**
antibiotics, 4, 13, 14, 26, 36, 37, 49, 56, 263, 422, **641–60**
   antineoplastic, 422
   cytoplasmic membrane function inhibitors, 649–50

antibiotics (cont'd)
   DNA synthesis and replication inhibitors, 649
   mechanism of action, 642–3
   peptidoglycan polymerization inhibitors, 645
   peptidoglycan synthesis inhibitors, 643–5
   resistance, **650**
   ribosomes, 647
   transcription inhibitors, 646
   translation inhibitors, 646–9
antibodies, 636
   monoclonal (mAbs), 17, 113, 115, 213–15, 233, **239,** 417, 427–8, **635,** 639, 695
   monoclonal humanized, 239
   monoclonal therapeutic, 239
   polyclonal, **635**
anticholinergic agents, 7, 34, 486, 498, 504, 507, 527, 680
antidepressants, 14, 15, 34, 35, 61, 259, 263, 338, 364, 366, 370–372, 481, 486, 511–12, **515,** 515–19
antidotes, 684
antiemetics, 65, 263, 338, 517
antiepileptic drugs, 51, **320,** 322–3, 325, 327, 369, 472, 530, 539
antifungal drugs, 338, **651–4**
   acting on cytoplasmic membrane, 652–3
   DNA and protein synthesis inhibitors, 653
   mechanism of action, 653
   mitotic inhibitors, 653
   resistance, **653**
antihistamines, 14, 263, 338, 483, 520, 527, 680, 684
antiinflammatory agents, 280, 518, 609, 611
antimalarial drugs, 338, 525–6
   amodiaquine, 526
   quinacrine, 526
antimania drugs, 338
antimetabolites, 262, 265, 415, **418–20,** 679
   antifolates, 418
   purine analogs, 419
   pyrimidine analogs, 419, 526
antineoplastic agents, 33, 214, **215, 415–30**
   conventional antitumor drugs, 33, **417–23**
   monoclonal antibodies, 33, **428–9**
   target-specific antitumor drugs, 33, **423–7**
antioxidants, 437, 439, 618
antipsychotics, 15, 65, 259, 264, 338, 482–3, 494, 507, 514, 517, 683
   atypical, 516
   typical, 517
antipyretics, 611, 638

antiviral drugs, **654–60**
   anti HBV, 659
   anti HCV, 658–9
      interferons, 659
      NS5B inhibitors, 658
      NS3/4 inhibitors, 658
   anti HIV, **659**
      integrase inhibitors, 660
      protease inhibitors, 393
      reverse transcriptase inhibitors, 659
      virus entry inhibitors, 660
   anti influenza virus, 656–7
   mechanism of action, **654–5**
   resistance, 656–60
anxiety, 14, 363, 481, 493–4, 511, 517, 529, 532–4
Apaf-1, **408,** 411
apamin, 339
apolipoproteins, 443, 448, 540
apomorphine, 489, 494, 540
apoptosis, 389, 558, 562–3
   regulators, 409
apoptosis-inducing factor (AIF), 411
apoptotic bodies, 407
   eat me signals, 408
   picnosis, 408
apoptotic pathway, 408–11
   extrinsic pathway, 408, 409
   intrinsic mitochondrial pathway, 408, 411
apraclonidine, 492
aprikalim, 335
aptamers, 286, 292
aquaporin, 339
arachidonic acid, 196, 321, 559, 586–7, **608–20,** 634
   enzymatic metabolism, 610–617
   metabolism, 610–617
   non enzymatic metabolism, 617–18
   release, 609
arachidonoylglycerol (2-AG), **587–8,** 595
area under the plasma concentration curve (AUC), **74–5**
arecoline, 504, 507
Arf, 154–7
argonaute proteins, **284–6**
   Argonaute2 (AGO2), **285**
aripiprazole, 516–17
aromatase, 65
aromatic amino acid decarboxylase, 513–14
arrestin (β-), 118, 178, 192, **199–200,** 218–20, 581
arsenic, 13, 338
artemisinin, 353, 581
aspartate, 178, 205, 346, 361–4, 366, 379–81, 385, 523, 542, 660
aspartate proteases, **381**
aspirin (acetylsalicylic acid, ASA), 4, 24, 34, 93, **611–13,** 710–711

association constant, **95**
astemizole, 337–8
asthma, 233, 264–5, 280, 290–291, 335, 339–40, 384, 491–3, 506–7, 524, 526, 561–4, 570, 573, 600, 615–16, 618, 620, 628, 636, 679
astrocyte, 362, 369, **466–7,** 592, 598, 603
ataxia telangiectasia, 286
atenolol, 27, 57, 89, 98, 325, 346, 392, 492, 493
atorvastatin, 71, 440–441, 450, 685
ATP binding cassette, 263, 345, 493
atracurium, 504, 506
atrial natriuretic peptide, 114
atropine, 57, 491, 498, 504–8, 680
   methylnitrate, 506, 507
autacoids *see* specific agents
autoimmunity, **626**
autonomic nervous system, 343, **477–95,** 553, 609, 619
avasimibe, **449**
avidity, **48,** 52, 81
azacytidine (Vidaza), 426
azathioprine, 62, 262, **626–31,** 635–6, 679
azimilide, 325, 337–8
azithromycin, 634, 642, 683–4
azoles, **651–4**
aztreonam, 642

bacitracin, 642
baclofen, 531, 533, 537–40
bacterial toxins, **159–63**
Bad, 413, 437
bafilomycin, 475
Bak, 411
bamifylline, 561, 563
barbiturates, 24, 30, 187, 324, **531–7,** 540, 549, 684
basal ganglia, 51, 481–2, 561–2, 585–6
basiliximab, 239, 628, **636,** 695
Batten disease, 303
Bax, 411
Bcl-2, 148–50, 210, 388, 390, 408, 409, **411,** 413, 425, 437
Bcl-XL, 411, 437
BCNU, 417
belladonna alkaloids, 8, 504
benzamides, 335, 426, 494, 516, 518
benzimidazoles, 335, 339–40, 351, 444
benzodiazepines, 320, 335, 531–2, 534–5, 537, 539
   discovery, 14
benzopyrenes, 335
benzothiadiazines, 335
benztropine, 506–7
bepridil, 338, 354–5
Bernard, C., 12
betaine, 364–9
betaxolol, 325, 492

betazole, 528
bethanechol, 504, 506–7
bevacizumab (Avastin), 214, 239, 417, **427–9**
bezafibrate, 440
BH3-only proteins, 410, **411**
    HRK, 411
    NOXA, 411
    PUMA, 411
biased signaling, **108**
bicuculline, 531, 533–4, 537, 539
Bid, 410, 411
    tBid, 411
biguanide, 459
biliary excretion, 58, 214
biliary secretion, **58**
bilirubin, 51
bimosiamose (TBC-1269), 233
binding, 97–103
    experiments, 97
    sites, 94–9, 102, **103**
    studies, **96**
Binz, C., 12
bioavailability, 32–3
bioequivalence, 214, 701, 705, **706**
bioinformatics, 692, **693**
biological response modifiers (BRMs), **625**
biological target, **692**
biotrasformation, 61, 65
biperiden, 506–7
bisoprolol, 492
bisphosphonates, 163, 437, 439–42, 680
blastocyst, 296
bleomycin, 380, 413, 422
blood-brain-barrier, 18, 24, **28–9**, 60, 263, 346, 454, 483, 484, 494, 504–7, 513, 521, 527, 541, 543, 569, 590, 592, 594, 700
bortezomib (PS341, Velcade), 387–9, 417, 425–6
bosutinib, 233
botulinum toxins A, B, 118, 160–161, 329, 381, 392, 475, 501, 505
Bovet, D., 13
Boyle, 11
brain-derived neurotrophic factor (BDNF), 157, 203, 205, 566
brefeldin A, 164, 474
bretylium, 49, 320, 325
bromocriptine, 458, 461, 483, 489, 490, 494
Buchheim, R., 12
bungarotoxins
    α-bungarotoxin, 95, 185–7, 501, 503
    β-bungarotoxin, 476
    n-bungarotoxin, 503
bunitrolol, 263
bupivacaine, 325, 337
buprenorphine, **128–9**, 574

bupropion, 71, **128–9**, 486
buspirone, 514, 516–17
butorphanol, 573
butylscopolamine, 506–7
butyrylcholinesterase (BChE), 498–500, 505–6
    inhibitor N1-phenylethyl-norcymserine, 499

CAAT box, **246**
cabozantinib, 214
cachexia, 564, 590–591
cactus, 124
cadherins, 114, 156, **226**, 232, 473
    signal transduction, **232**
caffeine, 12, 49, 143, 350, 553–4, 557, 562–3, 680, 682
calcineurin inhibitors, **630**
calcitonin, 38, 42, 190, 195, 199, 522, 566, 570
calcitonin gene related peptide (CGRP), 516, 566, 567, 570
calcium (Ca^{2+}), **139–45**
    ATPases, 141
    binding proteins, 140
    channels, 140–141
    diffusion in cytoplasm, 140
    distribution in cells, 139
    in endosomes, 145
    in the ER, 142
    fluxes in cells, 142
    homeostasis, **139–46**
    homeostasis in pathology, 145, 146
    intracellular concentration, 140
    intracellular pools, 145
    in mitochondria, 143, 435–6
    mitochondrial uniporter, 143–4
    in muscle function, 144
    in the nucleus, 145
    oscillations, 144–5
    pumps, 141–4
        endoplasmic reticulum, 142–4
        mitochondria, 143
        plasmamembranes, 141
    in *trans*-Golgi network (TGN), 145
    transporters, 141
calcium (Ca^{2+}) channels
    voltage gated (VGCC), **326–31**
        cellular organization, 326–7
        classification, 327
        drug action, 328, 331
        molecular structure, 327, 329
        pharmacology, 327
calcium-dependent activator protein for secretion (CAPS), 567
calmodulin, 113, 131, 132, 142, 145, 168, 340, 455, 581, 598, 604, **630**
calpain, 145, 380, 382–4, 387, **390–392**
calpastatin, 145, 380, 382–3, 391–2

cAMP, 195, 196, 523
cAMP Responsive Element Binding protein (CREB), 245, **251**, 252–3, 576, 581, 582
camptothecin, 413, 416, 421–2
canagliflozin, 460–461
cannabinoid receptors, **585–6, 593–5**
    antagonist CB1, 593
    antagonist/indirect agonist, 594
    antagonist/inverse agonist CB1, 593
    antagonist/inverse agonist, CB2, 594
    CB1 receptors (CB1R), 585
    CB2 receptors (CB2R), 586
    direct agonist CB1/CB2, 593
    direct agonist CB1 selective, 593
    direct agonist CB2 selective, 593
    GPR55 receptors, 586
    neutral receptor CB2, 594
    PPARα receptors, 586
cannabinoids, 124, 126, 190, 585–6, 590, 592–4
cannabis sativa, 585, **590**–591, 593
capecitabine, 429
capillary endothelium, 27–8
captopril, 386–7, 567
carbachol, 504, 506, 507
carbamazepine, 34, 49, 57, 72, 320, 323, 325, 679, 682
carbamic acid, derivatives of, 335
carbapenems, 642
carbidopa, 484
carbohydrate metabolism, 274
carboline (β), 532–4, 536, 539
carboplatin, 416
carboxylation, 136
carcinogenicity, **670**
cardiac arrhythmias, 35, 45, 78, 87, 320, 324, 325, 327–8, 330, **338,** 351, 493, 678, 681
cardiac conduction, 313, 323, 325
cardiac failure, 320, 325, 606, 631, 636
cardiac glycosides, 58, 141–2, 347–9, 354, 358
cardiac negative chronotropism, 560
cardiac negative inotropism, 560
cardiac positive chronotropism, 558, 560
cardiac positive inotropism, 558, 560
cardiomyocyte, 301
carfilzomib (PR-171), 387, 426
carnitine, 438
carnitine palmitoyl transferase I, 438
carteolol, 492
carvedilol, 263–5, 325, 492–3
caspase recruitment domain (CARD), 409
caspases, 380, **389–91**, 408–9, 549, 564
    inhibitors (IAPs), 411
    procaspases, 409
cassette dosing, **702**
catalase, 434

catecholamine receptors, 487
  principles of drug action, 490
catecholaminergic system, 477
  in the CNS, 480
catecholaminergic transmission, 477
  indirect sympathomimetic drugs, 485
catecholamines, 65, 177, 189–91, 265,
    359–61, 373, 453, **477–86,** 491,
    493, 495, 544, 555, 560, 562
  catabolism, 485
  release, 484
  reuptake, 485–6
  synthesis, 483–4
  vesicular storage, 484
catechol-O-methyltransferase (COMT),
    67, 485–6, 495
catenin (β-), 170
cathepsins, **383–4**
cationic channels, 339
  cyclic nucleotide-gated (CNG) channels,
    341–3
  hyperpolarization-activated cyclic
    nucleotide-gated (HCN)
    channels, 341–3
  modulated by cyclic nucleotides, 339–42
caveolae, 576
CD44, 226
  signal transduction, 229, **232–3**
Cdc42, 156
CD95/CD95R, 410, **411,** 413
cefaclor, 642
cefbuperazone, 642
cefotaxime, 642
cefoxitin, 642
ceftazidime, 642
ceftriaxone, 542, 642
celecoxib, 71, **611–13,** 682–3
celiprolol, 492–3
cell cycle, 405–7
  Cdc 25 phosphatase, 405, 406
  Cdk-activating kinase (CAK), 405, 406
  CDK inhibitor (CKI), 407
  checkpoints, **407**
  cyclin B, 406
  cyclin-dependent kinase (CDK), 405
  cyclins, 405
  E2F1, 406
  G0, 405
  G1, 405
  M, 405
  PEST sequence, 405
  Retinoblastoma binding protein (pRb),
    406, 409
  S phase, 405
cell death, 407
  apoptosis, 407, 558
  necroptosis, 413
  necrosis, 407, 412, 558, 562
  programmed cell death, 407

cell death defective protein (ced), 408
  Ced-3, 408
  Ced-4, 408
  Ced-9, 408
  Egl-1, 408
cell therapy, 296, **299–303**
cephalexin, 642
cephalosporins, 14, 56, 642, 645
cephamycins, 642
cerebral-spinal fluid (CSF), 29
cetuximab (Erbitux), 214–15, 417, **427–9**
cevimeline, 502, 504, 506
Chain, E., 14
channelopathies, **313–14**
chaperones, 117
chelating agents, 684
chemical antagonism, 685
chemicals, 675
  classification and labeling, 675
  documentations, 666
chemistry
  combinatorial, 690
  medicinal, 692–3
chemokine receptors, 219, 227
  Class I hematopoietin receptors, 217
  Class II hematopoietin receptors, 218
  Hematopoietin, 217–18
  The IL-1/IL-18 Receptor Family, 218
  signal transduction, **229**
  The TNF Family, 219
  tyrosine kinase receptors, 218
chemokines, 227, 229
chemoreceptor trigger zone (CTZ), 29,
    483, 494
chemotaxis, 560
chemotherapy, 5, **641–61**
chlorambucil, 417–18
chloramphenicol, 14, 49, 65, 69, 71, 642,
    647–8, 685
chlordiazepoxide, 14, 539
chloride (Cl⁻) channels, 532–40
chloromethylhistidine (α-), 525
chlorpheniramine, 527
chlorpromazine, 14, 49, 62, 69, 71, 338,
    494, 517
chocolate, 123, 562
cholecystokinin (CCK), 482, 497, 524,
    565, 567, 570
cholera toxin, 193–4
cholesterol, 15, 50, 62–4, 117, 163, 273,
    283, 288–91, 328, 345, 393,
    439–51, 535–6, 559–60, 603,
    614, 618, 633, 649, 689
  biosynthesis, **439–41**
  oligonucleotide conjugation with, 286–7
cholesterylester transfer protein (CEPT), 449
  inhibitors, **449–50**
choline, 475, **497–9,** 502, 504
choline acetyltransferase (ChAT), 498

choline acetyltransferase inhibitor,
    N-methyl-4-(1-naphtylvinyl)
    pyridinium, 499
cholinergic system, non-neuronal, 497
cholinergic transmission
  in CNS: in learning, memory and
    movement, 497–8
  in PNS: cholinergic sympathetic system,
    496–7
cholinesterases (ChEs), **499–500**
  inhibitors (ChEIs), 500, 505
  reactivators, 500
choline transporter 1 (CHT1), 499
chromanol, 337
chromatin, 426
chylomicron(s), 446, 448, 450
cimetidine, 15, 49, 57, 65, 71, 108, 527–8,
    680, 682
cinnoxicam, 612
ciprofloxacin, 71, 642, 649
cisapride, 71, 338, 518, 682
cisplatin, 416, **418**
citalopram, 71, 263, 366, 369, 370, 515,
    517, 682
clarithromycin, 71, 338, 642, 682–3
clathrin, 471
  coated pit, 471
  coated vesicles, 471
clavam, 642
clavulanic acid, 642
clearance, **54, 74,** 77–8, 84
  creatinine, 57–8, **84–8,** 448
  hepatic, **59**
  organ, **54**
  renal, **57**
clenbuterol, 492
clindamycin, 642
clinical investigation, 708–10
  ethical principles, 708
  phases, 709–10
  regulations, 708
clinical protocols, 709
  controls, 711
  design, 710–711
  end points, 711
  fast track, 712
  non inferiority, 711
  primary hypothesis, 711
  proof of concept, 712
  randomized, 711
  sample size calculation, 711–12
  sub group evaluation, 712
  validity, 711
clobenpropit, 527–8
clofibrate, 282, 440
clomipramine, 71, 517–18
clonazepam, 539
clonidine, 49, 78, 481, 487, 491–2
clopidogrel, 71, 261–3, 560–561, 711

clostridial toxin, 329, **475**
clozapine, 15, 65, 71, 264, 490, 494, 516–18, 682
c-myc, 408, 637
CNS depressants, 124
co-activators, 245, **246–50,** 254
coagulation factors, 116
cocaine, 12, 38, 123–4, 126, 321, 324, 338, 346, 361, 364, 366–8, 371–2, 374, 482–3, 486–7, 495, 512, 539, 544, 567, 584
codeine, 49, 57, 62, 260–261, 526, 573, 679, 684
coffee, 12, 562
cola, 562
colchicine, 94, 420, 474, 653
colistin, 642–3, 650
collagen, 42, 149, 205, 228, 300, 395, 399, 437, 597, 618
compound 48/80, 525
computer-aided drug design (CADD), 692
concanavalin A, 525
connexons, 465–6
conotoxin (α-), 502
constitutive androstane receptor, **272–83**
  drug metabolism, 283
contraception, 280
Contract Research Organization (CRO), 709
convertases, **386**
  endothelin convertase, 386
copy number variation (CNV), 257
co-regulators, **247–50**
co-repressors, **246–50**
CoREST, 249
corneal lesions, 559
corticotrophin releasing factor (CRF), 569, 570
  receptors, 569, 570
cortisol, 64, 273, 280, 452–3
cortisone, 14, 61, 280
cotransmission, 567, 569
crizotinib (Xalkori), 214
cromakalim, 320, 335
crosstalk, 168
curare, 14, 501, 503–4, 506, 508
cyanoguanidines, 335
cyclic polypeptides, 642
cyclizine, 527
cyclooxygenase (COX), 461, **610–613**
  COX-1, 610–613
  COX-2, 610–613
  gene, 610
  inhibition, 611–13
  protein, 610–611
cyclophilin A (CypA), 436
cyclophilin D (CypD), 436
cyclophosphamide, 417, **626–30,** 635
cyclopyrrolones, 534, 539

cycloserine, 642, 644–5
cyclosporin A (CsA), 15, 49, 71, 222, 253, 263, 353, 413, 436, 626–7, **629–30,** 679–80, 682–3, 685
  clinical use, 631
  pharmacokinetics, 631
cylexin (CY-1503), 233
cyproterone acetate, 281
cysteine proteases, **380**
cystic fibrosis, 136, 303, 306, 311, 314, 345–6, 387, 559–61, 600
cystine-glutamate exchanger, 544
cytarabine, 420
cytisine, 502–4
cytochrome c, 335, 384, 389–90, 409–11, 434–7, 596, 599–601, 605, 607
cytochromes p450 (CYPs), **62–5,** 258–62
  catalytic cycle, 63
  cellular localization, 63
  distribution, 68
  drug interaction, 682
  drug metabolism, 681
  enzymes, 62–5
  evolution, 64
  families, 64
  isoforms, 71
  polymorphism, 67, **258–62**
  toxicity, 62, 72
cytokine receptor, **217–24,** 235
  agonistic effect, 238
  antagonistic effect, 237
  soluble, 235
  soluble viral, 236
cytokines, 114, **220–224,** 236–8, **637–8**
  adaptive immunity, 222–3
  anti-inflammatory, 223
  classification, 217
  hematopoietic, 220
  innate immunity, 220–222
  pharmacology, 223–4
cytosine arabinoside (Ara-C), 419–20
cytoskeletal matrix, 117
cytoskeleton, 112, 117, 156, 159, 170, 198, 210, 227, 387, 473, 605
cytosolic 5' nucleotidase, 555

dacarbazine, 416–17
daclizumab, 239, 628, **636,** 695
dalcetrapib, **449–50**
danazol, 281
danger signals, 553
dantrolene, 321
dapsone, 642
daptomycin, 642–3, **650**
dasatinib, 233, 424
daunorubicin, 413, 416, 422–3
d-curarine, 525
death domain (DD), 410, 411
death effector domain, (DED), 409, 411

death receptors, 111, 115, 408
deltorphin, 574, 576
dementia, 86, 385, 483, 498, 504, 549, 550, 592
denufosol, 560–561
deoxyribo nucleic acid (DNA)
  polymorphisms, 266
  topoisomerase inhibitors, 421–3
dependence, **121–9**
  clinical phenomena, 125
  diagnostic criteria, 125
  physical, 580, 582, 584
  psychic, 7, **122–9,** 583, 584
  symptoms, 125
  therapy, **128**
deprenyl, 525–6
derived no effect level (DNEL), 674
dermorphin, 574, 576
desipramine, 35, 71, 338, 372
desogestrel, 280
dexamethasone, 49, 70, 187, 275, 280, 613, 630
dextran, 46, 525–6
dextromethorphan, 573
dextropropoxyphene, 574
diabetes insipidus, 192
diabetes mellitus, 14, 117, **150,** 152, 170, 192, 202, 248, 282, 295, 298, 314, 320, 332–5, 391, 438, 440, 443, 446–7, **452–61,** 695
diacylglycerol (DAG), 194, 196, 197
diacylglycerol acyltransferase, **445**
  inhibitors, **446**
  pharmacological/clinical activity, **446**
diadenosine polyphosphates (APxA), 559–60
diamine oxidase (DAO), 521
diazepam, 14, 49, 57, 61–2, 83, **534–5,** 539, 582
diazoxide, 320, 335, 437, 459
diclofenac, 71, 335, 611–13, 682, 710
diffusion, 23–6
  facilitated diffusion, 26
  simple diffusion, 24–5
  through membrane channels, 26
digoxin, 32, 34, 57, 77, 79, 86, 93, 263, **347–9,** 680–681, 683–6
dihydropyridines, 320–321, 326, 328, 330
diisopropyl fluorophosphate (DFP), 380, 500
dilazep, 556
diltiazem, 35, 49, 71, 263, 321, 325, 327, 330, 341, 343, 682–3
dimaprit, 527–8
dimethindene, 527
dipeptidyl peptidase-4 (DPP-4), 455
  inhibitors, **459–60**
diphenhydramine, 338, 506–7, 527
diphenoxylate, 573

dipivefrine, 492
dipyridamole, 556, 561
diquafosol, 559, 561
discovery
  alkaloid, 12
  benzodiazepine, 14
  glucosides, 12
  penicillin, 14
  vaccines, 13
disopyramide, 325, 338
dissociation constant, **96**
distribution, 21, 45, **46–53**
  redistribution, **83**
  volume, 46, **48–9, 74, 81,** 84
disulfide bonds, 136
disulfiram, **128–9,** 682
diuretics, 8, 15, 56, 100. 101, 280, 320–321, **347,** 356, 357, 461, 679–80, 685–6
dobutamine, 331, 492
dofetilide, 325, 337–8
dolasetron, 516–17
Domagk, G., 13
domoic acid, 549
donepezil, 500, 505
dopamine, **125–7,** 190, 492, 512, 514–15, 518, 582–3
dopaminergic receptors, 488
  agonists in cardiogenic shock, 493
  agonists in disorders of the CNS, 494
  dopaminergic agonist, 461
  dopaminergic antagonist, 494
  D_2 receptor, 190, 192, 195, 198
  drugs, 493–4
dopaminergic system, 481–3
  distribution, 481
  dopamine and behavioral disorders, 482
  dopamine on the ctz, hypothalamus and pituitary, 483
  dopaminergic nigrostriatal system and movement control, 482
  functions in the CNS, 482
dopamine transporter (DAT), 126, 360–361, 370–371
  pharmacology, 371
dose, **74**
  absorbed, 74–5, 78
  age, **84–7**
  attack dose, **80**
  dosage corrections, **84–7**
  dosing interval, **79**
  hepatic pathology, 86
  loading dose, **80**
  maximal tolerated dose (MTD), 669–70
  multiple, **77–8**
  renal pathology, **86–7**
dose-response curves, **99–102**
doxazosin, 492
doxepin, 338, 527

doxorubicin, 42–3, 263, 413, 416, 422–3
doxycycline, 642
droperidol, 338, 494
droxicam, 612
drug, 4, **93**
  acidity constant (pK_A), 24
  affinity, 6. 7
  agonist, 6
    allosteric, 6
  agonist inverse, 6, 109
  agonist partial, 6
  antagonist, 6
  biological, 5, 16, 694
  biosimilar, 5
  competition, 6, **99**
  concentration(s), 5, 27, 46, 48, 50, 59, 78–9, 86, 418
  dependence (*see* dependence)
  development, **689–99, 700–713**
    clinical, **708–13**
    diagnostic tests, 713
    failure rate, 709, 712
    process, 15, 16
  discovery, **700–703**
  generic, 4, **5,** 612, 691–2, 694, 706
  herbal, 5
  industries, 12, 13, 17
  interactions, 611, **678–86,** 705–6
    dietary supplements/components, 685
    grapefruit juice, 683
    herbal medicine, 685
    pharmacodynamic, 684
    pharmacokinetic *during absorption*, 679
    pharmacokinetic *during biotransformation*, 681
    pharmacokinetic *during distribution*, 681
    pharmacokinetic *during excretion*, 683
  metabolism, 24, 59, **61–72,** 258, 283, 673, 678, 681–2, **700–707**
    conjugation, 34, **58, 62,** 66–8, 258, 634, 703–4
    extrahepatic methabolism, 68
    genetic polimorphisms, **258**
    inducers, 70
    induction, 59, 65, **69–70,** 679, 681–5, 695
    inhibition, **71–2**
    inhibitors, 71
    intestinal flora, 69
    phase I, **62**
    phase II, 62, **66**
    phase II enzymes, 262–3
    phase II enzymes-polymorphisms, **262–3**
    phase II enzymes-polymorphisms-clinical effects, 264

    site, 68–9
    therapeutic relevance, 72
  names, 4
  orphan, 698
  potency, 7
  research (R), **691–2**
  selectivity, 7
  target, 93
  tolerance, 118
drug metabolism and pharmacokinetics (DMPK), 700–707
  in vitro, 700–707
  in vivo, 700–707
drug response, genetic variability, **256–7**
drugs approved per year, 709
drugs of abuse, 124
  entactogens, 124
  hallucinogens, 124
  psychostimulants, 124, 486–7
  sedative hypnotics, 124
dry eye syndrome, 561, 563
dynorphin, 567, 570, 572, 574–6, 578, 582, 584
dyslipidemia, 282, **439–51,** 591

echinocandins, 651
ecothiopate, 500
ecstasy (MDMA), 124, 126, 361, 366, 485, 515
ectonucleotidases, 555–6
  ecto-ADPase, 555
  ecto-ATPase, 555
edrophonium, 499–500, 505
efalizumab (Raptiva), 233, 628, 636
efavirenz, 655, 660
efficacy, **100,** 103, 108
  theory, **104**
EGFR *see* growth factor receptors
eicosanoids, 190, **608–20**
  epoxyeicosatrienoic acids, 608, 615–17
  hydroxyeicosatetraenoic acids, 608, 615–17
  isoeicosanoids, 608, 617–18
  isoprostanes, 608, 617–18
  leukotrienes, **608, 615–16,** 620
  prostanoids, **608, 610–614**
  receptors, **618–19**
eletriptan, 517
elimination, 22, **53**
  rate constant, 40, 53
  time constant, **74**
elongation factors, 646
empathogens, 124
enalapril, 387, 567
encainide, 325
endo-adenosine deaminase, 555
endocannabinoids (ECs), **586–95**
  transporter, 288

endocannabinoid system (ECS), **585–95**
  anxiety and depression, 591
  biological function, **588–93**
  cancer, 592
  eating disorders, 590
  immune system, 592
  neurodegenerative diseases, 592
  pain and inflammation, 591
  synaptic plasticity, 589
endocytosis, 23, 25–6, 116, 161, 198, 232, 360, 383, 439, 456, 471, 475, 582, 654
endomorphins, 574–6
endonuclease G, 411
endoplasmic reticulum (ER), 62, 63, 65, 66, 117, 135, 139, 142, 157, 339, 346, 352, 353, 380, 382, 383, 385, 409, 426, 442, 450, 484, 535, 538, 566, 568, 571, 610
endoprotease, 396
endorphins, 94, 567, 572
  β-endorphins, 567
endosomes, 25, 115, 145, 161, 199, 290, 360–361, 372, 381, 471
endothelin, 158, 381, 386–7
enhancer, **246**
enkephalins, 29, 94, 501, 510, 562, 565, 570, 572, 574–5, 578
entactogens, **124**
entecavir, **659**
enteramine, 509
enterochromaffin-like cells (ECL), 521
entero-hepatic cycle, **58**
enzymes generating lipid second messenger, 210–211
ephedrine, 485, 492
epibatidine, 502, 504
epigenetic modulators, 426
  demethylating agents, 426
  HDAC inhibitors, 426
  non-coding RNA, 426
epipodophyllotoxins, 416, 423
epirubicin, 416, 422–3
epoxide hydrolases (EH), 62, 64
epoxomicin, 426
equilibrium, **95–6**
  conditions, **96**
  constants, **95**
ERK, 147–8, 413, 581, 583
  cascade, 151–2
ERK5, 150
Erlich, 13
erlotinib (Tarceva), 152, 212, 214–15, 416, 424, 426
erythroidine (β-), 502–3
erythromycin, 14, 34, 71–2, 263, 338, 642, 647, 679, 682–5
escitalopram, 515, 517
esmolol, 325, 492

estradiol, 71, 167, **277–80,** 681
estrogen receptors, 272–80
  agonists, 280
  antagonists, 280
    indication, 281
  selective estrogen receptor modulators (SERMs), 281
estrogens, 42, 58, 264, 271, 272, 277, **280,** 446, 498, 523, 672, 679
estrone, 280
etanercept, 627, **636,** 695
etaracizumab (Abegrin), 233
ethacrynic acid, 56, 347, 357, 617
ethanol, 62, 64–5, 69–70, 100, 123–4, 126, 128, 350, 533–4, **537,** 539–40, 631, 682
ethanolamine, 530, 587
ethics committee, 709
ethinyl estradiol, 280, 681
ethosuximide, 320, 327, 331
ethylenediamine, 527
ethylephrine, 492
etodolac, 612
etofibrate, 440
etoposide, 416, 422–3
etoricoxib, 611–13
etretinate, 282
European Chemical Agency (ECHA), 675
European Food and Safety Authority (EFSA), 675
European Medicines Agency (EMA), 691, 708
excipient, 4, **32,** 36, 73
excitotoxicity, 146, 547–9
excretion, **55–9**
  hepatic, **58**
  renal, **55–8**
exocytosis, 23, 140, 168, 360, 411, 439, 454, 455, 467, **470–471,** 475, 514, 568
exon skipping, 286–7
extracellular matrix, 46, 114, 211, 225, 228, 384, 395, 466
extracellular protease, 395
  antagonists, 399
  inhibitors, 399
extraction index, 54–5

factor(s)
  general transcription (GTF), **243**
  inducible, **251–2**
  transcription, **243–6,** 251–2
FADH$_2$, 435
FAK, 230
famciclovir, 654
famotidine, 527–8
fampridine, 505
farnesoid X receptor, 272–82
farnesyl pyrophosphate synthase inhibitors, 441–2

farnesyltransferase, 157
Fas-associated protein with death domain (FADD), 410–411
Fas/FasL, 13, 410, **411**
fatty acid
  biosynthesis, 440, **443–4**
  desaturases inhibitors, 444
  synthase inhibitors, 443–4
    pharmacological activity, 444
feed-back loops, 170
felodipine, 327, 330
fenamates, 335, 612
fenfluramine (feh), 366, 370, 513
fenofibrate, 71, 440
fenoldopam, 104, 490
fenoprofen, 612
fenoterol, 492
fentanyl, 573, 582
fexofenadine, 263, 527
F_0F_1 ATP synthase, 436
fibrates, 70, 282, 438, 440, **447,** 679, 684
  pharmacological activity, **447**
fibrinogen, 228–9, 233
fibronectin, 115, 205, 217–18, 228–9, 233, 395–400
Fick's law, 24
first-pass effect, 35, 58
FK-506, 71, 222, 253, 263, 338, 627, **630–633,** 679–80, 682
flavin monooxygenases (FMO), 62–3
flecainide, 71, 325
Fleming, 14
fluconazole, 71, 682
flucytosine, 651, 654
fludarabine, 419
flufenamic acid, 612–13
fluoromethylhistidine (α-FMH), 521, 523, 525–6
fluoro-oxindoles, 335
fluorouracil (5-FU), 35, 262, 419–20, 653
fluoxetine, 71, 366, 370, 486, 514–15, 517
fluoxymesterone, 281
flupirtine, 320, 335, 339
flurbiprofen, 611–13
flutamide, 281
fluvastatin, 71, 440
fluvoxamine, 71, 514–15, 517, 682
flux-to-volume ratio, 27, 52
folic acid, 58, 288, 418–19, 541, 629, 642, 649
follicle-stimulating hormone (FSH), 42, 190–191, 481
fomivirsen, 654
Food and Drug Administration (FDA), 128, 132, 138, 152, 214, 256, 263, 299, 303, 388, 443, 451, 590, 592, 656, 658, 672, 691, **696–7,** 704, 706, 708–9
formoterol, 492

foscarnet, 656
frovatriptan, 517
furin, **386–7**, 393
furosemide, 49, 56–7, 101, 347, 357–9, 532
fusidanes, 642
fusidic acid, 642, 648

GABA *see* γ-aminobutyric acid (GABA)
GABA-α-ketoglutarate transaminase, 530
GABAergic transmission, 529–40
gabazine, 531, 533
galantamine, 500, 505
Galen, 8, 9
Galileo Galilei, 11
γ-aminobutyric acid (GABA), 190, 470, 512, **529–31**
  distribution, 529
  metabolism, 530
  receptor
    GABA$_A$, **531–5**
    GABA$_A$ extrasynaptic, **533–4**, 537, 539
    GABA$_A$ subtypes, **532–4**
    GABA$_B$, 190, 192, **537–9**
  release, 530–531
  reuptake, 530
  synthesis, 530
  transporters, 367
    pharmacology, 369
    structure, 368
ganaxolone, 536
ganciclovir, 655–6
ganglioplegic, 504, 506
gap junctions *see* junctions
GAPs, 158, 164
gastric parietal cell, 350
gastric secretion, 350
G-coupled cholecystokinin-B receptor (CCKB/gastrin receptor), 524
GDIs, 159
gefitinib (Iressa), 214–15, 416, 424
gemcitabine, 416, 419–20
gemfibrozil, 71, 282, 440, **447**, 682
gemtuzumab, 239, 695
gene delivery, 305
generic(s) *see* drug
gene therapy, 43, 262, 284, 295, 303–6, **303–6**
  ex-vivo, 304
  in-vivo, 304
genetic polymorphisms, **258**
  clinical effects, 258
  drug pharmacokinetics, 258
genomics, 694
gentamicin, 57, 642, 647, 685
gepirone, 516–18
geranylgeranyltransferase, 157
gimatecan (ST1481), 421–3

glial cells, 28, 202, 324, 362, 368, 466, 471, 472, 530, 535, 541, 543, 603
gliflozins, 458
glinides, 332–3, 455, 458–9
glomerular filtration, 82, **84–5**, 446, 479, 557, 603, 609, 632
glucagon, 106, 190, 343, **452–4**, 458–60, 563
glucagon-like peptide-1 (GLP-1), 351, 454, **458–60**
glucocorticoid receptor, 269–**278**
  agonists, 278
  regulated genes, 278
  selective agonists, 278
glucocorticoids, 110, 221–3, 269, 278, 370, 609, 618, 632, 679, 684, 6268
  mechanism of action, 632
  resistance, 280
  undesired effects, 276
glucose, 157, 162, 167, 169, 231, 280, 282, 321, 332–4, 350, 438, 440, 446–8, **452–61**, 467, 478, 480, 491, 493, 497, 499, 513, 541, 551, 559–60, 563, 586, 599, 605, 634, 652
  control, 452
  intestinal absorption, 460
  modulation of GLUT4 activity, 456
  phosphorylation, 454
  transport, 452
  transporters (GLUT), 453–4
  uptake, 452
glucosidase (α-), 460
glucosides, discovery, 12
glucuronidation, 66
glutamate (glutamic acid), 190–191, 470, 512, **541–52**
  metabolism, 541–2
  synapses, 543–5
  synthesis, 541–3
  transport, 542–4
glutamate decarboxylase (GAD), 529–30, 539–40, 542
glutamate dehydrogenase, 542
glutamate receptors, 544–9
  ionotropic, 545–6
    AMPA receptors, 469, 544–6, 550, 604
    kainate receptors, 43, 545–6
    NMDA receptors, 117, 129, 141, 469, 545–7, 549–50, 604
  ionotropic agonists, NMDA, 523
  metabotropic agonist, ACPD, 548
  metabotropic receptors (mGluRs), 190, 192, 198, 546–7
glutamate transporters (EAAT), **361–2**
  distribution, 352
  function, 352

pharmacology, 363
substrates, 364
glutaminase, 542
glutathione coniugation, 68
glutathione peroxidase, 434
glycemic control, **452–61**
  mechanisms, **452–7**
  pharmacology, **457–61**
glycine receptors, 111, 175
glycopeptides, 642, 645, 648
glycosylation, **135, 138**
  inhibitors, 138
glycosylphosphatidylinositols, 137
Golgi apparatus, 117, 135, 143, 156, 157, 163, 380, 383, 385, 566, 599
Good Clinical Practices (GCP), 696, 708–9
Good Laboratory Practices (GLP), 672, 696
Good Manufacturing Practices (GMP), 697
G-protein-coupled receptor kinase (GRK), 192, 194
G-protein-coupled receptors (GPCRs), **189–201**, 521, 555, 557, 559, 565, 568
  dimerization, 192
  effectors pathways, 195–8
  GPR17, 559
  G-protein-independent signaling, 199–200
  interacting proteins, 198–200
    AKAP, 198
    β-arrestin, 192, 199–200
    Homer, 198
    NHERF1, 200
    RAMP, 199
    RGS, 198
  ligand binding site, 191
  mutations, 192
  polymorphisms, 192
  structural organization, 190–192
G proteins, 192–5
  activation/inactivation cycle, 193
  βγ complex, 193–7
  effectors, 194–5
  regulators of G protein signaling (RGSs), 193
  α-subunits, 193–7
    modification by toxins, 194
    mutations, 194
Graft Versus Host Disease (GVHD), 626–7, 629, 631
granisetron, 338, 516–17
granulocyte colony-stimulating factor (G-CSF, filgrastim), 218, 220–221, 637–9, 694–5
granulocyte-macrophage colony-stimulating factor (GM-CSF), 217–8, 220–221, 237, 634, 637–9, 694
grapefruit juice, 682–3, 686

grepafloxacin, 338
growth factors, 203
  epidermal (EGF), 26, 113, 116, 151, 154–5, 168, 202–3, 205, 207, 218, 236, 275, 397, 428, 610, 637
  nerve growth factor (NGF), 113, 151, 167–8, 171, 202–3, 205–6, 498
  receptors, 202–16
    activation and signal transduction, 205–13
    epidermal growth factor receptor (EGFR), 203–5, 207, 214–15, 424, 427–8, 661, 691
    functional domains, 204–5
    molecular structure, 204–5
    target for anticancer drugs, 215
    vascular endothelial growth factor (VEGF), 150, 203, 207, 214–**15**, 254, 417, 428–9, 571
GSKβ3, 169, 170
guanabenz, 492
guanfacine, 492
guanine nucleotide exchange factors (GEFs), 158, 163
guanylyl cyclases, 604, 607

half-life ($t_{1/2}$), 38, **40, 53–4,** 75, 77
hallucinogens, 124
halofantrine, 338
haloperidol, 71, 108, 337–8, 494, 517
halothane, 62, 65, 321, 540, 682
Harvey, 11
hazard identification, **666,** 669, 673
heart failure, 117, 251, **264,** 269, 282, 342, 347–9, 492–3, 560, 602, 678
heart preconditioning, 560
heart rate, 196, 320, 342, 354, 479, 489, 491, 493, 507, 561, 579
hemicholinium (HC3), 499
heparin, 42, 49, 203, 228, 290, 521, 525, 585
hepatitis c, 429, **657**
heroin, 61, 124, 483
hexamethonium, 57, 112
hexokinase, 437, 454
high production volume (HPV), 675
high-temperature requirement A (HTRA), 388
high throughput screening (HTS), 690
Hippocrates, 8
histamine (2-(imidazol-4-yl)ethylamine), 28, 190, 520
  [^{3}H]-histamine, 526, 527
histamine-N-methyltransferase, 521
histamine receptors
  antagonists (antihistamines), 14, 263, 338, 483, 520, 527, 680, 684
  H_1R, 521
  H_2R, 520
  H_3R, 521

H_4R, 520–526
H_2R agonists, 528
H_4R agonists, 527, 528
H_4R antagonists, 527, 528
H_2R selective antagonists, 527, 528
histaminergic neurons, 523
histaminergic transmission, **520–528**
histidine-decarboxylase (HDC), 520
histone acetyltransferase (HAT), **248**
histone code, 249
histone deacetylase (HDAC), **248–9**
histones, 135, 148, 248, 274, 418, 425
HIV, 392
H^+/K^+ ATPase, **349–51,** 523
  gastric pump, 350
  inhibitors, 347, 351
  pharmacology, 350
  structure, 350
HMG-CoA reductase inhibitors *see* statins
homatropine, 506–7
hormone replacement therapy, 278
hormones, 190, 191
Human Genome Project, 691, 694
Huntington's chorea, 562
hydralazine, 34, 262
hydrocortisone, 38, 62, 187, 634–5
hydrogen peroxide, 433
hydrolysis, 65
hydromorphone, 573
hydroxitriptamine *see* serotonin
hydroxybutyric acid (γ-) (GHB), 539
hydroxyindolacetaldehyde, 513–14
hydroxyindoleacetic acid (5-HIAA), 511, 514
hydroxylation, 135
hydroxyl radical, 433
hydroxytryptophan (5-OH-tryptophan), 513
hydroxyzine, 338
hyperlipidemia, 444, 447, 459, 632
hypertension, 561
hypnotics, 56, 124
hypoglycemia, 453
  endocrine response, **453**
hypoglycemic drugs, 332, **458**
hypothalamic–pituitary–adrenal (HPA) axis, 513
hypoxia, 553–4, 562

ibuprofen, 71, 612–3
ibutilide, 325
ICAM-1, 636
idarubicin, 416, 422–3
If current, 196
imatinib mesylate (Gleevec), 212, 214, 233, 263, 416, 423–4
imetit, 527, 528
imidazole-N-methyltransferase, 525
imidazopyrazines, 335
imidazopyridine, 532–5, 539

imipenem, 642, 685
imipramine, 49, 57, 61, 71, 369–70, 486, 515, 682
immepip, 527–8
immune cells, 558, 564
immune system, pharmacological modulation 625
immunoadjuvants, 639
immunomodulators, 114, 630, **636,** 638, 657
immunostimulant drugs, 625, **637–9**
immunosuppressive drugs, 187, 222, 224, 296, 338, **625–37**
implitapide, 450
impromidine, 527–8
incretin effect, 454–5
incretins, **458–60**
indacaterol, 492
indinavir, 71, 263, 387, 393, 682–3, 685
indobufen, 612
indomethacin, 57, 71, 612
inflammation, 511
infliximab, 239, 627, 636
inosine, 555, 629
inositoltrisphosphate (IP3), 194, 196, 198
  receptor, 143–4
insulin, 16, 25, 31, 36, 42, 113, 117, 136, 150, 157, 170, 203–6, 275, 282, 320–321, 327, 331–3, 335, 389, 393, 444–5, 447–8, **454–61,** 493, 557, 559, 563, 566, 591, 600, 603, 684–5, 685
  exocytosis, 454
  IR tyrosine kinase, 457
  modulation of signaling, 459–60
  receptor (IR), 454–5
  resistance, **461**
  secretion, 455
integrase inibitors, 655, **660**
integration of receptor signaling, **166–71**
integrins, 114, **226–8**
  activation, **229**
  ligands, 290
  signal transduction, **229–32**
  FAK, **230**
  Src, 230
interactions
  drug–diet, 686
  drug–drug, 63, 108, 641, 678–86
  pharmacodynamics, 684
  pharmcokinetics, 678
interferons (IFNs), 114, 218, 236, **637–8, 657,** 694–5
  interferon α (IFNα), 17, 218, 236–7, 629, 638–9
  interferon β (IFNβ), 17, 42, 71, 114, 236–7, 280, 599, 637, 638, 657, 694–5
  interferon γ (IFNγ), 218, 222–4, 236, 597, 599, 605, 627, 630, 638–9

interleukins, 5, 114, 217, 389, 637, 639
　IL-1, 236–7, 290, 380, 409, 599, 609, 637
　IL-2, 17, 115, 243, 637, 694
　IL-3, 114, 217, 638
　IL-5, 217
　IL-6, 115, 217, 220–223, 237–9, 290, 560, 627–8, 636
　IL-7, 217–20, 237, 637
　IL-10, 114
　IL-18, 237
　IL-22, 114
intestinal natriuretic peptide, 114
intetumumab (CNTO-95), 233
intracellular receptors, 6, 110
　activating functions, 274–7
　classification, 268, 272
　DNA binding, 272, 274
　drugs and indications, 271, 278–83
　hormone responive element (HRE), 269
　ligand binding domain, 272
　ligand binding pocket, 272
　ligands, 274–7
　nomenclature, 272
　regulation of gene expression, 272–7
　specificity of action, 278
intrifiban (Integrilin, eptifibatide), 233
intrinsic activity, 104
inulin, 56, 85
iodofenpropit, 526–7
iodoprossifan, 526
iodoxuridine, 656
ion channels, **175–88, 311–59,** 468
　anionic channels, 339
　cationic channels, 339
　classification, 312
　function, 312
　gating mechanisms, 312
　　current-voltage relationship, 315
　ligand-gated, 111, 113, **175–88,** 312, 501, 568
　rectification process, 315
　structure, 316
　voltage-dependent anionic channels (VDAC), 411, 436
ionophores (A23187, X527A), 525
ipilimumab, 239
ipratropium, 506
ipsapirone, 514, 516–17
iPS cells, 298
irinotecan (SN38), 262–3, 266, 416, 421–2, 429
　UGT1A1 polymorphisms, **263**
ischemia, 555, 560, 562
isepamycin, 642, 647
isoflurane, 321, 336, 537, 540
isoniazid, 146, 262, 530, 682
isoprenaline, 57, 338
isoproterenol, 34–5, 57, 265, 487, 491–3

isosorbide, 597, 606
isotretinoin, 282
ivabradine, 196, 321, 341–3
ivermectin, 353

JAK2 tyrosine kinase, 200
JNJ7777120, 526–7
JNK, 147–8, 152–3, 413
josamycin, 642
junctional epidermiolysis bullosa (JEB), 306
junctions
　adherent, 56–7, 226, 232
　desmosomes, 232
　gap, 465–6
　neuromuscular, 111, 116–17, 176, 180, 328, 475, 497, 501–2, 508
　thigh, 32, 232

kanamycin, 49, 642, 647
kernicterus, 51
ketamine, 43, 124, 126, 549–50
ketanserin, 338, 491, 516–17
ketoconazole, 71, 263, 280, 338
ketoprofen, 612
ketorolac, 612
kidney, **55–8**
kinetic constants, **96**
kinetics
　first-order, 25, 38–41, 74
　multicompartmental, 81–82
　zero-order, 38
kiss and run, 471
kurtoxin, 327

labetalol, 35, 492
lactacystin, 387, 426
lactams (β-), 642, 645
laminin, 228
L-amino acid decarboxylase (AADC), 524
lamivudine, 655, 659
lamotrigine, 320, 325
Langmuir isotherm, 97
lapaquistat, 440, 443
lapatinib, 214–15
latrotoxin (α-), 475–6
Lavoisier, 11
law of mass action, 99
L-DOPA, 29, 34, 494, 681, 683
lead compound, 692
lead optimization, 701–3
lectins, 525
lenalidomide, 627, 630
Leroux, 12
leukocytes, 225, 227–9, 231
　recruitment, 225, 227, 233
leukotrienes, **608, 615–17,** 620
　LTA$_4$, 615, 616
　　metabolism, 616
　LTB$_4$, 615

LTC$_4$, 615
LTD$_4$, 615
leukotrienes receptors, 620
levcromakalim, 335
levobunolol, 492
levofloxacin, 263, 338, 642
levorphanol, 573
lidocaine, 35, 320, 325
lifetime, 40
ligand binding domain, 179, 272, 544–5
lincomycin, 642–3
lincosamids, 642, 648
linezolid, 642, 648, 650
linopirdine, 320, 337
lipid
　conjugation, 133, 288, 290
　lowering drugs, 439–51
　metabolism
　　transcriptional control genes, 446
　modifications, 136
　rafts, 117
　transfer proteins, 448
lipofectin, 287
lipooxygenase, **608–9, 615–17**
　inhibitors, 616
　5-lipooxygenase, 609, 615–16
　12-lipooxygenase, 609, 616
　15-lipooxygenase, 609, 616
　pathway, 615
lipopeptides, 642
liposomes, 43, 287
lithium, 14, 34, 57, 197–9, 338, 680, 684
liver, **58–9**
liver X receptors, 446–**8**
　regulated genes, 446–8
　synthetic ligands, 446–7
lomitapide, 450
long-term potentiation (LTP), 558, 561
loop diuretics, 347
loperamide, 263, 573
lopinavir, 387
loratadine, 527
lovastatin, 440
L-tryptophan, 513–14, 518
lumiracoxib, **611–13,** 710
luteinizing hormone (LH), 25, 126, 190–192, 481, 489
　receptor, 190–192, 195
lysergic acid (LSD), 124, 512, 516–17
lysophosphatidic acid, 190
lysosomal proteases, **382**
lysosomes, 117, 135, 198, 290, 361, 372, 380, 382–3

macrolides, 338, 642, 647
macrophages, 558
Magendie, 12
magic mushrooms, 124
Malpighi, 11

mapirocin, 642, 648
maprotiline, 366
margin of exposure, 668, 675
mastocytes, 558
MDR1/MDR2, 263, 283, 346
mecamylamine, 501, 504, 506
meclizine, 527
meclofen, 335
meclofenamate, 335, 612
mediator, **246**
medroxyprogesterone acetate, 280
mefenamic acid, 612–13
MEK, 152
melanocyte-stimulating hormones (MSH), 567–9
melatonin, 511, 513–14
meloxicam, 612
melphalan, 417–18
memantine 43, 549–50
membrane permeability, 701
membrane transporters, **345–58**
    clinical use, 347
memory, 122, 127
menopause, 279
meperidine, 65, 573
mepyramine, 526–7
mercaptopurine, 262, 416, 419
meropenem, 642
mesangioblast, 299
mescaline, 124
mesocorticolimbic system, 125–7, 583
meta-analysis, 7
metabolites
    metabolites in safety testing (MIST), 704
    reactive, 704
metalloprotease, **381, 399–401**
    cancer, 398
    cancer cell migration, 399
    cell migration, 398
    matrix metallo protease (MMP), 396, 400
    membrane metallo protease inhibitors, 399
    metastasis, 397, 399
    tissue inhibitors of metalloproteases (TIMP), 398
    tissue remodeling, 396, 399
metformin, 282, 438, 458–9
methacholine, 504, 506, 507
methadone, 124, **128–9**, 573
methamphetamine, 124, 370, 373, 485
methotrexate, 257, 416, 418–19, **629**, 636, 679–81, 684
methylation, 67, 136
methyldopa, 484, 492
methylhistamine, 525, 527
methyllycaconitine, 502
methylphenidate, 372, 485
methyltestosterone, 281

methylxanthine, 341, 343, 553, 562–3
methysergide, 517
metilfenidate, 366
metipranolol, 492
metoclopramide, 34, 483, 494, 517, 680
metoprolol, 34–5, 49, 108, 325, 492
metrifonate, 500
metronidazole, 643
mevalonate pathway, 437. 441
mexiletine, 320, 325
mianserin, 527–9
Michaelis–Menten equation, 97
microglial cells, 558
microRNA (miRNA), 285
microsomal triglyceride transfer protein (MPT), 450
    inhibitors, **450–459**
microtubule(s), 232, 413, 416, 420, 652
    acting drugs, 420
        taxanes, 420
        vinca alkaloids, 420
midodrine, 492
mifebradil, 320, 325, 338
mifepristone, 280
mineralcorticoid receptor, 272
    antagonists, 280
mineralcorticoids, 110, 272, 280
minocycline, 642
minoxidil, 320, 335
miokamycin, 642
mirabegron, 488, 492
miravirse, 658
mirtazapine, 486, 517
misoprostol, 615, 619
mitochondria, 143, 413, **433–8**
    permeability transition pores (PTPs), 411, 435–7
    reactive oxygen species (ROS), 411, **433–8**
mitochondrial drugs, **437–8**
    antineoplastic agents, 437
    antioxidant, 437
    mitochondrial dysfunction, 437
    mitochondrial metabolism, 437–8
mitochondrial membrane potential (MMP), 411, 433, 437
mitochondrial translocator protein, 437
Mitogen activated protein (MAP), 630
Mitogen activated protein kinases (MAPK), 133, 147–53, 195, 199, 576, 577, 581, 582, 583
    pharmacological inhibitors, 151–3
    specificity, 150–151
    subtypes, 147–50
mitoxantrone, 416, 422–3
molecular modeling, 692–3
monoamine oxidase (MAO), 370, 434, 521
    MAO A, 511, 513–4, 518
    MAO B, 525

monoamines transporters, 364–7, 371–2
    antidepressants, 372
    distribution, 366
    function, 372
    pharmacology, 366
    structure, 372
    substrates, 366
monobactams, 642
moricizine, 325
morphine, 57, 124, 126, 573
morpholines, 654
mosapride, 518
moxifloxacin, 642
mTOR, 169, 170
mucins, 225–8
multidrug resistant associated proteins, 26, 658
multiple sclerosis, 233, 280, 302, 320, 336, 505, 521, 539, 553, 590–599, 627, 635–7, 638
mupirocin, 642, 648
muromonab-CD3, 627, 635, 695
muscarine, 501, 504, 507
muscarinic toxins (MT), 503, 504
muscimol, 531, 537
muscular dystrophy, 295, 387, 563
    Becker muscular dystrophy (BMD), 301
    Duchenne muscular dystrophy (DMD), 286, 301, 600, 602, 605–6
myasthenia gravis, 117, 500, 505, 626–7
mycophenolate mofetil, 627, 629, 679
myocardial infarction, 437
myristoylation, 136

nabumetone, 612
Na^+/Ca^{2+} exchanger (NCX), 346, 353–6, **353–6**
    pharmacology, 354
    properties, 354
    structure /distribution, 353
N-acetyltransferases (NAT), 67
$NADH(H^+)$, 435
Na^+/H^+ exchanger (NHE), 346, **355–6**
    pharmacology, 356
    structure, 355
Na^+/H^+ exchanger regulatory factor 1 (NHERF1), 200
Na^+/K^+ ATPase, **346–9**
    operation mode, 349
    pharmacology, 348
    structure, 348
$Na^+/K^+/Cl^-$ cotransporter (NKCC), **356–8**
    pathology, 357
    pharmacology, 358
    structure, 357
nalbuphine, 573
nalidixic acid, 642, 649
naloxone, 129, 574, 580, 684
naltrexone, **128–9**, 574

namitecan (ST1968), 421
nanodrugs, 18
nanomedicine, 43
nanotube, 466
naphazoline, 492
naproxen, 71, 606, 611–13
naratriptan, 517
natalizumab (Tysabri), 233, 628, 636, 695
nebivolol, 492–3
necrostatin-1, 413
nelfinavir, 71, 387, 393, 693
neomycin, 14, 642, 647
neostigmine, 35, 49, 500, 503
neprilysin, 381, **386–7**
nerve growth factor (NGF) *see* growth factor
netilmycin, 642
neurexins, 473
neuroadaptation, **121–3, 126–7**
neuroligins, 473
neuromuscular blockers, 504, 506
neuromuscular junction *see* junctions
neuropeptides, **468–9, 565–71**
   colocalization, 367
   encoding genes, 566
   functions, 569
   neuropeptide Y (NPY), 567–70
   processing, 566–7
   receptors, 567–8
   secretion, 567–8
   storage, 567–8
   synthesis, 566–7
   therapeutic potential, 569
neuropharmacology, 474
neurosteroids, 533–4, 536, 540
   allopregnanolone, 536, 540
   tetrahydrodeoxycorticosterone (THDOC), 532, 540
neurotensin, 482, 525, 571
neurotoxins, 168, 475–6
neurotransmitters, 6, 57, 93, 94, 111, 113, 122, 175, 177, 186, 190, 194, 321, 331, 361–60, 417, 465–72, 475, 484, 500, 522, 538, 553, 561, 567–8, 604, 608
   receptors, 468
   transporters
      dependence from ions, 362
      family evolution, 367
      H^+-dependent, 372–3
      Na^+/Cl^--dependent, 364–72
      $Na^+ K^+$-dependent, 361–4
      structure and function, 362
neutral amino acid transporter, 513
nevirapine, 655, 660, 705
NF-kB, 251, 410, 428, 632
nicorandil, 335
nicotinamides, 335

nicotine, 124, 126, 501, 503–4
   replacement therapies, **128–9**
nifedipine, 35, 49, 71, 326–5, 682–3
nikkomycins, 651–2
nilotinib, 424
Nim 811, 436
nimesulide, 611–13
nimotuzumab, 239
nipecotic acid, 365, 368, 374
nitric oxide (NO), 497, 511, 516, 523, 561
   atheroscelrosis, 602
   biochemistry, **599–602**
   biosynthesis, **597–9**
      NO synthases, 597–9
      NO synthases control by NO, 599
   cardiovascular system, 602–3
   central and peripheral nervous systems, 603–4
   cytochrome c oxidase inhibition, 600–601
   guanylate cyclase activation, 599–600
   hypertension, 602
   and immune system, 605
   metabolic diseases, 603
   metabolic syndrome, 603
   microRNAs, 602
   nitrates tolerance, 606
   nitrate vasodilators, 606
   nitric esters, 606
   pharmacology, **606–7**
   respiratory system, 603
   skeletal muscle, 604–6
   S Nitros(yl)ation, 601–3
   stimulators of cGMP action, 606
   systemic and organ effects, **602–606**
nitrofurans, 643
nitrofurantoin, 643
nitroglycerin, 35, 38, 42, 49, 57, 597, 602, 606
nitroimidazoles, 643
nizatidine, 527, 528
NMDA receptors *see* glutamate receptors
N-methylhistamine, 521
N-methylimidazoleacetic acid, 521
N,N,diethyl-2-[4-(phenylmethyl) fenoxiletanamine] (DPPE), 523
nociceptin, 572, 574, 575
nociception, 558
nocodazole, 474
noncholinergic, nonadrenergic (NANC), 553
nonsteroidal anti-inflammatory drugs (NSAIDs), 611–13
   adverse effects, 611, 613
   coxibs, 611–13
   therapeutic activities, 611, 613
   traditional, 611–13
noradrenaline, 156, 360, 366, 371, 471, 478, 484–5, 492, 514, 517, 567

norfloxacin, 642, 680
norgestimate, 280
nortriptyline, 57, 259, 515, 517–18
Nuclear Factor of Activated T lymphocytes (NF-AT), 251
nucleosides, 553
nucleosome, **246**
nucleotides, 553
   cyclic, 176, 178, 179, 183, 317, 339, 341
nucleus basalis of Meynert, 495

observational studies, 710
occupancy
   theory, **101**
   threshold, **107**
ocular dryness, 559
oil-water partition coefficient, **24**, 32
olanzapine, 516–18
oligonucleotides, **285–91**
   antisense, **286**
   delivery, 287
   inhibition of miRNA, 286
   LNA, 286, 289
   2'-*O*-Alkyl-ribonucleotides, 289
   pharmacokinetics, 289
   pharmacotoxicology, 289–90
   phosphoroamidates, 289
   phosphorotioates, 289
   PNA, 289
   structure, 288
   toxicology, 289–90
omalizumab, 628, **636**
omeprazole, 15, 70, 71, 259, 347, **351**, 682, 685
ondansetron, 71, 263, 338, 517
operant conditioning, 123
opiates, 124, 126, 190, 573
   addiction, 583
   endogenous peptides, 572
   endogenous peptides distribution, 574
opioid receptors, **575**
   cardiovascular effects, 579
   distribution and effects, **576–80**
   effects on food intake and body temperature, 580
   effects on GI tract, 579
   effects on immune system, 580
   GPCR, 575–6, 580, 582, 584
   modulation of nociceptive transmission, 576
   NOP, 575, 584
   respiratory depression, 579
   signal transduction, 575
opioid system, **572–84**
organic nitrates, 35
organic phosphates, 338
ornithine transcarbamilase (OCT) deficiency, 304

orthosteric, 102–3
  binding sites, **103**
  ligands, **102**
oseltamivir, 656, 657
osteoblasts, 513, 563
osteoclasts, 384, 563
o-sulfate, 514
oxaliplatin, 416, 418
oxandrolone, 281
oxaprozin, 612
oxaxilidinones, 642, 648
oxidation, **62**
  P450-independent mixed oxidations, 65
oxidative phosphorylation, 437
oxotremorine, 507
oxprenolol, 492
oxycodone, 573
oxymetazoline, 492

p21, 406
p38, 150, 153, 413, 639
p53, 406–7, 411–13
p75, 115
paclitaxel, 263
pactimibe, **449**
pain, 12–13, 36, 38, 100, 123, 163, 260,
    313, 320, 324–5, 331, 339, 461,
    475, 510–511, 528, 547, 549,
    562, 565, 570–571, 573–4,
    578–9, 582, 584, 590–601, 612
  inflammatory, 562
  neuropathic, 562
palivizumab, 239, 695
palmitoylation, 137
palonosetron, 517
pancreatic cells, pharmacological
    stimulation of β-cell, **458–9**
pancuronium, 504, 506
panitumumab, 214, 695
para-aminohippuric (PAH), 56
para-aminosalicylic acid (PAS), 642
Paracelsus, 10
paracetamol, 62, 612, 638, 682
parachlorophenylalanine, 513–14
parathion, 500
Parkinson's disease, 480–486, 507, 560
  drugs, 494, 498, 506, 507
paroxetine, 366, 515
partition coefficient, 23–**6**, 29, **32**, 36,
    38–41, 46, 56
Pasteur, L., 12
patch clamp technique, 186, 372
patent, 691–2
pazopanib, 214
PC12 cell differentiation, 169
pefloxacin, 642, 649
pemoline, 485
penciclovir, 655–6
penicillin binding protein, 645

penicillin G, 36, 41, 642
penicillins, 56, 642, 645
  discovery, 14
pentamidine, 338
pentavalent antimonial, 338
pentazocine, 49, 573
peptide maturation, 386
peptidergic neurons, 568
peptidergic transmission, **568–9**
perhexilline, 438
peroxisome proliferator-activated receptors
    (PPAR), 272, **282,** 446–8
  pharmacological modulation, 447
  PPARα, **282,** 438
  PPARα agonists: fibrates, **447**
  PPARα/γ agonists, 283, 448
  PPARα regulated genes, **282**
  PPARβ, 282
  PPARγ, **282,** 438
  PPARγ agonists: thiazolidinediones,
    282, **447–8,** 460
  PPARγ regulated genes, **282**
personalized medicine, 712
pertussis toxin, 193–4
pertuzumab (Perjeta), 214
p-glycoprotein, 263, 683, 685
pH, 24
  gastric, 34
pharmaceutical form, 696
pharmaceutical interactions, 685
pharmacoeconomics, 5
pharmacogenetics, 5, **256–67**
  future trends, 266
pharmacogenomics, 5, 256
pharmacognosy, 5
pharmacokinetics, 5, 694–5, 700–707
  clinical, 704–5
  formulations, **706**
  generics, **706**
  in vitro, 700–707
  in vivo, 700–707
pharmacology
  arab, 9
  biotechnology, 15
  clinical, 5
  egiptian, 8
  hebrew, 8
  medieval, 9
  renaissance, 10
  romans, 9
  safety, 694, 696
  twelve century, 15
pharmacotherapy, modern, 14–15
phencyclidine, 124, 126
phenobarbital, 4, 49, 57, 69, 70, 85, 273,
    283, 680, 682
phenol-O-methyltransferase (POMT), 67
phenoxybenzamine, 486–7, 491–2
phentermine (phen), 513

phentolamine, 492
phenylalkylamines, 320, 328, 330
phenylbutazone, 56, 612
phenylephedrine, 492
phenyl histamine, 527
phenylpropanolamine, 485, 492
phenytoin, 51, 320, 325
phosphatidylcholine (lecithin), 499
phosphatidylinositol 3-kinase (PI3K),
    155, 194
phosphodiesterases, 196
  phosphodiesterase-4, 639
phospholipases, **609**
  phospholipase C, **196–8**
phosphomycin, 642, 645
photosensitization, 670
physiologically based pharmacokinetics
    (PBTK)/toxicokinetics (PBPK)
    modeling, 673
physostigmine (eserine), 187, 499, 505
picrotoxin, 531, 534
pigmentous retinitis, 192
pilocarpine, 504, 506, 507
pimozide, 338
pinacidil, 335
pindolol, 493
pioglitazone, 282, 440, **447,** 458, 460
piperacillin, 642
pirenzepine, 504, 506–7
piroxicam, 612
pituitary GH-secreting adenomas, 194
placebo, 5
placental barrier, 29–30
plasma concentration, **39,** 73–88
  fluctuations, **79–80**
  free drug, **50–51, 82**
  peak, 75
  protein bound, **50–51**
  steady state, **77–8**
  time course, **73–81**
  time-to-the-peak, 39, **74–5**
plasmamembrane Ca^{2+} ATPase, **351–2**
plasma proteins, **50–51,** 81
  binding to, **50, 59,** 81
plasminogen, **396–9**
plasminogen activator, 397–6
  inhibitor, 396
platelets, 226, 228–9, 232
  activation, **228–9**
platinum compounds, 418
p75^{NTR}, neurotrophin receptor, 167
poisoning, 9, 421, 500, 505–6, 584, 684
poly-(ADP-ribose)-polymerase
    (PARP), 549
polyamines, 653
polyenes, 654
polymixins, 525, 642, 650
polymorphisms, 182, 192, **257–66,** 363,
    584, 633

polypeptides, 642
posizolid, 642
post-synaptic density protein 95 (PSD-95), 473
posttranslational modifications, pharmacological modification, **130–138**
potassium (K$^+$) channels, **331–9**, 456
  2 transmembrane segments, (K$_{ATP}$), **332–5**
    composition, 332
    pharmacology, 332
    SUR composition, 332
  4 transmembrane segments, **336**
  6 transmembrane segments
    voltage gated (VGKC), **336–9**
      EAG type, 336
      ERG type, 336
      LQT type, 338
      Shaker type, 336
  7 transmembrane segments (7TM), 339
  Ca^{2+}-activated K$^+$ channels (BK channels), **340**
potassium conductance, 557, 560, 562
potency, **100**–101, 103–4, 106, 108
PP1 phosphatase, 406
pralidoxime, 500
pralnacasan, 387, 391
pramicidins, 651
prasugrel, 262, 561
pravastatin, 440
prazosin, 108, 491
precautionary principle, **676**
precocious puberty, 192
prednisolone, 187, 280
prednisone, 280, 628–9
pregnane X receptor, 272–83
  regulated genes, 283
pre-initiation complex (PIC), **243–6**, 249–50
preladenant, 561
prenylation, 136, 163
prepropeptide, 566
Priestley, J., 11
pristinamycin, 642
probenecid, 49, 56–7, 680, 684
probucol, 338
procainamide, 57, 262, 325, 338
procaine, 36, 65, 324
procyclidine, 506–7
prodrug, 61
progesterone, 263
progesterone receptor, 272–**81**
  antagonists, 278
progestins, 270, **280**
prohormone convertases, 567
prokineticin 2 (PK2), 569, 571

promoter, 110, **244**, 272, 600, 627, 632
  distal, **246**
  downstream element (DPE), 244
  minimal, **244**
  proximal elements, **244–6**
  upstream element (UPE), 246
proopiomelanocortin (POMC), 567, 570
propafenone, 71, 325
propeptide, 566
propranolol, 58, 325, 492
prostanoid receptors, 167, 618–20
  PGD$_2$ receptors, 618
  PGE$_2$ receptors, 619
  PGI$_2$ receptors, 619
  TXA$_2$ receptors, 619–20
prostanoids, **608–614**
  PGD$_2$, 609, 614
  PGE$_2$, 609, 614–15
  PGF$_2$, 609, 614
  TXA$_2$, 609, 613–14
prostanoid synthase, **613–15**
  prostaglandin D synthase, 614
  prostaglandin E synthase, 614–15
  prostaglandin F synthase, 614
  prostaglandin H synthase, **610–613**
  prostaglandin H synthase pathway, 610
  prostaglandin I synthase, 614
  thromboxane synthase, 613–14
protease domain, 397
protease inhibitors, 379, 382, 392–3, 399, 655, **657**, 660
proteases, **379–93**, 521, 525, 567, 658, 660
proteasome, **387–8**
  inhibitors, 425
protein kinases, **131–2**
  activation, 132
  5' AMP-activated protein kinase (AMPK), 459–60
  calcium/calmodulin-dependent protein kinase II, 131
  protein kinase A (PKA), **131–2**, 184, 196, 523
  protein kinase C (PKC), 184, 197
protein phosphatases, **132–3**
protein phosphorylation, **130–135**
protein tyrosine phosphatases, 434
proton pump inhibitors, 3, 15, 347, 351
prucalopride, 518
P-selectin, 523
pseudoephedrine, 485, 492
pseudomonic acid, 642
p66^{Shc}, 434
psilocybin, 124
psoriasis, 564
purinergic receptors, **553–64**
  P1, **556–64**
  P2, **556–64**

P2X, **557–8, 560–564**
P2Y, **557–63**
purines, **553–64**
pyridostigmine, 500, 505
pyridoxal-phosphate-6-azophenyl-2',4'-disulphonic-acid (PPADS), 557, 559
pyrimidines, 335, 553, 555
pyruvic dehydrogenase, 438

quantal release, 470
quantitative structure-activity relationship (QSAR), 673, 693
quetiapine, 338, 516–17
quinidine, 325, 338, 680–681
quinine, 338
quinolones, 338, 642, 649
quinpirole, 490
quinupristin-dalfopristin, 642

Rab, 156
Rac, 156
radezolid, 642
RAF, 152
raloxifene, 279, 281
R-(α)-methyl-histamine, 527
Ran, 154–7
ranitidine, 527–8
Rap, 156, 157
rapamycin, 263, 424, 627, **630**, 632–3
rapsyn, 176
Ras, 154–6
Ras-dependent transduction pathway, 209–10
Ras protein, 155–6, 163
REACH regulation, 673
reboxetine, 366, 486
receptor, 5, **93, 109–20**
  autoreceptor, 176–7, 482, 488, 491, 501, 512, 515–16, 520, 538
  cell membrane, 6, 110
  complex, **94**
  constitutively active, **107–8**
  cytoplasmic duality of receptors, 166
  density, **97**, 108
  desensitization, 7
  down regulation, 7, **117–18**
  endocytosis, 116
  frizzled receptors, 167
  heteroligomeric receptor, 177–8
  homomeric receptor, 177–9
  hypersensitivity, 117
  intracellular, 6, 110
  localization, **116–17**
  modulation of the activity, **109–19**
  nuclear, 110, 166
  nuclear classes 1-3, 272
  nuclear signal transduction, 110

orphan receptor, 191, 560
postsynaptic receptor, 175
presynaptic receptor, 175
reserve, **106**
spare receptors, **106**
subtypes, **99**
synthesis, 117
theory, **101–5**
traffic, **117**
two-state receptor model, **107**
upregulation, 118
receptor classes, **109–10**
   cytochines, 114
   death (DR), 115, 408
   decoy, 116, 236
      dominant negative, 237
   G protein coupled, 112
   guanylate ciclase activity, 114
   kinase activity, 113
   ligand gated, 111
   lipoprotein, 116
   toll-like, 115
   tumor necrosis factor (TNF), 115
receptor identification, **96**
   pharmacological profile, 96
   saturability, 96
   selectivity, 95–6
   specificity, 96
receptor-interacting protein-1 (RIP-1), 410, 413
receptor-interacting protein-3 (RIP-3), 410, 413
receptor interactions, **93–7**
   hydrogen bonds, 94
   hydrophobic interaction, 94
   ionic bonds, 94
   irreversible, 95
   reversible, 95
   van der Waals interactions, 94
receptor states
   active, **107**
   inactive, **107**
reduction, **65**
regenerative medicine, 296, **300–303**
regorafenib, 214
remifentanil, 573
renin, 330, 488, 489, 491, 493, 557, 559, 660
renin-angiotensin system, 330
repaglinide, 332–3, 458
repetitive administration, **75–80**
research stages, 692
research strategy, 691
reserpine, 373, 475, 514
retigabine, 320, 335, 339
retinoic acid receptor (RAR), 272–**82**
   translocation, **282**
retinoid acid receptor, 9-*cis* retinoic acid receptor (RXR), 272

retinoids, 282
   all trans retinoic acid, 282
Rheb, 156
rheumatoid arthritis, 564
Rho, 154–6, 194
rhodopsin, 189–90, 194
Rho family GTPases, 158–62
rifabutin, 642
rifamicins, 642, 646
rifampicin, 642
rifaximin, 642, 646
riluzole, 336, 549–50
rimonabant, 590–591
risk
   assessment, **673**
   characterization, **668**
risperidone, 494, 517
ritanserin, 517
ritodrine, 492
ritonavir, 263, 387, 683
rituximab, 239, 427, **635**
rivastigmine, 500, 505
rizatriptan, 516
RNA-induced silencing complex (RISC), 285, 292
RNA interference, **284–5**, 426
RNA polymerase II, **243**
rocuronium, 504
rofecoxib, 611–13
Roger, 12
rokitamycin, 642
roselipin, **446**
rosiglitazone, 282, 283, 440, **447**
rosuvastatin, 440–441
roxithromycin, 642
ryanodine receptor (RYR), 143–4

S-adenosylhomocysteine (SAH), 555–6
S-adenosyl-L-homocysteine, 525
S-adenosylmetionine (SAM), 555
S-adenosynil-L-methionine, 525
salbutamol, 492
salernitan medical School, 9
salicylates, 56, 461, 612, 681
salinosporamide A (NPI-0052), 426
salmeterol, 492
salsalate, 458, 461
Salvarsan, 13
saquinavir, 387, 393, 685
sarcoplasmic-endoplasmic reticulum Ca^{2+} ATPase (SERCA), 142–3, **352–3**
   distribution, 351
   pharmacology, 353
sarin, 500
sarpogrelate, 517
satellite cell, 301
scale-up, 697
Scatchard transformation, **98**

Scheele, K.W., 11
Schmiedeberg, O., 12
scopolamine, 42–3, 498, 504, **506–7**
secretory granules, 469
secretory vesicles, 168, 380, 385, **469**, 474
sedative hypnotics, 124
seizures, 562
selectin, 114, 225–227
   signal transduction, **231–2**
selective serotonin reuptake inhibitors (SSRI), 511, 513, 515, 518
selumetinib, 152
Semmola, G., 12
sensitization, **123**, 127
serine proteases, **380**
serotonin (5-hydroxytryptamine, 5-HT), 190, **509–19**
   cardiovascular system, 511–13
   drugs, 516–18
   gastrointestinal and genitourinal sistems, 513
   nervous system, 509–11
   release and reuptake, 514–15
   synthesis, 513–14
   transporter, 511, 513–15
serotonin receptors, 515–18
   5-HT1, 511–13, **515–17**
   5-HT2, 511–13, **515–17**
   5-HT3, 512, 513, **515–18**
   5-HT4, 511, 513, **516–18**
   5-HT5, 512, **516**
   5-HT6, 512, **516–18**
   5-HT7, 511–12, **516–18**
serotonin transporter (SERT), 369
   function, 370
   pharmacology, 370
sertindole, 338, 517
sertraline, 366, 515
setrons, 517
severe combined immunodeficiency (SCID), **304–5**
SH2 domains, 113, 158, 206–9, 211, 230
Shild graph, **103–4**
side effects, 7
signaling cascades, 168
signal transducer and activators of transcription (STAT), 208, 212–13, 251, **636**
signal transduction
   inside-out, **229**
   outside-in, **229–31**
sildenafil, 597, 606
silencer, **246**
simvastatin, 264, 440–441, 450
single nucleotide polymorphisms (SNP), 257
siRNA, **284–6**
   in clinical trials, 291

sirolimus, 263, 424, 627, **630,** 632–3
SIRT1, 249
sirtuins, **249**
skeletal muscle, 301, 332, 334, 340, 352, 446–7, 477, 488, 604
sleep–wake cycle, 509, 511, 513
SMAC, 411
small G proteins, **154–65**
   modulation, **159–64**
   physiological roles, 155
   posttranslational modification, 157, 163
   regulatory proteins, **157–9**
   structure, 154
   subcellular localization, 157
smooth muscle contraction, 118, 156, 158, 215, 320, 327, 510, 523–4, 527, 538, 558–9, 563, 579, 602–3, 609, 614, 617–19
snake venum toxin, 476
SNARE proteins, 329, 470–471, 475, 501, 505
sodium (Na^+) channels
   voltage gated (VGSC), **322–6**
      α-subunits, 322
      β-subunits, 322
      cellular organization, 322
      classification, 323
      functional properties, **322**
      molecular structure, **322**
      pharmacology, 323
sodium-glucose linked transporters (SGLTs), 453
   inhibitors, 460–461
   SGLT2, 458
solifenacin, 506–7
sorafenib, 152, 214–15, 426
sordarins, 653
sotalol, 320, 325, 338, 492
sparfloxacin, 338
specific flux, 27, **52**
spin-off, 689
spiramycin, 338, 642
spironolactone, 280, 676
squalene synthase inhibitors, 442–3
   pharmacological activity, 443
squalestatin, 440, 443
Src, 230
Src kinase, 199
stanozolol, 281
start-up, 787
statin-induced myopathy, 264
   OAT1B1 polymorphisms, **264**
statins, 15, 70, 137, 157, 162–3, 264, **441–3,** 679, 683–4, 691
   pharmacological activity, 441
STAT protein, 208, 212–13, 251, **636**
stem cells, **296–9**
   embryonic, 298

multipotent, 299
pluripotent, 298
steroids, **633–5**
   receptor, 166, 167
streptogramins, 642
Streptogramin, 640, 641, **645–46**
streptomycin, 642, 647
striatal interneurons, 562
substance P (SP), 525, 565, 567, 570
substances abuse, 124
subunit, **176–9**
   topology, 176–9
succinylcholine, 501, 504, 506
sufentanil, 573
sugammadex, 506
suicide inhibition (inactivation), 72
sulbactam, 642
sulfadiazine, 642
sulfamethoxazole, 642
sulfation, 67
sulfinpyrazone, 612
sulfonamides, 51, 642, 649
sulfonylureas, 320, **333,** 455, **458–9**
sulfotransferase, 514
sulpiride, 489, 490
sumatriptan, 516
sumo, proteins, 133
sumoylation, **133–4**
   conjugating enzymes, 133
   consensus motifs, 133
   SENP enzymes, 133
sunitinib, 214
superoxide anion, 433
superoxide dismutase, 63
suramin, 557, 559
survival factor inhibitors, 423
   HSP90 inhibitors, 425
   PI3K/akt/mTor inhibitors, 424
   raf/mek/erk inhibitors, 425
   Tyr-kinase inhibitors, 424
sushi regions, 538
sympathetic system, 478–9
   anatomical organization, 478
   cardiovascular effects, 479
   noncardiovascular effects, 479
sympathomimetic drugs, 485, 490
synapse, **465–76**
   active zone, 140, 465, 469
   chemical, 465
   cleft, 465
   electrical, 465
   excitatory synapse, 469, 547
   immunological, 116
   inhibitory synapse, 469
   large dense-core vesicles (LDCVs), 566–8
   plasticity, **472–4**

pruning, 473
small clear synaptic vesicles (SSVs), 567
tripartite, 466
vesicles, 465
synapsin, 471
synaptogenesis, **472**
synaptopathies, **468**
synaptotagmin, 471

T25, 675
tabun, 500
tachykinins, 565–7
tacrine, 500, 526
tacrolimus (FK-506), 71, 222, 253, 263, 338, 627, **630–633,** 679–80, 682
talavancin, 642, 650
tamoxifen, 71, 248, 261, 276, 279, 281, 426, 523, 682
tamsulosin, 492
TATA-box binding protein (TBP), **243–4**
taurine, 367
taxol, 413, 416
tazobactam, 642
t-butyl-bicycle-phosphorothionate (TBPS), 532
tea, 562
tedisamil, 325, 338
tedizolid, 642
tegaserod, 518
teicoplanin, 642
telavancin, 642
telbivudine, 659
tele-methylhistamine, 525
tele-methylimidazole acetic acid, 525
telomeres, telomerases, 426
temozolomide, 417–18
tenoxicam, 612
terazosin, 492
terbutaline, 492
terfenadine, 263, 338, 527
terodiline, 338
terpenic derivatives, 335
testosterone, 35, 192, 263
tetanus toxin, 329, 381, 475
tetrabenazine, 373, 514
tetra-chlorodibenzo-p-dioxin (TCDD), 666
tetracyclines, 642, 647
tetrahydrocannabinol (THC), 590
tetryzoline, 492
thalidomide, 14, 627, **630**
theobromine, 554, 557, 562
Theophrastus, 8, 10
theophylline, 554, 557, 561–2
therapeutic regimen, **83–7**
therapeutic window, 21, 39, 54, 80, 87, 705
therapy, personalized, 17
thiazide diuretics, 56, 347

thiazol ethylamine, 527
thiazolidinediones, 282, 438, **447–8,** 460
thiocarbamates, 652
thioformamides, 335
thioguanine, 262, 419
thioperamide, 526–7
thiopurine methyltransferase (TPMT), **262**
thiopurine-S-methyltransferase (TSMT), 67
thioridazine, 338
thiorphan, 386–7
threshold of toxicological concern (TTC), **675**
thrombin, 190–193, 195
thrombosis, 561
thromboxanes, 263, 605, 618
thyroid adenomas, 192, 194
thyroid hormones, 281
   receptors, 272–**81**
   triiodothyronine, 281
thyroid-stimulating hormone (TSH) receptor, 190–195
thyrotropin-releasing hormone (TRH), 190, 194, 481, 501, 567
ticagrelor, 560–561
ticarcillin, 642
ticlopidine, 71, 560–561, 682
tight junctions *see* junctions
timolol, 71, 325, 492
tiotidine, 526–8
tiotropium, 506
tirofiban (Aggrastat), 233
tissue plasminogen activator (tPA), **395–7**
TNFR1-associated death domain (TRADD), 410
TNF-related apoptosis-inducing ligand (TRAIL), 410, 413
tobramycin, 642, 647
tocainide, 320, 325–6
tocilizumab, 628, **636**
tolazoline, 492
tolbutamide, 71, 259, 332, 679, 681–2
tolerance, 7, **122–3,** 125, 127, 572, **580–583**
   adrenergic receptors, 493
   antibodies, 635
   GABA receptors, 534–5
   glucose, 455, 560
   immunodrugs, 630
   metabolism, 101
   nitrates, 606
   opioids, 572, **580–582**
toll-like receptors (TLR), 1–5, 115
tolmetin, 612
tolterodine, 506–7
topotecan, 421–2
torcetrapib, **449–50,** 711

toxic agents, 672–4
   annual production, 673
   mechanism of action, **672**
   mixtures, 674
toxicity
   dose response, **666**
   evaluation new chemicals, 669
   existing chemicals, **669**
   exposure assessment, **667**
   genotoxicity, 670
   irritation, 670
   phototoxicity, 670
   reproduction, **671**
   sensitization, 670
   test guidelines, **672**
toxicokinetics, **671**
toxicology, 5, **665–77,** 697–8
   environment, 674
   risk/benefit ratio, 7
   therapeutic index, 7
tramadol, 260, 573
tramazoline, 492
transcription
   regulation, **243**
   regulators, 275
   synthetic modulators of transcription activation, **254**
transcriptional control genes involved in lipid metabolism, **446**
transcription repressors, 244, 246, 276
transferrin, 26
translocator protein (TSPO), 535
transmission
   extra-synaptic, 568
   volume transmission, 468
   wiring transmission, 468
transplantation, 299
   allogeneic, 299
   autologous, 299
transporters
   ATP-dependent, **345–6**
   ATP-independent, **346**
   endocytosis, 361
   for excitatory amino acids, 361
   traffic, 361
trastuzumab (Herceptin), 214, 417, 427–8, 695
triamcinolone, 280
tricyclic antidepressants, 366, 370, 511, 515
trifluoromethylhistamine (α-), 525
triglyceride biosynthesis, 440, **444–5**
trimetaphan, 504, 506
trimethoprim, 642, 649
trimethoprim/sulfamethoxazole, 338
tripelennamine, 527
triprolidine, 527
triptans, 516

Trk neurotrophin receptors, 167
troglitazone, 440, 447
tropisetron, 517
tryptophan hydroxylase, 513
TSC2, 170
tubocurarine, 14, 49, 411, 502, 504, 506–7
tumor necrosis factor (TNF), 408, 410, 411, 413
   TNFα antagonist, **636**
   TNF receptor (TNFR), 115, 410, 411
tyramine, 65, 486, 680
tyrosine kinase, 207–15
   cytoplasmic, 211–12
   inhibitors, 213–15
   non-receptor, 211
   receptors, 207
   signal transducers, 208, 210
tyrosine phosphatases, 212
tyrphostin, 164

ubiquitination, **134**
   conjugating enzymes, 134
urapidil, 492
uric acid, 57
uridine, 556
uridine diphosphate (UDP), **553–64**
   galactose, **553–64**
   glucose, **553–64**
uridine-5'-triphosphate (UTP), **553–64**
urokinase (u-PA), 395–9
   pro-uPA, 398
   receptor (u-PAR), 399
ustekinumab, 628, **636,** 695

vagusstoff, 496
valdecoxib, 611–13
valproic acid, 49, 57, 325, 530, 682
vancomycin, 642–3, 645
vandetanib, 214
varenicline, **128–9,** 504, 507
vascular adhesion molecules (VCAM 1), 226, 228
vasoactive intestinal peptide (VIP), 569, 570
vasopressin, 158, 190, 482, 523, 569
vasopressin receptor, 190, 192
Vauquelin, L.-N., 12
vecuronium, 506
vedolizumab, 628, **636**
vemurafenib (PLX4032), 152, 417, 425
verapamil, 263, 325, 682
vernakalant, 336–7
vesamicol, 373–4, 475, 500
vesicular transporters, 361, **372–4**
   acetylcholine (VAChT), 374
   GABA (VGAT), 374
   glycine, 374
   monoamines (VMAT), 373

vidarabine, 656
viloxazine, 366
vinblastine, 263, 413, 420
vincristine, 263, 413, 420, 474
viral proteases, **392**
viramidine, 658
Virchow, R., 12
vitamin A, 282
vitamin B_6 (pyridoxal phosphate), 524
vitamin D, 281–**2**
   cholecalciferol, 281
   ergocalciferol, 281
vitamin D receptors, 272–**82**
   polymorphism, 281
vitronectin, 228

volociximab (M 200), 233
volume neurotransmission (VNT), 568
von Bezold-Jarisch reflex, 512
Von Linné, C., 11

warfarin, **51,** 71, 84, 260, 266, 679–82, 685, 705
whole body autoradiography, **703**
withdrawal, **123,** 125, 127, 574, 582
Wnt receptor agonists, 167, 170

xanomeline, 504, 506
xanthines, 557, 597
xantumol, **446**

xenobiotics, 61–2, 65, 71, 258, 263, 273, 283, 682–3
   activated receptors, 26
xylometazoline, 492

yohimbine, 492

zanamivir, 657
zatebradine, 321, 341
ziconotide, 320, 331
zidovudine, 655
ziv-aflibercept (Zaltrap), 214
zolmitriptan, 516
zotasetron, 516–17
zygote, 296

Printed in the USA/Agawam, MA
July 21, 2023

813399.013